LINEAR OPERATORS

PART I:
GENERAL THEORY

NELSON DUNFORD and **JACOB T. SCHWARTZ**

FORMER JAMES E. ENGLISH PROFESSOR OF MATHEMATICS YALE UNIVERSITY

PROFESSOR OF MATHEMATICS COURANT INSTITUTE NEW YORK UNIVERSITY

WITH THE ASSISTANCE OF

William G. Bade and **Robert G. Bartle**

PROFESSOR OF MATHEMATICS UNIVERSITY OF CALIFORNIA, BERKELEY

PROFESSOR OF MATHEMATICS UNIVERSITY OF ILLINOIS

Wiley Classics Library Edition Published 1988

A WILEY-INTERSCIENCE PUBLICATION

JOHN WILEY & SONS

New York • Chichester • Brisbane • Toronto • Singapore

This volume is dedicated to Virginia, Sandra, Eleanor and Doris

20 19 18 17 16

LIBRARY OF CONGRESS CATALOG CARD NUMBER
57-10545

ISBN 0 470 22605 6

ISBN 0-471-60848-3 (pbk.).

Printed in the United States of America

Preface

In the two parts of *Linear Operators* we endeavor to give a comprehensive survey of the general theory of linear operations, together with a survey of the application of this general theory to the diverse fields of more classical analysis. It has been our desire to emphasize the significance of the relationships between the abstract theory and its applications, that has set the general tone and determined the general structure of the present work. Thus, a very elaborate analysis (Chapter XIII) of the spectral theory of ordinary self-adjoint differential operators is presented, while on the other hand the theory of locally convex spaces is treated (Chapter V) rather briefly in its connection with the theory of *B*-spaces. Applications of the general theory are presented in two ways: as a part of the text, and as graded series of exercises. Thus Chapter VIII is devoted to ergodic theory and to the theory of semi-groups; Chapter XI to a variety of topics including the theories of integral equations, harmonic analysis, closure theorems of Wiener type, singular integral operators, and almost periodic functions; and Chapters XIII, XIV, XIX, and XX to various aspects of the spectral theory of differential operators. On the other hand, parts of the theory of summability of series and integrals are given as graded series of exercises in Chapters II and IV; the theory of orthogonal expansions as exercises in Chapter IV; the theory of inequalities in Chapter VI; the theory of Tauberian theorems of Hardy-Littlewood type in Chapter XI; etc. The exercises (of which there are approximately one thousand) have been chosen with considerable care. They are not normally routine drill problems but rather are designed to carry forward the theory presented in the text and to emphasize its interesting, and often surprising, applications. The reader is encouraged to read the exercises even though he may not care to work them out in all detail.

The division of the present work into two parts has been based on the following principle: all material relating to the topological theory of spaces and operators, and all material pertaining to the spectral theory of arbitrary operators into the first part; all material relating to the theory of completely reducible operators into the second part. Of course, we have occasionally found it convenient to violate this principle.

The present work is written for the student as well as for the mature mathematician. Much of the text has grown directly out of lectures given by the authors over many years, and the two parts are designed to form suitable texts for a variety of graduate courses. Thus, Chapters I, II, and selected topics from Chapters III and IV make a comprehensive one-year course in the theory of functions of a real variable. The material contained in Chapters VI, VII, IX, and X, with selections from V, VIII, and XI, has been used many times as the basis for a one-year set of lectures in operator theory. A one-year course in the spectral theory of self-adjoint differential operators and the associated boundary value problems may be based on Chapters IX, X, XII, and XIII. Many other topics, such as harmonic analysis, ergodic theory, the theory of semi-groups, and the general theory of completely reducible ("Spectral") operators in *B*-spaces developed in Chapters XV through XX are suitable for detailed study in a graduate seminar.

The present treatise is relatively self-contained, and nearly everything in it can be read by one who has studied the elementary algebraic and topological properties of the real and complex number systems, and those basic results of the theory of functions of a complex variable which center around the Cauchy integral theorem. At a very few isolated points, knowledge of certain less elementary results of algebra and analysis (e.g., determinant theory, the Weierstrass preparation theorem) is needed. Most of the notions and results from general topology and abstract algebra needed are presented and explained in the text, though in such a way as to require of the reader a considerable general mathematical maturity. It is desirable that he have as much familiarity with these two subjects as would be the normal fruit of one semester of graduate study of abstract algebra and complex variable theory.

To facilitate access to the very large amount of material collected in the present treatise, it has been furnished with a number of special features. A table graphically shows the interdependence of the sections of the various chapters. Tables of the properties of a number of special *B*-spaces and of operators mapping various of these spaces into each other are given in Chapters IV and VI. Many of the chapters end with sections entitled "Notes and Remarks", which have a dual purpose. On the one hand, they contain references to the original and subsequent papers in which the principal results of the chapter in question have been discussed in the literature. In addition, they contain references to many results related to but not included among those given in the text. In this capacity, the notes and remarks supplement the bibliography on the one hand, the exercises on the other, and furnish additional information for the research mathematician. To facilitate study, those results presented in the text which are particularly important for the subsequent development have been marked with a black arrow in the margin; such theorems and lemmas, some of which might otherwise have seemed relatively obscure, should be noted with particular care. We have tried to adhere to conventional terminology and notation except at those few points where the standard conventions seemed particularly unfortunate to us; at any rate, a general index and an index of notation have been provided to give guidance on this score. The theorems, lemmas, and definitions which compose the text are numbered serially in a single system that proceeds by sections. Lemma XI.5.4 is the fourth numbered item in the fifth section in the eleventh chapter. In the course of Chapter XI, this lemma is referred to as Lemma 5.4, and in the course of the fifth section of Chapter XI it is referred to as Lemma 4.

The general character of the present work may be indicated by a brief comparison with a number of the best-known books which have dealt with some of the topics treated here. The famous treatise of Banach is the inspiration and the prototype for Chapters IV, V, and VI. Stone's book on linear operators in Hilbert space contains substantially the material found in Chapters X and XII, though our treatment, which is based on the devices introduced by various Russian mathematicians and most notably by Gelfand, is quite different from that of Stone. The recent book of Riesz and Nagy is close in spirit to our work, and should be regarded as an excellent introduction to the much more extensive theory presented in the present Chapters III through XII. Naimark's recent book on linear differential operators is very close to Chapter XIII, and also discusses some points of the theory given in Chapter XIX.

Surveying in retrospect the theories presented in the following twenty chapters, it seems to the authors that the general theory of the first seven chapters, and the Hilbert space theory of self-adjoint operators given in Chapters IX, X, and XII, are theories which have now reached a relatively final form. The theory of semi-groups, the general harmonic analysis, and especially the theory of singular self-adjoint differential operators, though they have reached a considerable degree of maturity, should all still enjoy a substantial development. The new theory of spectral operators presented in Chapters XV through XVIII is, by comparison with the corresponding theory for self-adjoint operators, in an early and incomplete stage of development. Chapters XIX and XX give evidence that non-self-adjoint and non-normal spectral operators are of common enough occurrence among the interesting objects of mathematics to justify serious study. It is the authors' hope that the present treatise will indicate the location of the weak and of the strong spots in the edifice of theory built up till now, and thereby facilitate both the study of the theory as it exists and future research.

We have been especially fortunate in having the assistance of two of our colleagues. Without the patient and unstinting work of Professors Robert Bartle and William G. Bade, who revised and edited nearly every chapter and who contributed a number of sections, it is certain that nothing on the scale of the present treatise could have been completed. The great majority of the sections of "Notes and Remarks", in particular, are due to Professor Bartle.

We have received valuable advice and criticism from many other colleagues at Yale and at New York University. We have profited very much, and particularly in connection with our treatment of ergodic theory, from frequent contacts with Professor Shizuo Kakutani. We are indebted for many valuable suggestions concerning the theory of semi-groups to Professors Einar Hille and Ralph Phillips, who most generously made portions of their forthcoming treatise on this subject available to us. Innumerable contacts, both informal and in formal seminars, with Professors Berkowitz, Friedrichs, Friedman, Helson, Lax, Nirenberg, Rickart, and Wermer, and with Dr. Gian-Carlo Rota have been of immeasurable value to us, and we wish to thank all these friends for the help which they provided in the form of permission to quote manuscripts, help in revising our own manuscript, or/and written criticism. The final two sections of Chapter XIII, in particular, are due to Dr. Rota. Dr. Rota and Mr. David Mc Garvey edited many portions of the text, and, together with Drs. John Barry and Robert Christian, checked the accuracy of most of the problems in the text. We also wish to thank Dr. Alfred B. Wilcox for his help on Chapter IX, Dr. Marie Lesnick for editing Chapter V, and Mr. John Thompson for checking the calculations with hypergeometric and confluent hypergeometric functions made in Chapter XIII.

For most of the eight years that have gone into the writing of this volume the work was assisted by the support of the Office of Naval Research and particular thanks are due the administrators of its Mathematics Branch for their understanding and encouragement.

NELSON DUNFORD
JACOB SCHWARTZ

August, 1957.

La pensée n'est qu'un éclair au milieu de la nuit.
Mais c'est cet éclair qui est tout.

HENRI POINCARÉ

LINEAR OPERATORS IN THREE PARTS

PART I
General Theory

PART II
Spectral Theory, Self Adjoint Operators in Hilbert Space

PART III
Spectral Operators

Nelson Dunford and Jacob T. Schwartz

Contents

PART I. GENERAL THEORY

PART II. SPECTRAL THEORY

CHAPTER I

Preliminary Concepts

The study of linear operations requires a familiarity with certain basic concepts from the fields of set theory, topology and algebra. Chapter I discusses all of the concepts and results of these theories which will be used in this text. The exposition is complete, but brief; it contains results and proofs, with little accompanying illustrative or explanatory material. For the average reader, it will serve as a concise review of the topics treated, and, at the same time, as a collection of results available for easy reference. A reader familiar with set theory, metric spaces, and Hausdorff spaces, may prefer to start with Chapter II, and use Chapter I only for reference.

A. SET-THEORETIC PRELIMINARIES

1. Notation and Elementary Notions

Rather than list the undefined terms of set theory, the axioms relating them, and the logical postulates governing the manipulation of these axioms, we shall in this first section follow an intuitive approach. The theorems and their proofs will be carefully stated although in an informal manner.

Upper or lower case Latin or Greek letters will usually be used to denote *sets*, *collections*, *families*, or *classes*, as well as to denote *functions* or *mappings*. The symbol ϵ will indicate *membership* in a set; thus $x \epsilon A$ means that x is a member (or element) of the set A. If $P(x)$ is a proposition concerning x, the symbol $\{x|P(x)\}$ denotes the set of all x satisfying the proposition $P(x)$. The symbol $\{x, y, \ldots, z\}$ is used for the set whose elements are $x, y, \ldots, z$. Where no danger of confusion exists, we sometimes write x in place of $\{x\}$. In this notation then $\{x\} = \{y|y = x\}$. The *void set* is the set with no members; it is denoted by ϕ. If every element of a set A is also an element of a set B, then A is said to be *included* in B, or is said to be a *subset* of B, and B is said to *contain* A. This is denoted symbolically by $A \subseteq B$, or $B \supseteq A$. Two sets are the same if and only if they have the same elements, i.e., $A = B$ if and only if $A \subseteq B$ and $B \subseteq A$. The set A is said to be a *proper subset* of the set B if $A \subseteq B$ and $A \neq B$. The notation $A \subset B$, or $B \supset A$, will mean that A is a proper subset of B. The *complement of a set A relative to a set B* is the set whose elements are in B but not in A, i.e., the set $\{x | x \in B, x \notin A\}$. This set is sometimes denoted by $B—A$. In a discussion where the set B is clearly understood, we may simply employ the phrase *complement of A* and use the symbol A' for this complement; this is stated symbolically $A' = \{x | x \notin A\}$. If A is a set whose elements are sets a, the set of all x such that $x \in a$ for some $a \in A$ is called the *union*, or *sum*, of the sets a in A. This *union* is denoted by $\bigcup A$ or $\bigcup_{a \in A} a$. The *intersection*, or *product*, of the sets a in A is the set of all x in $\bigcup A$ which are elements of every $a \in A$. If $A = \{a, b, \ldots, c\}$ we will sometimes write the union $\bigcup A$ as $a \cup b \cup \ldots \cup c$, and the intersection $\bigcap A$ as $a \cap b \cap \ldots \cap c$, or simply as $ab \ldots c$. The operations of forming unions and intersections are *commutative* (i.e., $a \cup b = b \cup a$, $ab = ba$), and *associative* (i.e., $a \cup (b \cup c) = (a \cup b) \cup c$, $a(bc) = (ab)c$). Also, intersection is *distributive* with respect to union, and vice versa. This means that the following distributive laws hold:

$$x \bigcup_{a \in A} a = \bigcup_{a \in A} (xa), \qquad x \cup (\bigcap_{a \in A} a) = \bigcap_{a \in A} (x \cup a).$$

Moreover, there are identities, known as the *rules of De Morgan*, which relate the operations of complementation, taking unions, and taking intersections. These rules are expressed by the formulas

$$(\bigcup_{a \in A} a)' = \bigcap_{a \in A} a', \qquad (\bigcap_{a \in A} a)' = \bigcup_{a \in A} a',$$

where it is understood that all complements are taken relative to some set b which contains every element a of A.

Two sets are *disjoint* if their intersection is void. A family of sets is a *disjoint family* if every pair of distinct sets in the family is disjoint. A sequence $\{a_n\}$ of sets is a *sequence of disjoint sets* if $a_n \cap a_m = \phi$ for $n \neq m$. A set a is said to *intersect* a set b if $ab \neq \phi$.

The terms *function*, *mapping*, *transformation*, and *correspondence* will be used synonymously. The symbols $f : A \to B$ will mean that f is a function whose *domain* is A, and whose *range* is contained in B;

that is, for every $a \in A$, the function f assigns an element $f(a) \in B$. If $f : A \to B$ and $g : B \to C$, then the mapping $gf : A \to C$ is defined by the equation $(gf)(a) = g(f(a))$ for $a \in A$. If $f : A \to B$ and $C \subseteq A$, the symbol $f(C)$ is used for the set of all elements of the form $f(c)$ where $c \in C$. If $f : A \to B$ and $D \subseteq B$ then $f^{-1}(D)$ is defined as $\{x | x \in A, f(x) \in D\}$. The set $f(C)$ is called the *image* of C and the set $f^{-1}(D)$ is called the *inverse image* of D. If $f : A \to A$ and $C \subseteq A$, then C is said to be *invariant under* f in case that $f(C) \subseteq C$. The function f is said to map A *onto* B if $f(A) = B$ and *into* B if $f(A) \subseteq B$. The function f is said to be an *extension* of the function g and g a *restriction* of f if the domain of f contains the domain of g, and $f(x) = g(x)$ for x in the domain of g. The restriction of a function f to a subset A of its domain is sometimes denoted by $f|A$. If $f : A \to B$, and for each $b \in f(A)$ there is only one $a \in A$ with $f(a) = b$, f is said to have an *inverse* or to be *one-to-one*. The *inverse function* has domain $f(A)$ and range A; it is defined by the equation $a = f^{-1}(b)$. Thus the domain and range of f^{-1} are the range and domain, respectively, of f. The *characteristic function* χ_E of a set E is the real function defined by the equations $\chi_E(s) = 1$, $s \in E$, and $\chi_E(s) = 0$, $s \notin E$.

Sometimes, when the range of a transformation $f : A \to B$ is to be emphasized at the expense of f itself and its domain, we shall write $f(a)$ as b_a. Then $f(A)$ is said to be an *indexed set*, and A is said to be a *set of indices*. If B is a collection of sets, the union $\cup f(A)$ will sometimes be written as $\bigcup_{a \in A} b_a$, and $\cap f(A)$, as $\bigcap_{a \in A} b_a$.

A *relation* in (or on) a set A is a collection r of ordered pairs $[x, y]$ of elements of A. It is customary to write xry for $[x, y] \in r$. Other symbols for relations are $=$, $\leq$, $\subset$, $\subseteq$, $\sim$, and $\equiv$.

We presuppose a familiarity with the real and complex number systems. By the *extended real number system* we mean the real numbers with the symbols $+\infty$ and $-\infty$ adjoined; by the *extended complex number system* we mean the complex numbers with the single symbol ∞ adjoined. If A is a set of real numbers, then the *supremum*, or the *least upper bound*, of A is the smallest real number b such that $a \leq b$ for all a in A; if no such number exists, we take $+\infty$ as the supremum. In either case we write $\sup A$ or $\text{lub}\, A$ to denote the supremum. Similar definitions are given for the *infimum*, or the *greatest lower bound*, of a set A which we denote by the notation $\inf A$ or $\text{glb}\, A$.

If ϕ is the void subset of the real numbers, it is conventional to take $-\infty = \sup \phi$, $+\infty = \inf \phi$. If A is an infinite set of real numbers, then the symbol lim sup A denotes the infimum of all numbers b with the property that only a finite set of numbers in A exceed b; the definition of the symbol lim inf A is similar. In particular, if A is a sequence $\{a_n\}$ then lim sup A and lim inf A are usually denoted by

$$\limsup_{n\to\infty} a_n, \qquad \liminf_{n\to\infty} a_n,$$

respectively. If a and b are extended real numbers, then the symbol (a, b) denotes the *open interval* defined by $\{x|a < x < b\}$; the symbol $[a, b]$ denotes the *closed interval* defined by $\{x|a \leqq x \leqq b\}$; and the *semi-open intervals* $(a, b]$ and $[a, b)$ are given by $\{x|a < x \leqq b\}$ and $\{x|a \leqq x < b\}$, respectively. If f is a real function defined on an open interval containing zero then the equations

$$\limsup_{x\to 0} f(x) = \inf_{a>0} \sup f((-a, a)),$$

$$\liminf_{x\to 0} f(x) = \sup_{a>0} \inf f((-a, a)),$$

$$\limsup_{x\to 0^+} f(x) = \inf_{a>0} \sup f((0, a)),$$

$$\liminf_{x\to 0^+} f(x) = \sup_{a>0} \inf f((0, a)),$$

define the symbols on their left sides. Similar definitions hold for the symbols $\limsup_{x\to 0^-} f(x)$, $\liminf_{x\to 0^-} f(x)$. Finally if z is a complex number and $z = x + iy$, where x and y are real, then x and y are called the *real part* and the *imaginary part* of z and are denoted by $\mathscr{R}(z)$ and $\mathscr{I}(z)$, respectively.

2. Partially Ordered Systems

1 DEFINITION. A *partially ordered system* $(E, \leqq)$ (henceforth abbreviated p.o.s.) is a non-empty set E, together with a relation $\leqq$ on E, such that

(a) if $a \leqq b$ and $b \leqq c$ then $a \leqq c$,

(b) $a \leqq a$.

The relation $\leqq$ is called an *order relation* in E. The notation $y \geqq x$ is sometimes used in place of $x \leqq y$.

2 DEFINITION. A *totally ordered* subset F of a p.o.s. $(E, \leqq)$ is a subset of E such that for every pair $x, y \in F$ either $x \leqq y$ or $y \leqq x$.

3 DEFINITION. If F is a subset of a p.o.s. $(E, \leqq)$, then an element x in E is said to be an *upper bound* for F if every f in F has the property $f \leqq x$. An upper bound x for F is said to be a *least upper bound* of F if every upper bound g of F has the property $x \leqq g$. In a

similar fashion, the terms *lower bound* and *greatest lower bound* may be defined. As in the case of real numbers we denote the least upper bound of F by $\sup F$ or lub F and the greatest lower bound of F by $\inf F$ or glb F.

4 DEFINITION. An element x in E is said to be *maximal* if $x \leqq y$ implies $y \leqq x$.

The family A of all subsets of a set X affords an illustration of these concepts. The inclusion relation $\subseteqq$ between the sets contained in X makes the pair $(A, \subseteqq)$ a partially ordered system. An upper bound for a subfamily $B \subseteqq A$ is any set containing $\cup B$, and $\cup B$ is the only least upper bound of B. Similarly, $\cap B$ is the only greatest lower bound of B. The only maximal element of A is X. In general, when dealing with a family of subsets of a given set, it will be supposed that they are ordered by inclusion, unless some other ordering is explicitly defined.

The following proof contains the central argument of this section.

5 THEOREM. *Let $f : E \rightarrow E$ have the property that $f(x) \geqq x$ where $(E, \leqq)$ is a non-void partially ordered system with the additional properties:*

(α) If $a \leqq b$ and $b \leqq a$ then $a = b$;

(β) every totally ordered subset of E has a least upper bound.

Then there is a w in E with $w = f(w)$.

PROOF. Let a be an element of E which will remain fixed throughout the proof. A subset B of E with the following three properties will be called *admissible*.

(i) $a \in B$.

(ii) $f(B) \subseteqq B$.

(iii) Every least upper bound of a totally ordered subset of B is in B.

There exists a set, namely E itself, which has these properties. Also an intersection of admissible sets is admissible. Hence the intersection A of all such sets is a *minimal* admissible set. The set $\{x | x \in E, x \geqq a\}$ is an admissible subset of E, and thus

(iv) $a \leqq x, \; x \in A$.

Now let $P = \{x | x \in A; \; y \in A, \; y < x \text{ imply } f(y) \leqq x\}$ where $y < x$ means that $y \leqq x$ and $y \neq x$. It will be shown that

(v) $x \in P$, $z \in A$ imply either $z \leqq x$ or $z \geqq f(x)$.

Fix x in P, and define B to be the set of all z in A such that either $z \leqq x$ or $z \geqq f(x)$. The condition (iv) shows that B has the property (i). The set B has the property (ii). For if $z \geqq f(x)$, then $f(z) \geqq z \geqq f(x)$; if $z = x$ then $f(z) = f(x)$; and, finally, if $z < x$, then $f(z) \leqq x$

since $x \in P$. The set B has the property (iii), for if u is a least upper bound for the totally ordered subset F of B, then either $y \leqq x$ for every $y \in F$, in which case $u \leqq x$, or else $y \geqq f(x)$ for some $y \in F$, in which case $u \geqq f(x)$. Thus B is an admissible subset of A, and therefore $B = A$, which proves the assertion (v).

It will next be shown that P is admissible. The condition (i) is vacuously satisfied by P. To prove that P has the property (ii), let $x \in P$. It will be shown that if $z \in A$, and $z < f(x)$, then $f(z) \leqq f(x)$. From (v), either $z \geqq f(x)$, or $z \leqq x$, so that if $z < f(x)$, then $z \leqq x$. Then, since $x \in P$, if $z < x$, then $f(z) \leqq x \leqq f(x)$, and if $z = x$, then $f(z) = f(x)$. To verify that P has the property (iii), let v be a least upper bound for the totally ordered set $F \subseteqq P$. To show that $v \in P$, let $z \in A$, $z < v$. From (v), it is seen that every $x \in F$ satisfies either $z \leqq x$, or $x \leqq f(x) \leqq z$. The second alternative cannot be valid for every x in F, for then $v \leqq z$. Hence, for some x in F, $z \leqq x$. If $z < x$ then $f(z) \leqq x \leqq v$ by the definition of P. If $z = x$, then since $z \neq v$, there is a y in F with $z < y$, in which case $f(z) \leqq y \leqq v$. Thus, in both cases $f(z) \leqq v$, which proves that $v \in P$, and that P has the property (iii). Thus P is an admissible subset of A, and so $P = A$. Therefore, by (v) it is seen that for any two elements x, z in A, either $z \leqq x$, or $z \geqq f(x) \geqq x$, which shows that A is totally ordered. If w is a least upper bound for A, then, since $f(w) \in A$, $w \leqq f(w) \leqq w$, and $f(w) = w$. Q.E.D.

6 Theorem. (*Hausdorff maximality theorem*). *Every partially ordered system contains a maximal totally ordered subsystem.*

A more explicit statement of the theorem follows: Let the family $\mathscr{E}$ of totally ordered subsets of a p.o.s. $(E, \leqq)$ be made into a p.o.s. $(\mathscr{E}, \subseteqq)$, by using the relation of inclusion between the elements of $\mathscr{E}$ (which are subsets of E). Then $\mathscr{E}$ has a maximal element.

Proof. If $\mathscr{E}$ has no maximal element, then to every $A \in \mathscr{E}$ there corresponds an $f(A) \in \mathscr{E}$ containing A properly. Thus Theorem 5 is contradicted by the function $f : \mathscr{E} \to \mathscr{E}$. Q.E.D.

7 Theorem. (*Zorn's lemma*). *A partially ordered system has a maximal element if every totally ordered subsystem has an upper bound.*

Proof. Applying Theorem 6, let x be an upper bound of a maximal totally ordered subset E_0 of the p.o.s. $(E, \leqq)$. Let $x \leqq y$. Then, if $y \notin E_0$, the set $E_0 \cup \{y\}$ is a totally ordered set containing E_0 as a proper subset. Hence $y \in E_0$, so that $y \leqq x$. Q.E.D.

8 DEFINITION. A p.o.s. $(E, \leqq)$ is said to be *well-ordered* if

(i) $a \leqq b$ and $b \leqq a$ implies $a = b$.

(ii) Any non-void subset of E contains a lower bound for itself.

The positive integers in their usual order afford a familiar example of a well-ordered system.

9 THEOREM. (*Well-ordering theorem of Zermelo*) *Every set may be well-ordered.*

The theorem means that for every set E there is an order relation $\leqq$ in E such that the p.o.s. $(E, \leqq)$ is well-ordered.

PROOF. Consider the family $\mathscr{E}$ of all well-ordered sets $(E_0, \leqq_0)$ such that $E_0 \subseteqq E$. We define the ordering $\prec$ in $\mathscr{E}$ by placing $(E_0, \leqq_0) \prec (E_1, \leqq_1)$ if and only if

(i) $E_0 \subseteqq E_1$,

(ii) $x, y \in E_0$, $x \leqq_0 y$ imply $x \leqq_1 y$,

and

(iii) $x \in E_0$, $y \notin E_0$, $y \in E_1$ imply $x \leqq_1 y$.

Under this ordering every totally ordered subfamily $\mathscr{E}_0$ of $\mathscr{E}$ has an upper bound. Indeed, it will be seen that this upper bound may be defined as $(\cup \mathscr{E}_0, \leqq')$, where $x \leqq' y$, whenever x and y both belong to some subset $E_0 \in \mathscr{E}_0$, and $x \leqq_0 y$ in the ordering $\leqq_0$ of that E_0. It is clear that if $(\cup \mathscr{E}_0, \leqq')$ belongs to $\mathscr{E}$ it is an upper bound for $\mathscr{E}_0$. It will now be shown that it is well-ordered and hence belongs to $\mathscr{E}$. The statement $x \leqq' x$ for $x \in \cup \mathscr{E}_0$ is clear. If $x \leqq' y$ and $y \leqq' z$, then $x, y \in E_0 \in \mathscr{E}_0$, $y, z \in E_1 \in \mathscr{E}_0$, $x \leqq_0 y$, and $y \leqq_1 z$. Since $\mathscr{E}_0$ is totally ordered it may be supposed that $(E_0, \leqq_0) \prec (E_1, \leqq_1)$, and then it is seen that $x \leqq_1 z$, and thus that $x \leqq' z$. If $x \leqq' y$, and $y \leqq' x$, then $x, y \in E_0$, and $x, y \in E_1$, with $x \leqq_0 y$, and $y \leqq_1 x$. Then, supposing that $(E_0, \leqq_0) \prec (E_1, \leqq_1)$, it follows that $x = y$.

Now, let $F \subseteqq \cup \mathscr{E}_0$, and let F be non-void. Then, for some $E_0 \in \mathscr{E}_0$, $F \cap E_0 \neq \phi$. The p.o.s. $(E_0, \leqq_0)$ is well-ordered. Let $x_0 \in F \cap E_0$ be a lower bound for $F \cap E_0$, under the ordering $\leqq_0$. Then, if $y \in F$, $y \notin F \cap E_0$, it follows that $x_0, y \in E_1$, where $(E_0, \leqq_0) \prec (E_1, \leqq_1)$, so that $x_0 \leqq_1 y$. Thus, x_0 is a lower bound for F under the ordering $\leqq'$. Thus it has been shown that $\mathscr{E}_0$ has an upper bound.

Hence, by Theorem 7, there is a maximal well-ordered subset E_0 of E. Now $E_0 = E$ for if x is in E but not in E_0 the ordering $\leqq_0$ in E_0 may be extended to the set $E_0 \cup \{x\}$ by defining $y \leqq_0 x$ for $y \in E_0$. Q.E.D.

3. Exercises

1 If $(E, \leqq)$ is a p.o.s. with the property that every totally ordered set has a lower bound, then E contains a minimal element.

2 If a family $\mathscr{E}$ of subsets of a set has the property that $A \in \mathscr{E}$ if and only if every finite subset of A belongs to $\mathscr{E}$, then $\mathscr{E}$ contains a maximal element.

3 Derive Theorem 6 from the statement of Exercise 2.

4 Show that any one of the following implies any other one: Theorem 6, Theorem 7, Theorem 9, and Exercise 2.

5 Prove that if A and B are sets, there exists either a one-to-one mapping of A into B, or a one-to-one mapping of B into A. This is known as the *comparability theorem for cardinals.*

6 Show that there exists a one-to-one correspondence between any infinite set A and the set of all pairs (a, n), where $a \in A$ and n is an integer.

7 Let R be the set of real numbers. A subset $S \subseteq R$ is called a *Hamel basis* if every real number r can be uniquely represented as $r = \Sigma_1^n \alpha_i s_i$, where $s_i \in S$ and α_i is rational. (a) Show that there exists a Hamel basis. (b) Show that there exists a discontinuous real function of a real variable satisfying the identity $f(x+y) = f(x)+f(y)$.

8 A family $\mathscr{M}$ of subsets of a set X is said to have property (α) if

(α) X is not the union of a finite number of subsets in $\mathscr{M}$.

Show that if $\mathscr{M}$ has property (α), there exists a maximal family $\mathscr{N}$ of subsets of X with property (α) which contains $\mathscr{M}$. Also, show that any maximal $\mathscr{N}$ has the property

(β) If $A_i \subseteq X$, $i = 1, \ldots, n$, and $A_1 \cap \ldots \cap A_n \in \mathscr{N}$, then some $A_i \in \mathscr{N}$.

9 DEFINITION. A p.o.s. $(E, \leqq)$ is said to be *complete* if:

(i) $a \leqq b$ and $b \leqq a$ imply $a = b$,

(ii) Every non-void subset has a least upper bound and a greatest lower bound.

10 (Tarski) If $(E, \leqq)$ is a complete p.o.s., $f: E \rightarrow E$, and

$x \leqq y$ implies $f(x) \leqq f(y)$, then f has a fixed element x_0 ($f(x_0) = x_0$), and the set of all fixed elements contains its least upper bound and its greatest lower bound.

11 DEFINITION. For each x in a set X let A_x be a subset of a set A. Then the *Cartesian product* $P_{x \in X} A_x$ or PA_x is defined to be the set of all functions f on X to A for which

$$f(x) \in A_x, \qquad x \in X.$$

If X consists of a finite number of elements, $X = \{x_1, \ldots, x_n\}$, $\underset{x \in X}{P} A_x$ will sometimes be written $A_{x_1} \times A_{x_2} \times \ldots \times A_{x_n}$.

12 If X is the set $\{1, \ldots, n\}$, then $P_{x \in X} A_x$ can be regarded as the set of n-tuples of the form $[a_1, \ldots, a_n]$, where $a_i \in A_i$. If X is the set of integers, then PA_x may be regarded as the set of sequences $[a_1, a_2, \ldots]$, where $a_i \in A_i$.

13 Let Q_x, N_x, and M_x be subsets of A_x, for $x \in X$. Let $M = PM_x$, $N = PN_x$, and $Q = PQ_x$. Suppose that $N \neq \phi \neq Q$, $M = N \cup Q$, and $N \cap Q = \phi$. Then there exists a uniquely determined $x_0 \in X$, such that

(a) $M_{x_0} = N_{x_0} \cup Q_{x_0}$, $N_{x_0} \cap Q_{x_0} = \phi$,

(b) $M_x = N_x = Q_x$ if $x \neq x_0$, $x \in X$.

14 DEFINITION. If $Y \subseteqq X$ the mapping which takes each f in $P_{x \in X} A_x$ into its restriction $f|Y$ is called the projection of $P_{x \in X} A_x$ onto $P_{x \in Y} A_x$. This mapping is denoted by pr_Y. If $Y = \{x\}$ then pr_Y will be written as pr_x.

15 Let $X \neq \phi$.

(a) $\underset{x \in X}{P} M_x$ is void if and only if some M_x is void.

(b) If $M_x \subseteqq N_x$ for $x \in X$, $\underset{x \in X}{P} M_x \subseteqq \underset{x \in X}{P} N_x$.

(c) If $\underset{x \in X}{P} M_x \neq \phi$, the converse of (b) is true, and equality in the conclusion of (b) implies equality in the hypothesis.

(d) A set F is of the form $\underset{x \in X}{P} B_x$ with $B_x \subseteqq A_x$ if and only if $F = \underset{x \in X}{P} pr_x(F)$.

B. TOPOLOGICAL PRELIMINARIES

4. Definitions and Basic Properties

1 DEFINITION. A family τ of subsets of a set X is called a *topology* in X if τ contains the void set ϕ, the set X, the union of every one of its subfamilies, and the intersection of every one of its finite subfamilies. The pair (X, τ) is called a *topological space*; but sometimes if τ is understood, we refer to X as a topological space. If τ, τ_1 are two topologies in X, τ is said to be *stronger*, or *larger*, than τ_1 and τ_1 is said to be *weaker*, or *smaller*, than τ if $\tau \supseteqq \tau_1$. The sets in τ are called the *open sets* of (X, τ). A *neighborhood of the point* p is an open set containing p. A *neighborhood of the set* A is an open set containing A. If A is a subset of X, then a point p is a *limit point*, or a *point of accumulation*, of A provided every neighborhood of p contains at least one point $q \neq p$, with $q \in A$. The *interior* of a set in X is the union of its open subsets; a point in the interior of a set is called an *interior point* of the set.

2 LEMMA. *A set in a topological space is open if and only if it contains a neighborhood of each of its points.*

This lemma, and a number of those to follow, are immediate consequences of the definitions involved. Where this is the case, the proofs will be omitted.

3 DEFINITION. A set is said to be *closed* if its complement is open.

4 LEMMA. *The intersection of any family of closed sets is closed, the union of any finite family of closed sets is closed, and ϕ and X are closed.*

5 LEMMA. *If $\mathscr{F}$ is any family of subsets of X having the properties of Lemma 4, and τ is the family of complements of members of $\mathscr{F}$, then (X, τ) is a topological space, and $\mathscr{F}$ is the family of closed sets of this topology.*

6 DEFINITION. A family β of subsets of X is said to be a *base* for the topology τ if $\beta \subseteqq \tau$ and if every set in τ is the union of some subfamily of β. The family β is said to be a *subbase* for the topology τ if τ is the smallest topology containing β. If β is a collection of neighbor-

hoods of a set $A \subseteqq X$ and every neighborhood of A contains a set in β, then β is called a *fundamental family of neighborhoods for* A.

For example, the usual *topology of the real line* is the topology on $(-\infty, +\infty)$ which has as a base all open intervals (a, b), where a and b are arbitrary real numbers. Another base for this topology is obtained by taking a and b to be rational numbers. A subbase for this topology is given by all infinite intervals $(-\infty, a)$, $(b, +\infty)$, where a and b are either real or rational. The *topology of the extended real numbers* has as a base the sets $[-\infty, a)$, (a, b), $(b, +\infty]$, where a and b are real numbers. The complex numbers and the extended complex numbers are treated similarly.

7 Lemma. *If β is a family of subsets of X, and if τ is the family of all unions of subfamilies of β, then τ is a topology if and only if*

(i) *for every pair $U, V \in \beta$ and $x \in U \cap V$ there is a $W \in \beta$, such that $x \in W \subseteqq U \cap V$, and*

(ii) $X = \cup \beta$.

8 Lemma. *A family β is a subbase for the topology τ if and only if $\beta \subseteqq \tau$ and every open set is a union of finite intersections of sets in β.*

9 Definition. The *closure* $\bar{A}$ of a set A is the intersection of all closed sets containing A. The set of points in $\bar{A}$ and not in the interior of A is called the *boundary* of A.

10 Lemma. *The closure operation has the properties:*

(a) $\overline{A \cup B} = \bar{A} \cup \bar{B}$, (b) $\bar{A} \supseteqq A$,

(c) $\bar{\bar{A}} = \bar{A}$, (d) $\bar{A} = A$ *if* $A = \phi$,

(e) $p \in \bar{A}$ *if and only if every neighborhood $N(p)$ of p intersects A.*

Proof. Statement (b) is self-evident. Since $\bar{A}$ is closed, (Lemma 4), it follows that $\bar{\bar{A}} = \bar{A}$. The set $\overline{A \cup B}$ contains A and is closed. Thus $\overline{A \cup B} \supseteqq \bar{A}$, and similarly $\overline{A \cup B} \supseteqq \bar{B}$. Therefore $\overline{A \cup B} \supseteqq \bar{A} \cup \bar{B}$. Conversely, $\bar{A} \cup \bar{B}$ is closed (Lemma 4), so $\bar{A} \cup \bar{B} \supseteqq \overline{A \cup B}$, which proves (a). Statement (d) means simply that ϕ is closed. Statement (e) follows immediately from Definitions 3 and 9. Q.E.D.

11 Lemma. *Let $\mathscr{A}$ be the family of all subsets of X, and let $A \to \bar{A}$ be a mapping $\mathscr{A} \to \mathscr{A}$, which has the four properties* (a), . . ., (d) *of Lemma* 10. *Then the family $\mathscr{F} = \{A | A = \bar{A}\}$ has the properties*

listed in Lemma 5, so that the family of complements of elements of $\mathscr{F}$ *is a topology. The set* $\bar{A}$ *is the closure of* A *in this topology.*

PROOF. Statements 10(b), (d) show that $\phi, X \in \mathscr{F}$, and 10(a) shows that $A \cup B \in \mathscr{F}$ if $A, B \in \mathscr{F}$. It follows that any finite union of sets in $\mathscr{F}$ is in $\mathscr{F}$. From 10(a), it follows that $A \subset B$ implies $\bar{A} \subset \bar{B}$, and hence if $A_\alpha \in \mathscr{F}$, it is seen that

$$\overline{\cap_\alpha A_\alpha} \subseteq \bar{A}_\alpha = A_\alpha.$$

Thus

$$\cap_\alpha A_\alpha \subseteq \overline{\cap_\alpha A_\alpha} \subseteq \cap_\alpha A_\alpha,$$

and the family $\mathscr{F}$ has all of the properties listed in Lemma 5. From that lemma it is seen that the family of complements of elements of $\mathscr{F}$ form a topology. It remains to be shown only that $\bar{A}$ is the closure of A in this topology. Now if B is closed, and $B \supseteq \bar{A}$, then $\bar{B} \supseteq \bar{\bar{A}} = \bar{A}$, which shows that $\bar{A}$ is closed, and is, in fact, the smallest closed set containing A. Q.E.D.

12 DEFINITION. If $Y \subseteq X$, and τ is a topology for X, then the topology

$$\tau_Y = \{A \mid A = B \cap Y, B \in \tau\}$$

is called the *natural relative topology* of Y generated by τ. A subset of Y is said to be *relatively open* if it is in τ_Y; *relatively closed* if its complement relative to Y is relatively open. Other terms like *relative closure* of a set are defined analogously. A topological space X is said to be *connected* if it is not the union of two non-void disjoint closed sets.

13 LEMMA. *If* $Y \subseteq X$, *and* (X, τ) *is a topological space, then the relative closure of a subset* A *of* Y *is the intersection of the closure of* A *with* Y.

A subset of a topological space will always be regarded as a topological space with its relative topology, unless some other topology is given explicitly.

14 THEOREM. (*Lindelöf*) *Let* (X, τ) *be a topological space, and let* τ *have a countable base* β. *Then every family* $\sigma \subseteq \tau$ *contains a countable subfamily* σ_0 *with* $\cup \sigma = \cup \sigma_0$.

PROOF. Let $B_1, B_2, \ldots$ be an enumeration of β. Let β_0 be the family of elements of β which are contained in some subset of σ. If $B_n \in \beta_0$, let C_n be some set in σ which contains B_n, and let σ_0 be the family of all these C_n. Then it is evident that

$$\cup\sigma \supseteq \cup\sigma_0 \supseteq \cup\beta_0.$$

Since β is a basis, if $p \in A \in \sigma$, there is a $B_n \in \beta$ such that $p \in B_n \subseteq A$, and so $\cup\beta_0 \supseteq \cup\sigma$. Q.E.D.

15 DEFINITION. If (X, τ) and (Y, τ_1) are topological spaces, and $f : X \to Y$, then f is *continuous* if $f^{-1}(A) \in \tau$ for every A in τ_1. In other words a mapping between topological spaces is continuous if the inverse image of every open set is open. The function f is said to be *continuous at the point* x if to every neighborhood U of $f(x)$ there corresponds a neighborhood V of x with $f(V) \subseteq U$. If f is a continuous one-to-one map of X onto Y, and if the inverse function f^{-1} is also continuous, then f is called a *homeomorphism*, or a *topological isomorphism*. In this situation, the spaces X and Y are said to be *homeomorphic*.

16 LEMMA. *Let X, Y be topological spaces and let $f : X \to Y$. Then, each of the following statements is equivalent to the continuity of f:*

(a) *The function f is continuous at each point $x \in X$.*
(b) *The inverse image of a closed set is closed.*
(c) *If $A \subseteq Y$, then $f^{-1}(\bar{A}) \supseteq \overline{f^{-1}(A)}$.*
(d) *If $A \subseteq X$, then $f(\bar{A}) \subseteq \overline{f(A)}$.*
(e) *For every A in some subbase for the topology in Y the set $f^{-1}(A)$ is open.*

PROOF. If f is continuous and U is a neighborhood of $f(x)$, then $V = f^{-1}(U)$ is a neighborhood of x with $f(V) \subseteq U$. Thus f is continuous at each point x. Conversely, Lemma 2 shows that (a) implies the continuity of f. Since the inverse image of a complement is the complement of the inverse image, the continuity of f is equivalent to the statement (b).

Statement (b) implies (c), since if $f^{-1}(\bar{A})$ is closed, then $\overline{f^{-1}(A)} \subseteq f^{-1}(\bar{A})$. But (c) implies (b), since if A is closed, $f^{-1}(A) \supseteq \overline{f^{-1}(A)}$, so that $f^{-1}(A)$ is closed.

Statements (c) and (d) are equivalent, for if $f(\bar{A}) \subseteqq \overline{f(A)}$, then $f(\overline{f^{-1}(B)}) \subseteqq \bar{B}$, and hence $\overline{f^{-1}(B)} \subseteqq f^{-1}(\bar{B})$. On the other hand, if $f^{-1}(\bar{A}) \supseteqq \overline{f^{-1}(A)}$, then $f^{-1}(\overline{f(B)}) \supseteqq \overline{f^{-1}(f(B))} \supseteqq \bar{B}$, and hence $\overline{f(B)} \supseteqq f(\bar{B})$.

It is clear that (a) implies (e). Also, since the inverse image of an intersection (or union) is the intersection (or union) of the inverse images, (e) implies (a). Q.E.D.

17 LEMMA. *If X, Y, Z are topological spaces, and if $f : X \to Y$ and $g : Y \to Z$ are continuous, then the composite function fg is continuous.*

The term *scalar* will be used for a real or complex number; a *scalar function* is a real or complex valued function. The topology in the set of scalars is always assumed to be that determined by the base whose elements are neighborhoods of the form $\{\beta \mid |\beta - \alpha| < \varepsilon\}$.

18 LEMMA. *Let f, g be continuous scalar functions on a topological space X, and let α be a scalar. Then the functions given by the expressions*

$$|f(x)|, \qquad \alpha f(x), \qquad f(x)+g(x),$$

are continuous. If f, g are real, then the expressions

$$\max\,(f(x), g(x)), \qquad \min\,(f(x), g(x))$$

also define continuous functions.

5. Normal and Compact Spaces

Is there, on a given topological space, a continuous real function which is not a constant? If x and y are distinct points of a topological space X, is there a continuous real function on X with $f(x) \neq f(y)$? If the answer to this second question is in the affirmative for an arbitrary pair of distinct points, it is said that there are enough real continuous functions to distinguish between the points of the space. It is not clear from any of the preceding remarks whether or not a given topological space has this property. However, it will be shown presently that the normal and the compact Hausdorff spaces introduced in this section all have enough continuous real functions to distinguish between their points.

1 DEFINITION. A topological space X is a *Hausdorff* space if it has the properties (a), (b), listed below. It is a *regular* space if it has the properties (a), (c), a *normal* space, if it has the properties (a), (d).

(a) Sets consisting of single points are closed.

(b) For every pair of distinct points x and y, there are disjoint neighborhoods of x and y.

(c) For every closed set A, and every $x \notin A$, there are disjoint neighborhoods of A and x.

(d) For every pair of disjoint closed sets A and B, there are disjoint neighborhoods of A and B.

2 THEOREM. (*Urysohn*) *Let A and B be disjoint closed sets in a normal topological space X. Then there is a continuous real function f defined on X, such that $0 \leqq f(x) \leqq 1$, $f(A) = 0$, $f(B) = 1$.*

PROOF. Let $A_{1/2}$ and $B_{1/2}$ be disjoint open sets, containing A and B respectively. Then we have

$$A \subseteq A_{1/2} \subseteq \bar{A}_{1/2} \subseteq B'_{1/2}, \qquad B_{1/2} \supseteq B.$$

Then A and $A'_{1/2}$ are disjoint closed sets, and $B'_{1/2}$ and B are disjoint closed sets. Employing the hypothesis of normality again, we construct open sets $A_{1/4}$, and $A_{3/4}$, such that

$$A \subseteq A_{1/4} \subseteq \bar{A}_{1/4} \subseteq A_{1/2} \subseteq \bar{A}_{1/2} \subseteq A_{3/4} \subseteq \bar{A}_{3/4},$$

and $\bar{A}_{3/4} \cap B = \phi$. By induction, a family of open sets A_r may be defined for every dyadic rational r, $0 < r < 1$, such that

(i) $r < s$ implies $\bar{A}_r \subseteq A_s$

and

(ii) $A \subseteq A_r$, $B \cap \bar{A}_r = \phi$.

Let $f(x) = 0$, if x is in all of the sets A_r. Otherwise let

$$f(x) = \sup \{r | x \notin A_r\}.$$

To verify that f is continuous, let $c = f(x)$ be positive. Then, for some suitable arbitrarily small ε and $\eta < \varepsilon$, x is in the open set $A_{c+\varepsilon} \cap \bar{A}'_{c-\eta}$. If y is also in this open set, then $|f(x)-f(y)| < 2\varepsilon$. If $f(x) = 0$ an analogous proof of continuity holds. Q.E.D.

3 THEOREM. (*The Tietze extension theorem*) *If f is a bounded real continuous function defined on a closed set A of a normal space X, there*

is a continuous real function F defined on X with $F(x) = f(x)$ for x in A, and $\sup_{x \in X} |F(x)| = \sup_{x \in A} |f(x)|$.

PROOF. Since the theorem is obvious if f is identically zero on A, it will be assumed that this is not the case. Let $f_0(x) = f(x)$, $\mu_0 = \sup_{x \in A} |f_0(x)|$, $A_0 = \{x | x \in A, f_0(x) \leqq -\mu_0/3\}$, and $B_0 = \{x | x \in A, f_0(x) \geqq \mu_0/3\}$. Then $A_0 \cap B_0 = \phi$, and both are closed. Applying the previous theorem, we find a function $F_0(x)$ defined on all of X such that $F_0(A_0) = -\mu_0/3$, $F_0(B_0) = \mu_0/3$, $-\mu_0/3 \leqq F_0(x) \leqq \mu_0/3$. Let $f_1(x) = f_0(x) - F_0(x)$ for $x \in A$. Then f_1 is continuous, and $\mu_1 \leqq (2/3)\mu_0$ where $\mu_1 = \sup_{x \in A} |f_1(x)|$.

By applying to the pair f_1, μ_1 the procedure applied to f_0, μ_0, and then continuing inductively, one obtains a sequence F_i, $i = 1, 2, \ldots$, of real continuous functions on X, with the properties:

$$|f(x) - \sum_{i=0}^{n} F_i(x)| \leqq \left(\tfrac{2}{3}\right)^{n+1} \mu_0, \qquad x \in A,$$

and

$$\sup_{x \in X} |F_n(x)| \leqq \tfrac{1}{3}\left(\tfrac{2}{3}\right)^{n} \mu_0.$$

These properties show that the series $\sum_{n=0}^{\infty} F_n(x)$ converges and defines a function F on X which coincides with f on A. To verify that F is continuous, let $\varepsilon > 0$ and fix n so that $2\mu_0(2/3)^{n+1} < \varepsilon/2$. Then

$$\begin{aligned} |F(x) - F(y)| &\leqq |F(x) - \sum_{i=0}^{n} F_i(x)| \\ &\quad + |\sum_{i=0}^{n} F_i(x) - \sum_{i=0}^{n} F_i(y)| + |\sum_{i=0}^{n} F_i(y) - F(y)| \\ &\leqq 2\mu_0\left(\tfrac{2}{3}\right)^{n+1} + \sum_{i=0}^{n} |F_i(x) - F_i(y)| \\ &< \frac{\varepsilon}{2} + \sum_{i=0}^{n} |F_i(x) - F_i(y)|. \end{aligned}$$

The continuity of F will be shown, using Lemma 4.16(a), by proving that F is continuous at each point x. According to Lemmas 4.16 and 4.18, there is a neighborhood V of x such that

$$\sum_{i=0}^{n} |F_i(x) - F_i(y)| < \frac{\varepsilon}{2}, \qquad y \in V,$$

and thus $|F(x)-F(y)| < \varepsilon$ for y in V. Q.E.D.

4 Corollary. *A real continuous function defined on a closed subset of a normal space has a real continuous extension defined on the whole space.*

Proof. The only case where the theorem does not apply immediately is the case where f is unbounded. If f is real and continuous on the closed set A in the normal space X, the bounded function $\arctan f(x)$ has a continuous extension $\alpha(x)$ defined on X. The closed sets A and $B = \{x \mid |\alpha(x)| = \pi/2\}$ are disjoint, and hence there is a continuous function β with $0 \leqq \beta(x) \leqq 1$, which vanishes on B, and has the constant value 1 on A. Thus the function $\tan \beta(x)\alpha(x)$ is a continuous extension of f. Q.E.D.

5 Definition. A *covering* of a set A in a topological space X is a family of open sets whose union contains A. The space X is said to be *compact* if every covering of X contains a finite subset which is also a covering. A topological space X is said to be *locally compact* if every point has a neighborhood whose closure is compact. A family of sets has the *finite intersection property* if every finite subfamily has a non-void intersection. A subset of X is called *conditionally compact* if its closure is compact in its relative topology. It should be noted that a subset $A \subseteq X$ is compact in its relative topology if and only if every covering of A by open sets in X contains a finite subcovering.

A well-known example of a locally compact space is a closed set of real or complex numbers. Such a space is compact if and only if it is bounded. These statements constitute the Heine-Borel theorem.

6 Lemma. *A topological space is compact if and only if every family of closed sets with the finite intersection property has a non-void intersection.*

This lemma follows immediately from the rules of DeMorgan, and the next lemma follows readily from the definitions involved.

7 Lemma. (a) *A closed subset of a compact space is compact.*
(b) *A continuous image of a compact space is compact.*
(c) *A compact subset of a Hausdorff space is closed.*

8 LEMMA. *A continuous one-to-one map from a compact space to a Hausdorff space is a homeomorphism.*

PROOF. Let X be a compact space, Y a Hausdorff space, and f a one-to-one continuous function on X, with $f(X) = Y$. According to the preceding lemma, a closed set A in X is compact, its continuous image $f(A)$ is compact, and, since Y is a Hausdorff space, $f(A)$ is closed. Thus Lemma 4.16(b) shows that f^{-1} is continuous. Q.E.D.

9 THEOREM. *A compact Hausdorff space is normal.*

PROOF. Let A be a closed subset of the compact Hausdorff space X, and let $p \notin A$. Then, if $q \in A$, there is a neighborhood U_q of q and a neighborhood V_q of p, such that $U_q \cap V_q = \phi$. Since A is compact, a finite set $U_{q_1}, \ldots, U_{q_n}$ covers A and

$$(U_{q_1} \cup \ldots \cup U_{q_n}) \cap (V_{q_1} \cap \ldots \cap V_{q_n}) = \phi.$$

Thus any compact set A and any point $p \notin A$ have disjoint neighborhoods. Let A and B be closed and disjoint. Then if $p \in A$, there is a neighborhood U_p of p, and a neighborhood V_p of B, such that $U_p \cap V_p = \phi$. Then a finite set U_{p_i} will cover A, and the sets $U_{p_1} \cup U_{p_2} \cup \ldots \cup U_{p_m}$ and $V_{p_1} \cap V_{p_2} \cap \ldots \cap V_{p_m}$ are disjoint neighborhoods of A and B, respectively. Q.E.D.

10 LEMMA. *A real continuous function on a compact space attains its supremum and its infimum.*

PROOF. Let f be a real continuous function on the compact space X. By Lemma 7(b) the set $f(X)$ is compact and hence by Lemma 7(c) it is closed. Hence $f(X)$ is a bounded closed set of real numbers, and, therefore, it contains both its supremum and its infimum. Q.E.D.

6. Metric Spaces

1 DEFINITION. Let X be a set, and let ϱ be a real function on $X \times X$, with the properties:

(i) $\varrho(x, y) \geqq 0$,

(ii) $\varrho(x, y) = 0$ if and only if $x = y$,

(iii) $\varrho(x, y) = \varrho(y, x)$, and

(iv) $\varrho(x, y) \leqq \varrho(x, z) + \varrho(z, y)$.

Then ϱ is called a *metric*, or a *metric function* in X. Sets of the form

$$S(x, \varepsilon) = \{y \mid \varrho(x, y) < \varepsilon\}$$

are called *spheres* in X. The sphere $S(x, \varepsilon)$ has x for a *center*, and ε for a *radius*. The *metric topology* in X is the smallest topology containing the spheres. The set X, with its metric topology, is called a *metric space*. If X is a topological space such that there exists a metric function whose topology is the same as the original topology, we say that X is *metrizable*.

If A and B are subsets of a metric space, let $\varrho(A, B) = \inf_{a \in A,\, b \in B} \varrho(a, b)$. If A is a subset of a metric space, the *ε-neighborhood of A* is the set $S(A, \varepsilon) = \{x \mid \varrho(A, x) < \varepsilon\}$. The *diameter* of a set A, in symbols $\delta(A)$, is the number $\sup_{a, b \in A} \varrho(a, b)$.

2 LEMMA. *If ϱ is a metric function in the set X, the spheres form a base for the metric topology.*

PROOF. If $u \in S(x, \varepsilon)S(y, \varepsilon')$, choose $\delta > 0$, so that $\varrho(x, u) + \delta < \varepsilon$, $\varrho(y, u) + \delta < \varepsilon'$. Then if $v \in S(u, \delta)$,

$$\varrho(x, v) \leqq \varrho(x, u) + \varrho(u, v) < \varrho(x, u) + \delta < \varepsilon,$$

and

$$\varrho(y, v) \leqq \varrho(y, u) + \varrho(u, v) < \varrho(y, u) + \delta < \varepsilon',$$

which show that $S(u, \delta) \subseteqq S(x, \varepsilon)S(y, \varepsilon')$. The desired conclusion follows from Lemma 4.7. Q.E.D.

3 THEOREM. *A metric space is normal.*

PROOF. If $x \neq y$, $S(x, (1/2)\varrho(x, y))$ and $S(y, (1/2)\varrho(x, y))$ are disjoint neighborhoods of x and y respectively; therefore a metric space is a Hausdorff space. If A and B are disjoint closed sets, the sets $A_1 = \{x \mid \varrho(x, A) < \varrho(x, B)\}$ and $B_1 = \{x \mid \varrho(x, B) < \varrho(x, A)\}$, are disjoint neighborhoods of A and B, respectively. Q.E.D.

4 LEMMA. *In a metric space, any subset, with its relative topology, is also a metric space.*

PROOF. The restriction of the metric function to a subset determines a metric, the topology of which is the relative topology of the subset. Q.E.D.

5 DEFINITION. A sequence $\{a_n\}$ in a topological space is said to converge to a point a in the space if every neighborhood of a contains

all but a finite number of the points a_n. This notion is written symbolically $a_n \to a$, or $\lim_{n\to\infty} a_n = a$. A sequence $\{a_n\}$ is said to be *convergent* if $a_n \to a$ for some a. A sequence $\{a_n\}$ in a metric space is a *Cauchy sequence* if $\lim_{m,n} \varrho(a_m, a_n) = 0$. If every Cauchy sequence is convergent, a metric space is said to be *complete*.

The next three lemmas are immediate consequences of the definitions.

6 LEMMA. *In a metric space, a convergent sequence is a Cauchy sequence. A Cauchy sequence converges if and only if it has a convergent subsequence. A point a is in the closure of a set A in a metric space if and only if there is a sequence $\{a_n\}$ of points of A converging to a.*

7 LEMMA. *A closed subspace of a complete metric space is complete. A complete subspace of a metric space is closed.*

8 LEMMA. *A mapping $f: X \to Y$ between metric spaces is continuous at the point x if and only if $\{f(x_n)\}$ converges to $f(x)$, whenever $\{x_n\}$ converges to x.*

9 THEOREM. (*The Baire category theorem*) *If a complete metric space is the denumerable union of closed subsets, at least one of these closed subsets contains a non-void open set.*

PROOF. Let X be a complete metric space with metric $\varrho(x, y)$. Let $\{A_n\}$ be a sequence of closed sets whose union is X. For the purposes of an indirect proof, it is assumed that no A_n contains a non-void open set. Thus $A_1 \neq X$, and A_1' is open, and contains a sphere $S_1 = S(p_1, \varepsilon_1)$ with $0 < \varepsilon_1 < 1/2$. By assumption, the set A_2 does not contain the open set $S(p_1, \varepsilon_1/2)$; hence the non-void open set $A_2' \cap S(p_1, \varepsilon_1/2)$ contains a sphere $S_2 = S(p_2, \varepsilon_2)$ with $0 < \varepsilon_2 < 1/2^2$. By induction, a sequence $\{S_n\} = \{S(p_n, \varepsilon_n)\}$ of spheres is obtained with the properties

$$0 < \varepsilon_n < 1/2^n, \qquad S_{n+1} \subseteq S(p_n, \varepsilon_n/2), \qquad S_n A_n = \phi, \qquad n = 1, 2, \ldots.$$

Since, for $n < m$,

$$\varrho(p_n, p_m) \leqq \varrho(p_n, p_{n+1}) + \varrho(p_{n+1}, p_{n+2}) + \cdots + \varrho(p_{m-1}, p_m)$$
$$< \frac{1}{2^{n+1}} + \cdots + \frac{1}{2^m} < \frac{1}{2^n},$$

the centers p_m form a Cauchy sequence, and hence converge to a point p. Since

$$\begin{aligned}\varrho(p_n, p) &\leqq \varrho(p_n, p_m) + \varrho(p_m, p)\\ &< \frac{\varepsilon_n}{2} + \varrho(p_m, p) \to \frac{\varepsilon_n}{2}\end{aligned}$$

it is seen that $p \in S_n$ for every n. This implies that p is in none of the sets A_n, and hence, that it is not in their union. This contradicts the assumption that $X = \cup A_n$. Q.E.D.

10 **Definition.** A subset A of a topological space X is *sequentially compact,* if every sequence of points in A has a subsequence converging to a point of X.

11 **Definition.** A set is said to be *dense* in a topological space X, if its closure is X. It is said to be *nowhere dense* if its closure does not contain any open set. A space is *separable,* if it contains a denumerable dense set.

12 **Theorem.** *If a topological space has a countable base, it is separable. Conversely, if a metric space is separable, it has a countable base. Thus a subspace of a separable metric space is separable.*

Proof. Let $A_1, A_2, \ldots$ be a base for the topology in a topological space X, and let $p_n \in A_n$. If V is open, there is a base element A_n contained in V, and hence a point p_n in V. Thus the denumerable set $P = \{p_1, p_2, \ldots\}$ is dense in X. Conversely, let $p_1, p_2, \ldots$ be a countable dense set in a metric space. It will be shown that the denumerable set of spheres $S(p_n, r)$, with r rational, forms a base. If p is in the open set V, then $S(p, \varepsilon) \subseteq V$ for small ε. Some p_n is in $S(p, \varepsilon/4)$, and for such p_n

$$p \in S(p_n, r) \subseteq S(p, \varepsilon) \subset V,$$

where r is any rational number between $\varepsilon/4$ and $\varepsilon/2$. The final statement follows readily. Q.E.D.

13 **Theorem.** *A subset of a metric space is compact if and only if it is closed and sequentially compact.*

Proof. Let A be a compact subset of a metric space X. By Lemma 5.7(c), A is closed. If A is not sequentially compact, then some sequence $\{a_n\}$ contains no convergent subsequence. Hence each point in A has a neighborhood containing at most a finite number of a_n. Since a finite number of these neighborhoods cover A, the sequence $\{a_n\}$ consists of only a finite number of distinct points of A, and therefore most certainly does have a convergent subsequence. This contradiction proves that A is sequentially compact.

Conversely, suppose that A is sequentially compact and closed. It will first be shown that A is separable. Let p_0 be an arbitrary point of A, and let $d_0 = \sup_{p \in A} \varrho(p_0, p)$. The number d_0 is finite, for if $\varrho(p_0, q_n) \to \infty$, there is a convergent subsequence $q_{n_i} \to q$, and hence $\varrho(p_0, q) = \lim \varrho(p_0, q_{n_i}) = \infty$, an impossibility.

Now, inductively, let p_{i+1} be chosen so that $\min_{0 \leqq n \leqq i} \varrho(p_n, p_{i+1}) \geqq d_i/2$, where

$$d_i = \sup_{p \in A} \min_{0 \leqq n \leqq i} \varrho(p_n, p).$$

It is clear that $d_0 \geqq d_1 \geqq \ldots$. If $d_n \geqq \varepsilon > 0$ for all n, then no subsequence of $p_0, p_1, \ldots$ is a Cauchy sequence; by Lemma 6 no subsequence converges. Since this contradicts our hypothesis, it follows that $d_n \to 0$. Thus, for every p in A and every $\varepsilon > 0$, there is a p_n such that $\varrho(p_n, p) < \varepsilon$. Therefore, $\{p_0, p_1, \ldots\}$ is a countable dense set.

Now, as is seen from Theorems 12 and 4.14, to prove that A is compact, it suffices to show that every countable covering of A by open sets $G_1, G_2, \ldots$ contains a finite subcovering. If $\cup_{i=1}^n G_i \neq A$, let $x_n \notin \cup_{i=1}^n G_i$, $x_n \in A$. Let $x_{n_i} \to x$ be a convergent subsequence of $\{x_n\}$. Then, since the complement of $\cup_{i=1}^n G_i$ is closed, $x \notin \cup_{i=1}^\infty G_i$, so that $x \notin \cup_{i=1}^\infty G_i$; this contradicts the fact that $\cup_{i=1}^n G_i = A$. Q.E.D.

14 **Definition.** A subset K of a metric space is *totally bounded*, if for every $\varepsilon > 0$ it is possible to cover K by a finite number of spheres $S(k_i, \varepsilon)$, $i = 1, \ldots, n$, with centers in K.

15 **Theorem.** *If K is a set in a metric space X, the following statements are equivalent:*

(a) *K is sequentially compact:*

(b) *$\overline{K}$ is compact:*

(c) *K is totally bounded and $\overline{K}$ is complete.*

Furthermore, a compact metric space is complete and separable.

Proof. By the preceding theorem, (b) follows from (a) if we can show that $\bar{K}$ is sequentially compact. Let $\{p_n\}$ be a sequence in $\bar{K}$, and let $k_n \in K$ be such that $\varrho(p_n, k_n) < 1/n$, $n = 1, 2, \ldots$. Some subsequence of $\{k_n\}$ converges, and it is readily seen that the same subsequence of $\{p_n\}$ converges to the same point. This proves (b).

Suppose (b) is true and let $\varepsilon > 0$. Then, since K is dense in $\bar{K}$, the spheres $S(k, \varepsilon)$, $k \in K$, cover $\bar{K}$. Consequently, a finite number $S(k_1, \varepsilon), \ldots, S(k_n, \varepsilon)$ of these spheres cover $\bar{K}$ and hence K, proving that K is totally bounded. Let $\{p_n\}$ be a Cauchy sequence in $\bar{K}$; since $\bar{K}$ is sequentially compact (Theorem 13), some subsequence of $\{p_n\}$ converges to a point $p_0 \in \bar{K}$. It is readily seen that the entire sequence $\{p_n\}$ converges to p_0, so $\bar{K}$ is complete.

Let (c) be valid, and let $\{k_n\}$ be a sequence in K. By the total boundedness of K, for each number $1/n$, $n = 1, 2, \ldots$, there is a covering of K by a finite collection of spheres with centers in K and radius $1/n$. Thus some subsequence $\{k_{1,n}\}$ of $\{k_n\}$ is contained entirely in a sphere of radius one; a subsequence $\{k_{2,n}\}$ of $\{k_{1,n}\}$ is contained entirely in a sphere of radius $1/2$, etc. Continue this process and set $\bar{k}_n = k_{n,n}$. By construction, $\{\bar{k}_n\}$ is a Cauchy sequence; since $\bar{K}$ is complete, it is convergent. Hence K is sequentially compact, and the equivalence of (a), (b) and (c) is proved.

To prove the final statement, observe that if a space X is compact, then from the equivalence of (b) and (c) it follows that X is complete. The fact that X is separable was proved in the course of the proof of Theorem 13. Q.E.D.

The selection procedure used to establish part (c) in the preceding proof is known as the *Cantor diagonal process.*

16 DEFINITION. A mapping $f : X \to Y$ between metric spaces is said to be *uniformly continuous* on X, if to every $\varepsilon > 0$ there corresponds a $\delta > 0$ such that $\varrho(x, x') < \delta$ implies $\varrho(f(x), f(x')) < \varepsilon$.

The next theorem is of an elementary character, and will be used frequently throughout the text.

17 THEOREM. (*Principle of extension by continuity*) *Let X and Y be metric spaces, and let Y be complete. If $f : A \to Y$ is uniformly continuous on the dense subset A of X, then f has a unique continuous extension $g : X \to Y$. This unique extension is uniformly continuous on X.*

PROOF. If $x \in X$, there is a sequence of points $a_n \in A$ with $a_n \to x$. Since $\{a_n\}$ is a Cauchy sequence, and f is uniformly continuous, the sequence $\{f(a_n)\}$ is also a Cauchy sequence. Since Y is complete, there is a point $g(x) \in Y$ with $f(a_n) \to g(x)$. To verify that $g(x)$ depends only upon x, and not upon the particular sequence $a_n \to x$, let $\{b_n\}$ be another sequence in A with $b_n \to x$. Then $\varrho(a_n, b_n) \to 0$, and, since f is uniformly continuous, $\varrho(f(a_n), f(b_n)) \to 0$, and therefore $f(b_n) \to g(x)$. Now it is readily seen that $\varrho(x, x') < \delta$ implies $\varrho(g(x), g(x')) \leqq \varepsilon$; from this the uniform continuity of g follows. Finally, the uniqueness of g is obvious. Q.E.D.

In general, a continuous function is not uniformly continuous, but on a compact metric space these two notions coincide.

18 THEOREM. *If $f: X \to Y$ is a continuous mapping between metric spaces, and if X is compact, then f is uniformly continuous.*

PROOF. If f is not uniformly continuous, there exists an $\varepsilon > 0$ and two sequences $\{x_n\}$ and $\{z_n\}$ such that $\varrho(x_n, z_n) < 1/n$ and $\varrho(f(x_n), f(z_n)) > \varepsilon$ for $n = 1, 2, \ldots$. Since X is sequentially compact (Theorem 13), there exist subsequences $\{x_{n_k}\}$ and $\{z_{n_k}\}$ which converge, and it is evident that they converge to the same point. If f is continuous, then for sufficiently large k, we have $\varrho(f(x_{n_k}), f(z_{n_k})) < \varepsilon$ which contradicts the above inequality. Thus if f is not uniformly continuous, it is not continuous on X. Q.E.D.

The following result gives conditions under which the topology of a space can be given by a metric function.

19 THEOREM. (*Urysohn metrization theorem*) *A regular topological space with a countable base is metrizable. In particular, a compact Hausdorff space is metrizable if and only if it has a countable base.*

PROOF. We have seen in Theorem 15 that a compact metric space is separable and, by Theorem 12, has a countable base. On the other hand, a compact Hausdorff space is normal (5.9) and therefore regular (5.1), so the second statement follows from the first.

Let X be a regular space with a countable base $\{U_n\}$. It will first be shown that X is normal. Let A, B be closed and disjoint sets in X. Since X is regular there is for each point x in A a set $U \in \{U_n\}$ with $x \in U \subseteqq \bar{U} \subseteqq B'$. Thus if $\{V_n\}$ is the subsequence of $\{U_n\}$ consisting of all U_n with $\bar{U}_n \subseteqq B'$ we have $A \subseteqq \bigcup_{n=1}^{\infty} V_n$. Similarly if $\{W_n\}$ consists of those U_n with $\bar{U}_n \subseteq A'$ then $B \subseteq \cup_{n=1}^{\infty} W_n$. Now let $Y_1 = V_1$, $Z_1 = W_1 - \bar{Y}_1$ and, inductively let $Y_n = V_n - \cup_{j=1}^{n-1} \bar{Z}_j$, $Z_n = W_n - \cup_{j=1}^{n} \bar{Y}_j$. Thus $Y = \cup_{n=1}^{\infty} Y_n$ and $Z = \cup_{n=1}^{\infty} Z_n$ are open sets. It may be seen that Y and Z are disjoint by showing that $Y_n Z_m = \phi$ for all $n, m \geqq 1$. If $n \leqq m$ then $Z_m \subseteq W_m - \bar{Y}_n \subseteq W_m - Y_n$ and so $Y_n Z_m = \phi$. If $n > m$ then $Y_n \subseteq V_n - \bar{Z}_m \subseteq V_n - Z_m$ and so $Y_n Z_m = \phi$. To see that $A \subseteq Y$ let x be a point of A and choose m with x in V_m. Then, since $\bar{Z}_n \subseteq \bar{W}_n \subseteq A'$ for $n \geqq 1$, and since $x \in AV_m$, we have $x \in Y_m$ which proves that $A \subseteq Y$. Similarly $B \subseteq Z$ and thus X is a normal space.

Again let $U_1, U_2, \ldots$ be a base for the open sets in X. If $p \in U_m$, then there exists a U_n such that $p \in U_n \subseteq \bar{U}_n \subseteq U_m$. Thus there exist pairs (U_n, U_m) of sets selected from the base with the property that $\bar{U}_n \subseteq U_m$; but since there are only a countable number of sets in the base, there can only be a countable number of such pairs. Let

$$(U_{n_1}, U_{m_1}), \ldots, (U_{n_k}, U_{m_k}), \ldots$$

be an enumeration of such pairs, and for each $k = 1, 2, \ldots$, by Theorem 5.2 there exists a continuous function f_k with $f_k(\bar{U}_{n_k}) = 0$, $f_k(U'_{m_k}) = 1$, $0 \leqq f_k(x) \leqq 1$. Let ϱ be defined on $X \times X$ by

$$\varrho(x, y) = \sum_{k=1}^{\infty} 2^{-k} |f_k(x) - f_k(y)|, \qquad x, y \in X.$$

It is evident that ϱ satisfies (i), (iii) and (iv) of Definition 6.1. Suppose that $\varrho(x, y) = 0$ for some pair $x \neq y$; then $f_k(x) = f_k(y)$, $k = 1, 2, \ldots$. On the other hand there exists a set $U_{m(x)}$ from the base such that $x \in U_{m(x)}$, $y \notin U_{m(x)}$. By the regularity of the space there is some other set $U_{n(x)}$ from the base with $x \in U_{n(x)} \subseteq \bar{U}_{n(x)} \subseteq U_{m(x)}$, so that $(U_{n(x)}, U_{m(x)})$ is one of the pairs listed above. Hence $f_k(x) \neq f_k(y)$ for some choice of k, and this contradiction shows that ϱ is a metric function on X.

Let x be given and $\varepsilon > 0$. Then it is readily seen that $\varrho(x, .)$ is a continuous real valued function on X. Hence there exists a set U_m from the base with $x \in U_m$ such that if $y \in U_m$ then $\varrho(x, y) < \varepsilon$. This shows that the identity mapping of X with the given topology onto X with the metric topology defined by ϱ is a continuous function. On the other hand if $x \in U_m$ there is a U_n such that $x \in U_n \subseteq \bar{U}_n \subseteq U_m$. Hence (U_n, U_m) occurs in the enumeration of the pairs, say in the

k-th term. Then if $\varrho(x, y) < 2^{-k}$ we have $|f_k(y)| < 1$ and so $y \in U_m$. Thus $S(x, 2^{-k}) \subseteq U_m$, which shows that the identity mapping of X with the metric topology onto X with the given topology is continuous. Hence the identity is a homeomorphism and the space is metrizable. Q.E.D.

7. Convergence and Uniform Convergence of Generalized Sequences

The notion of convergence introduced in Definition 6.5 is not sufficiently general for all our purposes. We wish to indicate various ways in which it can be generalized.

If X and Y are topological spaces and $g : X \to Y$, the expression $\lim_{w \to x} g(w) = y$ is written to mean that for every neighborhood N_y of y, there exists a neighborhood N_x of x, such that $g(N_x) \subseteq N_y$. The following is a related, but more general notion: Let A be a set, and let X and Y be topological spaces. Let $f : A \to X$, and $g : A \to Y$. Then $\lim_{f(a) \to x} g(a) = y$ is written to mean that for every neighborhood N_y of y, there exists a neighborhood N_x of x, such that $g(f^{-1}(N_x)) \subseteq N_y$. For instance, the statement $\lim_{\varrho(x, x_0) \to 0} x = x_0$ is true in every metric space. Of course, if $A = X$, and f is the identity mapping, then $\lim_{f(a) \to x} g(a) = y$ if and only if $\lim_{a \to x} g(a) = y$.

The following definition gives a third important and interesting way in which the concept of convergence can be generalized:

1 DEFINITION. A partially ordered set $(D, \leqq)$ is said to be *directed*, if every finite subset of D has an upper bound. A map $f : D \to X$ of a directed set D into a set X is called a *generalized sequence* of elements in X, or simply a generalized sequence in X. If $f : D \to X$ is a generalized sequence in the topological space X, it is said to *converge to the point p in X*, if to every neighborhood N of p there corresponds a $d_0 \in D$, such that $d \geqq d_0$ implies $f(d) \in N$. In this case, it is also said that the *limit of f exists and is equal to p*, or, symbolically, $\lim f(d) = p$, or, if D is to be emphasized, $\lim_D f(d) = p$.

Each notion of convergence gives rise to a related notion of *uniform convergence*, For instance, let D be a directed set, A an arbitrary set, and X a metric space. Suppose that $f = f(d, a)$ maps

$D \times A$ into X. Then the statement $\lim_D f(d, a) = g(a)$ *uniformly* on A, or uniformly for $a \in A$, means that for every $\varepsilon > 0$ there exists a $d_0 \in D$, such that $\varrho(f(d, a), g(a)) < \varepsilon$ for $d > d_0$ and for every a in A.

If f and g are two generalized sequences of real or complex numbers defined on the same directed set D, then we write

$$f = O(g)$$

in case there exists an $A > 0$ such that $|f(d)| \leqq A|g(d)|$ whenever $d \geqq d_A$. Similarly the symbol

$$f = o(g)$$

indicates that for every $\varepsilon > 0$ there is a $d_\varepsilon \in D$ such that $|f(d)| \leqq \varepsilon|g(d)|$ whenever $d \geqq d_\varepsilon$.

In later chapters we will ordinarily denote a generalized sequence $f : D \to X$ by the notation $\{x_\alpha\}$, thus emphasizing the range of the function.

Generalized sequences may be used in a general topological space in much the same way as ordinary sequences are used in a metric space. Several of the following lemmas illustrate this statement.

2 LEMMA. *Let A be a set in a topological space. Then a point p is in the closure of A if and only if some generalized sequence in A converges to p.*

PROOF. If $p = \lim f(d)$, with $f(d)$ in A, then every neighborhood of p contains a point of A. Hence $p \in \bar{A}$. Conversely, let every neighborhood of p contain a point of A. Let the family $\{N\}$ of neighborhoods of p be directed by defining $N_1 \geqq N_2$ to mean that $N_1 \subseteqq N_2$. Let f be a function on $\{N\}$, whose value $f(N)$ is a point in NA. Then $p = \lim f(N)$. Q.E.D.

3 LEMMA. *A topological space is a Hausdorff space if and only if no generalized sequence of its elements has more than one limit.*

4 LEMMA. *If X and Y are topological spaces, and $f : X \to Y$, then f is continuous if and only if $\lim h(d) = x$ implies $\lim f(h(d)) = f(x)$ for every generalized sequence h of elements of X.*

The proofs of these lemmas are left to the reader.

Most of the notions related to the basic notion of convergence can be carried over from sequences to generalized sequences: Let

$f : D \to X$ be a generalized sequence of elements in a metric space X. We call f *a generalized Cauchy sequence* in X, if, for each $\varepsilon > 0$, there exists a $d_0 \in D$, such that $\varrho(f(p), f(q)) < \varepsilon$ if $p \geqq d_0$, $q \geqq d_0$.

5 **Lemma.** *If f is a generalized Cauchy sequence in a complete metric space X, there exists a $p \in X$ such that* $\lim f(d) = p$.

Proof. Let $d_n \in D$ be such that $c_1, c_2 \geqq d_n$ implies $\varrho(f(c_1), f(c_2)) < 1/n$. Let b_n be an upper bound for the finite set $\{d_1, d_2, \ldots, d_n\}$. Then it is evident that $f(b_n)$ is a Cauchy sequence. Hence there exists a $p \in X$ such that $\lim_{n\to\infty} f(b_n) = p$. Let $\varepsilon > 0$ be given, and choose n_0 such that $2n_0^{-1} < \varepsilon$ and such that $\varrho(f(b_{n_0}), p) < \varepsilon/2$. Then if $d \geqq b_{n_0}$, it is apparent that $\varrho(f(d), p) < \varepsilon$. Q.E.D.

The reader who reviews the preceding discussion of metric spaces will discover other results which can be carried over from sequences to generalized sequences.

The following important result on interchange of limit operations is due to E. H. Moore.

6 **Lemma.** *Let D_1 and D_2 be directed sets, and suppose that $D_1 \times D_2$ is directed by the relation $(d_1, d_2) \leqq (d_1', d_2')$, which is defined to mean that $d_1 \leqq d_1'$ and $d_2 \leqq d_2'$. Let $f : D_1 \times D_2 \to X$ be a generalized sequence in the complete metric space X. Suppose that:*

(a) *for each $d_2 \in D_2$, the limit* $g(d_2) = \lim_{D_1} f(d_1, d_2)$ *exists, and*

(b) *the limit* $h(d_1) = \lim_{D_2} f(d_1, d_2)$ *exists uniformly on D_1.*

Then the three limits

$$\lim_{D_2} g(d_2), \qquad \lim_{D_1} h(d_1), \qquad \lim_{D_1 \times D_2} f(d_1, d_2)$$

all exist and are equal.

Proof. Let $\varepsilon > 0$ be given. There is a $\delta_2 \in D_2$ such that $d_2 \geqq \delta_2$ implies $\varrho(f(d_1, d_2), h(d_1)) < \varepsilon/8$ for all $d_1 \in D_1$. It follows that

$$\varrho(f(d_1, d_2), f(d_1, \delta_2)) < \varepsilon/4$$

for $d_1 \in D_1$. Thus if $\delta_1 \in D_1$ is such that $d_1 \geqq \delta_1$ implies $\varrho(f(d_1, \delta_2), g(\delta_2)) < \varepsilon/8$, we have

$$\varrho(f(d_1, \delta_2), f(\delta_1, \delta_2)) < \varepsilon/4,$$

so that $\varrho(f(d_1, d_2), f(\delta_1, \delta_2)) < \varepsilon/2$. Hence, if $d_1, d_1' \geqq \delta_1$ and $d_2, d_2' \geqq \delta_2$, we have $\varrho(f(d_1, d_2), f(d_1', d_2')) < \varepsilon$. Thus, f is a generalized Cauchy sequence. By Lemma 5, $\lim_{D_1 \times D_2} f(d_1, d_2) = p$ exists. We have

$$\varrho(p, f(d_1', d_2')) = \varrho(\lim_{D_1 \times D_2} f(d_1, d_2), f(d_1', d_2')) \leqq \varepsilon, \quad d_1' \geqq \delta_1, \quad d_2' \geqq \delta_2.$$

Hence

$$\varrho(p, g(d_2')) = \varrho(p, \lim_{D_1} f(d_1', d_2')) \leqq \varepsilon, \qquad d_2' \geqq \delta_2,$$

so that $\lim_{D_2} g(d_2) = p$.

In the same way, we show that $\lim_{D_1} h(d_1) = p$. Q.E.D.

7 Corollary. *Let D be a directed set, let X and Y be topological spaces, and let Y be complete and metric. Let $f : D \times X \to Y$, so that $f(d, x)$ is a generalized sequence of functions on X, with values in Y Suppose that:*

(a) *for each $d_0 \in D$, the function $f(d_0, x)$ is continuous, and*

(b) *the limit $g(x) = \lim_D f(d, x)$ exists uniformly on X.*

Then $g(x)$ is continuous.

Proof. This follows readily from Lemmas 4 and 6. Q.E.D.

It is often convenient to have a characterization of compactness in terms of generalized sequences. We will find it convenient to use one dealing with the notion of a cluster point.

8 Definition. Let $f : D \to X$ be a generalized sequence in a topological space X. A point p is a *cluster point* of f if for each neighborhood U of p and $d_0 \in D$ there is a $d \geqq d_0$ such that $f(d) \in U$.

9 Lemma. *A space X is compact if and only if each generalized sequence in X has a cluster point.*

Proof. If $f : D \to X$ is a generalized sequence in a compact space X, for each $d \in D$ let

$$A_d = \{y | y = f(d'), d' \in D, d' \geqq d\}.$$

The collection $\{A_d\}$ and hence the collection $\{\bar{A}_d\}$ have the finite intersection property. By Lemma 5.6 there is a point p common to their closures. Since every neighborhood of p intersects each A_d, the point p is a cluster point of f.

Conversely, let each generalized sequence in X have a cluster point. Let $\mathscr{F}_1$ be a given family of closed sets with the finite intersection property, and let $\mathscr{F}$ be the set of all finite intersections of sets in $\mathscr{F}_1$. Then $\mathscr{F}$ has the finite intersection property and is directed by inclusion. Picking a point in each set of $\mathscr{F}$ we obtain a generalized sequence which therefore has a cluster point. This cluster point is clearly in each member of the family $\mathscr{F}$. Hence the sets in $\mathscr{F}_1$ have a common point and X is compact. Q.E.D.

There is another notion which is often of convenience in discussions of convergence.

10 **Definition.** A family $\mathscr{E}$ of subsets of a set is said to be a *filter* if it possesses the following properties:

(i) The void set ϕ is not in $\mathscr{E}$;

(ii) if $A \supseteq B$ and $B \in \mathscr{E}$, then $A \in \mathscr{E}$;

(iii) if $A, B \in \mathscr{E}$, then $A \cap B \in \mathscr{E}$.

If $\mathscr{E}$ and $\mathscr{E}'$ are filters of subsets of a set, we say that $\mathscr{E}$ *refines* $\mathscr{E}'$ if $\mathscr{E} \supseteq \mathscr{E}'$. A filter is called an *ultrafilter* if it is not refined by any filter but itself. A filter $\mathscr{E}$ of subsets of a topological space X *converges to a point* $p \in X$, if every neighborhood of p belongs to $\mathscr{E}$.

The reader will observe that the collection of all subsets of a topological space X which contain a neighborhood of a point p forms a filter $\mathscr{N}(p)$. Thus the filter $\mathscr{E}$ converges to p if and only if $\mathscr{E}$ refines $\mathscr{N}(p)$, i.e., $\mathscr{E} \supseteq \mathscr{N}(p)$. In applying the filter theory the following lemma, due to H. Cartan, is of considerable importance.

11 **Lemma.** *Every filter of subsets of a set X is refined by some ultrafilter of subsets of X.*

Proof. Let $\mathscr{E}_0$ be a filter in X, and let $\mathfrak{P}$ be the collection of all filters $\mathscr{E}$ in X with $\mathscr{E}_0 \subseteq \mathscr{E}$. Let $\mathfrak{P}$ be partially ordered by $\subseteq$. If $\mathfrak{G} = \{\mathscr{E}\}$ is a totally ordered subset of $\mathfrak{P}$, then $\mathscr{E}' = \{E | E \in \mathscr{E} \in \mathfrak{G}\}$ is readily seen to be a filter in $\mathfrak{P}$ and $\mathscr{E} \subseteq \mathscr{E}'$, for all $\mathscr{E} \in \mathfrak{G}$. By Zorn's lemma (2.7) there is a maximal element which is obviously an ultrafilter in X and which refines $\mathscr{E}_0$. Q.E.D.

It is evident that if $\mathscr{E}$ is an ultrafilter in X and if $E \subseteq X$, then precisely one of the sets E, E' is contained in $\mathscr{E}$.

12 **Lemma.** *A topological space X is compact if and only if every ultrafilter of subsets of X converges to a point in X.*

Proof. Let $\mathscr{E}$ be an ultrafilter in a compact space X. If $\mathscr{E}$ does not converge to $p \in X$, then some neighborhood N_p is not in $\mathscr{E}$, and thus its complement N'_p is in $\mathscr{E}$. If $\mathscr{E}$ does not converge to any point, then X is covered by N_p, $p \in X$, and hence by a finite number $N_1, \ldots, N_n$ of such sets. Therefore $\phi = N_1' \cap \ldots \cap N_n' = (N_1 \cup \ldots \cup N_n)' = X'$, being the intersection of a finite number of sets in $\mathscr{E}$ is also in $\mathscr{E}$. But this violates the definition of a filter.

Suppose that every ultrafilter of subsets of X converges. Let $\mathscr{F}$ be a family of closed subsets of X with the finite intersection property. Let $\mathscr{E}_1$ be the class of all sets which contain a finite intersection $F_1 \cap \ldots \cap F_n$ of sets $F_i \in \mathscr{F}$. It is evident that $\mathscr{E}_1$ is a filter. Let $\mathscr{E}$ be an ultrafilter refining $\mathscr{E}_1$, and let p be a point to which $\mathscr{E}$ converges. Then, every neighborhood of p has a non-void intersection with every set of $\mathscr{E}$. Hence, p is in the closure of every set of $\mathscr{E}$. In particular, $p \in F$ if $F \in \mathscr{F}$. Hence $p \in \cap \mathscr{F}$, so that $\cap \mathscr{F} \neq \phi$. The desired result follows from Lemma 5.6. Q.E.D.

8. Product Spaces

Given a Cartesian product (cf. Definition 3.11) $X = P_\alpha X_\alpha$ of topological spaces X_α, it is natural to seek to define a topology in X. For example, if X_1 and X_2 both represent the space of real numbers, the Cartesian plane $X_1 \times X_2$ may be topologized by using sets of the form $U \times V$, where U and V are open, as a base. With this topology, $X_1 \times X_2$ is two-dimensional Euclidean space. Abstracting a vital property of this example, we seek a topology τ for the general Cartesian product X, such that each projection $pr_\alpha : X \to X_\alpha$ is continuous. It is readily seen that the maps pr_α will be continuous if and only if each set of the form $U = P_\alpha U_\alpha$, where $U_\alpha \subseteq X_\alpha$ is open, and where $U_\alpha = X_\alpha$ except for some single $\alpha = \alpha_0$, is in τ. This property is not sufficient to characterize τ completely. Among the various topologies which make the maps pr_α continuous, we still have the three topologies determined by taking as a base the sets $U = P_\alpha U_\alpha$ with:

(a) all U_α open,

or

(b) all U_α open, $U_\alpha = X_\alpha$ for all but a *countable* set of α,

or

(c) all U_α open, $U_\alpha = X_\alpha$ for all but a *finite* set of α.

With a view to future applications, the topology (c), which is the weakest topology for X which will guarantee the continuity of the maps pr_α, is selected for further analysis. The formal definition follows:

1 DEFINITION. Let (X_α, τ_α) be an indexed family of topological spaces. Then the *product topology* is the topology τ in $X = P_\alpha X_\alpha$ obtained by taking the collection of all sets $U = P_\alpha U_\alpha$, where each U_α is open, and where $U_\alpha = X_\alpha$ except for a finite set of indices α, as a base. A *product space* is a Cartesian product of topological spaces with its product topology.

It is apparent that a generalized sequence f in the product space X converges if and only if the generalized sequence $pr_\alpha f$ converges in X_α for each α.

2 LEMMA. *A product of Hausdorff spaces is a Hausdorff space.*

3 LEMMA. *If $b \subseteqq a$, then the projection pr_b (cf. 3.14) of $P_{\alpha\epsilon a} X_\alpha$ onto $P_{\sigma\epsilon b} X_\alpha$ is continuous, and maps open sets into open sets.*

4 LEMMA. *A denumerable product of metric spaces X_n, $n = 1, 2, \ldots$, is a metric space. If ϱ_n is a metric for X_n, then a metric for the product space $X = P_n X_n$ is given by the formula*

$$\varrho(x, y) = \sum_{n=1}^{\infty} \frac{1}{2^n} \frac{\varrho_n(pr_n x, pr_n y)}{1+\varrho_n(pr_n x, pr_n y)}.$$

If each of the spaces X_n is complete, then the product space X with this metric is also complete.

The proofs of these lemmas are straightforward and elementary. They will be left to the reader.

5 THEOREM. (*Tychonoff*) *A Cartesian product of compact spaces is compact in its product topology.*

PROOF. The proof is based upon Lemma 7.12. Let X_α be compact, and let $\mathscr{E}$ be an ultrafilter of subsets of $X = P_\alpha X_\alpha$. Then the set $\mathscr{E}_\alpha$ of all sets of the form $pr_\alpha(E)$ with $E \in \mathscr{E}$ is an ultrafilter of sets in X_α. This is clear, since a proper refinement $\mathscr{E}'_\alpha$ of $\mathscr{E}_\alpha$ determines a proper refinement

$$\mathscr{E}' = \{E | E = pr_\alpha^{-1}(E_\alpha), E_\alpha \in \mathscr{E}'_\alpha\}$$

of $\mathscr{E}$. Lemma 7.12 shows that each $\mathscr{E}_\alpha$ converges to some point $x(\alpha) \in X_\alpha$. It follows readily from the definition of the product topology that $\mathscr{E}$ converges to the point $x = P_\alpha x(\alpha)$ in X. Q.E.D.

9. Exercises

1 Any non-void open interval of reals is homeomorphic to the set of all real numbers, but not homeomorphic to a half open or a closed interval. A closed interval is not homeomorphic to the set of complex numbers of absolute value one.

2 Let R be the space of real numbers, and let $R_x = R$ for each $x \in R$. Show that $Q = P_{x \in R} R_x$ contains a countable subset whose closure is Q, but that Q does not have a countable base.

3 Not every normal space is a metric space.

4 Let G_1 and G_2 be open sets in a normal space, and let F be a closed set with $F \subseteqq G_1 \cup G_2$. Show that F is the union of closed sets F_1, F_2 with $F_1 \subseteqq G_1$ and $F_2 \subseteqq G_2$.

5 Let S be an abstract set. Show that the collection $B(S)$ of all bounded real or complex functions on S is a complete metric space under the metric function ϱ defined by $\varrho(f, g) = \sup_{s \in S} |f(s) - g(s)|$. If S is a topological space, then the collection $C(S)$ of all bounded continuous real or complex functions on S is a closed subset of $B(S)$.

6 Establish the existence of continuous real functions of a real variable, such that

$$\limsup_{h \to 0+} \left| \frac{f(t+h) - f(t)}{h} \right| = \infty \text{ for all } t.$$

Hint: Let S be the real line and let $C(S)$ be the space of periodic real continuous functions metrized as in Exercise 5. For $m = 1, 2, \ldots$ let C_m be the set of all f in $C(S)$ for which

$$\left| \frac{f(t+h) - f(t)}{h} \right| \leqq m$$

for some t and all $h > 0$. Using the Baire category theorem, show that $C(S)$ is not the union of the sets C_m.

C. ALGEBRAIC PRELIMINARIES

The algebraic results used in this book will be introduced, and proved, as they are needed. This section will review basic notions, and give fundamental definitions.

10. Groups

A *group* is a set G, together with a mapping $\mu : G \times G \to G$, which has the properties (i), . . ., (iii) listed below. The binary operation μ is often written as $\mu(a, b) = ab$, and, when this notation is used, it is called *multiplication*. The element ab is called the *product* of a and b. The product ab is required to satisfy the following conditions:

(i) $a(bc) = (ab)c, \quad a, b, c \in G$;

(ii) there is an element e in G, called the *identity* or the *unit* of G, such that $ae = ea = a$ for every a in G;

(iii) to each a in G, there corresponds an element a^{-1}, called the *inverse* of a, such that $aa^{-1} = a^{-1}a = e$.

A *subgroup* of a group G is a subset of G which is itself a group with respect to the binary operation in G. A *proper subgroup* is a subgroup other than $\{e\}$ or G. *A commutative group*, or an *Abelian group*, is a group in which the commutative law $ab = ba$ holds. In the commutative case, the binary operation μ is sometimes written as $\mu(a, b) = a+b$, and, when this notation is used, it is called *addition*, $a+b$ is called the *sum* of a and b, and the group is called an *additive* group. The identity in an additive group is called the *zero element* of the group; its symbol is 0 instead of e. Moreover, in an additive group, $-a$ is written instead of a^{-1}, and $a-b$ instead of $a+(-b)$. The additive notation for a group will be used only if the group is commutative.

If A and B are subsets of a group G in which the group operation is multiplication, and if g and h are elements of G, the symbol A^{-1} is written for the set of elements of the form a^{-1} where $a \in A$, AB denotes the set of elements of the form ab with $a \in A$ and $b \in B$, gA denotes the set of elements ga with $a \in A$, and gAh denotes the set of elements of the form gah with $a \in A$. The corresponding notations $-A$, $A+B$, $g+A$ are adopted for additive groups.

The following frequently used assertions are immediate consequences of the above definitions: In a group G the identity e is unique; the inverse a^{-1} of a is unique; for every a and b in G, there are uniquely determined elements x and y in G for which $ax = b$, $ya = b$; the unit e in G belongs to every subgroup of G; and $(ab)^{-1} = b^{-1}a^{-1}$.

A mapping $h : A \to B$ between groups A and B is called a *homomorphism* if $h(ab) = h(a)h(b)$. A one-to-one homomorphism is called an *isomorphism.* If $h : A \to B$ is an isomorphism and if $h(A) = B$, then A and B are said to be *isomorphic*, or A is said to be *isomorphic with* B. An isomorphism of a group G with itself is called an *automorphism* of G. If a is an element of the group G, the transformation $h_a : G \to G$, defined by $h_a(x) = a^{-1}xa$, is an automorphism of G, and is called the *conjugation* of G by a. Conjugations are called *inner automorphisms*; all other automorphisms are *outer automorphisms.* The *group of automorphisms* of a group G is the set of all automorphisms taken together with the binary operation of composition; that is, the product of two automorphisms is defined by the equation $(hk)(x) = h(k(x))$. It is evident that the inner automorphisms form a subgroup of the group of automorphisms.

A subgroup A of a group G is said to be *invariant*, or *normal*, in G if $x^{-1}Ax = A$ for every x in G. If A is a subgroup of G, sets of the form Ax are called *right cosets* of A, and sets of the form xA are called *left cosets* of A. Thus, if A is normal, the right coset Ax is the same as the left coset xA. It is clear that the cosets of a normal subgroup A form a group under the operation $(Ax)(Ay) = Axy$. This group of cosets of A is called the *quotient group*, or *factor group* of G by A; it is denoted by G/A. If G is Abelian then every subgroup is normal.

11. Linear Spaces

A *ring* is an additive group R together with a mapping $r : R \times R \to R$ which has the properties (i), (ii) and (iii) listed below. The binary operation r is written as $r(a, b) = ab$, and is called *multiplication.* Multiplication is required to satisfy the following identities:

(i) $(ab)c = a(bc)$;
(ii) $a(b+c) = ab+ac$;
(iii) $(b+c)a = ba+ca$.

In a ring $a+b$ is called the *sum* of a and b, and ab their *product.* A ring is *commutative* if the identity $ab = ba$ is valid. A *field* is a commutative ring in which the non-zero elements form a group under multiplication. The unit of this group in a field will be written as 1 instead of e.

A *linear vector space, linear space,* or *vector space over a field* Φ is an additive group $\mathfrak{X}$ together with an operation $m : \Phi \times \mathfrak{X} \to \mathfrak{X}$, written as $m(\alpha, x) = \alpha x$, which satisfy the following four conditions:

$$\begin{array}{lll} \text{(i)} & \alpha(x+y) = \alpha x+\alpha y, & \alpha \in \Phi, \quad x, y \in \mathfrak{X}; \\ \text{(ii)} & (\alpha+\beta)x = \alpha x+\beta x, & \alpha, \beta \in \Phi, \quad x \in \mathfrak{X}; \\ \text{(iii)} & \alpha(\beta x) = (\alpha\beta)x, & \alpha, \beta \in \Phi, \quad x \in \mathfrak{X}; \\ \text{(iv)} & 1x = x, & x \in \mathfrak{X}. \end{array}$$

The elements of a vector space are called *vectors*. The elements of the coefficient field Φ are called *scalars*. The operations $x \to \alpha x$ and $x \to a+x$, where $\alpha \in \Phi$ and $a \in \mathfrak{X}$, are called *scalar multiplication by* α, and *translation by* a, respectively. A sum $\alpha x+\beta y+\ldots+\gamma z$, where $\alpha, \beta, \ldots, \gamma$ are scalars, is called a *linear combination* of the vectors $x, y, \ldots, z$. A *linear subspace, subspace* or *linear manifold* in a vector space, is a subset which contains all linear combinations of its vectors. The subspace *spanned* by, or *determined* by a given set E, is the set of all linear combinations of elements of E. This set is a linear manifold; it is the smallest linear manifold containing E.

A set A in a linear space over a field Φ is said to be *linearly independent* if the only linear combination $\alpha x+\beta y+\ldots+\gamma z$ of distinct elements $x, y, \ldots, z$ of A which vanishes is that for which $\alpha = \beta = \ldots = \gamma = 0$. It follows readily from Theorem 2.7 that in every linear vector space $\mathfrak{X}$ there is a maximal linearly independent set B. A subset B of the linear space $\mathfrak{X}$ is called a *Hamel basis* or a *Hamel base* for $\mathfrak{X}$ if every vector x in $\mathfrak{X}$ has a unique representation $x = \alpha_1 b_1+\ldots+\alpha_n b_n$ with $\alpha_i \in \Phi$ and $b_i \in B$. It is clear that a set B is a Hamel base if and only if it is a maximal linearly independent set. The cardinality of a Hamel basis is a number independent of the Hamel basis; it is called the *dimension* of the linear space. This independence is readily proved if there is a finite Hamel basis, in which case the space is said to be *finite dimensional.* In a finite dimensional space, a Hamel basis is usually called a *basis*.

A function T is said to be a *linear operator*, or a *linear transformation*, if its domain and range are linear spaces over the same field Φ, and if $T(x+y) = Tx+Ty$, $T(\alpha x) = \alpha Tx$ for every $\alpha \in \Phi$ and every pair x, y of vectors in the domain of T. Thus a linear transformation on a linear space $\mathfrak{X}$ is a homomorphism on the additive group $\mathfrak{X}$ which commutes with the operations of multiplication by scalars. If $T : \mathfrak{X} \to \mathfrak{Y}$ and $U : \mathfrak{Y} \to \mathfrak{Z}$ are linear transformations, and $\mathfrak{X}$, $\mathfrak{Y}$, $\mathfrak{Z}$ are linear spaces over the same field Φ, the *product* UT, defined by $(UT)x = U(Tx)$, is a linear transformation which maps $\mathfrak{X}$ into $\mathfrak{Z}$. If T is a linear operator on $\mathfrak{X}$ to $\mathfrak{X}$, it is said to be a *linear operator in* $\mathfrak{X}$. For such operators the symbol T^2 is used for TT, and, inductively, T^n for $T^{n-1}T$. The symbol I is used for the *identity operator*, $Ix = x$, and 0 for the *zero operator*, $0x = 0$. If $P(\lambda) = \alpha_0+\alpha_1\lambda+\ldots+\alpha_n\lambda^n$ is a polynomial, the symbol $P(T)$ is used for the operator $\alpha_0 I+\alpha_1 T+\ldots+\alpha_n T^n$. If T, U are both linear on $\mathfrak{X}$ to the linear space $\mathfrak{Y}$, the *sum* $T+U$, defined as $(T+U)x = Tx+Ux$, is also linear on $\mathfrak{X}$ to $\mathfrak{Y}$. If T is linear, and if α is a scalar, the operation αT defined by $(\alpha T)x=\alpha(Tx)$ is also linear. If $\mathfrak{X}$ and $\mathfrak{Y}$ are linear spaces over the same field Φ, the set of all linear operations from $\mathfrak{X}$ to $\mathfrak{Y}$ is a linear space over Φ under the operations $T+U$ and αT. If $\mathfrak{X} = \mathfrak{Y}$ this set is also a ring under the operations $T+U$ and TU. A linear operator E in $\mathfrak{X}$ is said to be a *projection* or a *projection in* $\mathfrak{X}$ if $E^2 = E$. A projection is sometimes called an *idempotent operator*. If $\mathfrak{X}$ is a vector space, if $A \subseteqq \mathfrak{X}$, and if α is a scalar, the symbol αA is written for the set of elements of the form αx with x in A. If $A, B \subseteqq \mathfrak{X}$, and $x \in \mathfrak{X}$, then, since $\mathfrak{X}$ is an additive group, the symbols $A+B$, $A-B$, and $x+A$ have already been assigned definite meanings. The vector space $\mathfrak{X}$ is said to be the *direct sum* of the vector spaces $\mathfrak{M}_i$, $i = 1, \ldots, n$; symbolically, $\mathfrak{X} = \mathfrak{M}_i \oplus \ldots \oplus \mathfrak{M}_n$, if the spaces $\mathfrak{M}_i$ are subspaces of $\mathfrak{X}$ with the property that every x in $\mathfrak{X}$ has a unique representation $x = m_1+\ldots+m_n$ with $m_i \in \mathfrak{M}_i$, $i = 1, \ldots, n$. The map $E_i : \mathfrak{X} \to \mathfrak{M}_i$, given by the equation $E_i x = m_i$, is a projection of $\mathfrak{X}$ onto $\mathfrak{M}_i$.

If $\mathfrak{X}_1, \ldots, \mathfrak{X}_n$ are vector spaces over the field Φ, the set $\mathfrak{X} = \mathfrak{X}_1 \times \mathfrak{X}_2 \times \ldots \times \mathfrak{X}_n$ is a vector space under the operations defined by the equations

$$[x_1, \ldots, x_n]+[y_1, \ldots, y_n] = [x_1+y_1, \ldots, x_n+y_n],$$
$$\alpha[x_1, \ldots, x_n] = [\alpha x_1, \ldots, \alpha x_n].$$

The space $\mathfrak{X}$ is the direct sum of the spaces $\mathfrak{M}_i$, where $\mathfrak{M}_i$ is the set of those vectors $[x_1, \ldots, x_n]$ in $\mathfrak{X}$ with $x_j = 0$ for $j \neq i$. Since there is a one-to-one linear map between the spaces $\mathfrak{M}_i$ and $\mathfrak{X}_i$, the space $\mathfrak{X}$ is often called the *direct sum* of the spaces $\mathfrak{X}_1, \ldots, \mathfrak{X}_n$.

If $\mathfrak{M}$ is a subspace of the vector space $\mathfrak{X}$ over the field Φ, the *factor space* $\mathfrak{X}/\mathfrak{M}$ is the set of cosets of $\mathfrak{M}$, i.e., the set of sets of the form $x+\mathfrak{M}$ with $x \in \mathfrak{X}$. The algebraic operations in $\mathfrak{X}/\mathfrak{M}$ are defined by the equations

$$(x+\mathfrak{M})+(y+\mathfrak{M}) = (x+y)+\mathfrak{M}, \qquad x, y \in \mathfrak{X},$$
$$\alpha(x+\mathfrak{M}) = \alpha x+\mathfrak{M}, \qquad \alpha \in \Phi, \quad x \in \mathfrak{X}.$$

With these operations the factor space $\mathfrak{X}/\mathfrak{M}$ is a linear vector space. The term *quotient space* is sometimes used instead of factor space. The mapping of $\mathfrak{X}$ onto $\mathfrak{X}/\mathfrak{M}$, defined by $x \rightarrow x+\mathfrak{M}$, is called the *natural homomorphism* of $\mathfrak{X}$ onto $\mathfrak{X}/\mathfrak{M}$. It is a linear transformation.

Unless otherwise stated, the coefficient field Φ for a linear space $\mathfrak{X}$ will be either the field of real numbers, in which case $\mathfrak{X}$ is called a *real linear space*, or the field of complex numbers, in which case $\mathfrak{X}$ is called a *complex linear space*. The term *linear functional* will be used for a linear transformation whose range is in the coefficient field.

12. Algebras

If R is a ring, then a subset $R_1 \subseteq R$ is called a *subring* if the elements in R_1 form a ring under the operations defined in R. A subring I of R is called a *right ideal* of R if it has the additional properties

(a) $Ix \subseteq I$, $x \in R$;

(b) $(0) \neq I \neq R$.

The definition of *left ideal* is similar. If I is both a right and a left ideal of R it is called a *two-sided ideal*. The subrings (0) and R are frequently called *trivial*, or *improper*, *ideals* and all other ideals *proper ideals*. Condition (a) above evidently shows that if R has a unit e, then e is not contained in any (proper) ideal. In addition, if R is a field, then R contains no proper ideals, for if I is an ideal and $a \in I$, $a \neq 0$, then I contains $a(a^{-1}x) = x$ for any $x \in R$. Conversely, if R is a commutative ring with unit e and if R has no proper ideals, then R is a field. To prove this let $a \in R$, $a \neq 0$. Then $I_a = \{ax|x \in R\}$ is a subring satisfying (a) which is not (0); therefore $I_a = R$, and so there exists an

$x \in R$ such that $e = ax = xa$. Thus every non-zero element of R has an inverse, so R is a field.

A right (left, or two-sided) ideal in a ring R is called a *maximal right* (*left*, or *two-sided*) *ideal*, if it is contained in no other ideal of the same type. If R contains a unit element e, than any right ideal I_0 is contained in a maximal right ideal. To see this, let $\mathscr{E}$ be the collection of all right ideals in R which contain I_0 and which do not contain e. Let $\mathscr{E}$ be ordered by $\subseteqq$; if $\mathscr{F}$ is a totally ordered subsystem of $\mathscr{E}$, then it is easy to see that $\cup \mathscr{F}$ is a right ideal in R which does not contain e. Zorn's lemma (2.7) applies to give a maximal element in $\mathscr{E}$, which is readily seen to be a maximal right ideal containing I_0.

If I is a two-sided ideal in R, let $x+I$ be defined as $x+I = \{x+y | y \in I\}$. If we define the operations

$$(x+I)+(y+I) = (x+y)+I,$$
$$(x+I)(y+I) = xy+I,$$

the sets $x+I$, $x \in R$, form a ring which is called the *quotient ring* and denoted by R/I. It is an easy exercise for the reader to verify that for the above equations to define uniquely the operations in R/I it is necessary and sufficient that I be a two-sided ideal. A mapping h of one ring R_1 into another ring R_2 is called a *homomorphism*, or a *ring homomorphism*, if

$$h(x+y) = h(x)+h(y)$$

and

$$h(xy) = h(x)h(y).$$

A one-to-one homomorphism is called an *isomorphism*. The *kernel* of a homomorphism is the set of elements which are mapped into zero. The mapping $x \to x+I$ is a homomorphism of R onto R/I and is called the *natural homomorphism*. The reader will observe that I is the kernel of this natural homomorphism. We notice that if J is a right ideal and I a two-sided ideal with $I \subseteqq J$, then the image of J under the natural homomorphism h of R onto R/I is a (possibly trivial) right ideal in R/I, which we denote by J/I. Conversely, if A is a right ideal in R/I, the set $J = h^{-1}(A)$ is readily seen to be a right ideal in R such that $I \subseteqq J$ and $A = J/I$. Further, J/I is a proper ideal if and only if $I \subset J \subset R$. This remark enables one to see that if R is a commutative ring with a unit, and I is an ideal in R, then R/I is a field if and only if

I is a maximal ideal. For if I is maximal, then R/I is a commutative ring with unit which has no proper ideals; by what we showed earlier R/I is a field. Conversely, if R/I is a field, it contains no ideals and hence R has no ideals properly containing I.

If R is a ring with unit e, then an element x in R is called (*right, left*) *regular* in R in case R contains a (right, left) inverse y for x, i.e., we have ($xy = e$, $yx = e$) $xy = yx = e$. If x is regular, its unique inverse is denoted by x^{-1}. An element which is not (right, left) regular is called (*right, left*) *singular*.

If Φ is a field, then a set X is said to be an *algebra* over Φ if X is a ring as well as a vector space over Φ and if

$$\alpha(xy) = (\alpha x)y = x(\alpha y), \qquad x, y \in X, \quad \alpha \in \Phi.$$

A *right* (*left, two-sided*) *ideal* in an algebra is a right (left, two-sided) ideal in the ring sense which is also closed under multiplication by scalars. If I is a two-sided ideal in an algebra X then the quotient ring X/I is an algebra, called the *quotient algebra*, under the operations defined above in the ring and the linear space cases. A mapping of an algebra X into another algebra over the same field is a *homomorphism*, or an *algebraic homomorphism*, if it is both a linear transformation and a ring homomorphism. If the homomorphism is one-to-one, it is called an *isomorphism*. If Φ is the field of complex numbers and if in the algebra X there is a single valued map $x \to x^*$ with $(x+y)^* = x^*+y^*$, $(\lambda x)^* = \bar{\lambda}x^*$, $(x^*)^* = x$, $(xy)^* = y^*x^*$ then we say that X is an *algebra with involution*, and x^* is called the *adjoint* of x.

An element x in a ring is said to be *idempotent* if $x^2 = x$ and to be *nilpotent* if $x^n = 0$ for some positive integer n. A *Boolean ring* is one in which every element is idempotent. The identity $x+x = 0$, or equivalently $x = -x$, holds in every Boolean ring. To see this, note that $2x = (2x)^2 = 4x^2 = 4x$ so that $2x = x+x = 0$. A Boolean ring is commutative, for $x+y = (x+y)^2 = x^2+xy+yx+y^2 = x+xy+yx+y$, so that $xy = -yx$ and thus $xy = yx$.

The smallest Boolean ring with unit consists of the integers modulo 2, i.e., the two numbers 0, 1. This Boolean ring, which we denote by Φ_2, is actually a field. Conversely, every Boolean ring with unit which is also a field is necessarily isomorphic to the field Φ_2. To prove this let 1 denote the unit and let x be a regular element. Then

$$1 = xx^{-1} = x^2x^{-1} = x(xx^{-1}) = x \,.\, 1 = x.$$

Any Boolean ring may be thought of as a commutative algebra over the field Φ_2, and in doing so we note that a set is a ring ideal if and only if it is an algebraic ideal. If I is an ideal in a Boolean ring R, then R/I is a Boolean ring. Further, from the above we see that if M is a maximal ideal in a Boolean ring R with unit, then R/M is isomorphic with the field Φ_2.

An important example of a Boolean ring with a unit is the ring of subsets of a fixed set. More precisely, let S be a set and let multiplication and addition of arbitrary subsets E and F of the set S be defined by

$$EF = E \cap F, \qquad E+F = (E \cap F') \cup (E' \cap F) = (E \cup F) \cap (E \cap F)'.$$

The reader may verify that the collection of all subsets of S forms a Boolean ring with S as unit element and the void set as zero. (The set $E+F$ defined above is usually called the *symmetric difference* of E and F; in later chapters we will denote it by $E \Delta F$). Other examples of Boolean rings will be given in later chapters, but the important fact is that every Boolean ring with unit may be represented as a Boolean ring of subsets of some set. This important result is due to Stone.

A topological space is said to be *totally disconnected* if its topology has a base consisting of sets which are simultaneously open and closed.

1 THEOREM. (*Stone representation theorem*) *Every Boolean ring with unit is isomorphic with the Boolean ring of all open and closed subsets of a totally disconnected compact Hausdorff space.*

PROOF. If the Boolean ring B contains only one element so that $e = 0$, the theorem is trivial; hence we suppose that $e \neq 0$. Let H be the set of all non-zero ring homomorphisms of B into the Boolean ring Φ_2. For each $x \in B$, let $H(x) = \{h | h \in H, h(x) = 1\}$. Throughout this proof, if $x \in B$, let $x' = e + x$; then $H(x') = H(x)'$ where the second prime denotes the complement in H of $H(x)$. The relations

$$H(xy) = H(x) \cap H(y)$$

and

$$H(x+y) = (H(x) \cap H(y)') \cup (H(x)' \cap H(y))$$

show that the mapping $x \to H(x)$ is a homomorphism of B into a collection of subsets of H.

As an aid for the remainder of the proof, we demonstrate the following statement: Let $B_1 \subseteqq B$ have the properties

(a) if $x, y \in B_1$, then $xy \in B_1$;
(b) if $x \in B_1$, then $x \neq 0$;

then there is a homomorphism $h_1 : B \to \Phi_2$ such that $h_1(x) = 1$ for $x \in B_1$. To prove this statement let I_1 be the set of all elements of the form ax' with $a \in B$, $x \in B_1$. To see that I_1 is an ideal, observe that

$$\begin{aligned}(ax'+by')(xy)' &= (ax'+by')(e+xy)\\ &= (ax'+by')+(ax'+by')xy\\ &= (ax'+by')+(a+ax+b+by)xy\\ &= (ax'+by')+axy+axy+bxy+bxy\\ &= ax'+by',\end{aligned}$$

so that the sum of two elements in I_1 is also in I_1. Since it is clear that I_1 is invariant under multiplications by elements in B, I_1 is an ideal. Further I_1 is a proper ideal, for if $ax' = e$, then

$$e = ax' = ax'x' = ex' = x'$$

from which it follows that $x = 0$ contradicting (b). Since B has a unit element, I_1 is contained in some maximal ideal M_1. Let h_1 be the natural homomorphism $h_1 : B \to B/M_1 = \Phi_2$. Now if $x \in B_1$, then $x' = e+x \in I_1 \subseteqq M_1$ and hence

$$h_1(e)+h_1(x) = h_1(e+x) = 0$$

showing that $h_1(x) = h_1(e) = 1$. This proves the statement made above.

We have seen that the mapping $x \to H(x)$ is a homomorphism. To show that it is an isomorphism, let $x_0 \neq y_0$. We will prove that there exists an $h_0 \in H$ with $h_0(x_0) \neq h_0(y_0)$. If $x_0 \neq y_0$, then either $x_0 \neq x_0y_0$ or $y_0 \neq x_0y_0$; we suppose the former is valid. Let $z_0 = x_0+x_0y_0 = x_0y_0'$, so that $z_0 \neq 0$. If $B_1 = \{z_0\}$, then by the preceding paragraph there is an $h_0 \in H$ with $h_0(z_0) = 1$. Now $z_0y_0 = 0$, so $h_0(y_0) = h_0(z_0)h_0(y_0) = h_0(z_0y_0) = h_0(0) = 0$; also $1 = h_0(z_0) = h_0(x_0+x_0y_0) = h_0(x_0)$. This proves that B is isomorphic to a Boolean ring of subsets of H.

It remains to prove that H may be topologized in such a way that it becomes a totally disconnected compact Hausdorff space in which the sets $H(x)$, $x \in B$, are precisely the collection of subsets of H which are both open and closed. We have seen above that $H(x) \cap H(y) = H(xy)$ and $H = H(e)$, so by Lemma 4.7 the collection $\{H(x) | x \in B\}$ is a base for a topology. Since $H(x)' = H(e+x)$, each set in the base is both open and closed, so H is totally disconnected. To prove that H is compact in this topology we will use Lemma 5.6. Since each closed set in H is the intersection of sets in $\{H(x) | x \in B\}$, it is sufficient to prove that if A_1 is a subset of B such that $\cap_{i=1}^{n} H(x_i) \neq \phi$ for each finite set $\{x_1, \ldots, x_n\} \subseteq A_1$, then $\bigcap_{x \in A_1} H(x) \neq \phi$. If B_1 is the set of all finite products of elements in A_1, then B_1 evidently satisfies conditions (a) and (b) above. Hence there exists an $h_1 \in H$ with $h_1(x) = 1$, $x \in A_1$, so that h_1 is in each $H(x)$, $x \in A_1$. Therefore H is compact.

Finally, if G is any set in H which is both open and closed in H, then since G is open we have $G = \cup_{\alpha} H(x_{\alpha})$. Since G is compact, a finite covering $G = H(x_1) \cup \ldots \cup H(x_n) = H((x_1' \ldots x_n')')$ can be extracted. Thus every open and closed set in H has the form $H(x)$ for some $x \in B$. This completes the proof of the theorem. Q.E.D.

It is convenient to give another formulation of this result in terms of order and lattice properties. A partially ordered set L is said to be a *lattice* if every pair x, y of elements in L has a least upper bound (2.3) and a greatest lower bound, denoted by $x \vee y$ and $x \wedge y$, respectively. The lattice L has a *unit* if there exists an element 1 such that $x \leqq 1$, $x \in L$, and a *zero* if there exists an element 0 such that $0 \leqq x$, $x \in L$. The lattice L is called *distributive* if

$$x \wedge (y \vee z) = (x \wedge y) \vee (x \wedge z), \qquad x, y, z \in L$$

and *complemented* if for every x in L there exists an x' in L such that

$$x \vee x' = 1, \qquad x \wedge x' = 0.$$

A lattice L is said to be *complete* if every subset bounded above has a least upper bound, or equivalently, every subset bounded below has a greatest lower bound. It is said to be *σ-complete* if this is true for denumerable subsets of L. A *Boolean algebra* is a lattice with unit and zero which is distributive and complemented.

Let B be a Boolean algebra, and define multiplication and addition as

$$xy = x \wedge y, \qquad x+y = (x \wedge y') \vee (x' \wedge y).$$

Then it may be verified that with these operations, B is a Boolean ring with 1 as unit. On the other hand, if B is a Boolean ring with unit denoted by 1, then if $x \leqq y$ is defined to mean $x = xy$, and $x' = 1+x$ then B is a Boolean algebra and

$$x \vee y = x+y+xy, \qquad x \wedge y = xy.$$

Thus the concepts of Boolean algebra and Boolean ring with unit are equivalent.

If B and C are Boolean algebras and $h : B \to C$, then h is said to be a *homomorphism*, or a *Boolean algebra homomorphism*, if

$$h(x \wedge y) = h(x) \wedge h(y), \qquad h(x \vee y) = h(x) \vee h(y), \qquad h(x') = h(x)'.$$

If h is one-to-one, it is called an *isomorphism*. If h is an isomorphism and $h(B) = C$, then we say that B is *isomorphic with* C, or that B and C are *isomorphic*. From the preceding paragraph, it will be seen that if h is a homomorphism of a Boolean algebra B, and if B is regarded as a Boolean ring with unit, then h is a ring homomorphism. The converse is also true.

As examples of Boolean algebras we mention the class $\Phi_2 = \{0, 1\}$, or the collection of all subsets of a given set where $\leqq$ is taken to be set inclusion, and $\wedge$, $\vee$ are taken as intersection and union, respectively.

In terms of these concepts, we can give the following formulation to the Stone representation theorem: *Every Boolean algebra is isomorphic with the Boolean algebra of all open and closed subsets of a totally disconnected compact Hausdorff space.*

13. Determinants

Let $\mathfrak{X}$ be a finite dimensional linear space, and let $x_1, \ldots, x_n$ be a basis for $\mathfrak{X}$. Suppose that T is a linear mapping of $\mathfrak{X}$ into itself. The coefficients (a_{ij}) in the formula

$$Tx_i = \sum_{j=1}^{n} a_{ij}x_j, \qquad i = 1, \ldots, n,$$

completely determine the linear operator T, and are collectively called the *matrix of T in terms of the basis $x_1, \ldots, x_n$*, or (if the basis is understood) simply the *matrix of T*. If i_k is an integer with $1 \leqq i_k \leqq n$, for $k = 1, \ldots, n$, let $\delta_{i_1, \ldots, i_n}$ be defined to be zero if $i_j = i_k$ for some pair with $j \neq k$; to be $+1$ if an even number of interchanges of adjacent terms brings $i_1, \ldots, i_n$ into the arrangement $1, \ldots, n$; and to be -1 if an odd number of such interchanges brings $i_1, \ldots, i_n$ into the arrangement $1, \ldots, n$. With this notation the number expressed by the sum

$$\sum_{i_1=1}^{n} \cdots \sum_{i_n=1}^{n} \delta_{i_1, \ldots, i_n} a_{i_1, 1}, \ldots, a_{i_n, n}$$

is called the *determinant of the matrix* (a_{ij}). If T is a linear operator, then it can be shown that the determinants of the matrices of T in terms of any two bases are equal, and so we may and shall call their common value the *determinant of* T, denoted by $\det(T)$. The determinant satisfies the important multiplicative relationship $\det(T_1 T_2) = \det(T_1)\det(T_2)$. A linear operator T in $\mathfrak{X}$ with a one-to-one inverse is said to be *non-singular*. An important theorem in the theory of determinants states that *a linear operator in a finite dimensional space is non-singular if and only if its determinant is not zero.*

If (a_{ij}) is an $n \times n$ matrix, then the *cofactor* of the element a_{ij} is the product of $(-1)^{i+j}$ with the $(n-1) \times (n-1)$ determinant obtained by deleting the ith row and the jth column in (a_{ij}). Expressed differently, the cofactor of a_{ij} is obtained by replacing the element a_{ij} by 1 and all the other elements of the ith row and jth column by 0 and calculating the resulting determinant. Denoting the cofactor of a_{ij} by A_{ij}, it may be seen that

$$[*] \qquad \det(a_{ij}) = \sum_{i=1}^{n} a_{ij} A_{ij} = \sum_{j=1}^{n} a_{ij} A_{ij}.$$

The first sum is an expansion of $\det(a_{ij})$ in terms of elements of the jth column and the second is in terms of elements of the ith row. Similarly, if $j \neq k$, then

$$0 = \sum_{i=1}^{n} a_{ij} A_{ik} = \sum_{i=1}^{n} a_{ji} A_{ki}.$$

Cramer's rule asserts that, if T is a nonsingular linear operator with matrix (a_{ij}), then the matrix (b_{ij}) of T^{-1}, relative to the same basis, may be calculated by the formula $b_{ij} = A_{ji}/\det(a_{ij})$.

In the second volume we will need the Laplace expansion for a determinant. Let (a_{ij}) be an $n \times n$ matrix, let p be an integer with

$1 \leqq p < n$, and let $i_1, \ldots, i_p$ and $j_1, \ldots, j_p$ be sets of indices with $1 \leqq i_1 < i_2 < \ldots < i_p \leqq n$ and $1 \leqq j_1 < j_2 < \ldots < j_p \leqq n$. Let $b(i_1, \ldots, i_p; j_1, \ldots, j_p)$ denote the $p \times p$ submatrix obtained from (a_{ij}) by retaining only the elements of the $i_1, \ldots, i_p$th rows and the $j_1, \ldots, j_p$th columns, and let $B(i_1, \ldots, i_p; j_1, \ldots, j_p)$ denote the determinant of $b(i_1, \ldots, i_p; j_1, \ldots, j_p)$. Then the *cofactor* of this submatrix is the product of $(-1)^{i_1+\ldots+i_p+j_1+\ldots+j_p}$ with the determinant of the $(n-p) \times (n-p)$ matrix obtained by suppressing the $i_1, \ldots, i_p$th rows and the $j_1, \ldots, j_p$th columns. Denoting this cofactor by $C(i_1, \ldots, i_p; j_1, \ldots, j_p)$, then the *Laplace expansion* of $\det(a_{ij})$ in terms of the elements of the $i_1, \ldots, i_p$th rows is given by the formula

$$\det(a_{ij}) = \sum_{i_1, \ldots, i_p} B(i_1, \ldots, i_p; j_1, \ldots, j_p) \times C(i_1, \ldots, i_p; j_1, \ldots, j_p),$$

where the summation is extended over all sets of p indices $1 \leqq j_1 < \ldots < j_p \leqq n$. Similarly, $\det(a_{ij})$ may be calculated in terms of elements of the $j_1, \ldots, j_p$th columns by summing over all sets of p indices $1 \leqq i_1 < \ldots < i_p \leqq n$. In case $p = 1$ the Laplace expansion reduces to the expansion by row or column given as [*] above.

14. Exercises

1 Show that there is a correspondence between non-zero linear functionals f on a vector space $\mathfrak{X}$ and subspaces $\mathfrak{M}$ such that $\mathfrak{X}/\mathfrak{M}$ is one-dimensional. This correspondence is defined by

$$\mathfrak{M} = \{x | f(x) = 0\}.$$

How can f be defined in terms of $\mathfrak{M}$? What class of f's corresponds to a single $\mathfrak{M}$?

2 Supply the details of the proof that every vector space has a Hamel basis. Prove that any two bases have the same cardinality. (The finite and infinite dimensional cases are to be handled separately. For the infinite dimensional use the theorem of Bernstein which states: If A and B are arbitrary sets, if the cardinality of A is at most equal to the cardinality of B and if the cardinality of B is at most equal to the cardinality of A, then these two sets have the same cardinality.)

3 If $\mathfrak{X}$ is a vector space over Φ, and Φ' is a subfield of Φ, show that $\mathfrak{X}$ can be regarded as a vector space over Φ'. What is the relation between the two corresponding values for the dimension of $\mathfrak{X}$?

4 The linear space $\mathfrak{X}$ is the direct sum of the subspaces $\mathfrak{M}_i$, $i = 1, \ldots, n$, if and only if there are projections E_i in $\mathfrak{X}$ with $E_iE_j=0$, $i \neq j$, $I = E_1+\ldots+E_n$, and $E_i\mathfrak{X} = \mathfrak{M}_i$.

5. Let T be a linear operator in a complex linear vector space. Let P, Q, R be polynomials with complex coefficients such that $P(\lambda)Q(\lambda) = R(\lambda)$ for all complex numbers λ. Show that $P(T)Q(T) = R(T)$.

6 The family $\mathscr{E}$ of projections in a linear space may be partially ordered by defining $A \leqq B$ to mean that $AB = BA = A$. Show that $(\mathscr{E}, \leqq)$ is a partially ordered set. For commuting projections A and B let $A \wedge B = AB$ and $A \vee B = A+B-AB$. Show that $A \wedge B$ is a projection which is the greatest lower bound of A and B and that its range is the intersection of the ranges of A and B. Show that $A \vee B$ is a projection which is the least upper bound of A and B and that its range is the linear manifold spanned by the ranges of A and B.

7 Let the Boolean ring $\Phi_2 = \{0, 1\}$ be topologized by taking all subsets to be open. For each x in a Boolean ring B with unit, let $\Phi_2(x) = \Phi_2$ and $P = \underset{x \in B}{P}\Phi_2(x)$. Let H be the family of all non-zero homomorphisms of B into Φ_2, and regard H as a subspace of P. Show that P is a totally disconnected compact Hausdorff space, and that H is a closed subset of P.

15. References

Since the discussion of many of the topics mentioned in this chapter is not complete, we cite here some texts to which the reader may wish to refer.

Set theory and logic. Gödel [1], Hausdorff [1, 2], Kamke [1, 2], Rosenbloom [1], Rosser [1], Wilder [1].

Topology. Alexandroff and Hopf [1; Chap. I], Bourbaki [5], Hausdorff [1, 2], Kelley [5], Lefschetz [1; Chap. I], Pontrjagin [1; Chap. 2].

Real variables. Carathéodory [1, 2], L. M. Graves [2], Hahn [4], Hausdorff [1, 2].

Complex variables. Ahlfors [1], Bieberbach [1], Knopp [1], Titchmarsh [1].

Algebra. Birkhoff and MacLane [1], Halmos [7], Jacobson [1], van der Waerden [1].

Lattice theory and Boolean algebras. Birkhoff [3], Stone [1, 9].

Determinants. Birkhoff and MacLane [1; Chap. X], Dresden [1; Chap. I], Kowalewski [1], and Veblen [1; Chap. I].

The treatment of set theory in Sections 1—2 is on the intuitive level; the reader who wishes to approach this theory axiomatically should refer to Gödel [1]. We do not explicitly mention Zermelo's "axiom of choice" (cf. Zermelo [1, 2]), since we do not mention any of the other axioms of logic or set theory. The reader will observe, however, that the use of this axiom is made in the proof of the Hausdorff maximality principle (Theorem 2.6) — it will be used in the text most frequently via Zorn's lemma (Theorem 2.7). The proof of Theorem 2.5 goes back to Zermelo's [2] second proof of the well-ordering theorem. This paper is also interesting for its polemical discussion of his axiom. The first occurrence of a maximum principle equivalent to the well-ordering principle (as in Theorem 2.6) is in Hausdorff [1; p. 140]. Zorn [1] gave a theorem essentially equivalent to Theorem 2.7. A similar theorem is due to R. L. Moore [1; p. 84]. For proofs of the equivalence of the well-ordering theorem and other theorems see Teichmüller [1] and Rosser [1]. Kelley [3] has proved that the well-ordering theorem is equivalent to Tychonoff's product theorem (Theorem 8.5).

In closing, we observe that while Gödel [2] has proved that if a system of logic is adequate for present day mathematics, then there can be no assurance that it does not contain a contradiction, he has also proved (Gödel [1]) that if the other axioms of set theory are consistent, then they remain consistent when the axiom of choice is added.

CHAPTER II

Three Basic Principles of Linear Analysis

In linear spaces with a suitable topology, one encounters three far-reaching principles concerning continuous linear transformations. These principles and their corollaries will be used repeatedly in the remaining chapters of the text. They furnish the foundation for many of the modern results in such fields of linear analysis as summability theory, the moment problem, ergodic theory, existence of invariant measures, and integration theory. The first of these principles is known as the *principle of uniform boundedness.* It establishes, among other things, that the limit of a sequence of continuous linear operators is continuous. The second is called the *interior mapping principle*; it asserts that continuous linear mappings between certain types of spaces map open sets onto open sets. The third, the *Hahn-Banach theorem,* is concerned with the existence of extensions of a linear functional. The Hahn-Banach theorem is the basis for several existence theorems that will be used frequently in later chapters.

1. The Principle of Uniform Boundedness

Throughout the text, *all linear vector spaces will be over the field of real or the field of complex numbers.* A *real vector space* is one for which the field Φ is the set of real numbers; a *complex vector space* is one for which Φ is the set of complex numbers. When a statement is made about a vector space, with no mention of the field, it is to be understood that the statement applies to both the real and the complex cases.

1 DEFINITION. A group G is said to be a *topological group* if:

(i) G is a Hausdorff space;

(ii) The mapping $(x, y) \to xy^{-1}$ of $G \times G$ into G is continuous.

A linear space $\mathfrak{X}$ is said to be a *linear topological space* if $\mathfrak{X}$ is a commutative topological group under addition, and if the mapping $(\alpha, x) \to \alpha x$ of $\Phi \times \mathfrak{X}$ into $\mathfrak{X}$ is continuous.

2 Lemma. (a) *In a topological group G, any algebraic combination of any number of variables $x_1, \ldots, x_n$ is continuous as a map of $G \times \ldots \times G$ into G.*

(b) *In a linear topological space $\mathfrak{X}$, all linear combinations of any number of scalars $\alpha_1, \ldots, \alpha_n$, and vectors $x_1, \ldots, x_n$, are continuous mappings of $\Phi \times \ldots \times \Phi \times \mathfrak{X} \times \ldots \times \mathfrak{X}$ into $\mathfrak{X}$.*

(c) *In a linear topological space $\mathfrak{X}$ (group G), the mapping(s) $x \to \alpha x (x \to x^{-1}, x \to ax$, or $x \to xa)$ where α is any scalar not zero, is a homeomorphism of $\mathfrak{X}$ onto itself (G onto itself).*

Proof. Statements (a) and (b) are easily proved, by induction, from the basic definitions. Statement (c) simply reflects the fact that the mapping $x \to \alpha x$ $(x \to x^{-1}$, $x \to ax$, or $x \to xa)$ has the inverse $x \to (1/\alpha)x$ $(x \to x^{-1}$, $x \to a^{-1}x$, or $x \to xa^{-1})$. Q.E.D.

3 Lemma. *The closure of a linear manifold in a linear topological space is a linear manifold.*

Proof. Let $\mathfrak{Z}$ be the closure of the linear manifold $\mathfrak{Y}$ in the linear topological space $\mathfrak{X}$. Let α, β be fixed scalars, and define the map

$$\xi : (u, v) \to \alpha u + \beta v$$

of $\mathfrak{X} \times \mathfrak{X}$ into $\mathfrak{X}$. Since $\mathfrak{Y}$ is linear,

$$\mathfrak{Y} \times \mathfrak{Y} \subseteq \xi^{-1}(\mathfrak{Y}) \subseteq \xi^{-1}(\mathfrak{Z}).$$

Lemma 2 shows that ξ is continuous, and thus $\xi^{-1}(\mathfrak{Z})$ is closed. Hence

$$\mathfrak{Z} \times \mathfrak{Z} \subseteq \xi^{-1}(\mathfrak{Z}),$$

which shows that $\mathfrak{Z}$ is linear. Q.E.D.

4 Definition. The subspace spanned by a set B in a linear space $\mathfrak{X}$ will be denoted by sp(B). If $\mathfrak{X}$ is a linear topological space, the closure of sp(B), denoted by $\overline{\text{sp}}(B)$, is called the *closed linear manifold determined by, or spanned by, B*. From Lemma 3, it is seen that $\overline{\text{sp}}(B)$ is a linear space. If $\overline{\text{sp}}(B) = \mathfrak{X}$, the set B is called *fundamental.*

5 Lemma. *The closed linear manifold determined by a denumerable set in a topological linear space is separable.*

PROOF. Let Φ_0 be a denumerable set dense in the scalar field Φ, and let B be a denumerable set in a linear topological space. Then the denumerable set of vectors of the form $\alpha x + \ldots + \beta y$ with $\alpha, \ldots, \beta$ in Φ_0 and $x, \ldots, y$ in B is dense in $\overline{\text{sp}}(B)$. Q.E.D.

6 LEMMA. *If a homomorphism of one topological group into another is continuous anywhere, it is continuous.*

PROOF. Let the homomorphism $f : G \to H$ be continuous at x, and let $y \in G$. If V is a neighborhood of $f(y)$, then, by Lemma 2(c), $Vf(y^{-1}x)$ is a neighborhood of $f(x)$. If U is a neighborhood of x such that $f(U) \subseteq Vf(y^{-1}x)$, then $Ux^{-1}y$ is a neighborhood of y such that $f(Ux^{-1}y) = f(U)f(x^{-1}y) \subseteq Vf(y^{-1}x)f(x^{-1}y) = V$. Therefore, f is continuous everywhere. Q.E.D.

7 DEFINITION. A set B in a linear topological space $\mathfrak{X}$ is *bounded* if, given any neighborhood V of the zero in $\mathfrak{X}$, there exists a positive real number ε such that $\alpha B \subseteq V$ providing $|\alpha| \leqq \varepsilon$.

8 LEMMA. *A compact subset of a linear topological space is bounded.*

PROOF. Let B be a compact set in the linear topological space $\mathfrak{X}$, and let V be a neighborhood of the origin in $\mathfrak{X}$. Since αx is continuous in both variables, there is a $\delta > 0$ and a neighborhood U of the origin in $\mathfrak{X}$ such that $\beta U \subset V$ if $|\beta| < \delta$. Now, since $x/n \to 0$, we have $\mathfrak{X} \subseteq \bigcup_{n=1}^{\infty} nU$, and since B is compact, $B \subset \bigcup_{n=1}^{m} nU$ for some m. Let $\varepsilon = \delta/m$. If $|\alpha| < \varepsilon$, then $|\alpha n| < \delta$ for $n = 1, \ldots, m$, and $\alpha B \subset \bigcup_{n=1}^{m} \alpha n U \subset V$. Q.E.D.

9 COROLLARY. *A convergent sequence in a linear topological space is bounded.*

PROOF. A convergent sequence together with its limit point is compact. Q.E.D.

10 DEFINITION. An *F-space*, or a space of *type F*, is a linear space $\mathfrak{X}$ which is also a metric space with the following properties:

(i) The metric ϱ in $\mathfrak{X}$ is *invariant*, i.e.,

$$\varrho(x, y) = \varrho(x-y, 0);$$

(ii) The mapping $(\alpha, x) \to \alpha x$ of $\Phi \times \mathfrak{X} \to \mathfrak{X}$ is continuous in α for each x, and continuous in x for each α;

(iii) The metric space $\mathfrak{X}$ is complete.

The symbol $|x|$, called the *norm* of x, is written for $\varrho(x, 0)$. In view of the invariance property stated in (i) above, it is readily seen that the properties $\varrho(x, y) \leqq \varrho(x, z)+\varrho(z, y)$, $\varrho(x, y) = 0$ if and only if $x = y$, and $\varrho(x, y) = \varrho(y, x)$, are equivalent to the properties $|x+y| \leqq |x|+|y|$, $|x| = 0$ if and only if $x = 0$, and $|-x| = |x|$, of the norm, respectively. Thus an F-space may also be defined as a linear space on which there is a non-negative function $|x|$ with these latter three properties and where, in addition, the properties (ii) and (iii) hold with reference to the metric function ϱ defined by $\varrho(x, y) = |x-y|$.

In this definition, it is not assumed that the operation $(\alpha, x) \to \alpha x$ of multiplication is a continuous function on the product space $\Phi \times \mathfrak{X}$; so it is not immediately apparent that an F-space is a linear topological space. This fact is established in Theorem 12. The notion of a bounded set in an F-space will be needed. It is defined just as it is in a linear topological space (cf. Definition 7).

The following theorem, a *principle of equi-continuity*, is the basic principle of this section. Because of the form which it assumes in the special case of B-spaces (cf. Section 3), it is known in the literature as the *principle of uniform boundedness.*

11 THEOREM. *For each a in the set A, let T_a be a continuous linear map of an F-space $\mathfrak{X}$ into an F-space $\mathfrak{Y}$.* If, *for each x in $\mathfrak{X}$, the set $\{T_a x | a \in A\}$ is bounded, then* $\lim_{x \to 0} T_a x = 0$ *uniformly for $a \in A$.*

The value of Theorem 11 is that it enables us to pass from two statements, in each of which one of the parameters a and x is fixed, to one in which they can both vary simultaneously. An application of the theorem will illustrate this statement. Let the set A of indices be the set of scalars of absolute value less than one. Let the corresponding maps be $T_\alpha : x \to \alpha x$. By Definition 10(ii), each of these mappings is linear and continuous. Moreover, for each fixed x_0, the set of all αx_0 is bounded. Indeed, if β is a small scalar, and if $|\alpha| < 1$, $\beta\alpha$ is a small scalar, and hence the required boundedness follows from Definition

10(ii). Theorem 11 establishes the existence of a $\delta(\varepsilon) > 0$ depending on ε such that $|\alpha x| < \varepsilon$ for $|\alpha| < 1$, and $|x| < \delta(\varepsilon)$. A moment's reflection shows us that this implies the continuity of the mapping $(\alpha, x) \to \alpha x$. Hence the following statement has been established:

12 THEOREM. *An F-space is a linear topological space.*

We return now to the proof of Theorem 11. Note that it is clearly a consequence of the following lemma, which is concerned with functions not necessarily linear, and which can be regarded as a refinement of Theorem 11. While Theorem 11 will suffice for the purposes of the present chapter, the more general form of the lemma will be used occasionally in later chapters.

13 LEMMA. *For each a in a set A, let V_a be a continuous map of an F-space $\mathfrak{X}$ into an F-space $\mathfrak{Y}$. Let V_a satisfy the following conditions:*

(i) $|V_a(x+y)| \leqq |V_a(x)|+|V_a(y)|, \qquad x, y \in \mathfrak{X};$

(ii) $|\alpha V_a(x)| = |V_a(\alpha x)|, \qquad \alpha \in \Phi,\ \alpha \geqq 0,\ x \in \mathfrak{X}.$

Then, if for each x in $\mathfrak{X}$, the set $\{V_a x | a \in A\}$ is bounded, the limit $\lim_{x\to 0} V_a(x) = 0$ exists uniformly for a in A.

PROOF. For given $\varepsilon > 0$, and for each positive integer k, consider

$$X_k = \left\{x \,\middle|\, \left|\frac{1}{k} V_a(x)\right| + \left|\frac{1}{k} V_a(-x)\right| \leqq \frac{\varepsilon}{2}, \quad a \in A\right\}.$$

Since V_a is continuous, it is seen that each set X_k is closed. Moreover, from our assumption of boundedness, $\bigcup_{k=1}^{\infty} X_k = \mathfrak{X}$. Hence, by the Baire category theorem (I.6.9), some X_{k_0} contains a sphere $S(x_0, \delta)$. That is, if $|x| < \delta$, then

$$\left|\frac{1}{k_0} V_a(x_0+x)\right| \leqq \frac{\varepsilon}{2}.$$

By assumptions (i) and (ii),

$$\left|\frac{1}{k_0} V_a(x)\right| \leqq \left|\frac{1}{k_0} V_a(x_0+x)\right| + \left|\frac{1}{k_0} V_a(-x_0)\right|.$$

Thus

$$\left|V_a\left(\frac{1}{k_0} x\right)\right| = \left|\frac{1}{k_0} V_a(x)\right| \leqq \varepsilon, \quad |x| < \delta, \quad a \in A.$$

Since the mapping $x \to x/k_0$ is a homeomorphism of $\mathfrak{X}$ with itself, by 10(ii), the proof is complete. Q.E.D.

We now are in a position to prove a number of basic results about F-spaces.

14 THEOREM. *A linear mapping of one F-space into another is continuous if and only if it maps bounded sets into bounded sets.*

PROOF. Let $\mathfrak{X}$, $\mathfrak{Y}$ be F-spaces, let $T : \mathfrak{X} \to \mathfrak{Y}$ be linear and continuous, and let $B \subseteq \mathfrak{X}$ be bounded. For every neighborhood V of the zero in Y, there is a neighborhood U of zero in $\mathfrak{X}$ such that $T(U) \subseteq V$. If α is a sufficiently small scalar, then $\alpha B \subseteq U$ and hence $\alpha T(B) = T(\alpha B) \subseteq V$.

Conversely, let T map bounded sets into bounded sets. In view of Lemma 6, it suffices to show that T is continuous at $x = 0$ to prove continuity. Suppose that $\lim_{i\to\infty} x_i = 0$; then $\lim_{i\to\infty} |x_i| = 0$, and there is a sequence of integers k_i such that $\lim_{i\to\infty} k_i = \infty$ while $\lim_{i\to\infty} k_i|x_i| = 0$. Now, $|k_i x_i| = |x_i + \ldots + x_i| \leqq k_i|x_i|$, so that $\lim_{i\to\infty} k_i x_i = 0$. The convergent sequence $\{k_i x_i\}$ is bounded (9 and 12); so by hypothesis the sequence $\{T(k_i x_i)\} = \{k_i T x_i\}$ is bounded. Therefore,

$$\lim_{i\to\infty} T(x_i) = \lim_{i\to\infty} \frac{1}{k_i} T(k_i x_i) = 0. \qquad \text{Q.E.D.}$$

The first corollary below follows from the first part of the proof of Theorem 14; the second from the second part.

15 COROLLARY. *Any continuous linear mapping from one linear topological space to another sends bounded sets into bounded sets.*

16 COROLLARY. *Any linear mapping from one F-space to another which sends sequences converging to zero into bounded sets is continuous.*

The following two convergence theorems have important applications, and will be used frequently.

17 THEOREM. *Let $\{T_n\}$ be a sequence of continuous linear maps of an F-space $\mathfrak{X}$ into an F-space $\mathfrak{Y}$, and let $Tx = \lim_{n\to\infty} T_n x$ exist for each*

x in $\mathfrak{X}$. Then $\lim_{x\to 0} T_n x = 0$ uniformly for $n = 1, 2, \ldots$, and T is a continuous linear map of $\mathfrak{X}$ into $\mathfrak{Y}$.

PROOF. The linearity of T is an immediate consequence of the linearity of the operators T_n. For each x, the sequence $\{T_n x\}$ is convergent, and hence bounded (9 and 12). By Theorem 11, for every $\varepsilon > 0$ there is a $\delta > 0$ such that $|T_n x| < \varepsilon$ when $|x| < \delta$ for $n = 1, 2, \ldots$ Hence $|Tx| \leqq \varepsilon$ for $|x| < \delta$, and the continuity of T follows from Lemma 6. Q.E.D.

18 THEOREM. *Let $T_\alpha : \mathfrak{X} \to \mathfrak{Y}$ be a generalized sequence of continuous linear maps between F-spaces. If $\lim_\alpha T_\alpha x$ exists for each x in a fundamental set, and if for each x in $\mathfrak{X}$ the set $\{T_\alpha x\}$ is bounded, then the limit $Tx = \lim_\alpha T_\alpha x$ exists for each x in $\mathfrak{X}$, and is continuous and linear.*

PROOF. Since T_α is linear, and $T_\alpha x$ converges for x in a fundamental set, it also converges for x in a dense set D. By Theorem 11, for every $\varepsilon > 0$ there is a $\delta > 0$ such that for all α we have $|T_\alpha z| < \varepsilon$ when $|z| < \delta$. Now, for an arbitrary $x \in \mathfrak{X}$, there is a $y \in D$ with $|x-y| < \delta$, and an $\alpha(\varepsilon)$ such that $|T_\alpha y - T_\beta y| < \varepsilon$ for $\alpha, \beta \geqq \alpha(\varepsilon)$. Therefore,

$$|T_\alpha x - T_\beta x| \leqq |T_\alpha(x-y)| + |T_\beta(y-x)| + |T_\alpha y - T_\beta y| < 3\varepsilon, \quad \alpha, \beta \geqq \alpha(\varepsilon).$$

Since $\mathfrak{Y}$ is complete, Lemma I.7.5 shows that $\lim_\alpha T_\alpha x$ exists for each $x \in \mathfrak{X}$. It is evident that T is linear. Theorem 11 and Lemma I.7.6 show that

$$\lim_{x\to 0} Tx = \lim_{x\to 0} \lim_{\alpha} T_\alpha x = \lim_{\alpha} \lim_{x\to 0} T_\alpha x = 0.$$

By Lemma 6, T is everywhere continuous. Q.E.D.

2. The Interior Mapping Principle

The interior mapping principle is stated in the following theorem.

1 THEOREM. *Under a continuous linear map of one F-space onto all of another, the image of every open set is open.*

PROOF. Let $\mathfrak{X}$, $\mathfrak{Y}$ be F-spaces, let $T : \mathfrak{X} \to \mathfrak{Y}$ be linear and continuous, and let $T\mathfrak{X} = \mathfrak{Y}$. It will first be shown that the closure $\overline{TG}$

of the image of any neighborhood G of the element 0 in $\mathfrak{X}$ contains a neighborhood of the element 0 in $\mathfrak{Y}$. Since $a-b$ is a continuous function of a and b, there is a neighborhood M of 0 such that $M-M \subseteq G$. For every $x \in \mathfrak{X}$, $x/n \to 0$, and so $x \in nM$ for large n. Thus

$$\mathfrak{X} = \bigcup_{n=1}^{\infty} nM, \qquad \mathfrak{Y} = T\mathfrak{X} = \bigcup_{n=1}^{\infty} nTM,$$

and, by the Baire category theorem (I.6.9), one of the sets $\overline{nTM}$ contains a non-void open set. Since the map $y \to ny$ is a homeomorphism in $\mathfrak{Y}$, $\overline{TM}$ contains a non-void open set V. Thus,

$$\overline{TG} \supseteq \overline{TM-TM} \supseteq \overline{TM}-\overline{TM} \supseteq V-V.$$

Since a map of the form $y \to a-y$ is a homeomorphism, the set $a-V$ is open. Since the set $V-V = \bigcup_{a \in V} (a-V)$ is the union of open sets, it is open, contains 0, and hence is a neighborhood of 0. Thus the closure of the image of a neighborhood of the origin contains a neighborhood of the origin.

For any $\varepsilon > 0$, let X_ε, Y_ε be the spheres in $\mathfrak{X}$, $\mathfrak{Y}$, respectively, with centers at their origins, and with radii ε. Let $\varepsilon_0 > 0$ be arbitrary, and let $\varepsilon_i > 0$ with $\sum_{i=1}^{\infty} \varepsilon_i < \varepsilon_0$. Then, according to the result stated in the preceding paragraph, there is a sequence $\{\eta_i,\ i = 0, 1, \ldots\}$ with $\eta_i > 0$, $\eta_i \to 0$, and such that

$$\text{(a)} \qquad \overline{TX_{\varepsilon_i}} \supset Y_{\eta_i}, \qquad i = 0, 1, \ldots.$$

Let $y \in Y_{\eta_0}$. It will be shown that there is an $x \in X_{2\varepsilon_0}$ such that $Tx = y$. From (a), with $i = 0$, it is seen that there is an $x_0 \in X_{\varepsilon_0}$ such that $|y-Tx_0| < \eta_1$. Since $y-Tx_0 \in Y_{\eta_1}$, from (a), with $i = 1$, there is an $x_1 \in X_{\varepsilon_1}$ with $|y-Tx_0-Tx_1| < \eta_2$. Continuing in this manner, a sequence $\{x_n\}$ may be defined for which $x_n \in X_{\varepsilon_n}$, and

$$\text{(b)} \qquad |y-T(\sum_{i=0}^{n} x_i)| < \eta_{n+1}, \qquad n = 0, 1, \ldots.$$

Let $z_m = x_0+\ldots+x_m$, so that for $m > n$, $|z_m-z_n| = |x_{n+1}+\ldots+x_m| < \varepsilon_{n+1}+\ldots+\varepsilon_m$. This shows that $\{z_n\}$ is a Cauchy sequence, and that the series $x_0+x_1+\ldots$ converges to a point x with

$$|x| = \lim_{n\to\infty} |z_n| \leqq \lim_{n\to\infty} (\varepsilon_0+\varepsilon_1+\ldots+\varepsilon_n) < 2\varepsilon_0.$$

Since T is continuous, it is seen from (b) that $y=Tx$. Thus it has been shown that an arbitrary sphere $X_{2\varepsilon_0}$ about the origin in $\mathfrak{X}$ maps onto a set $TX_{2\varepsilon_0}$ which contains a sphere Y_{η_0} about the origin in $\mathfrak{Y}$. Hence the image under T of a neighborhood of the origin in $\mathfrak{X}$ contains a neighborhood of the origin in $\mathfrak{Y}$.

Now let G be a non-void open set in $\mathfrak{X}$, and let N be a neighborhood of 0 in $\mathfrak{X}$ such that $x+N \subset G$. Let M be a neighborhood of 0 in $\mathfrak{Y}$ such that $TN \supseteq M$. Then

$$TG \supseteq T(x+N) = Tx+TN \supseteq Tx+M,$$

which shows that TG contains a neighborhood of every one of its points. Q.E.D.

2 THEOREM. *A continuous linear one-to-one map of one F-space onto all of another has a continuous linear inverse.*

PROOF. Let $\mathfrak{X}$, $\mathfrak{Y}$ be F-spaces and T a continuous linear one-to-one map with $T\mathfrak{X} = \mathfrak{Y}$. Since $(T^{-1})^{-1} = T$ maps open sets onto open sets (Theorem 1), the operator T^{-1} is continuous (I.4.15). Let $y_1, y_2 \in \mathfrak{Y}$, $x_1, x_2 \in \mathfrak{X}$, $Tx_1 = y_1$, $Tx_2 = y_2$, and $\alpha \in \Phi$. Then,

$$T(x_1+x_2) = Tx_1+Tx_2 = y_1+y_2, \qquad T\alpha x_1 = \alpha Tx_1 = \alpha y_1,$$

so that

$$T^{-1}(y_1+y_2) = x_1+x_2 = T^{-1}y_1+T^{-1}y_2,$$

and $T^{-1}(\alpha y_1) = \alpha x_1 = \alpha T^{-1}y_1$. These equations show that T^{-1} is linear. Q.E.D.

3 DEFINITION. Let T be a linear map whose domain $\mathfrak{D}(T)$ is a linear manifold in an F-space $\mathfrak{X}$, and whose range lies in an F-space $\mathfrak{Y}$. The *graph* of T is the set of all points in the product space $\mathfrak{X}\times\mathfrak{Y}$ of the form $[x, Tx]$ with $x \in \mathfrak{D}(T)$. The operator T is said to be *closed* if its graph is closed in the product space $\mathfrak{X}\times\mathfrak{Y}$. An equivalent statement is as follows: The operator T is closed if $x_n \in \mathfrak{D}(T)$, $x_n \to x$, $Tx_n \to y$ imply that $x \in \mathfrak{D}(T)$ and $Tx = y$.

➔ 4 THEOREM. (*Closed graph theorem*) *A closed linear map defined on all of an F-space, and with values in an F-space, is continuous.*

PROOF. Note first that the product $\mathfrak{X}\times\mathfrak{Y}$ of two F-spaces is an F-space, where the distance between two points, $[x, y]$ and $[x', y']$, is defined as $|x-x'|+|y-y'|$. The graph $\mathfrak{G}$ of T is a closed linear manifold in this product space; hence it is a complete metric space (I.6.7). Thus $\mathfrak{G}$ is an F-space. The map $pr_{\mathfrak{X}} : [x, Tx] \rightarrow x$ of $\mathfrak{G}$ onto $\mathfrak{X}$ is one-to-one, linear, and continuous (I.8.3). Hence, by Theorem 2, its inverse $pr_{\mathfrak{X}}^{-1}$ is continuous. Thus $T = pr_{\mathfrak{Y}} pr_{\mathfrak{X}}^{-1}$ is continuous (I.4.17). Q.E.D.

5 THEOREM. *If a linear space is an F-space under each of two metrics, and if one of the corresponding topologies contains the other, the two topologies are equal.*

PROOF. Let τ_1, τ_2 be metric topologies in the linear space $\mathfrak{X}$ for which the spaces $\mathfrak{X}_1 = (\mathfrak{X}, \tau_1)$, $\mathfrak{X}_2 = (\mathfrak{X}, \tau_2)$ are F-spaces. If $\tau_1 \subseteqq \tau_2$, then the one-to-one linear map $x \rightarrow x$ of $\mathfrak{X}_2$ onto $\mathfrak{X}_1$ is continuous. By Theorem 2, it is a homeomorphism, and so $\tau_1 = \tau_2$. Q.E.D.

6 DEFINITION. A family F of functions which map one vector space $\mathfrak{X}$ into another vector space $\mathfrak{Y}$ is called *total* if $x = 0$ is the only vector in $\mathfrak{X}$ for which $f(x) = 0$ for all f in F.

7 THEOREM. *Let $\mathfrak{X}$, $\mathfrak{Y}$,* and *$\mathfrak{W}$ be F-spaces and let F be a total family of continuous linear maps on $\mathfrak{X}$ to $\mathfrak{Y}$. Let T be a linear map from $\mathfrak{W}$ to $\mathfrak{X}$ such that fT is continuous for every f in F. Then T is continuous.*

PROOF. We shall show that T is closed, and apply Theorem 4. Let $\lim_{n\rightarrow\infty} w_n = w$, and let $\lim_{n\rightarrow\infty} Tw_n = x$. Then $\lim_{n\rightarrow\infty} f(Tw_n) = f(x)$ for each $f \in F$, since each $f \in F$ is continuous. On the other hand, $\lim_{n\rightarrow\infty} f(Tw_n) = f(Tw)$, since each of the functions fT is continuous. Therefore,

$$f(Tw) = f(x) \qquad \text{for } f \in F,$$

and, since F is total, $Tw = x$. This proves that T is closed, and Theorem 4 gives the desired conclusion. Q.E.D.

3. The Hahn-Banach Theorem

A number of important F-spaces are listed at the beginning of Chapter IV. The list includes spaces of continuous functions, functions of bounded variation, almost periodic functions, integrable functions, etc. Most of these spaces have a property not enjoyed by F-spaces in general: namely, the identity $|\alpha x| = |\alpha||x|$, which is satisfied by every scalar α and every vector x. An investigation into the consequences of this identity is the chief purpose of the present section.

1 DEFINITION. A linear space $\mathfrak{X}$ is a *normed linear space*, or a *normed space*, if to each $x \in \mathfrak{X}$ corresponds a real number $|x|$ called the *norm* of x which satisfies the conditions:

(i) $|0| = 0; \quad |x| > 0, \quad x \neq 0;$
(ii) $|x+y| \leqq |x|+|y|, \quad x, y \in \mathfrak{X};$
(iii) $|\alpha x| = |\alpha||x|, \quad \alpha \in \Phi, \quad x \in \mathfrak{X}.$

The zero element is called the *origin* of $\mathfrak{X}$; the *closed unit sphere* is the set $\{x||x| \leqq 1\}$.

The properties (i), (ii), and (iii) show that the function ϱ, defined by $\varrho(x, y) = |x-y|$, is an invariant metric in $\mathfrak{X}$. The metric topology in a normed linear space is sometimes called its *norm* or *strong* topology.

2 DEFINITION. A *complete normed linear space*, a space of *type B*, a *B-space*, or a *Banach space*, is a normed linear space which is complete in its norm topology.

The following is clearly an equivalent definition:
A B-space is an F-space in which the identity $|\alpha x| = |\alpha||x|$ *is satisfied.*

3 LEMMA. *A set B in a normed linear space is bounded if and only if* $\sup_{x \in B} |x| < \infty$.

PROOF. A neighborhood V of 0 contains an η-neighborhood $S_\eta = \{x||x| < \eta\}$ of 0. If $a = \sup_{x \in B} |x|$ is finite, and $\varepsilon = \eta/2a$, then $\alpha B \subseteq V$ when $|\alpha| \leqq \varepsilon$, which proves that B is bounded (cf. 1.7). Conversely, if B is bounded, there is an $\varepsilon > 0$ such that αB is contained in the unit sphere $S_1 = \{x||x| < 1\}$ for all $|\alpha| \leqq \varepsilon$. Thus for $x \in B$, $\varepsilon|x| = |\varepsilon x| < 1$; so $|x| < 1/\varepsilon$. Q.E.D.

→ 4 LEMMA. *For a linear map T between normed linear spaces the following statements are equivalent:*

(i) *T is continuous;*
(ii) *T is continuous at some point;*
(iii) *the supremum* $\sup_{|x| \leqq 1} |Tx|$ *is finite;*
(iv) *for some scalar M,* $|Tx| \leqq M|x|$ *for all x.*

PROOF. The equivalence of (i) and (ii) was proved in Lemma 1.6. If T is continuous at 0, there is an $\varepsilon > 0$ such that $|Tx| < 1$ if $|x| < \varepsilon$. For an arbitrary $x \neq 0$, the vector $y = (\varepsilon x)/2|x|$ has norm $|y| < \varepsilon$; so,

$$\frac{\varepsilon}{2|x|}\,|Tx| = |Ty| < 1, \qquad |Tx| < \frac{2}{\varepsilon}\,|x|.$$

This shows that (i) implies (iv). Statement (iv) clearly implies the continuity of T at 0; so (iv) implies (ii). This (i), (ii), and (iv) are equivalent. If $M = \sup_{|x| \leqq 1} |Tx|$ is finite, then for an arbitrary $x \neq 0$,

$$|Tx| = |x| \left| T\left(\frac{x}{|x|}\right) \right| \leqq M|x|.$$

This shows that (iii) implies (iv). It is obvious that (iv) implies (iii). Q.E.D.

5 DEFINITION. The *bound* or *norm* of a linear map T between normed linear spaces is the $\sup_{|x| \leqq 1} |Tx|$, denoted by $|T|$. The operator T is said to be *bounded* if $|T| < \infty$.

Thus, according to Lemma 4, *a linear map between normed linear spaces is continuous if and only if it is bounded.* This fact will be used frequently, and the terms "bounded" and "continuous" when applied to linear operators will be used interchangeably without reference to Lemma 4. Another frequently used consequence of Definition 5 is the inequality $|AB| \leqq |A||B|$, satisfied by linear maps between normed spaces, provided the domain of A contains the range of B.

→ 6 THEOREM. *Let $\mathfrak{X}$, $\mathfrak{Y}$ be B-spaces, and $\{T_n\}$ a sequence of bounded linear operators on $\mathfrak{X}$ to $\mathfrak{Y}$. Then the limit $Tx = \lim_{n \to \infty} T_n x$ exists for every x in $\mathfrak{X}$ if and only if*

(i) *the limit Tx exists for every x in a fundamental set, and*
(ii) *for each x in $\mathfrak{X}$ the supremum $\sup_n |T_n x| < \infty$.*

When the limit Tx exists for each x in $\mathfrak{X}$, the operator T is bounded, and

$$|T| \leqq \liminf_{n \to \infty} |T_n| \leqq \sup_n |T_n| < \infty.$$

PROOF. If Tx exists for each x, then, since a convergent sequence is bounded (1.9), statement (ii) follows from Lemma 3. Conversely, if

(i) and (ii) are satisfied, Lemma 3 shows that the hypotheses of Theorem 1.18 are satisfied. From that theorem it is seen that Tx exists for each x, and that T is continuous. Thus Lemma 4 shows that T is bounded. Now, when Tx exists,

$$|Tx| = \lim_{n\to\infty} |T_n x| \leqq \liminf_{n\to\infty} |T_n||x|,$$

and thus

$$|T| \leqq \liminf_{n\to\infty} |T_n|.$$

Finally, to verify that $\sup_n |T_n| < \infty$, if T is everywhere defined, we apply Theorem 1.11. According to that theorem, there is a $\delta > 0$ such that for all $n = 1, 2, \ldots, |T_n x| < 1$ if $|x| < \delta$. Hence $|T_n| \leqq 1/\delta$ for $n = 1, 2, \ldots$. Q.E.D.

7 Definition. If $\mathfrak{X}$ and $\mathfrak{Y}$ are linear topological spaces, the symbol $B(\mathfrak{X}, \mathfrak{Y})$ will be used for the linear space of all linear continuous maps of $\mathfrak{X}$ into $\mathfrak{Y}$. The symbol $B(\mathfrak{X})$ will be written for $B(\mathfrak{X}, \mathfrak{X})$, and $\mathfrak{X}^*$ for $B(\mathfrak{X}, \Phi)$. The linear space $\mathfrak{X}^*$ is called the *conjugate space, adjoint space,* or *dual space* of $\mathfrak{X}$. Thus the elements of $\mathfrak{X}^*$ are the continuous linear functionals on $\mathfrak{X}$.

8 Lemma. *If $\mathfrak{X}$ and $\mathfrak{Y}$ are normed linear spaces, and if $\mathfrak{Y}$ is complete, then the linear space $B(\mathfrak{X}, \mathfrak{Y})$, with the norm given in Definition 5, is a B-space.*

Proof. It is clear that $|T| = 0$ if and only if $T = 0$, and that $|\alpha T| = |\alpha||T|$. Since

$$|(T+U)x| \leqq |Tx|+|Ux| \leqq (|T|+|U|)|x|,$$

it follows that

$$|T+U| \leqq |T|+|U|.$$

To prove completeness, let $|T_n - T_m| < \varepsilon$ for $n, m \geqq n(\varepsilon)$. Then $Tx = \lim_{n\to\infty} T_n x$ exists for each x, and

$$|Tx - T_n x| \leqq |Tx - T_m x| + |T_m - T_n||x|.$$

Since the left side of this inequality is independent of m, it is seen by letting $m \to \infty$ that $|T - T_n| \leqq \varepsilon$ for $n \geqq n(\varepsilon)$. This shows that $|T| < \infty$, and that $|T - T_n| \to 0$. Q.E.D.

The fact that the field Φ is a B-space yields the following corollary.

9 Corollary. *The conjugate of a normed linear space is a B-space.*

This corollary suggests the question: are there any functionals other than 0 in the conjugate $\mathfrak{X}^*$ of a B-space $\mathfrak{X}$? As will be seen, the answer is in the affirmative; there are, in fact, enough functionals in $\mathfrak{X}^*$ to distinguish between the points of $\mathfrak{X}$. While this is not always the case for F-spaces, there are classes of linear topological spaces other than B-spaces for which it is true. These spaces are discussed in Chapter V. The following theorem is of basic importance in the analysis of questions concerning the existence of continuous linear functionals.

10 Theorem. (*Hahn-Banach*) *Let the real function p on the real linear space $\mathfrak{X}$ satisfy*

$$p(x+y) \leqq p(x)+p(y), \quad p(\alpha x) = \alpha p(x); \qquad \alpha \geqq 0, \quad x, y \in \mathfrak{X}.$$

Let f be a real linear functional on a subspace $\mathfrak{Y}$ of $\mathfrak{X}$ with

$$f(x) \leqq p(x), \qquad x \in \mathfrak{Y}.$$

Then there is a real linear functional F on $\mathfrak{X}$ for which

$$F(x) = f(x), \quad x \in \mathfrak{Y}; \qquad F(x) \leqq p(x), \quad x \in \mathfrak{X}.$$

Proof. Consider the family of all real linear extensions g of f for which the inequality $g(x) \leqq p(x)$ holds for x in the domain of g. The relation $h > g$, defined to mean that h is an extension of g, partially orders this family. Zorn's lemma (I.2.7.) implies the existence of a maximal linear extension g of f for which the inequality $g(x) \leqq p(x)$ holds for x in the domain of g. It remains to be shown that the domain $\mathfrak{Y}_0$ of g is equal to $\mathfrak{X}$.

For the purposes of an indirect proof, assume the existence of a vector y_1 in $\mathfrak{X}$ but not in $\mathfrak{Y}_0$. Every vector in the manifold $\mathfrak{Y}_1$ spanned by $\mathfrak{Y}_0$ and y_1 has a *unique* representation in the form $y+\alpha y_1$ with $y \in \mathfrak{Y}_0$. For any constant c, the function g_1 defined on $\mathfrak{Y}_1$ by the equation $g_1(y+\alpha y_1) = g(y)+\alpha c$ is a proper extension of g. The desired contradiction will be made and the proof finished when it is shown that c may be chosen so that $g_1(x) \leqq p(x)$ for x in $\mathfrak{Y}_1$. Let x, y be arbitrary points of $\mathfrak{Y}_0$; then the inequality

$$g(y)-g(x) = g(y-x) \leqq p(y-x) \leqq p(y+y_1)+p(-y_1-x)$$

yields

$$-p(-y_1-x)-g(x) \leqq p(y+y_1)-g(y).$$

Since the left side of this inequality is independent of y and the right side is independent of x, there is a constant c with

$$\text{(i) } c \leqq p(y+y_1)-g(y), \qquad y \in \mathfrak{Y}_0,$$
$$\text{(ii) } -p(-y_1-y)-g(y) \leqq c, \qquad y \in \mathfrak{Y}_0.$$

For $x = y+\alpha y_1$ in $\mathfrak{Y}_1$ the inequality

$$g_1(x) = g(y)+\alpha c \leqq p(y+\alpha y_1) = p(x),$$

which holds for $\alpha = 0$ by hypothesis, is obtained for $\alpha > 0$ by replacing y by y/α in (i), and for $\alpha < 0$ by replacing y by y/α in (ii). Q.E.D.

11 THEOREM. *Let $\mathfrak{Y}$ be a subspace of a normed linear space $\mathfrak{X}$. Then to every y^* in $\mathfrak{Y}^*$ corresponds an x^* in $\mathfrak{X}^*$ with*

$$|x^*| = |y^*|; \qquad x^*y = y^*y, \qquad y \in \mathfrak{Y}.$$

PROOF. If $\mathfrak{X}$ is a real space, this follows immediately from Theorem 10, by placing $p(x) = |y^*||x|$, and $f = y^*$ (cf. Lemma 4 and Definition 5). Now suppose that $\mathfrak{X}$ is a normed space over the field of complex numbers. For each y in $\mathfrak{Y}$, let $f_1(y)$, $f_2(y)$ be real numbers defined by the equation

$$y^*y = f_1(y)+if_2(y), \qquad y \in \mathfrak{Y}.$$

Then for α, β real and $x, y \in \mathfrak{Y}$,

$$f_1(\alpha x+\beta y) = \alpha f_1(x)+\beta f_1(y),$$
$$|f_1(y)| \leqq |y^*y| \leqq |y^*||y|.$$

Thus, regarding $\mathfrak{X}$ as a real linear space, and applying Theorem 10, a real linear function F_1 on $\mathfrak{X}$ is obtained for which

$$|F_1| \leqq |y^*|; \qquad F_1(y) = f_1(y), \qquad y \in \mathfrak{Y}.$$

Let the function x^* on the complex linear space $\mathfrak{X}$ be defined by the equation

$$x^*x = F_1(x)-iF_1(ix).$$

It will first be shown that x^* is linear. It is clearly additive, and

$x^*(\alpha x) = \alpha x^*x$ for α real. Also, $x^*(ix) = F_1(ix) - iF_1(-x) = ix^*x$; thus x^* is linear. It will next be shown that x^* is an extension of y^*. For y in $\mathfrak{Y}$,

$$f_1(iy) + if_2(iy) = y^*(iy) = iy^*y = if_1(y) - f_2(y),$$

which shows that $f_2(y) = -f_1(iy)$, and hence that

$$y^*y = f_1(y) - if_1(iy), \qquad y \in \mathfrak{Y}.$$

Thus x^* is an extension of y^*. Finally, let $x^*x = re^{i\theta}$, with $r > 0$ and θ real; then

$$|x^*x| = x^*(e^{-i\theta}x) = F_1(e^{-i\theta}x) \leq |y^*||e^{-i\theta}x| = |y^*||x|,$$

which proves that $|x^*| \leq |y^*|$. On the other hand, since x^* is an extension of y^*, $|x^*| \geq |y^*|$. Thus $|x^*| = |y^*|$. Q.E.D.

12 Lemma. *Let $\mathfrak{Y}$ be a subspace of the normed linear space $\mathfrak{X}$. Let $x \in \mathfrak{X}$ be such that*

$$\inf_{y \in \mathfrak{Y}} |y - x| = d > 0.$$

Then there is a point $x^* \in \mathfrak{X}^*$ with

$$x^*x = 1; \quad |x^*| = 1/d; \quad x^*y = 0, \qquad y \in \mathfrak{Y}.$$

Proof. Since $x \notin \mathfrak{Y}$, every point z in the linear manifold $\mathfrak{Z}$ spanned by $\mathfrak{Y}$ and x is uniquely representable in the form $z = y + \alpha x$ with $\alpha \in \Phi$ and $y \in \mathfrak{Y}$. For this z, let $z^*z = \alpha$. The function z^* is clearly linear on $\mathfrak{Z}$. For $\alpha \neq 0$,

$$|z| = |y + \alpha x| = |\alpha| \left| \frac{y}{\alpha} + x \right| \geq |\alpha| d,$$

and hence $|z^*z| \leq |z|/d$, $|z^*| \leq 1/d$. Let $y_n \in \mathfrak{Y}$ be such that $|x - y_n| \to d$. Then,

$$1 = z^*(x - y_n) \leq |z^*||x - y_n| \to |z^*|d,$$

and $1/d \leq |z^*|$. Thus $|z^*| = 1/d$. Theorem 11 gives the existence of the desired extension x^* of z^*. Q.E.D.

→ 13 Corollary. *Let x be a vector not in the closed subspace $\mathfrak{Y}$ of the normed linear space $\mathfrak{X}$. Then there is a functional x^* in $\mathfrak{X}^*$ with*

$$x^*x = 1; \quad x^*y = 0, \qquad y \in \mathfrak{Y}.$$

14 Corollary. *For every $x \neq 0$ in a normed linear space $\mathfrak{X}$, there is an $x^* \in \mathfrak{X}^*$ with $|x^*| = 1$* and $x^*x = |x|$.

Proof. Apply Lemma 12 with $\mathfrak{Y} = 0$. The x^* required in the present corollary may then be defined as $|x|$ times the x^* whose existence is established in Lemma 12. Q.E.D.

Corollary 14 shows that *there are enough functionals in the conjugate $\mathfrak{X}^*$ of a normed space $\mathfrak{X}$ to distinguish between the points of $\mathfrak{X}$*. The following result is an immediate consequence of Corollary 14.

15 Corollary. *For every x in a normed linear space $\mathfrak{X}$,*

$$|x| = \sup_{x^* \in S^*} |x^*x|,$$

where S^* is the closed unit sphere in the space $\mathfrak{X}^*$ conjugate to $\mathfrak{X}$.

16 Lemma. *If the conjugate $\mathfrak{X}^*$ of a normed linear space $\mathfrak{X}$ is separable, so is $\mathfrak{X}$.*

Proof. Let $\{x_n^*\}$ be dense in $\mathfrak{X}^*$, and let $x_n \in \mathfrak{X}$ be such that $|x_n| \leqq 1$, and $|x_n^* x_n| \geqq |x_n^*|/2$. The set L of finite linear combinations of the elements x_n with rational coefficients is a denumerable set. If L is not dense, there is, by Lemma 12, an $x^* \neq 0$ with $x^*L = 0$. Let $x_{n_i}^* \to x^*$. Since

$$|x^* - x_{n_i}^*| \geqq |(x^* - x_{n_i}^*)x_{n_i}| = |x_{n_i}^* x_{n_i}| \geqq |x_{n_i}^*|/2,$$

we have $x_{n_i}^* \to 0$, $x^* = 0$, a contradiction which proves the lemma. Q.E.D.

There is an important interpretation of Corollary 15. For every x in a normed linear space $\mathfrak{X}$, the scalar x^*x depends linearly and continuously upon x^*, and thus defines a continuous linear functional on $\mathfrak{X}^*$; i.e., each x in $\mathfrak{X}$ determines a unique point $\hat{x}$ in $(\mathfrak{X}^*)^*$ such that $\hat{x}x^* = x^*x$, $x \in \mathfrak{X}$. By definition, $|\hat{x}| = \sup_{|x^*| \leqq 1} |\hat{x}x^*|$, and thus, by Corollary 15, $|\hat{x}| = |x|$. These observations suggest the following definitions.

17 Definition. An *isomorphism* between two normed linear spaces $\mathfrak{X}$ and $\mathfrak{Y}$ is a one-to-one continuous linear map $T : \mathfrak{X} \to \mathfrak{Y}$ with $T\mathfrak{X} = \mathfrak{Y}$. When such an isomorphism exists, the spaces $\mathfrak{X}$ and $\mathfrak{Y}$ are called *equivalent*. An *isometric isomorphism* between two normed linear spaces $\mathfrak{X}$ and $\mathfrak{Y}$ is an isomorphism T between $\mathfrak{X}$ and $\mathfrak{Y}$ for which $|Tx| = |x|$. When such a T exists, the spaces are said to be *isometrically equivalent* or *isometrically isomorphic.*

18 **Definition.** Let $\mathfrak{X}$ be a normed linear space, and $\mathfrak{X}^{**}$ the conjugate of the B-space $\mathfrak{X}^*$. The mapping $\varkappa : x \to \hat{x}$ of $\mathfrak{X}$ into $\mathfrak{X}^{**}$, defined by $\hat{x}x^* = x^*x$ for x^* in $\mathfrak{X}^*$, is called the *natural embedding* of $\mathfrak{X}$ into $\mathfrak{X}^{**}$. The range of $\varkappa$ will be denoted by $\hat{\mathfrak{X}}$.

Thus, according to the remarks preceding Definition 17, the following theorem is a corollary of Corollary 15.

19 **Theorem.** *The natural embedding of a normed linear space $\mathfrak{X}$ into its second conjugate is an isometric isomorphism between $\mathfrak{X}$ and $\hat{\mathfrak{X}}$.*

In view of the property of the natural embedding given in Theorem 19, the map $\varkappa$ is sometimes called the *natural isometric isomorphism of* $\mathfrak{X}$ into $\mathfrak{X}^{**}$

20 **Theorem.** *Let x_α, $\alpha \in A$ be an indexed set of elements in a normed linear space $\mathfrak{X}$. If*

$$\sup_{\alpha \in A} |x^* x_\alpha| < \infty, \qquad x^* \in \mathfrak{X}^*,$$

then

$$\sup_{\alpha \in A} |x_\alpha| < \infty.$$

Proof. The functions $\hat{x}_\alpha$, $\alpha \in A$, satisfy the hypothesis of Theorem 1.11, which is to be applied to the conjugate space $\mathfrak{X}^*$. Then, by Theorem 1.11, there is a $\delta > 0$ such that $|\hat{x}_\alpha x^*| < 1$ if $|x^*| < \delta$. Hence $|\hat{x}_\alpha| \leqq 1/\delta$, $\alpha \in A$; and, by Theorem 19, $|x_\alpha| \leqq 1/\delta$, $\alpha \in A$. Q.E.D.

21 **Corollary.** *Let $\mathfrak{X}$, $\mathfrak{Y}$ be B-spaces, and let T_α, $\alpha \in A$, be bounded linear maps from $\mathfrak{X}$ to $\mathfrak{Y}$. Then the following statements are equivalent:*

(i) $\sup_{\alpha \in A} |T_\alpha| < \infty$;

(ii) $\sup_{\alpha \in A} |T_\alpha x| < \infty, \qquad x \in \mathfrak{X}$;

(iii) $\sup_{\alpha \in A} |y^* T_\alpha x| < \infty, \qquad x \in \mathfrak{X}, \qquad y^* \in \mathfrak{Y}^*$.

Proof. Theorem 20 shows that (iii) implies (ii). Now assume (ii). Lemma 3 shows that for each x the set $\{T_\alpha x | \alpha \in A\}$ is bounded, and Theorem 1.11 gives a $\delta > 0$ such that $|T_\alpha x| < 1$ if $|x| < \delta$. Thus $|T_\alpha| \leqq 1/\delta$, $\alpha \in A$. Q.E.D.

22 **Definition.** A B-space $\mathfrak{X}$ is *reflexive* if the natural embedding $\varkappa$ of Definition 18 maps $\mathfrak{X}$ onto all of $\mathfrak{X}^{**}$.

23 THEOREM. *A closed linear manifold in a reflexive B-space is reflexive.*

PROOF. Let $\mathfrak{Y}$ be a closed linear manifold in the reflexive space $\mathfrak{X}$. Let $\xi : x^* \to y^*$ be defined by the equation $\xi(x^*)y = x^*y$, $y \in \mathfrak{Y}$. Clearly $|\xi(x^*)| \leqq |x^*|$, so that $\xi : \mathfrak{X}^* \to \mathfrak{Y}^*$. Let $\eta : y^{**} \to x^{**}$ be defined by $\eta(y^{**})x^* = y^{**}(\xi(x^*))$, $x^* \in \mathfrak{X}^*$. It is evident that $|\eta(y^{**})x^*| \leqq |y^{**}||\xi(x^*)| \leqq |y^{**}||x^*|$; hence $\eta : \mathfrak{Y}^{**} \to \mathfrak{X}^{**}$. Let $\varkappa : x \to \hat{x}$ be the natural isometric isomorphism of $\mathfrak{X}$ into $\mathfrak{X}^{**}$. Since $\mathfrak{X}$ is reflexive, each $x^{**} = \hat{x}$ for some $x \in \mathfrak{X}$. It will be shown that $\varkappa^{-1}\eta(\mathfrak{Y}^{**}) \subseteqq \mathfrak{Y}$. If $x = \varkappa^{-1}\eta(y^{**}) \notin \mathfrak{Y}$ there is, by Corollary 13, an $x^* \in \mathfrak{X}^*$ with $x^*x \neq 0$, $x^*\mathfrak{Y} = 0$. Since $x^*\mathfrak{Y} = 0$, $\xi(x^*) = 0$. Thus,

$$0 = y^{**}\xi(x^*) = \eta(y^{**})x^* = \hat{x}x^* = x^*x,$$

a contradiction proving that $\varkappa^{-1}\eta(y^{**}) \subseteqq \mathfrak{Y}$. Now let $y_0^{**} \in \mathfrak{Y}^{**}$ and $x_0^{**} = \eta(y_0^{**})$. If $y^* \in \mathfrak{Y}^*$, let $x^* \in \mathfrak{X}^*$ be any extension of y^*. Then $y^* = \xi(x^*)$, and

$$y_0^{**}y^* = \eta(y_0^{**})x^* = x_0^{**}x^* = \hat{x}_0x^* = x^*x_0 = y^*x_0, \quad y^* \in \mathfrak{Y}^*,$$

since $x_0 = \varkappa^{-1}\eta(y_0^{**}) \in \mathfrak{Y}$. This shows that $\mathfrak{Y}$ is reflexive. Q.E.D.

24 COROLLARY. *A B-space is reflexive if and only if its conjugate space is reflexive.*

PROOF. Let $\mathfrak{X}$ be reflexive, and let $\varkappa$ be the natural isometric isomorphism of $\mathfrak{X}$ onto $\mathfrak{X}^{**}$. For an arbitrary point $x^{***} \in (\mathfrak{X}^*)^{**} = (\mathfrak{X}^{**})^*$, the functional $x^* = x^{***}\varkappa$ is in $\mathfrak{X}^*$, and

$$x^{**}x^* = \hat{x}x^* = x^*x = x^{***}\varkappa x = x^{***}\hat{x} = x^{***}x^{**}, \quad x^{**} \in \mathfrak{X}^{**},$$

which shows that $\mathfrak{X}^*$ is reflexive. Conversely, let $\mathfrak{X}^*$ be reflexive. Then $\mathfrak{X}^{**}$ is reflexive, and hence, by Theorem 23, the closed linear manifold $\hat{\mathfrak{X}}$ in $\mathfrak{X}^{**}$ is reflexive. From this it follows that $\mathfrak{X}$, which is isometrically equivalent to $\hat{\mathfrak{X}}$, is reflexive. Q.E.D.

25 DEFINITION. Let $\mathfrak{X}$ be a linear topological space. A generalized sequence $\{x_\alpha\}$ in $\mathfrak{X}$ is said to be *weakly convergent* if there is an x in $\mathfrak{X}$ with $x^*x = \lim_\alpha x^*x_\alpha$ for every $x^* \in \mathfrak{X}^*$. The point x is called a *weak limit* of the generalized sequence, and the generalized sequence $\{x_\alpha\}$ is said to *converge weakly* to x. A set $A \subseteqq \mathfrak{X}$ is said to be *weakly sequentially compact* if every sequence $\{x_n\}$ in A contains a subsequence

which converges weakly to a point in $\mathfrak{X}$. Every sequence $\{x_n\}$ such that $\{x^*x_n\}$ is a Cauchy sequence of scalars for each $x^* \in \mathfrak{X}^*$ is called a *weak Cauchy sequence.* The space $\mathfrak{X}$ is said to be *weakly complete* if every weak Cauchy sequence has a weak limit.

In Chapter V, a topology is introduced in certain linear spaces in such a way that they become Hausdorff spaces in which the notion of convergence of generalized sequences coincides with the notion of weak convergence as just defined.

26 LEMMA. *In a normed linear space, a weakly convergent generalized sequence has a unique limit.*

PROOF. If x and y are both weak limits of a generalized sequence, then for each $x^* \in \mathfrak{X}^*$, $x^*x = x^*y$, $x^*(x-y) = 0$ and $x = y$, by Corollary 14. Q.E.D.

27 LEMMA. *A weakly convergent sequence $\{x_n\}$ of points in a normed linear space is bounded. Its limit x is in the closed linear manifold determined by $\{x_n\}$, and $|x| \leqq \liminf_{n\to\infty} |x_n|$.*

PROOF. If, in Theorem 6, $\mathfrak{X}$, $\mathfrak{Y}$, T_n are replaced by $\mathfrak{X}^*$, Φ, $\hat{x}_n$, it is seen that $|\hat{x}| \leqq \liminf_{n\to\infty} |\hat{x}_n|$. Theorem 19 gives the inequality $|x| \leqq \liminf_{n\to\infty} |x_n|$, and Theorem 20 shows that $\sup_n |x_n| < \infty$. It follows from Corollary 13 that x is in the closed linear manifold determined by $\{x_n\}$. Q.E.D.

28 THEOREM. *A set in a reflexive space is weakly sequentially compact if and only if it is bounded.*

PROOF. Let $\{y_n\}$ be a bounded sequence in the reflexive space $\mathfrak{X}$ with $|y_n| \leqq K$, $n = 1, 2, \ldots$. Let $\mathfrak{Y}$ be the closed linear manifold determined by the sequence $\{y_n\}$. Then $\mathfrak{Y}$ is separable, and, by Theorem 23, reflexive. Thus $\mathfrak{Y} = \mathfrak{Y}^{**}$ is separable, and, by Lemma 16, $\mathfrak{Y}^*$ is separable. Let $\{y_n^*\}$ be dense in $\mathfrak{Y}^*$. Since the sequence $\{y_1^*y_p\}$ is bounded, it has a convergent subsequence $\{y_1^*y_{p_{1,i}}\}$. Since $\{y_2^*y_{p_{1,i}}\}$ is bounded, it has a convergent subsequence $\{y_2^*y_{p_{2,i}}\}$. In this manner one obtains, by induction, a sequence $\{p_{n,i}\}$, which is a subsequence of $\{p_{n-1,i}\}$, for which the sequence $\{y_n^*y_{p_{n,i}}\}$ is convergent. Thus the sequence $\{x_i\}$, defined by $x_i = y_{p_{i,i}}$, has the property that the limit

$\lim_{i\to\infty} y_n^* x_i = \lim_{i\to\infty} \hat{x}_i y_n^*$ exists for each $n = 1, 2, \ldots$. Since $\{y_n^*\}$ is dense in $\mathfrak{Y}^*$, and $|\hat{x}_i| \leq K$, Theorem 6 shows that there is a y^{**} in $\mathfrak{Y}^{**}$ with

$$\lim_i y^* x_i = y^{**} y^*, \qquad y^* \in \mathfrak{Y}^*.$$

Since $\mathfrak{Y}$ is reflexive, there is a point $y \in \mathfrak{Y}$ with $y^* x_i \to y^* y$ for every $y^* \in \mathfrak{Y}^*$. Now, any x^* in $\mathfrak{X}^*$ determines a y^* in $\mathfrak{Y}^*$ such that $x^* y = y^* y$ for y in $\mathfrak{Y}$; thus, since $x_n \in \mathfrak{Y}$, $x^* x_i \to x^* y$ for x^* in $\mathfrak{X}^*$, and $\{x_n\}$ converges weakly to y. Hence bounded sets are weakly sequentially compact. The converse follows from the preceding lemma. Q.E.D.

29 Corollary. *A reflexive space is weakly complete.*

Proof. If $\{x_n\}$ is a sequence in a reflexive space $\mathfrak{X}$ for which $\lim_{n\to\infty} x^* x_n$ exists for each x^* in $\mathfrak{X}^*$, then, by Theorem 20, $\{x_n\}$ is bounded. By Theorem 28, there is a subsequence $\{x_{n_i}\}$ which converges weakly to a point x in $\mathfrak{X}$. Thus,

$$\lim_n x^* x_n = \lim_i x^* x_{n_i} = x^* x, \qquad x^* \in \mathfrak{X}^*,$$

which proves that $\mathfrak{X}$ is weakly complete. Q.E.D.

The reader is referred to Chapter V for additional discussion of reflexive spaces and weak convergence. We conclude this section with a lemma on weak sequential compactness which has useful applications to ergodic theory.

30 Lemma. *Let $\mathfrak{X}$, $\mathfrak{Y}$ be B-spaces and let $\{T_n\}$ be a bounded sequence in $B(\mathfrak{X}, \mathfrak{Y})$. Then the set $\mathfrak{M}$ of all $x \in \mathfrak{X}$ for which $\{T_n x | n = 1, 2, .\}$ is weakly sequentially compact is a closed linear subspace of $\mathfrak{X}$.*

Proof. An obvious argument shows that $\mathfrak{M}$ is a linear space. It will be demonstrated here that $\mathfrak{M}$ is closed. Let $|T_n| \leq K$, $n = 1, 2, \ldots$ Let $x_n \in \mathfrak{M}$, $x_n \to x$. An arbitrary sequence $\{n_i\}$ of integers contains a subsequence $\{n_{1,i}\}$ for which $T_{n_{1,i}} x_1$ converges weakly to some point which will be denoted by y_1. Similarly, there is a point y_2 and a subsequence $\{n_{2,i}\}$ of $\{n_{1,i}\}$ such that $T_{n_{2,i}} x_2 \to y_2$ weakly. Proceeding by induction, a point y_m and a sequence $\{n_{m,i}\}$ are obtained for which $T_{n_{m,i}} x_m \to y_m$ weakly. The sequence $\{m_i\}$ defined by $m_i = n_{i,i}$ is a subsequence of $\{n_i\}$ for which $T_{m_i} x_k \to y_k$ weakly for all $k = 1, 2, \ldots$; the inequality

$$|y_i - y_j| \leqq K|x_i - x_j|, \qquad i, j = 1, 2, \ldots,$$

follows from Lemma 27. Since $\{x_j\}$ is a Cauchy sequence, $\{y_j\}$ is also. Let $y = \lim_{n\to\infty} y_n$. For $x^* \in Y^*$ and $p = 1, 2, \ldots$,

$$|x^*T_{m_i}x - x^*y| \leqq |x^*T_{m_i}x - x^*T_{m_i}x_p| + |x^*T_{m_i}x_p - x^*y_p| + |x^*y_p - x^*y|,$$

and hence

$$\limsup_{i\to\infty} |x^*T_{m_i}x - x^*y| \leqq K|x^*||x - x_p| + |x^*||y - y_p|.$$

By letting $p \to \infty$, it is seen that $T_{m_i}x \to y$ weakly. Thus every sequence $\{n_i\}$ of integers has a subsequence $\{m_i\}$ for which $T_{m_i}x \to y$ weakly. It follows that $T_n x \to y$ weakly. Q.E.D.

4. Exercises

1 A set C in a linear space is said to be *convex* if $\alpha x + (1-\alpha)y \in C$ whenever $x, y \in C$ and $0 \leqq \alpha \leqq 1$. Show that spheres in a B-space are convex. Show that the *closed sphere* $\{x \,|\, |x-y| \leqq \varepsilon\}$ is the closure of the open sphere $\{x \,|\, |x-y| < \varepsilon\}$. Show that the intersection of a decreasing sequence of closed spheres in a B-space is never void.

2 Find a decreasing sequence of non-void, bounded, closed, convex sets in some B-space whose intersection is void.

3 Let $\mathfrak{Y}$ and $\mathfrak{Z}$ be closed linear manifolds in the B-space $\mathfrak{X}$. Suppose that each x in $\mathfrak{X}$ has a unique representation in the form $x = y+z$, with $y \in \mathfrak{Y}$, $z \in \mathfrak{Z}$. Show that there is a constant K, with $|y| \leqq K|x|$, and $|z| \leqq K|x|$, for each x in $\mathfrak{X}$.

4 Let $\mathfrak{X}$, $\mathfrak{Y}$ and $\mathfrak{Z}$ be B-spaces, and let $z = (x, y)$ be a function on $\mathfrak{X} \times \mathfrak{Y}$ to $\mathfrak{Z}$ which is linear in x for each y, and linear in y for each x. Such a function is called *bilinear*. Suppose that, for each z^* in $\mathfrak{Z}^*$, the function $z^*(x, y)$ is continuous in y for each x, and continuous in x for each y. Show that there is a constant K such that

$$|(x, y)| \leqq K|x||y|.$$

5 Let $\mathfrak{Y}$ and $\mathfrak{Z}$ be B-spaces, and let $f : \mathfrak{Y} \to \mathfrak{Z}$ be such that $z^*f \in \mathfrak{Y}^*$ for each $z^* \in \mathfrak{Z}^*$. Show that $f \in B(\mathfrak{Y}, \mathfrak{Z})$.

6 Let $\mathfrak{X}$ be a complex B-space, and let G be an open subset of the complex plane. Let $f : G \to \mathfrak{X}$ be such that x^*f is analytic in G for each $x^* \in \mathfrak{X}^*$. Show that

$$\lim_{\mu\to\xi}\frac{f(\mu)-f(\xi)}{\mu-\xi}$$

exists for each $\xi \in G$.

7 DEFINITION. A sequence $\{x_n\}$ in an F-space $\mathfrak{X}$ is called a *basis* for $\mathfrak{X}$ if to each $x \in \mathfrak{X}$ there corresponds a unique sequence $\{\alpha_i\}$ of scalars such that

$$\lim_{n\to\infty} |x - \sum_{i=1}^{n} \alpha_i x_i| = 0.$$

8 Let $\{x_n\}$ be a basis in the F-space $\mathfrak{X}$, and let $\mathfrak{Y}$ be the vector space of all sequences $y = \{\alpha_i\}$ for which the series $\sum_{i=1}^{\infty} \alpha_i x_i$ converges. Show that with the definition

$$|y| = \sup_n |\sum_{i=1}^{n} \alpha_i x_i|,$$

$\mathfrak{Y}$ is an F-space, and that $\mathfrak{Y}$ is a B-space if $\mathfrak{X}$ is.

9 If $\{x_n\}$ is a basis in the F-space $\mathfrak{X}$, and for $x = \sum_{i=1}^{\infty} \alpha_i x_i$ we define $x_i^*(x) = \alpha_i$, $i = 1, 2, \ldots$, then the linear functional x_i^* is continuous. (Hint: Use the natural mapping between $\mathfrak{X}$ and $\mathfrak{Y}$ of Exercise 8.)

10 Show that no element of a basis in an F-space is in the closed linear manifold determined by the remaining elements of the basis.

11 DEFINITION. A pair of sequences $\{x_i\}$, $x_i \in \mathfrak{X}$, and $\{x_i^*\}$, $x_i^* \in \mathfrak{X}^*$, is called a *biorthogonal system* for a Banach space $\mathfrak{X}$ if $x_i^*(x_j) = \delta_{ij}$.

12 Let $\{x_i\}$, $\{x_i^*\}$ be a biorthogonal system for the B-space $\mathfrak{X}$.

(a) If $\sup_n |\sum_{i=1}^{n} x^*(x_i)x_i^*| < \infty$ for each $x^* \in \mathfrak{X}^*$, the representation

$$x = \sum_{i=1}^{\infty} x_i^*(x)x_i$$

is valid for each x in the closed linear span of the sequence $\{x_i\}$.

(b) If $\sup_n |\sum_{i=1}^{n} x_i^*(x)x_i| < \infty$ for each $x \in \mathfrak{X}$, the representation

$$x^* = \sum_{i=1}^{\infty} x^*(x_i)x_i^*$$

is valid for each x^* in the closed linear span of the sequence $\{x_i^*\}$.

13 Let $\mathfrak{X}$ be a B-space (or an F-space), and let $\mathfrak{Z}$ be a *closed*

linear manifold in $\mathfrak{X}$. Then $\mathfrak{X}/\mathfrak{Z}$ (cf. I.11) is a B-space (or an F-space) with the metric

$$|x+\mathfrak{Z}| = \inf_{z \in \mathfrak{Z}} |x+z|. .$$

(Hint: Given a Cauchy sequence in $\mathfrak{X}/\mathfrak{Z}$, define a subsequence for which $|x_k - x_{k+1} + \mathfrak{Z}| < 2^{-k}$, $k = 1, 2, \ldots$, and show that a Cauchy sequence in $\mathfrak{X}$ can be chosen which maps onto $\{x_k + \mathfrak{Z}\}$.)

14 Let $\mathfrak{X}$ be an F-space or a normed space. (a) The natural homomorphism $f : \mathfrak{X} \to \mathfrak{X}/\mathfrak{Z}$ defined by $f(x) = x + \mathfrak{Z}$ is continuous, maps open sets onto open sets, and has norm $|f| \leqq 1$.

(b) If $\mathfrak{X}/\mathfrak{Z}$ is given a topology actually stronger than the topology defined by the metric given above, then f will no longer be continuous.

(c) The function f maps the open unit sphere of $\mathfrak{X}$ onto the open unit sphere of $\mathfrak{X}/\mathfrak{Z}$.

15 (Banach). Let $\mathfrak{X}$ and $\mathfrak{Y}$ be B-spaces, and let T be a continuous linear mapping of $\mathfrak{X}$ onto all of $\mathfrak{Y}$. Show that there is a number $N > 0$ such that for every sequence $y_n \to y_0$ there is a sequence $x_n \to x_0$ with $|x_n| \leqq N|y_n|$ and $Tx_n = y_n$, $n = 0, 1, \ldots$.

16 Let $\mathfrak{X}$ be a normed linear space which is not assumed to be complete, and let $\mathfrak{Z}$ be a closed subspace of $\mathfrak{X}$. Then, if $\mathfrak{Z}$ and $\mathfrak{X}/\mathfrak{Z}$ are complete, show that $\mathfrak{X}$ is also.

17 Definition. If $\mathfrak{X}$ is a normed linear space, and $Z \subseteqq \mathfrak{X}$, the set $Z^{\perp} = \{x^* | x^* \in \mathfrak{X}^*, x^*(Z) = 0\}$ is called the *annihilator* or *orthogonal complement* of Z.

18 (a) If $\mathfrak{X}$ is a normed linear space, and $\mathfrak{Z}$ a linear manifold in $\mathfrak{X}$, the mapping $x^* + \mathfrak{Z}^{\perp} \to z^*$, where z^* is defined by $z^*z = x^*z$, $z \in \mathfrak{Z}$, is an isometric isomorphism of $\mathfrak{X}^*/\mathfrak{Z}^{\perp}$ onto all of $\mathfrak{Z}^*$.

(b) If $\mathfrak{Z}$ is a closed subspace of a B-space $\mathfrak{X}$, the mapping $x^* \to \bar{x}^*$ where $\bar{x}^*$ is defined by $\bar{x}^*(x+\mathfrak{Z}) = x^*(x)$, is an isometric isomorphism of $\mathfrak{Z}^{\perp}$ onto all of $(\mathfrak{X}/\mathfrak{Z})^*$.

(c) If $\mathfrak{X}$ is a reflexive B-space and $\mathfrak{Z}$ is a closed subspace of $\mathfrak{X}$, show that $\mathfrak{Z}^{\perp\perp} = \mathfrak{X}\mathfrak{Z}$. Is the result true if $\mathfrak{X}$ is not reflexive?

19 If $\mathfrak{X}$ is a reflexive B-space, and $\mathfrak{Z}$ is a closed subspace of $\mathfrak{X}$, show, using Exercise 18, that both $\mathfrak{Z}$ and $\mathfrak{X}/\mathfrak{Z}$ are reflexive.

20 If $\mathfrak{X}$ is a B-space, with $\mathfrak{Z} \subseteqq \mathfrak{X}$, and both $\mathfrak{Z}$ and $\mathfrak{X}/\mathfrak{Z}$ are reflexive, then $\mathfrak{X}$ is reflexive. (Hint: Given any $x_0^{**} \in \mathfrak{X}^{**}$, show that

there exists a $y_0 \in \mathfrak{X}$ such that $x_0^{**}(y^*) = y^*(y_0)$ for all $y^* \in \mathfrak{Z}^{\perp}$.)

21 Let m be the space of all bounded sequences $s = [s_1, s_2, \ldots]$. Show that if the norm of s be defined as

$$|s| = \operatorname*{lub}_{1 \leqq i < \infty} |s_i|,$$

m is a B-space.

22 (Banach limits) Let m_0 be the smallest closed subspace of m (cf. Exercise 21) containing all sequences of the form $t = [s_1, s_2 - s_1, s_3 - s_2, \ldots]$, where $s = [s_1, s_2, \ldots]$ is an element of m. Show that the sequence $e = [1, 1, \ldots]$ is not in m_0, and that there exists a continuous functional x^* on m such that $|x^*| = 1$, $x^*(e) = 1$, $x^*(x) = 0$ if $x \in m_0$. Letting $\operatorname*{LIM}_{n\to\infty} s_n = x^*(s)$, show that

(a) $\operatorname*{LIM}_{n\to\infty} s_n = \operatorname*{LIM}_{n\to\infty} s_{n+1}$;

(b) $\operatorname*{LIM}_{n\to\infty} (\alpha s_n + \beta t_n) = \alpha \operatorname*{LIM}_{n\to\infty} s_n + \beta \operatorname*{LIM}_{n\to\infty} t_n$;

(c) $\operatorname*{LIM}_{n\to\infty} s_n \geqq 0$ if $[s_n]$ is a non-negative sequence;

(d) $\liminf_{n\to\infty} s_n \leqq \operatorname*{LIM}_{n\to\infty} s_n \leqq \limsup_{n\to\infty} s_n$ if $[s_n]$ is a real sequence;

(e) If s_n is a convergent sequence, then

$$\operatorname*{LIM}_{n\to\infty} s_n = \lim_{n\to\infty} s_n.$$

23 Show that there exists a method of assigning a "generalized limit" $\operatorname*{LIM}_{s\to\infty}$ to every bounded complex valued function f of a real variable s in such a way that

(a) $\operatorname*{LIM}_{s\to\infty} f(s) = \operatorname*{LIM}_{s\to\infty} f(s+t)$ for each t;

(b) $\operatorname*{LIM}_{s\to\infty} \{\alpha f(s) + \beta g(s)\} = \alpha \operatorname*{LIM}_{s\to\infty} f(s) + \beta \operatorname*{LIM}_{s\to\infty} g(s)$;

(c) $\operatorname*{LIM}_{s\to\infty} f(s) \geqq 0$ if f is a non-negative function;

(d) $\liminf_{s\to\infty} f(s) \leqq \operatorname*{LIM}_{s\to\infty} f(s) \leqq \limsup_{s\to\infty} f(s)$ if f is real;

(e) $\operatorname*{LIM}_{s\to\infty} f(s) = \lim_{s\to\infty} f(s)$ if the right hand limit exists.

24 If $\mathfrak{Y}$ is a linear manifold which is dense in a B-space $\mathfrak{X}$, show that there is a natural isometric isomorphism between $\mathfrak{X}^*$ and $\mathfrak{Y}^*$.

25 Let $\mathfrak{N}$ be a separable linear manifold in $\mathfrak{X}^*$. Show that there is a separable subspace $\mathfrak{Z} \subseteq \mathfrak{X}$ such that $\mathfrak{N}$ is isometrically isomorphic to a subspace of $\mathfrak{Z}^*$.

26 If $\mathfrak{X}$ is a linear topological space, a linear functional on $\mathfrak{X}$ is continuous if and only if it is bounded on some neighborhood of the origin.

27 Definition. A B-space $\mathfrak{X}$ is *uniformly convex* if $|x_n| = |y_n| = 1$, $|x_n+y_n| \to 2$ implies $|x_n-y_n| \to 0$.

28 If $\mathfrak{X}$ is uniformly convex, $x_n \to x_0$ weakly, and $|x_n| \to |x_0|$, then $x_n \to x_0$ in the metric topology.

29 Let K be a closed convex set in a uniformly convex B-space. Then the norm function $f(x) = |x|$ assumes its minimum exactly once on K.

30 An infinite dimensional F-space never has a denumerable Hamel basis.

The next set of exercises forms a connected group pertaining to the summability theory of divergent series.

31 Let c denote the space of all convergent sequences $s = [s_1, s_2, \ldots]$ of scalars. Show that if the norm of s is defined as

$$|s| = \operatorname*{lub}_{1 \leqq i < \infty} |s_i|,$$

then c is a B-space.

32 Let l_1 denote the space of all sequences $s = [s_0, s_1, s_2, \ldots]$ such that $\sum_{i=0}^{\infty} |s_i| < \infty$. Show that if the norm of s' is defined as

$$|s| = \sum_{i=0}^{\infty} |s_i|,$$

then l_1 is a B-space.

33 For each $s = [s_0, s_1, s_2, \ldots] \in l_1$ and $t = [t_1, t_2, \ldots] \in c$, let

$$(gs)(t) = s_0 \lim_{n\to\infty} t_n + \sum_{n=1}^{\infty} s_n t_n .$$

Show that g is an isometric isomorphism of l_1 onto all of c^*.

34 Let T be a bounded linear map of c into c. Show that there is a double sequence $\{a_{ij}\}$, $i = 1, 2, \ldots$, $j = 0, 1, 2, \ldots$ such that

(a) $t_i = a_{i0} \lim_{j\to\infty} s_j + \sum_{j=1}^{\infty} a_{ij}s_j$, where $T[s_1, s_2, \ldots,] = [t_1, t_2, \ldots]$;

(b) $\operatorname{lub}_{1\leq i<\infty} \sum_{j=0}^{\infty} |a_{ij}| = M < \infty$;

(c) $\lim_{i\to\infty} a_{ij}$ exists for $j = 1, 2, \ldots,$

(d) $\lim_{i\to\infty} \sum_{j=0}^{\infty} a_{ij}$ exists;

(e) $|T| = M$.

Conversely, if $\{a_{ij}\}$ is a double sequence satisfying (b), (c) and (d), then (a) defines a bounded linear mapping T on c to c whose norm $|T| = M$.

35 DEFINITION. Suppose that a matrix (a_{ij}) defines a linear transformation T of c into itself by means of the formula

$$T[s_1, s_2, \ldots] = [t_1, t_2, \ldots] = [\sum_{j=1}^{\infty} a_{1j}s_j, \sum_{j=1}^{\infty} a_{2j}s_j, \ldots].$$

If T preserves limits of sequences (i.e. if $\lim_{i\to\infty} t_i = \lim s_i$ for every $[s_i] \in c$), then the matrix (a_{ij}) is said to define a *regular method of summability.*

36 (Silverman-Toeplitz). Show that a matrix (a_{ij}) defines a regular method of summability if and only if

(a) $\operatorname{lub}_{1\leq i<\infty} \sum_{j=1}^{\infty} |a_{ij}| = M < \infty$;

(b) $\lim_{i\to\infty} a_{ij} = 0$ for $1 \leq j < \infty$;

(c) $\lim_{i\to\infty} \sum_{j=1}^{\infty} a_{ij} = 1$.

37 Let $[p_k]$ be a sequence of positive numbers, and let $P_i = \sum_{k=1}^{i} p_k$. Show that the formula

$$t_i = P_i^{-1} \sum_{j=1}^{i} p_j s_j = \sum_{j=1}^{\infty} a_{ij} s_j$$

defines a regular method of summability if and only if $\Sigma_{k=1}^{\infty} p_k$ diverges. (If all $p_j = 1$, then this is called *Cesàro* $(C.1)$*-summability*, cf. Exercise 39 below.)

38 (Nörlund summability). Let $[p_k]$ be a sequence of positive

numbers, and let $P_i = \sum_{k=1}^{i} p_k$. Show that the formula $t_i = P_i^{-1} \sum_{j=1}^{i} p_{i-j+1} s_j$ defines a regular method of summability if and only if

$$\lim_{k\to\infty} P_k^{-1} p_k = 0.$$

39 (Cesàro (C, α)-summability). Show that the formula

$$t_n = (C_n^{\alpha+1})^{-1} \sum_{k=0}^{n} C_{n-k}^{\alpha} s_k$$

defines a regular method of summability for each complex number α such that $\mathscr{R}(\alpha) > 0$. Here $C_0^\beta = 1$ and $C_m^\beta = \beta(\beta+1)\ldots(\beta+m-1)/m!$ for $m > 0$.

40 Let $\sum_{n=0}^{\infty} p_n z^n$ be the power-series expansion of an entire function. Suppose that $p_n > 0$ for each n. Show that

$$\lim_{n\to\infty} s_n = \lim_{x\to\infty} \frac{\sum_{n=0}^{\infty} p_n s_n x^n}{\sum_{n=0}^{\infty} p_n x^n}$$

whenever the limit on the left exists.

41 Let $\sum_{n=0}^{\infty} p_n z^n$ be the power-series expansion of a function analytic in the circle $|z| < r$. Suppose that $p_n \geqq 0$ for each n and $\lim_{x\to r} \sum_{n=0}^{\infty} p_n x^n = \infty$, where $0 < x < r$. If $\lim_{n\to\infty} s_n$ exists, show that

$$\lim_{n\to\infty} s_n = \lim_{x\to r} \frac{\sum_{n=0}^{\infty} p_n s_n x^n}{\sum_{n=0}^{\infty} p_n x^n}$$

where the limit is taken through real values of x.

42 (Abel summability) Show that if $\lim_{n\to\infty} s_n = s$ exists, then $(1-z)\sum_{n=0}^{\infty} s_n z^n$ converges to s as z converges to 1 along any path inside the circle $C = \{z \mid |z| = 1\}$ not tangent to C.

43 (Abel) If the series $\sum_{n=0}^{\infty} a_n$ converges to a, if $\sum_{n=0}^{\infty} b_n$ converges to b, if $\sum_{n=0}^{\infty} c_n$ converges to c, and if $c_n = \sum_{j=0}^{n} a_{n-j} b_j$, show that $c = ab$. (Hint: Use Exercise 42).

44 The mapping $\sum_0^\infty a_n \to \sum_0^\infty \lambda_n a_n$ defined by the sequence $\{\lambda_n\}$ maps convergent series into convergent series if and only if $\sum_{n=0}^{\infty} |\lambda_n - \lambda_{n+1}| < \infty$.

45 A matrix (λ_{mn}), $m, n = 0, 1, 2, \ldots$, defines a transformation $\sum_{n=0}^{\infty} a_n \to \sum_{m=0}^{\infty} (\sum_{n=0}^{\infty} \lambda_{mn} a_n)$ which maps convergent series into convergent series if and only if

(1) $\sum_m \lambda_{mn}$ converges for every n;

(2) $\sum_n |\lambda_{mn} - \lambda_{m,n+1}|$ converges for each m;

(3) $\sup_M \sum_n |\sum_{m=0}^{M} (\lambda_{mn} - \lambda_{m,n+1})| < \infty.$

The transformation preserves sums of series (i.e. $\sum_{m=0}^{\infty} (\sum_{n=0}^{\infty} \lambda_{mn} a_n) = \sum_{n=0}^{\infty} a_n$) if and only if the equation

(1′) $\sum_m \lambda_{mn} = 1$

holds for each n, in addition to (2) and (3).

46 A matrix (λ_{mn}), $m, n = 0, 1, 2, \ldots$, defines a transformation $\sum_{n=0}^{\infty} a_n \to \sum_{n=0}^{\infty} \lambda_{mn} a_n = t_m$ which maps convergent series into convergent sequences if and only if

(1) $\lim_m \lambda_{mn}$ exists for each n;

(2) $\sup_m \sum_{n=0}^{\infty} |\lambda_{mn} - \lambda_{m,n+1}| < \infty.$

Further, $\lim_m t_m = \sum_{n=0}^{\infty} a_n$ if and only if the equation

(1′) $\lim_m \lambda_{mn} = 1$

holds for each n, in addition to (2).

47 Show that

$$\sum_{n=0}^{\infty} a_n = \lim_{m\to\infty} \{a_0 + \frac{m}{m+1} a_1 + \frac{m(m-1)}{(m+1)(m+2)} a_2 + \ldots\}$$

whenever the series on the left converges.

48 Show that

$$\sum_{n=1}^{\infty} a_n = \lim_{h\to 0} \sum_{n=1}^{\infty} a_n \left(\frac{\sin nh}{nh}\right)^k$$

whenever $k > 1$ and the series on the left converges.

49 (Schur-Mertens). Let $a = \{a_n\}$ and $b = \{b_n\}$ be two sequences of complex numbers, and let $c_m = \sum_{n=0}^{m} a_{m-n} b_n$. If $\sum |a_n|$ converges, then $\sum c_n$ converges whenever $\sum b_n$ converges. Conversely, if $\sum c_n$ converges whenever $\sum b_n$ converges, then $\sum |a_n|$ converges. (Compare this result with that of Exercise 43).

50 (a) If $\sum a_n$ converges and $\{\beta_n\}$ is a non-increasing sequence of positive numbers, then $\sum a_n \beta_n$ converges.

(b) If $\{\beta_{mn}\}$ is a sequence such that

1) $0 \leqq \beta_{mn} \leqq \beta_{m,n-1}$;
2) $\beta_{mn} \leqq M$ for some fixed $M < \infty$;
3) $\lim_{m\to\infty} \beta_{mn} = 1$ for each n;

then

$$\lim_{m\to\infty} \sum_{n=0}^{\infty} a_n\beta_{m,n} = \sum_{n=0}^{\infty} a_n$$

whenever the series on the right converges.

51 (Hardy-Littlewood) Suppose that $a(z) = \sum_{n=0}^{\infty} a_n z^n$ converges for $z = 1$. If $0 < x < 1$, show that

$$\sum_{k=0}^{\infty} \frac{a^{(k)}(x)(1-x)^k}{k!}$$

converges. (Hint: Use Exercise 50.)

52 Let u_n and v_n be elements of l_1 (cf. Exercise 32) defined by the formulae

$$u_n = [1, 1/2, 1/3, \ldots, 1/n, 0, 0, \ldots]$$
$$v_n = [(1-1/n), 1/2(1-1/n)^2, \ldots]$$

Show that $\{u_n - v_n\}$ is bounded, but does not converge to zero.

53 Suppose $\{a_n\}$ is a sequence of complex numbers for which $\{na_n\}$ is bounded, and let $a(z) = \sum_{n=0}^{\infty} a_n z^n$. Prove the following statements:

(a) $a(x)$ is bounded for $0 < x < 1$ if and only if the sequence $\{\sum_{k=0}^{n} a_k\}$ of partial sums of the series $\sum_0^{\infty} a_k$ is bounded;

(b) $a(z)$ is bounded in the circle $|z| < 1$ if and only if the sequence of partial sums of $\sum_{n=0}^{\infty} a_n z^n$ is uniformly bounded for $|z| = 1$. $|z| = 1$.

54 Suppose $\{a_k\}$ is a sequence such that $\{ka_k\}$ converges to zero. If $a(z) = \sum_{k=0}^{\infty} a_k z^k$, show that

$$\lim_{n\to\infty} \left[\sum_{k=0}^{n} a_k - a(1-1/n)\right] = 0.$$

Hence show that $\sum_{k=0}^{\infty} a_k$ converges if and only if $a(x)$ has a limit as $x \to 1$. (This famous result, due to Tauber, is the prototype of all "Tauberian" theorems. Theorems of this type state conditions under which "strong" (in this case Abel) summation reduces to ordinary summation. Compare the result of Exercise 54 with that of Exercise 42.)

5. Notes and Remarks

General remarks and references. Although the processes of differentiation and integration can be regarded as operations defined on a class of functions, before the end of the last century this viewpoint was employed as little more than a notational convenience. By the turn of the century, however, the work of Volterra and Fredholm on integral equations had emphasized the utility of these "operational" techniques. As integral equations and related theories were advanced by Hilbert, E. Schmidt, F. Riesz and others, and the similarities with corresponding algebraic problems emerged, the value of regarding a function as a vector or as a point in a "space" of functions became evident. This was recognized clearly by Schmidt [1], for example, whose geometrical notation and terminology in Hilbert space is substantially that in current use. At approximately the same time the foundations of modern point-set topology were being laid by Fréchet [1], Hausdorff [1], and F. Riesz [1], and it was only natural that these topological notions should be applied to algebra and analysis. Thus Kürschak [1] introduced topology into the theory of fields by defining a valuation, and thus the concrete Hilbert space and the L_p spaces soon led to the general concepts of normed and topological linear spaces.

Considering the extent to which Lie and his followers had developed the study of "continuous groups," it is somewhat surprising that the idea of a general topological group was so slow in taking form. For it was not until 1926 that Schreier [1] defined an abstract topological group and established some of its properties. (See also Leja [1].) A systematic treatment of the topologizing of abstract groups, rings, and fields was given by van Dantzig [1]. The concept of a normed algebra was first used by Michal and Martin [1], Nagumo [1] and Gelfand [1]. Fréchet [2, 3] appears to have originated the concept of an F-space.

References: Those interested in studying general topological groups in detail will want to consult the treatises of Loomis [1], Pontrjagin [1], and Weil [1]. The books of Chevalley [1, 2] are concerned with the more special Lie groups.

We particularly recommend the report in "Die Encyklopädie der Mathematischen Wissenschaften" by Hellinger and Toeplitz [3; particularly Sec. 24] as a remarkably readable and complete bibliographic account of the theory of integral equations, which served as the point of departure for much of linear space theory. Also of more than mere historical interest are the fundamental papers of Hilbert [1] on integral equations and the book of F. Riesz [6] on equations in infinitely many unknowns. The older "Encyklopädie" report of Pincherle [1], the volumes of E. H. Moore [1, 2], and the work of Volterra [1] are also related to much of this material. Finally, the book of Davis [1] cites many references to operational methods such as the Heaviside operational calculus, fractional differentiation and integration, etc.

Most intimately related to this present work is the classic treatise of Banach [1] and the recent books of F. Riesz and Sz.-Nagy [1] and of Zaanen [5]. The more special theory in Hilbert space is treated by Stone [3], Sz.-Nagy [3], Halmos [6], and Cooke [1]. The volume of Hille [1] also has many points of contact. The treatises of Nakano [1, 2], and Bourbaki [2] are more general in scope in that they are concerned principally with locally convex topological linear spaces. All of these books will be found useful as references.

Bounded sets. Definition 1.7 is essentially due to von Neumann [1]. An equivalent definition is: a set B is *bounded* if for every sequence $\{x_n\} \subseteq B$ and every sequence of scalars $\{\alpha_n\}$ with $\alpha_n \to 0$, it follows that $\alpha_n x_n \to 0$. This latter condition was used by Mazur and Orlicz [1]. In the case of B-spaces it is seen in Lemma 3.3 that a set is bounded if and only if it is contained in some sphere. An F-space, however, may have all of its spheres unbounded.

The principle of uniform boundedness. (The comments in this paragraph pertain to Theorems 1.11, 1.13, 1.17, and 1.18, and their applications to Theorems 3.6, 3.20, and 3.21. See also Gál [3].) There are many instances of results related to this theorem throughout analysis, although because of their frequent special character they are not always recognized as such, nor can they always be compared in strength. Some results of this nature were proved by Lebesgue [1] in his study of singular integrals, others by Hahn [1; p. 678], Steinhaus [1], and Saks and Tamarkin [1]. Certain consequences of Theorem 1.11 which are closely tied to the problems of representation of linear functionals and operators were proved in l_2 by Hellinger and Toeplitz [1, 2] in l_p, $p > 1$, by Landau [1], in c by Toeplitz [2], in

$C[0, 1]$ by Helly [1], and in L_p, $p > 1$, by F. Riesz [2; p. 457].

Theorems 1.11, 1.17, and 1.18 were proved for linear functionals on a general B-space by Hahn [2] who applied these results to a large number of special spaces. The first really general proofs of Theorems 1.11 and 1.13 were given by Hildebrandt [2] in the case of a B-space. Banach and Steinhaus [1; p, 53] observed that in the case of a B-space Theorem 1.11 remains valid if $\{T_a x | a \in A\}$ is bounded for x in a set of the second category. Banach and Steinhaus [1; p. 54] proved the following theorem, often called the theorem of condensation of singularities.

THEOREM. *If $\{U_{pq}\}$ is a double sequence of bounded linear operators between two B-spaces $\mathfrak{X}$ and $\mathfrak{Y}$ such that*

$$\limsup_{q \to \infty} |U_{pq}| = \infty, \qquad p = 1, 2, \ldots,$$

then there exists a set S of the second category in $\mathfrak{X}$ such that for each $x \in S$,

$$\limsup_{q \to \infty} |U_{pq}(x)| = \infty, \qquad p = 1, 2, \ldots.$$

Banach [3] proved Theorems 1.17 and 1.18 for B-spaces and gave [7] theorems related to these and the condensation theorem just mentioned for complete metric groups. (See also Banach [1; Chap. 1].) The extension of Theorems 1.11 and 1.13 to F-spaces was made by Mazur and Orlicz [1].

Pettis [5; p. 300] has given a very general extension of Theorem 1.13 to linear topological spaces. He has also shown that an extension of Theorem 1.17 is valid for a sequence of continuous homomorphisms between two topological groups.

Theorems 3.20 and 3.21 have been extended by Sargent [1, 2] so that they may be applied to a space of Denjoy integrable functions on [0, 1], which is of the first category.

A "uniform boundedness theorem" for a collection of real-valued functions on a metric space was proved by Goldstine [2]. Alexiewicz [1] has made a systematic study of circumstances under which theorems analogous to Theorems 1.17, 1.18, and the condensation theorem are valid for classes of continuous mappings between metric spaces. His conditions are such that they can be specialized to give those theorems or to give generalizations of some theorems of Saks [2, 3] on sequences of measures. Alexiewicz [I; II] also discussed these results in spaces where certain abstract notions of limit are available. Mazur and Orlicz [2] (see also Alexiewicz [1; I]) have extended these three theorems to polynomial operators in F-spaces. Alexiewicz [I; III] has also treated polynomial operators in a linear space in which various notions of limit are present. Orlicz [7] proves a theorem of the condensation type where the double sequence of operators also depends on a parameter in a complete metric space, which includes the following theorem.

THEOREM. *Let* U_n *be a continuous function on* $\mathfrak{X} \times [0, 1]$ *to* $\mathfrak{Y}$ *and such that for each* $t \in [0, 1]$, $U_n(., t) \in B(\mathfrak{X}, \mathfrak{Y})$. *If for each* $t \in [0, 1]$ *there exists an* x_t *such that* $\limsup_{n\to\infty} |U_n(x_t, t)| = \infty$, *then there exists an* x *such that* $\limsup_{n\to\infty} |U_n(x, t)| = \infty$ *for all* t *in a non-denumerable perfect set in* $[0, 1]$.

Gál [1, 2] has extended the uniform boundedness theorem and the condensation theorem to a class of non-linear homogeneous mappings between B-spaces, under certain hypotheses that compensate for the lack of linearity.

Let $\{T_\alpha | \alpha \in A\}$ be a generalized sequence of bounded linear operators from one B-space $\mathfrak{X}$ into another B-space $\mathfrak{Y}$. Day [9] has considered the question of when $\limsup_{\alpha\in A} |T_\alpha x| < \infty$, $x \in \mathfrak{X}$, implies that $\limsup_{\alpha\in A} |T_\alpha| < \infty$. Among other results, he showed that this is valid for any directed set A if and only if $\mathfrak{X}$ is finite dimensional.

A number of theorems of the uniform boundedness type together with applications are proved by Dunford [1; Chap. 1]. It is seen (Dunford [1; p. 308]) that the closed graph theorem (II.2.4) together with the Hahn-Banach theorem can be used to give a very simple proof of the uniform boundedness theorem for B-spaces.

Bourbaki [3] has demonstrated that Theorem 1.17 remains valid in a certain class of locally convex topological linear spaces. (See also Dieudonné and Schwartz [1; p. 73].)

Continuity of linear operations. (These remarks refer to Theorems 1.14—1.16 and 3.4.) Although special cases of Theorem 1.14 had been used before, it was first proved for a general B-space by Banach [3; p. 151]. Mazur and Orlicz [1; p. 153] extended this and 1.16 to

F-spaces. Wehausen [1; p. 161] observed that 1.16 remains true when the range is any topological linear space. He also derived necessary and sufficient conditions for the continuity of linear operators when the domain is a locally convex space. As may be surmised by analogy with the real case, additivity plus "measurability" of a function should imply continuity. That such results are valid in metric groups was shown by Banach [7] (see also Banach [1; Chap. 1] and Kuratowski [1]). Conditions of this nature are extended to polynomial operators by Mazur and Orlicz [2].

It is sometimes useful to define a notion of "continuity" of a linear mapping T between linear spaces $\mathfrak{X}$ and $\mathfrak{Y}$ if T maps sequences of $\mathfrak{X}$ which converge in some sense into convergent sequences in $\mathfrak{Y}$. This is the case, for example, when $\mathfrak{X}$ and $\mathfrak{Y}$ are concrete linear spaces having one or more natural notions of limit, or when $\mathfrak{X}$ and $\mathfrak{Y}$ have a notion of order. The reader will find such occasions in the references on partially ordered spaces. He may also compare Alexiewicz [1; II, III], Fichtenholz [1, 2], and Orlicz [5, 6].

The interior mapping principle. This result and the closed graph theorem are closely related to the notion of category. Theorem 2.2 was first proved by Banach [4; p. 238] for the case of a B-space; his proof was different than the one given. Schauder [7] gave a proof for B-spaces which is closer to the one in the text; he also stated Theorem 2.1. The validity of the theorems in Section 2 for F-spaces was proved in Banach's treatise by essentially the same arguments.

It is an interesting and important fact that Theorems 2.1—2.5 remain valid for homomorphic mappings of separable complete metric groups with left-invariant metrics. This fact was proved by Banach [7]. The hypothesis of separability is required in the group case, as is shown by the example of the identity map from the additive group of real numbers with its discrete topology into this group with its usual topology.

Pettis [5] has given several necessary and sufficient conditions that a homomorphism h between two topological groups X and Y be continuous, or that it be interior, i.e., map open sets into sets with an interior point. We mention the following result.

THEOREM. *If the topological group X is complete with respect to* *some right-invariant (possibly non-definite) metric, then a homomorphism* $h : X \to Y$ *is an interior map into* Y *and has a closed kernel,* $h^{-1}(0)$, *if and only if the graph of* h *is closed, and* h *maps each non-void open set onto a set whose closure contains an open set.*

Pettis [5] proved theorems extending some of the results of Banach [7], Freudenthal [2], and Lorentz [3], and which, when specialized to linear spaces, contain most of the results of this section.

The field of scalars can be any field with non-discrete absolute value (see Bourbaki [2; p. 34] for details). Further extensions have been made by Dieudonné and Schwartz [1; p. 72] and others (see Dieudonné [13; p. 504]) to the case where the domain space $\mathfrak{X}$ is the union of a countable number of expanding F-spaces $\mathfrak{X}_1 \subseteqq \mathfrak{X}_2 \subseteqq \ldots$, and the topology of $\mathfrak{X}$ is defined in a natural way, the range space $\mathfrak{Y}$ having certain completeness and neighborhood properties. See also Köthe [7, 10] and Pták [1].

Similar theorems may be proved in more general topological spaces, which do not necessarily have a group structure. For instance, Dunford [5] generalized the notion of category to derive conditions which are sufficient that a one-to-one continuous function send a non-void open set onto a set which contains an interior point. McShane [4] gave conditions for a set of the second category in a topological space to contain an interior point. His results were improved by Pettis [3, 5].

For systematic treatment of interior (or open) mappings in topological spaces, the reader is referred to Whyburn, [1, 2].

Schauder [4, 5] (see also Leray [3], and [1] for other references) has proved a theorem on the invariance of domain which is similar to 2.1 for a class of non-linear mappings in a B-space, but also closely related to certain fixed point theorems (see Sec. V. 10).

THEOREM. *Let* f *be a continuous mapping defined on the closure of a bounded open subset of a* B*-space* $\mathfrak{X}$ *with values in a compact subset of* $\mathfrak{X}$. *Then, if the mapping* $x \to x+f(x)$ *is one-to-one, the image of an open set is open.*

For a proof, the reader is referred to Nagumo [2]. L. M. Graves [5] has shown that a continuous non-linear function $F : \mathfrak{X} \to \mathfrak{Y}$ between B-spaces such that $F(x_0) = y_0$, maps a neighborhood of x_0 into a set containing a neighborhood of y_0, provided that F can be approximated near x_0 by a continuous linear operator mapping $\mathfrak{X}$ onto $\mathfrak{Y}$. The approximation is taken in a sense generalizing the notion of the differential. This extends theorems of Hildebrandt and Graves [1], where the approximating linear operator was taken to have an inverse. In this latter case, however, considerably more refined results can be obtained.

The following theorem, a special case of one due to Bartle and Graves [1], is a generalization of Theorem 2.1.

THEOREM. *Let $K(t)$ be a continuous linear map of a B-space $\mathfrak{X}$ onto a B-space $\mathfrak{Y}$ for $t \in [0, 1]$, and suppose that the map $t \to K(t)$ is continuous from $[0, 1]$ to $B(\mathfrak{X}, \mathfrak{Y})$ with the uniform operator topology. Then there exists a constant $N > 0$ such that if ψ is a continuous map of $[0, 1]$ into $\mathfrak{Y}$, there exists a continuous map $\varphi : [0, 1] \to \mathfrak{X}$ such that*

$$K(t)\varphi(t) = \psi(t), \qquad |\varphi(t)| \leqq N|\psi(t)|, \qquad t \in [0, 1].$$

An important special case of Theorem 2.4 is the case where T is a symmetric linear mapping on all of Hilbert space, i.e., where $(Tx, y) = (x, Ty)$ for all x, y. It follows readily that T is closed and hence continuous. In the terminology of bilinear forms in infinitely many variables this was proved by Hellinger and Toeplitz [1; p. 321—7], and in an abstract setting by von Neumann [7; p. 107]. See also Stone [3; p. 59].

Historical Comments. Axioms closely related to those of a normed linear space were introduced, in 1916, by Bennett [1] in a generalization of Newton's method for the location of roots. F. Riesz [4] extended much of the Fredholm theory of integral equations using the axioms of a complete normed linear space, and Lamson [1] proved an implicit function theorem for such spaces. In 1922, Banach [3], Hahn [2], and Wiener [1] published papers using the same or similar sets of axioms. Though Banach did not initiate the study of these spaces, his contributions were many and deep—for that reason many authors use the term *Banach space* to refer to a complete normed linear space. Throughout this work, we will adhere more closely to his own terminology and call them *B*-spaces.

The Hahn-Banach Theorem. Both Theorems 3.10 and 3.11 are called the Hahn-Banach theorem, but the reader should observe that the former applies to any linear space (topologized or not) and the latter is an application to normed spaces and yields the existence of *continuous* linear functionals. It will be seen in Chapter V that Theorem 3.10 yields the existence of many continuous linear functionals in any space when the topology can be defined by a family of *convex* neighborhoods of the origin. This plenitude of continuous linear functionals will be of utmost importance for much of what follows. On the other hand, Theorem 3.10 has other applications to the extension of measures, integrals, etc.—see the references given below.

Though some of the research of Helly [1, 2] and F. Riesz [2, 6] dealing with the problem of solving an infinite system of linear equations is very closely related to this theorem, the extension given by Theorem 3.11 was first proved by Hahn [3; p. 217] for a real B-space. Banach [4; pp. 212, 226] (see also Banach [1; pp. 28, 55]) proved both Theorem 3.11 and Theorem 3.10 and applied these results systematically to real B-spaces. The trick involved in the case of a complex space is due to Bohnenblust and Sobczyk [1] and, independently, to Soukhomlinoff [1] who treated the case of quaternion scalars as well.

The following theorem is due to Hahn [3; p. 216] and is a useful consequence of the Hahn-Banach theorem. This result contains corresponding theorems proved by F. Riesz [2; p. 470] [6; p. 61] for L_p and l_p, and by Helly [1; p. 271] for $C[0, 1]$.

THEOREM. *Let $\{c_n\}$ be a countable set of scalars, and $\{x_n\}$ a countable set of elements in a B-space $\mathfrak{X}$. Then there exists an $x^* \in \mathfrak{X}^*$ such that $x^*(x_n) = c_n$ for all n, and such that $|x^*| \leqq M$ if and only if*

$$\left| \sum \alpha_i c_i \right| \leqq M \left| \sum \alpha_i x_i \right|$$

for every finite collection of scalars $\{\alpha_i\}$.

A similar theorem dealing with the solution $x \in \mathfrak{X}$ of the infinite system

$$x_n^*(x) = c_n, \qquad n = 1, 2, \ldots,$$

is valid provided that the space $\mathfrak{X}$ is reflexive, but not in general. The following theorem, due to Helly [2; p. 73] is valid for general spaces. (See Kakutani [2] for an elementary proof.)

THEOREM. *Let $\mathfrak{X}$ be a normed linear space, $\{c_1, \ldots, c_n\}$ arbitrary* *scalars,* $\{x_1^*, \ldots, x_n^*\}$ *a finite set in* $\mathfrak{X}^*$, *and let* $M > 0$. *Then, for any* $\varepsilon > 0$ *there exists an* $x \in \mathfrak{X}$ *such that*

$$x_i^*(x) = c_i, \qquad i = 1, \ldots, n, \text{ and } \quad |x| < M+\varepsilon$$

if and only if

$$\left| \sum \alpha_i c_i \right| \leqq M \left| \sum \alpha_i x_i^* \right|$$

for every finite collection of scalars $\{\alpha_i\}$.

We cite a related theorem of Yamabe [1].

THEOREM. *Let* K *be a dense convex set in a normed linear space* $\mathfrak{X}$. *Let* $x \in \mathfrak{X}$, $\varepsilon > 0$, *and* $x_1^*, \ldots, x_n^* \in \mathfrak{X}^*$ *be arbitrary. Then there exists a* y *in* K *such that*

$$|y-x| < \varepsilon; \qquad x_i^*(y) = x_i^*(x), \qquad i = 1, \ldots, n.$$

Theorems of this nature are often useful in the theory of moments or of approximation.

The Hahn-Banach theorem can be sharpened to show the existence of certain invariant linear functionals. For example, a special case of a theorem of Agnew and Morse [1] reads:

THEOREM. *In addition to the hypotheses of Theorem* 3.10 *suppose that* G *is an Abelian (or a solvable) group of linear transformations of* $\mathfrak{X}$ *which map* $\mathfrak{Y}$ *into itself and that*

$$p(g(x)) = p(x), \qquad g \in G, \qquad x \in \mathfrak{X};$$
$$f(g(x)) = f(x), \qquad g \in G, \qquad x \in \mathfrak{Y}.$$

Then there exists a real linear functional F *on* $\mathfrak{X}$ *which extends* f *and such that*

$$F(x) \leqq p(x); \qquad F(g(x)) = F(x), \qquad g \in G, \qquad x \in \mathfrak{X}.$$

For further generalizations of this nature and applications to the extension of measures, etc., the reader should consult Agnew [1], Agnew and Morse [1], and Klee [5].

Let p be a positive function as in Theorem 3.10; then it is of interest to ask if there exists a real linear functional f defined on $\mathfrak{X}$ such that

$$f(x) < p(x), \qquad x \neq 0.$$

A necessary condition is that $p(x)+p(-x) \neq 0$, for $x \neq 0$. Aronszajn [5] showed that if $\mathfrak{X}$ is separable under the norm defined by

$$|x| = p(x)+p(-x),$$

then this condition is sufficient. Bonsall [1] showed that the separability condition cannot be dropped.

Ingleton [1] has given conditions for the Hahn-Banach theorem to hold when the field of scalars is non-Archimedean. (See also Fleischer [1] and Ono [1].)

The Hahn-Banach theorem may be used to give short proofs of the existence of Green's functions for the Laplace equation, and for other boundary value problems. We refer the reader to the papers of Garabedian [1], Garabedian and Shiffman [1], Lax [1] and Miranda [2], for details and other references.

Reflexivity. The fact that the natural isomorphism of a B-space $\mathfrak{X}$ into its second conjugate is isometric was proved by Hahn [3; p. 219] who was the first to formulate the notion of the conjugate space. He used the term *regular* to describe what we, following Lorch [2], have called *reflexive.* Examples of reflexive B-spaces are given in Chapter IV, and a necessary and sufficient condition for reflexivity is given in Theorem V.4.7. It is a consequence of Theorem V.6.1 that a necessary and sufficient condition for reflexivity is that spheres be weakly sequentially compact. This is due to Eberlein [1], and is the strongest criterion known.

Theorems 3.23 and 3.24 are due to Pettis [1]. Theorem 3.28 was proved by Schmidt [1] for L_2 and by F. Riesz [2; p. 467] for L_p; in these spaces it is sometimes called the "Theorem of Choice." In the case of abstract Hilbert space, it was proved by von Neumann [2; p. 381]; for a general reflexive B-space, Theorem 3.28 is due to Pettis [1].

It must be emphasized that reflexivity involves the isometric isomorphism of $\mathfrak{X}$ and $\mathfrak{X}^{**}$ under the natural mapping $\varkappa$ defined in 3.18. James [4, 5] has proved the following startling theorem.

THEOREM. *There exists a separable B-space which is isomorphic and isometric with its second conjugate space, but which is not reflexive.*

Factor spaces. Let $\mathfrak{X}$ be an F-space and $\mathfrak{M}$ a closed subspace of $\mathfrak{X}$. Let $\mathfrak{X}/\mathfrak{M}$ be the factor, or quotient, space as defined in section I.11. As is seen in Exercise II.4.13, $\mathfrak{X}/\mathfrak{M}$ becomes an F-space under the metric

$$|x+\mathfrak{M}| = \inf \{|x+m| \,|\, m \in \mathfrak{M}\}.$$

This useful device was introduced by Banach [6; pp. 47—9] and Hausdorff [3]. It will be fundamental in the discussion of B-algebras.

Completion of spaces. In the definitions of F- and B-spaces, we required the spaces to be complete in their metric topology. Occasionally it is necessary to consider metric linear spaces which are not complete. In such cases, the following theorem is often convenient.

THEOREM. *Let $\mathfrak{X}$ be a linear space satisfying properties (i) and (ii) of Definition* 1.10. *Then $\mathfrak{X}$ is isomorphic and isometric with a dense linear subspace of an F-space $\tilde{\mathfrak{X}}$. The space $\tilde{\mathfrak{X}}$ is uniquely determined up to isometric isomorphism. If $\mathfrak{X}$ is a normed linear space, then $\tilde{\mathfrak{X}}$ is a B-space.*

The proof of this theorem proceeds as in the Cantor completion of the rational numbers to obtain the real numbers. Let $\mathfrak{Y}$ be the linear space of all Cauchy sequences in $\mathfrak{X}$, whose vector addition and scalar multiplication in $\mathfrak{Y}$ are defined coordinatewise, and if $y = \{x_n\} \in \mathfrak{Y}$, let $|y| = \sup_n |x_n|$. Then $\mathfrak{Y}$ is an F- (or a B-) space. Let $\mathfrak{Z}$ be the closed subspace in $\mathfrak{Y}$ consisting of all Cauchy sequences in $\mathfrak{X}$ which converge to zero, and let $\tilde{\mathfrak{X}} = \mathfrak{Y}/\mathfrak{Z}$. The reader may verify that $\tilde{\mathfrak{X}}$ has the stated properties.

Direct sums and products. Let $\mathfrak{X}$ and $\mathfrak{Y}$ be two topological linear spaces over the same field of scalars. Let $\mathfrak{X} \oplus \mathfrak{Y}$ be the direct sum of the linear spaces $\mathfrak{X}$ and $\mathfrak{Y}$ in the sense of Section I.11, with the product topology of Section I.8. Then $\mathfrak{X} \oplus \mathfrak{Y}$ is readily seen to be a topological linear space. If $\mathfrak{X}$ and $\mathfrak{Y}$ are B- (or F-) spaces, then $\mathfrak{X} \oplus \mathfrak{Y}$ is a B- (or an F-) space under either of the norms

$$|[x, y]| = \max(|x|, |y|),$$
$$|[x, y]| = \{|x|^p + |y|^p\}^{1/p}, \qquad 1 \leqq p < \infty,$$

and these norms are equivalent to the product topology. The space $\mathfrak{X} \oplus \mathfrak{Y}$ so obtained is called the *direct sum* of $\mathfrak{X}$ and $\mathfrak{Y}$ (although the term *direct product* is used by some authors). The extension to any finite number of summands is immediate. By Lemma I.8.4, the direct sum of a denumerable number of B- (or F-) spaces can be made into an F-space, but in general not into a B-space. Finally, one can define the direct sum of an arbitrary family of topological linear spaces, but it is ordinarily not metric even if the summands are. It may be seen that in the case of B-spaces, if the norms are chosen appropriately, we have

$$\mathfrak{X}^* \oplus \mathfrak{Y}^* = (\mathfrak{X} \oplus \mathfrak{Y})^*.$$

For notational purposes, let us denote the couple $[x, y] \in \mathfrak{X} \oplus \mathfrak{Y}$ by the symbol $x \oplus y$. Then we have the linear relations:

$$x_1 \oplus y_1 + x_2 \oplus y_2 = (x_1+x_2) \oplus (y_1+y_2),$$
$$\alpha(x \oplus y) = \alpha x \oplus \alpha y.$$

We now ask if it is possible to construct a linear space from $\mathfrak{X}$ and $\mathfrak{Y}$ in such a way as to make valid the bilinear relations:

$$x \otimes (y_1+y_2) = x \otimes y_1 + x \otimes y_2,$$
$$(x_1+x_2) \otimes y = x_1 \otimes y + x_2 \otimes y,$$
$$\alpha\beta(x \otimes y) = (\alpha x) \otimes (\beta y)$$

This can be attained by taking the set $\mathfrak{X} \otimes \mathfrak{Y}$ of all finite formal sums $\sum x_i \otimes y_i$ with suitable identifications. Such a space is called the *direct product* (although the terms *tensor*, *cross*, and *Kronecker product* are also used). If $\mathfrak{X}$ and $\mathfrak{Y}$ are B-spaces, it is desired to define a topology on $\mathfrak{X} \otimes \mathfrak{Y}$ in such a manner that it becomes a B-space, and such that the relationship

$$\mathfrak{X}^* \otimes \mathfrak{Y}^* = (\mathfrak{X} \otimes \mathfrak{Y})^*$$

is valid. There are unexpected difficulties attached to this and associated problems. These questions have been treated by a number of authors; the reader may find the theory and further references in Schatten [1] for the case of B-spaces, and in Grothendieck [3] for general topological linear spaces.

Invariant metrics in groups. A metric ϱ on a (multiplicative) group G is said to be *left-invariant* if $\varrho(gx, gy) = \varrho(x, y)$ for all g in G. It is said to be *invariant* if $\varrho(gx, gy) = \varrho(x, y) = \varrho(xg, yg)$ for all g in G. If G is a topological group, we say it is *metrizable* if there is a metric whose topology is equivalent to the original topology.

G. Birkhoff [5] and Kakutani [12] proved that a topological group G admits a left-invariant metric if and only if the family of neighborhoods of the identity can be defined by a countable collection (i.e., G satisfies the first axiom of countability).

Klee [6] showed that if G is an Abelian topological group which admits an equivalent metric under which it is complete, then G admits an invariant metric and is complete under each invariant metric. Thus every complete linear metric space can be metrized to be an F-space. Further, a normed linear space is a B-space provided it is complete under some equivalent metric. See also van Dantzig [1], [2].

Norms in linear spaces. We have seen that in a normed linear space there are many continuous linear functionals. LaSalle [1] (see also Theorem V.2.8) showed that a non-zero continuous linear functional exists if and only if the space contains an open convex neighborhood of the origin which does not contain the entire space. However, Kolmogoroff [1] proved that a topological linear space is homeomorphic to a normed linear space if and only if there exists a *bounded* convex neighborhood of the origin. Wehausen [1] showed that if a topological linear space has a *bounded* neighborhood of the origin (not necessarily convex) then it has an equivalent invariant metric, but that an F-space may have no bounded sphere.

Eidelheit and Mazur [1] have proved that every F-space can be equipped with an invariant metric, equivalent to the original one, such that if $x \neq 0$, then the function $|\alpha x|$ is a monotone increasing function of the real variable α.

Isometry and linear dimension. Mazur and Ulam [1] (see also Banach [1; p. 166] and Aronszajn [1]) have proved the interesting result.

THEOREM. *Every mapping F of one real normed linear space onto another which is isometric, i.e., $|F(x)-F(y)| = |x-y|$, and such that $F(0) = 0$, is a linear mapping.*

A proof for finite dimensional F-space has been given by Charzyński [1].

In Theorem V.8.8 we will determine the form of the most general isometric isomorphism between two spaces of continuous functions. Banach [1; p. 174—180] has represented such mappings in other special B-spaces.

K. Borsuk (see Banach [1; p. 182—4)] showed that there is an algebraic isomorphism which is also a homeomorphism between each of the spaces L_p, l_p, $p \geqq 1$, c, c_0, $C[0, 1]$ and the direct sum of these spaces with themselves. Mazur [4] demonstrated that L_{p_1} and L_{p_2} are homeomorphic for $1 \leqq p_1 \leqq p_2$.

Banach [1; Chap. 12] defines $\mathfrak{X}$ to be of *smaller linear dimension* than $\mathfrak{Y}$ if there is a one-to-one continuous linear map of $\mathfrak{X}$ onto a closed linear manifold in $\mathfrak{Y}$, and writes $\dim_l \mathfrak{X} \leq \dim_l \mathfrak{Y}$. A number of results are obtained about the comparability of the L_p spaces in the sense of linear dimension. Banach and Mazur [1] have demonstrated that two separable B-spaces can be of equal linear dimension but not be topologically isomorphic.

Differential calculus in B-spaces. In Chapter III it will be seen that a satisfactory theory of integration can be constructed for functions with their domain in a measure space and their range in a B-space, and further that a theory of analytic functions of a complex variable with values in a complex B-space is also available. It is appropriate to observe that at least the fundamentals of a theory of the derivative exists for functions with both domain and range in B-spaces. In the case of complex B-spaces the theory is relatively complete and reminiscent of the theory of analytic functions. For this theory the reader is referred to Hille [1; Chap. 4]; the real case is somewhat more complicated. The central concept in both cases is that of the *Fréchet* (or *total*) *differential.*

DEFINITION. Let $\mathfrak{X}$ and $\mathfrak{Y}$ be B-spaces. Let D be an open set in $\mathfrak{X}$, and $F : D \rightarrow \mathfrak{Y}$. Then F is said to have a *Fréchet differential* at the point $a \in D$ if there is a linear operator $dF(a, .) \in B(\mathfrak{X}, \mathfrak{Y})$ such that

$$\lim_{|h|\rightarrow 0} |h|^{-1}|F(a+h)-F(a)-dF(a, h)| = 0.$$

L. M. Graves [3] has proved the validity of a generalization of Taylor's theorem with remainder. Kerner [1, 2] has extended Stokes' theorem and developed a form of differential geometry—for the latter see also Michal [1]. Differential equations have been treated by Michal and Elconin [1]. The implicit function theorem has been established by Hildebrandt and Graves [1]—other results along this line are due to Michal and Clifford [1], Cronin [1, 2] and Bartle [1].

For further references the reader should consult Hille [1] and the expository articles of L. M. Graves [1], Hyers [3], Michal [1], Rothe [4], and Taylor [10].

Convergence. A series $\sum_{i=1}^{\infty} x_i$ is said to be *unconditionally convergent* if every arrangement of its terms converges to the same element. It is clear that a sufficient condition for unconditional convergence is that the series be *absolutely convergent*, i.e., that $\sum_{i=1}^{\infty} |x_i|$ converges. Dvoretzky and Rogers [1] have shown that absolute convergence is equivalent to unconditional convergence if and only if the B-space is finite dimensional. The Orlicz-Banach theorem states: *a series is unconditionally convergent if and only if every partial sequence converges weakly to some element of the space.* (See Banach [1; p. 240], Dunford [1; p. 322]). In the case when the space is weakly complete, the series is unconditionally convergent if and only if $\sum_{i=1}^{\infty} |x^*x_i|$ is convergent for each x^* in $\mathfrak{X}^*$.

References on convergence: Dvoretzky and Rogers [1], Hildebrandt [1], Karlin [1], Macphail [1], Munroe [1], Nikodým [1], Orlicz [1, 2].

For special types of convergence in abstract linear spaces: Gagaev [1], Maddaus [1], Nachbin [1], Titov [1, 2], Vulich [3].

Orthogonality. At least four definitions of orthogonality of elements of a real normed linear space have been employed. Possibly the most fruitful is the one introduced by Birkhoff and extensively developed by James [2]: x is said to be *orthogonal* to y if and only if $|x| \leqq |x+ky|$ for all real k. This notion has been related to concepts of strict convexity, weak compactness, differentiability of the norm, and various properties of linear functionals. In terms of it, several necessary and sufficient conditions that an inner product can be defined have been given by James [2, 3].

References on orthogonality: Birkhoff [1], Fortet [1, 2], James [1, 2, 3], Roberts [1].

Bases. A sequence $\{x_i\}$ of elements of a B-space $\mathfrak{X}$ is called a *base* (or *basis*) if for every $x \in \mathfrak{X}$ there is a unique set $\{a_i\}$ of scalars such that

$$\lim_{n\to\infty} \left|x - \sum_{i=1}^{n} a_i x_i\right| = 0.$$

This notion was introduced by Schauder [1] and is convenient in extending results from finite to infinite dimensional spaces, whenever the base exists. A number of other types of bases have also been used by various authors. It can be seen that this concept is closely related to the problem of (biorthogonal) expansion of an arbitrary element.

It is clear that a B-space having a base in the sense defined above must be separable. The converse question as to whether every separable B-space possesses a base has not yet been solved. In the spaces c_0 or l_p, $1 \leqq p < \infty$, the vectors $\{x_i\}$, where $x_i = [\delta_{i1}, \delta_{i2}, \ldots]$, form a base. In c, these vectors with $x_0 = [1, 1, 1, \ldots]$ added form a base. Schauder [1] constructed a base in the space $C[0, 1]$, and proved (Schauder [3]) that the Haar orthogonal system is a base for L_p, $1 \leqq p < \infty$. In $L_2(0, 1)$ the trigonometric or Legendre polynomials can be normalized to form a base; in $L_2(0, \infty)$ the Laguerre functions, and in $L_2(-\infty, \infty)$ the Hermite functions, are sometimes used.

For addition results on bases and biorthogonal systems, see the following: Al'tman :[1, 2], Babenko [1], Bari [1], Banach [1; Chap. 7], Boas [2], Bohnenblust [3], Dieudonné [16], Dixmier [7], Frink [1], Gelbaum [1, 2, 3], Gelfand [6], Grinblyum [1, 2, 3, 4], Gurevič [1], James [4], Kaczmarz and Steinhaus [1], Karlin [1, 2], Kostyučenko and Skorohod [1], Kozlov [1, 2], Krein, Milman and Rutman [1], Lorch [1], Markouchevitch [1, 2, 3], Nikol'skiĭ [1], Orlicz [8], Schäfke [1, 2], Schauder [1, 3], Tseng [1], Vinokurov [1] and Wilansky [1].

CHAPTER III

Integration and Set Functions

1. Finitely Additive Set Functions

In contrast to the terms real function, complex function, etc., where the adjectives *real* and *complex* refer to the range of the function, the term *set function* is commonly used in mathematics for a function whose domain is a family of sets. The theory of the integral

$$\nu(E) = \int_E f(s)\mu(ds)$$

as developed in this chapter is based upon a set function μ. In some cases that will be encountered the values of μ are not scalars, but customarily where integration is used in this text μ is a scalar valued function and f a vector (or scalar) valued function. Thus, even if the integration process is defined with respect to a scalar valued set function μ, the resulting integral ν may be a vector valued set function. It is desirable therefore to formulate some of the basic and elementary concepts in such a way that vector as well as scalar valued set functions may be studied. A discussion of some of the deeper properties of vector valued set functions is contained in Section IV.10. It is desirable also to allow the set function μ to have its values in the extended real number system (which is not a vector space) but, since $\infty+(-\infty)$ is not defined and it will be necessary to add values in the range of μ, it will be stipulated that an extended real valued set function has at most one of the improper values ∞ and $-\infty$.

1 DEFINITION. A *set function* is a function defined on a family of sets, and having values either in a B-space, which may be the set of real or complex numbers, or in the extended real number system, in which case its range contains at most one of the improper values $+\infty$ and $-\infty$. A *positive set function* is a real valued or extended real valued set function which has no negative values.

2 DEFINITION. A set function μ defined on a family τ of sets is said to be *additive* or *finitely additive* if τ contains the void set ϕ, if $\mu(\phi) = 0$ and if

$$\mu(A_1 \cup A_2 \ldots \cup A_n) = \mu(A_1)+\mu(A_2)+\ldots+\mu(A_n),$$

for every finite family $\{A_1, \ldots, A_n\}$ of disjoint subsets of τ whose union is in τ.

For an example of a finitely additive set function, let $S = [0, 1)$ and let τ be the family of intervals $I = [a, b)$, $0 \leqq a < b < 1$, with $\mu(I) = b-a$.

We will ordinarily require that the domain of an additive set function be closed under the finite operations of union, intersection, and complementation. If A and B are subsets of a set S it is convenient to introduce the symbol $A-B$ to denote the set $A \cap B'$. However, this notation will be avoided whenever S is a group and there is possibility of confusion with the group operation. We will use the notation $A \Delta B$ for the *symmetric difference* $(A-B) \cup (B-A)$.

3 DEFINITION. Let S be a set. A *field of subsets* of S, or a *Boolean algebra of subsets* of S, is a non-empty family of subsets of S which contains the void set, the complement (relative to S) of each of its members, and the union of each finite collection of its members.

A field of sets clearly contains the difference and the symmetric difference of any two of its members. It follows from the rules of De Morgan that

$$A_1 \cap A_2 \cap \ldots \cap A_n = S-\{(S-A_1) \cup (S-A_2) \cup \ldots \cup (S-A_n)\}.$$

Thus a field of sets contains the intersection of each finite collection of its members.

In the example above, let Σ be the family of all finite unions of intervals $I = [a, b)$ in τ. Then Σ is a field. If $A \in \Sigma$, then $A = \cup_{i=1}^{n} I_i$, where the intervals I_i are disjoint. If $\mu(A) = \sum_{i=1}^{n} \mu(I_i)$, it can easily be seen that $\mu(A)$ depends only on A and not on its particular decomposition into intervals, and that μ is a finitely additive set function defined on the field Σ.

The next step in our analysis is to show that an arbitrarily given set function μ can be used to define a non-negative set function $v(\mu)$, called the *total variation* of μ. The set function $v(\mu)$ is defined so as to be equal to μ if μ itself is non-negative and additive, to be additive if μ is additive, and to be bounded if μ is bounded and additive. The total variation $v(\mu)$ of an additive set function μ is important because it dominates μ in the sense that $v(\mu, E) \geqq |\mu(E)|$ for $E \in \Sigma$; the reader should test his comprehension of Definition 4 below by proving that $v(\mu)$ is the smallest of the non-negative additive set functions λ such that $\lambda(E) \geqq |\mu(E)|$ for $E \in \Sigma$.

4 DEFINITION. Let μ be a set function defined on the field Σ of subsets of a set S. Then for every E in Σ the *total variation of μ on E*, denoted by $v(\mu, E)$, is defined as

$$v(\mu, E) = \sup \sum_{i=1}^{n} |\mu(E_i)|,$$

where the supremum is taken over all finite sequences $\{E_i\}$ of disjoint sets in Σ with $E_i \subseteqq E$. The set function μ is of *bounded variation* if $v(\mu, S) < \infty$, and it is of *bounded variation on a set* E in Σ if $v(\mu, E) < \infty$.

5 LEMMA. *If a real or complex valued additive set function defined on a field Σ of subsets of a set S is bounded, it is of bounded variation and* $v(\mu, S) \leqq 4 \sup_{E \in \Sigma} |\mu(E)|$.

PROOF. Let μ be an additive set function on Σ with $|\mu(E)| \leqq M$ for every E in Σ. If μ is real valued then for any finite sequence $\{E_1, \ldots, E_n\}$ of disjoint sets in Σ

$$\sum_{i=1}^{n} |\mu(E_i)| = \sum{}^{+} \mu(E_i) - \sum{}^{-} \mu(E_i) = \mu(\cup^{+} E_i) - \mu(\cup^{-} E_i)$$

where Σ^{+} and $\cup^{+}$ (Σ^{-} and $\cup^{-}$) are taken over those i for which $\mu(E_i) \geqq 0$ ($\mu(E_i) < 0$). Thus

$$v(\mu, S) = \sup_{A, B \in \Sigma} \{\mu(A) - \mu(B)\} \leqq 2M.$$

If μ is complex valued its real and imaginary parts are additive real set functions on Σ whose absolute values are bounded by M. Thus in this case $v(\mu, S) \leqq 4M$. Q.E.D.

If μ is understood we may write $v(E)$ instead of $v(\mu, E)$. If μ is non-negative and additive, $v(\mu, E) = \mu(E)$. It is often helpful to think of $v(E)$ as the limit of a generalized sequence in the following manner. Let the family of all finite sequences $\{E_i\}$ of disjoint sets in Σ with $E_i \subseteqq E$ be ordered by defining $\{E_i\} \subseteqq \{F_j\}$ to mean that each E_i

is the union of some of the sets F_j. Then $\Sigma|\mu(E_i)| \leqq \Sigma|\mu(F_j)|$ and we have $v(E) = \lim_{\{E_i\}} \Sigma|\mu(E_i)|$.

6 Lemma. *The total variation of an additive set function μ defined on a field Σ of subsets of a set S is also additive on Σ.*

Proof. Let $\{A_i\}$ be a finite sequence of disjoint sets in Σ with $A_i \subseteq E \cup F$ where $E, F \in \Sigma$ and $EF = \phi$. Let $E_i = EA_i$, $F_i = FA_i$, then

$$\sum|\mu(A_i)| \leqq \sum|\mu(E_i)|+\sum|\mu(F_i)| \leqq v(\mu, E)+v(\mu, F),$$

and hence

(i) $v(\mu, E \cup F) \leqq v(\mu, E)+v(\mu, F)$.

Thus if $v(\mu, E \cup F) = \infty$ it follows that $v(\mu, E \cup F) = v(\mu, E)+v(\mu, F)$. If $v(\mu, E \cup F) < \infty$ there are finite sequences $\{E_j\}$, $\{F_j\}$ of disjoint sets in Σ with $E_j \subseteq E$, $F_j \subseteq F$ and

$$v(\mu, E) \leqq \Sigma|\mu(E_j)|+\varepsilon, \quad v(\mu, F) \leqq \Sigma|\mu(F_j)|+\varepsilon$$

$$\begin{aligned} v(\mu, E)+v(\mu, F) &\leqq \Sigma|\mu(E_j)|+\Sigma|\mu(F_j)|+2\varepsilon \\ &\leqq v(\mu, E \cup F)+2\varepsilon. \end{aligned}$$

Since $\varepsilon > 0$ is arbitrary, $v(\mu, E)+v(\mu, F) \leqq v(\mu, E \cup F)$, which taken together with (i), shows that $v(\mu, E)$ is additive for E in Σ. Q.E.D.

In the next definition and theorem, we show how the total variation of a bounded additive real valued set function μ can be used to define the "positive" and "negative" parts of μ. The process is similar to the decomposition of a function $f(\cdot)$ into the difference of two non-negative functions: if we put $f^+(\cdot) = (1/2)(|f(\cdot)|+f(\cdot))$ and $f^-(\cdot) = (1/2)(|f(\cdot)|-f(\cdot))$, then f^+ and f^- are non-negative, and $f = f^+-f^-$.

7 Definition. Let μ be a bounded additive real set function defined on a field Σ of subsets of a set S. The *upper* or *positive variation* μ^+ and the *lower* or *negative variation* μ^- of μ are set functions defined on Σ by the equations

$$\mu^+(E) = \tfrac{1}{2}\{v(\mu, E)+\mu(E)\}, \qquad \mu^-(E) = \tfrac{1}{2}\{v(\mu, E)-\mu(E)\}.$$

8 Theorem. (*Jordan decomposition*) *If μ is a bounded additive real set function defined on a field Σ, then, for each E in Σ,*

$$\mu^+(E) = \sup_{F \subseteq E} \mu(F), \qquad \mu^-(E) = -\inf_{F \subseteq E} \mu(F),$$

where F is restricted to the domain Σ of μ. The set functions μ^+, μ^- are additive, non-negative, and for each E in Σ,

$$\mu(E) = \mu^+(E) - \mu^-(E), \quad v(\mu, E) = \mu^+(E) + \mu^-(E).$$

PROOF. If $F \subseteq E$, $E, F \in \Sigma$ then

$$\begin{aligned} 2\mu(F) &= \mu(F) + \mu(E) - \mu(E-F) \\ &\leqq \mu(E) + |\mu(F)| + |\mu(E-F)| \\ &\leqq \mu(E) + v(\mu, E) = 2\mu^+(E). \end{aligned}$$

Thus

$$\text{(i)} \qquad \sup_{F \subseteq E} \mu(F) \leqq \mu^+(E).$$

On the other hand let $\varepsilon > 0$ and let $E_1, \ldots, E_n$ be disjoint sets in Σ with $\cup E_i = E$ and $\Sigma|\mu(E_i)| > v(\mu, E) - \varepsilon$. Then, in the notation of Lemma 5,

$$\begin{aligned} 2\mu^+(E) - \varepsilon = v(\mu, E) + \mu(E) - \varepsilon &\leqq \Sigma|\mu(E_i)| + \mu(E) \\ &= \mu(\cup {}^+E_i) - \mu(\cup {}^-E_i) + \{\mu(\cup {}^+E_i) + \mu(\cup {}^-E_i)\} \\ &= 2\mu(\cup {}^+E_i) \leqq 2 \sup_{F \subseteq E} \mu(F). \end{aligned}$$

Since $\varepsilon > 0$ is arbitrary, $\mu^+(E) \leqq \sup_{F \subseteq E} \mu(F)$, which, with (i) above, shows that $\mu^+(E) = \sup_{F \subseteq E} \mu(F)$. Since $\mu^- = \{-\mu\}^+$ it follows that $\mu^-(E) = -\inf_{F \subseteq E} \mu(F)$. The remaining conclusions of the theorem follow readily from the definitions. Q.E.D.

In the theory of an additive set function μ certain non-empty sets are likely to occur which are in many ways negligible as far as μ is concerned. These are the μ-null sets as introduced in Definition 11 below. They may perhaps best be introduced by first considering a certain (not necessarily additive) extension μ^* of a positive set function μ which is defined as follows:

9 DEFINITION. Let μ be a positive extended real valued additive set function defined on a field Σ of subsets of a set S. For an arbitrary subset E of S the number $\mu^*(E)$ is defined by the equation

$$\mu^*(E) = \inf_{F \supseteq E} \mu(F),$$

where F is restricted to the domain Σ of μ.

10 Lemma. *Let μ be a positive extended real valued additive set function defined on a field Σ of subsets of a set S. Then*

(a) $\mu^*(E) = \mu(E)$, $\quad E \in \Sigma$;

(b) $\mu^*(A \cup B) \leqq \mu^*(A)+\mu^*(B)$, $\quad A, B \subseteq S$;

(c) $\mu^*(A) \leqq \mu^*(B)$, $\quad A \subseteq B \subseteq S$.

Proof. If $E \in \Sigma$, $F \in \Sigma$, and $F \supseteq E$, then $\mu(F) = \mu(E)+\mu(F-E)$, so that $\mu(F) \geqq \mu(E)$. Thus $\mu^*(E) \geqq \mu(E)$. On the other hand, since $E \supseteq E$, $\mu(E) \geqq \mu^*(E)$. This proves (a). To prove (b) let $\varepsilon > 0$ and let $A_1, B_1 \in \Sigma$ be such that $A_1 \supseteq A$, $B_1 \supseteq B$ and

$$\mu(A_1) \leqq \mu^*(A)+\varepsilon/2, \qquad \mu(B_1) \leqq \mu^*(B)+\varepsilon/2.$$

Then $A_1 \cup B_1 \supseteq A \cup B$ and

$$\begin{aligned} \mu^*(A \cup B) &\leqq \mu(A_1 \cup B_1) \\ &= \mu(A_1)+\mu(B_1-A_1) \\ &\leqq \mu(A_1)+\mu(B_1) \\ &\leqq \mu^*(A)+\mu^*(B)+\varepsilon. \end{aligned}$$

Since $\varepsilon > 0$ is arbitrary, this proves (b). Statement (c) follows immediately from the definition of μ^*. Q.E.D.

Definitions 4 and 9 allow us to introduce one of the most frequently occurring concepts in measure theory, namely, the concept of a null set. This is done in the following definition.

11 Definition. Let μ be an additive set function defined on a field of subsets of a set S. A subset N of S is said to be a *μ-null set* if $v^*(\mu, N) = 0$, where v^* is the extension of the total variation v of μ defined in Definition 9. It follows immediately from Lemma 10 that every subset of a μ-null set, and every finite union of μ-null sets, is a μ-null set. Any statement concerning the points of S is said to hold *μ-almost everywhere*, or, if μ is understood, simply *almost everywhere*, or for *almost all* s in S, if it is true except for those points s in a μ-null set. The phrase "almost everywhere" is customarily abbreviated "*a.e.*" Thus, if $\lim_n f_n(s) = f(s)$, $s \in S-N$, where N is a μ-null set, we say that the sequence $\{f_n\}$ converges to f *almost everywhere* on S. In addition to "μ-almost everywhere" there is another expression which is related to the notion of a μ-null set and which is used in connection with functions f on S where the emphasis is on the range of f

rather than on its domain S. Thus if there is a μ-null set N such that the restriction of f to $S-N$ is bounded, then f is said to be *μ-essentially bounded* or simply *essentially bounded.* The quantity

$$\inf_N \sup_{s \in S-N} |f(s)|,$$

where N ranges over the μ-null subsets of S is called the *μ-essential supremum* or *μ-essential least upper bound* of $|f(\cdot)|$ and is written as

$$\operatorname{ess\,sup}_{s \in S} |f(s)| \quad \text{or} \quad \operatorname{ess\,lub}_{s \in S} |f(s)|.$$

If for some null set N the restriction of f to $S-N$ has its values in a separable set, the function f is said to be *μ-essentially separably valued.* Expressions such as *essentially compact valued* are defined similarly.

In the example above, where Σ is generated by intervals $I = [a, b)$, $0 \leqq a < b < 1$, and $\mu(I) = b-a$, we note that any finite set of points as well as any convergent sequence of points in $[0, 1)$ is a μ-null set.

2. Integration

In this section and the following on Lebesgue spaces, we will be concerned with defining and developing the basic properties of the integral $\int f(s)\mu(ds)$. In these sections f will be a vector valued function defined on a set S and μ a finitely additive set function defined on a field of subsets of S. It will not be assumed that μ is bounded. Thus the basis for this material is a fixed set S, a field Σ of subsets of S, and a finitely additive complex or extended real valued set function μ on Σ. The functions to be integrated will have their values in a real or complex B-space $\mathfrak{X}$.

The first step in our analysis is to introduce a topology on the set of all functions defined on S with values in a B-space $\mathfrak{X}$. This topology will be defined by a certain metric function, which will be chosen so that two functions f and g are close together in terms of the metric if $f(s)$ is close to $g(s)$ except for those s in a set E in Σ for which $v(\mu, E)$ is small.

1 Definition. For every $E \subseteqq S$, every $\alpha \geqq 0$ and every function f on S to $\mathfrak{X}$ we define the subset $E(|f| > \alpha)$ of E by the equation

$$E(|f| > \alpha) = \{s | s \in E, |f(s)| > \alpha\},$$

and the norm $|f|$ of f by

$$|f| = \inf_{\alpha>0} \arctan \{\alpha + v^*(\mu, S(|f| > \alpha))\}.$$

It should be remarked that we take the principal value of the arctan; i.e., that value lying between 0 and $\pi/2$. The function arctan in Definition 1 is used to insure that $|f| < \infty$, even though $v^*(S) = \infty$. In fact, arctan could be replaced by any continuous increasing function φ such that $\varphi(0) = 0$, $\varphi(x_1+x_2) \leqq \varphi(x_1)+\varphi(x_2)$ for $x_1, x_2 \geqq 0$, and $\varphi(\infty) = \lim_{x\to\infty} \varphi(x)$ exists; e.g., $\varphi(x) = x(1+x)^{-1}$. If $v^*(S) < \infty$ we may take

$$|f| = \inf_{\alpha>0} [\alpha + v^*(\mu, S(|f| > \alpha))].$$

The reader should avoid confusing the norm of f with the norm $|f(s)|$ of the values $f(s)$ of the function f. If we wish to consider the function defined by $g(s) = |f(s)|$, we may write it on occasion as $|f(\cdot)|$, but *never* as $|f|$. If the reader keeps these notational conventions in mind, he will be spared considerable confusion.

Note also that, with the norm just defined, the set of all functions on S to $\mathfrak{X}$ is not in general a linear topological space, since ηf need not approach zero as η approaches zero (cf. Exercise 9.7).

2 Lemma. *If f and g are functions on S to $\mathfrak{X}$ then* $|f+g| \leqq |f|+|g|$.

Proof. Let $\alpha, \beta > 0$. Then $S(|f+g| > \alpha+\beta) \subseteqq S(|f| > \alpha) \cup S(|g| > \beta)$ and

$$\begin{aligned}
|f+g| &= \inf_{\alpha,\beta>0} \arctan \{\alpha+\beta+v^*(\mu, S(|f+g| > \alpha+\beta))\} \\
&\leqq \inf_{\alpha,\beta>0} \arctan \{\alpha+v^*(\mu, S(|f| > \alpha))+\beta+v^*(\mu, S(|g| > \beta))\} \\
&\leqq \inf_{\alpha>0} \arctan \{\alpha+v^*(\mu, S(|f| > \alpha))\} \\
&\quad + \inf_{\beta>0} \arctan \{\beta+v^*(\mu, S(|g| > \beta))\} \\
&= |f|+|g|. \quad \text{Q.E.D.}
\end{aligned}$$

If it were known that $f = 0$ whenever $|f| = 0$, the preceding lemma would show that the function $\varrho(f, g) = |f-g|$ is a metric in the space of all functions which map S into $\mathfrak{X}$ (see I.6.1). Unfortunately, this is rarely the case and so a slight detour will be made.

3 DEFINITION. The function f on S to $\mathfrak{X}$ is said to be a *μ-null function* or, when μ is understood, simply a *null function* if the set $S(|f| > \alpha)$ is a μ-null set for each $\alpha > 0$.

It is important to observe that a null function with respect to a finitely additive set function need not vanish almost everywhere. For an example of such a function, let $S = [0, 1)$ and Σ be the field of finite unions of intervals $I = [a, b), 0 \leqq a < b < 1$, with $\mu(I) = b-a$ as in Section 1. Let R denote the set of rational points in S. For $r = p/q \in R$ in lowest terms, we define $f(p/q) = 1/q$, and set $f(x) = 0$, $s \in S-R$. Since for each $\alpha > 0$, $S(|f| > \alpha)$ is a finite set, f is a μ-null function. However, $\mu^*(R) = 1$.

4 LEMMA. *A function f on S to $\mathfrak{X}$ is a μ-null function if and only if $|f| = 0$.*

PROOF. If $|f| = 0$, then, for each $\varepsilon > 0$, there is an $\alpha > 0$ such that $\alpha + v^*(\mu, S(|f| > \alpha)) < \varepsilon$. Thus $\alpha < \varepsilon$ so that $S(|f| > \alpha) \supseteqq S(|f| > \varepsilon)$ and $v^*(\mu, S(|f| > \varepsilon)) < \varepsilon$. Since $S(|f| > \delta) \subseteqq S(|f| > \varepsilon)$ for $\delta > \varepsilon$, we have $v^*(\mu, S(|f| > \delta)) < \varepsilon$ for $\delta > \varepsilon$, which proves that $v^*(\mu, S(|f| > \delta)) = 0$ for every $\delta > 0$. Conversely, it is clear that $|f| = 0$ provided that $v^*(\mu, S(|f| > \alpha)) = 0$ for each $\alpha > 0$. Q.E.D.

5 COROLLARY. *The null functions form a linear subspace of the space of all functions on S to $\mathfrak{X}$. If f is a null function and if $|g(s)| \leqq |f(s)|$ almost everywhere on S then g is a null function.*

Lemmas 2 and 4 show that the relationship between two functions f and g on S to $\mathfrak{X}$ which is expressed by the statement that $f-g$ is a null function, is an equivalence relation. The linear set of all functions on S to $\mathfrak{X}$ can therefore be divided into mutually exclusive equivalence classes. For an arbitrary function f on S to $\mathfrak{X}$ let $[f]$ denote the class of functions equivalent to f (i.e. all g such that $f-g$ is a μ-null function), and let $F(S, \Sigma, \mu, \mathfrak{X})$ denote the set of all such sets $[f]$. If the following equations are used to define their left hand members then Lemmas 2 and 4 show that $F(S, \Sigma, \mu, \mathfrak{X})$ is a linear vector space as well as a metric space with the distance function $\varrho([f], [g]) = |[f]-[g]|$:

$$[f]+[g] = [f+g];$$
$$\alpha[f] = [\alpha f];$$
$$|[f]| = |f|.$$

Since $|g| = |f|$ if $[g] = [f]$, it is clear that the above equation defines the norm uniquely. Moreover, just as in the general case for a factor space (see Section I.11) the addition and scalar multiplication of equivalence classes is well defined.

It is customary to speak of the elements of $F(S, \Sigma, \mu, \mathfrak{X})$ as if they were functions rather than sets of equivalent functions and this we shall ordinarily do. Thus, we shall write f instead of $[f]$ and think of $F(S, \Sigma, \mu, \mathfrak{X})$ as the set of all functions on S to $\mathfrak{X}$. No confusion should arise if it is remembered that two functions which differ only by a null function are considered to be the same. Thus a function ψ can not be considered as defined on the space $F(S, \Sigma, \mu, \mathfrak{X})$, unless $\psi(f) = \psi(g)$ whenever $f-g$ is a null function. Similarly, if a function f on S to $\mathfrak{X}$ is defined and is referred to as a point in the space $F(S, \Sigma, \mu, \mathfrak{X})$, what should be understood is the class of all functions g on S to $\mathfrak{X}$ which differ from f by a null function. Where the vector space $\mathfrak{X}$ is fixed throughout a discussion, the symbol $F(S, \Sigma, \mu)$ will sometimes be used instead of $F(S, \Sigma, \mu, \mathfrak{X})$. Likewise, if Σ and μ are clearly understood, the symbol $F(S)$ may be used in place of $F(S, \Sigma, \mu)$.

6 Definition. Convergence in the metric space $F(S)$ is called *convergence in μ-measure*, or simply *convergence in measure*. A sequence $\{f_n\}$ of functions on S to $\mathfrak{X}$ *converges in μ-measure to the function f* on S to $\mathfrak{X}$ if and only if

$$\lim_{n\to\infty} |f_n-f| = 0.$$

7 Lemma. *A sequence $\{f_n\}$ of functions on S to $\mathfrak{X}$ converges in measure to the function f on S to $\mathfrak{X}$ if and only if*

$$\lim_{n\to\infty} v^*(\mu, S(|f_n-f| > \varepsilon)) = 0$$

for every $\varepsilon > 0$.

Proof. The proof follows from the following elementary inequalities:

(a) if $|f_n-f| > \delta > 0$ and $0 < \varepsilon < (1/2) \tan \delta$, then $v^*(\mu, S(|f_n-f| > \varepsilon)) > (1/2) \tan \delta$;

(b) if $v^*(\mu, S(|f_n - f| > \varepsilon)) > \delta > 0$, $\varepsilon > 0$, then $|f_n - f| >$ min [arctan δ, arctan ε]. Q.E.D.

8 **Lemma.** *Let f and g be functions on S to $\mathfrak{X}$.*

(a) *The mapping $f, g \to f+g$ is a continuous map of $F(S) \times F(S)$ into $F(S)$.*

(b) *For a fixed scalar α the map $f \to \alpha f$ is a continuous map of $F(S)$ into itself.*

(c) *If f_n converges to f in μ-measure, then $|f_n(\cdot)|$ converges to $|f(\cdot)|$ in μ-measure.*

Proof. If $f_n \to f$ and $g_n \to g$, then, by Lemma 2, $|f_n + g_n - (f+g)| \leqq |f_n - f| + |g_n - g| \to 0$ and so $f+g$ is a continuous function of f and g. Also, since (b) is trivial if $\alpha = 0$, we can assume $\alpha \neq 0$, in which case

$$S(|\alpha f_n - \alpha f| > \varepsilon) = S(|f_n - f| > \varepsilon/|\alpha|),$$

which shows that $\alpha f_n \to \alpha f$ if $f_n \to f$. The last statement follows from the inequality

$$||f_n(s)| - |f(s)|| \leqq |f_n(s) - f(s)|. \qquad \text{Q.E.D.}$$

Various linear subspaces of $F(S)$ will be of particular importance in the development of the theory of the integral. We shall, in fact, first define the integral for functions of a very simple type, described in the next definition. The domain of definition of the integral will then be extended, by using its continuity properties, to a much larger class of functions.

9 **Definition.** Consider a function f on S to $\mathfrak{X}$ which has only a finite set $x_1, \ldots, x_n$ of values and for which

$$f^{-1}(x_i) = \{s | s \in S, f(s) = x_i\} \in \Sigma, \qquad i = 1, \ldots, n.$$

Any function g on S to $\mathfrak{X}$ which differs from such an f by a μ-null function is called a *μ-simple function.*

In accordance with our usual habit of saying "function" instead of "equivalence class of functions", we shall often treat a μ-simple function as if it were necessarily a function taking on only a finite number of values, and taking on these values on sets of Σ. If the reader remembers that we ordinarily make no distinction between two functions differing by a null function, this will cause no difficulty.

It follows from Corollary 5 that the μ-simple functions form a linear manifold in $F(S)$.

10 DEFINITION. The functions *totally μ-measurable on S*, or, if μ is understood, *totally measurable on S* are the functions in the closure $TM(S)$ in $F(S)$ of the μ-simple functions. If for every E in Σ with $v(\mu, E) < \infty$, the product $\chi_E f$ of f with the characteristic function χ_E of E is totally measurable, the function f is said to be *μ-measurable* or, if μ is understood, simply *measurable*. A *set A is measurable* if χ_A is measurable. Symbols for the set of measurable functions are $M(S, \Sigma, \mu, \mathfrak{X})$, $M(S, \Sigma, \mu)$, $M(S)$ and symbols besides $TM(S)$ for the set of totally measurable functions are $TM(S, \Sigma, \mu, \mathfrak{X})$ and $TM(S, \Sigma, \mu)$.

11 LEMMA. *The totally measurable functions as well as the measurable functions form a closed linear subspace of $F(S)$.*

PROOF. The set of μ-simple functions forms a linear manifold in $F(S)$ and hence, by Lemma 8, its closure $TM(S)$ does likewise. Since if $\{f_n\}$ converges in measure so does $\{f_n\chi_E\}$ for any set E in Σ, it follows that the set $M(S)$ is also a closed linear manifold in $F(S)$. Q.E.D.

12 LEMMA. Let f, β *be totally μ-measurable (μ-measurable) functions on S with β scalar valued and let g be a continuous function defined on the field of scalars. Then the functions βf, $|f(\cdot)|$, and $g(\beta(\cdot))$ are totally μ-measurable (μ-measurable). If one of β and f is μ-null, while the other is totally μ-measurable, βf is μ-null. Furthermore the mapping $\beta \to g(\beta(\cdot))$ is a continuous map of the space of totally measurable scalar functions into itself.*

PROOF. In view of the identities

$$\chi_E(s)\beta(s)f(s) = (\chi_E(s)\beta(s))(\chi_E(s)f(s)), \qquad \chi_E(s)|f(s)| = |\chi_E(s)f(s)|,$$
$$\chi_E(s)g(\beta(s)) = g(\chi_E(s)\beta(s)) - g(0)\chi_{E'}(s),$$

the statements made in the lemma concerning μ-measurable functions will follow from those concerning totally μ-measurable functions.

We now show that if β and f are totally μ-measurable so is their product. Let $\{f_n\}$ and $\{\beta_n\}$ be sequences of finitely valued μ-simple functions converging in measure to f and β respectively, and let ε be an arbitrary positive number. For n sufficiently large, the function β is approximated uniformly within ε on a set whose complement has measure less than ε by a bounded function β_n of the sequence, and a similar remark applies to f. Thus there is a constant M and a set A_0 such that

$$|\beta(s)| \leq M, \qquad |f(s)| \leq M, \qquad s \in A_0,$$

and $v^*(\mu, A_0') < \varepsilon$. Now, there exists an integer N_ε such that if $n \geqq N_\varepsilon$,

$$|f_n(s)-f(s)| < \frac{\varepsilon}{M}, \qquad |\beta_n(s)-\beta(s)| < \frac{\varepsilon}{M},$$

for all s in a set A_n for which $v^*(\mu, A_n') < \varepsilon$. Thus,

$$\begin{aligned} |f_n(s)\beta_n(s) - f(s)\beta(s)| &\leqq |f_n(s) - f(s)||\beta_n(s)| \\ &\quad + |\beta_n(s) - \beta(s)||f(s)| \\ &\leqq \frac{\varepsilon}{M}(\varepsilon+M) + \frac{\varepsilon}{M} M = 2\varepsilon + \frac{\varepsilon^2}{M} \end{aligned}$$

for $s \in A_0 \cap A_n$, where $v^*(\mu, (A_0 \cap A_n)') < 2\varepsilon$. This shows that the sequence $\{\beta_n f_n\}$ of μ-simple functions converges in μ-measure to βf. The fact that $|f(\cdot)|$ is totally μ-measurable follows from Lemma 8.

Now let g be a continuous function defined on the field of scalars and let β be totally μ-measurable. Let $\{\beta_n\}$ be a sequence of finitely valued μ-simple functions with $\beta_n \to \beta$ in μ-measure. Given $\varepsilon > 0$ there is, as before, a constant M and a set A_0 such that $v^*(\mu, A_0') < \varepsilon$ and $|\beta(s)| \leqq M$, $s \in A_0$. Let δ be a positive number less than one such that $|g(\alpha)-g(\gamma)| < \varepsilon$ if $|\alpha-\gamma| < \delta$ and $|\alpha|$, $|\gamma| \leqq M+1$. There is a sequence of sets $\{A_n\}$ and an integer N_ε such that if $n \geqq N_\varepsilon$,

$$|\beta_n(s)-\beta(s)| < \delta, \qquad s \in A_n,$$

and $v^*(\mu, A_n') < \varepsilon$. Thus for $n > N_\varepsilon$,

$$|g(\beta_n(s))-g(\beta(s))| < \varepsilon, \qquad s \in A_n \cap A_0,$$

and $v^*(\mu, (A_n \cap A_0)') < 2\varepsilon$. Thus $g(\beta_n(\cdot)) \to g(\beta(\cdot))$ in μ-measure and therefore $g(\beta(\cdot))$ is totally measurable. The argument of this paragraph may be used to prove the last statement of the lemma.

Finally, let f be totally measurable and let β be a scalar valued null function. Given $\varepsilon > 0$ we can, by our previous remarks, find a constant M and a set A_0 such that $|f(s)| \leqq M$, $s \in A_0$, $v^*(\mu, A_0') < \varepsilon$. Let $\delta > 0$ and define $B_0 = \{s | |\beta(s)f(s)| > \delta\}$. The set A_0B_0 is a null set since it is contained in the set of s for which $|\beta(s)| > \delta/M$, and $v^*(A_0'B_0) < \varepsilon$. Consequently, $v^*(B_0) < \varepsilon$, and βf is μ-null. The case in which β is totally μ-measurable and f is μ-null can be handled in exactly the same way. Q.E.D.

It is seen from Lemma 11 that the totally measurable functions form a linear space. Lemma 12 shows also that *the equivalence classes of scalar valued totally measurable functions form an algebra.*

13 DEFINITION. A μ-simple function is *μ-integrable* if it differs by a null function from a function of the form

$$f = \sum_{i=1}^{n} x_i \chi_{E_i},$$

where $E_i = f^{-1}(x_i)$, $i = 1, \ldots, n$, are disjoint sets in Σ with union S and where $x_i = 0$ if $v(\mu, E_i) = \infty$. The phrases "μ-integrable μ-simple function" and "μ-integrable simple function" will be used interchangeably. For an $E \in \Sigma$ the *integral over* E of a μ-integrable simple function h is defined by the equation

$$\int_E h(s)\mu(ds) = \int_E f(s)\mu(ds) = \sum_{i=1}^{n} x_i \mu(EE_i).$$

In this equation a term $x_i\mu(EE_i)$ with the form $0 \cdot \infty$ is defined to be zero.

To see that the integral is uniquely defined let

$$g = \sum_{j=1}^{m} y_j \chi_{A_j}$$

where $A_j = g^{-1}(y_j)$, $j = 1, \ldots, m$, are disjoint sets in Σ whose union is S and where $y_j = 0$ if $v(\mu, A_j) = \infty$, and suppose that g also differs from h by a null function. Then the function

$$f - g = \sum_{i=1}^{n} \sum_{j=1}^{m} (x_i - y_j)\chi_{E_i A_j}$$

is a null function and hence $x_i - y_j = 0$ if $v(\mu, E_i A_j) \neq 0$. Therefore

$$\sum_{i=1}^{n} x_i \mu(EE_i) - \sum_{j=1}^{m} y_j \mu(EA_j) = \sum_{i=1}^{n} \sum_{j=1}^{m} (x_i - y_j)\mu(EE_i A_j) = 0,$$

which proves that the definition given for $\int_E h(s)\mu(ds)$ is independent of the particular representing function f which is used. The symbol $\int_E h d\mu$ will sometimes be used in place of $\int_E h(s)\mu(ds)$. The above argument also shows that $\int_E h d\mu = \int_E k d\mu$ if h and k are both μ-integrable simple functions with $|h-k| = 0$ (see Lemma 4). Thus the integral may be regarded as defined on a subset of the metric space $F(S)$.

14 LEMMA. *The μ-integrable simple functions form a linear manifold in $F(S)$ and the integral $\int_E f d\mu$ is a linear mapping from this manifold into $\mathfrak{X}$. If f and μ are both non-negative, so is $\int_E f d\mu$.*

Proof. Suppose that f and g are μ-integrable simple functions having the form given in the penultimate paragraph. Then the values $z_1, \ldots, z_p$ of $f+g$ are found among the elements x_i+y_j, $1 \leqq i \leqq n$, $1 \leqq j \leqq m$ and

$$f+g = \sum_{k=1}^{p} z_k \chi_{B_k},$$

where B_k is the union of all the sets $E_i A_j$ for which $x_i+y_j = z_k$. If $z_k \neq 0$ and if $x_i+y_j = z_k$, then not both x_i and y_j are zero, and hence $v(\mu, B_k) < \infty$. Thus $f+g$ is a μ-integrable simple function. If P_k is the set of all pairs (i, j) with $x_i+y_j = z_k$ then $E_i A_j$ is void if (i, j) is in none of the sets P_k, $k = 1, \ldots, p$, and hence

$$\begin{aligned}
\int_E (f+g)d\mu &= \sum_{k=1}^{p} z_k \mu(EB_k) \\
&= \sum_{k=1}^{p} z_k \sum_{(i,j)\in P_k} \mu(EE_iA_j) \\
&= \sum_{i=1}^{n} \sum_{j=1}^{m} (x_i+y_j)\mu(EE_iA_j) \\
&= \sum_{i=1}^{n} \sum_{j=1}^{m} x_i\mu(EE_iA_j) + \sum_{i=1}^{n} \sum_{j=1}^{m} y_j\mu(EE_iA_j) \\
&= \sum_{i=1}^{n} x_i\mu(EE_i) + \sum_{j=1}^{m} y_j\mu(EA_j) \\
&= \int_E f d\mu + \int_E g d\mu.
\end{aligned}$$

The remaining conclusions of the lemma are immediate. Q.E.D.

15 **Lemma.** *If f is a μ-integrable simple function then*

$$\left| \int_E f(s)\mu(ds) \right| \leqq \int_E |f(s)| v(\mu, ds).$$

The set function $\lambda(E) = \int_E f(s)\mu(ds)$ is an additive set function on Σ whose total variation is

$$v(\lambda, E) = \int_E |f(s)| v(\mu, ds), \qquad E \in \Sigma.$$

Furthermore $\lim\limits_{v(\mu, E)\to 0} \int_E f(s)\mu(ds) = 0.$

Proof. Since $|\lambda(E)| \leqq v(\lambda, E)$ for $E \in \Sigma$, the second statement implies the first. To prove this second statement, let f have the distinct values $x_1, \ldots, x_n$ and let $E_i = f^{-1}(x_i)$. Since $x_i\mu(EE_i)$ is additive for E in Σ so is the integral $\lambda(E) = \sum_{i=1}^n x_i\mu(EE_i)$. Next, let $E \in \Sigma$ and let $A_j, j = 1, \ldots, m$, be disjoint sets in Σ with $A = \cup A_j \subseteq E$. Then, since the total variation $v(\mu, E)$ is additive in E (see 1.6),

$$\begin{aligned}\sum_{j=1}^m |\lambda(A_j)| &= \sum_{j=1}^m \Big| \sum_{i=1}^n x_i\mu(A_jE_i)\Big| \\ &\leqq \sum_{j=1}^m \sum_{i=1}^n |x_i| v(\mu, A_jE_i) \\ &= \sum_{i=1}^n |x_i| v(\mu, AE_i) \\ &\leqq \sum_{i=1}^n |x_i| v(\mu, EE_i) \\ &= \int_E |f(s)| v(\mu, ds)\end{aligned}$$

which shows that $v(\lambda, E) \leqq \int_E |f(s)| v(\mu, ds)$. Next let $E_j^m \in \Sigma$, $m = 1, \ldots, m_j$, be disjoint subsets of E_j with

$$\sum_{m=1}^{m_j} |\mu(EE_j^m)| > v(\mu, EE_j) - \frac{\varepsilon}{M}$$

where ε is a given positive number and $M = \sum_{j=1}^n |x_j|$. Then $|\lambda(EE_j^m)| = |x_j||\mu(EE_j^m)|$, and

$$\begin{aligned}v(\lambda, E) &\geqq \sum_{j=1}^n \sum_{m=1}^{m_j} |\lambda(EE_j^m)| \\ &= \sum_{j=1}^n |x_j| \sum_{m=1}^{m_j} |\mu(EE_j^m)| \\ &> \sum_{j=1}^n |x_j| v(\mu, EE_j) - \varepsilon \\ &= \int_E |f(s)| v(\mu, ds) - \varepsilon,\end{aligned}$$

which proves that $v(\lambda, E) \geqq \int_E |f(s)| v(\mu, ds)$ and thus completes the derivation of the formula for $v(\lambda, E)$. The final conclusion of the lemma follows from the inequality

$$\Big|\int_E f(s)\mu(ds)\Big| \leqq \sup_{1 \leqq i \leqq n} |x_i| v(\mu, E). \quad \text{Q.E.D.}$$

The next lemma provides us with the key "uniqueness theorem" needed in extending the definition of the integral to a class of functions larger than the class of simple integrable functions.

16 LEMMA. *If* $\{f_n^1\}$, $\{f_n^2\}$ *are sequences of* μ*-integrable simple functions both converging in* μ*-measure on* S *to the same limit and if*

$$\lim_{m,n} \int_S |f_n^i(s)-f_m^i(s)|v(\mu, ds) = 0, \qquad i = 1, 2,$$

then the limits $\lim_n \int_E f_n^i(s)\mu(ds)$, $i = 1, 2$, *exist uniformly with respect to* E *in* Σ *and are equal.*

PROOF. Using Lemma 15,

$$\left|\int_E f_n^i d\mu - \int_E f_m^i d\mu\right| \leqq \int_S |f_n^i(s) - f_m^i(s)|v(\mu, ds) \to 0, \quad i = 1, 2.$$

Thus, the desired limits exist uniformly with respect to E in Σ and it only remains to show that they are equal. In showing this, the following abbreviations will be used:

$$v(E) = v(\mu, E), \qquad p_n(s) = |f_n^1(s)-f_n^2(s)|, \qquad P_n(E) = \int_E p_n dv.$$

Since

$$|p_n(s) - p_m(s)| \leqq |f_n^1(s) - f_m^1(s)| + |f_n^2(s) - f_m^2(s)|$$

it is seen that $\lim_{n,m} \int_S |p_n(s)-p_m(s)|v(ds) = 0$, and hence it follows from the above argument that the limit $P(E) = \lim_n P_n(E)$ exists uniformly with respect to E in Σ. It will be shown that $P(E) = 0$ for every $E \in \Sigma$, and consequently, by Lemma 15,

$$\text{(i)} \left|\int_E f_n^1 d\mu - \int_E f_n^2 d\mu\right| \leqq P_n(E) \to 0, \qquad E \in \Sigma.$$

Since $\lim_{v(E)\to 0} P_n(E) = 0$ for each $n = 1, 2, \ldots$, (Lemma 15), it follows from I.7.6 that $\lim_{v(E)\to 0} P(E) = 0$. Thus for $\varepsilon > 0$, there is a $\delta > 0$ and an integer n_0 such that

(ii) $P(E) < \varepsilon$ for $v(E) < \delta$, and

(iii) $|P(E)-P_n(E)| < \varepsilon$ for $n \geqq n_0$, $\qquad E \in \Sigma$.

Since $p_{n_0}(s) = 0$ for s in the complement A' of a set $A \in \Sigma$ with $v(A) < \infty$, we have $P_{n_0}(A') = 0$, and so (iii) gives

$$\text{(iv)} \qquad P(A') < \varepsilon.$$

Now $p_n \to 0$ in measure on S by Lemma 8, and thus there is an integer $n_1 \geqq n_0$ and a set $B \in \Sigma$ with $v(B') < \delta$ such that

$$\text{(v)} \qquad p_{n_1}(s) < \frac{\varepsilon}{v(A)+1}, \qquad s \in B.$$

Thus from (iii) and (v)

$$\text{(vi)} \qquad P(AB) \leqq \int_{AB} p_{n_1} dv + \varepsilon \leqq 2\varepsilon.$$

Since $v(AB') \leqq v(B') < \delta$ it follows from (ii), (iv), and (vi) that

$$P(S) = P(AB)+P(AB')+P(A') < 4\varepsilon,$$

and since $\varepsilon > 0$ is arbitrary and $0 \leqq P(E) \leqq P(S)$, it follows that $P(E) = 0$ for every E in Σ, proving (i). This completes the proof of the lemma. Q.E.D.

17 Definition. A function f on S to $\mathfrak{X}$ is *μ-integrable on S* if there is a sequence $\{f_n\}$ of μ-integrable simple functions converging to f in μ-measure on S and satisfying in addition the equation

$$\lim_{m, n} \int_S |f_m(s) - f_n(s)| v(\mu, ds) = 0.$$

Such a sequence of μ-integrable simple functions will be said to *determine* f. For each $E \in \Sigma$, *the integral over* E, $\int_E f d\mu$, of a μ-integrable function f is defined in terms of such a sequence $\{f_n\}$ of μ-integrable simple functions by the equation

$$\int_E f(s)\mu(ds) = \lim_n \int_E f_n(s)\mu(ds), \qquad E \in \Sigma.$$

The preceding lemma shows that this limit exists and is independent of the particular sequence $\{f_n\}$ of μ-integrable simple functions. The set of all μ-integrable functions f on S to $\mathfrak{X}$ will be denoted by one of the symbols: $L(S, \Sigma, \mu, \mathfrak{X})$, $L(S, \Sigma, \mu)$, $L(S, \mu)$, or $L(S)$.

The remaining theorems of this section develop some of the basic properties of the integral.

18 LEMMA. *A function f on S is μ-integrable if and only if it is $v(\mu)$-integrable. If f is μ-integrable so is $|f(\cdot)|$. If $\{f_n\}$ is a sequence of μ-integrable simple functions determining f in accordance with the preceding definition, then the sequence $\{|f_n(\cdot)|\}$ determines $|f(\cdot)|$ and furthermore*

$$\lim_{n\to\infty} \int_S |f_n(s)-f(s)|v(\mu, ds) = 0.$$

PROOF. If $\{f_n(\cdot)\}$ determines f, it is clear that $\{|f_n(\cdot)|\}$ determines $|f(\cdot)|$ and, for a fixed m, $\{|f_n(\cdot)-f_m(\cdot)|\}$ determines $|f(\cdot)-f_m(\cdot)|$. Since convergence in μ-measure is the same as convergence in $v(\mu)$-measure, this lemma follows directly from Lemma 8 and Definition 17. Q.E.D.

➜ 19 THEOREM. *Let μ be a finitely additive function defined on a field Σ of subsets of a set S. Then*

(a) *the set $L(S)$ of μ-integrable functions on S is a linear space, and for each $E \in \Sigma$, the integral $\int_E f d\mu$ is linear on $L(S)$;*

(b) *if χ_A is the characteristic function of a set $A \in \Sigma$, and f is μ-integrable, then $f\chi_A$ is μ-integrable and*

$$\int_E f(s)\chi_A(s)\mu(ds) = \int_{EA} f(s)\mu(ds), \qquad E \in \Sigma;$$

(c) *if T is a bounded linear operator on $\mathfrak{X}$ to another B-space and f is μ-integrable, then $Tf(\cdot)$ is μ-integrable and*

$$\int_E Tf(s)\mu(ds) = T\int_E f(s)\mu(ds), \qquad E \in \Sigma.$$

PROOF. To prove statement (a) let $\{f_n\}$ and $\{g_n\}$ be sequences of μ-integrable simple functions defining the elements $f, g \in L(S)$ in accordance with Definition 17. By Lemma 8, $\alpha f_n+\beta g_n \to \alpha f+\beta g$ in measure on S, and if $h_n = \alpha f_n+\beta g_n$, then

$$\int_S |h_n(s)-h_m(s)|v(\mu, ds)$$

$$\leqq |\alpha| \int_S |f_n(s)-f_m(s)|v(\mu, ds)+|\beta| \int_S |g_n(s)-g_m(s)|v(\mu, ds) \to 0,$$

which proves that $L(S)$ is a linear space. By Lemma 14

$$\alpha\int_E f d\mu + \beta\int_E g d\mu = \lim_n (\alpha\int_E f_n d\mu + \beta\int_E g_n d\mu)$$

$$= \lim_n \int_E h_n d\mu = \int_E (\alpha f + \beta g) d\mu,$$

which proves that the integral is linear on $L(S)$.

Statement (b) is clear from the definitions.

Statement (c) is evident for μ-integrable simple functions; to prove it in the general case, let f_n be a sequence of finitely valued μ-simple functions which determine f. Then the functions $Tf_n(\cdot)$ are μ-integrable simple functions. Moreover, since for each $\varepsilon > 0$

$$v^*(\mu, S(|Tf-Tf_n| > \varepsilon)) \leqq v^*(\mu, S(|f-f_n| > \varepsilon/|T|)),$$

$Tf_n(\cdot) \to Tf(\cdot)$ in μ-measure. The inequality

$$\int_S |Tf_n(s) - Tf_m(s)|v(\mu, ds) \leqq |T| \int_S |f_n(s) - f_m(s)|v(\mu, ds)$$

shows that $\{Tf_n\}$ determines $\{Tf\}$. Statement (c) follows from these observations. Q.E.D.

The next theorem extends the results of Lemma 15 to arbitrary integrable functions.

→ 20 THEOREM. *Let g be μ-integrable and, for $E \in \Sigma$, let $G(E) = \int_E g(s)\mu(ds)$. Then*

(a) *$G(E)$ is additive on Σ and has total variation*

$$v(G, E) = \int_E |g(s)|v(\mu, ds), \qquad E \in \Sigma;$$

in particular, if g and μ are non-negative, the integral $G(E)$ is non-negative;

(b) $\lim_{v(\mu, E)\to 0} v(G, E) = 0;$

(c) *For each $\varepsilon > 0$ there are complementary sets A and A' in Σ with $v(\mu, A) < \infty$, $v(G, A') < \varepsilon$;*

(d) *$\int_S |g(s)|v(\mu, ds) = 0$ if and only if g is a μ-null function.*

PROOF. Let $\{g_n\}$ be a sequence of μ-integrable simple functions determining g in accordance with Definition 17. Since by Lemma 15 the set functions $G_n(E) = \int_E g_n(s)\mu(ds)$ are additive, it follows that $G(E)$ is additive. Let $E \in \Sigma$ be the union of disjoint sets $E_1, \ldots, E_k \in \Sigma$. Given $\varepsilon > 0$ there is by Lemma 18 an N_ε such that if $n > N_\varepsilon$

$$\left| \sum_{i=1}^{k} \left| \int_{E_i} g(s)\mu(ds) \right| - \sum_{i=1}^{k} \left| \int_{E_i} g_n(s)\mu(ds) \right| \right|$$
$$\leqq \sum_{i=1}^{k} \left| \int_{E_i} [g(s) - g_n(s)]\mu(ds) \right|$$
$$\leqq \int_E |g(s) - g_n(s)| \, v(\mu, ds) < \varepsilon,$$

independently of the choice of $E_1, \ldots, E_k$. Let us for the moment fix $n > N_\varepsilon$ and choose $A_1, \ldots, A_m$ such that

$$v(G, E) - \sum_{i=1}^{m} \left| \int_{A_i} g(s)\mu(ds) \right| < \varepsilon,$$

and $B_1, \ldots, B_p$ such that

$$v(G_n, E) - \sum_{i=1}^{p} \left| \int_{B_i} g_n(s)\mu(ds) \right| < \varepsilon.$$

Let $E_1, \ldots, E_k$ be the family of all intersections of sets A_i and sets B_j; then

$$v(G, E) - \sum_{i=1}^{k} \left| \int_{E_i} g(s)\mu(ds) \right| < \varepsilon,$$
$$v(G_n, E) - \sum_{i=1}^{k} \left| \int_{E_i} g_n(s)\mu(ds) \right| < \varepsilon.$$

Hence $|v(G, E) - v(G_n, E)| < 3\varepsilon$. Consequently,

$$\lim_{n \to \infty} v(G_n, E) = v(G, E), \qquad E \in \Sigma.$$

Since by Lemma 15 and Lemma 18

$$v(G_n, E) = \int_E |g_n(s)| \, v(\mu, ds) \to \int_E |g(s)| v(\mu, ds),$$

we have

$$v(G, E) = \int_E |g(s)| v(\mu, ds), \qquad E \in \Sigma.$$

Thus we have proved (a).

To prove (b) let ε be arbitrary and select a finitely valued μ-simple function g_ε such that

$$\int_S |g(s) - g_\varepsilon(s)| v(\mu, ds) < \varepsilon.$$

Then there is a set $A \in \Sigma$ with $v(\mu, A) < \infty$ and a constant M with $|g_\varepsilon(s)| < M$ for all $s \in S$ and $g_\varepsilon(s) = 0$ for $s \notin A$. Thus if $v(\mu, E) < \varepsilon/M$, $E \in \Sigma$,

$$\begin{aligned} v(G, E) &= \int_E |g(s)| v(\mu, ds) \\ &\leq \int_E |g(s) - g_\varepsilon(s)| v(\mu, ds) \\ &\qquad + \int_{AE} |g_\varepsilon(s)| v(\mu, ds) \\ &< \varepsilon + Mv(\mu, AE) < 2\varepsilon, \end{aligned}$$

proving (b). Since

$$v(G, A') \leq \int_{A'} |g(s) - g_\varepsilon(s)| v(\mu, ds) < \varepsilon$$

we have also proved statement (c).

Finally, to prove (d), let $\int_S |g(s)| v(\mu, ds) = 0$ and let $\{g_n\}$ be a sequence of functions, each assuming a finite set of values on sets of Σ, which determines g in accordance with Definition 17. Then

$$\lim_{n \to \infty} \int_S |g_n(s)| v(\mu, ds) = 0.$$

Let $\delta > 0$ and define

$$E_n(\delta) = \{s \,|\, |g_n(s)| > \delta\}, \qquad F_n(\delta) = \{s \,|\, |g(s) - g_n(s)| > \delta\}.$$

Then $E_n(\delta) \in \Sigma$ and

$$\int_S |g_n(s)| v(\mu, ds) \geq \int_{E_n(\delta)} |g_n(s)| v(\mu, ds) \geq \delta v(\mu, E_n(\delta)).$$

Consequently, $\lim_{n \to \infty} v(\mu, E_n(\delta)) = 0$. Since $g_n \to g$ in μ-measure, $\lim_{n \to \infty} v^*(\mu, F_n(\delta)) = 0$. But

$$|g(s)| \leq |g(s) - g_n(s)| + |g_n(s)| \leq 2\delta, \qquad s \notin E_n(\delta) \cup F_n(\delta).$$

Since $\lim_{n \to \infty} v^*(\mu, E_n(\delta) \cup F_n(\delta)) = 0$ and δ is arbitrary, it follows that g is a null function. Q.E.D.

21 Lemma. *Let f be a μ-measurable function, let g be a μ-integrable function, and suppose that $|f(s)| \leq |g(s)|$ almost everywhere. Then f is totally μ-measurable.*

Proof. Since g is μ-integrable there is, for each integer m, a μ-integrable simple function g_m assuming finitely many values on sets of Σ such that $|g(s)-g_m(s)| < 1/m$ except on a set E_m with $v(\mu, E_m) < 1/m$. Since g_m vanishes outside a set $F_m \in \Sigma$ with $v(\mu, F_m) < \infty$,

$$|g(s)| \leqq |g(s) - g_m(s)| + |g_m(s)| < \frac{1}{m}$$

for $s \notin A_m = E_m \cup F_m$, $v(\mu, A_m) < \infty$. Consequently if we define $f_m(s) = f(s)$ if $s \in A_m$ and $f_m(s) = 0$ if $s \notin A_m$, the sequence $\{f_m\}$ converges in measure to f. By Lemma 11, f is totally μ-measurable. Q.E.D.

22 Theorem. *Let μ be finitely additive on a field Σ of subsets of a set S. Then*

(a) *A μ-measurable function f is integrable if and only if the function $|f(\cdot)|$ is integrable.*

(b) *If g is a μ-integrable function on S to a B space $\mathfrak{Y}$ and f is a μ-measurable function on S to a B-space $\mathfrak{X}$, and $|f(s)| \leqq |g(s)|$ almost everywhere, then f is μ-integrable.*

Proof. The direct part of (a) follows from Lemma 18. Since the converse part of (a) clearly follows from (b), it is sufficient to prove (b).

By Lemma 21 it follows from the hypotheses of (b) that f is totally μ-measurable. Let $\{g_n\}$ be a sequence of μ-simple functions converging to f in μ-measure. It will first be shown that a sequence $\{f_n\}$ of μ-simple functions may be defined which converges to f in μ-measure and for which $|f_n(s)| \leqq 2|f(s)|$ for every s in S. There are sets $A_n \in \Sigma$, $n = 1, 2, \ldots$, for which $v(\mu, A_n) \to 0$ and a sequence of constants $\varepsilon_n \to 0$ such that

$$|g_n(s)-f(s)| < \varepsilon_n, \qquad s \notin A_n.$$

We now define f_n by the equations

$$f_n(s) = g_n(s) \text{ if } s \notin A_n \text{ and } |g_n(s)| > 2\varepsilon_n, \qquad f_n(s) = 0 \text{ otherwise.}$$

If $s \notin A_n$ and $|g_n(s)| > 2\varepsilon_n$ then $|f_n(s)-f(s)| < \varepsilon_n$, while if $s \notin A_n$ and $|g_n(s)| \leqq 2\varepsilon_n$ we have

$$|f(s)| \leqq |f(s)-g_n(s)|+|g_n(s)| < 3\varepsilon_n.$$

Thus

$$|f_n(s)-f(s)| < 3\varepsilon_n, \qquad s \notin A_n.$$

Consequently f_n approaches f in μ-measure. Now if $s \in A_n$ or if $|g_n(s)| \leqq 2\varepsilon_n$ then $f_n(s) = 0$ and so $|f_n(s)| \leqq 2|f(s)|$ while if $s \notin A_n$ and $|g_n(s)| > 2\varepsilon_n$ then

$$|f(s)| > |g_n(s)| - |g_n(s) - f(s)| > |g_n(s)| - \varepsilon_n > \tfrac{1}{2}|g_n(s)| = \tfrac{1}{2}|f_n(s)|.$$

Hence for all s in S we have $|f_n(s)| \leqq 2|f(s)|$.

It will next be shown that f_n is μ-integrable. Let x be a non-zero value of f_n and E the set of all s in S for which $f_n(s) = x$. Since $2|g| \geqq |f|$ we have $|g(s)| \geqq |x|/2$ on E. The argument of the preceding lemma then shows that $v(\mu, E) < \infty$.

$$|x| v(\mu, E) = \int_E |f_n(s)| v(\mu, ds) \leqq 2 \int_E |g(s)| v(\mu, ds) < \infty,$$

from which it follows that $v(\mu, E) < \infty$. This shows that the function f_n is a μ-integrable simple function. Since $|f_n(s)| \leqq 2|g(s)|$ we have

$$\int_E |f_n(s) - f_m(s)| v(\mu, ds) \leqq 4 \int_E |g(s)| v(\mu, ds), \qquad E \in \Sigma.$$

Let $\varepsilon > 0$ be given. By Theorem 20(c) there is a set $F \in \Sigma$ such that $v(\mu, F) < \infty$ and

$$\int_{S-F} |f_n(s) - f_m(s)| v(\mu, ds) < \varepsilon, \qquad n, m \geqq 1.$$

By Theorem 20(b) there is a $\delta > 0$ such that if $E \in \Sigma$ and $v(\mu, E) < \delta$,

$$\int_E |f_n(s) - f_m(s)| v(\mu, ds) < \varepsilon, \qquad n, m \geqq 1.$$

Since $f_n \to f$ in μ-measure, there is an integer N and a set $E_{n,m} \in \Sigma$ such that $v(\mu, E_{n,m}) < \delta$ and such that $|f_n(s) - f_m(s)| < \varepsilon / v(\mu, F)$ for $s \notin E_{n,m}$ and $n, m \geqq N$. Hence, if $n, m \geqq N$,

$$\int_S |f_n(s) - f_m(s)| v(\mu, ds) = \int_{S-F} + \int_{F-E_{nm}} + \int_{FE_{nm}} \leqq 3\varepsilon.$$

Thus the sequence $\{f_n\}$ determines f in accordance with Definition 17 and therefore f is integrable. Q.E.D.

For certain purposes, it is useful to extend the definition of measurability and integrability in the following ways.

First suppose that f is an extended real-valued function. Let $S^+ = f^{-1}(+\infty)$ and $S^- = f^{-1}(-\infty)$. Then f is said to be *μ-measurable* if

(a) there exist μ-null sets N^+ and N^- such that $S^+ \Delta N^+$ and $S^- \Delta N^-$ both belong to Σ, and

(b) the function g defined by $g(s) = f(s)$ if $s \notin S^+ \cup S^-$, $g(s) = 0$ if $s \in S^+ \cup S^-$, is μ-measurable.

Next suppose that we consider a function f (vector or extended real-valued) which is defined only on the complement of a μ-null set $N \subseteq S$. Then we say that f is μ-measurable if the function g defined by $g(s) = f(s)$ if $s \notin N$, $g(s) = 0$ if $s \in N$, is μ-measurable. A discussion similar to that preceding Definition 6 shows that by considering this somewhat more extended class of functions we make no change in $F(S, \Sigma, \mu, \mathfrak{X})$, or in any of the theorems or lemmas of this section.

Finally, suppose that f is a non-negative μ-measurable extended real valued function. If f is not μ-integrable, we write $\int_S f(s)v(\mu, ds) = +\infty$. It follows, from Theorem 22(b), that if $0 \leqq f_1(s) \leqq f_2(s)$ for almost all s, and if f_1 and f_2 are μ-measurable, we still have

$$\int_S f_1(s)v(\mu, ds) \leqq \int_S f_2(s)v(\mu, ds),$$

even though one or both of these integrals may be infinite.

3. The Lebesgue Spaces

The basis for this section is a finitely additive complex or extended real valued set function μ defined on a field Σ of subsets of a set S. The functions to be integrated with respect to μ will have their values in a real or complex B-space $\mathfrak{X}$. We shall define and discuss the properties of various linear spaces of μ-measurable functions.

1 Definition. Let $1 \leqq p < \infty$. Then $L_p^0(S, \Sigma, \mu, \mathfrak{X})$ (or, if $\mathfrak{X}$ is understood, $L_p^0(S, \Sigma, \mu)$), will denote the set of all μ-measurable functions f on S to $\mathfrak{X}$ such that the function $|f(\cdot)|^p$ is μ-integrable. By the *norm* $|f|$ of an element $f \in L_p^0(S, \Sigma, \mu)$ we mean the quantity

$$|f| = \left[\int_S |f(s)|^p v(\mu, ds)\right]^{1/p}.$$

When desirable for clarity, the symbol $|f|_p$ will be used for the norm of an element in $L_p^0(S, \Sigma, \mu)$.

➔ 2 Lemma. (*Hölder*) *Let f be a scalar valued function and g a vector valued function with $f \in L_p^0(S, \Sigma, \mu)$, $g \in L_q^0(S, \Sigma, \mu)$, where $p > 1$, $q > 1$, and $1/p+1/q = 1$. Then the function fg is μ-integrable, and*

$$\left|\int_S f(s)g(s)\mu(ds)\right| \leqq |f|_p|g|_q.$$

Proof. In case either $|f|_p$ or $|g|_q$ is zero the lemma follows from Lemmas 2.12, 2.21, and Theorem 2.20(d) and so we shall assume that neither of these norms is zero. The function $\varphi(t) = t^p/p+t^{-q}/q$ has a positive derivative for $t > 1$, and a negative derivative for $0 < t < 1$. Hence, its minimum value for $t > 0$ is $\varphi(1) = 1$. If we put $t = a^{1/q}b^{-1/p}$ we obtain the inequality $ab \leqq a^p/p+b^q/q$, valid for $a, b > 0$; hence, the inequality $|ab| \leqq |a|^p/p+|b|^q/q$ is valid for all scalars a, b. Putting $a = f(s)/|f|_p$, $b = g(s)/|g|_q$, we find that

$$|f(s)g(s)| \leqq \frac{1}{p}\, |f(s)|^p|f|_p^{1-p}|g|_q + \frac{1}{q}\, |g(s)|^q|g|_q^{1-q}|f|_p.$$

It follows from Lemma 2.12, Theorem 2.19(a), 2.20(a) and Theorem 2.22(b) that fg is integrable and that

$$\left|\int_S f(s)g(s)\mu(ds)\right| \leqq \left(\frac{1}{p}+\frac{1}{q}\right) |f|_p|g|_q. \quad \text{Q.E.D.}$$

➔ 3 **Lemma.** *Let* $1 \leqq p < \infty$, *let* f_1, f_2 *be in* $L_p^0(S, \Sigma, \mu)$, *and let* α *be a scalar. Then*

(a) *the function* αf_1 *lies in* $L_p^0(S, \Sigma, \mu)$, *and* $|\alpha f_1|_p = |\alpha||f_1|_p$;

(b) *the sum of* f_1 *and* f_2 *lies in* $L_p^0(S, \Sigma, \mu)$, *and* $|f_1+f_2|_p \leqq |f_1|_p+|f_2|_p$;

(c) $|f_1-f_2|_p = 0$ *if and only if* f_1-f_2 *is a null function.*

The inequality in part (b) is known as *Minkowski's inequality.*

Proof. Part (a) is obvious. Part (b) is obvious if $p = 1$; to prove part (b) for $p > 1$, we reason as follows: the function $(1+x)^p/(1+x^p)$ approaches 1 as $x \to 0$ or $x \to \infty$; consequently, it is bounded by some constant c in the whole range $0 \leqq x < \infty$. Putting $x = a/b$, it follows that we have $(a+b)^p \leqq c(a^p+b^p)$ for all $0 \leqq a, b < \infty$. Since this implies

$$|f_1(s)+f_2(s)|^p \leqq \{|f_1(s)|+|f_2(s)|\}^p \leqq c\{|f_1(s)|^p+|f_2(s)|^p\},$$

it follows from Theorem 2.22(b) that

$$f_1+f_2 \in L_p^0(S, \Sigma, \mu).$$

Now,

$$|f_1+f_2|_p^p = \int_S |f_1(s) + f_2(s)|^p v(\mu, ds)$$

$$\leqq \int_S \{|f_1(s)|+|f_2(s)|\}|f_1(s)+f_2(s)|^{p-1}v(\mu, ds)$$

$$= \int_S |f_1(s)||f_1(s)+f_2(s)|^{p-1}v(\mu, ds)$$

$$+ \int_S |f_2(s)||f_1(s)+f_2(s)|^{p-1}v(\mu, ds)$$

$$\leqq \{|f_1|_p + |f_2|_p\} \left\{ \int_S \{|f_1(s)| + |f_2(s)|\}^{q(p-1)} v(\mu, ds) \right\}^{1/q}$$

by Lemma 2, where $1/p+1/q = 1$. Since $q(p-1) = p$,

$$\{|f_1 + f_2|_p\}^p \leqq \{|f_1|_p + |f_2|_p\} \{|f_1 + f_2|_p\}^{p/q}.$$

Thus,

$$|f_1 + f_2|_p = \{|f_1 + f_2|_p\}^{p-p/q} \leqq |f_1|_p + |f_2|_p,$$

which proves (b).

Part (c) follows from Theorem 2.20(d). Q.E.D.

In view of Lemma 3(c) it is natural to consider the classes of functions in $L_p^0(S, \Sigma, \mu, \mathfrak{X})$ equivalent under the relation: f is equivalent to g if and only if $f-g$ is a null function. Denoting the equivalence class of $f \in L_p^0(S, \Sigma, \mu, \mathfrak{X})$ by $[f]$, it follows from Corollary 2.5 and Lemma 3(c) that the equivalence classes form a linear space in which $|[f]| = |f|_p$ is a norm. We refer the reader to the analogous discussion following Corollary 2.5 for the space $F(S, \Sigma, \mu, \mathfrak{X})$.

4 DEFINITION. The symbol $L_p(S, \Sigma, \mu, \mathfrak{X})$ will denote the set of equivalence classes $[f]$ of functions $f \in L_p^0(S, \Sigma, \mu, \mathfrak{X})$.

In view of the remarks above, we have the following theorem.

5 THEOREM. *The space $L_p(S, \Sigma, \mu, \mathfrak{X})$ is a normed linear space.*

As in the case of the space $F(S, \Sigma, \mu, \mathfrak{X})$ it is customary to refer to the elements of $L_p(S, \Sigma, \mu, \mathfrak{X})$ as if they were functions, rather than classes of equivalent functions. Thus where no confusion will result, we shall speak simply of "a function in L_p". In the sequel the symbol f rather than $[f]$ will be used for an element in L_p. We observe that the inequality of Minkowski and (in the case of scalar valued functions) the inequality of Hölder may be regarded as applying to the spaces L_p. This observation is obvious in the case of Minkowski's inequality. To see that it is justified in case of Hölder's inequality, note that Hölder's inequality implies that

$$|fg|_1 \leqq |f|_p|g|_q \text{ if } f \in L_p^0, \quad g \in L_q^0, \quad \frac{1}{p}+\frac{1}{q} = 1,$$

and at least one of f and g is scalar-valued. Thus, if one of f and g is μ-null, so is fg. Hence, if $f_1, f_2 \in L_p^0$, $g_1, g_2 \in L_q^0$, and $f_1 - f_2$ and $g_1 - g_2$ are μ-null functions, then $f_1 g_1 - f_2 g_2$ is a μ-null function.

The reader will more easily perceive the significance of the somewhat complicated conditions (ii) and (iii) in the following theorem, if he reads the statement and proof of Theorem 7 after the statement of Theorem 6 but before its proof.

6 **Theorem.** *Let* $1 \leqq p < \infty$; *let* $\{f_n\}$ *be a sequence of functions in* $L_p(S, \Sigma, \mu, \mathfrak{X})$ *and let* f *be a function on* S *to* $\mathfrak{X}$. *Then* f *is in* L_p *and* $|f_n - f|_p$ *converges to zero if and only if the three following conditions hold:*

(i) f_n *converges to* f *in measure;*

(ii) $\lim_{v(\mu, E) \to 0} \int_E |f_n(s)|^p v(\mu, ds) = 0$ *uniformly in* n;

(iii) *for each* $\varepsilon > 0$ *there is a set* E_ε *in* Σ *with* $v(\mu, E_\varepsilon) < \infty$ *and such that*

$$\int_{S-E_\varepsilon} |f_n(s)|^p v(\mu, ds) < \varepsilon, \qquad n = 1, 2, \ldots.$$

Proof. Assuming (i), (ii), and (iii) let $g_n = |f_n(\cdot)|^p$, $g = |f(\cdot)|^p$. We will first show that $\{g_n\}$ is a Cauchy sequence in $L_1(S)$. For $\varepsilon > 0$ there is, by (iii), a set E_ε with $v(\mu, E_\varepsilon) < \infty$ such that

$$|g_n - g_m|_1 \leqq 2\varepsilon + \int_{E_\varepsilon} |g_n(s) - g_m(s)| v(\mu, ds), \qquad n, m \geqq 1.$$

Thus to see that the sequence $\{g_n\}$ is a Cauchy sequence in $L_1(S)$ it will suffice to prove that it is a Cauchy sequence in $L_1(E_\varepsilon)$. Hence we may and shall assume that $v(\mu, S) < \infty$. By (ii) there is a $\delta > 0$ such that

$$\int_E g_n(s) v(\mu, ds) < \varepsilon, \qquad n \geqq 1,$$

provided that $v(\mu, E) < \delta$. By Lemma 2.12 $g_n \to g$ in measure, so there are sets E_{nm} in Σ with $v(\mu, E_{nm}) < \delta$ for all large values of n and m and with $|g_n(s) - g_m(s)| < \varepsilon$ for every s in $S - E_{nm}$. Thus, for sufficiently large n and m,

$$\begin{aligned} |g_n - g_m|_1 &\leqq 2\varepsilon + \int_{S-E_{nm}} |g_n(s) - g_m(s)| v(\mu, ds) \\ &\leqq \varepsilon(2 + v(\mu, S)), \end{aligned}$$

which proves that $\{g_n\}$ is a Cauchy sequence in $L_1(S)$. To see that g is integrable let h_n, $n \geqq 1$, be integrable simple functions with

$$\int_S |h_n(s) - g_n(s)| v(\mu, ds) < \frac{1}{n},$$

and $|h_n - g_n|_F < 1/n$, where the norm is that of the space $F(S, \Sigma, \mu)$. Since $\{g_n\}$ is a Cauchy sequence in $L_1(S)$ with $g_n \to g$ in measure it follows that $\{h_n\}$ is a Cauchy sequence in $L_1(S)$ which also converges to g in measure. By Definition 2.17, g is integrable and this proves that f is in L_p.

For convenience let us write $f_\infty = f$. In view of Theorem 2.20, conditions (ii) and (iii) hold uniformly for $1 \leqq n \leqq \infty$. For each $\varepsilon > 0$ there exists a set F_n such that $|f_n(s) - f_\infty(s)| < \varepsilon$ for $s \notin F_n$ and such that $v(\mu, F_n) \to 0$ as $n \to \infty$. Consequently

$$\begin{aligned}
|f_n - f|_p &= \left\{\int_S |f_n(s) - f(s)|^p v(\mu, ds)\right\}^{1/p} \\
&\leqq \left\{\int_{S-(E_\delta \cup F_n)} |f_n(s)|^p v(\mu, ds)\right\}^{1/p} \\
&\quad + \left\{\int_{S-(E_\delta \cup F_n)} |f_\infty(s)|^p v(\mu, ds)\right\}^{1/p} \\
&\quad + \left\{\int_{E_\delta - F_n} |f_n(s) - f_\infty(s)|^p v(\mu, ds)\right\}^{1/p} \\
&\quad + \left\{\int_{F_n} |f_n(s) - f_\infty(s)|^p v(\mu, ds)\right\}^{1/p} \\
&< 2\delta^{1/p} + \varepsilon[v(\mu, E_\delta)]^{1/p} \\
&\quad + \left\{\int_{F_n} \{|f_n(s)|^p v(\mu, ds)\right\}^{1/p} \\
&\quad + \left\{\int_{F_n} |f_\infty(s)|^p v(\mu, ds)\right\}^{1/p}.
\end{aligned}$$

Now let $\gamma > 0$. If we choose δ such that $2\delta^{1/p} < \gamma$, and then choose ε such that $\varepsilon[v(\mu, E_\delta)]^{1/p} < \gamma$ and $\{\int_F |f_n(s)|^p v(\mu, ds)\}^{1/p} < \gamma$, $1 \leqq n \leqq \infty$, whenever $v(\mu, F) < \varepsilon$, then $|f_n - f|_p < 4\gamma$ for $n > n_0$, which proves that $|f_n - f|_p \to 0$. This completes the proof of the sufficiency.

Next we will prove the necessity of (i). If $f \in L_p$ and $|f_n - f|_p \to 0$ then for each $\varepsilon > 0$ there is an integer n_0 such that $n \geqq n_0$ implies $\int_S g_n(s) v(\mu, ds) < \varepsilon$ where $g_n(\cdot) = |f_n(\cdot) - f(\cdot)|^p$. Now each g_n is the limit in μ-measure of a sequence $\{h_n^k\}$ of real valued simple functions, each assuming finitely many values on sets of Σ, such that

$$\lim_{k\to\infty} \int_S h_n^k(s)v(\mu, ds) = \int_S g_n(s)v(\mu, ds)$$

(cf. Lemma 2.18). Since we may replace $h_n^k(\cdot)$ by $|h_n^k(\cdot)|$, it may be supposed that $h_n^k(s) \geqq 0$. Consequently $\int_S h_n^k(s)v(\mu, ds) < 2\varepsilon$ for all sufficiently large k, and we may assume without loss of generality that $\int_S h_n^k(s)v(\mu, ds) < 2\varepsilon$ for all k. By the simple form of the functions h_n^k the set

$$E_n^k = \{s|h_n^k(s) > \gamma\}$$

is in Σ for each $\gamma > 0$. Moreover,

$$\gamma v(\mu, E_n^k) \leqq \int_S h_n^k(s)v(\mu, ds) < 2\varepsilon.$$

Consequently, $v(\mu, E_n^k) < 2\varepsilon/\gamma$. Since $\{h_n^k\}$ converges in μ-measure to g_n, for each sufficiently large k we can find a set $F_n^k \in \Sigma$ such that $v(\mu, F_n^k) < \varepsilon/\gamma$ and $|h_n^k(s) - g_n(s)| < \gamma$ for $s \notin F_n^k$. Thus $|f_n(s) - f(s)| < (2\gamma)^{1/p}$ for $s \notin F_n^k \cup E_n^k$. Given $\delta_1, \delta_2 > 0$, choose γ so small that $(2\gamma)^{1/p} < \delta_1$, and ε so small that $3\varepsilon/\gamma < \delta_2$. Then for $n \geqq n_0$, $|f_n(s) - f(s)| < \delta_1$ except when s lies in a set G_n such that $v(\mu, G_n) < \delta_2$. Thus $f_n \to f$ in measure and (i) is proved.

We now prove the necessity of (ii). By Theorem 2.20, for each $\varepsilon > 0$ there is a $\delta_1 > 0$ such that $v(\mu, E) < \delta_1$ implies $\left\{\int_E |f(s)|^p v(\mu, ds)\right\}^{1/p} < \varepsilon$. Let n_0 be such that $|f_n - f|_p < \varepsilon$ if $n \geqq n_0$, and let $\delta_2 < \delta_1$ be a positive quantity such that $v(\mu, E) < \delta_2$ implies

$$\left\{\int_E |f_n(s)|^p v(\mu, ds)\right\}^{1/p} < \varepsilon \text{ for } 1 \leqq n \leqq n_0.$$

Then,

$$\left\{\int_E |f_n(s)|^p v(\mu, ds)\right\}^{1/p} < 2\varepsilon \quad \text{for } 1 \leqq n \leqq \infty,$$

proving (ii).

Statement (iii) may be proved in the same way to complete the proof of the theorem. Q.E.D.

7 THEOREM. (*Dominated convergence theorem*) *Let* $1 \leqq p < \infty$, *let* $g \in L_p(S, \Sigma, \mu)$, *and let* $\{f_\alpha\}$ *be a generalized sequence of elements of*

$L_p(S, \Sigma, \mu)$ such that $|f_\alpha(s)| \leqq |g(s)|$ almost everywhere for each α. Then f_α converges to f in μ-measure if and only if f is in L_p and the norms $|f_\alpha - f|_p$ converge to zero.

Proof. Let us first consider the case when the generalized sequence is an ordinary sequence $\{f_n\}$. Since $|f_n(s)| \leqq |g(s)|$ almost everywhere, (ii) and (iii) of Theorem 6 hold automatically. Consequently the statements $f_n \to f$ in measure and $f_n \to f$ in L_p are equivalent by Theorem 6.

It will now be shown that the statement of the theorem for a generalized sequence follows from the case for sequences. We recall that the topologies of the spaces $F(S, \Sigma, \mu, \mathfrak{X})$ and $L_p(S, \Sigma, \mu, \mathfrak{X})$ are metric. Let $f_\alpha \to f$ in measure and suppose that f_α does not converge to f in L_p. Then there is an $\varepsilon > 0$, and for each α a $\beta_\alpha \geqq \alpha$, such that $|f_{\beta_\alpha} - f|_p > \varepsilon$. The generalized sequence $\{f_\gamma\}$, $\gamma = \beta_\alpha$, clearly converges to f in measure. Now from each sphere of radius $1/n$ about f in F select an element f_{γ_n}. Since $f_{\gamma_n} \to f$ in measure, it follows that $f_{\gamma_n} \to f$ in L_p, contradicting the inequality above. The proof of the converse is similar. Q.E.D.

8 Corollary. *Let $1 \leqq p < \infty$. The set of μ-simple μ-integrable functions is dense in $L_p(S, \Sigma, \mu, \mathfrak{X})$.*

Proof. Let $f \in L_p = L_p(S, \Sigma, \mu, \mathfrak{X})$ and $\varepsilon > 0$. By 2.20 we can find an $E \in \Sigma$ such that $v(\mu, E) < \infty$, and such that

$$\int_{S-E} |f(s)|^p v(\mu, ds) < \varepsilon^p.$$

Thus, if χ_E is the characteristic function of E, then $\chi_E f$ is an element of L_p and $|f - \chi_E f|_p < \varepsilon$. By Definition 2.10, $\chi_E f$ is the limit in μ-measure of a sequence $\{f_n\}$ of μ-simple functions. By using the argument presented in the proof of Theorem 2.22, we may and shall assume that $|f_n(s)| \leqq 2|\chi_E(s)||f(s)|$, $s \in S$. Since $f_n(s)$ vanishes for $s \notin E$, f_n is μ-integrable and μ-simple. By Theorem 7, $|f_n - \chi_E f|_p < \varepsilon$ for sufficiently large n. Hence $|f - f_n|_p < 2\varepsilon$ for sufficiently large n and the proof is complete. Q.E.D.

4. Countably Additive Set Functions

The basis for the present section is a countably additive set function defined on a σ-field of subsets of a set. In this case the results of the preceding sections can be considerably extended.

1 Definition. Let μ be a vector valued, complex valued, or extended real valued additive set function defined on a field Σ of subsets of a set S. Then μ is said to be *countably additive* if

$$\mu\left(\bigcup_{i=1}^{\infty} E_i\right) = \sum_{i=1}^{\infty} \mu(E_i)$$

whenever $E_1, E_2, \ldots$ are disjoint sets in Σ whose union also belongs to Σ.

2 Definition. A *σ-field* is a field Σ of subsets of a set S with the property that $\bigcup_{i=1}^{\infty} E_i \in \Sigma$ whenever $E_n \in \Sigma$, $n = 1, 2, \ldots$. In other words, a σ-field is a field which is closed under the operation of forming denumerable unions.

3 Definition. A triple (S, Σ, μ) consisting of a set S, a field Σ of subsets of S, and a complex valued or extended real valued additive set function μ defined on Σ is said to be a *measure space* if Σ is a σ-field, and μ is countably additive. Occasionally, S itself is called a measure space. The sets in Σ are called *measurable* sets. The measure space is said to be *finite* if μ does not take on either of the values $+\infty$ or $-\infty$, and to be *positive* if μ never takes on a negative value. A *measure* is a complex or extended real valued countably additive set function whose domain is a σ-field.

Throughout this section (S, Σ, μ) is assumed to be a measure space.

Suppose that $\{E_n\}$ is a sequence of sets. We define the *limit inferior* and the *limit superior* of $\{E_n\}$ by the equations

$$\liminf_n E_n = \bigcup_{n=1}^{\infty} \bigcap_{m=n}^{\infty} E_m, \qquad \limsup_n E_n = \bigcap_{n=1}^{\infty} \bigcup_{m=n}^{\infty} E_m.$$

If $\liminf_n E_n = \limsup_n E_n$, $\{E_n\}$ is said to be *convergent*, and we write the common value of the limit inferior and the limit superior as $\lim_n E_n$. A *non-decreasing* sequence $\{E_n\}$ is one for which $E_n \subseteq E_{n+1}$,

$n = 1, 2, \ldots$. It is evident that such a sequence has the limit $\bigcup_{n=1}^{\infty} E_n$. A *non-increasing* sequence $\{E_n\}$ is one for which $E_n \supseteq E_{n+1}$, $n = 1, 2, \ldots$. Such a sequence has the limit $\bigcap_{n=1}^{\infty} E_n$. A monotone sequence is one which is either non-decreasing or non-increasing. We note that the intersection, union, limit inferior, and limit superior of every sequence of measurable sets is measurable.

4 LEMMA. *The values of an extended real-valued measure which never assumes the value* $+\infty$ *have a finite upper bound.*

PROOF. Suppose that μ is not bounded above. Call a set $E_1 \in \Sigma$ unbounded if $\sup_{E \in \Sigma} \mu(EE_1) = +\infty$; otherwise call it bounded. Then either

(a) every unbounded set contains an unbounded set of arbitrarily large measure, or

(b) there exists an unbounded set $F \in \Sigma$ and an integer N such that F contains no unbounded set of measure greater than N.

In case (a), we can clearly apply induction, and find a decreasing sequence of unbounded sets such that $\mu(E_n) \geqq n$. Then

$$\mu\left(\bigcap_{i=1}^{\infty} E_i\right) + \sum_{i=n}^{\infty} \mu(E_i - E_{i+1}) = \mu(E_n);$$

since $\mu(E_n) \neq +\infty$, the series on the left converges to a finite quantity, and we have

$$\mu\left(\bigcap_{i=1}^{\infty} E_i\right) = \lim_{n\to\infty} \mu(E_n) = +\infty,$$

which contradicts our hypothesis.

In case (b) we let F_1 be a measurable subset of F such that $\mu(F_1) > N$. Then F_1 is not unbounded; since F is unbounded, $F - F_1$ is unbounded. Let A_1 be a measurable subset of $F - F_1$ such that $\mu(A_1) \geqq 1$. Then, since F contains no unbounded set of measure greater than N, $F_2 = F_1 \cup A_1$ is bounded, so $F - F_2$ is unbounded. Proceeding inductively, we construct a sequence of disjoint measurable sets $A_1, A_2, \ldots$ such that $\mu(A_k) \geqq 1$ for every k. It follows that $\mu(\bigcup_{i=1}^{\infty} A_i) = +\infty$, contrary to assumption. Q.E.D.

5 COROLLARY. *The set of values of a vector valued, countably additive set function defined on a* σ*-field of sets is bounded.*

Proof. Let μ be a set function defined on the σ-field Σ, and with values in the B-space X. Then, by Lemma 4, $\mathscr{R}x^*\mu(\Sigma)$ and $\mathscr{I}x^*\mu(\Sigma)$ are both bounded for each $x^* \in X^*$, so that by II.3.20, $\mu(\Sigma)$ is bounded. Q.E.D.

6 Corollary. *If (S, Σ, μ) is a finite measure space, μ is bounded.*

7 Lemma. *If (S, Σ, μ) is a finite measure space, the total variation $v(\mu)$ is countably additive and bounded. If μ is real valued, the upper and lower variations μ^+ and μ^- are also countably additive and bounded.*

Proof. The boundedness follows from Corollary 6 and Lemma 1.5, and only the countable additivity need be established. Since $\mu^+ = (1/2)(v(\mu)+\mu)$ and $\mu^- = (1/2)(v(\mu)-\mu)$, it is enough to show that $v(\mu)$ is countably additive. Let E_n be disjoint sets in Σ and let $\bigcup_{i=1}^{\infty} E_n = E$. Now, since $v(\mu)$ is non-negative and additive,

$$v(\mu, E) \geqq v(\mu, \bigcup_{n=1}^{m} E_n) = \sum_{n=1}^{m} v(\mu, E_n),$$

$$v(\mu, E) \geqq \sum_{n=1}^{\infty} v(\mu, E_n).$$

On the other hand, let $\{F_i\}$ be a finite sequence of disjoint measurable sets, with $F_i \subseteq E$. Then

$$\sum_{i=1}^{k} |\mu(F_i)| = \sum_{i=1}^{k} |\mu(\bigcup_{n=1}^{\infty} F_i E_n)| = \sum_{i=1}^{k} \left| \sum_{n=1}^{\infty} \mu(F_i E_n) \right|$$

$$\leqq \sum_{n=1}^{\infty} \sum_{i=1}^{k} |\mu(F_i E_n)| \leqq \sum_{n=1}^{\infty} v(\mu, E_n),$$

so that $\qquad v(\mu, E) \leqq \sum_{n=1}^{\infty} v(\mu, E_n)$. Q.E.D.

8 Lemma. *If (S, Σ, μ) is a finite positive measure space, and $\{E_n\}$ a sequence of measurable sets, then*

$$\mu(\liminf_n E_n) \leqq \liminf_n \mu(E_n) \leqq \limsup_n \mu(E_n) \leqq \mu(\limsup_n E_n).$$

Proof. Observe first that if $\{E_n\}$ is a non-decreasing sequence with limit E then $E = E_1 \cup (E_2 - E_1) \cup \ldots$, and $\mu(E) = \lim \mu(E_n)$. By taking complements it follows that this equation holds also for non-increasing sequences. Thus if μ is non-negative and $\{E_n\}$ is an arbitrary sequence in Σ,

$$\mu\left(\bigcup_{n=1}^{\infty} \bigcap_{m=n}^{\infty} E_m\right) = \lim_n \mu\left(\bigcap_{m=n}^{\infty} E_m\right) \leqq \liminf_n \mu(E_n),$$

$$\mu\left(\bigcap_{n=1}^{\infty} \bigcup_{m=n}^{\infty} E_m\right) = \lim_n \mu\left(\bigcup_{m=n}^{\infty} E_m\right) \geqq \limsup_n \mu(E_n). \quad \text{Q.E.D.}$$

9 COROLLARY. *If (S, Σ, μ) is a finite measure space, and $\{E_n\}$ a convergent sequence of measurable sets, then* $\lim_n \mu(E_n) = \mu(\lim_n E_n)$.

PROOF. If μ is positive, the conclusion follows from Lemma 8. The general case is treated by splitting μ into its real and imaginary parts, and then decomposing each of these, with the aid of the Jordan decomposition theorem (1.8), into the difference of two positive measures, and then applying Lemma 8. Q.E.D.

10 THEOREM. *(Hahn decomposition) To every extended real measure μ corresponds a measurable set E_0 such that μ is non-negative on measurable subsets of E_0 and non-positive on measurable subsets of E_0'.*

PROOF. Either μ or $-\mu$ fails to assume the value $+\infty$. Hence it may be assumed that $\mu(E) < \infty$ for every E in the σ-field Σ where μ is defined.

Let P consist of all sets $E \in \Sigma$ for which $\mu(AE) \geqq 0$, $A \in \Sigma$. Let $E_n \in P$ be such that

$$\mu(E_n) \to \sup_{E \in P} \mu(E).$$

Since μ is countably additive the set $E_0 = \bigcup_{n=1}^{\infty} E_n$ is in P and

$$\mu(E_0) = \sup_{E \in P} \mu(E).$$

It will now be shown by an indirect proof that $\mu(E_0'A) \leqq 0$ for every $A \in \Sigma$. Suppose that $A_0 \in \Sigma$, $A_0 \subseteqq E_0'$, and $\mu(A_0) > 0$. Partially order the family

$$Q = \{E | E \in \Sigma,\ E \subseteqq A_0,\ \mu(E) \geqq \mu(A_0)\}$$

by defining $E_1 \leqq E_2$ to mean that either $E_1 \supset E_2$ and $\mu(E_1) < \mu(E_2)$, or $E_1 = E_2$. Zorn's lemma (I.2.7) will be applied to show that Q contains a maximal element. Let Q_0 be a totally ordered subset of Q. If there is a $B_0 \in Q_0$ with $\mu(B_0) = \sup_{E \in Q_0} \mu(E)$ then clearly B_0 is an upper bound for Q_0. On the other hand, if

$$\mu(E) < \delta = \sup_{E \in Q_0} \mu(E), \qquad E \in Q_0,$$

there is a sequence $\{B_n\} \subset Q_0$ with $\mu(B_n) < \mu(B_{n+1}) \to \delta$. Since Q_0 is totally ordered, $B_n \supset B_{n+1}$, $n = 1, 2, \ldots$. Since $\mu(B_n) < \mu(B_{n+1}) < \infty$, all the quantities $\mu(B_n)$, $n \geqq 2$, are finite. If $C_n = B_n - B_{n+1}$, and $B_0 = \cap_{n=1}^{\infty} B_n$, then $B_n = B_0 \cup \bigcup_{k=n}^{\infty} C_k$, so that

$$\mu(B_n) = \mu(B_0) + \sum_{k=n}^{\infty} \mu(C_k), \qquad n \geqq 2.$$

Thus $\lim_{n\to\infty} \mu(B_n) = \mu(B_0)$, and the set $B_0 = \bigcap_{n=1}^{\infty} B_n$ is in Q, and $\delta = \mu(B_0)$. Let $E \in Q_0$; then $\mu(E) < \delta = \mu(B_0)$ and there is an n with $\mu(E) < \mu(B_n)$. Since Q_0 is totally ordered, $E \supset B_n \supset B_0$. Thus B_0 is an upper bound for Q_0. Since every totally ordered subset of Q has an upper bound, there is a maximal element M in Q. This maximal element M is in P, for otherwise there is a set $A \in \Sigma$, $A \subseteqq M$, $\mu(A) < 0$, $M - A \subseteqq A_0$, $\mu(M-A) = \mu(M) - \mu(A) > \mu(M) \geqq \mu(A_0)$, and thus $M - A \geqq M$.

Since $M \subseteqq E_0'$ and $M \in P$

$$\mu(M \cup E_0) = \mu(M) + \mu(E_0) > \sup_{E \in P} \mu(E),$$

a contradiction which proves that $\mu(E_0'A) \leqq 0$ for every $A \in \Sigma$. Q.E.D.

The Hahn decomposition allows us to extend the definition of the positive and negative variations to extended real valued measures.

11 COROLLARY. *Let* (S, Σ, μ) *be a measure space, and* μ *an extended real valued measure. Then there are non-negative measures* μ^+, μ^-, *one of which is finite and such that*

$$\mu = \mu^+ - \mu^-, \qquad v(\mu) = \mu^+ + \mu^-.$$

PROOF. If E_0 is the set whose existence is proved in the above theorem, then the functions μ^+, μ^- defined by

$$\mu^+(A) = \mu(E_0 A), \qquad \mu^-(A) = -\mu(E_0' A), \qquad A \in \Sigma,$$

clearly have the desired properties. Q.E.D.

It is clear that if μ is a bounded function the set functions μ^+ and μ^- coincide with those of the Jordan decomposition. We shall continue to call them, even if the μ is an extended real valued measure as in Corollary 11, the *positive* and *negative variations of* μ.

12 DEFINITION. Let λ, μ be finitely additive set functions defined on a field Σ. Then λ is said to be *continuous with respect to* μ or simply μ*-continuous*, if

$$\lim_{v(\mu, E)\to 0} \lambda(E) = 0.$$

The function λ is said to be μ*-singular* if there is a set $E_0 \in \Sigma$ such that

$$v(\mu, E_0) = 0; \qquad \lambda(E) = \lambda(EE_0), \qquad E \in \Sigma.$$

It is clear that the only set function which is both μ-singular and μ-continuous is the set function identically zero.

If λ, μ are both scalar valued additive set functions on a field Σ and if λ is μ-continuous then the inequality (III.1.5)

$$v(\lambda, E) \leqq 4 \sup_{A \subseteq E} |\lambda(A)|$$

shows that *the total variation of* λ *is also* μ*-continuous*. Thus *the positive and negative variations of a real* μ*-continuous set function are likewise* μ*-continuous*. It is also clear that the μ-continuous set functions as well as the μ-singular set functions form a linear vector space under the natural definitions of addition and scalar multiplication.

13 LEMMA. *Let* λ, μ *be countably additive set functions, complex or extended real valued, defined on the* σ*-field* Σ, *and let* λ *be finite. Then* λ *is* μ*-continuous if and only if* $v(\mu, E) = 0$ *implies that* $\lambda(E) = 0$.

PROOF. The necessity of the condition is obvious. To prove the sufficiency of the condition we observe first that a set function λ satisfies this condition if and only if the positive and negative variations of its real and imaginary parts satisfy the same condition. Thus, it may be assumed that λ is non-negative. If λ is not μ-continuous, there is an $\varepsilon > 0$ and sets $E_n \in \Sigma$, $n = 1, 2, \ldots$, with $\lambda(E_n) \geqq \varepsilon$ and $v(\mu, E_n) < 1/2^n$. Let $E_0 = \limsup_n E_n$. Then, for each $n = 1, 2, \ldots$,

$$v(\mu, E_0) \leqq v(\mu, \bigcup_{m=n}^{\infty} E_m) \leqq \sum_{m=n}^{\infty} 1/2^m,$$

which shows that $v(\mu, E_0) = 0$ and hence $\lambda(E_0) = 0$. On the other hand, by Lemma 8,

$$\lambda(E_0) \geqq \limsup_n \lambda(E_n) \geqq \varepsilon,$$

which is a direct contradiction, proving the lemma. Q.E.D.

This lemma shows that if each member of a generalized sequence $\{\lambda_\alpha\}$ of finite, countably additive measures is μ-continuous and if $\lim_\alpha \lambda_\alpha(E) = \lambda(E)$, $E \in \Sigma$, where λ is also a finite, countably additive measure, then λ is also μ-continuous.

It follows easily from Definition 12 that if $\lambda, \lambda_n, n = 1, 2, \ldots$, are finite countably additive measures with $\lambda_n(E) \to \lambda(E)$, $E \in \Sigma$, and if λ_n, $n = 1, 2, \ldots$ are μ-singular, then λ is also μ-singular.

14 THEOREM. (*Lebesgue decomposition*) *Let* (S, Σ, μ) *be a measure space. Then every finite countably additive measure* λ *defined on* Σ *is uniquely representable as a sum* $\lambda = \alpha + \beta$ *where* α *is* μ*-continuous and* β *is* μ*-singular.*

PROOF. The uniqueness of α and β is clear. In view of the Jordan decomposition, which may be applied to both the real and imaginary parts of λ if λ is complex valued, we may and shall assume that λ is non-negative. Partially order the collection N of all sets $E \in \Sigma$ with $v(\mu, E) = 0$ by defining $A \leqq B$ to mean that $\lambda(A) \leqq \lambda(B)$. If N_0 is a totally ordered subset of N and $\delta = \sup_{E \in N_0} \lambda(E)$ then either there is an upper bound E_0 of N_0 in the set N_0 itself or there is a sequence $\{E_n\}$ in N_0 with $\lambda(E_n) < \lambda(E_{n+1}) \to \delta$. In the latter case, $E_n \subset E_{n+1}$, and $E = \bigcup E_n$ is readily seen to be an upper bound for N_0. It follows from Zorn's lemma that N contains a maximal element E_0. The function β defined on Σ by $\beta(E) = \lambda(EE_0)$ is μ-singular. To see that the function α on Σ defined by the equation $\alpha(E) = \lambda(E) - \beta(E) = \lambda(EE_0')$ is μ-continuous, suppose that $E \in N$ and $\alpha(E) = \lambda(EE_0') > 0$. Then $E_0 \subsetneqq E_0 \cup EE_0' \in N$ which contradicts the maximality of E_0, and proves that α is μ-continuous. Q.E.D.

5. Extensions of Set Functions

A given countably additive set function defined on a field can be extended to a countably additive set function defined on a σ-field containing the given field. This extension theorem of Hahn and similar extension theorems, of importance in later applications, will be discussed in this section. Finally, it is shown how the extension theorems may be used to construct the classical measures of Borel, Lebesgue, and Stieltjes.

1 Definition. Let λ be a vector or extended real valued set function defined on a field Σ of sets in S for which $\lambda(\phi) = 0$. A set E is called a *λ-set* if $E \in \Sigma$ and if

$$\lambda(M) = \lambda(ME)+\lambda(ME'), \qquad M \in \Sigma.$$

2 Lemma. *Let λ be any vector valued or extended real valued set function defined on a field Σ of sets in S with $\lambda(\phi) = 0$. The family of λ-sets is a subfield of Σ upon which λ is additive. Furthermore, if E is the union of a finite sequence $\{E_n\}$ of disjoint λ-sets, then*

$$\lambda(ME) = \Sigma\lambda(ME_n), \qquad M \in \Sigma.$$

Proof. It is clear that the void set, the whole space, and the complement of any λ-set are λ-sets. It will now be shown that the product of two λ-sets A, B is also a λ-set. Let $M \in \Sigma$. Since A is a λ-set,

$$\text{(i)} \quad \lambda(MB) = \lambda(MBA)+\lambda(MBA'),$$

and since B is a λ-set,

$$\text{(ii)} \quad \lambda(M) = \lambda(MB)+\lambda(MB'),$$
$$\lambda(M(AB)') = \lambda(M(AB)'B)+\lambda(M(AB)'B'),$$

$$\text{(iii)} \quad \lambda(M(AB)') = \lambda(MBA')+\lambda(MB').$$

From (i) and (ii) it follows that

$$\lambda(M) = \lambda(MBA)+\lambda(MBA')+\lambda(MB'),$$

and from (iii) that

$$\lambda(M) = \lambda(MBA)+\lambda(M(AB)').$$

Thus AB is a λ-set. Since $\cup A_n = (\cap A_n')'$ we conclude that the λ-sets form a field. Now if E_1 and E_2 are disjoint λ-sets, it follows, by replacing M by $M(E_1 \cup E_2)$ in Definition 1, that

$$\lambda(M(E_1 \cup E_2)) = \lambda(ME_1)+\lambda(ME_2).$$

The final conclusion of the lemma follows from this by induction. Q.E.D.

3 Definition. An *outer measure* in S is a non-negative extended

real valued set function λ defined on a σ-field Σ of sets in S and satisfying

(i) $\lambda(\phi) = 0$;

(ii) $\lambda(A) \leqq \lambda(B), \qquad A \subseteqq B, \qquad A, B \in \Sigma$;

(iii) $\lambda(\bigcup_{n=1}^{\infty} E_n) \leqq \sum_{n=1}^{\infty} \lambda(E_n), \qquad \{E_n\} \subseteqq \Sigma$.

4 THEOREM. (*Carathéodory*) *If λ is an outer measure then the family of λ-sets is a σ-field upon which λ is countably additive.*

PROOF. Since the λ-sets form a field (Lemma 2), to see that they form a σ-field it will suffice to show that the union E of any sequence $\{E_n\}$ of disjoint λ-sets is a λ-set. It follows from Lemma 2 that, for $M \in \Sigma$,

$$\begin{aligned}\lambda(M) &= \lambda(M \bigcup_{n=1}^{k} E_n) + \lambda(M(\bigcup_{n=1}^{k} E_n)') \\ &= \sum_{n=1}^{k} \lambda(ME_n) + \lambda(M(\bigcup_{n=1}^{k} E_n)') \\ &\geqq \sum_{n=1}^{k} \lambda(ME_n) + \lambda(ME').\end{aligned}$$

Thus

$$\begin{aligned}\lambda(ME) + \lambda(ME') \geqq \lambda(M) &\geqq \sum_{n=1}^{\infty} \lambda(ME_n) + \lambda(ME') \\ &\geqq \lambda(ME) + \lambda(ME').\end{aligned}$$

This proves that E is a λ-set and shows also, by replacing M by ME, that

$$\lambda(ME) = \sum_{n=1}^{\infty} \lambda(ME_n). \qquad \text{Q.E.D.}$$

5 LEMMA. *Let μ be a non-negative, countably additive extended real valued set function defined on a field Σ of sets in S. For each $A \subseteqq S$ let*

$$\hat{\mu}(A) = \inf \sum_{n=1}^{\infty} \mu(E_n),$$

where the infimum is taken over all sequences $\{E_n\}$ of sets in Σ whose union contains A. Then $\hat{\mu}$ is an outer measure and every set in Σ is a $\hat{\mu}$-set. Furthermore, $\hat{\mu}(E) = \mu(E)$ for E in Σ.

Proof. The properties (i) and (ii) of Definition 3 are evident. Let E be the union of an arbitrary sequence $\{E_n\}$ of sets in S. Let $\varepsilon > 0$ and for each $n = 1, 2, \ldots$ let the sequence $\{E_{m,n}\}$ have the properties

$$E_{m,n} \in \Sigma, \qquad E_n \subseteqq \bigcup_{m=1}^{\infty} E_{m,n}, \qquad \sum_{m=1}^{\infty} \mu(E_{m,n}) \leqq \hat{\mu}(E_n)+\varepsilon/2^{n+1}.$$

Then $\bigcup_{m,n=1}^{\infty} E_{m,n} \supseteqq E$ and thus

$$\hat{\mu}(E) \leqq \sum_{n,m=1}^{\infty} \mu(E_{m,n}) \leqq \sum_{n=1}^{\infty} \hat{\mu}(E_n) + \varepsilon,$$

which proves, since $\varepsilon > 0$ is arbitrary, that $\hat{\mu}$ has the property (iii) of Definition 3. Thus $\hat{\mu}$ is an outer measure defined on the σ-field of all subsets of S.

Now let $E \in \Sigma$. Since $E \subseteqq E$, it follows that $\mu(E) \geqq \hat{\mu}(E)$. If $E_n \in \Sigma$, $n = 1, 2, \ldots$, and $E \subseteqq \bigcup_{n=1}^{\infty} E_n$ then the sets $A_1 = E_1$, $A_n = E_n(\bigcup_{j<n} E_j)'$, $n > 1$, are disjoint sets in Σ with $\bigcup A_n = \bigcup E_n$. Thus

$$\begin{aligned}\mu(E) = \mu(E \bigcup_{n=1}^{\infty} A_n) = \mu\Big(\bigcup_{n=1}^{\infty} EA_n\Big) = \sum_{n=1}^{\infty} \mu(EA_n) \\ \leqq \sum_{n=1}^{\infty} \mu(A_n) \leqq \sum_{n=1}^{\infty} \mu(E_n),\end{aligned}$$

which shows that $\mu(E) \leqq \hat{\mu}(E)$. Thus $\mu(E) = \hat{\mu}(E)$ if E is in Σ.

Finally, to show that every set E in Σ is a $\hat{\mu}$-set, let M be an arbitrary subset of S. Since $\hat{\mu}$ is an outer measure, $\hat{\mu}(ME)+\hat{\mu}(ME') \geqq \hat{\mu}(M)$. Thus to prove that E is a $\hat{\mu}$-set it suffices to show that

$$\hat{\mu}(M) \geqq \hat{\mu}(ME)+\hat{\mu}(ME').$$

For $\varepsilon > 0$ there are sets $E_n \in \Sigma$, $n = 1, 2, \ldots$, with $M \subseteqq \bigcup E_n$ and

$$\sum_{n=1}^{\infty} \mu(E_n) \leqq \hat{\mu}(M)+\varepsilon.$$

Now, since $ME \subseteqq \bigcup EE_n$, and $ME' \subseteqq \bigcup E'E_n$,

$$\begin{aligned}\varepsilon + \hat{\mu}(M) \geqq \sum_{n=1}^{\infty} \mu(E_n) = \sum_{n=1}^{\infty} \{\mu(E_nE) + \mu(E_nE')\} \\ \geqq \hat{\mu}(ME) + \hat{\mu}(ME'). \qquad \text{Q.E.D.}\end{aligned}$$

6 Lemma. *There is a uniquely determined smallest field and a uniquely determined smallest σ-field containing a given family of sets.*

Proof. There is at least one field, namely the field of all subsets of S, which contains a given family τ. The intersection of all fields containing τ is readily seen to be a field and hence it is the smallest field containing τ. A similar argument shows the existence of a smallest σ-field containing τ. Q.E.D.

The smallest field containing a family of sets will sometimes be called the *field determined by*, or *generated by* the family of sets, and the smallest σ-field containing a family of sets correspondingly called the *σ-field determined by*, or *generated by* the family.

7 Definition. Let Σ be a field of subsets of a set S, and μ an extended real valued function defined on Σ. Then μ is said to be *σ-finite on Σ* if S is the union of a sequence $\{E_n\}$ of sets in Σ such that $v(\mu, E_n) < \infty$, $n = 1, 2, \ldots$. A measure space (S, Σ, μ) is called *σ-finite* if μ is σ-finite on Σ.

8 Theorem. (*Hahn extension*) *Every countably additive non-negative extended real valued set function μ on a field Σ has a countably additive non-negative extension to the σ-field determined by Σ. If μ is σ-finite on Σ then this extension is unique.*

Proof. Theorem 4 and Lemma 5 show that the outer measure $\hat{\mu}$ is one non-negative countably additive extension of μ to the σ-field Σ_0 determined by Σ. Suppose that μ_1 is another such extension. If μ is σ-finite on Σ to prove the uniqueness of the extension, it will suffice to show that $\hat{\mu}(E) = \mu_1(E)$ for every set E in Σ_0 contained in a set F in Σ for which $\mu(F) < \infty$. Let $E_n \in \Sigma$ and $E \subseteq \bigcup E_n$. Then since $\mu_1(E) \leqq \Sigma \mu_1(E_n) = \Sigma \mu(E_n)$, it follows that $\mu_1(E) \leqq \hat{\mu}(E)$. Similarly $\mu_1(F - E) \leqq \hat{\mu}(F - E)$. Since

$$\mu_1(E) + \mu_1(F - E) = \mu_1(F) = \hat{\mu}(F) = \hat{\mu}(E) + \hat{\mu}(F - E)$$

the preceding inequalities show that $\mu_1(E) = \hat{\mu}(E)$. Q.E.D.

9 Corollary. *Every bounded, complex valued, countably additive set function on a field Σ has a unique countably additive extension to the σ-field determined by Σ.*

Proof. If μ is a bounded countably additive set function on the field Σ, the Jordan decomposition (1.8) and Lemma 4.7 show that its real and imaginary parts may be expressed as the difference of two non-negative countably additive set functions on Σ. The desired result follows from Theorem 8. Q.E.D.

The next results on the extension of measures make use of interesting relations between the topology of a space and certain measures which may be defined on it.

10 DEFINITION. The smallest σ-field $\mathscr{B}$ containing all the closed sets of a given topological space S is called the *Borel field of* S, and the sets in $\mathscr{B}$ are called the *Borel sets.*

11 DEFINITION. An additive set function μ defined on a field Σ of subsets of a topological space S is said to be *regular* if for each $E \in \Sigma$ and $\varepsilon > 0$ there is a set F in Σ whose closure is contained in E and a set G in Σ whose interior contains E such that $|\mu(C)| < \varepsilon$ for every C in Σ with $C \subseteqq G-F$.

For a complex or extended real valued additive set function μ, it is seen, from Lemma 1.5, that

$$\sup |\mu(C)| \leqq v(\mu, G-F) \leqq 4 \sup |\mu(C)|.$$

Thus for such functions the requirement for regularity, that $\sup |\mu(C)| < \varepsilon$, may be replaced by the equivalent condition: $v(\mu, G-F) < \varepsilon$.

12 LEMMA. *The total variation of a regular additive complex or extended real valued set function on a field is regular. Moreover, the positive and negative variations of a bounded regular real valued additive set function are also regular.*

PROOF. If μ is a regular additive complex or extended real valued set function on a field Σ of subsets of a set S, then the regularity of $v(\mu)$ is obvious from the remark following Definition 11. If μ is bounded and real valued, let $E \in \Sigma, \varepsilon > 0$, and let F and G be sets in Σ such that the closure of F is contained in E, E is contained in the interior of G, and $v(\mu, G-F) < \varepsilon$. Then since

$$v(\mu, G-F) = \mu^+(G-F)+\mu^-(G-F),$$

$\mu^+(G-F)$ and $\mu^-(G-F)$ are both less than ε, and hence μ^+ and μ^- are both regular. Q.E.D.

13 THEOREM. (*Alexandroff*) *Let* μ *be a bounded regular complex valued additive set function defined on a field* Σ *of subsets of a compact topological space* S. *Then* μ *is countably additive.*

PROOF. Let $\varepsilon > 0$. Let $\{E_n\}$ be a disjoint sequence of sets in Σ with union $E \in \Sigma$. There exists a set $F \in \Sigma$ such that $\bar{F} \subseteq E$ and such that $v(\mu, E-F) < \varepsilon$. Moreover there exists a set $G_n \in \Sigma$ such that E_n is contained in the interior G_n^0 of G_n, and such that $v(\mu, G_n - E_n) < \varepsilon/2^n$. Since $\bigcup_{n=1}^{\infty} G_n^0 \supseteq \bar{F}$, there is an integer m such that $\bigcup_{n=1}^{m} G_n^0 \supseteq F$. Hence

$$\sum_{n=1}^{\infty} v(\mu, E_n) \geqq \sum_{n=1}^{\infty} v(\mu, G_n) - \varepsilon \geqq \sum_{n=1}^{m} v(\mu, G_n) - \varepsilon$$
$$\geqq v(\mu, F) - \varepsilon \geqq v(\mu, E) - 2\varepsilon.$$

This proves that $\sum_{n=1}^{\infty} v(\mu, E_n) \geqq v(\mu, E)$. Since

$$v(\mu, E) \geqq v(\mu, \bigcup_{i=1}^{n} E_i) = \sum_{i=1}^{n} v(\mu, E_i), \qquad n = 1, 2, \ldots,$$

it is seen that $v(\mu, E) \geqq \sum_{i=1}^{\infty} v(\mu, E_i)$, and thus that $v(\mu)$ is countably additive.

Since μ is bounded, we see by Lemma 1.5 that $v(\mu, E) < \infty$. Consequently, $\sum_{i=1}^{\infty} v(\mu, E_i) < \infty$, and

$$v(\mu, \bigcup_{i=n}^{\infty} E_i) = \sum_{i=n}^{\infty} v(\mu, E_i) \to 0, \quad \text{as } n \to \infty.$$

Consequently,

$$|\mu(E) - \sum_{i=1}^{n-1} \mu(E_i)| = |\mu(\bigcup_{i=n}^{\infty} E_i)| \leqq v(\mu, \bigcup_{i=n}^{\infty} E_i) \to 0,$$

proving that $\mu(E) = \sum_{n=1}^{\infty} \mu(E_n)$. Q.E.D.

14 THEOREM. *Let* μ *be a bounded regular complex valued additive set function defined on a field* Σ *of subsets of a compact topological space* S. *Then* μ *has a unique regular countably additive extension to the* σ*-field determined by* Σ.

PROOF. Since μ is regular if and only if the positive and negative variations of its real and imaginary parts are regular, we may and shall assume that μ is non-negative. By Theorem 13, μ is countably additive in Σ. Thus the outer measure μ as defined in Lemma 5 furnishes, according to Theorem 4, Theorem 8, and Lemma 5, a countably additive extension to the σ-field Σ_1 generated by Σ. Thus for $E \in \Sigma_1$ and $\varepsilon > 0$, there are sets $E_n \varepsilon \Sigma$ with $\bigcup_{n=1}^{\infty} E_n \supseteqq E$ and

$$\hat{\mu}(\bigcup E_n - E) < \varepsilon/2.$$

By Definition 11 there is an open set G_n and a set $A_n \in \Sigma$ with $E_n \subseteqq G_n \subseteqq A_n$ and

$$\mu(A_n - E_n) < \varepsilon/2^{n+1}.$$

If $G = \bigcup G_n$ and $A = \bigcup A_n$, then G is open, $A \in \Sigma_1$, and

$$E \subseteqq G \subseteqq A, \qquad A - \bigcup_{n=1}^{\infty} E_n \subseteqq \bigcup_{n=1}^{\infty} (A_n - E_n),$$

$$\begin{aligned}\hat{\mu}(A - E) &\leqq \hat{\mu}(A - \bigcup E_n) + \hat{\mu}(\bigcup E_n - E)\\ &\leqq \sum_{n=1}^{\infty} \mu(A_n - E_n) + \varepsilon/2\\ &\leqq \sum_{n=1}^{\infty} \varepsilon/2^{n+1} + \varepsilon/2 = \varepsilon.\end{aligned}$$

Applying the same argument to E', we construct a set B in Σ_1 whose closure is contained in E, such that $\hat{\mu}(E - B) < \varepsilon$. This proves that $\hat{\mu}$ is regular on Σ_1. Q.E.D.

Theorem 14 may be used to construct many interesting examples of regular countably additive measures. One of the best known examples of such a measure is *Borel-Lebesgue*, or *Borel measure*. To construct Borel-Lebesgue measure on the compact interval $S = [a, b]$ of real numbers, consider intervals I having one of the two forms $[a, d]$ or $(c, d]$ where $a < c < d \leqq b$. For such intervals, place $\mu([a, d]) = d - a$, $\mu((c, d]) = d - c$. Let Σ consist of all finite unions of such intervals. It is clear that Σ is a field and that if a set $E \in \Sigma$ has the form

$$E = I_1 \cup I_2 \cup \ldots \cup I_n,$$

where I_j, $j = 1, \ldots, n$ are disjoint intervals of the type described, then $\mu(I_1) + \ldots + \mu(I_n)$ is independent of the particular family of disjoint intervals $I_1, \ldots, I_n$, whose union is E. The number $\mu(E)$ may thus be defined by the equation

$$\mu(E) = \mu(I_1)+\ldots+\mu(I_n).$$

The hypotheses of Theorem 14 are readily verified, and thus by that theorem there is a unique regular countably additive extension of μ to the σ-field of Borel subsets of S. This extension is known as *the Borel measure in* $[a, b]$.

The preceding construction can be generalized to several dimensions. It can also be extended by defining the measure differently on the basic field Σ. We shall illustrate this latter remark by describing here the construction of a Radon measure on an interval.

15 **Definition.** An *interval* is a set of points in the extended real number system which has one of the forms:

$$[a, b] = \{s|a \leqq s \leqq b\}, \qquad [a, b) = \{s|a \leqq s < b\},$$
$$(a, b] = \{s|a < s \leqq b\}, \qquad (a, b) = \{s|a < s < b\}.$$

The number a is called *the left end point* and b *the right end point* of any of these intervals. An interval is *finite* if both its end points are finite; otherwise it is an *infinite interval.* If f is a complex function on an interval I, the *total variation* of f on I is defined by the equation

$$v(f, I) = \sup \sum_{i=1}^{n} |f(b_i) - f(a_i)|,$$

where the supremum is taken over all finite sets of points a_i, $b_i \in I$ with $a_1 \leqq b_1 \leqq a_2 \leqq b_2 \leqq \ldots \leqq a_n \leqq b_n$. If $v(f, I) < \infty$, f is said to be of *bounded variation on I.*

16 **Lemma.** *Let f be a function of bounded variation on an interval I and let c be any point in I except the right end point. Then*

$$\lim_{\varepsilon \to 0+} v(f, (c, c+\varepsilon]) = 0.$$

Proof. If b is the right end point of I the function $v(f, (c, c+\varepsilon])$ is a non-decreasing function of ε in the interval $0 < \varepsilon < b-c$. We may therefore make an indirect proof by supposing that for some positive δ,

$$v(f, (c, c + \varepsilon]) > \delta, \qquad 0 < \varepsilon < b - c.$$

Thus if $0 < \varepsilon_1 < b-c$ there are n_1 points a_i, b_i in $(c, c+\varepsilon_1]$ with $c < a_{n_1} \leqq b_{n_1} \leqq \ldots \leqq a_1 \leqq b_1 \leqq c+\varepsilon_1$ and such that

$$\sum_{i=1}^{n_1} |f(b_i)-f(a_i)| > \delta.$$

Let $\varepsilon_2 = a_{n_1} - c$. Since $v(f, (c, c+\varepsilon_2]) > \delta$ there are points a_j, b_j, $j = n_1+1, \ldots, n_2$ in $(c, c+\varepsilon_2] = (c, a_{n_1}]$ with $c < a_{n_2} \leqq b_{n_2} \leqq \ldots \leqq a_{n_1+1} \leqq b_{n_1+1} \leqq c+\varepsilon_2 = a_{n_1}$ and such that

$$\sum_{i=n_1+1}^{n_2} |f(b_i)-f(a_i)| > \delta.$$

The argument may be repeated by defining $\varepsilon_3 = a_{n_2} - c$ and choosing appropriate points in the interval $(c, c+\varepsilon_3]$. By induction, it is clear that for every integer $k = 1, 2, \ldots$, there are points a_i, b_i with $c < a_{n_k} \leqq b_{n_k} \leqq \ldots \leqq a_1 \leqq b_1 \leqq c+\varepsilon_1$ such that

$$\sum_{i=1}^{n_k} |f(b_i)-f(a_i)| > (k-1)\delta, \qquad k = 1, 2, \ldots.$$

This contradicts the fact that f is of bounded variation on I and proves the lemma. Q.E.D.

Now let f be a function of bounded variation on the open interval $I = (a, b)$ which may be finite or infinite. It is assumed that

$$\text{(i)} \qquad f(s) = \lim_{\varepsilon \to 0} f(s+|\varepsilon|), \qquad s \in I;$$

i.e., f is continuous on the right at every point in the open interval I. The closed interval $\bar{I} = [a, b]$ is a compact subset of the extended real number system and we extend the domain of f to $\bar{I}$ by placing $f(a) = f(b) = 0$. Just as in the above construction of Borel measure, we let Σ be the field of all finite unions

$$\text{(ii)} \qquad E = I_1 \cup I_2 \cup \ldots \cup I_n$$

of intervals I_j, $j = 1, 2, \ldots, n$ where each I_j has one of the two forms: $[a, d]$ or $(c, d]$ with $a < c < d \leqq b$. If the intervals I_j, $j = 1, \ldots, n$ in (ii) are disjoint we define $\mu(E)$ for $E \in \Sigma$ by the equation

$$\text{(iii)} \qquad \mu(E) = \sum_{j=1}^{n} \mu(I_j)$$

where $\mu([a, d]) = f(d)-f(a)$ and $\mu((c, d]) = f(d)-f(c)$ for $a < c < d \leqq b$. It is readily seen that $\mu(E)$ is independent of the particular finite set $\{I_j\}$ used to represent E and that μ is additive on Σ. Since f is of bounded variation on I, μ is bounded. Moreover, if E consists of a single interval, then it follows readily from (i) that $v(\mu, E) = v(f, E)$. From this equation and the preceding lemma, the regularity of μ on Σ may be seen as follows: let E be given by (ii) where $I_j = (a_j, b_j]$, $a \leqq a_1 < b_1 \leqq a_2 < b_2 \ldots \leqq a_n < b_n \leqq b$, and let

$$E(\varepsilon) = \bigcup_{j=1}^{n} (a_j+\varepsilon, b_j], \qquad 0 < \varepsilon < \inf (b_j - a_j).$$

Then, by Lemma 16, $v(\mu, E-E(\varepsilon)) = \sum_{j=1}^{n} v(f, (a_j, a_j+\varepsilon]) \to 0$ which proves that μ is regular on Σ. (In the above expressions, $(a_1, b_1]$ and $(a_1, a_1+\varepsilon]$ are to be replaced by $[a, b_1]$ and $[a, a+\varepsilon]$ respectively if $a_1 = a$.)

It follows from Theorem 14 that μ has a regular countably additive extension to the σ-field of all Borel sets in $[a, b]$. The restriction of this extension to the σ-field of Borel subsets of (a, b) is called the *Radon* or *Borel-Stieltjes measure in* (a, b) *determined by the function* f.

The following extension theorem is elementary and independent of the preceding extension theorems. It establishes a general form of the relationship between the Borel-Stieltjes measures as just defined and the Lebesgue-Stieltjes measures to be defined presently.

17 THEOREM. *Let μ be a countably additive vector or extended real valued set function on the σ-field Σ. Let Σ^* consist of all sets of the form $E \cup N$ where E is in Σ and N is a subset of a set M in Σ with $v(\mu, M) = 0$. Then Σ^* is a σ-field and if the domain of μ is extended to Σ^* by defining $\mu(E \cup N) = \mu(E)$, the extended function is countably additive on Σ^*.*

PROOF. It will first be shown that the family Σ^* is a σ-field. Throughout the proof the symbol E with or without subscripts will denote a set in Σ, the symbol M with or without subscripts will denote a set in Σ for which $v(\mu, M) = 0$, and N with or without subscripts will denote a subset of a set M. To see that the complement of a set $E \cup N$ in Σ^* is also in Σ^*, let $N \subseteq M$ so that

$$(E \cup N)' = E'N' \supseteq E'M', \qquad E'N' - E'M' = E'(N'-M') \subseteq M.$$

Thus if $N_1 = E'N' - E'M'$ then $N_1 \subseteq M$ and $(E \cup N)' = (E'M') \cup N_1$. Hence Σ^* contains the complement of every one of its elements. Next let $\{E_n \cup N_n\} \subseteq \Sigma^*$ and $N_n \subseteq M_n$. Then, since

$$\bigcup_{n=1}^{\infty} (E_n \cup N_n) - \bigcup_{n=1}^{\infty} E_n \subseteq \bigcup_{n=1}^{\infty} N_n \subseteq \bigcup_{n=1}^{\infty} M_n = M,$$

it is seen that

$$\text{(i)} \qquad \bigcup_{n=1}^{\infty} (E_n \cup N_n) = (\bigcup_{n=1}^{\infty} E_n) \cup N$$

where

$$N = \bigcup_{n=1}^{\infty} (E_n \cup N_n) - \bigcup_{n=1}^{\infty} E_n.$$

Thus Σ^* is a σ-field. Now if $E_1 \cup N_1 = E_2 \cup N_2$ and $N_1 \subseteqq M_1$, $N_2 \subseteqq M_2$, let $M = M_1 \cup M_2$ so that $E_1 \cup M = E_2 \cup M$ and thus $\mu(E_1) = \mu(E_1 \cup M) = \mu(E_2)$. This shows that μ is uniquely defined on Σ^* and (i) shows that μ is countably additive on Σ^*. Q.E.D.

18 Definition. Let μ be a countably additive vector or extended real valued function on the σ-field Σ and let Σ^* be defined as in the preceding theorem. Then the function μ with domain Σ^* is known as the *Lebesgue extension of* μ. The σ-field Σ^* is known as the *Lebesgue extension* (*relative to* μ) of the σ-field Σ, and the measure space (S, Σ^*, μ) is the *Lebesgue extension* of the measure space (S, Σ, μ).

Expressions such as μ-simple function, totally μ-measurable function, μ-measurable function, μ-integrable function, etc., do not change meaning when μ is regarded as defined on Σ^*. This is because these expressions refer to equivalence classes of functions differing by a μ-null function rather than the functions themselves. Thus the use of the symbol μ for the measure on Σ as well as for its extension on Σ^* should cause no confusion.

The *measure of Lebesgue* in an interval $[a, b]$ of real numbers may be defined as the Lebesgue extension of the Borel measure in $[a, b]$. The *Lebesgue measurable sets in* $[a, b]$ are the sets in the Lebesgue extension (relative to Borel measure) of the σ-field of Borel sets in $[a, b]$. Similarly, the *Lebesgue-Stieltjes measure* determined by a function f of bounded variation on a finite or infinite interval is the Lebesgue extension of the Borel-Stieltjes measure determined by f.

If μ is either the Borel-Stieltjes or the Lebesgue-Stieltjes measure determined by the function f of bounded variation on the interval $I = (a, b)$ and if g is μ-integrable then the integral $\int_I g(s)\mu(ds)$ is often written $\int_a^b g(s)df(s)$. In the case where $f(s) = s$, i.e., if μ is Borel or Lebesgue measure, then this integral is sometimes written as $\int_a^b g(s)ds$.

The construction which we have given can be extended to monotone functions (defined on an open interval) which are not of bounded variation. Suppose that f is finite valued, real, monotone increasing, and continuous on the right in $(-\infty, +\infty)$. By our previous construction, we have already assigned a non-negative measure μ to every bounded Borel set B. Let $I_n = \{s | -n < s < +n\}$, and put $\mu(B) = \lim_{n\to\infty} \mu(I_n B)$ for every Borel set (the limit exists as an extended positive real number, since $\{\mu(I_n B)\}$ is an increasing sequence). To see that μ is countably additive, note that if B is written as the union of a sequence of disjoint Borel sets B_j, we have

$$\mu(B) \geqq \mu(BI_n) = \sum_{j=1}^{\infty} \mu(B_j I_n) \geqq \sum_{j=1}^{k} \mu(B_j I_n) \to \sum_{j=1}^{k} \mu(B_j),$$

so that

$$\mu(B) \geqq \sum_{j=1}^{\infty} \mu(B_j),$$

while

$$\mu(BI_n) = \sum_{j=1}^{\infty} \mu(B_j I_n) \leqq \sum_{j=1}^{\infty} \mu(B_j),$$

so that

$$\mu(B) \leqq \sum_{j=1}^{\infty} \mu(B_j).$$

In this case also we adopt the notation

$$\int_R g(s)\mu(ds) = \int_{-\infty}^{+\infty} g(s)df(s),$$

and speak of μ as *the measure determined by* f.

The reader will see without difficulty that the same construction can be carried out for an arbitrary open interval I provided only that f has finite real values, is monotone increasing, and is continuous on the right at each point of I.

6. Integration with Respect to a Countably Additive Measure

The basis for this section is a measure space (S, Σ, μ), i.e., a countably additive complex or extended real valued set function μ defined on a σ-field Σ of subsets of a set S. The measure space (S, Σ^*, μ)

is the Lebesgue extension of (S, Σ, μ). The functions f to be integrated with respect to μ will be extended real valued, or will have their values in a B-space $\mathfrak{X}$. It will be shown in this section that if (S, Σ, μ) is a measure space the various linear vector spaces of measurable and integrable functions we have encountered are *complete* metric spaces. Criteria for μ-measurability are given and almost everywhere convergence is discussed.

1 Definition. A sequence of functions $\{f_n\}$ defined on S with values in $\mathfrak{X}$ *converges μ-uniformly* if for each $\varepsilon > 0$ there is a set $E \in \Sigma$ such that $v(\mu, E) < \varepsilon$ and such that $\{f_n\}$ converges uniformly on $S-E$. The sequence $\{f_n\}$ *converges μ-uniformly to the function f* if for each $\varepsilon > 0$ there is a set $E \in \Sigma$ such that $v(\mu, E) < \varepsilon$ and such that $\{f_n\}$ converges uniformly to f on $S-E$.

It is clear that μ-uniform convergence of f_n to f implies convergence of f_n to f in μ-measure. The next lemma is a partial converse of this relation.

2 Lemma. *Let (S, Σ, μ) be a measure space. Let $\{f_n\}$ be a sequence of functions defined on S, and suppose that $\lim_{m, n\to\infty} (f_n-f_m) = 0$ in μ-measure. Then there exists a subsequence $\{f_{n_i}\}$ of $\{f_n\}$ and a function f such that $\{f_{n_i}\}$ converges μ-uniformly to f.*

Proof. If $\lim_{m, n\to\infty} (f_n-f_m) = 0$ in measure, we can find a subsequence $\{f_{n_i}\}$ and sets $E_i \in \Sigma$ such that $v(\mu, E_i) < 1/2^i$ and such that $|f_{n_i}(s)-f_{n_{i+1}}(s)| < 1/2^i$ if $s \notin E_i$. Then if $F_k = \bigcup_{i=k}^{\infty} E_i$, we have $v(\mu, F_k) < 1/2^{k-1}$, and, for $s \notin F_k$,

$$|f_{n_i}(s) - f_{n_j}(s)| \leqq \sum_{m=k}^{\infty} |f_{n_m}(s) - f_{n_{m+1}}(s)| < 1/2^{k-1},$$

where $j > i \geqq k$. Thus if $s \notin \bigcap_{k=1}^{\infty} F_k$ then $\{f_{n_i}(s)\}$ is a Cauchy sequence which converges to a function f uniformly on each of the sets $S-F_k$. This shows that $f_{n_i} \to f$ μ-uniformly. Q.E.D.

3 Corollary. *Let (S, Σ, μ) be a measure space. Let $\{f_n\}$ be a sequence of functions defined on S, and suppose that f_n converges to f in μ-measure. Then some subsequence converges to f μ-uniformly.*

4 Corollary. *If (S, Σ, μ) is a measure space, the space $F(S, \Sigma, \mu, \mathfrak{X})$ is complete.*

PROOF. Let $\{f_n\}$ be a Cauchy sequence in $F(S, \Sigma, \mu, \mathfrak{X})$. By Lemma 2, a subsequence $\{f_{n_i}\}$ of $\{f_n\}$ converges to a certain function f in μ-measure. Let $\varepsilon > 0$ be given. Then there exists an N such that $|f_n - f_m| < \varepsilon/2$ if $m, n \geqq N$, and an $n_i \geqq N$ such that $|f_{n_i} - f| < \varepsilon/2$. Consequently, $|f_n - f| \leqq |f_n - f_{n_i}| + |f_{n_i} - f| < \varepsilon$ for $n \geqq N$, so that $f_n \to f$ in μ-measure. Q.E.D.

5 COROLLARY. *If (S, Σ, μ) is a measure space, the spaces $TM(S, \Sigma, \mu, \mathfrak{X})$ and $M(S, \Sigma, \mu, \mathfrak{X})$ of totally measurable and measurable functions are complete.*

PROOF. By Lemma 2.11, TM and M are both closed subspaces of F. Hence, by I.6.7, these spaces are complete. Q.E.D.

→ 6 THEOREM. *If (S, Σ, μ) is a measure space, and $p \geqq 1$, then $L_p(S, \Sigma, \mu, \mathfrak{X})$ is complete and thus a B-space.*

PROOF. We observe first that a sequence $\{f_n\}$ in a metric space is a Cauchy sequence if and only if

$$\lim_{i \to \infty} (f_{n_i} - f_{n_{i+1}}) = 0$$

for every subsequence $\{n_i\}$. Thus if $\lim_{m, n \to \infty} |f_n - f_m|_p = 0$, it follows from Theorem 3.6 that $\lim_{m, n \to \infty} (f_m - f_n) = 0$ in μ-measure. Consequently, by Corollary 4, there exists an f such that $\lim_{n \to \infty} f_n = f$ in *μ-measure*. Let $\varepsilon > 0$, let N be so large that $|f_n - f_m|_p < \varepsilon$ for $n, m \geqq N$, and let δ be so small that

$$\left\{\int_E |f_n(s)|^p v(\mu, ds)\right\}^{1/p} < \varepsilon$$

for $v(\mu, E) < \delta$ and $1 \leqq n \leqq N$. Then

$$\left\{\int_E |f_n(s)|^p v(\mu, ds)\right\}^{1/p} < 2\varepsilon$$

for $v(\mu, E) < \delta$ and $1 \leqq n < \infty$, proving that

$$\lim_{v(\mu, E) \to 0} \int_E |f_n(s)|^p v(\mu, ds) = 0$$

uniformly in n. In the same way, we can show that for each $\varepsilon > 0$ there is a set $E \in \Sigma$ such that $v(\mu, E) < \infty$ and such that

$$\int_{S-E} |f_n(s)|^p v(\mu, ds) < \varepsilon, \qquad n = 1, 2, \ldots.$$

By Theorem 3.6, $f \in L_p$ and $\lim_{n\to\infty} |f_n - f|_p = 0$. Q.E.D.

7 LEMMA. *A set is a null set if and only if it is a subset of some measurable set F such that $v(\mu, F) = 0$.*

PROOF. If E is a null set, then $v^*(\mu, E) = 0$ and there are measurable sets E_n containing E with $v(\mu, E_n) < 1/n$. Thus the set $F = \cap E_n$ is a measurable set containing E and $v(\mu, F) = 0$. Q.E.D.

8 LEMMA. *If (S, Σ, μ) is a measure space, a function f defined on S to $\mathfrak{X}$ is a null function if and only if it vanishes almost everywhere. If f is μ-integrable then $\int_E f(s)\mu(ds) = 0$ for every E in Σ if and only if f vanishes almost everywhere.*

PROOF. It is clear that a function vanishing almost everywhere is a null function. Conversely, if f is a null function then for each $n = 1, 2, \ldots$, the set $E_n = \{s \mid |f(s)| > 1/n\}$ is a null set and the preceding lemma shows that the set $\cup_{n=1}^{\infty} E_n = \{s \mid f(s) \neq 0\}$ is contained in a set of measure zero. The final statement follows from parts (a) and (d) of Theorem 2.20. Q.E.D.

The next two results provide useful criteria for measurability.

9 LEMMA. *Let (S, Σ^*, μ) be the Lebesgue extension of the finite measure space (S, Σ, μ). Then a vector or extended real valued function f defined on S is μ-measurable if and only if*

(i) *f is μ-essentially separably valued, and*

(ii) *$f^{-1}(G)$ is in Σ^* for each open set G, or, equivalently,*

(ii′) *$f^{-1}(B)$ is in Σ^* for each Borel set B.*

PROOF. First assume that f is μ-measurable, and let $\{f_n\}$ be a sequence of simple functions converging to f in μ-measure. By Lemma 2, we can suppose that $\{f_n\}$ converges to f μ-uniformly. Let $E_n \in \Sigma$ be such that $v(\mu, E_n) < 1/n$ and $f_n(s) \to f(s)$ uniformly on $S - E_n$, and let $\cap_{n=1}^{\infty} E_n = E$. Then E is a null set, and $f_n(s) \to f(s)$ for $s \notin E$. The set $f_n(S-E)$ is finite and hence the closure of the union $\cup f_n(S-E)$ is a separable set containing $f(S-E)$ which proves (I.6.12) that $f(S-E)$ is separable.

Now let G be an open set, and let G_n be the set of x such that

$S(x, 1/n) \subseteq G$. Let $s \notin E$. Then $f(s) \in G$ if and only if $f_k(s)$ belongs to some set G_n for all sufficiently large k, i.e.,

$$f^{-1}(G) - E = \bigcup_{m,n=1}^{\infty} \bigcap_{k=m}^{\infty} f_k^{-1}(G_n) - E.$$

Since f_k is a simple function, we may suppose that the sets $f_k^{-1}(G_n)$ are in Σ. Thus, $f^{-1}(G) - E \in \Sigma$ and hence $f^{-1}(G) \in \Sigma^*$.

Since $f^{-1}(\bigcup_{n=1}^{\infty} B_n) = \bigcup_{n=1}^{\infty} f^{-1}(B_n)$, the family of sets B for which $f^{-1}(B) \in \Sigma^*$ is a σ-field. From this the equivalence of (ii) and (ii′) is evident.

Conversely, suppose that (i) and (ii′) hold. Let E be a null set and $\{x_n\}$ be a countable dense subset of $f(S-E)$. Let $\varepsilon > 0$ be given, and let A_n be the set of s not in E for which $|f(s)-x_n| < \varepsilon$, while $|f(s)-x_i| \geq \varepsilon$, $1 \leq i < n$. Then $A_n \in \Sigma^*$, and $E \cup \bigcup_{n=1}^{\infty} A_n = S$. Consequently, we can find an N so large that $v(\mu, \bigcup_{n=N}^{\infty} A_n) < \varepsilon$. Put $f_\varepsilon(s) = x_n$ if $s \in A_n$ and $n < N$, and let $f_\varepsilon(s) = 0$ otherwise. Then f_ε is a μ-simple function. Clearly, $f_\varepsilon \to f$ in μ-measure as $\varepsilon \to 0$. Consequently, f is μ-measurable. Q.E.D.

10 Theorem. *Let (S, Σ^*, μ) be the Lebesgue extension of the measure space (S, Σ, μ). Then a vector or extended real valued function f defined on S is μ-measurable if and only if for each measurable set F such that $v(\mu, F) < \infty$,*

(i) *f is μ-essentially separably valued on F;*
(ii) *$F \cap f^{-1}(G)$ is in Σ^* for each open set G; or, equivalently,*
(ii′) *$F \cap f^{-1}(B)$ is in Σ^* for each Borel set B.*

Proof. This result follows immediately from Lemma 9 and Definition 2.10. Q.E.D.

As a consequence of Theorem 10 we note the useful fact that an extended real valued function f is measurable if for each set $F \in \Sigma$ with $v(\mu, F) < \infty$, and real number c, the set $F \cap \{s|f(s) > c\}$ is in Σ^*. Equivalently, we may replace $\{s|f(s) > c\}$ by $\{s|f(s) \geq c\}$, $\{s|f(s) < c\}$, or $\{s|f(s) \leq c\}$. Each of these families of sets generates the Borel sets. If $\{f_n\}$ is a sequence of extended real valued measurable functions and $g = \sup_n f_n$, then

$$\{s|g(s) > c\} = \bigcup_{n=1}^{\infty} \{s|f_n(s) > c\}.$$

Thus $\sup f_n$ is measurable. Similarly the $\inf f_n$ is measurable, and consequently, $\limsup_{n\to\infty} f_n$, $\liminf_{n\to\infty} f_n$ and $\lim_{n\to\infty} f_n$ are measurable.

The next theorem applies specifically to vector valued functions, and provides a method for reducing the study of measurability of such functions to the scalar valued case. It is sometimes convenient to apply this theorem rather than Theorem 10.

11 **Theorem.** *If (S, Σ, μ) is a measure space, then a function f on S to a B-space $\mathfrak{X}$ is μ-measurable if and only if*

(i) *for every measurable set F with $v(\mu, F) < \infty$, the function f is μ-essentially separably valued on F, and*

(ii) *for every linear functional x^* in $\mathfrak{X}^*$ the scalar function x^*f on S is μ-measurable.*

Proof. The necessity of (i) is clear from Theorem 10. If f is measurable and $x^* \in \mathfrak{X}^*$, let $r(\cdot) = x^*f(\cdot)$. If H is an open set of scalars, $G = x^{*-1}(H)$ is open in $\mathfrak{X}$ and $r^{-1}(H) = f^{-1}(G) \in \Sigma^*$, proving (ii). To prove sufficiency we may suppose without loss of generality that (S, Σ, μ) is a finite measure space. Let $\{x_n\}$ be a countable dense subset of $f(S-E)$, where E is a null set. Let $\{x_n^*\}$ be a sequence of linear functionals which satisfy (cf. II.3.14) $|x_n^*| = 1$ and $x_n^*(x_n) = |x_n|$. Since $|f(s)| = \sup_n |x_n^* f(s)|$, $s \in S-E$, the function $|f(\cdot)|$ is measurable.

By the same argument the function $g_n(\cdot) = |f(\cdot)-x_n|$ is measurable. Let G be an open set in $\mathfrak{X}$. If $G = \phi$, $f^{-1}(G) = \phi \in \Sigma^*$. If $G \neq \phi$, let $\{y_n\}$ be that subset of $\{x_n\}$ that is in G and let ε_n be the radius of the largest open sphere $S(y_n, \varepsilon) \subseteq G$. If $G_n = S(x_n, \varepsilon_n)$ then $f^{-1}(G_n) = g_n^{-1}([0, \varepsilon_n)) \in \Sigma^*$, by applying Theorem 10 to g_n. Since $G = \cup_{n=1}^{\infty} G_n$, it follows that $f^{-1}(G) = \cup_{n=1}^{\infty} f^{-1}(G_n) \in \Sigma^*$, proving, by Theorem 10, that f is measurable. Q.E.D.

12 **Theorem.** *(Egoroff) If (S, Σ, μ) is a finite measure space, then a sequence $\{f_n\}$ of measurable functions defined on S with values in $\mathfrak{X}$ is μ-uniformly convergent to a function f if and only if $f_n(s)$ converges to $f(s)$ almost everywhere.*

Proof. Suppose that $f_n \to f$ μ-uniformly. Let $E_n \in \Sigma$ be such that $v(\mu, E_n) < 1/n$, and such that $f_m(s) \to f(s)$ uniformly for $s \notin E_n$. Then $E = \cap_{n=1}^{\infty} E_n$ is a null set such that $f_n(s) \to f(s)$ for $s \notin E$.

Conversely, suppose that E is a null set such that $f_n(s) \to f(s)$ for $s \notin E$. Let

$$E_{k,m} = \{s | s \notin E, |f_r(s) - f(s)| < 1/m \text{ for } r \geqq k\}.$$

Then $E_{k+1,m} \supseteqq E_{k,m}$, and, since $f_r(s) \to f(s)$ for each $s \in S - E$, $\bigcup_{k=1}^{\infty} E_{k,m} = S - E$ for all m. Hence, for each $\varepsilon > 0$, and each m we can find an integer k_m such that $v(\mu, S - E_{k_m, m}) < \varepsilon/2^m$. If we let $A_\varepsilon = \bigcap_{m=1}^{\infty} E_{k_m, m}$, then $v(\mu, S - A_\varepsilon) < \varepsilon$, while $|f_k(s) - f(s)| < 1/m$ for $k > k_m$ and $s \in A_\varepsilon$, proving that $f_k(s) \to f(s)$ uniformly in A_ε. Q.E.D.

It should be noted that the direct part of the proof does not use the finiteness of (S, Σ, μ), so that we have shown that if (S, Σ, μ) is a measure space, uniform μ-convergence implies convergence almost everywhere.

13 Corollary. (a) *If (S, Σ, μ) is a measure space, a sequence of measurable functions convergent in measure has a subsequence which converges almost everywhere.*

(b) *If (S, Σ, μ) is a finite measure space, an almost everywhere convergent sequence of measurable functions is convergent in μ-measure.*

14 Corollary. *If (S, Σ, μ) is a measure space, and $\{f_n\}$ is a sequence of measurable vector valued functions defined on S converging almost everywhere to a function f defined on S, f is measurable.*

Proof. This result follows from Theorem 12 and Definition 2.10. Q.E.D.

15 Theorem. (*Vitali convergence theorem*) *Let $1 \leqq p < \infty$, let (S, Σ, μ) be a measure space and let $\{f_n\}$ be a sequence of functions in $L_p(S, \Sigma, \mu, \mathfrak{X})$ converging almost everywhere to a function f. Then f is in $L_p(S, \Sigma, \mu, \mathfrak{X})$ and $|f_n - f|_p$ converges to zero if and only if*

$$\text{(i)} \quad \lim_{v(\mu, E) \to 0} \int_E |f_n(s)|^p v(\mu, ds) = 0 \text{ uniformly in } n;$$

(ii) *for each $\varepsilon > 0$, there is a set $E_\varepsilon \in \Sigma$ such that $v(\mu, E_\varepsilon) < \infty$ and such that*

$$\int_{S - E_\varepsilon} |f_n(s)|^p v(\mu, ds) < \varepsilon, \qquad n = 1, 2, \ldots.$$

Proof. The necessity of (i) and (ii) is an immediate consequence of Theorem 3.6. Conversely, suppose that (i) and (ii) hold. If E is a set such that $v(\mu, E) < \infty$, then

$$\int_E |f(s)|^p v(\mu, ds) < \infty \text{ and } \lim_{n\to\infty} \int_E |f_n(s)-f(s)|^p v(\mu, ds) = 0$$

follow easily from Corollary 13(b) and Theorem 3.6. Consequently,

$$\begin{aligned} \limsup_{m,n\to\infty} |f_n-f_m|_p &\leq \limsup_{m,n\to\infty} \left\{\int_{E_\varepsilon} |f_n(s)-f_m(s)|^p v(\mu, ds)\right\}^{1/p} + 2\varepsilon^{1/p} \\ &\leq \limsup_{m,n\to\infty} \left[\left\{\int_{E_\varepsilon} |f_n(s)-f(s)|^p v(\mu, ds)\right\}^{1/p}\right. \\ &\quad \left. + \left\{\int_{E_\varepsilon} |f_m(s)-f(s)|^p v(\mu, ds)\right\}^{1/p}\right] + 2\varepsilon^{1/p} \\ &= 2\varepsilon^{1/p}, \end{aligned}$$

so that $\lim_{m,n\to\infty} |f_n-f_m|_p = 0$. Since $L_p(S, \Sigma, \mu, \mathfrak{X})$ is complete, there exists a $g \in L_p$ such that $\lim_{n\to\infty} |f_n-g|_p = 0$. Then by Theorem 3.6 and Corollary 13(a), a subsequence of $\{f_n\}$ converges to g almost everywhere. Hence, $g = f$ almost everywhere, and consequently $f \in L_p$ and $|f_n-f|_p \to 0$. Q.E.D.

→ 16 Corollary. (*Lebesgue dominated convergence theorem*) *Let* $1 \leq p < \infty$, *let* (S, Σ, μ) *be a measure space, and let* $\{f_n\}$ *be a sequence of functions in* $L_p(S, \Sigma, \mu, \mathfrak{X})$ *converging almost everywhere to a function* f. *Suppose that there exists a function* g *in* $L_p(S, \Sigma, \mu, \mathfrak{X})$ *such that* $|f_n(s)| \leq |g(s)|$ *almost everywhere. Then* f *is in* $L_p(S, \Sigma, \mu, \mathfrak{X})$ *and* $|f_n-f|_p$ *converges to zero.*

Proof. This result follows immediately from Theorem 15, as conditions (i) and (ii) are satisfied by $\{f_n\}$. Q.E.D.

17 Corollary. *Let* (S, Σ, μ) *be a positive measure space, and let* $\{f_n\}$ *be a monotone increasing sequence of non-negative real measurable, but not necessarily integrable, functions converging almost everywhere to a function* f. *Then,*

$$\lim_n \int_S f_n(s)\mu(ds) = \int_S f(s)\mu(ds).$$

Proof. Since $0 \leq f_n(s) \leq f(s)$, our result follows immediately from Corollary 16 if $\int_S f(s)\mu(ds) < \infty$. Consequently, we have only to show that if $\int_S f_n(s)\mu(ds) \leq M < \infty$, then $\int_S f(s)\mu(ds) < \infty$. But if the sequence of integrals of $\{f_n\}$ has a finite upper bound, for each $\varepsilon > 0$ we can find an N such that

$$\int_S f_n(s)\mu(ds) \leqq \int_S f_N(s)\mu(ds) + \varepsilon, \qquad n = 1, 2, \ldots.$$

Consequently

$$\limsup_{\mu(E)\to 0} \int_E f_n(s)\mu(ds) \leqq \limsup_{\mu(E)\to 0} \int_E f_N(s)\mu(ds) + \varepsilon = \varepsilon$$

uniformly in n. This, since ε is arbitrary, implies that

$$\lim_{\mu(E)\to 0} \int_E f_n(s)\mu(ds) = 0 \text{ uniformly for } n \geqq 1.$$

In the same way, we can establish the existence of a set E of finite measure such that

$$\int_{S-E} f_n(s)\mu(ds) < \varepsilon, \qquad n \geqq 1.$$

Thus, by Theorem 15, $f \in L_1$. Q.E.D.

Corollaries 16 and 17 are historically important, and together with Theorem 6 are the principal results of the classical convergence theory for Lebesgue integrals.

18 COROLLARY. *If (S, Σ, μ) is a positive measure space and f is a non-negative measurable function, the function G defined by*

$$G(E) = \int_E f(s)\mu(ds), \qquad E \in \Sigma,$$

is an extended real valued set function which is countably additive.

PROOF. If $E \in \Sigma$ is the union of an increasing sequence of measurable sets E_n, let $f_n = \chi_{E_n} f$. Then

$$G(E) = \int_E f(s)\mu(ds) = \lim_n \int_E f_n(s)\mu(ds) = \lim_n G(E_n)$$

by Corollary 17. Q.E.D.

19 THEOREM. (*Fatou's lemma*) *Let (S, Σ, μ) be a positive measure space, and $\{f_n\}$ a sequence of non-negative measurable, but not necessarily integrable, functions. Then*

$$\int_S \liminf_{n\to\infty} f_n(s)\mu(ds) \leqq \liminf_{n\to\infty} \int_S f_n(s)\mu(ds).$$

PROOF. Let $g_n(s) = \inf_{k \geqq n} f_k(s)$. Then g_n is an increasing sequence of functions whose limit is $\liminf_{n\to\infty} f_n(s)$. Consequently, by Corollary 17,

$$\int_S \liminf_{n\to\infty} f_n(s)\mu(ds) = \lim_{n\to\infty} \int_S g_n(s)\mu(ds).$$

Since $g_n(s) \leqq f_n(s)$,

$$\lim_{n\to\infty} \int_S g_n(s)\mu(ds) \leqq \liminf_{n\to\infty} \int_S f_n(s)\mu(ds). \qquad \text{Q.E.D.}$$

We next show that an arbitrary closed operator commutes, in a certain sense, with the operation of integration. This supplements the similar result stated for bounded operators in Theorem 2.19.

20. THEOREM. *Let $\mathfrak{X}$, $\mathfrak{Y}$ be B-spaces and let T be a closed linear operator on a domain $\mathfrak{D} \subseteqq \mathfrak{X}$ and with range in $\mathfrak{Y}$. Let (S, Σ, μ) be a measure space and f a μ-integrable function with values in $\mathfrak{D}$. If Tf is also μ-integrable then $\int_S f(s)\mu(ds)$ is in $\mathfrak{D}$ and*

$$T\int_S f(s)\mu(ds) = \int_S Tf(s)\mu(ds).$$

PROOF. Consider the product space $\mathfrak{Z} = \mathfrak{X} \times \mathfrak{Y}$ with norm $|[x, y]| = |x| + |y|$. By hypothesis the graph $\mathfrak{G}$ of T is a closed linear manifold in $\mathfrak{Z}$ (cf. II.2.3). It will first be shown, using Theorem 11, that the function g on S to $\mathfrak{G}$ defined by the equation

$$g(s) = [f(s), Tf(s)], \qquad s \in S,$$

is μ-measurable. Let F be a measurable set with $v(\mu, F) < \infty$. Then since f and Tf are both μ-measurable there is a null set E such that $f(F-E)$ and $Tf(F-E)$ are both separable. Since

$$g(F-E) \subseteqq f(F-E) \times Tf(F-E),$$

g is μ-essentially separably valued on F. To prove that g is μ-measurable it will suffice, in view of Theorem 11, to show that g^*g is μ-measurable for every g^* in $\mathfrak{G}^*$. Let $z^* \in \mathfrak{Z}^*$ be an extension (see Theorem II.3.11) of g^* to $\mathfrak{X} \times \mathfrak{Y}$ and let $x^* \in \mathfrak{X}^*$, $y^* \in \mathfrak{Y}^*$ be defined by the equations

$$x^*x = z^*[x, 0], \qquad y^*y = z^*[0, y].$$

Then

$$g^*g(s) = z^*[f(s), Tf(s)] = x^*f(s) + y^*Tf(s)$$

which shows that g^*g is μ-measurable and this proves that g is μ-measurable. Since

$$|g(s)| = |f(s)|+|Tf(s)|,$$

it follows from Theorem 2.22 that g is μ-integrable. Since $g(s)$ is in $\mathfrak{G}$ for every s in S it follows that the integral

$$\int_S g(s)\mu(ds) = \left[\int_S f(s)\mu(ds), \int_S Tf(s)\mu(ds)\right]$$

is also in $\mathfrak{G}$. This means that $\int_S f(s)\mu(ds)$ is in $\mathfrak{D}$ and that

$$T\int_S f(s)\mu(ds) = \int_S Tf(s)\mu(ds). \qquad \text{Q.E.D.}$$

The convergence theory of this section may be used to prove a useful reciprocal relation between Stieltjes integrals with respect to two different functions of bounded variation, which is given by the famous formula for integration by parts. Before stating it, we prove a preliminary lemma.

21 LEMMA. *Let f be a function of bounded variation in the interval (a, b). Then $f(a+)$ and $f(b-)$ exist.*

PROOF. Since it is clear that f is of bounded variation in (a, b) if and only if its real and imaginary parts are of bounded variation on (a, b), we may and shall assume that f is real. If

$$\limsup_{\varepsilon\to 0} f(a+|\varepsilon|) > \liminf_{\varepsilon\to 0} f(a+|\varepsilon|)+\delta$$

where $\delta > 0$, there are decreasing sequences a_i, b_i, with $a_n > b_n > a_{n+1}$, $n = 1, 2, \ldots$, which converge to a, such that $f(a_n)-f(b_n) > \delta$ and this clearly contradicts the fact that f is of bounded variation on (a, b). Thus

$$\limsup_{\varepsilon\to 0} f(a+|\varepsilon|) = \liminf_{\varepsilon\to 0} f(a+|\varepsilon|) = \lim_{\varepsilon\to 0} f(a+|\varepsilon|) = f(a+).$$

In the same way we can show that $f(b-)$ exists. Q.E.D.

22 THEOREM. *Let $\alpha(x)$ and $\beta(x)$ be two functions of bounded variation in an interval (a, b). Suppose that one is continuous in (a, b), and that the other is continuous on the right. Then*

$$\int_a^b \alpha(x)d\beta(x) + \int_a^b \beta(x)d\alpha(x) = \alpha(b-)\beta(b-)-\alpha(a+)\beta(a+).$$

PROOF. Clearly, it suffices to prove the result for every finite subinterval of (a, b). For simplicity, and without loss of generality, we will suppose that $(a, b) = (0, 1)$. Put

$$\alpha_n(x) = \alpha\left(\frac{k+1}{n}\right), \qquad \frac{k}{n} \leqq x < \frac{k+1}{n}, \qquad 1 \leqq k \leqq n-2;$$

$$\alpha_n(x) = 0, \text{ if } 0 < x < \frac{1}{n} \text{ or } \frac{n-1}{n} \leqq x < 1.$$

$$\beta_n(x) = \beta\left(\frac{k}{n}\right), \qquad \frac{k}{n} \leqq x < \frac{k+1}{n}, \qquad 1 \leqq k \leqq n-2;$$

$$\beta_n(x) = 0, \text{ if } 0 < x < \frac{1}{n} \text{ or } \frac{n-1}{n} \leqq x < 1.$$

Then clearly, $\alpha_n(x) \to \alpha(x)$, $\beta_n(x) \to \beta(x)$, and

$$|\alpha_n(x)| \leqq \sup_{0<x<1} |\alpha(x)|,$$

$$|\beta_n(x)| \leqq \sup_{0<x<1} |\beta(x)|.$$

Consequently, by Corollary 16, both integrals are defined, and

$$\begin{aligned}
&\int_a^b \alpha(x)d\beta(x) + \int_a^b \beta(x)d\alpha(x) \\
&= \lim_{n\to\infty} \sum_{k=1}^{n-2} \alpha\left(\frac{k+1}{n}\right)\left\{\beta\left(\frac{k+1}{n}\right) - \beta\left(\frac{k}{n}\right)\right\} \\
&\quad + \lim_{n\to\infty} \sum_{k=1}^{n-2} \beta\left(\frac{k}{n}\right)\left\{\alpha\left(\frac{k+1}{n}\right) - \alpha\left(\frac{k}{n}\right)\right\} \\
&= \lim_{n\to\infty} \sum_{k=1}^{n-2} \left\{\alpha\left(\frac{k+1}{n}\right)\beta\left(\frac{k+1}{n}\right) - \alpha\left(\frac{k}{n}\right)\beta\left(\frac{k}{n}\right)\right\} \\
&= \lim_{n\to\infty} \left\{\alpha\left(1-\frac{1}{n}\right)\beta\left(1-\frac{1}{n}\right) - \alpha\left(\frac{1}{n}\right)\beta\left(\frac{1}{n}\right)\right\} \\
&= \alpha(1-)\beta(1-) - \alpha(0+)\beta(0+). \text{ Q.E.D.}
\end{aligned}$$

7. The Vitali–Hahn–Saks Theorem and Spaces of Measures

In this section a metric will be introduced in the σ-field Σ of a measure space (S, Σ, μ) in such a way that the corresponding metric space $\Sigma(\mu)$ is complete. The additive functions on Σ which are continuous on $\Sigma(\mu)$ are precisely the μ-continuous additive functions on Σ. A basic result concerned with a sequence $\{\nu_n\}$ of such functions is the Vitali-Hahn-Saks theorem. It states that if $\{\nu_n(E)\}$ converges for each E in Σ, then the continuity of ν_n on the metric space $\Sigma(\mu)$ is uniform with respect to $n = 1, 2, \ldots$. The reader is already familiar with a striking analogy to the Vitali-Hahn-Saks theorem—the theorem (II.1.17) asserting that continuity of the elements of a pointwise convergent sequence $\{T_n\}$ of continuous linear operators on an F-space is uniform in $n = 1, 2, \ldots$. It is enlightening to compare the proof of the Vitali-Hahn-Saks theorem given here (which is that of Saks) with the proof of the uniform boundedness principle in F-spaces (cf. II.1.13), for such comparison reveals a useful analogy between σ-fields and linear metric spaces. More direct relationships between set functions and certain B-spaces are explicitly discussed in the next chapter, where it is shown that the conjugate spaces of some familiar B-spaces may be represented in terms of set functions. In the present section it will be shown that the spaces of bounded additive set functions, and regular countably additive set functions are all B-spaces.

The metric space $\Sigma(\mu)$

Let (S, Σ, μ) be a measure space, and for each pair A, E in Σ define the equivalence relation $A \sim E$ by the equation $v(\mu, A \Delta E) = 0$, where $A \Delta E$ is the symmetric difference $(A \cup E) - AE$ of A and E. The fact that $\sim$ is an equivalence relation follows readily from the fact that the symmetric difference is a commutative and associative operation with $A \Delta A = \phi$. The set $\Sigma(\mu)$ of all equivalence classes $E(\mu) = \{A | A \sim E\}$ is a metric space with the distance function

$$\text{(i)} \qquad \varrho(E(\mu), F(\mu)) = \arctan v(\mu, E \Delta F).$$

For the sake of simplicity and convenience we shall speak of the sets E in Σ as elements of $\Sigma(\mu)$ just as we speak of the elements of the space $M(S, \Sigma, \mu)$ of measurable functions as functions on S rather than equivalence classes of such functions. Thus we will sometimes write, instead of (i), the equation

$$\text{(ii)} \qquad \varrho(E, F) = \arctan v(\mu, E \Delta F).$$

It should be noted however that a function λ on Σ cannot be regarded as defined on $\Sigma(\mu)$ unless $\lambda(E) = \lambda(F)$ whenever $v(\mu, E \Delta F) = 0$. We note that $A \sim E$ if and only if the characteristic functions χ_A and χ_B are equivalent as elements of $M(S, \Sigma, \mu)$, i.e., if and only if

χ_A and χ_B differ by a μ-null function. Thus the map $E \to \chi_E$ may be regarded as a homeomorphism of $\Sigma(\mu)$ onto a subset of $M(S, \Sigma, \mu)$. It is clear that this homeomorphism maps Cauchy sequences into Cauchy sequences. Thus if $\varrho(E_n, E_m) \to 0$, there is, by Corollary 6.5, a function χ in $M(S, \Sigma, \mu)$ with $\chi_{E_n} \to \chi$. By Corollary 6.13 some subsequence of $\{\chi_{E_n}\}$ converges almost everywhere to χ and thus for almost all s in S, $\chi(s)$ is either 0 or 1. Thus $\chi = \chi_E$ for some E in Σ. Since $\chi_{E_n} \to \chi_E$ in $M(S, \Sigma, \mu)$, equation (ii) shows that $E_n \to E$ in $\Sigma(\mu)$. The space $\Sigma(\mu)$ is therefore a complete metric space.

If λ is an additive vector or scalar valued function on Σ which is μ-continuous, then λ is defined and continuous on the metric space $\Sigma(\mu)$. To see this note first that the identities

$$v(\mu, E \Delta F) = v(\mu, E - EF) + v(\mu, F - EF),$$
$$\lambda(E) - \lambda(F) = \lambda(E - EF) - \lambda(F - EF),$$

show that $\lambda(E) = \lambda(F)$ whenever $v(\mu, E \Delta F) = 0$, and so a μ-continuous additive function λ on Σ is defined on the metric space $\Sigma(\mu)$. These same identities show that such a λ is continuous on $\Sigma(\mu)$, for if $E_m \to E$ in $\Sigma(\mu)$, then $v(\mu, E - EE_m) \to 0$ and $v(\mu, E_m - EE_m) \to 0$. Thus $\lambda(E - EE_m) \to 0$ and $\lambda(E_m - EE_m) \to 0$ and therefore $\lambda(E_m) \to \lambda(E)$. Thus a μ-continuous additive function λ is continuous on the metric space $\Sigma(\mu)$. Conversely, an additive function on Σ which is defined and continuous on $\Sigma(\mu)$ is μ-continuous. This follows immediately from the definitions.

It should also be observed that the binary operations $A \cup B$, AB, $A \Delta B$ are continuous maps of $\Sigma(\mu) \times \Sigma(\mu)$ into $\Sigma(\mu)$ and that the unary operation of complementation, $A \to A'$, is a continuous map of $\Sigma(\mu)$ into $\Sigma(\mu)$. These elementary but important facts are readily verified by an application of the inequality

$$v(\mu, E \Delta F) \leqq v(\mu, E) + v(\mu, F)$$

to the identities:

$$(A \cup B) \Delta (A_1 \cup B_1) = (A \Delta A_1) \Delta (B \Delta B_1) \Delta A(B \Delta B_1) \Delta B_1(A \Delta A_1),$$
$$(AB) \Delta (A_1 B_1) = A(B \Delta B_1) \Delta B_1(A \Delta A_1),$$
$$A' \Delta A_1' = A \Delta A_1,$$
$$(A \Delta B) \Delta (A_1 \Delta B_1) = (A \Delta A_1) \Delta (B \Delta B_1).$$

If $v(\mu, S) < \infty$ and if a sequence $\{E_n\}$ in Σ converges to E in the sense that

$$\text{(iii)} \qquad E = \liminf_n E_n = \limsup_n E_n,$$

then $\varrho(E_n, E) \to 0$, i.e., $E_n \to E$ in the metric space $\Sigma(\mu)$. For condition (iii) shows that $\chi_{E_n}(s) \to \chi_E(s)$ for every s in S and hence, by the dominated convergence theorem of Lebesgue (6.16), it is seen, using (ii), that

$$\varrho(E_n, E) = \arctan \int_S |\chi_{E_n}(s) - \chi_E(s)| v(\mu, ds) \to 0.$$

The above statements which relate the usual point set operations with the metric in $\Sigma(\mu)$ show immediately that, when $v(\mu, S) < \infty$, the closure in $\Sigma(\mu)$ of a subfield Σ_1 of Σ is itself a σ-field.

The remarks above are summarized in the following lemma.

1 **Lemma.** *If (S, Σ, μ) is a measure space, the σ-field Σ (or more precisely, the quotient set $\Sigma/\mathscr{N}$, where $\mathscr{N}$ is the ideal of all μ-null sets in Σ) is a complete metric space under the metric*

$$\varrho(A, B) = \arctan v(\mu, A \Delta B).$$

An additive vector or scalar valued function on Σ is defined and continuous on $\Sigma(\mu)$ if and only if it is μ-continuous. The operations $A \cup B$, AB, $A \Delta B$, A' are continuous functions of A and B. If $v(\mu, S) < \infty$ the closure in $\Sigma(\mu)$ of a subfield of Σ is a σ-field.

The following theorem is one of the most important results in the theory of set functions. The proof given here is that of Saks, and consists of applying the Baire category theorem to the space $\Sigma(\mu)$ in much the same way as it was applied to an F-space in the proof of the principle of uniform boundedness (cf. II.1.13).

2 **Theorem.** *(Vitali-Hahn-Saks) Let (S, Σ, μ) be a measure space and $\{\lambda_n\}$ a sequence of μ-continuous vector or scalar valued additive set functions on Σ. If the limit $\lim_n \lambda_n(E)$ exists for each E in Σ then*

$$\lim_{v(\mu, E) \to 0} \lambda_n(E) = 0,$$

uniformly for $n = 1, 2, \ldots$.

PROOF. By Lemma 1, λ_n is continuous on $\Sigma(\mu)$ and so for each $\varepsilon > 0$ the sets

$$\Sigma_{n,m} = \{E \mid E \in \Sigma, |\lambda_n(E) - \lambda_m(E)| \leqq \varepsilon\}, \qquad n, m = 1, 2, \ldots,$$

$$\Sigma_p = \bigcap_{n,m \geqq p} \Sigma_{n,m}, \qquad p = 1, 2, \ldots,$$

are closed in the complete metric space $\Sigma(\mu)$. Since the limit $\lim_n \lambda_n(E)$ exists for each $E \in \Sigma$ we have $\Sigma(\mu) = \bigcup_{p=1}^{\infty} \Sigma_p$. The Baire category theorem (I.6.9) shows that one of the sets Σ_p has an interior point. Thus there is an integer q, a positive number r and a set $A \in \Sigma$ such that

$$|\lambda_n(E) - \lambda_m(E)| \leqq \varepsilon, \qquad n, m \geqq q,$$

for every set E in the sphere

$$K = \{E \mid E \in \Sigma, v(\mu, A \Delta E) < r\}.$$

Let $0 < \delta < r$ be chosen so that

$$|\lambda_n(B)| < \varepsilon, \qquad n = 1, 2, \ldots, q,$$

for every $B \in \Sigma$ with $v(\mu, B) < \delta$. Now let $v(\mu, B) < \delta$ so that the sets $A \cup B$, $A - B$ are both in K. The identity

$$\begin{aligned}\lambda_n(B) &= \lambda_q(B) + \{\lambda_n(B) - \lambda_q(B)\} \\ &= \lambda_q(B) + \{\lambda_n(A \cup B) - \lambda_q(A \cup B)\} - \{\lambda_n(A - B) - \lambda_q(A - B)\},\end{aligned}$$

shows that $|\lambda_n(B)| < 3\varepsilon$ for all $n = 1, 2, \ldots$. Q.E.D.

3 COROLLARY. *Under the hypotheses of Theorem 2 and the additional hypothesis* $v(\mu, S) < \infty$, *the function* $\lambda(E) = \lim_n \lambda_n(E)$ *is countably additive on* Σ.

PROOF. The additivity of λ follows from that of λ_n and therefore to prove countable additivity it suffices to show that $\lambda(E_m) \to 0$ for every decreasing sequence $\{E_m\} \subset \Sigma$ with void intersection. Since $v(\mu, E_m) = \sum_{k=m}^{\infty} v(\mu, E_k - E_{k+1})$, it follows that for such sequences $v(\mu, E_m) \to 0$ and thus, by the theorem, for every $\varepsilon > 0$ there is an integer m_ε with

$$|\lambda_n(E_m)| < \varepsilon, \qquad m \geqq m_\varepsilon, \qquad n = 1, 2, \ldots$$

Hence

$$|\lambda(E_m)| \leqq \varepsilon, \qquad m \geqq m_\varepsilon. \qquad \text{Q.E.D.}$$

Corollary 3 can be strengthened, as in the following corollary, which is stated here only for scalar valued functions. Actually, the following result is true for vector valued functions, but the proof is postponed until Section IV.10, where a deeper investigation into the properties of vector valued set functions is made.

4 COROLLARY. (*Nikodým*) *Let* $\{\mu_n\}$ *be a sequence of countably additive scalar functions on the* σ*-field* Σ. *If* $\mu(E) = \lim_n \mu_n(E)$ *exists for each* E *in* Σ *then* μ *is countably additive on* Σ *and the countable additivity of* μ_n *is uniform in* $n = 1, 2, \ldots$.

PROOF. The final conclusion means that if $\{E_m\}$ is a decreasing sequence in Σ with void intersection then $\lim_m \mu_n(E_m) = 0$ uniformly for $n = 1, 2, \ldots$. To prove this we note first that since μ_n is finite on Σ the total variation $v(\mu_n, S)$ is also finite (Lemma 4.7). Now let $\lambda_n = 0$ if $\mu_n = 0$ and otherwise let

$$\lambda_n(E) = v(\mu_n, E)/v(\mu_n, S), \qquad \lambda = \sum_{n=1}^{\infty} \lambda_n/2^n$$

Then λ is countably additive on Σ and $\lambda(E_m) \to 0$. Since each μ_n is λ-continuous, the conclusions follow from Theorem 2 and Corollary 3. Q.E.D.

The spaces $ba(S, \Sigma, \mathfrak{X})$, $rba(S, \Sigma, \mathfrak{X})$, $ca(S, \Sigma, \mathfrak{X})$, $rca(S, \Sigma, \mathfrak{X})$.

It will now be shown that several familiar spaces of set functions are B-spaces.

Let Σ be a field of subsets of a set S, and let $\mathfrak{X}$ be a B-space. By $ba(S, \Sigma, \mathfrak{X})$ will be understood the family of all bounded finitely additive set functions with domain Σ and range in $\mathfrak{X}$. It is evident that the sum of two functions in $ba(S, \Sigma, \mathfrak{X})$ is again in $ba(S, \Sigma, \mathfrak{X})$, while if μ is in $ba(= ba(S, \Sigma, \mathfrak{X}))$ and α is a scalar, then $\alpha\mu$ is in ba. Thus ba is a linear vector space. If the norm of μ is defined by the equation

$$|\mu| = \sup_{E \in \Sigma} |\mu(E)|,$$

then ba becomes a normed linear space. To see that it is complete, let $\{\mu_n\}$ be a Cauchy sequence in ba. Clearly $\{\mu_n(E)\}$ is a Cauchy sequence in $\mathfrak{X}$ for each $E \in \Sigma$, so that $\mu(E) = \lim_{n\to\infty} \mu_n(E)$ defines an element

$\mu \epsilon ba$. For $\varepsilon > 0$ choose n_ε so that $|\mu_n - \mu_m| \leqq \varepsilon$ for $m, n \geqq n_\varepsilon$. Then $\mu(E) - \mu_n(E) = \lim_{m\to\infty} (\mu_m(E) - \mu_n(E))$, from which it follows that $|\mu - \mu_n| \leqq \varepsilon$ for $n \geqq n_\varepsilon$. Hence $\mu_n \to \mu$, which proves that $ba(S, \Sigma, \mathfrak{X})$ is complete. It follows, therefore, that $ba(S, \Sigma, \mathfrak{X})$ is a B-space.

If $\mathfrak{X}$ is the set of real or complex numbers, then according to Lemma 1.5,

$$\sup_{E \epsilon \Sigma} |\mu(E)| \leqq v(\mu, S) \leqq 4 \sup_{E \epsilon \Sigma} |\mu(E)|.$$

This shows that $v(\mu, S)$ is a norm in $ba(S, \Sigma, \mathfrak{X})$ which is equivalent to the norm $\sup_{E\epsilon\Sigma} |\mu(E)|$.

The set $ca(S, \Sigma, \mathfrak{X})$ of all countably additive set functions in $ba(S, \Sigma, \mathfrak{X})$ is clearly a linear manifold in $ba(S, \Sigma, \mathfrak{X})$. To see that it is a closed linear manifold, let $\mu \epsilon ba$, and $\{\mu_n\}$ be a sequence of elements of $ca(S, \Sigma, \mathfrak{X})$ converging to μ. Let $\{E_m\}$ be a sequence of disjoint sets in Σ whose union E is also in Σ. Then as $n \to \infty$,

$$\mu_n(E) - \sum_{j=1}^{m} \mu_n(E_j) = \mu_n \left(\bigcup_{j=m+1}^{\infty} E_j \right) \to \mu \left(\bigcup_{j=m+1}^{\infty} E_j \right) = \mu(E) - \sum_{j=1}^{m} \mu(E_j),$$

uniformly for $m = 1, 2, \ldots$. On the other hand, as $m \to \infty$, $\mu_n(E) - \sum_{j=1}^{m} \mu_n(E_j) \to 0$ for each $n = 1, 2, \ldots$. Thus, by I.7.6.

$$\mu(E) = \sum_{j=1}^{\infty} \mu(E_j),$$

which proves that μ is countably additive on Σ. Thus $ca(S, \Sigma, \mathfrak{X})$ is a closed linear manifold in $ba(S, \Sigma, \mathfrak{X})$ and is therefore a B-space.

If S is a topological space, $rba(S, \Sigma, \mathfrak{X})$ will denote the set of all regular set functions in $ba(S, \Sigma, \mathfrak{X})$. It is easily seen that $rba(S, \Sigma, \mathfrak{X})$ is a linear subspace of $ba(S, \Sigma, \mathfrak{X})$. Let $\mu \epsilon ba(S, \Sigma, \mathfrak{X})$, and let $\{\mu_n\}$ be a sequence of elements of $rba(S, \Sigma, \mathfrak{X})$ converging to μ. Let $\varepsilon > 0$ and $E \epsilon \Sigma$ be given, and fix n so large that $|\mu - \mu_n| \leqq \varepsilon$. Let $F, G \epsilon \Sigma$ be such that $\overline{F} \subseteqq E$, E is contained in the interior of G, and $|\mu_n(A)| \leqq \varepsilon$ for every A in Σ with $A \subseteqq G - F$. Then, for such A, we have $|\mu(A)| \leqq 2\varepsilon$, which proves that $\mu \epsilon rba(S, \Sigma, \mathfrak{X})$. Thus $rba(S, \Sigma, \mathfrak{X})$ is a closed linear manifold in $ba(S, \Sigma, \mathfrak{X})$ and therefore a B-space.

The set $rca(S, \Sigma, \mathfrak{X})$ of all regular countably additive measures in $ba(S, \Sigma, \mathfrak{X})$ is the intersection of $ca(S, \Sigma, \mathfrak{X})$ and $rba(S, \Sigma, \mathfrak{X})$. Consequently $rca(S, \Sigma, \mathfrak{X})$ is also a closed linear subspace of $ba(S, \Sigma, \mathfrak{X})$, and hence is also a B-space.

We note that if $\mathfrak{X}$ is the space of scalars (real or complex) it will usually be omitted from notations like $ca(S, \Sigma, \mathfrak{X})$. Thus $ca(S, \Sigma)$ is the space of all real or complex valued countably additive set functions defined on Σ.

We have just seen that the spaces $ba(S, \Sigma)$ and $ca(S, \Sigma)$ are B-spaces. For some purposes it is useful to know that they are also complete lattices. We close this section with some observations concerning their order properties and another decomposition theorem.

Let Σ be a field of subsets of a set S, then if μ is an additive set function such that $\mu(E) \geqq 0$, $E \in \Sigma$, we say that μ is *positive* and write $\mu \geqq 0$. If $\mu_1 - \mu_2 \geqq 0$ then we write $\mu_1 \geqq \mu_2$ or $\mu_2 \leqq \mu_1$. It is readily seen that this gives a partial ordering (I.2.1) on the collection of all real or complex valued additive set functions on Σ. The terms "upper bound", "least upper bound", etc., have the meanings given in I.2.3.

5 THEOREM. *Every subset of the partially ordered collection of additive scalar valued set functions on a field which has an upper bound (lower bound) has a least upper bound (greatest lower bound).*

PROOF. To prove the theorem it will clearly suffice to show that an indexed set $\{\mu_\alpha\}$ of positive elements has a greatest lower bound μ. We define

$$\mu(E) = \inf \{\mu_{\alpha_1}(E_1) + \ldots + \mu_{\alpha_n}(E_n)\}, \qquad E \in \Sigma,$$

where the infimum is taken over all finite subsets $\{\alpha_i\}$ of indices and all finite families of disjoint sets $\{E_i\}$ in Σ whose union is E. It will first be shown that μ is additive on Σ. Let E, F be disjoint sets in Σ and let $\varepsilon > 0$ be arbitrary. Let $E \cup F$ be partitioned into disjoint sets $A_1, \ldots, A_m$ in Σ with

$$\mu_{\alpha_1}(A_1) + \ldots + \mu_{\alpha_m}(A_m) \leqq \mu(E \cup F) + \varepsilon$$

for an appropriate choice of the indices $\alpha_1, \ldots, \alpha_m$. If $E_i = A_i E$ and $F_i = A_i F$ then

$$\mu(E) + \mu(F) \leqq \Sigma \mu_{\alpha_i}(E_i) + \Sigma \mu_{\alpha_i}(F_i) = \Sigma \mu_{\alpha_i}(A_i) \leqq \mu(E \cup F) + \varepsilon$$

which, since $\varepsilon > 0$ is arbitrary, proves that

$$\mu(E)+\mu(F) \leqq \mu(E \cup F).$$

Next let $\{B_i\}$, $\{C_i\}$ be finite partitionings of E and F into disjoint sets in Σ with

$$\Sigma\mu_{\alpha_i}(B_i) < \mu(E)+\varepsilon, \qquad \Sigma\mu_{\beta_j}(C_j) < \mu(F)+\varepsilon.$$

Then

$$\mu(E \cup F) \leqq \Sigma\mu_{\alpha_i}(B_i)+\Sigma\mu_{\beta_j}(C_j) < \mu(E)+\mu(F)+2\varepsilon,$$

and so

$$\mu(E \cup F) \leqq \mu(E)+\mu(F),$$

which completes the proof of the additivity of μ. To see that μ is the greatest lower bound of the set $\{\mu_\alpha\}$ let ν be an additive set function with $\nu \leqq \mu_\alpha$ for each α. Then for any finite partitioning $E_1, \ldots, E_n$ of the set E and any choice $\alpha_1, \ldots, \alpha_n$ of the indices we have $\nu(E_i) \leqq \mu_{\alpha_i}(E_i)$ and thus

$$\nu(E) = \Sigma\nu(E_i) \leqq \Sigma\mu_{\alpha_i}(E_i),$$

which proves that $\nu(E) \leqq \mu(E)$. Q.E.D.

We point out that the preceding theorem did not assume boundedness of the set functions.

6 Corollary. *The lattices $ba(S, \Sigma)$ and $ca(S, \Sigma)$ are complete.*

Proof. It suffices to show that if $\mu_\alpha \geqq 0$, then the greatest lower bound μ of the subset $\{\mu_\alpha\}$ has the property of boundedness or of countable additivity. If each μ_α is in $ba(S, \Sigma)$, then since $0 \leqq \mu(E) \leqq \mu_\alpha(E)$, $E \in \Sigma$, we have $\mu \in ba(S, \Sigma)$. If $\mu_\alpha \in ca(S, \Sigma)$, and if $\{E_n\} \subseteqq \Sigma$, $E_{n+1} \subseteqq E_n$, $\cap E_n = \phi$ then $\lim_n \mu_\alpha(E_n) = 0$. Since $0 \leqq \mu(E_n) \leqq \mu_\alpha(E_n)$, it follows that $\lim_n \mu(E_n) = 0$ and hence $\mu \in ca(S, \Sigma)$. Q.E.D.

We now show that if $\lambda \in ba(S, \Sigma)$, then λ can be uniquely decomposed into the sum of a countably additive set function and one which is finitely additive in a certain maximal sense.

7 Definition. If λ is in $ba(S, \Sigma)$ and if $\lambda \geqq 0$, we say that λ is *purely finitely additive* in case $0 \leqq \mu \leqq \lambda$ and $\mu \in ca(S, \Sigma)$ imply that $\mu = 0$.

8 Theorem. *If λ is in $ba(S, \Sigma)$ and $\lambda \geqq 0$ then there is a unique decomposition $\lambda = \lambda_1+\lambda_2$, $\lambda_i \geqq 0$, where λ_1 is countably additive and λ_2 is a purely finitely additive set function in $ba(S, \Sigma)$.*

Proof. Let C be the set of all μ in $ca(S, \Sigma)$ such that $0 \leqq \mu \leqq \lambda$. Let μ_n, $n = 1, 2, \ldots$, be chosen from C in such a way that $\lim_n \mu_n(S) = \sup_{\mu \in C} \mu(S) < \infty$. Since $\mu_i \leqq \sum_{j=1}^{n} \mu_j$, $i = 1, \ldots, n$, it follows from Corollary 6 that $\{\mu_1, \ldots, \mu_n\}$ has a least upper bound $\bar{\mu}_n$ in $ca(S, \Sigma)$. Clearly $\bar{\mu}_1 \leqq \bar{\mu}_2 \leqq \ldots \leqq \bar{\mu}_n \leqq \ldots$. Let Σ_1 be the σ-field determined by Σ and denote the unique extensions (5.8) of the measures $\{\bar{\mu}_n\}$ to Σ_1 by the same symbols. Since the extension of a non-negative set function on Σ to Σ_1 is non-negative, it follows that $\{\bar{\mu}_n(E)\}$ is a bounded non-decreasing set of real numbers for each $E \in \Sigma_1$. We define $\lambda_1(E) = \lim_n \bar{\mu}_n(E)$, $E \in \Sigma_1$. By Corollary 4, λ_1 is countably additive on Σ_1, and hence its restriction to Σ is in $ca(S, \Sigma)$. We define $\lambda_2(E) = \lambda(E) - \lambda_1(E)$, $E \in \Sigma$; clearly $\lambda_2 \geqq 0$. If λ_2 is not purely finitely additive, there is a non-zero $\lambda' \in ca(S, \Sigma)$ such that $\lambda' \leqq \lambda - \lambda_1$; hence $\lambda_1 \leqq \lambda_1 + \lambda' \leqq \lambda$ and $\sup_{\mu \in C} \mu(S) = \lambda_1(S) < \lambda_1(S) + \lambda'(S)$. This is a contradiction. To prove the uniqueness, suppose that $\lambda = \lambda_1 + \lambda_2 = \nu_1 + \nu_2$ where λ_1, ν_1 are in $ca(S, \Sigma)$ and λ_2, ν_2 are purely finitely additive non-negative set functions. Now $\lambda_1 - \nu_1 = \nu_2 - \lambda_2$ and so $\sup(\lambda_1 - \nu_1, 0) = \sup(\nu_2 - \lambda_2, 0)$. By Corollary 6, $\sup(\lambda_1 - \nu_1, 0) \in ca(S, \Sigma)$ and evidently $0 \leqq \sup(\lambda_1 - \nu_1, 0) \leqq \nu_2$ so that $\sup(\lambda_1 - \nu_1, 0) = 0$, i.e., $\lambda_1(E) \leqq \nu_1(E)$, $E \in \Sigma$. Similarly $\sup(\nu_1 - \lambda_1, 0) = 0$ and hence $\nu_1(E) \leqq \lambda_1(E)$, $E \in \Sigma$. Consequently $\lambda_i = \nu_i$, as desired. Q.E.D.

This result may be extended to complex valued set functions by decomposing them into real and imaginary and positive and negative parts by means of the Jordan decomposition (1.8). We leave this to the reader.

8. Relativization of Set Functions

Throughout this section, S is a set, Σ a field of subsets of S, and μ a finitely additive set function defined for $E \in \Sigma$. The field Σ_1 is a subfield of Σ, and μ_1 the restriction of μ from Σ to Σ_1. There are a number of elementary but useful relations between μ- and μ_1-measurability, μ- and μ_1-integrability, etc., which will be discussed in the present section.

First, if μ is non-negative, it is evident that $v^*(\mu, E) \leqq v^*(\mu_1, E)$ for $E \subseteqq S$. Thus, convergence in μ_1-measure implies convergence in μ-measure, a μ_1-null function is μ-null, and a μ_1-null set is μ-null.

Since a μ_1-simple function is clearly μ-simple, it follows immediately that a μ_1-measurable function is μ-measurable.

If f is a μ_1-integrable simple function, it is evident that f is also a μ-integrable simple function, and that $\int_E f(s)\mu_1(ds) = \int_E f(s)\mu(ds)$ for $E \in \Sigma_1$. Let f be a μ_1-integrable function and let $\{f_n\}$ be a sequence of μ_1-integrable simple functions converging to f in μ_1-measure, and such that

$$\lim_{m, n \to \infty} \int_S |f_n(s) - f_m(s)|\mu_1(ds) = 0.$$

Then $f_n \to f$ in μ-measure, and

$$\lim_{m, n \to \infty} \int_S |f_n(s) - f_m(s)|\mu(ds) = 0.$$

Consequently,

$$\int_E f(s)\mu_1(ds) = \lim_{n \to \infty} \int_E f_n(s)\mu_1(ds) = \int_E f(s)\mu(ds)$$

for each $E \in \Sigma_1$.

These remarks are summarized in the following lemma.

1 Lemma. *Let S be a set, Σ a field of subsets of S, and μ a non-negative finitely additive set function defined on Σ. Let Σ_1 be a subfield of Σ, and μ_1 the restriction of μ from Σ to Σ_1. Then*

(a) *convergence in μ_1-measure implies convergence in μ-measure;*
(b) *a μ_1-null function is a μ-null function;*
(c) *a μ_1-null set is a μ-null set;*
(d) *a μ_1-measurable function is μ-measurable;*
(e) *a μ_1-integrable function f is μ-integrable and*

$$\int_E f(s)\mu_1(ds) = \int_E f(s)\mu(ds), \qquad E \in \Sigma_1.$$

Let $\mathfrak{X}$ be an arbitrary B-space. It follows from (d) and (e) that a function $f \in L_p(S, \Sigma_1, \mu_1, \mathfrak{X})$ is also in $L_p(S, \Sigma, \mu, \mathfrak{X})$, and that its norm in both spaces is the same. Thus $L_p(S, \Sigma_1, \mu_1, \mathfrak{X})$ has a natural isometric embedding in $L_p(S, \Sigma, \mu, \mathfrak{X})$, and can be regarded as a subspace of $L_p(S, \Sigma, \mu, \mathfrak{X})$.

2 Lemma. *Under the hypotheses of Lemma 1 a totally μ_1-measurable function in $L_p(S, \Sigma, \mu, \mathfrak{X})$ is in $L_p(S, \Sigma_1, \mu_1, \mathfrak{X})$.*

PROOF. Let $\{f_n\}$ be a sequence of μ_1-simple functions converging in μ_1-measure to a function $f \in L_p(\mu) = L_p(S, \Sigma, \mu, \mathfrak{X})$. By the argument presented in the proof of Theorem 2.22 we may and shall assume that $|f_n(s)| \leqq 2|f(s)|$ for all s in S. By the dominated convergence theorem, $f_n \to f$ in $L_p(\mu)$, and hence $\{f_n\}$ is a Cauchy sequence in $L_p(\mu)$ and $\{|f_n(\cdot)|^p\}$ is a Cauchy sequence in $L_1(\mu)$. Since $|f_n(\cdot)|^p$ is a μ_1-simple function, its μ_1-integral coincides with its μ-integral and thus $\{|f_n(\cdot)|^p\}$ is a Cauchy sequence in $L_1(\mu_1)$. Since $|f_n(\cdot)|^p$ converges to $|f(\cdot)|^p$ in μ_1-measure, it follows from Definition 2.17 that $|f(\cdot)|^p$ is μ_1-integrable and thus that $f \in L_p(\mu_1)$. Q.E.D.

In addition to the restriction of μ to a subfield of Σ there is another type of restriction of common occurrence in integration theory. In the following discussion of this other type of restriction it is not necessary to assume that μ is non-negative.

Suppose that E is a set in Σ. If we put $\Sigma(E) = \{F \in \Sigma | F \subseteqq E\}$ it is clear that $\Sigma(E)$ is a field of subsets of E, and that $\Sigma(E)$ is the family of all sets AE, $A \in \Sigma$, and that if Σ is a σ-field, then $\Sigma(E)$ is a σ-field. $\Sigma(E)$ *is called the restriction of* Σ *to* E. If Σ_1 is a field, $E \in \Sigma_1$, and Σ is the σ-field generated by Σ_1, then we can easily show that $\Sigma(E)$ is the σ-field generated by $\Sigma_1(E)$. Indeed, $\Sigma(E)$ is a σ-field containing $\Sigma_1(E)$, while, conversely, if Σ_2 is a σ-field of subsets of E containing $\Sigma_1(E)$ it is evident that the family of unions of a set in Σ_2 with a set in $\Sigma(S-E)$ is a σ-field containing Σ_1, so that $\Sigma_2 \supseteqq \Sigma(E)$.

The restriction λ of μ to $\Sigma(E)$ is sometimes called *the restriction of* μ *to* E. It is clear that $(E, \Sigma(E), \lambda)$ is a measure space if (S, Σ, μ) is. The reader will have no difficulty in verifying that $v(\lambda, F) = v(\mu, F)$ if $F \subseteqq E$. The set of (vector or extended real valued) functions defined on E is in an obvious one-to-one correspondence with the set of functions defined on S and vanishing outside E; we have only to extend the domain of definition of a function f defined only on E to all of S in the natural way, by putting $f(s) = 0$ for $s \notin E$. Then a sequence of functions defined on E converges in λ-measure if and only if the sequence of extensions converges in μ-measure. In this case the original sequence of functions defined on E is said to *converge in* μ*-measure on* E. Similarly, if the domains of two functions f, g both contain the set E, the statement $f(s) = g(s)$ μ*-almost everywhere on* E means that there is a set $A \subseteqq E$ with $v^*(\mu, A) = 0$ and $f(s) = g(s)$

for every s in $E-A$. A function defined on E is λ-simple or λ-measurable if and only if its natural extension to S is μ-simple or μ-measurable, respectively. The concepts of λ-integrability and μ-integrability can be seen to be related in the same way. Thus, $L_p(E, \Sigma(E), \lambda, \mathfrak{X})$ is isometrically equivalent to the set of all functions in $L_p(S, \Sigma, \mu, \mathfrak{X})$ vanishing outside E, so that the former space may be regarded as a subspace of the latter.

3 **Lemma.** *Let (S, Σ, μ) be a positive measure space and μ_1 the restriction of μ to a subfield Σ_1 of Σ. If Σ is the σ-field generated by Σ_1 and if μ_1 is σ-finite on Σ_1, then the μ_1-integrable simple functions are dense in $L_p(S, \Sigma, \mu, \mathfrak{X})$, $1 \leqq p < \infty$.*

Proof. Since the μ-integrable simple functions are dense in $L_p(S, \Sigma, \mu, \mathfrak{X})$ (by Corollary 3.8), it suffices to prove that the characteristic function χ_E of a set E in Σ with $\mu(E) < \infty$ is the limit in $L_p(S, \Sigma, \mu)$ of a sequence χ_{E_n} with $E_n \in \Sigma_1$. Let A_m be an increasing sequence of sets in Σ_1, such that $\mu_1(A_m) < \infty$ and $\bigcup_{m=1}^{\infty} A_m = S$. Since $\mu(E) < \infty$ we have $|\chi_{EA_m} - \chi_E|_p = \{\mu(E-A_m)\}^{1/p} \to 0$, and it will therefore suffice to prove that χ_{EA_m} is the limit in $L_p(S, \Sigma, \mu)$ of a sequence χ_{E_n} with $E_n \in \Sigma_1$ for $m = 1, \ldots$. Since $\Sigma(A_m)$ is generated by $\Sigma_1(A_m)$, our whole argument can be confined to the set A_m. Thus, we may and shall assume without loss of generality that $\mu(S) < \infty$. By Lemma 7.1 the closure of Σ_1 in the metric space $\Sigma(\mu)$ is a σ-field containing Σ_1 and thus every set E in Σ is the limit of a sequence $\{E_n\}$ of sets in Σ_1. Since $\mu(S) < \infty$, the function identically equal to one is in $L_p(S, \Sigma, \mu)$ and dominates $\chi_{E_n}(s)$. Thus by Theorem 3.7, $\chi_{E_n} \to \chi_E$ in $L_p(S, \Sigma, \mu)$. Q.E.D.

4 **Lemma.** *The field generated by a countable family of sets is itself countable.*

Proof. Let $\mathscr{C} = \{E_n, n = 1, 2, \ldots\}$ be a countable collection of subsets of a set S. Let $\mathscr{C}_1$ be the family of all unions of finite subcollections of $\mathscr{C}$, $\mathscr{B}_1$ the family of all complements A' in S of sets A in $\mathscr{C}_1$, $\mathscr{C}_2$ the family of unions of finite subcollections of $\mathscr{B}_1$, $\mathscr{B}_2$ the family of all complements A' of sets A in $\mathscr{C}_2$, etc. It is clear that if A and B are in $\mathscr{C}_n$ then

$$A \cup B \in \mathscr{C}_n, \qquad A' \in \mathscr{C}_{n+1}, \qquad AB = (A' \cup B')' \in \mathscr{C}_{n+2}.$$

Thus the family $\Sigma = \cup_{n=1}^{\infty} \mathscr{C}_n$ is a field, and clearly the smallest field containing $\mathscr{C}$. An elementary induction shows that C_n, $n = 1, 2, \ldots$, is countable and thus Σ is a countable field. Q.E.D.

5 **Lemma.** *Let (S, Σ, μ) be a positive measure space and G a separable subset of $L_p(S, \Sigma, \mu, \mathfrak{X})$, where $1 \leqq p < \infty$. Then there is a set S_1 in Σ, a sub σ-field Σ_1 of $\Sigma(S_1)$, and a closed separable subspace $\mathfrak{X}_1$ of $\mathfrak{X}$ such that the restriction μ_1 of μ to Σ_1 has the following properties:*

(i) *the measure space (S_1, Σ_1, μ_1) is σ-finite;*

(ii) *the B-space $L_p(S_1, \Sigma_1, \mu_1, \mathfrak{X}_1)$ is separable;*

(iii) $G \subseteq L_p(S_1, \Sigma_1, \mu_1, \mathfrak{X}_1)$.

Proof. Let $\{f_n\}$ be dense in G and let $f_n^{(m)}$, $m, n = 1, 2, \ldots$, be μ-integrable simple functions with $\lim_{m \to \infty} |f_n^{(m)} - f_n|_p = 0$. Let X_0 be the countable set of non-zero values of the functions $f_n^{(m)}$ and $\mathfrak{X}_1$ the closed linear manifold in $\mathfrak{X}$ determined by X_0. By Lemma II.1.5, $\mathfrak{X}_1$ is separable. Let $\mathscr{C}$ be the countable collection of sets $E \in \Sigma$ which have the form $E = \{s | s \in S, f_n^{(m)}(s) = x_0\}$, where m, n are arbitrary positive integers and $x_0 \in X_0$. Let $S_1 = \cup \mathscr{C}$, let Σ_0 be the field of subsets of S_1 generated by $\mathscr{C}$ and Σ_1 be the σ-field of subsets of S_1 determined by Σ_0. Since every E in $\mathscr{C}$ has $\mu(E) < \infty$ and all the functions $f_n^{(m)}$ vanish on the complement of S_1, statements (i) and (iii) are immediate. If μ_0 is the restriction of μ to Σ_0, then it follows from Lemma 3 that the set of μ_1-integrable μ_0-simple functions is dense in $L_p(S_1, \Sigma_1, \mu_1, \mathfrak{X}_1)$. Since Σ_0 is countable (Lemma 4) and $\mathfrak{X}_1$ is separable, this set of functions is a separable set and hence its closure, i.e., $L_p(S_1, \Sigma_1, \mu_1, \mathfrak{X}_1)$, is separable. Q.E.D.

9. Exercises

In all the exercises below, unless the contrary is explicitly stated, S denotes a set, Σ a field of subsets of S, and μ a finitely additive (complex or extended real valued) set function on Σ. The letter f will denote a function defined on S with values in a B-space $\mathfrak{X}$.

1 Show that $f \in TM(S, \Sigma, \mu)$ if and only if for each $\varepsilon > 0$ there exists a set $E_\varepsilon \in \Sigma$ and a finite collection of disjoint sets $A_1, \ldots, A_n \in \Sigma$ such that $A_1 \cup \ldots \cup A_n = E'_\varepsilon$, $v(\mu, E_\varepsilon) < \varepsilon$, and $\sup_{s, t \in A_j} |f(s) - f(t)| < \varepsilon$, $j = 1, \ldots, n$.

2 Let $A \subseteq S$ be a μ-*null* set. Show that there need not exist any set $E \in \Sigma$ containing A such that $v(\mu, E) = 0$.

3 Suppose that S is the interval $(a, b]$, that Σ is the field of finite sums of intervals half open on the left, and that μ is the restriction of Lebesgue measure to Σ. Show that a real valued function f is μ-measurable if and only if it is continuous at every point not lying in a certain set E of Lebesgue measure zero, and μ-integrable if and only if it is μ-measurable and Lebesgue integrable.

4 Show that, even if μ is bounded, a uniformly bounded sequence $\{f_n\}$ of μ-measurable real valued functions defined on S can converge to zero everywhere without converging to zero in μ-measure.

5 Show that (i), (ii), and (iii) of Theorem 3.6 imply that f is in $L_p(S, \Sigma, \mu)$ and that $|f_n - f|_p$ converges to zero even if $\{f_n\}$ is a generalized sequence.

6 Let μ be bounded. Suppose that the field Σ is separable under the metric $\varrho(E, F) = v(\mu, E \Delta F)$. Show that $L_p(S, \Sigma, \mu, \mathfrak{X})$ is separable if the B-space $\mathfrak{X}$ is separable.

7 Show that $F(S)$ need not be a linear topological space if μ is not bounded, nor if μ is bounded but not countably additive.

8 Show that S can be the union of an increasing sequence of null sets even if μ is not identically zero.

9 Show that if f is defined on S and has values in a compact subset of $\mathfrak{X}$, and if $f^{-1}(G)$ is in Σ for each open subset G of $\mathfrak{X}$, then f is totally μ-measurable.

10 Find an example in which $L_1(S, \Sigma, \mu)$ is not complete. Show that if $TM(S, \Sigma, \mu)$ is complete, so is $L_1(S, \Sigma, \mu)$.

11 Show that if $f \in L_p(S, \Sigma, \mu)$ for some p with $1 \leqq p < \infty$, then $f \in TM(S, \Sigma, \mu)$.

12 Let Σ generate the σ-field Σ_1, and let μ be positive and countably additive on Σ. Show that if the measure space (S, Σ, μ) is not σ-finite, μ may have two distinct countably additive extensions to Σ_1.

13 Show that the map of $TM(S, \Sigma, \mu) \times TM(S, \Sigma, \mu)$ into $TM(S, \Sigma, \mu)$ defined by $[f, g] \to h$, where $h(s) = f(s)g(s)$, is contin-

uous. Show that this is no longer the case if TM is replaced by M.

14 Let Σ_1 be a subfield of Σ, and μ_1 the restriction of μ to Σ_1. Show that $v(\mu_1, E) \leqq v(\mu, E)$ holds for every $E \in \Sigma_1$, but that neither $v^*(\mu_1, E) \leqq v^*(\mu, E)$ nor $v^*(\mu, E) \leqq v^*(\mu_1, E)$ need hold for every subset E of S.

15 Show that Lemma 8.2 need no longer hold if we write "μ_1-measurable" for "totally μ_1-measurable".

16 Show that Σ forms an algebra in which $A^2 = A$ if we define

$$AB = A \cap B, \qquad A+B = A \Delta B$$

and that the μ-null sets are an ideal in this algebra.

17 Suppose that S is a normal topological space and that μ is regular and defined on the field Σ of Borel sets in S. Show that if $\mathfrak{X}$ is separable, the set of continuous functions in $TM(S, \Sigma, \mu, \mathfrak{X})$ is dense in $TM(S, \Sigma, \mu, \mathfrak{X})$. Show that for $1 \leqq p < \infty$, the set of continuous functions in $L_p(S, \Sigma, \mu, \mathfrak{X})$ is dense in $L_p(S, \Sigma, \mu, \mathfrak{X})$.

18 Let μ be a finite regular measure on the Borel subsets of a compact space S. Show that f is μ-measurable if and only if for each ε there exists an open set E_ε such that $v^*(\mu, E_\varepsilon) < \varepsilon$ and such that f is continuous on $S-E_\varepsilon$.

19 Find an example in which S is a metric space and μ is regular but not countably additive.

20 (Langlands) A regular complex valued additive set function defined on a field of sets in a compact space is countably additive.

21 Let S and S_1 be compact topological spaces with S_1 a Hausdorff space. Let $\varphi : S \to S_1$ be a continuous mapping. Let μ be a regular bounded additive set function defined on a field Σ containing the open subsets of S, and define the set function

$$\nu(E_1) = \mu(\varphi^{-1}E_1)$$

for each $E_1 \subseteqq S_1$ with $\varphi^{-1}(E_1) \in \Sigma$. Show that ν is a regular additive set function defined on the field $\{E_1 | \varphi^{-1}E_1 \in \Sigma\}$.

22 Let S be a topological space, and μ a bounded, countably additive measure. Call a set $E \in \Sigma$ regular if for each $\varepsilon > 0$ there exist sets F_1 and $F_2 \in \Sigma$, such that $\overline{F}_1 \subseteqq E$, $\overline{F}_2 \subseteqq S-E$, $v(\mu, S-F_1-F_2) < \varepsilon$. Show that the regular sets form a σ-field. Hence show that if S is a metric space and μ a bounded measure defined on the σ-field of Borel subsets of S, then μ is regular.

23 Suppose that S is a metric space, that Σ is a σ-field, and that μ is countably additive and bounded. Suppose that every continuous function is μ-measurable. Show that Σ^* contains every Borel set.

24 Suppose that S is a compact topological space, with the property that for every covering of S by a finite number of open sets $G_1, \ldots, G_n$ there exists a covering $E_1, \ldots, E_m$ of S by sets in Σ such that each E_j is contained in some G_i. Show that if f is continuous, f is μ-measurable.

25 Suppose that (S, Σ, μ) is a finite measure space. Let Σ_1 be a subfield of the σ-field Σ which generates Σ, and μ_1 the restriction of μ to Σ_1. Show that $v(\mu_1, E) = v(\mu, E)$ for $E \in \Sigma_1$.

26 Under the hypotheses of Exercise 25, let ν be a complex valued countably additive set function defined on Σ. Let ν_1 be the restriction of ν to Σ_1. Show that ν is μ-continuous if and only if ν_1 is μ_1-continuous.

27 Under the hypotheses of Exercise 26, $TM(S, \Sigma, \mu_1, \mathfrak{X})$ is dense in $TM(S, \Sigma, \mu, \mathfrak{X})$.

28 If (S, Σ, μ) is a measure space, $TM(S, \Sigma, \mu, \mathfrak{X})$ is an F-space, but $M(S, \Sigma, \mu, \mathfrak{X})$, though a complete metric space, is not necessarily an F-space.

29 Define $L_p(S, \Sigma, \mu)$ for $0 < p < 1$ as the set of all functions for which

$$|f| = \int_S |f(s)|^p \mu(ds) < \infty$$

Show that $|f_1+f_2| \leqq |f_1|+|f_2|$, while

$$|f_1+f_2|^{1/p} \geqq |f_1|^{1/p}+|f_2|^{1/p}$$

if f_1 and f_2 are both positive.

30 Let (S, Σ, μ) be a measure space, and let $0 < p < 1$. Show that the space $L_p(S, \Sigma, \mu)$ of Exercise 29 is an F-space.

31 Let $1 \leqq p < \infty$ and let f_1, f_2 be positive functions in L_p. Show that $|f_1+f_2|^p \geqq |f_1|^p+|f_2|^p$.

32 Find an example of a countably additive, bounded, vector valued function μ defined on a σ-field Σ of subsets of S for which $v(\mu, S) = +\infty$.

33 Show that (S, Σ, μ) can be a finite positive measure space, and $\{f_\alpha\}$ a uniformly bounded generalized sequence of non-negative

μ-simple functions defined on S converging to zero for each $s \in S$, but not converging to zero in μ-measure.

34 Let (S, Σ, μ) be a σ-finite positive measure space, and let Σ_1 be a subfield of the σ-field Σ, generating Σ and suppose that μ is σ-finite on Σ_1. Show that if $E \in \Sigma$,

$$\mu(E) = \inf \sum_{i=1} \mu(E_i),$$

where the infimum is taken over the family of all sequences $\{E_i\}$ of sets in Σ_1 such that $\cup_{i=1}^{\infty} E_i \supseteq E$.

35 Let (S, Σ, μ) be a positive measure space. If $f \in L_1(S, \Sigma, \mu)$, and $\{f_n\}$ is a sequence of real valued functions in $L_1(S, \Sigma, \mu)$, and $|f_n(s)| \leqq f(s)$ for $s \in S$, then

$$\begin{aligned}\int_S \liminf_{n\to\infty} f_n(s)\mu(ds) &\leqq \liminf_{n\to\infty} \int_S f_n(s)\mu(ds)\\ &\leqq \limsup_{n\to\infty} \int_S f_n(s)\mu(ds)\\ &\leqq \int_S \limsup_{n\to\infty} f_n(s)\mu(ds).\end{aligned}$$

Find a sequence $\{f_n\}$ of positive functions in $L_1(S, \Sigma, \mu)$ for which the preceding inequalities are not all valid.

36 Let S be the real axis, Σ the field of Lebesgue measurable subsets of S, and μ Lebesgue measure. Show that if $E \in \Sigma$, and a and b are real numbers, then $aE+b \in \Sigma$, and $\mu(aE+b) = |a|\mu(E)$. Show that if f is μ-integrable, then the function g defined by $g(s) = f(as+b)$ is μ-integrable, and

$$|a| \int_{-\infty}^{+\infty} f(as+b)ds = \int_{-\infty}^{+\infty} f(s)ds.$$

37 Let S be the real axis, Σ the field of Borel subsets of S, and μ any Borel-Stieltjes measure on Σ. Show that if the real function f is of bounded variation, or is continuous on the left, or is continuous on the right, then f is μ-measurable.

38 Let S be the closed unit interval, Σ the field of Borel sets of S, and μ Lebesgue measure. Find a μ-singular non-negative measure defined for $E \in \Sigma$ for which each point $p \in S$ has zero measure. (Hint: Use the Cantor perfect set.)

39 Let (S, Σ, μ) be as in Exercise 38. Find a continuous monotone increasing real function f defined on S such that f cannot be represented as

$$f(s) = \int_0^s g(t)dt \text{ with } g \text{ in } L_1(S, \Sigma, \mu).$$

40 Under the hypotheses of Exercise 36, let $0 < \alpha < 1$, and suppose that G is defined on $S \times S$, and continuous for all (s, t) such that $s \neq t$, and that

$$|G(s, t)| \leqq |s-t|^{-1-\alpha} |\sin(s-t)|.$$

Show that if h is a bounded, μ-measurable function defined on S, $G(s, \cdot)h(\cdot)$ is μ-integrable for each s, and

$$\int_{-\infty}^{+\infty} G(s, t)h(t)dt$$

is a continuous function of s.

41 Construct the Lebesgue extension of Borel measure by direct use of Theorem 5.4.

42 Let $1 \leqq p < \infty$, and let (S, Σ, μ) be a positive measure space. Let $f \in L_p$ and $g \in L_q$ $1/p+1/q = 1$, and let $h(s) = f(s)g(s)$. If $|\int_S h(s)\mu(ds)| = |f|_p|g|_q$, then $g(s) = A\overline{\text{sgn}}(f(s))|f(s)|^{p-1}$ μ-almost everywhere, where the function $\text{sgn}(re^{i\theta})$ of the complex number $z = re^{i\theta}$ is defined to be $e^{i\theta}$ if $r \neq 0$ and zero if $r = 0$. (Hint: Examine the cases of equality in the various steps of the proof of Lemma 3.2).

43 Let $1 < p < \infty$, and let (S, Σ, μ) be a measure space. Let $f, g \in L_p(S, \Sigma, \mu)$, and let $|f+g|_p = |f|_p+|g|_p$. Show that, if $g \neq 0$, then $f = \alpha g$ for some scalar α.

44 Let (S, Σ, μ) be a measure space. Let $\{f_n\}$ be a sequence of μ-measurable functions defined on S, with values in $\mathfrak{X}$, and let f be a function defined on S, with values in $\mathfrak{X}$. Suppose that for each $x^* \in \mathfrak{X}^*$, $x^*f_n(s) \to x^*f(s)$ for each $s \in S$. Show that f is μ-measurable.

45 Let (S, Σ, μ) be a measure space. Show that conditions (i) and (ii) in Theorem 6.15 may be replaced by the single condition

$$\lim_{m\to\infty} \int_{E_m} |f_n(s)|^p v(\mu, ds) = 0 \text{ uniformly in } n,$$

$\{E_m\}$ being an arbitrary decreasing sequence of sets with void intersection.

46 Show that Lemma 8.3 need no longer hold if we omit the hypothesis that μ_1 is σ-finite on Σ_1.

47 Let M be a metric space. Let $A \subseteq M$; let $\alpha, \varepsilon > 0$. Then define $\Lambda_\varepsilon(A, \alpha)$ as glb $\sum_{i=1}^{\infty}(\delta(A_i))^\alpha$, where $\{A_i\}$ is an arbitrary sequence of sets covering A whose diameters $\delta(A_i)$ are less than ε. Show that $\Lambda_\varepsilon(A, \alpha)$ is an outer measure, that $\Lambda_\varepsilon(A, \alpha)$ increases as ε decreases, and that $\Lambda(A, \alpha) = \lim_{\varepsilon \to 0} \Lambda_\varepsilon(A, \alpha)$ is an outer measure. (It can be shown that every Borel set is $\Lambda(A, \alpha)$ measurable; the Borel measure derived from $\Lambda(A, \alpha)$ is known as *Hausdorff α-measure.*)

48 Let $\{A_n\}$ and $\{B_n\}$ be sequences of subsets of S, and let $A_n \to A$, $B_n \to B$, both in the sense of the discussion following Definition 4.3. Show that $A_n B_n \to AB$, $A_n \cup B_n \to A \cup B$, $S - A_n \to S - A$.

10. The Radon–Nikodým Theorem

We have seen that the integral $\int_E f(s)\mu(ds)$ of a μ-integrable function is a μ-continuous set function. A partial converse to this statement is the important Radon-Nikodým theorem which asserts that if (S, Σ, μ) is a measure space then any finite scalar valued additive set function λ which is μ-continuous has the form $\lambda(E) = \int_E f(s)\mu(ds)$ for some μ-integrable function f. This theorem (Theorem 2) will be proved first under the assumption that μ is non-negative, and the complex valued case (Theorem 7) will be proved after some general results on change of measure have been established.

1 LEMMA. *Let (S, Σ, μ) be a finite positive measure space, and λ a finite positive μ-continuous measure defined on Σ. Then there exists a unique function f in $L_1(S, \Sigma, \mu)$ such that*

$$\lambda(E) = \int_E f(s)\mu(ds), \qquad E \in \Sigma.$$

Moreover, $v(\lambda, S) = |f|_1$.

PROOF. Let P be the set of non-negative integrable functions h such that

$$\int_E h(s)\mu(ds) \leq \lambda(E), \qquad E \in \Sigma.$$

The set P may be partially ordered by defining $h \leqq g$ to mean that $h(s) \leqq g(s)$ almost everywhere. We will show, using Zorn's lemma, that P contains a maximal element. Indeed, let Q be a totally ordered subfamily of P and $\alpha = \sup_{h \epsilon Q} \int_S h(s)\mu(ds)$. Then $0 \leqq \alpha \leqq \lambda(S)$. Let h_n be a sequence of elements of Q such that

$$\int_S h_n(s)\mu(ds) \leqq \int_S h_{n+1}(s)\mu(ds) \to \alpha.$$

Then, since Q is totally ordered, it follows that $h_n(s) \leqq h_{n+1}(s)$ almost everywhere and thus, without loss of generality, we may assume that $h_n(s) \leqq h_{n+1}(s)$ everywhere. By Corollary 6.17 $h(s) = \lim_n h_n(s)$ is integrable, $\int_S h(s)\mu(ds) = \alpha$, and $h_n \leqq h$, $n = 1, 2, \ldots$. To see that h is an upper bound for Q, let g be an arbitrary element of Q. Then either $g \leqq h_n$ for some n, in which case $g \leqq h$, or $g \geqq h_n$ for every n, in which case $g \geqq h$ and

$$0 \geqq \int_S g(s)\mu(ds) - \alpha = \int_S [g(s) - h(s)]\mu(ds) \geqq 0,$$

and this implies that $g(s) = h(s)$ almost everywhere. Thus h is an upper bound for Q and Zorn's lemma shows the existence of a maximal element f in P.

Let

$$\lambda_1(E) = \lambda(E) - \int_E f(s)\mu(ds), \qquad E \epsilon \Sigma.$$

Then λ_1 is a μ-continuous finite non-negative measure on Σ. To complete the proof we demonstrate that $\lambda_1(E) = 0$ for every E in Σ. If this is not so, then $\lambda_1(S) > 0$ and there is a positive number k such that

$$\text{(i)} \qquad \mu(S) - k\lambda_1(S) < 0.$$

By the Hahn decomposition (4.10) there is a set A in Σ such that

$$\mu(EA) - k\lambda_1(EA) \leqq 0, \qquad \mu(EA') - k\lambda_1(EA') \geqq 0, \qquad E \epsilon \Sigma,$$

and then, a fortiori,

$$\text{(ii)} \quad \mu(EA) - k\lambda_1(E) \leqq 0, \qquad \mu(EA') - k\lambda_1(EA') \geqq 0, \qquad E \epsilon \Sigma.$$

Therefore

$$\text{(iii)} \qquad \frac{1}{k}\mu(EA) - \lambda_1(E) \leqq 0, \qquad E \epsilon \Sigma.$$

If $\mu(A) = 0$ then $\lambda_1(A) = 0$ and $\mu(S) = \mu(A')$, $\lambda_1(S) = \lambda_1(A')$; thus, from (i) and (ii) we have

$$0 \leqq \mu(A') - k\lambda_1(A') = \mu(S) - k\lambda_1(S) < 0,$$

a contradiction. This proves that $\mu(A) > 0$.

Let g be defined by the equations:

$$g(s) = 1/k, \qquad s \in A;$$
$$g(s) = 0, \qquad s \notin A.$$

The inequality (iii) may then be written

$$\int_E g(s)\mu(ds) \leqq \lambda_1(E) = \lambda(E) - \int_E f(s)\mu(ds), \qquad E \in \Sigma,$$

which shows that

$$\int_E [f(s) + g(s)]\mu(ds) \leqq \lambda(E), \qquad E \in \Sigma.$$

Since $f+g > f$, this contradicts the maximality of f in P. Therefore $\lambda_1(E) = 0$ for every E in Σ.

The equation $v(\lambda, S) = |f|_1$ is simply Theorem 2.20(a). On the other hand, if f and g are two μ-integrable functions such that

$$\lambda(E) = \int_E f(s)\mu(ds) = \int_E g(s)\mu(ds),$$

then, by Lemma 6.8, f and g differ by a μ-null function. This establishes the uniqueness of f and completes the proof of the lemma. Q.E.D.

➔ 2 THEOREM. (*Radon-Nikodým*) *Let (S, Σ, μ) be a σ-finite positive measure space, and λ a finite μ-continuous measure defined on Σ. Then there exists a unique function f in $L_1(S, \Sigma, \mu)$ such that*

$$\lambda(E) = \int_E f(s)\mu(ds), \qquad E \in \Sigma.$$

Moreover, $v(\lambda, S) = |f|_1$.

PROOF. Once the existence of f is established, the equality $v(\lambda, S) = |f|_1$ follows from Theorem 2.20. Consequently, $\lambda(E) = 0$ if and only if $|f|_1 = 0$; i.e., if and only if f is a μ-null function. The uniqueness of f follows immediately from this. Thus, all that must be shown is the existence of f.

By separating λ into its real and imaginary parts, we find that it may be assumed that λ is real valued. A real valued set function can be represented as the difference of its positive and negative variations (4.11) and so we may also assume that λ is positive. Let, then, $\{E_n\}$ be a sequence of measurable sets such that $v(\mu, E_n) < \infty$, $E_n \subseteqq E_{n+1}$, and $\bigcup_{n=1}^{\infty} E_n = S$. By Lemma 1, for each $n = 1, 2, \ldots$ there is a non-negative integrable function f_n which vanishes on E_n' and for which

$$\int_E f_n(s)\mu(ds) = \lambda(E), \qquad E \subseteqq E_n.$$

By the uniqueness of f_n, we have $f_n(s) = f_{n+1}(s)$ almost everywhere in E_n; without loss of generality we may assume that $f_n(s) = f_{n+1}(s)$ everywhere in E_n. Then $f_n \in L_1(S, \Sigma, \mu)$, $f_n(s) \leqq f_{n+1}(s)$, and $\int_S f_n(s)\mu(ds) = \lambda(E_n) \leqq \lambda(S)$. Put $f(s) = \lim_{n\to\infty} f_n(s)$. Then, by Corollary 6.17,

$$\lambda(E) = \lim_{n\to\infty} \lambda(E_n E) = \lim_{n\to\infty} \int_E f_n(s)\mu(ds) = \int_E f(s)\mu(ds)$$

for every $E \in \Sigma$. Q.E.D.

The next results are useful complements to the Radon-Nikodým theorem, and may be used to extend that theorem. The key theorems in the set are 4 and 6, which give important "change of measure" statements.

3 Lemma. *Let (S, Σ, μ) be a measure space and let f be a μ-measurable function defined on S. Suppose that either*

(a) (S, Σ, μ) is positive and f is non-negative, or

(b) f is complex valued and integrable.

Let

$$\lambda(E) = \int_E f(s)\mu(ds), \qquad E \in \Sigma,$$

and let g be a function defined on S with values in a B-space $\mathfrak{X}$. Then g is λ-measurable if and only if fg is μ-measurable.

Proof. Let g be λ-measurable. To show that fg is μ-measurable, we shall show that $\chi_F fg$ is totally μ-measurable, where χ_F is the characteristic function of an arbitrary set F in Σ with $v(\mu, F) < \infty$. We observe first that it may be assumed that $v(\lambda, F) < \infty$. For, if f is integrable then $v(\lambda, F) < \infty$ for all $F \in \Sigma$; if f is only assumed to be positive and measurable we can put $F_n = \{s | s \in F, f(s) \leqq n\}$. Then $F = \bigcup_{n=1}^{\infty} F_n$ and $v(\lambda, F_n) < \infty$ for each n. Since

$$\chi_F(s)f(s)g(s) = \lim_n \chi_{F_n}(s)f(s)g(s), \qquad s \in S,$$

and the pointwise limit of a sequence of measurable functions is measurable, it will suffice then to show that $\chi_{F_n}fg$ is measurable. Thus we may and shall assume that $v(\mu, F) < \infty$ and $v(\lambda, F) < \infty$.

Since g is λ-measurable there is a sequence $\{g_n\}$ of simple functions converging to $g(s)$ for every s in F except on a set $E \subseteq F$ with $v(\lambda, E) = 0$ (by Corollary 6.13(a)). But

$$v(\lambda, E) = \int_E |f(s)| v(\mu, ds)$$

so that $f(s) = 0$ for s in E except in a set A with $v(\mu, A) = 0$. Thus

$$g_n(s)f(s) \to g(s)f(s), \qquad s \in F - A,$$

and Corollary 6.14 shows that $\chi_F fg$ is μ-measurable.

Conversely, let fg be μ-measurable, and let (S, Σ^*, μ) and (S, Σ_0^*, λ) be the Lebesgue extensions of (S, Σ, μ) and (S, Σ, λ) respectively. Then $\Sigma_0^* \supseteq \Sigma^*$. We note first that a given function is measurable with respect to a given measure if and only if it is measurable with respect to the Lebesgue extension of the measure. Now let $N = \{s | s \in S, f(s) = 0\}$, so that $\lambda(E) = 0$ for every $E \in \Sigma^*$ with $E \subseteq N$. Thus $g\chi_N$ is λ-measurable, and it need only be proved that $g\chi_{N'}$ is λ-measurable. Thus we may and shall assume that $f(s)$ never vanishes on S. Since

$$\{s | 1/f(s) \in G\} = \{s | f(s) = 1/z, z \in G\},$$

Theorem 6.10 shows that $1/f$ is μ-measurable, and thus it follows from Lemma 2.12 that g is μ-measurable. By Theorem 6.10, g is μ-essentially separably valued, and hence it is λ-essentially separably valued. Also for G open in $\mathfrak{X}$ we have, by 6.10, $g^{-1}(G) \in \Sigma^*$. Thus, since $\Sigma_0^* \supseteq \Sigma^*$, it is seen from Theorem 6.10 that g is λ-measurable. In the preceding statement we have implicitly used the fact that any $F \in \Sigma$ with

$$v(\lambda, F) = \int_F |f(s)| v(\mu, ds) < \infty$$

may be decomposed into a denumerable union of sets

$$F_n = \{s | s \in F, |f(s)| > 1/n\}$$

with $v(\mu, F_n) < \infty$. Q.E.D.

4 THEOREM. *Let (S, Σ, μ) be a positive measure space, f a non-negative μ-measurable function defined on S and*

$$\lambda(E) = \int_E f(s)\mu(ds), \qquad E \in \Sigma.$$

Let g be a non-negative λ-measurable function defined on S. Then fg is μ-measurable, and

$$\int_E g(s)\lambda(ds) = \int_E f(s)g(s)\mu(ds), \qquad E \in \Sigma.$$

PROOF. The μ-measurability of fg follows from Lemma 3. If we let H be the class of all non-negative λ-measurable functions h for which the equation

$$\int_E h(s)\lambda(ds) = \int_E f(s)h(s)\mu(ds), \qquad E \in \Sigma,$$

is valid, then H clearly contains all non-negative λ-simple functions. In view of Corollary 6.17 the class H is closed under the operation of taking limits of increasing sequences. Thus to prove the theorem, it will suffice to prove that a non-negative λ-measurable function g is the pointwise limit almost everywhere of an increasing sequence $\{g_n\}$ of simple functions.

To define such a sequence, decompose the set $E_n = \{s|g(s) < n\}$ into the n^2 disjoint parts

$$E(j, n) = \left\{s \,\middle|\, \frac{j-1}{n} \leqq g(s) < \frac{j}{n}\right\}, \qquad j = 1, \ldots, n^2.$$

Let

$$g_n(s) = \begin{cases} n\ , & s \notin E_n, \\ j/n, & s \in E(j, n), \end{cases}$$

so that $g_n(s)$ increases to $g(s)$ for each s in S. It follows from Theorem 6.10 that each g_n is a simple function relative to the Lebesgue extension of λ and thus the proof is complete. Q.E.D.

5 COROLLARY. *Let (S, Σ, μ) be a positive measure space, f a non-negative measurable function defined on S, and*

$$\lambda(E) = \int_E f(s)\mu(ds), \qquad E \in \Sigma.$$

Then a function g on S to the B-space $\mathfrak{X}$ is λ-integrable if and only if fg is μ-integrable, and in this case,

$$\int_E g(s)\lambda(ds) = \int_E f(s)g(s)\mu(ds), \qquad E \in \Sigma.$$

PROOF. Suppose that g is λ-integrable. By Lemma 3, fg is measurable. The μ-integrability of fg follows from Theorem 2.22 when it is observed (by Theorem 4) that $|f(\cdot)g(\cdot)|$ is μ-integrable. Theorem 4 applies in the same manner to prove the λ-integrability of g when it is assumed that fg is μ-integrable.

Since every real measurable function is the difference of two non-negative measurable functions, it follows from the theorem that

$$[*] \qquad \int_E g(s)\lambda(ds) = \int_E f(s)g(s)\mu(ds), \qquad E \in \Sigma,$$

for a real integrable function g. Since the real and imaginary parts of a complex measurable function are themselves measurable, it follows that [*] holds for complex valued functions g. If g takes its values in the B-space $\mathfrak{X}$, let x^* be a linear functional on $\mathfrak{X}$. Then using Theorem 2.19(c) and [*],

$$\begin{aligned} x^* \int_E g(s)\lambda(ds) &= \int_E x^*g(s)\lambda(ds) \\ &= \int_E x^*f(s)g(s)\mu(ds) = x^* \int_E f(s)g(s)\mu(ds), \end{aligned}$$

which, in view of Corollary II.3.15, proves [*] in the desired generality. Q.E.D.

6 COROLLARY. *Let (S, Σ, μ) be a measure space, f a complex valued μ-integrable function and*

$$\lambda(E) = \int_E f(s)\mu(ds), \qquad E \in \Sigma.$$

Then a function g on S to the B-space $\mathfrak{X}$ is λ-integrable if and only if fg is μ-integrable, and in this case we have

$$[*] \qquad \int_E g(s)\lambda(ds) = \int_E f(s)g(s)\mu(ds), \qquad E \in \Sigma.$$

PROOF. Let g be λ-integrable so that, by Lemma 3, fg is measurable. To show that fg is μ-integrable it will suffice, in view of Theorems 2.22 and 2.18, to show that $|f(\cdot)g(\cdot)|$ is $v(\mu)$-integrable. But since $|g(.)|$ is $v(\mu)$-integrable and $v(\lambda, E) = \int_E |f(s)|v(\mu, ds)$, this is a consequence of Theorem 4.

Conversely, suppose that fg is μ-integrable. Then, by Lemma 3, g

is measurable, and hence g is λ-integrable if and only if $|g(\cdot)|$ is $v(\mu)$-integrable. The $v(\mu)$-integrability of $|g(\cdot)|$, however, follows from Theorem 4 just as in the preceding argument. Thus only formula [*] remains to be verified. By the argument used in the proof of Corollary 5 it is seen that it will be sufficient to prove [*] in the case when g is a positive real function. It is also clear that [*] holds for λ-integrable simple functions. As in the proof of Theorem 4, there is an increasing sequence $\{g_n\}$ of non-negative simple functions converging pointwise to g. Since $g_n(s) \leqq g(s)$, we have $|g_n(s)f(s)| \leqq |g(s)f(s)|$ for every s in S. It follows from the dominated convergence theorem (6.16) that

$$\int_E |f(s)g_n(s) - f(s)g(s)|v(\mu, ds) \to 0,$$

and

$$\int_E |g_n(s) - g(s)|v(\lambda, ds) \to 0.$$

Since equation [*] holds for $g = g_n$ this shows that it holds also for g. Q.E.D.

The next result supplements Theorem 2 by allowing μ to be complex valued.

➔ 7 THEOREM. *Let (S, Σ, μ) be a finite measure space and λ a μ-continuous complex valued measure on Σ. Then there is a unique μ-integrable function f such that*

$$\lambda(E) = \int_E f(s)\mu(ds), \qquad E \in \Sigma.$$

PROOF. Using the Radon-Nikodým theorem (Theorem 2) we find $v(\mu)$-integrable functions g and h such that

$$\lambda(E) = \int_E g(s)v(\mu, ds), \qquad E \in \Sigma,$$

$$\mu(E) = \int_E h(s)v(\mu, ds), \qquad E \in \Sigma.$$

Since

$$v(\mu, E) = \int_E |h(s)|v(\mu, ds),$$

it follows that

$$\int_E \{1 - |h(s)|\}v(\mu, ds) = 0, \qquad E \in \Sigma,$$

from which it is seen that $|h(s)| = 1$ except on a set where $v(\mu)$ vanishes. Thus $f = g/h$ is μ-integrable, and by the preceding corollary,

$$\int_E f(s)\mu(ds) = \int_E g(s)v(\mu, ds) = \lambda(E). \quad \text{Q.E.D.}$$

Note: The unique μ-integrable function f of Theorems 2 and 7 is called the ***Radon-Nikodým derivative*** of λ with respect to μ and is often denoted by $d\lambda/d\mu$. Thus $d\lambda/d\mu$ is defined μ-almost everywhere by the formula

$$\lambda(E) = \int_E \left\{\frac{d\lambda}{d\mu}(s)\right\} \mu(ds), \qquad E \in \Sigma.$$

We close this section with a generalization of many "change of variable" theorems.

8 LEMMA. *Let S_1 and S_2 be sets and ϕ a mapping of S_1 into S_2. If Σ_2 is a field (resp. σ-field) of sets in S_2, then the family $\Sigma_1 = \{\phi^{-1}(E) | E \in \Sigma_2\}$ is a field (resp. σ-field) of sets in S_1. For an additive set function μ_2 on Σ_2 the equation $\mu_1(\phi^{-1}(E)) = \mu_2(E)$ defines an additive set function μ_1 on Σ_1. Moreover*

(a) *if μ_2 is countably additive, so is μ_1;*

(b) *if μ_2 is bounded, μ_1 is bounded;*

(c) $v(\mu_1, \phi^{-1}(E)) = v(\mu_2, E), \qquad E \in \Sigma_2;$

(d) *if a function f defined on S_2 is μ_2-measurable, then $f(\phi(\cdot))$ is μ_1-measurable;*

(e) *if μ_2 is non-negative and countably additive and a function f defined on S_2 is μ_2-measurable and non-negative, then*

$$\int_E f(s_2)\mu_2(ds_2) = \int_{\phi^{-1}(E)} f(\phi(s_1))\mu_1(ds_1);$$

(f) *if a function f defined on S_2 is μ_2-integrable, then $f(\phi(\cdot))$ is μ_1-integrable, and*

$$\int_E f(s_2)\mu_2(ds_2) = \int_{\phi^{-1}(E)} f(\phi(s_1))\mu_1(ds_1).$$

PROOF. Since $\phi(\phi^{-1}(E)) = E$ it is clear that μ_1 is well defined on Σ_1. Since $\phi^{-1}(EF) = \phi^{-1}(E)\phi^{-1}(F)$, $\phi^{-1}(E') = \{\phi^{-1}(E)\}'$, and $\phi^{-1}(\bigcup_{i=1}^{\infty} E_i) = \bigcup_{i=1}^{\infty} \phi^{-1}(E_i)$, it is clear that Σ_1 is a field and that μ_1 is additive on Σ_1. These identities also show that Σ_1 is a σ-field if Σ_2 is, and that μ_1 is countably additive if μ_2 is. Statement (b) is obvious. Since $v(\mu_1, \phi^{-1}(E))$ is a non-negative set function defined for $E \in \Sigma_2$ and

$$v(\mu_1, \phi^{-1}(E)) \geqq |\mu_1(\phi^{-1}(E))| = |\mu_2(E)|,$$

it follows readily from the definition of $v(\mu_2)$ that $v(\mu_1, \phi^{-1}(E)) \geqq v(\mu_2, E)$. Conversely, since $\phi(E_1 \cup E_2) = \phi(E_1) \cup \phi(E_2)$ and $\phi(E_1E_2) = \phi(E_1)\phi(E_2)$ if E_1, E_2 belong to Σ_1, $v(\mu_2, \phi(E))$ is a non-negative additive set function defined for $E \in \Sigma_1$. Since

$$v(\mu_2, \phi(E)) \geqq |\mu_2(\phi(E))| = |\mu_1(E)|,$$

it follows readily from the definition of $v(\mu_1)$ that $v(\mu_2, \phi(E)) \geqq v(\mu_1, E)$. As it has also been shown that $v(\mu_2, \phi(E)) \leqq v(\mu_1, E)$, we have $v(\mu_2, \phi(E)) = v(\mu_1, E)$ for $E \in \Sigma_1$, and therefore $v(\mu_2, E) = v(\mu_1, \phi^{-1}(E))$ for $E \in \Sigma_2$. This proves (c).

It is a consequence of (c) that if $\{f_n\}$ is a sequence of functions defined on S_2 and converging to a function f in μ_2-measure, then $\{f_n(\phi(\cdot))\}$ converges to $f(\phi(\cdot))$ in μ_1-measure. Since $g(\phi(\cdot))$ is μ_1-simple if g is μ_2-simple, (d) follows immediately.

Clearly, $g(\phi(\cdot))$ is a μ_1-integrable simple function if g is a μ_2-integrable simple function. From the definition of μ_1 it follows that for such a function g we have

$$\int_E g(s_2)\mu_2(ds_2) = \int_{\phi^{-1}(E)} g(\phi(s_1))\mu_1(ds_1), \qquad E \in \Sigma_2.$$

Thus (f) follows from the definition of integrability just as (d) from the definition of measurability. Finally, (e) is a consequence of Theorem 6.17 and the application of (f) to each member of an increasing sequence of μ_2-integrable simple functions which converge pointwise to the function f. Q.E.D.

11. Product Measures

Our principal objective in this section is to construct, from a given family (S_i, Σ_i, μ_i), $i = 1, \ldots, n$, of measure spaces, a measure defined in the Cartesian product space $S_1 \times \ldots \times S_n$, and to investigate the relations between integration in the product and integration in the component spaces. The most familiar example is, of course, the case of the Euclidean plane, the product of the real line with itself. For a rectangle E with sides parallel to the coordinate axes one defines the two dimensional measure of E to be the product of the lengths (one dimensional measure) of two adjacent sides. This measure function is then extended to the σ-field in the plane generated by such rectangles.

In the latter half of the section the theory of product measures and integration is extended to the product of an infinite family of measure spaces.

The following notation will be convenient in the statements of the theorems of this section. Suppose that for $i = 1, \ldots, n$, Σ_i is a field of subsets of a set S_i. We shall denote by $\mathcal{E}$ the family of all sets in the Cartesian product $S = S_1 \times \ldots \times S_n$ of the form $E = E_1 \times \ldots \times E_n$, where $E_i \in \Sigma_i$.

1 **Lemma.** *For $i = 1, \ldots, n$, let μ_i be a finitely additive complex valued set function defined on a field Σ_i of subsets of a set S_i. Then there is a unique additive set function μ defined on the smallest field Σ in $S = S_1 \times \ldots \times S_n$ containing $\mathcal{E}$ such that $\mu(E) = \prod_{i=1}^n \mu_i(E_i)$ for E in $\mathcal{E}$.*

Remark. The reader may imagine that this lemma is best proved by a direct construction. Such a proof can be given; but it is surprisingly cumbersome. For this reason, we give a proof depending on the theory of integration.

Proof. Let $\mathfrak{A}_0$ be the family of bounded complex valued functions f defined on S with the following property: For each fixed $[s_2, \ldots, s_n]$ the function $f(s_1, \ldots, s_n)$ is μ_1-integrable as a function of s_1, for each fixed $[s_3, \ldots, s_n]$ and E_1 in Σ_1 the function

$$\int_{E_1} f(s_1, \ldots, s_n)\mu_1(ds_1)$$

is μ_2-integrable as a function of s_2; for each fixed $[s_4, \ldots, s_n]$, $E_1 \in \Sigma_1$, and $E_2 \in \Sigma_2$, the function

$$\int_{E_2} \left[\int_{E_1} f(s_1, \ldots, s_n)\mu_1(ds_1) \right] \mu_2(ds_2)$$

is μ_3-integrable as a function of s_3, etc. We denote by $\mathfrak{A}_1$ the class of all $g \in \mathfrak{A}_0$ with the property that the product fg is in $\mathfrak{A}_0$ for every $f \in \mathfrak{A}_0$. Then $\mathfrak{A}_1$ is a linear class of functions which contains the product of any pair of its members. Now let Φ be the family of sets $F \subseteq S$ whose characteristic functions χ_F belong to $\mathfrak{A}_1$. It follows that Φ is closed under intersection and complementation and contains S. Thus Φ is a field of sets. Moreover, each set of $\mathcal{E}$ belongs to Φ and hence $\Phi \supseteq \Sigma$. If the set function μ is defined for $F \in \Phi$ by the formula

$$\mu(F) = \int_{S_n} \left[\cdots \left[\int_{S_1} \chi_F(s_1, \ldots, s_n)\mu_1(ds_1) \right] \cdots \right] \mu_n(ds_n),$$

the restriction of μ to Σ is an additive set function with the desired properties.

To prove the uniqueness of μ suppose that λ is an additive set function defined on Σ which agrees with μ on each set of $\mathscr{E}$. Let $\mathfrak{B}_0$ be the class of λ-integrable functions f in $\mathfrak{A}_0$ for which

$$\int_E f(s)\lambda(ds) = \int_{E_n}\left[\ldots\left[\int_{E_1} f(s_1, \ldots, s_n)\mu_1(ds_1)\right]\ldots\right]\mu_n(ds_n),$$

$E = E_1 \times \ldots \times E_n$ for arbitrary $E_i \in \Sigma_i$, $i=1, \ldots, n$. Let $\mathfrak{B}_1$ be the set of all $g \in \mathfrak{B}_0$ such that $fg \in \mathfrak{B}_0$ for all $f \in \mathfrak{B}_0$. Then $\mathfrak{B}_1$ is a linear class which is closed under products and contains the characteristic function of each set in $\mathscr{E}$. By the argument used before, it follows that

$$\begin{aligned}\lambda(F) &= \int_S \chi_F(s)\lambda(ds)\\ &= \int_{S_n}\left[\ldots\left[\int_{S_1} \chi_F(s_1, \ldots, s_n)\mu_1(ds_1)\right]\ldots\right]\mu_n(ds_n)\end{aligned}$$

for all $F \in \Sigma$. Q.E.D.

Remark. The field Σ consists of all finite disjoint unions of sets in $\mathscr{E}$. For, if Σ_0 is the collection of all finite disjoint unions, then Σ_0 is evidently closed under finite intersections. To prove that Σ_0 is a field it suffices to show that it is closed under complementation. If $n = 3$, this follows from the identity

$$(E_1 \times E_2 \times E_3)' = (E_1' \times S_2 \times S_3) \cup (E_1 \times E_2' \times S_3) \cup (E_1 \times E_2 \times E_3'),$$

and the general case is similar.

2 Theorem. *Let (S_i, Σ_i, μ_i), $i = 1, \ldots, n$, be finite measure spaces. Then there is a unique countably additive measure μ defined on the smallest σ-field in $S = S_1 \times \ldots \times S_n$ containing $\mathscr{E}$ and such that*

$$\mu(E_1 \times \ldots \times E_n) = \prod_{i=1}^{n} \mu(E_i)$$

for each set $E_1 \times \ldots \times E_n$ in $\mathscr{E}$.

Proof. Define $\mathfrak{A}_1$ as in Lemma 1. It follows from the countable additivity of the measures μ_i and from Theorem 6.16 that if $\{f_k\}$ is a uniformly bounded sequence in $\mathfrak{A}_1$ converging pointwise to a function f, then $f \in \mathfrak{A}_1$ and

$$\begin{aligned}\lim_{k\to\infty} &\int_{S_n}\left[\ldots\left[\int_{S_1} f_k(s_1, \ldots, s_n)\mu_1(ds_1)\right]\ldots\right]\mu_n(ds_n)\\ &= \int_{S_n}\left[\ldots\left[\int_{S_1} f(s_1, \ldots, s_n)\mu_1(ds)\right]\ldots\right]\mu_n(ds_n).\end{aligned}$$

Thus, the field Φ of Lemma 1 is a σ-field and the measure μ of Lemma 1 is countably additive on Φ. Consequently, the restriction of μ to the σ-field Σ generated by $\mathscr{E}$ is a measure with the desired properties.

To prove the uniqueness of μ, we note that it follows in the same way that the set $\mathfrak{B}_1$ of Lemma 1 contains the limit of a convergent uniformly bounded sequence of its elements, and hence contains the characteristic function of each set $F \in \Sigma$. Thus the argument used in proving uniqueness in Lemma 1 goes through without change in this case. Q.E.D.

3 **Definition.** The measure space (S, Σ, μ) constructed in Theorem 2 is called the *product measure space* of the measure spaces (S_n, Σ_n, μ_n). We write

$$\Sigma = \Sigma_1 \times \ldots \times \Sigma_n, \qquad \mu = \mu_1 \times \ldots \times \mu_n,$$
$$(S, \Sigma, \mu) = (S_1, \Sigma_1, \mu_1) \times \ldots \times (S_n, \Sigma_n, \mu_n),$$

and

$$(S, \Sigma, \mu) = \mathop{P}_{i=1}^{n} (S_i, \Sigma_i, \mu_i).$$

In the course of the proof of Theorem 2 we showed that the characteristic function of every set $E \in \Sigma = \Sigma_1 \times \ldots \times \Sigma_n$ is in $\mathfrak{A}_1$. That is, we have proved the following corollary, which is at the same time the first of a sequence of theorems on the relation between "multiple integrals" and "iterated integrals" to be proved in this section.

4 **Corollary.** *Let (S, Σ, μ) be the product of finite positive measure spaces (S_1, Σ_1, μ_1) and (S_2, Σ_2, μ_2). For each E in Σ and s_2 in S_2 the set $E(s_2) = \{s_1 | [s_1, s_2] \in E\}$ is μ_1-measurable. The function $\mu_1(E(s_2))$ is μ_2-integrable and*

$$\mu(E) = \int_{S_2} \mu_1(E(s_2))\, \mu_2(ds_2).$$

5 **Corollary.** *The product of finite positive measure spaces is a finite positive measure space.*

Proof. This fact follows immediately from the formula for $\mu(E)$ given in Corollary 4. Q.E.D.

It is now easy to extend the definition of the product measure to the case that the measure spaces (S_i, Σ_i, μ_i), $i = 1, \ldots, n$, are positive and σ-finite rather than finite.

6 Corollary. *If the measure spaces* (S_i, Σ_i, μ_i), $i = 1, \ldots, n$, *are positive and σ-finite, there is a unique measure μ defined on the σ-field generated by* $\mathscr{E}$ *such that* $\mu(E) = \prod_{i=1}^n \mu_i(E_i)$ *for* $E = P_{i=1}^n E_i$, $E_i \in \Sigma_i$.

Remark. It is understood that $\prod_{i=1}^n \mu_i(E_i) = +\infty$ if $\mu_i(E_i) = +\infty$ for some i and no $\mu_i(E_i) = 0$, while if some factor in a product is zero, we take the product to be zero.

Proof. For each $k = 1, \ldots, n$ let S_k be the union of an increasing sequence $\{E_k^{(j)}\}$, $j = 1, \ldots,$ of μ_k-measurable sets of finite μ_k-measure. If $E^{(j)} = P_{k=1}^n E_k^{(j)}$, then $\{E^{(j)}\}$ is an increasing sequence of subsets of S whose union is S. For each k, let $\Sigma_k^{(j)}$ be the field of μ_k-measurable subsets of $E_k^{(j)}$ and let $\mu_k^{(j)}$ be the restriction of μ_k to $E_k^{(j)}$. We have already seen how to form the finite measure spaces

$$(E^{(j)}, \Sigma^{(j)}, \mu^{(j)}) = \mathop{P}_{k=1}^{n} (E_k^{(j)}, \Sigma_k^{(j)}, \mu_k^{(j)}).$$

It follows from the uniqueness part of Theorem 2 that $\mu^{(j)}(F) = \mu^{(j+1)}(F)$ if $F \subseteq E^{(j)}$. Let Σ_0 be the family of all subsets E of S such that

$$E \cap E^{(j)} \in \Sigma^{(j)}$$

for each j. Since each $\Sigma^{(j)}$ is a σ-field, Σ_0 is itself a σ-field. Let Σ be the σ-field generated by $\mathscr{E}$. Then $\Sigma \supseteq \Sigma^{(j)}$ for each j, and thus $\Sigma \supseteq \Sigma_0$. Clearly $\Sigma \subseteq \Sigma_0$, and thus $\Sigma = \Sigma_0$.

If $F \in \Sigma$ define $\mu(F) = \lim_{j\to\infty} \mu^{(j)}(F \cap E^{(j)})$. Since this sequence is increasing, the limit exists but may equal $+\infty$. To see that μ is countably additive on Σ, let $F \in \Sigma$ be the union of a sequence $\{F_n\}$ of disjoint sets in Σ. Then for each j and k

$$\begin{aligned}\mu(F) &\geqq \mu^{(j)}(F \cap E^{(j)}) = \sum_{n=1}^{\infty} \mu^{(j)}(F_n \cap E^{(j)}) \\ &\geqq \sum_{n=1}^{k} \mu^{(j)}(F_n \cap E^{(j)}).\end{aligned}$$

Thus $\mu(F) \geqq \sum_{n=1}^k \mu(F_n)$, and hence $\mu(F) \geqq \sum_{n=1}^\infty \mu(F_n)$. On the other hand, for each j,

$$\begin{aligned}\sum_{n=1}^{\infty} \mu(F_n) &\geqq \sum_{n=1}^{\infty} \mu^{(j)}(F_n \cap E^{(j)}) \\ &= \mu^{(j)}(F \cap E^{(j)}),\end{aligned}$$

so $\sum_{n=1}^{\infty}\mu(F_n) \geqq \mu(F)$. It is clear that the measure has the property

$$\mu(\underset{i=1}{\overset{n}{P}} E_i) = \prod_{i=1}^{n} \mu_i(E_i), \qquad E_i \in \Sigma_i,$$

and it follows from the uniqueness part of Theorem 2 for each of the spaces $(E^{(j)}, \Sigma^{(j)}, \mu^{(j)})$, that μ is the unique countably additive measure on Σ with this property. Q.E.D.

As in the case of finite measure spaces we shall call the measure space (S, Σ, μ) constructed in Corollary 6 from the σ-finite measure spaces (S_i, Σ_i, μ_i) the *product measure space* and write $(S, \Sigma, \mu) = P_{i=1}^{n} (S_i, \Sigma_i, \mu_i)$.

The best known example of Theorem 2 and its Corollary 6 is obtained by taking (S_i, Σ_i, μ_i) to be the Borel-Lebesgue measure on the real line for $i=1, \ldots, n$. Then $S = P_{i=1}^{n} S_i$ is n-dimensional Euclidean space, and $\mu = \mu_1 \times \ldots \times \mu_n$ is known as *n-dimensional Borel-Lebesgue measure.* The Lebesgue extension of μ is known as *n-dimensional Lebesgue measure.* The characteristic property of this measure is then that the measure of an arbitrary "rectangle"

$$R = \{[s_1, \ldots, s_n] | a_1 \leqq s_1 \leqq b_1, \ldots, a_n \leqq s_n \leqq b_n\}$$

is the product $(b_1 - a_1) \ldots (b_n - a_n)$.

The next corollary is the σ-finite analogue of Corollary 4.

7 Corollary. *Let (S, Σ, μ) be the product of two positive σ-finite measure spaces (S_1, Σ_1, μ_1) and (S_2, Σ_2, μ_2). For each E in Σ and s_2 in S_2 the set $E(s_2) = \{s_1 | [s_1, s_2] \in E\}$ is μ_1-measurable. The function $\mu_1(E(s_2))$ is μ_2-measurable, and*

$$[*] \qquad \mu(E) = \int_{S_2} \mu_1(E(s_2))\mu_2(ds_2).$$

Proof. We shall use the notations introduced in the proof of Corollary 6. According to the proof of that corollary, $EE^{(j)}$ is in $\Sigma^{(j)}$ for each j, so that if $EE^{(j)}(s_2) = \{s_1 | [s_1, s_2] \in EE^{(j)}\}$ then $\mu_1(EE^{(j)}(s_2))$ is μ_2-measurable on $E_2^{(j)}$. Since $\mu_1(EE^{(j)}(s_2)) = 0$ for $s_2 \notin E_2^{(j)}$, $\mu_1(EE^{(j)}(s_2))$ is a μ_2-measurable function of s_2. Since $\{EE^{(j)}(s_2)\}$ is an increasing sequence of sets with union $E(s_2)$, $\mu_1(E(s_2)) = \lim_{j\to\infty} \mu_1(EE^{(j)}(s_2))$ for each $s_2 \in S_2$. Thus by Corollary 6.14, $\mu_1(E(s_2))$ is a μ_2-measurable function of s_2. According to Corollary 4,

$$\mu^{(j)}(EE^{(j)}) = \int_{E_2^{(j)}} \mu_1(EE^{(j)}(s_2))\mu_2(ds_2).$$

Since $\mu(E) = \lim_{j\to\infty} \mu^{(j)}(EE^{(j)})$ and

$$\int_{E_2^{(j)}} \mu_1(EE^{(j)}(s_2))\mu_2(ds_2) = \int_{S_2} \mu_1(EE^{(j)}(s_2))\mu_2(ds_2),$$

formula [*] follows immediately from Corollary 6.17. Q.E.D.

In proving the crucial Theorem 9 below, we will have use for the following easy consequence of Corollary 7.

8 COROLLARY. *Let (S, Σ, μ) be the product of two positive σ-finite measure spaces (S_1, Σ_1, μ_1) and (S_2, Σ_2, μ_2). Let $N \in \Sigma$ be a μ-null set. Then, for μ_1-almost all s_1 in S_1, the set $N(s_1) = \{s_2 | [s_1, s_2] \in N\}$ is a μ_2-null set.*

PROOF. This follows immediately from the formula given by Corollary 7, and from Lemma 6.8. Q.E.D.

At this point we make certain observations which will greatly simplify the later work of this section. Let (S_1, Σ_1, μ_1), (S_2, Σ_2, μ_2) and (S_3, Σ_3, μ_3) be measure spaces and $(S, \Sigma, \mu) = P_{i=1}^{3} (S_i, \Sigma_i, \mu_i)$. Thus S is the collection of points $[s_1, s_2, s_3]$, $s_i \in S_i$, and, strictly speaking, S should not be confused with the space $S_0 = (S_1 \times S_2) \times S_3$ whose elements have the form $[[s_1, s_2], s_3]$. However, it is reasonable and convenient to regard S and S_0 as identical, as there is clearly a natural one-to-one correspondence between their points. Moreover, as the reader will easily verify, this correspondence induces a one-to-one correspondence between the σ-fields Σ and Σ_0 which is measure preserving. Thus (S, Σ, μ) and (S_0, Σ_0, μ_0) may be identified as measure spaces. We may describe this situation by saying that the process of forming products of measure spaces is *associative*. It is easily seen to be *commutative* also. These remarks clearly extend to the product of any finite family of measure spaces. Thus if (S_i, Σ_i, μ_i), $i = 1, \ldots, n$, are measure spaces which are all finite or all positive and σ-finite, and α is any subset of $\{1, \ldots, n\}$, we may identify the measure spaces $P_{i=1}^{n} (S_i, \Sigma_i, \mu_i)$ and $(S_\alpha, \Sigma_\alpha, \mu_\alpha) \times (S_{\alpha'}, \Sigma_{\alpha'}, \mu_{\alpha'})$, where

$$(S_\alpha, \Sigma_\alpha, \mu_\alpha) = P_{i\in\alpha} (S_i, \Sigma_i, \mu_i),$$

and

$$(S_{\alpha'}, \Sigma_{\alpha'}, \mu_{\alpha'}) = P_{i\in\alpha'} (S_i, \Sigma_i, \mu_i).$$

The next theorems concern the relationship between integration in a product measure space and integration in the components of the product. It is a consequence of the associativity of the product measure that the discussion can be restricted without loss of generality to the case of the product (R, Σ_R, ϱ) of two measure spaces (S, Σ_S, μ) and (T, Σ_T, λ).

→ 9 THEOREM. (*Fubini*) *Let (S, Σ_S, μ) and (T, Σ_T, λ) be two positive σ-finite measure spaces. Let $(R, \Sigma_R, \varrho) = (S, \Sigma_S, \mu)\times(T, \Sigma_T, \lambda)$ and let f be in $L_1(R, \Sigma_R, \varrho, \mathfrak{X})$. Then for μ-almost all s in S, $f(s, \cdot)$ is in $L_1(T, \Sigma_T, \lambda, \mathfrak{X})$. Moreover, $\int_T f(\cdot, t)\lambda(dt)$ is in $L_1(S, \Sigma_S, \mu, \mathfrak{X})$, and*

$$\int_S \left\{\int_T f(s, t)\lambda(dt)\right\} \mu(ds) = \int_R f(r)\varrho(dr).$$

PROOF. Let $\mathfrak{A}_0$ be the subspace of ϱ-simple functions $g \in L_1(R, \Sigma_R, \varrho, \mathfrak{X})$. It follows from Corollary 7 that every $g \in \mathfrak{A}_0$ satisfies the conclusion of the theorem. By Lemma 2.18, we can find a sequence $\{g_n\}$ of elements of $\mathfrak{A}_0$ converging to f in the topology of $L_1(R, \Sigma_R, \varrho, \mathfrak{X})$. Then we have

$$\lim_{m, n\to\infty} \int_S \left\{\int_T |g_n(s, t) - g_m(s, t)|\lambda(dt)\right\} \mu(ds)$$
$$= \lim_{m, n\to\infty} \int_R |g_n(r) - g_m(r)|\varrho(dr) = 0.$$

Consequently, if we define the function G_n in

$$\tilde{L}_1 = L_1(S, \Sigma_S, \mu, L_1(T, \Sigma_T, \lambda, \mathfrak{X}))$$

by $G_n(s) = g_n(s, \cdot)$, we have

$$\lim_{m, n\to\infty} \int_S |G_n(s) - G_m(s)|\,\mu(ds)$$
$$= \lim_{m, n\to\infty} \int_S \left\{\int_T |g_n(s, t) - g_m(s, t)|\lambda(dt)\right\} \mu(ds) = 0.$$

By Corollary 6.6, there exists a $G \in \tilde{L}_1$ such that $G_n \to G$ in the norm of $\tilde{L}_1$. By Theorem 3.6(i) and Corollary 6.13(a), (and passing without loss of generality to a subsequence) we can find a μ-null set $N \in \Sigma_S$ such that $G_n(s) \to G(s)$ for $s \notin N$.

At the same time, passage to a subsequence allows us to assume that $g_n(r) \to f(r)$ for all r not belonging to a certain ϱ-null set M. By Corollary 8, this means that there exists a μ-null set N_1 such that $g_n(s, t) \to f(s, t)$ for λ-almost all t if $s \notin N_1$.

By Theorem 3.6(i) and Corollary 6.13(a), if $s_0 \notin N$ is given, we can find an increasing sequence of integers n_i such that $G_{n_i}(s_0)(t) \to G(s_0)(t)$ for λ-almost all t; i.e., such that

$$g_{n_i}(s_0, t) \to G(s_0)(t)$$

for λ-almost all t. Consequently, if $s_0 \notin N \cup N_1$, $f(s_0, t) = G(s_0)(t)$, so that $f(s_0, \cdot) \in L_1(T, \Sigma_T, \lambda, \mathfrak{X})$.

The equation $Uh = \int_T h(t)\lambda(dt)$ clearly defines a continuous linear mapping U of $L_1(T, \Sigma_T, \lambda, \mathfrak{X})$ into $\mathfrak{X}$. In terms of the mapping U we have

$$\int_T f(s_0, t)\lambda(dt) = \int_T G(s_0)(t)\lambda(dt) = UG(s_0), \qquad s_0 \notin N \cup N_1.$$

Hence, by Theorem 2.19(c), $\int_T f(\cdot, t)\lambda(dt) = UG$ is in $L_1(S, \Sigma, \mu, \mathfrak{X})$, and we have

$$[*] \qquad \int_S \left\{\int_T f(s, t)\lambda(dt)\right\} \mu(ds) = \int_S UG(s)\mu(ds)$$
$$= U \int_S G(s)\mu(ds)$$
$$= \lim_{n\to\infty} U \int_S G_n(s)\mu(ds)$$
$$= \lim_{n\to\infty} \int_S UG_n(s)\mu(ds)$$
$$= \lim_{n\to\infty} \int_S \left\{\int_T g_n(s, t)\lambda(dt)\right\} \mu(ds).$$

Since we have already observed that

$$\int_S \left\{\int_T g_n(s, t)\lambda(dt)\right\} \mu(ds) = \int_R g_n(r)\varrho(dr)$$

for the ϱ-simple functions g_n, and since $g_n \to f$ in the norm of $L_1(R, \Sigma_R, \varrho, \mathfrak{X})$, the limit on the right is $\int_R f(r)\varrho(dr)$, and the theorem is proved. Q.E.D.

Theorem 9 can easily be extended to the product of two arbitrary finite measure spaces. For this, the following three lemmas are convenient.

10 Lemma. *Let (R, Σ_R, ϱ) be the product of two σ-finite positive measure spaces (S, Σ_S, μ) and (T, Σ_T, λ), and let f be μ-measurable on S, with values in a B-space $\mathfrak{X}$. Then the function g on R to $\mathfrak{X}$, defined by the formula $g(s, t) = f(s)$ is ϱ-measurable.*

Proof. Let f be an arbitrary μ-measurable function; let E_n be an increasing sequence of sets in Σ_S such that $\cup E_n = S$ and $\mu(E_n) < \infty$. Let $f_n(s) = f(s)$ for $s \in E_n$ and $f_n(s) = 0$ otherwise. Let $g_n(s, t) = f_n(s)$. Then $g_n(r) \to f(r)$ for each r and hence, by Corollary 6.14, it is sufficient to show that each function g_n is ϱ-measurable. Thus, there is no loss in generality if we prove the lemma only for the case of the totally μ-measurable functions f_n.

Assume then that f is an arbitrary totally μ-measurable function. Using Corollary 6.13(a), let $\{f_n\}$ be a sequence of μ-simple functions converging to f μ-almost everywhere. If $g_n(s, t) = f_n(s)$ then $g_n(r) \to g(r)$ ϱ-almost everywhere. Hence, by Corollary 6.14, g is ϱ-measurable. Q.E.D.

11 **Lemma.** *Let (R, Σ_R, ϱ) be the product of the finite measure spaces (S, Σ_S, μ) and (T, Σ_T, λ). Then $v(\varrho) = v(\mu) \times v(\lambda)$.*

Proof. By the Radon-Nikodým theorem (10.2) there is a μ-integrable function g with

$$\mu(A) = \int_A g(s) v(\mu, ds), \qquad A \in \Sigma_S.$$

By Theorem 2.20, $v(\mu, A) = \int_A |g(s)| v(\mu, ds)$ and so $|g(s)| = 1$ for μ-almost all s. Without loss of generality we may and shall assume that $|g(s)| = 1$ for all s. Similarly, there is a λ-integrable function h with $|h(t)| = 1$ for all t and

$$\lambda(B) = \int_B h(t) v(\lambda, dt), \qquad B \in \Sigma_T.$$

For $r = [s, t]$ let $f(r) = g(s)h(t)$ so that $|f(r)| = 1$ for all r in R. Let $v = v(\mu) \times v(\lambda)$, so that for $A \in \Sigma_S$, $B \in \Sigma_T$, and $E = A \times B$ we have, from Theorem 9,

$$\begin{aligned}\int_E f(r) v(dr) &= \int_A \left\{\int_B g(s) h(t) v(\lambda, dt)\right\} v(\mu, ds) \\ &= \left\{\int_A g(s) v(\mu, ds)\right\} \left\{\int_B h(t) v(\lambda, dt)\right\} \\ &= \mu(A)\lambda(B).\end{aligned}$$

Hence, by the uniqueness assertion in Theorem 2, $\varrho(E) = \int_E f(r) v(dr)$. Since $|f(r)| = 1$ for all r we have, from Theorem 2.20,

$$v(\varrho, E) = \int_E |f(r)| v(dr) = v(E). \qquad \text{Q.E.D.}$$

12 LEMMA. *Let (R, Σ_R, ϱ) be the product of finite measure spaces (S, Σ_S, μ) and (T, Σ_T, λ). Let E be a ϱ-null set in R. Then for λ-almost all t, the set $E(t) = \{s|[s, t] \in E\}$ is a μ-null set.*

PROOF. By Lemma 11 it may be assumed that the measure spaces are positive, but in this case the desired conclusion was established in Corollary 8. Q.E.D.

13 THEOREM. *Let (S, Σ_S, μ) and (T, Σ_T, λ) be finite measure spaces and let (R, Σ_R, ϱ) be their product. Let $\mathfrak{X}$ be a B-space and F be a ϱ-integrable function on R to $\mathfrak{X}$. Then, for μ-almost all s in S, the function $F(s, \cdot)$ is λ-integrable on T and the function $\int_T F(\cdot, t)\lambda(dt)$ is μ-integrable on S. Moreover*

$$\int_S \left\{\int_T F(s, t)\lambda(dt)\right\} \mu(ds) = \int_R F(r)\varrho(dr).$$

PROOF. By Lemma 2.18, the function F is $v(\varrho)$-integrable and hence, by Theorem 9, $F(s, \cdot)$ is $v(\lambda)$-integrable for $v(\mu)$-almost all s in S. From this and Lemma 2.18 it follows that $F(s, \cdot)$ is λ-integrable for μ-almost all s in S. Now let the functions f, g, and h be defined as in the proof of Lemma 11. Then, by Corollary 10.6,

$$\int_T F(s, t)\lambda(dt) = \int_T F(s, t)h(t)v(\lambda, dt).$$

Using this equation, Theorem 9, Lemma 10, Lemma 2.18, and Theorem 2.22(a), it is seen that $\int_T F(\cdot, t)\lambda(dt)$ is μ-integrable on S. By Corollary 10.6, Theorem 9, and Lemma 11, we have

$$\begin{aligned} &\int_S \left\{\int_T F(s, t)\lambda(dt)\right\} \mu(ds) \\ &= \int_S \left\{\int_T F(s, t)g(s)h(t)v(\lambda, dt)\right\} v(\mu, ds) \\ &= \int_R F(r)f(r)v(\varrho, dr). \end{aligned}$$

Since, as was seen in the proof of Lemma 11,

$$\varrho(E) = \int_E f(r)v(\varrho, dr), \qquad E \in \Sigma_R,$$

it follows, from Corollary 10.6, that

$$\int_R F(r)f(r)v(\varrho, dr) = \int_R F(r)\varrho(dr).$$

Thus

$$\int_S \left\{\int_T F(s, t)\lambda(dt)\right\} \mu(ds) = \int_R F(r)\varrho(dr). \qquad \text{Q.E.D}$$

The next result is a useful complement to the Fubini theorem (Theorem 9).

➜ 14 THEOREM. (*Tonelli*) *Let* $(R, \Sigma_R, \varrho) = (S, \Sigma_S, \mu) \times (T, \Sigma_T, \lambda)$ *be the product of two positive* σ-*finite measure spaces. Let* f *be a positive* ϱ-*measurable function. Then, for* μ-*almost all* s *in* S, *the function* $f(s, \cdot)$ *is* λ-*measurable. Moreover, the (extended real valued) function* $\int_T f(\cdot, t)\lambda(dt)$ *is* μ-*measurable and*

$$[*] \qquad \int_S \left\{\int_T f(s, t)\lambda(dt)\right\} \mu(ds) = \int_R f(r)\varrho(dr),$$

irrespective of whether the integrals have finite or infinite values.

PROOF. Let $\{E_n\}$ be an increasing sequence of sets in Σ_R such that $R = \cup E_n$ and $\varrho(E_n) < \infty$. Put $f_n(r) = f(r)$ if $f(r) \leqq n$ and $r \in E_n$, $f_n(r) = 0$ otherwise. By Theorem 6.10, f_n is ϱ-measurable, and by Theorem 2.22(b), f_n is ϱ-integrable. If $\int_R f(r)\varrho(dr) < \infty$, our statement is merely Theorem 9. Hence, we have only to show that $\int_R f(r)\varrho(dr) = \infty$ implies that the iterated integral on the right of [*] is infinite. But this is obvious, since

$$\int_S \left\{\int_T f(s, t)\lambda(dt)\right\} \mu(ds) \geqq \int_S \left\{\int_T f_n(s, t)\lambda(dt)\right\} \mu(ds)$$

$$= \int_R f_n(r)\varrho(dr) \to \int_R f(r)\varrho(dr)$$

by Theorem 9 and Corollary 6.17. Q.E.D.

15 COROLLARY. *Under the hypothesis of Theorem* 14, *a* ϱ-*measurable vector valued function* g *for which*

$$\int_S \left\{\int_T |g(s, t)|\lambda(dt)\right\} \mu(ds) < \infty$$

is ϱ-*integrable on* R *and*

$$\int_R g(r)\varrho(dr) = \int_S \left\{\int_T g(s, t)\lambda(dt)\right\} \mu(ds).$$

PROOF. This is an immediate corollary of the theorems of Tonelli and Fubini and the fact that a ϱ-measurable function g is ϱ-integrable if the function $|g(\cdot)|$ is ϱ-integrable (2.22). Q.E.D.

Tonelli's theorem has the important consequence that a non-negative function measurable on the product of two positive σ-finite measure spaces yields the same value whether it be first integrated with respect to one variable and then with respect to the other, or vice versa. Indeed, according to Tonelli's theorem, both these integrals are equal to the integral of f with respect to the product measure (and we have already remarked that the product of measures is commutative.) Generalizing this statement, and making use of the commutativity and associativity of product measures, we can say that the integral of a non-negative function measurable on the product measure space of a finite collection of positive σ-finite measure spaces can be evaluated by "iterated" integration over the various factor spaces in any order. If either the function f or one of the measure spaces fails to be positive, we can no longer make this statement under the mere hypothesis that f is measurable on the product space. But, according to Theorems 9 and 14, if we assume that f is *integrable* on the product space, we can again evaluate its integral by "iterated" integration over the various factor spaces in any order. Thus the Fubini and Tonelli theorems provide us with very general results on "change of order of integration".

It follows that if f is a function defined on Euclidean n-dimensional space E and integrable with respect to n-dimensional Lebesgue measure λ, the "multiple" integral $\int_E f(s)\lambda(ds)$ is equal to the "iterated" integral $\int_{-\infty}^{+\infty} \{\ldots\{\int_{-\infty}^{+\infty} f(s_1, \ldots, s_n)ds_1\}\ldots\} ds_n$. Moreover, the order of integration in the iterated integral is immaterial. For this reason, both multiple and iterated n-dimensional Lebesgue integrals are commonly denoted by the incomplete notation

$$\int_{-\infty}^{+\infty} \cdots \int_{-\infty}^{+\infty} f(s_1, \ldots, s_n)\, ds_1 \ldots ds_n.$$

Correspondingly, if R is the "rectangle"

$$R = \{[s_1, \ldots, s_n] | a_1 \leqq s_1 \leqq b_1, \ldots, a_n \leqq s_n \leqq b_n\},$$

$\int_R f(s)\lambda(ds)$ is often written as

$$\int_{a_1}^{b_1} \cdots \int_{a_n}^{b_n} f(s_1, \ldots, s_n) ds_1 \ldots ds_n.$$

It should also be mentioned that just as in the one dimensional case, a specific symbol for n-dimensional Lebesgue measure is often omitted from the notation for n-dimensional Lebesgue integrals. Thus, when no confusion can arise, $\int_E f(s)\lambda(ds)$ is often written as $\int_E f(s)ds$, and $\int_R f(s)\lambda(ds)$ as $\int_R f(s)ds$.

Next we study the relation between the theory of product measures and the theory of vector valued integrals. In the application of the theory of vector valued integrals to concrete problems such as the representation of operators between Lebesgue spaces (cf. Section VI.8) one is faced with the following situation. Suppose that (S, Σ, μ) is a measure space and F is a μ-measurable function whose values are in $L_p(T, \Sigma_T, \lambda)$, $1 \leqq p < \infty$. For each s in S, $F(s)$ is an equivalence class of functions, any pair of whose members coincide λ-almost everywhere. If for each s we select a particular *function* $f(s, \cdot) \in F(s)$, the resulting function $f(s, t)$ defined on

$$(R, \Sigma_R, \varrho) = (S, \Sigma_S, \mu) \times (T, \Sigma_T, \lambda)$$

will be called a *representation* of the function F. It is important to know whether F has a ϱ-measurable representation, and, in case F is μ-integrable, whether $\int_S F(s)\mu(ds) = \int_S f(s, \cdot)\mu(ds)$. These questions are answered by Theorem 17 below.

16 Lemma. *Let (S, Σ_S, μ) and (T, Σ_T, λ) be measure spaces which are either both finite or both positive and σ-finite; let (R, Σ_R, ϱ) be their product and let $\mathfrak{X}$ be a B-space.*

(a) *If F is a function on S to $L_1(T, \Sigma_T, \lambda, \mathfrak{X})$ which is μ-integrable on S, then there is a ϱ-integrable function f on R to X such that $f(s, \cdot) = F(s)$ for μ-almost all s in S. Moreover, $\int_S f(s, t)\mu(ds)$ exists for λ-almost all t in T and, as a function of t, is equal to the element $\int_S F(s)\mu(ds)$ of $L_1(T, \Sigma_T, \lambda, \mathfrak{X})$.*

(b) *Let $1 \leqq p < \infty$, and let f be a ϱ-measurable function on (R, Σ_R, ϱ) to $\mathfrak{X}$ such that $F(s) = f(s, \cdot)$ is in $L_p(T, \Sigma_T, \lambda, \mathfrak{X})$ for μ-almost all s in S. Then the vector valued function F on (S, Σ_S, μ) to $L_p(T, \Sigma_T, \lambda, \mathfrak{X})$ is μ-measurable.*

Proof. For brevity we shall use the symbol $\tilde{L}_1$ for the space $L_1(S, \Sigma_S, \mu, L_1(T, \Sigma_T, \lambda, \mathfrak{X}))$. By Lemma 2.18 there is a sequence $\{F_n\}$ of μ-simple functions in $\tilde{L}_1$ with $F_n \to F$ in the norm of $\tilde{L}_1$. Each of the functions F_n is constant on each of a finite collection of disjoint sets $E_n^{(1)}, \ldots, E_n^{(m_n)}$ in Σ which together form a partition of S. Let $g_n^{(j)}$ be the value of F_n on $E_n^{(j)}$ and define the functions f_n, $n = 1, 2, \ldots$, on R by the equations

$$f_n(s, t) = g_n^{(j)}(t), \qquad s \in E_n^{(j)}.$$

It follows from Lemma 10 that f_n is ϱ-measurable. Moreover, it is clear that $f_n(s, \cdot) = F_n(s)$ for s in S. Also by Theorem 14

$$\int_R |f_n(r)|v(\varrho, dr) = \int_S \left\{\int_T |f_n(s, t)|v(\lambda, dt)\right\} v(\mu, ds)$$
$$= \int_S |F_n(s)|v(\mu, ds) < \infty,$$

and so (2.18) f_n is ϱ-integrable on R. Since $f_n(s, \cdot) = F_n(s)$ for s in S we have, by Theorem 9 and Lemma 11,

$$\lim_{m, n\to\infty} \int_R |f_n(r)-f_m(r)|v(\varrho, dr)$$
$$= \lim_{m, n\to\infty} \int_S \left\{\int_T \{|f_n(s, t)-f_m(s, t)|v(\lambda, dt)\right\} v(\mu, ds)$$
$$= \lim_{m, n\to\infty} \int_S |F_n(s)-F_m(s)|v(\mu, ds)$$
$$= \lim_{m, n\to\infty} |F_m-F_n| = 0.$$

Thus, by Corollary 6.6, there is a ϱ-integrable function f on R to $\mathfrak{X}$ such that

$$\lim_{n\to\infty} \int_R |f_n(r)-f(r)|v(\varrho, dr) = 0.$$

It is seen, by using Theorem 9, Lemma 11, and Lemma 2.18, that the function $G(s) = f(s, \cdot)$ is in $\tilde{L}_1$ and that

$$|F_n-G| = \int_S |F_n(s)-G(s)|v(\mu, ds)$$
$$= \int_S \left\{\int_T |f_n(s, t)-f(s, t)|v(\lambda, dt)\right\} v(\mu, ds)$$
$$= \int_S |f_n(r)-f(r)|v(\varrho, dr) \to 0.$$

Since $F_n \to F$ in $\tilde{L}_1$ we have $|F-G| = 0$ and therefore, (6.8), $F(s) = G(s) = f(s, \cdot)$ for μ-almost all s in S. Thus, the first assertion of part (a) is proved.

Since f is ϱ-integrable on R it follows from Theorems 9 and 13 that for λ-almost all t in T the function $f(\cdot, t)$ is μ-integrable on S and that $\int_S f(s, t)\mu(ds)$ is λ-integrable on T, proving the second assertion of part (a).

To prove the final assertion in (a) we define, for each E in Σ_T, the bounded linear operator U_E on $L_1(T, \Sigma_T, \lambda, \mathfrak{X})$ to $\mathfrak{X}$ by the equation

$$U_E g = \int_E g(t)\lambda(dt), \qquad g \in L_1(T, \Sigma_T, \lambda, \mathfrak{X}).$$

From Theorems 9, 13, and 2.19(c), it is seen that

$$U_E \int_S f(s, \cdot)\mu(ds) = \int_E \left\{\int_S f(s, t)\mu(ds)\right\} \lambda(dt)$$
$$= \int_{S\times E} f(r)\varrho(dr) = \int_S \left\{\int_E f(s, t)\lambda(dt)\right\}\mu(ds)$$
$$= \int_S U_E F(s)\mu(ds) = U_E \int_S F(s)\mu(ds).$$

The final assertion of part (a) therefore follows from Lemma 6.8.

It suffices to prove (b) under the additional assumption that $v(\mu, S)$ and $v(\lambda, T)$ are finite. Since f is ϱ-measurable, there is a sequence $\{f_n\}$ of ϱ-measurable simple functions converging in ϱ-measure to f. In virtue of Lemma 8.3 and the remark before Theorem 2 we may suppose that each f_n is a finite linear combination of characteristic functions of sets of the form $A \times B$, where $A \in \Sigma_S$, $B \in \Sigma_T$. In addition, we may assume that $|f_n(s, t)| \leqq |f(s, t)|$ ϱ-almost everywhere, and by Corollary 6.13(a) we may suppose that $\{f_n\}$ converges ϱ-almost everywhere to f. Put $F_n(s) = f_n(s, \cdot)$, $s \in S$, so that each F_n is a simple function on (S, Σ_S, μ) to $L_p(T, \Sigma_T, \lambda, \mathfrak{X})$. By Corollary 8 or Lemma 12, we conclude that for μ-almost all $s \in S$ the sequence $\{f_n(s, t)\}$ converges to $f(s, t)$ λ-almost everywhere on T. It follows from the Lebesgue dominated convergence theorem (6.16) that for μ-almost all $s \in S$, the sequence $\{F_n(s)\}$ converges to $F(s)$ in the norm of the space $L_p(T, \Sigma_T, \lambda, \mathfrak{X})$. The μ-measurability of F now follows from Corollary 6.14. Q.E.D.

In the following theorem the space $L_\infty(T, \Sigma_T, \lambda, \mathfrak{X})$ is understood to be the space of all λ-measurable, essentially bounded $\mathfrak{X}$-valued functions on T. The norm of such a function is its λ-essential supremum (cf. Definition 1.11).

17 Theorem. *Let (S, Σ_S, μ) and (T, Σ_T, λ) be measure spaces which are either both finite or both positive and σ-finite, and let (R, Σ_R, ϱ) be their product. Let $1 \leqq p \leqq \infty$ and let F be a μ-integrable function on S to $L_p(T, \Sigma_T, \lambda, \mathfrak{X})$ where $\mathfrak{X}$ is a real or complex B-space. Then there is a ϱ-measurable function f on R to $\mathfrak{X}$, which is uniquely determined except for a set of ϱ-measure zero, and such that $f(s, \cdot) = F(s)$ for μ-almost all s in S. Moreover $f(\cdot, t)$ is μ-integrable on S for λ-almost all t and the*

integral $\int_S f(s, t)\mu(ds)$, *as a function of* t, *is equal to the element* $\int_S F(s)\mu(ds)$ *of* $L_p(T, \Sigma_T, \lambda, \mathfrak{X})$.

PROOF. Let Σ_T be partitioned into a sequence $\{E_n\}$ of disjoint sets of finite λ-measure. For $1 \leqq p \leqq \infty$ let $L_p = L_p(T, \Sigma_T, \lambda, \mathfrak{X})$ and define the maps U_n, $n = 1, 2, \ldots$, of L_p into L_1 by the equations $(U_n g)(t) = g(t)\chi_{E_n}(t)$, where χ_{E_n} is the characteristic function of E_n. By Hölder's inequality (3.2),

$$|U_n g|_1 \leqq v(\lambda, E_n)^{1/q}|g|_p,$$

where $1/p+1/q = 1$, and thus U_n is a continuous linear map of L_p into L_1. By Theorem 2.19(c), the function $F_n(\cdot) = U_n F(\cdot)$ is a μ-integrable function on S to L_1. By applying Lemma 16(a) to F_n we obtain a ϱ-integrable function f_n on R to $\mathfrak{X}$ such that $f_n(s, \cdot) = F_n(s)$ for every s in S which is not in some μ-null set N_n. Furthermore, for λ-almost all t the function $f_n(\cdot, t)$ is μ-integrable on S and $\int_S f_n(s, t)\mu(ds)$ as a function of t defines the same element in L_1 as $\int_S F_n(s)\mu(ds)$. The function f on R to $\mathfrak{X}$ is now defined by the equation $f(s, t) = f_n(s, t)$ for t in E_n. Theorem 6.10 shows that f is a ϱ-measurable function on R to $\mathfrak{X}$. Since $F_n(s)(t) = F(s)(t)$ for t in E_n, it is clear that for each s not in the μ-null set $\bigcup_{n=1}^{\infty} N_n$, we have $f(s, t) = F(s)(t)$ for λ-almost all t in T. Thus $f(s, \cdot) = F(s)$ for μ-almost all s in S. Since $f_n(\cdot, t)$ is μ-integrable on S for λ-almost all t the same must be true for $f(\cdot, t)$. The integral $\int_S f_n(s, t)\mu(ds)$ as a function of t defines the same element in L_1 as

$$\int_S F_n(s)\mu(ds) = \int_S U_n F(s)\mu(ds) = U_n \int_S F(s)\mu(ds).$$

Since

$$\left\{U_n \int_S F(s)\mu(ds)\right\}(t) = \left\{\int_S \{F(s)\mu(ds)\}(t), \qquad t \in E_n,\right.$$

it follows that

$$\int_S f_n(s, t)\mu(ds) = \left\{\int_S F(s)\mu(ds)\right\}(t), \qquad t \in E_n$$

and thus that

$$\int_S f(s, t)\mu(ds) = \left\{\int_S F(s)\mu(ds)\right\}(t),$$

for λ-almost all t in T.

To complete the proof of the theorem it only remains to show that

f is uniquely determined up to a set of ϱ-measure zero. To prove this uniqueness it is evidently sufficient to show that a ϱ-measurable function h on R to $\mathfrak{X}$ for which $h(s, \cdot)$ is a λ-null function for μ-almost all s, must be a ϱ-null function. If h is such a function then, by Tonelli's theorem (14),

$$\int_R |h(r)| v(\varrho, dr) = \int_S \left\{ \int_T |h(s, t)| v(\lambda, dt) \right\} v(\mu, ds) = 0,$$

so that, by Lemma 6.8, $h(r) = 0$ for ϱ-almost all r in R. Q.E.D.

Infinite products. The theory of measure in product spaces will now be extended to the product of an infinite family $(S_\alpha, \Sigma_\alpha, \mu_\alpha)$, $\alpha \in A$, of measure spaces. We shall construct a measure space (S, Σ, μ) called the *product of the measure spaces* $(S_\alpha, \Sigma_\alpha, \mu_\alpha)$, and denoted by

$$(S, \Sigma, \mu) = \underset{\alpha \in A}{P} (S_\alpha, \Sigma_\alpha, \mu_\alpha),$$

which has the following properties: S is the Cartesian product $\underset{\alpha \in A}{P} S_\alpha$; Σ is the σ-field determined by all subsets of S of the form $E = \underset{\alpha \in A}{P} E_\alpha$ where $E_\alpha \in \Sigma_\alpha$ and, with a finite number of exceptions, $E_\alpha = S_\alpha$; and μ is a measure on Σ with the property that

$$\mu(E) = \prod_{\alpha \in A} \mu_\alpha(E_\alpha),$$

where E has the form just specified.

In order to avoid difficulties which might otherwise arise from the presence of infinite products such as $\prod_{\alpha \in A} \mu_\alpha(E_\alpha)$ it is convenient to assume that for all but a finite number of α, μ_α is non-negative and $\mu_\alpha(S_\alpha) = 1$. In this case the product $\prod_{\alpha \in A} \mu_\alpha(E_\alpha)$ is meaningful for the type of set $\underset{\alpha \in A}{P} E_\alpha$ mentioned above, since $E_\alpha = S_\alpha$ and hence $\mu_\alpha(S_\alpha) = 1$ for all but a finite number of α. We shall actually restrict ourselves somewhat more than this by assuming that, for all α, μ_α is non-negative and $\mu_\alpha(S_\alpha) = 1$. No generality is really lost by abandoning the possibility of including a finite number of factors in which μ_α is not necessarily positive and $\mu_\alpha(S_\alpha)$ is not necessarily one. In fact, we may apply the theory developed in the first part of this section to the finite number of "irregular" factors and the theory to be developed in what follows to the infinite family of regular factors and then, using only the theory developed in the first part of this section, form the product of these two separate parts.

Throughout the remaining part of this section the symbol A will be used for an arbitrary set of indices α, and S for the product $P_{\alpha \epsilon A} S_\alpha$. For a subset B of A we also write S_B for $P_{\alpha \epsilon B} S_\alpha$ so that $S = S_A$. The symbol π will be reserved for an arbitrary finite subset of A and π' will denote the complement $A - \pi$ of π in A. The symbol $\mathscr{E}_\pi$ will be used for the family of those sets in S_π which have the form $P_{\alpha \epsilon \pi} E_\alpha$ with E_α in Σ_α, and the symbol Σ_π used for the smallest field of sets in S_π which contains $\mathscr{E}_\pi$. There should be no confusion between the symbols Σ_π and Σ_α, for if π reduces to the single element α then the field Σ_π is the σ-field Σ_α. An *elementary set* in S is one of the form $P_{\alpha \epsilon A} E_\alpha$ where E_α is in Σ_α and, for all but a finite number of α, $E_\alpha = S_\alpha$. Equivalently, an elementary set in S is one having the form $S_{\pi'} \times E_\pi$ for some π and some E_π in $\mathscr{E}_\pi$. The family of elementary sets in S will be denoted by $\mathscr{E}$ and the field of sets in S determined by $\mathscr{E}$ will be denoted by Σ_1. The symbol Σ will be used for the σ-field of sets in S determined by Σ_1. The family of all sets in S of the form $S_{\pi'} \times E_\pi$ with $E_\pi \epsilon \Sigma_\pi$ will be denoted by Σ^π.

18 Lemma. *For each π the family Σ^π is a field of sets in S and $\Sigma_1 = \cup_\pi \Sigma^\pi$.*

Proof. The identities

$$(S_{\pi'} \times E_\pi) \cup (S_{\pi'} \times F_\pi) = S_{\pi'} \times (E_\pi \cup F_\pi),$$

and

$$(S_{\pi'} \times E_\pi)' = S_{\pi'} \times E'_\pi$$

show that Σ^π is a field. It is clear from its definition that Σ_1 contains all of the fields Σ^π. On the other hand, since $\Sigma^{\pi_1} \subseteqq \Sigma^{\pi_2}$ if $\pi_1 \subseteqq \pi_2$, the union $\bigcup_\pi \Sigma^\pi$ is a field. Thus $\Sigma_1 = \bigcup_\pi \Sigma^\pi$. Q.E.D.

19 Lemma. *For each α in A let $(S_\alpha, \Sigma_\alpha, \mu_\alpha)$ be a positive measure space with $\mu_\alpha(S_\alpha) = 1$. Then there is a unique additive set function μ on Σ_1 with the property that*

$$\mu(\underset{\alpha\in A}{P} E_\alpha) = \prod_{\alpha\in A} \mu_\alpha(E_\alpha),$$

for every elementary set $P_{\alpha\in A} E_\alpha$ *in* S.

PROOF. We will first show that μ is unique. Let λ be another additive set function on Σ_1 with the stated value on elementary sets. For each π let μ_π, λ_π be set functions on Σ_π defined by the formulas

$$\mu_\pi(E_\pi) = \mu(E_\pi \times S_{\pi'}), \qquad \lambda_\pi(E_\pi) = \lambda(E_\pi \times S_{\pi'}), \qquad E_\pi \in \Sigma_\pi.$$

Then

$$\mu_\pi(\underset{\alpha\in\pi}{P} E_\alpha) = \prod_{\alpha\in\pi} \mu_\alpha(E_\alpha) = \lambda_\pi(\underset{\alpha\in\pi}{P} E_\alpha),$$

and so, by Lemma 1, $\lambda(E) = \mu(E)$ for every E in Σ_π. This means that $\lambda(E) = \mu(E)$ for every E in Σ^π. Thus, by Lemma 18, $\lambda(E) = \mu(E)$ for every E in Σ_1, and μ is unique. This uniqueness argument suggests how we may prove the existence of μ. For each finite set π in A there is, by Lemma 1, a unique additive set function μ_π defined on Σ_π which has the property

$$\mu_\pi(\underset{\alpha\in\pi}{P} E_\alpha) = \prod_{\alpha\in\pi} \mu_\alpha(E_\alpha), \qquad E_\alpha \in \Sigma_\alpha.$$

Let μ^π be defined on Σ^π by the equation

$$\mu^\pi(E_\pi \times S_{\pi'}) = \mu_\pi(E_\pi), \qquad E_\pi \in \Sigma_\pi.$$

We note that if $\pi_1 \subseteq \pi$, then

$$\begin{aligned}
\mu^\pi(\underset{\alpha\in\pi_1}{P} E_\alpha \times S_{\pi_1'}) &= \mu_\pi(\underset{\alpha\in\pi_1}{P} E_\alpha \times \underset{\alpha\in\pi-\pi_1}{P} S_\alpha) \\
&= \{\prod_{\alpha\in\pi_1} \mu_\alpha(E_\alpha)\}\{\prod_{\alpha\in\pi-\pi_1} \mu_\alpha(S_\alpha)\} \\
&= \prod_{\alpha\in\pi_1} \mu_\alpha(E_\alpha) \\
&= \mu^{\pi_1}(\underset{\alpha\in\pi_1}{P} E_\alpha \times S_{\pi_1'}),
\end{aligned}$$

which shows that $\mu^{\pi_1}(E) = \mu^\pi(E)$ for every set E of the form $E \times S_{\pi_1'}$ with E in $\mathscr{E}_{\pi_1}$. It follows from the uniqueness argument presented at the beginning of the proof that $\mu^{\pi_1}(E) = \mu^\pi(E)$ for every E in Σ^π. Thus if π_1 and π_2 are arbitrary finite sets in A and if $E \in \Sigma^{\pi_1} \cap \Sigma^{\pi_2}$ then

$$\mu^{\pi_1}(E) = \mu^{\pi_1\cup\pi_2}(E) = \mu^{\pi_2}(E),$$

which shows, by Lemma 18, that the function

$$\mu(E) = \mu^{\pi}(E), \qquad E \in \Sigma^{\pi}$$

is uniquely defined on the field Σ_1. To see that μ is additive on Σ_1 let E_1, E_2 be disjoint sets in Σ_1. By Lemma 18, there are finite sets π_1, π_2 in A with $E_1 \in \Sigma^{\pi_1}$, $E_2 \in \Sigma^{\pi_2}$. Thus, if $\pi = \pi_1 \cup \pi_2$, we have $E_1, E_2 \in \Sigma^{\pi}$ and hence there are disjoint sets A_1, A_2 in Σ_{π} with

$$E_1 = A_1 \times S_{\pi'}, \qquad E_2 = A_2 \times S_{\pi'}$$

and

$$\begin{aligned}\mu(E_1 \cup E_2) &= \mu((A_1 \cup A_2) \times S_{\pi'}) = \mu_{\pi}(A_1 \cup A_2)\\ &= \mu_{\pi}(A_1) + \mu_{\pi}(A_2) = \mu(E_1) + \mu(E_2),\end{aligned}$$

which proves that μ is additive on Σ_1. Q.E.D.

20 Theorem. *For each α in A let $(S_{\alpha}, \Sigma_{\alpha}, \mu_{\alpha})$ be a positive finite measure space with $\mu_{\alpha}(S_{\alpha}) = 1$. Then there is a unique countably additive set function μ defined on the σ-field determined by the elementary sets in S with the property that*

$$\mu(\underset{\alpha \in A}{P}\, E_{\alpha}) = \prod_{\alpha \in A} \mu_{\alpha}(E_{\alpha})$$

for every elementary set $P_{\alpha \in A} E_{\alpha}$ in S.

Proof. Let μ be the set function on Σ_1 whose existence is asserted in Lemma 19. If it can be shown that μ is countably additive on Σ_1, then Corollary 5.9 will yield a unique extension of μ to Σ with the required properties. Let $\{E_i\}$ be a sequence of disjoint sets in Σ_1 whose union is also in Σ_1. Let $F_n = \cup_{i=n+1}^{\infty} E_i$; then $\cap_{n=1}^{\infty} F_n$ is void and it is required to show that $\lim_{n \to \infty} \mu(F_n) = 0$. By Lemma 18, there exists a sequence $\{\pi_n\}$ of finite subsets of A such that $F_n \in \Sigma^{\pi_n}$, $n \geqq 1$. Since $\Sigma^{\pi} \subseteqq \Sigma^{\tilde{\pi}}$ if $\pi \subseteqq \tilde{\pi}$ we may assume that $\pi_n \subseteqq \pi_{n+1}$. Thus there is a sequence $\{\alpha_i\}$ in A with $\pi_n = \{\alpha_1, \ldots, \alpha_{k_n}\}$. For each n, let μ_{π_n} be the set function on Σ_{π_n} defined by the formula

$$\mu_{\pi_n}(E_n) = \mu(E_n \times S_{\pi_n'}), \qquad E_n \in \Sigma_{\pi_n}.$$

Since F_n is in Σ^{π_n} there is an E_n in Σ_{π_n} with $F_n = E_n \times S_{\pi_n'}$. Thus, by the Fubini theorem,

$$\begin{aligned}(1) \qquad \mu(F_n) &= \mu_{\pi_n}(E_n)\\ &= \int_{S_{\alpha_1}} \left\{ \cdots \left\{ \int_{S_{\alpha_{k_n}}} \chi_{E_n}(s_{\alpha_1}, \ldots, s_{\alpha_{k_n}}) \mu_{\alpha_{k_n}}(ds_{\alpha_{k_n}}) \right\} \cdots \right\} \mu_{\alpha_1}(ds_{\alpha_1}).\end{aligned}$$

Since $\{F_n\}$ is a decreasing sequence so is $\mu(F_n)$. We shall make an indirect proof by assuming that $\mu(F_n) \geqq \delta > 0$ for all $n = 1, 2, \ldots$. By the Fubini theorem, the iterated integral

$$f_n(s_{\alpha_1})$$
$$= \int_{S_{\alpha_2}} \left\{ \cdots \left\{ \int_{S_{\alpha_{k_n}}} \chi_{E_n}(s_{\alpha_1}, \ldots, s_{\alpha_{k_n}}) \mu_{\alpha_{k_n}}(ds_{\alpha_{k_n}}) \right\} \cdots \right\} \mu_{\alpha_2}(ds_{\alpha_2})$$

is defined for μ_{α_1}-almost all s_{α_1} in S_{α_1}. Since $\mu(F_n) = \int_{S_{\alpha_1}} f_n(s_{\alpha_1}) \mu_{\alpha_1}(ds_{\alpha_1})$ does not converge to zero and since $0 \leqq f_n(s_{\alpha_1}) \leqq 1$ it follows from the dominated convergence theorem (6.16) that there is a point $s^0_{\alpha_1}$ in S_{α_1} for which $f_n(s^0_{\alpha_1})$ is defined for all n and for which the sequence $\{f_n(s^0_{\alpha_1})\}$ does not converge to zero. Thus, the sequence

$$(2) \qquad \int_{S_{\alpha_2}} \left\{ \cdots \left\{ \int_{S_{\alpha_{k_n}}} E_n(s^0_{\alpha_1}, s_{\alpha_2}, \ldots, s_{\alpha_{k_n}}) \mu_{\alpha_{k_n}}(ds_{\alpha_{k_n}}) \right\} \cdots \right\} \mu_{\alpha_2}(ds_{\alpha_2})$$

is defined and does not converge to zero. If we now apply to the sequence in (2) the argument just applied to the sequence (1) we may establish the existence of a point $s^0_{\alpha_2}$ in S_{α_2} for which the integral

$$\int_{S_{\alpha_3}} \left\{ \cdots \left\{ \int_{S_{\alpha_{k_n}}} E_n(s^0_{\alpha_1}, s^0_{\alpha_2}, s_{\alpha_3}, \ldots, s_{\alpha_{k_n}}) \mu_{\alpha_{k_n}}(ds_{\alpha_{k_n}}) \right\} \cdots \right\} \mu_{\alpha_3}(ds_{\alpha_3})$$

is defined for all n but does not converge to zero as n approaches infinity. By proceeding inductively in this fashion we arrive at a sequence $\{s^0_{\alpha_i}\}$ with $s^0_{\alpha_i} \in S_{\alpha_i}$, $i = 1, 2, \ldots$, and such that for each n and each $m < k_n$ the integral

(3)
$$\int_{S_{\alpha_m}} \left\{ \cdots \left\{ \int_{S_{\alpha_{k_n}}} \chi_{E_n}(s^0_{\alpha_1}, \ldots, s^0_{\alpha_{m-1}}, s_{\alpha_m}, \ldots, s_{\alpha_{k_n}}) \mu_{\alpha_{k_n}}(ds_{\alpha_{k_n}}) \right\} \cdots \right\} \mu_{\alpha_m}(ds_{\alpha_m})$$

exists and does not converge to zero as $n \to \infty$. If we apply this statement with $m = k_j+1$ it is seen from (3) that for some $n > j$ the number

$$\chi_{E_n}(s^0_{\alpha_1}, \ldots, s^0_{\alpha_{k_j}}, s_{\alpha_{k_j+1}}, \ldots, s_{\alpha_{k_n}})$$

is not zero for all $s_{\alpha_{k_j+1}}, \ldots, s_{\alpha_{k_n}}$. Thus for some t_j in S_{π_j}, the point $s^0_{\alpha_1} \times \ldots \times s^0_{\alpha_{k_j}} \times t_j$ is in F_n. Since $F_n \subseteqq F_j$, we have

$$(4) \qquad s^0_{\alpha_1} \times \ldots \times s^0_{\alpha_{k_j}} \times t_j \in F_j.$$

Since F_j is in Σ^{π_j} it has the form $F_j = E_j \times S_{\pi_j}$ and thus (4) is true for every t_j in S_{π_j}. We now let $\pi_\infty = \cup_{n=1}^{\infty} \pi_n$ and let s_α^0 be an arbitrary element of S_α for $\alpha \notin \pi_\infty$. It follows that

$$\underset{\alpha\in A}{P}\, s_\alpha^0 \in F_j, \qquad j = 1, 2, \ldots,$$

which contradicts the fact that the intersection of all the sets F_j, $j = 1, 2, \ldots$, is void. Q.E.D.

21 DEFINITION. The measure space of Theorem 20 is called the *product of the measure spaces* $(S_\alpha, \Sigma_\alpha, \mu_\alpha)$ and we write

$$(S, \Sigma, \mu) = \underset{\alpha\in A}{P}\, (S_\alpha, \Sigma_\alpha, \mu_\alpha).$$

Just as in the case of finite products the operation of forming infinite products of measure spaces is an associative operation. This property is explicitly formulated in the next lemma.

22 LEMMA. *For each α in the set A let $(S_\alpha, \Sigma_\alpha, \mu_\alpha)$ be a positive measure space with $\mu_\alpha(S_\alpha) = 1$. Let A be partitioned into a family of disjoint sets A_β, where β ranges over a set B. Then*

$$\underset{\alpha\in A}{P}\, (S_\alpha, \Sigma_\alpha, \mu_\alpha) = \underset{\beta\in B}{P}\ \underset{\alpha\in A_\beta}{P}\, (S_\alpha, \Sigma_\alpha, \mu_\alpha).$$

PROOF. We may suppose that B and A are disjoint and thus unambiguously define the measure spaces

$$(S_\beta, \Sigma_\beta, \mu_\beta) = \underset{\alpha\in A_\beta}{P}\, (S_\alpha, \Sigma_\alpha, \mu_\alpha), \qquad \beta \in B$$

$$(S_1, \Sigma_1, \mu_1) = \underset{\beta\in B}{P}\, (S_\beta, \Sigma_\beta, \mu_\beta),$$

$$(S, \Sigma, \mu) = \underset{\alpha\in A}{P}\, (S_\alpha, \Sigma_\alpha, \mu_\alpha).$$

With this notation it is required to show that $(S, \Sigma, \mu) = (S_1, \Sigma_1, \mu_1)$. Clearly $S = S_1$. For β_0 in B the family Σ^{β_0} of all sets of the form $E_{\beta_0} \times \underset{\beta\neq\beta_0}{P}\, S_\beta$ with E_{β_0} in Σ_{β_0} is readily seen to be the σ-field generated by sets of the form $\underset{\alpha\in A}{P}\, E_\alpha$ with E_α in Σ_α and $E_\alpha = S_\alpha$ except for a finite number of elements of A_{β_0}. Hence $\Sigma^{\beta_0} \subseteq \Sigma$ for each β_0 in B. Since Σ_1 is clearly the σ-field generated by the collection of all the families Σ^{β_0} with β_0 in B we have $\Sigma_1 \subseteq \Sigma$. On the other hand, if α_0 is an arbitrary element of A, α_0 is an element of some set A_{β_0}. Then each set of the form $E_{\alpha_0} \times P_{\alpha \neq \alpha_0} S_\alpha$, with E_{α_0} in Σ_{α_0}, is in Σ^{β_0}, and hence in Σ_1. Since the collection of all these sets generates the σ-field Σ, it is seen that $\Sigma \subseteq \Sigma_1$ and thus that $\Sigma = \Sigma_1$.

Now, let π be a finite subset of A, and let τ be a finite subset of B such that $\underset{\beta \epsilon \tau}{U} A_\beta \supseteq \pi$. Then

$$\begin{aligned}
\mu_1(\underset{\alpha\epsilon\pi}{P} E_\alpha \times \underset{\alpha\epsilon\pi'}{P} S_\alpha) &= \mu_1(\underset{\beta\epsilon\tau}{P}(\underset{\alpha\epsilon A_\beta\pi}{P} E_\alpha \times \underset{\alpha\epsilon A_\beta - \pi}{P} S_\alpha) \times \underset{\beta\epsilon\tau'}{P} S_\beta) \\
&= \prod_{\beta\epsilon\tau} \mu_1(\underset{\alpha\epsilon A_\beta\pi}{P} E_\alpha \times \underset{\alpha\epsilon A_\beta - \pi}{P} S_\alpha) \\
&= \prod_{\beta\epsilon\tau} \prod_{\alpha\epsilon A_\beta\pi} \mu_\alpha(E_\alpha) \\
&= \prod_{\alpha\epsilon\pi} \mu_\alpha(E_\alpha).
\end{aligned}$$

Thus it follows from the uniqueness assertion in Theorem 20 that $\mu = \mu_1$. Q.E.D.

23 Lemma. *Let (S_1, Σ_1, μ_1) and (S_2, Σ_2, μ_2) be two positive measure spaces with $\mu_1(S_1) = \mu_2(S_2) = 1$, and let (S, Σ, μ) be their product.* For *each f in $L_1(S, \Sigma, \mu, \mathfrak{X})$, where $\mathfrak{X}$ is a B-space, let Tf be the function on S defined by the formula*

$$(Tf)(s_1, s_2) = \int_{S_2} f(s_1, t_2)\mu_2(dt_2).$$

Then, if f is in $L_p(S, \Sigma, \mu, \mathfrak{X})$ with $1 \leq p < \infty$ the function Tf is also in $L_p(S, \Sigma, \mu, \mathfrak{X})$ and

$$|Tf|_p \leq |f|_p.$$

Proof. It follows from the Hölder inequality (3.2) that $L_p(S, \Sigma, \mu, \mathfrak{X})$ is contained in $L_1(S, \Sigma, \mu, \mathfrak{X})$ and thus the Fubini theorem (9) together with Lemma 10 shows that Tf is defined μ-almost everywhere and is μ-measurable. We have, by Theorem 14, and another application of the Hölder inequality,

$$\begin{aligned}
|Tf|_p^p &= \int_{S_2} \left\{ \int_{S_1} \left| \int_{S_2} f(s_1, t_2)\mu_2(dt_2) \right|^p \mu_1(ds_1) \right\} \mu_2(ds_2) \\
&\leq \int_{S_2} \left\{ \int_{S_1} \left\{ \int_{S_2} |f(s_1, t_2)|^p \mu_2(dt_2) \right\} \mu_1(ds_1) \right\} \mu_2(ds_2) \\
&= \int_{S_2} \left\{ \int_S |f(s)|^p \mu(ds) \right\} \mu_2(ds_2)
\end{aligned}$$

$$= \int_{S_2} |f|_p^p \mu_2(ds_2)$$
$$= |f|_p^p. \qquad \text{Q.E.D.}$$

24 THEOREM. (*Mean Fubini-Jessen theorem*) *Let* (S, Σ, μ) *be the product of the positive measure spaces* $(S_\alpha, \Sigma_\alpha, \mu_\alpha)$, $\alpha \in A$, *where* $\mu_\alpha(S_\alpha) = 1$. *Let the finite sets* π *in* A *be ordered by inclusion. Then, for any* f *in* $L_p(S, \Sigma, \mu, \mathfrak{X})$ *with* $1 \leqq p < \infty$, *the function* $f_{\pi'}$ *defined on* S *to the B-space* $\mathfrak{X}$ *by the formula*

$$f_{\pi'}(s_\pi \times s_{\pi'}) = \int_{S_\pi} f(s_\pi \times s_{\pi'})\mu_\pi(ds_\pi),$$

converges in the norm of $L_p(S, \Sigma, \mu, \mathfrak{X})$ *to the constant function whose value at each point of* S *is the integral*

$$\int_S f(s)\mu(ds).$$

Furthermore the function f_π *on* S *defined by*

$$f_\pi(s_\pi \times s_{\pi'}) = \int_{S_{\pi'}} f(s_\pi \times s_{\pi'})\mu_{\pi'}(ds_{\pi'})$$

converges in the norm of $L_p(S, \Sigma, \mu, \mathfrak{X})$ *to* f.

PROOF. By Lemma 23 and Theorem II.1.18 it is sufficient to restrict our attention to functions f in a fundamental set in $L_p(S, \Sigma, \mu, \mathfrak{X})$. Since Σ is the σ-field determined by the sets $E_{\alpha_0} \times P_{\alpha \neq \alpha_0} S_\alpha$ with E_{α_0} in Σ_{α_0}, it follows, from Lemma 8.3, that $L_p(S, \Sigma_1, \mu, \mathfrak{X})$ is dense in $L_p(S, \Sigma, \mu, \mathfrak{X})$ where Σ_1 is the field generated by these sets. Thus, it is sufficient to restrict our attention to the characteristic function of a set in Σ_1. By Lemma 18, $\Sigma_1 = \cup_\pi \Sigma^\pi$, so that any E in Σ_1 has the form $E = F \times S_{\pi_0'}$ for some finite set π_0 and some F in Σ_{π_0}. Let $f = \chi_E$ and let $\pi \supseteq \pi_0$. Then, since $f(s) = f(s_\pi \times s_{\pi'})$ is independent of $s_{\pi'}$ we have

$$f_{\pi'}(s) = \int_{S_\pi} \chi_E(s_\pi \times s_{\pi'})\mu_\pi(ds_\pi)$$
$$= \int_S f(s)\mu(ds),$$

and

$$f_\pi(s) = \int_{S_{\pi'}} \chi_E(s_\pi \times s_{\pi'})\mu_{\pi'}(ds_{\pi'})$$
$$= f(s). \qquad \text{Q.E.D.}$$

25 Corollary. *Under the hypothesis of Theorem 24, we have, for every f in a dense subset of $L_1(S, \Sigma, \mu)$, a finite set π_0 in A such that*

$$f_{\pi'} = \int_S f(s)\mu(ds), \qquad f_\pi = f, \qquad \pi \supseteq \pi_0.$$

We shall now prove a theorem analogous to the preceding one with the norm convergence of that theorem replaced by pointwise convergence almost everywhere. For this purpose the following lemma is needed.

26 Lemma. *Let (S, Σ, μ) be the product of the positive measure spaces (S_n, Σ_n, μ_n), $n = 1, 2, \ldots$, with $\mu_n(S_n) = 1$. Let $\pi_n = \{1, \ldots, n\}$ and for f in $L_1(S, \Sigma, \mu, \mathfrak{X})$ and s in S let $g_n(s)$ be either*

$$\left| \int_{S_{\pi_n'}} f(s_{\pi_n} \times s_{\pi_n'}) \mu_{\pi_n'}(ds_{\pi_n'}) \right|,$$

or

$$\left| \int_{S_{\pi_n}} f(s_{\pi_n} \times s_{\pi_n'}) \mu_{\pi_n}(ds_{\pi_n}) \right|.$$

For $\delta > 0$ let

$$A_\delta = \{s \mid \sup_{1 \leqq n < \infty} g_n(s) > \delta\}.$$

Then

$$\delta\mu(A_\delta) \leqq \int_{A_\delta} |f(s)|\mu(ds).$$

Proof. Since the measures are all positive, the norm of an integral is at most equal to the integral of the norm. Thus the set A_δ will not decrease if f is replaced by $|f(\cdot)|$ and we may, and shall, assume that f is real and positive. Let B_0 be void and

$$B_k = \{s \mid \sup_{1 \leqq n \leqq k} g_n(s) > \delta\}, \qquad C_k = B_k - \cup_{n=1}^{k-1} B_n, \qquad n = 1, 2, \ldots.$$

Then $\{C_k\}$ is a sequence of disjoint sets with union A_δ. We now suppose, for the sake of definiteness, that $g_n(s)$ is the first of the two alternative functions mentioned in the statement of the present lemma. Then $g_n(s)$ is independent of $s_{\pi_n'}$ and the same must therefore be true of the characteristic function of C_n. Consequently, by the Fubini theorem,

$$\int_{C_k} f(s)\mu(ds)$$

$$= \int_{S_{\pi_k}} \left\{ \int_{S_{\pi_k'}} \chi_{C_{\pi_k}}(s_{\pi_k} \times s_{\pi_k'}) f(s_{\pi_k} \times s_{\pi_k'}) \mu_{\pi_k'}(ds_{\pi_k'}) \right\} \mu_{\pi_k}(ds_{\pi_k})$$

$$= \int_{S_{\pi_k}} \chi_{C_k}(s_{\pi_k} \times s_{\pi_k'}) \left\{ \int_{S_{\pi_k'}} f(s_{\pi_k} \times s_{\pi_k'}) \mu_{\pi_k'}(ds_{\pi_k'}) \right\} \mu_{\pi_k}(ds_{\pi_k})$$

$$= \int_{S_{\pi_k}} \chi_{C_k}(s_{\pi_k} \times s_{\pi_k'}) g_k(s_{\pi_k} \times s_{\pi_k'}) \mu_{\pi_k}(ds_{\pi_k}),$$

and, since the integrand is independent of $s_{\pi_k'}$,

$$\int_{C_k} f(s)\mu(ds)$$

$$= \int_{S_{\pi_k'}} \left\{ \int_{S_{\pi_k}} \chi_{C_k}(s_{\pi_k} \times s_{\pi_k'}) g_k(s_{\pi_k} \times s_{\pi_k'}) \mu_{\pi_k}(ds_{\pi_k}) \right\} \mu_{\pi_k'}(ds_{\pi_k})$$

$$= \int_{C_k} g_k(s)\mu(ds) \geqq \delta\mu(C_k).$$

By summing on k we get

$$\int_{A_\delta} f(s)\mu(ds) \geqq \delta\mu(A_\delta).$$

A similar proof may be made for the other choice of $g_n(s)$. Q.E.D.

27 THEOREM. (*Pointwise Fubini-Jessen theorem*) *Let* (S, Σ, μ) *be the product of the positive measure spaces* (S_n, Σ_n, μ_n), $n = 1, 2, \ldots$, *with* $\mu_n(S_n) = 1$. *Let* $\pi_n = \{1, \ldots, n\}$ *and for* f *in* $L_1(S, \Sigma, \mu, \mathfrak{X})$ *and* s *in* S *let*

$$f_{\pi_n'}(s) = f_{\pi_n'}(s_{\pi_n} \times s_{\pi_n'}) = \int_{S_{\pi_n}} f(s_{\pi_n} \times s_{\pi_n'}) \mu_{\pi_n}(ds_{\pi_n})$$

and

$$f_{\pi_n} = f_{\pi_n}(s_{\pi_n} \times s_{\pi_n'}) = \int_{S_{\pi_n'}} f(s_{\pi_n} \times s_{\pi_n'}) \mu_{\pi_n'}(ds_{\pi_n}).$$

Then

$$\lim_n f_{\pi_n'}(s) = \int_S f(s)\mu(ds),$$

and

$$\lim_n f_{\pi_n}(s) = f(s),$$

for almost all s *in* S.

PROOF. For a given $\varepsilon > 0$, we may, by Corollary 25, write $f = g+h$ where $|h| < \varepsilon$ and where $g_{\pi_n} = g$ for all large n. If $r(s)$ is defined by the equation

$$r(s) = \lim_{q\to\infty} \sup_{m,n>q} |f_{\pi_n}(s) - f_{\pi_m}(s)|$$

we have

$$r(s) = \lim_{q\to\infty} \sup_{m,n>q} |h_{\pi_n}(s) - h_{\pi_m}(s)| \leqq 2 \sup_{1\leqq n<\infty} |h_{\pi_n}(s)|.$$

Thus, by Lemma 26,

$$\mu(\{s|r(s) > 2\delta\}) \leqq |h|/\delta \leqq \varepsilon/\delta.$$

Since $\varepsilon > 0$ is arbitrary we have $\mu(\{s|r(s) > 2\delta\}) = 0$ and since $\delta > 0$ is arbitrary it follows that $r(s) = 0$ almost everywhere on S. This means that the limit $f^*(s) = \lim_n f_{\pi_n}(s)$ exists almost everywhere on S. By Fatou's theorem (6.19) and the mean Fubini-Jessen theorem (24) we have

$$\int_S |f^*(s) - f(s)|\mu(ds) = \lim_n \int_S |f_{\pi_n}(s) - f(s)|\mu(ds) = 0,$$

and so (6.8) $f^*(s) = f(s)$ almost everywhere. The other conclusion may be proved similarly. Q.E.D.

12. Differentiation

In this section we will prove some of the fundamental theorems of the Lebesgue theory of the differentiation of set functions in Euclidean spaces, and a number of classical theorems of the form

$$f(s) = \lim_{n\to\infty} \int_{-\infty}^{+\infty} K_n(s, t)f(t)dt$$

where $\{K_n\}$ is a suitable sequence of kernels and convergence is almost everywhere. The first theorem, the covering theorem of Vitali, is the basic tool for the development of the differentiation theory.

1 LEMMA. *Let S be a compact metric space and A be an arbitrary subset of S. Suppose that $\mathscr{F}$ is a family of closed subsets of S with the property that for each point s in S there is a set F in $\mathscr{F}$ of arbitrarily small positive diameter $\delta(F)$ containing s. Then there exists a finite or denumerable family $\{F_n\}$ of disjoint sets of $\mathscr{F}$ such that $A \subseteqq \cup F_n$ if the family is finite, and*

$$[*] \qquad A \subseteqq F_1 \cup \ldots \cup F_n \cup \bigcup_{k=n+1}^{\infty} S(F_k, 3\delta(F_k))$$

for every n if the family is denumerable.

PROOF. We first define the family $\{F_n\}$. Let F_1 be chosen arbitrarily and suppose $F_1, \ldots, F_k$ already chosen. If $A \subseteq F_1 \cup \ldots \cup F_k$, the lemma is satisfied. Otherwise, let $\varepsilon_k = \text{lub}\, \delta(F)$, where F varies over all sets F in $\mathscr{F}$ satisfying

$$S(F, \delta(F))F_i = \phi, \qquad i = 1, \ldots, k.$$

It is clear, since S is a metric space and $\bigcup_{i=1}^k F_i$ is closed, that this collection is non-vacuous and that $\varepsilon_k > 0$. Let F_{k+1} be any set in $\mathscr{F}$ such that $S(F_{k+1}, \delta(F_{k+1}))F_i = \phi$, $i = 1, \ldots, k$ and satisfying $\delta(F_{k+1}) > 2\varepsilon_k/3$. We have thus defined the family $\{F_n\}$ by induction.

Suppose that [*] is false for some integer $n > 1$. If $p \in A - [F_1 \cup \ldots \cup F_n \cup \bigcup_{k=n+1}^\infty S(F_k, 3\delta(F_k))]$, let F be a fixed set in $\mathscr{F}$ containing p with $\delta(F) > 0$ and for which $S(F, \delta(F))F_i = \phi$, $i = 1, \ldots, n$. If it can be shown that $S(F, \delta(F))F_k \neq \phi$ for some $k > n$, this will involve us in the desired contradiction. For if $S(F, \delta(F))F_k \neq \phi$ for some $k > n$, let k_0 be the least such k. Then $\delta(F) < \varepsilon_{k_0-1}$ so that $\delta(F_{k_0}) > 2\delta(F)/3$. Since $k_0 > n$, by assumption, p is not in $S(F_{k_0}, 3\delta(F_{k_0}))$. If q is in $S(F, \delta(F))F_{k_0}$, then

$$\varrho(p, q) < 2\delta(F) < 3\delta(F_{k_0}).$$

This implies that $p \in S(q, 3\delta(F_{k_0})) \subseteq S(F_{k_0}, 3\delta(F_{k_0}))$, a contradiction.

Thus to prove that [*] holds for each $n > 1$, it remains to prove that under the assumptions above $S(F, \delta(F))F_k \neq \phi$ for some $k > n$. If this is not the case the intersection is void for every k, and by construction of $\{F_n\}$ we have $\delta(F) \leqq \varepsilon_k$ for every k. This implies $\delta(F_k) > 2\delta(F)/3 > 0$ for all $k > 1$. Let $\{p_k\}$ be any sequence of points in S such that $p_k \in F_k$. If $i < j$ then $S(F_j, \delta(F_j))F_i = \phi$. Since

$$S(p_j, 2\delta(F)/3) \subseteq S(F_j, \delta(F_j)),$$

we have $\varrho(p_i, p_j) > 2\delta(F)/3$. The sequence $\{p_k\}$ then clearly contains no convergent subsequence. This contradicts the compactness of S. We conclude that $S(F, \delta(F))F_k \neq \phi$ for some $k > n$, and thus [*] holds for each $n > 1$.

Finally, since

$$F_1 \cup \ldots \cup F_n \cup \bigcup_{k=n+1}^{\infty} S(F_k, 3\delta(F_k)) \subseteq F_1 \cup \bigcup_{k=2}^{\infty} S(F_k, 3\delta(F_k))$$

for every n, it follows that [*] holds for every n. Q.E.D.

2 **Definition.** Let μ be a finite positive measure defined on the σ-field of Borel sets of a compact metric space S. A set $A \subseteq S$ is said to be *covered in the sense of Vitali* by a family $\mathscr{F}$ of closed sets if each $F \in \mathscr{F}$ has positive μ-measure and there is a positive constant α such that each point of A is contained in sets $F \in \mathscr{F}$ of arbitrarily small positive diameter for which $\mu(S(F, 3\delta(F)))/\mu(F) \leqq \alpha$.

3 **Theorem.** (*Vitali covering theorem*) *Let μ be a finite positive measure defined on the Borel sets of a compact metric space S. If a family $\mathscr{F}$ of closed sets covers a set $A \subseteq S$ in the sense of Vitali, then there exists a sequence of disjoint sets $\{F_n\} \subseteq \mathscr{F}$ such that $A - \bigcup_{n=1}^{\infty} F_n$ is μ-null set.*

Proof. By Lemma 1 there is a sequence $\{F_n\}$ of disjoint sets in $\mathscr{F}$ such that $A - \bigcup_{k=1}^{n} F_k \subseteq \bigcup_{k=n+1}^{\infty} S(F_k, 3\delta(F_k))$ for each n. Now

$$\mu\left(\bigcup_{k=n+1}^{\infty} S(F_k, 3\delta(F_k))\right) \leqq \sum_{k=n+1}^{\infty} \mu(S(F_k, 3\delta(F_k)))$$

$$\leqq \alpha \sum_{k=n+1}^{\infty} \mu(F_k) \to 0$$

as $n \to \infty$ since $\sum_{k=1}^{\infty} \mu(F_k) \leqq \mu(S) < \infty$. Thus for each ε there is an n_ε such that

$$A - \bigcup_{k=1}^{\infty} F_k \subseteq A - \bigcup_{k=1}^{n_\varepsilon} F_k \subseteq \bigcup_{k=n_\varepsilon+1}^{\infty} S(F_k, 3\delta(F_k))$$

and $\mu(\bigcup_{k=n_\varepsilon+1}^{\infty} S(F_k, 3\delta(F_k))) < \varepsilon$. Thus $A - \bigcup_{n=1}^{\infty} F_n$ is a μ-null set. Q.E.D.

4 **Definition.** Let λ be a vector valued set function defined on all closed cubes contained in some open set G in real n-dimensional Euclidean space E^n and let μ be Lebesgue measure in E^n. Then λ is said to be *differentiable* at a point p in G if the limit

$$\frac{d\lambda}{d\mu}(p) = \lim_{\mu(C)\to 0} \frac{\lambda(C)}{\mu(C)}$$

exists, where C is a closed cube containing p. The function $d\lambda/d\mu$ is called the *derivative* of λ.

Theorem 6 below will show that the terminology and notation introduced in the preceding definition are consistent with those introduced in the note following Theorem 10.7.

5 Lemma. *Let λ be a finite positive measure defined on the Borel subsets of the closure of a bounded open set G of real Euclidean n-space E^n. Let $0 < r < \infty$.*

(a). *If for each p in a set $A \subseteq G$*

$$\liminf_{\mu(C)\to 0} \frac{\lambda(C)}{\mu(C)} < r,$$

where C is a closed cube containing p, then each neighborhood of A contains an open set Q such that $A-Q$ is a μ-null set and $\lambda(Q) < r\mu(Q)$.

(b) *If for each p in a set $A \subseteq G$*

$$\limsup_{\mu(C)\to 0} \frac{\lambda(C)}{\mu(C)} > r,$$

then each neighborhood of A contains a Borel set B such that $A-B$ is a μ-null set and $\lambda(B) > r\mu(B)$.

Proof. To prove statement (a) let U be an arbitrary open set such that $A \subseteq U \subseteq G$ and let $\mathscr{F}$ be the family of all closed cubes C contained in U satisfying $\lambda(C) < r\mu(C)$. Since every cube C satisfies $\mu(S(C, 3\delta(C))) \leqq (6\sqrt{n}+1)^n\mu(C)$, it follows that $\mathscr{F}$ covers A in the sense of Vitali. Hence by Theorem 3 there is a sequence of closed cubes $\{C_k\} \subseteq \mathscr{F}$ such that $A - \bigcup_{k=1}^{\infty} C_k$ is a μ-null set. For each k let D_k be the interior of C_k. Since the faces of a cube have zero μ-measure, $A - \bigcup_{k=1}^{\infty} D_k$ is a μ-null set and $\lambda(D_k) < r\mu(D_k)$. If $Q = \bigcup_{k=1}^{\infty} D_k$ then Q is open, $Q \subseteq U$, $A-Q$ is a μ-null set, and

$$\lambda(Q) = \sum_{j=1}^{\infty} \lambda(D_j) < r\sum_{j=1}^{\infty} \mu(D_j) = r\mu(Q),$$

proving (a).

To prove (b) let U be an arbitrary open set such that $A \subseteq U \subseteq G$ as before. Now the family $\mathscr{F}_1$, of closed cubes $C \subseteq U$ such that $\lambda(C) > r\mu(C)$, covers A in the sense of Vitali. If $\{C_k\} \subseteq \mathscr{F}_1$ is a sequence of disjoint cubes such that $A - \bigcup_{k=1}^{\infty} C_k$ is a μ-null set, let $B = \bigcup_{k=1}^{\infty} C_k$. Since

$$\lambda(B) = \sum_{j=1}^{\infty} \lambda(C_k) > r\sum_{j=1}^{\infty} \mu(C_j) = r\mu(B),$$

the set B satisfies the requirements of (b). Q.E.D.

6 THEOREM. *Let λ be a finite measure defined for the Borel subsets of an open set G in real Euclidean n-space E^n, and let μ denote Lebesgue measure in E^n. Then, for μ-almost all p in G,*

$$[*] \qquad \frac{d\lambda}{d\mu}(p) = \lim_{\mu(C)\to 0} \lambda(C)/\mu(C)$$

exists, where C denotes an arbitrary closed cube in G containing p. Moreover

(i) *$d\lambda/d\mu$ is μ-integrable;*

(ii) *$\lambda(B) = \int_B (d\lambda/d\mu)(p)\mu(dp)$ for every Borel subset B of G if and only if λ is μ-continuous;*

(iii) *$(d\lambda/d\mu)(p) = 0$ μ-almost everywhere if and only if λ is μ-singular.*

PROOF. Since G is the union of a denumerable family of open cubes whose closure is contained in G, it is clear that we can confine our attention entirely to subsets of the interior K_0 of some fixed cube K. By decomposing λ into its real and imaginary parts, and then decomposing each of these measures into the sum of a positive and negative measure by the Hahn decomposition theorem (4.10), it is also clear that we need only consider the case where λ is positive. Therefore, *we shall assume hereafter that λ is positive.*

First, we shall prove the existence of the limit. Let A be the set of points p in K_0 for which

$$\limsup_{\mu(C)\to 0} \frac{\lambda(C)}{\mu(C)} > \liminf_{\mu(C)\to 0} \frac{\lambda(C)}{\mu(C)},$$

where C is a closed cube containing p. Let

$$A_{mn} = \left\{p \middle| \limsup_{\mu(C)\to 0} \frac{\lambda(C)}{\mu(C)} > \frac{m+1}{n} > \frac{m}{n} > \liminf_{\mu(C)\to 0} \frac{\lambda(C)}{\mu(C)}\right\},$$

for every non-negative integer m and positive integer n. Suppose that A is not a μ-null set. Then since $A = \bigcup_{m,n=1}^{\infty} A_{mn}$, some set A_{ij} is not a μ-null set. If $\beta = \inf_{B \supseteq A_{ij}} \mu(B)$ where B is a Borel set, then $\beta > 0$. Now given $\varepsilon > 0$ and using the regularity of μ we can find an open set U such that $A_{ij} \subseteq U \subseteq K_0$ and $\mu(U) < \beta + \varepsilon$. By Lemma 5(a) there is an open set $Q \subseteq U$ such that $A_{ij} - Q$ is a μ-null set and

$$\lambda(Q) \leqq \frac{i}{j}\mu(Q) < \frac{i}{j}(\beta+\varepsilon).$$

By applying Lemma 5(b) to the set $A_{ij}Q$, we see there is a Borel set $B \subseteq Q$ such that $A_{ij}Q-B$ is a μ-null set and

$$\lambda(B) \geqq \frac{i+1}{j}\mu(B).$$

Since $A_{ij} \subseteq B \cup (A_{ij}-Q) \cup (A_{ij}Q-B)$, $\mu(B) \geqq \beta$. Thus

$$\frac{i}{j}(\beta+\varepsilon) > \lambda(Q) \geqq \lambda(B) > \frac{(i+1)}{j}\beta$$

which yields a contradiction for sufficiently small ε. Thus we have shown A is a μ-null set, and it follows that the limit [*] exists μ-almost everywhere in K.

We will show now that $d\lambda/d\mu$ is μ-measurable. Let $C(p, \alpha)$ denote the closed cube with center p and side length α. Let

$$\mu_m(p, \alpha) = 2m\int_{\alpha+1/2m}^{\alpha+1/m} \mu(C(p, \beta))d\beta,$$

$$\lambda_m(p, \alpha) = 2m\int_{\alpha+1/2m}^{\alpha+1/m} \lambda(C(p, \beta))d\beta.$$

Then $\lambda_m(p, \alpha)/\mu_m(p, \alpha)$ is a continuous function of p for each α and hence $(C(p, \alpha))/\mu(C(p, \alpha)) = \lim_{m\to\infty} \lambda_m(p, \alpha)/\mu_m(p, \alpha)$ is μ-measurable. Consequently,

$$\frac{d\lambda}{d\mu}(p) = \lim_{n\to\infty} \lambda\left(C\left(p, \frac{1}{n}\right)\right) / \mu\left(C\left(p, \frac{1}{n}\right)\right)$$

is μ-measurable.

It is convenient next to prove the converse assertion of (iii): if λ is μ-singular then $(d\lambda/d\mu)(p) = 0$ μ-almost everywhere. For, if λ is μ-singular, there is a μ-null Borel set N such that $\lambda(G-N) = 0$. Suppose that $D = \{p|(d\lambda/d\mu(p)) > 0\}$ is not a μ-null set. Then since $d\lambda/d\mu$ is measurable there is an $\varepsilon > 0$ and a Borel subset E of $D-N$ such that $\mu(E) > 0$ and $(d\lambda/d\mu)(p) > \varepsilon$, $p \in E$. Since μ is regular we can suppose that E is closed. By Lemma 5(b), every neighborhood of E has λ-measure greater than $\varepsilon\mu(E)$. Since E is the intersection of a sequence of its neighborhoods, $\lambda(D) \geqq \varepsilon\mu(D) > 0$, contradicting the fact that E is contained in the λ-null set $D-N$. We have thus shown $\mu(\{p|(d\lambda/d\mu)(p) > 0\}) = 0$. Since λ is non-negative, $d\lambda/d\mu \geqq 0$, and thus $(d\lambda/d\mu)(p) = 0$ μ-almost everywhere.

We proceed next to prove (i) and (ii). Using the result of the last paragraph and decomposing λ into the sum of a μ-continuous and μ-singular measure by Theorem 4.14, it is clear that we can assume, without loss of generality, that λ is μ-continuous. Then, by the Radon-Nikodým theorem (10.2), there exists a μ-integrable function f such that $\lambda(B) = \int_B f(p)\mu(dp)$ for μ-almost all p. If $\{p|f(p) > (d\lambda/d\mu)(p)\}$ is not a μ-null set, then there exist positive constants c and ε such that

$$A = \left\{p|f(p) > c+\varepsilon > c > \frac{d\lambda}{d\mu}(p)\right\}$$

is not a μ-null set, and by the regularity of μ, there exists a closed set $D \subseteqq A$ for which $\mu(D) \neq 0$. Then it follows that $\lambda(D) = \int_D f(p)\mu(dp) > (c+\varepsilon)\mu(D)$. However, by Lemma 5(a), any neighborhood of D contains a set Q such that $\lambda(D) \leqq \lambda(Q) < c\mu(Q)$. Since D is the intersection of a sequence of its neighborhoods, it follows that $\lambda(D) \leqq c\mu(D)$, and hence that $c\mu(D) > (c+\varepsilon)\mu(D)$, a contradiction. Thus we have shown, using Lemma 5(a), that $\mu(\{p|f(p) > (d\lambda/d\mu)(p)\}) = 0$. It follows similarly, using Lemma 5(b) that $\mu(\{p|f(p) < (d\lambda/d\mu)(p)\})=0$.

It remains to prove the direct assertion in (iii) that $(d\lambda/d\mu)(p)=0$ for μ-almost all p implies λ is μ-singular. We may write λ as the sum of a μ-singular positive measure λ_1 and a μ-continuous measure λ_2. From the definition of the derivative it is clear that $d\lambda/d\mu = d\lambda_1/d\mu + d\lambda_2/d\mu$, μ-almost everywhere. By the part of (iii) already proved $d\lambda_1/d\mu = 0$ μ-almost everywhere. Thus $d\lambda_2/d\mu = 0$ μ-almost everywhere. Applying (ii) to the μ-continuous measure λ_2, we see that

$$\lambda_2(B) = \int_B \frac{d\lambda_1}{d\mu}(p)\mu(dp) = 0$$

for each Borel set B. Q.E.D.

7 COROLLARY. *If f is a Lebesgue integrable function defined in an open set G of real Euclidean n-space then*

$$\lim_{h\to 0} \frac{1}{h^n}\int_{t_1}^{t_1+h} \cdots \int_{t_n}^{t_n+h} f(s_1, \ldots, s_n)ds_1, \ldots, ds_n = f(t_1, \ldots, t_n)$$

almost everywhere in G.

Note that we are not limited to the use of cubes in Theorem 6.

Many other families of closed sets (e.g., spheres) covering G in the sense of Vitali would do as well.

In the next result we extend the discussion to vector valued functions.

8 THEOREM. *Let f be a vector valued Lebesgue integrable function defined on an open set G in real n-dimensional Euclidean space and let $\alpha(E) = \int_E f(p)\mu(dp)$. Then*

$$\lim_{\mu(C)\to 0} \frac{1}{\mu(C)} \int_C |f(q)-f(p)|\mu(dq) = 0$$

for almost all p in G, where C is a closed cube containing p. In particular

$$\frac{d\alpha}{d\mu}(p) = \lim_{\mu(C)\to 0} \frac{1}{\mu(C)} \int_C f(q)\,\mu(dq) = f(p)$$

for almost all p in G.

PROOF. By Theorem 8.5 there is a μ-null set N_0 and a separable subspace $\mathfrak{Z}$ of $\mathfrak{X}$ such that $f(G-N_0) \subseteqq \mathfrak{Z}$. Let $\{z_n\}$ be a countable dense subset of $\mathfrak{Z}$. It follows from Theorem 6 that for each $n = 1, 2, \ldots$,

$$\lim_{\mu(C)\to \infty} \frac{1}{\mu(C)} \int_C |f(q)-z_n|\mu(dq) = |f(p)-z_n|$$

for p in the complement of a μ-null set N_n. The set $N = \cup_{i=0}^{\infty} N_i$ is a μ-null set. Let $p \in G-N$ and $\varepsilon > 0$. Select z_k such that $|f(p)-z_k| < \varepsilon$. Then

$$\begin{aligned}
&\limsup_{\mu(C)\to 0} \frac{1}{\mu(C)} \int_C |f(q)-f(p)|\mu(dq) \\
&\leqq \limsup_{\mu(C)\to 0} \int_C \{|f(q)-z_k|+|z_k-f(p)|\}\mu(dq) \\
&= |z_k-f(p)| + \limsup_{\mu(C)\to 0} \frac{1}{\mu(C)} \int_C |f(q)-z_k|\mu(dq) \\
&= 2|z_k-f(p)| < 2\varepsilon.
\end{aligned}$$

Since ε is arbitrary

$$\left| f(p) - \frac{1}{\mu(C)} \int_C f(q)\mu(dq) \right| < \frac{1}{\mu(C)} \int_C |f(q)-f(p)|\mu(dq) \to 0$$

as $\mu(C) \to 0$ for each $p \in G-N$. Q.E.D.

9 **Definition.** Let f be a vector valued Lebesgue integrable function defined on an open set in Euclidean n-space. The set of all points p at which

$$\lim_{\mu(C)\to 0} \frac{1}{\mu(C)} \int_C |f(q)-f(p)|\mu(dq) = 0$$

is called the ***Lebesgue set*** of the function f.

Clearly, the Lebesgue set of f contains each point of continuity of f.

Let us suppose that f is a vector valued Lebesgue integrable function of one real variable t, $-\infty < t < \infty$. Let

$$\begin{aligned} Q_n(t) &= \frac{n}{2}, \qquad |t| \leqq \frac{n^{-1}}{2}, \\ &= 0, \qquad |t| > \frac{n^{-1}}{2}. \end{aligned}$$

Theorem 8 states, in somewhat different notation, that

$$\lim_{n\to\infty} \int_{-\infty}^{\infty} Q_n(t-r)f(s)ds = \lim_{n\to\infty} \int_{-\infty}^{\infty} f(t-s)Q_n(s)ds = f(t)$$

for each t in the Lebesgue set of f. It is clear that the integrals above average the values of f in the neighborhoods $[t-1/n, t+1/n]$ of the point t. This interpretation of Theorem 8 as a theorem about "averages" is capable of great generalization. Consider the functions $Q_n^*(t) = (ne^{-n^2t^2})/\sqrt{\pi}$. Then $\int_{-\infty}^{\infty} Q_n^*(t)dt = 1$, $n = 1, 2, \ldots$. The functions Q_n^* have the familiar "bell-shaped" graph of the Gaussian density functions; for increasing n the peak at $t = 0$ becomes higher and narrower in such a way that for each $\varepsilon > 0$, $\lim_{n\to\infty} \int_{-\varepsilon}^{+\varepsilon} Q_n^*(t)dt = 1$.

In this localizing behavior the functions Q_n^* behave very much like the functions Q_n. It is reasonable to ask whether the "weighted averages" formed with the "density functions" Q_n^* behave like the ordinary averages formed with the functions Q_n, i.e., whether

$$\lim_{n\to\infty} \frac{n}{\sqrt{\pi}} \int_{-\infty}^{+\infty} e^{-n^2(t-s)^2} f(s)ds = f(t)$$

in the Lebesgue set of f as before. That this is true will follow from our Theorem 10 below. Other special consequences of Theorem 10 will be the limit relations

$$\lim_{n\to\infty}\int_{-\infty}^{\infty}\frac{\sin^2 n(t-s)}{n(t-s)}f(s)ds = f(t)$$

and

$$\lim_{n\to\infty} n\int_{t}^{\infty} e^{-n(t-s)}f(s)ds = f(t),$$

each holding in the Lebesgue set of f.

Instead of proving Theorem 10 directly we shall consider in Theorem 11 below the much more general question of when

$$\lim_{n\to\infty}\int_{-\infty}^{\infty} K_n(s, t)f(s)ds = f(t)$$

in the Lebesgue set of f, where the kernels $K_n(s, t)$ are not necessarily positive or even real. Theorem 10 will follow from Theorem 11 by taking $K_n(s, t) = Q_n(t-s)$.

10 THEOREM. *Let $\{Q_n\}$ be a sequence of non-negative real valued functions of the real variable t which satisfy*

(a) *$Q_n(t)$ is continuous on the right and is increasing for $t \leqq 0$ and decreasing for $t \geqq 0$;*

(b) $\lim_{|t|\to\infty} Q_n(t) = 0$;

(c) $\lim_{n\to\infty} Q_n(t) = 0$ *for* $t \neq 0$;

(d) $\lim_{n\to\infty}\int_{-1}^{1} Q_n(t)dt = 1.$

Then if f is a Lebesgue integrable function of t with values in a B-space $\mathfrak{X}$, the functions $Q_n(t-s)f(s)$ are integrable in s for each n and t and

$$\lim_{n\to\infty}\int_{-\infty}^{\infty} Q_n(t-s)f(s)ds = f(t)$$

for each t in the Lebesgue set of f.

11 THEOREM. *Let $\{K_n\}$, $\{R_n\}$ be sequences of scalar functions defined on the plane with R_n real valued and $|K_n(s,t)| \leqq R_n(s,t)$. Suppose that for each fixed s we have*

(a) *$R_n(s, t)$ is decreasing as a function of t when $t \geqq s$, is increasing when $t \leqq s$ and is continuous on the right;*

(b) $\lim_{|t|\to\infty} R_n(s, t) = 0$;

(c) *there is a constant $M(s)$ such that*

$$\int_{s-1}^{s+1} R_n(s, t)dt \leqq M(s), \qquad n = 1, 2, \ldots;$$

(d) $\lim_{n\to\infty} \int_{s-1}^{s+1} K_n(s, t)dt = 1;$

(e) $\lim_{n\to\infty} R_n(s, t) = 0$ *for each pair* (s, t) *with* $s \neq t$. *Then if* f *is a Lebesgue integrable function defined for* $-\infty < s < \infty$ *with values in a B-space* $\mathfrak{X}$,

$$\lim_{n\to\infty} \int_{-\infty}^{\infty} K_n(s, t)f(t)dt = f(s)$$

for every s *in the Lebesgue set of* f.

PROOF. Since $|K_n(s, t)| \leqq R_n(s, t) \leqq R_n(s, s)$ it follows from Theorem 2.22 that $K_n(s, t)f(t)$ is Lebesgue integrable for each n and s. Choose some fixed s_0 in the Lebesgue set of f and let $g(t) = f(t) -\chi_I(t)f(s_0)$ where $I = [s_0-1, s_0+1]$. In view of condition (d), Theorem 11 amounts to the assertion that

$$\lim_{n\to\infty} \int_{-\infty}^{\infty} K_n(s_0, t)g(t)dt = 0.$$

It is clearly sufficient to prove that

$$\lim_{n\to\infty} \int_{-\infty}^{\infty} R_n(s_0, t)|g(t)|dt = 0.$$

We shall prove

$$[*] \qquad \lim_{n\to\infty} \int_{s_0}^{\infty} R_n(s_0, t)|g(t)|dt = 0.$$

The corresponding assertion

$$\lim_{n\to\infty} \int_{-\infty}^{s_0} R_n(s_0, t)|g(t)|dt = 0$$

may be proved in an entirely analogous manner. The essence of the method is to integrate by parts; i.e., to rewrite the limit [*] in terms of a Lebesgue-Stieltjes integral (cf. the definitions and notation of Section 5.)

Defining $G(t) = \int_{s_0}^{t}|g(r)|dr$, we have, by Theorems 10.4 and 6.22,

$$[**] \qquad \int_{s_0}^{w} R_n(s_0, t)|g(t)|dt = G(w)R_n(s_0, w-) - \int_{s_0}^{w} G(t)dR_n(s_0, t).$$

As $w \to \infty$, $R_n(s_0, w-) \to 0$ by (b) while

$$G(w) \to \int_{s_0}^{\infty} |g(t)|dt < \infty.$$

Hence, letting $w \to \infty$ in [**] we find that

$$\lim_{w\to\infty} \int_{s_0}^{w} G(t)dR_n(s_0, t) = -\int_{s_0}^{\infty} R_n(s_0, t)|g(t)|dt.$$

Since G is non-negative, and since $R_n(s_0, \cdot)$ is decreasing so that $-R_n(s_0, \cdot)$ determines a non-negative measure, an application of Fatou's lemma (6.19) to the limit on the left shows that $\int_{s_0}^{\infty} G(t)dR_n(s_0, t)$ exists and

$$\int_{s_0}^{\infty} G(t)dR_n(s_0, t) = -\int_{s_0}^{\infty} R_n(s_0, t)|g(t)|dt.$$

Thus to prove [*] we need only show

$$[\dagger] \qquad \lim_{n\to\infty} \int_{s_0}^{\infty} G(t)dR_n(s_0, t) = 0.$$

Since s_0 is in the Lebesgue set of f,

$$\begin{aligned} 0 &= \lim_{h\to 0} h^{-1} \int_{s_0}^{s_0+h} |f(t)-f(s_0)|dt \\ &= \lim_{h\to 0} h^{-1} \int_{s_0}^{s_0+h} |g(t)|dt = \lim_{h\to 0} h^{-1}G(s_0+h). \end{aligned}$$

Thus given $\varepsilon > 0$ we can choose $\delta > 0$ such that $G(s_0+h) < \varepsilon h$ for $0 \leqq h \leqq \delta$. Then, integrating by parts (Theorem 6.22), we have

$$\begin{aligned} \int_{s_0}^{s_0+h} G(t)dR_n(s_0, t) &\leqq \varepsilon \int_{s_0}^{s_0+\delta} (t-s_0)dR_n(s_0, t) \\ &= \varepsilon\delta R_n(s_0, (s_0+\delta)-)-\varepsilon \int_{s_0}^{s_0+\delta} R_n(s_0, t)dt. \end{aligned}$$

Applying (b) and (e) it follows that

$$\limsup_{n\to\infty} \int_{s_0}^{s_0+\delta} G(t)dR_n(s_0, t) \leqq \varepsilon M(s_0).$$

On the other hand, since $\lim_{t\to\infty} G(t) < \infty$, $g(t)$ is bounded by some constant L. Thus

$$\int_{s_0+\delta}^{\infty} G(t)dR_n(s_0, t) \leqq L \int_{s_0+\delta}^{\infty} dR_n(s_0, t) = LR_n(s_0, s_0+\delta) \to 0$$

as $n \to \infty$ by (e). Consequently

$$\limsup_{n\to\infty} \int_{s_0}^{\infty} G(t)dR_n(s_0, t) < \varepsilon M(s_0).$$

Since ε is arbitrary, formula [†] is proved. Q.E.D.

The following special case of the previous two theorems is often useful.

12 COROLLARY. *Let Q be a non-negative function of the real variable t, $-\infty < t < \infty$. with the properties*

(a) *Q is increasing for $t \leqq 0$, decreasing for $t \geqq 0$, and is continuous on the right;*

(b) $\lim_{|t|\to\infty} tQ(t) = 0$;

(c) $\lim_{|t|\to\infty} \int_{-t}^{+t} Q(s)ds = 1.$

If f is a Lebesgue integrable function defined for $-\infty < t < \infty$ with values in a B-space $\mathfrak{X}$, then

$$Q(n(t-s))f(s)$$

is integrable for each n and t, and

$$\lim_{n\to\infty} \int_{-\infty}^{\infty} Q(n(t-s))f(s)ds = f(t)$$

for each t in the Lebesgue set of f.

PROOF. This result follows immediately on taking $Q_n(t) = nQ(nt)$ in Theorem 10. Q.E.D.

13. Exercises

1 Let (S, Σ, μ) be a finite measure space. Let α and β be bounded countably additive functions defined on Σ. Suppose that α is μ-continuous, and β is α-continuous. Show that β is μ-continuous, and that $(d\beta/d\mu)(s) = (d\beta/d\alpha)(s)(d\alpha/d\mu)(s)$ μ-almost everywhere.

2 Show that Theorem 10.2 fails without the hypothesis that (S, Σ, μ) is σ-finite.

3 Construct a bounded Lebesgue measurable function f for which $\int_0^t f(s)ds$ is non-differentiable on a given set of measure zero.

4 If f is a function with a derivative bounded on the finite interval (a, b), then f is of bounded variation, and

$$\int_a^b g(s)df(s) = \int_a^b g(s)f'(s)ds,$$

the integral on the left existing and the displayed formula being valid whenever the function gf' is Lebesgue integrable.

5 Let h be a function of bounded variation on the interval (a, b) and continuous on the right. Let g be a function defined on (a, b) such that the Lebesgue-Stieltjes integral $I = \int_a^b g(s)dh(s)$ exists. Let f be a continuous increasing function on an open interval (c, d), with $f(c)=a$, $f(d) = b$. Show that the Lebesgue-Stieltjes integral

$$\int_c^d g(f(s))\, dh(f(s))$$

exists and is equal to I.

6 Show that a monotone increasing function f defined on an interval is differentiable almost everywhere with respect to Lebesgue measure and that f' may vanish almost everywhere without f being a constant.

7 Let (S, Σ, μ) be the product of the regular measure spaces $(S_\alpha, \Sigma_\alpha, \mu_\alpha)$, $\alpha \in A$. If S is endowed with its product topology show that (S, Σ, μ) is also regular.

8 For each integer n, let (S_n, Σ_n, μ_n) be the measure space for which S_n is the set consisting of the two points 0 and 1, each of which is taken to have measure 1/2, and Σ_n is the family of all subsets of S_n. Let $(S, \Sigma, \mu) = P_n(S_n, \Sigma_n, \mu_n)$. Let I be the interval $[0, 1)$. Define a mapping of $I \to S$ as follows: for $0 \leqq s < 1$, let

$$s = \sum_{n=1}^{\infty} \frac{\varepsilon_n}{2^n}$$

be the unique dyadic expansion of s defined by the requirement that $\varepsilon_n = 0$ or 1 and, for infinitely many n, we have $\varepsilon_n = 0$, and let $\varphi(s) = [\varepsilon_1, \varepsilon_2, \varepsilon_3, \ldots]$. Show that the σ-field $\{\varphi^{-1}(E) | E \in \Sigma\}$ is the σ-field of Borel sets of I, and that if we put

$$\lambda(\varphi^{-1}(E)) = \mu(E),$$

λ is Borel-Lebesgue measure.

9 Let f be a Lebesgue integrable function defined on the unit interval $[0, 1)$. Put $f_n(s) = 2^n \int_{j/2^n}^{(j+1)/2^n} f(s)ds$ for $j/2^n \leqq s < (j+1)/2^n$, $0 \leqq j \leqq 2^n-1$. Show that $f_n(s) \to f(s)$ (Lebesgue) almost everywhere.

10 Let f be a Lebesgue integrable function defined on the unit interval $[0, 1)$. Put $f_n(s) = 2^{-n} \sum_{i=0}^{2^n-1} f(s+j/2^n)$ (where $s+t$ is to be interpreted as $s+t-1$ if $s+t \geqq 1$). Show that $f_n(s) \to \int_0^1 f(s)ds$ for (Lebesgue) almost all s.

11 Let (S, Σ, μ) be a σ-finite positive measure space, and $(R, \mathscr{B}, \lambda)$ the Borel-Lebesgue measure space on the real line R. Let $(S_1, \Sigma_1, \mu_1) = (S, \Sigma, \mu) \times (R, \mathscr{B}, \lambda)$. If f is a real function defined on S, its graph is the subset $P(f) = \{[s, f(s)] | s \in S\}$ of S_1. Show that f is μ-measurable if and only if its graph is a μ-null set.

12 Under the hypotheses of Exercise 11, let f be μ-measurable and non-negative. Show that

$$\int_S f(s)\mu(ds) = \mu_1\{[s, t] \in S_1 | 0 < t < f(s)\}.$$

13 Let $\{(S_\alpha, \Sigma_\alpha, \mu_\alpha)\}$ be a family of finite positive measure spaces each of which has $\mu_\alpha(S_\alpha) = 1$. Let $(S, \Sigma, \mu) = P_\alpha(S_\alpha, \Sigma_\alpha, \mu_\alpha)$, let $E_\alpha \in \Sigma_\alpha$, and let $E = P_\alpha E_\alpha$. Show that $E \in \Sigma$ if and only if $E_\alpha = S_\alpha$ for all but a countable number of indices α, and that in this case $\mu(E)$ can be expressed as the unconditionally convergent infinite product $\prod_\alpha \mu_\alpha(E_\alpha)$.

14. Functions of a Complex Variable

In some of the chapters to follow, and especially in Chapter VII, we shall use extensions of certain well-known results in the theory of analytic functions of a complex variable to the case where the functions are vector valued. It will be assumed that the reader is familiar with the elementary theory of complex valued analytic functions of one complex variable and this theory will be applied here to obtain the extensions which will be used later.

Throughout this section, $\mathfrak{X}$ will denote a *complex* B-space.

Definition. Let G be an open set in the space of n complex variables $z_1, \ldots, z_n$. (Such a set is often called a *domain.*) A function f defined on G and with values in $\mathfrak{X}$ is said to be *analytic on* G if f is continuous and the first partial derivatives $\partial f / \partial z_i$, $i = 1, \ldots, n$, exist at each point of G.

It is evident that if f is a vector valued analytic function of the complex variables $z_1, \ldots, z_n$, then $x^* f$ is a complex valued analytic function of $z_1, \ldots, z_n$ for each $x^* \in \mathfrak{X}^*$.

The theory of vector valued analytic functions, like that of complex valued analytic functions, can be developed most efficiently through the use of line integrals. These integrals may be defined as follows: Let $I = \{t | a \leqq t \leqq b\}$ be an interval of the real axis and let α be a complex valued function which is defined, continuous and of bounded variation on I. Then α is the parametrization of a *continuous rectifiable curve* $C = \alpha(I)$ in the complex plane. If $\alpha(t_1) \neq \alpha(t_2)$ unless $t_1 = t_2$ or t_1 and t_2 are the same as a and b, C is called a *simple Jordan curve.* If $\alpha(a) = \alpha(b)$, and C is a simple Jordan curve, then C is said to be *closed.*

If f is a vector valued function of a complex variable which is such that $f(\alpha(t))$ is defined for all $a \leqq t \leqq b$ and such that the Radon-Stieltjes integral $\int_a^b f(\alpha(t))d\alpha(t)$ is defined, we write

$$\int_a^b f(\alpha(t))d\alpha(t) = \int_C f(\alpha)d\alpha,$$

and call $\int_C f(\alpha)d\alpha$ the *line integral* of f over (or along) the curve C. It is easily seen that if we make a continuous monotone change of parameter $t = t(s)$, so that as s increases from a_1 to b_1, t increases from a to b, we have, putting $\alpha_1(s) = \alpha(t)s))$,

$$\int_{a_1}^{b_1} f(\alpha_1(s))d\alpha_1(s) = \int_a^b b(\alpha(t))d\alpha(t)$$

(cf. Theorem 10.8) In particular, if C is a closed Jordan curve, understood to be traversed in a certain direction, then $\int_C f(\alpha)d\alpha$ is independent of the particular parametrization of C, and thus, as the notation indicates, depends only on the point set C.

The fundamental theorem in the theory of complex line integrals is the *Cauchy integral theorem.* We may phrase this theorem as follows. Let U be a bounded open set in the complex plane, and let B denote the boundary of U. We suppose that B consists of a finite collection of disjoint closed rectifiable Jordan curves; that is, we suppose that B can be decomposed into the union $B = B_1 \cup B_2 \cup \ldots \cup B_k$ of disjoint closed sets B_j in such a way that each B_j is a closed rectifiable Jordan curve:

$$B_j = \alpha_j(I_j), \quad I_j = \{t | a_j \leqq t \leqq b_j\}.$$

We suppose in addition that the various curves B_j are oriented *in the positive sense customary in the theory of complex variables*: i.e., if the points in U lying close to B_j are inside B_j, we suppose that B_j is traversed in the counter-clockwise direction as t goes from a_j to b_j, while if the points in U lying close to B_j are outside B_j, we suppose that B_j is traversed in the clockwise direction as t goes from a_j to b_j (see figure). Let f be a function analytic in a neighborhood of $U \cup B$. By

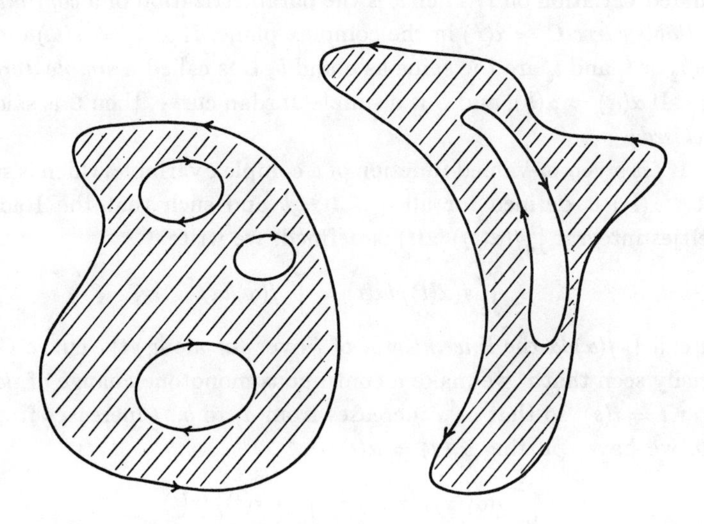

$\int_B f(\alpha)d\alpha$ we understood $\sum_{j=1}^k \int_{B_j} f(\alpha)d\alpha$. Then Cauchy's integral theorem states that

$$\int_B f(\alpha)d\alpha = 0.$$

The validity of this theorem for vector valued functions follows from its validity for complex functions: by Theorem 2.19(c)

$$x^* \int_B f(\alpha)d\alpha = \int_B x^* f(\alpha)d\alpha = 0, \qquad x^* \in \mathfrak{X}^*,$$

and thus, by Corollary II.3.15, $\int_B f(\alpha)d\alpha = 0$.

Just as in the familiar classical case of complex functions, the Cauchy integral theorem may be reworded in a somewhat more general form. To state this other form, let A be a compact set in the complex plane and V a neighborhood of A. Then V actually contains a neighborhood U of A which is bounded by a set B consisting of a finite number of closed Jordan curves. (To see this we may rule the complex plane into a sufficiently fine square mesh, and let U be the union of all open squares of the mesh whose closures intersect A, together with all open segments of the mesh which separate two such squares, and all vertices of the mesh which are common vertices of four such squares.) Let f be analytic in $V-A$. Then the second form of Cauchy's integral theorem states:

If we agree to orient the Jordan curves comprising B in the positive sense customary in the theory of complex variables, then $\int_B f(\alpha)d\alpha$ depends only on the function f and the set A, and is independent of any particular choice of the neighborhood U of A.

In other words, the integrals $\int_B$ and $\int_{B_1}$ are equal provided that B and B_1 bound domains U and U_1 which contain the same set of singularities of f. Here we are using the term singularity for any point where f is not defined and analytic. This statement may be proved by the use of linear functionals in precisely the same manner as in the proof of the first formulation.

The same use of functionals proves the *Cauchy integral formula*:

$$f(z) = \frac{1}{2\pi i}\int_B \frac{f(\alpha)}{\alpha-z}\,d\alpha,$$

where z is a point of the bounded open set U whose boundary B consists of a finite number of closed rectifiable Jordan curves oriented in the positive sense customary in complex variable theory, and f is a vector valued function analytic in a neighborhood of $U \cup B$. The Cauchy integral formula may also be proved directly from the Cauchy integral theorem just as in the case of complex valued analytic functions.

Let $U_1, \ldots, U_n$ be a finite family of bounded open sets, with each U_j having boundary B_j as above, and let f be a vector valued function of the complex variables $z_1, \ldots, z_n$, analytic in a neighborhood of $(U_1 \cup B_1)\times\ldots\times(U_n \cup B_n)$. Then, by applying the Cauchy integral formula to each of the variables $z_1, \ldots, z_n$ in succession, we readily derive the Cauchy integral formula in several variables:

$$f(z_1, \ldots, z_n) = \left(\frac{1}{2\pi i}\right)^n \int_{B_1}\cdots\int_{B_n} \frac{f(\alpha_1, \ldots, \alpha_n)}{(\alpha_1-z_1)\ldots(\alpha_n-z_n)}\,d\alpha_1\ldots d\alpha_n,$$

which is valid for every $[z_1, \ldots, z_n]$ in $U_1\times\ldots\times U_n$.

From this formula it follows readily, as in the classical one variable case, that f has continuous partial derivatives of all orders in

$U_1, \ldots, U_n$. Just as in the classical one variable case we can prove, using this Cauchy integral formula in several variables, the following convergence theorem of Weierstrass:

Let f_n be a uniformly bounded sequence of vector valued functions each defined and analytic on an open set U in the space of the complex variables $z_1, \ldots, z_n$. Let V be a bounded open subset of U whose closure is contained in U. If f_n converges at each point of U to a function f on U then f is analytic in U, and the partial derivatives of f_n of arbitrarily large order converge to the corresponding partial derivatives of f uniformly in V.

From this theorem, we can easily derive the useful fact that if f is analytic in the Cartesian product $U_1 \times \ldots \times U_n$ of a collection of open sets in the complex plane, and C is a continuous rectifiable curve lying wholly in U_n, then the function g defined by

$$g(z_1, \ldots, z_{n-1}) = \int_C f(z_1, \ldots, z_n) dz_n$$

is analytic in $U_1 \times \ldots \times U_{n-1}$.

If f is analytic in a neighborhood of the closure of the bounded open set U which is bounded by a finite collection B of closed disjoint rectifiable Jordan curves oriented in the customary positive sense then it follows from the Cauchy integral formula by differentiating p times that

$$f^{(p)}(z) = \frac{p!}{2\pi i} \int_B \frac{f(\alpha)}{(\alpha - z)^{p+1}} d\alpha, \qquad z \in U.$$

For such a function f the *Taylor expansion*

$$f(z) = \sum_{p=0}^{\infty} \frac{f^p(z_0)}{p!} (z - z_0)^p, \qquad z_0 \in U,$$

is valid and the series converges absolutely and uniformly for z in any closed set of the form $\{z \mid |z - z_0| \leqq r\}$ which is contained in U. This expansion may be proved from the formula for $f^{(p)}(z)$ by the same method that is used for complex functions.

Conversely, any power series

$$f(z) = \sum_{p=0}^{\infty} a_p (z - a_0)^p,$$

defines an analytic function in the open set $|z - z_0| < r$, where r is given by the formula

$$r = (\limsup_{p\to\infty} |a_p|^{1/p})^{-1}.$$

The series converges absolutely and uniformly on any set $|z-z_0| \leqq \alpha$ with $\alpha \leqq r$. Furthermore the series is uniquely defined by f, i.e.,

$$a_p = \frac{f^{(p)}(z_0)}{p!}, \qquad p = 0, 1, 2, \ldots.$$

These facts, as well as the following remarks about Laurent series may all be proved by the standard arguments used for complex functions.

A function f analytic in an annulus $\alpha < |z-z_0| < \beta$ has a unique *Laurent expansion*

$$f(z) = \sum_{p=-\infty}^{\infty} a_p(z-z_0)^p,$$

which converges uniformly and absolutely in every annulus $\alpha+\varepsilon \leqq |z-z_0| \leqq \beta-\varepsilon$ with $\varepsilon > 0$. The coefficients a_p are given by the formulas

$$a_p = \frac{1}{2\pi i}\int_C \frac{f(\alpha)}{(\alpha-z_0)^{p+1}}\, d\alpha, \qquad p = 0, \pm 1, \pm 2, \ldots,$$

where C is any closed rectifiable Jordan curve in the annulus $\alpha < |z-z_0| < \beta$ which separates the circles $|z-z_0| = \alpha$ and $|z-z_0| = \beta$ and which is traversed in the counter-clockwise direction. Conversely, an arbitrary series $\sum_{p=-\infty}^{\infty} a_p(z-z_0)^p$ converges in the annulus $\alpha < |z-z_0| < \beta$ where

$$\alpha = \limsup_{p\to-\infty} |a_p|^{1/p}, \qquad \beta^{-1} = \limsup_{p\to\infty} |a_p|^{1/p}.$$

Its sum f is an analytic function in this annulus and the series is the Laurent expansion of its sum. This annulus is the largest annulus with center z_0 in which an analytic function with the given Laurent expansion can be analytic.

If f is analytic in the (degenerate) annulus $0 < |z-z_0| < r$, but not analytic for $|z-z_0| < r$, then z_0 is said to be an *isolated singularity* of f. The Laurent expansion of $f(z) = \sum_{p=-\infty}^{+\infty} a_p(z-z_0)^p$ which converges in the annulus $0 < |z-z_0| < r$ is called the Laurent expansion of f about z_0. If an infinite number of coefficients a_p with $p < 0$ are non-zero, z_0 is said to be an *essential singularity* of f. If some, but only a finite number of a_p with $p < 0$ are non-zero, z_0 is said to be a *pole* of f.

The largest number n such that $a_{-|n|} \neq 0$ is called the *order* of the pole z_0. If no a_p with $p < 0$ is non-zero, and if we put $f(z_0) = a_0$, then f becomes analytic in $|z-z_0| < r$, so that the singularity at $z = z_0$ is *removable.* If $a_p = 0$ for $p \leqq 0$, z_0 is called a *zero* of f; thus z_0 is a zero of f if $f(z_0) = 0$. If, in this case, $a_p = 0$ for $p < n$ but $a_n \neq 0$, n is called the *order* of the zero z_0.

We recall that a set U in a topological space is said to be *connected* if it is not the union of two nonvoid disjoint sets, both open in the relative topology of U. Another useful criterion for connectivity is available in the space Z of n complex variables, for *a domain U in Z is connected if and only if every pair of its points lies on some simple Jordan curve contained in U.*

Let f be a function analytic on a domain U in the space of n complex variables, and let g be an analytic function defined on a domain V in this space. Then g is said to be an *analytic continuation of f* if V contains U properly, if $g(z_1, \ldots, z_n) = f(z_1, \ldots, z_n)$ for every point $z_1, \ldots, z_n$ in U, and if every point in V can be connected to a point of U by a continuous curve lying in V. If f admits no analytic continuations, U is said to be the *natural domain of existence* of f.

The well known maximum modulus principle is valid for vector valued functions. In particular, the following two forms of this principle will be used in the sequel.

Maximum modulus principle. Let f be an analytic function defined on a connected domain D in the complex plane and having its values in a complex B-space $\mathfrak{X}$. Then $|f(z)|$ does not have its maximum at any point of the domain D, unless $|f(z)|$ is identically constant.

To prove this an indirect argument may be used by assuming that, for some z_0 in D, $|f(z_0)| \geqq |f(z)|$ for every z in D. If C_r is a circle of small radius r with z_0 as center, then

$$f(z_0) = \frac{1}{2\pi i}\int_{C_r} \frac{f(z)dz}{z-z_0} = \frac{1}{2\pi i}\int_0^{2\pi} f(re^{i\theta}+z_0)d\theta.$$

Hence

$$|f(z_0)| \leqq \frac{1}{2\pi}\int_0^{2\pi} |f(re^{i\theta}+z_0)|d\theta,$$

so that

$$\int_0^{2\pi} \{|f(z_0)|-|f(re^{i\theta}+z_0)|\}d\theta \leqq 0.$$

On the other hand, since

$$|f(z_0)|-|f(re^{i\theta}+z_0)| \geqq 0,$$

we have $|f(z_0)| = |f(re^{i\theta}+z_0)|$ for almost all θ. Since f is continuous, this holds for all θ, which shows that $|f(z)| = |f(z_0)|$ for all z sufficiently close to z_0. The proof thus far shows that the set

$$\{z||f(z)| = |f(z_0)|\}$$

is open. Since this set is clearly closed and since D is connected, $|f(z)| = |f(z_0)|$ for all z in D.

Maximum modulus principle for a strip. Let $f(x+iy) = f(z)$ be an analytic function with values in a complex B-space $\mathfrak{X}$, defined and uniformly bounded on a strip $x_0 \leqq x \leqq x_1$, $-\infty < y < +\infty$. Suppose that

$$|f(x_0+iy)| \leqq M, \qquad |f(x_1+iy)| \leqq M.$$

Then $|f(x+iy)| \leqq M$ for $x_0 \leqq x \leqq x_1$.

To prove this we may, without loss of generality, suppose that $x_0 \geqq 1$. Then for each $\varepsilon > 0$, the function $z^{-\varepsilon}f(z)$ is analytic and uniformly bounded in the strip, tends to zero as $y \to \pm\infty$, and is bounded by M on the edges of the strip. Hence $|z^{-\varepsilon}f(z)|$ assumes its maximum value somewhere in the strip. By the maximum modulus principle, this maximum must be on one of the edges, and consequently $|z^{-\varepsilon}f(z)| \leqq M$ in the whole strip. As ε approaches zero, it is seen that $|f(z)| \leqq M$ everywhere in the strip.

If f is analytic in a connected open set U of the complex plane and not identically zero, U contains no point which is the limit of zeros of f. This fact can be concluded from the corresponding fact for complex functions as follows: if z_1 is the limit of zeros of f, then for each $x^* \in \mathfrak{X}^*$, z_1 is the limit of zeros of x^*f. Hence $x^*f = 0$ for each x^*, and by II.3.15, $f = 0$.

A function f defined and analytic in the whole complex plane is said to be *entire*. Liouville's theorem states that *a bounded entire function is constant.* To prove this, let f be bounded and entire, and define g by $g(z) = f(z)-f(0)$. Then g is bounded and entire, and $g(0) = 0$. For each $x^* \in \mathfrak{X}^*$, x^*g is bounded and entire, and $x^*g(0) = 0$. Hence by Liouville's theorem for the complex valued case, $x^*g = 0$, and by II.3.15, $g = 0$. Thus $f(z) = f(0)$, and f is constant.

Finally we will have need for the *Weierstrass preparation theorem* which we quote here for the reader's convenience in the form in which we will use it.

THEOREM. *Let $f(z, w)$ be complex valued and analytic in the two complex variables z, w for z in an open set U, and w in a neighborhood $|w| < \delta_1$ of the origin. Suppose that $f(z, 0)$ is not identically zero but has a zero at some point z_0 in U, and that the order of this zero is m. Then there exists a neighborhood V of z_0, a positive $\delta < \delta_1$, an integer $k \leqq m$ and an integer n, such that for each $|w| < \delta$, $w \neq 0$, $f(z, w)$ has exactly k distinct zeros $z_1(w), \ldots, z_k(w)$ in V, and these zeros are given by fractional power series*

$$z_j(w) = \sum_{p=0}^{\infty} a_{jp} w^{p/n}, \qquad j = 1, \ldots, k$$

i.e., by power series in $w' = w^{1/n}$.

15. Notes and Remarks

There are a number of expositions of the Lebesgue theory of integration of scalar valued functions with respect to a scalar measure. The treatises of Lebesgue [2] and Carathéodory [1, 2] are classical; more recent treatments are found in books on the theory of real variables and particularly in the treatises of Bourbaki [4], Hahn and Rosenthal [1], Halmos [5], McShane [2, 3], Munroe [2] and Saks [1]. Since the works of Hahn and Rosenthal [1] and Saks [1] contain excellent historical notes and further references, we shall content ourselves with a brief discussion of features of the treatment given here that differ from the standard approaches.

Vector integration. In this chapter we have been concerned with the integration of vector valued functions with respect to a scalar valued measure. The possibility of extending the integral to this realm was recognized by L. M. Graves [3], who discussed and applied the Riemann integral. A theory of the Lebesgue type was constructed by Bochner [2]. The Cauchy sequence method of the text was used by Dunford [4] to obtain an integral equivalent to the Bochner integral. Somewhat more general integrals have been obtained for B-space valued functions by Birkhoff [4], Gelfand [2], Pettis [4] and Price [1].

Integrals for functions with values in a locally convex topological space were obtained by Phillips [7] and Rickart [1]. All of these integrals are countably additive integrals. In Section IV.10 we will present a Lebesgue type theory of integration of a scalar function with respect to a countably additive vector valued measure. Instances in which both the function and the measure are vector valued have also been considered by Bochner and Taylor [1; pp. 915–917] and Gowurin [1] who gave Riemann type integrals, and by Day [9], Price [1], Rickart [1] and Bartle [3] who discussed Lebesgue integrals. For an excellent expository account of these integrals and their relations the reader should refer to Hildebrandt [4]. To be added to the references cited there are the works of Christian [1] and McShane [3] dealing with integrals defined by means of order; Birkhoff [6], Masani [1], Maslow [1] and Stewart [1] concerning multiplicative (as contrasted with additive) integration; and Monna [6] which discusses the integral of functions with range in a field with a non-Archimedean valuation.

A decomposition of a countably additive vector valued measure along the lines of the Lebesgue decomposition was obtained by Rickart [3] and Nakamura and Sunouchi [1].

Finitely additive set functions. The possibility of integrating bounded functions with respect to a finitely additive measure was demonstrated by Hildebrandt [3] and Fichtenholz and Kantorovitch [1]. Such an integral has been used by other authors, usually for bounded functions. Recently, Leader [1] has developed a theory of L_p spaces of finitely additive measures. He actually discusses the indefinite integrals of functions in L_p, which avoids certain limiting devices.

Alexandroff [1] has given an extensive account of the theory of bounded regular finitely additive measures on a "normal" topological space. In particular Theorem 5.13 is due to him (Alexandroff [1; p. 590]). Alexandroff [1; II. p. 618] has given conditions under which a bounded regular finitely additive measure can be decomposed into the sum of a countably additive and a finitely additive part. His decomposition differs from that presented in Theorem 7.8, which is due to Yosida and Hewitt [1; p. 52].

The Vitali-Hahn-Saks theorem. Fréchet [8] introduced a metric in the space of measurable functions on [0, 1] such that convergence with respect to the metric was equivalent to convergence in measure. When applied to characteristic functions, this yields the space $\Sigma(\mu)$ studied in Section 7. This metric space was specifically studied by Aronszajn and Nikodým [7, 8]; it is an important and powerful device.

Vitali [2; p. 147] showed that if $\{f_n\}$ is a sequence of Lebesgue integrable functions on [0, 1] which converge almost everywhere to f then

$$\int_0^1 f(s)ds, \qquad \lim_{n\to\infty} \int_0^1 f_n(s)ds$$

exist and are equal if and only if the indefinite integrals of the f_n are continuous uniformly with respect to Lebesgue measure. (This is essentially a special case of Theorem 6.15.) Hahn [2] proved that if $\{f_n\}$ is a sequence of Lebesgue integrable functions on [0, 1] and if $\lim \int_E f_n(s)ds$ exists for every measurable set E, then the indefinite integrals are continuous uniformly in n, and converge to a set function which is continuous with respect to Lebesgue measure. Another proof of this theorem was given by Banach [6; p. 152]. The important Theorem 7.2 is a generalization of this theorem and is due to Saks [3] in the case of scalar measures, although his proof is perfectly general. Actually, Saks proved that the theorem holds under somewhat weaker hypotheses.

Phillips [7; p. 125] and Rickart [1; p. 502] have observed that essentially the same proof is valid when the values of the indefinite integrals or the measures lie in a locally convex topological linear space. Other extensions have been given by Alexiewicz [1; I. pp. 15—20]. See also G. Sunouchi [1]. For other related results see Saks [2], Saks and Tamarkin [1], and Hahn and Rosenthal [1; pp. 56—60].

Corollary 7.3 in the case of scalar measures is due to Nikodým [6] who proved it previous to the publication of Theorem 7.2.

The Radon-Nikodým theorem. In 1904, Lebesgue [2; p. 129] gave a necessary and sufficient condition for a function on [0, 1] to be expressed as an indefinite integral. The next year, Vitali [1] characterized such functions as the now familiar absolutely continuous functions. These results were extended by Radon [2; p. 1349] for a Borel measure μ defined in Euclidean space. The general theorem is due to Nikodým [7], [8; p. 168]. Other proofs have been given, e.g., Yosida [2]—see also Hahn and Rosenthal [1; p. 171] for additional references.

An extension of the Radon-Nikodým theorem for finitely additive measures has been given by Bochner [3] and Bochner and Phillips [1]. This theorem will be given in IV.9.14.

Generalizations of the Radon-Nikodým theorem to the case of vector valued measures are discussed in Section IV.8 and additional remarks are found in Section IV.12.

Product measures. For a discussion of measures in finite products see Hahn and Rosenthal [1; Section 8] and Saks [1; Chap. 3], where many references are cited. Halmos [5; Chap. 7] also discusses infinite product measures. All of these treatments deal with scalar valued functions.

Jessen [2] was the first to extend the Fubini theorem to the case of infinitely many factors. The same problem, without the use of topology, was treated by von Neumann [4]. Theorems 11.24 and 11.27 were given originally by Jessen [1] and in the form stated here by Dunford and Tamarkin [1]. For other remarks on infinite product measures, see Dieudonné [12], Kakutani [14], and Sparre Andersen and Jessen [1, 2].

Differentiation. We refer the reader to Hahn and Rosenthal [1; Chap. V] and Saks [1; Chap. IV] for references concerning the theory of differentiation of scalar valued functions.

Differentiation in B-spaces. In this chapter the theory of the integral of a vector valued function has been considered at some length. There is also rather extensive literature on the theory of differentiation of a function on a linear interval to a B-space. We shall not discuss these results, but refer the reader to the following papers: Alaoglu [1], Alexiewicz [3], Alexiewicz and Orlicz [3], Birkhoff [4], Bochner [5], Bochner and Taylor [1], Clarkson [1], Dunford and Morse [1], Gelfand [2], L. M. Graves [3], Izumi [4], Munroe [3, 4], Pettis [1, 4, 7], Phillips [7], Sebastião e Silva [1].

CHAPTER IV

Special Spaces

1. Introduction

For concrete applications to analysis it is desirable to supplement the general theory of the text with a detailed investigation into the properties of special spaces. What, for example, is the analytical form of the general functional in the space conjugate to the Lebesgue space $L_1 = L_1(S, \Sigma, \mu)$? When does a sequence in L_1 converge weakly? What sets in L_1 are compact, or weakly compact? The answers to such questions, and similar ones for the spaces commonly encountered in mathematical analysis, will enhance the applicability of the general theory. This chapter is devoted to a systematic study of such concrete questions.

In Section 2 will be found a list of special spaces, mostly B-spaces, which have frequent occurrence in analysis. Each of these spaces is studied in an attempt to solve the eight problems listed below. The results of these studies are presented in tabular form in Section 14.

1 *Problem.* What is an analytical representation of the conjugate space $\mathfrak{X}^*$ of the given space $\mathfrak{X}$?

2 *Problem.* When does a sequence $\{x_n\}$ in $\mathfrak{X}$ have the property that the lim x^*x_n exists for each x^* in $\mathfrak{X}^*$?

3 *Problem.* When does a sequence converge weakly to a specified limit?

4 *Problem.* Is $\mathfrak{X}$ weakly complete?

5 *Problem.* Is $\mathfrak{X}$ reflexive?

6 *Problem.* Which subsets of $\mathfrak{X}$ are weakly sequentially compact?

7 *Problem.* Which subsets of $\mathfrak{X}$ are compact in the metric topology of $\mathfrak{X}$?

8 *Problem.* If $\{x_n\}$ is a sequence in $\mathfrak{X}$ and if $\mathfrak{X}$ is the conjugate of $\mathfrak{Y}$, when is the sequence $\{x_n\}$ $\mathfrak{Y}$-convergent in the sense that the limit $\lim_n x_n y$ exists for each y in $\mathfrak{Y}$?

During the course of the investigation a number of interesting special properties of the spaces to be considered and certain relations between them will appear. These will also be incorporated in the tabulation of Section 15.

2. A List of Special Spaces

We give below a list of various special B- and F-spaces. With the one exception of Hilbert space, each of them will consist of real or complex valued functions f, g defined on a specified domain S. Here, addition and multiplication are understood to be defined in the natural way, i.e., by the equations

$$(f+g)(s) = f(s)+g(s) \qquad (\alpha f)(s) = \alpha f(s).$$

Thus our zero vector will be the function which is identically zero. In the list below, we will not ordinarily specify whether the spaces are to consist of all real valued functions with the stated properties, or of all complex valued functions with these properties. Unless the contrary is specified, both possibilities are admitted. The first leads to a real B- or F-space, the second to a complex B- or F-space. Thus, for example, if S is a topological space, $C(S)$ can denote either the real B-space of all real valued bounded continuous functions on S or the complex B-space of all complex valued bounded continuous functions on S. We will only make a definite choice of the real or of the complex numbers when the real and the complex B-spaces under consideration actually require separate treatment.

The proof that a given space satisfies the required axioms will usually be found in the section where the space in question is discussed. The more elementary of such proofs are sometimes left as exercises for the reader.

1 The space E^n is the linear space of ordered n-tuples $x = [\alpha_1, \ldots, \alpha_n]$ of scalars $\alpha_1, \ldots, \alpha_n$. The norm is $|x| = (|\alpha_1|^2+\ldots+|\alpha_n|^2)^{1/2}$. If the field of scalars is the real number system E^n is called *n-dimensional Euclidean space*; if the field is the field of complex numbers E^n is called *n-dimensional unitary space*, or *n-dimensional Hilbert space.*

2 The space l_p^n is defined for positive integers n and real num-

bers p with $1 \leqq p < \infty$. It consists of ordered n-tuples $x = [\alpha_1, \ldots, \alpha_n]$ of scalars $\alpha_1, \ldots, \alpha_n$ and has the norm

$$|x| = \{\sum_{i=1}^{n} |\alpha_i|^p\}^{1/p}.$$

3. The space l_∞^n is the linear space of all ordered n-tuples $x = [\alpha_1, \ldots, \alpha_n]$ of scalars $\alpha_1, \ldots, \alpha_n$ with the norm

$$|x| = \sup_{1 \leqq i \leqq n} |\alpha_i|.$$

4. The space l_p is defined for $1 \leqq p < \infty$ as the linear space of all sequences $x = \{\alpha_n\}$ of scalars for which the norm

$$|x| = \{\sum_{n=1}^{\infty} |\alpha_n|^p\}^{1/p}$$

is finite.

5. The space l_∞ is the linear space of all bounded sequences $x = \{\alpha_n\}$ of scalars. The norm is given by the equation

$$|x| = \sup_n |\alpha_n|.$$

6. The space c is the linear space of all convergent sequences $x = \{\alpha_n\}$ of scalars. The norm is

$$|x| = \sup_n |\alpha_n|.$$

7. The space c_0 is the linear space of all sequences $x = \{\alpha_n\}$ converging to zero. The norm is

$$|x| = \sup_n |\alpha_n|.$$

8. The space bv is the linear space of all sequences $x = \{\alpha_n\}$ of scalars for which the norm

$$|x| = |\alpha_1| + \sum_{n=1}^{\infty} |\alpha_{n+1} - \alpha_n|$$

is finite.

9. The space bv_0 is the linear space of all sequences $x = \{\alpha_n\}$ of scalars with $\lim_n \alpha_n = 0$ and for which the norm

$$|x| = \sum_{n=1}^{\infty} |\alpha_{n+1} - \alpha_n|$$

is finite.

10. The space *bs* is the linear space of all sequences $x = \{\alpha_n\}$ of scalars for which the norm

$$|x| = \sup_n |\sum_{i=1}^n \alpha_i|$$

is finite.

11. The space *cs* is the linear space of all sequences $x = \{\alpha_n\}$ for which the series $\sum_{n=1}^\infty \alpha_n$ is convergent. The norm is

$$|x| = \sup_n |\sum_{i=1}^n \alpha_i|.$$

12. Let S be an arbitrary set, and Σ a field of subsets of S. The space $B(S, \Sigma)$ consists of all uniform limits of finite linear combinations of characteristic functions of sets in Σ. The norm in $B(S, \Sigma)$ is given by the formula

$$|f| = \sup_{s \in S} |f(s)|.$$

A scalar function f on S is *Σ-measurable* if $f^{-1}(A) \in \Sigma$ for every Borel set A in the range of f. It is clear that every bounded Σ-measurable function is in $B(S, \Sigma)$ and that such functions are dense in $B(S, \Sigma)$. It is evident that if we define the set function μ on Σ by placing $\mu(E) = \infty$ if $E \neq \phi$ and $\mu(\phi) = 0$, then a bounded function is Σ-measurable if and only if it is totally μ-measurable.

13. The space $B(S)$ is defined for an arbitrary set S and consists of all bounded scalar functions on S. The norm is given by

$$|f| = \sup_{s \in S} |f(s)|.$$

14. The space $C(S)$ is defined for a topological space S and consists of all bounded continuous scalar functions on S. The norm is

$$|f| = \sup_{s \in S} |f(s)|.$$

15. The space $ba(S, \Sigma)$ is defined for a field Σ of subsets of a set S and consists of all bounded additive scalar functions defined on Σ. The norm $|\mu|$ is the total variation of μ on S, i.e., $|\mu| = v(\mu, S)$.

16. The space $ca(S, \Sigma)$ is defined for a σ-field Σ of subsets of a set S and consists of all scalar functions which are defined and countably additive on Σ. The norm $|\mu|$ is the total variation $v(\mu, S)$.

17. The space $rca(S)$ is defined for a topological space S and consists of all regular countably additive scalar valued set functions

defined on the σ-field $\mathscr{B}$ of all Borel sets in S. The norm $|\mu|$ is the total variation $v(\mu, S)$.

18. The space $L_p(S, \Sigma, \mu)$ is defined for any real number p, $1 \leqq p < \infty$, and any positive measure space (S, Σ, μ). It consists of those μ-measurable scalar functions f on S for which the norm

$$|f| = \left\{\int_S |f(s)|^p \mu(ds)\right\}^{1/p}$$

is finite.

Remark. In Chapter III the space $L_p(S, \Sigma, \mu)$ was defined without the assumption that μ is non-negative. The space $L_p(S, \Sigma, \mu)$ is, in any case, the same as the space $L_p(S, \Sigma, v(\mu))$ and the variation $v(\mu)$ is non-negative. As indicated by the remarks after Corollary III.2.5 and Theorem III.3.5, the elements of $L_p(S, \Sigma, \mu)$ are not actually functions but equivalence classes of functions, two functions being equivalent if their difference is a μ-null function, (cf. III.6.8), i.e., if they are equal for μ-almost all $s \in S$. The same remarks apply to the spaces $L_\infty(S, \Sigma, \mu)$ and $TM(S, \Sigma, \mu)$ introduced below.

19. The space $L_\infty(S, \Sigma, \mu)$ is defined for a positive measure space (S, Σ, μ) and consists of all μ-essentially bounded μ-measurable scalar functions. (cf. Definition III.1.11) The norm is

$$|f| = \mu\text{-ess}\sup_{s \in S} |f(s)|.$$

Remark. In defining the following four spaces the term *interval* is used for a set of real numbers having any one of the following forms: $[a, b]$, $[a, b)$, $(a, b]$, or (a, b), where a is either real or $-\infty$ and b is either real or $+\infty$. In all four cases a is called the left *end point* of the interval.

20. The space $BV(I)$ is defined for an interval I and consists of all scalar functions on I which are of bounded variation (cf. III.5.15). If a is the left end point of I then

$$|f| = |f(a+)| + v(f, I),$$

$v(f, I)$ denoting, as usual, the total variation of f in I. (It has been shown in III.6.21 that the limit $f(a+)$ exists for every $f \in BV(I)$.)

21. The space $NBV(I)$ is defined for an interval I and consists of those functions f in $BV(I)$ which are normalized by the requirements that (1) f is continuous on the right at each interior point of I and (2) $f(a+) = 0$ where a is the left end point of I. The norm is given by the equation

$$|f| = v(f, I).$$

22. A function $f \in BV(I)$ is said to be *absolutely continuous* if for each $\varepsilon > 0$ there exists a $\delta > 0$ such that

$$\sum_{i=1}^{n} |f(b_i) - f(a_i)| < \varepsilon$$

whenever (a_i, b_i), $i = 1, 2, \ldots, n$, are non-overlapping subintervals of I with $\sum_{i=1}^{n} |b_i - a_i| < \delta$. The space $AC(I)$ is defined for an interval I and consists of all absolutely continuous functions on I. If a is the left end point of I the norm is

$$|f| = |f(a+)| + v(f, I).$$

23. The space $C^n(I)$ is defined for a compact interval I and a positive integer n as the family of those scalar functions on I having n bounded continuous derivatives. The norm is

$$|f| = \sum_{i=0}^{n} \sup_{s \in I} |f^{(i)}(s)|.$$

24. The space $A(D)$ is defined, for an open set D of complex numbers, as the family of those complex functions which are bounded and continuous on the closure of D and which are analytic on D. The norm is

$$|z| = \sup_{z \in D} |f(z)|.$$

The space $A(D)$ is a complex linear space with no obvious real analogue.

25. A function f of the real variable t is said to be *almost periodic* if for each $\varepsilon > 0$ there exists an $L > 0$ such that every interval of the real axis of length at least L contains some point x such that $|f(t) - f(t+x)| < \varepsilon$ for $-\infty < t < +\infty$. The space AP is the linear space of all continuous almost periodic functions of a real variable. The norm is

$$|f| = \sup_{-\infty < t < \infty} |f(t)|.$$

It will be shown in Section 7 that every almost periodic function is bounded.

26. *Hilbert space* is a linear vector space $\mathfrak{H}$ over the field Φ of complex numbers, together with a complex function $(\cdot, \cdot)$ defined on $\mathfrak{H} \times \mathfrak{H}$ with the following properties:

(i) $(x, x) = 0$ if and only if $x = 0$;

(ii) $(x, x) \geqq 0, \qquad x \in \mathfrak{H};$

(iii) $(x+y, z) = (x, z)+(y, z), \qquad x, y, z \in \mathfrak{H};$

(iv) $(\alpha x, y) = \alpha(x, y), \qquad \alpha \in \Phi, \qquad x, y \in \mathfrak{H};$

(v) $(x, y) = \overline{(y, x)};$

(vi) If $x_n \in \mathfrak{H}$, $n = 1, 2, \ldots$, and if $\lim_{n, m\to\infty} (x_n-x_m, x_n-x_m) = 0$, then there is an x in $\mathfrak{H}$ with $\lim_n (x_n-x, x_n-x) = 0$.

The function $(\cdot, \cdot)$ is called the *scalar* or *inner* product in $\mathfrak{H}$ and (x, y) is called the *scalar* or *inner product of x and y*. The *norm* in $\mathfrak{H}$ is $|x| = (x, x)^{1/2}$.

Remark. Hilbert space has been defined by a set of abstract axioms. It is noteworthy that some of the concrete spaces defined above satisfy these axioms, and hence are special cases of abstract Hilbert space. Thus, for instance, the n-dimensional unitary space E^n is a Hilbert space, if the inner product (x, y) of two elements $x = [\alpha_1, \ldots, \alpha_n]$ and $y = [\beta_1, \ldots, \beta_n]$ in E^n is defined by the formula

$$(x, y) = \sum_{i=1}^{n} \alpha_i \bar{\beta}_i.$$

In the same way, complex l_2 is a Hilbert space if the scalar product (x, y) of the vectors $x = \{\alpha_n\}$, $y = \{\beta_n\}$ is defined by the formula

$$(x, y) = \sum_{n=1}^{\infty} \alpha_n \bar{\beta}_n,$$

Also the complex space $L_2(S, \Sigma, \mu)$ is a Hilbert space with the scalar product

$$(f, g) = \int_S f(s)\overline{g(s)}\mu(ds) \qquad f, g \in L_2(S, \Sigma, \mu).$$

The final spaces on the list are F-spaces but not B-spaces.

27. The space $TM(S, \Sigma, \mu)$ is defined for a positive measure space (S, Σ, μ), and consists of all totally measurable functions f defined on S (cf. III.2.10). The metric function in $TM(S, \Sigma, \mu)$ is $\varrho(f, g) = |f-g|$, where

$$|f| = \inf_{\alpha>0} \{\alpha + \arctan \mu(S(|f| > \alpha))\},$$

$$S(|f| > \alpha) = \{s | s \in S, |f(s)| > \alpha\}$$

28. The space s is the space of all sequences $x = \{\alpha_n\}$ of scalars. The metric function in s is $\varrho(x, y) = |x-y|$ where

$$|x| = \sum_{n=1}^{\infty} \frac{1}{2^n} \frac{|\alpha_n|}{1+|\alpha_n|}.$$

3. Finite Dimensional Spaces

The space E^n, as will be seen presently, is the prototype of all n-dimensional normed linear spaces, and hence it should be observed first that E^n is a B-space. The Minkowski inequality (III.3.3) shows E^n to be a normed linear space. If $y = [\alpha_1, \ldots, \alpha_n] \in E^n$, then $|\alpha_i| \leqq |y|$ and hence the completeness of E^n follows from that of the field Φ of scalars. Thus E^n is a B-space. A bounded closed set in E^n is compact, for if $y^m = [\alpha_1^m, \ldots, \alpha_n^m]$ is bounded, the sequence $\{\alpha_i^m, m = 1, 2, \ldots\}$ is bounded in Φ, and consequently there is a subsequence $\{m_j\}$ of $\{m\}$ for which the limits $\lim_j \alpha_i^{m_j} = \alpha_i$, $i = 1, \ldots, n$, exist. Hence the sequence $\{y^{m_j}\}$ converges to the vector $[\alpha_1, \ldots, \alpha_n]$ in E^n.

1 Lemma. *A finite dimensional normed linear space is complete and hence it is a B-space.*

Proof. Let $\{b_1, \ldots, b_n\}$ be a Hamel basis for the finite dimensional normed linear space $\mathfrak{X}$. For each point $y = [\alpha_1, \ldots, \alpha_n]$ in E^n, let $Ty = x$ where $x = \sum_{i=1}^n \alpha_i b_i$. Then T is a one-to-one continuous map of E^n onto $\mathfrak{X}$. To see that $\mathfrak{X}$ is complete, it will first be shown that the inverse T^{-1} is continuous. If it is not continuous, there is a sequence $\{x^n\}$ in $\mathfrak{X}$ and an $\varepsilon > 0$ with $x^n \to 0$ and $|T^{-1}x^n| \geqq \varepsilon$, $n = 1, 2, \ldots$. If $y^n = (T^{-1}x^n)/|T^{-1}x^n|$ then $|y^n| = 1$ and $y^n = T^{-1}z^n$ where $z^n \to 0$. Since the sequence $\{y^n\}$ is in the compact set $\{y | y \in E^n, |y| = 1\}$ in E^n, there is a subsequence of $\{y^n\}$ converging to a point $y^0 = [\alpha_1^0, \ldots, \alpha_n^0]$ with $|y^0| = 1$. Since T is continuous, $0 = Ty^0 = \alpha_1^0 b_1 + \ldots + \alpha_n^0 b_n$, and, since the vectors $b_1, \ldots, b_n$ are linearly independent, $\alpha_1^0 = \ldots = \alpha_n^0 = 0$. This contradicts the fact that $|y^0| = 1$, and proves that T^{-1} is continuous. Thus, by Lemma II.3.4, there is a constant M with $|T^{-1}x| \leqq M|x|$, which shows that a Cauchy sequence $\{x^n\}$ in $\mathfrak{X}$ maps into a Cauchy sequence $y^n = T^{-1}x^n$ in E^n. If $y = \lim y^n$ then $x^n = Ty^n$ converges to Ty in $\mathfrak{X}$ and $\mathfrak{X}$ is complete. Q.E.D.

2 Corollary. *A finite dimensional linear manifold in a B-space is closed.*

Proof. This follows from the preceding lemma and Lemma I.6.7. Q.E.D.

3 Corollary. *An n-dimensional B-space is equivalent to* E^n.

4 Corollary. *Every linear operator on a finite dimensional normed linear space is continuous.*

Proof. Let $\{b_1, \ldots, b_n\}$ be a Hamel basis for the finite dimensional normed linear space $\mathfrak{X}$ so that every x in $\mathfrak{X}$ has a unique representation in the form $x = \alpha_1 b_1 + \ldots + \alpha_n b_n$. It was shown in the proof of Lemma 1 that α_i, $i = 1, \ldots, n$, depends continuously upon x. Thus if U is a linear operator on $\mathfrak{X}$ it is seen that $Ux = \alpha_1 Ub_1 + \ldots + \alpha_n Ub_n$ is also continuous in x. Q.E.D.

5 Theorem. *A normed linear space is finite dimensional if and only if its closed unit sphere is compact.*

Proof. Suppose that the sphere $S = \{x||x| \leqq 1\}$ in the normed linear space $\mathfrak{X}$ is compact. It will be shown that there is a finite set $\{x_1^*, \ldots, x_n^*\}$ in $\mathfrak{X}^*$ such that if $x_i^*(x) = 0$, $1 \leqq i \leqq n$, then $x = 0$. Suppose the contrary. Then for each finite set $A \subseteqq \mathfrak{X}^*$, the set

$$S(A) = \{x||x| = 1, x^*(x) = 0 \text{ for } x^* \in A\}$$

is a non-void closed subset of S. From our assumption it follows that any finite number of the sets $S(A)$ has a non-void intersection. Thus, by I.5.6, there must exist an x common to all $S(A)$. This vector x has the properties: $|x| = 1$, and $x^*(x) = 0$ for all x^*in $\mathfrak{X}^*$. This contradicts II.3.15 and proves the assertion. Let x_i^*, $i = 1, \ldots, n$, have the property that $x_i^* x = 0$, $i = 1, \ldots, n$, implies $x = 0$. Then the map $x \rightarrow [x_1^* x, \ldots, x_n^* x]$ of $\mathfrak{X}$ into E^n is linear, one-to-one, and has finite dimensional range. Thus $\mathfrak{X}$ has finite dimension.

The converse statement follows from Corollary 3. Q.E.D.

6 Lemma. *A normed linear space has finite dimension n if and only if its adjoint has dimension n.*

Proof. Let $\{b_1, \ldots, b_n\}$ be a Hamel basis for the normed linear space $\mathfrak{X}$. Then the functionals b_i^*, $i = 1, \ldots, n$, defined by

$$x = \sum_{i=1}^{n} b_i^*(x) b_i, \qquad x \in \mathfrak{X},$$

are, by 4, in $\mathfrak{X}^*$. Also, for any x^* in $\mathfrak{X}^*$ we have

$$x^* x = \sum_{i=1}^{n} b_i^*(x) x^*(b_i), \qquad x^* = \sum_{i=1}^{n} b_i^* x^*(b_i),$$

so that $sp\{b_i^*\} = \mathfrak{X}^*$. Now the vectors b_i^*, $i = 1, \ldots, n$ are linearly independent, for if $\sum_{i=1}^n \beta_i b_i^* = 0$ then $\beta_j = (\Sigma_{i=1}^n \beta_i b_i^*) b_j = 0$, $j = 1, \ldots, n$. Thus the dimension of $\mathfrak{X}^*$ is n. Conversely, let the dimension of $\mathfrak{X}^*$ be finite. Then the dimension of $\mathfrak{X}^{**}$ is finite, and, since $\mathfrak{X}$ is equivalent to a subspace of $\mathfrak{X}^{**}$ (II.3.19), the dimension of $\mathfrak{X}$ is finite. Hence, from the first part of this proof, $\mathfrak{X}$ and $\mathfrak{X}^*$ have the same dimension. Q.E.D.

In the preceding lemma it is assumed that the spaces are normed linear spaces, for there are examples (see the preliminary discussion in Section IV.11) of infinite dimensional F-spaces with zero dimensional conjugate spaces.

The following corollary was established during the first part of the preceding proof.

7 Corollary. *If* $\{b_1, \ldots, b_n\}$ *is a Hamel basis for the normed linear space* $\mathfrak{X}$ *then the functionals* b_i^*, $i = 1, \ldots, n$, *defined by the equations*

$$x = \sum_{i=1}^{n} b_i^*(x) b_i, \qquad x \in \mathfrak{X},$$

form a Hamel basis for $\mathfrak{X}^*$.

8 Corollary. *A finite dimensional normed linear space is reflexive.*

Proof. In the notation of Corollary 7,

$$x^*x = \sum_{i=1}^{n} b_i^*(x) x^*(b_i), \qquad x^* = \sum_{i=1}^{n} b_i^* x^*(b_i),$$

and thus for x^{**} in $\mathfrak{X}^{**}$ we have $x^{**}x^* = x^*x$, where $x = \sum_{i=1}^n x^{**}(b_i^*) b_i$. Q.E.D.

From Corollary 7 it follows that weak and strong convergence are the same in finite dimensional spaces and from Corollary 3 that a set is compact if and only if it is bounded and closed. Thus the Problems 2, . . ., 6 listed in Section 1 are readily solved for finite dimensional spaces.

According to Corollary 3 and Lemma 6 every finite dimensional normed linear space is equivalent to its conjugate, but this observation does not completely solve the problem of representing the conjugate of a given space. What is needed is an expression for the norm of the functional in terms of the scalars representing it. Thus, to complete the solutions of the problems listed in Section 1, it will be necessary to represent the conjugate spaces of E^n, l_p^n, and l_∞^n. Since $E_n = l_2^n$ the following theorem gives the desired results:

9 **Theorem.** *If* $1 \leqq p \leqq \infty$ *and* $p^{-1}+q^{-1} = 1$, *then the mapping* $x^* \longleftrightarrow [\alpha_1, \ldots, \alpha_n]$ *determined by the equation*

$$x^*x = \sum_{i=1}^{n} \alpha_i\beta_i, \qquad x = \{\beta_i\} \in l_p^n,$$

is an isometric isomorphism of $(l_p^n)^*$ *onto* l_q^n.

Proof. It is clear that the mapping is an isomorphism. To see that it is also an isometric map we suppose first that $1 < p < \infty$. Then, from Hölder's inequality (III.3.2),

$$|x^*x| \leqq \{ \sum_{i=1}^{n} |\alpha_i|^q\}^{1/q}\{ \sum_{i=1}^{n} |\beta_i|^p\}^{1/p},$$

which shows that $|x^*| \leqq \{\sum_{i=1}^{n} |\alpha_i|^q\}^{1/q}$. Now let $\beta_i = |\alpha_i|^q/\alpha_i$ if $\alpha_i \neq 0$ and $\beta_i = 0$ otherwise. Then

$$x^*x = \sum_{i=1}^{n} |\alpha_i|^q, \qquad |x| = \{\sum_{i=1}^{n} |\alpha_i|^{(q-1)p}\}^{1/p}$$

$$= \{\sum_{i=1}^{n} |\alpha_i|^q\}^{1/p},$$

which, together with the inequality $|x^*x| \leqq |x^*||x|$, gives $\{\sum_{i=1}^{n} |\alpha_i|^q\}^{1/q} \leqq |x^*|$. Thus $|x^*| = \{\sum_{i=1}^{n} |\alpha_i|^q\}^{1/q}$, which shows that the mapping is isometric. These steps, with obvious changes in notation, may be used to establish the isometry when $p = 1$ or ∞. Q.E.D.

4. Hilbert Space

Of the infinite dimensional B-spaces, Hilbert space is the most closely related, especially in its elementary geometrical aspects, to the Euclidean or finite dimensional unitary spaces. It is not immediate from the definition (2.26) that a Hilbert space is a B-space, but this fact is established in the following theorem. Throughout this discussion of Hilbert space the conditions (1), . . ., (vi) in Definition 2.26 will be used without explicit reference and the symbol $\mathfrak{H}$ will always be used for a Hilbert space.

1 THEOREM. *A Hilbert space $\mathfrak{H}$ is a complex B-space and*

$$|(x, y)| \leqq |x||y|, \qquad x, y \in \mathfrak{H}.$$

PROOF. The above inequality, known as the *Schwarz inequality*, will be proved first. It follows from the postulates for $\mathfrak{H}$ that the Schwarz inequality is valid if either x or y is zero. Hence suppose that $x \neq 0 \neq y$. For an arbitrary complex number α

$$\begin{aligned} 0 &\leqq (x+\alpha y, x+\alpha y) \\ &= |x|^2+|\alpha|^2|y|^2+\alpha(y, x)+\bar{\alpha}(x, y) \\ &= |x|^2+|\alpha|^2|y|^2+2\mathscr{R}(\alpha(y, x)), \end{aligned}$$

where the symbol $\mathscr{R}(\lambda)$ is used for the real part of λ. If $\alpha = re^{i\theta}$ and if θ is chosen properly, it follows from the above inequality that

$$|x|^2+r^2|y|^2 \geqq 2r|(x, y)|$$

for every positive r. Upon placing $r = |x|/|y|$ the Schwarz inequality follows.

To complete the proof of the theorem it will suffice to show that $|x+y| \leqq |x|+|y|$. First note that

$$(x, y)+(y, x) = 2\mathscr{R}(x, y) \leqq 2|x||y|$$

and hence that

$$\begin{aligned} |x+y|^2 &= |x|^2+|y|^2+(x, y)+(y, x) \\ &\leqq |x|^2+|y|^2+2|x||y| \\ &= (|x|+|y|)^2. \qquad \text{Q.E.D.} \end{aligned}$$

Remark. It should be noted that the above proofs of the Schwarz inequality and the triangle inequality $|x+y| \leqq |x|+|y|$ do not require that $\mathfrak{H}$ be complete or that (x, x) vanish only when $x = 0$.

2 LEMMA. *Let x be an element of $\mathfrak{H}$ and let K be a subset of $\mathfrak{H}$ with the property that $\frac{1}{2}(K+K) \subset K$. Let $\{k_i\}$ be a sequence in K with the property that*

$$\lim_i |x-k_i| = \inf_{k \in K} |x-k|.$$

Then $\{k_i\}$ is a convergent sequence.

PROOF. The identity

$$|x+y|^2+|x-y|^2 = 2|x|^2+2|y|^2, \qquad x, y \in \mathfrak{H},$$

called the *parallelogram identity*, follows immediately from the axioms. If $\delta = \inf_{k \in K} |x-k|$ the preceding identity shows that

$$\begin{aligned} |k_i-k_j|^2 &= 2|x-k_i|^2+2|x-k_j|^2-4|x-(k_i+k_j)/2|^2 \\ &\leqq 2|x-k_i|^2+2|x-k_j|^2-4\delta \to 0. \end{aligned} \qquad \text{Q.E.D.}$$

3 DEFINITION. Two vectors x, y in $\mathfrak{H}$ are said to be *orthogonal* if $(x, y) = 0$. Two manifolds $\mathfrak{M}$, $\mathfrak{N}$ in $\mathfrak{H}$ are *orthogonal manifolds* if $(\mathfrak{M}, \mathfrak{N}) = 0$. We write $x \perp y$ to indicate that x and y are orthogonal, and $\mathfrak{M} \perp \mathfrak{N}$ to indicate that $\mathfrak{M}$ and $\mathfrak{N}$ are orthogonal. The *orthocomplement* of a set $A \subseteq \mathfrak{H}$ is the set $\{x|(x, A) = 0\}$. It is sometimes denoted by $\mathfrak{H} \ominus A$, or, if $\mathfrak{H}$ is understood, by $A^\perp$.

➔ 4 LEMMA. *The orthocomplement* $\mathfrak{N}$ *of a closed linear manifold* $\mathfrak{M}$ *in* $\mathfrak{H}$ *is a closed linear manifold complementary to* $\mathfrak{M}$, i.e. $\mathfrak{H} = \mathfrak{M} \oplus \mathfrak{N}$.

PROOF. It follows from the linearity and the continuity of the scalar product (Theorem 1) that the orthocomplement of any set $\mathfrak{M}$ is a closed linear manifold. If $\mathfrak{M}$ is a closed linear manifold and if x is an arbitrary point in $\mathfrak{H}$ there is, by Lemma 2, an $m \in \mathfrak{M}$ such that $|x-m| = \delta = \inf_{m_1 \in \mathfrak{M}}|x-m_1|$. It will now be shown that the element $n = x-m$ is in $\mathfrak{N}$. For any complex number α and any m_1 in $\mathfrak{M}$ the vector $m+\alpha m_1$ is in $\mathfrak{M}$ and hence $|x-(m+\alpha m_1)| \geqq \delta$. Thus

$$\begin{aligned} 0 &\leqq |x-(m+\alpha m_1)|^2-|n|^2 = |n-\alpha m_1|^2-|n|^2 \\ &= -\alpha(m_1, n)-\bar{\alpha}(n, m_1)+|\alpha|^2|m_1|^2. \end{aligned}$$

Let $\alpha = \lambda(n, m_1)$ where λ is an arbitrary real number. Then

$$0 \leqq -2\lambda|(n, m_1)|^2+\lambda^2|(n, m_1)|^2|m_1|^2$$

which is possible only if $(n, m_1) = 0$. Thus $n \in \mathfrak{N}$. To complete the proof note that $x \in \mathfrak{M} \cap \mathfrak{N}$ implies $|x| = (x, x) = 0$. Thus $\mathfrak{M} \cap \mathfrak{N} = 0$ and $\mathfrak{H} = \mathfrak{M} \oplus \mathfrak{N}$. Q.E.D.

➔ 5 THEOREM. *Every* y^* *in* $\mathfrak{H}^*$ *uniquely determines a* y *in* $\mathfrak{H}$ *such that*

$$y^*x = (x, y), \qquad x \in \mathfrak{H}.$$

This map $\sigma : y^* \to y$ is a one-to-one isometric map of $\mathfrak{H}^*$ onto all of $\mathfrak{H}$ and $\sigma(y^*+z^*) = \sigma(y^*)+\sigma(z^*)$, $\sigma(\alpha y^*) = \bar{\alpha}\sigma(y^*)$.

PROOF. If $y^* = 0$ let $y = 0$. If $y^* \neq 0$ the set $\mathfrak{M} = \{x|y^*x = 0\}$ is a proper closed linear manifold in $\mathfrak{H}$ and its orthocomplement $\mathfrak{N}$ contains, by Lemma 4, a vector $y_1 \neq 0$. Let $y = \alpha y_1$ where $\bar{\alpha} = y^*(y_1)/|y_1|^2$. For an arbitrary vector x in $\mathfrak{H}$ the vector $x-(y^*x)/(y^*y_1)y_1$ is in $\mathfrak{M}$ so that $(x, y) = y^*x(y_1, y)/y^*y_1 = y^*x$, which proves the existence of the desired y. To see that y is unique, let y' be an element of $\mathfrak{H}$ such that $y^*x = (x, y')$ for every $x \in \mathfrak{H}$. Then $(x, y-y') = 0$ for every $x \in \mathfrak{H}$ and, in particular, $(y-y', y-y') = 0$, so $y = y'$. Thus σ is well defined. Since $|(x, y)| \leqq |x||y|$ it follows that $|y^*| \leqq |\sigma(y^*)|$, and since $(y, y) = |y|^2$ it follows that $|y^*| \geqq |\sigma(y^*)|$. Therefore σ is an isometry. The remaining properties to be proved for σ follow immediately from the postulated properties of the scalar product. Q.E.D.

6 COROLLARY. *The space $\mathfrak{H}^*$ is also a Hilbert space and $\mathfrak{H}$ is reflexive.*

PROOF. If the scalar product in $\mathfrak{H}^*$ is defined by

$$(x^*, y^*)_1 = (\sigma(y^*), \sigma(x^*)),$$

then $\mathfrak{H}^*$ is clearly a Hilbert space. For $y^{**} \in \mathfrak{H}^{**}$ there is, according to the theorem, an element $y^* \in \mathfrak{H}^*$ such that

$$y^{**}x^* = (x^*, y^*)_1 = (\sigma(y^*), \sigma(x^*)) = x^*y, \qquad x^* \in \mathfrak{H}^*,$$

where $y = \sigma(y^*)$. Q.E.D.

7 COROLLARY. *Hilbert space is weakly complete and a subset is weakly sequentially compact if and only if it is bounded.*

PROOF. This follows from 6, II.3.28, and II.3.29. Q.E.D.

8 DEFINITION. A set $A \subset \mathfrak{H}$ is called an *orthonormal* set if each vector in A has norm one and if every pair of distinct vectors in A is orthogonal. An orthonormal set is said to be *complete* if no non-zero vector is orthogonal to every vector in the set, i.e., A is complete if $\{0\} = \mathfrak{H} \ominus A$. We recall that a projection is a linear operator E with $E^2 = E$. A projection E in $\mathfrak{H}$ is called an *orthogonal projection* if the manifolds $E\mathfrak{H}$ and $(I-E)\mathfrak{H}$ are orthogonal.

It has been shown in Lemma 4 that

$$\mathfrak{H} = \mathfrak{M} \oplus (\mathfrak{H} \ominus \mathfrak{M})$$

where $\mathfrak{M}$ is an arbitrary closed linear manifold in $\mathfrak{H}$. If $x = y+z$ where $y \in \mathfrak{M}$ and $z \in \mathfrak{H} \ominus \mathfrak{M}$ let us define the transformation E in $\mathfrak{H}$ by placing $Ex = y$. It is clear that E is a projection, i.e., $E^2 = E$, and that E is an orthogonal projection. It is the uniquely determined orthogonal projection with $E\mathfrak{H} = \mathfrak{M}$. For if D is an orthogonal projection with $D\mathfrak{H} = \mathfrak{M}$ then $ED = D$ and, since $(I-D)\mathfrak{H} \subseteqq \mathfrak{H} \ominus \mathfrak{M}$, we see that $E(I-D) = 0$. Thus

$$D = ED+E(I-D) = E.$$

This uniquely determined orthogonal projection E with $E\mathfrak{H} = \mathfrak{M}$ is called *the orthogonal projection on* $\mathfrak{M}$ or sometimes simply *the projection on* $\mathfrak{M}$.

9 LEMMA. *If $\{y_i\}$ is an orthonormal sequence and $\{\alpha_i\}$ is a sequence of scalars, then the series $\Sigma\alpha_i y_i$ converges if and only if $\Sigma|\alpha_i|^2 < \infty$, and in this case*

$$|\Sigma\alpha_i y_i| = (\Sigma|\alpha_i|^2)^{1/2}.$$

When it converges, the series $\Sigma\alpha_i y_i$ is independent of the order in which its terms are arranged.

PROOF. For $m > n$

$$|\sum_{i=n}^{m} \alpha_i y_i|^2 = (\sum_{i=n}^{m} \alpha_i y_i, \sum_{j=n}^{m} \alpha_j y_j) = \sum_{i=n}^{m} \sum_{j=n}^{m} \alpha_i \bar{\alpha}_j (y_i, y_j) = \sum_{i=n}^{m} |\alpha_i|^2,$$

and so the one series converges if the other does. If, in the above equality, one puts $n = 1$ and allows m to increase indefinitely, the second conclusion of the lemma follows. Finally let $z = \sum_{n=1}^{\infty} \alpha_{i_n} y_{i_n}$ be a series obtained from $x = \sum \alpha_i y_i$ by a rearrangement of its terms. Then

$$|x-z|^2 = (x, x)-(x, z)-(z, x)+(z, z),$$

and a direct computation, similar to that above, shows that each of these scalar products is $\sum |\alpha_i|^2$. Thus $z = x$. Q.E.D.

→ 10 THEOREM. *Let A be an orthonormal set in $\mathfrak{H}$ and let x be an arbitrary vector in $\mathfrak{H}$. Then $(x, y) = 0$ for all but a countable number of y in A. The series*

$$Ex = \sum_{y \in A} (x, y)y, \qquad x \in \mathfrak{H}$$

converges and is independent of the order in which its non-zero terms are arranged. The operator E is the orthogonal projection on the closed linear manifold determined by A.

PROOF. Let $y_1, \ldots, y_n$ be distinct elements of A and let $y = \sum_{i=1}^n (x, y_i)y_i$ so that (by Lemma 9), $|y|^2 = \sum_{i=1}^n |(x, y_i)|^2$ and

$$0 \leqq |x-y|^2 = |x|^2-(x, y)-(y, x)+|y|^2,$$

$$(x, y) = \sum_{i=1}^{n} \overline{(x, y_i)}(x, y_i) = |y|^2,$$

$$(y, x) = \sum_{i=1}^{n} (x, y_i)\overline{(x, y_i)} = |y|^2.$$

Thus $|y|^2 \leqq |x|^2$, i.e.,

$$\sum_{i=1}^{n} |(x, y_i)|^2 \leqq |x|^2.$$

This shows that at most a finite number of vectors $y_1, \ldots, y_n$ in A can have $|(x, y_i)|$ greater than a preassigned positive number and proves that at most a countable number of the scalar products (x, y) with y in A fail to vanish. Since

$$\sum_{y \in A} |(x, y)|^2 \leqq |x|^2,$$

the preceding lemma shows that the series defining Ex converges and is independent of the order of its terms.

Now it is clear that E is a linear operator with $Ex = x$ for x in A. Thus $Ex = x$ for x in the closed linear manifold $\mathfrak{A}_1$ determined by A. Also $Ex = 0$ if x is orthogonal to A. Thus E is the orthogonal projection on $\mathfrak{A}_1$. Q.E.D.

11 DEFINITION. A set A is called an *orthonormal basis* for the linear manifold $\mathfrak{N}$ in $\mathfrak{H}$ if A is an orthonormal set contained in $\mathfrak{N}$ and if

$$x = \sum_{y \in A} (x, y)y, \qquad x \in \mathfrak{N}.$$

12 THEOREM. *Every closed linear manifold in $\mathfrak{H}$ contains an orthonormal basis for itself.*

PROOF. If the orthonormal sets in the closed linear manifold $\mathfrak{M}$ are ordered by inclusion, it is seen from Zorn's lemma (I.2.7) that there is a maximal one A which determines the closed linear manifold

$\mathfrak{A}_1 \subseteqq \mathfrak{M}$. Since A is maximal $\mathfrak{M} \ominus \mathfrak{A}_1 = 0$. But by Lemma 4, $\mathfrak{M} = \mathfrak{A}_1 \oplus (\mathfrak{M} \ominus \mathfrak{A}_1)$, and so $\mathfrak{M} = \mathfrak{A}_1$. The desired conclusion now follows from Theorem 10. Q.E.D.

13 THEOREM. *For an orthonormal set $A \subset \mathfrak{H}$ the following statements are equivalent*:

(i) *the set A is complete;*

(ii) *the set A is an orthonormal basis for $\mathfrak{H}$;*

(iii) $|x|^2 = \Sigma_{y \in A} |(x, y)|^2, \qquad x \in \mathfrak{H}$.

PROOF. The equivalence of statements (i) and (ii) is clear from Theorem 10. That either one of these implies (iii) follows from Theorem 10 and Lemma 9. Now assume (iii) and let x be an arbitrary vector in $\mathfrak{H}$. By Lemma 4, $x = u + v$ where $u \in \overline{sp}(A)$ and $v \in \mathfrak{H} \ominus \overline{sp}(A)$. Thus $|x|^2 = |u|^2 + |v|^2$. But, by Theorem 10 and Lemma 9, $|u|^2 = \Sigma_{y \in A} |(u, y)|^2$. Hence $|x|^2 = |u|^2$ and $v = 0$. This means that $\overline{sp}(A) = \mathfrak{H}$ from which (i) follows. Q.E.D.

The next result enables us to define the *dimension* of a Hilbert space.

14 THEOREM. *All orthonormal bases of a given Hilbert space $\mathfrak{H}$ have the same cardinality.*

PROOF. If $\mathfrak{H}$ is finite dimensional, the result is a well-known result in algebra. Suppose then that $\mathfrak{H}$ is infinite dimensional, and let $\{u_\alpha\}$ and $\{v_\beta\}$ be two orthonormal bases for $\mathfrak{H}$. We shall say that the vectors u_α and $u_{\alpha'}$ in the basis $\{u_\alpha\}$ are equivalent if there exists a finite chain

[*] $$u_\alpha, v_{\beta_1}, u_{\alpha_1}, \ldots, u_{\alpha_k}, v_{\beta_{k+1}}, u_{\alpha'},$$

of vectors in which the scalar product of any two successive vectors is non-zero and in which the terms alternate between vectors in $\{u_\alpha\}$ and $\{v_\beta\}$. The equivalence of two vectors v_β and $v_{\beta'}$ in $\{v_\beta\}$ is defined similarly. It follows immediately from Theorem 10 that any equivalence class of vectors is either finite or countable. An equivalence class U of vectors u_α will be said to correspond to an equivalence class V of vectors v_β if there is a pair of vectors, one from U and one from V, with a non-zero inner product. Suppose that U and V are corresponding equivalence classes and that $u_\alpha \in U$. Consider an arbitrary element v_β in the basis $\{v_\beta\}$ for which $(u_\alpha, v_\beta) \neq 0$. It will be shown that $v_\beta \in V$. Since U and V are corresponding classes there are elements

$u_{\alpha'}, v_{\beta'}$ in U, V respectively with $(u_{\alpha'}, v_{\beta'}) \neq 0$. Now since $u_{\alpha'} \in U$ there is a finite chain of the form given in [*] above in which successive vectors have non-zero scalar products. Thus by forming the chain $v_\beta, u_\alpha, \ldots, u_{\alpha'}, v_{\beta'}$ it is seen that v_β is equivalent to $v_{\beta'}$ and thus that v_β is in V. Since $\{v_\beta\}$ is a basis, the vector u_α has an expansion of the form $u_\alpha = \sum_\beta (u_\alpha, v_\beta) v_\beta$, so that u_α is in the closed linear manifold determined by those v_β with $(u_\alpha, v_\beta) \neq 0$. Since such v_β are in V we have $u_\alpha \in \overline{sp}[V]$ and thus $\overline{sp}[U] \subseteq \overline{sp}[V]$. Similarly $\overline{sp}[V] \subseteq \overline{sp}[U]$. It is thus seen that corresponding equivalence classes U and V determine the same closed linear manifold $\mathfrak{M}$. Hence, if one of U and V is finite, $\mathfrak{M}$ is finite dimensional, and therefore the other of U and V is finite and has the same cardinality. If neither of U and V is finite, both are countable. Thus $\{u_\alpha\}$ and $\{v_\beta\}$ break up into a disjoint union of corresponding pairs U, V of equivalence classes, each U having the same cardinality as the corresponding V. Consequently $\{u_\alpha\}$ and $\{v_\beta\}$ have the same cardinality. Q.E.D.

15 **Definition.** The cardinality of an arbitrary orthonormal basis of a Hilbert space $\mathfrak{H}$ is its *dimension*.

16 **Theorem.** *Two Hilbert spaces are isometrically isomorphic if and only if they have the same dimension.*

Proof. Let U be an isometric isomorphism of $\mathfrak{H}_1$ onto $\mathfrak{H}_2$. Then if x and y are orthogonal elements of $\mathfrak{H}_1$,

$$\begin{aligned}|U(x+\lambda y)|^2 = |x+\lambda y|^2 = |x|^2+|\lambda|^2|y|^2 &= |Ux+\lambda Uy|^2 \\ &= |Ux|^2+|\lambda|^2|Uy|^2+(Ux, \lambda Uy)+\overline{(Ux, \lambda Uy)} \\ &= |x|^2+|\lambda|^2|y|^2+(Ux, \lambda Uy)+\overline{(Ux, \lambda Uy)}.\end{aligned}$$

This shows that for arbitrary λ

$$0 = (Ux, \lambda Uy)+\overline{(Ux, \lambda Uy)},$$

and if we let $\lambda = (Ux, Uy)$ in this equation it is seen that $(Ux, Uy) = 0$. Thus U maps an orthonormal basis for $\mathfrak{H}_1$ onto an orthonormal basis for $\mathfrak{H}_2$, and consequently $\mathfrak{H}_1$ and $\mathfrak{H}_2$ have the same dimension.

Conversely, let $\mathfrak{H}_1$ and $\mathfrak{H}_2$ have the same dimension, and let $\{u_\alpha, \alpha \in A\}$ and $\{v_\alpha, \alpha \in A\}$ be orthonormal bases for $\mathfrak{H}_1$ and $\mathfrak{H}_2$ respectively. For each scalar function C on A with $C(\alpha) = 0$ for all but a countable set of indices α and such that $\sum |C(\alpha)|^2 < \infty$, let

$$U(\sum C(\alpha)u_\alpha) = \sum C(\alpha)v_\alpha.$$

It follows from Theorem 13 that U is an isometric isomorphism of $\mathfrak{H}_1$ onto $\mathfrak{H}_2$. Q.E.D.

Direct Sums of Hilbert Spaces

We recall (cf. I.11) that the direct sum

$$\mathfrak{X} = \mathfrak{X}_1 \oplus \ldots \oplus \mathfrak{X}_n$$

of the vector spaces $\mathfrak{X}_1, \ldots, \mathfrak{X}_n$ is the set $\mathfrak{X}_1 \times \mathfrak{X}_2 \times \ldots \times \mathfrak{X}_n$ in which addition and scalar multiplication are defined by the formulas

$$[x_1, \ldots, x_n] + [y_1, \ldots, y_n] = [x_1+y_1, \ldots, x_n+y_n],$$
$$\alpha[x_1, \ldots, x_n] = [\alpha x_1, \ldots, \alpha x_n].$$

The space $\mathfrak{X}_i$ is algebraically equivalent to the subspace $\mathfrak{M}_i$ of $\mathfrak{X}$ consisting of all vectors $[x_1, \ldots, x_n]$ in $\mathfrak{X}$ with $x_j = 0$ for $j \neq i$. It is sometimes convenient to refer to the space $\mathfrak{X}_i$ itself as a subspace of $\mathfrak{X}$ and, when such reference is made, it is the equivalent space $\mathfrak{M}_i$ that is to be understood. The map

$$[x_1, \ldots, x_n] \to [0, \ldots, x_i, \ldots, 0]$$

of $\mathfrak{X}$ onto $\mathfrak{M}_i$ is a projection and is sometimes called *the projection of* $\mathfrak{X}$ *onto* $\mathfrak{M}_i$. Equivalently, the map $[x_1, \ldots, x_n] \to x_i$ is called *the projection of* $\mathfrak{X}$ *onto* $\mathfrak{X}_i$. If each of the spaces $\mathfrak{X}_1, \ldots, \mathfrak{X}_n$ is a linear topological space, then the direct sum $\mathfrak{X}$, with the product topology (cf. I.8), is also a linear topological space in which the subspace $\mathfrak{M}_i$ is topologically as well as algebraically equivalent to $\mathfrak{X}_i$. If a topology in each of the summands $\mathfrak{X}_i$, $i = 1, \ldots, n$, is given by a norm $|\cdot|_i$, i.e., if each of the spaces $\mathfrak{X}_i$ is a normed linear space, then the space $\mathfrak{X}$ is a normed linear space. The norm in $\mathfrak{X}$ may be introduced in a variety of ways; for example, any one of the following norms defines the product topology in $\mathfrak{X}$.

(i) $|[x_1, \ldots, x_n]| = |x_1|_1 + |x_2|_2 + \ldots + |x_n|_n$,

(ii) $|[x_1, \ldots, x_n]| = \sup_{1 \leq i \leq n} |x_i|_i$,

(iii) $|[x_1, \ldots, x_n]| = (|x_1|_1^2 + \ldots + |x_n|_n^2)^{1/2}$.

Whenever the direct sum of normed linear spaces is used as a normed space, the norm will be explicitly mentioned. If, however, each of the spaces $\mathfrak{X}_1, \ldots, \mathfrak{X}_n$ is a Hilbert space then it will always be understood. sometimes without explicit mention, that $\mathfrak{X}$ is the uniquely determined Hilbert space with scalar product

$$\text{(iv)}\quad ([x_1, \ldots, x_n], [y_1, \ldots, y_n]) = \sum_{i=1}^{n} (x_i, y_i)_i,$$

where $(\cdot, \cdot)_i$ is the scalar product in $\mathfrak{X}_i$. Thus the norm in a direct sum of Hilbert spaces is always given by (iii). To summarize, we state the following definition.

17 Definition. For each $i = 1, \ldots, n$, let $\mathfrak{H}_i$ be a Hilbert space with scalar products $(\cdot, \cdot)_i$. The *direct sum of the Hilbert spaces* $\mathfrak{H}_1, \ldots, \mathfrak{H}_n$ is the linear space $\mathfrak{H} = \mathfrak{H}_1 \oplus \ldots \oplus \mathfrak{H}_n$ in which a scalar product is defined by (iv).

Let $\mathfrak{H} = \mathfrak{H}_1 \oplus \ldots \oplus \mathfrak{H}_n$ be the direct sum of the Hilbert spaces $\mathfrak{H}_1, \ldots, \mathfrak{H}_n$. Then for $i \neq j$ the manifolds $\mathfrak{H}_i$ and $\mathfrak{H}_j$ are orthogonal in $\mathfrak{H}$ and the projection of $\mathfrak{H}$ onto $\mathfrak{H}_i$ is the same as the orthogonal projection of $\mathfrak{H}$ onto $\mathfrak{H}_i$. Thus, for example, the subspace $\mathfrak{H}_2 \oplus \ldots \oplus \mathfrak{H}_n$ of $\mathfrak{H}$ is the orthocomplement of $\mathfrak{H}_1$.

The following definition generalizes Definition 17 to cover the case of an infinite family of direct summands.

18 Definition. For each ν in an index set A let $\mathfrak{H}_\nu$ be a Hilbert space. The *direct sum* $\sum \mathfrak{H}_\nu$ of the Hilbert spaces $\mathfrak{H}_\nu$ is defined to be the family of all functions $\{x_\nu\}$ on A such that for each ν, $x_\nu \in \mathfrak{H}_\nu$ and such that $\sum_{\nu \in A} |x_\nu|^2 < \infty$.

It is clear that $\sum \mathfrak{H}_\nu$ is a vector space if addition and scalar multiplication are defined by the formulas

$$\alpha\{x_\nu\} = \{\alpha x_\nu\}, \qquad \{x_\nu\}+\{y_\nu\} = \{x_\nu+y_\nu\}.$$

Moreover, one may define an inner product in $\sum \mathfrak{H}_\nu$ by the formula

$$(\{x_\nu\}, \{y_\nu\}) = \sum_\nu (x_\nu, y_\nu),$$

the series converging unconditionally since

$$\begin{aligned} \sum_\nu |(x_\nu, y_\nu)| &\leqq \sum_\nu |x_\nu||y_\nu| \\ &\leqq (\sum_\nu |x_\nu|^2)^{1/2} (\sum_\nu |y_\nu|^2)^{1/2}. \end{aligned}$$

Properties (i)—(v) of Definition 2.26 may readily be verified.

19 LEMMA. *If* $\{\mathfrak{H}_\nu\}$, $\nu \in A$, *is a family of Hilbert spaces, their direct sum* $\sum \mathfrak{H}_\nu$ *is a Hilbert space.*

PROOF. As remarked above, it only remains to prove the completeness of $\sum \mathfrak{H}_\nu$. If $\{x_\nu^n\}$, $n = 1, 2, \ldots$, is a Cauchy sequence in $\sum \mathfrak{H}_\nu$, it is clear that for fixed ν, $\{x_\nu^n\}$ is a Cauchy sequence in $\mathfrak{H}_\nu$ converging to some element x_ν^0. For any finite subset $\pi \subset A$ and any integer n,

$$\sum_{\nu \in \pi} |x_\nu^n - x_\nu^0|^2 = \lim_{m \to \infty} \sum_{\nu \in \pi} |x_\nu^n - x_\nu^m|^2$$
$$\leqq \limsup_{m \to \infty} |\{x_\nu^n\} - \{x_\nu^m\}|^2.$$

It follows that

$$\limsup_{n \to \infty} \sum\nolimits_\nu |x_\nu^n - x_\nu^0|^2 \leqq \limsup_{m, n \to \infty} |\{x_\nu^n\} - \{x_\nu^m\}|^2 = 0$$

showing that $\{x_\nu^0\}$ is in $\sum \mathfrak{H}_\nu$ and that $\{x_\nu^n\}$ converges to $\{x_\nu^0\}$. Q.E.D.

We conclude this section by listing, in the following lemma, a few useful properties of the orthocomplement.

→ 20 LEMMA. *Let* B *be a set in* $\mathfrak{H}$ *and* $\mathfrak{M}$ *a closed linear manifold in* $\mathfrak{H}$. *Then*

(i) $\mathfrak{H} = \mathfrak{M} \oplus (\mathfrak{H} \ominus \mathfrak{M})$;

(ii) $\mathfrak{M} = \mathfrak{H} \ominus (\mathfrak{H} \ominus \mathfrak{M})$;

(iii) $\overline{sp}\,(B) = \mathfrak{H} \ominus (\mathfrak{H} \ominus B)$.

PROOF. Equation (i) is merely a restatement of Lemma 4. Equation (ii) may be proved by replacing $\mathfrak{M}$ by $\mathfrak{H} \ominus \mathfrak{M}$ in equation (i). This shows that $\mathfrak{H} \ominus (\mathfrak{H} \ominus \mathfrak{M})$, as well as its closed subspace $\mathfrak{M}$, is a complementary manifold to $\mathfrak{H} \ominus \mathfrak{M}$. Thus $\mathfrak{M} = \mathfrak{H} \ominus (\mathfrak{H} \ominus \mathfrak{M})$. To prove (iii) note that for an arbitrary set $B \subseteqq \mathfrak{H}$ the condition $(B, x) = 0$ on an element x in $\mathfrak{H}$ is equivalent to the condition $(\overline{sp}\,(B), x) = 0$. Thus $\mathfrak{H} \ominus B = \mathfrak{H} \ominus \overline{sp}(B)$ and (iii) follows by placing $\mathfrak{M} = \overline{sp}\,(B)$ in (ii). Q.E.D.

5. The Spaces $B(S, \Sigma)$ and $B(S)$

In this section it is observed that $B(S)$ and $B(S, \Sigma)$ are B-spaces. Their conjugate spaces are represented and a condition for compactness is given. Weak compactness and weak convergence in $B(S)$ will be discussed at the end of Section 6.

It is clear that $B(S)$ is a normed linear space. To see that it is complete and hence a B-space, suppose that $\{f_n\}$ is a Cauchy sequence in $B(S)$. Then for each $\varepsilon > 0$ there is an $m(\varepsilon)$ with $|f_n - f_m| < \varepsilon$ for $n, m \geqq m(\varepsilon)$. Let $f(s) = \lim_n f_n(s)$ for each s in S. For each s in S there is a $p \geqq m(\varepsilon)$ with $|f(s) - f_p(s)| < \varepsilon$ and thus

$$|f(s)-f_n(s)| \leqq |f(s)-f_p(s)|+|f_p(s)-f_n(s)| < 2\varepsilon,$$
$$n \geqq m(\varepsilon), \qquad s \in S.$$

This shows that f_n converges to f in $B(S)$ and proves that $B(S)$ is a B-space. To see that $B(S, \Sigma)$ is also a B-space it suffices to note that it has been defined as the closure in $B(S)$ of the set of all finite linear combinations of characteristic functions of sets in Σ. Thus $B(S, \Sigma)$ is a closed linear manifold in $B(S)$ and is therefore a B-space. Since the product of two characteristic functions of sets in Σ is also a characteristic function of a set in Σ, it follows from the definition of $B(S, \Sigma)$ that it contains the product of any two of its elements. Thus $B(S, \Sigma)$ *is a closed subalgebra of* $B(S)$, a fact that will be useful later.

1 Theorem. *There is an isometric isomorphism between* $B^*(S, \Sigma)$ *and* $ba(S, \Sigma)$, *determined by the identity*

$$[*] \qquad x^*f = \int_S f(s)\mu(ds).$$

Thus, for each x^* *in* $B^*(S, \Sigma)$ *there is a unique* μ *in* $ba(S, \Sigma)$ *such that* [*] *holds; for each* μ *in* $ba(S, \Sigma)$ *there is a unique* x^* *such that* [*] *holds; and the correspondence between* x^* *and* μ *is linear and isometric.*

Proof. It is clear from the definition of a μ-integrable function that for any $\mu \in ba(S, \Sigma)$ we have $B(S, \Sigma) \subseteqq L_1(S, \Sigma, \mu)$ and that for f in $B(S, \Sigma)$

$$\left| \int_S f(s)\mu(ds) \right| \leqq \sup_{s \in S} |f(s)| v(\mu, S).$$

Thus for each μ in $ba(S, \Sigma)$, equation [*] defines a point x^* in $B^*(S, \Sigma)$ with $|x^*| \leqq |\mu|$. To see that $|x^*| \geqq |\mu|$, let $\varepsilon > 0$ and let $E_1, \ldots, E_n$ be a partition of S into disjoint sets in Σ such that $\sum_{i=1}^n |\mu(E_i)| > |\mu| - \varepsilon$. Then if we let θ_j satisfy $0 < \theta_j \leqq 2\pi$ and $e^{i\theta_j}\mu(E_j) = |\mu(E_j)|$

for $j = 1, \ldots, n$, and put $f(s) = e^{i\theta_j}$ for $s \in E_j$, then $|f| \leqq 1$, $f \in B(S,\Sigma)$, and

$$|x^*| \geqq \int_S f(s)\mu(ds) = \sum_{i=1}^{\infty} |\mu(E_i)| > |\mu| - \varepsilon.$$

Since ε is arbitrary, $|x^*| \geqq |\mu|$. Thus, $|x^*| = |\mu|$. To show that every $x^* \in B^*(S, \Sigma)$ is given by some $\mu \in ba(S, \Sigma)$, we have only to let χ_E be the characteristic function of a set $E \in \Sigma$, and define $\mu(E) = x^*(\chi_E)$. It is then evident that μ is additive, that $|\mu(E)| \leqq |x^*|$, and that [*] holds for every function f which is a linear combination of characteristic functions of sets in Σ. Such functions are dense in $B(S, \Sigma)$ and therefore, since both sides of equation [*] are continuous in f, it follows that [*] holds for all f in $B(S, \Sigma)$. Since the linearity of the correspondence between x^* and μ is clear, the theorem is proved. Q.E.D.

2 Definition. If Σ is the family of all subsets of a set S, we abbreviate $ba(S, \Sigma)$ as $ba(S)$.

Since $B(S) = B(S, \Sigma)$ if Σ is the family of all subsets of S, we have the following corollary.

3 Corollary. *There is an isometric isomorphism between $B^*(S)$ and $ba(S)$ determined by the identity*

$$x^*f = \int_S f(s)\mu(ds), \qquad f \in B(S).$$

Next we wish to give conditions on a subset of $B(S, \Sigma)$ equivalent to compactness. Since a set in a complete metric space is compact if and only if it is closed and conditionally compact (I.6.15), it will be sufficient and convenient for our present purposes to state conditions for conditional compactness. The following elementary lemma is useful here as well as in later discussions of compactness in other B-spaces.

4 Lemma. *Let $\{U_a\}$ be a uniformly bounded generalized sequence of linear operators in the B-space $\mathfrak{X}$. If $\lim_a U_a x = x$ for every x in $\mathfrak{X}$, then this limit exists uniformly on any compact set. Conversely, if $\lim_a U_a x = x$ uniformly for x in a bounded set K, and if, in addition, $U_a(\{x \mid |x| \leqq 1\})$ is conditionally compact for each a, then K is conditionally compact.*

Proof. Let K be compact and $\varepsilon > 0$. There is a finite set $\{k_1, \ldots, k_n\}$ of elements of K such that $\inf_{1 \leq i \leq n} |k-k_i| < \varepsilon$ for every k in K (I.6.15). Let a_ε be such that $|U_a k_i - k_i| < \varepsilon$ for $a \geq a_\varepsilon$ and for $1 \leq i \leq n$. Then, if $|U_a| \leq M$ for all a,

$$|U_a k - k| \leq \inf_{1 \leq i \leq n} [|U_a(k-k_i)| + |U_a k_i - k_i| + |k_i - k|] \leq (M+2)\varepsilon,$$

for all k in K and all $a \geq a_\varepsilon$. Thus $U_a k \to k$ uniformly for k in K.

Conversely, let $\varepsilon > 0$, and let a be such that

$$|U_a k - k| < \varepsilon, \qquad k \in K.$$

Since $U_a K$ is conditionally compact there are elements $k_i \in K$ with

$$\inf_{1 \leq i \leq n} |U_a k - U_2 k_i| < \varepsilon, \qquad k \in K.$$

Thus $\inf_{1 \leq i \leq n} |k-k_i| < 2\varepsilon$ for every k in K. Q.E.D.

5 **Corollary.** *If $\{x_i\}$ is a basis for $\mathfrak{X}$ then the expansion $x = \sum_{i=1}^{\infty} \alpha_i x_i$ converges uniformly for x in a bounded set K if and only if K is conditionally compact.*

6 **Theorem.** *A bounded set K in $B(S, \Sigma)$ is conditionally compact if and only if for every $\varepsilon > 0$ there is a finite collection $\{E_1, \ldots, E_n\}$ of disjoint sets in Σ with union S, and points s_i in E_i, $i = 1, \ldots, n$, such that*

$$\sup_{s \in E_i} |f(s_i) - f(s)| < \varepsilon, \qquad f \in K, \qquad i = 1, \ldots, n.$$

Proof. Let A be the set each of whose elements $a = \{E_1, \ldots, E_n; s_1, \ldots, s_n\}$ consists of a finite collection $\{E_1, \ldots, E_n\}$ of disjoint sets in Σ with union S, and points $s_1, \ldots, s_n$ with $s_i \in E_i$, $i = 1, \ldots, n$. Let A be ordered by defining $a \leq a'$ to mean that each set in a is a union of sets in a'. For f in $B(S, \Sigma)$ and $a = \{E_1, \ldots, E_n; s_1, \ldots, s_n\}$ in A, let $U_a f = f_a$ where

$$f_a = \sum_{i=1}^{n} f(s_i) \chi_{E_i},$$

and χ_E is the characteristic function of E. Clearly

$$|U_a| = 1, \qquad U_{a'} U_a = U_a \text{ for } a' \geq a.$$

Now if f_0 is a finite linear combination of characteristic functions of sets in Σ it is clear that there is an element a_0 in A such that f_0 is constant on each set in a_0. Thus $U_a f_0 = f_0$ for $a \geqq a_0$. For such f_0 therefore, $\lim_a U_a f_0 = f_0$. Since these functions are dense in $B(S, \Sigma)$ it follows from II.1.18 that $\lim_a U_a f = f$ for every f in $B(S, \Sigma)$. The conclusion now follows from Lemma 4 and Theorem 3.5. Q.E.D.

6. The Space $C(S)$

In this section, it will be assumed at first only that S is a normal topological space. The space $C(S)$ is the set of all bounded continuous real or complex functions defined on S. The norm in $C(S)$ is given by the formula

$$|f| = \sup_{s \in S} |f(s)|.$$

It follows from Lemma I.4.18 and Corollary I.7.7 that $C(S)$ is a B-space. We shall begin our discussion by representing the conjugate space, $C^*(S)$.

1 Definition. If S is a topological space, the space $rba(S)$ is the linear space of regular bounded additive set functions defined on the field generated by the closed sets. The norm $|\mu|$ of μ is its total variation. The space $rba(S)$ is partially ordered by defining $\mu \geqq \lambda$ to mean that $\mu(E) \geqq \lambda(E)$ for every E in the field determined by the closed sets. Similarly, the space $C(S)$ is partially ordered by defining $f \geqq g$ to mean that $f(s) \geqq g(s)$ for all s in S. Finally, $C^*(S)$ is partially ordered by defining $x^* \geqq y^*$ to mean that $x^*f \geqq y^*f$ for every $f \in C(S)$ with $f \geqq 0$.

Before representing the conjugate space $C^*(S)$ we first observe that every f in $C(S)$ is integrable with respect to every μ in $rba(S)$. To see this, cover $f(S)$ with open sets $G_1, \ldots, G_n$ with diameters less than a given positive ε. Let

$$A_1 = G_1, \qquad A_j = G_j - \bigcup_{i=1}^{j-1} G_i, \qquad j = 2, \ldots, n,$$

and, if A_j is not void, choose a point $\alpha_j \in A_j$. If A_j is void, let $\alpha_j = 0$. Since G_j is open, $f^{-1}(G_j)$ is also open, and thus the set $B_j = f^{-1}(A_j)$ is in the domain of μ. Then the function

$$f_\varepsilon = \sum_{j=1}^{n} \alpha_j \chi_{B_j}$$

is a μ-simple function with $\sup_s |f_\varepsilon(s) - f(s)| < \varepsilon$. The function f is therefore the uniform limit of μ-simple functions and, since $v(\mu, S) < \infty$, f is μ-integrable. Since the integral $\int_S f(s)\mu(ds)$ satisfies the inequality

$$\left| \int_S f(s)\mu(ds) \right| \leqq \sup_s |f(s)| v(\mu, S),$$

it is clear that the integral is a continuous linear functional on $C(S)$. The following theorem is a converse to this statement.

2 **Theorem.** *If S is normal, there is an isometric isomorphism between $C^*(S)$ and $rba(S)$ such that corresponding elements x^* and μ satisfy the identity*

$$[*] \qquad x^* f = \int_S f(s)\mu(ds), \qquad f \in C(S).$$

Furthermore, this isomorphism preserves order.

Proof. It has just been observed that every $\mu \in rba(S)$ determines a functional $x^* \in C^*(S)$ by the formula [*] and that $|x^*| \leqq |\mu|$. To show that $|x^*| = |\mu|$, let $\varepsilon > 0$, and let $E_1, \ldots, E_n$ be disjoint sets in the domain of μ such that $\sum_{i=1}^{n} |\mu(E_i)| \geqq |\mu| - \varepsilon = v(\mu, S) - \varepsilon$. Let C_i be a closed subset of E_i such that $v(\mu, E_i - C_i) \leqq \varepsilon/n$ and let $\{G_1, \ldots, G_n\}$ be a family of disjoint open sets containing the disjoint closed sets $C_1, \ldots, C_n$. Since μ is regular, it is clear that we may assume that $v(\mu, G_i - C_i) \leqq \varepsilon/n$. By I.5.2, there exists a set $\{f_1, \ldots, f_n\}$ of continuous functions with $0 \leqq f_i(s) \leqq 1$, and such that $f_i(s) = 0$ if $s \notin G_i$ and $f_i(s) = 1$ if $s \in C_i$. Let $\alpha_1, \ldots, \alpha_n$ be complex constants of modulus one such that $\alpha_i \mu(E_i) = |\mu(E_i)|$ and put $f_0 = \sum_{i=1}^{n} \alpha_i f_i$. Then $\big| x^*(f_0) - |\mu| \big| \leqq 3\varepsilon$, so that $\sup_{|f| \leqq 1} |x^*(f)| = |\mu|$.

We next consider order. It is clear that $\int_S f(s)\mu(ds) \geqq 0$ if μ is a non-negative set function and f a non-negative continuous function. Conversely, let $\int_S f(s)\mu(ds) \geqq 0$ for each non-negative $f \in C(S)$. If there exists a μ-measurable set E such that $\mu(E) < -\varepsilon < 0$ we can find a closed subset $F \subseteq E$ such that $v(\mu, E - F) < \varepsilon/2$, and an open set $G \supseteq E$ such that $v(\mu, G - F) < \varepsilon/4$. If $g \in C(S)$ is chosen by I.5.2 to satisfy $0 \leqq g(s) \leqq 1$, $g(s) = 0$ if $s \notin G$ and $g(s) = 1$ if $s \in F$, then

$|\int_S g(s)\mu(ds) - \mu(E)|$ is less than $3\varepsilon/4$, so that $\int_S g(s)\mu(ds) \geqq 0$ is impossible.

Thus all that remains to be shown is that every continuous functional x^* on $C(S)$ can be represented in the form [*], where $\mu \in rba(S)$. By II.3.11, x^* can be extended to a continuous functional y^* on $B(S)$ and by Corollary 5.3, there exists an element $\lambda \in ba(S)$ such that $y^*f = \int_S f(s)\lambda(ds)$ for $f \in B(S)$. By the Jordan decomposition (III.1.8), λ can be written in the form $\lambda = \lambda_1 - \lambda_2 + i(\lambda_3 - \lambda_4)$, where $\lambda_i \geqq 0$, $i = 1, \ldots, 4$, and so it suffices to consider the case in which λ is non-negative, and to find a $\mu \in rba(S)$ such that $\int_S f(s)\mu(ds) = \int_S f(s)\lambda(ds)$ for every $f \in C(S)$.

Let F represent the general closed subset, G the general open subset, and E the general subset of S. Define set functions μ_1 and μ_2 by putting

$$\mu_1(F) = \inf_{G \supseteq F} \lambda(G), \qquad \mu_2(E) = \sup_{F \subseteq E} \mu_1(F).$$

It is clear that these set functions are non-negative and non-decreasing. Let G_1 be open and F_1 be closed. Then, if $G \supseteq F_1 - G_1$, it follows that $G_1 \cup G \supseteq F_1$ and $\lambda(G_1 \cup G) \leqq \lambda(G_1) + \lambda(G)$ so that $\mu_1(F_1) \leqq \lambda(G_1) + \lambda(G)$. Since G is an arbitrary open set containing $F_1 - G_1$ we have

$$\mu_1(F_1) \leqq \lambda(G_1) + \mu_1(F_1 - G_1).$$

If F is a closed set it follows from this inequality, by allowing G_1 to range over all open sets containing FF_1, that

$$\mu_1(F_1) \leqq \mu_1(FF_1) + \mu_2(F_1 - F).$$

If E is an arbitrary subset of S and F_1 ranges over the closed subsets of E it follows from the preceding inequality that

$$\text{(i)} \qquad \mu_2(E) \leqq \mu_2(EF) + \mu_2(E - F).$$

It will next be shown that for an arbitrary set E in S and an arbitrary closed set F in S we have

$$\text{(ii)} \qquad \mu_2(E) \geqq \mu_2(EF) + \mu_2(E - F).$$

To see this let F_1 and F_2 be disjoint closed sets. Since S is normal there are disjoint neighborhoods G_1 and G_2 of F_1 and F_2 respectively. If G is an arbitrary neighborhood of $F_1 \cup F_2$ then $\lambda(G) \geqq \lambda(GG_1) + \lambda(GG_2)$ so that

$$\mu_1(F_1 \cup F_2) \geqq \mu_1(F_1)+\mu_1(F_2).$$

We now let E and F be arbitrary sets in S with F closed and let F_1 range over the closed subsets of EF while F_2 ranges over the closed subsets of $E-F$. The preceding inequality then proves (ii). From (i) and (ii) we have

$$\text{(iii)} \quad \mu_2(E) = \mu_2(EF)+\mu_2(EF'), \qquad E \subseteq S, \qquad F \text{ closed}.$$

The function μ_2 is defined on the field of all subsets of S and it follows from (iii) that every closed set F is a μ_2-set in the sense of Definition III.5.1. If μ is the restriction of μ_2 to the field determined by the closed sets, it follows from Lemma III.5.2 that μ is additive on this field. It is clear from the definition of μ_1 and μ_2 that $\mu_1(F) = \mu_2(F) = \mu(F)$ if F is closed and thus $\mu(E) = \sup_{F \subseteq E} \mu(F)$. This shows that μ is regular, and since $\mu(S) < \infty$, we have $\mu \in rba(S)$. All that remains to be shown is that

$$\text{(iv)} \qquad \int_S f(s)\lambda(ds) = \int_S f(s)\mu(ds), \qquad f \in C(S).$$

It will clearly suffice to prove (iv) for real f and, since a real function in $C(S)$ is the difference of two non-negative functions in $C(S)$, it is sufficient to prove (iv) for non-negative f. Finally, since every f in $C(S)$ is bounded, we may and shall restrict the proof of (iv) to the case where $0 \leqq f(s) \leqq 1$.

Suppose then that f is continuous on S and $0 \leqq f(s) \leqq 1$. For $\varepsilon > 0$, let $E_1, \ldots, E_n$ be a partitioning of S into disjoint sets in the domain of μ such that

$$\sum_{i=1}^{n} a_i\mu(E_i)+\varepsilon \geqq \int_S f(s)\mu(ds),$$

where $a_i = \inf_{s \in E_i} f(s)$. Since μ is regular, there are closed sets $F_i \subseteq E_i$ such that

$$\sum_{i=1}^{n} a_i\mu(F_i)+2\varepsilon \geqq \int_S f(s)\mu(ds).$$

It follows from the normality of S and the continuity of f that there are disjoint open sets $G_1, \ldots, G_n$ with $G_i \supseteq F_i$, $i = 1, \ldots, n$ and such that

$$b_i = \inf_{s \in G_i} f(s) \geqq a_i - \frac{\varepsilon}{n|\mu|},$$

and hence

$$\sum_{i=1}^{n} b_i \mu(G_i) + 3\varepsilon \geqq \int_S f(s)\mu(ds).$$

Now it is clear that $\mu_1(F) = \mu_2(F) = \mu(F)$ for a closed set F, and if the open set G contains F then $\mu(F) \leqq \lambda(G)$. Thus, since μ is regular, $\mu(G) \leqq \lambda(G)$ for G open. We therefore have

$$\sum_{i=1}^{n} b_i \mu(G_i) \leqq \sum_{i=1}^{n} b_i \lambda(G_i) \leqq \int_S f(s)\lambda(ds),$$

and so

$$\text{(v)} \qquad \int_S f(s)\lambda(ds) \geqq \int_S f(s)\mu(ds).$$

Since $\mu(S) = \lambda(S)$ it follows from (v) that

$$\int_S (1-f(s))\lambda(ds) \leqq \int_S (1-f(s))\mu(ds).$$

But, since $0 \leqq 1-f(s) \leqq 1$, the function f may be replaced by $1-f$ in the inequality (v) and this, together with the preceding inequality, shows that $\int (1-f)d\lambda = \int (1-f)d\mu$ and this proves (iv). Q.E.D.

→ 3 **Theorem.** (*Riesz representation theorem*) *If S is a compact Hausdorff space there is an isometric isomorphism between $C^*(S)$ and $rca(S)$ such that the corresponding elements x^* and μ satisfy the identity*

$$[*] \qquad x^*f = \int_S f(s)\mu(ds), \qquad f \in C(S).$$

Furthermore, this isomorphism preserves order.

Proof. The previous proof shows that each $\mu \in rca(S)$ determines an $x^* \in C^*(S)$ by the formula [*], that $|x^*| = |\mu|$, and that the correspondence between x^* and μ is linear and preserves order. Thus, in order to prove our present theorem, we have only to show that each $\lambda \in rba(S)$ determines a $\mu \in rca(S)$ such that $\int_S f(s)\lambda(ds) = \int_S f(s)\mu(ds)$ for $f \in C(S)$. This, however, follows from III.5.14 and III.8.1(e). Q.E.D.

4 **Corollary.** *If $f_n, f \in C(S)$, $n = 1, 2, \ldots$, where S is a compact Hausdorff space, then the sequence $\{f_n\}$ converges weakly to f if and only if it is bounded and $f(s) = \lim_n f_n(s)$ for each s in S.*

PROOF. If $f_n \to f$ weakly then (II.3.20) $\sup_n |f_n| < \infty$. Also, for a fixed s in S the number $g(s)$ is continuous and linear in g. Thus, $f_n(s) \to f(s)$ for each s in S. The converse statement follows from the preceding theorem and the dominated convergence theorem of Lebesgue (III.6.16). Q.E.D.

5 THEOREM. *Let S be an arbitrary topological space and K a bounded subset of $C(S)$. Then K is conditionally compact if and only if for every $\varepsilon > 0$ there is a finite collection $\{E_1, \ldots, E_n\}$ of sets with union S, and points s_i in E_i, $i = 1, \ldots, n$, such that*

$$\sup_{f \in K} \sup_{s \in E_i} |f(s_i) - f(s)| < \varepsilon, \qquad i = 1, \ldots, n.$$

PROOF. If Σ is the field of all sets in S then $C(S)$ is, by Corollary I.7.7, a closed linear manifold in $B(S, \Sigma)$ and the theorem follows from Theorem 5.6. Q.E.D.

6 DEFINITION. A subset $K \subseteq C(S)$ is said to be *equicontinuous* if to every $\varepsilon > 0$ and every $s \in S$ there corresponds a neighborhood $N = N(s)$ of s with

$$\sup_{f \in K} \sup_{t \in N} |f(s) - f(t)| < \varepsilon.$$

An equivalent condition is as follows: if $\{s_\alpha\}$ is a convergent generalized sequence with $s_\alpha \to s$, then $f(s_\alpha) \to f(s)$ uniformly with respect to f in K.

7 THEOREM. (*Arzelà-Ascoli*) *If S is compact, then a set in $C(S)$ is conditionally compact if and only if it is bounded and equicontinuous.*

PROOF. Let K be an equicontinuous and bounded set in $C(S)$ and let $\varepsilon > 0$. From among the neighborhoods whose existence is asserted in the preceding definition, choose a finite number $N_1, \ldots, N_m$ which cover S. Then

$$\sup_{f \in K} \sup_{s \in N_i} |f(s_i) - f(s)| < \varepsilon,$$

and so, by Theorem 5, K is conditionally compact.

Conversely, let K be conditionally compact and hence totally bounded (I.6.15), and bounded (II.1.8). If $\varepsilon > 0$ there are functions $f_1, \ldots, f_n$ in K such that every $f \in K$ has distance less than $\varepsilon/3$ from one of the functions $f_1, \ldots, f_n$. For $s \in S$ choose the neighborhood $N = N(s)$ of s such that

$$|f_i(s)-f_i(t)| < \varepsilon/3, \qquad t \in N, \qquad i = 1, \ldots, n.$$

Then, for each f in K, each t in N, and some $i \leqq n$,

$$|f(s)-f(t)| \leqq |f(s)-f_i(s)|+|f_i(s)-f_i(t)|+|f_i(t)-f(t)| < \varepsilon,$$

and so K is equicontinuous. Q.E.D.

➜ 8 Corollary. *Let S be a compact metric space and K be a bounded set in $C(S)$. Then K is conditionally compact if and only if for every $\varepsilon > 0$ there is a $\delta > 0$ such that*

$$\sup_{f \in K} |f(s)-f(t)| < \varepsilon, \qquad \varrho(s, t) < \delta.$$

Proof. If the condition is satisfied then K is an equicontinuous family in $C(S)$ and, by Theorem 7, K is conditionally compact.

Conversely, let K be conditionally compact and suppose the condition is not satisfied. Then there is an $\varepsilon > 0$ and sequences $\{s_n\}, \{t_n\} \subseteqq S$, $\{f_n\} \subseteqq K$ with $|f_n(s_n)-f_n(t_n)| > \varepsilon$, $\varrho(s_n, t_n) \to 0$. Since S and $\overline{K}$ are compact metric spaces, it may be supposed that the sequences $\{s_n\}$ and $\{f_n\}$ are convergent. If $f_n \to f$ in $C(S)$ and $s_n \to s$, then $t_n \to s$ and, by Lemma I.7.6, $0 = |f(s)-f(s)| \geqq \varepsilon > 0$, a contradiction which proves the corollary. Q.E.D.

9 Corollary. *Let S be a compact subset of a topological group G and let K be a bounded set in $C(S)$. Then K is conditionally compact if and only if for every $\varepsilon > 0$ there is a neighborhood U of the identity in G such that $|f(t)-f(s)| < \varepsilon$ for every f in K and every pair s, t in S with t in Us.*

Proof. If the condition is satisfied, the family K is clearly equicontinuous and hence, by Theorem 7, conditionally compact. Conversely, if K is conditionally compact then, by Theorem 7, it is equicontinuous and for every $\varepsilon > 0$ and $s \in S$ there is a neighborhood V_s of the unit in G such that

$$\text{(i)} \qquad |f(t)-f(s)| < \varepsilon/2, \qquad f \in K,$$

if t is a point in S with $t \in V_s s$. Since the group operations are continuous, there is a neighborhood U_s of the unit of G such that $U_s^{-1}U_s \subset V_s$. Since S is compact, a finite set $U_{s_1}s_1, \ldots, U_{s_n}s_n$ covers S. Let $U = \cap_{i=1}^n U_{s_i}$ and let $s, t \in S$ with $t \in Us$. Then, for some j and $u_j \in U_{s_j}$ and some $u \in U$, $t = us$, $t = u_j s_j$, and

$$s = u^{-1}u_j s_j \in U^{-1}U_{s_j}s_j \subseteq U_{s_j}^{-1}U_{s_j}s_j \subset V_{s_j}s_j.$$

Thus it follows from (i) that

$$|f(s)-f(s_j)| < \varepsilon/2, \qquad f \in K.$$

Since

$$|f(t)-f(s)| \leqq |f(t)-f(s_j)| + |f(s_j)-f(s)|,$$

and since $t \in U_{s_j}s_j \subset V_{s_j}s_j$, it follows that

$$|f(t)-f(s)| < \varepsilon \quad \text{for } f \in K. \qquad \text{Q.E.D.}$$

10 Definition. A generalized sequence $\{f_\alpha\}$ of functions on a set S is said to be *quasi-uniformly convergent on* S if there exists a function f_0 on S with $f_\alpha(s) \to f_0(s)$ for every s in S, and such that for every $\varepsilon > 0$ and α_0 there exists a finite number of indices $\alpha_1, \ldots, \alpha_n \geqq \alpha_0$ such that for each $s \in S$,

$$\min_{1 \leqq i \leqq n} |f_{\alpha_i}(s) - f_0(s)| < \varepsilon.$$

11 Theorem. (*Arzelà*) *If S is a compact Hausdorff space and $\{f_\alpha\}$ is a generalized sequence in $C(S)$ which converges at each point of S to a function f_0, then f_0 is continuous if and only if $\{f_\alpha\}$ converges quasi-uniformly on S.*

Proof. If $f_0 \in C(S)$, then given α_0, $\varepsilon > 0$ and any $t \in S$, there is an $\alpha(t) \geqq \alpha_0$ such that $|f_{\alpha(t)}(t) - f_0(t)| < \varepsilon$. Let $N(t) = \{s \,|\, |f_{\alpha(t)}(s) - f_0(s)| < \varepsilon\}$; since f_0 is continuous, $N(t)$ is an open set containing t. By the compactness of S, only a finite number of $N(t)$ are required to cover S, which proves the quasi-uniform convergence of $\{f_\alpha\}$.

Conversely, suppose that $\{f_\alpha\}$ converges quasi-uniformly to f_0. We will show that the continuity of f_0 at a point s_0 follows from that of the f_α. For, given $s_0 \in S$, $\varepsilon > 0$, there exists an α_0 such that if $\alpha \geqq \alpha_0$, then $|f_\alpha(s_0) - f_0(s_0)| < \varepsilon$. Select $\alpha_1, \ldots, \alpha_n \geqq \alpha_0$ as described in the definition of quasi-uniform convergence, and let

$$N_i(s_0) = \{s \,|\, |f_{\alpha_i}(s) - f_{\alpha_i}(s_0)| < \varepsilon\} \quad \text{for } i = 1, \ldots, n.$$

Since the f_α are continuous, the sets $N_i(s_0)$ are open; hence $N(s_0) = \bigcap_{i=1}^{n} N_i(s_0)$ is open and contains s_0. Now for the proper choice of i and for any $s \in N(s_0)$ we have

$$|f_0(s) - f_0(s_0)| \leqq |f_0(s) - f_{\alpha_i}(s)| + |f_{\alpha_i}(s) - f_{\alpha_i}(s_0)| + |f_{\alpha_i}(s_0) - f_0(s_0)| < 3\varepsilon.$$

Thus f_0 is continuous at the point s_0. Q.E.D.

12 COROLLARY. *A sequence in $C(S)$ is weakly convergent if and only if it is bounded and quasi-uniformly convergent on S.*

13 DEFINITION. A family $F = \{f\} \subseteq C(S)$ is said to be *quasi-equicontinuous on S* if $s_\alpha \to s_0$ implies that the convergence $f(s_\alpha) \to f(s_0)$ is quasi-uniform on F. That is, given any $\varepsilon > 0$ and α_0 there exists a finite set of indices $\alpha_i \geqq \alpha_0$, $i = 1, \ldots, n$, such that for each $f \in F$,

$$\min_{1 \leqq i \leqq n} |f(s_{\alpha_i}) - f(s_0)| < \varepsilon.$$

We recall that if F is a family of functions from a set S to the scalar field Φ, then F can be regarded as a subset of the product space $P_{s \in S} \Phi$ under the one-to-one embedding $f \to P_{s \in S} f(s)$. The relative *product topology* on F is the topology generated by the neighborhoods $N(f_0; A, \varepsilon) = \{f | f \in F, |f(s) - f_0(s)| < \varepsilon, s \in A\}$ where A is a finite subset of S. It is clear that convergence of a generalized sequence $\{f_\alpha\}$ in the product topology is equivalent to the convergence of the scalars $\{f_\alpha(s)\}$ for each $s \in S$. The relative product topology on F is seen to be the weakest topology on F in which each $s \in S$ generates a *continuous* function $\hat{s}$ on F defined by $\hat{s}(f) = f(s)$, $f \in F$.

The *weak topology* of $C(S)$ is the topology generated by the neighborhoods $N(f_0; B, \varepsilon) = \{f | f \in C(S), |x^*(f - f_0)| < \varepsilon, x^* \in B\}$ where B is a finite set in $C^*(S)$. Since every point in S gives rise to a continuous linear functional on $C(S)$, it is clear that the product topology of $C(S)$ is weaker than the weak topology of $C(S)$.

The next theorem indicates a close relation between weak sequential compactness of a set of continuous functions on a compact Hausdorff space, compactness in the product topology, compactness in the weak topology, and quasi-equicontinuity.

14 THEOREM. *Let S be a compact Hausdorff space and $F \subseteq C(S)$. Then the following conditions are equivalent.*

(1) *The closure of F in the weak topology of $C(S)$ is weakly compact.*

(2) *F is bounded and its closure in the product topology is a compact set of continuous functions in this topology.*

(3) *F is bounded and quasi-equicontinuous on S.*

(4) *F is bounded and if F_0 is a denumerable subset of F and*

$\{s_0, s_1, s_2, \ldots\}$ is a sequence in S for which $f(s_n) \to f(s_0)$, $f \in F_0$, then $s_n \to s_0$ quasi-uniformly on F_0.

(5) *F is weakly sequentially compact.*

Proof. If (1) holds, then, since the product topology of $C(S)$ is weaker than the weak topology, the weak closure of F in $C(S)$ is also compact in the product topology. Also, since a continuous scalar function on a compact set is bounded (I.5.10), the set x^*F is bounded for each x^* in $\mathfrak{X}^*$ and thus (II.3.20) the set F is bounded in the norm topology of $\mathfrak{X} = C(S)$. Hence (1) implies (2).

To see that (2) implies (3), let $\overline{F}$ be the closure of F in the product topology. Since each $f \in \overline{F}$ is a continuous function, if $\{s_\alpha\}$ is a generalized sequence converging to $s_0 \in S$, then $f(s_\alpha) \to f(s_0)$. On the other hand, each $s \in S$ gives rise to a continuous function $\hat{s}$ on the compact Hausdorff space $\overline{F}$. Thus $\hat{s}_\alpha(f) \to \hat{s}_0(f)$ for every $f \in \overline{F}$, and since $\hat{s}_0$ is continuous, it follows from Theorem 11 that this convergence must be quasi-uniform on $\overline{F}$. But this implies that F is quasi-equicontinuous in the sense of Definition 13.

Let (3) be true and suppose that $s_0, s_1, s_2, \ldots$ is a sequence in S such that $f(s_n) \to f(s_0)$ for f in some denumerable subset F_0 of F. Let $\mathscr{E}_0$ be the collection of all subsets of S which contain some set $E_n = \{s_n, s_{n+1}, \ldots\}$, $n = 1, 2, \ldots$. Clearly $\mathscr{E}_0$ is a filter on S (cf. Definition I.7.10); let $\mathscr{E} = \{K_\alpha\}$, $\alpha \in A$, be an ultrafilter which refines $\mathscr{E}_0$. Then every set $K_\alpha \in \mathscr{E}$ contains some point s_n with $n \geqq 1$, for otherwise $\phi = E_1 \cap K_\alpha \in \mathscr{E}$, which is a contradiction. For each K_α in $\mathscr{E}$, let $t_\alpha = s_n$, where n is such that s_n is in $E_1 \cap K_\alpha$. If the set A is ordered by the requirement that $\alpha \leqq \beta$ means that $K_\alpha \supseteq K_\beta$, then it is clear that A is a directed set. It follows from Lemma I.7.12 that the ultrafilter $\mathscr{E}$ converges to a unique point $t_0 \in S$, and hence that the generalized sequence $\{t_\alpha\}$ converges to t_0. Thus the hypothesis in (3) implies that the convergence of $\{f(t_\alpha)\}$ to $f(t_0)$ is quasi-uniform for $f \in F$ and *a fortiori* in F_0, and it is readily seen that $f(s_0) = f(t_0)$ for $f \in F_0$, since $E_n \in \mathscr{E}$ for each n.

Now let $\varepsilon > 0$ and n_0 be given, and let α_0 be the index corresponding to E_{n_0}. Then by the quasi-uniformity of the convergence, there exist $\alpha_1, \ldots, \alpha_r \geqq \alpha_0$ such that

$$\min_{1 \leqq j \leqq r} |f(t_{\alpha_j}) - f(t_0)| < \varepsilon, \qquad f \in F_0.$$

But $t_{\alpha_j} = s_{n_j}$, $j = 1, \ldots, r$, where each $n_j \geqq n_0$. Hence we have shown that there exists $n_1, \ldots, n_r \geqq n_0$ such that

$$\min_{1 \leqq j \leqq r} |f(s_{n_j}) - f(s_0)| < \varepsilon, \qquad f \in F_0,$$

which proves the quasi-uniform convergence of $f(s_n)$ to $f(s_0)$. Thus (3) implies (4).

We now show that (4) implies (5). Let $F_0 = \{f_1, f_2, \ldots\}$ be a denumerable set of functions in F. Let

$$\varrho(s, t) = \sum_{i=1}^{\infty} \frac{1}{2^i} \frac{|f_i(s) - f_i(t)|}{1 + |f_i(s) - f_i(t)|}, \qquad s, t \in S.$$

Then ϱ defines a metric on the set S which may identify points in S; we denote this metric space by S_ϱ. It is clear that the natural mapping of S onto S_ϱ is continuous (but perhaps not one-to-one); thus S_ϱ is a compact metric space. By I. 6. 15, S_ϱ is separable. Let $T = \{t_1, t_2, \ldots\}$ be a denumerable dense subset of S_ϱ. By a diagonal process, one can select a subsequence $\{g_n\}$ of $\{f_n\}$ such that g_n converges on T to a limit f_0 defined on T.

It will next be shown that f_0 has a continuous extension $\tilde{f}_0$ to all of S_ϱ and that $\{g_n\}$ converges at every point of S_ϱ to $\tilde{f}_0$. If $\{t_r\} \subseteqq T$ and $t_r \to s_0$ in S_ϱ, then from (4), the convergence of every subsequence of $\{t_r\}$ is quasi-uniform on F_0. We now show that if a subsequence $\{h_i\}$ of $\{g_n\}$ has the property that $h_i(s_0) \to L$, then $f_0(t_r) \to L$. For if $h_i(s_0) \to L$, then given $\varepsilon > 0$ there is an i_0 such that if $i \geqq i_0$ then $|h_i(s_0) - L| < \varepsilon$. Let $\{t_r'\}$ be a subsequence of $\{t_r\}$, so that given any r_0 there is a finite set of indices $r_1, \ldots, r_m \geqq r_0$ such that if $f \in F_0$ there is some index j with $1 \leqq j \leqq m$ such that $|f(t_{r_j}') - f(s_0)| < \varepsilon$. Since $\{h_i\}$ is a subsequence of $\{g_n\}$, it converges on T. Thus there exists a j_0 such that if $i \geqq j_0$ then

$$|h_i(t_{r_j}') - f_0(t_{r_j}')| < \varepsilon, \qquad j = 1, \ldots, m.$$

Now take a fixed $i \geqq i_0, j_0$; then for some j we have

$$|f_0(t_{r_j}') - L| \leqq |f_0(t_{r_j}') - h_i(t_{r_j}')| + |h_i(t_{r_j}') - h_i(s_0)| + |h_i(s_0) - L| < 3\varepsilon.$$

This shows that every subsequence $\{t_r'\}$ of $\{t_r\}$ has a subsequence $\{t_r''\}$ such that $f_0(t_r'') \to L$, which implies that $f_0(t_r) \to L$. Now if some other subsequence of $\{g_n\}$ converges at s_0 to a limit L', then the argument ust given shows that $f_0(t_r) \to L'$. Therefore $L' = L$.

Thus we conclude that $g_n(s_0)$ converges to a limit L and that if $t_r \to s_0$ in S_ϱ then $f_0(t_r) \to L$. It follows that f_0 has a unique continuous extension $\bar{f}_0$ to S_ϱ and that $\{g_n\}$ converges to $\bar{f}_0$ on S_ϱ. Since the mapping of S into S_ϱ is continuous, and weak convergence of a bounded sequence in $C(S)$ is implied by pointwise convergence, we conclude that F is weakly sequentially compact.

The fact that (5) implies (1) is due to Eberlein and is proved in Chapter V (cf. V.6.1). Q.E.D.

Remark. It should be observed that we can require that the points $s_1, s_2, \ldots$ in condition (4) be contained in a preassigned dense subset of S. This fact will be used in the proof of Theorem 29.

We continue our analysis of the space $C(S)$ with a discussion of certain important special properties related to its structure as an algebra. One of these properties is a well known approximation theorem of Weierstrass, which asserts that a scalar function continuous on a compact interval of real numbers is the uniform limit on the interval of polynomials. This important theorem has had a number of far-reaching generalizations; notable among these is one due to M. H. Stone which will be discussed here. We note first that $C(S)$ is an *algebra*, for if f and g are in $C(S)$, then the product fg, defined by $(fg)(s) = f(s)g(s)$, is also in $C(S)$ (cf. I.4.18). The algebra $C(S)$ has a unit e, i.e., $ef = f$ for f in $C(S)$. This unit is defined by the equation $e(s) = 1, s \in S$. A *closed subalgebra* of $C(S)$ is a closed linear manifold in $C(S)$ which contains the product of every pair of its elements.

15 Definition. A set A in $B(S)$ is said to *distinguish between the points of* S if for every pair s, t of distinct points of S there is a function f in A with $f(s) \neq f(t)$.

Stone's generalization of the Weierstrass theorem may be stated in terms of these concepts as follows:

16 Theorem. *Let S be a compact Hausdorff space and $C(S)$ be the algebra of all real continuous functions on S. Let $\mathfrak{A}$ be a closed subalgebra of $C(S)$ which contains the unit e. Then $\mathfrak{A} = C(S)$ if and only if $\mathfrak{A}$ distinguishes between the points of S.*

Proof. Let $\mathfrak{A} = C(S)$ and s and t be distinct points of S. Since a compact Hausdorff space is normal (I.5.9), there is, by Urysohn's lemma (I.5.2), a function $f \in \mathfrak{A}$ with $f(s) = 1$ and $f(t) = 0$. To prove the converse we define the functions $f \vee g$, $f \wedge g$, and $\phi(f)$ as follows:

$$(f \vee g)(s) = \max \{f(s), g(s)\},$$
$$(f \wedge g)(s) = \min \{f(s), g(s)\},$$
$$\phi(f)(s) = |f(s)|.$$

Now by the classical Weierstrass theorem there is a sequence P_n of polynomials with

$$||\lambda| - P_n(\lambda)| \leqq 1/n, \qquad -n \leqq \lambda \leqq n.$$

Thus

$$||g(s)| - P_n(g)(s)| = ||g(s)| - P_n(g(s))| \leqq 1/n,$$

provided that $-n \leqq g(s) \leqq n$. This shows that $\phi(g) \in \mathfrak{A}$ if $g \in \mathfrak{A}$. Since

$$f \vee g = (f+g)/2 + \phi(f-g)/2,$$

and

$$f \wedge g = (f+g)/2 - \phi(f-g)/2,$$

we see that $\mathfrak{A}$ is closed under the lattice operations $\vee$ and $\wedge$. Next, we note that for an arbitrary $F \in C(S)$ and an arbitrary pair $s, t \in S$ there is an $f_{s,t} \in \mathfrak{A}$ with $f_{s,t}(s) = F(s)$ and $f_{s,t}(t) = F(t)$. To see this, let $g \in \mathfrak{A}$ and $g(s) \neq g(t)$. Then real numbers α and β may be found so that

$$\alpha g(s) + \beta = F(s),$$
$$\alpha g(t) + \beta = F(t).$$

Now if $t \in S$, then for each $s \in S$ there is a neighborhood U_s of s such that $f_{s,t}(u) > F(u) - \varepsilon$ for $u \in U_s$. Suppose that $U_{s_1}, \ldots, U_{s_p}$ cover S and define

$$f_t = f_{s_1,t} \vee \ldots \vee f_{s_p,t}.$$

Thus $f_t(u) > F(u) - \varepsilon$ for $u \in S$. Since $f_{s_i,t}(t) = F(t)$, we have $f_t(t) = F(t)$, and hence there is a neighborhood V_t of t with

$$f_t(u) < F(u) + \varepsilon, \quad u \in V_t.$$

Let $V_{t_1}, \ldots, V_{t_q}$ cover S, and define

$$f = f_{t_1} \wedge \ldots \wedge f_{t_q}.$$

Since $f_{t_i}(u) > F(u) - \varepsilon$, $u \in S$, we have also

$$f(u) > F(u) - \varepsilon, \qquad u \in S.$$

On the other hand, for an arbitrary $u \in S$, say $u \in V_{t_i}$, we have

$$f(u) \leqq f_{t_i}(u) < F(u)+\varepsilon,$$

and so

$$|f(u)-F(u)| < \varepsilon, \qquad u \in S.$$

Since $f \in \mathfrak{A}$ and $\varepsilon > 0$ is arbitrary, the theorem is proved. Q.E.D.

➡ 17 THEOREM. *Let S be a compact Hausdorff space and $C(S)$ be the algebra of all complex continuous functions on S. Let $\mathfrak{A}$ be a closed subalgebra of $C(S)$ which contains the unit e and contains, with f, its complex conjugate $\bar{f}$ defined by $\bar{f}(s) = \overline{f(s)}$. Then $\mathfrak{A} = C(S)$ if and only if $\mathfrak{A}$ distinguishes between the points of S.*

PROOF. The necessity of the condition is proved as it was in Theorem 16. To prove the converse, let $\mathfrak{A}_r$ consist of the real functions in $\mathfrak{A}$. Then $\mathfrak{A}_r$ is a closed subalgebra containing the unit of the real algebra $C_r(S)$ of all real continuous functions on S. If $f \in \mathfrak{A}$ and $f = f_1+if_2$ with f_1, f_2 real, then $f_1 = (f+\bar{f})/2$ and $f_2 = (f-\bar{f})/2i$ are in $\mathfrak{A}$ and thus in $\mathfrak{A}_r$. Hence if $f(s) \neq f(t)$, then either $f_1(s) \neq f_1(t)$ or $f_2(s) \neq f_2(t)$. This shows that $\mathfrak{A}_r$ distinguishes between the points of S if $\mathfrak{A}$ does. The preceding theorem gives $C_r(S) = \mathfrak{A}_r \subset \mathfrak{A}$. Since every function f in $C(S)$ is a linear combination $f = f_1+if_2$ of real functions f_1, f_2 in $C_r(S)$, it follows that $\mathfrak{A} = C(S)$. Q.E.D.

Theorem 17 can be used to establish an intimate relation between the space $B(S)$ and the space of continuous functions on a certain compact Hausdorff space. The space $B(S)$ is an algebra under the natural definition of multiplication, $(fg)(s) = f(s)g(s)$. Also, $B(S)$ has a unit e defined by $e(s) = 1$, $s \in S$. As in the case of the algebras of continuous functions, a *closed subalgebra of* $B(S)$ is defined to be a closed linear manifold in $B(S)$ which contains the product of every pair of its elements.

18 THEOREM. *Let $\mathfrak{A}$ be a closed subalgebra of the complex algebra $B(S)$ containing the unit e as well as the complex conjugate of each of its elements. Then there exists a compact Hausdorff space S_1 and an isometric algebraic isomorphism U between the algebras $C(S_1)$ and $\mathfrak{A}$. Furthermore, U maps real functions into real functions, positive functions into positive functions, and complex conjugate functions into complex conjugate functions,* i.e., $U\bar{f} = \overline{Uf}$ *for every f in $\mathfrak{A}$. Moreover, if β is an arbitrary contin-*

uous complex function of a complex variable and if f is in $\mathfrak{A}$, then $\beta(f)$ is in $\mathfrak{A}$ and $U(\beta(f)) = \beta(U(f))$.

Proof. Let S_1 consist of those non-zero continuous linear functionals in the closed unit sphere of $\mathfrak{A}^*$ for which $x^*(fg) = (x^*f)(x^*g)$ and $x^*(\bar{f}) = \overline{x^*(f)}$. Every $s \in S$ defines an $x_s^* \in S_1$ by the equation $x_s^*f = f(s)$. Clearly

$$\text{(i)} \qquad \sup_s |x_s^*f| = |f|.$$

If $x^* \in S_1$ there is an $f \in \mathfrak{A}$ with $x^*f \neq 0$ and hence, since $x^*f = x^*(fe) = (x^*f)(x^*e)$, it is seen that $x^*e = 1$ and $|x^*| \geqq 1$. If $|f| \leqq 1$ then $|x^*f|^n = |x^*(f^n)| \leqq |x^*|$ and so $1 \leqq |x^*| \leqq 1$. This fact combined with (i) gives

$$\text{(ii)} \qquad \sup_{x^* \in S_1} |x^*f| = |f|, \qquad f \in \mathfrak{A}.$$

Since

$$x^*f \in I_f = \{\lambda | \lambda \in \Phi, |\lambda| \leqq |f|\}, x^* \in S_1, f \in \mathfrak{A},$$

the topology in the product space $P\, I_f$ taken over $f \in \mathfrak{A}$ induces a relative topology in S_1. Since $P\, I_f$ is a compact Hausdorff space(I.8.2, I.8.5), S_1 is also a compact Hausdorff space if it is closed (I.5.7). Let $\lambda \in \bar{S}_1$; then (I.7.2) some generalized sequence $\{x_\alpha^*\} \subseteqq S_1$ converges to λ. This means that $x_\alpha^*f \to \lambda f$ for f in $\mathfrak{A}$. It follows that

$$\lambda(fg) = \lim_\alpha x_\alpha^*(fg) = \lim_\alpha (x_\alpha^*f)(x_\alpha^*g) = (\lambda f)(\lambda g).$$

It may be shown similarly that λ is linear and that $|\lambda f| \leqq |f|$. Thus $\lambda \in S_1$ and S_1 is compact. Define the map $U : f \to f_1$ of $\mathfrak{A}$ into $C(S_1)$ by placing $f_1 x^* = x^*f$. Then U is linear and it follows from (ii) that U is an isometry. Clearly $U\mathfrak{A}$ distinguishes between the points of S_1, and since $x^*e = 1$ for x^* in S_1, $U\mathfrak{A}$ contains the unit in $C(S_1)$. Since $|Uf| = |f|$, the algebra $U\mathfrak{A}$ is closed in $C(S_1)$. Thus Theorem 17 shows that $U\mathfrak{A} = C(S_1)$.

It is clear that U maps products into products, complex conjugate functions into complex conjugate functions, and therefore real functions into real functions. Thus if α is a polynomial in two variables, then

$$U(\alpha(f, \bar{f})) = \alpha(Uf, \overline{Uf}).$$

By the Weierstrass theorem there is a sequence $\{\alpha_n\}$ of such polynomials for which $\alpha_n(\lambda, \bar{\lambda})$ converges to $\beta(\lambda)$ uniformly for λ in the range of f. Thus

$$\beta(f(s)) = \lim_n \alpha_n(f(s), \overline{f(s)})$$

uniformly for s in S. Consequently $\beta(f)$ is in $\mathfrak{A}$ and $U(\beta(f)) = \beta(U(f))$. If we consider the function $\beta(\lambda) = |\lambda|$ it is clear that U maps positive functions into positive functions. Q.E.D.

19 COROLLARY. *Suppose, in addition to the hypotheses of Theorem* 18, *that the functions of $\mathfrak{A}$ distinguish between the points of S. Then there exists a compact Hausdorff space S_1 and a one-to-one embedding of S as a dense subset of S_1 such that each f in $\mathfrak{A}$ has a unique continuous extension f_1 to S_1, and such that the correspondence $f \longleftrightarrow f_1$ is an isometric isomorphism of $\mathfrak{A}$ and $C(S_1)$.*

PROOF. We use the notations of the proof of Theorem 18. It is clear that the map $s \to x_s^*$ is a one-to-one embedding of S in S_1, and to prove the corollary it will suffice to show that S is dense in S_1. If this is not the case, there exists, by I.5.2, an $f \in C(S_1)$ such that $f \neq 0$ and $f(s) = 0$ for $s \in S$. If $g \in \mathfrak{A}$ is such that $Ug = f$, then $g(s) = x_s^* f = 0$ for $s \in S$, i.e., $g = 0$, which contradicts the fact that $0 \neq f = Ug$. Q.E.D.

20 THEOREM. *Let the closed subalgebra $\mathfrak{A}$ of the real algebra $B(S)$ contain the unit. Then there is a compact Hausdorff space S_1 and an isometric isomorphism between the real algebras $\mathfrak{A}$ and $C(S_1)$.*

PROOF. The proof follows the lines of that of Theorem 18, except that Theorem 16 is used in place of Theorem 17. Q.E.D.

21 DEFINITION. A topological space S is *completely regular* if points are closed and, given any point s_0 in S and any closed set F not containing s_0, there is a function f defined and continuous on S and satisfying $0 \leq f(s) \leq 1$, $f(s_0) = 0$, and $f(s) = 1$ for s in F.

This class of spaces contains, for example, all normal and all compact Hausdorff spaces (cf. I.5.2 and I.5.9).

22 THEOREM. *If S is completely regular, it is homeomorphic with a dense subset of a compact Hausdorff space S_1 such that every bounded*

continuous complex function on S has a unique extension to a continuous function to S_1.

PROOF. Since the algebra $\mathfrak{A} = C(S)$ satisfies the hypotheses of Corollary 19, the present theorem follows from that corollary if it is shown that the one-to-one embedding of S in S_1 is a homeomorphism. In the notation introduced in the proof of Theorem 18 this means that it will suffice to show that the one-to-one correspondence $s \longleftrightarrow x_s^*$ between S and a subset S_0 of S_1 is a topological one. Here we are using the symbol S_0 for the set of all points x^* in S_1 having the form $x^* = x_s^*$ for some s in S. Let $\varepsilon > 0$, $s_0 \in S$, and $f_1, \ldots, f_n \in C(S)$. Then the set $\{s \mid |f_i(s) - f_i(s_0)| < \varepsilon, i = 1, \ldots, n\}$ is open and contains s_0. Its correspondent in S_0 is the set

$$\{x_s^* \mid |x_s^* f_i - x_{s_0}^* f_i| < \varepsilon, i = 1, \ldots, n\},$$

which is, by definition, the general neighborhood in S_0. Thus to complete the proof it will suffice to show that neighborhoods in S of the form $\{s \mid |f_i(s) - f_i(s_0)| < \varepsilon, i = 1, \ldots, n\}$ form a basis for the topology in S. This may be done by using the complete regularity of S in the following way. Let s_0 be in the open set G in S. Then there is a function f in $C(S)$ with $0 \leqq f(s) \leqq 1$, $f(s_0) = 0$, and $f(s) = 1$ for s in the complement of G. The set

$$B = \{s \mid |f(s) - f(s_0)| < 1/2\}$$

is a neighborhood of s_0 which is contained in G. This completes the proof of the theorem. Q.E.D.

To see the remarkable nature of the preceding theorem suppose that S is the half open interval $0 < s \leqq 1$ of real numbers. The space S is clearly a dense subset of the compact interval $0 \leqq s \leqq 1$, but the functions $\sin(1/s)$ on S has no continuous extension to this interval. Even in this simple case the compact space S_1 of the preceding theorem has no simple or familiar representation.

23 DEFINITION. A linear functional x^* on an algebra $\mathfrak{A}$ will be called *multiplicative* if $x^*(fg) = (x^*f)(x^*g)$ for every f and g in $\mathfrak{A}$.

24 LEMMA. *Let $\mathfrak{A}$ be a closed subalgebra of the algebra $B(S)$ and let $\mathfrak{A}$ contain the unit e. Then any non-zero multiplicative linear functional on $\mathfrak{A}$ is continuous and has norm one.*

Proof. Suppose that x^* is a multiplicative linear functional on $\mathfrak{A}$ and that f is an element of $\mathfrak{A}$ with $x^*f \neq 0$. Then $x^*f = x^*(ef) = (x^*e)(x^*f)$ so that $x^*e = 1$ and hence $|x^*| \geqq 1$. Next let g be an arbitrary element of $\mathfrak{A}$ with $|g| \leqq 1$ and let λ be a scalar with $|\lambda| > 1$. Then the series $\sum_{n=0}^{\infty} g^n/\lambda^{n+1}$ converges to an element h in $\mathfrak{A}$ and

$$(\lambda e - g)h = \sum_{n=0}^{\infty} \left(\frac{g^n}{\lambda^n} - \frac{g^{n+1}}{\lambda^{n+1}} \right) = e.$$

Thus

$$1 = x^*e = (\lambda - x^*g)x^*h,$$

which shows that $x^*g \neq \lambda$. Since λ is an arbitrary scalar with $|\lambda| > 1$ it follows that $|x^*g| \leqq 1$. Since g is an arbitrary element of $\mathfrak{A}$ with $|g| \leqq 1$ it follows that $|x^*| \leqq 1$. Therefore $|x^*| = 1$. Q.E.D.

25 **Lemma.** *Let S be a compact Hausdorff space and x^* a non-zero multiplicative linear functional on $C(S)$. In the complex case it is also assumed that $x^*\bar{f} = \overline{x^*f}$. Then there is a point s in S such that*

$$x^*f = f(s), \qquad f \in C(S).$$

Proof. By the preceding lemma x^* is a point in the space S_1 defined in the proof of Theorem 18. By Theorem 22, S is homeomorphic to a dense subset S_0 in S_1. Since S is compact, S_0 is compact and thus closed (I.5.7). Therefore $S_0 = S_1$ and $x^* \in S_0$. According to the definition of S_0 as given in the proof of Theorem 22, this means that for some s in S we have $x^*f = f(s)$ for every f in $C(S)$. Q.E.D.

26 **Theorem.** *Let H be an algebraic homomorphism of $C(S)$ into $C(T)$, where S and T are compact Hausdorff spaces. If the algebras $C(S)$ and $C(T)$ are over the field of complex numbers, it is also assumed that $H\bar{f} = \overline{Hf}$. Then H is continuous and has the form*

$$(Hf)(t) = f(h(t)), \qquad t \in T, \qquad f \in C(S),$$

where h is a continuous map of T into S. If H is an isomorphism of $C(S)$ onto $C(T)$ then h is a homeomorphism of T onto S.

Proof. For each t in T define the multiplicative linear functional x_t^* on $C(T)$ by the equation

$$x_t^*f = f(t), \qquad f \in C(T).$$

For each t in T let $y^*(t)$ be the functional on $C(S)$ defined by the equation

$$y^*(t)f = x_t^* Hf, \qquad f \in C(S).$$

Since x_t^* and H are both multiplicative, it is clear that $y^*(t)$ is multiplicative. Thus by the preceding lemma there is a point $s = h(t)$ in S such that $y^*(t)f = f(s)$. Hence

$$\text{(i)} \qquad (Hf)(t) = f(h(t)), \qquad t \in T, \qquad f \in C(S),$$

which shows that $|Hf| \leqq |f|$ and proves the H is continuous. To see that h is continuous, let N be a neighborhood of the point $s_0 = h(t_0)$. By Theorems I.5.2 and I.5.9 there is a continuous function f on S with $f(s_0) = 1$ and $f(s) = 0$ for every s in the complement of N. Since $f(h(t)) = (Hf)(t)$ is continuous in t, the set

$$U = \{t | f(h(t)) \neq 0\}$$

is a neighborhood of t_0. If t is in U then $f(h(t)) \neq 0$. This shows that $h(t)$ is in N. Thus $h(U) \subseteq N$ and h is continuous. If H is an isomorphism with $HC(S) = C(T)$ the result already proved yields a continuous function h_1 on S to T with

$$\text{(ii)} \qquad (H^{-1}f)(s) = f(h_1(s)), \qquad s \in S, \qquad f \in C(T).$$

This, combined with (i), shows that

$$\text{(iii)} \qquad f(s) = f(h(h_1(s))), \qquad s \in S, \qquad f \in C(S).$$

Since there are sufficiently many functions in $C(S)$ to distinguish between points of S, it follows that $s = h(h_1(s))$. Similarly, $t = h_1(h(t))$ for every t in T. Since h and h_1 are continuous, the proof is complete. Q.E.D.

27 COROLLARY. *If S and T are compact Hausdorff spaces such that the real algebras $C(S)$ and $C(T)$ are algebraically equivalent, then the spaces S and T are homeomorphic.*

Corollary 27 shows that the compact Hausdorff space S_1 associated with a given completely regular space S in the manner described in Theorem 22 is unique. It is called the *Stone-Čech compactification* of S.

We conclude this section by showing how Theorem 14 may be used to derive conditions for weak sequential compactness in $B(S)$. First, however, a few introductory remarks must be made.

Let $\mathscr{E} = \{E_\alpha\}$ be an ultrafilter of sets on the set S, and for each α, let s_α be some point in E_α. If $\{\alpha\}$ is ordered by the requirement that $\alpha \leqq \beta$ means that $E_\alpha \supseteqq E_\beta$, then it is evident that $\{s_\alpha\}$ is a generalized sequence of points in S. We will say that the generalized sequence $\{s_\alpha\}$ is *generated by the ultrafilter* $\mathscr{E} = \{E_\alpha\}$. Further, for each $f \in B(S)$ the generalized sequence $\{f(s_\alpha)\}$ of scalars has a limit, which we will denote by $f(\mathscr{E})$. Though it is obvious that each ultrafilter generates many generalized sequences, it is easily seen that if $\{s'_\alpha\}$ is any other generalized sequence generated by $\mathscr{E}$, then $\{f(s'_\alpha)\}$ also converges to $f(\mathscr{E})$, so that this notation is unambiguous.

28 Definition. *A* set $F \subseteqq B(S)$ is said to be *quasi-equicontinuous on* S if every generalized sequence $\{s_\alpha\}$ in S which is generated by an ultrafilter $\mathscr{E}$ has the property that $f(s_\alpha)$ converges to $f(\mathscr{E})$ quasi-uniformly on F. Note that in this definition we make no assumption as to the nature of the set S. If S is a compact Hausdorff space, it can be proved that this definition is equivalent to the statement that $F \subseteqq C(S)$ and is quasi-equicontinuous on S in the sense of Definition 13.

29 Theorem. *Let S be an arbitrary set and $F \subseteqq B(S)$. Then the following statements are equivalent.*

(1) *F is bounded and quasi-equicontinuous on S.*

(2) *F is bounded and if F_0 is a denumerable subset of F and $\{s_1, s_2, \ldots\}$ is a sequence in S for which $\{f(s_n)\}$ converges for each $f \in F_0$, then the convergence is quasi-uniform on F_0.*

(3) *F is weakly sequentially compact.*

Proof. That (1) implies (2) can be proved in a manner similar to that used in Theorem 14 to show that condition (3) of that theorem implies (4).

From Corollary 19 it follows that S may be embedded as a dense subset of a compact Hausdorff space S_1 in such a way that each $f \in B(S)$ has a unique extension f_1 to $C(S_1)$, and so that the correspondence $f \longleftrightarrow f_1$ is an isometric isomorphism of $C(S)$ with $C(S_1)$. That (2) implies (3) then follows by virtue of our observation after Theorem 14 that it is sufficient to select the sequence of points of Theorem 14(4) from a set S dense in S_1. The relation between $B(S)$ and $C(S_1)$ also shows that the implication that (3) implies (1) follows from the implication that (5) implies (3) in Theorem 14. Q.E.D.

30 LEMMA. *Let A be a dense subset of a compact Hausdorff space S, and suppose that a sequence $\{f_n\}$ of continuous functions converges at every point of A to a continuous limit f_0. Then $\{f_n\}$ converges to f_0 at every point of S if and only if $\{f_n\}$ and every subsequence of $\{f_n\}$ converges to f_0 quasi-uniformly on A.*

PROOF. Theorem 11 implies that the condition is necessary. To prove the sufficiency, suppose that $f_n(s_0)$ does not converge to $f_0(s_0)$. Then there exists an ε_0 and a subsequence $\{g_k\}$ such that $|g_k(s_0)-f_0(s_0)| > \varepsilon_0$, $k = 1, 2, \ldots$. Let $k_1, \ldots, k_r$ be the indices corresponding to ε_0 and $k = 1$ guaranteed by the quasi-uniform convergence of $\{g_k\}$. Then

$$U_i = \{s \,|\, |g_{k_i}(s)-f_0(s)| > \varepsilon_0\}$$

is an open set containing s_0 for $i = 1, \ldots, r$. Since A is dense in S, there exists a point $s \in A \cap U_1 \cap \ldots \cap U_r$ for which $|g_{k_i}(s)-f_0(s)| > \varepsilon_0$, $i = 1, \ldots, r$, contradicting the quasi-uniform convergence of $\{g_k\}$. Q.E.D.

31 THEOREM. *Let S be an arbitrary set. A sequence $\{f_n\}$ in $B(S)$ converges weakly to f_0 if and only if it is bounded and, together with every subsequence, converges to f_0 quasi-uniformly on S.*

PROOF. The sequence $\{f_n\}$ in $B(S)$ converges weakly to f_0 if and only if the corresponding sequence of continuous functions $\{\tilde{f}_n\}$ in $C(S_1)$ converges weakly to f_0. (See Corollary 19, which shows that S may be identified with a dense subset of a compact Hausdorff space S_1.) Making this identification, we can write $\tilde{f}_n(s) = f_n(s)$, $s \in S$, and the theorem follows from Lemma 30 and Corollary 4. Q.E.D.

7. The Space AP

Harald Bohr's elegant theory of almost periodic functions characterizes those complex functions of a real variable which may be uniformly approximated over the whole real number system by trigonometric polynomials of the form

$$s(x) = \sum_{n=1}^{m} \alpha_n e^{i\lambda_n x}$$

where $\alpha_1, \ldots, \alpha_n$ are arbitrary complex numbers and $\lambda_1, \ldots, \lambda_n$ are arbitrary real numbers. In other words, the theory gives an intrinsic characterization, without reference to trigonometric polynomials, of those complex functions $f(x)$, $-\infty < x < \infty$ with the property that for every $\varepsilon > 0$ there is a trigonometric polynomial s with the above form such that

$$|f(x)-s(x)| < \varepsilon, \qquad -\infty < x < \infty.$$

The principal result in Bohr's theory is that the class of such complex functions of a real variable is precisely the class AP of almost periodic functions. In this section the class AP will be defined, it will be shown that AP is a B-space, and a criterion equivalent to almost-periodicity due to Bochner will be given. Other important results of the theory (in particular the result which identifies each function in AP with the uniform limit of trigonometric polynomials) will be presented later, when some of the basic tools of spectral theory have been developed.

1 Definition. For a complex function f defined on the class R of real numbers and a positive number ε the set $T(\varepsilon, f) \subseteq R$ is defined as

$$T(\varepsilon, f) = \{t \mid |f(x+t)-f(x)| < \varepsilon, \qquad x \in R\}.$$

Any number $t \in T(\varepsilon, f)$ is called a *translation number of f corresponding to ε*. When there can be no confusion the symbol $T(\varepsilon)$ may be used in place of $T(\varepsilon, f)$. It is clear that $T(\varepsilon) \subseteq T(\delta)$ if $\varepsilon < \delta$ and that $-t \in T(\varepsilon)$ whenever $t \in T(\varepsilon)$. The function f is said to be *almost periodic* if it is continuous and if for every $\varepsilon > 0$ there is an $L = L(\varepsilon) > 0$ such that every interval in R of length L contains at least one point of $T(\varepsilon)$.

It is clear that a periodic function is almost periodic. It also follows immediately from the definition that if $f \in AP$, if α is an arbitrary complex number, and if λ is an arbitrary real number then the functions defined by the expressions

$$\alpha f(t), \qquad f(t+\lambda), \qquad \overline{f(t)}, \qquad t \in R,$$

are also in AP. Since

$$||f(t+\lambda)|-|f(t)|| \leqq |f(t+\lambda)-f(t)|$$

it is clear that $|f(\cdot)| \in AP$ whenever $f \in AP$. It is also helpful to note that unless the almost periodic function f is actually periodic the admissible numbers $L(\varepsilon)$ are unbounded for ε near zero. For if $L(\varepsilon) \leqq K$, $\varepsilon > 0$ then there is a sequence $\{t_n\}$ with $0 \leqq t_n \leqq K$ and

$$|f(x+t_n)-f(x)| < 1/n, \qquad x \in R,$$

and thus any point of accumulation of the sequence $\{t_n\}$ is a period for f.

For $\lambda \in R$ the *translate* f_λ of f is defined by the equation $f_\lambda(x) = f(x+\lambda)$.

2 THEOREM. (*Bochner*) *A function in* $C(R)$ *is almost periodic if and only if the set of its translates is conditionally compact.*

PROOF. The proof will require the following two lemmas.

3 LEMMA. *An almost periodic function is bounded.*

PROOF OF LEMMA 3. Since an almost periodic function f is continuous, the function $|f(s)|$ has a maximum K on the interval $0 \leqq x \leqq L(1)$. Let x be an arbitrary real number and choose $t \in T(1, f)$ in the interval $-x \leqq t \leqq -x+L(1)$. Thus $0 \leqq t+x \leqq L(1)$ and

$$|f(x+t)| \leqq K,$$
$$|f(x)| \leqq |f(x+t)|+|f(x)-f(x+t)| \leqq K+1. \quad \text{Q.E.D.}$$

4 LEMMA. *An almost periodic function is uniformly continuous.*

PROOF OF LEMMA 4. Since an almost periodic function f is continuous, for each $\varepsilon > 0$ there is a δ with $0 < \delta < 1$ such that $|f(s)-f(t)| < \varepsilon/3$ for every pair s, t with $-1 \leqq s, t \leqq L(\varepsilon/3)+1$ and $|s-t| < \delta$. Now suppose that x, y are arbitrary real numbers with $|x-y| < \delta$. Choose $u \in T(\varepsilon/3)$ satisfying $-x \leqq u \leqq -x+L(\varepsilon/3)$. Then $0 \leqq x+u \leqq L(\varepsilon/3)$ and $-1 < y+u < L(\varepsilon/3)+1$. Thus

$$\begin{aligned}|f(x)-f(y)| \leqq |f(x)-f(x+u)|+|f(x+u)-f(y+u)| \\ +|f(y+u)-f(y)| < \varepsilon. \quad \text{Q.E.D.}\end{aligned}$$

PROOF OF THEOREM 2. Using Lemma 4, for each $\varepsilon > 0$ there is a $\delta > 0$ such that $|f_\lambda(x)-f_\lambda(y)| < \varepsilon$ for all λ when $|x-y| < \delta$. Thus it is seen (5.8) that an arbitrary sequence $\{f_{\lambda_i}\}$ contains a subsequence $\{f_{1,i}\}$ uniformly convergent on the interval $|x| \leqq 1$. By the same reasoning $\{f_{1,i}\}$ contains a subsequence $\{f_{2,i}\}$ convergent uniformly on the interval $|x| \leqq 2$. In this way successive subsequences are chosen so that the limit $\lim_i f_{n,i}(x)$ exists uniformly on the interval $|x| \leqq n$. The sequence $\{g_n = f_{n,n}, n = 1, 2, \ldots\}$ then is a subsequence of $\{f_{\lambda_i}\}$ which converges uniformly on any finite interval. Now let $\varepsilon > 0$ and choose $n(\varepsilon)$ so that

$$|g_n(x)-g_m(x)| < \frac{\varepsilon}{3},$$

for all n, m, and x with $0 \leqq x \leqq L(\varepsilon/3)$, and $n, m \geqq n(\varepsilon)$. For every real number x choose a $y \in T(\varepsilon/3, f)$ such that $-x \leqq y \leqq -x+L(e/3)$ and note that, since $g_n = f_{\mu_n}$ for some μ_n, the number y is also in $T(\varepsilon/3, g_n) = T(\varepsilon/3, f)$ for every $n = 1, 2, \ldots$. Thus for $n, m \geqq n(\varepsilon)$,

$$\begin{aligned}|g_n(x)-g_m(x)| \leqq |g_n(x)-g_n(x+y)|+|g_n(x+y)-g_m(x+y)|\\ +|g_m(x+y)-g_m(x)| < \varepsilon.\end{aligned}$$

We have shown that an arbitrary sequence $\{f_{\lambda_i}\}$ of translates of f contains a subsequence which is uniformly convergent on the whole real number system R. Thus if f is an almost periodic function the set $\{f_\lambda, \lambda \in R\}$ is conditionally compact in $C(R)$.

Conversely let f be a bounded continuous complex function on the real number system R whose translates $\{f_\lambda, \lambda \in R\}$ form a conditionally compact set in $C(R)$. The set $\{f_\lambda, \lambda \in R\}$ is totally bounded (I.6.15) and thus for every $\varepsilon > 0$ there are numbers $\lambda_1, \ldots, \lambda_m$ such that each translate f_λ satisfies one of the inequalities $|f_\lambda - f_{\lambda_i}| < \varepsilon$, $i = 1, \ldots, m$. That is, for every $\lambda \in R$ and $x \in R$,

$$|f(x+\lambda)-f(x+\lambda_i)| < \varepsilon$$

for some integer $i \leqq m$. Thus for every $\lambda \in R$ and every $x \in R$,

$$|f(x)-f(x+\lambda-\lambda_i)| < \varepsilon$$

for some $i \leqq m$. If $k = \max_{1 \leqq i \leqq m} |\lambda_i|$ it follows that any interval of length $2k$ contains a point in $T(\varepsilon, f)$. Thus f is almost periodic. Q.E.D.

5 Theorem. *The space AP of all complex periodic functions of a real variable is a complex B-space under the norm*

$$|f| = \sup_{-\infty < x < \infty} |f(x)|.$$

Proof. It has already been observed that the product αf of a scalar α and an almost periodic function f is again almost periodic. If f and g are both almost periodic, it is seen from Theorem 2 that any sequence $(f+g)_{\lambda_i} = f_{\lambda_i} + g_{\lambda_i}$ contains a subsequence $\mu_n = \lambda_{i_n}$ such that $\{f_{\mu_n}\}$ and $\{g_{\mu_n}\}$ are both uniformly convergent. Thus the sequence $\{(f+g)_{\mu_n}\} = \{f_{\mu_n} + g_{\mu_n}\}$ is also uniformly convergent and Theorem 2 shows that $f+g$ is an almost periodic function. Finally let $\{f_n\}$ be a uniformly convergent sequence of almost periodic functions and let f be the limit of f_n in $C(R)$. Fix n_0 so that the function $g = f_{n_0}$ satisfies the inequality $|g-f| < \varepsilon/3$. Using Theorem 2 (and I.6.15) choose $\lambda_1, \ldots, \lambda_m$ so that for every λ, $|g_\lambda - g_{\lambda_i}| < \varepsilon/3$ for some $i \leqq m$. Then for every λ,

$$|f_\lambda - f_{\lambda_i}| \leqq |f_\lambda - g_\lambda| + |g_\lambda - g_{\lambda_i}| + |g_{\lambda_i} - f_{\lambda_i}| < \varepsilon$$

for some $i \leqq m$. Thus the set $\{f_\lambda, \lambda \in R\}$ is totally bounded in $C(R)$ and hence (I.6.15) is conditionally compact in $C(R)$. Theorem 2 then shows that f is almost periodic. Q.E.D.

6 THEOREM. *The space AP of all complex almost periodic functions of a real variable contains, with f, g, also the functions fg and* $\bar{f}$ *defined by* $(fg)(t) = f(t)g(t)$, $\bar{f}(t) = \overline{f(t)}$. *There is an isometric algebraic (i.e., multiplication and conjugation preserving) isomorphism between AP and the algebra C(S) of all complex continuous functions on some compact Hausdorff space S.*

PROOF. It has already been observed that $\bar{f}$ belongs to AP whenever f does. The argument in the preceding proof which was used to show that the sum $f+g$ of two almost periodic functions is almost periodic, also shows that the product fg is almost periodic. Thus the present theorem is a corollary of Theorem 6.18. Q.E.D.

8. The Spaces $L_p(S, \Sigma, \mu)$

The spaces $L_p(S, \Sigma, \mu)$, $1 \leqq p < \infty$, have already been studied in Chapter III. In particular it was shown in Theorem III.6.6 that they are B-spaces. In this section the study of these spaces will be continued with a view to solving the problems listed in Section 1. In addition we shall study the space $L_\infty(S, \Sigma, \mu)$ of Definition 2.19. It is clear from Corollary III.6.14 that this space is a B-space. Since $L_p(S, \Sigma, \mu) = L_p(S, \Sigma, v(\mu))$ we may and shall assume throughout the present section that (S, Σ, μ) is a *positive* measure space.

The reader should note that the space l_p is just the space $L_p(S,\Sigma,\mu)$ in which S is the set of all integers, Σ the family of all subsets of S, and $\mu(E)$ the number (finite or infinite) of elements of E. Thus the discussion of the present section covers the space l_p as well as the space $L_p(S, \Sigma, \mu)$.

→ 1 **Theorem.** *If $1 < p < \infty$ and $1/p+1/q = 1$, there is an isometric isomorphism between $L_p^*(S, \Sigma, \mu)$ and $L_q(S, \Sigma, \mu)$ in which corresponding vectors x^* and g are related by the identity*

$$x^*f = \int_S g(s)f(s)\mu(ds), \qquad f \in L_p(S, \Sigma, \mu).$$

Proof. For $1 < p < \infty$ let $L_p = L_p(S, \Sigma, \mu)$ and let $|f|_p$ be the norm of f as an element of L_p. Let $x^* \in L_p^*$ and assume, for the present, that $\mu(S) < \infty$. If χ_E is the characteristic function of the set $E \in \Sigma$, then, if $\{E_j\}$ is a disjoint sequence of measurable subsets of S and $\bigcup_{j=1}^{\infty} E_j = E_0$, it follows from III.6.16 that $\sum_{j=1}^{\infty} \chi_{E_j} = \chi_{E_0}$, the series converging in the norm of L_p. Hence $x^*\chi_{E_0} = \sum_{j=1}^{\infty} x^*\chi_{E_j}$, so that $x^*\chi_E$ is a countably additive set function. Since $|\chi_E|_p \to 0$ as $\mu(E) \to 0$, $x^*\chi_E$ is μ-continuous, and the Radon-Nikodým theorem (III.10.2) yields a $g \in L_1$ with $x^*\chi_E = \int_S g(s)\chi_E(s)\mu(ds)$. Thus for simple functions f

$$\text{(i)} \qquad x^*f = \int_S g(s)f(s)\mu(ds).$$

If $\{f_n\}$ is a sequence of μ-integrable simple functions converging μ-almost everywhere and in L_p to f (by III.3.8, III.3.6, and III.6.13, such a sequence always exists), then $gf_n \to gf$ μ-almost everywhere. Since $x^*(f_n\chi_E)$ converges, it is seen from the Vitali-Hahn-Saks theorem (III.7.2) that

$$\lim_{\mu(E)\to 0} \int_E g(s)f_n(s)\mu(ds) = 0$$

uniformly in n. Since it is assumed for the present that $\mu(S) < \infty$ it is seen from Theorem III.6.15 that fg is in L_1 and that equation (i) is valid for every f in L_p. It will next be shown that g is in L_q. For a complex number z let $\alpha(z) = e^{-i\theta}$ if $z = re^{i\theta}$, $\alpha(0) = 0$. Then $\alpha(g(\cdot))$ is μ-measurable by III.6.9, so that, by III.2.12, $g_1(\cdot) = |g(\cdot)|^{1/p}\alpha(g(\cdot))$ is μ-measurable and hence is in L_p. Thus

$$\begin{aligned}
\int_S |g(s)|^{1+\frac{1}{p}} \mu(ds) &= x^*(g_1) \leqq |x^*|\,|g_1|_p \\
&= |x^*| \left\{ \int_S |g(s)|\mu(ds) \right\}^{\frac{1}{p}} \\
&= |x^*| \{x^*\alpha(g)\}^{\frac{1}{p}} \\
&\leqq |x^*|^{1+\frac{1}{p}} \mu(S)^{\frac{1}{p}}.
\end{aligned}$$

From this we see that $|g(\cdot)|^{1+1/p} \in L_1$ so that the function $g_2(\cdot) = |g(\cdot)|^{(1+1/p)/p}\alpha(g(\cdot))$ is in L_p, and moreover,

$$|g_2|_p = \left[\int_S |g(s)|^{1+\frac{1}{p}}\mu(ds)\right]^{\frac{1}{p}}$$

$$\leqq |x^*|^{\frac{1}{p}+\frac{1}{p^2}}\mu(S)^{\frac{1}{p^2}}.$$

Thus

$$\int_S |g(s)|^{1+\frac{1}{p}+\frac{1}{p^2}}\mu(ds) = x^*(g_2) \leqq |x^*||g_2|_p \leqq |x^*|^{1+\frac{1}{p}+\frac{1}{p^2}}\mu(S)^{\frac{1}{p^2}}.$$

Proceeding inductively by defining

$$g_n(\cdot) = |g(\cdot)|^{\frac{1}{p}+\frac{1}{p^2}+\cdots+\frac{1}{p^n}}\alpha(g(\cdot))$$

it is seen that

$$\text{(ii)} \quad \int_S |g(s)|^{1+\frac{1}{p}+\cdots+\frac{1}{p^n}}\mu(ds) \leqq |x^*|^{1+\frac{1}{p}+\cdots+\frac{1}{p^n}}\mu(S)^{\frac{1}{p^n}},$$

$$n = 1, 2, \ldots.$$

Since $p > 1$, $\sum_{n=0}^{\infty} 1/p^n = q$, and Fatou's lemma (III.6.19) gives $|g|_q \leqq |x^*|$. On the other hand, Hölder's inequality (III.3.2) gives $|x^*| \leqq |g|_q$. Thus $|x^*| = |g|_q$.

The mapping $x^* \to g$ is then a one-to-one isometric map of L_p^* into L_q. It is evident from the Hölder inequality that any $g \in L_q$ determines an $x^* \in L_p^*$ satisfying (i), so that the mapping $x^* \longleftrightarrow g$ is a one-to-one isometric map of L_p^* onto L_q. Since the linearity of the map is evident, the theorem is proved for the case $\mu(S) < \infty$.

Now let (S, Σ, μ) be an arbitrary positive measure space, and let Σ_1 consist of those sets $E \in \Sigma$ with $\mu(E) < \infty$. For $E \in \Sigma_1$ let $L_p(E)$ be the closed linear manifold in L_p consisting of those functions which vanish on the complement of E. Let x_E^* be the restriction of x^* to $L_p(E)$ so that if $E \in \Sigma_1$ there is, by what we have already shown, a $g_E \in L_q(E)$ with

$$x_E^* f = \int_S g_E(s)f(s)\mu(ds), \qquad f \in L_p(E),$$

$$|g_E|_q = |x_E^*| \leqq |x^*|, \qquad E \in \Sigma_1.$$

It is clear from the uniqueness of the function g_E that if $A, B \in \Sigma_1$ then $g_A(s) = g_B(s) = g_{AB}(s)$ for μ-almost all s in AB. Thus $|g_E|_q^q = \int_E |g_E(s)|^q \mu(ds)$ is a non-negative additive set function defined for

$E \in \Sigma_1$. Consequently, there is a non-decreasing sequence $\{E_n\} \subset \Sigma_1$ with

$$|x^*_{E_n}| \to \sup_{E \in \Sigma_1} |x^*_E| \leqq |x^*|.$$

Since $g_{E_n}(s) = g_{E_{n+1}}(s)$ μ-almost everywhere on E_n the limit $g(s) = \lim_n g_{E_n}(s)$ exists μ-almost everywhere and vanishes on the complement of $F = \bigcup_{n=1}^{\infty} E_n$. It follows (III.6.17) that

$$|g|_q = \lim_n |g_{E_n}|_q = \sup_{E \in \Sigma_1} |x^*_E| \leqq |x^*|.$$

If $E \in \Sigma_1$ and $EF = \phi$, then $EE_n = \phi$ for each n, and $|g_{E \cup E_n}|_q^q = |g_E|_q^q + |g_{E_n}|_q^q$. Since $|g_{E_n}|_q^q \to \sup_{E \in \Sigma_1} |g_E|_q^q$, we must have $|g_E| = 0$ if $EF = \phi$. Hence, if $f \in L_p(E)$ for some $E \in \Sigma_1$, we have

$$\begin{aligned} x^*f = x^*_E f &= \int_E g_E(s)f(s)\mu(ds) \\ &= \int_{E-F} g_E(s)f(s)\mu(ds) + \int_{EF} g_E(s)f(s)\mu(ds) \\ &= \int_{E-F} g_{E-F}(s)f(s)\mu(ds) + \int_{EF} g_{EF}(s)f(s)\mu(ds) \\ &= \int_{EF} g_{EF}(s)f(s)\mu(ds). \end{aligned}$$

Since $g_{EF}(s) = g_{E_n}(s)$ μ-almost everywhere in $EE_nF = EE_n$, $g_{EF}(s) = g(s)$ μ-almost everywhere in EF. Hence if $f \in L_p(E)$ for $E \in \Sigma_1$,

$$x^*f = \int_{EF} g(s)f(s)\mu(ds) = \int_S g(s)f(s)\mu(ds).$$

Since $\bigcup_{E \in \Sigma_1} L_p(E)$ is dense in L_p by III.3.8 and the right and left sides of the last equation are both continuous in f,

$$x^*f = \int_S g(s)f(s)\mu(ds), \qquad f \in L_p.$$

It has been shown that $|g|_q \leqq |x^*|$. On the other hand it follows from Hölder's inequality that $|x^*| \leqq |g|_q$. Thus $x^* \to g$ is an isometric map of L^*_p into L_q. The remainder of the proof is identical with that given above for the case $\mu(S) < \infty$. Q.E.D.

2 Corollary. *If $1 < p < \infty$, the space $L_p(S, \Sigma, \mu)$ is reflexive.*

PROOF. Let $x^{**} \in (L_p^*)^*$. By Theorem 1, L_p^* is isometrically isomorphic to L_q, so that there is a functional $y^* \in L_q^*$ such that

$$x^{**}(x^*) = y^*(g)$$

when g and x^* are connected, as in Theorem 1, by the formula

$$x^*f = \int_S f(s)g(s)\mu(ds), \qquad f \in L_p.$$

Applying Theorem 1 once more, this time to L_q^* and L_p, we find there exists an $h \in L_p$ such that

$$y^*f = \int_S h(s)f(s)\mu(ds), \qquad f \in L_q.$$

Thus $x^{**}(x^*) = y^*(g) = \int_S g(s)h(s)\mu(ds) = x^*(h)$; i.e., for each $x^{**} \in L_p^{**}$ there is an $h \in L_p$ such that $x^{**}(x^*) = x^*(h)$. Thus L_p is reflexive. Q.E.D.

3 COROLLARY. *If $1 < p < \infty$ the space $L_p(S, \Sigma, \mu)$ is weakly complete.*

PROOF. This follows from Corollary 2 and Corollary II.3.29. Q.E.D.

4 COROLLARY. *If $1 < p < \infty$, a set in $L_p(S, \Sigma, \mu)$ is weakly sequentially compact if and only if it is bounded.*

PROOF. This follows from Corollary 2 and Theorem II.3.28. Q.E.D.

Next we consider the problem of representing the conjugate space L_p^* when $p = 1$. A result analogous to that of Theorem 1 may be obtained by assuming that (S, Σ, μ) is a σ-finite measure space.

➔ 5 THEOREM. *If (S, Σ, μ) is a positive σ-finite measure space, there is an isometric isomorphism between $L_1^*(S, \Sigma, \mu)$ and $L_\infty(S, \Sigma, \mu)$ in which corresponding vectors x^* and g are related by the identity*

$$x^*f = \int_S g(s)f(s)\mu(ds), \qquad f \in L_1(S, \Sigma, \mu).$$

PROOF. First assume $\mu(S) < \infty$. Then the steps in the proof of Theorem 1 apply without change through the point where we obtained the inequality (ii) in that proof. For $p = 1$ this inequality becomes

$$\int_S |g(s)|^n \mu(ds) \leqq |x^*|^n \mu(S), \qquad n = 1, 2, \ldots.$$

If we let E_m be the set where $|g(s)| \geqq m$, this shows that $m^n \mu(E_m)$

$\leqq |x^*|^n\mu(S)$; i.e., $[\mu(E_m)/\mu(S)]^{1/n} \leqq |x^*|/m$. Letting $n \to \infty$, we find that $\mu(E_m) = 0$ if $m > |x^*|$. Thus $|g|_\infty \leqq |x^*|$. It is obvious that $|x^*| \leqq |g|_\infty$, and so $|x^*| = |g|_\infty$.

Now suppose that (S, Σ, μ) is σ-finite, and let E_n be an increasing sequence of measurable sets of finite measure whose union is S. Using the theorem for $L_1(E_n) = L_1(E_n, \Sigma(E_n), \mu)$, we obtain a sequence $\{g_n\}$ of functions in L_∞ such that $|g_n|_\infty \leqq |x^*|$, $g_n(s) = g_{n+1}(s)$ for μ-almost all s in E_n, and

$$x^*f = \int_{E_n} g_n(s)f(s)\mu(ds),$$

for every f in L_1 which vanishes outside E_n. Let $g(s) = \lim_n g_n(s)$ so that g is defined almost everywhere, $|g|_\infty \leqq |x^*|$, and

$$x^*(\chi_{E_n}f) = \int_S g(s)\chi_{E_n}(s)f(s)\mu(ds), \quad f \in L_1, \; n = 1, 2, \ldots,$$

where χ_{E_n} is the characteristic function of the set E_n. Since, by III.6.16 $|f-\chi_{E_n}f|_1 \to 0$ as $n \to \infty$, it is seen that

$$x^*f = \int_S g(s)f(s)\mu(ds), \qquad f \in L_1.$$

It is then clear that $|x^*| \leqq |g|_\infty$, and since it has been shown that $|g|_\infty \leqq |x^*|$, there is an isometric correspondence between L_1^* and L_∞. The linearity of this correspondence being evident, the theorem is proved. Q.E.D.

Remark. It is a matter of some importance in the study of locally compact groups that the preceding theorem remains valid if S is the union of a (possibly uncountable) collection of disjoint subsets $\{S_\alpha\}$ in Σ each of which is σ-finite and such that if $E \in \Sigma$ and $\mu(E) < \infty$, then E has non-void intersection with at most a countable number of the sets S_α, and such that $\mu(K) = 0$ for every set K in Σ for which $\mu(KS_\alpha) = 0$ for all α. The preceding argument shows that a function the uniqueness has been demonstrated for any σ-finite space, the uniqueness on S follows from the fact that any set of finite measure is contained in at most a countable union of the S_α, which is a σ-finite set.

6 Theorem. *The space $L_1(S, \Sigma, \mu)$ is weakly complete.*

Proof. Let $\{f_n\}$ be a weak Cauchy sequence in L_1. By III.8.5, there exists a σ-finite set $E \in \Sigma$ such that all the functions f_n belong to the closed subspace $L_1(E) = L_1(E, \Sigma(E), \mu)$ of $L_1(S, \Sigma, \mu)$ consisting of all functions vanishing outside E. If we can find an $f \in L_1(E)$ to which $\{f_n\}$ converges weakly, we will have shown that L_1 is weakly complete. Since $\{f_n\}$ is a weak Cauchy sequence in $L_1(E)$ (by II.3.11) the whole argument may be made in the space $L_1(E)$. Thus, without loss of generality we may assume that (S, Σ, μ) is σ-finite.

By II.3.20 the sequence $\{f_n\}$ is bounded in L_1. Since the characteristic function of any $E \in \Sigma$ is in $L_\infty(S, \Sigma, \mu)$, the number $\lambda(E) = \lim_{n\to\infty} \int_E f_n(s)\mu(ds)$ exists for each E in Σ. It follows from III.7.2 that λ is μ-continuous. Thus, by the Radon-Nikodým theorem (III.10.2), there is a function f in $L_1(S, \Sigma, \mu)$ such that

$$\lim_{n\to\infty} \int_E f_n(s)\mu(ds) = \int_E f(s)\mu(ds), \qquad E \in \Sigma.$$

Hence

$$\text{(i)} \qquad \lim_{n\to\infty} \int_S f_n(s)h(s)\mu(ds) = \int_S f(s)h(s)\mu(ds)$$

for every μ-simple function h. Since, by II.3.20, the sequence $\{f_n\}$ is a bounded sequence in L_1, it will follow from Theorem II.1.18 that equation (i) holds for every h in L_∞ as soon as it is shown that the μ-simple functions are dense in L_∞.

Thus let $\varepsilon > 0$ and let h be an arbitrary element of L_∞. We may suppose without loss of generality that h is bounded. Let $A_1, \ldots, A_n$ be a finite collection of disjoint Borel sets in the field of scalars, each with diameter less than ε, and such that $h(S) \subset \cup A_i$. Let $\alpha_i \in A_i$, $B_i = f^{-1}(A_i)$, and $h_\varepsilon(s) = \alpha_i$ if $s \in B_i$. Then by III.6.10, and III.5.17, h_ε is a μ-simple function and $|h-h_\varepsilon|_\infty < \varepsilon$. Thus equation (i) holds for every h in L_∞ and the desired conclusion follows from Theorem 5. Q.E.D.

7 **Theorem.** *A sequence $\{f_n\}$ in $L_1(S, \Sigma, \mu)$ is a weak Cauchy sequence if and only if it is bounded and the limit $\lim_n \int_E f_n(s)\mu(ds)$ exists for every E in Σ. The sequence $\{f_n\}$ converges weakly to an element f in L_1 (S, Σ, μ) if and only if it is bounded and*

$$\int_E f(s)\mu(ds) = \lim_n \int_E f_n(s)\mu(ds), \qquad E \in \Sigma.$$

PROOF. In the final two paragraphs of the proof of Theorem 6 it was shown that a bounded sequence in $L_1(S, \Sigma, \mu)$ converges weakly to an element f in $L_1(S, \Sigma, \mu)$ provided that $\int_E f_n(s)\mu(ds) \to \int_E f(s)\mu(ds)$ for every E in Σ. To prove the necessity of the conditions, let $\{f_n\}$ be a weak Cauchy sequence in $L_1(S, \Sigma, \mu)$. In view of Theorem 6, the sequence $\{f_n\}$ converges weakly to a function f in $L_1(S, \Sigma, \mu)$. Since for each E in Σ the integral $\int_E f_n(s)\mu(ds)$ depends linearly and continuously on f, we have $\int_E f_n(s)\mu(ds) \to \int_E f(s)\mu(ds)$. The boundedness of $\{f_n\}$ follows from the principle of uniform boundedness (II.3.27). Q.E.D.

8 LEMMA. *Let Σ be a σ-field of sets, Σ_1 a subfield of Σ determining the σ-field Σ, and $\{\mu_n\}$ a sequence of countably additive set functions defined on Σ with values in a B-space $\mathfrak{X}$. Suppose that the countable additivity of μ_n is uniform in n, and that* $\lim_{n\to\infty} \mu_n(E)$ *exists for $E \in \Sigma_1$. Then* $\lim_{n\to\infty} \mu_n(E)$ *exists for $E \in \Sigma$.*

PROOF. Let Σ_2 be the family of all sets E in Σ for which $\lim_{n\to\infty} \mu_n(EF)$ exists for each $F \in \Sigma_1$, and let Σ_3 be the family of all sets E in Σ_2 for which $EF \in \Sigma_2$ for each $F \in \Sigma_2$. It is clear that if F_1 and F_2 are elements of Σ_3, then $F_1F_2 \in \Sigma_3$. It is also clear that if $F_1 \in \Sigma_3$, then $S-F_1 \in \Sigma_3$, and that if $F_1, F_2 \in \Sigma_3$ with $F_1F_2 = \phi$, then $F_1 \cup F_2 \in \Sigma_3$. It follows that Σ_3 is a field. If $\{F_k\}$ is a sequence of disjoint elements of Σ_3 with union F and if $E_1 \in \Sigma_1$ and $E \in \Sigma_2$, then, by hypothesis,

$$\lim_{m\to\infty} \sum_{k=1}^{m} \mu_n(F_k E E_1) = \mu_n(FEE_1)$$

uniformly in n. Since $\lim_{n\to\infty} \mu_n(F_k E E_1)$ exists for each k, it follows by Lemma I.7.6 that $\lim_{n\to\infty} \mu_n(FEE_1)$ exists. Thus Σ_3 is a σ-field. Since it is clear that $\Sigma_3 \supseteq \Sigma_1$, we conclude that $\Sigma_3 \supseteq \Sigma$, from which the desired result follows. Q.E.D.

9 THEOREM. *A subset K of $L_1(S, \Sigma, \mu)$ is weakly sequentially compact if and only if it is bounded and the countable additivity of the integrals $\int_E f(s)\mu(ds)$ is uniform with respect to f in K.*

PROOF. The statement that the countable additivity of the integrals $\int_E f(s)\mu(ds)$ is uniform with respect to f in K means that for each decreasing sequence $\{E_n\}$ in Σ with void intersection the limit

$$\lim_n \int_{E_n} f(s)\mu(ds) = 0$$

is uniform for f in K. Assume now that K is weakly sequentially compact. It follows from Lemma II.3.27 that K is bounded. If the integrals $\int_E f(s)\mu(ds)$ are not countably additive uniformly with respect to f in K there is a number $\varepsilon > 0$, a decreasing sequence $E_n \in \Sigma$ with a void limit, and functions f_n in K such that $|\int_{E_n} f_n(s)\mu(ds)| \geqq \varepsilon$ for $n = 1, 2, \ldots$. Since K is weakly sequentially compact it may be assumed that $\{f_n\}$ is weakly convergent. The limit $\lim_{n\to\infty} \int_E f_n(s)\mu(ds)$ then exists for each $E \in \Sigma$ and Corollary III.7.4 is contradicted.

Conversely, suppose that K is bounded and that the integrals $\int_E f(s)\mu(ds)$ are countably additive uniformly with respect to f in K. Let $f_n \in K$ and suppose that $|f_n| \leqq C$ for $n = 1, 2, \ldots$. By Lemma III.8.5 there is a σ-finite set S_1 in Σ and a sub σ-field Σ_1 of $\Sigma(S_1)$ which is generated by a denumerable field $\Sigma_0 = \{E_n\}$ such that the functions all vanish outside of S_1 and $\{f_n\} \subset L_1(S_1, \Sigma_1, \mu)$. We now, by the Cantor diagonal process, choose a subsequence $\{g_n\}$ of $\{f_n\}$ such that the limit

$$\lambda(E) = \lim_{n\to\infty} \int_E g_n(s)\mu(ds)$$

exists for every E in Σ_0. By Lemma 8 the limit $\lambda(E)$ exists for each E in Σ_1. Thus, by Theorems 6 and 7, the sequence $\{g_n\}$ is weakly convergent in $L_1(S_1, \Sigma_1, \mu)$. Since $L_1(S_1, \Sigma_1, \mu)$ is a linear subspace of $L_1(S, \Sigma, \mu)$ the sequence $\{g_n\}$ is weakly convergent in $L_1(S, \Sigma, \mu)$. Q.E.D.

10 Corollary. *If $\{f\}$ is a family of functions in $L_1(S, \Sigma, \mu)$ which is weakly sequentially compact, then the family $\{|f(\cdot)|\}$ is also weakly sequentially compact.*

Proof. First, $\{|f(\cdot)|\}$ is clearly a bounded set. By Theorem 9, it is sufficient to show that if E_n is a decreasing sequence of sets with void intersection, then $\lim_{n\to\infty} \int_{E_n} |f(s)|\mu(ds) = 0$ uniformly for f in the family $K = \{f\}$. If this is not the case, then one of the statements

$$\lim_{n\to\infty} \int_{E_n} |\mathscr{R}f(s)|\mu(ds) = 0, \quad \text{uniformly for } f \in K,$$

and

$$\lim_{n\to\infty}\int_{E_n}|\mathscr{I}f(s)|\mu(ds)=0, \quad \text{uniformly for } f\in K$$

is false. We may suppose, without loss of generality, that the former does not hold. If $E\in\Sigma$, $f\in K$, and we define $E^+=\{s\in E|\mathscr{R}f(s)\geqq 0\}$ and $E^-=\{s\in E|\mathscr{R}f(s)<0\}$, then

$$\int_E|\mathscr{R}f(s)|\mu(ds)=\int_{E^+}\mathscr{R}f(s)\mu(ds)-\int_{E^-}\mathscr{R}f(s)\mu(ds).$$

Since there exists an $\varepsilon>0$ such that for each n there is some $f_n\in K$ such that $\int_{E_n}|\mathscr{R}f_n(s)|\mu(ds)\geqq\varepsilon$, we can find a subset A_n of E_n such that

$$\left|\int_{A_n}f_n(s)\mu(ds)\right|\geqq\varepsilon/2.$$

Since K is weakly sequentially compact, a subsequence of $\{f_n\}$ converges weakly, and it may be assumed without loss of generality that $\{f_n\}$ itself converges weakly. By Theorem 7, $\{\int_E f_n(s)\mu(ds)\}$ converges for each $E\in\Sigma$.

Let

$$\lambda_n(E)=\int_E|f_n(s)|\mu(ds), \qquad \lambda(E)=\sum_{n=1}^{\infty}\frac{1}{2^n}\frac{\lambda_n(E)}{\lambda_n(S)}.$$

Then, since $ca(S,\Sigma)$ is a B-space (cf. Section III.7), λ is in $ca(S,\Sigma)$. Since each set function $\int_E f_n(s)\mu(ds)$ is clearly λ-continuous, it follows from III.7.2 that $\lim_{\lambda(E)\to 0}\int_E f_n(s)\mu(ds)=0$ uniformly in n. Since E_n is a decreasing sequence with void intersection, $\lambda(E_n)\to 0$, and since $A_n\subseteqq E_n$, it follows that $\lambda(A_n)\to 0$. Hence $\lim_{m\to\infty}\int_{A_m}f_n(s)\mu(ds)=0$ uniformly in n, contradicting the inequality $|\int_{A_n}f_n(s)\mu(ds)|\geqq\varepsilon/2$. Q.E.D.

11 Corollary. *If a set K in $L_1(S,\Sigma,\mu)$ is weakly sequentially compact then*

$$\lim_{\mu(E)\to 0}\int_E f(s)\mu(ds)=0,$$

uniformly for f in K. If $\mu(S)<\infty$ then conversely this condition is sufficient for a bounded set K to be weakly sequentially compact.

Proof. Let K be a weakly sequentially compact set in $L_1(S,\Sigma,\mu)$. If the μ-continuity of integrals $\int_E f(s)\mu(ds)$ is not uniform for f in K there is an $\varepsilon>0$, a sequence $\{E_n\}$ in Σ, and a sequence $\{f_n\}$ in K with

$\mu(E_n) \to 0$ and $\int_{E_n} f_n(s)\mu(ds) \geqq \varepsilon$. We may and shall assume that $\{f_n\}$ is weakly convergent and thus that the sequence $\{\int_E f_n(s)\mu(ds)\}$ converges for each E in Σ. Thus III.7.2 gives the desired contradiction. Conversely, if $\mu(S) < \infty$ and if the μ-continuity of the integrals $\int_E f(s)\mu(ds)$ is uniform for f in a bounded set K then the countable additivity of the integrals is uniform with respect to f in K and, by Theorem 9, K is weakly compact. Q.E.D.

12 THEOREM. *Let f_n converge weakly to f in $L_1(S, \Sigma, \mu)$. Then f_n converges strongly to f if and only if f_n converges in μ-measure to f on every measurable set of finite measure.*

PROOF. If $|f_n - f| \to 0$ then Theorem III.3.6 shows that f_n converges to f in μ-measure. To prove the converse, observe first that since f_n approaches f weakly the sequence $\{f_n - f\}$ is weakly compact and thus the sequence $\{|f_n(\cdot) - f(\cdot)|\}$ is, by Corollary 10, also weakly compact. Then, by Corollary 11, the μ-continuity of the integrals $\int_E |f_n(s) - f(s)|\mu(ds)$ is uniform in n. Since $f_n - f$ converges to zero in μ-measure on every measurable set of finite measure it follows from Theorem III.3.6 that $\int_E |f_n(s) - f(s)|\mu(ds) \to 0$ for every set E of finite μ-measure. Now, since the functions $f, f_n, n = 1, 2, \ldots$, are μ-integrable they all vanish outside of some σ-finite set E. Let $\{E_m\}$ be an increasing sequence of sets with union E and with $\mu(E_m) < \infty$. Since the sequence $\{|f_n(\cdot) - f(\cdot)|\}$ is weakly sequentially compact there is, for each $\varepsilon > 0$, an m (Theorem 9) such that

$$\int_{S-E_m} |f_n(s) - f(s)|\mu(ds) \leqq \varepsilon/2, \qquad n = 1, 2, \ldots.$$

Since $\int_{E_m} |f_n(s) - f(s)|\mu(ds) \to 0$ as $n \to \infty$ there is an $n(\varepsilon)$ with

$$|f_n - f| = \int_{S-E_m} |f_n(s) - f(s)|\mu(ds) + \int_{E_m} |f_n(s) - f(s)|\mu(ds) < \varepsilon, \qquad n \geqq n(\varepsilon),$$

and thus $|f_n - f| \to 0$. Q.E.D.

13 COROLLARY. *If every point has non-zero measure then weak and strong convergence of sequences in $L_1(S, \Sigma, \mu)$ are the same.*

PROOF. Let f_n converge weakly to f in $L_1(S, \Sigma, \mu)$. Since f_n, f are integrable, $f_n(s) = f(s) = 0$ for any point s with $\mu(\{s\}) = \infty$. If $0 < \mu(\{s\}) < \infty$ then

$$\mu(\{s\})f_n(s) = \int_{\{s\}} f_n(s)\mu(ds) \to \int_{\{s\}} f(s)\mu(ds) = \mu(\{s\})f(s),$$

and so $f_n(s) \to f(s)$ for every point s satisfying $0 < \mu(\{s\}) < \infty$. Thus (III.6.13(b)) f_n converges in measure to f on every set of finite measure. The desired conclusion now follows from Theorem 12. Q.E.D.

14 Corollary. *Weak and strong convergence of sequences in* l_1 *are the same.*

15 Definition. Let (S, Σ, μ) be a positive measure space. Let Σ^* be its Lebesgue extension, and let Σ_1 be the family of all sets $E \subseteq S$ for which $AE \in \Sigma^*$ for every set $A \in \Sigma$ with $\mu(A) < \infty$. Clearly Σ_1 is a σ-field containing Σ. The function μ_1 on Σ_1 defined by the equations

$$\begin{aligned} \mu_1(E) &= \mu(E), & E &\in \Sigma^*, \\ &= \infty, & E &\in \Sigma_1 - \Sigma, \end{aligned}$$

is a countably additive extension of μ from Σ^* to Σ_1. The space $ba(S, \Sigma, \mu)$ is the space of those bounded additive functions on Σ which vanish on sets of μ-measure zero. The norm of an element in $ba(S, \Sigma, \mu)$ is its total variation.

It should be noted that if (S, Σ, μ) is a σ-finite measure space then $(S, \Sigma^*, \mu) = (S, \Sigma_1, \mu_1)$. If S is the set of integers, Σ the set of all subsets of S, and $\mu(E)$ the cardinality of the set $E \in \Sigma$, then $L_\infty(S, \Sigma, \mu)$ is simply l_∞. The theorems concerned with L_∞ to follow therefore apply to l_∞ as well.

16 Theorem. *There is an isometric isomorphism between* $L^*_\infty(S, \Sigma, \mu)$ *and* $ba(S, \Sigma_1, \mu_1)$ *determined by the identity*

$$[*] \qquad x^*f = \int_S f(s)\lambda(ds), \qquad f \in L_\infty(S, \Sigma, \mu).$$

Proof. Let $f \subset L_\infty(S, \Sigma, \mu)$. Then, there exists a μ-null set N such that $f(S-N)$ is bounded. The set $f(S-N)$ is therefore contained in the union of disjoint Borel sets $A_1, \ldots, A_n$ in the field of scalars, each of diameter less than a prescribed positive number ε. By Theorem III.6.10, the set $E_i = f^{-1}(A_i)$ is in Σ_1. If we let $\alpha_i \in A_i$ and put $f_\varepsilon = \Sigma \alpha_i \chi_{E_i}$, then $|f(s) - f_\varepsilon(s)| < \varepsilon$ for $s \in S-N$. Since, for $\lambda \in ba(S_1, \Sigma_1, \mu_1)$, each μ_1-null set is a λ-null set, it follows that f is λ-measurable. Since f is λ-essentially bounded, it is λ-integrable. It follows from Theorem

III.2.20(a) that $|\int_S f(s)\lambda(ds)| \leqq |f||\lambda|$. Thus, equation [*] does define an element $x^* \in L_\infty^*(S, \Sigma, \mu)$ with $|x^*| \leqq |\lambda|$.

If E_i, $i = 1, 2, \ldots, n$ are disjoint sets in Σ_1 with $\sum_{i=1}^n |\lambda(E_i)| > |\lambda| - \varepsilon$, let $\alpha_1, \ldots, \alpha_n$ be scalars with $|\alpha_i| = 1$, $\alpha_i \lambda(E_i) = |\lambda(E_i)|$, and define $f = \sum_{i=1}^n \alpha_i \chi_{E_i}$, where χ_E is the characteristic function of E. Then $f \in L_\infty(S, \Sigma, \mu)$, $|f| = 1$ and $|x^*| \geqq x^* f \geqq |\lambda| - \varepsilon$. Thus $|x^*| = |\lambda|$.

It is clear that the map $\lambda \to x^*$ of $ba(S, \Sigma_1, \mu_1)$ into $L_\infty^*(S, \Sigma, \mu)$ defined by [*] is one-to-one. To see that an arbitrary x^* in $L_\infty^*(S, \Sigma, \mu)$ is the correspondent of some λ in $ba(S, \Sigma_1, \mu_1)$, let $\lambda(E) = x^* \chi_E$ for $E \in \Sigma_1$, and let $f \in L_\infty(S, \Sigma, \mu)$. Then $|f_\varepsilon - f| < \varepsilon$, while $x^* f_\varepsilon = \int_S f_\varepsilon(s) \lambda(ds)$ is evident. Thus, allowing ε to approach zero, $\int_S f(s)\lambda(ds) = x^* f$. Q.E.D.

Next we turn to the study of compactness in L_p spaces. There are various types of criteria for compactness that can be given. First, we give a quite general criterion based on Lemma IV.5.4, which, however, is somewhat difficult to apply to specific cases.

17 Definition. Let (S, Σ, μ) be a positive measure space and let Π be the set of all finite collections $\pi = \{E_1, \ldots, E_n\}$ of disjoint sets in Σ with finite positive measure. Let Π be ordered by defining $\pi \leqq \pi_1$ to mean that every set in π is, except for a set of measure zero, a union of sets in π_1. For every $\pi = \{E_1, \ldots, E_n\} \in \Pi$, and every function f on S which is integrable on every set of finite measure, define the function f_π by the equations

$$f_\pi(s) = 0 \qquad \text{if } s \notin \bigcup_{i=1}^n E_i,$$

$$f_\pi(s) = \frac{1}{\mu(E_i)} \int_{E_i} f(s)\mu(ds) \qquad \text{if } s \in E_i,$$

and let $U_\pi f = f_\pi$.

18 Theorem. *Let $1 \leqq p < \infty$ and let U_π be the map defined in the preceding definition. Then a bounded set K in $L_p(S, \Sigma, \mu)$ is conditionally compact if and only if $\lim_\pi U_\pi f = f$ uniformly on K. If $\mu(S) < \infty$ the criterion is also valid in $L_\infty(S, \Sigma, \mu)$.*

Proof. Let $1 \leqq p < \infty$, let χ_E be the characteristic function of E, and let $f \in L_p(S, \Sigma, \mu)$. Then for $\pi = \{E_1, \ldots, E_n\}$, we have

$$U_\pi f = \sum_{i=1}^{n} \left\{ \mu(E_i)^{-1} \int_{E_i} f(s)\mu(ds) \right\} \chi_{E_i},$$

$$|U_\pi f| \leqq \sum_{i=1}^{n} \int_{E_i} |f(s)|\mu(ds)\mu(E_i)^{-1+\frac{1}{p}}.$$

Thus, from the Hölder inequality (III.3.2)

$$|U_\pi f| \leqq \sum_{i=1}^{n} \left[\int_{E_i} |f(s)|^p \mu(ds) \right]^{1/p} \leqq |f|.$$

Since U_π has a finite dimensional range it maps bounded sets into conditionally compact sets. If $\pi = \{E_1, \ldots, E_n\}$ and $f = \sum_{i=1}^{n} \alpha_i \chi_{E_i}$, then $U_{\pi_1} f = f$ for $\pi_1 \geqq \pi$. Thus $U_\pi f \to f$ for simple functions. Since $|U_\pi| \leqq 1$ it follows by III.3.8 that $U_\pi f \to f$ for all f in $L_p(S, \Sigma, \mu)$. Hence if $1 \leqq p < \infty$, the theorem follows from Lemma IV.5.4. If $\mu(S) < \infty$ a similar argument proves the desired result in $L_\infty(S, \Sigma, \mu)$. Q.E.D.

If S is n-dimensional Euclidean space, Σ the field of Borel sets in S and μ Lebesgue measure, more applicable conditions may be given. For this purpose we need the following lemma.

19 **Lemma.** *If μ is a regular finitely additive set function on a field Σ of subsets of a normal space S, then, for $1 \leqq p < \infty$, the functions in $L_p(S, \Sigma, \mu)$ which are bounded and continuous are dense in $L_p(S, \Sigma, \mu)$.*

Proof. Since μ-integrable simple functions are dense in $L_p(S, \Sigma, \mu)$, (III.3.8), it suffices to prove that the characteristic function of a set $E \in \Sigma$ with $v(\mu, E) < \infty$ may be approximated by bounded continuous functions in $L_p(S, \Sigma, \mu)$. Let F_1, F_2 be sets in Σ with $\overline{F}_1 \subseteq E$, $\overline{F}_2 \subseteq E'$, and

$$v(\mu, E - F_1) < \varepsilon, \qquad v(\mu, E' - F_2) < \varepsilon.$$

By Urysohn's theorem (I.5.2) there is a continuous function f on S with $0 \leqq f(s) \leqq 1$ and with $f(s) = 1$ for s in $\overline{F}_1$ and $f(s) = 0$ for s in $\overline{F}_2$. Thus

$$\int_S |f(s) - \chi_E(s)|^p v(\mu, ds) \leqq 2\varepsilon. \qquad \text{Q.E.D.}$$

20 **Theorem.** *Let S be the real axis, $\mathscr{B}$ the field of Borel subsets of S, and μ the Lebesgue measure of sets in $\mathscr{B}$. Suppose $1 \leqq p < \infty$. Then a subset K of $L_p(S, \mathscr{B}, \mu)$ is conditionally compact if and only if it is bounded and*

(a) $\lim_{x\to 0} \int_{-\infty}^{+\infty} |f(x+y)-f(y)|^p dy = 0$ *uniformly for* $f \in K$, *and*

(b) $\lim_{A\to\infty} \int_A^{\infty} + \int_{-\infty}^{-A} |f(y)|^p dy = 0$ *uniformly for* $f \in K$.

PROOF. Suppose K is conditionally compact. Then K is bounded. Let $\varepsilon > 0$ be given. By I.6.15 and III.3.8 we can find a finite set of μ-integrable simple functions $g_1, \ldots, g_N$ such that for each $f \in K$ there exists some j with $|f-g_j| < \varepsilon$. By III.7.1, we may suppose that each g_j is a linear combination of characteristic functions of intervals. Supposing this, it follows that all the functions $g_1, \ldots, g_N$ vanish outside some sufficiently large interval $[-A_0, +A_0]$, so that

$$\int_A^{\infty} + \int_{-\infty}^{-A} |f(y)|^p dy = \int_A^{\infty} + \int_{-\infty}^{-A} |f(y)-g_j(y)|^p dy$$
$$\leqq |f-g_j|^p \leqq \varepsilon^p$$

for $A \geqq A_0$, proving (b).

To prove (a) we note first that it is evident that

$$\lim_{x\to 0} \int_{-\infty}^{+\infty} |\chi(x+y)-\chi(y)|^p dy = 0$$

if χ is the characteristic function of a finite interval. Thus $\lim_{x\to 0} \int_{-\infty}^{+\infty} |g_j(x+y)-g_j(y)|^p dy = 0$ for each function g_j; and hence

$$\limsup_{x\to 0} \int_{-\infty}^{+\infty} |f(x+y)-f(y)|^p dy \leqq \limsup_{x\to 0} \int_{-\infty}^{+\infty} |f(x+y)-g_j(x+y)|^p dy$$
$$+ \limsup_{x\to 0} \int_{-\infty}^{+\infty} |f(y)-g_j(y)|^p dy$$
$$+ \limsup_{x\to 0} \int_{-\infty}^{+\infty} |g_j(x+y)-g_j(y)|^p dy$$
$$= 2|f-g_j|^p$$
$$\leqq 2\varepsilon^p,$$

uniformly for $f \in K$. Thus (a) is proved.

To prove the converse, we argue as follows. The measure μ is the unique Borel measure defined by the condition $\mu([a, b]) = b-a$ for each interval $[a, b]$. Hence $\mu(x+E) = \mu(E)$ for each Borel set E. It follows by III.10.8 that the operator T_x defined by the equation $(T_x f)(y) = f(y+x)$ defines a map of L_p into L_p, and that $|T_x f| = |f|$. Since $\lim_{x\to y} |T_x f - T_y f| = 0$ is evident if f is the characteristic function of a bounded interval, it follows by III.7.1 and III.3.8 that $\lim_{x\to y} |T_x f - T_y f| = 0$ for all $f \in L_p$. Thus, for each fixed f, $T_x f$ is a continuous function of the real variable x. Assumption (a) states that $\lim_{x\to 0} T_x f = f$ uniformly for f in K. Let

$$J_a f = \frac{1}{2a}\int_{-a}^{+a} T_y f dy;$$

then, by III.2.20(a),

$$|J_a f - f| \leqq \sup_{-a \leqq x \leqq +a} |T_x f - f|,$$

so that $J_a f \to f$ uniformly for $f \in K$ as $a \to \infty$.

Since $T_x f$ is continuous in x, the function h_n defined by $h_n(x) = T_{ja/n} f$ for $ja/n \leqq x < (j+1)a/n$, $j = 0, \pm 1, \pm 2, \ldots$, converges to $T_x f$ uniformly on every finite interval of x. Hence

$$J_a f = \lim_{n\to\infty} \frac{1}{2a}\int_{-a}^{+a} h_n(x)dx = \lim_{n\to\infty} \frac{1}{2an}\sum_{j=-n}^{n=1} T_{\frac{ja}{n}} f,$$

where the limits are to be taken in the metric of L_p. By III.3.6 and III.6.13, on passing to a subsequence $\{n_i\}$ of 1, 2, . . ., we have

$$(J_a f)(x) = \lim_{i\to\infty} \frac{1}{2an}\sum_{j=-n_i}^{n_i-1} f\left(\frac{ja}{n} + x\right)$$

for μ-almost all x. If f is continuous, the limit on the right is evidently

$$\frac{1}{2a}\int_{-a}^{+a} f(y+x)dy = \frac{1}{2a}\int_{x-a}^{x+a} f(y)dy.$$

Thus if we define the element $\phi^*_{a,x} \in L^*_p$ by putting $\phi^*_{a,x}(g) = (1/2a)\int_{x-a}^{x+a} g(y)dy$ for $g \in L_p$, then by Theorem 1 $|\phi^*_{a,x}| = (2a)^{-1/p}$, and we have $(J_a f)(x) = \phi^*_{a,x} f$ for $-\infty < a < +\infty$, f continuous and belonging to L_p, and μ-almost all x. By the preceding lemma there is a sequence $\{f_n\}$ of continuous functions converging in L_p to f. Then $J_a f_n \to J_a f$ and $\phi^*_{a,x} f_n \to \phi^*_{a,x} f$. If we choose a subsequence by III.6.13, we can also have $(J_a f_n)(x) \to (J_a f)(x)$ for almost all x. It follows that $(J_a f)(x) = \phi^*_{a,x} f$ for almost all x. Since the function on the left of this equation is only defined modulo null sets, we can take this equation to hold for all a and x; i.e.,

$$(J_a f)(x) = \phi^*_{a,x} f, \qquad -\infty < a, x < +\infty, \qquad f \in L_p.$$

Since $T_x T_y f = T_{x+y} f = T_y T_x f$, it follows, from III.2.19(c), that

$$T_y J_a f = T_y \left[\frac{1}{2a} \int_{-a}^{+a} T_x f dx \right] = \frac{1}{2a} \int_{-a}^{+a} T_x (T_y f) dx = J_a(T_y f)$$

for $f \in L_p$. If $\delta_1 > 0$ is given, we can find δ_2 so small that $|T_y f - f| < \delta_1$ for $|y| < \delta_2$ and $f \in K$. Then for each fixed $a > 0$,

$$|(J_a f)(x+y) - (J_a f)(x)| = |\{J_a(T_y f - f)\}(x)| = |\phi^*_{a,x}(T_y f - f)| < \delta_1 (2a)^{-1/p}.$$

Therefore, for each a, the set of functions $J_a f$, $f \in K$, is uniformly equicontinuous.

Let $\varepsilon > 0$ be given. Choose a so small that $|J_a f - f| < \varepsilon$ for $f \in K$. Choose A so large that

$$\int_{-\infty}^{-A} + \int_{A}^{+\infty} |f(x)|^p dx < \varepsilon, \qquad \text{for } f \in K.$$

Then, using Theorem 6.17, and I.6.15, choose a finite set of continuous functions $g_1, \ldots, g_N$ defined on the interval $[-A, +A]$ such that $|g_j(x) - J_a f(x)| < \varepsilon A^{-1/p}$ for $-A \leqq x \leqq A$. Then, if we put $v_j(x) = g_j(x)$ for $x \in [-A, +A]$ and $v_j(x) = 0$ otherwise, it is evident that $v_j \in L_p$, and that $|f - v_j| < 3\varepsilon$. Thus the conditional compactness of K follows from I.6.15. Q.E.D.

Theorem 20 may readily be generalized to Euclidean n-space. We will state the generalization, leaving the appropriate modification of the details of the proof of Theorem 20 to the reader.

21 THEOREM. *Let S be Euclidean n-space, $\mathscr{B}$ the field of Borel subsets of S, and μ the Lebesgue measure of sets in $\mathscr{B}$. Then a subset K of $L_p(S, \mathscr{B}, \mu)$ is conditionally compact if and only if it is bounded and the following limits are uniform for f in K.*

$$\text{(a)} \quad \lim_{x_1 \to 0 \ldots x_n \to 0} \int_{-\infty}^{+\infty} \cdots \int_{-\infty}^{+\infty} |f(x_1+y_1, \ldots, x_n+y_n) - f(y_1, \ldots, y_n)|^p = 0,$$

and

$$\text{(b)} \quad \lim_{A \to \infty} \int_{S - C_A} |f(y)|^p dy = 0,$$

where C_A is the cube $-A \leqq x_1, \ldots, x_n \leqq A$.

The ultimate generalization of this line of thought is the generalization to arbitrary groups with invariant measures, a topic which is discussed in Chapter XI.

We now append some properties of the space L_p which spring from its natural ordering, and which will be useful later. We say that $f \in L_p(S, \Sigma, \mu)$ is *positive* and write $f \geqq 0$ if $f(s) \geqq 0$ for μ-almost all $s \in S$. If f_1 and f_2 are two real or complex valued functions in $L_p(S, \Sigma, \mu)$ such that $f_1 - f_2 \geqq 0$ we write $f_1 \geqq f_2$ or $f_2 \leqq f_1$. This clearly is a partial ordering (I.2.1) in L_p. We now establish completeness (I.12) with respect to this ordering.

22 THEOREM. *Let (S, Σ, μ) be a positive measure space. Then the real partially ordered space $L_p(S, \Sigma, \mu)$, $1 \leqq p < \infty$, is a complete lattice.*

PROOF. It is evidently sufficient to show that if $\{f_\alpha\}$ is a set of functions in L_1 such that $0 \leqq f_\alpha \leqq g_0$ for some $g_0 \in L_1$, then $\sup_\alpha \{f_\alpha\}$ exists in L_1. Further, since g_0 is integrable it vanishes μ-almost everywhere outside of a set of σ-finite measure. We may therefore suppose that (S, Σ, μ) is a σ-finite positive measure space. Now the mapping $f_\alpha \to \lambda_\alpha$ defined by $\lambda_\alpha(E) = \int_E f_\alpha(s)\mu(ds)$, $E \in \Sigma$, is a one-to-one order preserving map of $L_1(S, \Sigma, \mu)$ into the space $ca(S, \Sigma)$ with its natural ordering (III.7). Thus if g_0 is mapped into $\lambda_0 \in ca(S, \Sigma)$, then $0 \leqq \lambda_\alpha \leqq \lambda_0$. Using Corollary III.7.6, let $\nu = \sup \{\lambda_\alpha\}$, i.e., ν is the least upper bound of the set $\{\lambda_\alpha\}$. Hence $\nu \in ca(S, \Sigma)$ and $0 \leqq \lambda_\alpha \leqq \nu \leqq \lambda_0$. Since λ_0 is μ-continuous so is ν and the Radon-Nikodým theorem (III.10.7) implies that there is an $h \in L_1$ such that $\nu(E) = \int_E h(s)\mu(ds)$. The fact that $h = \sup_\alpha \{f_\alpha\}$ follows from the relation $\nu = \sup_\alpha \{\lambda_\alpha\}$ and the order preserving character of the map of L_1 into ca. Q.E.D.

23 THEOREM. *If (S, Σ, μ) is a σ-finite positive measure space, then the real partially ordered space $L_\infty(S, \Sigma, \mu)$, is a complete lattice.*

PROOF. Let $S_n \in \Sigma$ be such that $\mu(S_n) < \infty$, $S_n \subseteqq S_{n+1}$ and $S = \cup S_n$. Let $f_\alpha, g_0 \in L_\infty$ be such that $f_\alpha \leqq g_0$. Since $L_\infty(S_n, \Sigma, \mu) \subseteqq L_1(S_n, \Sigma, \mu)$ the preceding theorem implies that there exists an $h_n \in L_1(S_n, \Sigma, \mu)$ such that h_n is the supremum of $\{f_\alpha\}$ regarded in $L_1(S_n, \Sigma, \mu)$. In particular, for each α, $f_\alpha(s) \leqq h_n(s) \leqq g_0(s)$ for μ-almost all $s \in S_n$, and so $h_n \in L_\infty(S_n, \Sigma, \mu)$. We may take h_n to vanish outside S_n and obtain an $h_n \in L_\infty(S, \Sigma, \mu)$. Now for μ-almost all $s \in S$, the sequence $\{h_n(s)\}$ is an increasing sequence of real numbers and we take $h(s) = \lim_n h_n(s)$. The function h is measurable and since $h(s) \leqq g_0(s)$ for almost all s, it is essentially bounded. By construction h is an upper bound for the set $\{f_\alpha\}$. If it is not the least upper bound, there is a measurable set of finite positive measure $E \subseteqq S_{n_0}$ and a measurable function h' such that for each α, $f_\alpha(s) \leqq h'(s) < h(s) = h_{n_0}(s)$ for almost all s in E. But this violates the construction of h_{n_0}. Q.E.D.

Some additional results will be found in Section 11. We observe that L_1^{**} has been proved in Theorem 16 to be a space of set functions, and so by III.7.6 it has the property that subsets which possess upper bounds have least upper bounds. Now L_1 can be regarded as being embedded, by the natural mapping $\varkappa$, in the space L_1^{**}. For a later application it is important to know that the least upper bound of a subset in L_1 is the same as its least upper bound when regarded as a subset of L_1^{**}.

24 THEOREM. *Let (S, Σ, μ) be a σ-finite positive measure space. If $\varkappa$ is the natural isometric isomorphism of $L_1 = L_1(S, \Sigma, \mu)$ into L_1^{**}, and if $\{f_\alpha\}$ has an upper bound in the partially ordered space L_1, then* $\varkappa(\sup_\alpha \{f_\alpha\}) = \sup_\alpha \{\varkappa f\}$.

PROOF. Since (S, Σ, μ) is σ-finite it follows that $L_1^{**}(S, \Sigma, \mu) = ba(S, \Sigma^*, \mu)$, the space of all bounded additive functions on the Lebesgue extension Σ^* of Σ which vanish on sets from Σ^* of μ-measure zero. If $f \in L_1$, it is readily seen that $\varkappa f$ is the set function defined by $(\varkappa f)(E) = \int_E f(s)\mu(ds)$, $E \in \Sigma^*$. In particular $\varkappa f$ is countably additive and μ-continuous. Suppose that $0 \leqq f_0 = \sup_\alpha \{f_\alpha\}$. Since $\varkappa$ preserves order it is clear that $0 \leqq \lambda = \sup_\alpha \{\varkappa f_\alpha\} \leqq \varkappa f_0$, the least upper bound being taken in the partially ordered space $ba(S, \Sigma^*, \mu)$. It follows that λ is countably additive and μ-continuous, so by the Radon-Nikodým theorem there is a function $g \in L_1(S, \Sigma^*, \mu)$ such that $\lambda(E) = \int_E g(s)\mu(ds)$, $E \in \Sigma^*$. Since $\varkappa g = \lambda \leqq \varkappa f_0$ we have $g \leqq f_0$. On the other hand, since $\varkappa f_\alpha \leqq \varkappa g$, $f_\alpha \leqq g$ for each α. Thus $f_0 = \sup_\alpha f_\alpha \leqq g$ so that $f_0 = g$. Q.E.D.

We now show that the space of linear mappings between the L_p spaces also is a partially ordered space. We say that a linear mapping

$T : L_p \to L_q$ is *positive* and write $T \geqq 0$ when $f \in L_p$ and $f \geqq 0$ imply that $Tf \geqq 0$. Similarly $T_1 \geqq T_2$ or $T_2 \leqq T_1$ means that $T_1 - T_2 \geqq 0$. It is readily seen that if $T \geqq 0$, then T maps real functions into real functions.

25 Lemma. *Let (S, Σ, μ) be a positive measure space and let $1 \leqq p \leqq \infty$. Suppose $f_1 + f_2 = \sum_{k=1}^n g_k$, where f_j, g_k are positive elements in $L_p(S, \Sigma, \mu)$. Then there are positive elements h_{jk} in $L_p(S, \Sigma, \mu)$, $j = 1, 2$, $k = 1, \ldots, n$, such that*

$$f_j = \sum_{k=1}^n h_{jk}, \quad g_k = h_{1k} + h_{2k}, \quad j = 1, 2, \quad k = 1, \ldots, n.$$

Proof. Suppose $n = 2$. Since $0 \leqq f_1, g_1 \leqq f_1 + g_1$ the bounds $\inf\{f_1, g_1\}$ and $\sup\{f_1, g_1\}$ exist in $L_p(S, \Sigma, \mu)$. We take $h_{11} = \inf\{f_1, g_1\}$, $h_{12} = f_1 - h_{11}$, $h_{21} = g_1 - h_{11}$ and $h_{22} = (f_1 + f_2) - \sup\{f_1, g_1\}$. The statement follows from these choices for the h_{jk} and the observation that

$$\inf\{f_1, g_1\} + \sup\{f_1, g_1\} = f_1 + g_1.$$

The statement for arbitrary choice of n follows by induction and the result for $n = 2$ by grouping $f_1 + f_2 = \sum_{k=1}^{n-1} g_k + (g_n + g_{n+1})$. Q.E.D.

26 Theorem. *Let (S, Σ, μ) be a positive measure space and let $1 \leqq p \leqq \infty$ and $1 \leqq q < \infty$. Then the partially ordered space of linear mappings from the real space $L_p(S, \Sigma, \mu)$ to the real space $L_q(S, \Sigma, \mu)$ is a complete lattice. If (S, Σ, μ) is σ-finite, the result also holds if $q = \infty$.*

Proof. It is sufficient to show that if $\{T_\alpha\}$ is a set of positive linear mappings, then they have a greatest lower bound T_0 which is also a linear mapping. We first define T_0 on the positive elements f of $L_p(S, \Sigma, \mu)$. To do this, let $f = \sum_{i=1}^n g_i$ be a decomposition of f into a finite sum of positive functions g_i in $L_p(S, \Sigma, \mu)$. Set

$$T_0 f = \inf\{\sum_{i=1}^n T_{\alpha_i} g_i\}$$

where the infimum is taken over all such finite decompositions $f = \sum g_i$ of f and arbitrary choices of T_{α_i}, $i = 1, \ldots, n$. It is evident that $0 \leqq T_0 f \leqq T_\alpha f$ for any α. Also if S is any lower bound for $\{T_\alpha\}$, then

$$Sf = \sum_{i=1}^{n} Sg_i \leqq \sum_{i=1}^{n} T_{\alpha_i} g_i$$

for any decomposition and hence $Sf \leqq T_0 f$. We now show that T_0 is additive on positive elements. Let $f_j \in L_p(S, \Sigma, \mu)$, $f_j \geqq 0$, $j = 1, 2$, and let $f_1 = \sum_{i=1}^n g_{1i}$, $f_2 = \sum_{i=1}^n g_{2i}$ be decompositions of f_1 and f_2 into positive functions. Then $\sum g_{1i} + \sum g_{2i}$ is a decomposition of f_1+f_2, and so

$$T_0(f_1+f_2) \leqq T_0(f_1)+T_0(f_2).$$

To show the opposite inequality, let $f_1+f_2 = \sum_{k=1}^n g_k$ be a decomposition of f_1+f_2 into a sum with $g_k \geqq 0$. By the preceding lemma there is a set h_{jk}, $j = 1, 2$, $k = 1, \ldots, n$, of positive elements in $L_p(S, \Sigma, \mu)$ with $f_j = \sum_{k=1}^n h_{jk}$ and $g_k = h_{1k}+h_{2k}$. Thus

$$T_0(f_1)+T_0(f_2) \leqq \sum_{k=1}^{n} T_{\alpha_k}(h_{1k}) + \sum_{k=1}^{n} T_{\alpha_k}(h_{2k}) = \sum_{k=1}^{n} T_{\alpha_k}(g_k),$$

and since $T_0(f_1+f_2)$ is the infimum of sums of the last kind, we conclude that

$$T_0(f_1)+T_0(f_2) \leqq T_0(f_1+f_2).$$

Hence T_0 is additive on positive functions, and its homogeneity with respect to positive scalars is evident.

If f_1, f_2, g_1, g_2 are positive and $f_1-f_2 = g_1-g_2$ then $f_1+g_2 = g_1+f_2$, $T_0 f_1+T_0 g_2 = T_0 g_1+T_0 f_2$ and $T_0 f_1-T_0 f_2 = T_0 g_1-T_0 g_2$. Thus T_0 may be defined for a real function f by placing $T_0 f = T_0 f_1-T_0 f_2$ where f_1, f_2 are positive and $f = f_1-f_2$. This extended function T_0 is a linear map, and it is clear from the way in which it was defined that it is the greatest lower bound of the maps T_α. Q.E.D.

9. Spaces of Set Functions

In this section we investigate special properties of $ba(S, \Sigma)$, the space of bounded additive scalar functions on a field of sets, and $ca(S, \Sigma)$, the space of countably additive measures defined on a σ-field. The first theorems identify the weakly compact sets of $ca(S, \Sigma)$.

1 THEOREM. *A set $K \subset ca(S, \Sigma)$ is weakly sequentially compact if and only if it is bounded and the countable additivity of μ on Σ is uniform with respect to μ in K.*

PROOF. If K is weakly sequentially compact, then II.3.27 shows that K is bounded. If the countable additivity of μ is not uniform for μ in K there is a decreasing sequence of sets E_n in Σ with void intersection, a sequence $\{\mu_n\} \subseteqq K$, and a number $\varepsilon > 0$ such that

$$|\mu_n(E_n)| > \varepsilon, \qquad n = 1, 2, \ldots.$$

The functions μ_n are all continuous with respect to the measure defined by

$$\lambda(E) = \sum_{n=1}^{\infty} \frac{1}{2^n} \frac{v(\mu_n, E)}{1+v(\mu_n, E)}, \qquad E \in \Sigma,$$

and thus all belong to the subspace $ca(S, \Sigma, \lambda)$ consisting of all λ-continuous functions in $ca(S, \Sigma)$. According to the Radon-Nikodým theorem (III.10.2) the formula

$$\mu(E) = \int_E f(s)\lambda(ds)$$

establishes an isometric isomorphism between $ca(S, \Sigma, \lambda)$ and $L_1(S, \Sigma, \lambda)$. Thus the set K' in $L_1(S, \Sigma, \lambda)$ corresponding to K is sequentially weakly compact. Theorem 8.9 shows that the countable additivity of $\mu(E) = \int_E f(s)\lambda(ds)$ is uniform with respect to f in K' and hence uniform with respect to μ in K.

Conversely, suppose that the set $K \subset ca(S, \Sigma)$ satisfies the two conditions and let $\mu_n \in K$, $n = 1, 2, \ldots$. Using the measure λ defined above we have functions $f_n \in L_1(S, \Sigma, \lambda)$ such that

$$\mu_n(E) = \int_E f_n(s)\lambda(ds), \qquad |\mu_n| = |f_n|, \qquad n = 1, 2, \ldots.$$

Thus (by 8.9) the sequence $\{f_n\}$ has a subsequence converging weakly in $L_1(S, \Sigma, \lambda)$. Since the spaces $ca(S, \Sigma, \lambda)$ and $L_1(S, \Sigma, \lambda)$ are equivalent the sequence $\{\mu_n\}$ contains a subsequence weakly convergent in $ca(S, \Sigma, \lambda)$, and hence in $ca(S, \Sigma)$. Q.E.D.

Another useful criterion for weak sequential compactness in $ca(S, \Sigma)$ is given in the following theorem.

2 THEOREM. *A subset $K \subset ca(S, \Sigma)$ is weakly sequentially compact if and only if it is bounded and, for some positive λ in $ca(S, \Sigma)$, the limit $\lim_{\lambda(E)\to 0} \mu(E) = 0$ is uniform with respect to μ in K.*

PROOF. The sufficiency of the conditions as well as the necessity of the boundedness condition follows from Theorem 1. Thus let K be compact and let M be a bound for the norms $|\mu|$ of elements μ in K. It will first be shown that for each $\varepsilon > 0$ there exists a finite set $\mu_1, \ldots, \mu_n$ in K and a $\delta > 0$ such that $|\mu(E)| < \varepsilon$ for every μ in K provided that $v(\mu_i, E) < \delta$, $i = 1, \ldots, n$. If this is not true, there is an $\varepsilon > 0$ such that for every $\mu_1 \in K$ there is a set $E_1 \in \Sigma$ and a $\mu_2 \in K$ such that

$$v(\mu_1, E_1) < \tfrac{1}{2}, \qquad |\mu_2(E_1)| \geqq \varepsilon.$$

There also exists a set $E_2 \in \Sigma$ and an element μ_3 in K such that

$$v(\mu_1, E_2) < \frac{1}{2^2}, \qquad v(\mu_2, E_2) < \frac{1}{2^2}, \qquad |\mu_3(E_2)| \geqq \varepsilon.$$

In this fashion sequences $\{\mu_n\} \subseteqq K$, $\{E_n\} \subseteqq \Sigma$ are defined for which

$$v(\mu_i, E_n) < \frac{1}{2^n}, \qquad i = 1, \ldots, n,$$

$$|\mu_{n+1}(E_n)| \geqq \varepsilon.$$

Since K is weakly sequentially compact a subsequence of $\{\mu_n\}$ is weakly convergent. In order to simplify notation we suppose that the sequence $\{\mu_n\}$ itself is weakly convergent. Let $\lambda_0 = \sum_{j=1}^{\infty} 2^{-j} v(\mu_j)$ so that each μ_n is λ_0-continuous. Since $\{\mu_n\}$ is weakly convergent, $\lim \mu_n(E)$ exists for each E in Σ. It follows from the Vitali-Hahn-Saks theorem (III.7.2) that $\lim_{\lambda_0(E)\to 0} \mu_n(E) = 0$ uniformly with respect to n. Now

$$\lambda_0(E_n) \leqq \sum_{j=1}^{n} \frac{1}{2^j}\frac{1}{2^n} + \sum_{j=n+1}^{\infty} \frac{1}{2^j} M < \frac{1+M}{2^n},$$

and therefore $\lim_n \mu_m(E_n) = 0$ uniformly with respect to $m = 1, 2, \ldots$. But this contradicts the fact that $|\mu_{n+1}(E_n)| \geqq \varepsilon > 0$ and proves the original assertion. Now let $\delta_n > 0$ and let $\mu_1^{(n)}, \ldots, \mu_{m_n}^{(n)}$ be elements of K such that $|\mu(E)| < 1/n$ for every μ in K and every $E \in \Sigma$ for which $v(\mu_i^{(n)}, E) < \delta_n$, $i = 1, \ldots, m_n$. If λ is defined by

$$\lambda = \sum_{n=1}^{\infty} \sum_{i=1}^{m_n} \frac{1}{2^n}\frac{1}{2^i} v(\mu_i^{(n)}),$$

it follows from Lemma III.4.13 that every μ in K is λ-continuous, i.e., $K \subset ca(S, \Sigma, \lambda)$. The Radon-Nikodým theorem (III.10.2) shows that the formula

$$\mu(E) = \int_E f(s)\lambda(ds)$$

establishes an equivalence between $ca(S, \Sigma, \lambda)$ and $L_1(S, \Sigma, \lambda)$ and thus the present theorem follows from Corollary 8.11. Q.E.D.

3 COROLLARY. *Under the hypothesis of Theorem 2, λ may be chosen so that*

$$\lambda(E) \leqq \sup_{\mu \in K} |\mu(E)|, \qquad E \in \Sigma.$$

PROOF. In view of Lemma III.1.5 and the formula defining λ, the measure $\lambda/4$ has the desired properties. Q.E.D.

4 THEOREM. *The space $ca(S, \Sigma)$ is weakly complete.*

PROOF. If $\{\mu_n\}$ is a weak Cauchy sequence in $ca(S, \Sigma)$ then the limit $\lim \mu_n(E)$ exists for every E in Σ and, by II.3.27, the sequence $\{\mu_n\}$ is bounded. According to Corollary III.7.4 the countable additivity of $\mu_n(E)$ is uniform with respect to $n = 1, 2, \ldots$ and thus, since $\{\mu_n\}$ is a weak Cauchy sequence, Theorem 1 shows that it is a weakly convergent sequence. Q.E.D.

5 THEOREM. *A sequence $\{\mu_n\}$ in $ca(S, \Sigma)$ converges weakly (to μ) if and only if it is bounded and the limit $\lim_n \mu_n(E)$ exists (and is equal to $\mu(E)$) for every E in Σ.*

PROOF. Let λ be defined as in the proof of Theorem 1. Then $ca(S, \Sigma, \lambda)$ is equivalent to $L_1(S, \Sigma, \lambda)$ and the sequence

$$\mu_n(E) = \int_E f_n(s)\lambda(ds), \qquad n = 1, 2, \ldots$$

is weakly convergent in $ca(S, \Sigma, \lambda)$ (and thus in $ca(S, \Sigma)$) if and only if the sequence $\{f_n\}$ is weakly convergent in $L_1(S, \Sigma, \lambda)$. The desired conclusion follows from Theorem 8.7. Q.E.D.

6 DEFINITION. Let (S, Σ, μ) be a measure space. A set $E \in \Sigma$ is called an *atom* if $\mu(E) \neq 0$, and if $F \in \Sigma$, $F \subseteqq E$, then either $\mu(F) = \mu(E)$ or $\mu(F) = 0$.

It is clear that if E_1 and E_2 are atoms, then either $\mu(E_1E_2) = 0$ or $\mu(E_1 \triangle E_2) = 0$. Also a finite positive measure space can have at most a countable family of disjoint atoms.

7 LEMMA. (*Saks*) *Let (S, Σ, μ) be a finite positive measure space. If $\varepsilon > 0$ then S is the union of a finite sequence of disjoint sets $E_1, \ldots, E_n \in \Sigma$ such that each E_i is either an atom or $\mu(E_i) \leqq \varepsilon$.*

PROOF. Let ε be an arbitrary positive number. Since $\mu(S) < \infty$ there is at most a finite set $\{E_1, \ldots, E_p\}$ of disjoint atoms with measure greater than ε. The set $A = S - \cup_{i=1}^{p} E_i$ then contains no atoms having measure greater than ε. It will now be shown that every measurable subset B of A contains a set F with $0 < \mu(F) \leqq \varepsilon$. Suppose, on the contrary, that some such set B contains no set of Σ with positive measure at most ε. Then B is not an atom and thus contains a set G_1 in Σ with $0 < \mu(G_1) < \mu(B)$. The set $B - G_1$ also contains no set in Σ with positive measure at most ε. Thus there is a set G_2 in Σ with $G_2 \subseteqq B - G_1$ and with $0 < \mu(G_2) < \mu(B - G_1)$. Continuing inductively we obtain a sequence $\{G_n\}$ of disjoint sets of positive measure. Since $\Sigma\mu(G_i) < \infty$ we must have $\mu(G_n) < \varepsilon$ for sufficiently large n. This contradiction proves the existence of a set $F \subseteqq B$ with $0 < \mu(F) \leqq \varepsilon$.

For every set E in Σ let $\beta(E) = \sup \mu(H)$ where H varies over all measurable subsets of E with $\mu(H) \leqq \varepsilon$. Then, by the preceding argument, $0 < \beta(G) \leqq \varepsilon$ for every measurable subset G of A with $\mu(G) > 0$. By induction we determine a sequence $\{F_n\}$ of disjoint measurable subsets of A with

$$\tfrac{1}{2}\beta(A - \bigcup_{i=1}^{n} F_i) \leqq \mu(F_{n+1}) \leqq \varepsilon, \qquad n = 1, 2, \ldots.$$

Let $F_0 = A - \cup_{i=1}^{\infty} F_i$ so that

$$\beta(F_0) \leqq \beta(A - \bigcup_{i=1}^{n} F_i) \leqq 2\mu(F_{n+1}), \qquad n = 1, 2, \ldots.$$

However $\sum \mu(F_i) \leqq \mu(A) < \infty$ and consequently $\lim \mu(F_n) = 0$. Thus, the above inequality shows that $\beta(F_0) = 0$. Consequently $\mu(F_0) = 0$. Let r be an integer such that $\sum_{i=r+1}^{\infty} \mu(F_i) < \varepsilon$ and let $E_{m+1} = F_1, \ldots, E_{m+r} = F_r$, and $E_{m+r+1} = \cup_{i=r+1}^{\infty} F_i \cup F_0$. The sets $E_1, \ldots, E_p$ with $p = m+r+1$ satisfy the requirements of the lemma. Q.E.D.

The next theorem is a striking improvement of the principle of uniform boundedness in the space $ca(S, \Sigma)$.

8 THEOREM. (*Nikodým*) *If M is a set in $ca(S, \Sigma)$ and if for each E in Σ there is an $N(E) < \infty$ such that*

$$|\mu(E)| < N(E), \qquad \mu \in M,$$

then there exists a number $N < \infty$ *such that*

$$|\mu(E)| < N, \qquad \mu \in M, \qquad E \in \Sigma.$$

PROOF. If the conclusion is false, then for each integer n there exist measures $\mu_n \in M$ and sets $G_n \in \Sigma$ such that $|\mu_n(G_n)| > n$. Let $\lambda \in ca(S, \Sigma)$ be defined by

$$\lambda(E) = \sum_{n=1}^{\infty} \frac{1}{2^n} \frac{v(\mu_n, E)}{v(\mu_n, S)}, \qquad E \in \Sigma,$$

and consider the complete metric space $\Sigma(\lambda)$ discussed in Section III.7. Let

$$H_m = \{E \in \Sigma(\lambda) \,|\, |\mu_n(E)| \leqq m,\ n = 1, 2, \ldots\},$$

so that H_m is a closed set in $\Sigma(\lambda)$ and $\Sigma(\lambda) = \bigcup_{m=1}^{\infty} H_m$. The Baire category theorem (I.6.9) asserts that there exists a set $B_0 \in \Sigma(\lambda)$ and a number $\varepsilon > 0$ such that if $\lambda(E \,\Delta\, B_0) \leqq \varepsilon$, then $|\mu_n(E)| \leqq m_0$ for some integer m_0 and all integers $n = 1, 2, \ldots$.

Let A be an atom of λ. If $F \subseteqq A$, $F \in \Sigma$, and $0 < v(\mu_n, F) < v(\mu_n, A)$ for some n, then $0 < \lambda(F) < \lambda(A)$. Consequently, for each n, either $v(\mu_n, A) = 0$ or A is an atom of $v(\mu_n)$. If A is an atom of $v(\mu_n)$ then if $F \subseteqq A$, $F \in \Sigma$, and $\mu_n(F) \neq 0$, we must have $v(\mu_n, A-F) = 0$, from which it follows that $\mu_n(F) = \mu_n(A)$; i.e., A is an atom of μ_n. Now let $\varepsilon > 0$ and let $E_1, \ldots, E_m$ be a decomposition of S with respect to the measure λ as described in Lemma 7. Let $E \in \Sigma$ and $F_k = E \cap E_k$, $k = 1, \ldots, m$. If $E_1, \ldots, E_p$ are the atoms of the family $\{E_1, \ldots, E_m\}$ for which $\lambda(E_i) \geqq \varepsilon$, then by our remarks above,

$$|\mu_n(F_k)| \leqq |\mu_n(E_k)| < N(E_k), \qquad k = 1, \ldots, p.$$

Now $\lambda(F_k) \leqq \varepsilon$ for $k = p+1, \ldots, m$. Write

$$F_k = (B_0 \cup F_k) - (B_0 - F_k).$$

Since $B_0 \Delta (B_0 \cup F_k) = F_k - B_0$ and $B_0 \Delta (B_0 - F_k) = B_0 \cap F_k$, it follows that $\lambda(B_0 \Delta (B_0 - F_k))$ and $\lambda(B_0 \Delta (B_0 \cup F_k))$ are at most equal to $\lambda(F_k) \leqq \lambda(E_k) \leqq \varepsilon$. Consequently

$$|\mu_n(E_k)| \leqq 2m_0, \qquad k = p+1, \ldots, m.$$

Thus

$$|\mu_n(E)| \leqq \sum_{k=1}^{p} N(E_k) + 2m_0(m-p)$$

for any $E \in \Sigma$ and all $n = 1, 2, \ldots$. The right hand side is independent of n, contradicting the supposition that there was a set $G_n \in \Sigma$ with $|\mu_n(G_n)| > n$ for each integer n. Q.E.D.

Next we turn to an investigation of the space $ba(S, \Sigma)$.

9 Theorem. *The space* $ba(S, \Sigma)$ *is weakly complete. If* S *is a topological space, then* $rba(S)$ *is also weakly complete.*

Proof. Consider the closed subspace $B(S, \Sigma)$ of $B(S)$. According to Theorems 6.18 and 6.20 there is a compact Hausdorff space S_1 such that $B(S, \Sigma)$ is equivalent to $C(S_1)$. Theorem 5.1 shows that there is an isometric isomorphism $x^* \longleftrightarrow \mu$ between $B^*(S, \Sigma)$ and $ba(S, \Sigma)$, which is determined by the equation $x^*\chi_E = \mu(E)$, $E \in \Sigma$. Thus, since $B(S, \Sigma)$ is equivalent to $C(S_1)$, $ba(S, \Sigma)$ is equivalent to $rca(S_1)$ (Theorem 6.3). But $rca(S_1)$, being a closed subspace of the space of countably additive measures on the Borel sets in S_1, is weakly complete by Theorem 4. Therefore $ba(S, \Sigma)$ is also weakly complete.

Since $rba(S)$ is a closed subspace of $ba(S, \Sigma)$, where Σ is the field generated by the closed subsets of the topological space S, $rba(S)$ is weakly complete. Q.E.D.

The essential tool of the proof of Theorem 9 was the isometric isomorphism of $B(S, \Sigma)$ with $C(S_1)$ where S_1 is a particular compact Hausdorff space. We shall now investigate what properties the existence of this isomorphism implies for S_1 and investigate in more detail the isomorphism of $ba(S, \Sigma)$ onto $rca(S_1)$. This information will be used to obtain further properties of the space $ba(S, \Sigma)$. Let H denote the isomorphism of $B(S, \Sigma)$ onto $C(S_1)$ and let $E \in \Sigma$. We observe that if χ_E is the characteristic function of E, then $H(\chi_E)$ is continuous on S_1 and $H(\chi_E)^2 = H(\chi_E^2) = H(\chi_E)$. Thus $H(\chi_E)(s_1)$ is either zero or one for each $s_1 \in S_1$, i.e., $H(\chi_E)$ is the characteristic function of a set $E_1 \subseteq S_1$. In view of the continuity of $H(\chi_E)$, E_1 must be both open and closed. One may ask whether the converse is true: If E_1 is an open and closed set in S_1, is $H^{-1}(\chi_{E_1})$ the characteristic function χ_E of a set $E \in \Sigma$? Stated differently, does the existence of the isomorphism H of $B(S, \Sigma)$ onto $C(S_1)$ imply the existence of an isomorphic mapping τ of Σ onto the field Σ_1 of all open and closed sets of S_1?

That this is so will be proved in Lemma 10 below. Moreover, it will be seen that S_1 is totally disconnected, i.e., the open and closed sets form a basis for the topology.

10 LEMMA. *Let S_1 be a compact Hausdorff space such that $B(S, \Sigma)$ is isometrically isomorphic with $C(S_1)$. Then the space S_1 is uniquely determined to within a homeomorphism and is totally disconnected. The correspondence $\chi_E \to \chi_{E_1}$ establishes an isomorphism τ of the field Σ onto the field Σ_1 of all open and closed sets in S_1, i.e., $\tau(E \cup F) = \tau(E) \cup \tau(F)$, $\tau(EF) = \tau(E)\tau(F)$, and $\tau(E') = \tau(E)'$ for all $E, F \in \Sigma$.*

PROOF. As before, let H be the isomorphism of $B(S, \Sigma)$ onto $C(S_1)$ given by Theorems 6.18 and 6.20. We have shown that $\tau(E)$ is open and closed for $E \in \Sigma$. Conversely, let E_1 be any open and closed set in S_1. We wish to show that the set $E = \tau^{-1}(E_1)$ is in Σ. Since χ_E is Σ-measurable, there exists a finite decomposition of S into disjoint sets $A_1, \ldots, A_n$ in Σ, and scalars $\alpha_1, \ldots, \alpha_n$ such that $|\chi_E - \sum_{i=1}^n \alpha_i \chi_{A_i}| < 1/4$. We observe first that $E \subseteq \bigcup_{i=1}^n A_i$, since if s belongs to $E - \bigcup_{i=1}^n A_i$, then $|\chi_E(s) - \sum_{i=1}^n \alpha_i \chi_{A_i}(s)| = 1 > 1/4$. We will show that if $EA_i \neq \phi$, then $A_i \subseteq E$, from which it will follow that E is the union of all the sets A_j which it contains, so that $E \in \Sigma$. Indeed, if $s \in EA_i$ for some i, then $|\chi_E(s) - \sum_{i=1}^n \alpha_i \chi_{A_i}(s)| = |1 - \alpha_i| < 1/4$. Thus $|\alpha_i| > 3/4$. However, if $t \in A_i$ and $\chi_E(t) = 0$, then $|\chi_E(t) - \sum_{i=1}^n \alpha_i \chi_{A_i}(t)| = |\alpha_i| < 1/4$, which is impossible. This shows that $A_i \subseteq E$, and hence that $E \in \Sigma$. The fact that τ is an isomorphism of Σ onto Σ_1 follows from the equations $H(1 - \chi_E) = 1 - H(\chi_E)$, $H(\chi_E \chi_F) = H(\chi_E)H(\chi_F)$, and $H(\chi_E + \chi_F - \chi_{EF}) = H(\chi_E) + H(\chi_F) - H(\chi_E)H(\chi_F)$.

It will be shown next that S_1 is totally disconnected. If G_1 is a non-void open set in S_1 let $t_0 \in G_1$ and let f_1 be a continuous function such that $f_1(t_0) = 1$, and $f_1(t) = 0$ for t in G_1' (cf. I.5.2). As before, we select a simple function $g = \sum_{i=1}^n \alpha_i \chi_{A_i}$ in $B(S, \Sigma)$ where $A_1, \ldots, A_n \in \Sigma$, such that $|g - H^{-1}(f_1)| < 1/4$. If $g_1 = H(g) = \sum_{i=1}^n \alpha_i \chi_{\tau(A_i)}$, let $U_1 = \{s_1 \,|\, |g_1(s_1)| > 1/2\}$. Then U_1 is open and closed and $t_0 \in U_1 \subseteq G_1$.

The uniqueness of S_1 follows from Theorem 6.26. Q.E.D.

11 LEMMA. *Employing the notation of Lemma 10, let $B(S, \Sigma)$ be isometrically isomorphic with $C(S_1)$.*

(a) *There is an isometric isomorphism T of $ba(S, \Sigma)$ onto $ba(S_1, \Sigma_1)$ determined by the correspondence $(T\mu)(E_1) = \mu(\tau^{-1}(E_1))$ for μ in $ba(S, \Sigma)$ and E_1 in Σ_1.*

(b) *Each* $\mu_1 \in ba(S_1, \Sigma_1)$ *has a unique extension to a regular countably additive measure* μ_2 *in* $ca(S_1, \Sigma_2)$ *where* Σ_2 *is the* σ*-field generated by* Σ_1. *Each* μ_2 *in* $ca(S_1, \Sigma_2)$ *is regular. The correspondence* $U : \mu_1 \to \mu_2$ *is an isometric isomorphism of* $ba(S_1, \Sigma_1)$ *onto* $ca(S_1, \Sigma_2)$.

(c) *If* E_1 *is in* Σ_1 *then* $v(\mu_1, E_1) = v(U(\mu_1), E_1)$ *for all* μ_1 *in* $ba(S_1, \Sigma_1)$.

PROOF. Recalling that τ is an isomorphism of Σ onto Σ_1, it is clear that the mapping T is an isometric isomorphism of $ba(S, \Sigma)$ onto $ba(S_1, \Sigma_1)$, since

$$|T\mu| = \sup \sum_{i=1}^{n} |(T\mu)(\tau E_i)| = \sup \sum_{i=1}^{n} |\mu(E_i)| = |\mu|,$$

where $\{E_1, \ldots, E_n\}$ is an arbitrary partition of S. This proves (a).

Since each $\mu_1 \in ba(S_1, \Sigma_1)$ is clearly regular, it follows from Theorem III.5.13 that every μ_1 in $ba(S_1, \Sigma_1)$ is countably additive. Hence by III.5.14 every $\mu_1 \in ba(S_1, \Sigma_1)$ has a unique regular extension to a $\mu_2 \in ca(S_1, \Sigma_2)$. On the other hand the restriction of each $\mu_2 \in ca(S_1, \Sigma_2)$ to Σ_1 is regular. Thus each $\mu_2 \in ca(S_1, \Sigma_2)$ is regular and the mapping U is an algebraic isomorphism of $ba(S_1, \Sigma_1)$ onto $ca(S_1, \Sigma_2)$. We shall show that U is isometric. By Theorem 6.3

$$|U\mu_1| = \sup_{|f|=1} \left| \int_{S_1} f(s_1)(U\mu_1)(ds_1) \right|, \qquad f \in C(S_1).$$

Since, by Lemma 10, the field Σ_1 is a basis for the topology in S_1, it follows from the Stone-Weierstrass Theorem 6.16 (or 6.17) that functions of the form $\sum_{i=1}^{n} \alpha_i \chi_{E_i}$ are dense in $C(S_1)$, where $\alpha_1, \ldots, \alpha_n$ are scalars and $E_1, \ldots, E_n$ are disjoint sets in Σ_1. Moreover,

$$\int_{S_1} \{\sum \alpha_i \chi_{E_i}(s_1)\}(U\mu_1)(ds_1) = \int_{S_1} \{\sum \alpha_i \chi_{E_i}(s_1)\}\mu_1(ds_1).$$

Thus if H is the isometric isomorphism of $B(S, \Sigma)$ onto $C(S_1)$,

$$\begin{aligned} |U\mu_1| &= \sup_{|f|=1} \left| \int_{S_1} f(s_1)\mu_1(ds_1) \right| \\ &= \sup_{|H^{-1}(f)|=1} \left| \int_S (H^{-1}f)(s)(T^{-1}\mu_1)(ds) \right| \\ &= |T^{-1}\mu_1| = |\mu_1| \end{aligned}$$

by (a) and Theorem 5.1.

To prove (c) let $\mu_1 \in ba(S_1, \Sigma_1)$ and $E \in \Sigma_1$. Let λ_1 be defined by the equation $\lambda_1(F) = \mu_1(EF)$, $F \in \Sigma_1$. Then $U(\lambda_1)(G) = (U\mu_1)(GE)$ for G in Σ_2. By part (b),

$$v(U(\mu_1), E) = |U(\lambda_1)| = |\lambda_1| = v(\mu_1, E). \qquad \text{Q.E.D.}$$

12 THEOREM. *A subset K of $ba(S, \Sigma)$ is weakly sequentially compact if and only if there exists a non-negative μ in $ba(S, \Sigma)$ such that*

$$\lim_{\mu(E)\to 0} \lambda(E) = 0$$

uniformly for $\lambda \in K$.

PROOF. Let $K \subseteq ba(S, \Sigma)$ be weakly sequentially compact and let $V = UT$ be the isometric isomorphism of $ba(S, \Sigma)$ onto $ca(S_1, \Sigma_2)$. Then $VK \subseteq ca(S_1, \Sigma_2)$ is weakly sequentially compact. By Theorem 2 there exists a non-negative $\mu_2 \in ca(S_1, \Sigma_2)$ such that $\lim_{\mu_2(E)\to 0} \lambda_2(E) = 0$ uniformly for $\lambda_2 \in VK$. It is then clear that $\mu = V^{-1}(\mu_2)$ is a non-negative element of $ba(S, \Sigma)$ such that $\lim_{\mu(E)\to 0} \lambda(E) = 0$ uniformly for $\lambda \in K$.

To prove the converse, suppose there exists a non-negative $\mu \in ba(S, \Sigma)$ such that $\lim_{\mu(E)\to 0} \lambda(E) = 0$ uniformly for λ in K. Then $\mu_1 = T(\mu)$ is a non-negative measure in $ba(S_1, \Sigma_1)$ such that $\lim_{\mu_1(E)\to 0} \lambda_1(E) = 0$, $E \in \Sigma_1$, uniformly for $\lambda_1 \in K_1 = T(K)$. Let $\mu_2 = U\mu_1$ and $K_2 = UK_1$. We shall show that $\lim_{\mu_2(E)\to 0} \lambda_2(E) = 0$, $E \in \Sigma_2$, uniformly for $\lambda_2 \in K_2$, from which it will follow that K_2 is weakly sequentially compact in $ca(S_1, \Sigma_2)$, whence K is weakly sequentially compact in $ba\,(S, \Sigma)$.

For $A \subseteq S_1$ let $\hat{\mu}_1(A) = \inf \sum_{i=1}^{\infty} \mu_1(E_i)$, where the infimum is to be taken over all sequences $\{E_i\}$ of sets in Σ_1 such that $\bigcup_{i=1}^{\infty} E_i \supseteq A$. By the uniqueness assertion of Theorem III.5.8, and by Theorem III.5.4 and Lemma III.5.5, $\hat{\mu}_1(A) = \mu_2(A)$ for A in Σ_2. Let $\varepsilon > 0$ be given, and let $\delta > 0$ be such that $\mu_1(E) < \delta$, with $E \in \Sigma_1$, implies $|\lambda_1(E)| < \varepsilon$ for all λ_1 in K_1, so that $v(\lambda_1, E) < 4\varepsilon$, (cf. III.1.5). Then if $A \in \Sigma_2$ and $\mu_2(A) < \delta/2$, there is a sequence $\{E_i\}$ of sets in Σ_1 such that $\bigcup_{i=1}^{\infty} E_i \supseteq A$, and $\sum_{i=1}^{\infty} \mu_1(E_i) < \delta$. Since μ_1 is positive, we may and shall assume that the E_i are disjoint. Then, since $\mu_1(\bigcup_{i=1}^{n} E_i) < \delta$ for each n, $v(\lambda_1, \bigcup_{i=1}^{n} E_i) < 4\varepsilon$ for each n and each λ_1 in K_1. Consequently, by Lemma 11(c), $v(\lambda_2, \bigcup_{i=1}^{n} E_i) < 4\varepsilon$ for each n and λ_2 in K_2. Letting $n \to \infty$, we find that $v(\lambda_2, \bigcup_{i=1}^{\infty} E_i) \leqq 4\varepsilon$ for λ_2 in K_2. Hence $\mu_2(A) < \delta/2$ implies $|\lambda_2(A)| < 4\varepsilon$ for $\lambda_2 \in K_2$, so we have proved $\lim_{\mu_2(E)\to 0} \lambda_2(E) = 0$, $E \in \Sigma_2$, uniformly for $\lambda_2 \in K_2$. Q.E.D.

Our next lemma is a corollary of the proof of Theorem 12.

13 COROLLARY. *Let S_1 be a set, Σ_1 a field of subsets of S_1, and let μ_1, λ_1 be in $ba(S_1, \Sigma_1)$. Suppose that Σ_2 is the σ-field generated by Σ_1, and that μ_1 and λ_1 have countably additive extensions μ_2 and λ_2 to Σ_2. Then λ_1 is μ_1-continuous if and only if λ_2 is μ_2-continuous.*

PROOF. Clearly if λ_2 is μ_2-continuous, λ_1 is μ_1-continuous. To prove the converse we recall (cf. the remarks following Definition III.4.12) that it is sufficient to show that if $v(\lambda_1)$ is $v(\mu_1)$-continuous, then $v(\lambda_2)$ is $v(\mu_2)$-continuous. However, it follows from the last paragraph of the proof of Theorem 12, that, given ε and $\delta > 0$, if δ is such that $v(\mu_1, E) < \delta$ implies $v(\lambda_1, E) < \varepsilon$, $E \in \Sigma_1$, then $v(\mu_2, A) < \delta/2$ implies $v(\lambda_2, A) \leqq 4\varepsilon$, $A \in \Sigma_2$. Q.E.D.

We shall now prove a generalization, due to Bochner, of the Radon-Nikodým theorem.

14 THEOREM. *Let μ be a non-negative element of $ba(S, \Sigma)$. Let $\lambda \in ba(S, \Sigma)$ be μ-continuous. Then for each $\varepsilon > 0$ there is a μ-integrable simple function f_ε such that the function F defined by the equation $F(E) = \int_E f_\varepsilon(s)\mu(ds)$, $E \in \Sigma$, satisfies the inequality $|\lambda - F| = v(\lambda - F, S) < \varepsilon$.*

PROOF. Let U and T be as in Lemma 11 so that $V = UT$ is an isometric isomorphism of $ba(S, \Sigma)$ onto $ca(S_1, \Sigma_2)$. Since λ is μ-continuous, Lemma 11(a) shows that $T\lambda$ is $T\mu$-continuous and thus Corollary 13 shows that the function $\lambda_2 = V\lambda$ is continuous with respect to the function $\mu_2 = V\mu$. Hence, by the Radon-Nikodým theorem (III.10.7), there exists a μ_2-integrable function g such that

$$\lambda_2(E) = \int_E g(s_1)\mu_2(ds_1), \qquad E \in \Sigma_2.$$

If $\mu_1 = U^{-1}\mu_2$, then by Lemma III.8.3 and Lemma 11(c) there exists a μ_1-integrable, simple function h_ε such that $|\lambda_1 - F_1| < \varepsilon$ where

$$F_1(E) = \int_E h_\varepsilon(s_1)\mu_1(ds_1).$$

Let $E_1, \ldots, E_n$ be a partition of S_1 and let $\alpha_1, \ldots, \alpha_n$ be a set of scalars such that $E_i \in \Sigma_1$ and $\alpha_i = h_\varepsilon(s_i)$ where $s_i \in E_i$. Put $G_i = \tau^{-1}(E_i) \in \Sigma$ and define $f_\varepsilon(s) = \alpha_i$ for s in G_i. It is evident that f_ε is a μ-integrable simple function and that if

$$F(G) = \int_G f_\varepsilon(s)\mu(ds), \qquad G \in \Sigma,$$

then $F_1 = T(F)$. Since T and U are isometries we have $|\lambda - F| < \varepsilon$. Q.E.D.

This section will be concluded by giving one solution to Problem 1.8 in the case where $\mathfrak{X} = C^*(S)$ and $\mathfrak{Y} = C(S)$. The following theorem is stated in a somewhat more general form, for it applies to set functions in $ba(S, \Sigma)$ and not just to those in $C^*(S) = rba(S)$.

15 THEOREM. (*Alexandroff*) *Let* $\mu, \mu_n, n = 1, 2, \ldots$, *be a bounded sequence in* $ba(S, \Sigma)$ *where* Σ *is a field containing the open sets in the topological space* S. *In order that*

$$\text{(i)} \qquad \int_S f(s)\mu_n(ds) \to \int_S f(s)\mu(ds), \qquad f \in C(S),$$

it is sufficient that

$$\text{(ii)} \qquad \mu_n(G) \to \mu(G)$$

for every open set G *with* $\mu(\bar{G}) = \mu(G)$. *If* S *is normal,* μ *is regular, and* μ, μ_n, $n = 1, 2, \ldots$ *are non-negative, the condition is also necessary.*

PROOF. It will be sufficient to prove (i) with f real valued. Thus suppose that $-M < f(s) < M$, $s \in S$, and choose $\alpha_0, \ldots, \alpha_m$ with

$$-M = \alpha_0 < \alpha_1 < \alpha_2 < \ldots < \alpha_m = M,$$

$$\alpha_i - \alpha_{i-1} < \varepsilon, \qquad i = 1, \ldots, m,$$

and

$$\mu(\{s | f(s) = \alpha_i\}) = 0, \qquad i = 1, \ldots, m.$$

This last condition may be achieved since sets of the form $F_\alpha = \{s | f(s) = \alpha\}$ are disjoint and thus there are at most a countable number of values of α for which $\mu(F_\alpha) \neq 0$. If $G_i = \{s | f(s) < \alpha_i\}$, then $\bar{G}_i \subseteq \{s | f(s) \leqq \alpha_i\}$. From (ii) it is seen that $\mu_n(G_i) \to \mu(G_i)$ and hence that

$$\text{(iii)} \qquad \mu_n(G_i - G_{i-1}) \to \mu(G_i - G_{i-1}).$$

Let χ_i be the characteristic function of $G_i - G_{i-1}$ and let $f_\varepsilon = \sum_{i=1}^m \alpha_i \chi_i$. Then $|f - f_\varepsilon| < \varepsilon$ and (iii) shows that

$$\lim_{n\to\infty} \int_S f_\varepsilon(s)\mu_n(ds) = \int_S f_\varepsilon(s)\mu(ds).$$

Thus, if $|\mu_n| \leqq K$ and $|\mu| \leqq K$, then

$$\left|\int_S f(s)\mu_n(ds) - \int_S f(s)\mu(ds)\right|$$

$$\leqq \left|\int_S (f-f_\varepsilon)(s)(\mu_n-\mu)(ds)\right| + \left|\int_S f_\varepsilon(s)(\mu_n-\mu)(ds)\right|$$

$$\leqq 2\varepsilon K + \left|\int_S f_\varepsilon(s)(\mu_n-\mu)(ds)\right|,$$

and

$$\limsup_{n\to\infty} \left|\int_S f(s)\mu_n(ds) - \int_S f(s)\mu(ds)\right| \leqq 2\varepsilon K.$$

This proves (i).

We shall now prove the converse, assuming that S is normal, that μ is regular, and that all the set functions μ, μ_n, $n = 1, 2, \ldots$, are non-negative. Let G be a fixed open set in S with $\mu(\bar{G}) = \mu(G)$. Let $\varepsilon > 0$ and choose a closed set F and an open set H such that

$$F \subseteq G \subseteq \bar{G} \subseteq H,$$

$$\mu(H-F) < \varepsilon.$$

Let f, h be continuous functions with

$$0 \leqq f(s),\ h(s) \leqq 1, \qquad s \in S;$$

$$f(s) = \begin{cases} 1, & s \in F; \\ 0, & s \notin G; \end{cases} \qquad h(s) = \begin{cases} 0, & s \notin H; \\ 1, & s \in G. \end{cases}$$

Then

$$\mu(F) \leqq \int_S f(s)\mu(ds) \leqq \mu(G)$$

$$= \mu(\bar{G}) \leqq \int_S h(s)\mu(ds) \leqq \mu(H).$$

Thus

$$0 \leqq \int_S \{h(s)-f(s)\}\mu(ds) \leqq \mu(H-F) < \varepsilon.$$

Consequently, we have

$$\int_S f(s)\mu_n(ds) \leqq \mu_n(G) \leqq \int_S h(s)\mu_n(ds),$$

$$\int_S f(s)\mu(ds) \leqq \mu(G) \leqq \int_S h(s)\mu(ds) \leqq \int_S f(s)\mu(ds)+\varepsilon.$$

Thus the hypothesis (i) shows that

$$\limsup_n |\mu_n(G)-\mu(G)| < \varepsilon.$$

and this proves (ii). Q.E.D.

10. Vector Valued Measures

The theorems on spaces of set functions proved in the last section permit us to develop a more satisfactory theory of vector valued countably additive set functions (briefly, *vector valued measures*) than we were able to develop in Chapter III. In particular, we will now be able to add to the integration theory of Chapter III a satisfactory theory of integration of scalar valued functions with respect to a vector valued measure.

Throughout this section, we shall suppose that S is a fixed set, that Σ is a σ-field of subsets of S, and μ is an additive set function defined on Σ with values in a B-space $\mathfrak{X}$. We suppose in addition that μ is *weakly countably additive*; that is, $\sum_{i=1}^{\infty} x^*\mu(E_i) = x^*\mu(\bigcup_{i=1}^{\infty} E_i)$ for each x^* in $\mathfrak{X}^*$ and each sequence of disjoint sets E_n in Σ.

1 THEOREM. (*Pettis*) *A weakly countably additive vector valued set function μ defined on a σ-field Σ is countably additive. In addition, if λ is a finite positive measure on Σ and if μ vanishes on λ-null sets, then μ is λ-continuous.*

PROOF. If $\{E_n\}$ is a sequence of sets in Σ, let Σ_0 be the field generated by $\{E_n\}$ and let Σ_1 be the σ-field generated by Σ_0. By Lemma III.8.4, Σ_0 is countable and so $\mathfrak{X}_1 = \overline{\text{sp}}\{\mu(E)/E \in \Sigma_0\}$ is a separable subspace of $\mathfrak{X}$. We assert that $\mu(F) \in \mathfrak{X}_1$ if $F \in \Sigma_1$. If this is not the case for some $F \in \Sigma_1$, then by Corollary II.3.13 there is an $x^* \in \mathfrak{X}^*$ such that $x^*\mu(F) \neq 0$, and $x^*\mu(E) = 0$, $E \in \Sigma_0$, which contradicts the uniqueness assertion of Corollary IV.5.9.

If μ is not countably additive, then for some $\varepsilon > 0$ there exists a decreasing sequence $\{E_n\}$ in Σ with void intersection such that $|\mu(E_n)| > \varepsilon$, $n = 1, 2, \ldots$. By Corollary II.3.14 there is a sequence $x_n^* \in \mathfrak{X}^*$ such that $|x_n^*| = 1$ and $x_n^*\mu(E_n) = |\mu(E_n)| > \varepsilon, n = 1, 2, \ldots$. Form Σ_1 and $\mathfrak{X}_1$ as in the preceding paragraph and let $\{x_k\}$ be a countable dense set in $\mathfrak{X}_1$. Using a Cantor diagonal procedure, there is a subsequence $\{y_m^* = x_{n_m}^*\}$ of $\{x_n^*\}$ such that $\lim_{m\to\infty} y_m^* x_k$ exists for all k. Since $|y_m^*| = 1$, $m = 1, 2, \ldots$, it follows from Theorem II.3.6 that $\lim_{m\to\infty} y_m^* x$ exists for every $x \in \mathfrak{X}_1$, and, in particular, $\lim_{m\to\infty} y_m^* \mu(E)$ exists for each $E \in \Sigma_1$. By Corollary III.7.4, the set $\{y_m^* \mu\}$ of scalar valued measures is uniformly countably additive on Σ_1, which contradicts the assumption that $y_m^* \mu(E_{n_m}) > \varepsilon$, $m = 1, 2, \ldots$. This proves the first assertion.

If μ is not λ-continuous, then for some $\varepsilon > 0$ there is a sequence $\{E_n\}$ in Σ such that $\lambda(E_n) < 1/n$ and $|\mu(E_n)| > \varepsilon$, $n = 1, 2, \ldots$. Proceeding exactly as before, a sequence $\{y_m^*\} \subset \mathfrak{X}^*$ is obtained such that $\lim_{m\to\infty} y_m^* \mu(E)$ exists for $E \in \Sigma_1$ and $y_m^* \mu(E_{n_m}) > \varepsilon$, $m = 1, 2, \ldots$. Since $y_m^* \mu(E) = 0$ whenever $\lambda(E) = 0$, by Lemma III.4.13 each measure $y_m^* \mu$ is λ-continuous, and it follows from the Vitali-Hahn-Saks theorem (III.7.2) that the λ-continuity is uniform for $m = 1, 2, \ldots$. But this contradicts the assumption that $\lambda(E_{n_m}) < 1/n_m$ and $y_m^* \mu(E_{n_m}) > \varepsilon$, $m = 1, 2, \ldots$. This shows that μ is λ-continuous, and completes the proof. Q.E.D.

2 Corollary. *A vector measure is bounded, and the set* $\{x^*\mu | x^* \in \mathfrak{X}^*, |x^*| \leq 1\}$ *of numerical measures is weakly sequentially compact as a subset of* $ca(S, \Sigma)$.

Proof. Since $|x^*\mu(E)| \leq v(x^*\mu, S)$ for each x^* and $E \in \Sigma$, it follows from Theorem II.3.20 that there is a constant M such that $|\mu(E)| \leq M$, $E \in \Sigma$. Since

$$|x^*\mu| = v(x^*\mu, S) \leq 4 \sup_{E \subseteq S} |x^*\mu(E)| \leq 4M, \quad |x^*| \leq 1,$$

the set of measures is bounded in $ca(S, \Sigma)$. Let $\{E_n\}$ be a decreasing sequence of sets in Σ with void intersection. Since μ is countably additive, $\lim_{n\to\infty} \mu(E_n) = 0$. Thus, $\lim_{n\to\infty} x^*\mu(E_i) = 0$ uniformly for $|x^*| \leq 1$. The desired conclusion follows from Theorem 9.1. Q.E.D.

In contrast to the case of complex valued measures, the total variation of a vector valued measure (cf. Definition III.1.4) need not be finite. Our next step is to construct a finite positive set function which can replace the total variation in the vector valued case.

3 DEFINITION. The *semi-variation* of the vector valued measure μ is defined by

$$\|\mu\|(E) = \sup \left| \sum_{i=1}^{n} \alpha_i \mu(E_i) \right|, \qquad E \in \Sigma,$$

where the supremum is taken over all finite collections of scalars with $|\alpha_i| \leqq 1$ and all partitions of E into a finite number of disjoint sets in Σ.

A number of elementary properties of the semi-variation are listed in the next lemma.

4 LEMMA. *Let μ be a vector valued measure. Then*

(a) $\|\mu\|(E) \geqq |\mu(E)| \geqq 0, \qquad E \in \Sigma;$

(b) $\|\mu\|(E) \leqq 4 \sup_{F \subseteq E} |\mu(F)| < \infty, \qquad E \in \Sigma;$

(c) $\|\mu\|(F) \leqq \|\mu\|(E)$ *if* $F \subseteqq E$;

(d) *if* $\{E_i\}$ *is a sequence of sets in* Σ *then* $\|\mu\|(\cup_{i=1}^{\infty} E_i) \leqq \sum_{i=1}^{\infty} \|\mu\|(E_i)$.

Remark. Even if the sets E_i in (d) are disjoint, the inequality may be strict; that is, $\|\mu\|$ need not be an additive function. It is easy to see that $\|\mu\|$ is additive if and only if $\|\mu\| = v(\mu)$; so that if $v(\mu, S) = \infty$, $\|\mu\|$ *cannot* be additive.

PROOF. Statements (a) and (c) are obvious. To prove (b), observe that we have

$$\begin{aligned}\|\mu\|(E) &= \sup \left| \sum_{i=1}^{n} \alpha_i \mu(E_i) \right| = \sup_{|x^*| \leqq 1} \sup \left| \sum_{i=1}^{n} \alpha_i x^* \mu(E_i) \right| \\ &\leqq \sup_{|x^*| \leqq 1} \sup \sum_{i=1}^{n} |\alpha_i| v(x^*\mu, E_i) \leqq \sup_{|x^*| \leqq 1} v(x^*\mu, E) \\ &\leqq 4 \sup_{|x^*| \leqq 1} \sup_{F \subseteq E} |x^*\mu(F)| = 4 \sup_{F \subseteq E} |\mu(F)| < \infty\end{aligned}$$

by Corollary II.3.15 and Corollary 2.

In proving (d) we shall assume that the sets E_i are disjoint. Note that if $F_1, \ldots, F_k$ is a disjoint partition of $E = \cup_{i=1}^{\infty} E_i$, then $E_i F_i, \ldots, E_i F_k$ is a disjoint partition of E_i for each i. Thus if $|\alpha_j| \leqq 1$, $j = 1, \ldots, k$, we have

$$|\sum_{j=1}^{k} \alpha_j \mu(F_j)| = |\sum_{j=1}^{k} \sum_{i=1}^{\infty} \alpha_j \mu(E_i F_j)| \leqq \sum_{i=1}^{\infty} ||\mu||(E_i),$$

so that

$$||\mu||(E) \leqq \sum_{i=1}^{\infty} ||\mu||(E_i). \qquad \text{Q.E.D.}$$

The next lemma is crucial for the development of the integral for scalar functions with respect to μ.

5 LEMMA. *There exists a finite positive measure λ defined on Σ such that*

(a) $\lambda(E) \leqq ||\mu||(E), \qquad E \in \Sigma;$

(b) $\lim_{\lambda(E) \to 0} ||\mu||(E) = 0.$

PROOF. By Corollary 2, Theorem 9.2, and Corollary 9.3, we can find a positive measure λ such that $\lambda(E) \leqq \sup_{|x^*| \leqq 1} |x^*\mu(E)| = |\mu(E)|$, and such that $\lim_{\lambda(E) \to 0} x^*\mu(E) = 0$ uniformly for $|x^*| \leqq 1$. Thus $\lambda(E) \leqq ||\mu||(E)$ by Lemma 4(a), and $\lim_{\lambda(E) \to 0} ||\mu||(E) = \lim_{\lambda(E) \to 0} |\mu(E)| = 0$ by Lemma 4(b). Q.E.D.

Lemma 5 enables us to prove a result, promised in Chapter III.7, which generalizes a theorem of Nikodým to the case of vector valued measures.

6 THEOREM. *Let $\{\mu_n\}$ be a sequence of vector valued measures defined on the σ-field Σ. If $\mu(E) = \lim_{n \to \infty} \mu_n(E)$ exists for each E in Σ, μ is a vector measure on Σ and the countable additivity of μ_n is uniform in $n = 1, 2, \ldots$.*

PROOF. For each n let λ_n be a positive finite measure corresponding to μ_n by Lemma 5. Let λ be the measure defined by the formula

$$\lambda(E) = \sum_{n=1}^{\infty} \frac{1}{2^n} \frac{\lambda_n(E)}{1 + \lambda_n(S)}.$$

Then each μ_n is λ-continuous. Corollary III.7.3 shows that μ is countably additive. If $\{E_m\}$ is a decreasing sequence of sets in Σ with void intersection, $\lim_{m \to \infty} \lambda(E_m) = 0$. By the Vitali-Hahn-Saks theorem (III.7.2), $\lim_{m \to \infty} \mu_n(E_m) = 0$ uniformly for $n = 1, 2, \ldots$. Q.E.D.

For the rest of this section, λ will be a finite positive measure related to μ as in Lemma 5.

We now proceed to develop a theory of integration of scalar functions with respect to the vector measure μ. A *μ-null set* is a subset of a set $E \in \Sigma$ such that $\|\mu\|(E) = 0$; by Lemma 5, this is the same as a λ-null set. The term *μ-almost everywhere* refers to the complement of a μ-null set, and is hence synonymous with the term λ-almost everywhere. The symbol Σ^* denotes the Lebesgue extension of Σ. Thus (III.5.17) Σ^* is the σ-field of unions $E \cup N$, $E \in \Sigma$, N a μ-null set. A scalar function f defined on S is *μ-measurable* if for every Borel set B of scalars, $f^{-1}(B) \in \Sigma^*$; by III.6.9, this is the case if f is λ-measurable. A scalar valued function f defined on S is *μ-simple* if it is a finite linear combination of characteristic functions of sets in Σ^*; this is evidently the case if and only if f is *λ-simple.* It follows from Corollaries III.6.13 and III.6.14 that f is μ-measurable if and only if it is the limit μ-almost everywhere of a sequence of μ-simple functions. By III.6.14, the limit of a sequence of μ-measurable functions converging μ-almost everywhere is μ-measurable.

If f is the μ-simple function $\sum_{i=1}^{n} \alpha_i \chi_{E_i}$, where $E_i, \ldots, E_n$ are sets in Σ, then the *integral* of f over a set $E \in \Sigma$ is defined by the equation

$$\int_E f(s)\mu(ds) = \sum_{i=1}^{n} \alpha_i \mu(E \cap E_i).$$

It follows just as in the case in which μ is complex valued that the integral of f is independent of its particular representation as a linear combination of characteristic functions (cf. the paragraph following Definition III.2.13).

Obviously, integration of simple functions over E is a linear operation. Also, the integral of a simple function is a countably additive set function with values in $\mathfrak{X}$. If f is a simple function such that $|f(s)| \leqq M$ for each $s \in \Sigma$, then

$$\left|\int_E f(s)\mu(ds)\right| = \left|\sum_{i=1}^{n} \alpha_i \mu(E \cap E_i)\right| = M \left|\sum_{i=1}^{n} \left(\frac{\alpha_i}{M}\right) \mu(E \cap E_i)\right|$$
$$\leqq M\|\mu\|(E);$$

hence

$$[*] \qquad \left|\int_E f(s)\mu(ds)\right| \leqq \{\sup_{s \in E} |f(s)|\}\|\mu\|(E).$$

If f is an arbitrary measurable function, we define the *μ-essential supremum* of f on E to be the infimum of those numbers A for which $\{s \in E \,|\, |f(s)| > A\}$ is a μ-null set. If $\mu\text{-ess}\sup_{s \in E} |f(s)| < \infty$, we say that f is *μ-essentially bounded* on the set $E \in \Sigma$.

It is clear that $\mu\text{-ess}\sup_{s \in E} |f(s)| = \lambda\text{-ess}\sup_{s \in E} |f(s)|$, and that f is μ-essentially bounded if and only if f is λ-essentially bounded. Equation [*] may be written somewhat more generally as

$$\left| \int_E f(s)\mu(ds) \right| \leqq \{\mu\text{-ess}\sup_{s \in E} |f(s)|\} \|\mu\|(E),$$

if f is μ-simple.

7 Definition. A scalar valued measurable function f is said to be *integrable* if there exists a sequence $\{f_n\}$ of simple functions such that

(i) $f_n(s)$ converges to $f(s)$ μ-almost everywhere;

(ii) the sequence $\{\int_E f_n(s)\mu(ds)\}$ converges in the norm of $\mathfrak{X}$ for each $E \in \Sigma$.

The limit of this sequence of integrals is defined to be *the integral of f with respect to μ over the set $E \in \Sigma$*, in symbols:

$$\int_E f(s)\mu(ds).$$

8 Theorem. (a) *If $E \in \Sigma$ and f is scalar valued and μ-integrable, the integral of f with respect to μ over E is an unambiguously defined element of $\mathfrak{X}$;*

(b) *if f and g are scalar valued and μ-integrable, if α and β are scalars, and if $E \in \Sigma$, then*

$$\int_E \{\alpha f(s) + \beta g(s)\}\mu(ds) = \alpha \int_E f(s)\mu(ds) + \beta \int_E g(s)\mu(ds);$$

(c) *if f is a μ-measurable and scalar valued function which is μ-essentially bounded on E, then f is μ-integrable and*

$$\left| \int_E f(s)\mu(ds) \right| \leqq \{\mu\text{-ess}\sup_{s \in E} |f(s)|\}\{\|\mu\|(E)\};$$

(d) *if f is scalar valued and μ-integrable, then $\int_E f(s)\mu(ds)$ is a countably additive function on Σ to $\mathfrak{X}$;*

(e) *if f is scalar valued and μ-integrable, then*

$$\lim_{\|\mu\|(E)\to 0} \int_E f(s)\mu(ds) = 0;$$

(f) *if U is a bounded linear operator from $\mathfrak{X}$ into a Banach space $\mathfrak{Y}$, then $U\mu$ is a vector measure from Σ to $\mathfrak{Y}$, and for any μ-integrable scalar valued function f and $E \in \Sigma$, we have*

$$U\left\{\int_E f(s)\mu(ds)\right\} = \int_E f(s)U\mu(ds).$$

PROOF. To prove (a), let $\{f_n\}$ and $\{g_n\}$ be two sequences of simple functions as in Definition 7. We are required to show that the two sequences of integrals approach the same limit. We define $h_n(s) = 0$ if s is a point at which either $\{f_n(s)\}$ or $\{g_n(s)\}$ fails to converge to $f(s)$, and set $h_n(s) = f_n(s)-g_n(s)$ otherwise. It is evident that $\{h_n\}$ converges to zero everywhere, and that $\{\int_E h_n(s)\mu(ds)\}$ converges in the norm of $\mathfrak{X}$ for $E \in \Sigma$. We must show that this sequence of integrals converges to the zero element of $\mathfrak{X}$.

Let λ be a positive measure related to μ as in Lemma 5. Clearly, since each h_n is a simple function,

$$(*) \qquad \lim_{\lambda(E)\to 0} \int_E h_n(s)\mu(ds) = 0, \qquad n = 1, 2, \ldots;$$

further the sequence of integrals $\{\int_E h_n(s)\mu(ds)\}$ converges for each $E \in \Sigma$, so by the Vitali-Hahn-Saks theorem (III.7.2), the limit in (*) is uniform in n. Consequently, for each $\varepsilon > 0$ there exists a $\delta=\delta(\varepsilon)>0$ such that if $A \in \Sigma$ and $\lambda(A) < \delta$ then

$$\left|\int_A h_n(s)\mu(ds)\right| < \varepsilon, \qquad n = 1, 2, \ldots.$$

By Egoroff's theorem (III.6.12) there exists a set $A \in \Sigma$ with $\lambda(A) < \delta$ such that $\{h_n(s)\}$ converges to zero uniformly for $s \in S-A$. Having specified ε and chosen $\delta = \delta(\varepsilon)$ as above, there exists an $N = N(\varepsilon)$ such that if $n \geqq N$, then $|h_n(s)| < \varepsilon$ for $s \in S-A$. Hence if $n \geqq N$,

$$\begin{aligned}\left|\int_E h_n(s)\mu(ds)\right| &\leqq \left|\int_{E-A} h_n(s)\mu(ds)\right| + \left|\int_{E\cap A} h_n(s)\mu(ds)\right| \\ &\leqq \varepsilon\|\mu\|(S)+\varepsilon\end{aligned}$$

uniformly for $E \in \Sigma$. Thus the integral is well defined.

Statement (b) follows from its validity for simple functions and the additivity of the limit operation.

To prove (c), let f be μ-measurable with μ-essential bound B on E, and let $\varepsilon > 0$. Let $F_1, \ldots, F_n$ be a covering of $f(E)$ by disjoint Borel sets of scalars, and define $E_j = f^{-1}(F_j)$. Let $\alpha_j \in F_j$, and define $f_\varepsilon(s) = \alpha_j$ for $s \in F_j$. Then f_ε (modified if need be on a μ-null set) is a μ-simple function, and $\mu\text{-ess sup}_{s \in E} |f_\varepsilon(s) - f(s)| < \varepsilon$. Let $\varepsilon_n \to 0$. Then $\lim_{m,n\to\infty} \mu\text{-ess sup}_{s \in E} |f_{\varepsilon_n}(s) - f_{\varepsilon_m}(s)| = 0$. Since we have already observed that (c) holds for each μ-simple function, we can immediately conclude that

$$\lim_{m,n\to\infty} \left| \int_E f_{\varepsilon_n}(s)\mu(ds) - \int_E f_{\varepsilon_m}(s)\mu(ds) \right| = 0,$$

so that $\{\int_E f_{\varepsilon_m}(s)\mu(ds)\}$ converges for each $E \in \Sigma$, and we have

$$\int_E f(s)\mu(ds) = \lim_{n\to\infty} \int_E f_{\varepsilon_n}(s)\mu(ds), \qquad E \in \Sigma.$$

Since $\mu\text{-ess sup}_{s \in E} |f_{\varepsilon_n}(s)| \leqq B + \varepsilon_n$, the general validity of (c) now follows from its validity for μ-simple functions.

We have already seen that (d) and (e) are true for simple functions. Let f be an arbitrary μ-integrable function and $\{f_n\}$ be a sequence of simple functions as in Definition 7. The Vitali-Hahn-Saks theorem (III.7.2) implies (d), and the fact that

$$\lim_{\lambda(E)\to 0} \int_E f(s)\mu(ds) = 0,$$

from which (e) follows immediately by virtue of Lemma 4(a).

The first assertion in (f) is obvious, while the second follows by an evident limiting argument from its validity for μ-simple functions. Q.E.D.

9 THEOREM. *Let $\{f_n\}$ be a sequence of μ-integrable functions which converge μ-almost everywhere to f. Then f is μ-integrable if*

$$\lim_{\|\mu\|(E)\to 0} \int_E f_n(s)\mu(ds) = 0$$

uniformly for $n = 1, 2, \ldots$. In this case we have

$$\int_E f(s)\mu(ds) = \lim_{n\to\infty} \int_E f_n(s)\mu(ds).$$

PROOF. Let k be a positive integer and let $\delta_k > 0$ be such that if $E \in \Sigma$ and $\|\mu\|(E) < \delta_k$, then

$$\left|\int_E f_n(s)\mu(ds)\right| < 2^{-k}, \qquad n = 1, 2, \ldots. \tag{1}$$

Evidently we may assume that $\delta_k \leqq 2^{-k}$. Let $\eta_k > 0$ be such that if $A \in \Sigma$ and $\lambda(A) < \eta_k$ then $\|\mu\|(A) < \delta_k$. The sequence $\{f_n\}$ converges λ-almost everywhere, and, by an application of Egoroff's theorem (III.6.12), we can select a set $A \in \Sigma$ such that $\lambda(A) < \eta_k$ and such that the convergence of $\{f_n\}$ is uniform on $S-A$. Thus there is an N_k such that if $E \in \Sigma$ and $n, m > N_k$, we have

$$\begin{aligned} \left|\int_E \{f_n(s)-f_m(s)\}\mu(ds)\right| &\leqq \left|\int_{E-A} \{f_n(s)-f_m(s)\}\mu(ds)\right| \\ &+ \left|\int_{E\cap A} f_n(s)\mu(ds)\right| + \left|\int_{E\cap A} f_m(s)\mu(ds)\right| < 2^{-k}\{\|\mu\|(S)+2\}. \end{aligned} \tag{2}$$

But since k is an arbitrary positive integer, this proves that the sequence $\{\int_E f_n(s)\mu(ds)\}$ converges in the norm of $\mathfrak{X}$ for any $E \in \Sigma$.

We now prove that f is μ-integrable. Let δ_k and η_k have the same meaning as in the previous paragraph. Since each f_k is μ-integrable it follows from Egoroff's theorem (III.6.12) and the second paragraph of the proof of Theorem III.2.22 that there is a simple function g_k and a set $A_k \in \Sigma$ with $\lambda(A_k) < \eta_k$ such that

$$|f_k(s)-g_k(s)| < 2^{-k}, \qquad s \in S-A_k, \tag{3}$$

$$|g_k(s)| \leqq 2|f_k(s)|, \qquad s \in S. \tag{4}$$

Let $B_k = \bigcup_{i=k}^{\infty} A_i$ so that $B_k \in \Sigma$ and $\{B_k\}$ decreases to the set $B = \bigcap_{k=1}^{\infty} B_k$. Since

$$\|\mu\|(B_k) \leqq \sum_{i=k}^{\infty} \|\mu\|(A_i) < \sum_{i=k}^{\infty} \delta_i \leqq \sum_{i=k}^{\infty} 2^{-i} = 2^{-(k-1)},$$

it follows that $\|\mu\|(B) = 0$. Now

$$|f(s)-g_k(s)| \leqq |f(s)-f_k(s)|+|f_k(s)-g_k(s)|.$$

If $s \in S-B$, then $s \in S-B_k$ for k greater than some integer $K(s)$, and so (3) holds provided that $k > K(s)$. Since f_k was assumed to converge μ-almost everywhere to f, we conclude that the sequence $\{g_k\}$ converges μ-almost everywhere to f.

It remains to show that the integrals $\{\int_E g_n(s)\mu(ds)\}$ converge for $E \in \Sigma$. But

$$\left|\int_E \{f_k(s)-g_k(s)\}\mu(ds)\right| \leqq \left|\int_{E-A_k} \{f_k(s)-g_k(s)\}\mu(ds)\right| + \left|\int_{E\cap A_k} f_k(s)\mu(ds)\right| + \left|\int_{E\cap A_k} g_k(s)\mu(ds)\right|.$$

The integral over $E-A_k$ is at most $2^{-k}\|\mu\|(S)$ by (3). Since $\|\mu\|(E\cap A_k) < \delta_k$, we have seen in (1) that the second term on the right is at most 2^{-k}. To estimate the last term let $x^* \in \mathfrak{X}^*$, $|x^*| \leqq 1$, $F \subseteqq E$, $F \in \Sigma$, and $\|\mu\|(E) < \delta_k$. Then by (1)

$$\left|\int_F f_n(s)x^*\mu(ds)\right| < 2^{-k},$$

so that by Theorem III.2.20 and the remark following Definition III.4.12,

$$(5) \qquad \int_E |f_n(s)|v(x^*\mu, ds) < 4 \,.\, 2^{-k}, \qquad n = 1, 2, \ldots,$$

if $\|\mu\|(E) < \delta_k$. Thus by (4) and (5)

$$\begin{aligned}\left|\int_{E\cap A_k} g_k(s)\mu(ds)\right| &= \sup_{|x^*|\leqq 1} \left|\int_{E\cap A_k} g_k(s)x^*\mu(ds)\right| \\ &\leqq \sup_{|x^*|\leqq 1} \int_{E\cap A_k} |g_k(s)|v(x^*\mu, ds) \\ &\leqq 2 \sup_{|x^*|\leqq 1} \int_{E\cap A_k} |f_k(s)|v(x^*\mu, ds) \leqq 8 \cdot 2^{-k}.\end{aligned}$$

Combining these statements we conclude that

$$(6) \qquad \left|\int_E \{f_k(s)-g_k(s)\}\mu(ds)\right| \leqq 2^{-k}\{\|\mu\|(S)+9\}$$

which, since $\{\int_E f_k(s)\mu(ds)\}$ converges, shows that $\{\int_E g_k(s)\mu(ds)\}$ converges, and that f is μ-integrable.

To obtain the last conclusion of the theorem, observe that since $\{\int_E f_k(s)\mu(ds)\}$ converges, it follows by the Vitali-Hahn-Saks theorem (III.7.2) that for each $x^* \in \mathfrak{X}^*$ we have $\lim_{v(x^*\mu, E)\to 0} \int_E f_k(s)x^*\mu(ds) = 0$ uniformly in k. Thus, by Theorem III.2.20 and the remark following Definition III.4.12,

$$\lim_{v(x^*\mu, E)\to 0} \int_E |f_k(s)|v(x^*\mu, ds) = 0$$

uniformly in k. It follows from Theorem III.6.15 that

$$\lim_{n\to\infty} x^* \int_E f_k(s)\mu(ds) = x^* \int_E f(s)\mu(ds)$$

for each $x^* \in \mathfrak{X}^*$. Thus $x^*(\int_E f(s)\mu(ds) - \lim_{n\to\infty} \int_E f_n(s)\mu(ds)) = 0$ for $x^* \in \mathfrak{X}^*$, so that by Corollary II.3.13,

$$\lim_{n\to\infty} \int_E f_n(s)\mu(ds) = \int_E f(s)\mu(ds). \qquad \text{Q.E.D.}$$

We now show that the theorem on dominated convergence is valid for vector measures.

10 THEOREM. *If $\{f_n\}$ is a sequence of μ-integrable functions which converges μ-almost everywhere to f and if g is a μ-integrable function such that $|f_n(s)| \leqq g(s)$ μ-almost everywhere, $n = 1, 2, \ldots$, then f is μ-integrable and*

$$\int_E f(s)\mu(ds) = \lim_{n\to\infty} \int_E f_n(s)\mu(ds), \qquad E \in \Sigma.$$

PROOF. By the preceding theorem, it suffices to show that

$$\lim_{\|\mu\|(E)\to 0} \int_E f_n(s)\mu(ds) = 0$$

uniformly for $n = 1, 2, \ldots$. Given $\varepsilon > 0$, we may, by Theorem 8(e), choose a $\delta > 0$ such that if $E \in \Sigma$ and $\|\mu\|(E) < \delta$, then

$$\left|\int_E g(s)\mu(ds)\right| < \varepsilon.$$

Hence if $F \subseteq E$, $F \in \Sigma$, and $\|\mu\|(E) < \delta$, we have

$$\left|\int_F g(s)x^*\mu(ds)\right| < \varepsilon, \qquad |x^*| \leqq 1.$$

By III.2.20 and the remark following III.4.12 it follows that

$$\int_E g(s)v(x^*\mu, ds) < 4\varepsilon, \qquad |x^*| \leqq 1.$$

Consequently, if $\|\mu\|(E) < \delta$, then

$$\begin{aligned} \left|\int_E f_n(s)\mu(ds)\right| &\leqq \sup_{|x^*|\leqq 1} \int_E |f_n(s)|v(x^*\mu, ds) \\ &\leqq \sup_{|x^*|\leqq 1} \int_E g(s)v(x^*\mu, ds) \leqq 4\varepsilon, \qquad n = 1, 2, \ldots, \end{aligned}$$

from which the conclusion follows. Q.E.D.

11. The Space $TM(S, \Sigma, \mu)$

We shall be concerned here with a set S, a σ-field Σ of its subsets, and a scalar valued countably additive set function μ on Σ. The symbol $TM(S, \Sigma, \mu)$ will be used for the set of all scalar valued functions on S which are totally μ-measurable (cf. Definition III.2.10). More precisely, as in Chapter III, the elements of $TM(S, \Sigma, \mu)$ are equivalence classes where two totally measurable functions are defined to be equivalent if their difference is a null function. A number of properties of $TM(S, \Sigma, \mu)$ have been established in Chapter III. In particular $TM(S, \Sigma, \mu)$ is a linear vector space which is a metric space under the distance function $\varrho(f, g) = |f-g|$, the norm $|f|$ being defined by

$$|f| = \inf_{\alpha>0} \arctan\left(\alpha+v^*(\mu, S(|f| > \alpha))\right),$$

where $S(|f| > \alpha) = \{s | s \in S, |f(s)| > \alpha\}$. This metric space is complete (III.6.5) and convergence of a sequence $\{f_n\} \subseteq TM(S, \Sigma, \mu)$ is equivalent to convergence in measure of the functions f_n on S.

It is easily seen that $TM(S, \Sigma, \mu)$ is an F-space. To see this we note that, in view of Lemma III.2.8(b), it suffices to prove that $\lim_{\alpha\to 0} \alpha f = 0$, for each $f \in TM(S, \Sigma, \mu)$. Let $f \in TM(S, \Sigma, \mu)$ and let $\varepsilon > 0$. Let g_ε be a μ-simple function such that $|f(s)-g_\varepsilon(s)| < \varepsilon$ for s in the complement of a set $E_\varepsilon \in \Sigma$ such that $\mu(E_\varepsilon) < \varepsilon$. If $M = \operatorname{ess\,sup}_{s\in S} |g_\varepsilon(s)|$ and $|\alpha| < \varepsilon/(M+\varepsilon)$, then $|\alpha f(s)| < \varepsilon$ for $s \notin E_\varepsilon$. Thus, by Lemma III.2.7, $\lim_{\alpha\to 0} \alpha f = 0$.

Note that $\lim_{\alpha\to 0} \alpha f = 0$ in μ-measure is not necessarily true for every measurable function. For example, if $S = (-\infty, \infty)$ and μ is Lebesgue measure, let f be the function defined by $f(s) = s$. Then f/n does not approach zero in measure since $\mu(E(|f/n| > \alpha)) = \infty$ for all $n = 1, 2, \ldots$ and all $\alpha > 0$. Thus the space $M(S, \Sigma, \mu)$ of all measurable functions need not be an F-space.

The F-space $TM(S, \Sigma, \mu)$ need not admit any continuous linear functionals. To see this, consider the case of Lebesgue measure μ on the set $S = [0, 1]$. If $0 \neq x^* \in TM^*(S, \Sigma, \mu)$, i.e., if x^* is a continuous linear functional on $TM(S, \Sigma, \mu)$ which is not identically zero, then, since linear combinations of characteristic functions of intervals are dense in $TM(S, \Sigma, \mu)$, there is a subinterval Δ_n of $[0, 1]$ of length less than $1/n$ and such that x^* does not vanish on the characteristic function χ_n of Δ_n. Let $x^*\chi_n = \delta_n \neq 0$ and let $f_n = \chi_n/\delta_n$ so that $f_n \to 0$ and $x^*f_n = 1$, $n = 1, 2, \ldots$, which is a contradiction. Thus the conjugate space of the space of totally measurable functions may consist of the zero vector only. This is not always the case; for if the measure of the set $\{s_0\}$ consisting of the single point s_0 is not zero, then $f(s_0)$ depends linearly and continuously on f and $f(s_0) = g(s_0)$ for every pair f, g of equivalent functions.

Thus all but one of the problems described in Section 1 fail to have significance for some spaces of measurable functions. The one problem of the list which has significance in all cases is the seventh one, i.e., to determine which subsets of $TM(S, \Sigma, \mu)$ are compact in the metric topology of $TM(S, \Sigma, \mu)$. One answer to this question is contained in the following theorem.

1 **Theorem.** *Let μ be a complex or extended real valued countably additive set function on a σ-field Σ of subsets of a set S. Then a subset A of $TM(S, \Sigma, \mu)$ is conditionally compact if and only if for every $\varepsilon > 0$ there are sets $E_1, \ldots, E_n$ in Σ, a constant K, and, corresponding to each $f \in A$, a set E_f in Σ such that*

(1) $E_1 \cup E_2 \cup \ldots \cup E_n = S, \qquad E_i E_j = \phi, \qquad i \neq j;$

(2) $v^*(\mu, E_f) < \varepsilon, \qquad f \in A;$

(3) $|f(s)| < K, \qquad f \in A, \qquad s \notin E_f;$

(4) $\sup_{s, t \in E_i - E_f} |f(s) - f(t)| < \varepsilon, \qquad 1 \leq i \leq n.$

Proof. Since μ is countably additive on the σ-field Σ, the space $TM(S, \Sigma, \mu)$ of all totally measurable functions on S is a complete metric space (III.6.5) and thus (I.6.15) a set $A \subseteq TM$ is conditionally compact if and only if A is totally bounded. (It is worth noting that the preceding argument is the only place in the proof where the countable additivity of μ is used, i.e., the conditions of the theorem are necessary and sufficient for total boundedness even if Σ is just a field and μ is only an additive, not necessarily bounded, function on Σ.)

First suppose that A is totally bounded so that given $\varepsilon > 0$ there are functions $f_1, \ldots, f_q$ in A with

$$\inf_{1 \leq i \leq q} |f_i - f| < \varepsilon/4, \qquad f \in A.$$

From the definition of total measurability (III.2.10) it follows that there are simple functions $g_1, \ldots, g_q$ such that

$$|g_i - f_i| < \varepsilon/4, \qquad i = 1, \ldots, q,$$

$$\inf_{1 \leq i \leq q} |g_i - f| < \varepsilon/2, \qquad f \in A.$$

Thus there are sets $E_1, \ldots, E_n$ satisfying (i) and constants $\alpha_i^{(j)}$, $j = 1, \ldots, n$; $i = 1, \ldots, q$ with

$$g_i(s) = \sum_{j=1}^{n} \alpha_i^{(j)} \chi_{E_j}(s), \qquad i = 1, \ldots, q,$$

where χ_{E_j} is the characteristic function of E_j.

For each f in A there is an i for which $|g_i - f| < \varepsilon/2$, and hence an $\alpha > 0$ such that

$$\alpha + v^*(\mu, S(|g_i - f| > \alpha)) < \varepsilon/2,$$

where the set $S(|g_i - f| > \alpha)$ is defined by

$$S(|g_i - f| > \alpha) \equiv \{s | s \in S, |g_i(s) - f(s)| > \alpha\}.$$

Since $\alpha \leqq \varepsilon/2$, it follows that

$$S(|g_i - f| > \varepsilon/2) \subseteq S(|g_i - f| > \alpha),$$

$$v^*(\mu, S(|g_i - f| > \varepsilon/2)) \leqq v^*(\mu, S(|g_i - f| > \alpha)) < \varepsilon/2.$$

If $E_f = S(|g_i - f| > \varepsilon/2)$, $K = \varepsilon + \text{ess sup}_{i,s} |g_i(s)|$, (2) and (3) follow immediately. Since $g_i(s) = g_i(t)$ if s and t both lie in some set E_j, and since $|g_i(s) - f(s)| < \varepsilon/2$ for $s \notin E_f$, (4) is also evident.

Now conversely suppose that A satisfies the conditions (1), . . ., (4). Let $\varepsilon > 0$, $\alpha = 2^{-1} \tan \varepsilon$,

$$-K = \alpha_0 < \alpha_1 < \ldots < \alpha_p = K$$

$$\alpha_i - \alpha_{i-1} < \alpha, \qquad i = 1, \ldots, p,$$

and consider the set of all simple functions which are constant on each of the sets $E_1, \ldots, E_n$ and whose values are in the set $\{\alpha_i, i=0, 1, \ldots, p\}$. There are a finite number $g_1, \ldots, g_m$ of such functions, and in view of

(2), (3), and (4) we have for every f in A an $i \leqq m$ and a set $E_f \in \Sigma$ with

$$v^*(\mu, E_f) < \alpha, \qquad |g_i(s)-f(s)| < \alpha, \qquad s \notin E_f.$$

Thus

$$S(|g_i-f| > \alpha) \subseteqq E_f$$

and hence

$$\alpha + v^*(\mu, S(|g_i-f| > \alpha)) < 2\alpha = \tan \varepsilon,$$

which shows that $|g_i-f| < \varepsilon$. This proves that every f in A is within distance ε of one of the functions $g_1, \ldots, g_m$ and proves that A is totally bounded. Q.E.D.

We know from the general theory of continuous linear maps between F-spaces $\mathfrak{X}$ and $\mathfrak{Y}$ that a sequence $\{T_n\}$ of such maps with the properties that (i) $\{T_n x\}$ is bounded for each x and (ii) $\{T_n x\}$ is convergent for each x in a dense set, also has the property that $\{T_n x\}$ is convergent for each x in the domain space (see Theorem II.1.18). If the range space $\mathfrak{Y}$ is the space $TM(S, \Sigma, \mu)$ where (S, Σ, μ) is a positive σ-finite measure space, there is an analogous theorem in which the notions of boundedness and convergence are taken in the sense of holding almost everywhere on S. If T is a mapping from an F-space $\mathfrak{X}$ into $TM(S, \Sigma, \mu)$ we shall write $T(x, s)$ for the value of Tx (i.e., the value of any one of the functions in the equivalence class determined by Tx) at the point s.

2 THEOREM. (*Banach*) *Let* $\{T_n\}$ *be a sequence of continuous linear maps from an* F*-space* $\mathfrak{X}$ *into the space* $TM(S, \Sigma, \mu)$ *of all real or complex totally measurable functions on a positive* σ*-finite measure space* (S, Σ, μ). *Suppose that for each* x *in* $\mathfrak{X}$ *we have* $\sup_n |T_n(x, s)| < \infty$ *for almost all* s *in* S. *Suppose also that for each* x *in a dense set in* $\mathfrak{X}$, *the limit* $\lim_n T_n(x, s)$ *exists for almost all* s *in* S. *Then for every* x *in* $\mathfrak{X}$ *the limit* $\lim_n T_n(x, s)$ *exists almost everywhere on* S.

We shall occasionally need a generalization of this result which is given in the following theorem. Theorem 2 is seen to be a special case of the next theorem by taking the set A_k to be the set of all integers $n \geqq k$.

3 THEOREM. *Let* (S, Σ, μ) *be a* σ*-finite positive measure space.*

Let $A_1 \supseteq A_2 \supseteq \ldots$ be denumerable sets and for each $a \in A_1$ let T_a be a continuous linear map from an F-space $\mathfrak{X}$ into the space $TM(S, \Sigma, \mu)$ of real or complex valued totally measurable functions. Suppose that

(i) *for each x in $\mathfrak{X}$*

$$\sup_{a \in A_1} |T_a(x, s)| < \infty,$$

almost everywhere on S; and

(ii) *for each x in a set dense in $\mathfrak{X}$*

$$\lim_{p \to \infty} \sup_{a, b \in A_p} |T_a(x, s) - T_b(x, s)| = 0,$$

almost everywhere on S. Then the equation appearing in (ii) is valid for every x in $\mathfrak{X}$.

PROOF. First we observe that it is sufficient to prove the theorem for the case of a finite measure space. For, let $S = \cup_{n=1}^{\infty} S_n$ where $\{S_n\}$ is a sequence of disjoint measurable sets of finite μ-measure. Then if μ_1 is the measure defined by

$$\mu_1(E) = \sum_{n=1}^{\infty} \frac{\mu(ES_n)}{2^n[1+\mu(S_n)]}, \qquad E \in \Sigma,$$

then μ_1 is a finite measure and its null sets are the same as those of μ. Consequently we suppose that μ is a finite measure, and so $TM(S, \Sigma, \mu) = M(S, \Sigma, \mu)$.

Let $a_1, a_2, \ldots$ be an enumeration of the elements of A_1. Let the maps $W, V, V_n, n = 1, 2, \ldots$ of the space $\mathfrak{X}$ into $TM(S, \Sigma, \mu)$ be defined by the equations

$$V_n(x, s) = \sup_{1 \leq m \leq n} |T_{a_m}(x, s)|, \qquad V(x, s) = \sup_{a \in A_1} |T_a(x, s)|,$$

$$W(x, s) = \lim_{p \to \infty} \sup_{a, b \in A_p} |T_a(x, s) - T_b(x, s)|.$$

It is easily seen that V_n is a continuous map of $\mathfrak{X}$ into $TM(S, \Sigma, \mu)$ satisfying $|V_n(x+y)| \leq |V_n(x)|+|V_n(y)|$ and $|V_n(\alpha x)| = |\alpha V_n(x)|$ for every pair x, y in $\mathfrak{X}$ and every scalar α. Condition (i) assures that V maps $\mathfrak{X}$ into $TM(S, \Sigma, \mu)$, and it follows from their definitions and the fact just proved that $|\alpha V_n(x)| \leq |\alpha V(x)|$. For each x, $\alpha V(x) \to 0$ as $\alpha \to 0$ and so the set $\{V_n(x) | n = 1, 2, \ldots\}$ is a bounded set (II.1.7) in the space $TM(S, \Sigma, \mu)$. By Lemma II.1.13 we conclude that

$\lim_{x\to 0} V_n(x) = 0$ uniformly for $n \geqq 1$. By Corollary III.6.13(b), $V_n(x) \to V(x)$ and so it follows that V is continuous at $x = 0$. Since $|W(x, s)| \leqq 2V(x, s)$, it follows that $|W(x)| \leqq 2|V(x)|$ and hence W is continuous at $x = 0$. Now it may be readily shown that

$$|W(x, s)-W(y, s)| \leqq W(x-y, s),$$

for almost all s, and hence it follows that

$$|W(x)-W(y)| \leqq |W(x-y)|, \qquad x, y \in \mathfrak{X}.$$

This proves that W is continuous at every point of $\mathfrak{X}$. By (ii), W vanishes on a set which is dense in $\mathfrak{X}$; consequently W vanishes identically. Q.E.D.

We shall now consider some order properties of the space $M(S, \Sigma, \mu)$. If f is in $M(S, \Sigma, \mu)$ and if $f(s) \geqq 0$ for μ-almost all $s \in S$, we say that f is *positive* and write $f \geqq 0$. If f_1 and f_2 are real or complex valued functions in $M(S, \Sigma, \mu)$ and if $f_1-f_2 \geqq 0$, we write $f_1 \geqq f_2$ or $f_2 \leqq f_1$. This makes $M(S, \Sigma, \mu)$ into a partially ordered system in the sense of Definition I.2.1. We note that the space $TM(S, \Sigma, \mu)$ is also partially ordered by the same ordering.

4 LEMMA. *If Σ is a field of subsets of S and μ a finitely additive set function, then the real spaces $M(S, \Sigma, \mu)$ and $TM(S, \Sigma, \mu)$ are lattices. If (S, Σ, μ) is a positive measure space then the real space $M(S, \Sigma, \mu)$ is a σ-complete lattice.*

PROOF. The definitions of a lattice and of σ-completeness were given near the end of Section I.12. To prove the first statement it suffices to observe that

$$\sup\{f, g\} = \tfrac{1}{2}[|f(\cdot)+g(\cdot)|+|f(\cdot)-g(\cdot)|],$$
$$\inf\{f, g\} = \tfrac{1}{2}[|f(\cdot)+g(\cdot)|-|f(\cdot)-g(\cdot)|].$$

The fact that these functions are in $M(S, \Sigma, \mu)$ or in $TM(S, \Sigma, \mu)$ follows from Lemmas III.2.11 and III.2.12. It is also clear that the value of $\sup\{f, g\}$ at the point $s \in S$ is almost everywhere equal to the number $\sup\{f(s), g(s)\}$. The second statement of the lemma was proved following Theorem III.6.10. Q.E.D.

In order to establish another type of completeness for the space $M(S, \Sigma, \mu)$, it will be convenient to have a general lemma on σ-complete lattices.

5 **Lemma.** *Let L be a σ-complete lattice in which every set of elements of L which is well-ordered under the partial ordering of L is at most countable. Then L is complete and every subset A of L has a least upper bound which is the least upper bound of a countable subset of A.*

Proof. Let A be a subset of L which has an upper bound and let A_1 be the collection of least upper bounds of all countable subsets of A. Consider the family W of subsets of A_1 which are well-ordered under the ordering inherited from L. We shall order W by defining the relation $a \leqq b$ between elements a, b in W to mean that $a \subseteqq b$ and that each element x which is in b but not a is an upper bound for a. It will first be shown that W satisfies the hypothesis of Zorn's lemma. To do this we let W_0 be a totally ordered subset of W (I.22) and let $c \subseteqq \bigcup W_0$. Then, for some $a \in W_0$, $c \cap a$ is not void. Let x be the smallest element of $c \cap a$ and let y be any other element of c. If $y \in b \in W_0$, we have either $b \leqq a$ or $a \leqq b$. If $b \leqq a$, then, since $y \in b$, we have also $y \in a$ and thus $y \in c \cap a$ and $y \geqq x$. If $a \leqq b$ and $y \in a$ then $y \in c \cap a$ and $y \geqq x$. Finally, if $a \leqq b$ and $y \notin a$ then $y \geqq x$ by the definition of the ordering in W. Thus x is the smallest element of c. This shows that $\bigcup W_0$ is well-ordered and is thus an upper bound in W for W_0. Zorn's lemma shows that W contains a maximal element b_0. By hypothesis, b_0 is at most countable and hence $y_0 = \sup b_0$ exists and is in A_1. We now prove that $y \leqq y_0$ for any $y \in A_1$. For if this were not true for some $y_1 \in A_1$, then $\sup\{y_0, y_1\} > y_0$; but $y_2 = \sup\{y_0, y_1\} \in A_1$, $b_0 \leqq b_0 \cup \{y_2\}$, and so b_0 is not maximal. Therefore $y \leqq y_0$, $y \in A_1$, and since $y_0 \in A_1$ we conclude that $y_0 = \sup A_1 = \sup A$. It may be shown in a similar manner that the infimum of a set with a lower bound exists. This proves that L is complete; the final statement of the lemma follows from the observation that y_0 is the least upper bound of a sequence from A_1 and hence of a sequence from A. Q.E.D.

6 **Theorem.** *Let (S, Σ, μ) be a positive σ-finite measure space. Then the partially ordered space of real measurable functions on S, where $f \geqq g$ means that $f(s) \geqq g(s)$ for μ-almost all s in S, is a complete lattice. Furthermore the least upper bound of every bounded set B in this lattice is the least upper bound of a suitably chosen countable subset of B.*

Proof. Let B be a bounded set in the lattice $M(S, \Sigma, \mu)$ of real measurable functions on S. We may and shall suppose that the functions in B are positive and bounded above by the measurable function g. We shall first show that the lattice L of all f in $M(S, \Sigma, \mu)$ with $0 \leqq f \leqq g$ has the property that every well-ordered subset is countable. To do this let (R, Σ_0, λ) be the real line under Lebesgue measure; if we put $(T, \Sigma_1, \theta) = (S, \Sigma, \mu) \times (R, \Sigma_0, \lambda)$ then the space (T, Σ_1, θ) is σ-finite (III.11.6). To each $f \in L$ we associate the measurable set $A(f) = \{[s, r] \in S \times R | 0 \leqq r \leqq f(s)\}$. Note that if $f_1, f_2 \in L$ and if $f_1 \leqq f_2$, then $A(f_1) \subseteqq A(f_2)$ and $0 \leqq \theta(A(f_2)) - A(f_1)) = \int_S \{f_2(s) - f_1(s)\} \mu(ds) \leqq \infty$. This implies that f_1 and f_2 are equal μ-almost everywhere if and only if $\theta(A(f_1))$ and $\theta(A(f_2))$ are equal. Now the σ-finite space (T, Σ_1, θ) can contain at most a countable number of disjoint measurable sets of non-zero θ-measure, and so any well-ordered subset of L can contain only a countable number of non-equivalent functions. The theorem then follows from Lemmas 4 and 5. Q.E.D.

The natural ordering just used for functions in $M(S, \Sigma, \mu)$ has already been employed for functions in the spaces $L_p(S, \Sigma, \mu)$, $1 \leqq p \leqq \infty$. In theorems 8.22 and 8.23 we proved that if a subset of one of these spaces is bounded above by some function then there exists a least upper bound in the same space. Each of these spaces is, of course, a subspace of $M(S, \Sigma, \mu)$, and it is useful to know that the least upper bounds may be taken in either the space $L_p(S, \Sigma, \mu)$ or in $M(S, \Sigma, \mu)$. We state this result formally.

7 COROLLARY. *Let (S, Σ, μ) be a positive σ-finite measure space. Then the real spaces $L_p(S, \Sigma, \mu)$, $1 \leqq p \leqq \infty$, and $M(S, \Sigma, \mu)$ are complete lattices and the supremum in $L_p(S, \Sigma, \mu)$ of a set bounded in this lattice is the same as its supremum in the lattice $M(S, \Sigma, \mu)$. Furthermore any set B which is bounded in the lattice $L_p(S, \Sigma \mu)$ contains a countable subset having the same supremum as B.*

PROOF. It follows from the fact that $L_p(S, \Sigma, \mu) \subseteqq M(S, \Sigma, \mu)$ that $M-\sup_\alpha \{f_\alpha\} \leqq L_p-\sup_\alpha \{f_\alpha\}$; since $L_p-\sup_\alpha \{f_\alpha\}$ is in the space $L_p(S, \Sigma, \mu)$ it follows from Theorem III.2.22(b) that $M-\sup\{f_\alpha\}$ is also in $L_p(S, \Sigma, \mu)$ and hence the two suprema coincide. The final statement follows from the preceding theorem. Q.E.D.

12. Functions of Bounded Variation

Let I be an interval with end points a and b. Let Σ be the field of sets in I consisting of all unions of intervals $(c, d]$, half open on the left unless c is the left end point a of I, and $a \in I$, in which case we take the closed interval $[a, d]$. If $f \in BV(I)$ is given, define $\mu_f \in ba(I, \Sigma)$ by $\mu_f([a, d]) = f(d)-f(a)$, $\mu_f((c, d]) = f(d)-f(c)$ if $c \neq a$. It is then evident that $v(\mu_f, I) = v(f, I)$. Thus, $ba(S, \Sigma)$ is isometrically isomorphic with the closed subspace $BV_0(I)$ of all $f \in BV(I)$ such that $f(a+) = 0$. If N is the one-dimensional space of constant functions, it is evident that $BV(I) = BV_0(I) \oplus N$. Thus $BV(I)$ is isometrically isomorphic to the direct sum of $ba(I, \Sigma)$ and a one-dimensional space. From this, the following theorem is evident (cf. 9.9).

1 THEOREM. *The space $BV(I)$ is a weakly complete B-space.*

It is easily seen that the isomorphism defined above can be used to give answers valid for the space $BV(I)$ to many other of the questions asked in Section 1. We leave the details to the reader as an exercise.

It was shown in Chapter III.5 (cf. the discussion following Lemma III.5.16) that if f is in NBV, then μ_f is regular; and the converse is equally evident. Thus $f \longleftrightarrow \mu_f$ determines an isometric isomorphism between $NBV(I)$ and $rba(I, \Sigma)$. Using Theorem 9.9, we obtain the following result.

2 THEOREM. *The space $NBV(I)$ is a weakly complete B-space.*

Again we find that we are considering a space isometrically isomorphic to a space studied earlier; and again this isometric isomorphism can be used to give answers to the problems listed in Section 1. We leave the details to the reader as an exercise.

Finally, let $f \in AC(I)$, and suppose that $f(a+) = 0$, so that $f \in NBV(I)$. Let Σ_1 be the σ-field determined by Σ; i.e., the field of all Borel subsets of I. Let λ denote Borel-Lebesgue measure, and λ_1 its restriction to the field Σ. It is then clear that μ_f is λ_1-continuous. Conversely, if $f \in NBV(I)$ and μ_f is λ_1-continuous, it is clear that $f \in AC(I)$.

If $\hat{\mu}_f$ denotes the unique countably additive extension of μ_f to Σ_1, (which exists, according to the discussion following III.5.16) then, by Lemma 9.13, $\hat{\mu}_f$ is λ-continuous if and only if μ_f is λ_1-continuous.

According to the Radon-Nikodým theorem (III.10.7), $\hat{\mu}_f$ is representable as an integral in the form

$$\hat{\mu}_f(E) = \int_E g(x)dx, \qquad g \in L_1(I, \Sigma, \lambda).$$

Consequently, an arbitrary function $f \in AC(I)$ can be written in the form

$$f(x) = f(a+) + \int_0^x g(y)dy,$$

where g is an element in $L_1(S, \Sigma, \lambda)$. Conversely, every function f of this form clearly belongs to $AC(I)$. It is readily seen that

$$|f| = |f(a+)| + \int_a^b |g(y)|dy.$$

These facts establish the following result.

3 THEOREM. *The space $AC(I)$ is isometrically isomorphic to the direct sum of $L_1(I, \Sigma, \lambda)$ and a one-dimensional space. Consequently, $AC(I)$ is weakly complete.*

Solutions of the problems about $AC(I)$ raised in Section 1 can easily be given by using the isometric isomorphism noted in Theorem 3. Details are left to the reader as exercises.

13. Exercises

A. Exercises which complete the table of Section 15.

1 Show that all possible topologies of a finite dimensional linear topological space $\mathfrak{X}$ are equivalent. Show hence that a subset of $\mathfrak{X}$ is conditionally compact if and only if it is bounded. Show that if $\{x_1, \ldots, x_n\}$ is a basis for the n-dimensional linear topological space $\mathfrak{X}$, and $\{y_m\}$ is a sequence in $\mathfrak{X}$, then y_m converges to an element y if and only if $\alpha_m^{(i)} \to \alpha^{(i)}$, where $y_m = \sum_{i=1}^n \alpha_m^{(i)} x_i$, $y = \sum_{i=1}^n \alpha^{(i)} x_i$. If $\mathfrak{X}$ is a B-space the sequence $\{y_m\}$ is a weak Cauchy sequence if and only if $\lim_{m\to\infty} a_m^{(i)}$ exists for each i, $1 \leqq i \leqq n$.

2 Show that no space $L_1(S, \Sigma, \mu)$ is reflexive unless it is finite-dimensional.

3 Show that a subset K of l_p, $p \geqq 1$, is conditionally compact if and only if it is bounded and $\lim_{n\to\infty} \sum_{n=i}^{\infty} | a_i|^p = 0$ uniformly for

$[a_1, a_2, \ldots] \in K$. Show that in l_1, strong conditional compactness and weak sequential compactness are the same.

4 If $\infty > p > 1$ a sequence $x^{(n)} = \{\xi_i^{(n)}, i = 1, 2, \ldots\}$ is a weak Cauchy sequence in l_p if and only if it is bounded and the limits $\xi_i = \lim_n \xi_i^{(n)}$, $i = 1, 2, \ldots$ all exist, and that such a sequence converges weakly to the element $x = \{\xi_i\}$. Show that if $p = 1$, the same condition describes c_0-convergence in $l_1 = c_0^*$.

5 Show that no space $B(S, \Sigma)$ is weakly complete unless Σ is finite and that no space $B(S)$ is weakly complete unless S is finite. Show that $ba(S, \Sigma)$ is not reflexive unless it is finite dimensional.

6 Let $f, f_n \in L_\infty(S, \Sigma, \mu)$ then $\lim_n \int_S f_n(s)g(s)\mu(ds) = \int_S f(s)g(s)\mu(ds)$ for every g in $L_1(S, \Sigma, \mu)$ if and only if the sequence $\{f_n\}$ is bounded in $L_\infty(S, \Sigma, \mu)$ and

$$\lim_n \int_e f_n(s)\mu(ds) = \int_e f(s)\mu(ds)$$

for every e in some field which determines the σ-field Σ.

7 Using 6.3, represent the spaces c_0^* and c^*. (cf. Exercise II.4.33).

8 Show that c_0 and c are not weakly complete and not reflexive.

9 A set K in c or in c_0 is conditionally compact if and only if it is bounded and the limit $\lim_n \xi_n$ exists uniformly for $x = \{\xi_n\}$ in K. A set K in c or c_0 is weakly sequentially compact if and only if it is bounded and $\lim_n \xi_n$ exists quasi-uniformly for $x = \{\xi_n\}$ in K.

10 Let $x_n = \{\xi_i^{(n)}\}$, $x = \{\xi_i\}$ be vectors in c or in c_0. Then $x_n \to x$ weakly if and only if $\{x_n\}$ is bounded and

$$\lim_n \lim_i \xi_i^{(n)} = \lim_i \xi_i$$

$$\lim_n \xi_i^{(n)} = \xi_i, \qquad i = 1, 2, \ldots.$$

The sequence $\{x_n\}$ is a weak Cauchy sequence if and only if $\lim_n \xi_i^{(n)}$ and $\lim_n \lim_i \xi_i^{(n)}$ exist.

11 Show that the space bv_0 is isometrically isomorphic to the space l_1, and that bv is the direct sum of bv_0 and a one dimensional subspace. Using this isomorphism, the results obtained for l_1 all carry over to these spaces. What are the detailed results obtained?

12 Show that bv may be interpreted naturally as cs^*, and that

bv_0 has a similar interpretation as the conjugate of the subspace of *cs* consisting of series with sum equal to zero. Use these facts to extend Exercise 4 to *bv* and bv_0.

13 Show that the space *bs* is isometrically isomorphic with the space l_∞. Show how this isomorphism can be used to solve all the problems indicated in the table for *bs*. What are the detailed forms of the results obtained?

14 The space *cs* is isometrically isomorphic with the space *c*. On this basis, solve all the problems indicated in the table for this space.

15 Let S be a compact Hausdorff space. Show that $C(S)$ is weakly complete if and only if it is finite dimensional, which is the case if and only if S is finite. Using this result and 6.19, show that the same statements may be made for any normal space.

16 Let S be a completely regular topological space. Show that $C(S)$ is separable if and only if S is compact and metric.

17 Show that a sequence $\{\lambda_n\}$ of elements of $ba(S, \Sigma)$ converge weakly to an element $\lambda \in ba(S, \Sigma)$ if and only if there exists a non-negative $\mu \in ba(S, \Sigma)$ such that $\lim_{\mu(E)\to 0} |\lambda_n(E)| = 0$ uniformly in n, and

$$\lim_{n\to\infty} \lambda_n(E) = \lambda(E) \quad \text{for } E \in \Sigma.$$

18 Let $\mu_n \in ba(S, \Sigma)$. Then there is a $\mu \in ba(S, \Sigma)$ such that

$$\lim_n \int_S f(s)\mu_n(ds) = \int_S f(s)\mu(ds), \qquad f \in B(S, \Sigma)$$

if and only if $\{|\mu_n|\}$ is bounded and the limit $\lim_n \mu_n(E)$ exists for every set $E \in \Sigma$.

19 Let μ be a non-negative element of $ba(S, \Sigma)$ and let $ba(S,\Sigma,\mu)$ be the subspace of $ba(S, \Sigma)$ consisting of those λ which are μ-continuous. For each $a = (e_1, \ldots, e_n)$ in the partially ordered set A defined in 5.6 let the operator U_a in $ba(S, \Sigma, \mu)$ be defined by the equation

$$(U_a\lambda)(e) = \sum_{i=1}^{n} \frac{\lambda(e_i)}{\mu(e_i)} \mu(e_ie), \qquad e \in \Sigma.$$

Show that a set $K \subset ba(S, \Sigma)$ is conditionally compact if and only if

(i) K is bounded.

(ii) There is a non-negative μ in $ba(S, \Sigma)$ with respect to which every λ in K is continuous.

(iii) $\lim_a U_a \lambda = \lambda$ uniformly with respect to $\lambda \in K$.

20 Let $\Sigma = \{E_n\}$ be a countable field of subsets of a set S, and let Σ_1 be the σ-field generated by Σ. Let μ be a non-negative finite countably additive measure defined on Σ_1. In the set A_μ of μ-measurable functions f with $\mu\text{-ess}\sup_{s \in S} |f(s)| \leqq 1$ we introduce the metric

$$\varrho_\mu(f, g) = \sum_{n=1}^{\infty} \frac{1}{2^n} \left| \int_{E_\mu} \{f(s) - g(s)\} \mu(ds) \right|.$$

Show that A_μ is a compact metric space. Show that a bounded set K in $ca(S, \Sigma_1)$ is conditionally compact if and only if there exists a non-negative μ in $ca(S, \Sigma_1)$ such that the continuity of

$$\int_S f(s) \lambda(ds)$$

for f in A_μ is uniform for λ in K.

21 Show that $ca(S, \Sigma)$ is reflexive only when it is finite dimensional.

22 Let S be a normal topological space and $rca(S)$ the regular countably additive set functions on the field of Borel sets in S. Prove that

(i) $rca(S)$ is weakly complete.

(ii) A sequence $\{\mu_n\}$ in $rca(S)$ is a weak Cauchy sequence if and only if it is bounded and the limit $\lim_n \mu_n(e) = \mu(e)$ exists for every Borel set e, in which case $\mu \in rca$ and $\mu_n \to \mu$ weakly.

(iii) A subset $K \subseteqq rca(S)$ is weakly sequentially compact if and only if it is bounded and, for some positive λ in $rca(S)$, the limit $\lim_{\lambda(e)=0} \mu(e) = 0$ uniformly with respect to μ in K.

(iv) A set $K \subseteqq rca(S)$ is weakly sequentially compact if and only if it is bounded and the countable additivity of $\mu(e)$ for e in the field of Borel sets is uniform with respect to μ in K.

(v) $rca(S)$ is reflexive only if it is finite dimensional.

23 Let $\infty > p > 1$, $f_n, f \in L_p[0, 1]$ (Lebesgue measure is understood). Then $f_n \to f$ weakly if and only if

(i) $$\sup_n \int_0^1 |f_n(t)|^p \, dt < \infty$$

(ii) $$\int_0^x f_n(t)dt \to \int_0^x f(t)dt, \qquad 0 \leqq x \leqq 1.$$

The sequence $\{f_n\}$ is a weak Cauchy sequence if and only if (i) holds and $\lim_n \int_0^x f_n(t)\,dt$ exists for each $0 \leqq x \leqq 1$.

24 Let $f, f_n \in L_p(S, \Sigma, \mu)$ where $1 < p < \infty$, and let Σ_0 be a family of sets of finite measure whose characteristic functions form a fundamental set in $L_p(S, \Sigma, \mu)$. Then the sequence $\{f_n\}$ converges to f weakly if and only if it is bounded and

$$\lim_n \int_E f_n(s)\mu(ds) = \int_E f(s)\mu(ds), \qquad E \in \Sigma_0.$$

The sequence $\{f_n\}$ is a weak Cauchy sequence if and only if it is bounded and $\lim_n \int_E f_n(s)\mu(ds)$ exists for each $E \in \Sigma_0$.

25 Let $f_n, f \in L_1(S, \Sigma, \mu)$. Then $f_n \to f$ weakly if and only if $\lim_n \int_E f_n(s)\mu(ds) = \int_E f(s)\mu(ds)$ for each $E \in \Sigma$. Moreover, $\{f_n\}$ is a weak Cauchy sequence if and only if $\lim_n \int_E f_n(s)\mu(ds)$ exists for all $E \in \Sigma$.

26 Show that a bounded subset $K \subseteq L_\infty(S, \Sigma, \mu)$ is compact if and only if for each $\varepsilon > 0$ there exists a partition π of S into a finite number of measurable sets such that

$$\operatorname*{ess\,sup}_{\substack{s \in A \\ t \in A}} |f(s) - f(t)| < \varepsilon$$

for each $A \in \pi$.

27 Let $g, g_n \in L_\infty(0, 1)$. Then $\int_0^1 g_n(t)f(t)dt \to \int_0^1 g(t)f(t)dt$ for every $f \in L_1(0, 1)$ if and only if

(i) $\sup_n \operatorname*{ess\,sup}_t |g_n(t)| \leqq \infty$

(ii) $\int_0^x g_n(t)dt \to \int_0^x g(t)dt$.

28 Show that $BV(I)$, $NBV(I)$ and $AC(I)$ are not reflexive.

29 If $L_\infty(I)$ is the space of essentially bounded Lebesgue mea-

surable functions on I and if the space $\Phi + L_\infty(I)$ is normed by the equation

$$|a+g| = |a|+|g| = |a| + \operatorname{ess\,sup}_{s \in I} |g(s)|,$$

then the equation

$$x^*f = af(a) + \int_I g(s)f'(s)ds, \qquad f \in AC(I)$$

establishes an isometric isomorphism between $AC^*(I)$ and $\Phi + L_\infty(I)$.

30 A bounded subset K of $NBV(I)$ (or, $BV(I)$) is sequentially weakly compact if and only if there is a $g \in NBV$ (or, BV) such that for $\varepsilon > 0$ we have a $\delta > 0$ with

$$\sum_{i=1}^{n} |g(s_i) - g(t_i)| < \delta \text{ implies } \sum_{i=1}^{n} |f(s_i) - f(t_i)| < \varepsilon$$

for each f in K.

A sequence $\{f_n\}$ in $BV(I)$ or $NBV(I)$ converges weakly to an element f if and only if the set $\{f_n\}$ is weakly sequentially compact and we have $f_n(s) \to f(s)$ for s in I.

31 A bounded subset K of $AC(I)$ is sequentially weakly compact if and only if for each $\varepsilon > 0$ there is a $\delta > 0$ such that

$$\sum_{i=1}^{n} |s_i - t_i| < \delta \text{ implies } \sum_{i=1}^{n} |f(s_i) - f(t_i)| < \varepsilon$$

for each f in K.

A sequence $\{f_n\}$ in $AC(I)$ converges weakly to an element f if and only if $\{f_n\}$ is weakly sequentially compact and $f_n(s) \to f(s)$ for all s.

32 Let $K \subseteq AC(I)$ and if I is not $(-\infty, \infty)$ let $AC(I)$ be embedded in $AC((-\infty, \infty))$ by defining each $f \in AC(I)$ to be constant outside of I. The set K is conditionally compact in $AC(I)$ if and only if

(i) K is bounded;

(ii) $\lim_{n\to\infty} v(f, I_n^+ \cup I_n^-) = 0$ uniformly for f in K, where $I_n^+ = [n, \infty)$ and $I_n^- = (-\infty, n]$;

(iii) $\lim_{\varepsilon\to\infty} |f - f_\varepsilon| = 0$ uniformly for f in K, where $f_\varepsilon(s) = f(s+\varepsilon)$.

33 On the closed interval I, let Σ be the field of sets generated by the family of subintervals with rational end points, and let $\{E_n\}$ be an enumeration of Σ. Let μ be Lebesgue measure. In the set A of

Lebesgue measurable functions such that $\operatorname{ess\,sup}_{s \in I} |f(s)| \leqq 1$ we introduce the metric

$$\varrho(f, g) = \sum_{n=1}^{\infty} \frac{1}{2^n} \left| \int_{E_n} \{f(s) - g(s)\} \mu(ds) \right| .$$

Show that A is a compact metric space. Show that a bounded set K in $AC(I)$ is conditionally compact if and only if the continuity of $\int_I f(s) dg(s)$ for f in A is uniform for f in K.

34 Under the same assumptions and with the same notation as in the preceding exercise, let K be a bounded subset of $NBV(I)$. Using the notation of Exercise 20 and of Section 12, show that K is conditionally compact if and only if there exists a function g in $NBV(I)$ such that the continuity of $\int_I h(s) \mu_f(ds)$ for h in A_{μ_g} is uniform for f in K.

35 (a) Let I be a closed interval. Show that the formula

$$x_g^*(f) = \int_I f(s) dg(s), \qquad f \in C(I)$$

determines an isometric isomorphism between $C(I)^*$ and $NBV(I)$. Show that if $\{g_n\}$ is a sequence of functions in $NBV(I)$, we have $\lim_{n \to \infty} \int_I f(s) dg_n(s) = \int_I f(s) dg_0(s)$ for all f in $C(I)$ if and only if

(i) $v(g_n, I)$ is uniformly bounded;

(ii) $g_n(s) \to g_0(s)$ at each point s of continuity of g_0.

(b) Using the relationship between $BV(I)$ and the space $ba(I, \Sigma)$ discussed in the first paragraph of Section 12, show how $BV(I)$ may be regarded as the conjugate space of the direct sum $\mathfrak{X}$ of $B(S, \Sigma)$ with a one dimensional space, and give a necessary and sufficient condition for $\mathfrak{X}$-convergence of a sequence of elements in $BV(I)$.

36 Show that the space $C^p[a, b]$ is decomposable into the direct sum of a finite dimensional space and a space isomorphic to $C[a, b]$. Show that

(a) every element x^* in the conjugate space of C^p is uniquely representable in the form

$$x^*(f) = \sum_{i=0}^{p-1} \alpha_i f^{(i)}(a) + \int_a^b f^{(p)}(s) \mu(ds),$$

where μ is a regular Borel measure.

(b) A sequence f_n is a weak Cauchy sequence (converges weakly to f)

if and only if it is bounded and $f_n^{(j)}(s)$ converges (to $f^{(j)}(s)$) for each s in $[a, b]$ and each $j = 0, 1, \ldots, p$.
(c) C^p is not weakly complete, and not reflexive.
(d) A subset $A \subseteq C^p$ is conditionally compact if and only if it is bounded and for each $\varepsilon > 0$ there is a $\delta > 0$ such that $|s-t| < \delta$ implies $|f^{(p)}(s)-f^{(p)}(t)| < \varepsilon$ for $s, t \in [a, b]$.
(e) A subset $A \subseteq C^p$ is weakly sequentially compact if and only if it is bounded and the set $\{f^{(p)}\}$, $f \in A$, is quasi-equicontinuous.

37 Let D be a bounded domain. Show that a sequence of functions in $A(D)$ is a weak Cauchy sequence (a sequence converging weakly to f in $A(D)$) if and only if it is uniformly bounded and converges at each point of the boundary of D (to f). Show that a subset of $A(D)$ is conditionally compact if and only if it is bounded and the functions in it are uniformly equicontinuous. Show that a subset of $A(D)$ is weakly sequentially compact if and only if it is bounded and quasi-equicontinuous on the boundary of D. Show that $A(D)$ is never weakly complete and never reflexive. Show that $A(D)$ is a closed subspace of $C(\bar{D})$.

38 Show that AP is not weakly complete and not reflexive.

39 Show that a bounded subset K of the space AP is conditionally compact if and only if
(i) there exists a function l_ε defined for $\varepsilon > 0$ such that the set

$$M_\varepsilon = \{t \,|\, |f(s+t)-f(s)| < \varepsilon \quad \text{for } -\infty < s < +\infty \text{ and } f \text{ in } K\}$$

has a non-void intersection with every interval of length l_ε;
(ii) for every $\varepsilon > 0$ there is a $\delta > 0$ such that

$$|f(s)-f(t)| < \varepsilon \quad \text{if } |s-t| < \delta \text{ and } f \text{ is in } K.$$

40 Let S be a compact Hausdorff space. Show that a sequence $\{f_n\} \in C(S)$ is a weak Cauchy sequence if and only if $|f_n|$ is uniformly bounded and $\lim_{n\to\infty} f_n(s)$ exists for each $s \in S$.

41 Show that a sequence $\{f_n\}$ of elements of AP is a weak Cauchy sequence if and only if $|f_n|$ is uniformly bounded and each sequence $\{s_m\}$ of real numbers contains a subsequence $\{s_{m_i}\}$ such that $\lim_{n\to\infty} \lim_{i\to\infty} f_n(s_{m_i})$ exists. Show that a sequence $\{f_n\}$ of elements of AP converges weakly to f in AP if and only if each sequence

$\{s_n\}$ of real numbers contains a subsequence $\{s_{m_i}\}$ such that $\lim_{n\to\infty} \lim_{i\to\infty} f_n(s_{m_i}) = \lim_{i\to\infty} f(s_{m_i})$.

42 Let S be a normal topological space. Show that a sequence $\{f_n\} \in C(S)$ is a weak Cauchy sequence if and only if $|f_n|$ is uniformly bounded and each sequence $\{s_m\}$ of points in S contains a subsequence $\{s_{m_i}\}$ such that $\lim_{n\to\infty} \lim_{i\to\infty} f_n(s_{m_i})$ exists.

43 Show that a sequence $\{f_n\}$ of elements of $B(S, \Sigma)$ is a weak Cauchy sequence if and only if $|f_n|$ is uniformly bounded and each sequence $\{s_m\}$ of points in S contains a subsequence $\{s_{m_i}\}$ such that $\lim_{n\to\infty} \lim_{i\to\infty} f_n(s_{m_i})$ exists.

44 Show that a sequence $\{x_n\}$ of elements of Hilbert space is a weak Cauchy sequence (with weak limit x) if and only if $|x_n|$ is uniformly bounded and $\lim_{n\to\infty} (x_n, y_\alpha)$ exists ($\lim_{n\to\infty} (x_n, y_\alpha) = (x, y_\alpha)$) for each element y_α in an orthonormal basis.

45 Let $\{y_\alpha\}$ be an orthonormal basis for Hilbert space $\mathfrak{H}$. Show that a bounded subset A of $\mathfrak{H}$ is conditionally compact if and only if there exists, for each $\varepsilon > 0$, a finite collection $y_{\alpha_1}, \ldots, y_{\alpha_n}$ of the orthonormal basis $\{y_\alpha\}$ such that $\sum_{\alpha \neq \alpha_1, \ldots, \alpha_n} |(x, y_\alpha)|^2 < \varepsilon$, all x in A.

46 Show that every continuous linear functional on the space s is of the form

$$f(x) = \sum_{i=1}^{n} \alpha_i \xi_i,$$

where $x = \{\xi_i\} \in s$, and $[\alpha_1, \ldots, \alpha_n]$ is a finite collection of complex numbers.

47 Show that a collection $\{x_\alpha\} = \{\xi_i^{(\alpha)}\}$ of elements of s is conditionally compact if and only if $\xi_i^{(\alpha)}$ is bounded in α for each fixed i.
each fixed i.

48 Show that $BV(I)$ is decomposed as the direct sum of its subspace $NBV(I)$ and the subspace consisting of all functions in BV which vanish except for a denumerable set of points. Prove that this latter space is isometrically isomorphic to a space L_1. Use this fact and Exercise 34 to give a characterization of the family of conditionally compact subsets of $BV(I)$.

B. Miscellaneous Exercises.

49 (a) Show that if strong convergence and weak convergence of sequences coincide in $L_1(S, \Sigma, \mu)$, then every set of positive measure in S is an atom.

(b) Show that $L_1(S, \Sigma, \mu)$ is equivalent to l_1 if and only if there exists a countable collection of atoms of finite measure $\{E_n\}$ in Σ such that every measurable subset of $S - \bigcup_{n=1}^{\infty} E_n$ is either an atom of infinite measure or a null set.

50 Show that no space equivalent to a space $C(S)$ is equivalent to a closed subspace of a space $ba(S, \Sigma)$ unless both are finite dimensional.

51 Show that no space $L_p(S, \Sigma, \mu)$, $1 < p < \infty$, is equivalent either to a space $C(S)$ or a space $L_1(S_1, \Sigma_1, \mu_1)$, unless it is finite dimensional.

52 Show that if $\mathfrak{X}_1$ is a finite dimensional subspace of a B-space $\mathfrak{X}$, $\mathfrak{X}_1$ is closed, and there exists a second closed subspace $\mathfrak{X}_2$ of $\mathfrak{X}$ such that $\mathfrak{X} = \mathfrak{X}_1 \oplus \mathfrak{X}_2$.

53 Let (S, Σ, μ) be a measure space which is not necessarily positive. Let $v(E) = v(\mu, E)$ for $E \in \Sigma$. Show that the spaces $L_p(S, \Sigma, \mu)$ and $L_p(S, \Sigma, v)$ are isometrically isomorphic in a natural way for $1 \leqq p \leqq \infty$. Rephrase all the theorems of Section 7 to obtain versions valid in this slightly more general setting.

54 Show that a bounded subset K of $L_1(S, \Sigma, \mu)$ is weakly sequentially compact if and only if

(i) $\lim_{\mu(E)\to 0} \int_E f(s)\mu(s) = 0$ uniformly for $f \in K$;

(ii) there exists a sequence of sets $\{A_n\}$, $A_n \in \Sigma$, $\mu(A_n) < \infty$, such that

$$\lim_{n\to\infty} \int_{S-A_n} f(s)\mu(ds) = 0 \text{ uniformly for } f \in K.$$

55 Let $\mathfrak{X}$ be a B-space, and let (S, Σ, μ) be a measure space. Suppose that $f : S \to \mathfrak{X}$, and that $x^*f(\cdot)$ is μ-integrable for each x^* in $\mathfrak{X}^*$. Show that there exists an x^{**} in $\mathfrak{X}^{**}$ such that

$$x^{**}(x^*) = \int_S x^*f(t)\mu(dt), \qquad x^* \in \mathfrak{X}^*.$$

56 (Gelfand). Let $\mathfrak{X}$ be a B-space, and (S, Σ, μ) a measure space. Let $f^* : S \to \mathfrak{X}^*$, and suppose that $f^*(\cdot)x$ is μ-integrable for each x in $\mathfrak{X}$. Show that there exists an x^* in $\mathfrak{X}^*$ such that

$$x^*x = \int_S f^*(s)x\mu(ds), \qquad x \in \mathfrak{X}.$$

57 Let $\mathfrak{X}$ be a B-space, and let K be a real valued function defined on $\mathfrak{X} \times [0, 1]$ which is linear in x for each fixed t in $[0, 1]$, and which lies in $L_q[0, 1]$ for each fixed x in $\mathfrak{X}$ (Lebesgue measure understood.) Suppose that

$$\lim_{x \to 0} \int_0^s K_x(t)dt = 0, \qquad 0 \leqq s \leqq 1.$$

Prove that

$$\lim_{x \to 0} \int_0^1 |K_x(t)|^q dt = 0 \qquad \text{if } 1 \leqq q < \infty,$$

and that

$$\lim_{x \to 0} \operatorname*{ess\,sup}_{0 \leqq t \leqq 1} |K_x(t)| = 0 \qquad \text{if } q = \infty.$$

58 Let $\mathfrak{X}$ be a Banach space, and K a complex valued function on $\mathfrak{X} \times [0, 1]$ which is linear in x for each fixed t in $[0, 1]$ and belongs to $NBV[0, 1]$ for each fixed x. Suppose that $\lim_{x \to 0} K_x(t) = 0$ for $0 \leqq t \leqq 1$. Prove that $\lim_{x \to 0} v(K_x, [0, 1]) = 0$.

59 Let $1 \leqq p \leqq \infty$ and let $\{\alpha_n\}$ be a sequence of complex numbers such that $\sum \alpha_n \xi_n$ converges for every $\{\xi_n\}$ in l_p. Then $\{\alpha_n\}$ is in $l_{p'}$, where $1/p + 1/p' = 1$.

60 If $\{\alpha_n\}$ is a sequence of complex numbers such that $\sum \alpha_n \xi_n$ converges for every $x = \{\xi_n\}$ in c, show that $\sum |\alpha_n| < \infty$.

61 Let $1 \leqq p < \infty$, $1 < q \leqq \infty$. Let (α_{ij}) be an infinite matrix such that the series $\eta_i = \sum_{j=1}^{\infty} \alpha_{ij}\xi_j$ all converge for every $\xi = [\xi_1, \xi_2, \ldots]$ in l_p, and such that $\eta = [\eta_1, \eta_2, \ldots]$ lies in l_q. Show that if $q < \infty$, there exists a constant K such that

$$\Big(\sum_{i=1}^{\infty} |\eta_i|^q\Big)^{1/q} \leqq K\Big(\sum_{j=1}^{\infty} |\xi_j|^p\Big)^{1/p} \quad \text{for every } \xi = [\xi_j] \text{ in } L_p$$

and if $q = \infty$ there exists a constant K such that

$$\sup_{1 \leqq i < \infty} |\eta_i| \leqq K\Big(\sum_{j=1}^{\infty} |\xi_j|^p\Big)^{1/p} \text{ for every } \xi = [\xi_j] \text{ in } L_p.$$

62 Let $1 \leqq p < \infty$ and $1 \leqq q < \infty$. Let (S, Σ, μ) and (S_1, Σ_1, μ_1) be measure spaces, and let K be a complex valued $\mu \times \mu_1$-

measurable function defined on $S \times S_1$ such that for each f in $L_p(S_1, \Sigma_1, \mu_1)$, the integral

$$g(s) = \int_{S_1} K(s, s_1) f(s_1) \mu_1(ds_1)$$

exists μ_1-almost everywhere and lies in L_q. Show that there exists a constant $M < \infty$ such that $|g|_q \leqq M|f|_p$.

63 If f and g are complex valued functions defined on the closed interval $[a, b]$, we say that the integral $\int_a^b f(s)dg(s)$ exists in the Riemann-Stieltjes sense if for each $\varepsilon > 0$, there exist points $s_1, \ldots, s_n$ in $[a, b]$ such that if $t_1, \ldots, t_n$ and $u_1, \ldots, u_l$ are two increasing sequences of points in $[a, b]$, both containing all the points $s_1, \ldots, s_n$, we have

$$\left| \sum_{i=1}^{n-1} f(t_i)(g(t_{i+1}) - g(t_i)) - \sum_{i=1}^{l-1} f(u_i)(g(u_{i+1}) - g(u_i)) \right| \leqq \varepsilon.$$

Show that if g is a function defined on $[a, b]$ such that the integral $\int_a^b f(s)dg(s)$ exists in the Riemann-Stieltjes sense for each f in $C[a, b]$ then g is in $BV[a, b]$, and conversely.

64 Show that every continuous complex valued function $f(s)$ of the real variable s which satisfies $f(s) = f(s+2\pi)$ can be approximated by a finite linear combination of terms e^{ins}, $-\infty < s < +\infty$.

65 Show that every continuous function on the Cartesian product $S \times T$ of two compact Hausdorff spaces S and T can be approximated uniformly by finite linear combinations of terms $f(s) \cdot g(t)$.

66 Let S be a compact Hausdorff space, $x, x_n \in C(S)$, and

(a) $\lim_n x_n(s) = x(s), \qquad s \in S;$

(b) $|x_n(s)| \leqq M < \infty, \qquad s \in S, \quad n = 1, 2, \ldots.$

Then for every $\varepsilon > 0$ there is an integer n and complex numbers α_i, $i = 1, \ldots, n$, with

$$\left| x(s) - \sum_{i=1}^{n} \alpha_i x_i(s) \right| < \varepsilon \qquad s \in S.$$

67 Show that Hilbert space is isometrically equivalent to a space $L_2(S, \Sigma, \mu)$ where Σ consists of all subsets of S and where μ is a countably additive function assuming the value one for each set consisting of one point.

68 Let (S, Σ, μ) be a finite measure space, and let K be a bound-

ed subset of $L_\infty(S, \Sigma, \mu)$. Show that K is weakly sequentially compact when regarded as a subset of $L_1(S, \Sigma, \mu)$.

69 Show that the family of characteristic functions of sets in Σ is fundamental in $L_\infty(S, \Sigma, \mu)$.

70 The set of all vectors $x = \{\eta_n\}$ with $|\eta_n| \leqq 1/n$ in real l_2 is called the *Hilbert cube*. Show that this set is compact in l_2.

71 Let (S, Σ, μ) be a measure space. Let f be a complex valued function defined on S such that fg is in $L_1(S, \Sigma, \mu)$ for every g in $L_p(S, \Sigma, \mu)$, $1 < p \leqq \infty$. Show that f is in $L_q(S, \Sigma, \mu)$, where $1/p+1/q = 1$. Show that if S is σ-finite, the same holds for $p = 1$.

72 An operator A in Hilbert space $\mathfrak{H}$ is said to be Hermitian if $(Ax, y) = (x, Ay)$ for each x, y in $\mathfrak{H}$.

(a) Show that if A is Hermitian, $A = 0$ and $(Ax, x) = 0$ for all x in $\mathfrak{H}$ are equivalent.

(b) Show that if A_n is Hermitian for $n \geqq 0$, $\lim_{n\to\infty} (A_n x, y) = (A_0 x, y)$ for all x, y in $\mathfrak{H}$ if and only if $\lim_{n\to\infty} (A_n x, x) = (A_0 x, x)$ for all x in $\mathfrak{H}$.

73 A real function defined on an interval I belongs to $BV(I)$ if and only if it is the difference of two bounded monotone increasing functions defined on I.

74 Let (S, Σ, μ) be a finite positive measure space. Show that an equivalent norm in the F-space $TM(S, \Sigma, \mu)$ is given by the formula

$$|f| = \int_S \frac{|f(s)|}{1+|f(s)|}\, \mu(ds).$$

75 Let S be a topological space, Σ a σ-field of subsets of S, and $\mathfrak{X}$ a B-space. Let μ be a set function defined on Σ and with values in $\mathfrak{X}$. Suppose that for each x^* in $\mathfrak{X}^*$ the set function $x^*\mu$ is regular and countably additive. Then there exists a non-negative regular countably additive set function v in $ca(S, \Sigma)$ such that $\lim_{v(E)\to 0} \mu(E) = 0$.

76 Introduce an ordering into $B(S, \Sigma)$ by putting $f \geqq g$ if $f(s) \geqq g(s)$ for all s in S, and an ordering into $ba(S, \Sigma)$ by putting $\mu \geqq \lambda$ if $\mu(E) \geqq \lambda(E)$ for E in Σ. Show that $\mu \geqq \lambda$ if and only if

$$\int_S f(s)\mu(ds) \geqq \int_S f(s)\lambda(ds) \qquad \text{for all } f \geqq 0,$$

and that $f \geqq g$ if and only if

$$\int_S f(s)\mu(ds) \geqq \int_S g(s)\mu(ds) \qquad \text{for all } \mu \geqq 0.$$

77 Introduce an ordering into $L_\infty(S, \Sigma, \mu)$ by putting $f \geqq g$ if $f(s) \geqq g(s)$ for μ-almost all s in S, and an ordering into $ba(S, \Sigma_1, \mu_1)$ (cf. Definition 8.15 for notation) by putting $\mu \geqq \lambda$ if $\mu(E) \geqq \lambda(E)$ for E in Σ_1. Show that $\mu \geqq \lambda$ if and only if

$$\int_S f(s)\mu(ds) \geqq \int_S f(s)\lambda(ds) \qquad \text{for all } f \geqq 0,$$

and that $f \geqq g$ if and only if

$$\int_S f(s)\mu(ds) \geqq \int_S g(s)\mu(ds) \qquad \text{for all } \mu \geqq 0.$$

C. *Summability Exercises for Integrals*

The following exercises are "continuous" analogues of exercises II.4.31—II.4.54. The reader can compare the solutions of the corresponding exercises in Section II.4 to construct solutions to the exercises in the present set. The reader should observe that except where otherwise noted Exercises 78—99 remain valid if the spaces M_0, M_1 and M_2 defined below are interpreted as spaces of $\mathfrak{X}$-valued functions, where $\mathfrak{X}$ is a B-space.

78 Let R denote the positive real axis, $\mathscr{B}$ the field of Borel subsets of R, and λ the Lebesgue measure of sets in $\mathscr{B}$. Let M_0 be the family of bounded λ-measurable functions f on R such that $\lim_{x\to\infty} f(x)$ exists. Show that M_0 is a B-space, if we put

$$|f| = \operatorname*{lub}_{0 \leqq x < \infty} |f(x)|.$$

79 Let $K(x, y)$ be a $\lambda \times \lambda$ measurable function on $R \times R$. Suppose that $K(x, y)$ is λ-integrable for each fixed x. Show that $\int_0^\infty K(x, y)f(y)dy$ is in M_0 whenever f is in M_0 if and only if

(a) $\int_0^\infty |K(x, y)|dy$ is bounded in x;

(b) $\lim_{x\to\infty} \int_0^A K(x, y)dy$ exists for all $0 < A < \infty$;

(c) For each $\varepsilon > 0$ and $A > 0$ there exists a $\delta > 0$ and $N > 0$ such that

$$\left|\int_E K(x, y)dy\right| < \varepsilon \quad \text{if } E \subseteqq (0, A],\ \lambda(E) < \delta,\ x > N;$$

(d) $\lim_{x\to\infty} \int_0^\infty K(x, y)dy$ exists.

Show that $\int_0^\infty K(x, y)f(y)dy$ has the same limit at $x = \infty$ as $f(x)$ if and only if we have

(b′) $\lim_{x\to\infty} \int_0^A K(x, y)dy = 0$ for all $0 < A < \infty$

and

(d′) $\lim_{x\to\infty} \int_0^\infty K(x, y)dy = 1.$

80 Let k be a non-negative function of a real variable which is Lebesgue integrable over every finite interval. Put $K(x) = \int_0^x k(y)dy$. Show that we have

$$\lim_{x\to\infty} f(x) = \lim_{x\to\infty} \frac{1}{K(x)} \int_0^x f(y)k(y)dy$$

for all f in M_0 if and only if $\lim_{x\to\infty} K(x) = \infty$.

81 Let k be an increasing non-negative function of a real variable which is integrable over every finite interval. Put $K(x) = \int_0^x k(y)dy$. Show that if $\lim_{x\to\infty} k(x)/K(x) = 0$, then

$$\lim_{x\to\infty} \frac{1}{K(x)} \int_0^x f(y)k(x-y)dy = \lim_{x\to\infty} f(x)$$

for all f in M_0.

82 Let μ be a non-negative measure defined for all Borel subsets of the positive real axis. Suppose that $\int_0^\infty e^{-s_0 t}\mu(dt) = \infty$, but that $\int_0^\infty e^{-st}\mu(dt) = N(s) < \infty$ for $s > s_0$. Put $N(f, s) = \int_0^\infty e^{-st}f(t)\mu(dt)$ for $s > s_0$ and f in M_0. Show that

$$\lim_{s\to s_0} N(s)^{-1}N(f, s) = \lim_{x\to\infty} f(x) \text{ for } f \text{ in } M_0.$$

83 Show that if f is in M_0

(a) (Cesàro) $\lim_{x\to\infty} \frac{1}{x} \int_0^x f(y)dy = \lim_{x\to\infty} f(x)$;

(b) if $\alpha > 0$, $\lim_{x\to\infty} \frac{\alpha}{x^\alpha} \int_0^x (x-y)^{\alpha-1} f(y)dy = \lim_{x\to\infty} f(x)$;

(c) (Abel) $\lim_{x\to 0} x \int_0^\infty e^{-xy} f(y)dy = \lim_{x\to\infty} f(x)$;

(d) (Gauss) $\lim_{x\to 0} \frac{2x}{\sqrt{\pi}} \int_0^\infty e^{-(xy)^2} f(y)dy = \lim_{x\to\infty} f(x)$;

(e) $\lim_{x\to 0} \frac{2}{\pi x} \int_0^\infty \left(\frac{\sin xy}{y}\right)^2 f(y)dy = \lim_{x\to\infty} f(x)$.

84 Let (R, Σ, λ) be as in Exercise 78. Let M_1 be the space of all bounded λ-measurable functions f defined on R such that $\lim_{x\to 0} f(x)$ exists. Putting

$$|f| = \operatorname*{lub}_{0 \leqq x < \infty} |f(x)|,$$

show that M_1 is a B-space.

85 Let $K(x, y)$ be a $\lambda \times \lambda$ measurable function on $R \times R$. Suppose that $K(x, y)$ is λ-integrable for each fixed x. Show that $\int_0^\infty K(x, y)f(y)dy$ is in M_1 whenever f is in M_1 if and only if

(a) $\int_0^\infty |K(x, y)|dy$ is bounded in x;

(b) $\lim_{x\to 0} \int_A^\infty K(x, y)dy$ exists for all $0 < A < \infty$;

(c) For each $\varepsilon > 0$ and $A > 0$ there exists a $\delta > 0$ and $N > 0$ such that $|\int_E K(x, y)dy| < \varepsilon$ if $E \subseteqq [A, \infty]$, $\lambda(E) < \delta$, and $x < N$;

(d) for each $\varepsilon > 0$ there exists an $M > 0$ and an $N > 0$ such that $\int_M^\infty |K(x, y)|dy < \varepsilon$ for $x < N$;

(e) $\lim_{x\to 0} \int_0^\infty K(x, y)dy$ exists.

Show that $f(x)$ has the same limit at $x = 0$ as $\int_0^\infty K(x, y)f(y)dy$ if and only if we have

(b′) $\lim_{x\to 0} \int_A^\infty K(x, y)dy = 0$ for $0 < A < \infty$;

(e′) $\lim_{x\to 0} \int_0^\infty K(x, y)dy = 1$.

86 Show that if f is in M_1 then

(a) $\lim_{x\to 0} \frac{1}{x} \int_0^x f(y)dy = \lim_{x\to 0} f(x)$;

(b) $\lim_{x\to 0} \frac{\alpha}{x^\alpha} \int_0^x (x-y)^{\alpha-1} f(y)dy = \lim_{x\to 0} f(x)$;

(c) $\lim_{x\to\infty} x \int_0^\infty e^{-xy} f(y)dy = \lim_{x\to 0} f(x)$;

(d) $\lim_{x\to\infty} \dfrac{2x}{\sqrt{\pi}} \int_0^\infty e^{-(xy)^2} f(y)dy = \lim_{x\to 0} f(x);$

(e) $\lim_{x\to\infty} \dfrac{2}{\pi x} \int_0^\infty \left(\dfrac{\sin xy}{y}\right)^2 f(y)dy = \lim_{x\to 0} f(x).$

87 Show that if f is in $L_\infty(R, \mathscr{B}, \lambda)$, then $z^2 \int_0^\infty e^{-zx} \int_0^x f(y)dy\, dx = z \int_0^\infty e^{-zy} f(y)dy$, and hence that if $\lim_{x\to\infty} (1/x) \int_0^x f(y)dy$ exists, then $\lim_{z\to 0} z \int_0^\infty e^{-zy} f(y)dy$ exists and has the same value. Show that if f is in $L_\infty(R, \mathscr{B}, \lambda)$, $\alpha > 0$, and $\lim_{x\to\infty} (\alpha/x^\alpha) \int_0^x (x-y)^{\alpha-1} f(y)dy$ exists, then $\lim_{z\to 0} z \int_0^\infty e^{-zy} f(y)dy$ exists and has the same value.

88 Show that if f is in $L_\infty(R, \mathscr{B}, \lambda)$ and if $\lim_{x\to\infty} xf(x) = 0$ then $\lim_{x\to 0} \int_0^\infty e^{-xy} f(y)dy = A$ if and only if $\lim_{x\to\infty} (1/x) \int_0^x (x-y)f(y)dy = A$. (Hint: Compare exercise II.4.54.)

89 Let M_2 be the set of all λ-measurable functions f such that,

(a) f is λ-integrable on each finite interval $[0, A]$

(b) $\lim_{A\to\infty} \int_0^A f(x)dx$ converges.

Put $|f| = \sup_{0\leq A<\infty} |\int_0^A f(x)dx| + \sum_{n=1}^\infty I_n/(2^n(1+I_n))$, where $I_n = \int_0^n |f(x)|dx$.

Show that M_2 is a separable F-space. If M_2 is interpreted as a space of $\mathfrak{X}$-valued functions where $\mathfrak{X}$ is a B-space then M_2 is separable if and only if $\mathfrak{X}$ is separable.

90 Show that if β is a function of bounded variation defined on R, then $f\beta$ is in M_2 if f is in M_2.

91 Show that if $\beta(x, y)$ is a function defined on $R \times R$ such that

(a) the variation $v(\beta(x, \cdot), (0, \infty))$ is uniformly bounded;

(b) $|\beta(x, 0+)|$ is uniformly bounded;

(c) $\lim_{x\to\infty} \int_0^A \beta(x, y)dy = A$ for each $0 < A < \infty$;

then

$$\lim_{x\to\infty} \lim_{A\to\infty} \int_0^A \beta(x, y)f(y)dy = \lim_{A\to\infty} \int_0^A f(y)dy, \qquad f \text{ in } M_2.$$

92 Show that if f is in M_2, then

(a) $\lim_{x\to 0} \lim_{A\to\infty} \int_0^A e^{-xy} f(y)dy = \lim_{A\to\infty} \int_0^A f(y)dy;$

(b) $\lim_{x\to 0} \lim_{A\to\infty} \int_0^A e^{-(xy)^2} f(y)dy = \lim_{A\to\infty} \int_0^A f(y)dy;$

(c) $\lim_{x\to\infty} x^{-\alpha} \int_0^x (x-y)^\alpha f(y)dy = \lim_{A\to\infty} \int_0^A f(y)dy, \qquad \alpha \geqq 0;$

(d) $$\lim_{x\to 0} \lim_{A\to\infty} \int_0^A \left(\frac{\sin xy}{xy}\right)^2 f(y)dy = \lim_{A\to\infty} \int_0^A f(y)dy.$$

93 If $\{\alpha\}$ is a set of indices,

(a) $\beta_\alpha(0)$ is uniformly bounded in α;

(b) $v(\beta_\alpha, [0, \infty))$ is uniformly bounded in α.

Then if f is in M_2,

$$\lim_{A\to\infty} \int_0^A f(x)\beta_\alpha(x)dx$$

exists uniformly in α.

94 Let $f(x)$ be a λ-measurable function which is λ-integrable over every finite interval $[0, A]$. Suppose that for some complex number $\lim_{A\to\infty} \int_0^A e^{-zx} f(x)dx$ exists. Then

$$\lim_{A\to\infty} \int_0^A e^{-z'x} f(x)dx$$

exists for $\mathscr{R}(z') > \mathscr{R}(z)$ and is analytic in this half-plane. (Hint: use Exercise 93).

95 If λ_n is a sequence of positive real numbers monotonically approaching ∞, and if the series (generalized Dirichlet series) $\sum_{n=0}^\infty a_n \lambda_n^{-z}$ converges for some complex number z_0, then the series converges in the half plane $\mathscr{R}(z) > \mathscr{R}(z_0)$ and is analytic there. (Hint. cf. Exercises 93 and 94.)

96 If f is integrable over any finite interval, if $(1/x) \int_0^x |f(t)|dt \leqq B$, and if $\lim_{x\to 0} \int_0^x f(t)dt = A$, then

$$\lim_{x\to\infty} \frac{8}{3\pi x^2} \int_0^\infty \frac{\sin^3 xt}{t^3} f(t)dt = A.$$

97 Let K be the Laplace transform of a function g such that $g(s)/s$ belongs to L_1 and $\int_0^\infty (g(s)/s)ds = 1$. If f is integrable over any finite interval, if

$$\lim_{t\to 0} t \int_0^\infty e^{-tz} f(z)dz = A,$$

and if

$$t \int_0^\infty e^{-tz} |f(z)| dz \leqq B,$$

then

$$\lim_{t \to 0} t \int_0^\infty K(tz) f(z) dz = A.$$

98 If f is integrable over any finite interval, if $\lim_{t \to 0} t \int_0^\infty e^{-tz} f(z) dz \to A$, and $t \int_0^\infty e^{-tz} |f(z)| dz \leqq B$, then

$$\lim_{t \to 0} \int_0^\infty \frac{at}{(tz+a)^2} f(z) dz = A$$

for $a > 0$.

99 Let $1 \leqq p < \infty$ and for each f in $L_p(R, \mathscr{B}, \lambda)$ let $f_y(x) = f(x+y)$. Then f_y is a continuous function of y with values in $L_p(R, \mathscr{B}, \lambda)$.

100 For every s in R, let K_s be an element of $L_1(R, \mathscr{B}, \lambda)$. Suppose that

(a) $\int_0^\infty |K_s(x)| dx \leqq M < \infty, \qquad s \in R;$

(b) $\lim_{s \to \infty} \int_A^\infty K_s(x) dx = 0, \qquad 0 < A < \infty;$

(c) for each $\varepsilon > 0$ there exists an $N \geqq 0$ and a $K > 0$ such that

$$\int_K^\infty |K_s(x)| dx < \varepsilon, \qquad s \geqq N;$$

(d) for each $\varepsilon > 0$ and $A > 0$ there exists a $\delta > 0$ and an $N > 0$ such that

$$\left| \int_E K_s(x) dx \right| < \varepsilon$$

if $E \subseteqq [A, \infty)$, $\lambda(E) < \delta$, and $s \geqq N$.

For $1 \leqq p < \infty$ show that if f is in $L_p(R, \mathscr{B}, \lambda)$, the function f_s defined by

$$f_s(x) = \int_0^\infty K_s(y) f(x+y) dy$$

is defined λ-almost everywhere for $s \in R$, belongs to $L_p(R, \mathscr{B}, \lambda)$ and that the limit relation

$$\lim_{s \to \infty} f_s = f$$

holds in the norm of L_p.

101 Let f be Lebesgue measurable on the whole real axis, and let $\int_{-\infty}^{+\infty} |f(x)|^p dx < \infty$, where $1 \leqq p < \infty$. Show that

(a) $\lim_{\varepsilon\to 0}\int_{-\infty}^{+\infty}\left|\left\{\frac{1}{\varepsilon}\int_{x}^{x+\varepsilon}f(y)dy\right\}-f(x)\right|^{p}dx=0;$

(b) $\lim_{\varepsilon\to 0}\int_{-\infty}^{+\infty}\left|\left\{\frac{\alpha}{\varepsilon^{\alpha}}\int_{x}^{x+\varepsilon}(\varepsilon+x-y)^{\alpha-1}f(y)dy\right\}-f(x)\right|^{p}dx=0;$

(c) $\lim_{n\to\infty}\int_{-\infty}^{+\infty}|\{n\int_0^\infty e^{-ny}f(x-y)dy\}-f(x)|^p dx=0;$

(d) $\lim_{n\to\infty}\int_{-\infty}^{+\infty}\left|\left\{\frac{n}{\sqrt{\pi}}\int_{-\infty}^{+\infty}e^{-(ny)^2}f(x-y)dy\right\}-f(x)\right|^{p}dx=0;$

(e) $\lim_{n\to\infty}\int_{-\infty}^{+\infty}\left|\left\{\frac{1}{\pi n}\int_{-\infty}^{+\infty}\left(\frac{\sin ny}{y}\right)^{2}f(x-y)dy\right\}-f(x)\right|^{p}dx=0.$

14. Exercises on Orthogonal Series and Analytic Functions

The following set of exercises is concerned with the application of linear space methods to the theory of orthogonal series. The most important special case of this theory is that of Fourier series in which we consider the expansion of arbitrary functions on the interval $[0, 2\pi]$ in series of the form $\sum_{-\infty}^{+\infty} a_n e^{inx}$. In view of the importance of this special case, (and more out of convenience than out of necessity) we shall take $[0, 2\pi]$ as a standard interval in all our problems. Then, by $C^{(k)}$ we shall mean $C^{(k)}[0, 2\pi]$; by AC and BV, $AC[0, 2\pi]$ and $BV[0, 2\pi]$; by L_p, $1 \leq p < \infty$, the space of functions f defined and Lebesgue measurable on the interval $[0, 2\pi]$ such that $|f|_p = \{\int_0 |f(x)|^p dx\}^{1/p} < \infty$, etc. We shall also find it convenient to use the symbols C^∞ for $\bigcap_{n=1}^{\infty} C^{(n)}$ and CBV for $C \cap BV$.

In the specific case of Fourier series it will be assumed that our functions are periodic: i.e., $f(0)=f(2\pi)$ for f in C, AC, BV, or CBV; $f^{(k)}(0) = f^{(k)}(2\pi)$, $0 \leq k \leq n$, for $f \in C^{(n)}$, and $f^{(k)}(0)=f^{(k)}(2\pi)$ for all $k \geq 0$ if $f \in C^{(\infty)}$. Of course, a restriction like $f(0) = f(2\pi)$ is meaningless in the space L_p.

1 Definition. By a closed orthonormal (c.o.n.) system we mean a two-sided sequence of functions ϕ_n, $n = 0, \pm 1, \pm 2, \ldots$ in $C^{(\infty)}$ such that

(i) linear combinations of ϕ_n are dense in every space $C^{(k)}$, $0 \leq k < \infty$.

(ii) $$\int_0^{2\pi} \phi_n(x)\overline{\phi_m(x)}dx = 0, \qquad n \neq m,$$

$$\int_0^{2\pi} |\phi_n(x)|^2 dx = 1, \qquad -\infty < n < \infty.$$

By f_n, the nth coefficient of a function $f \in L_1$, we understand the quantity $\int_0^{2\pi} f(x)\overline{\phi_n(x)}dx$. By $S_n f$, the nth partial sum of the orthogonal expansion of f, we understand the sum $\sum_{j=-n}^{+n} f_n \phi_n(x)$.

2 Let $\{\phi_n\}$ be a c.o.n. system. Let $E_n(x, y) = \sum_{-n}^{n} \phi_i(x)\overline{\phi_i(y)}$. Show that $S_n f$ is given by the formula

$$(S_n f)(x) = \int_0^{2\pi} E_n(x, y) f(y) dy,$$

and is a projection S_n in each of the spaces L_p, BV, CBV, AC, $C^{(k)}$, $1 \leqq p \leqq \infty$, $k = 0, 1, 2, \ldots, \infty$. Show that the range of S_n lies in $C^{(\infty)}$.

3 Show that $S_n \to I$ strongly in any one of the spaces $C^{(k)}$, $k < \infty$, AC, L_p, $1 \leqq p < \infty$ if and only if $|S_n| \leqq K$, where $|S_n|$ denotes the operator norm of S_n in the space. Show that in L_2, $S_n \to I$ strongly for all c.o.n. systems.

4 Show that for a given c.o.n. system, $S_n \to I$ strongly in L_1 if and only if $S_n \to I$ strongly in C.

5 Show that $S_n \to I$ strongly in L_p if and only if $S_n \to I$ strongly in L_q, $(1/p + 1/q = 1)$.

6 Show that $(S_n f)(x)$ converges to $f(x)$ uniformly on $[0, 2\pi]$ for all $f \in C^{(k)}$ if and only if the operator norm of S_n as an operator from $C^{(k)}$ into C is uniformly bounded as $n \to \infty$.

7 Show that $(S_n f)(x)$ converges to $f(x)$ uniformly on $[0, 2\pi]$ for each f in AC if and only if $|\int_0^y E_n(x, z)dz| \leqq M$ for all y and x.

8 Show that $S_n \to I$ strongly in C if and only if

$$\int_0^{2\pi} |E_n(x, y)| dy \leqq M, \qquad 0 \leqq x \leqq 2\pi, \ n > 0.$$

9 Show that $\int_0^{2\pi} \phi_n(x) f(x) dx \to 0$ as $n \to \infty$ for all f in L_1 if and only if the ϕ_n are uniformly bounded.

10 Show that $\int_0^{2\pi} \phi_n(x) f(x) dx \to 0$ for all f in L_p, $1 < p < \infty$, if and only if $\int_0^{2\pi} |\phi_n(x)|^q dx < M$, $1/p + 1/q = 1$.

11 Show that the functions $\{(2\pi)^{-1/2} e^{inx}\}_{-\infty}^{\infty}$ form a c.o.n. system.

12 Show that

$$E_n(x, y) = \frac{1}{2\pi} \frac{\sin\{[n+(1/2)](x-y)\}}{\sin(1/2)(x-y)}$$

for $\phi_n(x) = e^{inx}/\sqrt{2\pi}$.

For f in L_1,

$$f_n = (2\pi)^{-1/2} \int_0^{2\pi} f(x)e^{-inx}\,dx$$

is called the nth *Fourier coefficient of* f and the formal series

$$(2\pi)^{-1/2} \sum_{-\infty}^{\infty} f_n e^{inx}$$

is called the *Fourier series of* f.

13 Show that if $\{c_n\}$ is a sequence with $\sum_{-\infty}^{\infty} |c_n|^2 < \infty$, then there is some f in L_2 whose nth Fourier coefficient is c_n. Show that, conversely, if f is in L^2, and f_n is the nth Fourier coefficient of f, then $\sum_{-\infty}^{\infty} |f_n|^2 < \infty$.

14 The nth Fourier coefficient of a function in L_1 approaches zero as n approaches infinity.

15 Show that the Fourier series of some continuous function fails to converge uniformly on $[0, 2\pi]$. Show that there exists a continuous function whose Fourier series diverges at 0.

16 Show that there are functions in L_1 whose Fourier series diverges in L_1.

17 Show that if $f \in AC$ and hence if $f \in C^{(k)}$, $k \geqq 1$, then the Fourier series of f converges uniformly on $[0, 2\pi]$.

18 Show that there exists an f in AC such that $\sum_{n=0}^{\infty} f_n e^{inx}$ fails to converge uniformly.

19 Define the operator E by

$$E\left(\sum_{n=-N}^{N} a_n e^{inx}\right) = \sum_{n=0}^{N} a_n e^{inx}.$$

It will be shown later that for $p > 1$, E may be extended to be a bounded operator from L_p to L_p. Deduce from this that $S_n \to I$ strongly in L_p, $1 < p < \infty$.

20 Show that there are functions f in C having Fourier series $\sum_{n=-\infty}^{\infty} f_n e^{inx}$ such that no g in C has Fourier series $\sum_{n=0}^{\infty} f_n e^{inx}$.

The convergence of the partial sums $S_n f$ with respect to a given closed orthonormal system is said to be *localized* if for every f in L_1

which vanishes in a neighborhood of a point p the sequence $(S_n f)(x)$ converges to zero uniformly for x in some neighborhood of p.

21 Show that if $|\int_0^y E_n(x, z)dz| \leqq M$, then the convergence of $S_n f$ for a given c.o.n. system is localized if and only if $\max_{|x-y|\geqq\varepsilon} |E_n(x, y)| \leqq M_\varepsilon < \infty$ for each $\varepsilon > 0$.

22 Suppose that $(S_n f)(x) \to f(x)$ uniformly for every f in AC. Show that there exists a finite constant K such that for f in CBV,

$$|(S_n f)(x)| \leqq K(v(f, [0, 2\pi]) + \sup_{0\leqq x\leqq 2\pi} |f(x)|), \quad 0 \leqq x \leqq 2\pi.$$

23 Suppose that

(i) $(S_n f)(x) \to f(x)$ uniformly in x for f in AC.

(ii) The convergence of $S_n f$ is localized. Show that for any f in CBV, $(S_n f)(x) \to f(x)$ uniformly in x.

24 Suppose (i), (ii) of the last exercise. Show that if f is in BV then $(S_n f)(x) \to f(x)$ at each point x where f is continuous.

25 Suppose that $(S_n f)(x) \to (1/2)(f(x+)+f(x-))$ for each step function f. Show that under hypotheses (i) and (ii) of Exercise 23, $(S_n f)(x) \to (1/2)(f(x+)+f(x-))$ for every $f \in BV$ and each x.

26 Show that convergence of Fourier series is localized.

27 Show that for Fourier series

(a) $(S_n f)(x) \to f(x)$ uniformly in x for f in CBV.

(b) $(S_n f)(x) \to f(x)$ if f is in BV and if f is continuous at x.

(c) $(S_n f)(x) \to (1/2)(f(x+)+f(x-))$ for all x if f is in BV.

28 (Dini) Let ϕ_n be a c.o.n. system such that $(S_n f)(x) \to f(x)$ uniformly for all f in AC. Let $0 \leqq x_0 \leqq 2\pi$. Show that the condition $(y-x_0)^{-1}(f(y)-f(x_0)) \in L_1$ is sufficient to insure that $(S_n f)(x_0) \to f(x_0)$ if and only if there exists a finite constant M such that $|E_n(x_0, y)(x_0-y)| \leqq M$ for $0 \leqq y \leqq 2\pi$.

29 Show that the Fourier series of a function in L_1 converges to the value of the function at every point where the function is differentiable.

30 Show that if $f \in C^{(1)}$, the Fourier series of f converges absolutely. Show that this is no longer the case for each $f \in AC$.

31 Show that not every sequence $\{f_n, -\infty < n < \infty\}$ with $\lim_{|n|\to\infty} f_n = 0$, is the sequence of Fourier coefficients of a function in L_1.

32 Let $F^{(n)}$ be a sequence of functions on $[0, 2\pi]$ with absolutely

convergent Fourier series, let $g \in L_1$. Suppose that

(i) $\lim_{n\to\infty} F^{(n)}(x) = g(x)$ for all x;

(ii) $F^{(n)}(x) = \sum_{\nu=-\infty}^{\infty} F_\nu^{(n)} e^{i\nu x}$ and $\sum_{\nu=-\infty}^{\infty} |F_\nu^{(n)}| \leqq K$, $\quad n = 1, 2, \ldots$.

Show that then the Fourier series of $g(x)$ converges absolutely and that g is equivalent to a continuous function. (A. Beurling, "Sur les intégrales de Fourier absolument convergentes", 9th Scandinavian Mathematics Congress, 1938.)

33 Let (α, β) be any subinterval of $[0, 2\pi]$. Show that not every continuous function f on (α, β) can be represented on (α, β) as an absolutely convergent Fourier series $f(x) = \sum_{n=-\infty}^{\infty} c_n e^{inx}$, $\alpha \leqq x \leqq \beta$, $\sum_{n=-\infty}^{\infty}|c_n| < \infty$. (Hint: Use Exercise 32 and the open mapping principle to show that in the case excluded by our statement there exists a discontinuous function with an absolutely convergent Fourier series.)

In the following set of problems, we suppose that we have given a summation procedure which maps convergent series into convergent sequences (cf. Exercise II.4.46). That is, we have a collection of real numbers $\{\lambda_{mn}\}$, $1 \leqq m \leqq \infty$, $-\infty < n < +\infty$, such that if the series $\sum_{n=-\infty}^{+\infty} C_n$ converges to a limit C, then each of the series $\sum_{n=-\infty}^{+\infty} \lambda_{mn} C_n$, $m = 1, \ldots$, converges, and we have $\lim_{m\to\infty} \sum_{n=-\infty}^{+\infty} \lambda_{mn} C_m = C$. We make the additional standing hypotheses

(i) $\sum_{n=-\infty}^{+\infty} |\lambda_{mn}| < \infty$, $\quad m = 1, 2, \ldots$

(ii) The c.o.n. system ϕ_n is uniformly bounded.

This being the case, $\sum_{n=-\infty}^{+\infty} \lambda_{mn} f_n \phi_n(x)$ converges uniformly in x for all $m > 1$, for each f in L_1. (f_n is the nth coefficient of $f : f_n = \int_0^{2\pi} f(x)\overline{\phi_n(x)}dx$). We shall write the sum of this series as $T_m f$, i.e., we put

$$(T_m f)(x) = \sum_{n=-\infty}^{+\infty} \lambda_{mn} f_n \phi_n(x), \qquad m \geqq 1.$$

34 Show that there exists a continuous function $K_m(x, y)$ of two real variables such that

$$(T_m f)(x) = \int_0^{2\pi} f(y) K_m(x, y) dy, \qquad m \geqq 1, f \in L_1.$$

35 Show that $T_m f \to f$ strongly as $m \to \infty$ in any one of the spaces $C^{(k)}$, AC or L_p if and only if T_m maps the given space into itself

and $|T_m| < K$, where $|T_m|$ is the operator norm of T_m regarded as an operator in the given space.

36 Show that convergence of $T_m f$ in the space C or in the space L_1 is equivalent to the statement that

$$\int_0^{2\pi} |K_m(x, y)| dy \leqq M, \qquad 0 \leqq x \leqq 2\pi.$$

37 Show that if $\phi_0(x) = (2\pi)^{-1/2}$ and if $K_m(x, y) \geqq 0$ for all m, then T_m converges in C and L_1.

38 Given a bounded sequence of numbers $\{a_n\}$, $-\infty < n < +\infty$. Under the hypotheses of Exercise 37, show that there exists f in C with $a_n = \int_0^{2\pi} f(x)\overline{\phi_n(x)}dx$ if and only if the functions $\sum_{n=-\infty}^{\infty} \lambda_{mn} a_n \phi_n(x)$, $m \geqq 1$, are uniformly bounded and equicontinuous.

39 Let $\{a_n\}$, $-\infty < n < +\infty$, be a bounded sequence of numbers. Show that under the hypotheses of Exercise 37 there exists a complex-valued regular measure μ in C^* with $a_n = \int_0^{2\pi} \overline{\phi_n(y)}\mu(dy)$ if and only if $\int_0^{2\pi} |\sum_{n=-\infty}^{\infty} \lambda_{mn} a_n \phi_n(x)| dx < K$, for all $m \geqq 1$. Show that there exists an f in L_1 with $a_n = \int_0^{2\pi} \overline{\phi_n(y)} f(y) dy$ if and only if in addition to the preceding hypotheses, $\lim_{\lambda(E)\to 0} \int_E |\sum_{n=-\infty}^{\infty} \lambda_{mn} a_n \phi_n(x)| dx = 0$, uniformly in n, E denoting a Borel set and $\lambda(E)$ its Lebesgue measure.

40 Let $\{a_n\}$, $-\infty < n < +\infty$, be a bounded sequence of numbers and let $1 < p < \infty$. It is assumed that the summation procedure is such that $T_m f$ converges to f in L_p for every f in L_p. Show that there exists an $f \in L_p$ such that $a_n = \int_0^{2\pi} f(x)\phi_n(x)dx$ if and only if there exists an $M < \infty$ such that

$$\int_0^{2\pi} |\sum_{n=-\infty}^{\infty} \lambda_{mn} a_n \phi_n(x)|^p dx \leqq M, \qquad m \geqq 1.$$

41 Let $\{a_n\}$, $-\infty < n < +\infty$, be a bounded sequence of numbers and let $1 < p < \infty$. Show that there exists an f in L_p such that $a_n = \int_0^{2\pi} f(x) e^{-inx} dx$ if and only if there exists an $M < \infty$ such that

$$\int_0^{2\pi} |\sum_{n=-m}^{+m} a_n e^{inx}|^p dx < M, \qquad \text{for } m = 1, 2, \ldots.$$

42 Let $\{a_n\}$, $-\infty < n < +\infty$, be a bounded sequence of numbers. Let $1 < p < \infty$ and let $1/p + 1/q = 1$. Show that there exists an $f \in L_p$ such that $a_n = \int_0^{2\pi} f(x) e^{-inx} dx$, if and only if the series

$\sum_{n=-\infty}^{+\infty} a_n b_n$ converges for each sequence $\{b_n\}$ of the form $b_n = \int_0^{2\pi} g(x) e^{-inx} dx$, with $g \in L_q$.

43 Let $\{a_n\}$, $-\infty < n < +\infty$, be a bounded sequence of numbers. Show that, under the hypotheses of Exercise 40, there exists an $f \in L_p$ such that $a_n = \int_0^{2\pi} f(x)\overline{\phi_n(x)}dx$ if and only if $\lim_{m\to\infty} \sum_{n=-\infty}^{+\infty} \lambda_{mn} a_n b_n$ exists for each sequence $\{b_n\}$ of the form $b_n = \int_0^{2\pi} \phi_n(x) g(x) dx$, with $g \in L_q$ where $1/p + 1/q = 1$.

44 (Cesaro Summability) Let

$$\begin{aligned} \lambda_{mn} &= 1-(|n|/m), \quad n = 0, \pm 1, \pm 2, \ldots, \pm(m-1) \\ \lambda_{mn} &= 0, \qquad\qquad\quad |n| \geqq m. \end{aligned}$$

Find the corresponding kernels $K_m(x, y)$. Show that for $f \in L_1$,

$$T_m f = \frac{1}{m} \sum_{j=0}^{m-1} S_j f$$

where S_j and T_m have their earlier meanings.

45 Show that if $\{\phi_n\} = \{(2\pi)^{-1/2} e^{inx}\}$ and $\{\lambda_{mn}\}$ is as in Exercise 44, then

(a) $K_m(x, y) \geqq 0$

(b) $T_m f \to f$ in C and in L_1.

46 Let $\{\phi_n\}$ be any uniformly bounded c.o.n. system. For $f \in L_1$, set $f_n = \int_0^{2\pi} f(y)\phi_n(y)dy$, and set

$$(T_r f)(x) = \sum_{n=-\infty}^{\infty} r^{|n|} f_n \phi_n(x), \qquad 0 < r < 1.$$

Find a kernel $P_r(x, y)$ so that

$$(T_r f)(x) = \int_0^{2\pi} f(y) P_r(x, y) dy.$$

47 (Poisson-Summability). Let $\{\phi_n\} = \{(2\pi)^{-1/2} e^{inx}\}$. Show that here the kernel of Exercise 46 is

$$P_r(x, y) = \frac{1}{2\pi} \frac{1-r^2}{1+r^2-2r\cos(x-y)}.$$

Show that $T_r f \to f$ in C and L_1.

48 For $f \in L_1$ and $-\infty < t < +\infty$ define $U_t f$ by the formula $(U_t f)(x) = f(x+t)$, $0 \leqq x \leqq 2\pi$ (we define $f(x)$ for $x < 0$ by periodicity; i.e., by requiring that $f(x) = f(x+2\pi)$, $-\infty < x < +\infty$).

Show that if $f \in L_p$ or AC or $C^{(n)}$ (being in each case subject to the periodicity requirements set down in the discussion preceding Definition 1) $U_t f$ is a vector-valued function of t (with values lying in the given space L_p or AC or $C^{(n)}$, as the case may be) which varies continuously with t, and that $|U_t f| = |f|$.

49 Show that in case $\phi_n(x) = (2\pi)^{-1/2}e^{inx}$, the operator T_m is given by the integral

$$T_m f = \int_0^{2\pi} (U_t f)K_m(t)dt,$$

where $K_m(t)$ is $K_m(0, t)$ in the notation of Exercise 34. (Here f is in L_p, AC, or $C^{(n)}$).

50 Show that in case $\phi_n(x) = (2\pi)^{-1/2}e^{inx}$, $T_m f \to f$ in the norm of L_p (or AC, or $C^{(n)}$) for $f \in L_p$ (or AC, or $C^{(n)}$) if $T_m f \to f$ in the norm of C for each $f \in C$. (Hint: Use Exercises 35 and 49.)

51 Show that if T_m is the operator of Exercise 45, $(T_m f)(x) \to f(x)$ at every point of the Lebesgue set of a function $f \in L_1$. Show that the same holds for the operator T_r of Exercise 47. (Hint. Use Theorem III.12.11.)

52 (Hardy) Let $f(z)$ be analytic in $|z| < 1$. Show that for $p \geqq 1$,

$$N_r(f, p) = \left\{ \int_0^{2\pi} |f(re^{i\theta})|^p d\theta \right\}^{1/p}$$

is an increasing function of r, $0 \leqq r \leqq 1$. (Hint. Apply the maximum modulus principle to the vector-valued function F defined by $(F(z))(\theta) = f(ze^{i\theta})$.)

53 Let the class H_p consist of all functions $f(z)$ analytic in $|z| < 1$ such that

$$N(f, p) = \sup_{r<1} N_r(f, p) < \infty,$$

where $N_r(f, p)$ is defined as in Exercise 52. Assume $p > 1$. Show that if $f(z) = \sum_{n=0}^{\infty} f_n z^n \in H_p$, then there exists a function $\tilde{f}$ in L_p with

$$\begin{aligned} \int_0^{2\pi} \tilde{f}(\theta)e^{-in\theta} d\theta &= 0, \qquad && n < 0 \\ &= 2\pi f_n, && n \geqq 0. \end{aligned}$$

54 Suppose that F is in L_p for some $p > 1$ and is such that $\int_0^{2\pi} F(\theta)e^{in\theta} d\theta = 0$, for all $n < 0$. Let $f(z) = \sum_{n=0}^{\infty} F_n z^n$ where $F_n = (1/2\pi)\int_0^{2\pi} F(\theta)e^{-in\theta} d\theta$ for $n \geqq 0$. Show that

(a) $f(re^{i\theta}) = \frac{1}{2\pi}\int_0^{2\pi} F(t)P_r(t, \theta)dt,$

(b) $\{\int_0^{2\pi}|f(re^{i\theta})|^p d\theta\}^{1/p} \leqq |F|_p.$

Show that the map: $f \to \tilde{f}$ defined in Exercise 53 maps H_p in a linear one-one manner onto the closed subspace of L_p consisting of those F all of whose negative Fourier coefficients vanish.

55 Using the notations of Exercises 53 and 54, show that if $f \in H_p$ and if $(U_r f)(\theta) = f(re^{i\theta})$, then $|U_r f - \tilde{f}|_p \to 0$ as $r \to 1$. $(p > 1)$.

56 Using the notations of Exercises 53 and 54, show that if $f \in H_p$, $p > 1$, then for $|z| < 1$,

$$f(z) = \frac{1}{2\pi i}\int_{|\xi|=1} \frac{\tilde{f}(\xi)d\xi}{\xi - z}$$

57 Show that if we norm H_p by setting $|f| = \sup_{r<1}\{\int_0^{2\pi}|f(re^{i\theta})|^p d\theta\}^{1/p}$, H_p becomes a reflexive Banach space.

58 Using Exercise 19, show that if $F \in L_p$, $G \in L_q$, $1/p + 1/q = 1$, then $\sum_{n=-\infty}^{\infty} F_n \bar{G}_n$ converges and

$$\frac{1}{2\pi}\int_0^{2\pi} F(t)\overline{G(t)}dt = \sum_{n=-\infty}^{\infty} F_n \bar{G}_n$$

where $F_n = (2\pi)^{-1}\int_0^{2\pi} F(t)e^{-int}dt$, $G_n = (2\pi)^{-1}\int_0^{2\pi} G(t)e^{-int}dt$.

59 Let $\{a_n\}_{-\infty}^{\infty}$ be a bounded sequence of numbers. Show that if

$$u(r, \theta) = \sum_{n=-\infty}^{\infty} a_n r^{|n|} e^{in\theta} \geqq 0, \qquad r > 1, \qquad 0 \leqq \theta < 2\pi,$$

then there exists a positive measure μ in C^* with $\int_0^{2\pi} e^{-int} d\mu(t) = a_n$, $+\infty > n > -\infty$. (Hint: cf. Exercise 39.)

60 If $P_r(t, \theta)$ is defined as in Exercise 47, and $u(r, \theta)$ and μ as in Exercise 59, show that

$$u(r, \theta) = 2\pi \int_0^{2\pi} P_r(t, \theta)d\mu(t), \qquad r < 1, \qquad 0 \leqq \theta < 2\pi.$$

61 (Herglotz) Let $f(z)$ be analytic in $|z| < 1$ and $\mathscr{R}f(z) \geqq 0$. Show that there exists a positive measure $\mu \in C^*$ and a real constant v_0 such that

$$f(z) = \int_0^{2\pi} \frac{e^{it}+z}{e^{it}-z}\,\mu(dt) + iv_0, \qquad |z| < 1.$$

(Hint. Use previous exercise; in $|z| < 1$ an analytic function is determined by its real part up to an imaginary constant.)

62 Let $\{f_n\}$ be a sequence of monotone increasing functions which is uniformly bounded. Let f be a monotone increasing function and assume that

$$\lim_{k\to\infty} \int_0^{2\pi} e^{-inx} f_k(x)dx = \int_0^{2\pi} e^{-inx} f(x)dx,$$

for all n. Show that $f_k(x)$ converges to $f(x)$ at each point of continuity of f in $(0, 2\pi)$.

63 Let $\{\phi_n\}$ be a c.o.n. system and for f in L_1 set $a_n(f) = \int_0^{2\pi} \phi_n(x)f(x)dx$. A set E of integers is called *p, q lacunary* if $f \in L_p$ and $f_n = 0$ for $n \notin E$ implies $\{f_n\}_{-\infty}^{\infty} \in l_q$. Let $1 \leqq p < \infty$ and $1 \leqq q < \infty$. Let p', q' be so that $1/p+1/p' = 1$, $1/q+1/q' = 1$. Show that E is p, q lacunary if and only if whenever $\{\alpha_n\}_{-\infty}^{\infty}$ is in $l_{q'}$ there exists an f in $L_{p'}$ such that $\alpha_n = f_n$ for $n \in E$.

Let $\{\phi_n\}$ be a uniformly bounded c.o.n. system and let $\{\lambda_n\}$, $-\infty < n < +\infty$, be a sequence of numbers. Let $\mathfrak{A}$ be any of the function spaces on $[0, 2\pi]$ which we have been considering. We say that $\{\lambda_n\}$ is a *factor-sequence of type* $(\mathfrak{A}, \mathfrak{A})$, or briefly $\{\lambda_n\}$ is $(\mathfrak{A}, \mathfrak{A})$, if for each $f \in \mathfrak{A}$ there exists some $f^* \in \mathfrak{A}$ with $f_n^* = \lambda_n f_n$. If $rca = C^*$ denotes the space of regular measures, we say $\{\lambda_n\}$ is (rca, rca) if for each $\mu \in rca$, there exists μ^* in rca with

$$\int_0^{2\pi} \overline{\phi_n(x)}\mu^*(dx) = \lambda_n \int_0^{2\pi} \overline{\phi_n(x)}\mu(dx).$$

64 Show that if each λ_n is real, the sequence $\{\lambda_n\}$ is (L_1, L_1) if and only if it is (L_∞, L_∞), (C, C) or (rca, rca). (Hint: $C \subset L_\infty$, rca is the conjugate space of C, $L_1 \subset rca$, L_∞ is the conjugate space of L_1.)

65 Let $\{\lambda_n\}$ be a given sequence. Show that if

$$\int_0^{2\pi} \left| \sum_{j=-(n-1)}^{n-1} \lambda_j \left(1 - \frac{|j|}{n}\right) \phi_j(x)\overline{\phi_j(y)} \right| dy$$

is bounded for all n, x, then $\{\lambda_n\}$ is a factor sequence of type (C, C). Show that if we have Cesàro summability in C, (i.e., if the operators T_m of Exercise 44 converge strongly to I) then the condition is also necessary.

66 Show that a factor sequence $\{\lambda_n\}$ is in (C, C) with respect to the system $(2\pi)^{-1/2}e^{inx}$ if and only if there exists a regular measure μ with $\lambda_n = \int_0^{2\pi} e^{-inx}\mu(dx)$.

Multiple Fourier Series

Let A denote the Cartesian product, with its product topology, of n replicas of the interval $[0, 2\pi]$. The symbol L_p will be used for $L_p(A, \mathscr{B}, \lambda)$ where $\mathscr{B}$ is the field of Borel sets in A and λ is the Lebesgue measure on $\mathscr{B}$. In the next exercise a notion of a closed orthonormal system, which is somewhat less restrictive than that introduced in Definition 1, will be used. It is given in the following definition.

67 DEFINITION. By a *closed orthonormal* (c.o.n.) *system* of functions on A will be understood an indexed set $\{\phi_\alpha\}$ of functions in $C(A)$ such that

(i) linear combinations of the functions ϕ_α are dense in $C(A)$;
(ii) $\int_A \phi_\alpha(x_1, \ldots, x_n)\overline{\phi_\beta(x_1, \ldots, x_n)}dx_1 \ldots dx_n = 0, \qquad \alpha \neq \beta;$
(iii) $\int_A |\phi_\alpha(x_1, \ldots, x_n)|^2 dx_1 \ldots dx_n = 1, \qquad$ for all α.

68 Let $\{\phi_m\}$ be a c.o.n. system on $[0, 2\pi]$ (in the sense of Definition 1). Show that the set of all products of the form

$$\phi_{m_1}(x_1)\phi_{m_2}(x_2) \ldots \phi_{m_n}(x_n)$$

form a c.o.n. system of functions on A (in the sense of Definition 67).

A *multiple Fourier series* is a series whose general term is a constant multiple of the function $\exp\{i(m_1x_1+\ldots+m_nx_n)\}$. It will be assumed in the following exercises on multiplicative Fourier series that $C(A)$ is the space of all scalar valued continuous functions f on A which are *multiply periodic* in the sense that

$$f(0, x_2, \ldots, x_n) = f(2\pi, x_2, \ldots, x_n); \ldots; f(x_1, \ldots, x_{n-1}, 0) = \\ = f(x_1, \ldots, x_{n-1}, 2\pi).$$

69 Show that the family of functions $\exp\{i(m_1x_1+\ldots+m_nx_n)\}$ with $-\infty < m_j < \infty$, $j = 1, \ldots, n$ is dense in $C(A)$, as well as in $L_p(A)$ for each p with $1 \leq p < \infty$.

70 Let K be a continuous function on $A \times A$ of the form

$$K(x_1, \ldots, x_n; y_1, \ldots, y_n) = K_1(x_1, y_1) \ldots K_n(x_n, y_n),$$

where $K_1, \ldots, K_n$ are continuous functions on $[0, 2\pi] \times [0, 2\pi]$. Express the norm of the operator $\hat{K}$ in $C(A)$ defined by the equation

$$(\hat{K}f)(y_1, \ldots, y_n) = \int_A K(y_1, \ldots, y_n; x_1, \ldots, x_n)f(x_1, \ldots, x_n)dx_1 \ldots dx_n$$

in terms of the norms of the operators $\hat{K}_1, \ldots, \hat{K}_n$ in $C[0, 2\pi]$ defined similarly in terms of the kernels $K_1, \ldots, K_n$. How are the corresponding norms of $\hat{K}$ and $\hat{K}_1, \ldots, \hat{K}_n$ as operators in $L_p(A)$ and $L_p(0, 2\pi)$ related?

71 For each f in $L_1(A)$ let

$$f_{m_1} \cdots {}_{m_n} = \int_A f(x_1, \ldots, x_n) \exp\{-i(m_1x_1 + \ldots + m_nx_n)\}dx_1 \ldots dx_n.$$

If f is in $L_p(A)$ with $1 < p < \infty$, then as $R_1 \to \infty, \ldots, R_n \to \infty$,

$$f = \lim \left(\frac{1}{2\pi}\right)^n \sum_{m_1=-R_1}^{R_1} \cdots \sum_{m_n=-R_n}^{R_n} f_{m_1 \ldots m_n} \exp\{i(m_1x_1 + \ldots + m_nx_n)\}$$

in the norm of $L_p(A)$. This is not true if $p = 1$.

72 With the notation introduced in Exercise 71 we have, as $r_1 \to 1, \ldots, r_n \to 1$,

$$f = \lim \left(\frac{1}{2\pi}\right)^n \sum_{m_1=-\infty}^{\infty} \cdots \sum_{m_n=-\infty}^{\infty} f_{m_1 \ldots m_n} r_1^{|m_1|} \ldots r_n^{|m_n|} \exp\{i(m_1x_1 + \ldots + m_nx_n)\}$$

where for each f in $C(A)$ the limit exists in the norm of $C(A)$ and for each f in $L_p(A)$ with $1 \leq p < \infty$ the limit exists in the norm of $L_p(A)$.

73 With the notation introduced in Exercise 71 we have, as $N_1 \to \infty, \ldots, N_n \to \infty$,

$$f = \lim \left(\frac{1}{2\pi}\right)^n \sum_{m_1=-N_1}^{N_1} \cdots \sum_{m_n=-N_n}^{N_n} \left(1 - \frac{|m_1|}{N_1}\right) \cdots \left(1 - \frac{|m_n|}{N_n}\right) f_{m_1} \cdots {}_{m_n} \exp\{i(m_1x_1 + \ldots + m_nx_n)\}$$

where for each f in $C(A)$ the limit exists in the norm of $C(A)$ and for each f in $L_p(A)$ with $1 \leq p < \infty$ the limit exists in the norm of $L_p(A)$.

Extremal Methods for Polynomials and Classes H_p

74 Let $\mathfrak{X}$ be a finite-dimensional subspace of a B-space $\mathfrak{Y}$, and let f be a linear functional on $\mathfrak{X}$ whose norm is $|f|$. Then there exists an x_0 in $\mathfrak{X}$, and a linear functional f^* on $\mathfrak{Y}$, such that

(a) $f^*(x) = f(x), \quad x \in \mathfrak{X}$.

(b) $f^*(x_0) = |f^*| = |f|$.

(c) $|x_0| = 1$.

75 Let P_n be the space of all polynomials of order not more

than n. Regard P_n as a subspace of $C[-1, 1]$. Let f be a linear functional on P_n. Then there exists an x_0 in P_n and a measure μ in $rca[-1, 1]$ such that

(a) $f(x) = \int_{-1}^{1} x(t)\mu(dt), \qquad x \in P_n;$

(b) $\max_{-1 \leqq t \leqq 1} |x_0(t)| = 1;$

(c) if C is the set of t in the interval $-1 \leqq t \leqq 1$ where $|x_0(t)|=1$, then $v(\mu, C') = 0$.

Consequently, unless $x_0(t)$ is a constant of absolute value 1, C is a set of at most $n+1$ points $-1 \leqq t_1 \leqq t_2 \ldots < t_k \leqq 1$, and there are constants $c_1, \ldots, c_k$ with $\sum_{i=1}^{k} |c_i| = |f|$, and in terms of which we may write an "interpolation formula"

$$f(x) = \sum_{j=1}^{k} c_j x(t_j), \qquad x \in P_n.$$

76 Let τ_n be a real polynomial of degree n, not identically constant, such that

$$\max_{-1 \leqq t \leqq 1} |\tau_n(t)| = 1,$$

and such that $|\tau_n(t)|$ takes on its maximum value 1 at $n+1$ distinct points. Then τ_n satisfies the differential equation

$$(t^2-1)(\tau_n'(t))^2 = n^2(\tau_n(t)^2-1),$$

and is consequently identical (up to sign) with the nth *Tchebicheff polynomial for the interval* $[-1, +1]$, i.e.,

$$\tau_n(t) = \cos(n \operatorname{arc} \cos t).$$

77 Let P_n and f be as in Exercise 75, and τ_n be as in Exercise 76. Suppose that $\max_{x \in P_n |x| \leqq 1} |f(x)|$ is not reached when x is the constant function $x(t) \equiv 1$. Then unless there exist k points $t_1 < \ldots < t_k$, with $k \leqq n$, such that the values $f(x)$, $x \in P_n$ are determined by the values $x(t_1), \ldots, x(t_k)$, we have

$$\max_{x \in P_n, |x| \leqq 1} |f(x)| = |f(\tau_n)|.$$

78 If $a_n(x)$ is the leading coefficient of the polynomial x of degree n, then

$$|a_n(x)| \leqq 2^{n-1} \max_{-1 \leqq t \leqq 1} |x(t)|,$$

and the equality is attained only by scalar multiples of the Tchebicheff polynomial τ_n.

79 The inequality

$$|x'(1)| \leqq n^2 \max_{-1 \leqq t \leqq 1} |x(t)|$$

holds for each x in P_n.

80 If n is odd, the inequality

$$|x'(0)| \leqq n \max_{-1 \leqq t \leqq +1} |x(t)|$$

holds for each x in P_n. If n is even, the inequality

$$|x'(0)| \leqq (n-1) \max_{-1 \leqq t \leqq +1} |x(t)|$$

holds for each x in P_n.

(Hint: The function x_0 of Exercise 75 may be taken in the present case to be odd.)

81 (Bernstein) The inequality

$$|x'(1)| \leqq n \max_{|z| \leqq 1} |x(z)|$$

holds for every polynomial of degree n.

(Hint: The function x_0 of Exercise 75 is of constant modulus in the present case.)

82 (H. S. Shapiro) Let Π_n denote the space of polynomials of degree n in the variable z, with the norm

$$|x| = \max_{|z| \leqq 1} |x(z)|.$$

Let f be a linear functional on Π_n, and define a linear mapping F by putting

$$F(x)(\zeta) = f(x_\zeta),$$

where $x_\zeta(z) = x(\zeta z)$. Then F maps Π_n into itself, and has norm equal to $|f|$. More generally, for $1 \leqq p < \infty$ and x in Π_n,

$$\left\{\frac{1}{2\pi}\int_0^{2\pi} |F(x)(e^{i\theta})|^p \, d\theta\right\}^{1/p} \leqq |f| \left\{\frac{1}{2\pi}\int_0^{2\pi} |x(e^{i\theta})|^p \, d\theta\right\}^{i/p}.$$

(Hint: Find a measure like that of Exercise 75.)

83 If $1 \leqq p < \infty$ and if x is in Π_n, then

$$\left\{\frac{1}{2\pi}\int_0^{2\pi} |x'(e^{i\theta})|^p \, d\theta\right\}^{1/p} \leqq n \left\{\frac{1}{2\pi}\int_0^{2\pi} |x(e^{i\theta})|^p \, d\theta\right\}^{1/p}.$$

84 Let (S, Σ, μ) be a positive measure space, and let $L_p = L_p(S, \Sigma, \mu)$ with $p \geqq 1$. Let f and g be two elements of L_p such that $|f+\lambda g| \geqq |f|$ for each scalar λ. Then

$$\int_S |f(s)|^{p-1} \operatorname{sgn}(f(s))\overline{g(s)}\mu(ds) = 0,$$

where sgn z is defined by the equations

$$\begin{aligned} \operatorname{sgn} z &= \frac{z}{|z|}, && z \neq 0 \\ &= 0, && z = 0. \end{aligned}$$

85 (Blaschke Product) Let $p > 1$ and let f be a function in H_p which does not vanish identically. Then there is a function g in H_∞ with $|g(e^{i\theta})| = 1$ for almost all θ and which has the same zeros as f in the interior of the unit disc. (Hint: Suppose that $f(z_0) \neq 0$. From among all functions h with $f(z_0) = h(z_0)$ and for which $h(z) = 0$ if $f(z) = 0$ choose one with smallest norm.)

86 Let p, f be as in Exercise 85. Then $f(e^{i\theta}) \neq 0$ for almost all θ.

87 Let $p > 1$ and let f be a function in H_p. Then there exists a function g in H_∞ such that $g(e^{i\theta}) = 1$ for almost all θ,

$$\frac{f}{g} \in H_p, \qquad \left|\frac{f}{g}\right| = |f|,$$

and such that f/g has no zeros.

(Hint: Generalize the argument of Exercise 85 to apply to zeros on the boundary of the unit disc.)

88 Show that Exercise 87 is valid even if $p = 1$.

89 Every function f in H_1 can be written as a product gh, where g and h are in H_2. (Hint: Use Exercise 88.)

90 Show that Exercises 55, 54, and 56 remain valid for $p = 1$. (Hint: Use Exercise 89.)

91 (F. Riesz-M. Riesz). A Borel measure μ on the interval $[0, 2\pi]$ such that

$$\int_0^{2\pi} e^{in\theta}\mu(d\theta) = 0, \qquad n \geqq 0,$$

is absolutely continuous.

(Hint: Use Exercise 90 and the method of Exercise 60.)

15. A Tabulation of Results

In this section we will give a summary of what is known in the way of an answer to the eight questions of Section 1 as applied to the twenty-eight spaces listed in Section 2. The information is presented in Table IV.A (pp. 374–379) cross-indexing questions and spaces, and giving references to the appropriate theorems and exercises in this chapter, in previous chapters, or, in a small number of cases, in chapters to follow.

Definite "yes or no" statements on the problems of weak completeness and reflexivity are made in the table; these should be understood as covering the "general cases" of the spaces under consideration. For instance, in writing "no" in the table as the answer to the question "Is $B(S)$ weakly complete?", we have in mind the fact, stated in Exercise 13.5, that $B(S)$ is weakly complete if and only if S is finite.

Symbols like (**F3**) in parentheses refer to footnotes collected immediately after the table.

16. Notes and Remarks

Finite dimensional and Hilbert spaces. The notion of a finite dimensional space, is of course, an algebraization of ordinary geometrical concepts. The axioms for finite dimensional Euclidean space were stated explicitly, e.g., in Weyl [3; p. 15—25]. The study of norms other than the Euclidean norm, is primarily due to Minkowski. The spaces l_2 and L_2 were studied in detail by Hilbert and others, but the abstract axiomatization of Hilbert space is due to von Neumann [8; p. 15—17], [7] in the separable case, and Löwig [1] and Rellich [3] in general.

Tychonoff [1; p. 769] proved that any finite dimensional topological linear space is equivalent to an Euclidean space. This, in particular, implies the completeness. Theorem 3.5, which characterizes locally compact B-spaces is due to F. Riesz [4].

The "Schwarz inequality" (Theorem 4.1) goes back very far. For $n = 3$, it is a consequence of a well-known identity due to Lagrange [1; p. 662—3] proved in 1773. For a finite sum, it was proved in 1821 by Cauchy [2; p. 373]. It was proved for integrals in 1859 by Buniakowsky [1; p.4] and in 1885 by Schwarz [1; p. 251]. Of course, it is a special case of the Hölder inequality (III.3.2) obtained for sums by Hölder [2; p. 44] and for integrals by F. Riesz [2; p. 456].

Lemma 4.2 is due to F. Riesz [8; p. 36] and was also used by Sz.-Nagy [5]. A similar argument may be applied to obtain this result in any uniformly convex B-space. This generalizes and abstracts a result proved for closed linear manifolds in $L_2[0, 1]$ by E. Fischer [2].

The fact that a linear manifold which is not dense in the entire space has a non-zero orthogonal complement (proved in 4.4) was proved without the assumption of separability by F. Riesz [8]. His proof follows the lines of an argument employed by Levi [1; Sec. 7] in a study of the Dirichlet problem.

The fact that every continuous linear functional on $L_2[0, 1]$ arises from an element of L_2 was proved independently by Fréchet [4] [5; p. 439] and F. Riesz [9]. The proof of Theorem 4.5 given here is due to F. Riesz [8]. Corollary 4.7 was proved for $L_p[0, 1]$ by F. Riesz [2; p. 466].

The circle of ideas in 4.9—4.13 express in abstract form the results usually known as the Riesz-Fischer theorem, which was proved independently by E. Fischer [1] and F. Riesz [5], although their names are also given to the theorem asserting that L_2 is complete. The equation in Theorem 4.13 is classically known as "Parseval's equality", and the related inequality in the proof of Theorem 4.10 is called "Bessel's inequality".

The fact that all orthonormal bases of a Hilbert space have the same cardinality was shown by Löwig [1; p. 31] and Rellich [3; p. 355]. Theorem 4.16 is due to Löwig [1; p. 27].

The spaces $B(S, \Sigma)$ *and* $B(S)$. The results in 5.1 and 5.3 were proved independently by Hildebrandt [3] and by Fichtenholz and Kantorovitch [1]. (See also Yosida and Hewitt [1].) Lemma 5.4 and Corollary 5.5 are due to Phillips [3; p. 526]. Theorem 5.6 was proved by Veress [1], [2; p. 184].

The space $C(S)$. Since the results in Section 6 fall into several groupings, we will break our comments into various subheadings.

The Riesz representation theorem (6.1—6.3). This name is frequently given to Theorems 2 and 3, since this fundamental integral representation for a continuous linear functional on $C[0, 1]$ was first discovered by F. Riesz [7]. (Text continues on p. 380.)

TABLE IV.A

Space $\mathfrak{X}$	1. E^n	2. l_p^n	3. l_∞^n	4.A. l_p $1 < p < \infty$	4.B. l_1
Conjugate Space $\mathfrak{X}^*$	3.9	3.9	3.9	8.1	8.5
Weak Completeness	yes II.3.29	yes II.3.29	yes II.3.29	yes II.3.29	yes 8.6
Reflexivity	yes 3.8	yes 3.8	yes 3.8	yes 8.2	no 13.2
Strongly Compact Sets	13.1	13.1	13.1	13.3	13.3
Weakly Compact Sets	II.3.28	II.3.28	II.3.28	II.3.28	13.3
Weak Cauchy Sequences	13.1 (F3)	13.1 (F3)	13.1 (F3)	13.4 13.24 (F3)	8.14 13.25 (F3)
Weak Convergence to x	13.1	13.1	13.1	13.4 13.24	8.14 13.25
$\mathfrak{Y}$ Convergence in $\mathfrak{X} = \mathfrak{Y}^*$	13.1	13.1	13.1	13.4	13.4
Miscellaneous Properties	3.3, 3.4, 3.6, 3.7, 13.1 13.50 13.51	3.3, 3.4, 3.6, 3.7, 3.9 13.1	3.3, 3.4, 3.6, 3.7, 3.9 13.1	13.59 13.61 13.70 (F6)	8.14, 13.49, 13.59 13.60 13.61 (F5)

Footnotes

(F1) No completely satisfactory representation for the conjugate space of $ba(S, \Sigma)$, $ca(S, \Sigma)$, or $rca(S, \Sigma)$, or for the conjugate spaces of the spaces $NBV(I)$ and $BV(I)$, which are isometrically isomorphic to spaces of measures, seems to be known. However, various sorts of representations have been considered in the literature. References are given in the notes and comments section below.

TABLE IV.A (*Continued*)

Space $\mathfrak{X}$	5. l_∞	6. c	7. c_0	8. bv	9. bv_0
Conjugate Space $\mathfrak{X}^*$	8.16	13.7	13.7	13.11	13.11
Weak Completeness	no 13.5	no 13.8	no 13.8	yes 13.11	yes 13.11
Reflexivity	no II.3.29	no 13.8	no 13.8	no 13.11	no 13.11
Strongly Compact Sets	5.6	13.9	13.9	13.11	13.11
Weakly Compact Sets	6.29	13.9	13.9	13.11	13.11
Weak Cauchy Sequences	13.43	13.10	13.10	13.11	13.11
Weak Convergence to x	6.31	13.10	13.10	13.11	13.11
$\mathfrak{Y}$ Convergence in $\mathfrak{X} = \mathfrak{Y}^*$	13.6	meaningless	meaningless	13.11	13.11
Miscellaneous Properties	6.18 (F4) 13.59	(F4) 13.60	(F4)	13.11 13.12 (F5)	13.11 13.12 (F5)

Table continued

(F2) A representation for $L_1^*(S, \Sigma, \mu)$ in terms of a space of measures, valid in the non σ-finite case, was given in J. Schwartz [1].

(F3) Note that in a weakly complete space a weak Cauchy sequence converges weakly to some definite element.

Footnotes continued

TABLE IV.A (*Continued*)

Space $\mathfrak{X}$	10. *bs*	11. *cs*	12. $B(S, \Sigma)$	13. $B(S)$	14. $C(S)$
Conjugate Space $\mathfrak{X}^*$	13.13	13.14	5.1	5.3	6.2 6.3
Weak Completeness	no 13.13	no 13.14	no 13.5	no 13.5	no 13.15
Reflexivity	no 13.13	no 13.14	no II.3.29	no II.3.29	no 13.15
Strongly Compact Sets	13.13	13.14	5.6	5.6	6.5, 6.7, 6.8, 6.9
Weakly Compact Sets	13.13	13.14	6.29	6.29	6.14
Weak Cauchy Sequences	13.13	13.14	13.43	13.43	13.40
Weak Convergence to x	13.13	13.14	6.31	6.31	6.4, 6.12, 6.31
$\mathfrak{Y}$ Convergence in $\mathfrak{X} = \mathfrak{Y}^*$	13.13	13.14	meaning-less	meaning-less	meaning-less
Miscellaneous Properties	13.13 (F4)	13.14 (F4)	6.18 6.19 (F4) 13.76	6.18 6.19 (F4)	6.16, 6.17, 6.26 V.8.8 (F4) 13.50 13.51 13.63 13.64 13.65 13.66

(F4) In a space $\mathfrak{X}$ isometrically ismorphic to spaces of continuous functions, an interesting order relation can be introduced, which makes $\mathfrak{X}$ into a vector lattice of M-space type. We may also introduce a multiplication of functions which makes $\mathfrak{X}$ into a B^*-algebra. References for these two points are given in the notes and comments section below.

TABLE IV.A (*Continued*)

Space $\mathfrak{X}$	15. $ba(S, \Sigma)$	16. $ca(S, \Sigma)$	17. $rca(S, \Sigma)$	18.A. $L_p(S,\Sigma,\mu)$ $1 < p < \infty$	18.B. $L_1(S, \Sigma, \mu)$
Conjugate Space $\mathfrak{X}^*$	(F1)	(F1)	(F1)	8.1	8.5 (F2)
Weak Completeness	yes 9.9	yes 9.4	yes 13.22	yes II.3.29	yes 8.6
Reflexivity	no 13.5	no 13.21	no 13.22	yes 8.2	no 13.2
Strongly Compact Sets	13.19	13.19 13.20	13.19 13.20	8.18 8.20	8.18 8.20 13.68
Weakly Compact Sets	9.12	9.1 9.2	13.22	II.3.28	8.9 8.11 13.54
Weak Cauchy Sequences	13.17 (F3)	9.5 (F3)	13.22 (F3)	13.23 13.24 (F3)	13.25 (F3)
Weak Convergence to x	13.17	9.5	13.22	13.23 13.24	13.25
$\mathfrak{Y}$ Convergence in $\mathfrak{X} = \mathfrak{Y}^*$	13.18	meaningless	9.15	13.24	meaningless
Miscellaneous Properties	9.11(b) III.7.5 III.7.6 13.50 13.76 13.77 (F5)	III.7.5 III.7.6. 9.8 (F5)	(F5)	13.51, 8.22 13.53, 8.26 13.62, 11.7 13.71 (F6)	13.49, 8.10 13.51, 8.13 13.53, 8.14 13.54, 8.22 13.62, 8.24 13.68, 8.26 13.71, 11.7 (F5)

Table continued

(F5) Spaces of measures, L_1-spaces, and spaces isometrically isomorphic to such spaces can all be made into vector lattices of L-space type by the introduction of a suitable order relation. Compare the remarks in the notes and comments section below.

(F6) The spaces l_2 and $L_2(S, \Sigma, \mu)$, as Hilbert spaces, have numerous special properties.

Footnotes continued

TABLE IV.A (*Continued*)

Space $\mathfrak{X}$	19. $L_\infty(S,\Sigma,\mu)$	20. $BV(I)$	21. $NBV(I)$	22. $AC(I)$	23. $C^n(I)$
Conjugate Space $\mathfrak{X}^*$	8.16	(F1)	(F1)	13.29	13.36
Weak Completeness	no V.11.2	yes 12.1	yes 12.2	yes 12.3	no 13.36
Reflexivity	no V.11.2	no 13.28	no 13.28	no 13.28	no 13.36
Strongly Compact Sets	6.26	13.48	13.34	13.32 13.33	13.36
Weakly Compact Sets	(F8)	13.30	13.30	13.31	13.36
Weak Cauchy Sequences	V.11.2	13.30 (F3)	13.30 (F3)	13.31 (F3)	13.36
Weak Convergence to x	V.11.2	13.30	13.30	13.31	13.36
$\mathfrak{Y}$ Convergence in $\mathfrak{X} = \mathfrak{Y}^*$	13.27	13.35(B)	13.35(A)	meaning-less	meaning-less
Miscellaneous Properties	11.7, V.8.11 13.53, 8.23 13.69, 8.26 13.77 13.71 (F4)	13.48 13.63 13.73 (F5)	13.35 (F5)	12.3 (F5)	13.36

(F7) Note, however, the result given in 13.46.

(F8) Using the isomorphism between $L_\infty(S, \Sigma, \mu)$ and a space of continuous functions established in V.8.11, a condition for weak compactness can be obtained from 6.14 (Compare V.11.2, where the corresponding result is given for weak convergence.) Unfortunately, this condition can only be stated in an excessively cumbersome form.

TABLE IV.A (*Continued*)

Space $\mathfrak{X}$	24. $A(D)$	25. AP	26. Hilbert Space	27. $TM(S,\Sigma,\mu)$	28. s
Conjugate Space $\mathfrak{X}^*$		(F9)	4.5	meaningless	meaningless (F7)
Weak Completeness	no 13.37	no 13.38	yes 4.7	meaningless	meaningless
Reflexivity	no 13.37	no 13.38	yes 4.6	meaningless	meaningless
Strongly Compact Sets	13.37	13.39	13.45	11.1	13.47
Weakly Compact Sets	13.37	6.29	4.7	meaningless	meaningless
Weak Cauchy Sequences	13.37	13.41	13.44	meaningless	meaningless
Weak Convergence to x	13.37	13.41	13.44	meaningless	meaningless
$\mathfrak{Y}$ Convergence in $\mathfrak{X} = \mathfrak{Y}^*$	meaningless	meaningless	13.44	meaningless	meaningless
Miscellaneous Properties		7.6 Section XI.2 (F4)	4.1, 4.4, 4.5, 4.9, 4.10, 4.12, 4.15, 4.16, 13.67 13.70 13.72	11.2 11.4 11.6 11.7 13.74	

(F9) Using Theorem 7.6, we can, of course, represent AP^* as $rca(S)$, S being the compact space of that theorem, which is sometimes known as the *Bohr Compactification* of the real line. No more concrete representation of AP^* seems to be known. The result in this direction announced by E. Hewitt (Hewitt [6]) does not seem to be completely proved.

Earlier, however, certain representations for linear functionals on $C[0, 1]$ had been given, but they all suffered from various defects, such as the lack of uniqueness. For example, Hadamard [1] showed that each $x^* \in C^*[0, 1]$ has the form

$$x^*f = \lim_{m\to\infty} \int_0^1 k_m(s)f(s)ds$$

where $k_m \in C[0, 1]$. Fréchet [5; I] gave a different proof of this fact, and observed [5; II] that the k_m may be taken to vanish at 0 and 1, or may be taken to be polynomials (but not both). In particular, there exists a double sequence of constants b_{nm}, $0 \leqq n \leqq m$, $1 \leqq m < \infty$, such that

$$x^*f = \lim_{m\to\infty} \sum_{n=1}^{m} b_{nm} f\left(\frac{n}{m}\right).$$

Fréchet gave other applications of these results.

Riesz announced his result in 1909 (F. Riesz [7]). Since then he has given several different proofs of this result (Riesz [3, 10, 11]). Another proof, still for $C[0, 1]$ was supplied by Helly [1]. Radon [2; p. 1333] extended the theorem to a compact set in E^n, and cast it in the dress of an integral with respect to a regular measure, rather than a Stieltjes integral. See also the paper of C. A. Fischer [2]. Hildebrandt [8] and Hildebrandt and Schoenberg [1] showed that the representation theorem implies and is implied by the theorem of Hausdorff on moments. A further extension of the theorem was made, in 1937, when Banach (in Note II of Saks [1]) proved it for $C(S)$, where S is a compact metric space. In the same framework, the theorem was proved by Saks [4]. In 1941, Kakutani [9; p. 1009] extended the theorem to compact Hausdorff spaces by modifying a process used in some unpublished notes of von Neumann.

Somewhat earlier, in 1938, the first attempt to extend the result to non-compact spaces was made by Markov [2], who considered the bounded continuous functions on a space which satisfies the normal separation axiom except that points are not assumed to be closed. He showed that positive functionals correspond to positive regular finitely additive measures, and considered certain invariant functionals, i.e., such that $x^*(fg) = x^*(f)$, for all $f \in C(S)$. A similar investigation was made by Alexandroff [1] for "spaces" which satisfy the normal separation axiom but in which uncountable unions of open sets are not necessarily open. For such spaces, Alexandroff [1; II p. 577] proved Theorem 2, and related the additivity properties to the compactness properties of the space [II. p. 587] and the convergence properties of the functional [II. p. 593].

There are also a number of recent papers related to this result. For example, Halmos [5; Chap. 10], Hewitt [3] and Edwards [1] consider functionals on the space of continuous functions on a locally compact Hausdorff space which vanish outside of compact sets. Arens [3] discussed certain sub-algebras of continuous functions on a topological space such that $\{s \mid |f(s)| > c > 0\}$ is compact for each $c > 0$. In each of these cases the representation is effected by a regular countably additive measure on a certain field of subsets. Glicksberg [1] showed that if S is completely regular, then every non-negative functional on $C(S)$ may be represented by means of a countably additive measure if and only if any one of the following conditions hold: (1) $f_n \in C(S)$, $n = 0, 1, 2, \ldots$ and $f_n(s) \uparrow f_0(s)$, $s \in S$ imply that $f_n \to f_0$ uniformly on S; (2) every continuous real valued function on S is bounded; (3) every continuous real valued function on S assumes its maximum; (4) every bounded equicontinuous family in $C(S)$ is conditionally compact.

Up to now we have discussed linear functionals on the *bounded* continuous functions. Hewitt [1] showed that if S is completely regular, then a positive linear functional on the space of all continuous functions on S may be represented by means of a countably additive measure μ defined on a certain σ-field, and that every function is bounded except on a μ-null set. This extends a result announced for $C(-\infty, \infty)$ by Wehausen [1; p. 164]. In this paper, Hewitt also treated linear functionals continuous in certain topologies (e.g., compact-open, and product topology).

An inspection of the proof of Theorem 2 indicates that the problem of representation may be viewed as one of extending the linear function, originally defined on $C(S)$, to be defined on a larger collection of functions on S such as the bounded Borel functions. Once this extension is achieved the measure is readily obtained, and the functional represented as an integral. This is the point of view of the Daniell [1] approach to the theory of integration, modifications of which have been presented by Bourbaki [4], Loomis [1] and Stone [6]. Hewitt [2] and Loomis [2] have applied this view to discuss linear functionals. The Daniell attack has been extended greatly to order preserving maps between ordered spaces by McShane [3] and to positive mappings of $C(S)$ into ordered spaces by Christian [1].

Strong and weak compactness in $C(S)$ (6.5—6.14). Ascoli [2; p. 545] defined the notion of equicontinuity of a set of continuous functions (of a real variable) at a fixed point; this concept was also used by Arzelà [5] about the same time. Definition 6.6 merely applies his definition at every point of the domain. This fundamental concept can readily be extended to functions with range in a metric space, or still more general functions (see Bourbaki [5; Chap. 10]).

The important Theorem 4.7 is known as the Arzelà-Ascoli theorem, although many authors use only one of these names. In the case of $C[0, 1]$, Ascoli [2; 545—549] used a construction which essentially amounts to the sufficiency of the condition for compactness. Arzelà [2] proved the necessity of this condition. Both of these papers are phrased in geometrical terms and it is difficult to resurrect this result from them. However, an extremely clear presentation of this and related theorems was given by Arzelà [3; p. 56—60]. The extension to a case in which the domain is a space with a notion of limit (say, a metric space) was done by Fréchet [1]. It is not difficult to extend it to a case where the range is taken to be a metric space instead of the real or complex numbers, although in this case we must also suppose that $\{f(s) | f \in K\}$ is conditionally compact for each $s \in S$.

In the case that S is not necessarily compact or that the range of the functions is not necessarily a metric space, similar compactness criteria have been established by Arens [5], Gale [1], and Myers [1] (see also Bourbaki [5; Chap. 10] and Kelley [5]). In these cases one usually employs the "compact-open" topology of functions. A different type of criterion for compactness in $C[0, 1]$ was given by Izumi [2].

It is perhaps appropriate to insert at this place a few historical comments concerning the notion of uniform convergence—the mode of convergence in this space. The importance of this mode of convergence is now fully appreciated, but it has not always been so. Even the mighty Cauchy stumbled on this score and asserted, in 1821, that the sum of a convergent series of continuous functions is itself a continuous function (see Cauchy [1; p. 120]). That this was false was pointed out, in 1826, by Abel [1; p. 316]. There the matter rested for some years. In 1847 Stokes [1; p. 562], in 1848 Seidel [1], and in 1853 Cauchy [1; p. 30—36] independently demonstrated that uniform convergence is sufficient to assure the continuity of the limit function. (It is interesting that Weierstrass [1; p. 67, 70] had employed this notion of convergence in some unpublished manuscripts written in 1841—he even used the term "gleichmässig".) It is to Seidel's credit that he observed that he was unable to prove the necessity as well; Stokes went astray on this point, and Cauchy remained silent. Conditions that are necessary as well as sufficient were not to come for a number of years.

The notion of the quasi-uniform convergence of a sequence of functions was introduced in 1884 by Arzelà [1] to whom Theorem 6.11 is due in the classical case of a sequence in $C[0, 1]$. In 1878, however, Dini [1; p. 107—109] had given necessary and sufficient conditions for continuity of the limit function at a point, and announced [p. 110—112] his classical theorem concerning monotone convergence of continuous functions.

The concept of quasi-equicontinuity (which is related to quasi-uniform convergence in exactly the same way ordinary equicontinuity is related to uniform convergence) was introduced by Sirvint [2, 3] for $C[0, 1]$, and by Bartle [2] for $C(S)$. In the case of $C[0, 1]$, the equivalence of statements (1), (2) and (3) of Theorem 6.14 was proved by Sirvint [2, 3; p. 76, 82] and by Bourgin [1; p. 601]. In a general framework, Grothendieck [2; p. 180—182] proved the equivalence of statements (1), (2) and (5). The general equivalence of (1), (3) and (4) was proved by Bartle [2], to whom this proof is due. Grothendieck [2] considered the relationship between compactness, countable compactness and sequential compactness in the case where the range of the continuous functions is in a more general topological space. Some of Bartle's [2] results are valid in this more general case, as well.

The Stone-Weierstrass and related theorems (6.15—6.27). The classical theorem on the approximation of continuous functions by polynomials was given by Weierstrass [2; p. 5]. Many proofs of this theorem have been given; see, e.g., L. M. Graves [1, 4], Hobson [1; II, p. 228—234] and Widder [1; p. 152—153]. (The proof in Graves

[4] shows that if $f \in C^n[0, 1]$, then the polynomials may be chosen so that all derivatives including those of order n converge uniformly to the corresponding derivatives of f.)

Of the many extensions of the Weierstrass theorem for $[0, 1]$ we mention the striking theorem of Müntz [1] which asserts that the set $\{1, x^{\alpha_1}, \ldots, x^{\alpha_n}, \ldots\}$ is fundamental in $C[0, 1]$ if and only if the series $\sum \alpha_i^{-1}$ diverges. This theorem was sharpened and simplified by Szász [1] and Clarkson and Erdös [1].

Stone's extension of the Weierstrass theorem for real functions was made in Stone [1; p. 466—469], the discussion being concerned principally with the continuous functions as an algebra. The complex case was treated by Gelfand and Šilov [1]. Kakutani [9] discussed a similar question, basing his work on the lattice properties of the space. In [4], Stone considered both of these aspects and presented a complete and elementary approach to this beautiful and important generalization. In this latter paper, Stone investigates a number of related aspects of this problem, and makes many applications of this result. A different type of proof, based on semi-group theory, was given by Dunford and Segal [1].

In the case of real scalars, Stone's theorem asserts that algebraic combinations of a family $D \subset C(S)$ are dense in $C(S)$ if and only if the functions in D separate the points of S. Hewitt [4] showed that if S is only assumed to be completely regular and not compact, then the theorem is no longer valid. Hewitt also proved that a stronger separation property (concerning closed sets rather than points) suffices to establish the theorem in this case.

Arens [4] has given a generalization of the Stone-Weierstrass theorem in the case where the range is in an Abelian group with certain lattice and topological properties. Kaplansky [1; p. 228—233] proved a version of this theorem in an instance where the space S is compact Hausdorff, but the functions at a point $s \in S$ take their values in a C^*-algebra A_s which depends on s. This is a "non-commutative" generalization of the theorem. Even in the ordinary case, his proof is different from the one presented here. Kaplansky [2] has also presented a theorem of the Stone type where the scalars are taken to be in a division ring with a valuation.

All of the abstract theorems mentioned above deal with establishing conditions that are sufficient to insure that a class of functions generates the entire space. Wermer [6, 9] has given sufficient conditions that two functions generate $C(S)$ when S is the unit circle. If f is one-to-one, or the function $f(\lambda) = \lambda^2$, then necessary and sufficient conditions are given on a function g in order that f and g generate $C(S)$. This is closely related to classifying the closed subalgebras B which are maximal in the sense that if B' is a closed subalgebra of $C(S)$ and if $B \subseteqq B'$, then B' is either B or all of $C(S)$. Additional results concerning maximal subalgebras and related topics are found in Wermer [10, 11], Helson and Quigley [1, 2] and Rudin [1].

We now comment briefly on the Theorems 6.18—6.27. They are essentially due, at least in the real case, to Stone [1], although his terminology and proofs often differ from that given here. It should be mentioned that Theorem 6.22 was proved independently by Čech [1] only slightly later. (See also Stone [5] for an elementary treatment.) Lemma 6.25 was proved in Stone [1; p. 465]—extensions of this result are also found in Hewitt [5] and Kaplansky.

There are several theorems closely related to Theorems 6.26 and 6.27—see for example Theorem V.8.8 where it is shown that if $C(S)$ is isometrically isomorphic with $C(T)$, where S and T are compact Hausdorff spaces, then S and T are homeomorphic. The theorem just mentioned was proved for real scalars and compact metric spaces by Banach [1; p. 170] and for compact Hausdorff spaces by Stone [1; p. 469]. (Related results have been given by Eilenberg [1], Arens and Kelley [1] and Hewitt [5]). It may be summarized as asserting "the Banach space $C(S)$ characterizes S." Šilov [1] showed that $C(S)$ as a topological ring characterizes S, provided S is compact metric. Gelfand and Kolmogoroff [1] proved the sharper theorem that if S is compact Hausdorff, then the ring $C(S)$ characterizes S. This includes Corollary 6.27. Stone [7; II] showed that as a lattice-ordered group $C(S)$ characterizes S. Finally, Kaplansky [3; I] proved that merely as a lattice, $C(S)$ characterizes S. Hewitt [5], Nagata [1], and Shirota [1] have obtained some results when S is not required to be compact. Kadison [2] has obtained results of this nature for non-commutative C^*-algebras.

The preceding paragraph makes it appear that any reasonable topological or algebraic property of $C(S)$ characterizes S, provided that S is compact. That this breaks down if $C(S)$ is regarded as a topological linear space is demonstrated by the fact that $C[0, 1]$ may be mapped in a linear and homeomorphic manner on the space $C([0, 1] \cup 2)$. (See Banach [1; p. 184]).

Weak compactness in $B(S)$. Definition 6.27 is equivalent to a definition given by Bartle [2], to whom Theorem 6.28 is due. Lemma 6.29 is essentially due to Bourgin [1; p. 600], the proof given here and its application to 6.30 is due to Bartle [2]. Theorem 6.30 was first proved by Sirvint [3; p. 80], who based his proof on some related theorems in Banach [1; p. 217—225]. This theorem had been announced earlier without proof by Fichtenholz and Kantorovitch [2].

The space AP. The theory of almost periodic functions was created by Harald Bohr [1], although it has been considerably extended by many other mathematicians. Although it is a natural and beautiful generalization of the theory of periodic functions, Bohr was led to it by the study of Dirichlet series. For a most readable expository account of some different proofs of the principal theorems of the theory, the reader is referred to Bohr [4]. The book of Bohr [2] gives a very enjoyable account of the elementary theory, and that of Besicovitch [1] considers certain generalizations as well as analytic almost periodic functions. (See also Maak [1].)

Theorem 7.2 has been used as a definition of almost periodic functions by Bochner [4]. This led von Neumann [9] and Bochner and von Neumann [1] to extend the theory to an arbitrary abstract group. They required G to be an abstract group; a function $f \in B(G)$ is said to be *right almost periodic* if the set $\{f_a | a \in G\}$, where $f_a(x) = f(xa)$, is a conditionally compact set in $B(G)$. For an expository paper on this facet of the theory, see Bohr [3]. If the strong topology is replaced by the weak topology of $B(G)$, then we have a *weakly almost periodic* function in the sense of Eberlein [3].

Certain aspects of these generalizations will be discussed in a later chapter. The reader will find other references in Loomis [1] and Weil [1].

The space S of Theorem 7.6 is called the *Bohr compactification* of the real line. Weil [2, 1; p. 124—139] has shown that a similar result may be obtained for any locally compact Abelian group and has given applications of this fact. Anzai and Kakutani [1] showed, among other things, that the Bohr compactification of a locally compact Abelian group G may be obtained by taking the character group $\hat{G}$ of G, equipping $\hat{G}$ with its discrete topology, and then taking the character group of this discrete group. Hewitt [6] has used the Bohr compactification to study the adjoint space of AP. (See also Artemenko [2] and Krein [5] for certain positive linear functionals.)

The space L_p. In the case of $L_2[0, 1]$, Theorem 8.1 was announced independently and simultaneously in 1907 by Fréchet [4] and F. Riesz [9]. A detailed proof was given by Fréchet [5; III. p. 441]. The theorem for $L_p[0, 1]$, $1 < p < \infty$, was demonstrated by F. Riesz [2; p. 475]. In the case of a finite measure space the theorem was established by Nikodým [9; p. 132] and later by Dunford [1; p. 338] by essentially the same method used here. A proof based on the property of uniform convexity was given by McShane [1] for a completely arbitrary measure space. A different proof in this context was given by J. Schwartz [1]. Generalizations of this theorem to the spaces of Orlicz were obtained by Zaanen [1, 5; p. 138].

In the case $L_p[0, 1]$, $1 < p < \infty$, Corollaries 8.3 and 8.4 were proved by F. Riesz [2; p. 467].

Steinhaus [2] proved Theorem 8.5 for [0, 1], and obtained Corollary 8.6. Dunford [1; p. 338] extended 8.5 to a finite measure space. An example due to T. Botts (given in J. Schwartz [1]) shows that this theorem is not valid without the assumption of σ-finiteness. However, if S is locally compact and μ is a regular measure, then $L_1^* = L_\infty$. (See Dieudonné [7; p. 83].) J. Schwartz [1] has given a representation of L_1^* in the case of a general measure. Before the Steinhaus paper, Fréchet [5; III] gave a representation for the linear functionals F on $L_1[0, 1]$ which are continuous in the sense that if $f_n(s) \to f(s)$ almost everywhere, then $F(f_n) \to F(f)$. An extension of the Steinhaus theorem has been obtained by San Juan [1].

Theorem 8.9 was proved in the case of finite measure by Dunford [9; p. 643], and in the σ-finite case by Dunford and Pettis [1; p. 376]. If S is locally compact and μ is a regular positive measure on the Borel sets Σ, then Dieudonné [7; p. 93] has shown that a set $K \subset L_1(S, \Sigma, \mu)$ is weakly compact if and only if (1) K is bounded; (2) for any $\varepsilon > 0$ there is a $\delta > 0$ such that if $\mu(E) < \delta$ then

$\int_A |f(s)| \mu(ds) < \varepsilon$, $f \in K$; (3) for any $\varepsilon > 0$ there is a compact set $C \subseteq S$ such that $\int_{S-C} |f(s)| \mu(ds) < \varepsilon$, $f \in K$.

Dieudonné [7] has extended these results to a certain class of functions closely related to L_1, but which do not form a B-space. The reader is referred to his paper for the details.

Corollary 8.13 is a classical result due to Schur [1]. There is a similar theorem valid in L_p, $1 < p < \infty$, which was proved by Radon [2; p. 1363] and F. Riesz [13].

THEOREM. *If* $1 < p < \infty$ *then a sequence* $\{f_n\}$ *converges strongly to* f *in* $L_p(S, \Sigma, \mu)$ *if and only if it converges weakly and* $|f_n| \to |f|$. This theorem remains valid in any uniformly convex B-space.

Theorem 8.15 was proved independently by Hildebrandt [3; p. 875] and Fichtenholz and Kantorovitch [1; p. 76]. Earlier, Steinhaus [2] showed that the general continuous linear functional on the essentially bounded measurable functions with the L_1-norm is given by an element of L_∞.

Fréchet [9; p. 308] gave necessary and sufficient conditions for a set in l_2 to be compact. Similar results for $L_2[0, 1]$ were given (see Fréchet [7; p. 118]) but they involved a preassigned orthonormal set. Other results for $L_2[0, 1]$ were obtained by Veress [2]. Theorem 8.17 is essentially due to Kolmogoroff [2] in the case that $1 < p < \infty$ and S is a bounded set in finite dimensional Euclidean space. The extension to unbounded sets was made by Tamarkin [1]. Tulajkov [1] showed that the Tamarkin theorem was valid for $p = 1$ as well. M. Riesz [2] gave a different proof of this condition. Takahashi [1] showed that the same result is valid in the generalized L_p spaces of Orlicz. The form of Theorem 8.17 presented here and its validity for $p = \infty$ was given by Phillips [3; p. 527]. A different criterion was given for L_p, $p \geqq 1$, by Fréchet [7]. Nicolescu [1] has also discussed the Kolmogoroff result. Conditions of the type introduced by Kolmogoroff were employed by Izumi [2] to find compactness criteria for $C[0, 1]$ and $TM(S, \Sigma, \lambda)$.

Theorems 8.18 and 8.20 are due to M. Riesz [2]. An extension to the case $0 < p < 1$ was given by Tsuji [2]. Lemma 8.25 and Theorem 8.26 are due to F. Riesz [23], who was concerned with a somewhat more abstract situation.

Since each element of L_∞ is an equivalence class of functions, it is natural to inquire whether bounded representatives may be chosen from these equivalence classes in such a manner as to preserve sums and scalar products. It is a striking fact, due to von Neumann [21], that such a selection can be made in $L_\infty(0, 1)$ so as to preserve polynomial identities.

THEOREM. *Let (S, Σ, μ) be a positive finite measure space such that $L_1(S, \Sigma, \mu)$ is separable. Then there exists a linear mapping T of $L_\infty(S, \Sigma, \mu)$ into $B(S)$ with $|T| \leqq 1$ such that Tf and f are contained in the same equivalence class for all $f \in L_\infty(S, \Sigma, \mu)$.*

For $S = [0, 1]$, this is a special case of a theorem of von Neumann. A short proof of this result, using a theorem of Halmos and von Neumann [1] has been given by Dieudonné [9], who has shown that the above result is intimately related with some theorems to be discussed in Section VI.8.

Spaces of set functions. The fact that uniform countable additivity of a set in $ca(S, \Sigma)$ is equivalent to equicontinuity with respect to a fixed positive measure was proved by Dubrovskiĭ [2]. Dubrosvkiĭ [1] also proved in the case of a compact cube in E^n, either assumption plus boundedness implies that every sequence has a subsequence which converges for each Borel set. For related results, see also Cafiero [3, 4, 5] and Dubrovskiĭ [3, 4, 5, 6]. A proof of Theorems 9.1—9.3 different from that found here is found in Bartle, Dunford and Schwartz [1].

Similar weak compactness criteria have been proved for regular measures by Grothendieck [4; p. 146]. We cite his result.

THEOREM. *Let M be locally compact space, and let $K \subset rca(M)$. Then the following statements are equivalent.*

(1) *K is weakly sequentially compact;*

(2) *if $\{f_n\}$ is a uniformly bounded sequence of continuous functions on M which converge at each point to zero, then*

$$\lim_{n\to\infty} \int_M f_n(s)\mu(ds) = 0$$

uniformly with respect to μ in K;

(3) *for every sequence $\{G_n\}$ of pairwise disjoint open sets* $\lim \mu(G_n) = 0$ *uniformly for μ in K;*

(4) (a) *for every compact set* $C \subseteqq M$ *and* $\varepsilon > 0$ *there exists an open neighborhood* U *of* C *such that* $v(\mu; U - C) \leqq \varepsilon$ *for every* μ *in* K,

(b) *given* $\varepsilon > 0$ *there exists a compact set* $C \subseteqq M$ *such that* $v(\mu, M - C) \leqq \varepsilon$ *for every* μ *in* K.

In view of the Radon-Nikodým theorem, Theorem 9.5 may be regarded as a generalization of a theorem due to Lebesgue [1; p. 57]. Theorem 9.4 is a special case of this result and a theorem of Nikodým [6], namely Theorem III.7.3.

Theorem 9.8, which strengthens the uniform boundedness theorem in a special case, is due to Nikodým [5]. The proof given here, as well as Lemma 9.7, is due to Saks [3], where a slightly stronger version of this result is proved. (For other results related to the atomic structure of a measure space, see Hahn and Rosenthal [1; pp.45–53] and Halmos [5; pp. 165–174].)

Theorem 9.14 is due to Bochner [3]. Bochner and Phillips [1] also gave a proof of this result.

Theorem 9.15 was proved by Alexandroff [1; III. p. 182] to whom a number of similar results are due (cf. Alexandroff [1; III]). The following comments are pertinent to this theorem.

Some remarks on $\mathfrak{X}$*-convergence in* $\mathfrak{X}^*$. One of the problems set in Section 1 of this chapter is the determination of concrete conditions for $\mathfrak{X}$-convergence in $\mathfrak{X}^*$; i.e., to describe in terms of the spaces $\mathfrak{X}$ and $\mathfrak{X}^*$ what is meant by the statement $x_n^* x \to x^* x, x \in \mathfrak{X}$. Theorem 9.15 is an instance of such a theorem. Similar questions were considered in a classic paper of Lebesgue [1]. He derived several necessary and sufficient conditions guaranteeing that

$$\int_0^1 f(s)\phi_n(s)dx \to 0, \qquad f \in F.$$

Where the functions f and $\{\phi_n\}$ are taken to be in certain preassigned classes, e.g., L_1, L_2, L_∞, C, BV, etc. Some of these results are weak convergence, some $\mathfrak{X}$-convergence in $\mathfrak{X}^*$, and some are neither. Camp [1] extended a number of Lebesgue's results to several variables. An extensive discussion of such conditions was given by Hahn [2], who considered a wide variety of spaces.

Corresponding problems suggest themselves for measures—that is, to determine conditions on $\{\mu_n\}$ such that

$$[*] \qquad \int f(s)\mu_n(ds) \to 0, \qquad f \in F,$$

where F is a specified collection of functions. Here the integral may be in the Stieltjes sense. In this connection we mention explicitly a classical theorem due to Helly [1; p. 268] and Bray [1; p. 180].

THEOREM. *If* $\{\alpha_n\}$ *is a sequence of functions of uniformly bounded variation, and if there is a function* $\alpha \in BV[0, 1]$ *such that* $\alpha_n(x) \to \alpha(x)$ *for* x *in a dense subset of* $[0, 1]$ *containing* 0 *and* 1 *then*

$$\int_0^1 f(s)\alpha_n(ds) \to \int_0^1 f(s)\alpha(ds), \qquad f \in C[0, 1].$$

The condition in this theorem is sufficient, but not necessary; a necessary and sufficient condition was given by Hildebrandt [9]. For other results the reader may refer to Glivenko [1; Chap. 7] or L. M. Graves [2; p. 281—292]. H. M. Schwartz [1] has also obtained a number of results in this direction when f is not assumed to be continuous.

Dieudonné [11] considered the convergence in [*] where S is a compact Hausdorff space, the μ_n are regular measures, and F is one of the following classes: (1) continuous functions; (2) "Riemann integrable functions"; (3) semi-continuous functions; (4) bounded Borel measurable functions. Probably his most striking result is

THEOREM. *If* S *is compact metric and* $\mu_n \in rca(S)$, *then if* $\mu_n(G) \to \mu(G)$ *for every open set* $G \subseteq S$, *then* $\mu_n(E) \to \mu(E)$ *for every Borel set* E.

Vector valued measures. The important Theorem 10.1 is due to Pettis [4; p. 283) who proved it for an indefinite integral by means of a theorem of Orlicz and Banach, although this proof is valid for the general case. The following year, Pettis [6] announced the general result. Kunisawa [1] gave the first published proof of the general theorem. (See also Nakamura and Sunouchi [1].)

The semi-variation, as defined in 10.3, reduces in the scalar valued cases to the ordinary total variation. In an abstract case, it was employed by Gowurin [1] who developed a Riemann-type integration theory where the function and measure take values in two vector spaces on the product of which there is defined a vector valued bilinear function.

The Lebesgue-type theory of integration of a scalar valued function with respect to a vector valued measure which is presented here is the one given in Bartle, Dunford and Schwartz [1]. A similar procedure has been employed by Bartle [3] to obtain a Lebesgue-type integration theory where both function and measure are vector valued.

The space $TM(S, \Sigma, \mu)$. Although the notion of convergence in measure was introduced in 1909 by F. Riesz [12], it was Fréchet [8; p. 199] who introduced a metric in the space of measurable functions in such a way that metric convergence is equivalent to convergence in measure. Fréchet [6] gave necessary and sufficient conditions for a set of measurable functions on [0, 1] to be compact under this metric. (Earlier, Veress [1] had given conditions assuring that a sequence of measurable functions has a uniformly convergent subsequence.) Fréchet's arguments were simplified by Hanson [1]. Somewhat different conditions have been given by Izumi [2] and Medvedev [1]. Theorem 11.1 is a generalization of Fréchet's results and is due to Šmulian [14]. For another generalization, see Cafiero [2].

Theorem 11.2 is a generalization of a theorem due to Banach [8; p. 37]. In its form given here it is a generalization of a theorem of Dunford and Miller [1; p. 542]. The proof given here is essentially that of Mazur and Orlicz [1; p. 157]. Alexiewicz [1; IV] treated polynomial operators when the domain space is not assumed to be a B-space, but satisfies certain limit conditions. For related theorems and applications see Saks [2, 5].

We have seen that the space $TM(S, \Sigma, \mu)$ is not a B-space, and hence the existence of non-zero continuous linear functions is in doubt. Nikodým [9; p. 141] showed that if $\mu(S) < \infty$ then a necessary and sufficient condition that there exists a non-zero continuous linear functional on $TM(S, \Sigma, \mu)$ is that there exists an atom for μ.

Functions of bounded variation. The concept of function of bounded variation was introduced, in 1881, by Jordan [1] and that of an absolutely continuous function, in 1905, by Vitali [1]. While these classes of functions are of considerable importance in much of analysis, their study has largely merged into the more general modern concept of measure. This merger was effected principally by Radon [2].

Representations of linear functionals on BV have been given by Hildebrandt [7] and by Artemenko [1] and Grosberg [1]. In somewhat more generality this problem has been treated by Šreĭder [1], but none of these results are entirely natural. Adams [1] and Adams and Morse [1, 2] have discussed the space BV and certain of its subspaces with a different metric under which it is not a B-space. They have been able to give natural representations for the most general functionals which are continuous and uniformly continuous under this metric.

There are a number of definitions of bounded variation and absolute continuity for functions of two variables. The relation between these definitions and the properties of the associated notions have been examined in detail by Adams and Clarkson [1, 2, 3, 4].

Characterization of Hilbert space. Jordan and von Neumann [1] proved that in a normed linear space $\mathfrak{X}$ of two or more dimensions for which the "parallelogram identity"

$$|x+y|^2+|x-y|^2 = 2\{|x|^2+|y|^2\}$$

is valid for arbitrary $x, y \in \mathfrak{X}$, an inner product can be defined such that $|x|^2 = (x, x)$. Thus the parallelogram identity is characteristic of inner product spaces. (See also Rubin and Stone [1].) The same conclusion has been obtained from other identities or inequalities assumed for the norm by Birkhoff [1], Day [7], Ficken [1], James [2], Lorch [3, 4], and Schoenberg [1].

Characterizing conditions based upon certain properties of linear functionals, hyperplanes, and types of orthogonality have been discussed by Birkhoff [1] and James [1, 2, 3].

Kakutani and Mackey [1] showed that a real B-space $\mathfrak{X}$ can be given an equivalent norm under which it is a real Hilbert space if there is a mapping $T \to T^*$ of the ring $B(\mathfrak{X})$ of bounded linear operators in $\mathfrak{X}$ which satisfies $(T_1T_2)^* = T_2^*T_1^*$, $(T_1+T_2)^* = T_1^*+T_2^*$, $T^{**} = T$ and $T \neq 0$ implies that $T^*T \neq 0$. They showed that the same conclusion is true if there is a mapping $\mathfrak{M} \to \mathfrak{M}'$ of the lattice $\mathscr{L}$ of closed linear manifolds of $\mathfrak{X}$ for which $\mathfrak{M}'' = \mathfrak{M}$, $\mathfrak{M}' \cap \mathfrak{M} = 0$, and $\mathfrak{M}_1 \subset \mathfrak{M}_2$ implies $\mathfrak{M}_1' \supset \mathfrak{M}_2'$. In this case the space $\mathfrak{M}'$ can be identified with the orthogonal complement of $\mathfrak{M}$.

Kakutani [6] proved that if $\mathfrak{X}$ is a normed linear space of three or more dimensions, then the norm can be given by an inner product if and only if every two dimensional subspace is the range of a projection of norm 1. (See also Bohnenblust [2], Phillips [2], Sobczyk [1].)

For a related problem, see Blumenthal [1].

References. Aronszajn [1], Birkhoff [1], Blumenthal [1], Bohnenblust [2], Day [7], Ellis [1], Ficken [1], James [1, 2, 3], Jordan and von Neumann [1], Kakutani [6], Kakutani and Mackey [1], Lorch [3, 4], Nagumo [3], Ōhira [1, 2], Phillips [2], Rubin and Stone [1], Schoenberg [1], Sobczyk [1].

Ordered spaces. There is a vast literature dealing with vector spaces which are also assumed to possess an additional structure of order. For example, a *partially ordered vector space* is a vector space $\mathfrak{V}$ in which there is defined a relation $x \geqq y$ for some pairs of elements in $\mathfrak{V}$ for which

(i) $x \geqq 0$ and $-x \geqq 0$ imply $x = 0$;
(ii) $x \geqq y$ and $y \geqq z$ imply $x \geqq z$;
(iii) $x \geqq 0$ and λ a non-negative real number imply $\lambda x \geqq 0$;
(iv) $x \geqq y$ implies $x+z \geqq y+z$.

If, for every two elements in $\mathfrak{V}$ there is a least upper bound $x \vee y$ and a greatest lower bound $x \wedge y$, one says that $\mathfrak{V}$ is a *vector lattice.*

Frequently there are other relations between the ordering and the algebraic structure (as in the case of ordered algebras) or between the ordering and the topological or metric structures. This field lends itself to a wide variety of possibilities. We will confine our remarks to representation theorems for abstract L- and M-spaces.

A real B-space is said to be an *abstract L-space* if it is a vector lattice in which

(l) $\quad x \geqq 0$ and $y \geqq 0$ imply $|x+y| = |x|+|y|$.

Such B-spaces were introduced axiomatically by G. Birkhoff [2] as abstractions from the concrete B-spaces of Lebesgue integrable functions on a measure space. (See also Freudenthal [1], Kakutani [3, 7, 8, 9], and Smiley [1].)

Kakutani [8] showed that every abstract L-space can be provided with an equivalent norm which satisfies (l) and also the condition

(m) $\quad x \wedge y = 0$ implies $|x+y| = |x-y|$.

An abstract L-space is said to have a *unit* if there exists an element e for which $x > 0$ implies $e \wedge x > 0$. Extending a result of Freudenthal, Kakutani [8] proved:

THEOREM. *Given any abstract L-space satisfying (m) and possessing a unit, there exists a totally disconnected compact topological space S and* *a countably additive measure μ defined on the Borel field Σ of S such that the abstract L-space is isometric and lattice isomorphic to the real B-space $L_1(S, \Sigma, \mu)$.*

Kakutani [3, 9] (and Bohnenblust and Kakutani [1]) define a real B-space to be an *abstract M-space* if it is a vector lattice for which

(v) $x_n \geqq y_n$, $x_n \to x_0$, $y_n \to y_0$ implies $x_0 \geqq y_0$.

(m*) $x \wedge y = 0$ implies that $|x+y| = |x-y|$ and $|x \vee y| = \max(|x|, |y|)$.

We say that an abstract M-space has a *unit* if there exists an element e for which $e \geqq 0$, $|e| = 1$ and $|x| \leqq 1$ implies $x \leqq e$. Independently, Kakutani [8] and M. and S. Krein [1, 2] showed:

THEOREM. *For any abstract M-space with unit, there exists a compact Hausdorff space S such that the abstract space is isometric and lattice isomorphic to the real B-space C(S).*

In the case where the abstract space is not assumed to possess a unit, a similar representation theorem can be obtained in which there may be linear relations between the values of the functions at pairs of points.

The following result is also due to Kakutani.

THEOREM. *The conjugate space of an abstract M-space is an abstract L-space. The conjugate of an abstract L-space is an abstract M-space with unit.*

Representation theorems for non-normed spaces, similar to those cited above, have been obtained by Kuller [1].

The following is an incomplete list of works primarily concerned with various aspects of the theory of ordered spaces. Books: Birkhoff [3], Nakano [2], Kantorovič, Vulih and Pinsker [1]. Papers: Berri [1], Birkhoff [2], Bochner [1], Bochner and Fan [1], Bochner and Phillips [1], Bohnenblust [1], Bohnenblust and Kakutani [1], Dieudonné [4, 5, 6], Freudenthal [1], Fan [1, 2], Grosberg and Krein [1], Kadison [1], Kakutani [3, 7, 8, 9], Kantorovitch [1, 2, 3], Kantorovič, Vulih, and Pinsker [2], Krein [2, 3, 4], Krein and Krein [1, 2], Krein and Rutman [1], Mikusiński [1], Nachbin [2], Nakamura [1, 2], Nakano [2, 3, 4, 5, 6, 14, 15, 16], Ogasawara [1, 2, 3, 4, 5, 6], Ogasawara and Maeda [1, 2], Orihara [1], Pierce [1], Pinsker [1, 2, 3, 4, 5, 6, 7], F. Riesz [23], Širohov [1], Smiley [1], Šmulian [11], Sz.-Nagy [1],

Tagamlitzki [1], Udin [1], Vulich [1, 2, 3, 4, 5, 6, 7, 8, 9, 10, 11], Wassilkoff [1, 2, 3, 4], Yosida [1, 2].

Characterization of L_1 and L_p. In the preceding paragraph we have discussed the characterization of the space L_1 as a concrete representation of the abstract L-spaces. Bohnenblust [1] has given a very interesting characterization of the L_p spaces, $1 \leqq p < \infty$. We comment very briefly on his results. If $\mathfrak{B}$ is a partially ordered real B-space, let the *absolute value* of $x \in \mathfrak{B}$ be defined as $||x|| = x \vee (-x)$. Two elements x, y in $\mathfrak{B}$ are said to be orthogonal if $||x|| \wedge ||y|| = 0$. A *unit* here denotes the same kind of unit as for an L-space. For convenience we say that $\mathfrak{B}$ has property P in case:

(P) if $x = x_1 + x_2$, where x_1 and x_2 are orthogonal, and $y = y_1 + y_2$, where y_1 and y_2 are orthogonal, and if

$$|x_1| = |y_1|, \qquad |x_2| = |y_2|,$$

then it follows that $|x| = |y|$.

We may now state Bohnenblust's theorem.

THEOREM. *Any separable, partially ordered real B-space with unit which is a σ-complete lattice, having at least three dimensions and possessing property P is equivalent to one of the spaces l_p^n, l_∞^n, l_p, L_p, $L_p \oplus l_p^n$, $L_p \oplus l_p$, $1 \leqq p < \infty$, or c_0. Further, the isomorphism preserves norm and order.*

Additional conditions can be given to distinguish between the various possibilities. Another characterization of L_1, related to the work of Clarkson [2], has been given by Fullerton [5]. Another characterization of L_p spaces has been given by Nakano [14, 15, 16].

Characterization of C. There have been many characterizations of the space of real continuous functions. In a preceding paragraph we have already mentioned the theorem of Kakutani [9] and M. and S. Krein [2] representing M-spaces as $C(S)$, for some compact Hausdorff S. Similar results employing norm and lattice properties were obtained by Stone [7; II] and Yosida [1]. Stone [7; I] also gave a characterization in terms of algebra, norm and order properties. Further results based only on algebra and norm properties were developed by Gelfand [1] in the complex case and were further studied by Gelfand and Neumark [1], Arens [6, 7], Arens and Kaplansky [1] and Segal [1]. We will deal with these aspects in a subsequent chapter on B-algebras.

Most of the characterizations arising from algebraic considerations are for complex scalars. The lattice and order properties generally give rise to real $C(S)$ spaces. Arens [4], however, has given conditions under which a real B-algebra is the space of quaternion valued continuous functions on a compact Hausdorff space.

Characterization on the basis of order and linear properties were made by Fan [2] and Kadison [1]. Kadison's work is sufficiently general so that he is able to encompass most of the preceding theories. Nachbin [2] employed order, norm and linear properties.

Results characterizing $C(S)$ among real B-spaces $\mathfrak{X}$ have been given by Arens and Kelley [1], Clarkson [2], Jerison [1] and Myers [2, 3, 4]. These results are concerned with certain special properties possessed by the unit sphere of $\mathfrak{X}$ or $\mathfrak{X}^*$. For example, Clarkson [2; p. 847] found that a real B-space is $C(S)$ for some S if and only if (1) there exists a point v, with $|v| = 1$, and such that any element on $\{x| \, |x| = 1\}$ may be connected with either v or $-v$ by a line segment lying entirely in this surface, and (2) the half-cone of lines from v to every point in the solid unit sphere has the property that the intersection of two of its translates is itself a translate. See also Fullerton [1, 5].

Arens and Kelley [1] showed that $\mathfrak{X}$ is isometrically equivalent with $C(S)$ if and only if (1) the extremal points of the unit sphere U of $\mathfrak{X}^*$ are contained in two supporting hyperplanes, and (2) any collection of extremal points of U whose closure contains no antipodal points lies entirely in some hyperplane supporting U. They gave another condition which was extended by Jerison.

Several other interesting conditions were obtained by Myers. The reader is referred to his papers for details—Myers [4] is an expository article.

Abdelhay [1, 2] has used ring and lattice properties to characterize c_0, and the space of continuous functions vanishing at a point. Jerison [1] has treated the lattice question purely by B-space methods.

Special $C(S)$ spaces. If S is assumed to have special properties, then they are often reflected by the space $C(S)$. For example, M. and S. Krein [1] showed that if S is completely regular, then $C(S)$ is separable if and only if S is compact metric. (See also Myers [3].) That the space $C(S)$ can be decomposed as a direct sum when and only when S is not connected, was proved by Eilenberg [1].

A completely regular space is *extremally disconnected* if the closure of every open set is itself open. Compact extremally disconnected spaces are frequently called *Stone spaces*, since Stone [1] proved that every complete Boolean algebra is isomorphic (as a Boolean algebra) to the Boolean algebra of all open and closed sets of such a space. Such disconnectedness is reflected in the fact that the real space $C(S)$ is a complete lattice in the natural order. (See Stone [8] for additional results of this sort.) We will see in Chapter V that the real space L_∞ is isometrically isomorphic to a real $C(S)$ where S is a Stone space.

Let S be a Stone space and $\mathfrak{X} = C(S)$ the real continuous functions. Grothendieck [4; p. 168] showed that every $\mathfrak{X}$-convergent sequence in $\mathfrak{X}^*$ is actually $\mathfrak{X}^{**}$-convergent, and that if $\mathfrak{Y}$ is a separable B-space, then any operator in $B(\mathfrak{X}, \mathfrak{Y})$ is weakly compact. Goodner [1; p. 103] proved that the unit sphere in $\mathfrak{X}$ is the closed convex hull of its extremal points. Kelley [2] completed a result of Nachbin [3] and Goodner [1] to the effect that if $\mathfrak{X} = C(S)$ is a closed linear manifold of a B-space $\mathfrak{Z}$, then there is a projection of norm one mapping $\mathfrak{Z}$ onto $\mathfrak{X}$. In addition, this property is characteristic in that any B-space possessing it is isometrically equivalent to $C(S)$ where S is a Stone space. Nachbin [3] also proved that if $\mathfrak{Y}$ is a real B-space whose unit sphere contains an extremal point and such that every collection of spheres, every two of which intersect, has non-void intersection, then $\mathfrak{Y}$ is isometrically equivalent to $C(S)$, S a Stone space.

Even more special spaces S (the *hyper-Stone spaces*) are found to be useful in the study of algebras of operators. The reader should refer to Dixmier [3] for a discussion of them.

Other special spaces. In addition to the spaces listed in Section 2, there are many other spaces which have been studied by various authors. Some of these are B-spaces, some only topological linear spaces. We mention here only a few that have been investigated.

Day [1] has considered the space L_p, $0 < p < 1$, which is a linear space of functions on which there is a "norm" defined which does not satisfy the triangle inequality, although there are inequalities which partially compensate for this lack. In particular, Day showed that the only continuous linear functional on this space is the zero functional. On the other hand the space $L_p[0, 2\pi]$ contains the subspace H_p of all functions regular in the unit circle and such that

$$|f|_p = \sup_{0 \leqq r < 1} \left[\frac{1}{2\pi}\int_0^{2\pi} |f(re^{i\theta})|^p \, d\theta\right]^{1/p} < \infty.$$

Even though the space L_p, $0 < p < 1$, has no non-trivial continuous linear functions, Walters [1] has shown that H_p, $0 < p < 1$, has sufficiently many functionals to separate between functions in the space. (See also Livingston [1] and Walters [2].)

Arens [2] has introduced the space $L_\omega[0, 1]$, which consists of all functions f for which $|f|_1, |f|_2, \ldots$ are all finite. He showed that the inclusions $L_\infty \subset L_\omega \subset L_p$ are proper, and although L_ω is a locally convex topological linear algebra, the topology cannot be given by a norm. In particular if U is a convex open set containing 0 for which $UU \subset U$, then $U = L_\omega$. For other remarks about the spaces L_p, $0 < p < 1$, and L_ω, see LaSalle [3].

The B-space H_p, $p \geqq 1$, with the norm described above (or with the norm given as a subspace of L_p) has been studied by many authors. See, in particular, the work of Taylor [4, 5, 6, 7]. For the study of integral functions, see Iyer [1]. An extensive theory of analytic functions from the linear space view, going back to the work of Fantappiè and Volterra, has been developed. In this connection we mention only the papers of Grothendieck [5], Sébastião e Silva [2, 3] and da Silvas Dias [1], where addition references may be found. Spaces consisting of functions which are almost periodic in various senses have been investigated by Bohr and Følner [1] (see also Hartman and Wintner [1]).

Köthe and Toeplitz [1] introduced a class of topological vector spaces, called "perfect spaces" ("vollkommene Räume"), formed by sequences $\{x_n\}$ of real or complex numbers which satisfy a set of conditions of the form

$$\sum |a_n^{(\alpha)} x_n| < \infty.$$

These spaces include some of the classical sequences spaces, admit a theory of duality (Köthe [1—9], Toeplitz [1]) and have applications to the solutions of systems of equations with infinitely many unknowns. Generalizations and other results concerning this type of space (particularly the "Stufenräume" introduced by Köthe [5]) have been presented by Dieudonné [7], Dieudonné and Gomes [1], and Dieudonné and Schwartz [1]. Dieudonné and Schwartz [1], see also Dieudonné [13], have studied a class of spaces which are the union of a collection of locally convex F-spaces, the topology being defined in a suitable manner, and have shown that many of the basic results on B-spaces remain valid in this realm of generality. (For example, the space C of continuous function on $(-\infty, \infty)$ may be considered as a union of $C[-n, n]$, $n = 1, 2, \ldots$, the topology in C becoming uniform convergence on compact sets.) The same type of spaces have been investigated independently by Mazur and Orlicz [3], who have developed an extensive theory.

Vector spaces in which the scalars are taken from non-Archimedean fields have been discussed by I. S. Cohen [1] and Monna [1—19]. See also Fleischer [1], Ingleton [1] and Ono [1].

Birnbaum and Orlicz [1], Orlicz [3, 4] and Zygmund [1; Chap. 4] have discussed a generalization of the L_p spaces of the following type. We let M be a continuous, convex function on $[0, \infty)$, vanishing only for $u = 0$, and for which $u^{-1}M(u)$ approaches 0 and ∞ with u. The space $L_M(0, 1)$ is defined to be the set of measurable functions f on $(0, 1)$ for which

$$\int_0^1 M(|f(x)|)dx < \infty.$$

These spaces are a class of B-spaces which include the spaces L_p, $1 < p < \infty$, and possess many analogous results. For example, there exists a function N having the same properties as M, which plays the role of the conjugate function. In particular if there exists a $k > 0$ such that $M(2u) \leqq kM(u)$, then $(L_M)^*$ is equivalent to the space L_N (see Zaanen [1], where a definition is given that includes the spaces L_1 and L_∞ as well). Conditions for conditional compactness in L_M are exactly similar to those in L_p (Takahashi [1]). Further conditions that an integral operator in L_M be compact are quite analogous to conditions in L_p (Zaanen [2]). These spaces are discussed in detail in the book of Zaanen [5]. See also Krasnosel'skiĭ and Rutickiĭ [1, 2, 3].

For other generalizations of the Lebesgue spaces, see Ellis and Halperin [1], Halperin [1, 3, 4] and Lorentz [1, 2]. Spaces of Denjoy-integrable functions have been studied by Sargent [1, 2].

A function defined in a subset E of Euclidean space for which

$$|f| = \sup_{x, y \in E} \frac{|f(x)-f(y)|}{|x-y|^\alpha} < \infty$$

is said to satisfy a *Hölder condition of exponent* α *in* E, or to be *Hölder continuous*. It is clear that the family of all such functions, with the indicated norm, forms a B-space. This B-space of functions plays an important role in connection with the study of certain singular integral operators, particularly those which arise in the theory of partial differential operators. More detailed indications on this score are given in the section of Notes and Remarks at the end of Chapter XI of Part II. For an account of the principal inequalities connecting Hölder continuous functions and singular integral operators, see Friedrichs [11]. For an illustration of the application of this type of inequality in a connection other than the theory of partial differential operators, see Friedrichs [1]. Very little seems to be known about the B-space properties of the B-space of all functions satisfying a Hölder condition of exponent α, and additional information on this score would be useful.

An interesting variety of locally convex linear topological space (see the next chapter for the general definition of such spaces) has been introduced into consideration by Laurent Schwartz, and seems destined to play a conspicuous part in many branches of analysis. Let I be an open interval of the real axis, and let $C_0^\infty(I)$ denote the linear space of all complex-valued functions defined in I which are infinitely often differentiable and vanish outside a compact subset of I. Let r be an arbitrary positive integer, and $\phi_0, \ldots, \phi_r$ be an arbitrary set of $r+1$ everywhere positive functions defined in I. Then, if we put

$$N(g; \phi_0, \ldots, \phi_r) = \{f \in C_0^\infty(I) \mid |f^{(j)}(t) - g^{(j)}(t)| < \phi_j(t), \quad t \in I, t \in I, \quad j = 0, \ldots, r\}$$

for each $g \in C_0^\infty(I)$, the set of neighborhoods $N(g; \phi_0, \ldots, \phi_r)$ define a topology for $C_0^\infty(I)$ in terms of which the linear space becomes a (locally convex) linear topological space. The space of all continuous linear functionals defined on this linear topological space, which we shall denote by the symbol $D(I)$, is called the space of *distributions defined in* I. The mapping $f \to F_f$, where

$$F_f(g) = \int_I f(t)g(t)dt,$$

imbeds $C_0^\infty(I)$ (and even $L_1(I)$) into $D(I)$, and enables us to regard

$D(I)$ as forming a space of generalized functions. Very many of the generalized functions which have been considered from time to time in heuristic analysis (such as the famous Dirac δ-function and its derivatives) can then be identified with rigorously defined elements of $D(I)$. If a suitable weak topology is defined in $D(I)$, it becomes possible to define such analytic operations as differentiation as continuous mappings in $D(I)$, and, in this sense, to define a suitable generalized derivative for every function in, say, $L_1(I)$. In his groundbreaking book [5], L. Schwartz treats all these questions in detail, and also studies the theory of Fourier expansions, integral transforms, convolutions, etc. of distributions. In addition, he gives a variety of applications of this theory to various branches of analysis. J. Dieudonné and L. Schwartz in [1] give a general theory of a variety of linear topological space including the spaces $C_0^\infty(I)$ and $D(I)$. We shall reproduce parts of the theory of distributions in Chapter XIV of Part II, in connection with the study of partial differential operators to be made there.

The Gauss-Wiener Integral in Hilbert Space

Let E^n be n-dimensional (real) Euclidean space. Let μ_n denote the Borel measure in E^n defined by

$$\mu_n(e) = \pi^{-n/2} \int_e e^{-|x|^2} dx,$$

the integral being an n-dimensional Lebesgue integral, and $|x|$ denoting the norm of the vector x in E^n. Since

$$\int_{-\infty}^{+\infty} e^{-t^2} dt = \sqrt{\pi},$$

$\mu_n(E^n) = 1$. Also, the measure μ_n is easily seen to have the following fundamental property of *rotational invariance*:

$$\mu_n(e) = \mu_n(Ue),$$

where U is any linear and norm-preserving mapping of E^n onto all of itself. The space E^{n+m} may evidently be regarded as the direct sum of $E^n \oplus E^m$ in the sense of Section IV.4. Then $(|x| \oplus |y|)^2 = |x|^2 \oplus |y|^2$, and hence $\exp(-(|x| \oplus |y|)^2) = \exp(-|x|^2) \exp(-|y|^2)$ for all x in E^n, and y in E^m. It follows readily that if e and f are Borel sets in

E^n and E^m, respectively, and if $e \oplus f$ denotes the set $\{x \oplus y | x \in e, y \in f\}$, then $\mu_{n+m}(e \oplus f) = \mu_n(e)\mu_m(f)$. In particular,

$$[1] \qquad \mu_{n+m}(e \oplus E^m) = \mu_n(e).$$

It follows from the rotational invariance of μ_n that μ_n may be regarded as a measure defined intrinsically in any n-dimensional real Hilbert space $\mathfrak{H}$, without reference to any particular coordinate system in that space. This measure will be called the *Gauss measure* in $\mathfrak{H}$. In particular, let $\mathfrak{H}_0$ be such a space, and $\mathfrak{H}_1$ a subspace of $\mathfrak{H}_0$. Let E be the orthogonal projection of $\mathfrak{H}_0$ onto $\mathfrak{H}_1$. Suppose that $\mu_{\mathfrak{H}_0}$ and $\mu_{\mathfrak{H}_1}$ denote the Gauss measure in $\mathfrak{H}_0$ and $\mathfrak{H}_1$, respectively. Let e be a Borel subset of $\mathfrak{H}_0$ such that $x \in e$ if and only if $Ex \in e$. Then it follows easily from (1) that

$$(2) \qquad \mu_{\mathfrak{H}_1}(e\mathfrak{H}_1) = \mu_{\mathfrak{H}_0}(e).$$

Equation (2) enables us to define a *Gauss measure* in real *Hilbert space* $\mathfrak{H}$ as follows. Call a Borel subset e of $\mathfrak{H}$ a *cylinder set* if there exists an orthogonal projection E of $\mathfrak{H}$ onto a finite dimensional subspace $\mathfrak{H}_1$ of $\mathfrak{H}$ such that $x \in e$ if and only if $Ex \in e$. If this is the case, put

$$(3) \qquad \mu(e) = \mu_{\mathfrak{H}_1}(e\mathfrak{H}_1).$$

Then (2) shows that the left side of (3) is independent of $\mathfrak{H}_1$; i.e., that the definition given in (3) is univalent and hence legitimate.

It follows very easily from the corresponding properties of finite dimensional Gauss measures that

(i) μ is a non-negative, finitely additive set function defined on the field Σ of cylinder sets in $\mathfrak{H}$.

(ii) Let U be a linear mapping of $\mathfrak{H}$ into itself which is norm-preserving. Then, for every e in Σ, UE is also in Σ and $\mu(e) = \mu(Ue)$. This is the property of *rotational invariance in Hilbert space.*

(iii) $\mu(\mathfrak{H}) = 1$.

Using (i) and the general theory of Sections III.1, III.2, and III.3, we may now establish an integration theory (finitely additive) for functions on $\mathfrak{H}$. It turns out, however, that this integration theory is not sufficiently inclusive for the customary applications of the Gauss integral. Hence, it is well to extend this theory somewhat, as follows.

(a) First establish the integration theory of Sections III.1, III.2, and III.3, which gives among other things a definition of μ-measurable and μ-integrable functions.

(b) Let $\mathscr{E}$ be the set of all projections E of $\mathfrak{H}$ onto its finite dimensional subspaces, and let $\mathscr{E}_1$ be the set of all bounded linear maps in $\mathfrak{H}$ which have finite dimensional ranges. Put

$$|f| = \inf_{\substack{E \in \mathscr{E} \\ \varepsilon > 0}} \sup_{\substack{F \in \mathscr{E}_1 \\ |E-FE| < \varepsilon}} \int_{\mathfrak{H}} |f(Fx)|\mu(dx)$$

if this quantity is finite; otherwise put $|f| = \infty$. Then $|f|$ defines a norm on a certain linear subspace Φ of the set of functions on $\mathfrak{H}$. It is easy to see that Φ includes the space Φ_0 of all μ-integrable functions, and that the linear functional S

$$S : f \to \int_{\mathfrak{H}} f(x)\mu(dx), \qquad f \in \Phi_0$$

is uniformly continuous on Φ_0 relative to the topology induced by the norm $|f|$. Hence, by Theorem I.6.17, S may be extended uniquely to a continuous linear functional defined on all of $\overline{\Phi}_0 = \Phi_1$. We say that a function f in Φ_1 is μ-integrable *in the extended sense*, and write $\int_{\mathfrak{H}} f(x)\mu(dx)$ for $S(f)$ if f is in Φ_1. This "extended sense" theory has many of the properties of ordinary finitely additive integration theory. In addition to these, it is rotationally invariant, i.e., if *U is a linear norm-preserving mapping of $\mathfrak{H}$ into itself*, and if $f(\cdot)$ is in Φ_1, then $f(U(\cdot))$ is in Φ_1, *and*

$$\int_{\mathfrak{H}} f(x)\mu(dx) = \int_{\mathfrak{H}} f(Ux)\mu(dx).$$

Suppose that $\mathfrak{H}$ is separable, and that we realize it as l_2, i.e., we map it in a one-to-one, linear, and norm-preserving way on l_2 (say, by use of Theorem 4.16). Since μ is intrinsic to $\mathfrak{H}$, we obtain a Gauss measure μ in l_2, and it is easy to see from the definitions that this measure satisfies the equation

$$\mu(\{[x_i] \in l_2 | a_i \leqq x_i \leqq b_i, \quad i = 1, \ldots, n\})$$
$$= \pi^{-n/2} \int_{a_1}^{b_1} \int_{a_2}^{b_2} \cdots \int_{a_n}^{b_n} \exp\,[-(y_1^2 + \cdots + y_n^2)] dy_1 \cdots dy_n.$$

Consequently, it is very closely related with the infinite product measure $\mu_\infty = \mu_1 \times \mu_1 \times \ldots$ defined on the infinite product s of countably many replicas of the real axis, μ_1 being as above the one-dimensional Gauss measure on the real line. The space s is, of course, the space of all real sequences $x = [x_1]$ as distinct from l_2, which is the subspace of s determined by the condition $\sum_{1=1}^{\infty} x_i^2 < \infty$. The measure μ_∞ is countably additive; the subset l_2 of s is of μ_∞-measure zero; the measure μ is not countably additive. In fact, using the relation between μ_∞ and μ, it is easy to see that l_2 is the union of a countable family of μ-null sets. These relationships may be summed up in the following heuristic formula: by passing from the full infinite product space s to its subset l_2, we lose countable additivity, but gain in return an important property of rotational invariance.

As a general reference for the above remarks, see Friedrichs [12], pp. 52—63.

There is yet another method, introduced by Norbert Wiener, for obtaining a countably additive theory at the cost of abandoning rotational invariance. This may be described as follows. Represent $\mathfrak{H}$ as the real space $L_2(-\infty, +\infty)$. Then the mapping $f \to Tf$ defined by

$$(Tf)(x) = \int_0^x f(t)dt, \qquad f \in L_2(-\infty, +\infty),$$

maps L_2 into a (rather "thin") subset of the space $\hat{C}$ of all continuous functions defined on the real axis. This enables one to define a field $\hat{\Sigma}$ of subsets of $\hat{C}$ and an additive set function $\hat{\mu}$ on $\hat{\Sigma}$ as follows:

$$\hat{\Sigma} = \{e \in \hat{C} | x_T - 1_e \in \Phi_1\}$$

$$\hat{\mu}(e) = \int_{\mathfrak{H}} \chi_T - 1_e(x)\mu(dx), \qquad e \in \hat{\Sigma},$$

χ_{e_0} denoting the characteristic function of the set e_0, as usual. This measure $\hat{\mu}$ on $\hat{C}$ is substantially the same as the measure of Wiener.

More specifically, note that every set of the form $\{f \in \hat{C} | f(t_0) \in e_0\}$, where $-\infty < t_0 < \infty$ and e_0 is a Borel subset of the field of scalars, belongs to $\hat{\Sigma}$. Let $\hat{\Sigma}_0$ be the subfield of $\hat{\Sigma}$ generated by the family of sets of this form. Then it may be shown that $\hat{\mu}$ is countably additive on $\hat{\Sigma}_0$. Hence, by Theorem III.5.8, $\hat{\mu}$ may be extended to a countably additive measure μ_w on the σ-field Σ_w of subsets of $\hat{C}$ generated by $\hat{\Sigma}_0$. The measure space $(\hat{C}, \Sigma_w, \mu_w)$ is the Wiener measure space.

The theory of this measure space has been studied quite elaborately. The fundamental early papers are Wiener [6] and [7], and Paley, Wiener, and Zygmund [1]. Most of the subsequent progress is due to

R. H. Cameron, W. T. Martin, and their students. Cameron and Martin [3], [5], [7], and Cameron and Fagan [1], study the effect of various sorts of mappings in $\hat{C}$ on the measure μ_w, establishing generalizations of the Jacobian determinant law governing changes of variable in finite dimensional multiple integrals. Cameron and Martin [6] evaluate various Wiener integrals by the use of a general method involving "changes of variable" in the space $\hat{C}$, and making use of certain Sturm-Liouville differential equations. Cameron and Martin [2], Cameron and Hatfield [1], [2] establish a complete orthonormal set of "Fourier-Hermite functionals" in $L_2(\hat{C}, \Sigma_w, \mu_w)$, and study the theory of expansion of arbitrary elements of $L_1(\hat{C}, \Sigma_w, \mu_w)$ in these functionals. Cameron [4] and Cameron and Martin [9] provide a related theory of Fourier integrals. Cameron and Martin [1], [4], Cameron and Shapiro [1], Cameron, Lindgren, and Martin [1] show how the solutions of certain non-linear integral equations can be expressed "explicitly" in terms of Wiener integrals. Cameron [2] and Owchar [1] study formulas for Wiener integrals related to the formula

$$\frac{d}{dt}\int^{t} f(x)dx = f(t)$$

for functions of a real variable. Cameron [1] gives a "Simpson's rule" for the numerical evaluation of Wiener integrals.

See also Kac [2], Marumaya [1], Cameron [8], Cameron and Graves [1], Cameron [5], Wiener [8], Paley and Wiener [1], Chapter IX.

A most interesting development of the theory begins with the thesis of the physicist Feynman [1], and continues in the studies of Kac [1] and [3], Rosenblatt [1], Tingley [1], Fortet [4], Cameron [3], Montroll [1], Steinberg [1]. The formula

$$[*] \qquad f(t, x) = \frac{1}{\sqrt{\pi t}}\int_{-\infty}^{+\infty} e^{-(x-y)^2/t} f(y)dy$$

expresses, as is well-known, the solution of the partial differential initial value problem of "heat flow"

$$\frac{\partial}{\partial t} f(x, t) = \frac{1}{4}\frac{\partial^2}{\partial x^2} f(x, t), \qquad t \geqq 0; \qquad f(x, 0) = f(x).$$

It is easy to see from the definition of Wiener's measure that this may be written as

$$f(x, t) = \int_G f(\varphi(t) + x)\mu_w(d\varphi).$$

There exists a corresponding Wiener-integral formula for the solution F of the more general parabolic initial value problem.

$$[\dagger] \qquad \frac{\partial}{\partial t} F(x, t) = \frac{1}{4}\frac{\partial^2}{\partial x^2} F(x, t) + V(x, t)F(x, t), \qquad t \geqq 0;$$

$$F(x, 0) = f(x),$$

V being a given coefficient function. This formula is

$$[\dagger\dagger] \quad F(x, t) = \int_{\hat{G}} f(\varphi(t)+x) \exp\left\{\int_0^t V\big(t-s, \ \varphi(s)+x\big)ds\right\} \mu_w(d\varphi).$$

The authors cited above have investigated various aspects of the relationship between the partial differential problem [†] and the formula [††]. Cameron [3] in particular gives very detailed results under very mild analytical restrictions. It should also be remarked that the connection of [†] and [††] is related to a more general relationship, known from the probability theory of Markoff processes, between a Markoff process in its "transition probability" and its "sample function" representations. For applications of formula [††] to the theory of probability, see Kac [1] and Erdos and Kac [1]. Gelfand and Yaglom [1] give an excellent review of the theory of the Wiener integral with special emphasis on formula [†].

For various other, quite different, approaches to the problem of integration in Hilbert space, see K. Löwner [1]; E. R. Lorch [12], [13], [14]; Friedrichs [12], pp. 121–132; P. Levy [1], pp. 209–355.

CHAPTER V

Convex Sets and Weak Topologies

In Chapter II we have seen that the continuous linear functionals provide important tools in the study of B-spaces; the present chapter continues these investigations in more general spaces. We begin by studying the notion of convexity in a general linear space, and by proving a fundamental lemma equivalent to the Hahn-Banach theorem, which relates linear functionals to convex sets. These results are examined in Section 2 under the additional assumption that the space is a linear topological space. In Section 3 it is shown how certain classes of linear functionals determine topologies in a linear space. In particular for B-spaces a topology, called the weak topology, may be introduced such that weak convergence of elements, as defined in II.3 is equivalent to convergence in the weak topology.

Sections 4–6 continue the study of various topologies for B-spaces determined by linear functionals. In particular, an investigation is made of compactness properties, relations to metric topologies, reflexivity, bounded and unbounded sets, and sequential properties. Many of these results are valid for locally convex linear topological spaces.

Sections 8–10 discuss the topics of extremal points, tangent planes, and fixed point theorems. The ideas of these sections, while highly interesting in themselves, will be applied much less frequently in later chapters. Additional results and examples will be found in the exercises of Sections 7 and 11.

1. Convex Sets in Linear Spaces

In this section, $\mathfrak{X}$, $\mathfrak{Y}$, etc. will denote linear vector spaces, and p, q, x, y, . . ., points in these spaces. The symbols α, β, . . . will denote real or complex numbers, a, b, . . ., real numbers.

1 DEFINITION. A set $K \subseteq \mathfrak{X}$ is *convex* if $x, y \in K$, and $0 \leqq a \leqq 1$, imply $ax+(1-a)y \in K$.

The following lemma is an obvious consequence of Definition 1.

2 LEMMA. *The intersection of an arbitrary family of convex subsets of the linear space $\mathfrak{X}$ is convex.*

As examples of convex subsets of $\mathfrak{X}$, we note the subspaces of $\mathfrak{X}$, and the subsets of $\mathfrak{X}$ consisting of one point.

3 LEMMA. *Let $x_1, \ldots, x_n$ be points in the convex set K and let $a_1, \ldots, a_n$ be non-negative scalars with $a_1+\ldots+a_n = 1$. Then $a_ix_1+\ldots+a_nx_n$ is in K.*

PROOF. If $n = 2$, the statement is true by definition. Suppose that the lemma is true for $n = m$. Then, setting $b = a_2+\ldots+a_{m+1}$, and

$$y = (a_2/b)x_2+\ldots+(a_{m+1}/b)x_{m+1},$$

it follows from our induction hypothesis that $y \in K$. Since $a_1+b = 1$,

$$\sum_{i=1}^{m+1} a_ix_i = a_1x_1+by \in K.$$

Q.E.D.

4 LEMMA. *Let $\mathfrak{X}$ be a linear space. If $K_1, K_2 \subseteq \mathfrak{X}$ are convex, then βK_1 and $K_1 \pm K_2$ are convex.*

PROOF. If $x, y \in \beta K_1$, we have $x = \beta x'$, $y = \beta y'$, for some $x', y' \in K_1$. Then, if $0 \leqq a \leqq 1$, $ax+(1-a)y = \beta\{ax'+(1-a)y'\} \in \beta K_1$, since K_1 is convex. The second part of the lemma may be proved in the same way. Q.E.D.

The proof of the following lemma employs the same idea.

5 LEMMA. *If T is a linear map from $\mathfrak{X}$ to $\mathfrak{Y}$, and K is a convex set in $\mathfrak{X}$, then TK is convex.*

6 DEFINITION. If M is a subset of the linear space $\mathfrak{X}$, then a point $p \in M$ is called an *internal point* of M if, for each $x \in \mathfrak{X}$, there exists an $\varepsilon > 0$ such that $p + \delta x \in M$ for $|\delta| \leqq \varepsilon$. A point $p \in \mathfrak{X}$ is called a *bounding* point of M if p is neither an internal point of M nor of the complement of M.

7 DEFINITION. Let K be a convex set in a linear space $\mathfrak{X}$ and let the origin 0 of $\mathfrak{X}$ be an internal point of K. For each $x \in \mathfrak{X}$, let $I(x)$

$= \{a | a > 0, a^{-1}x \in K\}$, and $\mathfrak{k}(x) = \inf_{a \in I(x)} a$. The function $\mathfrak{k}(x)$ is called the *support function* of K.

For example, if K is the unit sphere of a B-space $\mathfrak{X}$, then $\mathfrak{k}(x) = |x|$.

8 LEMMA. *Let K be a convex set in $\mathfrak{X}$ containing the origin as an internal point, and let $\mathfrak{k}$ be its support function. Then*

(a) $\mathfrak{k}(x) \geqq 0$;

(b) $\mathfrak{k}(x) < +\infty$;

(c) $\mathfrak{k}(ax) = a\mathfrak{k}(x)$ *for* $a \geqq 0$;

(d) *if* $x \in K$, *then* $\mathfrak{k}(x) \leqq 1$;

(e) $\mathfrak{k}(x+y) \leqq \mathfrak{k}(x)+\mathfrak{k}(y)$;

(f) *the set of internal points of K is characterized by the condition* $\mathfrak{k}(x) < 1$, *and the set of bounding points by the condition* $\mathfrak{k}(x) = 1$.

PROOF. Statement (a) is obvious. Statement (b) follows from the fact that the origin is an internal point of K. Statements (c) and (d) are self-evident. To prove (e), we note that if $c > \mathfrak{k}(x)+\mathfrak{k}(y)$, then $c = a+b$ with $a > \mathfrak{k}(x)$, $b > \mathfrak{k}(y)$. It follows from the convexity of K that the point

$$\frac{x+y}{c} = \frac{x+y}{a+b} = \frac{a(a^{-1}x)+b(b^{-1}y)}{a+b}$$

is in K, since $a^{-1}x$ and $b^{-1}y$ are both in K. Hence $\mathfrak{k}(x+y) \leqq c$.

If x is an internal point of K, then $x+\varepsilon x = (1+\varepsilon)x$ is in K for some sufficiently small $\varepsilon > 0$, so that $\mathfrak{k}(x) \leqq (1+\varepsilon)^{-1}$. Conversely, if $\mathfrak{k}(x) < 1$, let $\varepsilon = 1-\mathfrak{k}(x)$. To complete the proof of (f) we shall suppose that $\mathfrak{X}$ is real, leaving the details of the complex case to the reader. Let $|\delta|(\mathfrak{k}(y)+\mathfrak{k}(-y)) < \varepsilon$. Then

$$\mathfrak{k}(x+\delta y) < (1-\varepsilon)+\varepsilon = 1,$$

irrespective of whether δ is positive or negative, so that $1 \cdot (x+\delta y) = x + \delta y \in K$. Therefore, x is an internal point of K. In the same way it follows that $\mathfrak{k}(x) > 1$ characterizes the internal points of the complement of K. Thus, $\mathfrak{k}(x) = 1$ characterizes the boundary points of K. Q.E.D.

9 DEFINITION. If $\mathfrak{X}$ is a vector space, and M and N are subsets of $\mathfrak{X}$, a functional f on $\mathfrak{X}$ is said to *separate* M and N if there exists a real constant c with $\mathscr{R}f(M) \geqq c$, $\mathscr{R}f(N) \leqq c$.

10 LEMMA. *The linear functional f separates the subsets M and N of $\mathfrak{X}$ if and only if it separates the subsets $M-N$ and $\{0\}$ of $\mathfrak{X}$.*

The proof is elementary, and is left to the reader.

In dealing with subspaces, it is often convenient to make use of the following lemma:

11 LEMMA. *Let f be a linear functional on the vector space $\mathfrak{X}$, and let $\mathfrak{Y}$ be a subspace of $\mathfrak{X}$. If $f(\mathfrak{Y})$ is not the whole field of scalars, then $f(\mathfrak{Y}) = 0$.*

PROOF. Suppose that there exists a $y \in \mathfrak{Y}$ with $f(y) \neq 0$. Then $f(\alpha y/f(y)) = \alpha$, so that every scalar is in $f(\mathfrak{Y})$. Q.E.D.

12 THEOREM. (*Basic separation theorem*) *Let M and N be disjoint convex subsets of a linear space $\mathfrak{X}$, and let M have an internal point. Then there exists a non-zero linear functional f which separates M and N.*

PROOF. Suppose that $\mathfrak{X}$ is a real vector space. If m is an internal point of M, then the origin 0 of $\mathfrak{X}$ is an internal point of $M-m$. It is easily seen that a functional separates M and N if and only if it separates $M-m$ and $N-m$. Thus it suffices to prove the theorem under the additional assumption that 0 is an internal point of M.

Let p be any point of N, so that $-p$ is an internal point of $M-N$, and 0 an internal point of $K = M-N+p$. Since M and N are disjoint, the set $M-N$ does not contain 0; hence, K does not contain p. Let $\mathfrak{k}$ be the support function of K, so $\mathfrak{k}(p) \geqq 1$. If we put $f_0(ap) = a\mathfrak{k}(p)$, then f_0 is a linear functional defined on the one dimensional subspace of $\mathfrak{X}$ which consists of real multiples of p. Moreover, $f_0(ap) \leqq \mathfrak{k}(ap)$ for all real a, since for $a \geqq 0$ we have $f_0(ap) = \mathfrak{k}(ap)$, while for $a < 0$ we have $f_0(ap) = af_0(p) < 0 \leqq \mathfrak{k}(ap)$. By the Hahn-Banach theorem (II.3.10), f_0 can be extended to a linear functional f such that $f(x) \leqq \mathfrak{k}(x)$ for all $x \in \mathfrak{X}$. It follows that $f(K) \leqq 1$, while $f(p) \geqq 1$. Thus f separates K and $\{p\}$; by Lemma 10, f separates $M-N$ and $\{0\}$, and, again by Lemma 10, f separates M and N. Thus the theorem is proved for real spaces.

If the space $\mathfrak{X}$ is complex, we can still regard it as a vector space over the subfield of real scalars. By the proof given above, we can construct a real-valued function f on $\mathfrak{X}$, such that $f(x+y) = f(x)+f(y)$, $f(\alpha x) = \alpha f(x)$ for α real, and such that $f(M)$ and $f(N)$ belong to non-overlapping intervals. Then, the function $F(x) = f(x)-if(ix)$ is a non-zero linear functional on the complex space $\mathfrak{X}$, which separates the sets M and N. Q.E.D.

2. Linear Topological Spaces

In this section, the results of Section 1 are applied to linear topological spaces. Statements 1—6 of this section are elementary results. Statements 7—12 are applications of the fundamental Theorem 1.12.

1 THEOREM. (a) *The closure and the interior of a convex set in a linear topological space are convex.*

(b) *An interior point of a set in a linear topological space is an internal point of the set.*

(c) *If a convex set K in a linear topological space has at least one interior point, then a point p is an internal point of K if and only if it is an interior point; it is a bounding point if and only if it is a boundary point. Moreover, the interior of K is dense in K.*

PROOF. Let $\mathfrak{X}$ be a linear topological space, let K be a subset of $\mathfrak{X}$ and let I be the closed unit interval. Then K is convex if and only if the mapping

$$\psi : [x, y, a] \rightarrow ax+(1-a)y$$

of $\mathfrak{X}\times\mathfrak{X}\times I$ into $\mathfrak{X}$, sends $K\times K\times I$ into K. But, since ψ is continuous, and $\overline{K\times K\times I} = \overline{K}\times\overline{K}\times I$, we have

$$\psi(\overline{K}\times\overline{K}\times I) = \psi(\overline{K\times K\times I}) \subseteq \overline{\psi(K\times K\times I)} \subseteq \overline{K}$$

whenever K is convex. Thus $\overline{K}$ is convex if K is convex.

Next we show that if p is an interior point of K, and q a point in $\overline{K}$, then $ap+(1-a)q$ is an interior point of K for $0 < a < 1$. Indeed, there exists a neighborhood U of the origin such that $p+U \subseteq K$, and a point $q_1 \in K$ in the neighborhood $a(a-1)^{-1}U+q$ of q. Now, since K is convex, the open set $U_1 = a(p+U)+(1-a)q_1$ lies in K. Since $(1-a)(q-q_1) \in aU$, $ap+(1-a)q = ap+(1-a)q_1+(1-a)(q-q_1) \in U_1$, and hence $ap+(1-a)q$ is interior to K.

The second part of (a), and the last part of (c), follow immediately from what we have just proved. Statement (b) is an immediate consequence of the definition of a linear topological space. It follows from

(b) that a bounding point of K is a boundary point of K; it remains to show that if the convex set K has at least one interior point p, an internal point q_1 is an interior point, and a boundary point q_2 is a bounding point.

Since q_1 is internal, the vector $r = -\varepsilon p+(1+\varepsilon)q_1$ belongs to K for some sufficiently small positive ε. It follows from the above that for some sufficiently small positive ε, $q_1 = r/(1+\varepsilon)+\varepsilon p/(1+\varepsilon)$ is an interior point of K. Since q_2 is a boundary point, it is not an internal point of K. But we have shown that $ap+(1-a)q_2 \in K$ for $0 < a < 1$; hence q_2 is not an internal point of the complement of K. Therefore q_2 is a bounding point of K. Q.E.D.

2 DEFINITION. If A is a subset of the linear space $\mathfrak{X}$, the set co (A), called the *convex hull* of A, is the intersection of all convex sets containing A; if $\mathfrak{X}$ is a linear topological space, the set $\overline{\text{co}}\,(A)$, called the *closed convex hull* of A, is the intersection of all closed convex sets containing A.

It is readily seen that $\text{co}(A)$ is the set of all linear combinations $\sum_{i=1}^{n} a_i x_i$ of elements $x_i \in A$ in which $0 \leqq a_i \leqq 1$, and $\sum_{i=1}^{n} a_i = 1$. Such linear combinations are sometimes called *convex combinations*; hence $\text{co}(A)$ is the set of all convex combinations of points of A.

The operations $\text{co}(A)$ and $\overline{\text{co}}(A)$ map sets into sets. Some of the elementary properties of these operations are given in Lemma 4 below. The proof of Lemma 4 requires the following lemma on topological groups.

3 LEMMA. *If A and K are closed subsets of an additive topological group G, with K compact, then $A+K$ is closed.*

PROOF. Let $p \in \overline{A+K}$. For each neighborhood U of p, let $K_U = \{k | k \in K,\ k \in U-A\}$. Since $p \in \overline{A+K}$, each K_U is non-void. It is evident that if $U_1 \subseteqq U_2$, then $K_{U_1} \subseteqq K_{U_2}$. It follows that the closed sets $\overline{K}_U$ have the finite intersection property. We can now apply Lemma I.5.6, and conclude that there is a point $k_0 \in K$ common to all the $\overline{K}_U$. Thus, if N is any neighborhood of the identity,

$$(N+k_0) \cap (N+p-A) \neq \phi.$$

This means that $(N-N+k_0) \cap (p-A) \neq \phi$. If M is any neighborhood of the identity, there is a neighborhood N of the identity such that

$N-N \subseteqq M$. Thus, any neighborhood of k_0 intersects $p-A$. Since A is closed, $p-A$ is closed. Hence $k_0 \in p-A$, and thus $p \in A+k_0 \subseteqq A+K$. Q.E.D.

Since the commutativity of the group G is not essential to the proof, the same result holds for non-Abelian topological groups.

4 Lemma. *For arbitrary sets A, B in a linear space $\mathfrak{X}$:*

(i) $\text{co}(\alpha A) = \alpha\, \text{co}(A)$, $\text{co}(A+B) = \text{co}(A)+\text{co}(B)$.

If $\mathfrak{X}$ is a linear topological space, then

(ii) $\overline{\text{co}}(A) = \overline{\text{co}(\overline{A})}$,

(iii) $\overline{\text{co}}(\alpha A) = \alpha\, \overline{\text{co}}(A)$,

(iv) *If $\overline{\text{co}}(A)$ is compact, then $\overline{\text{co}}(A+B) = \overline{\text{co}}(A)+\overline{\text{co}}(B)$.*

Proof. The first part of statement (i) follows in an elementary fashion from Lemma 1.4. Lemma 1.4 also shows that $\text{co}(A+B) \subseteqq \text{co}(A)+\text{co}(B)$. Now, if $y \in B$ and $x = \sum_{i=1}^{n} a_i x_i \in \text{co}(A)$, then $x+y = \sum_{i=1}^{n} a_i(x_i+y)$, so $\text{co}(A)+y = \text{co}(A+y)$, and thus $\text{co}(A)+B \subseteqq \text{co}(A+B)$. The same argument shows that $\text{co}(A)+\text{co}(B) \subseteqq \text{co}(\text{co}(A)+B)$. Thus $\text{co}(A)+\text{co}(B) \subseteqq \text{co}(\text{co}(A)+B) \subseteqq \text{co}(\text{co}(A+B)) = \text{co}(A+B)$. This completes the proof of (i). To prove (ii), note that $\overline{\text{co}}(A)$ is closed and contains $\text{co}(A)$. Thus $\overline{\text{co}}(A) \supseteqq \overline{\text{co}(A)}$. By Theorem 1, the closure of a convex set is convex; so $\overline{\text{co}(A)}$ is convex, and includes A. Thus $\overline{\text{co}(A)} \supseteqq \overline{\text{co}}(A)$, which completes the proof of (ii). Statement (iii) follows from (i) and (ii). We now prove (iv). Statement (i) and Lemma 3 show that $\overline{\text{co}}(A)+\overline{\text{co}}(B)$ is convex and closed, so that $\overline{\text{co}}(A+B) \subseteqq \overline{\text{co}}(A)+\overline{\text{co}}(B)$. Now, since $x+y$ is a continuous function of x and y, $\overline{X}_1+\overline{Y}_1 \subseteqq \overline{X_1+Y_1}$ for arbitrary subsets X_1, Y_1 of $\mathfrak{X}$. Hence, from (i) and (ii),

$$\overline{\text{co}}(A+B) = \overline{\text{co}(A)+\text{co}(B)} \supseteqq \overline{\text{co}}(A)+\overline{\text{co}}(B).$$

This completes the proof of (iv). Q.E.D.

5 Lemma. *Let A, B be sets in a linear topological space. If the closed convex hulls of A and B are compact, then $\overline{\text{co}}(A \cup B) = \text{co}(\overline{\text{co}}(A) \cup \overline{\text{co}}(B))$.*

Proof. The inclusion $\text{co}(\overline{\text{co}}(A) \cup \overline{\text{co}}(B)) \subseteqq \overline{\text{co}}(A \cup B)$ is clear from the definitions involved. Let $K_1 = \overline{\text{co}}(A)$, $K_2 = \overline{\text{co}}(B)$. The map

$$\psi : (A, p, q) \to ap+(1-a)q,$$

is a continuous mapping of the compact space $K = [0, 1]\times K_1 \times K_2$

into $\text{co}(K_1 \cup K_2)$; therefore $\psi(K)$ is compact and hence closed. But $A \cup B \subseteq K_1 \cup K_2 \subseteq \psi(K)$. If $\psi(K)$ is also convex, then $\overline{\text{co}}(A \cup B) \subseteq \psi(K) \subseteq \text{co}(K_1 \cup K_2)$. But this fact is shown by the following elementary calculation.

If $0 \leqq a_1, a_2, b \leqq 1$, then

$$b\{a_1p_1+(1-a_1)q_1\}+(1-b)\{a_2p_2+(1-a_2)q_2\}$$
$$=\{ba_1+(1-b)a_2\}\left\{\frac{ba_1}{ba_1+(1-b)a_2}p_1+\frac{(1-b)a_2}{ba_1+(1-b)a_2}p_2\right\}$$
$$+\{b(1-a_1)+(1-b)(1-a_2)\}\left\{\frac{b(1-a_1)}{b(1-a_1)+(1-b)(1-a_2)}q_1\right.$$
$$\left.+\frac{(1-b)(1-a_2)}{b(1-a_1)+(1-b)(1-a_2)}q_2\right\}.$$

Q.E.D.

6 **Theorem.** (*Mazur*) *Let $\mathfrak{X}$ be a B-space, and let $A \subseteq \mathfrak{X}$ be compact. Then $\overline{\text{co}}(A)$ is compact.*

Proof. As a closed subset of a complete space, the set $\overline{\text{co}}(A)$ is complete. Hence, by Theorem I.6.15, it suffices to show that $\overline{\text{co}}(A)$ is totally bounded. Let $\varepsilon > 0$. Since A is totally bounded, there is a finite subset $\{z_1, \ldots, z_n\} \subseteq A$ such that $A \subseteq S(\{z_1, \ldots, z_n\}, \varepsilon/4)$. Let $K = \text{co}(\{z_1, \ldots, z_n\})$. Now $\overline{\text{co}}(A) \subset S(\text{co}(A), \varepsilon/4)$. But if $y \in \text{co}(A)$, $y = \sum_{i=1}^m a_i y_i$, where $y_i \in A$, $a_i \geqq 0$ and $\sum_{i=1}^m a_i = 1$. Let v be a function on A to $\{1, \ldots, n\}$, such that if $x \in A$, $|x - z_{v(x)}| < \varepsilon/4$. Then,

$$\left|y - \sum_{i=1}^m a_i z_{v(y_i)}\right| = \left|\sum_{i=1}^m a_i(y_i - z_{v(y_i)})\right| < \frac{\varepsilon}{4},$$

and thus $\overline{\text{co}}(A) \subset S(K, \varepsilon/2)$. Now,

$$K = \{k \mid k = \sum_{i=1}^n a_i z_i,\ a_i \geqq 0,\ \sum_{i=1}^n a_i = 1\}.$$

The mapping,

$$\psi : (a_1, \ldots, a_n, z_1, \ldots, z_n) \to \sum_{i=1}^n a_i z_i,$$

is a continuous mapping of the compact set $[0, 1] \times \ldots \times [0, 1] \times \{z_1\} \times \ldots \times \{z_n\}$ onto K. Thus K is compact, and hence totally bounded. There is, then, a finite subset $\{k_1, \ldots, k_m\}$ of K, such that $K \subset$

$S(\{k_1, \ldots, k_m\}, \varepsilon/2)$. But then $\overline{\text{co}}(A) \subset S(\{k_1, \ldots, k_m\}, \varepsilon)$. Q.E.D.

7 LEMMA. *If a linear functional on a linear topological space separates two sets, one of which has an interior point, then the functional is continuous.*

PROOF. Let $\mathfrak{X}$ be a linear topological space, and $A_1, A_2 \subseteqq \mathfrak{X}$. Let h be a linear functional separating A_1 and A_2, and let p be an interior point of A_1. If f, g are the real and imaginary parts of h, then $g(x) = -f(ix)$, and hence, to show that h is continuous, it suffices to prove the continuity of f. Let N be a neighborhood of the origin such that $p+N \subseteqq A_1$. Then $f(N) \subseteqq f(A_1)-f(p)$, and $f(N)$ is contained in a proper subinterval $[-a, \infty)$, or $(-\infty, a]$, of the real axis, where $a > 0$. Let $M = N \cap (-N)$; then $M = -M$, and M is a neighborhood of the origin such that $f(M)$ is contained in the interval $[-a, a]$. In this case, $f(\varepsilon a^{-1}M)$ is contained in the interval $[-\varepsilon, \varepsilon]$. Since $\varepsilon a^{-1}M$ is a neighborhood of the origin, f is continuous at 0. By Lemma II.1.3, f is continuous. Q.E.D.

Lemma 7 and Theorem 1.12 yield the following result.

8 THEOREM. *In a linear topological space, any two disjoint convex sets, one of which has an interior point, can be separated by a non-zero continuous linear functional.*

9 DEFINITION. A linear topological space is said to be *locally convex* if it possesses a base for its topology consisting of convex sets.

10 THEOREM. *If K_1 and K_2 are disjoint closed convex subsets of a locally convex linear topological space $\mathfrak{X}$, and if K_1 is compact, then there exist constants c and ε, $\varepsilon > 0$, and a continuous linear functional f on $\mathfrak{X}$, such that*

$$\mathscr{R}f(K_2) \leqq c-\varepsilon < c \leqq \mathscr{R}f(K_1).$$

PROOF. By Lemma 3, the set K_1-K_2 is closed; by Lemma 1.4, it is convex. Since K_1-K_2 does not contain 0, there is a convex neighborhood U of 0, which does not intersect K_1-K_2. By Theorem 8, some continuous non-zero linear functional f separates U and K_1-K_2; i.e., there exists a real constant d, such that $\mathscr{R}f(K_1-K_2) \geqq d$, $\mathscr{R}f(U) \leqq d$. Now, since f is non-zero, there exists an $x \in \mathfrak{X}$ with $f(x) = 1$. It follows that $f(\alpha x) = \alpha$. On the other hand, $\alpha x \in U$ for α sufficiently small. Thus, there exists an $\varepsilon > 0$ such that $f(U)$ contains every scalar of modulus less than ε. Hence $\mathscr{R}f(K_1)-\mathscr{R}f(K_2) = \mathscr{R}f(K_1-K_2) \geqq d \geqq \varepsilon$, so that every number in $\mathscr{R}f(K_1)$ is at least ε greater than any number in $\mathscr{R}f(K_2)$. The desired conclusion follows by placing $c = \inf \mathscr{R}f(K_1)$. Q.E.D.

→ 11 COROLLARY. *If K_1, K_2 are disjoint closed, convex subsets of a locally convex linear topological space $\mathfrak{X}$, and if K_1 is compact, then some non-zero continuous linear functional on $\mathfrak{X}$ separates K_1 and K_2.*

12 COROLLARY. *If K is a closed convex subset of a locally convex linear topological space, and $p \notin K$, then some non-zero continuous linear functional separates K and p.*

13 COROLLARY. *If p and q are distinct points of a locally convex linear topological space $\mathfrak{X}$, there is a continuous linear functional f defined on $\mathfrak{X}$ such that $f(p) \neq f(q)$.*

As a last corollary we state a result which will have important applications in later sections.

14 COROLLARY. *Let a linear space $\mathfrak{X}$ be given two locally convex topologies τ_1 and τ_2. If $(\mathfrak{X}, \tau_1)$ and $(\mathfrak{X}, \tau_2)$ have the same continuous linear functionals, then a convex set is closed in $(\mathfrak{X}, \tau_1)$ if and only if it is closed in $(\mathfrak{X}, \tau_2)$.*

PROOF. Let K be a convex set closed in $(\mathfrak{X}, \tau_1)$ and let $p \notin K$. Theorem 10 yields a continuous linear functional f on $(\mathfrak{X}, \tau_1)$ and real numbers c and ε, with $\varepsilon > 0$, such that

$$\mathscr{R}f(K) \leqq c < c+\varepsilon \leqq \mathscr{R}f(p).$$

Since f is also continuous on $(\mathfrak{X}, \tau_2)$, the neighborhood $\{x \mid |f(x)-f(p)| < \varepsilon\}$ of p in $(\mathfrak{X}, \tau_2)$ does not meet K. Thus K is closed in $(\mathfrak{X}, \tau_2)$. Q.E.D.

3. Weak Topologies.
Definitions and Fundamental Properties

1 DEFINITION. If $\mathfrak{X}$ is a linear vector space, $\mathfrak{X}^+$ is the space of all linear functionals on $\mathfrak{X}$.

A linear subspace Γ of $\mathfrak{X}^+$ is called *total* (cf. II.2.6) if $f(x) = 0$ for all $f \in \Gamma$ implies $x = 0$. The space Γ is often called a *total space of functionals on* $\mathfrak{X}$.

2 DEFINITION. Let $\mathfrak{X}$ be a linear vector space, and let Γ be a total subspace of $\mathfrak{X}^+$. Then the *Γ topology* of $\mathfrak{X}$ is the topology obtained by taking as base all sets of the form

$$N(p; A, \varepsilon) = \{q \,|\, |f(p) - f(q)| < \varepsilon, \ f \in A\},$$

where $p \in \mathfrak{X}$, A is a finite subset of Γ, and $\varepsilon > 0$.

The terms Γ-open and Γ-closed subsets of $\mathfrak{X}$, Γ-continuous maps, etc. will refer to the Γ topology of $\mathfrak{X}$.

The following lemma is a consequence of Definition 2.

3 LEMMA. *If Γ is a total linear space of linear functionals on $\mathfrak{X}$, $\mathfrak{X}$ is a locally convex linear topological space in its Γ topology.*

Note that $\mathfrak{X}$ may already be a linear topological space, possessing notions of open and closed subsets, continuous maps, etc. These concepts have to be distinguished from the corresponding concepts in the Γ topology. Thus, if $\mathfrak{X}$ is a B-space, $\mathfrak{X}$ has a natural metric topology defined by its norm. This topology is often referred to as the *strong*, or *metric*, *topology*. When reference is made to a closed subset of $\mathfrak{X}$, or a continuous map of $\mathfrak{X}$, without qualification, it is to be understood that the topology in question is the strong topology.

The Γ topology of a linear space $\mathfrak{X}$ is related to the product space topology of Definition I.8.1. Let $\mathfrak{X}$ be a linear space over the field Φ, and let Γ be a total subspace of $\mathfrak{X}^+$. Let $\Phi_f = \Phi$ for each $f \in \Gamma$, and let $\Psi = \underset{f \in \Gamma}{P} \Phi_f$. Let the map T on $\mathfrak{X}$ to Ψ be defined by

$$T(x) = \underset{f \in \Gamma}{P} f(x).$$

Since Γ is total, T is a one-to-one embedding of $\mathfrak{X}$ in Ψ, consequently $\mathfrak{X}$ may be regarded as a subset of Ψ. It is then evident, from Definition 2 and Definition I.8.1, that the Γ topology of $\mathfrak{X}$ is identical with the relative topology of $\mathfrak{X}$ as a subset of the product space Ψ. In the next section this remark will enable us to prove a number of interesting theorems on the Γ topology of the linear space $\mathfrak{X}$.

There are two particularly important instances of locally convex topologies defined by total sets of linear functionals. If $\mathfrak{X}$ is a B-space, or, more generally, any locally convex linear topological space, and $\Gamma = \mathfrak{X}^*$ is the set of all continuous linear functionals on $\mathfrak{X}$ (which exist by Corollary 2.13), then the Γ topology is the *$\mathfrak{X}^*$ topology*, or the *weak topology*, of $\mathfrak{X}$. A generalized sequence $\{x_\alpha\}$ will converge to x

in the $\mathfrak{X}^*$ topology if and only if $\lim_\alpha x_\alpha = x$ weakly in the sense of Definition II.3.25. On the other hand, if $\mathfrak{X}$ is a subspace of $\mathfrak{Y}^+$, then each element $y \in \mathfrak{Y}$ determines the linear functional f_y on $\mathfrak{X}$ defined by

$$f_y(x) = x(y), \qquad x \in \mathfrak{X},$$

and the subspace $\Gamma = \{f_y | y \in Y\} \subseteq \mathfrak{X}^+$ is obviously total. The Γ topology of $\mathfrak{X}$ is often called the $\mathfrak{Y}$ *topology* of $\mathfrak{X}$. It is clear that a generalized sequence $\{x_\alpha\}$ converges to x in this topology if and only if $\lim\ x_\alpha(y) = x(y)$ for each $y \in \mathfrak{Y}$.

The most important case of this latter type of topology for a space of functionals occurs when $\mathfrak{Y}$ is a linear topological space and $\mathfrak{X} = \mathfrak{Y}^*$. In this case, what is known as the $\mathfrak{Y}$ *topology* of $\mathfrak{Y}^*$ is obtained.

The reader will have observed that in certain cases a number of different topologies have been defined for the same space $\mathfrak{X}$. For instance, if $\mathfrak{X}$ is a B-space, there is both a metric and an $\mathfrak{X}^*$ topology for $\mathfrak{X}$. If $\mathfrak{Y}$ is a B-space, its conjugate space $\mathfrak{X} = \mathfrak{Y}^*$ has a metric, a $\mathfrak{Y}$, and a $\mathfrak{Y}^{**}$ (or $\mathfrak{X}^*$) topology. In addition, various fragmentary topological notions, such as weak sequential compactness have already been defined in Chapter II (cf. II.3.25).

The next few sections will investigate the relations between the various topologies that have been defined.

4 LEMMA. *The topology of a locally convex space $\mathfrak{X}$ is stronger than the $\mathfrak{X}^*$ topology of $\mathfrak{X}$.*

5 COROLLARY. *The metric topology of a B-space $\mathfrak{X}$ is stronger than its weak topology.*

6 LEMMA. *Let $\mathfrak{X}$ be a linear space, and let Γ_1 and Γ_2 be two total subspaces of $\mathfrak{X}^+$. If $\Gamma_1 \subseteq \Gamma_2$, then the Γ_1 topology of $\mathfrak{X}$ is weaker than the Γ_2 topology of $\mathfrak{X}$.*

7 COROLLARY. *If $\mathfrak{X}$ is a B-space, the $\mathfrak{X}$ topology of $\mathfrak{X}^*$ is weaker than the $\mathfrak{X}^{**}$ topology of $\mathfrak{X}^*$.*

The proofs of 4–7 are elementary, and are left to the reader.

8 LEMMA. *Let $\mathfrak{X}$ be a linear space, and let Γ be a total subspace of $\mathfrak{X}^+$. Then the Γ topology of $\mathfrak{X}$ is the weakest topology in which every functional in Γ is continuous.*

The proof is elementary, and is left to the reader. Lemma 8 has the following important converse.

9 THEOREM. *Let $\mathfrak{X}$ be a linear space, and let Γ be a total subspace of $\mathfrak{X}^+$. Then the linear functionals on $\mathfrak{X}$ which are continuous in the Γ topology are precisely the functionals in Γ.*

The proof of Theorem 9 will be based on the following lemma.

10 LEMMA. *If $g, f_1, \ldots, f_n$ are any $n+1$ linear functionals on a linear space $\mathfrak{X}$, and if $f_i(x) = 0$ for $i = 1, \ldots, n$, implies $g(x) = 0$, then g is a linear combination of the f_i.*

PROOF. Consider the linear map $T : \mathfrak{X} \to E^n$, defined by

$$T(x) = [f_1(x), \ldots, f_n(x)].$$

On the linear subspace $T(\mathfrak{X})$ of E^n, define the mapping ψ by

$$\psi[T(x)] = \psi[f_1(x), \ldots, f_n(x)] = g(x).$$

The map ψ is well-defined, since $T(x) = T(y)$ implies that $T(x-y)=0$, so that $g(x) = g(y)$. It is obvious that ψ is a linear functional on the subspace $T(\mathfrak{X})$ of E^n. By II.3.11, it can be extended to a linear functional ψ_1 on E^n. By IV.3.7, ψ_1 has the form

$$\psi_1[y_1, \ldots, y_n] = \sum_{i=1}^{n} \alpha_i y_i.$$

Hence

$$g(x) = \sum_{i=1}^{n} \alpha_i f_i(x), \qquad x \in \mathfrak{X}.$$

Q.E.D.

PROOF OF THEOREM 9. Every functional in Γ is Γ-continuous, by Lemma 8.

Conversely, let $g \neq 0$ be a linear functional on $\mathfrak{X}$ which is Γ-continuous. There exists a Γ-neighborhood $N(0; f_1, \ldots, f_n; \varepsilon)$ which is mapped by g into the unit sphere of Φ. For $f \in \mathfrak{X}^+$ let $\mathfrak{H}_f = \{x | f(x)=0\}$, and suppose that $x_0 \in \bigcap_{i=1}^{n} \mathfrak{H}_{f_i}$. Then $x_0 \in N(0; f_1, \ldots, f_n, \varepsilon)$, and hence $|g(x_0)| < 1$. Since $\bigcap_{i=1}^{n} \mathfrak{H}_{f_i}$ is a linear space, it contains mx_0 for every integer m. Hence $m|g(x_0)| = |g(mx_0)| < 1$, from which we conclude that $g(x_0) = 0$. That is, $g(x_0) = 0$, whenever $f_i(x_0) = 0$ for $i = 1, \ldots, n$. It follows, from Lemma 10, that g is a linear combination of the f_i. Hence $g \in \Gamma$. Q.E.D.

11 Corollary. *Let f be a linear functional on the linear space $\mathfrak{X}$, and let Γ be a total subspace of $\mathfrak{X}^+$. Then the following statements are equivalent:*

(i) *f is in Γ;*

(ii) *f is Γ-continuous;*

(iii) *$\mathfrak{H}_f = \{x | f(x) = 0\}$ is Γ-closed.*

Proof. By Theorem 9, (i) is equivalent to (ii). Clearly, (ii) implies (iii); we now show that (iii) implies (i). Suppose that $f \neq 0$, and let $p \in \mathfrak{X}$ be such that $f(p) \neq 0$. By Theorem 2.10, there exists a non-zero linear Γ-continuous functional g, and a real constant c, such that $\mathscr{R}g(\mathfrak{H}_f) \leqq c$. By Lemma 1.11, $g(\mathfrak{H}_f) = 0$; i.e., $f(x) = 0$ implies $g(x)=0$. It follows from Lemma 10 that $g = \alpha f$ for some non-zero scalar α. By Theorem 9, g is in Γ; thus f is in Γ. Q.E.D.

12 Corollary. *Let $\mathfrak{X}$ be a linear vector space, and let Γ be a total subspace of $\mathfrak{X}^+$. A linear subspace $\mathfrak{Y}$ of $\mathfrak{X}$ is Γ-closed if and only if for x not in $\mathfrak{Y}$ there exists an f in Γ with $f(\mathfrak{Y}) = 0$, $f(x) = 1$.*

Proof. If $\mathfrak{Y}$ is not Γ-closed, let $x \in \mathfrak{Y}' \cap \overline{\mathfrak{Y}}$, where the closure is taken in the Γ topology. By continuity, if $f \in \Gamma$, and $f(\mathfrak{Y}) = 0$, then $f(\overline{\mathfrak{Y}}) = 0$, and thus $f(x) = 0$. Conversely, if $\mathfrak{Y}$ is Γ-closed and $x \notin \mathfrak{Y}$, then, by Corollary 2.12, there is a Γ-continuous f_0 and a constant c such that $\mathscr{R}f_0(\mathfrak{Y}) \leqq c$, $f_0(x) \neq 0$. By Lemma 1.11, $f_0(\mathfrak{Y}) = 0$; by Theorem 9, $f_0 \in \Gamma$. Put $f = f_0/f_0(x)$, and the corollary is proved. Q.E.D.

13 Theorem. *A convex subset of a locally convex linear topological space is $\mathfrak{X}^*$-closed if and only if it is closed.*

Proof. This result follows from Theorem 9 and Corollary 2.14. Q.E.D.

14 Corollary. *If $\mathfrak{X}$ is a* B-space, *and $\{x_n\}$ is a sequence of elements of $\mathfrak{X}$, converging weakly to x, then some sequence of convex combinations of the elements x_n converges to x in the metric topology.*

Proof. Let $A = \overline{\text{co}}\{x_n\}$. By the theorem just proved, A is closed in the weak topology, hence $x \in A$. The corollary now follows readily from Lemma 2.4(ii) and Lemma I.6.6. Q.E.D.

15 Theorem. *Let T be a linear mapping of a B-space $\mathfrak{X}$ into a B-space $\mathfrak{Y}$. Then T is continuous with respect to the metric topologies in $\mathfrak{X}$ and $\mathfrak{Y}$ if and only if it is continuous with respect to the weak topologies.*

PROOF. Suppose that T is continuous with respect to the metric topologies. Let $N(0; y_1^*, \ldots, y_n^*, \varepsilon)$ be a neighborhood of zero in $\mathfrak{Y}$. For each y_i^*, define x_i^* by

$$x_i^*(x) = y_i^*(Tx).$$

Then $|x_i^*| \leqq |y_i^*| \cdot |T|$, and so $x_i^* \in \mathfrak{X}^*$. If $x \in N(0; x_i^*, \ldots, x_n^*, \varepsilon)$ then $|x_i^*(x)| < \varepsilon$. Hence $|y_i^*(Tx)| < \varepsilon$, so that $Tx \in N(0; y_1^*, \ldots, y_n^*, \varepsilon)$. Therefore, T is weakly continuous at the origin, and hence at every point.

Conversely, suppose that T is weakly continuous, and $y^* \in \mathfrak{Y}^*$. Then $y^* T$ is a linear functional on $\mathfrak{X}$ which is $\mathfrak{X}^*$-continuous. Hence, by Theorem 9, $y^* T \in \mathfrak{X}^*$ for $y^* \in \mathfrak{Y}^*$, so that $y^* T$ is continuous in the metric topology for $y^* \in \mathfrak{Y}^*$. Then, by Theorem II.2.7, T is continuous in the metric topology. Q.E.D.

4. Weak Topologies. Compactness and Reflexivity

The following two sections are devoted to the study of the $\mathfrak{X}$ topology for the conjugate space $\mathfrak{X}^*$ of a B-space. The next basic lemma is nothing more than a simple consequence of Tychonoff's theorem (I.8.5).

1 LEMMA. *Let $\mathfrak{X}$ be a linear space, and let c be a real valued function on $\mathfrak{X}$. Then the set*

$$K = \{f | f \in \mathfrak{X}^+, |f(x)| \leqq c(x)\}$$

is compact in the $\mathfrak{X}$ topology of $\mathfrak{X}^+$.

PROOF. For each $x \in \mathfrak{X}$, let $I(x)$ be the set of scalars α such that $|\alpha| \leqq c(x)$, and let $I = P_{x\in\mathfrak{X}} I(x)$. Define the map $\tau : K \to I$ by $\tau(f) = P_{x\in\mathfrak{X}} f(x)$. Let $\mathfrak{X}^+$ have the $\mathfrak{X}$ topology, K the relative topology as a subset of $\mathfrak{X}^+$, and I the product topology. Then, as is evident from the definitions and the discussion following Lemma 3.3, τ is a homeomorphism. By Theorem I.8.5, I is compact. Hence, by Lemma I.5.7(a), it remains to show that τK is a closed subset of I.

It is easy to verify that τK is the set of all $g \in I$ which lie in all the sets $A(x, y) = \{g | pr_{x+y} g = pr_x g + pr_y g\}$, and in all the sets $B(\alpha, x)$

$= \{g | \alpha pr_x g = pr_{\alpha x} g\}$. Since each projection is a continuous map, each o the sets $A(x, y)$ and $B(\alpha, x)$ is closed. Hence $\tau K = \bigcap_{x, y \in \mathfrak{X}} A(x, y) \cap \bigcap_{\alpha \in \Phi, x \in \mathfrak{X}} B(\alpha, x)$ is also closed. Q.E.D.

2 THEOREM. (*Alaoglu*) *The closed unit sphere in the conjugate space $\mathfrak{X}^*$ of the B-space $\mathfrak{X}$ is compact in the $\mathfrak{X}$ topology of $\mathfrak{X}^*$.*

PROOF. By Definition II.3.5, the unit sphere in $\mathfrak{X}^*$ is the set $\{f | f \in \mathfrak{X}^+, |f(x)| \leqq |x|\}$, and thus the theorem follows from Lemma 1. Q.E.D.

3 COROLLARY. *If $\mathfrak{X}$ is a B-space, a subset of $\mathfrak{X}^*$ is compact in the $\mathfrak{X}$ topology if and only if it is closed in the $\mathfrak{X}$ topology, and bounded in the metric topology.*

4 COROLLARY. *If $\mathfrak{X}$ is a B-space, then $\mathfrak{X}$ is isometrically isomorphic to a closed subspace of the space $C(\Lambda)$ of continuous functions on some compact Hausdorff space Λ.*

PROOF. Let Λ be the closed unit sphere of $\mathfrak{X}^*$. Then, by Theorem 2, Λ is a compact Hausdorff space in its $\mathfrak{X}$ topology. Let $\varkappa$ be the natural embedding of $\mathfrak{X}$ into $\mathfrak{X}^{**}$. Then, by Lemma 3.8, for each $x \in \mathfrak{X}$ the restriction of $\varkappa x = \hat{x}$ to Λ is a continuous function on Λ. Moreover,

$$\sup_{x^* \in \Lambda} |\hat{x}(x^*)| = |x|$$

by II.3.15. Thus $\varkappa$ determines an isometric isomorphism of $\mathfrak{X}$ with a subspace $\mathfrak{X}_1$ of $C(\Lambda)$. Since $\mathfrak{X}$ is complete, $\mathfrak{X}_1$ is complete and therefore a closed subset of $C(\Lambda)$. Q.E.D.

Since the natural embedding $\varkappa : \mathfrak{X} \to \mathfrak{X}^{**}$ is an isometry, it maps metrically closed subsets of $\mathfrak{X}$ onto metrically closed subsets of $\mathfrak{X}^{**}$. However, if the metric topology of $\mathfrak{X}^{**}$ is replaced by its $\mathfrak{X}^*$ topology, the situation is quite different, as the next theorem shows.

5 THEOREM. (*Goldstine*) *Let $\varkappa$ be the natural embedding of the B-space $\mathfrak{X}$ into its second conjugate space $\mathfrak{X}^{**}$ and let S, S^{**} be the closed unit spheres in $\mathfrak{X}$, $\mathfrak{X}^{**}$, respectively. Then $\varkappa S$ is $\mathfrak{X}^*$-dense in S^{**}.*

PROOF. Let S_1 be the $\mathfrak{X}^*$ closure of $\varkappa(S)$. Since S^{**} is $\mathfrak{X}^*$-closed (Theorem 2), $S_1 \subseteq S^{**}$. Also S_1 is convex, by Theorem 2.1. It will be shown that $S_1 = S^{**}$. If there is an element $x^{**} \in S^{**}$ but not in S_1, then, by Theorem 2.10, there exist an $\mathfrak{X}^*$-continuous functional f on

X^{**}, and constants c and ε, $\varepsilon > 0$, such that $\mathscr{R}f(S_1) \leqq c$, $\mathscr{R}f(x^{**}) \geqq c+\varepsilon$. By Theorem 3.9, there is an element $x^* \in \mathfrak{X}^*$ such that $f(x^{**}) = x^{**}x^*$ for $x^{**} \in \mathfrak{X}^{**}$. Since $\varkappa(S) \subseteqq S_1$, it follows that $\mathscr{R}x^*(x) \leqq c$ for $x \in S$. But $x \in S$ implies $\alpha x \in S$ for $|\alpha| = 1$, hence $|x^*(x)| \leqq c$ for $x \in S$. Thus, $|x^*| \leqq c$, and $|x^{**}(x^*)| \leqq c|x^{**}| \leqq c$, contradicting $\mathscr{R}x^{**}(x^*) \geqq c+\varepsilon$. Therefore every $x^{**} \in S^{**}$ is in S_1. Q.E.D.

6 Corollary. *If $\varkappa$ is the natural embedding of a B-space $\mathfrak{X}$ into $\mathfrak{X}^{**}$, then $\varkappa\mathfrak{X}$* is *$\mathfrak{X}^*$-dense in $\mathfrak{X}^{**}$.*

Proof. The $\mathfrak{X}^*$-closure of $\varkappa(\mathfrak{X})$ is a subspace of $\mathfrak{X}^{**}$, which, by Theorem 5, contains the unit sphere of $\mathfrak{X}^{**}$. It follows immediately that it contains every point in $\mathfrak{X}^{**}$. Q.E.D.

Theorems 2 and 5 lead to an important result on reflexive spaces.

7 Theorem. *A B-space is reflexive if and only if its closed unit sphere is compact in the weak topology.*

Proof. Let $\mathfrak{X}$ be a reflexive B-space, and let $\varkappa$ be the natural embedding of $\mathfrak{X}$ onto $\mathfrak{X}^{**}$. Then $\varkappa$ and $\varkappa^{-1}$ are isometries, and $\varkappa$ maps the closed unit sphere S of $\mathfrak{X}$ onto the closed unit sphere S^{**} of $\mathfrak{X}^{**}$. It is clear from the definitions of the two topologies that $\varkappa$ is a homeomorphism between S with its $\mathfrak{X}^*$ topology, and S^{**} with its $\mathfrak{X}^*$ topology. Theorem 2 now establishes that S is weakly compact.

Conversely, let the closed unit sphere S of $\mathfrak{X}$ be weakly compact. Since $\varkappa$ is a homeomorphism between S and $\varkappa(S)$ with the $\mathfrak{X}^*$ topology on both S and $\varkappa(S)$, it follows that $\varkappa(S)$ is compact. By Lemma I.5.7, $\varkappa(S)$ is closed in the $\mathfrak{X}^*$ topology. By Theorem 5, $\varkappa(S)$ is dense in S^{**}. It follows that $\varkappa(S) = S^{**}$. Thus $\varkappa(\mathfrak{X}) = \mathfrak{X}^{**}$, and $\mathfrak{X}$ is reflexive. Q.E.D.

8 Corollary. *A bounded weakly closed set in a reflexive B-space is weakly compact. Conversely, this property characterizes reflexive spaces.*

5. Weak Topologies.
Metrizability. Unbounded Sets

This section continues the discussion of the $\mathfrak{X}$ topology of the conjugate space $\mathfrak{X}^*$ of a B-space.

1 THEOREM. *If $\mathfrak{X}$ is a B-space, then the $\mathfrak{X}$ topology of the closed unit sphere S^* of $\mathfrak{X}^*$ is a metric topology if and only if $\mathfrak{X}$ is separable.*

PROOF. If $\mathfrak{X}$ is separable, let $\{x_n\}$ be a countable dense subset of $\mathfrak{X}$, and define

$$\varrho(x^*, y^*) = \sum_{n=1}^{\infty} \frac{1}{2^n} \frac{|(x^*-y^*)x_n|}{1+|(x^*-y^*)x_n|}.$$

It is easy to verify that the topology of S^* defined by this metric is weaker than the $\mathfrak{X}$ topology of S^*. Hence, by Theorem 4.2 and Lemma I.5.8, the metric topology for S^* defined by ϱ is the same as the $\mathfrak{X}$ topology of S^*.

Conversely, if the $\mathfrak{X}$ topology of S^* is a metric topology, there is a denumerable sequence $\{U_n^*\}$ of $\mathfrak{X}$-neighborhoods of the origin 0 of $\mathfrak{X}^*$, such that $\bigcap_{n=1}^{\infty} U_n^* = \{0\}$. We may suppose that

$$U_n^* = \{x^* | x^* \in S^*, |x^*(x)| < \varepsilon_n, x \in A_n\},$$

where A_n is a finite subset of $\mathfrak{X}$, and $\varepsilon_n > 0$. Let $A = \bigcup_{n=1}^{\infty} A_n$; if $x^*(A) = 0$, then $x^* \in U_n^*$ for each n, and consequently $x^* = 0$. Let $\mathfrak{X}_1 = \overline{\text{sp}}(A)$. It follows from Lemma II.1.17 that $\mathfrak{X}_1$ is separable, and from Corollary II.3.13 that $\mathfrak{X}_1 = \mathfrak{X}$. Q.E.D.

2 THEOREM. *If $\mathfrak{X}$ is a B-space, then the $\mathfrak{X}^*$ topology of the closed unit sphere of $\mathfrak{X}$ is a metric topology if and only if $\mathfrak{X}^*$ is separable.*

PROOF. Let $\mathfrak{X}^*$ be separable, and let $\varkappa : \mathfrak{X} \to \mathfrak{X}^{**}$ be the natural embedding. By Theorem 1, the $\mathfrak{X}^*$ topology of the closed unit sphere S^{**} of $\mathfrak{X}^{**}$ is a metric topology. If S is the closed unit sphere in $\mathfrak{X}$, the mapping $\varkappa : S \to S^{**}$ is a homeomorphism of S and $\varkappa(S)$ in their $\mathfrak{X}^*$ topologies. It follows from Lemma I.6.4 that the $\mathfrak{X}^*$ topology of S is a metric topology.

Conversely, let the $\mathfrak{X}^*$ topology of S be a metric topology. Then there exists a sequence $\{U_n\}$ of $\mathfrak{X}^*$-neighborhoods of the origin 0 of $\mathfrak{X}$, such that every $\mathfrak{X}^*$-neighborhood of 0 contains some neighborhood U_n. We may suppose that

$$U_n = \{x | x \in S, |x^*(x)| < \varepsilon_n, x^* \in A_n^*\},$$

where A_n^* is a finite subset of $\mathfrak{X}^*$ and $\varepsilon_n > 0$.

Let $A^* = \bigcup_{n=1}^{\infty} A_n^*$, and let $X_1^* = \overline{\text{sp}}(A^*)$. By Lemma II.1.17, $\mathfrak{X}_1^*$ is separable, and it remains to show that $\mathfrak{X}^* = \mathfrak{X}_1^*$.

If $\mathfrak{X}^* \neq \mathfrak{X}_1^*$, we can choose $y^* \in \mathfrak{X}^*$, $y^* \notin \mathfrak{X}_1^*$. Let

$$d = \inf_{x_1^* \in \mathfrak{X}_1^*} |y^* - x_1^*|.$$

Then $d > 0$, and, by II.3.12, there is an $x^{**} \in \mathfrak{X}^{**}$ with $|x^{**}| = 1/d$ such that $x^{**}(\mathfrak{X}_1^*) = 0$, $x^{**}(y^*) = 1$. The set $V = \{x | x \in S, |y^*(x)| < d/2\}$ is an $\mathfrak{X}^*$-neighborhood of 0, and hence $V \supseteq U_n$ for some n. Since $dx^{**} \in S^{**}$, there is, by Theorem 4.5, an $x_1 \in S$ such that

$$|x^*(x_1)| = |dx^{**}(x^*) - x^*(x_1)| < \varepsilon_n, \qquad x^* \in A_n^*,$$

$$|d - y^*(x_1)| = |dx^{**}(y^*) - y^*(x_1)| < \frac{d}{2}.$$

Thus

$$|y^*(x_1)| > d/2; \qquad |x^*(x_1)| < \varepsilon_n, \qquad x^* \in A_n^*.$$

But this means that $x_1 \in U_n$, and $x_1 \notin V$, a contradiction which proves that $\mathfrak{X}_1^* = \mathfrak{X}^*$. Q.E.D.

The remaining theorems in this section concern convex sets which are not necessarily bounded. We may remark that the analogues for the $\mathfrak{X}^*$ topology of $\mathfrak{X}$ of Statements 3–6 are all trivial consequences of Theorem 3.13.

3 Definition. Let $\mathfrak{X}$ be a B-space. The *bounded $\mathfrak{X}$ topology*, or *BX topology*, for $\mathfrak{X}^*$ is the strongest topology which coincides with the $\mathfrak{X}$ topology on each set $aS^* = \{x^* | x^* \in \mathfrak{X}^*, |x^*| \leqq a\}$. Thus, a set $U \subseteq \mathfrak{X}^*$ is BX-open if and only if $U \cap aS^*$ is a relatively $\mathfrak{X}$-open subset of aS^* for every $a \geqq 0$, and a set $K \subseteq \mathfrak{X}^*$ is BX-closed if and only if $K \cap aS^*$ is $\mathfrak{X}$-closed for every $a \geqq 0$.

4 Lemma. *Let $\mathfrak{X}$ be a B-space. A fundamental system of neighborhoods of the origin for the bounded $\mathfrak{X}$ topology of $\mathfrak{X}^*$ consists of the sets $\{x^* | |x^*(x)| < 1, x \in A\}$, where $A = \{x_i\}$ is a sequence of elements of $\mathfrak{X}$ converging to zero.*

Proof. Let S^* be the closed unit sphere of $\mathfrak{X}^*$. If A is a sequence of elements converging to zero, then $\{x^* | |x^*(x)| < 1, x \in A\} \cap aS^* = \{x^* | |x^*(x)| < 1, x \in A_1\} \cap aS^*$, where A_1 is the finite set of elements $x \in A$ of norm $|x| \geqq 1/a$. Thus, $\{x^* | |x^*(x)| < 1, x \in A\} \cap aS^*$ is a relatively $\mathfrak{X}$-open subset of aS^*.

In proving the converse assertion of the lemma, the notation $A^0 = \{x^* | |x^*(x)| \leqq 1, x \in A\}$ for subsets A of $\mathfrak{X}$ is convenient. Let

U be a BX-neighborhood of the origin. Then, by definition of the BX topology, there is a finite set $A_1 \subseteq \mathfrak{X}$ such that $A_1^0 \cap S^* \subseteq U$. Suppose that for some integer n we have defined a finite set $A_n \subseteq \mathfrak{X}$ such that $A_n^0 \cap nS^* \subseteq U$. It will be shown that there is a finite set of elements $B_n \subseteq \mathfrak{X}$ such that $|B_n| \leqq 1/n$ and such that $(A_n \cup B_n)^0 \cap (n+1)S^* \subseteq U$.

If this is not the case, it is clear that the family of sets of the form $(A_n \cup B)^\circ \cap (n+1)S^* \cap U'$ where B is finite and $|B| \leqq 1/n$, has the finite intersection property. Since U' is BX-closed, all these sets are $\mathfrak{X}$-closed, and, since $(n+1)S^*$ is $\mathfrak{X}$-compact, by Theorem 4.2, it follows from I.5.6 that there exists an $x^* \in (n+1)S^* \cap U' \cap A_n^0$ such that $|x^*(x)| \leqq 1$ for every $x \in \mathfrak{X}$ with $|x| \leqq 1/n$. Thus $|x^*| \leqq n$, so that $x^* \in nS^* \cap A_n^0 \cap U'$, contradicting the fact that $nS^* \cap A_n^0 \subseteq U$.

Defining $A_{n+1} = A_n \cup B_n$, we have established an inductive construction of a sequence of finite sets $A_n \subset \mathfrak{X}$ such that $A_n^0 \cap nS^* \subseteq U$, and such that any enumeration of $A = \bigcup_{i=1}^{\infty} A_n$ is a sequence of elements of $\mathfrak{X}$ tending to zero. Hence $\{x^* | x^* \in \mathfrak{X}^*, |x^*(x)| < 1, x \in A\}$ is a set of the desired form which is contained in U. Q.E.D.

The reader will have no difficulty in applying Lemma 3 to prove the next corollary.

5 Corollary. *Let $\mathfrak{X}$ be a B-space. Then $\mathfrak{X}^*$, with its bounded $\mathfrak{X}$ topology, is a locally convex linear topological space.*

The next theorem gives a fundamental property of the BX topology.

6 Theorem. *A functional θ on $\mathfrak{X}^*$ is continuous in the $\mathfrak{X}$ topology if and only if it is continuous in the bounded $\mathfrak{X}$ topology.*

Proof. By definition the BX topology is stronger than the $\mathfrak{X}$ topology so that an $\mathfrak{X}$-continuous functional is BX-continuous. Conversely, let the linear functional θ on $\mathfrak{X}^*$ be continuous in the BX topology. Then there is a sequence $\{x_i\}$, with $\lim_{i\to\infty} x_i = 0$ such that

$$|\theta(x^*)| \leqq 1 \quad \text{if} \quad |x^*(x_i)| < 1, \qquad i = 1, 2, \ldots.$$

Let $T: x^* \to [x^*(x_i)]$ map $\mathfrak{X}^*$ into the B-space c_0. Since $x^*(x_i) = 0$, $i = 1, 2, \ldots$, implies $\theta(x^*) = 0$, the functional $h(Tx^*) = \theta(x^*)$ on $T\mathfrak{X}^*$ is well-defined. It is clearly continuous, and by II.3.11 can be extended to the B-space $c = C(S)$, where $S = \{0, 1/n, n \geqq 1\}$. Thus, by

IV.6.3, there is a sequence $[\alpha_0, \alpha_1, \ldots]$ with $\sum_{i=0}^{\infty} |\alpha_i| < \infty$ such that $h(\xi) = \sum_{i=1}^{\infty} \alpha_i \xi_i$ for $\xi = [\xi_1, \xi_2, \ldots] \in c_0$. (cf. IV.13.7). Thus

$$\theta(x^*) = \sum_{i=1}^{\infty} \alpha_i x^*(x_i) = x^*\left(\sum_{i=1}^{\infty} \alpha_i x_i\right), \qquad x^* \in \mathfrak{X}^*.$$

That is, θ has the form $\theta(x^*) = x^*(x)$, where $x = \sum_{i=1}^{\infty} \alpha_i x_i \in \mathfrak{X}$. Hence, by Theorem 3.9, θ is $\mathfrak{X}$-continuous. Q.E.D.

7 Theorem. (*Krein-Šmulian*) *A convex set in $\mathfrak{X}^*$ is $\mathfrak{X}$-closed if and only if its intersection with every positive multiple of the closed unit sphere of $\mathfrak{X}^*$ is $\mathfrak{X}$-closed.*

Proof. This follows from the preceding theorem and Corollary 2.14. Q.E.D.

8 Corollary. *If $\mathfrak{X}$ is a B-space, a linear subspace $\mathfrak{Y} \subseteq \mathfrak{X}^*$ is $\mathfrak{X}$-closed if and only if there exists an $\mathfrak{X}$-closed bounded subset K of $\mathfrak{X}^*$ containing a non-void metrically open subset of $\mathfrak{Y}$.*

Proof. If $\mathfrak{Y}$ is $\mathfrak{X}$-closed, the closed unit sphere S^* of $\mathfrak{X}^*$ is $\mathfrak{X}$-closed, and we may take $K = \mathfrak{Y} \cap S^*$.

Conversely, let K be a bounded $\mathfrak{X}$-closed subset of $\mathfrak{Y}$, and $K \supseteq \mathfrak{Y} \cap S^*(p^*, \delta)$. If $a > 0$, the mapping $x^* \to a(x^* - p^*)/\delta$ is an $\mathfrak{X}$-homeomorphism of $\mathfrak{X}^*$ with itself. Thus $a(K-p^*)/\delta$ is $\mathfrak{X}$-closed. Since $\mathfrak{Y} \cap \delta S^* \subseteq K - p^*$, $\mathfrak{Y} \cap aS^* \subseteq a(K-p^*)/\delta$, and hence $\mathfrak{Y} \cap aS^* = aS^* \cap a(K-p^*)/\delta$ is $\mathfrak{X}$-closed. The desired conclusion follows from Theorem 7. Q.E.D.

9 Corollary. *Let $\mathfrak{X}$ be a B-space, and let K be a convex $\mathfrak{X}$-closed subset of $\mathfrak{X}^*$. Let $\mathfrak{Y}$ be the linear space spanned by K. Then $\mathfrak{Y}$ is closed in the metric topology of $\mathfrak{X}^*$, if and only if it is $\mathfrak{X}$-closed.*

Proof. If $\mathfrak{Y}$ is $\mathfrak{X}$-closed, then it is closed in the metric topology by Corollaries 3.5 and 3.7.

Conversely, let $\mathfrak{Y}$ be closed in the metric topology. We shall assume that $\mathfrak{X}$ is real, leaving the details of the complex case to the reader. Let S^* be the closed unit sphere in $\mathfrak{X}^*$, and $K_n = K \cap nS^*$. Let $\tilde{K} = \text{co}(K \cup -K)$, and $\tilde{K}_n = \text{co}(K_n \cup -K_n)$. Then $\tilde{K}_n$ is $\mathfrak{X}$-closed by Lemmas 2.5 and 4.3; hence, by Corollaries 3.5 and 3.7, $\tilde{K}_n$ is closed in the metric topology. Now each $y \in \mathfrak{Y}$ can be written as $y = \sum_{i=1}^{p} a_i x_i - \sum_{i=p+1}^{m} a_i x_i$, where $x_i \in K$, and all the a_i are positive.

It follows that $y \in a\tilde{K}$ with $a = \sum_{i=1}^{m} a_i$; since $0 \in \tilde{K}$, $a\tilde{K} \subseteqq b\tilde{K}$ if $0 \leqq a \leqq b$; hence any element $y \in \mathfrak{Y}$ is in $n\tilde{K}$ for all sufficiently large integers n. Since $\tilde{K} = \bigcup_{n=1}^{\infty} \tilde{K}_n$, it follows that $\mathfrak{Y} = \bigcup_{n=1}^{\infty} n\tilde{K}_n$. Since $\mathfrak{Y}$ is a closed subspace of $\mathfrak{X}^*$, it is complete by Lemma I.6.7. Hence, by the Baire category theorem, I.6.9, some set $\tilde{K}_n$ contains a non-void metrically open subset of $\mathfrak{Y}$. Our result now follows from Corollary 8. Q.E.D.

6. Weak Topologies. Weak Compactness

We have already introduced and employed the concepts of weak sequential compactness (II.3.25) and compactness in the $\mathfrak{X}^*$ (or weak) topology. There is at least one other type of weak compactness that is occasionally of use. It is a remarkable and important fact that these three concepts are equivalent.

1 **Theorem.** (*Eberlein-Šmulian*) *Let A be a subset of a B-space $\mathfrak{X}$. Then the following statements are equivalent:*

(i) *A is weakly sequentially compact — i.e., any sequence in A has a subsequence which converges weakly to an element of $\mathfrak{X}$;*

(ii) *every countably infinite subset of A has a weak limit point in $\mathfrak{X}$—i.e., a point such that every weak neighborhood contains an element in the infinite subset;*

(iii) *the closure of A in the $\mathfrak{X}^*$ topology is $\mathfrak{X}^*$-compact.*

Proof. We observe first that each of the three conditions implies that A is bounded in the metric topology, for $x^*(A)$ is a bounded set of scalars for each $x^* \in \mathfrak{X}^*$ and we can apply II.3.20. It is readily seen that (iii) implies (ii). The other implications we desire are decidedly non-trivial; we will complete the proof by showing first that (ii) implies (i), and then that (i) implies (iii).

Proof that condition (ii) implies (i). Let $\{x_n\}$ be an arbitrary sequence in A and let $\mathfrak{X}_0 = \overline{\text{sp}}\{x_n\}$, so that, by Lemma II.1.5, $\mathfrak{X}_0$ is separable. By Theorem 5.1 and Corollary 4.2, the unit sphere in $\mathfrak{X}_0^*$ is separable in the $\mathfrak{X}_0$ topology. Since $\mathfrak{X}_0^*$ is the union of a sequence of multiples of its unit sphere, $\mathfrak{X}_0^*$ is separable in the $\mathfrak{X}_0$ topology. Let H_0 be a denumerable dense set in $\mathfrak{X}_0^*$. Clearly H_0 is a total set on $\mathfrak{X}_0$ and each element of H_0 can be extended to a linear functional on all of $\mathfrak{X}$. Taking one such extension for each element of H_0 we get a denumerable subset H of $\mathfrak{X}^*$.

Since A is bounded, by a diagonal process we may extract a subsequence $\{y_m\}$ of $\{x_n\}$ such that $\lim_{m\to\infty} x^* y_m$ exists for each $x^* \in H$. By condition (ii) there is a point $y_0 \in \mathfrak{X}$ such that every weak neighborhood of y_0 contains at least one y_m. Since $\{y_m\} \subset \mathfrak{X}_0$ and $\mathfrak{X}_0$ is $\mathfrak{X}^*$-closed, it follows that $y_0 \in \mathfrak{X}_0$. It is clear that

$$x^* y_0 = \lim_{m\to\infty} x^* y_m, \qquad x^* \in H,$$

and it remains to show that this is true for every $x^* \in \mathfrak{X}^*$. If this is not true, there exists an $x_0^* \in \mathfrak{X}^*$, an $\varepsilon > 0$, and a subsequence $\{y_{m_k}\}$ such that

$$(*) \qquad |x_0^*(y_{m_k} - y_0)| > \varepsilon, \qquad k = 1, 2, \ldots.$$

Applying condition (ii) to $\{y_{m_k}\}$ we obtain a point $y_0' \in \mathfrak{X}$ such that every weak neighborhood of y_0' contains at least one y_{m_k}. Just as above we show that $y_0' \in \mathfrak{X}_0$ and

$$(**) \qquad x^* y_0' = \lim_{k\to\infty} x^* y_{m_k}, \qquad x^* \in H.$$

Hence $x^* y_0 = x^* y_0'$ for all $x^* \in H$, and since H is total on $\mathfrak{X}_0$, $y_0 = y_0'$. But this fact and (**) contradict (*). Hence the arbitrary sequence $\{x_n\}$ in A contains a weakly convergent subsequence $\{y_m\}$, and so A is weakly sequentially compact.

Proof that condition (i) implies (iii). Let $\bar{A}$ be the closure of A in the $\mathfrak{X}^*$ topology of $\mathfrak{X}$; we must show that (i) implies that $\bar{A}$ is $\mathfrak{X}^*$-compact. Since the natural embedding $\varkappa : \mathfrak{X} \to \mathfrak{X}^{**}$ is a homeomorphism between $\mathfrak{X}$ and $\varkappa(\mathfrak{X})$ in their $\mathfrak{X}^*$ topologies, we have $\varkappa(\bar{A}) = \overline{\varkappa(A)} \cap \varkappa(\mathfrak{X})$, and moreover $\bar{A}$ is $\mathfrak{X}^*$-compact if and only if $\varkappa(\bar{A})$ is $\mathfrak{X}^*$-compact. Since $\bar{A}$ is bounded, by Corollary 4.3, $\varkappa(\bar{A})$ is $\mathfrak{X}^*$-compact if and only if $\varkappa(\bar{A})$ is an $\mathfrak{X}^*$-closed subset of $\mathfrak{X}^{**}$. Since $\varkappa(\bar{A}) \subseteqq \overline{\varkappa(A)}$, the bar denoting closure in the $\mathfrak{X}^*$ topology, it will suffice to show that $\overline{\varkappa(A)} \subseteqq \varkappa(\bar{A})$, but this will be true by virtue of $\varkappa(\bar{A}) = \overline{\varkappa(A)} \cap \varkappa(\mathfrak{X})$ once it is shown that $\overline{\varkappa(A)} \subseteqq \varkappa(\mathfrak{X})$. Let $x^{**} \in \mathfrak{X}^{**}$ be an element in the $\mathfrak{X}^*$-closure of $\varkappa(A)$; we will show that $x^{**} \in \varkappa(\mathfrak{X})$. This amounts to showing that there exists an $x \in \mathfrak{X}$ such that $x^{**}x^* = x^*x$, $x^* \in \mathfrak{X}^*$.

First, however, we prove the weaker assertion: if $\{x_1^*, \ldots, x_n^*\}$ is an arbitrary finite subset of $\mathfrak{X}^*$, then there exists a $z \in \bar{A}$ such that $x^{**} x_i^* = x_i^* z$, $i = 1, \ldots, n$. To see this, let m be an arbitrary integer; since x^{**} is in the $\mathfrak{X}^*$-closure of $\varkappa(A)$, there is an element $z_m \in A$ such that

$$|x_i^*(z_m) - x^{**}(x_i^*)| < 1/m, \qquad i = 1, \ldots, n.$$

Since A is weakly sequentially compact, a subsequence of $\{z_m\}$ will converge weakly to an element z which is certainly in $\bar{A}$, since the sequential closure of A is contained in $\bar{A}$, and hence $x_i^*(z) = x^{**}(x_i^*)$, $i = 1, \ldots, n$.

The remainder of the proof is concerned with showing that $x^{**} \in \varkappa(\mathfrak{X})$. By Corollary 3.11 this is true if and only if the subspace $\mathfrak{Q}$ of $\mathfrak{X}^*$ defined by $\mathfrak{Q} = \{x^* \in \mathfrak{X}^* | x^{**} x^* = 0\}$ is $\mathfrak{X}$-closed. If S^* is the closed unit sphere of $\mathfrak{X}^*$, then, by Corollary 5.8, $\mathfrak{Q}$ is $\mathfrak{X}$-closed if $\mathfrak{Q} \cap S^*$ is $\mathfrak{X}$-closed. Thus we must show that if y_0^* is in the $\mathfrak{X}$-closure of $\mathfrak{Q} \cap S^*$, then $y_0^* \in \mathfrak{Q} \cap S^*$. To do this, we choose an arbitrary $\varepsilon > 0$ and will construct three sequences $\{z_n\} \subseteqq \bar{A}$, $\{x_n\} \subseteqq A$, and $\{y_n^*\} \subseteqq \mathfrak{Q} \cap S^*$, in the following manner: From the remark in the preceding paragraph there exists a $z_1 \in \bar{A}$ such that $y_0^*(z_1) = x^{**}(y_0^*)$. Since z_1 is in the $\mathfrak{X}^*$-closure of A, there exists an $x_1 \in A$ such that $|y_0^*(x_1) - y_0^*(z_1)| < \varepsilon/4$. Since y_0^* is in the $\mathfrak{X}$-closure of $\mathfrak{Q} \cap S^*$, there exists a $y_1^* \in \mathfrak{Q} \cap S^*$ such that $|y_1^*(x_1) - y_0^*(x_1)| < \varepsilon/4$.

By induction, if elements with subscripts less than n have been chosen, we select $z_n \in \bar{A}$, $x_n \in A$, and $y_n^* \in \mathfrak{Q} \cap S^*$ in such a way that

(a) $y_m^*(z_n) = x^{**}(y_m^*) = 0, \qquad m = 0, \ldots, n-1;$

(b) $|y_m^*(x_n) - y_m^*(z_n)| < \varepsilon/4, \qquad m = 0, \ldots, n-1;$

(c) $|y_n^*(x_i) - y_0^*(x_i)| < \varepsilon/4, \qquad i = 1, \ldots, n.$

In this way we construct three sequences. From the construction we have

(d) $|y_m^*| \leqq 1, \qquad m = 1, 2, \ldots;$

(e) $y_m^*(z_n) = 0, \qquad m = 1, 2, \ldots, n-1.$

Let $m = 0$ in (a) and (b) and combine with (c) to get

(f) $|x^{**}(y_0^*) - y_n^*(x_i)| < \varepsilon/2, \qquad i = 1, \ldots, n.$

Since $\{x_n\} \subseteqq A$ and A is weakly sequentially compact, there is a subsequence of $\{x_n\}$ which converges weakly to an element $x \in \bar{A}$. To avoid changing notation we suppose (without loss of generality) that the entire sequence $\{x_n\}$ converges weakly to x. Now from (b) and (e) we have

$$|y_m^*(x_n)| < \varepsilon/4, \qquad m = 1, \ldots, n-1,$$

and hence it follows that

$$\text{(g)} \quad |y_m^*(x)| \leqq \varepsilon/4, \qquad m = 1, 2, \ldots.$$

By Corollary 3.14, there is a convex combination $w = \sum_{i=1}^N a_i x_i$, with $a_i \geqq 0$, $\sum_{i=1}^N a_i = 1$, such that $|x-w| < \varepsilon/4$. In (f) let $n = N$ so that

$$\text{(h)} \quad |x^{**}(y_0^*)-y_N^*(w)| \leqq \sum_{i=1}^N a_i|x^{**}(y_0^*)-y_N^*(x_i)| < \varepsilon/2.$$

It follows that

$$|x^{**}(y_0^*)| \leqq |x^{**}(y_0^*)-y_N^*(w)|+|y_N^*(w)-y_N^*(x)|+|y_N^*(x)|.$$

But the first term is at most $\varepsilon/2$ by (h), the second less than $\varepsilon/4$ by (d) and the fact that $|w-x| < \varepsilon/4$, and by (g) the final term is dominated by $\varepsilon/4$. Since ε was chosen arbitrarily it follows that

$$x^{**}(y_0^*) = 0.$$

Hence $y_0^* \in \mathfrak{Q}$; since S^* is $\mathfrak{X}$-closed, $y_0^* \in \mathfrak{Q} \cap S^*$. This shows that $\mathfrak{Q}$ is $\mathfrak{X}$-closed, so that $x^{**} \in \varkappa(\mathfrak{X})$ and the proof of the theorem is complete. Q.E.D.

2 THEOREM. (*Šmulian*) *A convex subset K of a B-space $\mathfrak{X}$ is weakly compact if and only if every decreasing sequence of non-void closed convex subsets of K has a non-empty intersection.*

PROOF. The necessity of the condition is immediate from I.5.6. For the sufficiency, we observe that the condition implies that K is bounded. For otherwise, there exists an $x_1^* \in \mathfrak{X}^*$ such that $x_1^*(K)$ is an unbounded convex set of scalars, and hence there exists an $x^* \in \mathfrak{X}^*$ such that $x^*(K)$ contains a segment $[N, \infty)$. If $K_n = \{x \in K | x^*x \geqq N+n\}$, then the sequence $\{K_n\}$ violates the assumed condition. Further, the condition implies that K is closed, for if $x_n \to x_0$, $x_n \in K$, we set $K_n = K \cap \overline{\text{co}}\{x_n, x_{n+1}, \ldots\}$. Clearly $\{x_0\} = \bigcap_{n=1}^\infty \overline{\text{co}}\{x_n, x_{n+1}, \ldots\}$, so the condition implies that $x_0 \in K$.

To prove that K is weakly compact take an arbitrary sequence $\{x_n\}$

and proceed as in the first part of the proof of the preceding theorem to construct a subsequence $\{y_m\}$ of $\{x_n\}$ such that $\lim_{m\to\infty} x^*y_m$ exists for each x^* in the set H of that proof. Let $K_m = \overline{\text{co}}\{y_m, y_{m+1}, \ldots\}$ and let y_0 be an arbitrary point in $\bigcap_{m=1}^{\infty} K_m$. For each $x^* \in \mathfrak{X}^*$, we have

$$x^* y_0 \in \bigcap_{m=1}^{\infty} x^*(K_m).$$

This fact readily implies that

$$x^* y_0 = \lim_{m\to\infty} x^* y_m, \qquad x^* \in H.$$

To prove that this is true for all $x^* \in \mathfrak{X}^*$ we proceed as in the proof of the first part of Theorem 1, replacing condition (ii) of that theorem with our present hypothesis. We thus conclude that K is weakly sequentially compact, but since K is a closed convex set and therefore (3.13) $\mathfrak{X}^*$-closed, the preceding theorem applies to show that K is $\mathfrak{X}^*$-compact. Q.E.D.

3 Theorem. *The weak topology of a weakly compact subset A of a separable B-space is a metric topology.*

Proof. Suppose that $\mathfrak{X} = \overline{\text{sp}}\{x_1, x_2, \ldots\}$. Construct $H \subseteq \mathfrak{X}^*$ as in the proof of Theorem 1. Let $H = \{x_n^*\}$. The metric

$$\varrho(x, y) = \sum_{n=1}^{\infty} \frac{1}{2^n} \frac{|x_n^*(x-y)|}{1+|x_n^*(x-y)|}$$

defines a topology for A which is weaker than the $\mathfrak{X}^*$ topology of A. Hence, by I.5.8 these two topologies are the same on the set A. Q.E.D.

It should be noted that the analogue of Theorem 3 for the $\mathfrak{X}$ topology of $\mathfrak{X}^*$ follows trivially from Corollary 4.3 and Theorem 5.1. The analogue of Theorem 1 for the $\mathfrak{X}$ topology of $\mathfrak{X}^*$ is not true. While it follows immediately from Theorem 4.3 that an $\mathfrak{X}$-sequentially compact subset of $\mathfrak{X}^*$ has an $\mathfrak{X}$-compact $\mathfrak{X}$-closure, a subset of $\mathfrak{X}^*$ can be $\mathfrak{X}$-compact without being $\mathfrak{X}$-sequentially compact. A method for constructing an example of this phenomenon is given in Exercise 7.32.

As an application of Theorem 1, we prove:

4 Theorem. (*Krein-Šmulian*) *The closed convex hull of a weakly compact subset of a B-space is itself weakly compact.*

PROOF. Let A be a weakly compact subset of the B-space $\mathfrak{X}$. Since $\overline{\text{co}}(A)$ is $\mathfrak{X}^*$-closed by Theorem 3.13, it follows from Theorem 1 that it suffices to show that co(A) is weakly sequentially compact. Let $\{p_n\}$ be a sequence of points in co(A); then p_n is a convex combination of a finite set B_n of points of A. Let $B_0 = \bigcup_{i=1}^{\infty} B_n$, and let $\mathfrak{X}_0 = \overline{\text{sp}}(B_0)$; by II.1.5, $\mathfrak{X}_0$ is separable. Let $A_0 = A\mathfrak{X}_0$; by Theorem 3.13 and I.5.7(a), A_0 is weakly compact. Since $\{p_n\} \subseteqq \text{co}(A_0)$ the theorem will be proved if we show $\overline{\text{co}}(A_0)$ is weakly compact.

By Theorem 1, A_0 is weakly sequentially compact; by Theorem II.3.20 there exists a constant K with $|A_0| \leqq K$. Further, in the relative $\mathfrak{X}^*$ topology A_0 is a compact Hausdorff space. Let $C(A_0)$ be the space of continuous functions on A_0, and $C^*(A_0)$ the conjugate space of $C(A_0)$, so from Theorem IV.6.3, $C^*(A_0)$ is isometrically isomorphic to the space of all regular measures on A_0. Let S^* be the closed unit sphere of $C^*(A_0)$.

Define a linear map $\psi : C^*(A_0) \to \mathfrak{X}$ by

$$\psi(f^*) = \int_{A_0} a\mu_{f^*}(da),$$

where $f^* \in C^*(A_0)$, and μ_{f^*} is the regular measure corresponding to f^*. Since the set A_0 is separable, and since $|a| \leqq K$ for $a \in A_0$ this integral is defined (cf. III.6.9.). Define rx^* for $x^* \in \mathfrak{X}^*$ as the restriction of the linear functional x^* to A_0. Then, by 3.8, $rx^* \in C(A_0)$. Moreover, it is seen from the definition of ψ that $x^*\psi f^* = f^* rx^*$ for $f^* \in C^*(A_0)$ and $x^* \in \mathfrak{X}^*$. Hence, if J is an arbitrary finite subset of $\mathfrak{X}^*$, ψ maps the $C(A)$-neighborhood $N(f^*; rJ, \varepsilon)$ of f^* into the $\mathfrak{X}^*$-neighborhood $N(\psi f^*; J, \varepsilon)$ of ψf^* (cf. Definition 3.1). Thus $\psi : C^*(A_0) \to X$ is continuous if $C^*(A_0)$ is given its $C(A_0)$ topology, and $\mathfrak{X}$ its $\mathfrak{X}^*$ topology. By Lemma 1.5, the set $\psi(S^*)$ is convex; and by Theorem 4.2 and Lemma I.5.7(b), this set is $\mathfrak{X}^*$-compact. If we put $f_a^*(g) = g(a)$ for $a \in A_0$ and $g \in C(A_0)$, then it is easily seen that $\psi(f_a^*) = a$ and hence $\psi(S^*) \supseteqq A_0$. Since $\psi(S^*)$ is closed in the metric topology by Lemma I.5.7(c) and Lemma 3.4, it follows that $\psi(S^*) \supseteqq \overline{\text{co}}(A_0)$. By Theorem 3.13 and Lemma I.5.7(a), $\overline{\text{co}}(A_0)$ is weakly compact. Q.E.D.

It should be noted that the analogue of Theorem 4 for $\mathfrak{X}$-compact subsets of $\mathfrak{X}^*$ follows trivially from Corollary 4.3.

7. Exercises

1 Let $\mathfrak{X}$ be a linear vector space. Show that $\mathfrak{X}^+$ is a total space of linear functionals on $\mathfrak{X}$.

2 If $\mathfrak{X}$ is an infinite dimensional B-space then $\mathfrak{X}^+ \neq \mathfrak{X}^*$.

3 Let $\mathfrak{X}$ be a linear topological space. Then $\mathfrak{X}$ has a non-zero continuous linear functional if and only if the origin is interior to some convex proper subset of $\mathfrak{X}$.

4 If a convex set in a linear topological space has an interior point, it has the same interior as its closure.

5 Let Γ be a total set of linear functionals on the space $\mathfrak{X}$. Show that if $\mathfrak{X}$ contains a non-void Γ-open bounded (II.1.7) set, then $\mathfrak{X}$ must be finite dimensional.

6 Let $\mathfrak{X}$ be a B-space, and $\mathfrak{X}_1$ a subspace of $\mathfrak{X}$. Show that the $\mathfrak{X}_1^*$ topology of $\mathfrak{X}_1$ is the same as the relative $\mathfrak{X}^*$ topology of $\mathfrak{X}_1$.

7 Let $\mathfrak{X}$ be a linear space, and Γ a total subspace of $\mathfrak{X}^+$. Show that a set $A \subseteq \mathfrak{X}$ is Γ-bounded if and only if $f(A)$ is a bounded set of scalars for each $f \in \Gamma$.

8 Let $\mathfrak{X}$ be a B-space. Show that a subset of $\mathfrak{X}$ is $\mathfrak{X}^*$-bounded if and only if it is metrically bounded, and that a subset of $\mathfrak{X}^*$ is $\mathfrak{X}$-bounded if and only if it is metrically bounded.

9 Show that the weak and the metric topologies of the unit sphere of a normed space are the same if and only if the space is finite dimensional.

10 Let $\mathfrak{X}$ be a linear space, and let Γ_1 and Γ_2 be two total subspaces of $\mathfrak{X}^+$. Show that if Γ_1 and Γ_2 determine the same topology for $\mathfrak{X}$, then $\Gamma_1 = \Gamma_2$.

11 Prove II.3.28, II.3.29, and II.3.24 using Theorems 4.8 and 6.1.

12 Show that a B-space is separable if and only if it is isometrically isomorphic to a closed subspace of $C(S)$, where S is a compact metric space.

13 Show that there exists a continuous mapping of the Cantor perfect set onto an arbitrary compact metric space S. (Hint. Construct a covering of $S = \bigcup_{i=1}^{2n} C_i^n$ by a sequence of closed sets C_i^n such that $C_i^n = C_{2i}^{n+1} \cup C_{2i-1}^{n+1}$); and such that the diameter of $C_j^n \to 0$ as $n \to \infty$.

14 Show that a B-space is separable if and only if it is isometrically isomorphic to a closed subspace of $C(P)$, where P is the Cantor perfect set.

15 If $\mathfrak{X}$ is a separable linear topological space and A is an $\mathfrak{X}$-compact subset of $\mathfrak{X}^*$, then the $\mathfrak{X}$-topology of A is a metric topology.

16 If $\mathfrak{X}$ is a separable B-space, a convex subset A of $\mathfrak{X}^*$ is $\mathfrak{X}$-closed if and only if $x_n^* \in A$, and $\lim_{n \to \infty} x_n^*(x) = x^*(x)$, $x \in \mathfrak{X}$, imply $x^* \in A$.

17 If S is a normal space and $C(S)$ is separable, then S is a compact metric space, and conversely.

18 If S is a normal space, S is homeomorphic to a subset of the unit sphere of the conjugate of $C(S)$ with the $C(S)$ topology.

19 Every normal space S is homeomorphic to a dense subset S_1 of a compact Hausdorff space C_1, such that every bounded continuous function on S_1 has a unique extension to a continuous function on C_1.

20 If $\mathfrak{X}$ is a B-space, then a convex set $K \subseteqq \mathfrak{X}$ is weakly closed if and only if its intersection with every bounded weakly closed set is weakly closed.

21 If S is a compact Hausdorff space, and $\{f_n\}$ is a sequence of continuous functions on S such that $\sup |f_n| < \infty$ and $f_n(s) \to f(s)$ for each $s \in S$, and f is continuous, then some sequence of convex combinations of the f_n converges uniformly to f.

22 Let $\mathfrak{X}$ be a B-space, and $A \subseteqq \mathfrak{X}$. If every separable subspace of $\mathfrak{X}$ intersects A in a weakly compact set, then the weak closure of A is weakly compact.

23 Show that every neighborhood N of the identity of a linear topological space contains a neighborhood M such that $\alpha M \subseteqq M$ for $|\alpha| \leqq 1$.

24 Show that if the linear space $\mathfrak{X}$ is finite dimensional, $K \subseteqq \mathfrak{X}$ is convex, and $p \notin K$, some functional on $\mathfrak{X}$ separates K and p.

25 Find a convex set K and a point $p \notin K$ in a linear space such that no non-zero functional separates K and p. (Hint. Let $\mathfrak{X}$ be a space with a denumerable Hamel basis $\{x_n\}$, and consider the set K of vectors of the form $\sum_{i=1}^n a_i x_i$ with $a_n > 0$.)

26 Find a convex set K with an internal point and a point $p \notin K$ in a B-space such that no non-zero continuous functional separates K and p.

27 Every infinite dimensional B-space is the union of two disjoint dense convex subsets.

28 Find two closed subsets A_1 and A_2 of a topological group such that A_1+A_2 is not closed.

29 Find two closed convex sets A_1, A_2 in a linear topological space such that $\text{co}(A_1 \cup A_2) \neq \overline{\text{co}}(A_1 \cup A_2)$, and two sets B_1, B_2 such that $\overline{\text{co}}(B_1+B_2) \neq \overline{\text{co}}(B_1) + \overline{\text{co}}(B_2)$.

30 A linear homeomorphic mapping of a locally convex space onto some normed space exists if and only if some open set of the locally convex space is bounded.

31 If K is a convex subset of a linear topological space, and the origin is an internal point of K, show that the support function of K is continuous if and only if the origin is an interior point of K.

32 Show that a compact space may contain a sequence having no convergent subsequence.

33 Show that if a subset of the conjugate $\mathfrak{X}^*$ of a B-space $\mathfrak{X}$ is $\mathfrak{X}$-sequentially compact, its $\mathfrak{X}$-closure is $\mathfrak{X}$-compact, but that the converse is not necessarily true.

34 If $\mathfrak{X}$ is a linear space and Γ a total subspace of $\mathfrak{X}^+$, then the Γ-topology of $\mathfrak{X}$ is a metric topology if and only if Γ has a countable Hamel basis.

35 If the weak topology of the unit sphere of a B-space is a metric topology, $\mathfrak{X}^*$ contains a countable total set.

36 If $\mathfrak{X}$ is reflexive, and $\mathfrak{X}^*$ contains a countable total set, $\mathfrak{X}^*$ is separable.

37 Let S be the unit sphere $S = \{x \mid |x| < 1\}$ of the F-space $L_p(0, 1)$ where $0 < p < 1$. The norm in this space is given by the equation $|x| = \int_0^1 |x(t)|^p dt$. Show that $\text{co}(S)$ is the whole space $L_p(0, 1)$. Show that there is no non-zero continuous linear functional on $L_p(0, 1)$.

38 (von Neumann) Let A be the subset of l_2 consisting of vectors $\{x_{mn} \mid 1 \leqq m < n < \infty\}$, where the m-th coordinate of x_{mn} is one, the n-th coordinate is m, and all other coordinates are zero. Show that the origin is in the weak closure of A, but that no sequence of elements of A converges weakly to zero.

39 Show that if a point p in l_2 is in the weak closure of a bounded set $A \subseteq l_2$, then p is the weak limit of a sequence of elements of A.

40 Let $\mathfrak{Z}$ be a dense linear manifold of a B-space $\mathfrak{X}$. Show that $\mathfrak{Z}^*$ and $\mathfrak{X}^*$ are isometrically isomorphic, but that the isometric iso-

morphism between them is not a homeomorphism with the $\mathfrak{Z}$ topology of $\mathfrak{Z}^*$ and the $\mathfrak{X}$ topology of $\mathfrak{X}^*$ unless $\mathfrak{X} = \mathfrak{Z}$.

41 Let $\mathfrak{X}$ be a locally convex linear topological space and let $\mathfrak{G}$ be a linear subspace of $\mathfrak{X}^*$. Then $\mathfrak{G}$ is $\mathfrak{X}$-dense in $\mathfrak{X}^*$ if and only if $\mathfrak{G}$ is a total set of functionals on $\mathfrak{X}$.

42 Let $\mathfrak{X}$ be a locally convex linear topological space, and $\mathfrak{X}_1$ a subspace of $\mathfrak{X}$. Let $x_1^* \in \mathfrak{X}_1^*$. Then there exists an element $x^* \in \mathfrak{X}^*$ such that $x^*(x_1) = x_1^*(x_1)$ for $x_1 \in \mathfrak{X}_1$.

43 Let $\{x_n\}$ be a bounded sequence in a reflexive B-space $\mathfrak{X}$ and let $K_n = \overline{\text{co}}\{x_n, x_{n+1}, \ldots\}$. Then $x_n \to x_0$ weakly if and only if $\{x_0\} = \bigcap_{n=1}^{\infty} K_n$. Show that this theorem is not true if we omit the assumption of reflexivity.

44 Let $\mathfrak{X}$ be a complex locally convex linear topological space. Show that the weak topology of $\mathfrak{X}$ is the same whether we take it as a complex space or regard it as a vector space over the field of real scalars.

8. Extremal Points

1 Definition. Let K be a subset of a real or complex linear vector space $\mathfrak{X}$. A non-void subset $A \subseteqq K$ is said to be an *extremal subset* of K if a proper convex combination $ak_1+(1-a)k_2$, $0 < a < 1$, of two points of K is in A only if both k_1 and k_2 are in A. An extremal subset of K consisting of just one point is called an *extremal point* of K.

For example, in three dimensional Euclidean space the surface of a solid sphere is an extremal subset of the sphere, and every point on the surface is an extremal point. The vertices, sides, and faces of a solid cube form an extremal subset of the cube, but only the eight vertices are extremal points of the cube, the remaining points on the sides and faces are neither extremal nor internal points. A convex set may have no extremal point at all, as in the case of an open sphere.

2 Lemma. *A non-void compact set in a locally convex linear topological space has extremal points.*

Proof. Let K be a compact subset of the locally convex linear topological space $\mathfrak{X}$. Let $\mathscr{A}$ be the non-void family of closed extremal subsets of K. Order $\mathscr{A}$ by inclusion. It is readily seen that if $\mathscr{A}_1$ is a totally ordered subfamily of $\mathscr{A}$, the non-void set $\cap \mathscr{A}_1$ is a closed extremal subset of K which furnishes a lower bound for $\mathscr{A}_1$. It follows by Zorn's lemma that $\mathscr{A}$ contains a minimal element A_0. Suppose that A_0 contains two distinct points p and q. Then, by 2.13, there is a functional $x^* \in \mathfrak{X}^*$ such that $\mathscr{R}x^*(p) \neq \mathscr{R}x^*(q)$. This implies that $A_1 = \{x | x \in A_0, \mathscr{R}x^*(x) = \inf_{y \in A_0} \mathscr{R}x^*(y)\}$ is a proper subset of A_0. On the other hand, if k_1 and k_2 are points of K such that $ak_1+(1-a)k_2 \in A_1$ for some $0 < a < 1$, then $k_1, k_2 \in A_0$ since A_0 is extremal. It follows from the definition of A_1 that $k_1, k_2 \in A_1$. Hence A_1 is a proper closed extremal subset of A_0. This contradiction shows that A_0 contains only one point which is therefore an extremal point of K. Q.E.D.

3 LEMMA. *Let K be a subset of a linear space, let A_1 be an extremal subset of K, and A_2 an extremal subset of A_1. Then A_2 is an extremal subset of K.*

The proof is elementary, and is left to the reader.

4 THEOREM. (*Krein-Milman*) *If K is a compact subset of a locally convex linear topological space $\mathfrak{X}$ and E is the set of its extremal points, then $\overline{\text{co}}(E) \supseteq K$. Consequently $\overline{\text{co}}(E) = \overline{\text{co}}(K)$, and $\overline{\text{co}}(E) = K$ if K is convex.*

PROOF. Let $k \in K$ and $k \notin \overline{\text{co}}(E)$. Then, by 2.10, we can find an $x^* \in \mathfrak{X}^*$ and real constants c and ε, $\varepsilon > 0$, such that $\mathscr{R}x^*(k) \leq c$, $\mathscr{R}x^*(\overline{\text{co}}(E)) \geq c+\varepsilon$. Let $K_1 = \{x | x \in K, \mathscr{R}x^*(x) = \inf_{y \in K} \mathscr{R}x^*(y)\}$. Then K_1 is a closed extremal subset of K, and $K_1 E = \phi$. By Lemma 3, K_1 has no extremal points, but this contradicts Lemma 2. Thus the first assertion of Theorem 4 is proved and the remaining assertions follow trivially. Q.E.D.

5 LEMMA. *Let Q be a compact set in a locally convex linear topological space $\mathfrak{X}$ whose closed convex hull is compact. Then the only extremal points in $\overline{\text{co}}(Q)$ are points in Q.*

PROOF. Let p be an extremal point of $\overline{\text{co}}(Q)$ which is not in Q. Since Q is closed, we can find a neighborhood U_0 of the origin of $\mathfrak{X}$ such that $(p+U_0) \cap Q = \phi$, and can then find a convex neighborhood U of the origin such that $U - U \subseteq U_0$. It follows that $(p+U) \cap (Q+U) = \phi$, so that $p \notin \overline{Q+U}$. The family of sets $\{q+U\}$, $q \in Q$, is an open covering of Q; let $\{q_i+U\}$, $i = 1, \ldots, n$, be a finite subcovering. Put $K_i = \overline{\text{co}}((\overline{q_i+U}) \cap Q) \subseteqq \overline{q_i+U}$. Then K_i is a closed, and hence a compact, subset of $\overline{\text{co}}(Q)$. Hence

$$\overline{\text{co}}(Q) = \overline{\text{co}}(K_1 \cup \ldots \cup K_n) = \text{co}(K_1 \cup \ldots \cup K_n),$$

by an easy induction on Lemma 2.5. It follows readily that p has the form $p = \sum_{i=1}^n a_i k_i$, $a_i \geqq 0$, $\sum_{i=1}^n a_i = 1$, $k_i \in K_i$; and, since p is extremal, that $k_i = p$ if $a_i > 0$. Hence $p \in \bigcup_{i=1}^n K_i \subseteqq \bigcup_{i=1}^n \overline{(q_i+U)} \subseteqq \overline{Q+U}$. This contradiction proves the lemma. Q.E.D.

6 Lemma. *Let $\mathfrak{X}$ be a closed linear manifold in the B-space $C(Q)$ of all real (or complex) continuous functions on the compact Hausdorff space Q. For each q in Q let x_q^* in $\mathfrak{X}^*$ be defined by*

$$x_q^* f = f(q), \qquad f \in \mathfrak{X}.$$

Then every extremal point of the closed unit sphere S^* of $\mathfrak{X}^*$ is of the form αx_q^* with $|\alpha| = 1$ and q in Q. If $\mathfrak{X} = C(Q)$ the converse holds, i.e., every element of the form αx_q^* with $|\alpha| = 1$ and q in Q is an extremal point S^*.

Proof. Let A be the set of all points in $\mathfrak{X}^*$ of the form αx_q^* with $|\alpha| = 1$ and $q \in Q$, so that $A \subseteqq S^*$. Let $\mathfrak{X}^*$ have its $\mathfrak{X}$ topology; since S^* is convex and $\mathfrak{X}$-closed, by V.2.4 the $\mathfrak{X}$-closure $\overline{\text{co}(A)} = \overline{\text{co}}(A) \subseteqq S^*$. If $x^* \notin \overline{\text{co}}(A)$, then by 2.12 and 3.9, there is an $x \in \mathfrak{X}$ and real constants c and ε, $\varepsilon > 0$, with

$$\mathscr{R}x^*(x) \geqq c; \qquad \mathscr{R}\alpha x(q) \leqq c-\varepsilon, \qquad q \in Q, \qquad |\alpha| = 1.$$

Hence $|x| \leqq c-\varepsilon$, so that $|x^*| > 1$. Thus $\overline{\text{co}}(A) \supseteqq S^*$, so that $\overline{\text{co}}(A) = S^*$. It then follows from Lemma 5 and Theorem 4.2 that every extremal point of S^* lies in A.

Conversely, let $\mathfrak{X} = C(Q)$ and let $q \in Q$ be such that $x_q^* = ay^*+(1-a)z^*$, where $0 < a < 1$ and $y^*, z^* \in S^*$. It will be shown that $y^* = z^* = x_q^*$. Let $x_0 \in C(Q)$, $|x_0| \leqq 1$, and $x_0(p) = 0$ for p in some neighborhood N of q. By Theorem I.5.3, there exists a $y \in C(Q)$ such that $|y| \leqq 1$, $y(q) = 1$, $y(p) = 0$ for $p \notin N$. Then $ay^*(y)+(1-a)z^*(y) = x_q^*(y) = 1$ and $|y^*(y)| \leqq 1$, $|z^*(y)| \leqq 1$. Hence $y^*(y) = z^*(y) = 1$. In the same way it follows that $y^*(x_0+y) = z^*(x_0+y) = 1$. Thus $y^*(x_0) = z^*(x_0) = 0$. Now let $x_1 \in C(Q)$, $|x_1| \leqq 1$, and $x_1(q) = 0$. Then for each integer n there exists a neighborhood N_n of q such that $|x_1(p)| < 1/n$ for $p \in N_n$. Let M_n be a neighborhood of q

such that $\overline{M}_n \subseteqq N_n$; by Theorem I.5.3, let $g_n \in C(Q)$ be such that $|g_n| \leqq 1/n$, $g_n(p) = 0$ for $p \notin N_n$, and $g_n(p) = x_1(p)$ for $p \in M_n$. Then $x_1 - g_n \to x_1$, $|x_1 - g_n| \leqq 1$, and $x_1 - g_n$ vanishes in M_n. It follows that $y^*(x_1) = z^*(x_1) = 0$. If $x \in C(Q)$ is such that $x(q) = 0$, then $|x/n| \leqq 1$ for some sufficiently large integer n, so that $y^*(x) = z^*(x) = 0$. Now Lemma 3.10 can be applied to show that scalars α and β exist such that $y^* = \alpha x_q^*$, $z^* = \beta x_q^*$. Since $y^*, z^* \in S^*$, $|\alpha| \leqq 1$, $|\beta| \leqq 1$. Since $x_q^* = (a\alpha + (1-a)\beta)x_q^*$, $\alpha = \beta = 1$. This proves that x_q^* is an extremal point of S^*; a similar argument shows that αx_q^* with $|\alpha| = 1$ is also extremal. Q.E.D.

7 Lemma. *If Q is a compact Hausdorff space, and $C(Q)$ is the B-space of all real or complex continuous functions on Q, then the mapping $\lambda : q \to x_q^*$ of Q into a subset $\hat{Q}$ of the extremal points of S^* is a homeomorphism where $C^*(Q)$ has its $C(Q)$ topology.*

Proof. The mapping $\lambda : q \to x_q^*$ is clearly one-to-one. By Lemma 6 it sends Q into a subset of the extremal points of S^*. To show that λ is continuous, let $N(x_q^*; x_0, \varepsilon) = \{x^* \mid |(x^* - x_q^*)x_0| < \varepsilon\}$ be a subbasic neighborhood of x_q^*. Since $x_0 \in C(Q)$, the set $N(q) = \{p \mid |x_0(p) - x_0(q)| < \varepsilon\}$ is open in Q, and it is clear that $\lambda(N(q)) \subseteqq N(x_q^*; x_0, \varepsilon)$. Hence $\hat{Q} = \lambda(Q)$ is compact, and λ is a homeomorphism. Q.E.D.

8 Theorem. *(Banach-Stone) Let Q and R be compact Hausdorff spaces, and let T be an isometric isomorphism between $C(Q)$ and $C(R)$. Then there exists a homeomorphism τ between R and Q, and a function α in $C(R)$ with $|\alpha(r)| = 1$, $r \in R$, such that*

$$[*] \qquad (Tx)(r) = \alpha(r)x(\tau(r)); \qquad x \in C(Q), \qquad r \in R.$$

Proof. The linear map $T^* : C^*(R) \to C^*(Q)$, defined by $(T^*y^*)x = y^*(Tx)$; $y^* \in C^*(R)$, $x \in C(Q)$, is readily seen to be an isometric mapping. Furthermore T^* maps onto $C^*(Q)$; for given an arbitrary functional $x^* \in C^*(Q)$, let $y^* = x^*T^{-1}$. Thus if S_Q^* and S_R are the unit spheres in $C^*(Q)$ and $C^*(R)$, $T^*(S_R^*) = S_Q^*$. Since T^* is linear and isometric, it is clear that it maps the extremal points E_R of S_R^* onto the extremal points E_Q of S_Q^* in a one-to-one fashion.

It is elementary to show that T^* is continuous with the $C(R)$ topology in $C^*(R)$ and the $C(Q)$ topology in $C^*(Q)$. Since, by Lemma 7, $\hat{R}$ is compact in the $C(R)$ topology, T^* is a homeomorphism between

$\hat{R}$ and a subset of E_Q in these topologies (I.5.8). From Lemma 6, to each $r \in R$, $T^* y_r^* = \alpha(r) x_{\tau(r)}^*$, where $|\alpha(r)| = 1$ and $\tau(r) \in Q$. From the fact that T^* maps E_R onto E_Q, and from Lemma 6, it follows that $t : y_r^* \to x_{\tau(r)}^*$ maps $\hat{R}$ onto $\hat{Q}$, and since T^* is one-to-one, t is also. Thus t is a homeomorphism between $\hat{R}$ and $\hat{Q}$. But since, by Lemma 7, $\lambda : Q \to \hat{Q}$ and $\mu : R \to \hat{R}$ are homeomorphisms, $\tau = \lambda^{-1} t \mu$ is a homeomorphism from R to Q. Thus, for any $x \in C(Q)$ and $r \in R$, we have

$$(Tx)(r) = y_r^*(Tx) = (T^* y_r^*)(x) = \alpha(r) x_{\tau(r)}^*(x) = \alpha(r) x(\tau(r)),$$

which is equation [*]. It remains to show that α is continuous. In the formula above, substitute the function $x_0 \in C(Q)$ which is identically one. This yields $\alpha = Tx_0 \in C(R)$. Q.E.D.

Just as we were able to prove a number of interesting results about the space $C(S)$ by considering the extremal points of the unit sphere of $C^*(S)$, so we shall see that consideration of the extremal points of $L_\infty^*(S, \Sigma, \mu)$ leads to interesting facts about $L_\infty(S, \Sigma, \mu)$.

9 LEMMA. *Let (S, Σ, μ) be a positive measure space. A functional x^* in $L_\infty^*(S, \Sigma, \mu)$ is an extremal point of the closed unit sphere S^* of $L_\infty^*(S, \Sigma, \mu)$ only if it is of the form $x^* = \alpha y^*$, where $|\alpha| = 1$, and where y^* is non-zero and multiplicative:*

$$y^*(fg) = y^*(f) y^*(g),$$

where $(fg)(s) = f(s)g(s)$ for μ-almost all s in S.

PROOF. We will first show that the norm of an extremal point x^* in S^* is equal to one. Clearly $x^* \neq 0$ so that $y^* = x^*/|x^*|$ is in S^* and

$$x^* = |x^*| y^* + (1 - |x^*|)0.$$

Thus, by Definition 1, $1 - |x^*| = 0$.

By Theorem IV.8.16, there is a λ in $ba(S, \Sigma_1, \mu_1)$ with $|\lambda| = 1$ and

$$x^* f = \int_S f(s) \lambda(ds), \qquad f \in L_\infty(S, \Sigma, \mu).$$

It will next be shown that λ vanishes on at least one set in every pair of disjoint sets in Σ_1. To see this, assume that there exist two disjoint sets $E_1, E_2 \in \Sigma_1$ with $\lambda(E_1) \neq 0$ and $\lambda(E_2) \neq 0$. If $\lambda_1(E) = \lambda(EE_1)$ and $\lambda_2(E) = \lambda(E(S - E_1))$ for $E \in \Sigma_1$ it is clear that $\lambda_1, \lambda_2 \in ba(S, \Sigma_1, \mu_1)$, that $v(\lambda_1, E) = v(\lambda, EE_1)$, and that $v(\lambda_2, E) = v(\lambda, E(S - E_1))$ for

$E \in \Sigma_1$. Hence, $1 = |\lambda| = |\lambda_1|+|\lambda_2|$. Since $\lambda_1 \neq 0$, $\lambda_2 \neq 0$, we may put $v_1 = \lambda_1/|\lambda_1|$ and $v_2 = \lambda_2/|\lambda_2|$. Then $v_1, v_2 \in S^*$, and $\lambda = |\lambda|_1 v_1 + (1-|\lambda_1|)v_2$. Consequently $v_1 = v_2 = \lambda$, and thus $0 \neq \lambda(E_1) = v_2(E_1) = 0$, which is the desired contradiction.

Since λ vanishes on one set in each complementary pair we have

$$\lambda(E)(\lambda(S)-\lambda(E)) = 0, \qquad E \in \Sigma_1,$$

which shows that the function $m = \lambda/\lambda(S)$ assumes only the values 0 and 1. Thus

$$m(AB) = m(A)m(B), \qquad A, B \in \Sigma_1, \tag{1}$$

for, if either A or B is an m-null set so is AB, while if $m(A)=1=m(B)$ then, since $A-AB$ and $B-AB$ are disjoint, one of them is an m-null set and $m(AB) = 1$.

If y^* is defined by the equation

$$y^* f = \int_S f(s)m(ds),$$

for f in L_∞, then $|y^*| = |m| = 1$ and $y^* = \alpha x^*$ where $|\alpha| = 1/|\lambda(S)|=1$. Equation (1) shows that $y^*(fg) = y^*(f)y^*(g)$ if f and g are both characteristic functions of sets in Σ_1. For any g in L_∞ the manifold

$$\mathfrak{M}_g = \{f | f \in L_\infty, y^*(fg) = y^*(f)y^*(g)\}$$

is clearly a closed linear manifold in L_∞, and it follows from the preceding remarks that $\mathfrak{M}_g = L_\infty$ if g is a characteristic function. (It was shown in the proof of Theorem IV.8.16 that characteristic functions form a fundamental set in L_∞.) This means that if f is an arbitrary function in L_∞ the manifold $\mathfrak{M}_f$ contains all characteristic functions and thus $\mathfrak{M}_f = L_\infty$. Thus $y^*(fg) = y^*(f)y^*(g)$ for all f and g in L_∞. Q.E.D.

10 Corollary. *Let (S, Σ, μ) be a positive measure space. Let M be the collection of all non-zero x^* in the closed unit sphere of $L_\infty^*(S, \Sigma, \mu)$ such that $x^*(h) = x^*(f)x^*(g)$ whenever $h(s) = f(s)g(s)$ μ-almost everywhere. Then*

$$\sup_{x^* \in M} |x^*(f)| = |f|, \qquad f \in L_\infty(S, \Sigma, \mu).$$

Proof. It is clear that $\sup_{x^* \in M} |x^*(f)| \leqq |f|$ for $f \in L_\infty(S, \Sigma, \mu)$. Suppose that for some $f_0 \in L_\infty(S, \Sigma, \mu)$, $\sup_{x^* \in M} |x^*(f_0)| \leqq |f_0|-\varepsilon$, where $\varepsilon > 0$. Then, according to Lemma 9, the set

$$A = \{x^* \mid |x^*| \leqq 1, \; |x^*(f_0)| \leqq |f_0| - \varepsilon\}$$

contains all the extremal points of the closed unit sphere of $L_\infty^*(S, \Sigma, \mu)$. On the other hand, A is evidently convex and $L_\infty(S, \Sigma, \mu)$-closed. By Theorem 4.2 and Theorem 4, A contains the entire unit sphere of L_∞^*. That is

$$\sup_{|x^*| \leqq 1} |x^*(f_0)| \leqq |f_0| - \varepsilon.$$

But, by the Hahn-Banach theorem, I.3.15, this is impossible. Q.E.D.

11 Theorem. *Let (S, Σ, μ) be a positive measure space. Then there exists a compact Hausdorff space S_1 and an isometric isomorphism Λ between $L_\infty(S, \Sigma, \mu)$ and $C(S_1)$. The isomorphism Λ maps real functions (i.e., functions real μ-almost everywhere) into real functions, positive functions into positive functions, and complex conjugate functions into complex conjugate functions. Moreover, Λ is an algebraic isomorphism in the sense that if $h(s) = f(s)g(s)$ μ-almost everywhere, $\Lambda h = \Lambda f \cdot \Lambda g$. If β is an arbitrary continuous function of a complex variable, and f is in $L_\infty(S, \Sigma, \mu)$, then $\Lambda(\beta(f)) = \beta(\Lambda(f))$.*

Proof. In view of Corollary 10, we may take the proof to be word for word the same as the proof of Theorem IV.6.18. Q.E.D.

9. Tangent Functionals

We begin this section with a study of functionals tangent to a convex set K. Since many of our results depend on the assumption that K has an internal point, it is convenient to observe that p is an internal point of K if and only if the origin is an internal point of $K - p$. Consequently we will consider only subsets which contain the origin as an internal point. The reader will have no difficulty in carrying over our statements and definitions to the slightly more general case of sets not necessarily containing the origin.

1 Lemma. *Let K be a convex set in a linear space $\mathfrak{X}$, and let the origin be an internal point of K. Let $\mathfrak{k}$ be the support function of K. Then $(\mathfrak{k}(x + ay) - \mathfrak{k}(x))/a$ is an increasing function of the positive real variable*

a for all x, y in $\mathfrak{X}$. The limit

$$\tau(x, y) = \lim_{a \to 0+} \frac{1}{a}(\mathfrak{k}(x+ay) - \mathfrak{k}(x))$$

exists for each x, y in $\mathfrak{X}$.

Proof. Let $a_1 \geqq a_2 > 0$; then

$$\mathfrak{k}(a_1 x + a_1 a_2 y) \leqq \mathfrak{k}(a_2 x + a_1 a_2 y) + \mathfrak{k}((a_1 - a_2)x)$$

by 1.8(e). Thus

$$a_1\{\mathfrak{k}(x+a_2 y) - \mathfrak{k}(x)\} \leqq a_2\{\mathfrak{k}(x+a_1 y) - \mathfrak{k}(x)\}$$

by 1.8(c), so that

$$\frac{1}{a_2}\{\mathfrak{k}(x+a_2 y) - \mathfrak{k}(x)\} \leqq \frac{1}{a_1}\{\mathfrak{k}(x+a_1 y) - \mathfrak{k}(x)\}.$$

Thus the function

$$\frac{1}{a}\{\mathfrak{k}(x+ay) - \mathfrak{k}(x)\}$$

decreases as a decreases. Since $\mathfrak{k}(x+ay) + \mathfrak{k}(-ay) \geqq \mathfrak{k}(x)$, $\{\mathfrak{k}(x+ay) - \mathfrak{k}(x)\}/a \geqq -\mathfrak{k}(-y)$, so that $\{k(x+ay) - \mathfrak{k}(x)\}/a$ is bounded below. This proves the lemma. Q.E.D.

2 Definition. Let K be a convex set in a linear space $\mathfrak{X}$ and let K contain the origin as an internal point. If $\mathfrak{k}$ is the support function of K, then the real valued function τ defined for $x, y \in \mathfrak{X}$ by

$$\tau(x, y) = \lim_{a \to 0+} \frac{1}{a}\{\mathfrak{k}(x+ay) - \mathfrak{k}(x)\}$$

is called the *tangent function* of K.

3 Lemma. *Let K be a convex set in a linear space, and let the origin be an internal point of K. Let $\mathfrak{k}$ be the support function and τ the tangent function of K. Then*

(a) $\tau(x, y) \leqq \mathfrak{k}(y)$;
(b) $\tau(x, y_1+y_2) \leqq \tau(x, y_1) + \tau(x, y_2)$;
(c) $\tau(x, ay) = a\tau(x, y)$ *for* $a \geqq 0$;
(d) $-\tau(x, -y) \leqq \tau(x, y)$;
(e) $\tau(x, ax) = a\mathfrak{k}(x)$ *for a real.*

PROOF. By 1.8(c) and 1.8(e), we have

$$\frac{1}{a}\{\mathfrak{k}(x+ay)-\mathfrak{k}(x)\} \leqq \mathfrak{k}(y),$$

which implies (a). Statement (b) follows from the inequality

$$2\mathfrak{k}(x+\frac{a}{2}(y_1+y_2)) \leqq \mathfrak{k}(x+ay_1)+\mathfrak{k}(x+ay_2).$$

Statement (c) is trivial. Statement (d) follows from the inequality

$$\tau(x, -y)+\tau(x, y) \geqq \tau(x, 0) = 0 \cdot \tau(x, y) = 0.$$

Statement (e) is trivial. Q.E.D.

4 DEFINITION. If A is a subset of a linear space $\mathfrak{X}$, and x is a bounding point of A, a functional $f \in \mathfrak{X}^+$ is said to be *tangent* to A at x if there exists a real constant c such that

$$\mathscr{R}f(A) \leqq c, \qquad \mathscr{R}f(x) = c.$$

We note that if f is tangent to A at x, so is every real multiple of f. Conversely, if every tangent to A at x is a multiple of f, we say that *A has a unique tangent at x.*

Observe that if $\mathfrak{X}$ is a linear topological space and A has an interior point, then it follows from Lemma 2.7 that any functional tangent to A is continuous.

The next theorem gives a criterion for the existence of linear functionals tangent to a convex set in terms of its tangent function.

5 THEOREM. *Let $\mathfrak{X}$ be a linear space, K a convex subset of $\mathfrak{X}$, and suppose that the origin is an internal point of K. Let τ be the tangent function of K. Then if x is a bounding point of K, a functional f in $\mathfrak{X}^+$ with $f(x) = 1$ is tangent to A at x if and only if*

$$-\tau(x, -y) \leqq \mathscr{R}f(y) \leqq \tau(x, y) \text{ for all } y \in \mathfrak{X}.$$

Conversely, if x is a bounding point of K and y is any point in $\mathfrak{X}$ for which

$$-\tau(x, -y) \leqq c \leqq \tau(x, y),$$

then there is a functional f tangent to K at x with $f(x) = 1$ and $\mathscr{R}f(y) = c$.

PROOF. If the functional f is tangent to K at x and $f(x) = 1$, then by 1.8(f), $\mathfrak{k}(y) < 1$ implies $\mathscr{R}f(y) \leqq 1$. It then follows readily from

1.8(c) that $\mathfrak{k}(y) \geqq \mathscr{R}f(y)$. Consequently, since $f(x) = \mathfrak{k}(x) = 1$,

$$\mathscr{R}f(y) = \mathscr{R}\frac{1}{a}\{f(x+ay)-f(x)\} \leqq \frac{1}{a}\{\mathfrak{k}(x+ay)-\mathfrak{k}(x)\},$$

so that $\mathscr{R}f(y) \leqq \tau(x, y)$. Replacing y by $-y$, we get $-\mathscr{R}f(y) \leqq \tau(x, -y)$, so that $-\tau(x, -y) \leqq \mathscr{R}f(y) \leqq \tau(x, y)$. On the other hand if f satisfies this inequality, 3(a) shows f is tangent to K at x.

To prove the converse, we regard $\mathfrak{X}$ as a linear space over the reals. Observe that the functions $\mathfrak{k}$ and τ are unchanged. Let $y \in \mathfrak{X}$ and $-\tau(x, -y) \leqq c \leqq \tau(x, y)$. Every element z of $\mathfrak{S} = \text{sp}\{x, y\}$ has the form $z = ax+by$; define f_0 on $\mathfrak{S}$ by $f_0(z) = a+bc$. If $y = dx$, then it follows from 3(e) that $c = d$; so that $z = 0$ implies $f_0(z) = 0$. If y is not a multiple of x, the representation $z = ax+by$ is unique. Thus f_0 is a well defined linear functional on $\mathfrak{S}$.

We wish to show that $f_0(z) \leqq \mathfrak{k}(z)$ for $z \in \mathfrak{S}$. Since this is trivial when y is a scalar multiple of x, we can assume that every $z \in \mathfrak{S}$ has a unique representation $z = ax+by$.

Case 1. $a > 0$. Since $f_0(y) = c$, it suffices to show that

$$\mathfrak{k}(x+\frac{b}{a}y) \geqq f_0(x)+\frac{bc}{a};$$

or, letting $b/a = a_1$, that

$$\mathfrak{k}(x+a_1y)-\mathfrak{k}(x) \geqq a_1c,$$

since $f_0(x) = \mathfrak{k}(x) = 1$. If $a_1 = 0$ this is obvious. If $a_1 > 0$, $\mathfrak{k}(x+a_1y)-\mathfrak{k}(x) \geqq a_1\tau(x, y) \geqq a_1c$ by Lemma 1. If $a_1 < 0$, put $a_2 = -a_1$, and $y_1 = -y$. Then $\mathfrak{k}(x+a_2y_1)-\mathfrak{k}(x) \geqq a_2\tau(x, -y) \geqq -a_2c$ by Lemma 1, so that $\mathfrak{k}(x+a_1y)-\mathfrak{k}(x) \geqq a_1c$ in this case also.

Case 2. $a \leqq 0$. We wish to show that

$$\mathfrak{k}(ax+by)-a\mathfrak{k}(x) \geqq bc,$$

i.e., that

$$\mathfrak{k}(ax+by)+\mathfrak{k}(-ax) \geqq bc.$$

If $b \geqq 0$, we have

$$\mathfrak{k}(ax+by)+\mathfrak{k}(-ax) \geqq \mathfrak{k}(by) = b\mathfrak{k}(y) \geqq b\tau(x, y) \geqq bc$$

by 1.8(e) and 3(a). If $b < 0$,

$$\begin{aligned}\mathfrak{k}(ax+by)+\mathfrak{k}(-ax) &\geqq \mathfrak{k}(by) = -b\mathfrak{k}(-y)\\ &\geqq -b\tau(x, -y) \geqq (-b)(-c) = bc.\end{aligned}$$

Hence $f_0(z) \leqq \mathfrak{k}(z)$ for $z \in \mathfrak{S}$. By the Hahn-Banach theorem (II.3.10), f_0 can be extended to a linear functional f defined on all of $\mathfrak{X}$ such that $f(x) \leqq \mathfrak{k}(x)$ for $x \in \mathfrak{X}$. It follows by 1.8(d) that $f(K) \leqq 1$, so that f is tangent to K at x, and $f(y) = c$. In the case that $\mathfrak{X}$ is a linear space over the complex numbers, the functional $f(x)-if(ix)$ is the required tangent functional. Q.E.D.

6 COROLLARY. *If a convex set K in a linear space has the origin as an internal point, K has a non-zero tangent functional at each of its bounding points. There is a unique tangent functional at x if and only if $\tau(x, y) = -\tau(x, -y)$ for all $y \in \mathfrak{X}$.*

In the case of a linear topological space, we have a converse result.

7 THEOREM. *Let $\mathfrak{X}$ be a linear topological space, and let A be a closed subset of $\mathfrak{X}$, having interior points. Suppose that A has a non-zero tangent functional at each point of a dense subset of its boundary. Then A is convex.*

PROOF. Denoting the interior of A by A_1, we will show first that there is no non-zero tangent to A at any point of $aA_1+(1-a)A$, for $0 < a < 1$. Let $p \in A_1$, $q \in A$, and let $x = ap+(1-a)q$ where $0 < a < 1$. If f is a functional tangent to A at x, there is a real number c such that $\mathscr{R}f(x) = c$, $\mathscr{R}f(A) \leqq c$. Since $f(x) = af(p)+(1-a)f(q)$, it follows that $\mathscr{R}f(p) = \mathscr{R}f(q) = c$. Let N be a neighborhood of the origin such that $p+N \subseteq A$, and let M be a neighborhood of the origin with $M \cup (-M) \subseteq N$. Then $\mathscr{R}f(N+p) \leqq c$, and thus $\mathscr{R}f(N) \leqq 0$. Hence, $0 \leqq \mathscr{R}f(M) \leqq 0$, i.e., $\mathscr{R}f(M) = 0$, from which it follows that $f = 0$.

Let B_1 be the dense subset of the boundary B of A at which there are non-zero functionals tangent to A. We have seen that $(aA_1+(1-a)A) \cap B_1 = \phi$ for $0 < a < 1$. Since $aA_1+(1-a)A$ is open, $(aA_1+(1-a)A) \cap B = \phi$ for $0 < a < 1$. Let $p \in A_1$ and $q \in A$; since p is interior to A, $(1-a)p + aq$ is in A for all sufficiently small positive a. If d is the least upper bound of the set

$$\{a | 0 < a < 1,\ (1-a)p+aq \in A\},$$

then it is seen that $(1-d)p+dq \in B$, so from the above, $d = 1$. Consequently $(1-a)p+aq \in A$ for $0 < a < 1$, and $(1-a)A_1+aA \subseteq A$

for $0 < a < 1$. Since $(1-a)A_1+aA$ is open for $0 < a < 1$, $(1-a)A_1+aA_1 \subseteqq A_1$ for $0 < a < 1$; therefore A_1 is convex.

Since A is closed, $A \supseteqq \bar{A}_1$. On the other hand, if $p \in A_1$ and if $q \in A$, $q = \lim_{a\to 0}\{(1-a)q+ap\}$, so $A \subseteqq \bar{A}_1$. The convexity of $A = \bar{A}_1$ follows from that of A_1 and Theorem 2.1(a). Q.E.D.

While it is clear from elementary examples that a convex set need not have a unique tangent functional at every bounding point, the next theorem gives a strong result in this direction.

8 **Theorem.** *If a convex subset of a separable B-space $\mathfrak{X}$ has an interior point, it has a unique tangent at each point of a dense subset of its boundary.*

Proof. Let K be the convex set. It will be shown that $-\tau(x, y) = \tau(x, -y)$, $y \in \mathfrak{X}$, for x in a dense subset Z of $\mathfrak{X}$. The set K contains some sphere $S(0, 1/N)$ about the origin. This clearly implies that the support function $\mathfrak{k}$ of K satisfies $\mathfrak{k}(x) \leqq N|x|$. By Lemma 1.8(e),

[*] $$|\mathfrak{k}(y_1)-\mathfrak{k}(y_2)| \leqq N|y_1-y_2|).$$

Also

$$\begin{aligned}\tau(x, y_1)-\tau(x, y_2) &\leqq \tau(x, y_1-y_2)\\ &\leqq \mathfrak{k}(y_1-y_2) \leqq N|y_1-y_2|.\end{aligned}$$

Consequently,

$$|\tau(x, y_1)-\tau(x, y_2)| \leqq N|y_1-y_2|.$$

Thus, if we let $\{y_n\}$ be a denumerable dense subset of $\mathfrak{X}$, and put

$$Z_n = \{x|x \in \mathfrak{X}, \tau(x, y_n) = -\tau(x, -y_n)\},$$

then $x \in \bigcap_{n=1}^{\infty} Z_n$ is equivalent to $\tau(x, y) = -\tau(x, -y)$ for all $y \in \mathfrak{X}$. It remains to prove, therefore, that $Z = \bigcap_{n=1}^{\infty} Z_n$ is dense in $\mathfrak{X}$.

Now, Z_n is defined by the condition

$$\lim_{a\to 0+} \frac{1}{a}\{\mathfrak{k}(x+ay_n)-\mathfrak{k}(x)\} = \lim_{a\to 0+} -\frac{1}{a}\{\mathfrak{k}(x-ay_n)-\mathfrak{k}(x)\},$$

and since these limits exist, Z_n is the set of x for which

$$\lim_{a\to 0+} \frac{1}{a}\{\mathfrak{k}(x+ay_n)-2\mathfrak{k}(x)+\mathfrak{k}(x-ay_n)\} = 0.$$

Since the function

$$g(a, x, y) = \frac{1}{a}\{\mathfrak{f}(x+ay)-2\mathfrak{f}(x)+\mathfrak{f}(x-ay)\}$$

is (by Lemma 1) the sum of two monotone increasing functions of a it has this same property for all $x, y \in \mathfrak{X}$. Consequently, letting

$$Z_{n,i,j} = \left\{ x \middle| x \in \mathfrak{X},\ j\{\mathfrak{f}(x+y_n/j)-2\mathfrak{f}(x)+\mathfrak{f}(x-y_n/j)\} < \frac{1}{i} \right\},$$

we have $Z_n = \bigcap_{i=1}^{\infty} \bigcup_{j=1}^{\infty} Z_{n,i,j}$. If we put $Z_{n,i} = \bigcup_{j=1}^{\infty} Z_{n,i,j}$, then $Z_{n,i}$ is open in $\mathfrak{X}$, and $Z_n = \bigcap_{i=1}^{\infty} Z_{n,i}$.

We wish to prove that $Z = \bigcap_{n=1}^{\infty} Z_n = \bigcap_{n=1}^{\infty} \bigcap_{i=1}^{\infty} Z_{n,i}$ is dense in $\mathfrak{X}$. Suppose that Z is not dense in $\mathfrak{X}$, and that $p \notin \overline{Z}$. Then some sphere $S(p, \varepsilon)$ does not intersect Z. If $S = \overline{S(p, \varepsilon/2)}$, then $SZ = \phi$. Hence $\bigcup_{n=1}^{\infty} \bigcup_{i=1}^{\infty} SZ'_{n,i} = S$. It follows from Theorem I.6.9 that some set $Z'_{n,i}$ contains an open set, i.e., that some set $Z_{n,i}$ is not dense in $\mathfrak{X}$; since $Z_{n,i} \supseteq Z_n$, this implies that some set Z_n is not dense. We will show that this is impossible.

Suppose that $x_1 \in \mathfrak{X}$, $x_1 \notin \overline{Z}_n$, then $\tau(x, y_n) \neq -\tau(x, -y_n)$ for x in some relative neighborhood of x_1. Thus, the function $\mathfrak{f}(x_1+ay_n)$ has no derivative at any point a in a neighborhood of zero. But [*] shows that $\varphi(a) = \mathfrak{f}(x_1+ay_n)$ is continuous and of bounded variation. Thus by the remarks preceding Theorem III.5.17, there exists a Borel-Stieltjes measure μ such that $\mu(c, d) = \varphi(c)-\varphi(d)$. Thus, from III.11.6 it follows that $\mathfrak{f}(x_1+ay_n)$ has a derivative almost everywhere. This contradiction proves that Z_n is dense in $\mathfrak{X}$. Hence it follows that Z is dense in $\mathfrak{X}$. Since $\tau(ax, ay) = a\tau(x, y)$ for $a > 0$, $ax \in Z$ if $x \in Z$ and $a > 0$. Consequently, the continuous mapping $x \to x/\mathfrak{f}(x)$ of $\{x \in \mathfrak{X} | x \neq 0\}$ onto the boundary B of K maps Z into a subset of itself which, since Z is dense in $\mathfrak{X}$, is clearly dense in B. This shows that $Z \cap B$ is dense in B. The present theorem now follows immediately from Corollary 6. Q.E.D.

There are interesting connections between the tangents to a set in a linear space $\mathfrak{X}$ and certain special convex subsets of $\mathfrak{X}$, called cones.

9 Definition. If $\mathfrak{X}$ is a vector space, a convex set $K \subseteq \mathfrak{X}$ is called a *cone with vertex* p, if $p+x \in K$ implies that $p+rx \in K$ for $r \geq 0$. The *cone K with vertex p generated by A* is the intersection of all

cones with vertex p which contain the set A. It is easy to see that $K = \{z|z = r(q-p)+p, q \in A, r \geqq 0\}$, if A is convex.

10 THEOREM. *If K is a closed cone with vertex p in a real locally convex linear topological space $\mathfrak{X}$, and $K \neq \mathfrak{X}$, then there exists a non-zero continuous linear functional tangent to K at p. If A is a subset of $\mathfrak{X}$, and p is in A, then there exists a non-zero continuous linear functional tangent to A at p if and only if the cone B with vertex p generated by A is not dense in $\mathfrak{X}$.*

PROOF. If $q \notin K$, then, by 2.12 we can find a functional f and a real constant c with $\mathscr{R}f(K) \leqq c \leqq \mathscr{R}f(q)$. Let $\mathscr{R}f(p) = a$. Then if $z \in K$ and $\mathscr{R}f(z) > a$, we have

$$\mathscr{R}f(z-p) \geqq \varepsilon > 0, \text{ and } \mathscr{R}f(r(z-p)+p) \geqq r\varepsilon+a,$$

which contradicts $\mathscr{R}f(K) \leqq c$ for r sufficiently large. Hence $\mathscr{R}f(K) \leqq \mathscr{R}f(p)$, which proves the first part of the theorem.

It is clear that B and $\bar{B}$ are cones. If $\bar{B} \neq \mathfrak{X}$, then any non-zero continuous linear functional tangent to $\bar{B}$ at p is a tangent to A at p.

Conversely, if f is a non-zero continuous linear functional tangent to A at p, then $\mathscr{R}f(p) = c$, $\mathscr{R}f(A) \leqq c$. It follows that $\mathscr{R}f(a(q-p)+p) \leqq c$ for $a \geqq 0$, so that $\mathscr{R}f(B) \leqq c$, and since $\bar{B} = \mathfrak{X}$ Lemma 1.11 implies $f(\mathfrak{X}) = 0$; since $f \neq 0$, this is impossible. Q.E.D.

11 THEOREM. *Let $\mathfrak{X}$ be a normed linear space, and let K be a convex subset of $\mathfrak{X}$. If there are points p, q with p in K, q not in K, and*

$$|q-p| = \inf_{z \in K} |q-z|,$$

then there exists a non-zero continuous linear functional tangent to K at p.

PROOF. Let $d = |q-p|$ and $S = S(q, d)$. Then S is an open set and $S \cap K = \phi$. By Theorem 2.8, there exists a non-zero continuous linear functional f such that $\mathscr{R}f(K) \leqq \mathscr{R}f(S)$. Since p is in the closure of both K and S, it is evident that f is tangent to K at p. Q.E.D.

12 COROLLARY. *Let $\mathfrak{X}$ be a normed linear space and K a convex subset of $\mathfrak{X}$. Suppose that for each q not in K there exists a p in K such that $|p-q| = \inf_{z \in K} |z-q|$. Then K has non-zero continuous linear tangent functionals at each point of a dense subset of its boundary.*

PROOF. Let $z \in K$ be in the boundary of K; let $q_n \notin K$, $q_n \to z$. Then, if $p_n \in K$ is such that $|p_n - q_n| = \inf_{z \in K} |z - q_n|$, it follows from Theorem 11 that there exists a non-zero functional tangent to K at p_n. But, clearly, $p_n \to z$. Q.E.D.

10. Fixed Point Theorems

1 DEFINITION. A topological space R is said to have the *fixed point property*, if for every continuous mapping $T : R \to R$, there exists a $p \in R$ with $p = T(p)$.

The famous Brouwer fixed point theorem states: *The closed unit sphere of E^n has the fixed point property.*

This section is devoted to proving a striking generalization (Theorem 5) of Brouwer's theorem. We shall use the Brouwer theorem in proving Theorem 5; however, since the Brouwer theorem is quite well-known, we do not give the details of its proof in this section but we refer the reader to the notes at the end of the chapter where a proof and some additional references are given.

The *Hilbert cube* C is the set in the B-space l_2 consisting of all sequences $[\xi_n]$ with $|\xi_n| \leqq 1/n$, $n = 1, 2, \ldots$. It is seen that C is a compact subset of l_2 (e.g. see Corollary IV.5.5).

2 LEMMA. *The Hilbert cube has the fixed point property.*

PROOF. Let $T : C \to C$ be continuous and let $P_n : C \to C$ be the map defined by

$$P_n([\xi_1, \ldots, \xi_n, \xi_{n+1}, \ldots]) = [\xi_1, \ldots, \xi_n, 0, 0, \ldots].$$

The set $C_n = P_n(C)$ is homeomorphic to the closed unit sphere in E^n. Since the mapping $P_nT : C_n \to C_n$ is continuous, the Brouwer theorem implies that it has a fixed point $y_n \in C_n \subset C$, so that

$$|y_n - T(y_n)| \leqq \sqrt{\sum_{i=n+1}^{\infty} 1/i^2}.$$

Since C is compact, $\{y_n\}$ has a convergent subsequence. The limit of this sequence is clearly a fixed point of T. Q.E.D.

3 LEMMA. *Any convex closed subset K of the Hilbert cube C has the fixed point property.*

Proof. To each point $p \in C$ there corresponds a unique nearest point $N(p) \in K$. To see this, note that Lemma IV.4.2 implies that if $\{k_i\}$ is a sequence in K such that $\lim_{i\to\infty} |p-k_i| = \inf_{k\in K}|p-k|$ then $\{k_i\}$ converges, say, to $q \in K$. If $\{k_i'\}$ is another such minimizing sequence converging to $q' \in K$, then IV.4.2 implies that $\{k_1, k_1', k_2, k_2', \ldots\}$ is also convergent. Hence $q = q'$ and q is the desired nearest point $N(p)$.

Now, $N(p)$ is a continuous function of p. For, if $p_n \to p$, and $N(p_n) \not\to N(p)$, then since K is compact, $\{N(p_n)\}$ has a convergent subsequence $\{N(p_{n_i})\}$, converging to an element q in K different from $N(p)$. Thus

$$|p_{n_i}-N(p_{n_i})| \leqq |p_{n_i}-N(p)| \leqq |p_{n_i}-p|+|p-N(p)|;$$

so that $|p-q| \leqq |p-N(p)|$; by the first part of the proof, this implies $N(p) = q$. We note that $N(C) \subseteq K$, while $p \in K$ implies that $N(p) = p$. Now, if $T : K \to K$ is continuous, $TN : C \to K$ is continuous and by the preceding lemma has a fixed point. This fixed point is in K; it is therefore a fixed point of the mapping T. Q.E.D.

4 **Lemma.** *Let K be a compact convex subset of a locally convex linear topological space $\mathfrak{X}$. Let $T : K \to K$ be continuous. If K contains at least two points, there exists a proper closed convex subset $K_1 \subset K$ such that $T(K_1) \subseteq K_1$.*

Proof. We may suppose that K has the $\mathfrak{X}^*$ topology, since the identity mapping from $\mathfrak{X}$ with its original topology to $\mathfrak{X}$ with its $\mathfrak{X}^*$ topology is continuous. Since a continuous and one-to-one mapping of K is a homeomorphism, (cf. Lemma I.5.8), changing to the $\mathfrak{X}^*$ topology does not affect the hypotheses of the lemma.

We will say that a set of continuous linear functionals $F = \{f\}$ is determined by another set $G = \{g\}$, if for each $f \in F$ and $\varepsilon > 0$, there exists a neighborhood $N(0; \gamma, \delta) = \{x \,|\, |g(x)| < \delta,\ g \in \gamma\}$, where γ is a finite subset of G, with the property that if $p, q \in K$ and $p-q \in N(0;\gamma,\delta)$, then $|f(Tp)-f(Tq)| < \varepsilon$. It is clear that if F is determined by G, then $g(p) = g(q)$, $g \in G$, implies that $f(Tp) = f(Tq)$, $f \in F$.

Each continuous linear functional f is determined by some denumerable set of functionals $G = \{g_m\}$. For, by IV.6.9, the scalar function $f(Tp)$ is uniformly continuous on the compact set K. Hence for each integer n there is a neighborhood $N(0; \gamma_n, \delta_n)$ of the origin in

$\mathfrak{X}$, given by a finite set γ_n of continuous linear functionals, and a $\delta_n > 0$, such that if $p, q \in K$ and $p-q \in N(0, \gamma_n, \delta_n)$, then $|f(Tp)-f(Tq)| < 1/n$. Let $G = \bigcup_{n=1}^{\infty} \gamma_n$; then f is determined by G. It follows that if F is a denumerable subset of $\mathfrak{X}^*$, there exists a denumerable subset G_F of $\mathfrak{X}^*$ such that each $f \in F$ is determined by G_F.

We can even assert that each continuous linear functional f can be included in a denumerable self-determined set G of functionals. For, if f is determined by the denumerable set G_1, let each functional in G_1 be determined by the denumerable set G_2, each functional in G_2 by the denumerable set G_3, etc. Put $G = \{f\} \cup \bigcup_{i=1}^{\infty} G_i$.

Now, suppose that K contains two distinct points, p and q, and let $f \in \mathfrak{X}^*$ be such that $f(p) \neq f(q)$. Let $G = \{g_i\}$ be a denumerable self-determined set of continuous linear functionals containing f. Since K is compact, $g_i(K)$ is a bounded set of scalars for each i, and since we can multiply g_i by a suitable constant, we may suppose that $|g_i(K)| \leqq 1/i$. In this case, the mapping $H : K \to l_2$, defined by $H(k) = [g_i(k)]$, is a continuous mapping of K onto a compact convex subset K_0 of the Hilbert cube which contains at least two points. Consider the mapping $T_0 = HTH^{-1} : K_0 \to K_0$; since G is self-determined, T_0 is single-valued. To see that T_0 is continuous, let $b_0 \in K_0$ and $0 < \varepsilon < 1$. Choose N such that $\sum_{i=N+1}^{\infty} 1/i^2 < \varepsilon$. Then since G is self-determined, there is a $\delta > 0$ and an m such that if $|g_j(p)-g_j(q)| < \delta$, $j = 1, \ldots, m$, then

$$[*] \qquad |g_i(Tp)-g_i(Tq)| < \sqrt{\frac{\varepsilon}{N}}, \qquad i = 1, \ldots, N.$$

Thus if $|b-b_0| < \delta$, and if p and q are any points in K with $b = [g_i(p)]$ and $b_0 = [g_i(q)]$, then [*] holds and

$$\begin{aligned}|T_0(b)-T_0(b_0)|^2 &= |HTH^{-1}(b)-HTH^{-1}(b_0)|^2 \\ &\leqq \sum_{i=1}^{N} |g_i(Tp)-g_i(Tq)|^2 + 2 \sum_{i=N+1}^{\infty} 1/i^2 < 3\varepsilon.\end{aligned}$$

Hence T_0 is a continuous mapping of K_0 into K_0. It follows from Lemma 3 that T_0 has a fixed point k_0. Thus $TH^{-1}(k_0) \subseteqq H^{-1}T_0(k_0) = H^{-1}(k_0)$. Setting $K_1 = H^{-1}(k_0)$, we note that K_1 is a proper closed subset of K, and that $T(K_1) \subseteqq K_1$. The linearity of H implies that K_1 is convex. Thus the lemma is proved. Q.E.D.

5 THEOREM. (*Schauder-Tychonoff*) *A compact convex subset of a locally convex linear topological space has the fixed point property.*

PROOF. By Zorn's lemma there exists a minimal convex subset K_1 of K with the property that $TK_1 \subseteq K_1$. By the previous lemma, this minimal subset contains only one point. Q.E.D.

The importance of Theorem 5 lies in the fact that it applies to non-linear mappings. If we consider only linear mappings, we can prove a stronger result by more elementary means.

6 THEOREM. (*Markov-Kakutani*) *Let K be a compact convex subset of a linear topological space $\mathfrak{X}$. Let $\mathfrak{F}$ be a commuting family of continuous linear mappings which map K into itself. Then there exists a point p in K such that $Tp = p$ for each T in $\mathfrak{F}$.*

PROOF. Let n be a positive integer, and $T \in \mathfrak{F}$. Put $T_n = n^{-1}(I+T+\ldots+T^{n-1})$. Let $\mathscr{K}$ be the family of all sets $T_n(K)$ for $n \geqq 1$ and $T \in \mathfrak{F}$. Then, each set in $\mathscr{K}$ is convex by Lemma 1.5, and compact by I.5.7(b); since K is convex $T_n(K) \subseteq K$. Since T_n and S_m commute, it follows that $T_nS_m(K) \subseteq T_n(K) \cap S_m(K)$, $T, S \in \mathfrak{F}$. Thus any finite subfamily of $\mathscr{K}$ has a non-void intersection. By I.5.6, there is a $p \in \bigcap \mathscr{K}$.

If $T \in \mathfrak{F}$ and $Tp \neq p$, there is a neighborhood U of the origin of $\mathfrak{X}$, such that $Tp-p \notin U$. If n is an arbitrary positive integer, since $p \in T_n(K)$, there exists a $q \in K$ such that $p = n^{-1}(I+T+\ldots+T^{n-1})q$. Hence $Tp-p = n^{-1}(T^n-I)q \notin U$. Since $T^nq \in K$, it follows that $n^{-1}(K-K)$ is not a subset of U for any positive integer n. But $K-K = \phi(K\times K)$, where $\phi(x, y) = x-y$, and thus $K-K$ is compact. But this contradicts II.1.8. Q.E.D.

Examination of the above proof will show that the only property of the mappings T besides continuity that was used was the property

$$T(ax+(1-a)y) = aT(x)+(1-a)Ty.$$

for $x, y \in \mathfrak{X}$ and $0 \leqq a \leqq 1$. Mappings of K which have this property are often called *affine* mappings.

7 DEFINITION. A family $\mathfrak{G}$ of linear transformations on a linear topological space $\mathfrak{X}$ is said to be *equicontinuous on a subset K* of $\mathfrak{X}$ if for every neighborhood V of the origin in $\mathfrak{X}$ there is a neighborhood U of

the origin such that if $k_1, k_2 \in K$ and $k_1-k_2 \in U$ then $\mathfrak{G}(k_1-k_2) \subseteqq V$; that is, $T(k_1-k_2) \in V$ for $T \in \mathfrak{G}$.

8 THEOREM. (*Kakutani*) *Let K be a compact convex subset of a locally convex linear topological space $\mathfrak{X}$, and let $\mathfrak{G}$ be a group of linear mappings which is equicontinuous on K and such that $\mathfrak{G}(K) \subseteqq K$. Then there exists a point $p \in K$ such that $\mathfrak{G}(p) = p$.*

PROOF. By Zorn's lemma K contains a minimal non-void compact convex subset K_1 such that $\mathfrak{G}K_1 \subseteqq K_1$. If K_1 contains just one point, our proof is complete. If this is not the case, the compact (II.1.2) set K_1-K_1 contains some point other than the origin, and, consequently, there exists a neighborhood V of the origin such that $\bar{V}$ does not contain K_1-K_1. There is a convex neighborhood V_1 of the origin such that $\alpha V_1 \subseteqq V$ for $|\alpha| \leqq 1$. By the equicontinuity of $\mathfrak{G}$ on the set K_1, there is a neighborhood U_1 of the origin such that if $k_1, k_2 \in K_1$ and $k_1-k_2 \in U_1$ then $\mathfrak{G}(k_1-k_2) \subseteqq V_1$. Put $U_2 = \text{co}(\mathfrak{G}U_1)$, since $\mathfrak{G}$ is a group $\mathfrak{G}U_2 = U_2$, and from Lemma I.4.16(d) we see that $\mathfrak{G}\bar{U}_2 = \bar{U}_2$. Let $\delta = \inf \{a | a>0, aU_2 \supseteqq K_1-K_1\}$, and $U = \delta U_2$. It follows readily that for each ε with $0 < \varepsilon < 1$ the set K_1-K_1 is not contained in $(1-\varepsilon)\bar{U}$, while $K_1-K_1 \subseteqq (1+\varepsilon)U$. The family of open sets $\{2^{-1}U+k\}$, $k \in K_1$, is a covering of K_1. Let $\{2^{-1}U+k_1, \ldots, 2^{-1}U+k_n\}$ be a finite subcovering and let $p = (k_1+\ldots+k_n)/n$. If k is any point in K_1, then $k_i-k \in 2^{-1}U$ for some i between 1 and n. Since $k_i-k \in (1+\varepsilon)U$ for $1 \leqq i \leqq n$ and $\varepsilon > 0$, we have $p \in n^{-1}(2^{-1}U+(n-1)(1+\varepsilon)U)+k$; substituting $\varepsilon = 1/4(n-1)$, we have $p \in (1-1/4n)U+k$ for each $k \in K_1$. Now let $K_2 = K_1 \cap \bigcap_{k \in K_1} ((1-1/4n)\bar{U}+k) \neq \phi$. Since $(1-1/4n)\bar{U}$ does not contain K_1-K_1, $K_2 \neq K_1$. The closed set K_2 is clearly convex. Further, since $T(a\bar{U}) \subseteqq a\bar{U}$ for $T \in \mathfrak{G}$, we have $T(a\bar{U}+k) \subseteqq a\bar{U}+Tk$ for $T \in \mathfrak{G}$, $k \in K_1$. Since $\mathfrak{G}$ is a group and $TK_1 \subseteqq K_1$, $T \in \mathfrak{G}$, it follows that $TK_1 = K_1$ for $T \in \mathfrak{G}$. This implies that $\mathfrak{G}K_2 \subseteqq K_2$, which contradicts the minimality of K_1. Q.E.D.

11. Exercises

1 Let C be a closed convex separable set in a B-space, and let p be an extremal point of C. Let μ be a positive measure defined for the Borel subsets of C, such that $\mu(C) = 1$. Show that if $p = \int_C x\mu(dx)$, then $\mu(p) = 1$.

2 Let (S, Σ, μ) be a positive measure space. Show that $L_\infty(S, \Sigma, \mu)$ is not weakly complete or reflexive unless it is finite dimensional. Let $\{f_n\}$ be a sequence in $L_\infty(S, \Sigma, \mu)$, and $f \in L_\infty(S, \Sigma, \mu)$. Let M be the set of all λ in $ba(S, \Sigma_1, \mu_1)$ (cf. Theorem IV.8.16 for notation) which take on only the values 0 and 1. Show that $\{f_n\}$ is a weak Cauchy sequence if and only if it is uniformly bounded and $\lim_{n\to\infty} \int_S f_n(s)\lambda(ds)$ exists for each λ in M. Show that $f_n \to f$ weakly if and only if $\{f_n\}$ is uniformly bounded and $\lim_{n\to\infty} \int_S f_n(s)\lambda(ds) = \int_S f(s)\lambda(ds)$ for each λ in M.

3 Show that the closed unit sphere of c_0 contains no extremal points.

4 Let (S, Σ, μ) be a σ-finite positive measure space. Show that $L_1(S, \Sigma, \mu)$ is isometrically isomorphic to the conjugate $\mathfrak{X}^*$ of a B-space $\mathfrak{X}$ if and only if S can be written as a denumerable union $S = \bigcup_{i=1}^\infty S_i$ of measurable subsets S_i such that $\mu(S_i) < \infty$, and such that every measurable subset A of S_i satisfies either $\mu(A) = \mu(S_i)$ or $\mu(A) = 0$.

5 If the closed unit sphere of an infinite dimensional B-space $\mathfrak{X}$ contains only a finite number of extremal points, then $\mathfrak{X}$ is not isometrically isomorphic to the conjugate of any B-space.

6 Let S be a topological space, and let $C(S)$ be the B-space of real bounded continuous functions on S. How many extremal points are there in the closed unit sphere of $C(S)$?

7 Show that every boundary point of the closed unit sphere S of a B-space $\mathfrak{X}$ is an extremal point if and only if $|x+y| \neq |x|+|y|$ unless x and y are linearly dependent. A space with this property is called *strictly convex*. Show that $\mathfrak{X}$ is strictly convex if and only if for each x^* in $\mathfrak{X}^*$, $\sup_{x\in S} \mathscr{R}x^*x$ is attained by at most one x in S.

8 Let $\mathfrak{X} = c$ and $\mathfrak{Y} = c \times c$, where the norm in $\mathfrak{Y}$ is given by $|[y_1, y_2]| = \max(|y_1|, |y_2|)$. Show that $\mathfrak{X}^*$ and $\mathfrak{Y}^*$ are isometrically isomorphic, but that $\mathfrak{X}$ and $\mathfrak{Y}$ are not.

9 (Klee) Let K be the set of sequences $x = [\xi_i] \in l_p$, $1 \leqq p < \infty$, for which $\xi_i \geqq 0$, $i = 1, 2, \ldots$. Then K is a closed convex set, each point of which is a boundary point. But K has a continuous tangent functional at $[\xi_i]$ if and only if $\xi_m = 0$ for some $m \geqq 0$.

10 Calculate the tangent function of the closed unit sphere of Hilbert space.

11 Let (S, Σ, μ) be a positive measure space, and let $p > 1$. Calculate the tangent function of the closed unit sphere of $L_p(S, \Sigma, \mu)$. Show that the closed unit sphere of $L_p(S, \Sigma, \mu)$ has a unique tangent functional at each of its boundary points. Is this true if $p = 1$?

12 Let $\mathfrak{X}$ be a B-space, and let K be a weakly compact convex subset of $\mathfrak{X}$. Show that K has a continuous tangent functional at each point of a dense subset of its boundary.

13 Let $\mathfrak{X}$ be a B-space, and let K^* be a bounded $\mathfrak{X}$-closed convex subset of $\mathfrak{X}^*$. Show that K^* has a continuous tangent functional at each point of a dense subset of its boundary.

14 Definition. A point p of a subset A of a metric space is called *diametral* if $\sup_{x \in A} \varrho(p, x) = \sup_{x, y \in A} \varrho(x, y)$. A subset of a metric space M is said to be *admissible* if it is an intersection of closed spheres $\{x | \varrho(x, y_0) \leqq c_0\}$, $y_0 \in M$, $0 < c_0 \leqq \infty$. A metric space has *normal structure* if each admissible subset which contains more than one point contains a non-diametral point.

15 A compact convex subset of a B-space has normal structure.

16 Let M be a compact metric space, and let A be a closed subset of M. Let $T : M \to M$ be such that $\varrho(Tx, Ty) \geqq \varrho(x, y)$ for $x, y \in M$. Then, if $TA \supseteqq A$, $TA = A$, and if $TA \subseteqq A$, $TA = A$.

17 Show that if a metric space M with normal structure has at least two points, it contains a proper subset A such that $TA \subseteqq A$ for every T mapping M onto M such that $\varrho(Tx, Ty) \leqq \varrho(x, y)$, $x, y \in M$. Show that if in addition M is compact then M contains a point p such that $Tp = p$ for every T mapping M onto M such that $\varrho(Tx, Ty) \leqq \varrho(x, y)$.

18 Let M be a compact metric space with normal structure. Show that there is a point $p \in M$ such that $Tp = p$ for every map $T : M \to M$ such that $\varrho(Tx, Ty) \geqq \varrho(x, y)$, $x, y \in M$.

19 Let M be a complete metric space, and let $T : M \to M$ be such that $\varrho(Tx, Ty) \leqq a\varrho(x, y)$ for $x, y \in M$, where $0 < a < 1$. Show that there is exactly one point $p \in M$ such that $Tp = p$.

20 Let T be a non-linear map of the reflexive B-space $\mathfrak{X}$

into itself which is continuous with the $\mathfrak{X}^*$ topology in $\mathfrak{X}$. Let $\lim_{|x|\to\infty}|Tx|/|x| = 0$. Show that $(I+T)\mathfrak{X} = \mathfrak{X}$.

21 Let S be a compact Hausdorff space, and let μ be a finite regular measure on S. Let Φ be the real or complex numbers, and let K be an element of $C(S\times S\times\Phi)$. Show that the non-linear integral equation

$$f(s) = g(s) - \int_S K(s, t, g(t))\mu(dt)$$

has a solution $g \in C(S)$ for each $f \in C(S)$.

22 DEFINITION. Let G be a topological group. Then a regular measure μ on G is said to be *left-invariant* if $\mu(sE)=\mu(E)$ for all $s \in G$, $E \in \Sigma$. Such a measure is often called *Haar measure.*

23 (Haar) Let G be a compact topological group, and let Σ be the Borel sets of G. Show that there exists a regular measure on Σ which is left-invariant and is not identically zero, and that any two left-invariant measures differ only by a scalar factor. Show that a left-invariant measure also satisfies $\mu(Es) = \mu(E)$ and $\mu(E^{-1})=\mu(E)$.

12. Notes and Remarks

Convex sets and linear topological spaces. The proof of Theorem 1.12 is due essentially to Mazur [1; p. 73], who proved that a convex set with interior point in a real normed linear space can be separated (in the sense of Definition 1.9) from any non-interior point. This generalized a similar result proved for separable spaces by Ascoli [1; p. 206]. Eidelheit [1] extended this result to include the case of two convex sets which have interior points, but have no such points in common. Simpler proofs of Eidelheit's theorem were constructed by Kakutani [1] and Botts [1].

The first statement of Theorem 1.12 in its full generality is due to Dieudonné [1], but his proof was somewhat different from ours. Dieudonné observed that although the hypothesis concerning the existence of an internal point can be weakened to the assumption that M has an internal point relative to the least vector subspace containing M, it cannot be eliminated entirely. Another approach to this result was given by Stone [2] and summarized by Klee [3; p. 455].

Tukey [1; p. 96] proved that separation of an open convex set and an arbitrary convex set is possible, provided they are disjoint (compare Theorem 2.8). He also proved that a convex set K which is compact in the $\mathfrak{X}^*$ topology of the normed linear space $\mathfrak{X}$, can be separated from an arbitrary closed convex set disjoint from K (compare Theorem 2.10). Tukey observed that this result implies that in a reflexive B-space two disjoint closed convex sets, one of which is bounded, can be separated by a hyperplane.

Tukey demonstrated that these results cannot be extended very far by giving examples of:

(i) Two disjoint bounded convex sets, one of which is closed, which cannot be separated.

(ii) Two disjoint closed convex sets of a Hilbert space which cannot be separated.

In this spirit, Dieudonné [2] exhibited:

(iii) Two disjoint closed bounded convex sets in l_1 which cannot be separated.

This result was extended by Klee [4; p. 881] as follows:

(iv) Every non-reflexive separable B-space contains two disjoint closed bounded convex sets which cannot be separated.

In view of these examples, the following theorem of Tukey [1; p. 99] seems remarkable: If A and B are closed convex sets of a normed linear space, and if A is bounded, then the difference set $A-B = \{a-b|a \in A, b \in B\}$ contains every sphere in which it is dense.

Tukey showed that every infinite dimensional normed linear space $\mathfrak{X}$ is the union of two complementary dense convex sets. That the number 2 can be replaced by any cardinal number not greater than the cardinality of $\mathfrak{X}$, was demonstrated by Klee [2]. Klee [3; p. 454] also extended Tukey's result to a more general type of space.

Lemmas 2.4 and 2.5 were proved in the important special case of subsets of a conjugate space $\mathfrak{X}^*$ taken with its $\mathfrak{X}$ topology by Krein and Šmulian [1].

Theorem 2.6 is due to Mazur [2].

Weak topologies. The notion of weak convergence of a sequence in $L_2[0, 1]$ was used by Hilbert, and in $L_p[0, 1]$ by F. Riesz, but the use of the weak neighborhoods in defining a true topology was introduced by von Neumann [2; p. 380], who demonstrated (cf. Exercise 7.38) that it is not sufficient to employ sequential notions alone in using this topology. Bourgin [1; p. 608] amplified this by exhibiting a set which is compact in the $\mathfrak{X}$ topology of the conjugate of a B-space $\mathfrak{X}$, and whose only convergent sequences are constant after some index. Wehausen [1] showed that the weak topology is equivalent to a norm topology if and only if the space is finite dimensional.

It may be mentioned that some authors refer to the $\mathfrak{X}$ topology of $\mathfrak{X}^*$ as the *weak** (or w*) *topology* of $\mathfrak{X}^*$, but we shall not use this term.

Theorem 3.9 is due to Phillips [7; p. 116]. Dieudonné [3; p. 109] proved 3.9 after establishing Lemma 3.10 by induction. The case of Theorem 3.9 in which $\mathfrak{X} = \mathfrak{Y}^*$ and $\Gamma = \mathfrak{Y}$, was proved in the separable case by Banach [1; p. 131] and in full generality by Alaoglu [1; p. 256]. Michael [1] gave an extension of Lemma 3.10 to a case where linear operators are considered.

The equivalence of weak and strong closure for subspaces of a normed linear space (a special case of Theorem 3.13) appears in Banach [1; pp. 58, 134]. Mazur [1; p. 80] showed that a strongly closed convex set in a normed linear space contains all the weak limits of sequences in the set. His proof requires only minor changes to yield Theorem 3.13. Banach and Saks [1] proved that every weakly convergent sequence $\{x_n\}$ in $L_p(0, 1)$ or l_p, $p > 1$, contains a subsequence $\{y_i\}$ which is $(C, 1)$-summable in norm; i.e., $\{n^{-1} \sum_{i=0}^{n-1} y_i\}$ converges strongly. That this stronger version of Corollary 3.14 is not true in $C[0, 1]$ was demonstrated by a counterexample due to Schreier [1]. By studying general convex combinations, Zalcwasser [1] and Gillespie and Hurwitz [1; p. 538] independently proved the validity of Corollary 3.14 in the space $C[0, 1]$. Kakutani [22] gave a proof of the Banach-Saks theorem valid in any uniformly convex space.

The fact that the ordinary (metric) continuity of a linear transformation between B-spaces implies weak continuity was observed by Banach [1; p. 143]. The converse implication in Theorem 3.15 was proved by Dunford [1; p. 317]. A number of other results in this direction have been obtained by Dieudonné [3; pp. 122, 131—137].

In addition to the topological notions of closure which we have studied, various special definitions of closure for subspaces or convex subsets of a linear topological space have been considered. In his treatise, Banach introduced the notions of *regular* and *transfinite closure* for linear manifolds in the conjugate of a general B-space, and showed that these concepts are equivalent (see Banach [1; p. 121]). Banach [1; p. 124] also proved that in the case of a separable space these notions coincide with that of closure in the $\mathfrak{X}$ topology of $\mathfrak{X}^*$. Alaoglu [1; p. 256] and Kakutani [2; p. 170] independently established the equivalence of these types of closure without any separability assumptions. In a similar vein, Krein and Šmulian [1] introduced a definition of a *regularly convex* set in $\mathfrak{X}^*$. It is not difficult to establish that the regularly convex sets in $\mathfrak{X}^*$ are merely the convex sets which are closed in the $\mathfrak{X}$ topology of $\mathfrak{X}^*$.

Weak topologies and reflexivity. A sequential form of the important Theorem 4.2 was demonstrated by Banach [1; p. 123] for separable spaces; Alaoglu [1; p. 255] proved it in the form stated. (This theorem had been announced earlier in full generality, but without proof, by Alaoglu [2], Bourbaki [1] and Kakutani [3; p. 63]). Theorem 4.5 is due to Goldstine [1; p. 128], although he expressed this result in other terms, and gave quite a different proof. Other proofs, more closely related to the one presented here were offered by Kakutani [2; p. 171], Day [2; p. 764], and Dieudonné [3; p. 137].

Theorem 4.7 was proved for separable spaces by Banach [1; p. 189], and a number of generalizations to arbitrary spaces have been made. The form in which we state Theorem 4.7 is the same as that employed by Bourbaki [1; p. 1703], whose argument was not presented in detail. The method of our proof is very similar to that suggested by Kakutani [3; p. 64] and [2; p. 171]. This theorem was also proved by Šmulian [2; p. 471] independently. The stronger result that reflexivity is equivalent to weak sequential compactness of the unit sphere of a B-space is due to Eberlein [1].

Let $\mathfrak{X}$ be a linear space, and let Γ be a total subspace of $\mathfrak{X}^*$. The notion of Γ-compactness is related to various other notions of compactness which have been introduced for convex sets. The theorem below, which makes use of the theory of cardinal and ordinal numbers, is typical of the results which have been proved in this direction. Results of this sort, which relate compactness and convexity to various types of transfinite processes, are studied in Banach's treatise, but much of the recent progress along these lines is due to the work of such Russian mathematicians as Šmulian [1, 5, 8, 10], Gantmacher and Šmulian [1], and Milman [1]. See also Phillips [1].

THEOREM. (*Šmulian*) *Let $\mathfrak{X}$ be a linear space and let K be a convex subset of $\mathfrak{X}$. Let Γ be a total subspace of $\mathfrak{X}^*$. Then statements* (1)–(4) *are equivalent.*

(1) *K is compact in the Γ topology.*

(2) *If x_ξ, $\xi < \theta$, θ a limit ordinal, is a transfinite sequence in K, then there is an $x_0 \in K$ with*

$$\liminf_{\xi\to\theta} \mathscr{R}f(x_\xi) \leqq f(x_0) \leqq \limsup_{\xi\to\theta} \mathscr{R}f(x_\xi),$$

for $f \in \Gamma$.

(3) *For every $f \in \Gamma$, $\sup_{x\in K} \mathscr{R}f(x) < \infty$; moreover, each linear functional Φ on Γ which satisfies*

$$\inf_{x\in K} \mathscr{R}f(x) \leqq \mathscr{R}\Phi(f) \leqq \sup_{x\in K} \mathscr{R}f(x), \qquad f \in \Gamma$$

also satisfies

$$\Phi(f) = f(x_0), \qquad f \in \Gamma$$

for some $x_0 \in K$.

(4) *If K_ξ, $\xi < \theta$, θ a limit ordinal, is a monotone decreasing transfinite sequence of Γ-closed convex sets, all intersecting K, then*

$$K \cap (\cap K_\xi) \neq \Phi.$$

PROOF. That (1) implies (4) is clear. Statement (4) implies statement (2); for if we define

$$K_\xi = \overline{\mathrm{co}}(\bigcup_{\eta<\xi} x_\eta),$$

the closure being taken in the Γ topology, then K_ξ satisfies the requirements of (4); and if $x_0 \in K \cap (\cap K_\xi)$, x_0 is clearly as desired in (2).

Now, (2) implies (3); for, assuming (2), let Φ be as in (3). Suppose (3) is false, and let $\aleph$ be the least cardinal for which there exists a subset Γ' of Γ with cardinality $\aleph$ such that we do not have

$$\Phi(f) = f(x), \qquad f \in \Gamma'$$

for any $x \in K$. Let θ be the least ordinal of cardinality $\aleph$, and well-order Γ' as f_ξ, $\xi < \theta$. If $\aleph$ is infinite, θ is a limit ordinal. In this case we can find a transfinite sequence $\{x_\xi\}$ of elements of K, such that

$$(*) \qquad \Phi(f_\eta) = f_\eta(x_\xi) \quad \text{for} \quad \eta \leqq \xi.$$

Then, by (2), we can find $\bar{x} \in K$ with

$$\liminf_{\xi \to \theta} \mathscr{R}f(x_\xi) \leqq \mathscr{R}f(\bar{x}) \leqq \limsup_{\xi \to \theta} \mathscr{R}f(x_\xi), \qquad f \in \Gamma.$$

Since, by (*),

$$\lim_{\xi \to \theta} \mathscr{R}f(x_\xi) = \mathscr{R}\Phi(f) \quad \text{for} \quad f \in \Gamma',$$

we have $\mathscr{R}\Phi(f) = \mathscr{R}f(\bar{x})$ for $f \in \Gamma'$, from which $\Phi(f) = f(\bar{x})$, $f \in \Gamma'$ follows readily. If $\aleph$ is a finite cardinal n, then $\Gamma' = \{f_1, \ldots, f_n\}$, and we can argue as follows. The set $K' = \{[f_1(x), \ldots, f_n(x)]\}$, $x \in K$, is clearly a convex subset of n-dimensional Euclidean space. The set K' is easily seen to be closed, by (2). If we put $p_i = \Phi(f_i)$ and $p = [p_1, \ldots, p_n]$, we see that $p \notin K'$. Thus, by Theorem 2.10, we can find scalars $\alpha_1, \ldots, \alpha_n$ and real numbers c and ε, $\varepsilon > 0$, such that

$$\mathscr{R}\left(\sum_{i=1}^{n} \alpha_i p_i\right) > c+\varepsilon > c > \mathscr{R}\left(\sum_{i=1}^{n} \alpha_i f_i(x)\right), \qquad x \in K.$$

Thus $\sum_{i=1}^{n} \alpha_i f_i = f \in \Gamma$, and

$$\mathscr{R}\Phi(f) > c+\varepsilon > c > \mathscr{R}f(x), \qquad x \in K,$$

which contradicts the hypothesis in (3).

Finally, (3) implies (1); for, let (3) hold. To each $x \in K$ let there correspond the functional Φ on Γ defined by

$$\Phi(f) = f(x).$$

It follows from (3) that K is mapped onto the set

$$\{\Phi | \Phi \in \Gamma^+, \quad \inf \mathscr{R}f(K) \leqq \mathscr{R}\Phi(f) \leqq \sup \mathscr{R}f(K)\}.$$

This mapping is easily seen to be a homeomorphism if both $\mathfrak{X}$ and Γ^+ are taken with their Γ topology. The compactness of K' now follows readily from Lemma 4.1. Q.E.D.

Weak topologies and compactness. The direct assertions in Theorem 5.1 and Theorem 5.2 are due to Banach [1; p. 185].

The special case of Theorem 5.7 dealing with subspaces of the conjugate space of a B-space was announced by Bourbaki [1], and a proof given by Dieudonné [3; p. 129]. The general case of Theorem 5.7 is due to Krein and Šmulian [1].

The definition of the BX-topology, and the proof of Lemma 5.4, are due to Dieudonné [8]. For an extension of Theorem 5.7 to more general spaces, see Köthe [10].

The fundamental Theorem 6.1 developed gradually. Šmulian [8] proved that (ii) implies (i) and also that the weak sequential closure of a weakly sequentially compact set is itself weakly sequentially compact. (See Krein and Šmulian [1] for the latter.) Eberlein [1] showed that a set is weakly compact if and only if it is weakly closed and weakly sequentially compact. The proof given here is an ingenious modification of the proofs of Šmulian and Eberlein and is due to Brace [1, 2]. Generalizations of these results to more general spaces have been given by Grothendieck [1, 2], Collins [1], Dieudonné and Schwartz [1; p. 89], and Pták [1, 2, 3].

Theorem 6.2 is due to Šmulian [5]. Another proof was given by Klee [4], and an extension to more general spaces by Dieudonné [15].

In the case of a separable B-space, Theorem 6.4 was proved by Krein [1], and the general case by Krein and Šmulian [1; p. 581]; see also Phillips [1].

Extremal points. Theorem 8.4 is due essentially to Krein and Milman [1]. An improved version of Theorem 8.4 was subsequently proved by Milman and Rutman [1]. We have given the proof of Theorem 8.4 due to Kelley [1]. See also Hotta [1], Yosida and Fukamiya [1].

Lemma 8.6 is due to Arens and Kelley [1], who used it to prove Theorem 8.8. Milman [2, 3, 4] gives related results concerning extremal points. Theorem 8.8 itself is due to Banach [1; p. 173] in a special case, and to Stone [1; p. 469] in the general case.

Tangent functionals. Lemmas 9.1 and 9.3 and Theorem 9.5 are due to Ascoli [1; pp. 53–56, 205] in the case of a separable normed space, and to Mazur [1; pp. 75–78] in the case of an arbitrary normed space. Theorem 9.8 is also due to Mazur [1]. Moskovitz and Dines [1; p. 526] proved Theorem 9.7 for Hilbert space, but their proof needs no modification for a general space.

Other proofs that a convex set with an interior point has a tangent plane at every boundary point were given by Moskovitz and Dines [2], and Klee [1, 3; p. 457], to whom 9.10–9.12 are due. Moskovitz and Dines [1; p. 531], and Klee [1; p. 771], give examples to show that it is not possible to drop the condition that an interior point exists and still have a tangent plane at *every* boundary point.

A number of other results concerning convex sets in very general spaces are contained in Klee [3].

Fixed point theorems. A proof of the Brouwer fixed point theorem, using a minimum of homology theory, is given in Graves [2; p. 149]. Alexandroff and Hopf [1; 377] give another proof of this theorem together with a discussion of various other fixed point theorems that can be obtained by homological methods. See also Hurewicz and Wallman [1; p. 40] or Lefschetz [1; p. 318 ff], [2; p. 117].

Before we prove the Brouwer fixed point theorem, we observe that the case of complex scalars is a consequence of the case of real scalars. This follows from the fact that the complex space E^n is isometric with the natural space E^{2n}, and the unit spheres in these spaces correspond in a natural way. Thus we restrict our attention to real Euclidean space. We will need the following lemma.

Lemma. *Let f be an infinitely differentiable function of $n+1$ variables $(x_0, \ldots, x_n)$ with values in E^n. Let D_i denote the determinant whose columns are the n partial derivatives $f_{x_0}, \ldots, f_{x_{i-1}}, f_{x_{i+1}}, \ldots, f_{x_n}$. Then*

$$[*] \qquad \sum_{i=0}^{n} (-1)^i \frac{\partial}{\partial x_i} D_i = 0.$$

Proof. For every pair i, j of unequal integers between 0 and n, let C_{ij} denote the determinant whose first column is $f_{x_i x_j}$ and whose remaining columns are $f_{x_0}, \ldots, f_{x_n}$ arranged in order of increasing indices, and where f_{x_i} and f_{x_j} are omitted from the enumeration. Clearly $C_{ij} = C_{ji}$, and by the laws governing differentiation of determinants and interchange of columns in them we have

$$\frac{\partial}{\partial x_i} D_i = \sum_{j<i} (-1)^j C_{ij} + \sum_{j>i} (-1)^{j-1} C_{ij}.$$

Hence

$$(-1)^i \frac{\partial}{\partial x_i} D_i = \sum_{j=0}^{n} (-1)^{i+j} C_{ij} \sigma(i, j),$$

where $\sigma(i, j) = 1$ if $j < i$, $\sigma(i, j) = 0$ if $i = j$, and $\sigma(i, j) = -1$ if $j > i$. Thus

$$\sum_{i=0}^{n} (-1)^i \frac{\partial}{\partial x_i} D_i = \sum_{i,j=0}^{n} (-1)^{i+j} C_{ij} \sigma(i, j).$$

Interchanging the dummy indices i, j in this latter expression and using the fact that $\sigma(i, j) = -\sigma(j, i)$, we see that

$$\sum_{i,j=0}^{n} (-1)^{i+j} C_{ij}\sigma(i, j) = \sum_{i,j=0}^{n} (-1)^{j+i} C_{ji}\sigma(j, i)$$
$$= (-1) \sum_{i,j=0}^{n} (-1)^{i+j} C_{ij}\sigma(i, j).$$

Thus all the three equal quantities in the formula just written must be zero, and formula [*] is proved. Q.E.D.

THEOREM. (*Brouwer*) *If ϕ is a continuous mapping of the closed unit sphere $S = \{x \in E^n | |x| \leqq 1\}$ of Euclidean n-space into itself, then there is a point y in S such that $\phi(y) = y$.*

PROOF. We have remarked that it suffices to consider real Euclidean space. Further, the Weierstrass approximation theorem for continuous functions of n variables implies that every continuous map ϕ of S into itself is the uniform limit of a sequence $\{\phi_k\}$ of infinitely differentiable mappings of S into itself. Suppose that the theorem were proved for infinitely differentiable maps. Then, for each integer k there is a point $y_k \in S$ such that $\phi_k(y_k) = y_k$. Since S is compact, some subsequence $\{y_{k_i}\}$ converges to a point y in S. Since $\lim_{i\to\infty} \phi_{k_i}(x) = \phi(x)$ uniformly on S, $\phi(y) = \lim_{i\to\infty} \phi_{k_i}(y_{k_i}) = \lim_{i\to\infty} y_{k_i} = y$. This shows that it is sufficient to consider the case that ϕ is infinitely differentiable.

We suppose that ϕ is an infinitely differentiable map of S into itself and that $\phi(x) \neq x$, $x \in S$. Let $a = a(x)$ be the larger root of the quadratic equation $|x+a(x-\phi(x))|^2 = 1$, so that

$$\begin{aligned} 1 &= (x+a(x-\phi(x)), x+a(x-\phi(x))) \\ &= |x|^2+2a(x, x-\phi(x))+a^2|x-\phi(x)|^2. \end{aligned}$$

By the quadratic formula

$$[+] \qquad \begin{aligned} |x-\phi(x)|^2 a = {} & (x, \phi(x)-x) \\ & +\{(x, x-\phi(x))^2+(1-|x|^2)|x-\phi(x)|^2\}^{1/2}. \end{aligned}$$

Since $|x-\phi(x)| \neq 0$ for $x \in S$, the discriminant $(x, x-\phi(x))^2 +(1-|x|^2)|x-\phi(x)|^2$ is positive when $|x| \neq 1$. Also if $|x| = 1$, then

$(x, x-\phi(x)) \neq 0$, for otherwise $(x, \phi(x)) = 1$ and the inner product of two vectors with length at most 1 can be equal to 1 only when they are equal. Thus the discriminant is never zero for x in S. Since the function $t^{1/2}$ is an infinitely differentiable function of t for $t > 0$, and since $|x-\phi(x)| \neq 0$, $x \in S$, it follows from formula [+] that the function $a(x)$ is an infinitely differentiable function of $x \in S$. Moreover, it follows from [+] that $a(x) = 0$ for $|x| = 1$. Now, for each real number t, put $f(t; x) = x+ta(x)(x-\phi(x))$. Then f is an infinitely differentiable function of the $n+1$ variables $t, x_1, \ldots, x_n$ with values in E^n. Since $a(x) = 0$ for $|x| = 1$, we have $f_t(t; x) = 0$ for $|x| = 1$. Also $f(0; x) = x$, and from the definition of a we have $|f(1; x)| = 1$ for all $x \in S$.

Denote the determinant whose columns are the vectors $f_{x_1}(t; x)$, $\ldots, f_{x_n}(t; x)$ by $D_0(t; x)$ and consider the integral

$$I(t) = \int \underset{S}{\cdots} \int D_0(t; x) dx_1 \ldots dx_n.$$

It is clear that $I(0)$ is the volume of S and hence $I(0) \neq 0$. Since $f(1; x)$ satisfies the non-trivial functional dependence $|f(1; x)| = 1$, it follows that the Jacobian determinant $D_0(1; x)$ is identically zero, hence $I(1) = 0$. The desired contradiction will be obtained if we can show that $I(t)$ is a constant; i.e., that $I'(t) = 0$. To prove this, differentiate under the integral sign and employ [*] to conclude that $I'(t)$ is a sum of integrals of the form

$$[++] \qquad \pm \int \underset{S}{\cdots} \int \frac{\partial}{\partial x_i} D_i(t; x) dx_1 \ldots dx_n,$$

where $D_i(t; x)$ is the determinant whose columns are the vectors

$$f_t(t; x), f_{x_1}(t; x), \ldots, f_{x_{i-1}}(t; x), f_{x_{i+1}}(t; x), \ldots, f_{x_n}(t; x).$$

Let S_i denote the unit sphere in the space of variables $x_1, \ldots, x_{i-1}$, $x_{i+1}, \ldots, x_n$. Let x_i^+ denote the positive square root $\{1-(x_1^2+\ldots+x_{i-1}^2+x_{i+1}^2+\ldots+x_n^2)\}^{1/2}$, and x_i^- denote the corresponding negative square root; let p_i^+ denote the point whose j-th coordinate is x_j if $j \neq i$ and is x_i^+ if $j = i$, and let p_i^- denote the point whose j-th coordinate is x_j if $j \neq i$ and is x_i^- if $j = i$. Then the integral in [++] is

$$\pm \int_{S_i} \cdots \int D_i(t; p_i^+) dx_1 \ldots dx_{i-1} dx_{i+1} \ldots dx_n$$

$$\pm \int_{S_i} \cdots \int D_i(t; p_i^-) dx_1 \ldots dx_{i-1} dx_{i+1} \ldots dx_n.$$

But $|p_i^+| = |p_i^-| = 1$ and since $f_i(t; x) = 0$ for $|x| = 1$, it follows from the definition of D_i that these integrals are zero. This completes the proof. Q.E.D.

Birkhoff and Kellogg [1] were the first to make an extension to infinite dimensional vector spaces, by proving that compact convex sets in $C^n[0, 1]$ and $L_2[0, 1]$ have the fixed point property, and by applying these results to differential and integral equations. Schauder extended this theorem, first to compact convex sets in a B-space with a basis [1], and then to arbitrary B-spaces [2]. It remained for Tychonoff [1] to make the generalization to locally convex topological linear spaces, in which form the theorem is applicable to the weak topologies as well as the strong topology in B-spaces.

Of other generalizations of fixed point theorems, we mention the following due to Rothe [1]: A continuous mapping of the solid unit sphere S of a B-space $\mathfrak{X}$ into a conditionally compact subset of $\mathfrak{X}$, which sends the boundary $\{x||x| = 1\}$ into S, has at least one fixed point.

Most of the papers mentioned above have applications to differential equations. Discussions of fixed point and other abstract approaches to existence theorems for differential equations can be found in the expository articles of L. M. Graves [1] and Leray [1]. The work of Miranda [1] and Niemytsky [1] will also be useful; the former is particularly recommended because of its extensive bibliography. It might be mentioned that although fixed points and other topological methods generally yield only existence theorems, Aronszajn [2] and Rothe [2], among others, have indicated how uniqueness theorems can also be obtained.

Applying Corollary 4.7, it follows that every weakly continuous (i.e., continuous in the weak topology) mapping of the unit sphere of a reflexive B-space into itself always has a fixed point. Kakutani [5] gave an example to show that the corresponding statement for strongly continuous maps is not true even for a homeomorphism of the sphere of Hilbert space onto itself. Dugundji [1] showed that the unit sphere of a B-space with the strong topology has the fixed point property if and only if the space is finite dimensional.

Studies have been made of the "nature" and the "stability" of fixed points of functions in a suitable neighborhood of a given function whose fixed points are known. For questions of this nature see Fort [1], Kinoshita [1], and Wehausen [2].

Markov [1] proved Theorem 10.6 by using the Tychonoff theorem. The proof given here is essentially that of Kakutani [4], who made several applications of this result, and to whom Theorem 10.8 is due. (See also Peck [1].)

The material in 11.14—11.18 is taken from Brodskiĭ and Milman [1].

Finite dimensional spaces. This chapter has been confined to the treatment of convex sets in infinite dimensional spaces. There is a very extensive theory of the special properties of convex sets in finite dimensional spaces. For this, the reader is referred to the work of Minkowski [1] and the treatise of Bonnesen and Fenchel [1]. The latter contains a great wealth of results and applications, and a large bibliography.

Locally convex spaces. We have discussed only a small portion of the very extensive theory of locally convex linear topological spaces. The theory of such spaces has been extended considerably further than we have indicated. Moreover, many results have been obtained for topological linear spaces which are not assumed to have the property of local convexity. The interested reader should consult the treatises of Bourbaki [2] and Nakano [1], and the expository article of Hyers [3].

Uniform convexity and differentiability of norms. The tangent plane theory of Section 9 has been considerably extended in various directions. One of the most interesting lines of investigation is concerned with the derivability of the support function $\mathfrak{k}(x)$ of a convex body in a stronger sense than that of 9.2.

DEFINITION. Let $\mathfrak{k}$ be the support function of a convex body containing the origin as an interior point, and let τ be its tangent function. If

$$\lim_{|y|\to 0} \frac{1}{|y|} \{\mathfrak{f}(x+y) - \mathfrak{f}(x) - \tau(x, y)\} = 0,$$

we say that $\mathfrak{f}$ is *strongly differentiable* at the point x.

Banach [1; p. 168] showed that the norm in $C[0, 1]$ is strongly differentiable at $x_0 \in C[0, 1]$ if and only if the function x_0 achieves its maximum at exactly one point. Mazur [1; p. 78–79] proved that the same condition holds in $B(S)$, that the norm in L_p, $p > 1$ is strongly differentiable at every point but the origin, and gave conditions for strong differentiability of the norm in the space L_1. He also showed that in the F-space of measurable functions on $[0, 1]$, the norm is not differentiable at any point.

Mazur [3] proved that in a reflexive space in which the norm is strongly differentiable at every non-zero point, a bounded closed convex set is the intersection of all the closed spheres which contain it.

Šmulian [4, 6, 7, 9] proved a number of results concerning differentiability and other properties. In the first two papers, necessary and sufficient conditions for a weaker type of differentiability of the norms in $\mathfrak{X}$ and $\mathfrak{X}^*$ are given, together with results relating differentiability, tangent planes to the unit sphere, reflexivity and various other geometrical properties.

Šmulian [7] obtains two interesting necessary and sufficient conditions for strong differentiability of the norm.

THEOREM. *In order that the norm is strongly differentiable at a point x in a B-space $\mathfrak{X}$, it is necessary and sufficient that every sequence of elements $x_n^* \in \mathfrak{X}^*$ satisfying $|x_n^*| \leqq 1$ and $x_n^*(x) \to |x|$, is convergent.*

THEOREM. *In order that the norm is strongly differentiable at a point x^* in the conjugate $\mathfrak{X}^*$ of a B-space $\mathfrak{X}$, it is necessary and sufficient that every sequence of elements $x_n \in \mathfrak{X}$ satisfying $|x_n| \leqq 1$ and $x^*(x_n) \to |x^*|$ is convergent.*

In [7], Šmulian gives results relating strong differentiability of the norm with reflexivity, weak completeness, and uniform convexity (see below). Šmulian [9] pursues this study further, deriving conditions for strong differentiability in l_∞, c, c_0, $C(S)$ and L_1, and proving a few results concerning differentiability of norms in Banach algebras.

James [2] applies differentiability concepts to derive a number of results concerning a type of "orthogonality" in normed linear spaces.

The two results of Šmulian, cited above, add interest to the following definition, due to Clarkson [1].

DEFINITION. A B-space $\mathfrak{X}$ is said to be *uniformly convex* if whenever $x_n \in \mathfrak{X}$, $y_n \in \mathfrak{X}$, $|x_n| \leqq 1$, $|y_n| \leqq 1$, and $|x_n+y_n| \to 2$, then $|x_n-y_n| \to 0$.

Clarkson [1] showed that the spaces L_p, $p > 1$, are uniformly convex (see also Boas [1]). It has been shown by Milman [1] and Pettis [2] (see also Kakutani [2]) that every uniformly convex B-space is reflexive, but that there are reflexive spaces which are not equivalent to any uniformly convex space (Day [3]).

Day [5] showed that uniform convexity in the neighborhood of a point implies uniform convexity, and (Day [4, 6]) demonstrated that under certain conditions combinations also yield uniformly convex spaces. A dual notion of "uniform flattening" is related to that of uniform convexity by Day [6].

For other results on uniform convexity see the papers of Šmulian already cited, and also Fortet [1, 2, 3], James [2], Kračkovskiĭ and Vinogradov [1], and Ruston [1].

Bibliography

Convex sets: Ascoli [1], Botts [1], Brodskiĭ and Milman [1], Dieudonné [1, 2], Eberlein [2], Eidelheit [1], Halperin [2], Klee [1, 2, 3, 4], Krein [1], Krein and Šmulian [1], Mazur [1, 2, 3], Moskovitz and Dines [1, 2], Schoenberg [2], Šmulian [8, 10], Stone [2], Tagamlitzki [2], Tukey [1], Yosida and Fukamiya [1].

Weak topologies and reflexivity: Alaoglu [1], Arens [1], Bourbaki [1], Bourgin [1], Dieudonné [3], Dieudonné and Schwartz [1], Day [2, 3], Eberlein [1], Gantmacher and Šmulian [1, 2], Goldstine [1], James [4], Kakutani [2, 3], Klee [7], Krein and Šmulian [1], Mackey [1, 2], Milman [1], von Neumann [2], Pettis [1, 2], Phillips [1], Ruston [4], Šmulian [1—10, 12, 13], Taylor [2, 3].

Extremal points: Arens and Kelley [1], Hotta [1], Jerison [2], Kelley [1], Krein and Milman [1], Milman [2, 3, 4, 7], Milman and Rutman [1], Tomita [1], Yosida and Fukamiya [1].

Fixed point theorems: Birkhoff and Kellogg [1], Brouwer [1], Fort [1], Hukuhara [1], Inaba [1], Kakutani [4, 5] Kinoshita [1], Krein and Šmulian [1], Leray [1], Markov [1], Miranda [1], Niemytsky [1], O'Neill [1], Peck [1], Rothe [1, 2], Schauder [1, 2], Tychonoff [1], Wehausen [2], Yood [3].

Topological linear spaces: Arens [1], Bourgin [2], Dieudonné [3], Dieudonné and Schwartz [1], Donoghue and Smith [1], Grothendieck [1, 2], Hyers [1, 2], Klee [3, 4], Kolomogoroff [1], Mackey [1, 2], von Neumann [1], Tychonoff [1], Wehausen [1].

Uniform convexity: Boas [1], Clarkson [1], Day [3, 4, 5, 6], Fortet [1, 2, 3], James [2], Kakutani [22], Kračkovskiĭ and Vinogradov [1], Lovaglia [1], Milman [1], Pettis [2], Ruston [1], Šmulian [4, 6, 7, 9].

Differentiability of the norm: Banach [1], James [2], Mazur [1], Šmulian [4, 6, 7, 9].

CHAPTER VI

Operators and Their Adjoints

This chapter continues the study of linear maps between B-spaces begun in Chapter II. Various topologies are introduced in the space $B(\mathfrak{X}, \mathfrak{Y})$ of bounded linear maps between the B-spaces $\mathfrak{X}$ and $\mathfrak{Y}$. The notions of adjoint operators, projection operators, weakly compact operators and compact operators are introduced, and the basic properties of these operators are studied. Representations for various general classes of operators in the spaces of continuous and integrable functions are given. Other similar representation theorems are contained in the exercises. In addition, the important convexity theorem of M. Riesz is discussed. *Throughout the chapter, the symbols* $\mathfrak{X}$, $\mathfrak{Y}$, $\mathfrak{Z}$ *will denote B-spaces*, unless the contrary is explicitly stated.

We recall that if T is in $B(\mathfrak{X})$, we shall frequently say that T is an operator in $\mathfrak{X}$.

1. The Space $B(\mathfrak{X}, \mathfrak{Y})$

A B-space $\mathfrak{X}$ has at least two important topologies: the strong, or metric, topology and the weak, or $\mathfrak{X}^*$, topology. If $\mathfrak{X}$ is the conjugate space of $\mathfrak{Y}$, there is also the $\mathfrak{Y}$ topology in $\mathfrak{X}$.

The linear space $B(\mathfrak{X}, \mathfrak{Y})$ of continuous linear mappings $T : \mathfrak{X} \to \mathfrak{Y}$ has a correspondingly more complicated set of topologies. The three most commonly used topologies in $B(\mathfrak{X}, \mathfrak{Y})$ are the *uniform*, *strong*, and *weak operator topologies*, as given in the following definitions.

1 Definition. The *uniform operator topology in* $B(\mathfrak{X}, \mathfrak{Y})$ is the metric topology of $B(\mathfrak{X}, \mathfrak{Y})$ induced by its norm,

$$|T| = \sup_{|x| \leq 1} |Tx|.$$

2 Definition. The *strong operator topology in* $B(\mathfrak{X}, \mathfrak{Y})$ is the topology defined by the basic set of neighborhoods

$$N(T; A, \varepsilon) = \{R | R \in B(\mathfrak{X}, \mathfrak{Y}), |(T-R)x| < \varepsilon, x \in A\}$$

where A is an arbitrary finite subset of $\mathfrak{X}$, and $\varepsilon > 0$ is arbitrary. Thus, in the strong topology, a generalized sequence $\{T_\alpha\}$ converges to T if and only if $\{T_\alpha x\}$ converges to Tx for every x in $\mathfrak{X}$.

3 DEFINITION. The *weak operator topology* in $B(\mathfrak{X}, \mathfrak{Y})$ is the topology defined by the basic set of neighborhoods

$$N(T; A, B, \varepsilon) = \{R | R \in B(\mathfrak{X}, \mathfrak{Y}), |y^*(T-R)x| < \varepsilon, \quad y^* \in B, x \in A\},$$

where A and B are arbitrary finite sets of elements in $\mathfrak{X}$ and $\mathfrak{Y}^*$, respectively, and $\varepsilon > 0$ is arbitrary. Thus, in the weak topology, a generalized sequence $\{T_\alpha\}$ converges to T if and only if $\{y^*T_\alpha x\}$ converges to y^*Tx for every x in $\mathfrak{X}$ and y^* in $\mathfrak{Y}^*$.

These topologies are by no means the only interesting topologies of $B(\mathfrak{X}, \mathfrak{Y})$. Other topologies in $B(\mathfrak{X}, \mathfrak{Y})$ can be introduced in such a way that the convergence of a generalized sequence $\{T_\alpha\}$ to a limit T means any one of the following:

(i) for each x in $\mathfrak{X}$, the generalized sequence $\{T_\alpha x\}$ converges to Tx uniformly for x in any compact subset of $\mathfrak{X}$;

(ii) for each y^* in $\mathfrak{Y}^*$, the generalized sequence $\{y^*T_\alpha x\}$ converges to y^*Tx uniformly for x in any compact (or bounded) subset of $\mathfrak{X}$;

(iii) the generalized sequence $\{T_\alpha\}$ converges to T as defined in (i) or (ii), with "compact" replaced by "weakly compact";

(iv) if $\mathfrak{Y}$ is the conjugate space of $\mathfrak{Z}$, then in the $\mathfrak{Z}$ topology of $\mathfrak{Y}$, $\lim_\alpha T_\alpha x = Tx$ for any $x \in \mathfrak{X}$, or the same limit holding uniformly in compact, weakly sequentially compact, or bounded subsets of $\mathfrak{X}$.

As a final example, we mention the topology whose basic neighborhoods are

$$N(T; x_1, x_2, \ldots, \varepsilon) = \{R | R \in B(\mathfrak{X}, \mathfrak{Y}), \sum_{i=1}^{\infty} |(R-T)x_i| < \varepsilon\},$$

where $\varepsilon > 0$ is arbitrary, and $x_1, x_2, \ldots$ is an arbitrary sequence of elements in $\mathfrak{X}$ with $\sum_{i=1}^{\infty} |x_i| < \infty$.

With each of these topologies, $B(\mathfrak{X}, \mathfrak{Y})$ is a locally convex linear topological space. In this text, interest centers in the uniform, the strong, and the weak operator topologies in $B(\mathfrak{X}, \mathfrak{Y})$. It is evident that the uniform operator topology is stronger than the strong operator topology, and that the strong operator topology is stronger than the weak operator topology.

With its uniform operator topology, $B(\mathfrak{X}, \mathfrak{Y})$ becomes a B-space; as such, it has a weak topology which should not be confused with the weak operator topology in $B(\mathfrak{X}, \mathfrak{Y})$.

4 THEOREM. *A linear functional on* $B(\mathfrak{X}, \mathfrak{Y})$ *is continuous with respect to the weak operator topology if and only if it is continuous with respect to the strong operator topology.*

PROOF. Since the strong operator topology is stronger than the weak operator topology, a functional continuous in the weak topology is continuous in the strong topology. Conversely, let F be a functional on $\mathfrak{B} = B(\mathfrak{X}, \mathfrak{Y})$ which is continuous in the strong topology. Then there exists a finite subset $\{x_1, \ldots, x_m\}$ of $\mathfrak{X}$ and an $\varepsilon > 0$ such that $|Tx_i| < \varepsilon$, $T \in \mathfrak{B}$, $i = 1, \ldots, n$ implies $|F(T)| < 1$.

Consider the B-space $\mathfrak{Y}_n = \mathfrak{Y} \oplus \ldots \oplus \mathfrak{Y}$ of all n-tuples $\mathfrak{y} = [y_1, \ldots, y_n]$ with $y_i \in \mathfrak{Y}$, $i = 1, \ldots, n$; the norm in $\mathfrak{Y}_n$ is $|[y_1, \ldots, y_n]| = \max_{1 \leq i \leq n} |y_i|$. Define $H : \mathfrak{B} \to \mathfrak{Y}_n$ by $HT = [Tx_1, \ldots, Tx_n]$, and put $f(\mathfrak{y}) = F(T)$ if $\mathfrak{y} \in H(\mathfrak{B})$ and $\mathfrak{y} = HT$. Since $|H(T)| < \delta\varepsilon$ implies $|F(T)| < \delta$, f is well-defined and continuous on $H(\mathfrak{B})$. By II.3.11, f has a continuous linear extension f_1 defined on all of $\mathfrak{Y}_n$. It is easily seen that any such functional must have the form

$$f_1[y_1, \ldots, y_n] = \sum_{i=1}^{n} y_i^*(y_i),$$

where $y_i^* \in \mathfrak{Y}^*$, $i = 1, \ldots, n$. Consequently, $F(T) = f_1(HT)$ has the form $F(T) = \sum_{i=1}^{n} y_i^* Tx_i$, and it is then obvious that F is continuous in the weak operator topology. Q.E.D.

5 COROLLARY. *A convex set in* $B(\mathfrak{X}, \mathfrak{Y})$ *has the same closure in the weak operator topology as it does in the strong operator topology.*

PROOF. This follows immediately from Theorem 4 and Corollary V.2.14. Q.E.D.

2. Adjoints

1 Definition. The *adjoint* T^* of a linear operator T in $B(\mathfrak{X}, \mathfrak{Y})$ is the mapping from $\mathfrak{Y}^*$ to $\mathfrak{X}^*$ defined by $T^*y^* = y^*T$.

2 Lemma. *The mapping $T \to T^*$ is an isometric isomorphism of $B(\mathfrak{X}, \mathfrak{Y})$ into $B(\mathfrak{Y}^*, \mathfrak{X}^*)$.*

Proof. The linear functional y^*T is continuous (I.4.17), and hence $T^*y^* \in \mathfrak{X}^*$. The map $T \to T^*$ is clearly linear. From II.3.15, $|Tx| = \sup_{|y^*|\leq 1} |y^*Tx|$, and thus

$$\begin{aligned} |T^*| &= \sup_{|y^*|\leq 1} |T^*y^*| = \sup_{|y^*|\leq 1} \sup_{|x|\leq 1} |y^*Tx| \\ &= \sup_{|x|\leq 1} \sup_{|y^*|\leq 1} |y^*Tx| = \sup_{|x|\leq 1} |Tx| = |T|, \end{aligned}$$

which shows that the map $T \to T^*$ is an isometric isomorphism. Q.E.D.

3 Lemma. *The adjoint T^* of an operator T in $B(\mathfrak{X}, \mathfrak{Y})$ is a continuous mapping of $\mathfrak{Y}^*$ into $\mathfrak{X}^*$ if these spaces have the $\mathfrak{Y}$ and $\mathfrak{X}$ topologies, respectively.*

The proof is trivial, and is left to the reader.

4 Lemma. *If T is in $B(\mathfrak{X}, \mathfrak{Y})$, and U is in $B(\mathfrak{Y}, \mathfrak{Z})$, then $(UT)^* = T^*U^*$. The adjoint of the identity in $B(\mathfrak{X})$ is the identity in $B(\mathfrak{X}^*)$.*

Proof. For $z^* \in \mathfrak{Z}^*$, $x^* \in \mathfrak{X}^*$

$$(UT)^*z^* = z^*UT = (U^*z^*)T = T^*(U^*z^*) = (T^*U^*)z^*,$$

$$I^*x^* = x^*I = x^*.$$

Q.E.D.

Thus the mapping $T \to T^*$ of the ring $B(\mathfrak{X})$ into the ring $B(\mathfrak{X}^*)$ is an "anti-isomorphism."

5 Definition. Let $\hat{\mathfrak{X}}$, $\hat{\mathfrak{Y}}$ be the images under the natural embedding of $\mathfrak{X}$, $\mathfrak{Y}$ into $\mathfrak{X}^{**}$, $\mathfrak{Y}^{**}$, respectively. For $T \in B(\mathfrak{X}, \mathfrak{Y})$, define $\hat{T} \in B(\hat{\mathfrak{X}}, \hat{\mathfrak{Y}})$ by the equation $\hat{T}\hat{x} = \hat{y}$, where $y = Tx$. A function U whose domain is a set in $\mathfrak{X}^{**}$ containing $\hat{\mathfrak{X}}$ is said to be an *extension of* T, if $U\hat{x} = \hat{T}\hat{x}$ for $\hat{x}$ in $\hat{\mathfrak{X}}$. Thus the notion of an extension of T is applied as though $\mathfrak{X}$ and $\hat{\mathfrak{X}}$ were identified. Similarly, if $\hat{\mathfrak{X}} = \mathfrak{X}^{**}$, the equation $U = T$ is understood to mean that $U\hat{x} = \hat{T}\hat{x}$ for $\hat{x}$ in $\hat{\mathfrak{X}}$.

6 LEMMA. *If T is in $B(\mathfrak{X}, \mathfrak{Y})$, the second adjoint $T^{**} : \mathfrak{X}^{**} \to \mathfrak{Y}^{**}$ is an extension of T. If $\mathfrak{X}$ is reflexive then $T^{**} = T$.*

PROOF. Let $x \in \mathfrak{X}$, $y^* \in \mathfrak{Y}^*$. Then,

$$(T^{**}\hat{x})y^* = \hat{x}T^*y^* = (T^*y^*)x = y^*Tx = (\hat{T}\hat{x})y^*.$$

Q.E.D.

7 LEMMA. *A linear operator T in $B(\mathfrak{X}, \mathfrak{Y})$ has a bounded inverse T^{-1} defined on all of $\mathfrak{Y}$ if and only if its adjoint T^* has a bounded inverse $(T^*)^{-1}$ defined on all of $\mathfrak{X}^*$. When these inverses exist, $(T^{-1})^* = (T^*)^{-1}$.*

PROOF. If T^{-1} exists and is in $B(\mathfrak{Y}, \mathfrak{X})$, then, by Lemma 4, $(TT^{-1})^* = (T^{-1})^*T^*$ is the identity in $\mathfrak{Y}^*$ and $(T^{-1}T)^* = T^*(T^{-1})^*$ is the identity on $\mathfrak{X}^*$. Thus $(T^*)^{-1}$ exists, is in $B(\mathfrak{X}^*, \mathfrak{Y}^*)$, and equals $(T^{-1})^*$. Conversely, if $(T^*)^{-1}$ exists and is in $B(\mathfrak{X}^*, \mathfrak{Y}^*)$, then, by what has already been proved, $(T^{**})^{-1}$ exists and is in $B(\mathfrak{Y}^{**}, \mathfrak{X}^{**})$. Thus T^{**} is a homeomorphism; by Lemma 6, it is an extension of T. Hence T is one-to-one, and $T\mathfrak{X}$ is closed. It only remains to be shown that $T\mathfrak{X} = \mathfrak{Y}$. If $y \in \mathfrak{Y}$ and $y \notin T\mathfrak{X}$, there is (II.3.13) a y^* in $\mathfrak{Y}^*$ with $y^* \neq 0$, $y^*T = T^*y^* = 0$. This contradicts the assumption that T^* is one-to-one, and proves the lemma. Q.E.D.

The final argument in the above proof also establishes the following lemma.

8 LEMMA. *If T is in $B(\mathfrak{X}, \mathfrak{Y})$, then the closure in $\mathfrak{Y}$ of $T\mathfrak{X}$ is the set of vectors y such that $y^*y = 0$ for each y^* which satisfies the equation $T^*y^* = 0$.*

For operators in a Hilbert space, a slightly different notion of the adjoint is customary. Let $\mathfrak{H}$ be a Hilbert space, and let $T \in B(\mathfrak{H})$. Then the adjoint operator T_1 of T is in $B(\mathfrak{H}^*)$. However, since $\mathfrak{H}$ and $\mathfrak{H}^*$ are intimately related, as is shown by Theorem IV.4.5, it is customary to refer to the operator $T_2 = \sigma T_1 \sigma^{-1}$, where $\sigma : \mathfrak{H}^* \to \mathfrak{H}$ is the map given by Theorem IV.4.5, as the adjoint of T, if $T \in B(\mathfrak{H})$. This has the advantage that T_2 is in $B(\mathfrak{H})$ rather than in $B(\mathfrak{H}^*)$. We call T_2 the *Hilbert space adjoint* of T. Let us give a formal definition.

9 DEFINITION. Let $\mathfrak{H}$ be a Hilbert space, and let $T \in B(\mathfrak{H})$. There is a uniquely determined operator $T^* \in B(\mathfrak{H})$, called the *Hilbert space adjoint* of T, which satisfies the identity

$$(Tx, y) = (x, T^*y), \qquad x, y \in \mathfrak{H}.$$

Unless the contrary is explicitly stated, *the term adjoint, and the symbol T^*, when applied to an operator in Hilbert space, will refer to the Hilbert space adjoint.* An operator T in Hilbert space is called *self-adjoint* if $T^* = T$.

The preceding lemmas concerning adjoints take the following form in Hilbert space.

➔ 10 Lemma. *If T and U are bounded linear operators in Hilbert space, then*

(a) $(T+U)^* = T^*+U^*$;

(b) $(TU)^* = U^*T^*$;

(c) $(\alpha T)^* = \bar{\alpha}T^*$;

(d) $I^* = I$;

(e) $T^{**} = T$;

(f) $|T^*| = |T|$;

(g) *if either T^{-1} or $(T^*)^{-1}$ exists and is in $B(\mathfrak{H})$, then the other is also, and* $(T^{-1})^* = (T^*)^{-1}$.

These statements follow from Lemmas 2, 4, 6, and 7. Q.E.D.

3. Projections

A projection in an arbitrary linear space $\mathfrak{X}$ has been defined in Section I.11 as a linear operator E for which $E^2 = E$. If $\mathfrak{X}$ is a linear topological space, we shall require, from this point on, that E be continuous.

1 Definition. A *projection* in $\mathfrak{X}$ is an operator $E \in B(\mathfrak{X})$ with $E^2 = E$.

➔ If E is a projection in $\mathfrak{X}$, then every element x can be written uniquely as a sum $x = x_1+x_2$ where $Ex_1 = x_1$, $Ex_2 = 0$. Indeed, $x = Ex+(I-E)x$ is such a decomposition. Conversely, given two closed linear subspaces $\mathfrak{X}_1$ and $\mathfrak{X}_2$ of $\mathfrak{X}$ such that every x can be written uniquely as $x = x_1+x_2$, $x_1 \in \mathfrak{X}_1$ and $x_2 \in \mathfrak{X}_2$, then the function E defined by $E(x) = x_1$ is a closed, and hence, by Theorem II.2.4, a bounded linear transformation of $\mathfrak{X}$ into $\mathfrak{X}$ with $E^2 = E$. Thus there is a one-to-one correspondence between projections in $\mathfrak{X}$ and direct sum decompositions of $\mathfrak{X}$ into two closed subspaces $\mathfrak{X}_1$ and $\mathfrak{X}_2$. The subspaces corresponding to E are $E\mathfrak{X}$ and $(I-E)\mathfrak{X}$, characterized respectively by $Ex = x$ and $Ex = 0$.

Some useful elementary properties of projections are summarized in the following lemmas. Further properties will be found in the exercises.

2 Lemma. *Let E_1 and E_2 be commuting projections in a B-space $\mathfrak{X}$. If $\mathfrak{M}_i = E_i\mathfrak{X}$, $\mathfrak{N}_i = (I-E_i)\mathfrak{X}$, $i = 1, 2$, then*

(a) *$E_1E_2 = E_2$ if and only if $\mathfrak{M}_2 \subseteqq \mathfrak{M}_1$, or equivalently, if and only if $\mathfrak{N}_1 \subseteqq \mathfrak{N}_2$;*

(b) *if $E = E_1+E_2-E_1E_2$, then E is a projection with $E\mathfrak{X} = \text{sp}\{\mathfrak{M}_1 \cup \mathfrak{M}_2\}$ and $(I-E)\mathfrak{X} = \mathfrak{N}_1 \cap \mathfrak{N}_2$;*

(c) *if $E = E_1E_2$, then E is a projection with $E\mathfrak{X} = \mathfrak{M}_1 \cap \mathfrak{M}_2$ and $(I-E)\mathfrak{X} = \text{sp}\{\mathfrak{N}_1 \cup \mathfrak{N}_2\}$;*

(d) *if $E = E_1-E_2$, then E is a projection if and only if $E_1E_2 = E_2$. In this case $E\mathfrak{X} = \mathfrak{M}_1 \cap \mathfrak{N}_2$, $(I-E)\mathfrak{X} = \mathfrak{N}_1 \oplus \mathfrak{M}_2$.*

The proofs are immediate consequences of the definitions and are left to the reader. (cf. Exercise I.14.6.)

3 Lemma. *If E is a projection in $\mathfrak{X}$, E^* is a projection in $\mathfrak{X}^*$ and*

$$E^*\mathfrak{X}^* = \{x^* | x^*x = 0, \ x \in (I-E)\mathfrak{X}\}$$
$$(I-E^*)\mathfrak{X}^* = \{x^* | x^*x = 0, \ x \in E\mathfrak{X}\}.$$

Proof. Clearly $E^{*2} = E^*$ so E^* is a projection. If $x \in \mathfrak{X}$ and $E^*y^* = y^*$, then $y^*x = E^*y^*x = y^*Ex$ and $y^*(x-Ex) = 0$. Conversely, let $y^*(x-Ex) = 0$ for every x in $\mathfrak{X}$. Then $y^*x = y^*Ex$, $y^* = y^*E$, and $y^* = E^*y^*$. The second statement follows from the first by substituting $I-E^*$ for E^*. Q.E.D.

There is an interesting way in which projections can be ordered.

4 Definition. Two projections are said to be ordered in their *natural order* $E_1 \leqq E_2$ if $E_1E_2 = E_2E_1 = E_1$.

This amounts to requiring $E_1\mathfrak{X} \subseteqq E_2\mathfrak{X}$, $(I-E_1)\mathfrak{X} \supseteqq (I-E_2)\mathfrak{X}$.

The reader will have no difficulty in proving the next lemma.

5 Lemma. *The natural ordering $\leqq$ of projections satisfies*

(i) *$E_1 \leqq E_1$;*

(ii) *if $E_1 \leqq E_2$ and $E_2 \leqq E_3$ then $E_1 \leqq E_3$;*

(iii) *if $E_1 \leqq E_2$ and $E_2 \leqq E_1$ then $E_1 = E_2$.*

Thus, the family of all projections in $B(\mathfrak{X})$ form a partially ordered set. Any two commuting projections, E_1 and E_2, will have the least upper bound

$$E_1 \vee E_2 = E_1 + E_2 - E_1 E_2,$$

and the greatest lower bound

$$E_1 \wedge E_2 = E_1 E_2.$$

These two operations will be of fundamental importance in Volume II where we will introduce and discuss the concept of a Boolean algebra of projections in a B-space.

We conclude this section with a few remarks on projections in Hilbert space. Let E be a projection in Hilbert space $\mathfrak{H}$, and let E^* be its Hilbert space adjoint. Then the identity

$$(Ex, (I-E)y) = (Ex, y) - (E^*Ex, y)$$

shows that the complementary manifolds $E\mathfrak{H}$ and $(I-E)\mathfrak{H}$ are orthogonal if and only if $E = E^*E$, or, equivalently, $E = E^*$. Thus a self-adjoint projection is sometimes called an *orthogonal* or *perpendicular projection*. Since closed complementary manifolds $\mathfrak{X}$ and $\mathfrak{Y}$ determine uniquely a projection E with $E\mathfrak{H} = \mathfrak{X}$, $(I-E)\mathfrak{H} = \mathfrak{Y}$, the above identity shows that there is a one-to-one correspondence between self-adjoint projections and complementary orthogonal manifolds. It should be recalled (IV.4.4) that every closed linear manifold in $\mathfrak{H}$ determines uniquely an ortho-complement; hence every closed linear manifold $\mathfrak{X}$ in $\mathfrak{H}$ determines uniquely a self-adjoint projection E in $\mathfrak{H}$ with $E\mathfrak{H} = \mathfrak{X}$.

4. Weakly Compact Operators

1 Definition. Let $T \in B(\mathfrak{X}, \mathfrak{Y})$, and let S be the closed unit sphere in $\mathfrak{X}$. The operator T is said to be *weakly compact* if the weak closure of TS is compact in the weak topology of $\mathfrak{Y}$. Thus, by Theorem V.6.1, *an operator is weakly compact if and only if it maps bounded sets into weakly sequentially compact sets.*

2 Theorem. *A linear operator T in $B(\mathfrak{X}, \mathfrak{Y})$ is weakly compact if and only if $T^{**}\mathfrak{X}^{**}$ is contained in the natural embedding $\varkappa\mathfrak{Y}$ of $\mathfrak{Y}$ into $\mathfrak{Y}^{**}$.*

Proof. The symbols S, S^{**} will be used for the closed unit spheres in $\mathfrak{X}$, $\mathfrak{X}^{**}$, respectively. Note that T^{**} is continuous with the $\mathfrak{X}^*$, $\mathfrak{Y}^*$ topologies in $\mathfrak{X}^{**}$, $\mathfrak{Y}^{**}$, respectively (cf. 2.3). Hence, since T^{**} is an extension of T (cf. 2.6),

$$\text{(i)}\quad T^{**}(S_1) \subseteq \overline{T^{**}(\varkappa S)} = \overline{\varkappa(TS)} \subseteq \overline{\varkappa(\overline{TS})},$$

where S_1 is the $\mathfrak{X}^*$-closure of $\varkappa S$, and where the bars denote closures in the $\mathfrak{Y}^*$ topology. If T is weakly compact, then $\overline{TS}$ is compact in the $\mathfrak{Y}^*$ topology of $\mathfrak{Y}$ and thus $\varkappa(\overline{TS})$ is compact and hence closed in the $\mathfrak{Y}^*$ topology of $\mathfrak{Y}^{**}$. Thus if T is weakly compact, (i) yields

$$T^{**}(S_1) \subseteq \varkappa(\overline{TS}).$$

According to Theorem V.4.5, $S_1 = S^{**}$, and so $T^{**}S^{**} \subseteq \varkappa(\overline{TS})$, which proves that

$$\text{(ii)}\qquad T^{**}\mathfrak{X}^{**} \subseteq \varkappa\mathfrak{Y}.$$

Conversely, let the operator $T \in B(\mathfrak{X}, \mathfrak{Y})$ satisfy (ii). Since, by Lemma 2.3, T^{**} is continuous with the $\mathfrak{X}^*$, $\mathfrak{Y}^*$ topologies in $\mathfrak{X}^{**}$, $\mathfrak{Y}^{**}$, respectively, and S^{**} is $\mathfrak{X}^*$-compact in $\mathfrak{X}^{**}$ (Theorem V.4.2), it follows that $T^{**}S^{**} \subseteq \varkappa\mathfrak{Y}$ is $\mathfrak{Y}^*$-compact (I.5.7). Thus the $\mathfrak{Y}^*$-homeomorphic image $\varkappa(TS)$ of TS is a subset of a $\mathfrak{Y}^*$-compact subset of $\varkappa\mathfrak{Y}$. Hence, the $\mathfrak{Y}^*$-closure of $\varkappa(TS)$ is a $\mathfrak{Y}^*$-compact subset of $\varkappa\mathfrak{Y}$, and the $\mathfrak{Y}^*$-closure of TS is a $\mathfrak{Y}^*$-compact subset of $\mathfrak{Y}$. Q.E.D.

3 **Corollary.** *If either $\mathfrak{X}$ or $\mathfrak{Y}$ is reflexive, every operator in $B(\mathfrak{X}, \mathfrak{Y})$ is weakly compact.*

Proof. Let $T \in B(\mathfrak{X}, \mathfrak{Y})$. If $\mathfrak{Y}$ is reflexive, then

$$T^{**}\mathfrak{X}^{**} \subseteq \mathfrak{Y}^{**} = \varkappa\mathfrak{Y},$$

and, if $\mathfrak{X}$ is reflexive,

$$T^{**}\mathfrak{X}^{**} = T^{**}\varkappa\mathfrak{X} = \varkappa T\mathfrak{X} \subseteq \varkappa\mathfrak{Y}.$$

Thus, in either case, Theorem 2 shows that T is weakly compact. Q.E.D.

4 **Corollary.** *The set of weakly compact operators is closed in the uniform operator topology of $B(\mathfrak{X}, \mathfrak{Y})$.*

Proof. If $T_n \to T$ in $B(\mathfrak{X}, \mathfrak{Y})$, then, by Lemma 2.2, $|T_n^{**} - T^{**}| \to 0$. If T_n is weakly compact, then for each x^{**} in $\mathfrak{X}^{**}$, $T_n^{**}x^{**} \in \varkappa\mathfrak{Y}$

(Theorem 2), and, since $\varkappa\mathfrak{Y}$ is closed in the metric topology of $\mathfrak{Y}^{**}$, $T^{**}x^{**} \in \varkappa\mathfrak{Y}$. Hence $T^{**}\mathfrak{X}^{**} \subseteqq \varkappa\mathfrak{Y}$, and Theorem 2 gives the desired conclusion. Q.E.D.

5 THEOREM. *Linear combinations of weakly compact operators are weakly compact. The product of a weakly compact linear operator and a bounded linear operator is weakly compact.*

PROOF. Let $T, U \in B(\mathfrak{X}, \mathfrak{Y})$, $W \in B(\mathfrak{Y}, \mathfrak{Z})$, $V \in B(\mathfrak{Z}, \mathfrak{X})$, $\alpha, \beta \in \Phi$, and let T, U be weakly compact. The following inclusions then follow from Theorem 2, Lemma 2.2 and Lemma 2.3.

$$(\alpha T+\beta U)^{**}\mathfrak{X}^{**} = (\alpha T^{**}+\beta U^{**})\mathfrak{X}^{**} \subseteqq \varkappa\mathfrak{Y},$$

$$(TV)^{**}\mathfrak{Z}^{**} = T^{**}V^{**}\mathfrak{Z}^{**} \subseteqq T^{**}\mathfrak{X}^{**} \subseteqq \varkappa\mathfrak{Y},$$

$$(WT)^{**}\mathfrak{X}^{**} = W^{**}T^{**}\mathfrak{X}^{**} \subseteqq W^{**}\varkappa\mathfrak{Y} = W\mathfrak{Y} \subseteqq \varkappa\mathfrak{Z}.$$

Theorem 5 follows from these inclusions and Theorem 2. Q.E.D.

6 COROLLARY. *In the uniform operator topology of $B(\mathfrak{X})$, the weakly compact operators form a closed two-sided ideal.*

We have already observed (2.3) that the adjoint T^* of any T in $B(\mathfrak{X}, \mathfrak{Y})$ is continuous relative to the $\mathfrak{X}$, $\mathfrak{Y}$ topologies in $\mathfrak{X}^*$, $\mathfrak{Y}^*$, respectively. The following result shows that, if T is weakly compact, its adjoint T^* has a stronger continuity property.

7 LEMMA. *An operator in $B(\mathfrak{X}, \mathfrak{Y})$ is weakly compact if and only if its adjoint is continuous with respect to the $\mathfrak{X}^{**}$, $\mathfrak{Y}$ topologies in $\mathfrak{X}^*$, $\mathfrak{Y}^*$, respectively.*

PROOF. Let T be weakly compact. By Theorem 2, for each x^{**} in $\mathfrak{X}^{**}$, there is a y in $\mathfrak{Y}$ with

$$x^{**}(T^*y^*) = (T^{**}x^{**})y^* = y^*y, \qquad y^* \in \mathfrak{Y}^*.$$

Thus, if $y^*_\alpha \to y^*$ in the $\mathfrak{Y}$-topology of $\mathfrak{Y}^*$, then $T^*y^*_\alpha \to T^*y^*$ in the $\mathfrak{X}^{**}$ topology of $\mathfrak{X}^*$. Hence T^* has the stated continuity property (I.7.4). Conversely, let T^* be continuous relative to the $\mathfrak{X}^{**}$, $\mathfrak{Y}$ topologies in $\mathfrak{X}^*$, $\mathfrak{Y}^*$, respectively and let $x_0^{**} \in \mathfrak{X}^{**}$. If $y^*_\alpha y \to y^*y$, $y \in \mathfrak{Y}$, then $x_0^{**}T^*y^*_\alpha = T^{**}x_0^{**}y^*_\alpha \to T^{**}x_0^{**}y^*$. Thus $T^{**}x_0^{**}$ in $\mathfrak{Y}^{**}$ is a continuous linear functional on $\mathfrak{Y}^*$ with its $\mathfrak{Y}$ topology. Theorem V.3.9 shows that $T^{**}x_0^{**} \in \varkappa\mathfrak{Y}$, so $T^{**}\mathfrak{X}^{**} \subset \varkappa\mathfrak{Y}$. The desired conclusion follows from Theorem 2. Q.E.D.

8 THEOREM. (*Gantmacher*) *An operator in* $B(\mathfrak{X}, \mathfrak{Y})$ *is weakly compact if and only if its adjoint is weakly compact.*

PROOF. Let T be weakly compact. Since the closed unit sphere S^* of $\mathfrak{Y}^*$ is $\mathfrak{Y}$-compact (V.4.2), it follows from Lemma 7 and Lemma I.5.7 that T^*S^* is compact in the $\mathfrak{X}^{**}$ topology of $\mathfrak{X}^*$. Hence T^* is weakly compact.

Conversely, if T^* is weakly compact, it follows from Lemma 7 that T^{**} is continuous relative to the $\mathfrak{X}^*$, $\mathfrak{Y}^{***}$ topologies in $\mathfrak{X}^{**}$, $\mathfrak{Y}^{**}$, respectively. If S, S^{**} are the closed unit spheres in $\mathfrak{X}$, $\mathfrak{X}^{**}$, respectively, and if $\varkappa$ is the natural embedding of $\mathfrak{X}$ into $\mathfrak{X}^{**}$, then by Theorem V.4.5, $\varkappa S$ is $\mathfrak{X}^*$-dense in S^{**}, and so, from the continuity of T^{**} we see that $T^{**}S^{**}$ is contained in the $\mathfrak{Y}^{***}$-closure of $T^{**}\varkappa S = \varkappa TS$. According to Theorem V.3.13 the $\mathfrak{Y}^{***}$-closure of the convex set $\varkappa TS$ is the same as its strong closure. Thus $T^{**}S^{**} \subseteqq \varkappa\mathfrak{Y}$, and Theorem 2 gives the desired conclusion. Q.E.D.

5. Compact Operators

1 DEFINITION. Let $T \in B(\mathfrak{X}, \mathfrak{Y})$, and let S be the closed unit sphere in $\mathfrak{X}$. The operator T is said to be *compact* if the strong closure of TS is compact in the strong topology of $\mathfrak{Y}$.

2 THEOREM. (*Schauder*) *An operator in* $B(\mathfrak{X}, \mathfrak{Y})$ *is compact if and only if its adjoint is compact.*

PROOF. Let S, S^* be the closed unit spheres in $\mathfrak{X}$, $\mathfrak{Y}^*$, respectively. Let T be compact, and let $\{y_n^*\}$ be an arbitrary sequence in S^*. Since $|y_n^*y - y_n^*z| \leqq |y-z|$, $n = 1, 2, \ldots$, it follows, from Theorem IV.6.7, that some subsequence $y_{n_i}^*y$ converges uniformly for y in the compact set $\overline{TS}$. Hence $y_{n_i}^*Tx = (T^*y_{n_i}^*)x$ converges uniformly for x in S. It follows that $T^*y_{n_i}^*$ converges in the strong topology of $\mathfrak{X}^*$. Thus T^* is compact.

Conversely, let T^* be compact. Then, by the point just proved, T^{**} is compact; hence if S^{**} is the closed unit sphere in $\mathfrak{X}^{**}$, $T^{**}S^{**}$ is totally bounded (I.6.15). Thus, since $\varkappa TS \subseteqq T^{**}S^{**}$, $\varkappa TS$ is totally bounded; hence TS is totally bounded. Therefore $\overline{TS}$ is compact, (I.6.15) and T is compact. Q.E.D.

3 **Lemma.** *The set of compact operators is closed in the uniform operator topology of* $B(\mathfrak{X}, \mathfrak{Y})$.

Proof. Let S be the closed unit sphere in $\mathfrak{X}$, let T_n be compact, and let $|T_n - T| \to 0$. Then, for $\varepsilon > 0$, there is an n with $|T_n - T| < \varepsilon/3$. Since T_n is compact, there are, by Theorem I.6.15, points $x_1, \ldots, x_p$ in S, with

$$\inf_{1 \leq i \leq p} |T_n x - T_n x_i| < \varepsilon/3, \qquad x \in S.$$

It follows that

$$\inf_{1 \leq i \leq p} |Tx - Tx_i| < \varepsilon, \qquad x \in S.$$

Thus TS is totally bounded, and $\overline{TS}$ is compact (I.6.15). Q.E.D.

4 **Theorem.** *Linear combinations of compact linear operators are compact operators, and any product of a compact linear operator and a bounded linear operator is a compact linear operator.*

Proof. The conclusions of Theorem 4 follow readily from the fact that a set in a metric space is compact if and only if it is sequentially compact (I.6.13). Q.E.D.

5 **Corollary.** *In the uniform operator topology of* $B(\mathfrak{X})$, *the compact operators form a closed two-sided ideal.*

Proof. This corollary follows immediately from 3 and 4. Q.E.D.

The reader will observe the analogy between Lemma 4.7 and the next result.

6 **Theorem.** *An operator in* $B(\mathfrak{X}, \mathfrak{Y})$ *is compact if and only if its adjoint sends bounded generalized sequences which converge in the* $\mathfrak{Y}$ *topology of* $\mathfrak{Y}^*$ *into generalized sequences which converge in the metric of* $\mathfrak{X}^*$.

Proof. Let S and S^* denote the unit spheres of $\mathfrak{X}$ and $\mathfrak{Y}^*$, respectively, and let $T \in B(\mathfrak{X}, \mathfrak{Y})$. We recall that a generalized sequence in $\mathfrak{X}^*$ converges in the metric of $\mathfrak{X}^*$ if and only if it converges uniformly on S. It is seen from the proof of V.4.4 that $T(S)$ is isometric to a bounded subset of $C(S^*)$, where S^* has the $\mathfrak{Y}$ topology. The stated condition is equivalent to the assertion that if $\{y_\alpha^*\}$ is bounded and $y_\alpha^* \to y_0^*$ in the $\mathfrak{Y}$ topology, then $y_\alpha^*(Tx) \to y_0^*(Tx)$, uniformly for x

in S. Hence, by IV.6.6, the condition is equivalent to the statement that $T(S)$ is an equicontinuous subset of $C(S^*)$. It follows from Theorem IV.6.7, that $T(S)$ is conditionally compact in the metric of $\mathfrak{Y}$ if and only if the condition is satisfied. Q.E.D.

6. Operators with Closed Range

It was observed in Lemma 2.8 that the closure of the range of an operator $U \in B(\mathfrak{X}, \mathfrak{Y})$ consists of those vectors y such that $y^*U\mathfrak{X} = 0$ implies $y^*y = 0$. Or, in other words,

$$\overline{U\mathfrak{X}} = \{y | U^*y^* = 0 \text{ implies } y^*y = 0\}.$$

The dual of this theorem states the following: if $U \in B(\mathfrak{X}, \mathfrak{Y})$, then

$$\overline{U^*\mathfrak{Y}^*} = \{x^* | Ux = 0 \text{ implies } x^*x = 0\},$$

and this statement is, in general, false. This will be seen in an exercise. However the dual statement is true if the range $U\mathfrak{X}$ is closed, in which case the range $U^*\mathfrak{Y}^*$ is also closed. Dually, if $U^*\mathfrak{Y}^*$ is closed, so is $U\mathfrak{X}$. These results are contained in the next two theorems. Additional information along these lines is to be found in the set of exercises in Section 9.

1 Lemma. *If the range of an operator U in $B(\mathfrak{X}, \mathfrak{Y})$ is closed, there is a constant K such that to each y in $U\mathfrak{X}$ corresponds an x with $Ux = y$ and $|x| \leqq K|y|$.*

Proof. It is seen from the interior mapping principle, (II.2.1), that the unit sphere S in $\mathfrak{X}$ maps onto a set US which contains some relative sphere $\{y | y \in U\mathfrak{X}, |y| < \delta\}$ with $\delta > 0$. Thus, for $0 \neq y \in U\mathfrak{X}$, the vector $\delta y/2|y|$ is the image under U of a vector z with $|z| \leqq 1$. Hence, if $x = 2|y|z/\delta$, we have $Ux = y$, and $|x| \leqq (2/\delta)|y|$. Q.E.D.

2 Theorem. *If the range of the operator U in $B(\mathfrak{X}, \mathfrak{Y})$ is closed, then the range of its adjoint is the set of x^* in $\mathfrak{X}^*$ such that $Ux = 0$ implies $x^*x = 0$.*

Proof. Let x^* satisfy this condition, and define a (possibly discontinuous) linear functional y_0^* on $\mathfrak{Y}_0 = U\mathfrak{X}$ by the formula $y_0^*(Ux) = x^*(x)$. Because of the condition imposed on x^*, this defines y_0^* uniquely. By the preceding lemma, there exists a constant K such that to each $y \in \mathfrak{Y}_0$ there corresponds an x with $|x| \leqq K|y|$, $Ux = y$. Hence, $|y_0^*(y)| \leqq K|x^*||y|$. It follows from Theorem II.3.11 that y_0^* can be extended to a continuous linear functional y^* on all of $\mathfrak{Y}$, and now $U^*y^* = x^*$.

It follows from the definition of U^* that every element in its range satisfies the stated condition. Q.E.D.

3 LEMMA. *If the adjoint of an operator U in $B(\mathfrak{X}, \mathfrak{Y})$ is one-to-one and has a closed range, then $U\mathfrak{X} = \mathfrak{Y}$.*

PROOF. Let $0 \neq y \in \mathfrak{Y}$ and define

$$\Gamma = \{y^* | y^* \in \mathfrak{Y}^*,\ y^*y = 0\}.$$

Then Γ is $\mathfrak{Y}$-closed in $\mathfrak{Y}^*$.

Suppose, for the moment, that $U^*\Gamma$ is $\mathfrak{X}$-closed and different from $U^*\mathfrak{Y}^*$. From Corollary V.3.12 it is seen that there is an $x \in \mathfrak{X}$ with $\hat{x}U^*\mathfrak{Y}^* \neq 0$, $\hat{x}U^*\Gamma = 0$. This means that $Ux \neq 0$, and $y^*Ux = 0$, for every $y^* \in \Gamma$. By Lemma V.3.10, for some non-zero scalar α, $Ux = \alpha y$. Therefore $y \in U\mathfrak{X}$ and $U\mathfrak{X} = \overline{\mathfrak{Y}}$ as desired.

It remains to be shown that $U^*\Gamma$ is $\mathfrak{X}$-closed, but not equal to $U^*\mathfrak{Y}^*$. Since $y \neq 0$, Γ is a proper subset of $\mathfrak{Y}^*$. Hence, since U^* has an inverse, $U^*\Gamma$ is a proper subset of $U^*\mathfrak{Y}^*$. Finally, to show that $U^*\Gamma$ is $\mathfrak{X}$-closed, it suffices, in view of Theorem V.5.7 (or V.5.8), to show that $(U^*\Gamma) \cap S^*$ is $\mathfrak{X}$-closed, where S^* is the closed unit sphere in $\mathfrak{X}^*$. By Lemma 1, $(U^*)^{-1}S^*$ is bounded. Thus, $(U^*)^{-1}S^*$ is contained in some multiple nS_1^* of the closed unit sphere S_1^* of $\mathfrak{Y}^*$. Theorem V.4.2 shows that nS_1^* is compact in the $\mathfrak{Y}$-topology of $\mathfrak{Y}^*$. Since, by Lemma 2.3, U^* is a continuous mapping from $\mathfrak{Y}^*$ with its $\mathfrak{Y}$-topology to $\mathfrak{X}^*$ with its $\mathfrak{X}$-topology, the image under U^* of the compact set $\Gamma \cap nS_1^*$ is closed. Hence $(U^*\Gamma) \cap S^* = S^* \cap U^*(\Gamma \cap nS_1^*)$ is $\mathfrak{X}$-closed. Q.E.D.

4 THEOREM. *If the adjoint of an operator U in $B(\mathfrak{X}, \mathfrak{Y})$ has a closed range, then the range of U is closed and consists of those vectors y in $\mathfrak{Y}$ for which $U^*y^* = 0$ implies $y^*y = 0$.*

PROOF. Consider the map U_1 from $\mathfrak{X}$ to $\mathfrak{Z} = \overline{U(\mathfrak{X})}$, defined by $U_1(x) = U(x)$. Then, since U_1 has a dense range, U_1^* is one-to-one. If $x^* \in \mathfrak{X}^*$ is in the closure of $U_1^*\mathfrak{Z}^*$, then $x^* = \lim_{n\to\infty} U_1^*z_n^*$, where $z_n^* \in \mathfrak{Z}^*$. If y_n^* is an extension to $\mathfrak{Y}$ of z_n^* (II.3.11), then $x^* = \lim_{n\to\infty} U^*y_n^*$, and, since the range of U^* is closed, $x^* = U^*y^*$ for some $y^* \in \mathfrak{Y}^*$. If z^* is the restriction of y^* to $\mathfrak{Z}$, then $x^* = U_1^*z^*$. Hence, the range of U_1^* is also closed. It follows from the previous lemma that $U_1\mathfrak{X} = U\mathfrak{X} = \mathfrak{Z}$. Hence, U has a closed range. Q.E.D.

5 THEOREM. *If U is in $B(\mathfrak{X}, \mathfrak{Y})$ and maps bounded closed sets onto closed sets, then U has a closed range.*

PROOF. Let $y = \lim_n Ux_n$ be a point in the closure of $U\mathfrak{X}$, and let $\mathfrak{M} = \{x|Ux = 0\}$. Let d_n be the distance between x_n and $\mathfrak{M}$, and let $w_n \in \mathfrak{M}$ be such that

$$d_n \leqq |x_n - w_n| \leqq 2d_n.$$

If the sequence $\{x_n - w_n\}$ contains a bounded subsequence $\{x_{n_i} - w_{n_i}\}$, then, by hypothesis, the closure of this subsequence is mapped onto a closed set containing $y = \lim_i U(x_{n_i} - w_{n_i})$, and so $y \in U\mathfrak{X}$. Thus, to complete the proof, it will be sufficient to show that the assumption $|x_n - w_n| \to \infty$ leads to a contradiction.

If $|x_n - w_n| \to \infty$, then, since $U(x_n - w_n) \to y$, we have $U((x_n - w_n)/|x_n - w_n|) \to 0$, and hence, by hypothesis, $\mathfrak{M}$ contains a point w of the closure of the bounded sequence $\{(x_n - w_n)/|x_n - w_n|\}$. If n is fixed so that

$$\left| \frac{x_n - w_n}{|x_n - w_n|} - w \right| < \frac{1}{3},$$

then

$$|x_n - w_n - w|x_n - w_n|| < \tfrac{1}{3}|x_n - w_n| < \tfrac{2}{3} d_n,$$

which contradicts the definition of d_n. Q.E.D.

6 COROLLARY. *If U is in $B(\mathfrak{X})$ and maps bounded closed sets onto closed sets, the ranges of all its iterates are closed.*

7. Representation of Operators in $C(S)$

A detailed knowledge of the specific analytical form of the most general compact or weakly compact linear map between a given pair of B-spaces is often useful in applying the theorems of Sections 4 and 5. In certain cases fairly complete information is easily obtained. It will be seen in this section that this is so if the range of the operator is in a space of continuous functions. In some other cases very little is known. For example, while it is easy to see that the general continuous linear map from $L_p[0, 1]$, $p > 1$, to $L_q[0, 1]$ has the form

$$g(s) = \frac{d}{ds}\int_0^1 K(s, t)f(t)dt,$$

no satisfactory expression for the norm of T is known. No conditions on $K(s, t)$ are known which are equivalent to the compactness of T. Of course, conditions on $K(s, t)$ which are *sufficient* to insure the compactness of T may be given. Such conditions will be found in the exercises.

The present section, which is concerned with a compact Hausdorff space S, is divided into two parts. We first consider operators with range in $C(S)$ and then operators defined on $C(S)$.

1 THEOREM. *Let S be a compact Hausdorff space and let T be a bounded linear operator from a B-space $\mathfrak{X}$ into $C(S)$. Then there exists a mapping $\tau : S \to \mathfrak{X}^*$ which is continuous with the $\mathfrak{X}$ topology in $\mathfrak{X}^*$ such that*

(1) $Tx(s) = \tau(s)x, \ x \in \mathfrak{X}, \ s \in S;$

(2) $|T| = \sup_{s \in S} |\tau(s)|.$

Conversely, if such a map τ is given, then the operator T defined by (1) *is a bounded linear operator from $\mathfrak{X}$ into $C(S)$ with norm given by* (2). *The operator T is weakly compact if and only if τ is continuous with the $\mathfrak{X}^{**}$ topology in $\mathfrak{X}^*$. The operator T is compact if and only if τ is continuous with the norm topology in $\mathfrak{X}^*$.*

PROOF. Let T be a bounded linear map of $\mathfrak{X}$ into $C(S)$. Its conjugate T^* maps $C^*(S)$ into $\mathfrak{X}^*$. Furthermore the map $\pi : S \to C^*(S)$, defined by the equation

$$\pi(s)(f) = f(s), \qquad f \in C(S), \qquad s \in S$$

is a homeomorphism of S into a compact subset of $C^*(S)$ in the $C(S)$ topology (see V.8.7). By Lemma 2.3, T^* is continuous with the $C(S)$ topology in $C^*(S)$ and the $\mathfrak{X}$ topology in $\mathfrak{X}^*$ so the map $\tau : S \to \mathfrak{X}^*$ defined by $\tau = T^*\pi$ is continuous with the $\mathfrak{X}$ topology in $\mathfrak{X}^*$. Equations (1) and (2) are easily seen to be valid. Conversely if $\tau : S \to \mathfrak{X}^*$

is continuous with the $\mathfrak{X}$ topology in $\mathfrak{X}^*$, then equation (1) clearly defines a linear map from $\mathfrak{X}$ into $C(S)$ whose norm is

$$\begin{aligned} |T| &= \sup_{|x|\leq 1} \sup_{s\in S} |\tau(s)x| \\ &= \sup_{s\in S} \sup_{|x|\leq 1} |\tau(s)x| \\ &= \sup_{s\in S} |\tau(s)|. \end{aligned}$$

This completes the proof of the first part of the theorem.

If T is weakly compact, Lemma 4.7 implies that T^* is continuous with the $C(S)$ topology in $C^*(S)$ and the $\mathfrak{X}^{**}$ topology in $\mathfrak{X}$, so the asserted continuity of τ is guaranteed. Conversely, if τ is continuous in the $\mathfrak{X}^{**}$ topology, and if $s_\alpha \to s_0$ in S, then $\tau(s_\alpha) \to \tau(s_0)$ in the $\mathfrak{X}^{**}$ topology of $\mathfrak{X}^*$. Now $\tau(s_\alpha)$ and $\tau(s_0)$ are in $C(B_2)$, where B_2 is the solid unit sphere in $\mathfrak{X}^{**}$ endowed with its $\mathfrak{X}^*$ topology. By Arzelà's theorem (IV.6.11) the convergence is quasi-uniform on B_2 and hence on B, the unit sphere in $\mathfrak{X}$. From this fact and equation (1) we conclude that the bounded set $T(B)$ is a quasi-equicontinuous collection of functions in $C(S)$. It follows from Theorem IV.6.14 that $T(B)$ is conditionally weakly compact, so that T is a weakly compact operator. This completes the proof of the assertion concerning weakly compact operators.

Finally if T is compact, then it follows from 5.6 and the fact that $\pi(S)$ is bounded in $C^*(S)$ that τ is continuous with the norm topology in $\mathfrak{X}^*$. Conversely, let τ be continuous with the norm topology in $\mathfrak{X}^*$. Then given $\varepsilon > 0$ and $s_0 \in S$, there is a neighborhood N of s_0 such that if $s \in N$ then

$$\sup_{x\in B} |Tx(s)-Tx(s_0)| = |\tau(s)-\tau(s_0)| < \varepsilon.$$

Hence if s is in N, then $|Tx(s)-Tx(s_0)| < \varepsilon$ for all x in B. It follows from Definition IV.6.6 that $T(B)$ is an equicontinuous set in $C(S)$. Since $T(B)$ is bounded, it is a consequence of Theorem IV.6.7 that it is conditionally strongly compact. Thus T is a compact operator and the theorem is proved. Q.E.D.

The theorem just proved yields rather complete information concerning operators with range in $C(S)$. We now turn to the question of representing operators T defined on $C(S)$. Motivated by the Riesz representation theorem (IV.6.3) for linear functionals on $C(S)$, we are led to hope that the representation of operators will be effected by a vector measure whose values lie in $\mathfrak{X}$. This turns out to be the case for weakly compact operators; for the general operator the measure has its values in $\mathfrak{X}^{**}$.

In the following, $\mathscr{B}$ denotes the field of Borel sets in S, i.e., the σ-field generated by the closed sets of S. If μ is a function on $\mathscr{B}$ with values in a B-space, then as in Definition IV.10.3, the symbol $\|\mu\|(E)$ denotes the semi-variation of μ over $E \in \mathscr{B}$ and is defined as

$$\|\mu\|(E) = \sup \left| \sum_{i=1}^{n} \alpha_i \mu(E_i) \right|,$$

where the supremum is taken over all finite collections of disjoint Borel sets in E and all finite sets of scalars $\alpha_1, \ldots, \alpha_n$ with $|\alpha_i| \leq 1$.

We are now prepared to represent the general operator.

2 THEOREM. *Let S be a compact Hausdorff space and let T be an operator on $C(S)$ to $\mathfrak{X}$. Then there exists a unique set function μ, defined on the Borel sets in S and having values in $\mathfrak{X}^{**}$, such that*

(a) *$\mu(\cdot)x^*$ is in $rca(S)$ for each x^* in $\mathfrak{X}^*$;*

(b) *the mapping $x^* \to \mu(\cdot)x^*$ of $\mathfrak{X}^*$ into $rca(S)$ is continuous with the $\mathfrak{X}$ and $C(S)$ topologies in these spaces respectively;*

(c) $x^*Tf = \int_S f(s)\mu(ds)x^*, \quad f \in C(S), \quad x^* \in \mathfrak{X}^*$;

(d) $|T| = \|\mu\|(S)$.

Conversely, if μ is a set function on the Borel sets in S to $\mathfrak{X}^{**}$ which satisfies (a) *and* (b), *then equation* (c) *defines a linear map T of $C(S)$ into $\mathfrak{X}$ whose norm is given by* (d), *and such that $T^*x^* = \mu(\cdot)x^*$.*

PROOF. If $E \in \mathscr{B}$, let ϕ_E be the element in $C^{**}(S)$, the second adjoint of $C(S)$, defined by

$$\phi_E(\lambda) = \lambda(E), \qquad \lambda \in rca(S).$$

Define the set function $\mu : \mathscr{B} \to \mathfrak{X}^{**}$ by the equation

$$\mu(E) = T^{**}(\phi_E), \qquad E \in \mathscr{B}.$$

It follows from the Riesz representation theorem (IV.6.3) that T^*x^* is a measure $\lambda_{x^*} \in rca(S)$. Since

$$\lambda_{x^*}(E) = \phi_E(\lambda_{x^*}) = \phi_E(T^*x^*) = T^{**}\phi_E(x^*) = \mu(E)x^*,$$

it follows that (a) and (c) are valid. This equation also shows that $T^*x^* = \mu(\cdot)x^*$ from which (b) follows. Equality (d) is readily checked.

Conversely, if (a) and (b) are satisfied for the mapping which sends x^* into $\mu(\cdot)x^*$, then it follows that for each fixed $f \in C(S)$ the mapping

$$x^* \to \int_S f(s)\mu(ds)x^*$$

is continuous in the $\mathfrak{X}$ topology of $\mathfrak{X}^*$ and therefore (V.3.9) is generated by some element $x_f \in \mathfrak{X}$. Thus the mapping $T : f \to x_f$ defined by (c) is a linear mapping of $C(S)$ into $\mathfrak{X}$. It is easy to verify that it is bounded and has the stated properties. Q.E.D.

In the case of a weakly compact operator the measure takes its values in $\mathfrak{X}$, and not merely in $\mathfrak{X}^{**}$.

3 **Theorem.** *Let S be a compact Hausdorff space and let T be a weakly compact operator from $C(S)$ to $\mathfrak{X}$. Then there exists a vector measure μ defined on the Borel sets in S and having values in $\mathfrak{X}$ such that*

(a) $x^*\mu$ *is in* $rca(S)$, $\qquad x^* \in \mathfrak{X}^*$;

(b) $Tf = \int_S f(s)\mu(ds)$, $\qquad f \in C(S)$;

(c) $|T| = \|\mu\|(S)$,

(d) $T^*x^* = x^*\mu$, $\qquad x^* \in \mathfrak{X}^*$.

Conversely if μ is a vector measure on the Borel sets in S to the B-space $\mathfrak{X}$ which satisfies (a) *then the operator T defined by* (b) *is a weakly compact operator from $C(S)$ to $\mathfrak{X}$ whose norm is given by* (c) *and whose adjoint is given by* (d).

Proof. If T is weakly compact, then by Theorem 4.2, T^{**} maps $C^{**}(S)$ into $\varkappa(\mathfrak{X})$, the natural embedding of $\mathfrak{X}$ into $\mathfrak{X}^{**}$. Hence by the construction in Theorem 2, μ is defined on the Borel sets Σ and has its values in $\varkappa(\mathfrak{X})$. Thus we may and shall regard μ as a mapping into $\mathfrak{X}$. Since $x^*\mu$ is in $rca(S)$ for every x^* in $\mathfrak{X}^*$, it follows from Theorem IV.10.1 that μ is a strongly countably additive vector measure. Thus from Theorem IV.10.8(c) the integral in equation (b) exists and is in $\mathfrak{X}$. From equation (c) in Theorem 2 we conclude that

$$Tf = \int_S f(s)\mu(ds), \qquad f \in C(S).$$

The validity of equations (c) and (d) are consequences of the corresponding results in Theorem 2.

Conversely, let μ be an $\mathfrak{X}$-valued measure defined on the Borel sets in S with the property that $x^*\mu$ is in $rca(S)$ for every x^* in $\mathfrak{X}^*$. It is clear that the operator T, defined by (b), is a bounded linear operator on $C(S)$ to $\mathfrak{X}$ whose adjoint T^* is given by (d). From IV.10.2 we conclude that T^* maps the unit sphere of $\mathfrak{X}^*$ into a conditionally weakly compact set of $rca(S)$, and therefore T^* is a weakly compact operator. By Theorem 4.8 this implies that T is a weakly compact operator. Q.E.D.

4 THEOREM. *If T is a weakly compact operator from $C(S)$ to $\mathfrak{X}$, then T sends weak Cauchy sequences into strongly convergent sequences. Consequently, T maps conditionally weakly compact subsets of $C(S)$ into conditionally strongly compact subsets of $\mathfrak{X}$.*

PROOF. If $\{f_n\}$ is a weak Cauchy sequence in $C(S)$, it is (II.3.20) bounded. Clearly the limit $f_0(s) = \lim f_n(s)$ exists for each $s \in S$. Although the limit function f_0 may not be in $C(S)$, it is certainly bounded and measurable. Let μ be the vector measure corresponding to T as in the preceding theorem. Then

$$Tf_n = \int_S f_n(s)\mu(ds), \qquad n = 0, 1, 2, \ldots.$$

Thus by Theorem IV.10.10 it follows that $\{Tf_n\}$ is a convergent sequence and this proves the first assertion. The second conclusion follows immediately from the first. Q.E.D.

5 COROLLARY. *The product of two weakly compact operators in $C(S)$ is compact; in particular, the square of a weakly compact operator in $C(S)$ is compact.*

PROOF. This fact follows directly from Theorem 4. Q.E.D.

It was proved in Corollary 4.3 that an arbitrary bounded operator with range in a reflexive B-space is weakly compact. The next theorem asserts that this remains true for a weakly complete space provided the domain is $C(S)$.

6 THEOREM. *An arbitrary bounded linear operator from $C(S)$ into a weakly complete B-space $\mathfrak{X}$ is weakly compact.*

PROOF. We shall first prove the theorem under the assumption that S is a compact metric space. In this case let μ be the set function given by Theorem 2. To prove the theorem it will, by Theorem 3, suffice to show that $\mu(E)$ is in $\varkappa(\mathfrak{X})$ for every Borel set E.

Let $\mathscr{B}$ be the Borel sets in S. Then the space $B(S, \mathscr{B})$ (cf. IV.2.12) is the space of bounded Borel measurable functions on S. We let $\mathfrak{B}_0$ be the intersection of all linear manifolds $\mathfrak{B} \subseteq B(S, \mathscr{B})$ with the two properties:

(i) $C(S) \subseteq \mathfrak{B}$,

(ii) if $\{f_n\}$ is a uniformly bounded sequence in $\mathfrak{B}$ and if $f(s) = \lim_{n\to\infty} f_n(s)$ for every s in S, then $f \in \mathfrak{B}$.

It is evident that $\mathfrak{B}_0$ possesses these two properties. We now prove that $\mathfrak{B}_0$ is an algebra under the natural product $(fg)(s) = f(s)g(s)$, $s \in S$. To see this, let h be a fixed element of $C(S)$ and let $\mathfrak{B}(h) = \{f \in \mathfrak{B}_0 | fh \in \mathfrak{B}_0\}$. Evidently $\mathfrak{B}(h) \subseteq \mathfrak{B}_0$, and it is clear that $\mathfrak{B}(h)$ satisfies properties (i) and (ii) above so that $\mathfrak{B}(h) = \mathfrak{B}_0$. This proves that if $h \in C(S)$ and $f \in \mathfrak{B}_0$, then $fh \in \mathfrak{B}_0$. Now let f be a fixed element of $\mathfrak{B}_0$, and let $\mathfrak{B}(f) = \{g \in \mathfrak{B}_0 | fg \in \mathfrak{B}_0\}$. We have just proved that $C(S) \subseteq \mathfrak{B}(f)$, and it is clear that (ii) is satisfied. As before, this shows that $\mathfrak{B}_0 = \mathfrak{B}(f)$, so that $\mathfrak{B}_0$ is an algebra.

Denoting the characteristic function of a set E by χ_E, we let $\mathscr{B}_0 = \{E \in \mathscr{B} | \chi_E \in \mathfrak{B}_0\}$. Since $\mathfrak{B}_0$ is an algebra and satisfies condition (ii), it is readily seen that $\mathscr{B}_0$ is a σ-field contained in $\mathscr{B}$. We now show that $\mathscr{B}_0 = \mathscr{B}$ by proving that $\mathscr{B}_0$ contains all the closed sets.

Let F be an arbitrary closed set in S. Since S is assumed to be a compact metric space, there is an increasing sequence of open sets $\{G_n\}$ in S with $F' = \bigcup_{n=1}^{\infty} G_n$ and $\overline{G_n} \cap F = \phi$. By the Urysohn theorem (I.5.2) there is an $f_n \in C(S)$ with $|f_n| = 1$ and such that $f_n(F) = 1$, $f_n(\overline{G_n}) = 0$. Clearly $f_n(s) \to \chi_F(s)$, $s \in S$, and so $\chi_F \in \mathfrak{B}_0$.

We now show that $\mu(E) \in \mathfrak{X}$ for $E \in \mathscr{B}$. Consider the collection $\mathfrak{B}_1$ of all $f \in B(S, \mathscr{B})$ such that

$$\int_S f(s)\mu(ds) \in \mathfrak{X}.$$

The set $\mathfrak{B}_1$ forms a linear manifold which contains $C(S)$. That $\mathfrak{B}_1$ satisfies (ii) follows from the relation

$$\left(\int_S f(s)\mu(ds)\right) x^* = \lim_{n\to\infty} x^* \int_S f_n(s)\mu(ds), \qquad x^* \in \mathfrak{X}^*,$$

and the weak completeness of $\mathfrak{X}$. Hence $\mathfrak{B}_0 \supseteq \mathfrak{B}_1$. Since

$$\mu(E) = \int_E \chi(s)\mu(ds) \in \mathfrak{X}$$

for $E \in \mathscr{B} = \mathscr{B}_0$, it follows that μ maps $\mathscr{B}$ into $\mathfrak{X}$. This proves the theorem under the additional assumption that S is a compact metric space.

To complete the proof, let S be a compact Hausdorff space. Let $\{f_n\}$ be an arbitrary sequence in the unit sphere of $C(S)$. Let S_0 be the set of equivalence classes of S under the relation: $s \sim s'$ if and only if $f_n(s) = f_n(s')$, $n = 1, 2, \ldots$. We make S_0 into a metric space by defining the metric

$$\varrho(s, t) = \sum_{n=1}^{\infty} 2^{-n}|f_n(s)-f_n(t)|, \qquad s, t \in S_0.$$

Let $\pi : s \to s_0$ be the canonical map of a point into its equivalence class. From the continuity of the f_n we see that π is continuous. Consequently, S_0 is a compact metric space. Consider the space $C(S_0)$; we note that if $\varphi \in C(S_0)$ then the function f defined by $f(s) = \varphi(\pi(s))$ is in $C(S)$, and we define $T_0 : C(S_0) \to \mathfrak{X}$ by $T_0\varphi = Tf$. Clearly T_0 is a linear map and $|T_0| \leqq |T|$. We also note that each f_n gives a well-defined function $\varphi_n \in C(S_0)$ such that $f_n(s) = \varphi_n(\pi(s))$. Now from the first part of the proof we have that T_0 is weakly compact, so there is a subsequence $\{\varphi_{n_k}\}$ such that $\{T\varphi_{n_k}\}$ converges weakly in $\mathfrak{X}$. Since $Tf_n = T_0\varphi_n$, we conclude that the subsequence $\{Tf_{n_k}\}$ converges weakly, so that T is weakly compact, as was to be proved. Q.E.D.

Since a compact operator is weakly compact, it may be represented by a vector measure as in Theorem 3. As would be expected this measure has a special nature.

7 Theorem. *An operator $T : C(S) \to \mathfrak{X}$ is compact if and only if the vector measure $\mu : \mathscr{B} \to \mathfrak{X}$ corresponding to it as in Theorem 3 takes its values in a compact subset of $\mathfrak{X}$.*

Proof. If T is compact, then it follows from Theorems 4.2 and 5.2 that T^{**} is a compact operator from $C^{**}(S)$ to $\mathfrak{X}$, and it follows from the construction of μ that it has its values in a compact set of $\mathfrak{X}$.

To prove the converse, it is evidently sufficient to show that the set K of all sums of the form

$$\sum_{i=1}^{n} \alpha_i \mu(E_i),$$

where the $\{E_i\}$ are disjoint and $|\alpha_i| \leqq 1$, is a totally bounded set in $\mathfrak{X}$. Let $\varepsilon > 0$ be given and let M be the semi-variation of μ on S. Select a set $\{\beta_1, \ldots, \beta_p\}$ of complex numbers with $|\beta_i| \leqq 1$ such that if $|\alpha| \leqq 1$ then there exists a $\beta_i = \beta(\alpha)$ with $|\beta(\alpha) - \alpha| < \varepsilon/2M$. Let $\mathscr{B}$ be the Borel sets in S and let $\{F_1, \ldots, F_q\} \subset \mathscr{B}$ be such that if $E \in \mathscr{B}$, there is an $F_k = F(E)$ with $|\mu(F(E)) - \mu(E)| < \varepsilon/2p$. Then from the definition of the semi-variation

$$\begin{aligned} \Big|\sum_{i=1}^{n} \alpha_i \mu(E_i) - \sum_{i=1}^{n} \beta(\alpha_i)\mu(E_i)\Big| &= \Big|\sum_{i=1}^{n} \{\alpha_i - \beta(\alpha_i)\}\mu(E_i)\Big| \\ &\leqq \frac{\varepsilon}{2M} M = \frac{\varepsilon}{2}. \end{aligned}$$

Now $\sum_{i=1}^{n} \beta(\alpha_i)\mu(E_i)$ can be written as a sum $\sum_{j=1}^{p} \beta_j \mu(E_j^+)$, with $\{E_j^+\}$ a disjoint family in $\mathscr{B}$. Thus

$$\begin{aligned} \Big|\sum_{j=1}^{p} \beta_j \mu(E_j^+) - \sum_{j=1}^{p} \beta_j \mu(F(E_j^+))\Big| &\leqq \sum_{j=1}^{p} |\mu(E_j^+) - \mu(F(E_j^+))| \\ &< p \cdot \frac{\varepsilon}{2p} = \frac{\varepsilon}{2}. \end{aligned}$$

Thus it has been shown that each element in K can be approximated within an arbitrary positive distance ε by sums of the form $\sum_{j=1}^{p} \beta_j \mu(F_{k_j})$. This proves that K is totally bounded. Hence the operator T given by

$$Tf = \int_S f(s)\mu(ds), \qquad f \in C(S)$$

is compact. Q.E.D.

We have already seen (IV.6.18, IV.7.6 and V.8.11) that the spaces $B(S)$, $(B(S, \Sigma)$, AP, and $L_\infty(S, \Sigma, \mu)$ are isometrically isomorphic to a space $C(S_1)$ where S_1 is some compact Hausdorff space. The same is true (IV.6.22) for the space of bounded continuous functions on a completely regular space, and for the spaces c and l_∞. Consequently any operator with one of these spaces as domain and with range in a weakly complete space is automatically weakly compact. Further, any weakly compact operator defined on one of these spaces maps a weak Cauchy sequence into a strongly convergent sequence, and the square of such an operator in these spaces is compact.

Additional information and special cases will be found in the exercises of Section 9.

8. The Representation of Operators in a Lebesgue Space

This section parallels the preceding one. At first the general operator and the weakly compact operator from an arbitrary B-space to an L_1 space are represented. Then the problem of representing an operator whose domain is an L_1 space is considered. The compact and weakly compact operators are given a kernel representation and the topological properties of the operator are stated in equivalent form in terms of the kernel. The property of mapping a weakly convergent sequence into a strongly convergent sequence, which is enjoyed by the weakly compact operators on $C(S)$ (Theorem 7.4), is a property also shared by the weakly compact operators on L_1.

1 THEOREM. *Let* (S, Σ, μ) *be a measure space and let* T *be a continuous linear map of the* B-space $\mathfrak{X}$ *into* $L_1(S, \Sigma, \mu)$. *Then there is a uniquely determined function* $x^*(\cdot)$ *on* Σ *to* $\mathfrak{X}^*$ *such that*

(i) *for each* x *in* $\mathfrak{X}$ *the set function* $x^*(\cdot)x$ *is* μ*-continuous and countably additive on* Σ, *and*

$$\text{(ii)} \quad Tx = \frac{dx^*(\cdot)x}{d\mu}, \qquad x \in \mathfrak{X}.$$

The norm of T *satisfies the relations*

$$\text{(iii)} \quad \sup_{E \in \Sigma} |x^*(E)| \leqq |T| \leqq 4 \sup_{E \in \Sigma} |x^*(E)|.$$

Conversely, if $x^*(\cdot)$ *on* Σ *to* $\mathfrak{X}^*$ *satisfies* (i) *then* (ii) *defines an operator* T *on* $\mathfrak{X}$ *to* $L_1(S, \Sigma, \mu)$ *whose norm satisfies* (iii).

Furthermore T *is weakly compact if and only if* $x^*(\cdot)$ *is countably additive on* Σ *in the strong topology of* $\mathfrak{X}^*$.

PROOF. If, for E in Σ, the functional $x^*(E)$ in $\mathfrak{X}^*$ is defined by the equation

$$x^*(E)x = \int_E (Tx)(s)\mu(ds),$$

then statements (i) and (ii) are immediate. By Theorems III.2.20(a) and III.1.5, we have, for each E in Σ,

$$\begin{aligned}|x^*(E)| &= \sup_{|x|\leqq 1} |x^*(E)x| \leqq \sup_{|x|\leqq 1} \int_S |(Tx)(s)|v(\mu, ds) \\ &= |T| = \sup_{|x|\leqq 1} v(x^*(\cdot)x, S) \\ &\leqq 4 \sup_{|x|\leqq 1} \sup_{E\epsilon\Sigma} |x^*(E)x| = 4 \sup_{E\epsilon\Sigma} |x^*(E)|.\end{aligned}$$

Conversely, if $x^*(\cdot)$ on Σ to $\mathfrak{X}^*$ has the property stated in (i), then the operator T defined by (ii) (cf. the note following Theorem III.10.7) is clearly linear. To see that it is continuous it will therefore suffice to show that it is closed (II.2.4). If $x_n \to x$ and $Tx_n \to f$, then, for each E in Σ,

$$\begin{aligned}\int_E f(s)\mu(ds) &= \lim_n \int_E (Tx_n)(s)\mu(ds) \\ &= \lim_n x^*(E)x_n = x^*(E)x = \int_E (Tx)(s)\mu(ds),\end{aligned}$$

which shows that $Tx = f$ and proves that T is closed and thus continuous. To prove the final statement of the theorem consider the subspace $ca(S, \Sigma, \mu)$ of $ca(S, \Sigma)$ which consists of all μ-continuous functions in $ca(S, \Sigma)$. By the general Radon-Nikodým theorem (III.10.7), and Theorem III.2.20(a), the space $ca(S, \Sigma, \mu)$ is equivalent to the space $L_1(S, \Sigma, \mu)$. Under this equivalence T determines the operator $x \to x^*(\cdot)x$ on $\mathfrak{X}$ to $ca(S, \Sigma, \mu)$. Thus T is weakly compact if and only if the set of all set functions $x^*(\cdot)x$ with $|x| \leqq 1$ is conditionally weakly compact in $ca(S, \Sigma, \mu)$. According to Theorem IV.9.1 this is the case if and only if the countable additivity of $x^*(\cdot)x$ is uniform with respect to $|x| \leqq 1$. Thus T is weakly compact if and only if $x^*(\cdot)$ is countably additive in the strong topology of $\mathfrak{X}^*$. Q.E.D.

2 THEOREM. *Let (S, Σ, μ) be a σ-finite positive measure space and let T be a continuous linear map of $L_1(S, \Sigma, \mu)$ into a linear topological space $\mathfrak{X}$. Let T map the closed unit sphere in $L_1(S, \Sigma, \mu)$ onto a set whose closure K is compact and has a countable base. Then there is a function $x(\cdot)$ on S to K such that $x^*x(\cdot)$ is in $L_\infty(S, \Sigma, \mu)$ for each x^* in $\mathfrak{X}^*$ and such that*

$$\int_S x^*x(s)f(s)\mu(ds) = x^*Tf, \qquad f \in L_1(S, \Sigma, \mu).$$

The theorem will first be proved under the assumption that $\mu(S) < \infty$. The proof will be based upon the following three lemmas whose statements require the introduction of the following notation: If $\pi = \{E_1, \ldots, E_n\}$ and $\pi' = (F_1, \ldots, F_m\}$ are two partitions of S into disjoint sets in Σ of positive measure we write $\pi' \geqq \pi$ if and only if for each $F_j \in \pi'$ there is an $E_k \in \pi$ such that $\mu(F_j - E_k) = 0$. It is straightforward to verify that the collection of partitions of S forms a directed set (I.7.1) under this ordering. If $\pi = \{E_1, \ldots, E_n\}$ is a partition of S and h is a bounded measurable function on S, we define the function h_π by

$$h_\pi(t) = \sum_{i=1}^{n} \left\{\frac{1}{\mu(E_i)} \int_{E_i} h(s)\mu(ds)\right\} \chi_{E_i}(t).$$

We define

$$x_\pi(t) = \sum_{i=1}^{n} \frac{T(\chi_{E_i})}{\mu(E_i)} \chi_{E_i}(t).$$

3 Lemma. *Let (S, Σ, μ) be a finite positive measure space and let h be a bounded measurable function on S. Then given $\varepsilon > 0$, there is a partition π_0 of S such that if $\pi \geqq \pi_0$, then*

$$|h_\pi(s) - h(s)| < \varepsilon$$

for all s not contained in a null set $E(\pi)$ depending on π.

Proof. Let the range of h be written as the union of disjoint Borel sets $A_1, \ldots, A_k$ of diameter less than ε. At least one of the subsets $G_i = h^{-1}(A_i)$, $i = 1, \ldots, k$ of S has positive measure, and we suppose it to be G_1. The sets $E_1, \ldots, E_n$ of the partition π_0 are obtained as follows. We adjoin to G_1 those sets G_j of measure zero, obtaining E_1. The remaining sets G_j of positive measure are the sets $E_2, \ldots, E_n$. Thus there is a null set E_0 such that

$$|h(s) - h(t)| < \varepsilon, \qquad s, t \in E_i - E_0, \qquad i = 1, \ldots, n.$$

Now if $\pi \geqq \pi_0$, $\pi = (F_1, \ldots, F_m)$, there is a null set F_0 depending on π such that each of the sets $F_i - F_0$ is contained in some one of the sets E_j. Thus for $s \notin E_0 \cup F_0$

$$|h_\pi(s) - h(s)| = \frac{1}{\mu(F_i)} \left| \int_{F_i} \{h(s) - h(t)\}\mu(dt) \right|$$

$$= \frac{1}{\mu(E_j \cap F_i)} \left| \int_{E_j \cap F_i} \{h(t) - h(s)\}\mu(dt) \right| < \varepsilon.$$

Q.E.D.

4 LEMMA. *If the compact Hausdorff space S has a countable base the space $C(S)$ is separable.*

PROOF. By Theorems I.6.19 and I.6.12, S is a separable metric space. If $\varrho(x, y)$ is a metric in S let $f_n(x) = \varrho(x, x_n)$ where $\{x_n\}$ is a denumerable dense set in S. By I.5.10 f_n is a bounded function and hence an element of $C(S)$. By IV.6.16 (or IV.6.17) the closed algebra generated by the denumerable set $\{f_n\}$ is all of $C(S)$. Thus $C(S)$ is separable. Q.E.D.

Let U be the closed unit sphere in $L_1(S, \Sigma, \mu)$. Then, by the preceding lemma, since $K = \overline{TU}$ is compact and has a countable base the space $C(K)$ is separable. Since a subspace of a separable metric space is separable (I.6.12) there is a sequence $\{x_i^*\} \subseteqq \mathfrak{X}^*$ such that for any x^* in $\mathfrak{X}^*$ and $\varepsilon > 0$

$$\text{(i)} \qquad |x^*(k) - x_j^*(k)| < \varepsilon, \qquad k \in K,$$

for some integer j.

Since we are assuming that $\mu(S) < \infty$, the set function $x^*T\chi_E$ is, for each x^* in $\mathfrak{X}^*$, μ-continuous and thus, by Theorem III.10.2, there is a function $\alpha(\cdot, x^*)$ in $L_1(S, \Sigma, \mu)$ such that

$$\text{(ii)} \qquad x^*T\chi_E = \int_E \alpha(t, x^*)\mu(dt), \qquad E \in \Sigma.$$

Since x^*T is a continuous linear functional on $L_1(S, \Sigma, \mu)$ we have

$$|x^*T\chi_E| \leqq |x^*T|\mu(E), \qquad E \in \Sigma$$

from which it follows that $\alpha(\cdot, x^*)$ is μ-essentially bounded. Thus $\int_S \alpha(t, x^*)f(t)\mu(dt)$ and x^*Tf are both defined and continuous for f in $L_1(S, \Sigma, \mu)$. Since they are equal, by (ii), for characteristic functions they must coincide everywhere on $L_1(S, \Sigma, \mu)$, i.e.,

$$\text{(iii)} \qquad x^*Tf = \int_S \alpha(t, x^*)f(t)\mu(dt), \qquad f \in L_1(S, \Sigma, \mu).$$

5 LEMMA. *There is a sequence $\{\pi_n\}$ of partitions of S into sets of positive measure and a null set E_0 such that*

$$\lim_{n\to\infty} x_i^* x_{\pi_n}(t) = \alpha(t, x_i^*), \qquad i = 1, 2, \ldots,$$

uniformly for t in $S - E_0$.

PROOF. We first observe that for any partition $\pi = \{E_1, \ldots, E_n\}$

$$\alpha_\pi(t, x^*) = \sum_{j=1}^{n} \left[\frac{1}{\mu(E_j)} \int_{E_j} \alpha(s, x^*)\mu(ds) \right] \chi_{E_j}(t)$$

$$= \sum_{j=1}^{n} \frac{x^*T(\chi_{E_j})}{\mu(E_j)} \chi_{E_j}(t) = x^*x_\pi(t).$$

The partition π_n will be chosen by induction. By Lemma 3 there is a partition π_1 and a set E_1 with $\mu(E_1) = 0$ such that

$$|x_1^*x_{\pi_1}(t) - \alpha(t, x_1^*)| < 1, \qquad t \in S - E_1.$$

Now suppose that for $i \leqq k$ the partition π_i and the set E_i with $\mu(E_i) = 0$ have been chosen in such a manner that

$$|x_i^*x_{\pi_k}(t) - \alpha(t, x_i^*)| < \frac{1}{k}, \quad 1 \leqq i \leqq k, \quad t \in S - E_k.$$

By applying Lemma 3 in turn to each of the functions $\alpha(t, x_i^*)$, $i = 1, \ldots, k+1$, with ε replaced by $1/(k+1)$ a partition $\pi_{k+1} \geqq \pi_k$ is obtained such that

$$|x_i^*x_{\pi_{k+1}}(t) - \alpha(t, x_i^*)| < \frac{1}{k+1}, \quad 1 \leqq i \leqq k+1, \quad t \in S - E_{k+1},$$

for some μ-null set E_{k+1}. The conclusion of the lemma then follows by letting $E_0 = \cup_{k=1}^{\infty} E_k$. Q.E.D.

Proof of Theorem 2. We note that for each t_0 in S the sequence $\{x_{\pi_n}(t_0)\}$ lies in K. Since K is a compact Hausdorff space with a countable base it is a metric space (I.6.19) and thus a subsequence converges to a vector $x(t_0)$ in K (I.6.13). Let $x(t)$ be arbitrarily defined for each t in S as the limit of such a subsequence. Then for $i = 1, 2, \ldots$, we have

$$\lim_{n\to\infty} x_i^*x_{\pi_n}(t) = \alpha(t, x_i^*) = x_i^*x(t), \qquad t \in S - E_0,$$

so that the functions $x_i^*x(\cdot)$ are bounded and measurable and

$$x_i^*Tf = \int_S x_i^*x(t)f(t)\mu(dt), \qquad i = 1, 2, \ldots, \qquad f \in L_1(S, \Sigma, \mu).$$

Since $x(S) \subseteqq K$ we see from (i) that for every x^* in $\mathfrak{X}^*$ and every $\varepsilon > 0$ there is an integer j such that

$$|x^*x(t) - x_j^*x(t)| < \varepsilon, \qquad t \in S.$$

Since $x^*x(\cdot)$ is the uniform limit of bounded measurable functions it is bounded and measurable. Moreover, if $|f| \leqq 1$, then

$$\begin{aligned} &|x^*Tf - \int_S x^*x(t)f(t)\mu(dt)| \\ &\leqq |x^*Tf - x_j^*Tf| + \int_S |x^*x(t) - x_j^*x(t)||f(t)|\mu(dt) \\ &< 2\varepsilon, \end{aligned}$$

which proves the theorem under the assumption that $\mu(S) < \infty$.

In case $S = \cup_{n=1}^{\infty} S_n$ where $\mu(S_n) < \infty$ and $S_n \subset S_{n+1}$, we may apply the result already proved to the space $L_1(S_n, \Sigma_n, \mu)$ where Σ_n consists of those sets in Σ which are subsets of S_n. The function $x(\cdot)$ on S to K is defined in the natural way as the limit of the sequence $x_n(\cdot)$ obtained for (S_n, Σ_n, μ). Q.E.D.

In the following we shall give several applications of Theorem 2. The first makes use of the compactness Theorem V.4.3 to give the general form of a mapping from $L_1(S, \Sigma, \mu)$ to the conjugate of a separable space.

6 THEOREM. *Let (S, Σ, μ) be a σ-finite positive measure space and let T be a continuous linear map of $L_1(S, \Sigma, \mu)$ into the conjugate $\mathfrak{X}^*$ of a separable B-space $\mathfrak{X}$. Then there is a μ-essentially unique function $x^*(\cdot)$ on S to $\mathfrak{X}^*$ such that $x^*(\cdot)x$ is μ-essentially bounded for each x in $\mathfrak{X}$ and*

$$[*] \qquad (Tf)x = \int_S x^*(s)xf(s)\mu(ds), \quad f \in L_1(S, \Sigma, \mu), \quad x \in \mathfrak{X}.$$

Moreover, $|T| = \operatorname{ess\,sup}_{s\in S} |x^*(s)|$. Conversely, if $x^*(\cdot)$ is any function on S to $\mathfrak{X}^*$ such that $x^*(\cdot)x$ is measurable for each x in $\mathfrak{X}$, and such that

$$\operatorname{ess\,sup}_{s\in S} |x^*(s)| = M < \infty,$$

then equation [*] defines a continuous linear map T of $L_1(S, \Sigma, \mu)$ into $\mathfrak{X}^*$ whose norm is M.

PROOF. If T is a bounded operator from $L_1(S, \Sigma, \mu)$ to $\mathfrak{X}^*$ and U is the unit sphere of $L_1(S, \Sigma, \mu)$, then $K = \overline{TU}$ is compact in the $\mathfrak{X}$ topology (V.4.3) and has a countable base in this topology (V.5.1). By Theorem 2 there is a map $s \to x^*(s)$ of S into $K \subset \mathfrak{X}^*$ such that [*] holds, and where $|x(s)| \leqq |T|$. On the other hand it follows from [*] that $|T| \leqq \operatorname{ess\,sup} |x^*(s)|$, so $|T| = \operatorname{ess\,sup} |x^*(s)|$. We now prove the μ-essential uniqueness of the function $x^*(s)$. If $x^*(s)x \in L_\infty(S, \Sigma, \mu)$ for each $x \in \mathfrak{X}$ and $\int x^*(s)xf(s)\mu(ds) = 0$ for each $x \in \mathfrak{X}$ and $f \in L_1(S, \Sigma, \mu)$, then by the uniqueness part of IV.8.5, there exists, for each $x \in \mathfrak{X}$, a measurable set E_x of measure zero such that $x^*(s)x = 0$ for $s \in S$, $s \notin E_x$. Let X_1 be a countable dense subset of $\mathfrak{X}$, and let $E = \bigcup_{x \in X_1} E_x$. Then it is clear that $x^*(s)x = 0$ for $x \in \mathfrak{X}$, $s \in S$, $s \notin E$. Hence $x^*(s) = 0$ for $x \notin E$.

Conversely, let $x^*(\cdot)$ be a map of S into $\mathfrak{X}^*$ such that ess sup $|x^*(s)| = M < \infty$ and $x^*(\cdot)x$ is measurable for each $x \in \mathfrak{X}$. Define T on L_1 by [*]. Since $|(Tf)x| \leqq M|f||x|$, it is evident that $Tf \in \mathfrak{X}^*$. Hence $T \in B(L_1(S, \Sigma, \mu), \mathfrak{X}^*)$. The statement concerning the norm of T follows from the first part of the theorem. Q.E.D.

7 COROLLARY. *Theorem 6 remains valid if the hypothesis that $\mathfrak{X}^*$ is the conjugate space of a separable space is replaced by the hypothesis that $\mathfrak{X}^*$ is the conjugate space of an arbitrary B-space and T has a separable range.*

The proof of the corollary will require, besides Theorem 6, the following lemma.

8 LEMMA. *Let $\mathfrak{X}^*$ be the conjugate space of a B-space $\mathfrak{X}$. Let $\mathfrak{M} \subseteqq \mathfrak{X}^*$ be a separable linear manifold. Then there exists a closed separable subspace $\mathfrak{Y}$ of $\mathfrak{X}$ such that $\mathfrak{M}$ is equivalent to a subspace of $\mathfrak{Y}^*$.*

PROOF. Let $\{x_n^*\}$ be a countable dense subset of $\mathfrak{M}$, and let $\{x_{mn}\}$, $m, n = 1, 2, \ldots$, belong to $\mathfrak{X}$ and satisfy the relations $|x_{mn}| = 1$ and

$$|x_n^*(x_{mn})| \geqq \left(1 - \frac{1}{m}\right) |x_n^*|.$$

If $\mathfrak{Y} = \overline{\text{sp}}\, \{x_{mn}\}$, then $\mathfrak{Y}$ is separable. The mapping $V : \mathfrak{X}^* \to \mathfrak{Y}^*$ defined by

$$Vx^*(y) = x^*(y), \qquad y \in \mathfrak{Y}, \quad x^* \in \mathfrak{X}^*$$

satisfies $|V| \leqq 1$. However, since

$$|Vx_n^*| \geqq \sup_m |x_n^*(x_{mn})| = |x_n^*|, \qquad n = 1, 2, \ldots,$$

the mapping V defines an isometric mapping of $\mathfrak{M}$ into $\mathfrak{Y}^*$. Q.E.D.

PROOF OF COROLLARY 7. Let $\mathfrak{M}$ be the closure of the range of T. By the lemma, $\mathfrak{M}$ is equivalent to a subspace of the conjugate $\mathfrak{Y}^*$ of a separable subspace $\mathfrak{Y}$ of $\mathfrak{X}$. By Theorem 6 there is a μ-essentially unique function $x^*(\cdot)$ on S to $\mathfrak{M}$ with $|T| = \text{ess sup } |x^*(s)|$ and such that for each x in $\mathfrak{Y}$ the function $x^*(\cdot)x$ is μ-measurable and

$$\text{(i)} \qquad \int_S x^*(s)xf(s)\mu(ds) = (Tf)x, \qquad f \in L_1(S, \Sigma, \mu).$$

To complete the proof of the corollary it must be shown that for an arbitrary x in $\mathfrak{X}$ the function $x^*(\cdot)x$ is μ-measurable and equation (i) holds. Consequently, let x_0 be a fixed element of $\mathfrak{X}$ and let $\mathfrak{Y}_0$ be a closed subspace spanned by x_0 and $\mathfrak{Y}$. Then, as before, there is a μ-essentially bounded function $x_0^*(\cdot)$ on S to $\mathfrak{M}$ such that $x_0^*(\cdot)x$ is μ-measurable for every x in $\mathfrak{Y}_0$ and for such x

$$\text{(ii)} \qquad \int_S x_0^*(s)xf(s)\mu(ds) = (Tf)x, \qquad f \in L_1(S, \Sigma, \mu).$$

Thus, since $\mathfrak{Y} \subseteq \mathfrak{Y}_0$, it is seen from (i) that

$$\int_E \{x_0^*(s) - x^*(s)\}x\mu(ds) = 0, \qquad E \in \Sigma$$

for x in $\mathfrak{Y}$. Thus there is a μ-null set E_x such that $x_0^*(s)x = x^*(s)x$ for x in $\mathfrak{Y}$ and $s \notin E_x$. Since $\mathfrak{Y}$ is separable there is a μ-null set E such that for all x in $\mathfrak{Y}$

$$\{x_0^*(s) - x^*(s)\}x = 0, \qquad s \notin E.$$

Since the only vector in $\mathfrak{M}$ which vanishes on $\mathfrak{Y}$ is the zero vector and since $x_0^*(s) - x^*(s)$ is in $\mathfrak{M}$, it follows from the preceding equation that $x_0^*(s) = x^*(s)$ for μ-almost all s in S. Equation (ii) holds for x_0 and therefore equation (i) does likewise. Q.E.D.

In the following theorem it is shown that if the range of an operator on $L_1(S, \Sigma, \mu)$ is another L_1 space we may give a specific scalar kernel representation of T rather than represent it by means of a vector valued integral. In the statement of this theorem the derivative notation $d\lambda/d\mu$ introduced in connection with the Radon-Nikodým theorem (III.10.7) is used. We recall that for a μ-continuous complex valued measure λ there is a unique μ-integrable function $d\lambda/d\mu$ with

$$\lambda(E) = \int_E \left(\frac{d\lambda}{d\mu}\right)(s)\mu(ds), \qquad E \in \Sigma.$$

9 Corollary. *Let (S, Σ, μ) be a σ-finite positive measure space and let λ be a positive regular measure on the Borel sets $\mathscr{B}$ in the compact*

Hausdorff space W. Let T be a continuous linear map of $L_1(S, \Sigma, \mu)$ into $L_1(W, \mathscr{B}, \lambda)$. It is assumed that either (a), *W is a metric space, or* (b), *T has a separable range. Then there is a scalar function K on $\mathscr{B} \times S$ with the following properties.*

(i) *For each s in S, $K(\cdot, s)$ is a regular measure on $\mathscr{B}$ with*

$$\operatorname{ess\,sup}_s v(K(\cdot, s), S) = M < \infty.$$

(ii) *For each E in $\mathscr{B}$, $K(E, \cdot)$ is a measurable function on S.*

(iii) *For each A in Σ with $\mu(A) < \infty$ the measure $\int_A K(E, s)\mu(ds)$ defined for E in $\mathscr{B}$ is regular and λ-continuous.*

$$\text{(iv)}\quad Tf = \frac{d}{d\lambda}\int_S K(\cdot, s)f(s)\mu(ds), \qquad f \in L_1(S, \Sigma, \mu).$$

(v) $|T| = M$.

Conversely, if the function K on $\mathscr{B} \times S$ satisfies (i), (ii), *and* (iii) *then* (iv) *defines a continuous linear map of $L_1(S, \Sigma, \mu)$ into $L_1(W, \mathscr{B}, \lambda)$ whose norm is given by* (v).

Proof. Let $\mathfrak{X} = C(W)$ so that $\mathfrak{X}^*$ is the space of regular measures on W (Theorem IV.6.3). The space $L_1(W, \mathscr{B}, \lambda)$ is isometrically isomorphic to the set of regular measures which are λ-continuous (cf. III.10.7 and III.2.20(a)). Thus $L_1(W, \mathscr{B}, \lambda)$ may be regarded as a subspace of $\mathfrak{X}^*$. If condition (a) is satisfied $\mathfrak{X} = C(W)$ is separable by Lemma 4 and we may apply Theorem 6 to obtain the function K. In the case of condition (b) we apply Corollary 7. Q.E.D.

We return now to the case where the range of T is in an arbitrary B-space. It will be shown that if T is weakly compact and has a separable range its integral representation may be accomplished by a vector valued kernel which has many properties not enjoyed by the kernels of Theorems 2, 6, and 7. In this case, the kernel is a bounded μ-measurable function $x(\cdot)$ and thus the vector function $x(\cdot)f(\cdot)$ is μ-integrable for every f in $L_1(S, \Sigma, \mu)$. The representation of T in this situation is

$$Tf = \int_S x(s)f(s)\mu(ds), \qquad f \in L_1(S, \Sigma, \mu),$$

which is easily stated without the use of functionals x^* in $\mathfrak{X}^*$. Since bounded sets in a reflexive space are weakly sequentially compact it may be noted that the next theorem applies to every continuous linear map of $L_1(S, \Sigma, \mu)$ into a separable reflexive space.

10 THEOREM. *Let (S, Σ, μ) be a σ-finite positive measure space, and let T be a weakly compact operator on $L_1(S, \Sigma, \mu)$ to a separable subset of the B-space $\mathfrak{X}$. Then there exists a μ-essentially unique bounded measurable function $x(\cdot)$ on S to a weakly compact subset of $\mathfrak{X}$ such that*

$$[*] \qquad Tf = \int_S x(s)f(s)\mu(ds), \qquad f \in L_1(S, \Sigma, \mu),$$

and $|T| = \text{ess sup } |x(s)|$.

Conversely, if $x(\cdot)$ is a measurable, μ-essentially bounded function on S with almost all of its values in a weakly compact subset of $\mathfrak{X}$, then equation [*] defines a weakly compact mapping of $L_1(S, \Sigma, \mu)$ into $\mathfrak{X}$ with separable range.

PROOF. Let U be the closed unit sphere of $L_1(S, \Sigma, \mu)$. If T is weakly compact and has a separable range, the existence of a function $x(\cdot)$ on S such that all the values $x(s)$ lie in the separable weakly compact set $\overline{T(U)}$ follows from V.6.3 and Theorem 2. By III.6.11, $x(s)$ is measurable, so that [*] holds. Since $x(s) \in \overline{TU}$, we have $\sup_s |x(s)| \leqq |T|$. On the other hand, [*] shows that $|T| \leqq \text{ess sup}_s |x(s)|$ so $|T| = \text{ess sup } |x(s)|$.

Under the hypotheses of the converse part of our theorem it is clear we can assume without loss of generality there is a weakly compact set K such that $x(s) \in K$ for $s \in S$. Let

$$K_1 = \overline{\text{co}}(K \cup -K \cup iK \cup -iK).$$

Then, by V.6.4, K_1 is weakly compact. Since it is easily seen that $Tf \in K_1$ whenever f is a finitely-valued function in $L_1(S, \Sigma, \mu)$ such that $|f| \leqq 1$, it follows readily that $TU \subseteqq K$. Thus T is weakly compact. By III.6.11, there exists a separable subspace $\mathfrak{X}_1$ of $\mathfrak{X}$ and a null set $F \in \Sigma$ such that $x(s) \in \mathfrak{X}_1$ if $s \in S - F$. Since $Tf \in \mathfrak{X}_1$ for finitely valued functions f, T has a separable range. Finally, the uniqueness of $x(\cdot)$ follows immediately from Lemma III.6.8. Q.E.D.

Since every compact operator is weakly compact and has a separable range, Theorem 10 yields the following corollary.

11 COROLLARY. *The operator T of Theorem 10 is compact if and*

only if there is a μ-null set E and a compact set K in $\mathfrak{X}$ with $x(s)$ in K for every s in the complement of E.

PROOF. Let U be the unit sphere of $L_1(S, \Sigma, \mu)$. Then if T is compact, $K = TU$ is a compact set in the separable space $\overline{T(L_1)}$, and it follows from the proof of Theorem 10 that $x(s) \in TU$ for $s \in S$.

Conversely, let K be compact and let E be a subset of S of measure zero such that $x(s) \in K$ for $s \notin E$. It follows from the proof of Theorem 10 that

$$x(s) \in \overline{TU} \subseteqq \overline{\text{co}}(K \cup -K \cup iK \cup -iK).$$

Hence, it follows from V.2.6 that T is compact. Q.E.D.

Remark. The following statement follows readily from Theorem 10 if we use Theorem III.11.17.

Let (S, Σ, μ) and $S_1, \Sigma_1, \mu_1)$ be σ-finite positive measure spaces. Let T be a weakly compact mapping of $L_1(S, \Sigma, \mu)$ into a separable subspace of $L_1(S_1, \Sigma_1, \mu_1)$. Then there exists a $\mu \times \mu_1$-essentially unique, $\mu \times \mu_1$-measurable function K defined on $S \times S_1$, such that

$$\mu\text{-ess sup} \int_{S_1} |K(s, s_1)| \mu_1(ds_1) < \infty$$

and

$$(Tf)(s_1) = \int_S K(s, s_1) f(s) \mu(ds), \qquad f \in L_1(S, \Sigma, \mu).$$

Moreover, we have

$$\mu\text{-ess sup} \int_{S_1} |K(s, s_1)| \mu_1(ds_1) = |T|.$$

On the basis of Theorem 7.4 it was observed at the end of the preceding section that a weakly compact operator on any of the spaces $B(S)$, $B(S, \Sigma)$, AP, $C(S)$, or $L_\infty(S, \Sigma, \mu)$ maps every weak Cauchy sequence into a strongly convergent sequence. The next theorem shows that the Lebesgue spaces $L_1(S, \Sigma, \mu)$ also have this property.

12 THEOREM. *Let (S, Σ, μ) be a positive measure space and let T be a weakly compact linear map of $L_1(S, \Sigma, \mu)$ into a B-space. Then T maps weak Cauchy sequences into strongly convergent sequences. Consequently, T maps conditionally weakly compact sets into conditionally strongly compact sets.*

PROOF. Let $\{f_n\}$ be a weak Cauchy sequence in $L_1(S, \Sigma, \mu)$. By Lemma III.8.5 there is a set $S_1 \in \Sigma$ and a σ-field $\Sigma_1 \subseteqq \Sigma(S_1)$ such that the restriction μ_1 of μ to Σ_1 has the property that the measure space (S_1, Σ_1, μ_1) is σ-finite and the space $L_1(S_1, \Sigma_1, \mu_1)$ is a separable subspace of $L_1(S, \Sigma, \mu)$ which contains the sequence $\{f_n\}$. Thus, by Theorem 10, there is μ_1-measurable, μ_1-essentially bounded function $x(\cdot)$ on S_1 to the separable closed linear manifold $\mathfrak{X}_1$ spanned by the set $TL_1(S_1, \Sigma_1, \mu_1)$ and such that

$$Tf_n = \int_{S_1} x(s) f_n(s) \mu_1(ds).$$

Let $\{z_k\}$ be dense in $\mathfrak{X}_1$ and let

$$B_1 = S(z_1, \varepsilon), \qquad B_k = S(z_k, \varepsilon) - \bigcup_{i=1}^{k-1} B_i, \qquad k > 1.$$

Then the sets $A_k = x^{-1}(B_k)$ are disjoint and the function $x_\varepsilon(\cdot)$ defined by the equation $x_\varepsilon(s) = z_k$ for x in A_k has the property that

$$\text{(i)} \qquad |x_\varepsilon(s) - x(s)| \leq \varepsilon, \qquad s \in S_1.$$

For each $k = 1, 2, \ldots$, we let $x_k(s) = x_\varepsilon(s)$ for s in $A_1 \cup \ldots \cup A_k$, and $x_k(s) = 0$ elsewhere. By Theorem III.6.10 the sets A_k are in the Lebesgue extension Σ_1^* of Σ_1. We may and shall assume that $\Sigma_1^* = \Sigma_1$ so that the functions x_ε and x_k are μ_1-measurable. Now

$$\text{(ii)} \qquad \begin{aligned} &\left| \int_{S_1} (x_k(s) - x_\varepsilon(s)) f_n(s) \mu_1(ds) \right| \\ &= \left| \int_{C_{k+1}} x_\varepsilon(s) f_n(s) \mu_1(ds) \right| \\ &\leq M \int_{C_{k+1}} |f_n(s)| \mu_1(ds), \end{aligned}$$

where $M = \text{ess sup } |x(s)|$ and $C_k = \bigcup_{i=k}^{\infty} A_i$. Since $L_1(S_1, \Sigma_1, \mu_1)$ is weakly complete (IV.8.6) the sequence $\{f_n\}$ is weakly compact and thus the sequence $\{|f_n(\cdot)|\}$ is also weakly compact (IV.8.9). Hence Theorem IV.8.8 shows that

$$\lim_{k \to \infty} \int_{C_{k+1}} |f_n(s)| \mu_1(ds) = 0$$

uniformly with respect to $n = 1, 2, \ldots$. From (ii) it follows that

$$\text{(iii)} \qquad \lim_{k \to \infty} \int_{S_1} x_k(s) f_n(s) \mu_1(ds) = \int_{S_1} x_\varepsilon(s) f_n(s) \mu_1(ds),$$

uniformly in $n = 1, 2, \ldots$. Since $\{f_n\}$ is weakly convergent, the limit $\lim_n \int_E f_n(s) \mu_1(ds)$ exists for each E in Σ_1 and thus for the simple functions x_k the limits

$$\lim_n \int_{S_1} x_k(s) f_n(s) \mu_1(ds), \qquad k = 1, 2, \ldots$$

exist. Since the limit in (iii) is uniform in $n = 1, 2, \ldots$, Lemma I.7.6 shows that the limit

$$\lim_{n \to \infty} \int_{S_1} x_\varepsilon(s) f_n(s) \mu_1(ds), \qquad \varepsilon > 0$$

exists. Since $\{f_n\}$ is bounded (II.3.27) the inequality (i) shows that

$$\lim_{\varepsilon \to 0} \int_{S_1} x_\varepsilon(s) f_n(s) \mu_1(ds) = \int_S x(s) f_n(s) \mu_1(ds)$$

uniformly with respect to $n = 1, 2, \ldots$. Thus Lemma I.7.6 may be applied again to yield the existence of the limit

$$\lim_n \int_{S_1} x(s) f_n(s) \mu_1(ds) = \lim_n T f_n. \qquad \text{Q.E.D.}$$

13 Corollary. *Let (S, Σ, μ) be a positive measure space. Then the product of two weakly compact operators in $L_1(S, \Sigma, \mu)$ is compact. In particular, the square of a weakly compact operator in $L_1(S, \Sigma, \mu)$ is compact.*

It should be observed that the only place in the preceding proof where the hypothesis of weak compactness is used is to show that the operator T has the form $\int_S x(s) f(s) \mu(ds)$ where $x(\cdot)$ is μ-measurable and bounded. The fact that the kernel $x(\cdot)$ representing a weakly compact operator has its values in a weakly compact set was not used. Thus the preceding proof has also demonstrated the following theorem.

14 Theorem. *Let (S, Σ, μ) be a positive measure space and let x be a μ-measurable bounded function on S to the B-space $\mathfrak{X}$. Then the operator*

$$Tf = \int_S x(s) f(s) \mu(ds), \qquad f \in L_1(S, \Sigma, \mu),$$

maps weak Cauchy sequences into strongly convergent sequences.

It has been mentioned in Section IV.16 that the abstract L-spaces are equivalent to the concrete L-spaces and so Theorem 12 and Corollary 13 remain valid for these spaces. The equivalence needed to prove this statement for each of the L-spaces on the list in Section IV.2 may readily be made on the basis of the analysis presented in

Chapter IV or indicated in the exercises of Section IV.13. Thus Theorem 12 and Corollary 13 are true if the space L_1 is replaced by any one of the spaces l_1, bv, bv_0, $ba(S, \Sigma)$, $rba(S, \Sigma)$, $ca(S, \Sigma)$, $rca(S, \Sigma)$, $BV(I)$, or $AC(I)$. These spaces together with those spaces which are equivalent modulo a finite dimensional space to spaces $C(S)$ of continuous functions, and for which it is known that Theorem 12 is true (7.4), practically exhaust the list in Section IV.2. In fact the only remaining B-space on the list besides the reflexive spaces $\mathfrak{H}$, l_p, L_p, $1 < p < \infty$, is the space $A(D)$ of analytic functions.

9. Exercises

1 The space $B(\mathfrak{X}, \mathfrak{Y})$ is algebraically isomorphic to a subspace of the product $P_{x \in \mathfrak{X}} \mathfrak{Y}_x$, where $\mathfrak{Y}_x = \mathfrak{Y}$, under the mapping $T \to P_{x \in \mathfrak{X}} Tx$. Show that the strong topology of operators in $B(\mathfrak{X}, \mathfrak{Y})$ is identical with the usual product topology where the strong topology is taken in each $\mathfrak{Y}_x$, and that the weak operator topology is that where each factor is taken to have the $\mathfrak{Y}^*$ topology.

2 A set $A \subset B(\mathfrak{X}, \mathfrak{Y})$ is compact in the weak operator topology if and only if it is closed in the weak operator topology and the weak closure of Ax is weakly compact for each $x \in \mathfrak{X}$. A set $A \subset B(\mathfrak{X}, \mathfrak{Y})$ is compact in the strong operator topology if and only if it is closed in the strong operator topology and Ax is conditionally compact for each $x \in \mathfrak{X}$.

3 Let $A \subset B(\mathfrak{X}, \mathfrak{Y})$. Then if A is compact in the weak operator topology, so is the weak operator closure of $\text{co}(A)$. If A is compact in the strong operator topology, so is the strong operator closure of $\text{co}(A)$.

4 If a set $A \subset B(\mathfrak{X}, \mathfrak{Y})$ is sequentially compact in the weak operator topology, its weak operator closure is compact in the weak operator topology. If A is sequentially compact in the strong operator topology, then its strong operator closure is compact in the strong operator topology.

5 If U is a subset of a topological space, let $\tilde{U}$ be the set of all sequential limit points of U. If $\mathfrak{X}$ is separable, a set $A \subseteq B(\mathfrak{X}, \mathfrak{Y})$ is sequentially compact in the strong operator topology if and only if $\tilde{A}$

is compact in the strong operator topology. If $\mathfrak{Y}$ is also separable, A is sequentially compact in the weak operator topology if and only if $\tilde{A}$ is compact in the weak operator topology.

6 If $\mathfrak{Y}$ is reflexive, then the closed unit sphere of $B(\mathfrak{X}, \mathfrak{Y})$ is compact in the weak operator topology. Conversely, if the closed unit sphere of $B(\mathfrak{X}, \mathfrak{Y})$ is compact in the weak operator topology, $\mathfrak{Y}$ is reflexive.

7 Define the BWO topology for $B(\mathfrak{X}, \mathfrak{Y})$ to be the strongest topology which coincides with the weak topology on every positive multiple aS of the closed unit sphere S of $B(\mathfrak{X}, \mathfrak{Y})$. Show by the method of proof of V.5.4 that if $\mathfrak{Y}$ is reflexive then a fundamental set of neighborhoods of zero in the BWO topology is given by the sets

$$\{T | T \in B(\mathfrak{X}, \mathfrak{Y}), |y_i^* T x_i| < \varepsilon, \quad i = 1, 2, \ldots\},$$

where $y_i^* \in \mathfrak{Y}^*$, and $y_i^* \to 0$; $x_i \in \mathfrak{X}$, and $x_i \to 0$.

8 Let $\mathfrak{Y}$ be reflexive. Show that a continuous linear functional on $B(\mathfrak{X}, \mathfrak{Y})$ which we take with its BWO topology must have the form $T \to \sum_{i=0}^{\infty} \alpha_i y_i^* T x_i$, where $\{\alpha_i\}$ is an absolutely convergent series of scalars, $\{y_i^*\}$ is a bounded sequence in $\mathfrak{Y}^*$, and $\{x_i\}$ a bounded sequence in $\mathfrak{X}$. Conversely, show that a functional on $B(\mathfrak{X}, \mathfrak{Y})$ with the given form is continuous in the BWO topology.

9 Define the BSO topology of $B(\mathfrak{X}, \mathfrak{Y})$ to be the strongest topology which coincides with the strong operator topology on each positive multiple aS of the unit sphere S of $B(\mathfrak{X}, \mathfrak{Y})$. Show that a convex subset of $B(\mathfrak{X}, \mathfrak{Y})$ is BWO-closed if and only if it is BSO-closed and that a linear functional is BSO-continuous if and only if it is BWO-continuous.

10 Show that every BWO-continuous linear function on $B(\mathfrak{X},\mathfrak{Y})$ is continuous in the weak operator topology if and only if either $\mathfrak{X}$ or $\mathfrak{Y}$ is finite dimensional.

11 Let $A, B \in B(\mathfrak{X})$. Then the mapping $(A, B) \to AB$ is a continuous function in each variable separately when $B(\mathfrak{X})$ is taken to have the uniform, strong or weak topologies. This mapping is continuous in both variables with the uniform topology, and also in the strong topology provided that A is restricted to a bounded set in $B(\mathfrak{X})$. If $\mathfrak{X} = l_2$ and $\{A_n\}$ is defined by $A_n[x_1, \ldots, x_n, \ldots] = [x_{n+1}, \ldots]$, show that both $\{A_n\}$ and $\{A_n^*\}$ converge to zero in the weak operator

topology, but that $\{A_n A_n^*\}$ does not converge to zero in this topology even though $\{A_n^* A_n\}$ converges to zero in the strong operator topology.

12 If $\mathfrak{H}$ is a Hilbert space, the mapping $T \to T^*$ of $B(\mathfrak{H})$ into itself is continuous with either the uniform or weak operator topology. By considering the sequence $\{A_n\}$ defined in Exercise 11, show that this mapping is not continuous in the strong operator topology.

13 If $U : \mathfrak{Y}^* \to \mathfrak{X}^*$ is a linear mapping which is continuous with the $\mathfrak{Y}$ topology in $\mathfrak{Y}^*$ and the $\mathfrak{X}$ topology in $\mathfrak{X}^*$, then there exists a bounded linear operator $T : \mathfrak{X} \to \mathfrak{Y}$ such that $T^* = U$.

14 Let T be a linear, but not necessarily continuous, mapping between B-space $\mathfrak{X}$ and $\mathfrak{Y}$. Let T^* be defined on the set $\mathfrak{D}(T^*)$ of those $y^* \in \mathfrak{Y}^*$ for which y^*Tx is continuous in x by setting $T^*y^* = y^*T$. Then the following statements are equivalent.

(i) T^* is defined on all of $\mathfrak{Y}^*$.

(ii) T is continuous with the strong topologies in $\mathfrak{X}$ and $\mathfrak{Y}$.

(iii) T is continuous with the $\mathfrak{X}^*$ and $\mathfrak{Y}^*$ topologies in $\mathfrak{X}$ and $\mathfrak{Y}$.

15 Let $\mathfrak{X}$, $\mathfrak{Y}$ be B-spaces, and U a continuous linear operator from $\mathfrak{X}$ to $\mathfrak{Y}$.

(i) If U has a continuous inverse, the range of U is closed.

(ii) The range of U is closed if there exists a constant K such that for any y in the range there exists a solution of $y = Ux$ such that $|x| \leq K|y|$.

(iii) U is one-to-one if the range of U^* is dense in $\mathfrak{X}^*$.

(iv) U^* is one-to-one if and only if the range of U is dense in $\mathfrak{Y}$.

(v) If U maps onto $\mathfrak{Y}$, then U^* has a continuous inverse.

(vi) If U^* maps onto $\mathfrak{X}^*$, then U has a continuous inverse.

16 If $\mathfrak{Y}$ is a closed subspace of a B-space and $\mathfrak{N}$ is a finite dimensional subspace, then $\mathfrak{Y} \oplus \mathfrak{N}$ is a closed subspace. If $\mathfrak{Y} \oplus \mathfrak{N}$ is a closed subspace, and $\mathfrak{N}$ is finite dimensional, it does not follow that $\mathfrak{Y}$ is closed.

17 Let $\mathfrak{X}$ be a B-space and suppose that $\mathfrak{X} = \mathfrak{Y} \oplus \mathfrak{N}$, where $\mathfrak{Y}$ is closed and $\mathfrak{N}$ is finite dimensional. Let $T : \mathfrak{X} \to \mathfrak{Z}$ be a bounded linear operator mapping into a second B-space. Then $T(\mathfrak{X})$ is closed if and only if $T(\mathfrak{Y})$ is closed.

18 Every non-null, finite dimensional, proper subspace of a B-space has infinitely many projections mapping onto it.

19 If E is a projection with n dimensional range, then E^* is a projection with n dimensional range.

20 A projection has finite dimensional range if and only if it is compact.

21 A linear mapping E such that $E^2 = E$ is a projection (i.e., is bounded) if and only if the ranges of E and $I-E$ are closed.

22 Let E be a projection in the B-space $\mathfrak{X}$, and let $\mathfrak{M} = E(\mathfrak{X})$ and $\mathfrak{N} = (I-E)(\mathfrak{X})$. Let $\mathfrak{M}^\perp = \{x^* | x^*x = 0,\ x \in \mathfrak{M}\}$ and similarly for $\mathfrak{N}^\perp$. Show that $\mathfrak{X}^* = \mathfrak{M}^\perp \oplus \mathfrak{N}^\perp$ and that $E^*(\mathfrak{X}^*) = \mathfrak{N}^\perp$, $(I-E)^*(\mathfrak{X}^*) = \mathfrak{M}^\perp$.

23 The letters $\mathfrak{M}$ and $\mathfrak{N}$ have the same meaning for a projection as in the preceding exercise.

(i) If E_1 and E_2 are projections, $E_1E_2 = E_2$ if and only if $\mathfrak{M}_2 \subseteq \mathfrak{M}_1$. Also $E_2E_1 = E_2$ if and only if $\mathfrak{N}_1 \subseteq \mathfrak{N}_2$.

(ii) If the E_i are perpendicular projections in Hilbert space, then $E_1E_2 = 0$ if and only if $E_2E_1 = 0$.

(iii) If $E_1, \ldots, E_n$ are projections, then $E = E_1 + \ldots + E_n$ is a projection if $E_iE_j = 0$ for $i \neq j$. In this case $\mathfrak{M} = \mathfrak{M}_1 \oplus \ldots \oplus \mathfrak{M}_n$ and $\mathfrak{N} = \mathfrak{N}_1 \cap \ldots \cap \mathfrak{N}_n$.

(iv) If the projections in (iii) are perpendicular projections in a Hilbert space, E is a projection if and only if $E_iE_j = 0$ for $i \neq j$.

24 (i) $E = E_1 - E_2$ is a projection if and only if $E_1E_2 = E_2E_1 = E_2$. In this case $\mathfrak{M} = \mathfrak{M}_1 \cap \mathfrak{N}_2$ and $\mathfrak{N} = \mathfrak{N}_1 \oplus \mathfrak{M}_2$.

(ii) E_1E_2 is a projection if and only if $E_1(\mathfrak{M}_2) \subseteq \mathfrak{M}_2 \oplus \mathfrak{N}_1 \cap \mathfrak{N}_2$.

(iii) If $E = E_1E_2 = E_2E_1$, then E is a projection and $\mathfrak{M} = \mathfrak{M}_1 \cap \mathfrak{M}_2$, $\mathfrak{N} = \text{sp}\,\{\mathfrak{N}_1 \cup \mathfrak{N}_2\}$, which is therefore closed.

(iv) $E_1E_2 = E_2E_1$ if and only if $\mathfrak{M}_2 = \mathfrak{M}_2 \cap \mathfrak{M}_1 \oplus \mathfrak{M}_2 \cap \mathfrak{N}_1$ and $\mathfrak{N}_2 = \mathfrak{N}_2 \cap \mathfrak{M}_1 \oplus \mathfrak{N}_2 \cap \mathfrak{N}_1$.

(v) If $E_1E_2 = E_2E_1$, then $E_1 + E_2 - E_1E_2$ is a projection with range $\text{sp}\,\{\mathfrak{N}_1 \cup \mathfrak{N}_2\}$ and null manifold $\mathfrak{N}_1 \cap \mathfrak{N}_2$.

25 A weakly compact projection in a space $C(S)$ or a space $B(S)$ has a finite dimensional range.

26 Show that a linear operator T in a finite dimensional space may be represented as a matrix. In terms of this matrix, what is the representation of the adjoint T^*?

27 In this exercise let $\mathfrak{X}$ be n-dimensional Euclidean space and let $e_1, \ldots, e_n$ be the basis of coordinate vectors.

(i) If E is a perpendicular projection, show that its matrix representation is $(a_{ij}) = (\sum_{k=1}^{r} \bar{p}_{ik} p_{jk})$, where $g_k = \sum_{i=1}^{n} p_{ik} e_k$, $k = 1, \ldots, r$ is any orthonormal basis of $\mathfrak{M} = E(\mathfrak{X})$.

(ii) Let $\mathfrak{M}$ and $\mathfrak{N}$ have bases $\{a_1, \ldots, a_r\}$ and $\{b_1, \ldots, b_{n-r}\}$, respectively. Let T be the transformation defined by

$$Te_i = a_i, \qquad i = 1, \ldots, r;$$

$$Te_{i+r} = b_i, \qquad i = 1, \ldots, n-r.$$

If $T = (t_{ij})$ and $T^{-1} = (s_{ij})$, show that $E = (\sum_{k=1}^{r} s_{ik} t_{kj})$.

28 Define the trace of a matrix $A = (a_{ij})$ by $\text{tr}(A) = \sum_{i=1}^{n} a_{ii}$. Show that $\text{tr}(A+B) = \text{tr}(A)+\text{tr}(B)$ and $\text{tr}(AB) = \text{tr}(BA)$. If E is a projection in Euclidean space, show that $\text{tr}(E) = \dim(\mathfrak{M})$.

29 Let $E_1, \ldots, E_p$ be projections in a finite dimensional Euclidean space. If $E = E_1 + \ldots + E_p$ is a projection, show that the manifolds $\{\mathfrak{M}_i\}$ are linearly independent, that $\mathfrak{M} = \mathfrak{M}_1 \oplus \ldots \oplus \mathfrak{M}_p$, and that $E_i E_j = 0$ for $i \neq j$. (Hint: Use Exercise 28.)

30 If $T \in B(\mathfrak{X}, \mathfrak{Y})$ is compact then it maps every weakly convergent sequence onto a strongly convergent sequence in $\mathfrak{Y}$. If $\mathfrak{X}$ is reflexive, this property implies that T is a compact operator.

31 Let $T \in B(\mathfrak{X})$ be compact and $\lambda \neq 0$. Then the equation $(\lambda I - T)x = y$ has a unique solution for every y in $\mathfrak{Y}$ if and only if the equation $(\lambda I - T)x = 0$ has no solution other than $x = 0$.

32 If $\mathfrak{Y}$ has a basis, every compact operator in $B(\mathfrak{X}, \mathfrak{Y})$ is the limit in the uniform operator topology of a sequence of operators with finite dimensional ranges.

33 If (S, Σ, μ) is a positive measure space, weakly compact projections in $L_1(S, \Sigma, \mu)$ have finite dimensional ranges.

34 A continuous linear mapping from a reflexive space to l_1 is compact.

35 A continuous linear mapping from c to a weakly complete space is compact.

36 Let ϕ be a function mapping a set S into itself. If T is the bounded linear operator in $B(S)$ defined by $Tf(s) = f(\phi(s))$, represent the adjoint T^*.

37 Let ϕ be a continuous function defined on a compact Hausdorff space S to S. Let T be the bounded linear operator in $C(S)$ defined by $Tf(s) = f(\phi(s))$. Represent the adjoint T^*.

38 (Markov) Let S be a non-void set and ϕ a function on S to S. A function μ defined on the family of subsets of S is said to be ϕ-invariant in case $\mu(E) = \mu(\phi^{-1}E)$, $E \subset S$, where $\phi^{-1}E = [s|\phi s \in E]$. Show that there is a non-negative bounded additive function μ defined for all subsets of S which is not identically zero and is ϕ-invariant.

39 Let S be a compact Hausdorff space and ϕ a continuous function on S to S. Show that there is a regular countably additive non-negative measure μ defined for all Borel sets in S with the properties that μ is not identically zero and μ is ϕ-invariant.

40 Let S be a non-void set and G a family of functions ϕ on S to S. Suppose that $\phi_1(\phi_2(s)) = \phi_2(\phi_1(s))$, $s \in S$, $\phi_1, \phi_2 \in G$. Show that there is a non-negative additive set function μ defined for all subsets of S and with the properties that μ is not identically zero and μ is ϕ-invariant for every $\phi \in G$.

41 Let S be a compact Hausdorff space and G a family of continuous functions ϕ on S to S with $\phi_1\phi_2 s = \phi_2\phi_1 s$, $s \in S$, $\phi_1, \phi_2 \in G$. Show that there is a countably additive non-negative measure defined on the Borel sets of S which is not identically zero and is ϕ-invariant for every $\phi \in G$.

42 Show that in Exercise 38 the set function μ is unique up to a positive constant factor if and only if $n^{-1} \sum_{i=0}^{n-1} f(\phi^i(s))$ converges uniformly to a constant for each $f \in B(S)$.

43 Show that in Exercise 39 the measure μ is unique up to a positive constant factor if and only if $n^{-1} \sum_{i=0}^{n-1} f(\phi^i(s))$ converges uniformly to a constant for each $f \in C(S)$.

44 Let S be a compact metric space, and $\phi : S \to S$ a mapping such that $\varrho(\phi(x), \phi(y)) \leqq \varrho(x, y)$. Suppose that there exists an s_0 in S such that the set $\{\phi^n s_0 | n \geqq 0\}$, is dense. Show that there exists a unique regular measure μ defined for the Borel subsets of S such that $\mu(\phi^{-1}(E)) = \mu(E)$ for each Borel subset of S, and such that $\mu(S) = 1$. (Hint. Use Exercise 43 and the fact that only constant functions can satisfy the equation $f(\phi(s)) = f(s)$.)

45 Let S be a compact Hausdorff space, K be a continuous function on $S \times S$, and $\mu \in rca(S)$. Define the operator $g = Tf$ by

$$g(s) = \int_S K(s, t) f(t) \mu(dt).$$

Show that T is a compact operator on $C(S)$ to $C(S)$ and that its adjoint is represented by the formula

$$T^*x^*(E) = \int_E \left[\int_S K(s, t)x^*(ds) \right] \mu(dt),$$

for every Borel set E.

46 Let $C = C[0, 1]$ and $T \in B(C, C)$. Then there is a scalar function K on $[0, 1] \times [0, 1]$ such that

(i) For each $s \in [0, 1]$ the function $K(s, \cdot)$ is a normalized function of bounded variation on $[0, 1]$;

(ii) $T(f, s) = \int_0^1 f(t)K(s, dt), \quad 0 \leqq s \leqq 1, \quad f \in C$;

(iii) $K(s, 1)$ and $\int_0^r K(s, t)dt$ are continuous in s for each $r \in [0, 1]$;

(iv) $\sup_s \operatorname{var}_{0 \leqq t \leqq 1} K(s, t) = M < \infty$;

(v) $M = |T|$.

Conversely if K is a function satisfying (i), (iii), and (iv) then (ii) defines an operator $T \in B(C, C)$ whose norm is given by (v).

47 State and prove theorems similar to that of the preceding exercise for the representation of the general linear operators in $B(L_p, C)$, $B(B(S), C)$, $B(c, C)$, and $B(C, c)$ where $C = C[0, 1]$ and $L_p = L_p[0, 1]$.

48 Represent the general compact operator in $B(L_p, C)$, $B(B(S), C)$, $B(c, C)$ where $C = C[0, 1]$ and $L_p = L_p[0, 1]$.

49 Show that the adjoint of the operator T in Exercise 46 is given by the formula

$$T^*(g, t) = \int_0^1 K(s, t)dg(s), \qquad g \in NBV[0, 1].$$

50 Let $1 \leqq p, q < \infty$ and let $T \in B(L_p[0, 1], L_q[0, 1])$. Show that there is a function K on $[0, 1] \times [0, 1]$ such that

$$T(f, s) = \frac{d}{ds} \int_0^1 K(s, t)f(t)dt, \qquad f \in L_p[0, 1].$$

51 Let (S, Σ, μ) be a positive measure space, and K a measurable function on S to the B-space $\mathfrak{X}$. Let

$$\left(\int_S |K(s)|^{p'} \mu(ds) \right)^{1/p'} = M < \infty,$$

where $1 < p < \infty$ and $(1/p) + (1/p') = 1$. Then the operator T on $L_p(S, \Sigma, \mu)$ to $\mathfrak{X}$ defined by

$$Tf = \int_S K(s)f(s)ds$$

is compact and has norm at most M.

52 Let $1 < p < \infty$, $(1/p)+(1/p') = 1$. Let (S, Σ, μ) be a positive measure space and K a scalar valued measurable function on $S \times S$ which satisfies the condition

$$\left[\int_S \left(\int_S |K(s, t)|^p \mu(ds)\right)^{p'/p} \mu(dt)\right]^{1/p'} = M < \infty.$$

Let $g = Tf$ be defined by

$$g(s) = \int_S K(s, t)f(t)dt.$$

Then T is a compact operator in $L_p(S, \Sigma, \mu)$ whose norm is at most M.

53 Let $1 < p < \infty$, $(1/p)+(1/p') = 1$, and let (S, Σ, μ) be a positive finite measure space. Let K be a measurable function on $S \times S$ with

$$\sup_s \int_S |K(s, t)|^{p'} dt < \infty.$$

Then the operator $g = Tf$ defined by

$$g(s) = \int_S K(s, t)f(t)dt$$

is a compact operator in $L_p(S, \Sigma, \mu)$.

54 Let (S, Σ, μ) be a positive measure space, $L_p = L_p(S, \Sigma, \mu)$, $1 \leqq p \leqq \infty$, and K a measurable function on $S \times S$ with

$$\operatorname*{ess\,sup}_s \int_S |K(s, t)|\mu(dt) \leqq M,$$

$$\operatorname*{ess\,sup}_t \int_S |K(s, t)|\mu(ds) \leqq M.$$

Then the operator $g = Tf$ defined by

$$g(s) = \int_S K(s, t)f(t)dt$$

is a continuous linear map of L_p into L_p and $|T| \leqq M$.

55 Let K be a Lebesgue measurable scalar function with period 2π and

$$|K| = \int_0^{2\pi} |K(s)|ds < \infty.$$

Then for $1 \leqq p < \infty$ the operator $g = Tf$ defined by

$$g(s) = \int_0^{2\pi} K(s-t)f(t)dt$$

is a compact linear operator in $L_p(0, 2\pi)$ whose norm is at most $|K|$.

56 Let (S, Σ, μ) be a finite measure space whose space S is a compact Hausdorff space. If $1 < p < \infty$, show that any continuous linear operator from $L_p(S, \Sigma, \mu)$ into $C(S)$ is a compact operator when it is regarded as mapping into $L_p(S, \Sigma, \mu)$.

57 Let (S, Σ, μ) be a positive finite measure space and let $K(s, t)$ be bounded and measurable on $S \times S$. Let T in $B(L_1(S, \Sigma, \mu))$ be defined by

$$T(f, s) = \int_S K(s, t) f(t) \mu(dt).$$

Show that T^* in $L_\infty(S, \Sigma, \mu)$ is given by

$$T^*(g, t) = \int_S g(s) K(s, t) \mu(ds).$$

Show that the operator T, when regarded as acting in $L_\infty(S, \Sigma, \mu)$, is weakly compact and that its square is compact.

58 Let (S, Σ, μ) be a positive finite measure space and K an essentially bounded measurable function on $S \times S$. Let ν be a bounded additive function on Σ which vanishes on sets of μ-measure zero. Show that

$$\lambda(E) = \int_S \left[\int_E K(s, t) \mu(ds) \right] \nu(dt)$$

is countably additive for $E \in \Sigma$. If the measure space is σ-finite instead of finite, what restrictions on K are equivalent to the countable additivity of λ?

59 Let (S, Σ, μ) and (S', Σ', μ') be positive σ-finite measure spaces. Let $L = L(S, \Sigma, \mu)$ and $L_p = L_p(S', \Sigma', \mu')$. If $p > 1$ and T is a separable bounded linear map of L into L_p there is a scalar function K on $S \times S'$ measurable with respect to $\Sigma \times \Sigma'$ and such that

(i) $T(f, s) = \int_S K(s, t) f(s) \mu(ds), \qquad f \in L;$

(ii) $\operatorname{ess\,sup}_s \left(\int_{S'} |K(s, t)|^p \mu'(ds) \right)^{1/p} = M < \infty;$

(iii) $|T| = M.$

Conversely, if K is a measurable function satisfying (ii) then (i) defines an operator $T \in B(L, L_p)$ whose norm is given by (iii).

60 Let L be the space of Lebesgue integrable functions on $[0, 1]$. Represent the general linear operators in $B(L, L)$, $B(L, l_1)$, $B(l_1, L)$, and $B(l_1, l_1)$. In each case give the norm of the operator in terms of the kernel.

10. The Riesz Convexity Theorem

In this section we prove the deep and important convexity theorem of M. Riesz by the method due to G. O. Thorin. The proof is a particularly beautiful instance of the application of complex variable theory to a seemingly unrelated problem in the theory of linear spaces.

Throughout this section $\mathfrak{B}$ denotes a *complex* B-space, and E^k denotes k-dimensional *real* Euclidean space. The reader should observe that the same letter M is repeatedly used for different functions, each explicitly defined in the statement of a given theorem or in its proof.

1 Definition. Let C be a convex subset of E^k, and let M be a function defined on C and having values which are either real numbers or $+\infty$. We say that M is *convex* if for any $u, v \in C$,

$$M(\alpha u+(1-\alpha)v) \leqq \alpha M(u)+(1-\alpha)M(v),$$

whenever $0 \leqq \alpha \leqq 1$.

2 Lemma. *Let $\{M_\pi\}$ be a family of convex functions defined on a convex set $C \subseteq E^k$. Then the function M defined by*

$$M(x) = \sup_\pi M_\pi(x)$$

is a convex function.

Proof. For each π we have

$$M_\pi(\alpha u+(1-\alpha)v) \leqq \alpha M_\pi(u)+(1-\alpha)M_\pi(v),$$

whenever $u, v \in C$, and $0 \leqq \alpha \leqq 1$. Taking the supremum over all π, we have the desired assertion. Q.E.D.

The next result is sometimes called the "three lines theorem".

3 Theorem. *Let f be an analytic function of a complex variable $z = x+iy$ with values in $\mathfrak{B}$. Suppose that f is defined and bounded in the strip $x_0 \leqq x \leqq x_1$, $-\infty < y < +\infty$. Let M be defined for $x_0 \leqq x \leqq x_1$ by*

$$M(x) = \sup_{-\infty<y<\infty} |f(x+iy)|.$$

Then $\log M(x)$ *is a convex function of the real variable x.*

Proof. Let $x_0 \leqq a \leqq c \leqq x_1$ and let $b = \alpha a+(1-\alpha)c$ for some α with $0 \leqq \alpha \leqq 1$. For any real number r, the function $e^{rz}f(z)$ is analytic and bounded in the strip $a \leqq x \leqq c$. By the maximum modulus principle for a strip (cf. III.14), we have

$$e^{rb}M(b) \leqq \max \{e^{ra}M(a), e^{rc}M(c)\}.$$

We now choose r such that $e^{ra}M(a) = e^{rc}M(c)$; that is,

$$r = \frac{\log M(c) - \log M(a)}{a-c}.$$

Substituting this value of r into

$$rb + \log M(b) \leqq rc + \log M(c)$$

and using the fact that $c-b = -\alpha(a-c)$, we find after simplification that

$$\log M(b) \leqq \alpha \log M(a) + (1-\alpha) \log M(c).$$

Thus $\log M(x)$ is convex as stated. Q.E.D.

We now extend this result to k variables.

4 COROLLARY. *Let f be a bounded analytic function of the complex variables $z_1, \ldots, z_k$, $z_j = x_j + iy_j$, which is defined for $x = [x_1, \ldots, x_n]$ lying in a convex set $C \subseteqq E^k$ and $y = [y_1, \ldots, y_n]$ unrestricted. Suppose f has values in $\mathfrak{W}$, and let M be defined on C by*

$$M(x) = \sup \{|f(x+iy)| \mid -\infty < y_j < \infty,\ j = 1, \ldots, k\}.$$

Then $\log M(x)$ is a convex function of the vector variable x in C.

PROOF. Let $x', x'' \in C$ be held fixed, and let $y = [y_1, \ldots, y_k]$ be arbitrary. Let $\lambda = \alpha + i\beta$ be a complex variable, and define g by

$$g(\lambda) = f(\lambda x' + (1-\lambda)x'' + iy) = f(u_1, \ldots, u_k),$$

where

$$u_j = \alpha x_j' + (1-\alpha)x_j'' + i[\beta x_j' + (1-\beta)x_j'' + y_j], \quad j = 1, \ldots, k.$$

We now apply Theorem 3 to the function g on the strip $0 \leqq \alpha \leqq 1$, $-\infty < \beta < +\infty$, obtaining

$$\begin{aligned} \log |g(\alpha)| &\leqq \log \{\sup_\beta |g(\alpha + i\beta)|\} \\ &\leqq \alpha \log \{\sup_\beta |g(1+i\beta)|\} + (1-\alpha) \log \{\sup_\beta |g(0+i\beta)|\} \\ &\leqq \alpha \log M(x') + (1-\alpha) \log M(x''). \end{aligned}$$

We have shown that for arbitrary choice of $y = [y_1, \ldots, y_k]$,

$$\begin{aligned} \log |f(\alpha x' + (1-\alpha)x'' + iy)| &= \log |g(\alpha)| \\ &\leqq \alpha \log M(x') + (1-\alpha) \log M(x''). \end{aligned}$$

Taking the supremum over all choices of y we conclude that

$$\log M(\alpha x'+(1-\alpha)x'') \leqq \alpha \log M(x')+(1-\alpha) \log M(x''),$$

which proves the statement. Q.E.D.

5 DEFINITION. If $\mathfrak{Z}$ is a complex vector space, a function $G : \mathfrak{Z} \to \mathfrak{W}$ is said to be *analytic* if $G(\lambda_1 z_1+\ldots+\lambda_p z_p)$ is an analytic function of the complex variables $\lambda_1, \ldots, \lambda_p$ in the sense of Section III.14 for every finite collection $\{z_1, \ldots, z_p\} \subset \mathfrak{Z}$. If $\mathfrak{Z}_1, \ldots, \mathfrak{Z}_k$ are complex vector spaces, let $\mathfrak{Z}$ be their direct sum (cf. I.11); then the definition just given also gives a concept of analyticity for a function $G : \mathfrak{Z} = \mathfrak{Z}_1 \oplus \ldots \oplus \mathfrak{Z}_k \to \mathfrak{W}$.

To facilitate the statement of the next result it will be convenient to introduce the following notation.

6 DEFINITION. Let (S_j, Σ_j, μ_j) be a positive measure space for each $j = 1, \ldots, k$. Let $\mathfrak{Z}_j$ be the space of all complex valued μ_j-integrable simple functions on S_j. If $a = [a_1, \ldots, a_k] \in E^k$ and $a_j > 0$, we define $A(a) = A(a_1, \ldots, a_k)$ to be the set of all collections $f = [f_1, \ldots, f_k]$ with $f_j \in \mathfrak{Z}_j$ and

$$[*] \qquad \int_{S_j} |f_j(s)|^{1/a_j} \mu_j(ds) \leqq 1, \qquad j = 1, \ldots, k.$$

If $a_j = 0$, then condition [*] is replaced by the requirement that

$$|f_j(s)| \leqq 1 \qquad \mu_j\text{-almost everywhere,}$$

for that value of j.

7 LEMMA. *With the terminology of the preceding definitions, let G be an analytic function on $\mathfrak{Z} = \mathfrak{Z}_1 \oplus \ldots \oplus \mathfrak{Z}_k$ with values in $\mathfrak{W}$. Let*

$$M(a) = \sup_{A(a)} |G(f)|$$

Then $\log M(a)$ *is a convex function of $a = [a_1, \ldots, a_k]$ for $a_j \geqq 0$, $j = 1, \ldots, k$.*

PROOF. Let $A^+(1)$ denote the set of all $f = [f_1, \ldots, f_k]$ in $A(1, \ldots, 1)$ such that each $f_j \geqq 0$. Let $a = [a_1, \ldots, a_k]$ with $a_j \geqq 0$, then if $f \in A^+(1)$ and $g \in A(0) = A(0, \ldots, 0)$, the collection

$$f^a g = [f_1^{a_1} g_1, \ldots, f_k^{a_k} g_k]$$

is an element of $A(a)$. It is readily seen that an arbitrary element of

$A(a)$ can be obtained in this way. Further, if $b = [b_1, \ldots, b_k] \in E^k$ is arbitrary, then

$$f^{a+ib}g = [f_1^{a_1+ib_1}g_1, \ldots, f_k^{a_k+ib_k}g_k]$$

is in $A(a)$. Consequently,

$$\begin{aligned} M(a) = \sup_{A^+(1), A(0)} |G(f^a g)| &= \sup_{b} \sup_{A^+(1), A(0)} |G(f^{a+ib}g)| \\ &= \sup_{A^+(1), A(0)} \sup_{b} |G(f^{a+ib}g)|. \end{aligned}$$

Let $\lambda_j = a_j + ib_j$, $a_j \geqq 0$; then since f_j and g_j are simple functions, $f_j^{\lambda_j}g_j$ can be written as a sum $\sum_{m=1}^{n_j} p_{jm}^{\lambda_j} z_{jm}$, where $z_{jm} \in Z_j$ and p_{jm} is a positive real number. Since G is assumed to be analytic, it follows that $G(f^\lambda g) = G(f^{a+ib}g)$ is an analytic function of the complex variables $\lambda = [\lambda_1, \ldots, \lambda_k]$, and is bounded on each strip $0 \leqq \mathscr{R}(\lambda_j) \leqq c_j < \infty$, $j = 1, \ldots, k$. By Corollary 4, Lemma 2, and the increasing nature of the logarithm we conclude that

$$\log M(a) = \sup_{A^+(1), A(0)} \log \{\sup_{b} |G(f^{a+ib}g)|\}$$

is a convex function of the vector $a = [a_1, \ldots, a_k]$ for $a_j \geqq 0$. Q.E.D.

8 Theorem. *Let (S, Σ, μ) be a positive measure space and L_0 the space of all μ-integrable simple functions. Let T be a linear mapping of L_0 into a complex B-space $\mathfrak{X}$. If T can be extended to a bounded linear transformation of $L_p(S, \Sigma, \mu)$ into $\mathfrak{X}$, let $|T|_p$ be the norm of this extension; if no such extension exists, let $|T|_p = +\infty$. Then log $|T|_{1/a}$ is a convex function of a for $0 \leqq a \leqq 1$.*

Proof. Since L_0 is dense in $L_p(S, \Sigma, \mu)$ for $1 \leqq p < \infty$, it is clear that if $0 < a \leqq 1$,

$$[*] \qquad |T|_{1/a} = \sup \{|Tf| \mid f \in L_0, \ |f|_{1/a} \leqq 1\}.$$

It is also evident that

$$|T|_\infty \geqq \sup \{|Tf| \mid f \in L_0, \ |f|_\infty \leqq 1\}.$$

Thus, if $M(a)$, $0 \leqq a \leqq 1$, denotes the supremum in [*] and N is defined by $N(0) = |T|_\infty$, $N(a) = 0$ for $0 < a \leqq 1$, then $|T|_{1/a} = \max \{M(a), N(a)\}$. Since a linear mapping is obviously analytic, the theorem follows from Lemmas 7 and 2 and the observation that the logarithm of the maximum is the maximum of the logarithms. Q.E.D.

9 Lemma. *Let (S, Σ, μ) be a positive measure space, and let f be a μ-measurable function defined on S. If $f \in L_p(S, \Sigma, \mu)$, let $|f|_p$ denote its norm as an element of this space; otherwise put $|f|_p = +\infty$. Then $\log |f|_{1/a}$ is a convex function of a, $0 \leqq a \leqq 1$.*

Proof. If $|f|_{1/a} = +\infty$ for all a, $0 < a \leqq 1$, the conclusion is trivial, so we suppose that $f \in L_{p_0}(S, \Sigma, \mu)$ for some p_0. By Lemma III.8.5, we may suppose that S is σ-finite. Let L_0 denote the class of μ-integrable simple functions and $g \in L_0$, then Hölder's inequality (III.3.2) implies that fg is μ-integrable. Let

$$M(a) = \sup \left\{ \left| \int_S f(s)g(s)\mu(ds) \right| \; g \in L_0, \quad |g|_{1/1-a} \leqq 1 \right\}.$$

If f is in $L_{1/a}(S, \Sigma, \mu)$, then the Hölder inequality asserts that $M(a) \leqq |f|_{1/a}$. Conversely, we shall show that if $M(a) < \infty$, $0<a<1$, then f is in $L_{1/a}(S, \Sigma, \mu)$ and $|f|_{1/a} \leqq M(a)$. This will prove the lemma, since if we put $N(1) = |f|_1$, $N(0) = |f|_\infty$, $N(a) = 0$ for $0 < a < 1$ then $|f|_{1/a} = \max \{M(a), N(a)\}$ and the conclusion follows from Lemmas 7 and 2.

Suppose that $M(a) < \infty$ for some $a \in (0, 1)$. Since L_0 is dense in $L_{1/1-a}(S, \Sigma, \mu)$, it follows that

$$x^*(g) = \int_S f(s)g(s)\mu(ds)$$

can be extended to be a continuous linear functional of norm $M(a)$ on all of $L_{1/1-a}$. By Theorem IV.8.1, there exists an $\bar{f} \in L_{1/a}(S, \Sigma, \mu)$ such that $|\bar{f}|_{1/a} = M(a)$ and

$$\int_S f(s)g(s)\mu(ds) = \int_S \bar{f}(s)g(s)\mu(ds), \qquad g \in L_0.$$

In particular, if $\mu(E) < \infty$, then

$$\int_E f(s)\mu(ds) = \int_E \bar{f}(s)\mu(ds).$$

By Lemma III.6.8, it follows that $f(s) = \bar{f}(s)$ μ-almost everywhere on each set in Σ of finite measure. Since we have taken S to be σ-finite, $f(s) = \bar{f}(s)$ μ-almost everywhere. Hence $f \in L_{1/a}(S, \Sigma, \mu)$ and $|f|_{1/a} = M(a)$. Q.E.D.

The next theorem is a companion result to Theorem 8.

10 Theorem. *Let (S, Σ, μ) be a positive measure space, let $\mathfrak{X}$ be a complex B-space and let $M = M(S, \Sigma, \mu)$ be the space of all complex*

valued μ-measurable functions defined on S. Let T be a linear mapping of $\mathfrak{X}$ into M. If T is a bounded linear mapping of $\mathfrak{X}$ into $L_p(S, \Sigma, \mu)$, let $|T|_p$ be the norm of this operator, otherwise let $|T|_p = +\infty$. Then $\log |T|_{1/a}$ *is a convex function of a for $0 \leqq a \leqq 1$.*

PROOF. If $x \in \mathfrak{X}$, the preceding lemma shows that $\log |Tx|_{1/a}$ is a convex function of a, $0 \leqq a \leqq 1$. It follows from Lemma 2 that

$$\log |T|_{1/a} = \log \sup_{|x| \leqq 1} |Tx|_{1/a} = \sup_{|x| \leqq 1} \log |Tx|_{1/a}$$

is also a convex function of a in this range. Q.E.D.

→ 11 THEOREM. (*Riesz convexity theorem*) *Let (S_1, Σ_1, μ_1) and (S_2, Σ_2, μ_2) be positive measure spaces. Let L_0 be the set of all complex valued μ_1-integrable simple functions, let M be the set of all complex valued μ_2-measurable functions, and let T be a linear mapping of L_0 into M. If for a given pair (p, q), T has an extension to a bounded linear mapping of $L_p(S_1, \Sigma_1, \mu_1)$ into $L_q(S_2, \Sigma_2, \mu_2)$, let $|T|_{p,q}$ denote the norm of this extension; if no such extension exists, let $|T|_{p,q} = +\infty$. Then* $\log |T|_{1/a,1/b}$ *is a convex function of (a, b) in the rectangle $0 \leqq a, b \leqq 1$.*

PROOF. If $|T|_{p,q} = +\infty$ for (p, q) in the range $1 \leqq p \leqq \infty$, $1 \leqq q \leqq \infty$, then the assertion follows from Theorem 8 with $\mathfrak{X} = L_\infty(S_2, \Sigma_2, \mu_2)$. Hence we may suppose that there is some pair (p_0, q_0) in this range with $|T|_{p_0,q_0}$ finite. Hence if $f_1 \in L_0^{(1)} = L_0$, Tf_1 is in some space $L_{q_0}(S_2, \Sigma_2, \mu_2)$. It follows from Hölder's inequality, that if f_2 is any function in the space $L_0^{(2)}$ of all μ_2-integrable simple functions, then the integral

$$[*] \qquad G(f_1, f_2) = \int_{S_2} (Tf_1)(s) f_2(s) \mu_2(ds)$$

exists. Thus, under the assumption that there is some $|T|_{p_0,q_0}$ which is finite, then the integral in [*] exists for any $f_1 \in L_0^{(1)}$ and any $f_2 \in L_0^{(2)}$.

Let $a = 1/p$, $b = 1/q$, $1/q + 1/q' = 1$, and put $M(a, b) = \sup |G(f_1, f_2)|$, where the supremum is taken over all $f_1 \in L_0^{(1)}$, $|f_1|_p \leqq 1$, and over all $f_2 \in L_0^{(2)}$, $|f_2|_{q'} \leqq 1$. It is clear that $M(a, b) \leqq |T|_{p,q}$.

Conversely, suppose that $0 < a, b < 1$ and $M(a, b) < \infty$. We shall show that $|T|_{p,q} \leqq M(a, b)$. To see this, observe that for each $f_1 \in L_0^{(1)}$, equation [*] determines a linear functional on a dense subset of $L_{q'}(S_2, \Sigma_2, \mu_2)$ with bound at most equal to $M(a, b)|f_1|_p$. Consequently, it can be uniquely extended to all of $L_{q'}$, and by Theorem IV.8.1 there is a $g \in L_q(S_2, \Sigma_2, \mu_2)$ with $|g|_q \leqq M(a, b)|f_1|_p$ such that

$$\int_{S_2} (Tf_1)(s)f_2(s)\mu_2(ds) = \int_{S_2} g(s)f_2(s)\mu_2(ds), \qquad f_1 \in L_0^{(1)}, \quad f_2 \in L_0^{(2)}.$$

Applying Lemmas III.6.8 and III.8.5, we conclude that $Tf_1(s) = g(s)$ μ_2-almost everywhere; hence $Tf_1 \in L_q(S_2, \Sigma_2, \mu_2)$ and

$$|Tf_1|_q \leqq M(a, b)|f_1|_p.$$

Since this holds on the dense subset $L_0^{(1)}$ of $L_p(S_1, \Sigma_1, \mu_1)$, T may be extended to a mapping of this space into $L_q(S_2, \Sigma_2, \mu_2)$ with norm at most $M(a, b)$. Hence $|T|_{p,q} \leqq M(a, b)$.

Now the function G is a complex valued bilinear function defined on $L_0^{(1)} \times L_0^{(2)}$, and so is analytic. By Lemma 7, $\log M(a, b)$ is a convex function on the rectangle. In addition we have shown that $M(a, b) \leqq |T|_{1/a,1/b}$ for all a, b and that if $0 < a, b < 1$, then $M(a, b) = |T|_{1/a, 1/b}$. Let N be defined on the rectangle by $N(a, b) = |T|_{1/a,1/b}$ if either a or b is equal to 0 or 1 and $N(a, b) = 0$ for all other pairs. Theorems 8 and 10 imply that $\log N(a, b)$ is a convex function. Since $|T|_{1/a,1/b} = \max \{M(a, b), N(a, b)\}$, the conclusion of the theorem follows from Lemma 2. Q.E.D.

We conclude this section by giving a very simple application of the Riesz convexity theorem which will be used later.

12 Corollary. *Let (S, Σ, μ) be a positive measure space and let T be a linear mapping which sends each complex $L_p(S, \Sigma, \mu)$ space into itself, $1 \leqq p \leqq \infty$. If T is known to be continuous for $p = 1$ and $p = \infty$, with norm at most C, then $T : L_p \to L_p$ is continuous for all p, $1 \leqq p \leqq \infty$, and has norm at most C.*

The proof of this statement follows readily from Theorem 11 and will be left to the reader.

11. Exercises on Inequalities

The convexity methods developed in the previous section, if taken together with a few elementary devices, can be used to derive a large number of the most familiar and important inequalities of analysis. In the present section, we shall give a connected series of exercises on inequalities, designed to illustrate this statement.

A. *Inequalities obtainable from convexity theorem and knowledge of extreme cases.*

1 Let (S, Σ, μ) be a positive measure space. Show, using a convexity argument, that the mapping $[f, g] \to h$ defined by $h(s) = f(s)g(s)$ is a continuous map from $L_p(S, \Sigma, \mu) \times L_q(S, \Sigma, \mu)$ to $L_r(S,\Sigma,\mu)$, where $1/p+1/q = 1/r$, and that $|h|_r \leqq |f|_p |g|_q$, thereby obtaining Hölder's inequality as the special case $1/p+1/q = 1$. $(p, q, r \geqq 1.)$

2 Generalize Exercise 1 to show that if $h(s) = f_1(s)\ldots f_n(s)$, $f_i \in L_{p_i}(S, \Sigma, \mu)$, and $\sum_{i=1}^n 1/p_i = 1/r$, then h is in $L_r(S, \Sigma, \mu)$ and $|h|_r \leqq |f_1|_{p_1} \cdots |f_n|_{p_n}$. $(p_1, \ldots, p_n, r \geqq 1.)$

3 Let (S_1, Σ_1, μ_1) and (S_2, Σ_2, μ_2) be two positive measure spaces. Let K be a $\mu_1 \times \mu_2$-measurable function defined on $S_1 \times S_2$, and suppose that $\int_{S_1} |K(s_1, s_2)| \mu_1(ds_1) \leqq M_1 < \infty$ for μ_2-almost all s_2 in S_2, and $\int_{S_2} |K(s_1, s_2)| \mu_2(ds_2) \leqq M_2 < \infty$ for μ_1-almost all s_1 in S_1. Show that if f is in $L_p(S_2, \Sigma_2, \mu_2)$, then the integral

$$\int_{S_2} K(s_1, s_2) f(s_2) \mu(ds_2)$$

exists for μ_1-almost all s_1, and defines a function in $L_p(S_1, \Sigma_1, \mu_1)$ of norm at most $M_2^{1/p} M_1^{1-1/p} |f|_p$. $(1 \leqq p \leqq \infty)$

4 Let (S_1, Σ_1, μ_1) and (S_2, Σ_2, μ_2) be two positive measure spaces. Let K be a $\mu_1 \times \mu_2$-measurable function defined on $S_1 \times S_2$, and suppose that $\int_{S_2} |K(s_1, s_2)|^{p_2} \mu_2(ds_2) \leqq M_2 < \infty$ for μ_1-almost all s_1, and $\int_{S_1} |K(s_1, s_2)|^{p_1} \mu_1(ds_1) \leqq M_1 < \infty$ for μ_2-almost all s_2. Show that for each f in $L_q(S_2, \Sigma_2, \mu_2)$, the integral

$$\int_{S_2} K(s_1, s_2) f(s_2) \mu_2(ds_2)$$

exists for μ_1-almost all s_1, and defines a function in $L_r(S_1, \Sigma_1, \mu_1)$ of norm at most

$$|f|_q M_1^{(1/p_1)-(p_2/pp_1)} M_2^{1/p}.$$

Here $1 \leqq p_1, p_2 \leqq \infty$, $1/p_2+1/q_2 = 1$, $1/p+1/q = 1$, $1 \leqq q \leqq q_2$, and $r = pp_1(p-p_2)^{-1}$.

5 Let (S_1, Σ_1, μ_1) and (S_2, Σ_2, μ_2) be two positive measure spaces. Let K be a $\mu_1 \times \mu_2$-measurable function defined on $S_1 \times S_2$, and suppose that

$$\int_{S_1} \left\{ \int_{S_2} |K(s_1, s_2)|^{p_1} \mu_2(ds_2) \right\}^{r_1} \mu_1(ds_1) = M_1 < \infty,$$

$$\int_{S_2} \left\{ \int_{S_1} |K(s_1, s_2)|^{p_2} \mu_1(ds_1) \right\}^{r_2} \mu_2(ds_2) = M_2 < \infty.$$

Show that for each $f \in L_q(S_2, \Sigma_2, \mu_2)$,

$$\int_{S_2} K(s_1, s_2) f(s_2) \mu_2(ds_2)$$

exists for μ_1-almost all s_1, and defines a function in $L_t(S_1, \Sigma_1, \mu_1)$ of norm at most $|f|_q M^{\alpha/r_1p_1} M_2^{(1-\alpha)/r_2p_2}$. Here q is between q_2 and $r_2p_2(r_2p_2-1)^{-1}$,

$$\alpha = \frac{qr_2p_2-q-r_2p_2}{q-qr_2},$$

and

$$t = \frac{r_1s_1p_1}{\alpha(p_1+r_1s_1)+r_1s_1p_1-r_1s_1}.$$

Similar inequalities might easily be given for multilinear forms of the type

$$\int_{S_1} \cdots \int_{S_n} K(s_1, \ldots, s_n) f_1(s_1) \ldots f_n(s_n) \mu_1(ds_1) \ldots \mu_n(ds_n),$$

but we leave the development of this line of thought to the reader.

In VIII.1.24 we will show that if f is a function on the real line which is Lebesgue measurable, the function g of two real variables defined by

$$g(x, y) = f(x-y)$$

is measurable with respect to two dimensional Lebesgue measure. In the following exercises on integrals of the "convolution" type, free use may be made of this fact. (In the next four exercises, L_p denotes the L_p space formed with respect to the Lebesgue measure on the real line.)

6 Let f be in L_1, g be in L_p, $p \geqq 1$. Then the integral $h(x) = \int_{-\infty}^{+\infty} f(x-y)g(y)dy$ exists for almost all x, and defines a function in L_p. Moreover, $|h|_p \leqq |g|_p |f|_1$.

7 Let f be in L_p, and g be in L_r, where $1/p+1/r \geqq 1$, $p \geqq 1$, and $r \geqq 1$. Show that the integral $h(x) = \int_{-\infty}^{+\infty} f(x-y)g(y)dy$ exists for almost all x and defines a function in L_s, where $s^{-1} = r^{-1}+p^{-1}-1$. Moreover, $|h|_s \leqq |f|_p |g|_r$.

8 Let f_i be in L_{p_i}, $p_i \geqq 1$, $i = 1, \ldots, n$, and suppose that $\sum_{i=1}^n 1/p_i \geqq n-1$. Show that the integral

$$h(x) = \int_{-\infty}^{+\infty} \cdots \int_{-\infty}^{+\infty} f_1(x-y_1) f_2(y_1-y_2) \cdots f_{n-1}(y_{n-2}-y_{n-1}) f_n(y_{n-1}) \, dy_1 \ldots dy_{n-1}$$

exists for almost all x, and defines a function h in L_r, $1/r = \sum_{i=1}^{n} p_i^{-1} - n + 1$. Moreover, $|h|_r \leq |f_1|_{p_1} \cdots |f_n|_{p_n}$.

9 (Hausdorff-Young) Let k be an integer ≥ 1, and let $p = (2k+1)/2k$. Let f be in L_p. Show that the multiple integral

$$h(x) = \int_{-\infty}^{+\infty} \cdots \int_{-\infty}^{+\infty} f(x-y_1)f(y_1-y_2)\ldots f(y_{n-2}-y_{n-1})f(y_{n-1})dy_1 \ldots dy_{n-1}$$

exists for almost all x, and defines a function in L_2. Show that $|h|_2 \leq |f|_p^k$.

10 Show that if $\{a_n\}$ and $\{b_n\}$ are (two-sided) sequences of complex numbers such that $\sum_{n=-\infty}^{+\infty} |a_n| < \infty$ and $\sum_{n=-\infty}^{+\infty} |b_n|^p$ converge, $p \geq 1$, then $c_n = \sum_{m=-\infty}^{+\infty} a_{n-m} b_m$ is defined for each n, and

$$\{\sum_{n=-\infty}^{+\infty} |c_n|^p\}^{1/p} \leq \sum_{n=-\infty}^{+\infty} |a_n| \{\sum_{n=-\infty}^{+\infty} |b_n|^p\}^{1/p}.$$

11 Let $\{a_n\}$ and $\{b_n\}$ be (two-sided) sequences of complex numbers such that $\sum_{n=-\infty}^{+\infty} |a_n|^p < \infty$, $\sum_{n=-\infty}^{+\infty} |b_n|^r < \infty$, $1 \leq p < \infty$, $1 \leq r < \infty$, where $1/p + 1/r \geq 1$. Show that $c_n = \sum_{m=-\infty}^{+\infty} a_{n-m} b_m$ is defined for all n, and that

$$\{\sum_{n=-\infty}^{+\infty} |c_n|^s\}^{1/s} \leq \{\sum_{n=-\infty}^{+\infty} |a_n|^p\}^{1/p} \{\sum_{n=-\infty}^{+\infty} |b_n|^r\}^{1/r},$$

where $1/r + 1/p = (1/s) + 1$.

12 Let $\{a_n^{(i)}\}$, $-\infty < n < +\infty$, $i = 1, \ldots, k+1$, be a family of sequences (two-sided) of complex numbers, such that $\sum_{n=-\infty}^{+\infty} |a_n^{(i)}|^{p_i} < \infty$, $p_i \geq 1$, $i = 1, \ldots, k+1$, and $\sum_{i=1}^{k} 1/p_i \geq n-1$. Show that the multiple series

$$c_n = \sum_{m_1=-\infty}^{+\infty} \cdots \sum_{m_k=-\infty}^{+\infty} a_{n-m_1}^{(1)} a_{m_1-m_2}^{(2)} \cdots a_{m-m_k}^{(k)} a_{m_k}^{(k+1)}$$

converges absolutely for each n, and that

$$\{\sum_{n=-\infty}^{+\infty} |c_n|^r\}^{1/r} \leq \prod_{i=1}^{k+1} \{\sum_{n=-\infty}^{+\infty} |a_n^{(i)}|^{p_i}\}^{1/p_i}.$$

B. *Generalizations of Hölder's and Minkowski's inequalities.*

Often a familiar inequality, when applied to vector valued functions, will take on an interesting, but less familiar form. Minkowski's inequality leads immediately to the formula

$$\int_S |f(s)|\mu(ds) \geq \left| \int_S f(s)\mu(ds) \right|$$

for functions with values in the B-space L_p. In more analytical terms, we obtain the following inequalities:

13 Let (S_1, Σ_1, μ_1) and (S_2, Σ_2, μ_2) be positive measure spaces, and let K be a $\mu_1 \times \mu_2$-integrable function on $S_1 \times S_2$. Then, for $p \geqq 1$,

$$\int_{S_1} \left\{ \int_{S_2} |K(s_1, s_2)|^p \mu_2(ds_2) \right\}^{1/p} \mu_1(ds_1) \geqq \left\{ \int_{S_2} \left[\int_{S_1} |K(s_1, s_2)| \mu_1(ds_1) \right]^p \mu_2(ds_2) \right\}^{1/p}.$$

14 (Jessen) Under the hypotheses of Exercise 13, if $r \geqq s > 0$,

$$\left\{ \int_{S_1} \left[\int_{S_2} |K(s_1, s_2)|^r \mu_2(ds_2) \right]^{s/r} \mu_1(ds_1) \right\}^{1/s} \geqq \left\{ \int_{S_2} \left[\int_{S_1} |K(s_1, s_2)|^s \mu_1(ds_1) \right]^{r/s} \mu_2(ds_2) \right\}^{1/r}.$$

(Hint. Put $p = r/s$ in Exercise 13.)

In the same way, Hölder's inequality, when applied to vector valued functions, may lead to unfamiliar looking inequalities.

15 Let (S, Σ, μ) be a positive measure space, $\mathfrak{X}$ and $\mathfrak{Y}$ B-spaces, f a μ-measurable function with values in $\mathfrak{X}$, g a μ-measurable function with values in $B(\mathfrak{X}, \mathfrak{Y})$. Show that if $p, q \geqq 1$, $1/p+1/q = 1$ and $f \in L_p(S, \Sigma, \mu, \mathfrak{X})$, $g \in L_q(S, \Sigma, \mu, B(\mathfrak{X}, \mathfrak{Y}))$, then the function h defined by $h(s) = g(s)f(s)$ is in $L_1(S, \Sigma, \mu, \mathfrak{Y})$, and we have $|h|_1 \leqq |f|_p |g|_q$.

16 Let (S_1, Σ_1, μ_1) and (S_2, Σ_2, μ_2) be positive measure spaces. Let K_1 and K_2 be $\mu_1 \times \mu_2$-measurable functions defined on $S_1 \times S_2$. Suppose that $p_1, p_2 \geqq 1$, $1/p_1+1/q_1 = 1$, $1/p_2+1/q_2 = 1$. Show that if

$$\left\{ \int_{S_2} \left[\int_{S_1} |K_1(s_1, s_2)|^{p_1} \mu_1(ds_1) \right]^{p_2/p_1} \mu_2(ds_2) \right\}^{1/p_2} = M_1 < \infty,$$

and

$$\left\{ \int_{S_2} \left[\int_{S_1} |K_2(s_1, s_2)|^{q_1} \mu_1(ds_1) \right]^{q_2/q_1} \mu_2(ds_2) \right\}^{1/q_2} = M_2 < \infty,$$

then

$$\int_{S_1} \int_{S_2} K_1(s_1, s_2) K_2(s_1, s_2) \mu_1(ds_1) \mu_2(ds_2)$$

exists, and is in absolute value no greater than $M_1 M_2$.

17 Let (S, Σ, μ) be a measure space, let $p \geqq 1$ and $1/p+1/q = 1$, and let $\{K_n\}$ and $\{L_n\}$ be sequences of μ-measurable functions defined on S. Show that

$$|\sum_{n=0}^{\infty}\int_S K_n(s)L_n(s)\mu(ds)|$$
$$\leqq \Big[\int_S \{\sum_{n=0}^{\infty}|K_n(s)|^p\}^{1/p}\mu(ds)\Big]\Big[\int_S \{\sum_{n=0}^{\infty}|L_n(s)|^q\}^{1/q}\mu(ds)\Big],$$

the series on the left being well-defined and absolutely convergent whenever the quantities on the right are finite.

18 Let f and g be Lebesgue measurable functions of two real variables. Let $p \geqq 1$, $1/p+1/q = 1$. Show that

$$\Big\{\int_{-\infty}^{+\infty}\Big[\int_{-\infty}^{+\infty}\Big|\int_{-\infty}^{+\infty} f(x, y-z)g(x, z)dz\Big|dx\Big]^p dy\Big\}^{1/p}$$
$$\leqq \Big\{\int_{-\infty}^{+\infty}\int_{-\infty}^{+\infty}|g(x, y)|^p\, dxdy\Big\}^{1/p}\int_{-\infty}^{+\infty}\Big[\int_{-\infty}^{+\infty}|f(x, y)|^q\, dx\Big]^{1/q} dy,$$

the quantity on the left being defined if the quantities on the right are finite. (Hint. This is the vector form of Exercise 6.)

C. *Inequalities of Hardy-Hilbert type.*

The next set of inequalities are all variations on the elementary theme given in Exercise 15, which lends itself to surprisingly manifold ramifications. As an introductory example, we give:

19 Let f be a Lebesgue-measurable function defined on the real axis, and let f be in L_p. Define the mapping $T_t : L_p \to L_p$ by putting $(T_t f)(x) = f(x-t)$. Show that $T_t f$ is a continuous function of t with values in L_p for each f in L_p, and that $|T_t f| = |f|$. Show that if g is in L_1,

$$\Big(\int_{-\infty}^{+\infty}(T_t f)g(t)dt\Big)(x) = \int_{-\infty}^{+\infty} f(x-t)g(t)dt$$

for almost all x, and hence derive the result of Exercise 6 from Exercise 15. (Or from the consideration mentioned at the beginning of B.)

In the following exercises we deal once more with Lebesgue measure, this time on the positive real axis. By p we will always denote a quantity between 1 and ∞.

20 (Hardy-Littlewood-Polya) Let K be Lebesgue measurable. Show that the map T defined by

$$(Tf)(x) = \int_0^{\infty} K(y)f(xy)dy = \frac{1}{x}\int_0^{\infty} K(y/x)f(y)dy, \qquad x > 0,$$

is a map of $L_p(0, \infty)$ into itself of norm at most $\int_0^{\infty}|K(y)|y^{-1/p}dy$. Show that if $K(x) \geqq 0$ for almost all x, this gives the exact norm.

21 Show that the map T defined by

(a) (Hardy) $(Tf)(x) = \frac{1}{x}\int_0^x f(y)dy$

is a map in $L_p(0, \infty)$ of norm $p/(p-1)$, $p > 1$,

(b) (Hilbert, Schur, Hardy, M. Riesz)

$$(Tf)(x) = \int_0^\infty \frac{f(y)}{x+y}\,dy$$

is a map in $L_p(0, \infty)$ of norm $\pi(\sin \pi/p)^{-1}$, $p > 1$,

(c) (Hardy, Littlewood, Polya)

$$(Tf)(x) = \int_0^\infty \frac{f(y)}{\max(x, y)}\,dy$$

is a map in $L_p(0, \infty)$ of norm $p^2(p-1)^{-1}$, $p > 1$.

22 Show that the mapping $T : \{a_n\} \to \{b_n\}$ of sequences defined by

(a) $$b_n = \frac{1}{n}\sum_{k=1}^{n} a_k$$

is a map in l_p of norm $p/(p-1)$, $p > 1$,

(b) $$b_n = \sum_{j=1}^{\infty} \frac{a_j}{n+j}$$

is a map in l_p of norm $\pi(\sin \pi/p)^{-1}$, $p > 1$,

(c) $$b_n = \sum_{j=1}^{\infty} \frac{a_j}{\max(j, n)}$$

is a map in l_p of norm $p^2(p-1)^{-1}$, $p > 1$.

23 Let K be Lebesgue measurable, and let $\int_0^\infty |K(y)|y^{-1/p}\,dy < \infty$. Let $p \geqq 1$, $1/p+1/q = 1$. Show that the adjoint of the map $T : L_p(0, \infty) \to L_p(0, \infty)$ of Exercise 20 is the map $S : L_q(0. \infty) \to L_q(0, \infty)$ defined by the formula

$$(Sg)(x) = \int_0^\infty K(x/y)g(y)y^{-1}dy.$$

If $K(x) \geqq 0$ for all x, what is the norm of S?

24 (a) (Hardy) Show that the map S defined by

$$(Sf)(x) = \int_x^\infty f(y)y^{-1}dy$$

is a map in $L_p(0, \infty)$ of norm p.

(b) (Hardy) Show that the map $S : \{a_n\} \to \{b_n\}$ defined by

$$b_n = \sum_{k=n}^{\infty} a_k/k$$

is a map in l_p of norm p.

25 Let K be a function of n real variables defined for positive values of all these variables and measurable with respect to n-dimensional Lebesgue measure. Let $p_i \geqq 1$, $i = 1, \ldots, n$, and $q \geqq 1$. Let $q^{-1} = \sum_{i=1}^{n} p_i^{-1}$. Suppose that

$$\int_0^\infty \cdots \int_0^\infty |K(x_1, \ldots, x_n)| x_1^{-1/p_1} \ldots x_n^{-1/p_n} dx_1 \ldots dx_n = c < \infty.$$

Show that if f_i is in $L_{p_i}(0, \infty)$, the integral

$$I(x) = \int_0^\infty \cdots \int_0^\infty K(y_1/x, \ldots, y_n/x) f_1(y_1) \ldots f_n(y_n) dy_1 \ldots dy_n$$

exists for almost all $x \geqq 0$, and that we have

$$\left\{ \int_0^\infty \left| \frac{1}{x^n} I(x) \right|^q dx \right\}^{1/q} \leqq c |f_1|_{p_1} \ldots |f_n|_{p_n}.$$

Show that if K is non-negative, the constant c in this inequality is the best possible.

26 Let $\mathfrak{H}_1$ and $\mathfrak{H}_2$ be Hilbert spaces. Let T_1 map a dense subset D_1 of $\mathfrak{H}_1$ into $\mathfrak{H}_2$, and let T_2 map a dense subset D_2 of $\mathfrak{H}_2$ into $\mathfrak{H}_1$. Suppose that T_1 and T_2 are adjoint in the sense that $(T_1 x, y) = (x, T_2 y)$ for x in D_1 and y in D_2. Show that T_2, $T_1 T_2$, $T_2 T_1$, and T_1 are all extendable to everywhere defined bounded operators if any one of them is. How are the norms of the extensions related?

27 (Hardy) Consider the map T defined by

$$(Tf)(x) = \int_0^\infty e^{-xt} f(t) dt.$$

Show that T is a bounded operator in $L_2(0, \infty)$, of norm $\sqrt{\pi}$, and that TT^* is given by the formula

$$(TT^* f)(x) = \int_0^\infty f(t)(x+t)^{-1} dt.$$

Show that if $1 \leqq p \leqq 2$, $1/p + 1/q = 1$, then $|Tf|_q \leqq \pi^{1/q} |f|_p$ for f in $L_p(0, \infty)$.

28 Show that the map T which sends the Lebesgue integrable

function f defined on the interval $(0, 1)$ into the sequence $\{a_n\}$ defined by

$$a_n = \int_0^1 x^n f(x)dx, \qquad n \geqq 0,$$

is a bounded map of L_2 into l_2 such that TT^* is the map $\{a_n\} \to \{b_n\}$ defined by

$$b_n = \sum_{j=0}^{\infty} \frac{a_j}{n+j+1}.$$

Show that T has norm $\sqrt{\pi}$. Show that the map S of $L_2(0, 1)$ into itself defined by

$$(Sf)(x) = \int_0^1 f(y)(1-xy)^{-1}\,dy$$

has norm π.

29 Let K be a Lebesgue measurable function defined on the positive real axis. Show that the map T defined by

$$(Tf)(x) = \int_0^\infty K(xy)y^{(2/p)-1}f(y)dy$$

is a map in $L_p(0, \infty)$ of norm at most $\int_0^\infty |K(x)|x^{(1/p)-1}dx$, and that this is the exact norm if $K(x) \geqq 0$ for all $x > 0$. Show that the adjoint map is

$$(T^*f)(x) = x^{(2/p)-1}\int_0^\infty K(xy)f(y)dy.$$

(Hint. Put $y = y_1^{-1}$.)

D. *General properties of L_p Norms.*

In this set of problems, let (S, Σ, μ) be a positive measure space, and let $L_p(S, \Sigma, \mu)$ be defined for $0 < p < \infty$ as the set of μ-measurable functions satisfying the requirement:

$$|f|_p = \left\{\int_S |f(s)|^p \mu(ds)\right\}^{1/p} < \infty.$$

$L_\infty(S, \Sigma, \mu)$ has a known meaning; in the course of this section we will also see how to define $L_0(S, \Sigma, \mu)$.

For $1 \leqq p \leqq \infty$, L_p is a B-space, but for $0 < p < 1$, L_p is merely an F-space (cf. Exercise III.9.30, where, however, the symbol $|f|_p$ has a slightly different meaning for $0 < p < 1$). Nevertheless, there are certain interesting properties of the function $|f|_p$ which hold in the extended range $0 < p \leqq \infty$.

30 If $\infty \geqq p_1 \geqq p \geqq p_2 > 0$, and f is in $L_{p_1}(S, \Sigma, \mu) \cap L_{p_2}(S, \Sigma, \mu)$, then f is in $L_p(S, \Sigma, \mu)$, and $\log |f|_p$ is a continuous convex function of $1/p$. (Hint. Prove the equivalent fact that $p \cdot \log |f|_p$ is a continuous convex function of p.)

31 If $\mu(S) = 1$ then $L_{p_1}(S, \Sigma, \mu)$ is contained in $L_{p_2}(S, \Sigma, \mu)$ if $0 < p_1 \leqq p_2$, and $|f|_p$ is an increasing function of p.

32 Suppose that f is in $L_p(S, \Sigma, \mu)$ for some $p > 0$. Show that unless f vanishes outside a certain set E in Σ such that $\mu(E) \leqq 1$, $|f|_p \to \infty$ as $p \to 0$. Show that in the contrary case, $|f|_p$ is a decreasing function converging to

$$\exp\left\{\int_E \log |f(x)|\mu(dx)\right\} = |f|_0.$$

33 Suppose that $\mu(S) = 1$. Show that the set $L_0(S, \Sigma, \mu)$ of all functions f such that

$$-\infty \leqq \int_S \log |f(x)|\mu(dx) < \infty$$

form a linear space, and the result of Exercise 31 holds in the extended range $0 \leqq p \leqq \infty$.

34 (Inequality of arithmetic and geometric means) If $a_1, \ldots, a_n$ are complex numbers, show that

$$|a_1 \ldots a_n|^{1/n} \leqq (1/n)(|a_1| + \ldots + |a_n|).$$

35 Assume $\mu(S) = 1$. Put

$$\exp\left\{\int_S \log |f(s)|\mu(ds)\right\} = |f|_0$$

for f in $L_0(S, \Sigma, \mu)$. Show that if f and g are non-negative functions in $L_0(S, \Sigma, \mu)$, then $|f+g|_0 \geqq |f|_0 + |g|_0$. (Hint. Use Exercise III.9.29.)

36 Let (S, Σ, μ) and (S_1, Σ_1, μ_1) be positive measure spaces. Assume $\mu(S) = 1$. Show that if K is a $\mu \times \mu_1$-measurable function defined on $S \times S_1$,

$$\int_{S_1} \exp\left\{\int_S \log |K(s, s_1)|\mu(ds)\right\}\mu_1(ds_1)$$
$$\leqq \exp\left\{\int_S \left[\log\left(\int_{S_1} |K(s, s_1)|\mu_1(ds_1)\right)\right]\mu(ds)\right\}.$$

(Hint. Use Exercise 14.)

37 (a) Show that if f and K are Lebesgue measurable functions

defined on the positive real axis, and if K is non-negative and has $\int_0^\infty K(x)dx = 1$, then

$$\int_0^\infty \exp\left\{\int_0^\infty K(x)\log|f(xy)|dx\right\}dy \leqq \exp\left\{-\int_0^\infty K(x)\log x\, dx\right\}\cdot\int_0^\infty |f(y)|dy.$$

(b) (Knopp) Show that if f is a Lebesgue integrable function defined on the positive real axis, then

$$\int_0^\infty \exp\left\{\frac{1}{x}\int_0^x \log|f(y)|dy\right\}dx \leqq e\int_0^\infty |f(y)|dy,$$

the constant e being the best possible.

(c) (Carleman) Show that $\sum_{n=1}^\infty |a_1 \ldots a_n|^{1/n} \leqq e\sum_{n=1}^\infty |a_n|$ for each sequence $\{a_n\}$, the constant e being the best possible.

E. *Extensions of the convexity theorem.*

38 Let (S, Σ, μ) be a σ-finite measure space, and let f be a μ-measurable function defined on S with values in a B-space $\mathfrak{X}$, and suppose that f is μ-integrable over every set E in Σ such that $\mu(E) < \infty$. Show that if $1 \leqq p \leqq \infty$ then f is in $L_p(S, \Sigma, \mu, \mathfrak{X})$ if and only if

$$[*] \qquad \left|\int_S g(s)f(s)\mu(ds)\right| < \infty, \qquad g \in \mathfrak{L}, \quad |g|_q \leqq 1,$$

where $\mathfrak{L}$ is the subspace of μ-simple functions in $L_q(S, \Sigma, \mu, \mathfrak{X}^*)$ and $1/p+1/q = 1$. Show that in this case, $|f|_p$ is equal to the supremum in [*].

39 Show that Lemma 10.7, Theorem 10.8, Lemma 10.9, Theorem 10.10, and Theorem 10.11 all hold even if the spaces of complex valued functions are replaced by the corresponding spaces of vector valued functions (i.e., if $L_p(S, \Sigma, \mu, \mathfrak{X})$ replaces $L_p(S, \Sigma, \mu)$, etc.).

40 Let (S_1, Σ_1, μ_1) and (S_2, Σ_2, μ_2) be positive measure spaces. Let $z = x+iy$ be a complex parameter ranging over a strip $c_1 \leqq x \leqq c_2$. Suppose that for each z, $T(z)$ is a linear mapping of the space $L^{(1)}$ of all μ_1-integrable simple functions into the space $L^{(2)}$ of all μ_2-measurable functions integrable over every set of finite measure, and that for each f_1 in $L^{(1)}$ and E in Σ_2 such that $\mu(E) < \infty$, $\int_E (T(z)f_1)(s)\mu_2(ds)$ is analytic and uniformly bounded. Let $|T(z)|_{p,q}$ be defined for each z as in the statement of Theorem 11. Show that

$\max_{-\infty<y<+\infty} \log |T(x+iy)|_{1/a,\,1/b}$ is a convex function of $[x, a, b]$ in the region $c_1 \leqq x \leqq c_2$, $0 \leqq a \leqq 1$, $0 \leqq b \leqq 1$.

41 Let (S_1, Σ_1, μ_1) and (S_2, Σ_2, μ_2) be positive measure spaces, and K a non-negative $\mu_1 \times \mu_2$-measurable function defined $S_1 \times S_2$. Suppose that $\int_{E_1}\int_{E_2} K(s_1, s_2)\mu_1(ds_1)\mu_2(ds_2) < \infty$ if E_1 is in Σ_1, $i = 1, 2$ and $\mu_i(E_i) < \infty$, $i = 1, 2$. Show that the norm of the map T of $L_p(S_2, \Sigma_2, \mu_2) \to L_q(S_1, \Sigma_1, \mu_1)$ defined by the formula

$$[*] \qquad (Tf)(s_1) = \int_{S_2} \{K(s_1, s_2)\}^\alpha f(s_2)\mu_2(ds_2)$$

(in the sense that the integral is to exist μ_1-almost everywhere for each f in $L_p(S_2, \Sigma_2, \mu_2)$ and lie in $L_q(S_1, \Sigma_1, \mu_1)$) is a convex function of α, $1/p$, $1/q$. (Of course, if [*] does not define a bounded map of $L_p \to L_q$, we put the norm equal to $+\infty$).

42 Let f be Lebesgue measurable on $(0, \infty)$, $p > 1$, $0 \leqq \lambda \leqq 1$, and $1/p+1/q = 1$. Show that if f is in $L_{q/q-\lambda}(0, \infty)$, the integral $\int_0^\infty f(x)/(x+y)^\lambda dx$ exists for almost all y, and determines a function in $L_{p/\lambda}(0, \infty)$. Give an upper bound for the norm of this function.

F. *Inequalities from the theory of orthonormal series.*

In the next set of exercises, we adopt the notion and definitions of IV.14 and in particular, of Definition 1 of that section.

43 Let $\{\phi_n\}$, $-\infty < n < +\infty$ be a uniformly bounded c.o.n. system and let $1 \leqq p \leqq 2$. Show that if $f \in L_p$, and $a_n = \int_0^{2\pi} f(x)\overline{\phi_n(x)}dx$, then $\sum_{-\infty}^{+\infty} |a_n|^q < \infty$, where $1/p+1/q = 1$.

44 Let $\{\phi_n\}$, $-\infty < n < +\infty$, be a uniformly bounded c.o.n. system. Suppose that $1 \leqq p \leqq 2$, $1/p+1/q = 1$, and $\sum_{-\infty}^{+\infty} |a_n|^p < \infty$. Show that there exists an f in L_q such that $a_n = \int_0^{2\pi} f(x)\overline{\phi_n(x)}dx$.

45 Show, under the hypotheses of Exercise 43, and the additional assumption that $\{\beta_n\}$ is a sequence of complex numbers such that $\sum_{-\infty}^{+\infty} |\beta_n| < \infty$, that $\sum_{-\infty}^{+\infty} |\beta_n|^{2-p}|a_n|^p < \infty$ for all f in L_p.

46 Let T_n be the operator defined in the paragraph preceding Exercise IV.14.34. Show that if $T_n f \to f$ in the norm of C for every $f \in C$, then $T_n f \to f$ in the norm of L_p for every $f \in L_p$.

47 Let $\{\lambda_n\}$ be a real factor sequence which is of type (C, C) in the sense of the definition preceding Exercise IV.14.64. Show that $\{\lambda_n\}$ is also of type (L_p, L_p) for $1 \leqq p \leqq \infty$.

G. *Some miscellaneous convexity inequalities.*

48 (Hadamard three circles theorem) Let f be an analytic function defined in the annular domain $a < |z| < b$ and having values in a B-space $\mathfrak{X}$. Show that if $M(r) = \max_{|z|=r} |f(z)|$, $\log M(r)$ is a convex function of $\log r$, $a < r < b$.

49 (Hardy) Let f be a complex-valued analytic function defined in the annular domain $a < |z| < b$. Let $1 \leqq p < \infty$. Show that if $M_p(r) = \{\int_0^{2\pi} |f(re^{i\theta})|^p d\theta\}^{1/p}$, $\log M_p(r)$ is a convex function of $\log r$, $a < r < b$. (Hint. Apply Exercise 48 to the function $F(z)$ defined by $(F(z))(\theta) = f(ze^{i\theta})$).

50 Let f be a complex valued analytic function defined in an annular domain $a < |z| < b$. Let $0 < \alpha < 1$, and let

$$M_\alpha(r) = \max_{|z_1|=|z_2|=r} \frac{|f(z_1)-f(z_2)|}{|z_1-z_2|^\alpha}.$$

Show that $\log M_\alpha(r)$ is a convex function of $\log r$, $a < r < b$.

12. Notes and Remarks

Topologies, adjoints and projections. The strong and weak operator topologies for bounded operators on Hilbert space were introduced and employed systematically by von Neumann [2]. The notions of strong and weak convergence of sequences of operators had been used earlier, however, (see Hilbert [1], Riesz [6; pp. 107, 111]). The final topology mentioned in Section 1 is sometimes called the "strongest" operator topology, and was introduced by von Neumann [5]. The question of finding the form of the linear functionals on $B(\mathfrak{H})$ which are continuous in these and other topologies was considered by Dixmier [2]. He proved Theorem 1.4 for this case; the general form is due to Bade [3]. Related results were obtained by Michael [1].

The formal concept of the adjoint operator goes back to matrix theory and to the theories of differential and integral equations. In the spaces L_p, $p > 1$, and l_2, Riesz [2; p. 478, 6; p. 85] made use of this notion and proved the validity of Lemma 2.2. Banach [4; p. 235] introduced the adjoint operator in a general B-space and proved Lemmas 2.7 and 2.8 in this context. The fundamental idea of a projection operator and its importance was clearly recognized by E. Schmidt [1], to whom the geometrical terminology in linear space theory is due.

Compact and weakly compact operators. The concept of a compact (or completely continuous) operator is essentially due to Hilbert [1; IV] where it was defined for bilinear forms in l_2. In terms of operators, Hilbert requires the operator to map weakly convergent sequences into strongly convergent sequences. In reflexive spaces this is equivalent to Definition 5.1, which is due to F. Riesz [4] who made a penetrating study of these operators.

The study of weakly compact (or weakly completely continuous) operators was initiated by Kakutani [13] and Yosida [4] in connection with ergodic theory. Gantmacher [1] proved Theorem 4.8, and also Theorems 4.2 and 4.7 for separable spaces. These facts are proved without the assumption of separability by Nakamura [3].

Theorem 5.2 is due to Schauder [6] and a sequential form of Theorem 5.6, valid in separable spaces to Gelfand [2; p. 269]. Kakutani [11] has given a symmetric proof of Theorem 5.2; a similar treatment for weakly compact operators is found in Bartle [2].

Operators with closed range. Some particular cases of these theorems were proved by Hellinger and Toeplitz [1] for l_2 and by F. Riesz [2, 6] for L_p, l_p, $p > 1$. Related abstractly-phrased results were obtained by Hahn [3]. In the form presented they are essentially due to Banach [4; p. 234–9, 1; p. 145–152] although his proofs are different. For additional results of this nature the reader may consult Hausdorff [3] and Dieudonné [3].

Representation of operators in C. The representation of the general operator with domain and range in $C[0, 1]$ was given by Radon [1]. Both C. Fischer [1] and Radon [1] treated the compact operator from $C[0, 1]$ to itself. Gelfand [2] represented the general and compact operators with range in $C[0, 1]$ but domain an arbitrary B-space. Sirvint [2, 3] gave a similar representation for weakly compact operators mapping $\mathfrak{X}$ into $C[0, 1]$. Bartle [2] showed that this same representation is valid for any of these three types of operators mapping $\mathfrak{X}$ into the B-space of bounded continuous functions on an arbitrary topological space.

In the case that $C[0, 1]$ is the domain and $\mathfrak{X}$ the range, Gelfand [2] represented by means of Stieltjes integrals both the compact operator and the operator mapping into a weakly complete space $\mathfrak{X}$. Weakly compact operators from $C[0, 1]$ to $\mathfrak{X}$ were treated by Sirvint [3]. A very incisive discussion of weakly compact operators with domain $C(S)$ was given by Grothendieck [4] who proved Theorems 7.4—7.6 by other means. Grothendieck showed there is a one-to-one correspondence between weakly compact operators on $C(S)$ to $\mathfrak{X}$ and certain vector measures but did not use this correspondence to represent them. This integral representation was employed by Bartle, Dunford and Schwartz [1] to prove these theorems essentially as discussed here.

In Section 7 and Section 8, it is shown that an arbitrary weakly compact operator on a B-space $\mathfrak{X}$ to a B-space $\mathfrak{Y}$ has the property that it sends weakly convergent sequences into strongly convergent sequences, provided that $\mathfrak{X}$ is C or L_1 or one of the many spaces isometrically equivalent to a C or L_1 space. An abstract study of this property of the space $\mathfrak{X}$ was made independently by Brace [1] and Grothendieck [4]. Brace employed the following condition on a B-space $\mathfrak{X}$: if $\{x_n\}$ converges weakly to x_0 and $\{x_n^*\}$ converges weakly to x_0^* then $\{x_n^* x_n\}$ converges to $x_0^* x_0$. Brace [1; p. 18] proved that if this condition is satisfied in $\mathfrak{X}$, then every weakly compact operator $T : \mathfrak{X} \to \mathfrak{Y}$ maps weakly convergent sequences in $\mathfrak{X}$ into strongly convergent sequences in $\mathfrak{Y}$. The converse of this statement was proved by Grothendieck [4; p. 138]. For other related results consult these two works.

Representation of operators in a Lebesgue space. The form of the representation in Theorem 8.1 for the general operator was given by Kantorovitch and Vulich [1; p. 138]. The bound for norm of the operator was found in the case that $\mathfrak{X} = L_p$ by Fullerton [3]. Related theorems have been obtained by Bochner and Taylor [1; p. 941—943].

The general operator from $L_1[0, 1]$ to a uniformly convex space $\mathfrak{X}$, or a space with a certain kind of basis was represented by Dunford [8]; he also treated general and compact operators from $L_1[0, 1]$ to $L_p[0, 1]$. In a similar vein, Gelfand [2] represented the compact operator from $L_1[0, 1]$ to a general B-space, and also the general operator from $L_1[0, 1]$ to a space which is reflexive or a separable conjugate space. These last results were extended by Dunford and Pettis [1] to the case of a measure space. When S is a finite or infinite Euclidean interval, they gave a concrete representation for the weakly compact and compact operators from $L_1(S)$ to an arbitrary space $\mathfrak{X}$. Phillips [3] treated the same case for an arbitrary σ-finite measure space $\mathfrak{X}$ obtaining a representation of the general operator as an integral with respect to a Lipschitzean vector measure, and obtaining kernel representations for the weakly compact and strongly compact operators. See also Phillips [1].

Sirvint [1] gave an example of a (weakly compact) operator in $L_1[0, 1]$ which is not compact but whose square is compact. An almost exactly similar example was given by Yosida, Mimura and Kakutani [1]. If S is Euclidean, Dunford and Pettis [1] showed that the square of any weakly compact operator in $L_1(S)$ is strongly compact. This was proved for a measure space by Phillips [3].

Closely related to the problem of representing operators on $L_1(S)$ is the vector form of the Radon-Nikodým theorem. Dunford and Pettis [1] showed that if (S, Σ, α) is a σ-finite positive measure space and if μ is a measure on Σ with values in a separable conjugate space $\mathfrak{X}$ such that $|\mu(E)| \leqq K\alpha(E)$, $E \in \Sigma$, then there exists a measurable function x on S to $\mathfrak{X}$ such that ess sup $|x(s)| \leqq K$ and $\mu(E) = \int_E x(s)\alpha(ds)$. A similar theorem is due to Phillips [1] where $\mathfrak{X}$ is now an arbitrary B-space and the set

$$\{\mu(E)/\alpha(E) | E \in \Sigma, 0 < \alpha(E) < \infty\}$$

is weakly compact. Dieudonné [9, 10, 14] has given other extensions of the Dunford-Pettis theorem. A more general Radon-Nikodým theorem is due to Rickart [2]; in his case the function to be integrated is not single-valued, in general.

Riesz convexity theorem. The principal result in this section is due to M. Riesz [1], and has a number of important applications (see, for example, Hardy, Littlewood and Polya [1; Chapter VIII] and Zygmund [1; Chapter IX]). The proof given here is essentially due to Thorin [1], although his discussion is principally concerned with the spaces $l_{p,n}$. Caldéron and Zygmund [1, 2] have extended this theorem to the strip $0 \leqq a \leqq 1$, $0 \leqq b < \infty$, thereby including the spaces L_p, $0 < p < 1$, and enabling them to make some applications to the spaces H_p. The theorem may also be extended to multilinear functions.

In addition to the above references, the reader will find other proofs (sometimes only for $l_{p,n}$) and applications in Paley [1, 2] (completed by Zygmund [1]), Salem [1], Salem and Zygmund [1], Tamarkin and Zygmund [1], and Young [1].

We have discussed the case of complex spaces. In the real case, the convexity is only valid in the triangle $0 \leqq a, b \leqq 1, a+b \geqq 1$. The reader should consult the above references, e.g. Thorin [1] for a discussion of this case.

As is indicated in the Section 11 of exercises, many of the most important inequalities of analysis may be readily derived from the Riesz convexity theorem and from a few elementary ideas on vector functions. There is a class of deeper inequalities which will be discussed in the text of Chapter XI and in the section of *Notes and Remarks* appended to that chapter.

Representation of linear operators. In the application of the abstract theory of linear operators to concrete spaces it is of some importance to know the form of the operator in terms of the properties of the domain and range spaces. For the spaces C and L_1, concrete representations were given in Sections 7—8. For the convenience of the reader, we have tabulated here references to the literature where such representation theorems are explicitly stated. There are four tables; one for the general operator, one for each of the compact and weakly compact operators, and one for representations concerning various order properties.

In a sense the results may be regarded as complete for a given space when the form of the operator mapping an arbitrary B-space into that space, and the form of the operator mapping that space into an unrestricted B-space are both known. Actually, however, it frequently happens that when it is a question of two specific spaces, these general results may be improved or phrased in terms more easily verified.

The user of these tables will find some duplications and different sets of conditions. Sometimes, for example, more detailed information is available under the assumptions of separability, reflexivity, etc.; such instances are noted. No attempt has been made to distinguish between theorems stated for $L_p[0, 1]$ and $L_p(S)$, where S is a σ-finite or an arbitrary measure space, for example, although not all of these results stated for $[0, 1]$ generalize.

Further, in most of the spaces there are other notions of conver-

TABLE VI

Representation of Operators

Legend

The rows denote a fixed domain space, and the columns denote a fixed range space. The papers listed below are entered into the tables by their code letters and usually the page number for a particular theorem is noted unless the paper is short. The following abbreviations are used:

b = some basis restriction,
c = conjugate space,
p = positive operator,
r = reflexive space,
s = separable space,
u = uniformly convex space,
w = weakly complete space.

[A] A. Alexiewicz [2].

[B] R. G. Bartle [2].
[BDS] R. G. Bartle, N. Dunford, and J. Schwartz [1].
[BT] S. Bochner and A. E. Taylor [1].

[Ch] R. R. Christian [1].
[C] L. W. Cohen [1].
[CD] L. W. Cohen and N. Dunford [1].

[D] N. Dunford [8].
[DP] N. Dunford and B. J. Pettis [1].

[E] I. A. Ezrohi [1].

[F] G. Fichtenholz [3].
[Fi] G. Fichtenholz [2].
[FK] G. Fichtenholz and L. Kantorovitch [1].
[Fs] C. A. Fischer [1].
[Ft] R. E. Fullerton [2].
[Fu] R. E. Fullerton [3].
[Ful] R. E. Fullerton [4].

[G] I. Gelfand [2].
[Go] M. Gowurin [1].
[Gr] A. Grothendieck [4].

[HT] E. Hille and J. D. Tamarkin [2].

[IS] S. Izumi and G. Sunouchi [1].

[K] L. Kantorovitch [3].
[Ka] L. Kantorovitch [4].
[KV] L. Kantorovitch and B. Vulich [1].
[KaV] L. Kantorovitch and B. Vulich [2].

[L] G. Lorentz [4]

[M] I. Maddaus [2].

[O] W. Orlicz [5].

[Pe] B. J. Pettis [1].
[P] R. S. Phillips [3].
[Ph] R. S. Phillips [1].

[Ra] J. Radon [1].
[R] C. E. Rickart [2].

[S] G. Sirvint [3].
[Sm] F. Smithies [2].

[V] B. Vulich [12].
[Vu] B. Vulich [13].
[Vh] B. Vulich [14].

TABLE VI.A

Representation of General Linear Operator

	General B-space	C	L_1	L_p $1 < p < \infty$
General B-space	[CD;689]b [E]b	[B] [IS] [BT;943]s [G;267]	[BT;943] [KV;138]	[BT;941] [KV;138]
C	[BDS] [G;280]w [P;531]	[BDS] [F] [F]p [Ra]	[S;93]	
L_1	[D;482,5]b [G;275]r [D;482]u [G;276]sc [DP;369]r [PE;428]r [DP;345-6]sc [R;65]	[V;279]	[D;483] [DP;358] [KV;146] [V;286] [Ful]	[D;483.5] [Ful] [DP;347] [KaV] [DP;367] [K;264] [DP;379] [Vh]
L_p $1 < p < \infty$	[D;476] [D;482]b [P;528]	[G;267]	[D;483] [Fu] [Ful]	[D;483] [Ful] [K;275] [Vh]
L_∞	[D;482]b [K;238] [P;528] [A;146]	[FK;90]	[D;483]	[D;483]
l_1	[E] [G;270] [P;530]			
l_p $1 < p < \infty$	[E] [P;530]			
l_∞	[P;530]			
c	[E] [G;273]			
c_0	[Gr;168]*			
$B(S)$	[A;146] [Go] [K;238]	[FK;90] [V;295]	[V;300]	
$B(S, \Sigma)$	[A;146] [Go]	[FK;90] [V;295]	[V;300]	

TABLE VI.A (*Continued*)

	L_∞	l_1	l_p $1 < p < \infty$	l_∞	c
General B-space	[BT;942-3] [G;279]	[E] [KV;128] [CD;698]b [P;531] [G;271] [Pe;424]r [IS]	[E] [IS] [KV;128] [P;531]	[IS]	[E] [IS]
C					
L_1	[DP;348-50] [G;279] [V;288]	[D;486]	[D;486] [Vu;42]		
L_p $1 < p < \infty$		[Pe;425]	[KV;131]		
L_∞					
l_1		[G;270] [L;85]	[CD;697]		
l_p $1 < p < \infty$		[CD;699] [Pe;425]	[K;272] [Vh]		
l_∞				[C;334]	
c					
c_0		[CD;699]			
$B(S)$	[V;300]				
$B(S, \Sigma)$	[V;300]				

Table continued

TABLE VI.A (*Continued*)

	$B(S)$	$B(S, \Sigma)$	BV	C^n
General B-space	[B] [IS] [P;538]			[Ft;270] [Ft;272]r
C				
L_1		[V;278]	[DP;352-7] [V;279] [G;278] [K;262] [Ka;106]	[Ft;277]
L_p				[Ft;277]
L_∞				
l_1				[Ft;274]
l_p $1 < p < \infty$				[Ft;274]
l_∞				
c				[Ft;274]
c_0				
$B(S)$			[V;296]	
$B(S, \Sigma)$		[V;294]	[V;296]	

TABLE VI.B

Representation of Compact Operator

	General B-space	C	L_1	L_p $1 < p < \infty$	L_∞
General B-space	[CD;693]b [M]b	[B] [G;267]			
C	[BDS] [G;282] [P;537]	[BDS] [Fs] [Ra]			
L_1	[DP;369]r [DP;369] [G;277] [P;529] [P;534] [P;536]		[DP;370] [DP;379] [G;278]	[D;490] [DP;369-70] [DP;379-80] [G;278]	[DP;370] [DP;379] [DP;381]
L_p $1 < p < \infty$	[DP;383] [P;529] [P;536]	[G;268]	[D;487] [DP;384]	[D;487] [DP;384] [HT;446] [Sm]	[D;492] [DP;384]
L_∞	[P;529] [P;537]		[D;487]	[D;487]	
l_1	[G;270] [P;530] [P;536]				
l_p $1 < p < \infty$	[P;530] [P;536]			[Sm]	
l_∞	[P;530] [P;536]				
c					
c_0					

Table continued

TABLE VI.B (*Continued*)

	l_1	l_p $1 < p < \infty$	l_∞	c	c_0	BV	C^n
General B-space	[G;271] [P;531] [Pe;424]r	[P;531]				[G;284]	[Ft;272]
C							
L_1	[CD;700]	[CD;700]					[Ft;280]
L_p $1<p<\infty$	[Pe;425]	[Sm]					[Ft;280]
L_∞							
l_1	[C;327-9] [G;271] [L;85]	[C;327] [CD;697]			[CD;697]		
l_p $1<p<\infty$		[C;327-8] [CD;694-5] [Sm]					[Ft;275]
l_∞							
c				[CD;692]	[C;329]		[Ft;275]
c_0							

TABLE VI.C

Representation of Weakly Compact Operator

	General *B*-space	C	L_1	L_∞	
General *B*-space		[B] [S;87]			
C	[BDS] [Gr;167-8]* [Gr;173]* [S;93]s	[BDS] [S;88]	[S;93]		
L_1	[DP;368-9] [DP;375]s [P;534] [Ph;131]	[S;88]	[DP;378]	[DP;381]	
L_∞	[Gr;140]* [Gr;155]*				
l_1	[DP;368]	[S;88]			

TABLE VI.D

Representation of Order-continuous Linear Operators

	General B-space	Ordered Space	C	L_1	L_p $1 < p < \infty$
Ordered Space				[KV;138-9]	[IS] [KV;138-9] [KV;151]s
C		[Ch]p [K;241]		[KV;154]	[KV;154]
L_1		[Ka;103] [K;255-6] [KV;140]	[KV;154]	[K;259] [KV;154] [V;286]	[K;259] [K;264]
L_p $1 < p < \infty$		[K;273] [Ka;104]		[KV;153]	[K;275] [KV;153]
L_∞	[A;147]	[Ka;104] [K;247] [K;250]		[K;252-3] [KV;148]	[K;252]
l_1		[Ka;102]			
l_p $1 < p < \infty$		[K;269] [Ka;102]			
$B(S)$	[A;147] [K;254]	[K;237]			
$B(S, \Sigma)$	[A;147]	[K;237]			

TABLE VI.D (*Continued*)

	L_∞	l_1	l_p $1 < p < \infty$	$B(S)$	BV
Ordered Space	[KV;157] [KV;158] [KV;161]		[KV;128] [KV;130]		
C	[KV;163]	[KV;133]	[KV;133]		
L_1			[K;258] [Ka;105]		[K;262] [V;279]
L_p $1 < p < \infty$	[KV;162]				[Ka;106]
L_∞	[Fi;218] [K;245] [KV;158] [O;76]				
l_1					
l_p $1 < p < \infty$		[KV;132]	[K;272] [KV;132]		
$B(S)$				[K;245]	
$B(S, \Sigma)$					

gence, such as pointwise convergence, convergence in measure, etc. the final table refers to theorems which represent operators enjoying various order properties.

While some care has been exercised in the compilation of these tables, the authors are aware that they are not complete and not entirely accurate; the authors welcome additions and corrections.

Of particular importance are the expressions of the norm of the operator in terms of the objects used in the representation. The user will observe that this representation of the norm is not always known.

ADDITIONAL REMARKS.

1. A question related to the representation of linear operators is that dealing with the representation of linear functions defined on certain spaces with values in a B-space. Gowurin gave a representation for the general bounded linear functional on the space of continuous functions on $[a, b]$ with values in $\mathfrak{X}$. Bochner and Taylor treated this space and also similar spaces corresponding to L_p, $1 \leqq p < \infty$.

2. The paper of Grothendieck is not specifically concerned with representation; however, he proves a number of deep theorems concerning weakly compact operators on C and L_∞.

3. Izumi and Sunouchi represent operators mapping an arbitrary B-space into a B-space of functions satisfying specified conditions.

4. Maddaus gave conditions which are sufficient for a compact operator to be the limit of a sequence of operators with finite-dimensional range.

5. The explicit expression of the norm is difficult to obtain for maps between L_p spaces. A lower bound for this quantity has been obtained by Fullerton [4].

Bilinear functionals. It is easily seen that the collection of all scalar valued bilinear functions on a product $\mathfrak{X} \times \mathfrak{Y}$ of two B-spaces is in one-to-one correspondence with the collection of linear mappings of $\mathfrak{X}$ into $\mathfrak{Y}^*$ (or of $\mathfrak{Y}$ into $\mathfrak{X}^*$). In this way problems concerning linear operators can be replaced in terms of bilinear functionals, and conversely.

A concrete representation for bilinear functionals on $C \times C$ was given many years ago by Fréchet [10]. Recently, Morse and Transue have made a comprehensive study of bilinear functions on very general function spaces, including c, C, l_p, L_p, $1 \leqq p < \infty$, and others. (We refer only to their papers [1, 2, 3]; additional references may be found in Morse [1] and Morse and Transue [4].) It is seen in Morse and Transue [2; I] that under certain conditions the bilinear functionals can be represented in a canonical manner by repeated Lebesgue-Stieltjes integrals.

In their later papers extensive applications are made, among other things, to the convergence of double Fourier series.

Ideals of operators. Let $\mathfrak{X}$ be a B-space. Let $\mathfrak{F} \subseteqq B(\mathfrak{X})$ be the set of all operators with finite dimensional range, let $\mathfrak{C} \subseteq B(\mathfrak{X})$ be the set of compact operators, $\mathfrak{W}$ the set of weakly compact operators, and $\mathfrak{P}$ the operators which map weakly convergent sequences into strongly convergent sequences. All of these four classes are seen to be two-sided ideals; that $\mathfrak{C}$ and $\mathfrak{W}$ are closed in the uniform topology of $B(\mathfrak{X})$ was established in Corollaries 5.4 and 4.6, and the reader can easily show that $\mathfrak{P}$ is also closed. It is always true that

$$\mathfrak{F} \subseteq \mathfrak{C} \subseteq \mathfrak{P}, \qquad \mathfrak{F} \subseteq \mathfrak{C} \subseteq \mathfrak{W}.$$

In a finite dimensional space all these sets coincide with $B(\mathfrak{X})$. In a reflexive space, $\mathfrak{C} = \mathfrak{P}$ and $\mathfrak{W} = B(\mathfrak{X})$. In Hilbert space, or a B-space with a basis, $\mathfrak{C}$ is the closure of $\mathfrak{F}$. Calkin [1; p. 841] has shown that in Hilbert space $\mathfrak{C}$ is a maximal two-sided ideal. We have seen in Theorems 7.4 and 8.12 that in the spaces C and L_1,

$$\mathfrak{F} \subset \mathfrak{C} \subset \mathfrak{W} \subseteq \mathfrak{P}.$$

Grothendieck [4; p. 153] proved that in C, the ideals $\mathfrak{W}$ and $\mathfrak{P}$ coincide. This is not the case in L_1, however, for since L_1 is not reflexive there exist subsets which are bounded but not weakly compact. The fact that $\mathfrak{W} \subset \mathfrak{P}$ follows from this observation and Theorems 8.10 and 8.14.

Another ideal closely related to the compact operators was introduced by Kleinecke [1] and is briefly discussed in VII.11.4.

Complements and projections. Two linear manifolds $\mathfrak{M}$ and $\mathfrak{N}$ in a B-space $\mathfrak{X}$ are said to be *complementary* in case $\mathfrak{M} \cap \mathfrak{N} = 0$, $\mathfrak{M} \oplus \mathfrak{N} = \mathfrak{X}$. Either one is said to be a *complement* of the other. We have commented on the relation between *closed* complements and the existence of bounded projections. In the case of finite dimensional submanifolds there are always infinitely many projections, and hence closed complements. In the case of infinite dimensional subspaces this is not always true, as was shown by Murray [1] for certain subspaces of l_p, $1 < p \neq 2$. Other subspaces which have no complements have been constructed by Komatuzaki [1, 2], Phillips [3] and Sobczyk [1, 2]; see also the exercises in Section 9.

If we require only that $\mathfrak{M} \oplus \mathfrak{N}$ be dense in $\mathfrak{X}$, then we say that $\mathfrak{N}$ is a *quasi-complement* of $\mathfrak{M}$. In this case the "projection" on $\mathfrak{M}$ is not a bounded operator. That every closed subspace has (infinitely many) quasi-complements was shown by Murray [2] when $\mathfrak{X}$ is a separable reflexive space, and by Mackey [3] without the assumption of reflexivity.

References. Bohnenblust [3, 4], Dunford [2], Goodner [1], Kober [1], Komatuzaki [1, 2], Lorch [2], Mackey [3], Murray [1, 2], Phillips [3], Sobczyk [1, 2, 3].

Extension of Linear Transformation. Taylor [1] studied conditions under which the extension of linear functionals will be a uniquely defined operation. Kakutani [6] showed that this operation is linear and isometric for each closed linear manifold if and only if the space is a Hilbert space.

It is clear that the closed subspace $\mathfrak{X}_0 \subset \mathfrak{X}$ has the property that every bounded operator $T : \mathfrak{X}_0 \to \mathfrak{Y}_T$ has an extension (with the same norm) $\overline{T} : \mathfrak{X} \to \mathfrak{Y}_T$ if and only if there is a projection (of norm 1) from $\mathfrak{X}$ onto $\mathfrak{X}_0$. Thus Kakutani [6] proved that $\mathfrak{X}$ is a Hilbert space if and only if an arbitrary bounded operator defined on a subspace of $\mathfrak{X}$ has an extension to all of $\mathfrak{X}$ with the same norm.

By allowing the range space to increase, Phillips [3] and Sobczyk [3] showed that it is always possible to find an extension without increasing the norm. Sobczyk [3] proved that if $\mathfrak{X}_0$ is closed in $\mathfrak{X}$ and T is a one-to-one bounded linear operator from $\mathfrak{X}_0$ to $\mathfrak{Y}$, then there exists a space $\mathfrak{W} \supset \mathfrak{X}$ and an extension $\overline{T} : \mathfrak{X} \to \mathfrak{W}$ which is also one-to-one and has the same norm as T.

Kelley [2] has considered a reciprocal problem of characterizing real B-spaces $\mathfrak{X}$ with the property that a bounded linear operator on a closed subspace of any B-space $\mathfrak{Y}$ to $\mathfrak{X}$ has a norm-preserving extension to all of $\mathfrak{Y}$. He shows that $\mathfrak{X}$ is equivalent to $C(S)$, where S is an extremally disconnected compact Hausdorff space. This extends a previous result of Nachbin [3] and Goodner [1].

References. Akilov [1, 2], Goodner [1], Kakutani [6], Kelley [2], Nachbin [3], Phillips [3], Sobczyk [3], Taylor [1].

CHAPTER VII

General Spectral Theory

In the first section of this chapter we shall see that a study of the polynomials in an operator on a finite dimensional unitary space leads to a rather complete description of the analytic behavior of the operator and, at the same time, furnishes a clear geometric picture of the manner in which the operator transforms the unitary space upon which it operates. In attempting to make a corresponding study of an operator T on an infinite dimensional complex B-space, one is immediately confronted with the necessity of introducing an algebra larger than that consisting of the polynomials in T. The development of the theory in a finite dimensional space suggests that a useful definition of a function $f(T)$ of an operator T is given by the Cauchy formula

$$f(T) = \frac{1}{2\pi i}\int_C \frac{f(\lambda)}{\lambda - T}\, d\lambda,$$

in which f is an analytic scalar function and C is an appropriately chosen contour. In giving a meaning to this formula we are naturally led to the study of questions concerning the existence of, and properties of the function $(\lambda I - T)^{-1}$. It will be seen that $(\lambda I - T)^{-1}$ is defined and analytic everywhere in the λ-plane except for λ in a certain compact set, which is known as the *spectrum* of T. The general concepts and methods introduced in this chapter all center around the concept of the spectrum of an operator. It is for this reason that the term "spectral theory" is used in the title of the chapter.

The present chapter is a fundamental turning point in our investigations. Until now our study has been directed towards the topological aspect of operator theory. The function-theoretic and algebraic mechanisms introduced here will make possible our future detailed analyses of operators.

1. Spectral Theory in a Finite Dimensional Space

Throughout this section, $\mathfrak{X}$ will denote a finite dimensional complex B-space, and T a linear operator in $B(\mathfrak{X})$. Our aim is to study the algebraic and topological properties of T. This is achieved, here and in the more general theory of the next sections, by studying a class of functions of the operator T. The symbol I will denote the identity operator in $\mathfrak{X}$; we will interpret T^0 as I.

1 DEFINITION. If $P(\lambda) = \sum_{i=0}^{n} \alpha_i \lambda^i$ is a polynomial with complex coefficients, then by $P(T)$ is meant the sum $\sum_{i=0}^{n} \alpha_i T^i = \alpha_n T^n + \ldots + \alpha_1 T + \alpha_0 I$.

Our first task is to discover when two polynomials determine the same function of T.

2 DEFINITION. The *spectrum* $\sigma(T)$ of an operator T in a finite dimensional B-space is the set of complex numbers λ such that $\lambda I - T$ is not one-to-one. The *index* $\nu(\lambda)$ of a complex number λ is the smallest non-negative integer ν such that $(\lambda I - T)^{\nu} x = 0$ for every vector x for which $(\lambda I - T)^{\nu+1} x = 0$.

It follows that if $\lambda_0 \in \sigma(T)$, there exists an $x_0 \neq 0$ such that $(T - \lambda_0 I)x_0 = 0$. The number λ_0 is frequently called an *eigenvalue* of T, and any corresponding x_0 is called an *eigenvector*.

For each non-negative integer n and complex number λ, define the linear manifold $\mathfrak{N}_\lambda^n = \{x | (T - \lambda I)^n x = 0\}$. Then the index $\nu(\lambda)$ is the least integer ν such that $\mathfrak{N}_\lambda^{\nu+1} = \mathfrak{N}_\lambda^{\nu}$. Observe that $\mathfrak{N}_\lambda^n = \mathfrak{N}_\lambda^{\nu(\lambda)}$ for $n \geqq \nu(\lambda)$. Since $\mathfrak{X}$ has finite dimension, there can be proper inclusion for at most a finite number of terms in the sequence $\mathfrak{N}_\lambda^0 \subseteqq \mathfrak{N}_\lambda^1 \subseteqq \mathfrak{N}_\lambda^2 \subseteqq \ldots$, and thus $\nu(\lambda) \leqq \dim \mathfrak{X}$ for every λ. We note that $\lambda \in \sigma(T)$ if and only if $\nu(\lambda) > 0$. For example, the operator T in two dimensional space given by the matrix $\left(\begin{smallmatrix} 0 & 1 \\ 0 & 0 \end{smallmatrix}\right)$ has $\sigma(T) = \{0\}$ and $\nu(0) = 2$.

3 THEOREM. *If P and Q are polynomials, then $P(T) = Q(T)$ if and only if $P - Q$ has a zero of order $\nu(\lambda)$ at each point λ in $\sigma(T)$.*

PROOF. It is clearly sufficient to consider the case $Q = 0$. Let $\mathfrak{X}$ be k-dimensional, with basis $\{x_1, \ldots, x_k\}$. Then the $k+1$ vectors $x_1, Tx_1, \ldots, T^k x_1$ must be linearly dependent, so there exists a non-zero polynomial S_1 such that $S_1(T)x_1 = 0$. In the same way, there exist non-zero polynomials S_i, $i = 2, \ldots, k$ such that $S_i(T)x_i = 0$. If $R = S_1 \cdot S_2 \cdots S_k$, then $R(T)x_i = 0$, and consequently $R(T)x = 0$ for all $x \in \mathfrak{X}$. Thus a non-zero polynomial R exists such that $R(T) = 0$. Let R be factored as $R(\lambda) = \beta \prod_{i=1}^{m} (\lambda - \lambda_i)^{\alpha_i}$. If $\lambda_i \notin \sigma(T)$, then $(T - \lambda_i I)x = 0$ implies $x = 0$. Consequently, the product R_1 of all the factors $(\lambda - \lambda_i)^{\alpha_i}$ in R such that $\lambda_i \in \sigma(T)$, still satisfies $R_1(T) = 0$. In the same way, the product R_2 of all the factors $(\lambda - \lambda_i)^{\beta_i}$, where $\beta_i = \min(\alpha_i, \nu(\lambda_i))$, satisfies $R_2(T) = 0$. Since any polynomial P having a zero of order $\nu(\lambda)$ at each $\lambda \in \sigma(T)$ is divisible by R_2, any such polynomial satisfies $P(T) = 0$.

To prove the converse statement, suppose that $P(T) = 0$, where $P(\lambda) = \beta \prod_{i=1}^{p} (\lambda - \lambda_i)^{\alpha_i}$. By the argument used in the first part of the theorem we may suppose each $\lambda_i \in \sigma(T)$. We now show $\sigma(T) = \{\lambda_1, \ldots, \lambda_p\}$. For if $\lambda_0 \in \sigma(T)$, then $Ty = \lambda_0 y$ for some vector $y \neq 0$. Then $P(T)y = P(\lambda_0)y$, and, since $P(T) = 0$, it follows that $P(\lambda_0) = 0$. To prove that $\alpha_1 \geqq \nu(\lambda_1)$, we suppose on the contrary that $\alpha_1 < \nu(\lambda_1)$, so that there exists an $x_1 \neq 0$ with $(T - \lambda_1 I)^{\alpha_1 + 1} x_1 = 0$ and $y_1 = (T - \lambda_1 I)^{\alpha_1} x_1 \neq 0$. Let $P(\lambda) = (\lambda - \lambda_1)^{\alpha_1} Q(\lambda)$, where $Q(\lambda_1) \neq 0$. Since $Ty_1 = \lambda_1 y_1$, $P(T)x_1 = Q(T)y_1 = Q(\lambda_1)y_1 \neq 0$, contradicting the fact $P(T) = 0$. Similarly $\alpha_i \geqq \nu(\lambda_i)$, $i = 2, \ldots, p$, so that P has a root of order $\nu(\lambda)$ for each $\lambda \in \sigma(T)$. Q.E.D.

4 Corollary. *The spectrum of an operator in a finite dimensional space is a non-void finite set of points.*

Proof. This follows from Theorem 3. We have seen that T always satisfies a non-zero polynomial equation $P(T) = 0$, where the roots of $P(\lambda)$ are the spectral points of T. Q.E.D.

Theorem 3 enables us to define the operator $f(T)$ for f in a class of functions larger than that consisting of the polynomial functions. Let $\mathscr{F}(T)$ be the class of all functions of the complex variable λ which are analytic in some open set containing $\sigma(T)$. The open set need not be connected, and can vary with the function in $\mathscr{F}(T)$. If $f \in \mathscr{F}(T)$, let P be a polynomial such that $f^{(m)}(\lambda) = P^{(m)}(\lambda)$, $m \leqq \nu(\lambda) - 1$, for each $\lambda \in \sigma(T)$. We define $f(T) = P(T)$. It follows from Theorem 3 that this definition of $f(T)$ is unambiguous. The following theorem follows immediately from the corresponding results for polynomial functions.

5 THEOREM. *If f, g are in $\mathscr{F}(T)$ and α, β are complex numbers, then:*

(a) *$\alpha f+\beta g$ is in $\mathscr{F}(T)$ and $(\alpha f+\beta g)(T) = \alpha f(T)+\beta g(T)$;*
(b) *$f \cdot g$ is in $\mathscr{F}(T)$ and $(f \cdot g)(T) = f(T) \cdot g(T)$;*
(c) *if $f(\lambda) = \sum_{n=0}^{m} \alpha_n \lambda^n$ then $f(T) = \sum_{n=0}^{m} \alpha_n T^n$;*
(d) *$f(T) = 0$ if and only if*

$$f^{(m)}(\lambda) = 0, \qquad \lambda \in \sigma(T), \quad 0 \leqq m \leqq \nu(\lambda)-1.$$

Note that (b) implies $f(T)g(T) = g(T)f(T)$ for all $f, g \in \mathscr{F}(T)$.

If λ_0 is a complex number, let $e_{\lambda_0}(\lambda)$ be identically equal to one in a neighborhood of λ_0, and identically equal to zero in a neighborhood of each point of $\sigma(T) \cap \{\lambda_0\}'$. Put $E(\lambda_0) = e_{\lambda_0}(T)$. The next theorem follows immediately from Theorem 5.

6 THEOREM. (a) *$E(\lambda_0) \neq 0$ if and only if λ_0 is in $\sigma(T)$.*
(b) *$E(\lambda_0)^2 = E(\lambda_0)$ and $E(\lambda_0)E(\lambda_1) = 0$ for $\lambda_0 \neq \lambda_1$.*
(c) *$I = \sum_{\lambda \epsilon \sigma(T)} E(\lambda)$.*

Let $\{\lambda_1, \ldots, \lambda_k\}$ be an enumeration of $\sigma(T)$, and let $\mathfrak{X}_i = E(\lambda_i)\mathfrak{X}$. It follows from (b) and (c) of Theorem 6 that

$$\mathfrak{X} = \mathfrak{X}_1 \oplus \ldots \oplus \mathfrak{X}_k.$$

Moreover, since $TE(\lambda_i) = E(\lambda_i)T$, it follows that $T\mathfrak{X}_i \subseteq \mathfrak{X}_i$, $i = 1, \ldots, k$. Thus, to the decomposition of the spectrum $\sigma(T)$ into k points there corresponds a direct sum decomposition of $\mathfrak{X}$ into k subspaces, each of which is mapped into itself by the operator T. Thus the study of the action of T on $\mathfrak{X}$ may be reduced to the study of the action of T on each of the subspaces $\mathfrak{X}_i$. Since $(\lambda_i-\lambda)^{\nu(\lambda_i)} e_{\lambda_i}(\lambda)$ has a zero of order $\nu(\lambda)$ at each point λ in $\sigma(T)$, the operator $(\lambda_i I-T)^{\nu(\lambda_i)}E(\lambda_i) = 0$. Thus, in each space $\mathfrak{X}_i$, the operator T is the sum of a scalar multiple $\lambda_i I$ of the identity and a nilpotent operator $T-\lambda_i I$. Such a direct sum decomposition of $\mathfrak{X}$ can be enormously advantageous in many investigations of the properties of T. Theorem 7 clarifies the relation between the index $\nu(\lambda)$ of a point in $\sigma(T)$ and the corresponding projection $E(\lambda)$.

7 THEOREM. *If λ is in $\sigma(T)$, then*

$$E(\lambda)\mathfrak{X} = \mathfrak{N}_\lambda^{\nu(\lambda)} = \{x \mid (T-\lambda I)^{\nu(\lambda)}x = 0\}.$$

PROOF. The preceding paragraph demonstrated that if $\lambda \in \sigma(T)$, then $(\lambda I - T)^{\nu(\lambda)}E(\lambda) = 0$. This shows that $E(\lambda)\mathfrak{X} \subseteq \mathfrak{N}_\lambda^{\nu(\lambda)}$. Since $\sum_{\lambda\in\sigma(T)} E(\lambda) = I$, to show the reverse inclusion it is sufficient to prove that $\mathfrak{N}_\lambda^{\nu(\lambda)} \cap \mathfrak{N}_\mu^{\nu(\mu)} = \{0\}$, for $\lambda \neq \mu$, $\lambda, \mu \in \sigma(T)$.

Suppose that there is an $x \neq 0$ with $x \in \mathfrak{N}_\lambda^{\nu(\lambda)} \cap \mathfrak{N}_\mu^{\nu(\mu)}$. Let α, $0 \leqq \alpha < \nu(\lambda)$, be the greatest integer such that $z = (T-\lambda I)^\alpha x \neq 0$. Then $Tz = \lambda z$, so $(T-\mu I)^{\nu(\mu)}z = (\lambda-\mu)^{\nu(\mu)}z \neq 0$. On the other hand, $(T-\mu I)^{\nu(\mu)}z = (T-\mu I)^{\nu(\mu)}(T-\lambda I)^\alpha x = (T-\lambda I)^\alpha(T-\mu I)^{\nu(\mu)}x = 0$, since $x \in \mathfrak{N}_\mu^{\nu(\mu)}$. Hence x must be zero. Q.E.D.

The projections $E(\lambda)$ define a very useful direct sum decomposition of $\mathfrak{X}$, and permit us to give a simple analytical expression for functions of T.

8 THEOREM. *If f is in $\mathscr{F}(T)$, then*

$$f(T) = \sum_{\lambda\in\sigma(T)} \sum_{i=0}^{\nu(\lambda)-1} \frac{(T-\lambda I)^i}{i!} f^{(i)}(\lambda)E(\lambda).$$

PROOF. This formula follows immediately from Theorem 5, since f and the function $g \in \mathscr{F}(T)$ defined by

$$g(\mu) = \sum_{\lambda\in\sigma(T)} \sum_{i=0}^{\nu(\lambda)-1} \frac{(\mu-\lambda)^i}{i!} f^{(i)}(\lambda)e_\lambda(\mu)$$

satisfy the relations $f^{(m)}(\lambda) = g^{(m)}(\lambda)$, $m \leqq \nu(\lambda)-1$, for $\lambda \in \sigma(T)$. Q.E.D.

Theorem 8 gives us a powerful method for explicit calculation of functions of T, and has a number of interesting theoretical implications.

9 THEOREM. *Let f_n be in $\mathscr{F}(T)$. Then $\{f_n(T)\}$ converges if and only if the sequences $\{f_n^{(m)}(\lambda)\}$, $0 \leqq m \leqq \nu(\lambda)-1$, converge for λ in $\sigma(T)$. If f is in $\mathscr{F}(T)$, then $f_n(T) \to f(T)$ if and only if $f_n^{(m)}(\lambda) \to f^{(m)}(\lambda)$, $0 \leqq m \leqq \nu(\lambda)-1$, for λ in $\sigma(T)$.*

PROOF. The sufficiency of the first condition follows immediately from Theorem 8. Conversely, suppose that $\{f_n(T)\}$ converges, and let $\lambda \in \sigma(T)$. Since $(T-\lambda I)^{\nu(\lambda)-1}E(\lambda) \neq 0$, we can find an x such that $(T-\lambda I)^{\nu(\lambda)-1}E(\lambda)x \neq 0$. Let $y = E(\lambda)x$, and $y_k = (T-\lambda I)^k y$, for $0 \leqq k \leqq \nu(\lambda)-1$. Put $\nu(\lambda)-1 = r$. Then $f_n(T)y_r = f_n(\lambda)y_r$, so that $\{f_n(\lambda)\}$ converges. In the same way, $f_n(T)y_{r-1} = f_n(\lambda)y_{r-1} + f_n'(\lambda)y_r$, so that $\{f_n'(\lambda)\}$ converges. Proceeding in this inductive fashion, we establish that $\{f_n^{(m)}(\lambda)\}$ converges for each $m \leqq \nu(\lambda)-1$. The second statement is proved in the same manner. Q.E.D.

10 THEOREM. *Let f in $\mathscr{F}(T)$ be analytic in a domain containing the closure of an open set U containing $\sigma(T)$, and suppose that the boundary B of U consists of a finite number of closed rectifiable Jordan curves, oriented in the positive sense customary in the theory of complex variables. Then $f(T)$ may be expressed as a Riemann contour integral over B by the following formula:*

$$f(T) = \frac{1}{2\pi i}\int_B f(\lambda)(\lambda I - T)^{-1} d\lambda.$$

PROOF. Let $\lambda \notin \sigma(T) = \{\lambda_1, \ldots, \lambda_k\}$, and let $r(\xi) = (\lambda-\xi)^{-1}$. Then, by Theorems 5 and 8,

$$(\lambda I - T)^{-1} = r(T) = \sum_{j=1}^{k} \sum_{\nu=0}^{\nu(\lambda_j)-1} \frac{(T-\lambda_j I)^{\nu}}{(\lambda-\lambda_j)^{\nu+1}} E(\lambda_j).$$

Thus, if $f \in \mathscr{F}(T)$,

$$\frac{1}{2\pi i}\int_B f(\lambda)(\lambda I - T)^{-1} d\lambda = \sum_{j=1}^{k} \sum_{\nu=0}^{\nu(\lambda_j)-1} (T-\lambda_j I)^{\nu} \frac{f^{(\nu)}(\lambda_j)}{\nu!} E(\lambda_j) = f(T).$$

Q.E.D.

A few examples will illustrate applications of the preceding theorems.

(a) Put $f_n(\lambda) = n^{-1} \sum_{m=0}^{n-1} \lambda^m$ and $g_n(\lambda) = \lambda^n/n$. Then $\{f_n(\lambda)\}$ converges if and only if $|\lambda| \leqq 1$, while if $j > 0$, $\{f_n^{(j)}(\lambda)\}$ converges if and only if $|\lambda| < 1$. It is apparent that the restrictions on λ in order that $g_n(\lambda) \to 0$, and $g_n^{(j)}(\lambda) \to 0$, for $j > 0$, are exactly the same. Hence, it follows from Theorem 9 that $\{n^{-1} \sum_{m=0}^{n-1} T^m\}$ converges if and only if $\{T^n/n\}$ converges to 0, and that this is the case if and only if $|\sigma(T)| \leqq 1$ and $\nu(\lambda_0) = 1$ for those $\lambda_0 \in \sigma(T)$ with $|\lambda_0| = 1$.

(b) Let a machine shuffle a deck of cards in such a way that there is a definite probability p_{ij} that a card originally in the i-th place be found in the j-th place after the shuffle. Suppose that this probability p_{ij} is independent of the preceding shuffles. Let $p_{ij}^{(n)}$ be the probability that the card originally in the i-th place be found in the j-th place after n shuffles, and let $p_{ij}^{(0)} = \delta_{ij}$. We show the existence of the limit

$$\lim_{n\to\infty} \frac{1}{n} \sum_{\nu=0}^{n-1} p_{ij}^{(\nu)}, \qquad i, j = 1, 2, \ldots, 52.$$

There are 52 different ways in which the card originally in the i-th place may transfer to the j-th place in two shuffles; it may go to first place after the first shuffle and then from first place to the j-th place, or it may go to second place and then to j-th place, etc. Thus

$$p_{ij}^{(2)} = \sum_{i=1}^{52} p_{ij} p_{jk}.$$

A similar inductive argument shows that $p_{ij}^{(n+1)} = \sum_{k=1}^{52} p_{ik}^{(n)} p_{kj}$ for $n = 0, 1, \ldots$. The matrix $P = (p_{ij})$ defines a linear transformation in E^{52} whose n-th power is $P^n = (p_{ij}^{(n)})$. Since $0 \leqq p_{ij}^{(n)} \leqq 1$, we have $P^n/n \to 0$. The preceding example shows that $n^{-1} \sum_{\nu=0}^{n-1} p_{ij}^{(\nu)}$ converges.

(c) For an arbitrary scalar t, the function $e^{\lambda t}$ is entire in λ, and hence belongs to $\mathscr{F}(T)$. Furthermore,

$$\lim_{h\to 0} \frac{d^n}{d\lambda^n} \frac{e^{\lambda(t+h)} - e^{\lambda t}}{h} = \frac{d^n}{d\lambda^n} (\lambda e^{\lambda t}), \qquad n = 0, 1, 2, \ldots.$$

Thus, by Theorem 9, $(d/dt)e^{tA} = Ae^{tA}$ for all matrices A in E^m. Since $e^{-tA}e^{tA} = I$, the columns of the matrix e^{tA} form a set of m linearly independent solutions of the equation $dy/dt = Ay$.

(d) If T is such that $\nu(\lambda) = 1$ for $\lambda \in \sigma(T)$, Theorem 8 yields

$$f(T) = \sum_{i=1}^{k} f(\lambda_i) E(\lambda_i).$$

We shall show that $\nu(\lambda) = 1$ for $\lambda \in \sigma(T)$, if T is an *Hermitian matrix* in the finite dimensional Hilbert space E^n; i.e., if T satisfies the identity $(Tx, y) = (x, Ty)$ for $x, y \in E^n$. For, if T is Hermitian, and $(T-\lambda I)x = 0$ for some $x \neq 0$, then $(Tx, x) = \lambda(x, x)$. Since $(Tx, x) = (x, Tx) = \overline{(Tx, x)}$, it follows that λ is real; hence $\sigma(T)$ is a subset of the real axis. Now let $\lambda \in \sigma(T)$, and let $(T-\lambda I)^2 y = 0$. Since λ is real, $((T-\lambda I)^2 y, y) = ((T-\lambda I)y, (T-\lambda I)y) = |(T-\lambda I)y|^2 = 0$, so that $(T-\lambda I)y = 0$. Consequently, $\nu(\lambda) = 1$, for any $\lambda \in \sigma(T)$.

2. Exercises

1 Let T be a matrix, and let $\Delta(\lambda)$ be the determinant of the matrix $\lambda I - T$. Show that $\lambda \in \sigma(T)$ if and only if λ is a root of the poly-

nomial equation $\Delta(\lambda) = 0$. The polynomial $\Delta(\lambda)$ is called the *characteristic polynomial* of the matrix T.

2 If T is a linear operator in a finite dimensional space, show that any two matrix representations of T have the same characteristic polynomial. This will permit us to speak without ambiguity of the characteristic polynomial of a linear operator.

3 If T is a linear operator in a finite dimensional space and $\mu \in \sigma(T)$, let T_μ be the restriction of T to the subspace $\mathfrak{N}_\mu^{\nu(\mu)}$. Let $\Delta_\mu(\lambda)$ be the characteristic polynomial of T_μ. Show that

(a) $\Delta_\mu(\lambda) = (\lambda-\mu)^{n(\mu)}$, where $n(\mu) = \dim \mathfrak{N}_\mu^{\nu(\mu)}$;

(b) $\nu(\mu) \leqq n(\mu)$;

(c) $\Delta(\lambda) = \prod_{\mu \in \sigma(T)} \Delta_\mu(\lambda)$.

4 (Hamilton-Cayley) Show that any linear transformation T in a finite dimensional space satisfies its characteristic equation, i.e., $\Delta(T) = 0$.

5 Let

$$T = \begin{pmatrix} 1 & 0 & 3 \\ 2 & 1 & 2 \\ 0 & 0 & 2 \end{pmatrix}.$$

Find matrices A and B for which

$$A^4 = T^3, \qquad T = \sum_{n=0}^{\infty} \frac{B^n}{n!}.$$

How many matrices can you find?

6 Let $\sigma(T) = \{\lambda_1, \ldots, \lambda_k\}$. Show that if $\nu(\lambda_i) = 1$, $i = 1, \ldots, k$, and $f \in \mathscr{F}(T)$, then

$$f(T) = \sum_{i=1}^{k} \frac{f(\lambda_i)}{m'(\lambda_i)} \prod_{j \neq i} (T-\lambda_j I),$$

where $m(\lambda) = (\lambda-\lambda_1)\ldots(\lambda-\lambda_k)$.

7 Put $\mathfrak{N}_\lambda^j = \{x | (T-\lambda I)^j x = 0\}$. Show that $\mathfrak{X} = \mathfrak{N}_\lambda^j \oplus (T-\lambda I)^j \mathfrak{X}$ if and only if $j \geqq \nu(\lambda)$ or $j = 0$.

8 If $E = \lim_n n^{-1} \sum_{i=0}^{n-1} T^i$ show that $E^2 = E$, $ET = TE$, that $E(\mathfrak{X})$ is the set $\{x | x \in \mathfrak{X}, Tx = x\}$, and that $(I-E)(\mathfrak{X})$ is the set $\{x | x \in \mathfrak{X}, x = (I-T)y\}$.

9 Show that $\qquad |(\lambda-\lambda_0)^j (\lambda I-T)^{-1}|$

is bounded in a neighborhood of λ_0 if and only if $j \geqq \nu(\lambda_0)$.

10 Let $f \in \mathscr{F}(T)$, $g \in \mathscr{F}(f(T))$, and $F(\xi) = g(f(\xi))$. Then $F \in \mathscr{F}(T)$, and $F(T) = g(f(T))$.

11 Let $f \in \mathscr{F}(T)$ and let $f(\lambda) = \sum_{n=0}^{\infty} \alpha_n \lambda^n$. Find conditions which are necessary and sufficient in order that $f(T) = \sum_{n=0}^{\infty} \alpha_n T^n$.

12 If, for some real number $r < 2$, the elements of the matrix $\lambda^r(\lambda I - T)^{-1}$ are bounded in some deleted neighborhood $0 \neq |\lambda| < \varepsilon$ of the origin, what can be said about the existence of the limit $\lim_{\lambda \to 0} \lambda(\lambda I - T)^{-1}$?

13 The spectra $\sigma(T)$ and $\sigma(T^*)$ are identical. Moreover $f(T^*) = f(T)^*$ for $f \in \mathscr{F}(T)$.

14 An operator T in the finite-dimensional Hilbert space is said to be *normal* if $TT^* = T^*T$, where T^* is the Hilbert space adjoint defined by the identity $(Tx, y) = (x, T^*y)$. Show that, if T is normal, then $\nu(\lambda) = 1$ for $\lambda \in \sigma(T)$.

15 If $\lambda \notin \sigma(T)$, the domain of $(\lambda I - T)^{-1}$ is $\mathfrak{X}$.

16 Let $\mathfrak{X}$ have dimension n. If T satisfies $T^n = 0$, show that there is a basis $\{x_1, \ldots, x_n\}$ for $\mathfrak{X}$, and a set of integers $n_0, \ldots, n_k$, $1 \leqq n_0 < n_1 < \ldots < n_k = n$, such that $Tx_i = x_{i+1}$ unless i is one of the integers n_j, in which case $Tx_i = 0$.

17 (Jordan) Let $\mathfrak{X}$ have dimension n. Show that there is a basis $\{x_1, \ldots, x_n\}$ for $\mathfrak{X}$, a set of integers $n_0, \ldots, n_k$, such that $0 = n_0 < n_1 < \ldots < n_k = n$, and an enumeration (with possible repetitions) $\lambda_1, \ldots, \lambda_k$ of $\sigma(T)$, such that $Tx_i = \lambda_j x_i + x_{i+1}$ for $n_{j-1} < i < n_j$, $Tx_{n_j} = \lambda_j x_{n_j}$.

18 Let B be the Boolean algebra of sets in the complex plane and let B_1 be the Boolean algebra of projections generated by the projections $E(\lambda_1), \ldots, E(\lambda_k)$. Let a map $E : B \to B_1$ be defined by placing $E(\sigma) = 0$ if $\sigma \cap \sigma(T)$ is void and otherwise defining $E(\sigma)$ as the sum of all $E(\lambda_i)$ with $\lambda_i \in \sigma$. Show that

(i) E is a homomorphism and $E(\sigma(T)) = I$;

(ii) $E(\sigma)T = TE(\sigma)$, $\quad \sigma \in B$;

(iii) the spectrum of T when considered as an operator in $E(\sigma)\mathfrak{X}$ is contained in σ.

In addition, show that no other map E of B into B_1 satisfies the conditions (i), (ii), and (iii).

19 Find the most general solution of the differential equation $y' = Ty$ where T is the matrix of Exercise 5.

The next ten exercises refer to the stability theory of systems of n linear homogeneous differential equations, $dy(t)/dt = A(t)y$. Here $A(t) = (a_{ij}(t))$ is an $n \times n$ complex valued matrix, continuous in the real variable $t \in I$: $-\infty \leqq \alpha < t < \beta \leqq +\infty$, and a solution is a complex valued (column) vector $y(t) \in E^n$, differentiable and satisfying the differential system for each $t \in I$. The theory of differential equations assures us that for each $t_0 \in I$ and each $y_0 \in E^n$ there exists a unique solution $y(t)$ such that $y(t_0) = y_0$.

20 Show that the solutions of $dy/dt = A(t)y$ form an n-dimensional complex linear vector space.

21 Consider the matrix differential equation $dY/dt = A(t)Y$ where a solution is a $n \times n$ complex valued matrix $Y(t)$, differentiable and satisfying the differential equation for each $t \in I$. Show that for each $t_0 \in I$ and each complex (constant) matrix Y_0 there exists a unique solution matrix $Y(t)$ such that $Y(t_0) = Y_0$. Show that a set of solution vectors $y_1(t), \ldots, y_n(t)$ of $dy/dt = A(t)y$ form a basis for the solution space if and only if they form the columns of a solution matrix $Y(t)$ of $dY/dt = A(t)Y$ which corresponds to a *non-singular* initial matrix Y_0.

22 Show that if $\int_{t_0}^t A(s)ds$ and $A(t)$ commute for each $t \in I$, then the matrix solution of $dY/dt = A(t)Y$ with $Y(t_0) = Y_0$ is $Y(t) = [\exp \int_{t_0}^t A(s)ds]Y_0$. Show that if $A(t)$ is either constant or diagonal (i.e. all entries off the principal diagonal are zero), then the above formula for $Y(t)$ is correct.

23 The differential system $dy/dt = A(t)y$, where $\alpha < t < \beta = \infty$, is called *stable* if every solution remains bounded as $t \to \infty$; that is, $\limsup_{t\to\infty} |y(t)| < \infty$. Show that, if $A(t) = A$ is constant, the system is stable if and only if no points of $\sigma(A)$ have positive real parts, while all pure imaginary $\lambda \in \sigma(A)$ have $\nu(\lambda) = 1$. Show that $\limsup_{t\to\infty} |y(t)| = 0$ for every solution if and only if $\mathscr{R}\sigma(A) < 0$.

24 Let $Y(t)$ be a solution matrix of $dY/dt = A(t)Y$ and show that

$$(d/dt)\,[\det Y(t)] = [\det Y(t)]\,[\operatorname{tr} A(t)]$$

at each point $t \in I$ where $Y(t)$ is non-singular and $\operatorname{tr} A(t)$ is the sum of the entries on the principal diagonal. Show that a solution matrix

$Y(t)$ is non-singular everywhere on I if it is non-singular at one point on I.

25 Let $Y(t)$ be a solution matrix of $dY/dt = A(t)Y$ which is non-singular. Show that the set of all non-singular matrix solutions are precisely the matrices $Y(t)C$ where C is any $n \times n$ constant, non-singular matrix.

26 Let $A(t)$ have period $p > 0$; that is, $A(t+p) = A(t)$ for all t on $-\infty < t < \infty$. If $Y(t)$ is a non-singular solution matrix of $dY/dt = A(t)Y$ then show that $Y(t+p) = Y(t)C$ for some non-singular constant matrix C. Show that the correspondence $y(t) \to y(t+p)$ defines a non-singular linear transformation T of the solution space of $dy/dt = A(t)y$ onto itself and that the matrix of T, relative to the basis formed by the columns of $Y(t)$, is just C.

27 Let A, Y, and G be as in Exercise 26. Show that $X(t)$ is a non-singular solution matrix of $dX/dt = KX$ if and only if $Y(t) = P(t)X(t)$ is a non-singular solution matrix of $dY/dt = A(t)Y$; where $P(t) = Y(t) \exp(-t/p \log C)$ and $A(t)$ have period $p > 0$, $K = p^{-1} \log C$, and C is the constant period-shift matrix for $Y(t)$. Show that the real parts of the points in $\sigma(K)$ are independent of the determination of $\log C$. These values $\mathscr{R}\sigma(K)$ are called the characteristic exponents of $A(t)$. Show that the system $dy/dt = A(t)y$ is stable if and only if the characteristic exponents of $A(t)$ are non-positive while those pure imaginary $\lambda \in \sigma(K)$ have $\nu(\lambda) = 1$. Also every solution $y(t) \to 0$ if and only if the characteristic exponents of $A(t)$ are all negative.

28 Suppose $A(t)$ is continuous and $\sup |a_{ij}(t)| < B$ for all $\alpha < t < \infty$ and some constant B. Show that for every solution vector $y_i(t)$ of $dy/dt = A(t)y$, $\limsup_{t\to\infty} t^{-1} \log |y_i(t)| = \lambda_i$, $|\lambda_i| < Bn$. The real numbers λ_i which arise in this way are called the generalized characteristic exponents of $A(t)$. Show that there are at most n distinct generalized characteristic exponents and that the differential system is stable whenever all the λ's are negative.

29 If $A(t)$ is either a constant or a periodic matrix show that the real parts of $\sigma(A)$ or the characteristic exponents of $A(t)$, respectively, are precisely the generalized characteristic exponents of $A(t)$.

3. Functions of an Operator

Throughout the remainder of this chapter, $\mathfrak{X}$ will denote a complex B-space, and T a bounded linear operator in $\mathfrak{X}$. *We exclude the trivial case* $\mathfrak{X} = \{0\}$.

If $\mathfrak{X}$ is infinite dimensional, T need not satisfy any non-trivial polynomial equation $P(T) = 0$. Nevertheless, the procedure suggested by Theorem 1.10 will enable us to generalize many of the results of Section 1 to the infinite dimensional case. We begin by studying the function $(\lambda I - T)^{-1}$.

1 DEFINITION. The *resolvent set* $\varrho(T)$ of T is the set of complex numbers λ, for which $(\lambda I - T)^{-1}$ exists as a bounded operator with domain $\mathfrak{X}$. The *spectrum* $\sigma(T)$ of T is the complement of $\varrho(T)$. The function $R(\lambda; T) = (\lambda I - T)^{-1}$, defined on $\varrho(T)$, is called the *resolvent function* of T, or simply the *resolvent* of T.

→ 2 LEMMA. *The resolvent set $\varrho(T)$ is open. The function $R(\lambda; T)$ is analytic in $\varrho(T)$.*

PROOF. Let λ be a fixed point in $\varrho(T)$, and let μ be any complex number with $|\mu| < |R(\lambda; T)|^{-1}$. It will be shown that $\lambda + \mu \in \varrho(T)$. Heuristic considerations, based upon an analogy with geometric series, suggest that if $(\lambda+\mu)I - T = \mu I + (\lambda I - T)$ has an inverse, it is given by the series

$$S(\mu) = \sum_{k=0}^{\infty} (-\mu)^k (\lambda I - T)^{-(k+1)} = \sum_{k=0}^{\infty} (-\mu)^k R(\lambda; T)^{k+1}.$$

Since $|\mu R(\lambda; T)| < 1$, this series converges. Since $S(\mu)$ commutes with T, and

$$\begin{aligned}((\lambda+\mu)I - T)S(\mu) &= (\lambda I - T)S(\mu) + \mu S(\mu) \\ &= \sum_{k=0}^{\infty} \{(-\mu R(\lambda; T))^k - (-\mu R(\lambda; T))^{k+1}\} \\ &= I,\end{aligned}$$

it follows that $\lambda + \mu \in \varrho(T)$, and that $R(\lambda+\mu; T) = S(\mu)$ is analytic at the point $\mu = 0$. Q.E.D.

3 COROLLARY. *If $d(\lambda)$ is the distance from λ to the spectrum $\sigma(T)$, then*

$$|R(\lambda; T)| \geqq \frac{1}{d(\lambda)}, \qquad \lambda \in \varrho(T).$$

Thus $|R(\lambda; T)| \to \infty$ *as* $d(\lambda) \to 0$, *and the resolvent set is the natural domain of analyticity of* $R(\lambda;$ T).

Proof. We have seen in the proof of Lemma 2 that if $|\mu| < |R(\lambda; T)|^{-1}$, then $\lambda+\mu \in \varrho(T)$. Hence $d(\lambda) \geqq |R(\lambda; T)|^{-1}$, from which the statements follow. Q.E.D.

➔ 4 Lemma. *The closed set* $\sigma(T)$ *is bounded and non-void. Moreover* $\sup |\sigma(T)| = \lim_{n\to\infty} \sqrt[n]{|T^n|} \leqq |T|$. *For* $|\lambda| > \sup |\sigma(T)|$ *the series* $R(\lambda; T) = \sum_{n=0}^{\infty} T^n/\lambda^{n+1}$ *converges in the uniform operator topology.*

Proof. Let $f(\lambda) = \sum_{n=0}^{\infty} T^n/\lambda^{n+1}$. From III.14 it is seen that the series $f(\lambda)$ has domain of convergence $D = \{\lambda | \, |\lambda| > \limsup_{n\to\infty} \sqrt[n]{|T^n|}\}$. As in the proof of Lemma 2, we verify that $f(\lambda)(\lambda I-T) = (\lambda I-T)f(\lambda) = I$, for $\lambda \in D$, so that $\varrho(T) \supseteqq D$, and consequently $\sigma(T)$ is bounded. Since $\varrho(T)$ is the natural domain of analyticity of $R(\lambda; T)$ by Corollary 3, the Laurent series for $R(\lambda; T)$ will have the domain of convergence $|\lambda| > \sup |\sigma(T)|$. Thus $\sup |\sigma(T)| = \limsup \sqrt[n]{|T^n|}$. It will next be shown that $\sup |\sigma(T)| \leqq \liminf \sqrt[n]{|T^n|}$. Note that if λ is an arbitrary point of $\sigma(T)$, then $\lambda^n \in \sigma(T^n)$; for the factorization

$$(\lambda^n I - T^n) = (\lambda I - T)P_n(T) = P_n(T)(\lambda I - T)$$

shows that if $(\lambda^n I - T^n)$ has a bounded inverse, so will $(\lambda I - T)$. Thus $|\lambda|^n \leqq |T^n|$, and hence

$$\sup |\sigma(T)| \leqq |T^n|^{1/n}.$$

It remains to show that the spectrum is not void. If $\sigma(T) = \phi$, then $R(\lambda; T)$ is entire, and, since $R(\lambda; T)$ is readily seen from its Laurent expansion to be analytic at infinity, it follows from Liouville's theorem (III.14) that $R(\lambda; T)$ is a constant. Hence, the coefficient of λ^{-1} in the Laurent expansion of $R(\lambda; T)$ vanishes, so that $I = 0$, which contradicts the assumption $\mathfrak{X} \neq \{0\}$. Q.E.D.

5 Definition. The quantity

$$r(T) = \sup |\sigma(T)| = \lim_{n\to\infty} \sqrt[n]{|T^n|}$$

is called the *spectral radius* of T.

6 LEMMA. *The following identity, known as the resolvent equation, is valid for every pair* λ, μ *in* $\varrho(T)$:

$$R(\lambda; T)-R(\mu; T) = (\mu-\lambda)R(\lambda; T)R(\mu; T).$$

PROOF. The lemma follows by multiplying both sides of the equation

$$\begin{aligned}(\mu I-T)(\lambda I-T)\{R(\lambda; T)-R(\mu; T)\} &= (\mu I-T)-(\lambda I-T)\\ &= (\mu-\lambda)I\end{aligned}$$

by $R(\lambda; T)R(\mu; T)$. Q.E.D.

7 LEMMA. *The spectrum of the adjoint* T^* *is identical with the spectrum of* T. *Further,* $R(\lambda; T^*) = R(\lambda; T)^*$ *for* λ *in* $\varrho(T) = \varrho(T^*)$.

PROOF. This follows immediately from VI.2.7. Q.E.D.

8 DEFINITION. By $\mathscr{F}(T)$, we denote the family of all functions f which are analytic on some neighborhood of $\sigma(T)$. (The neighborhood need not be connected, and can depend on $f \in \mathscr{F}(T)$.)

9 DEFINITION. Let $f \in \mathscr{F}(T)$, and let U be an open set whose boundary B consists of a finite number of rectifiable Jordan curves, oriented in the positive sense customary in the theory of complex variables. Suppose that $U \supseteq \sigma(T)$, and that $U \cup B$ is contained in the domain of analyticity of f. Then the operator $f(T)$ is defined by the equation

$$f(T) = \frac{1}{2\pi i}\int_B f(\lambda)R(\lambda; T)d\lambda.$$

It follows from Lemma 2, and from the Cauchy integral theorem, that $f(T)$ depends only on the function f, and not on the domain U.

→ 10 THEOREM. *If* f, g *are in* $\mathscr{F}(T)$, *and* α, β *are complex numbers, then*

(a) $\alpha f+\beta g$ *is in* $\mathscr{F}(T)$ *and* $\alpha f(T)+\beta g(T) = (\alpha f+\beta g)(T)$;

(b) $f \cdot g$ *is in* $\mathscr{F}(T)$ *and* $f(T) \cdot g(T) = (f \cdot g)(T)$;

(c) *if* f *has the power series expansion* $f(\lambda) = \sum_{k=0}^{\infty} \alpha_k \lambda^k$, *valid in a neighborhood of* $\sigma(T)$, *then* $f(T) = \sum_{k=0}^{\infty} \alpha_k T^k$;

(d) f *is in* $\mathscr{F}(T^*)$ *and* $f(T^*) = f(T)^*$.

PROOF. Statement (a) is obvious. It is clear that $f \cdot g \in \mathscr{F}(T)$; let U_1 and U_2 be two neighborhoods of $\sigma(T)$ whose boundaries B_1 and B_2 consist of a finite number of rectifiable Jordan curves, and suppose that $U_1 \cup B_1 \subseteqq U_2$. Suppose also that $U_2 \cup B_2$ is contained in a common region of analyticity of f and g. Then

$$
\begin{aligned}
f(T)g(T) &= -\frac{1}{4\pi^2}\left\{\int_{B_1} f(\lambda)R(\lambda; T)d\lambda\right\}\left\{\int_{B_2} g(\mu)R(\mu; T)d\mu\right\} \\
&= -\frac{1}{4\pi^2}\int_{B_1}\int_{B_2} f(\lambda)g(\mu)R(\lambda; T)R(\mu; T)d\mu d\lambda \\
&= -\frac{1}{4\pi^2}\int_{B_1}\int_{B_2} \frac{f(\lambda)g(\mu)(R(\lambda; T)-R(\mu; T))}{(\mu-\lambda)} d\mu d\lambda \\
&= -\frac{1}{4\pi^2}\int_{B_1} f(\lambda)R(\lambda; T)\left\{\int_{B_2}\frac{g(\mu)}{\mu-\lambda} d\mu\right\} d\lambda \\
&\qquad\qquad + \frac{1}{4\pi^2}\int_{B_2} g(\mu)R(\mu; T)\left\{\int_{B_1}\frac{f(\lambda)}{\mu-\lambda} d\lambda\right\} d\mu \\
&= \frac{1}{2\pi i}\int_{B_1} f(\lambda)g(\lambda)R(\lambda; T)d\lambda \\
&= (f\cdot g)(T),
\end{aligned}
$$

by Lemma 6 and the Cauchy integral formula. This proves (b). To prove (c), we note that the power series $\sum_{k=0}^{\infty}\alpha_k\lambda^k$ converges uniformly on the circle $C = \{\lambda \mid |\lambda| \leqq r(T)+\varepsilon\}$ for ε sufficiently small. Consequently,

$$
\begin{aligned}
f(T) &= \frac{1}{2\pi i}\int_C \left\{\sum_{k=0}^{\infty}\alpha_k\lambda^k\right\} R(\lambda; T)d\lambda \\
&= \frac{1}{2\pi i}\sum_{k=0}^{\infty}\alpha_k\int_C \lambda^k R(\lambda; T)d\lambda \\
&= \frac{1}{2\pi i}\sum_{k=0}^{\infty}\alpha_k\int_C \lambda^k\{\sum_{j=0}^{\infty} T^j/\lambda^{j+1}\}\, d\lambda \\
&= \sum_{k=0}^{\infty}\alpha_k T^k,
\end{aligned}
$$

by Lemma 4 and Cauchy's integral formula. Statement (d) follows immediately from Lemma 7. Q.E.D.

➜ 11 THEOREM. (*Spectral mapping theorem.*) *If f is in $\mathscr{F}(T)$, then* $f(\sigma(T)) = \sigma(f(T))$.

Proof. Let $\lambda \in \sigma(T)$, and define the function g in the domain of definition of f by

$$g(\xi) = (f(\lambda)-f(\xi))/(\lambda-\xi).$$

By Theorem 10, $f(\lambda)I-f(T) = (\lambda I-T)g(T)$. Hence, if $f(\lambda)I-f(T)$ had a bounded everywhere defined inverse A then $g(T)A$ would be a bounded everywhere defined inverse for $\lambda I-T$. Consequently $f(\lambda) \in \sigma(f(T))$.

Conversely, let $\mu \in \sigma(f(T))$, and suppose that $\mu \notin f(\sigma(T))$. Then the function $h(\xi) = (f(\xi)-\mu)^{-1}$ belongs to $\mathscr{F}(T)$. By Theorem 10, $h(T)(f(T)-\mu I) = I$, which contradicts the assumption that $\mu \in \sigma(f(T))$. Q.E.D.

➔ 12 **Theorem.** *Let f be in $\mathscr{F}(T)$, g be in $\mathscr{F}(f(T))$, and $F(\xi) = g(f(\xi))$. Then F is in $\mathscr{F}(T)$, and $F(T) = g(f(T))$.*

Proof. The statement $F \in \mathscr{F}(T)$ follows immediately from Theorem 11. Let U be a neighborhood of $\sigma(f(T))$ whose boundary B consists of a finite number of rectifiable Jordan arcs, and suppose that $U \cup B$ is contained in the domain of analyticity of g. Let V be a neighborhood of $\sigma(T)$ whose boundary C consists of a finite number of rectifiable Jordan arcs, and suppose that $V \cup C$ is contained in the domain of analyticity of f. Suppose, moreover, that $f(V \cup C) \subseteqq U$.

By Theorem 10, the operator

$$A(\lambda) = \frac{1}{2\pi i}\int_C \frac{R(\xi; T)}{\lambda-f(\xi)}\, d\xi$$

satisfies the equations $(\lambda I-f(T))A(\lambda) = A(\lambda)(\lambda I-f(T)) = I$. Thus $A(\lambda) = R(\lambda; f(T))$. Consequently,

$$\begin{aligned} g(f(T)) &= \frac{1}{2\pi i}\int_B g(\lambda)R(\lambda; f(T))d\lambda \\ &= -\frac{1}{4\pi^2}\int_B\int_C \frac{g(\lambda)R(\xi; T)}{\lambda-f(\xi)}\, d\xi d\lambda \\ &= \frac{1}{2\pi i}\int_C R(\xi; T)g(f(\xi))d\xi = F(T), \end{aligned}$$

by Cauchy's integral formula. Q.E.D.

The elementary algebraic rules of operation given by Theorems

10 and 12 will be used in the rest of this chapter without explicit reference to these theorems.

13 Lemma. *Let f_n be in $\mathscr{F}(T)$, $n = 1, \ldots$, and suppose that all the functions f_n are analytic in a fixed neighborhood V of $\sigma(T)$. Then, if f_n converges uniformly to f on V, $f_n(T)$ converges to $f(T)$ in the uniform topology of operators.*

Proof. Let U be a neighborhood of $\sigma(T)$ whose boundary B consists of a finite number of rectifiable Jordan arcs, and such that $U \cup B \subseteq V$. Then $f_n \to f$ uniformly on B, and consequently

$$\frac{1}{2\pi i}\int_B f_n(\lambda)R(\lambda; T)d\lambda$$

converges in the uniform topology of operators to

$$\frac{1}{2\pi i}\int_B f(\lambda)R(\lambda;T)d\lambda. \qquad \text{Q.E.D.}$$

The next lemma can be proved in the same way.

14 Lemma. *Let V be a neighborhood of $\sigma(T)$, and let U be an open set in the complex plane. Suppose that f is an analytic function of the two complex variables λ, μ for $[\lambda, \mu]$ in $V \times U$. Then $f(T, \mu)$ is a $B(\mathfrak{X})$-valued function which is analytic for $\mu \in U$.*

15 Definition. A point $\lambda_0 \in \sigma(T)$ is said to be an *isolated* point of $\sigma(T)$, if there is a neighborhood U of λ_0 such that $\sigma(T) \cap U = \{\lambda_0\}$. An isolated point λ_0 of $\sigma(T)$ is called a *pole of T*, or simply a *pole*, if $R(\lambda; T)$ has a pole at λ_0. By the *order* $\nu(\lambda_0)$ of a pole λ_0 is meant the order of λ_0 as a pole of $R(\lambda; T)$.

The following theorem, which is the analogue of Theorem 1.3, takes on a somewhat different form in the general case.

16 Theorem. (*Minimal equation*) *Let f, g be in $\mathscr{F}(T)$. Then $f(T) = g(T)$ if and only if $f(\lambda) = g(\lambda)$ identically on an open set including all of $\sigma(T)$ except for a finite number of poles $\{\lambda_1, \ldots, \lambda_k\}$, and for each i with $1 \leq i \leq k$, $f-g$ has a zero of order at least $\nu(\lambda_i)$ at λ_i.*

Proof. It is clearly sufficient to consider the case $g = 0$. Let f vanish identically on an open set V containing all of $\sigma(T)$ but the poles $\lambda_1, \ldots, \lambda_k$. Then it is clear from Definition 9 that

$$f(T) = \sum_{i=1}^{k} \frac{1}{2\pi i} \int_{C_k} f(\lambda)R(\lambda; T)d\lambda,$$

where C_k is a sufficiently small circle about λ_k. If $f(\lambda)$ has a zero of order at least $\nu(\lambda_i)$ at $\lambda = \lambda_i$, then, since $R(\lambda; T)$ has a pole of order $\nu(\lambda_i)$ at $\lambda = \lambda_i$, it follows that $f(\lambda)R(\lambda; T)$ is regular inside C_k. Thus, by Cauchy's integral formula, $f(T) = 0$.

Conversely, let $f(T) = 0$; then, by Theorem 11, $f(\sigma(T)) = 0$. Let f be analytic on a neighborhood U of $\sigma(T)$. For each $\alpha \in \sigma(T)$, there is an $\varepsilon(\alpha) > 0$ such that the sphere $S(\alpha, \varepsilon(\alpha)) \subseteq U$. Since $\sigma(T)$ is compact, a finite set of spheres $S(\alpha_1, \varepsilon(\alpha_1)), \ldots, S(\alpha_n, \varepsilon(\alpha_n))$ will cover $\sigma(T)$. If some sphere $S(\alpha_i, \varepsilon(\alpha_i))$ contains an infinite number of points of $\sigma(T)$, it follows from the theory of complex variables that f vanishes identically on $S(\alpha_i, \varepsilon(\alpha_i))$.

Thus, if U_1 is the union of those spheres $S(\alpha_1, \varepsilon(\alpha_i))$ which contain an infinite number of points of $\sigma(T)$, then f vanishes identically on U_1. Hence, U_1 contains all but a finite number of isolated points of $\sigma(T)$, which we suppose to be the points $\{\lambda_1, \ldots, \lambda_r\}$.

Suppose that f does not vanish identically in any neighborhood of λ_1. Then, since $f(\sigma(T)) = 0$, f has a zero of finite order n at λ_1. Consequently, the function g_1, defined by $g_1(\xi) = (\lambda_1 - \xi)^n / f(\xi)$, is analytic in a neighborhood of λ_1. Let e be a function identically one in a neighborhood of λ_1 and identically zero in a neighborhood of every other point of $\sigma(T)$, and let $g = g_1 e$. Then $(\lambda_1 I - T)^n e(T) = f(T)g(T) = 0$. The Laurent expansion of $R(\xi; T)$ in a neighborhood $0 < |\xi - \lambda_1| < \varepsilon$ is given by

$$R(\xi; T) = \sum_{m=-\infty}^{\infty} A_m(\lambda_1 - \xi)^m,$$

where

$$A_{-m} = -\frac{1}{2\pi i} \int_{C_1} (\lambda_1 - \xi)^{m-1} R(\xi; T)d\xi = -(\lambda_1 I - T)^{m-1} e(T),$$

and where C_1 is a small circle about λ_1. Thus $A_{-(m+1)} = -(\lambda_1 I - T)^m e(T) = 0$ for $m \geq n$, and therefore λ_1 is a pole of order at most n. In the same way we see that either f vanishes identically in a neighborhood of λ_i, $i = 2, \ldots, r$, or λ_i is a pole of $\sigma(T)$ and f has a zero of order at least $\nu(\lambda_i)$ at λ_i. Q.E.D.

17 **Definition.** A subset of $\sigma(T)$ which is both open and closed in $\sigma(T)$ is called a *spectral set*.

It is evident that the spectral sets form a Boolean algebra of subsets of $\sigma(T)$. If σ is a spectral set, there is a function $f \in \mathscr{F}(T)$ which is identically one on σ and which vanishes on the rest of $\sigma(T)$. We put $E(\sigma; T) = f(T)$. If the operator T is understood, we may write $E(\sigma; T)$ simply as $E(\sigma)$. It is clear from Theorem 16 that $E(\sigma)$ depends only on σ, and not on the particular $f \in \mathscr{F}(T)$ chosen to define it. If the spectral set σ consists of the single point λ, the symbol $E(\lambda; T)$ may sometimes be used instead of $E(\{\lambda\}; T)$. It will occasionally be convenient to use the symbol $E(\sigma)$ for any set σ of complex numbers for which $\sigma \cap \sigma(T)$ is a spectral set. In this case, we put

$$E(\sigma) = E(\sigma \cap \sigma(T)).$$

Thus $E(\sigma) = 0$ if $\sigma \cap \sigma(T)$ is void.

A comparison of Theorem 16 with Theorem 1.3 suggests that there is a relationship between the order of a pole and its index as defined in Definition 1.2. The following theorem gives such a relationship. Though the notion of index was defined for operators in a finite dimensional space, the definition given has a general application.

18 THEOREM. *If λ is a pole of T of order ν, then λ has index ν. Furthermore, an isolated point λ in the spectrum of T is a pole of order ν if and only if*

$$(\lambda I - T)^{\nu} E(\lambda; T) = 0, \quad (\lambda I - T)^{\nu-1} E(\lambda; T) \neq 0.$$

PROOF. It was shown during the course of the proof of Theorem 16 that the Laurent expansion of $R(\xi; T)$, in the neighborhood $0 < |\xi - \lambda| < \varepsilon$ of an isolated point λ, is given by

$$R(\xi; T) = \sum_{n=-\infty}^{\infty} A_n (\lambda - \xi)^n,$$

with

$$A_{-(m+1)} = -(\lambda I - T)^m E(\lambda; T).$$

Thus λ is a pole of order ν if and only if

$$(\lambda I - T)^{\nu} E(\lambda; T) = 0, \quad (\lambda I - T)^{\nu-1} E(\lambda; T) \neq 0.$$

To prove the first conclusion of the theorem let λ be a pole of order ν. Then there is a vector x with

$$(\lambda I - T)^{\nu} x = 0, \quad (\lambda I - T)^{\nu-1} x \neq 0,$$

which shows that the index of λ is at least ν. Now, suppose that n is the index of λ. Then for some x, $(\lambda I-T)^n x = 0$, $(\lambda I-T)^{n-1}x \neq 0$. Since

$$R(\xi; T) = -\sum_{j=0}^{\infty} \frac{(\lambda I-T)^j}{(\lambda-\xi)^{j+1}}, \quad |\lambda-\xi| > |\lambda I-T|,$$

as is seen by multiplying both sides of this equation by $(\xi I-T) = (\xi-\lambda)I+(\lambda I-T)$, the function

$$R(\xi; T)x = -\sum_{j=0}^{n-1} \frac{(\lambda I-T)^j x}{(\lambda-\xi)^{j+1}}$$

is regular over the entire plane, except perhaps at the point $\xi = \lambda$. Thus, if K is any rectifiable contour surrounding $\sigma(T)$, and C is a small circle about λ,

$$x = \frac{1}{2\pi i}\int_K R(\xi; T)x d\xi = \frac{1}{2\pi i}\int_C R(\xi; T)x d\xi = e(T)x.$$

This shows that $(\lambda I-T)^\nu x = (\lambda I-T)^\nu e(T)x = 0$, and proves that the index $n \leqq \nu$. Q.E.D.

19 **Theorem.** *Let f be in $\mathcal{F}(T)$, and let τ be a spectral set of $f(T)$. Then $\sigma(T)\cap f^{-1}(\tau)$ is a spectral set of T, and*

$$E(\tau; f(T)) = E(f^{-1}(\tau); T).$$

Proof. Let $e_\tau(\mu) = 1$ for μ in a neighborhood of τ, and let $e_\tau(\mu) = 0$ for μ in a neighborhood of the rest of $\sigma(f(T))$. Then $e_\tau(f(T)) = E(\tau; f(T))$. If τ' is the complement of τ in $\sigma(f(T))$, Theorem 11 shows that $\sigma(T)$ is the union of the disjoint sets $f^{-1}(\tau)$ and $f^{-1}(\tau')$. Since f is continuous, these two sets are both open and closed. It follows that $\sigma = \sigma(T)\cap f^{-1}(\tau)$ is a spectral set of T. If $e_\sigma(\lambda) = e_\tau(f(\lambda))$, then $E(\sigma; T) = e_\sigma(T)$, and Theorem 12 shows that

$$E(\tau; f(T)) = E(\sigma; T) = E(f^{-1}(\tau); T). \qquad \text{Q.E.D.}$$

For any set σ for which $E(\sigma)$ is defined we place $\mathfrak{X}_\sigma = E(\sigma)\mathfrak{X}$. Then $T\mathfrak{X}_\sigma \subseteq \mathfrak{X}_\sigma$, and the restriction of T to $\mathfrak{X}_\sigma$ will be denoted by T_σ.

20 **Theorem.** *Let σ be a spectral set of $\sigma(T)$. Then $\sigma(T_\sigma) = \sigma$. If f is in $\mathcal{F}(T)$, then f is in $\mathcal{F}(T_\sigma)$, and $f(T)_\sigma = f(T_\sigma)$. A point λ in σ is a pole of T of order ν if and only if it is a pole of T_σ of order ν.*

PROOF. Let $\lambda \in \sigma$, and suppose that $\lambda \notin \sigma(T_\sigma)$. Then there exists a bounded linear operator A in the space $\mathfrak{X}_\sigma$ such that $(\lambda I-T)Ax = A(\lambda I-T)x = x$ for $x \in \mathfrak{X}_\sigma$. Let the function g be equal to zero for μ in a neighborhood of σ, and equal to $(\lambda-\mu)^{-1}$ for μ in a neighborhood of the remaining points of $\sigma(T)$. Then $g(T)(\lambda I-T) = (\lambda I-T)g(T) = I-E(\sigma)$. If we define $A_1 : \mathfrak{X} \to \mathfrak{X}$ by $A_1 x = AE(\sigma)x$, then

$$(\lambda I-T)(A_1+g(T)) = (A_1+g(T))(\lambda I-T) = I.$$

Consequently $\lambda \in \varrho(T)$, contradicting $\lambda \in \sigma$. This shows that $\sigma \subseteq \sigma(T_\sigma)$.

Conversely, let $\lambda \notin \sigma$. Then define h to be equal to $(\lambda-\mu)^{-1}$ for μ in a neighborhood of σ not containing λ, and to be identically zero in a neighborhood of the remainder of $\sigma(T)$. We have $h(T)(\lambda I-T) = (\lambda I-T)h(T) = E(\sigma)$. Consequently, the restriction $h(T)_\sigma$ of $h(T)$ to $\mathfrak{X}_\sigma$ satisfies $h(T)_\sigma(\lambda I_\sigma - T_\sigma) = (\lambda I_\sigma - T_\sigma)h(T)_\sigma = I_\sigma$, so that $\lambda \notin \sigma(T_\sigma)$. This proves that $\sigma(T_\sigma) \subseteq \sigma$, and that $R(\lambda; T_\sigma) = R(\lambda; T)_\sigma$.

If $f \in \mathscr{F}(T)$, let U be a neighborhood of $\sigma(T)$ whose boundary B consists of a finite number of rectifiable Jordan arcs, and such that $U \cup B$ is included in the domain of analyticity of f. Then

$$\begin{aligned} f(T)_\sigma &= \frac{1}{2\pi i}\left\{\int_B f(\lambda)R(\lambda; T)d\lambda\right\}_\sigma \\ &= \frac{1}{2\pi i}\int_B f(\lambda)R(\lambda; T)_\sigma d\lambda \\ &= \frac{1}{2\pi i}\int_B f(\lambda)R(\lambda; T_\sigma)d\lambda = f(T_\sigma). \end{aligned}$$

By Theorem 18, λ is a pole of order ν for T if and only if

$$(\lambda I-T)^\nu E(\lambda) = 0, \qquad (\lambda I-T)^{\nu-1}E(\lambda) \neq 0.$$

Since $\lambda \in \sigma$ we have $E(\lambda)E(\sigma) = E(\lambda)$ and thus

$$(\lambda I-T)^m E(\lambda) = (\lambda I_\sigma - T_\sigma)^m E(\lambda), \qquad m = 1, 2, \ldots.$$

Hence λ is a pole of T of order ν if and only if it is a pole of T_σ of order ν. Q.E.D.

21 COROLLARY. *The map $\sigma \to E(\sigma)$ is an isomorphism of the Boolean algebra of spectral sets onto the Boolean algebra of all projections of the form $E(\sigma)$, with σ a spectral set.*

PROOF. It follows from Theorem 10 that the map $\sigma \rightarrow E(\sigma)$ is a homeomorphism. To verify that it is an isomorphism, it will suffice to show that $E(\sigma) = 0$ only when σ is the void set ϕ. Now if $E(\sigma) = 0$, then $\mathfrak{X}_\sigma = \{0\}$ and $\sigma(T_\sigma) = \phi$. It follows from Theorem 20 that $\sigma = \sigma(T_\sigma) = \phi$. Q.E.D.

22 THEOREM. *Let $\lambda_1, \ldots, \lambda_k$ be poles of T, let $\nu_1, \ldots, \nu_k$ be their orders, and let $\sigma = \{\lambda_1, \ldots, \lambda_k\}$. Then, for f in $\mathscr{F}(T)$,*

$$f(T)E(\sigma) = \sum_{i=1}^{k} \sum_{m=0}^{\nu_i - 1} \frac{f^{(m)}(\lambda_i)}{m!} (T-\lambda_i I)^m E(\lambda_i).$$

PROOF. If

$$g(\lambda) = \sum_{i=1}^{k} \sum_{m=0}^{\nu_i - 1} \frac{f^{(m)}(\lambda_i)}{m!} (\lambda-\lambda_i)^m,$$

it follows that $g^{(m)}(\lambda_i) = f^{(m)}(\lambda_i)$, for $1 \leqq i \leqq k$ and $0 \leqq m < \nu_i$. Thus the equation $f(T)E(\sigma) = g(T)E(\sigma)$ follows from Theorem 16. Q.E.D.

23 COROLLARY. *Let the functions f, f_n, $n = 1, 2, \ldots$, be in $\mathscr{F}(T)$ and assume that $f_n(T)$ converges to $f(T)$ in the weak operator topology. Then, for every pole λ of T of order ν we have*

$$f^{(m)}(\lambda) = \lim_{n\to\infty} f_n^{(m)}(\lambda), \qquad 0 \leqq m < \nu.$$

PROOF. If T_λ is the restriction of T to $E(\lambda)\mathfrak{X}$, then, by Theorem 20, λ is a pole of order ν for T_λ. Theorem 18 shows that T_λ has index ν. Thus the corollary follows from Theorem 22 by the same argument used to derive Theorem 1.9 from Theorem 1.8. Q.E.D.

24 THEOREM. *Let λ be a pole of order ν for T and let σ_1 be the complement in $\sigma(T)$ of the set $\{\lambda\}$. Then*

$$\mathfrak{X}_\lambda = \{x | (T-\lambda I)^\nu x = 0\},$$

and

$$\mathfrak{X}_{\sigma_1} = (T-\lambda I)^\nu \mathfrak{X}.$$

PROOF. By Theorem 20, $(T-\lambda I)^\nu \mathfrak{X}_{\sigma_1} = \mathfrak{X}_{\sigma_1}$. By Theorem 18,

$$\text{(i)} \qquad (T-\lambda I)^\nu \mathfrak{X}_\lambda = 0,$$

and thus

$$(T-\lambda I)^\nu \mathfrak{X} = (T-\lambda I)^\nu \{\mathfrak{X}_\lambda \oplus \mathfrak{X}_{\sigma_1}\} = \mathfrak{X}_{\sigma_1}.$$

In view of (i) we have

(ii) $$\mathfrak{X}_\lambda \subseteqq \{x|(T-\lambda I)^\nu x = 0\}.$$

Also if $(T-\lambda I)^\nu x = 0$ then, by (i),

$$0 = (T-\lambda I)^\nu\{E(\lambda)x+E(\sigma_1)x\} = (T-\lambda I)^\nu E(\sigma_1)x.$$

Since by Theorem 20, $(T-\lambda I)$ is one-to-one on $\mathfrak{X}_{\sigma_1}$ it follows that $E(\sigma_1)x = 0$ and thus that $E(\lambda)x = x$. This proves that $\{x|(T-\lambda I)^\nu x = 0\} \subseteqq \mathfrak{X}_\lambda$ which, when combined with (ii), completes the proof of the theorem. Q.E.D.

4. Spectral Theory of Compact Operators

It has been observed in some of the exercises in Section VI.9 that a number of linear operators of importance in analysis either are compact, or have compact squares. The structure of the spectrum of such an operator is particularly simple. The spectral theory of compact operators, as presented in this section, was initiated by F. Riesz, and constitutes a generalization of some of the work of Fredholm on the theory of linear integral equations.

1 Lemma. *Let T be a compact operator, and let λ be a non-zero complex number. If $\lambda I-T$ is one-to-one, the range of $\lambda I-T$ is closed.*

Proof. Let $y = \lim_{n\to\infty} y_n$, where $y_n = (\lambda I-T)x_n$. If the sequence $\{x_n\}$ contains a bounded subsequence, then, since T maps bounded sets into conditionally compact sets, $\{x_n\}$ contains a sequence $\{x_{n_i}\}$ such that $\{Tx_{n_i}\}$ converges. Since $x_{n_i} = (y_{n_i}+Tx_{n_i})/\lambda$, $\{x_{n_i}\}$ converges to some element $x \in \mathfrak{X}$, and $y = (\lambda I-T)x$.

If $\{x_n\}$ contains no bounded subsequence, then $|x_n| \to \infty$. Put $z_n = x_n/|x_n|$, so that $(\lambda I-T)z_n \to 0$ and $|z_n| = 1$. Since T is compact, there exists a subsequence $\{z_{n_i}\}$ of $\{z_n\}$ such that $\{Tz_{n_i}\}$ converges. Since $z_{n_i}-\lambda^{-1}Tz_{n_i} \to 0$, it follows that $\{z_{n_i}\}$ converges. Let $z = \lim z_{n_i}$. Then $|z| = 1$, and $(\lambda I-T)z = 0$; consequently, contrary to hypothesis, $\lambda I-T$ is not one-to-one. Q.E.D.

2 Corollary. *Let T be a compact operator, and let $0 \neq \lambda$ be in $\sigma(T)$. Then there exists either a non-zero x in $\mathfrak{X}$ such that $Tx = \lambda x$, or a non-zero x^* in $\mathfrak{X}^*$ such that $T^*x^* = \lambda x^*$.*

Proof. If $\lambda I-T$ is one-to-one, and if $(\lambda I-T)\mathfrak{X}$ is dense in $\mathfrak{X}$, then, by Lemma 1, $(\lambda I - T)\mathfrak{X} = \mathfrak{X}$, and $\lambda I - T$ has an everywhere defined inverse. By Theorem II.2.2, this inverse is bounded; therefore $\lambda \in \varrho(T)$, contradicting the assumption that $\lambda \in \sigma(T)$. It follows that $(\lambda I - T)\mathfrak{X}$ is not dense in $\mathfrak{X}$. By Corollary II.2.13, there is an $x^* \neq 0$ in $\mathfrak{X}^*$ with $x^*(\lambda I - T)\mathfrak{X} = 0$. Hence $(\lambda x^* - T^* x^*)\mathfrak{X} = 0$, and $\lambda x^* = T^* x^*$. Q.E.D.

It will follow from Theorem 5 of this section that if $0 \neq \lambda \in \sigma(T)$, neither the operator $(\lambda I - T)$ nor the operator $(\lambda I - T^*)$ is one-to-one.

3 LEMMA. *Let $\mathfrak{A}$ and $\mathfrak{B}$ be closed subspaces of $\mathfrak{X}$, and suppose that $\mathfrak{A}$ is a proper subspace of $\mathfrak{B}$. Then, for every $\varepsilon > 0$, there is a y in $\mathfrak{B}$, with $|y| = 1$ and $|x - y| > 1 - \varepsilon$, for each x in $\mathfrak{A}$.*

PROOF. Let $b \in \mathfrak{B}$, $b \notin \mathfrak{A}$, and $\delta = \inf_{a \in \mathfrak{A}} |b - a|$. Since $\mathfrak{A}$ is closed, $\delta > 0$, and there is an $a_0 \in \mathfrak{A}$ with $|b - a_0| < \delta(1+\varepsilon)$. If $b_1 = b - a_0$, then

$$(1+\varepsilon) \inf_{a \in \mathfrak{A}} |b_1 - a| = (1+\varepsilon)\delta > |b_1|.$$

If $y = b_1/|b_1|$, then $|y| = 1$, and

$$|y - a| = \left| \frac{b_1}{|b_1|} - a \right| = \frac{1}{|b_1|} |b_1 - a'|$$

$$> \frac{1}{\delta(1+\varepsilon)} |b_1 - a'| \geqq \frac{1}{1+\varepsilon} > 1 - \varepsilon. \qquad \text{Q.E.D.}$$

4 LEMMA. *Let T be a compact operator, $\{\lambda_n\}$ a sequence of distinct scalars, and $\{x_n\}$ a sequence of non-zero vectors such that $(T - \lambda_n I)x_n = 0$ for each $n = 1, 2, \ldots$. Then λ_n approaches zero.*

PROOF. If λ_n does not approach zero, there is an $\varepsilon > 0$ and a subsequence $\{\lambda_{n_i}\}$ with $|\lambda_{n_i}| \geqq \varepsilon$. We may suppose that $\lambda_{n_i} = \lambda_i$. Let $\mathfrak{A}_n = \text{sp}\,\{x_1, \ldots, x_n\}$. By Corollary IV.3.2 $\mathfrak{A}_n$ is a closed subspace of $\mathfrak{X}$. To verify that $\mathfrak{A}_n$ is properly contained in $\mathfrak{A}_{n+1}$, it will be shown that for any n the vectors $x_1, \ldots, x_n$ are linearly independent. Suppose that $x_1, \ldots, x_{n-1}$ are linearly independent, but that $x_n = \alpha_1 x_1 + \ldots + \alpha_{n-1} x_{n-1}$. Then

$$0 = (T - \lambda_n I)x_n = \alpha_1(\lambda_1 - \lambda_n)x_1 + \ldots + \alpha_{n-1}(\lambda_{n-1} - \lambda_n)x_{n-1},$$

and, since $\lambda_i - \lambda_n \neq 0$ for $i \neq n$, we have $\alpha_i = 0$, $i = 1, \ldots, n-1$, and $x_n = 0$, a contradiction. Thus $\mathfrak{A}_n \subset \mathfrak{A}_{n+1}$ properly, and, by Lemma

3, there is a y_n in $\mathfrak{A}_n$ with $|y_n| = 1$ and $|y_n - x| > 1/2$ for x in $\mathfrak{A}_{n-1}$. The vector y_n is of the form $\alpha_1 x_1 + \ldots + \alpha_n x_n$, so that $(T - \lambda_n I) y_n \in \mathfrak{A}_{n-1}$. Thus, if $n > m$, the vector $z_{n,m} = (y_n - \lambda_n^{-1} T y_n) + \lambda_m^{-1} T y_m$ is in $\mathfrak{A}_{n-1}$, and therefore

$$\left| T\left(\frac{1}{\lambda_n} y_n\right) - T\left(\frac{1}{\lambda_m} y_m\right) \right| = |y_n - z_{n,m}| > \frac{1}{2}.$$

Hence no subsequence of $\{T(y_n/\lambda_n)\}$ converges. Since $|y_n/\lambda_n| \leqq 1/\varepsilon$, this contradicts the compactness of T. Q.E.D.

→ 5 THEOREM. *If T is a compact operator, its spectrum is at most denumerable and has no point of accumulation in the complex plane except possibly $\lambda = 0$. Every non-zero number in $\sigma(T)$ is a pole of T and has finite positive index. For such a number λ, the projection $E(\lambda)$ has a non-zero finite dimensional range given by the formula*

$$E(\lambda)\mathfrak{X} = \{x | (T - \lambda I)^\nu x = 0\}$$

where ν is the order of the pole.

PROOF. Since $\sigma(T)$ is compact, the conclusions stated in the first sentence of the theorem will follow as soon as it is known that every non-zero $\lambda \in \sigma(T)$ is isolated. But, if $\lambda \neq 0$ is not isolated, we can find a sequence of distinct points $\lambda_n \in \sigma(T)$, such that $\lambda_n \to \lambda$. By Lemma 4, only a finite number of the maps $\lambda_n I - T$ are not one-to-one. Then, by Corollary 2, only a finite number of the maps $\lambda_n I^* - T^*$ are one-to-one. However, T^* is compact (by VI.5.2), and Lemma 4, applied to T^*, yields a contradiction. This proves that every non-zero $\lambda \in \sigma(T)$ is isolated.

Let T_λ be the restriction of T to $\mathfrak{X}_\lambda = E(\lambda)\mathfrak{X}$. It is seen from Theorem 3.20 that $0 \neq \lambda = \sigma(T_\lambda)$, and, hence, that T_λ has a bounded inverse. Thus if S is the closed unit sphere in $\mathfrak{X}_\lambda$, then $T_\lambda^{-1} S$ is bounded, and, since T_λ is compact, $S = T_\lambda T_\lambda^{-1} S$ is compact. Theorem IV.3.5 then shows that $E(\lambda)\mathfrak{X}$ is finite dimensional. According to Theorem 1.3, there is an integer ν with $(T_\lambda - \lambda I_\lambda)^\nu = 0$. By Theorem 3.18, λ is a pole of T_λ, and by Theorem 3.20, λ is a pole of T. Thus Theorem 3.18 shows that the index of λ is a positive finite number. The final conclusion follows from Theorem 3.24. Q.E.D.

6 THEOREM. *All of the conclusions of Theorem 5 remain valid if it is assumed only that T^n is compact for some positive integer n.*

PROOF. By Theorem 3.11, $\{\sigma(T)\}^n = \sigma(T^n)$ and so Theorem 5 shows that $\sigma(T^n)$ is either finite or else a denumerable sequence converging to zero. Let $\lambda \neq 0$ be a point in $\sigma(T)$ and let $\lambda_1, \ldots, \lambda_k$ be the set of all points in $\sigma(T)$ for which $\lambda_i^n = \lambda^n$. By Theorem 3.19

$$E(\lambda^n; T^n) = \sum_{i=1}^{k} E(\lambda_i; T)$$

which shows that

$$E(\lambda; T)\mathfrak{X} \subseteqq E(\lambda^n; T^n)\mathfrak{X}.$$

By Theorem 5, $E(\lambda^n; T^n)$ has a finite dimensional range and the above inclusion shows that $E(\lambda; T)$ also has a finite dimensional range. The proof of Theorem 6 may now be completed by the reasoning used in the proof of Theorem 5. Q.E.D.

The weakly compact operators in the spaces $C(S)$, AP, $B(S)$, $L_1(S)$, $rca(S)$, $ca(S)$, $ba(S)$, $rba(S)$ etc. which have been discussed in Sections 7 and 8 of Chapter VI provide instances of operators to which Theorem 6 may be applied even though Theorem 5 is not always applicable.

5. Exercises

It is convenient for many purposes to introduce a rough classification of the points of $\sigma(T)$.

1 DEFINITION. (a) The set of $\lambda \in \sigma(T)$ such that $\lambda I - T$ is not one-to-one is called the *point spectrum* of T, and is denoted by $\sigma_p(T)$. Thus, $\lambda \in \sigma_p(T)$ if and only if $Tx = \lambda x$ for some non-zero $x \in \mathfrak{X}$.

(b) The set of all $\lambda \in \sigma(T)$ for which $\lambda I - T$ is one-to-one, and $(\lambda I - T)\mathfrak{X}$ is dense in $\mathfrak{X}$, but such that $(\lambda I - T)\mathfrak{X} \neq \mathfrak{X}$ is called the *continuous spectrum* of T, and denoted $\sigma_c(T)$.

(c) The set of all $\lambda \in \sigma(T)$ for which $\lambda I - T$ is one-to-one, but such that $(\lambda I - T)\mathfrak{X}$ is not dense in $\mathfrak{X}$ is called the *residual spectrum* of T, and is denoted by $\sigma_r(T)$.

2 Prove that $\sigma_r(T)$, $\sigma_c(T)$, and $\sigma_p(T)$ are disjoint and $\sigma(T) = \sigma_r(T) \cup \sigma_c(T) \cup \sigma_p(T)$.

3 Determine the point, residual, and continuous spectra for the operator $y = Tx$ in $C[0, 1]$ defined by $y(s) = \int_0^s x(t)dt$. Does your answer change if T is regarded as an operator in $L[0, 1]$, or in $C_0[0, 1]$,

the space of all continuous functions on $[0, 1]$ which vanish at 0?

4 Let $y = Tx$ be the operator in $C[0, 1]$ defined by $y(t) = tx(t)$. Determine $\sigma_r(T)$, $\sigma_p(T)$, $\sigma_c(T)$. Find $f(T)$ for $f \in \mathscr{F}(T)$.

5 Let $\{\alpha_i\}$ be a bounded sequence of complex numbers. Let T be the map in l_2 defined by $T[\xi_i] = [\alpha_i \xi_i]$. Find $\sigma_p(T)$, $\sigma_c(T)$, and $\sigma_r(T)$.

6 Show that any compact set in the plane can be the spectrum of an operator.

7 Let T be the map in l_p, $1 \leqq p \leqq \infty$, defined by $T[\xi_1, \xi_2, \ldots] = [\xi_2, \xi_3, \ldots]$. Find $\sigma_p(T)$, $\sigma_r(T)$, $\sigma_c(T)$.

8 If E is a projection operator, find the resolvent $R(\lambda; E)$ explicitly in terms of E and λ. What is the spectrum $\sigma(E)$? Find $f(E)$ for $f \in \mathscr{F}(E)$.

9 For any bounded linear operator T show that

$$\sigma_r(T) \subseteqq \sigma_p(T^*) \subseteqq \sigma_r(T) \cup \sigma_p(T).$$

10 If there is a number λ on the circle $|\lambda| = |T|$ which is in the point spectrum of T, then λ is also in the point spectrum of T^*.

11 Show that if T_1 and T_2 in $B(\mathfrak{X})$ satisfy $T_1T_2 = T_2T_1$, then $r(T_1+T_2) \leqq r(T_1)+r(T_2)$, but that if $T_1T_2 \neq T_2T_1$, this need not be the case even if $\mathfrak{X}$ is two-dimensional.

12 Definition. The operator T is called *quasi-nilpotent* if $\lim_{n\to\infty} \sqrt[n]{|T^n|} = 0$.

13 Show that T is quasi-nilpotent if and only if $\sigma(T) = \{0\}$.

14 Let $\sigma = \{\lambda_n\}$ be a denumerable compact set in the complex plane such that $\lambda_n \to 0$. Show that there exists a compact operator T in some B-space with $\sigma(T) = \sigma$.

15 Let (S, Σ, μ) be a positive σ-finite measure space and h be μ-essentially bounded on S. In $L_p(S, \Sigma, \mu)$, $1 \leqq p \leqq \infty$, let T be the transformation defined by $Tx(t) = h(t)x(t)$. Show that $\sigma(T)$ is the set of all points each of whose neighborhoods M satisfies the inequality $\mu(h^{-1}(M)) > 0$. Find $f(T)$ for $f \in \mathscr{F}(T)$.

16 Show that e^{tT} is a differentiable operator valued function of the real variable t, and that

$$\frac{d}{dt} e^{tT} = Te^{tT}.$$

Let $y \in \mathfrak{X}$. Solve the differential equation

$$\frac{dy(t)}{dt} = Ty(t), \qquad y(0) = y,$$

and show that the solution is unique.

17 The operator T satisfies a non-trivial polynomial equation $P(T) = 0$ if and only if $\sigma(T)$ consists of a finite set of poles.

18 Let α be a spectral subset of $\sigma(T)$. Then $\sigma_p(T_\alpha) = \alpha \cap \sigma_p(T)$, $\sigma_c(T_\alpha) = \alpha \cap \sigma_c(T)$, $\sigma_r(T_\alpha) = \alpha \cap \sigma_r(T)$.

19 Let $f \in \mathscr{F}(T)$, and let σ be a spectral subset of $\sigma(T)$. Let $\alpha = \sigma(T) \cap \sigma'$, and let f be non-zero on σ. If $f(T)x = 0$ for some $x \in \mathfrak{X}$, then x is in $\mathfrak{X}_\alpha$.

20 Let λ be a pole of $\sigma(T)$. Then

$$\mathfrak{X} = (\lambda I - T)^k \mathfrak{X} \oplus \{x | x \in \mathfrak{X}, (\lambda I - T)^k x = 0\}$$

if and only if $k \geqq \nu(\lambda)$.

21 Let $f \in \mathscr{F}(T)$ and μ be a complex number, and suppose that $f^{-1}(\mu) \cap \sigma(T)$ is finite. Show that μ is a pole of $\sigma(f(T))$ if and only if each point of $f^{-1}(\mu) \cap \sigma(T)$ is a pole of $\sigma(T)$.

22 Let $f \in \mathscr{F}(T)$, and suppose that $f^{-1}(0) \cap \sigma(T)$ is a finite set $\{\lambda_1, \ldots, \lambda_k\}$ of complex numbers. Show that the following statements are equivalent:

(a) $f(T)\mathfrak{X}$ is closed and $\mathfrak{X} = f(T)\mathfrak{X} \oplus \{x | x \in \mathfrak{X}, f(T)x = 0\}$.

(b) Each point λ_i, $i = 1, \ldots, k$, is a pole of $\sigma(T)$ of order less than the order of λ_i as a zero of f.

(c) The set $\sigma = \{\lambda_1, \ldots, \lambda_k\}$ is a spectral set, $f(T)\mathfrak{X} = E(\sigma(T) \cap \sigma')\mathfrak{X}$, and $\{x | x \in \mathfrak{X}, f(T)x = 0\} = E(\sigma)\mathfrak{X}$.

23 Let T be an operator in the B-space $\mathfrak{X}$, and let T^* be its adjoint. Let σ_1 and σ_2 be disjoint spectral sets of T. Show that $x_2^*(x_1) = 0$ for $x_2^* \in \mathfrak{X}^*_{\sigma_2}$ and $x_1 \in \mathfrak{X}_{\sigma_1}$.

24 (Riesz and Sz.-Nagy) If $|T| \leqq r' < r$ and if f is analytic for $|\lambda| \leqq r$ and $|f(\lambda)| \leqq R$ on $|\lambda| = r$, show that

$$|f(T)| \leqq \frac{rR}{r - r'}.$$

25 (Riesz and Sz.-Nagy) Let σ be a spectral set of T lying inside the circle $|\lambda - \lambda_0| = r$. Let $\sigma(T) \cap \sigma'$ lie outside the circle. Show that $x \in \mathfrak{X}_\sigma$ if and only if

$$\limsup_{n \to \infty} |(\lambda_0 I - T)^n x|^{1/n} < r.$$

26 Let $\lambda \in \sigma_c(T)$.

(a) Show there exists a sequence $\{x_n\}$ such that $|x_n| = 1$ and $|(\lambda I - T)x_n| \to 0$.

(b) If σ is a spectral set and $\lambda \in \sigma$ show that for the sequence in (a),

$$\lim_{n\to\infty} |x_n - E(\sigma)x_n| = 0.$$

27 Let $\mathfrak{X} = C[0, 1]$ and let T be the operator in $\mathfrak{X}$ defined by

$$Tx(t) = \int_0^t x(s)ds.$$

Calculate $R(\lambda; T)$ and show that the differential equation

$$x^{(n)}(t) + a_1 x^{(n-1)}(t) + \ldots + a_n x(t) = y(t),$$
$$x(0) = x'(0) = \ldots = x^{n-1}(0) = 0$$

has the solution

$$x = \frac{1}{2\pi i}\int_C \frac{\lambda^n R(\lambda; T) y d\lambda}{1 + a_1\lambda + \ldots + a_n\lambda^n}$$

where C is a circle about the origin excluding the roots of the denominator, or (setting $z = 1/\lambda$)

$$x(t) = \frac{1}{2\pi i}\int_0^t y(s)ds \int_K \frac{e^{z(t-s)}\, dz}{z^n + a_1 z^{n-1} + \ldots + a_n}$$

where K encloses the roots of the denominator.

28 Let $\sigma(T)$ lie in the half plane $\mathscr{R}(\lambda) < 0$. Show that

$$\exp(\xi T) = \lim_{w\to\infty} \frac{1}{2\pi i}\int_{-iw}^{+iw} e^{\lambda\xi} R(\lambda; T)d\lambda, \qquad \xi > 0.$$

29 Assuming that $\sigma(T)$ lies in $\mathscr{R}(\lambda) < 0$, show that

$$R(\lambda; T) = \int_0^\infty e^{-\lambda\xi} \exp(\xi T)d\xi,$$

where the integral converges for $\mathscr{R}(\lambda) > 0$.

30 Show

$$\sup_{\lambda\in\sigma(T)} \mathscr{R}(\lambda) = \lim_{\xi\to\infty} \frac{1}{\xi} \log|\exp(\xi T)|.$$

31 Let T be an operator in the B-space $\mathfrak{X}$, and suppose that $\sigma(T)$ does not intersect the ray $re^{i\theta}$, $r > 0$. If $|R(re^{i\theta}; T)| = 0(r^{-3/2+\varepsilon})$

for some $\varepsilon > 0$ as $r \to 0$, show that there is a sequence $\{f_n\} \subset \mathscr{F}(T)$ and an operator U such that $f_n(T) \to U$ and $U^2 = T$.

32 Let $f_n \in \mathscr{F}(T)$, and suppose that $f_n(T)$ converges in the uniform topology of operators. Then f_n converges uniformly on $\sigma(T)$.

33 Let U be an open set in the complex plane. Suppose that $f(\lambda, \cdot) \in \mathscr{F}(T)$ for each $\lambda \in U$ and that $f(\lambda, T)$ is an analytic function with values in $B(\mathfrak{X})$. Then $f(\lambda, \mu)$ is an analytic function of λ for each $\mu \in \sigma(T)$.

34 If σ is a spectral subset of $\sigma(T)$ such that $\mathfrak{X}_\sigma$ is finite-dimensional, then σ is a finite set of poles.

35 If $\lambda \in \sigma(T)$ is isolated and $\mathfrak{X}_\lambda$ is finite-dimensional, then $(\lambda I - T)^k \mathfrak{X}$ is the set of $x \in \mathfrak{X}$ such that $y^*(x) = 0$ for every $y^* \in \mathfrak{X}^*$ satisfying $(\lambda I^* - T^*)^k y^* = 0$, and $(\lambda I^* - T^*)^k \mathfrak{X}^*$ is the set of $x^* \in \mathfrak{X}^*$ such that $x^*(y) = 0$ for every $y \in \mathfrak{X}$ satisfying $(\lambda I - T)^k y = 0$. The sets $\{x | x \in \mathfrak{X}, (\lambda I - T)^k x = 0\}$ and $\{x^* | x^* \in \mathfrak{X}^*, (\lambda I^* - T^*)^k x^* = 0\}$ have the same dimension.

6. Perturbation Theory

Let $\mu \to T(\mu)$ be an operator valued function of a complex parameter μ which is continuous (or analytic) in the uniform operator topology. The purpose of this section is to investigate how the spectrum and the resolvent operator change with a small change in μ. The basic result of this section is a theorem, essentially due to Rellich, describing how certain isolated points of the spectrum of $T(0)$ vary when $T(\mu)$ depends analytically on μ.

1 **Lemma.** *The set G of elements in $B(\mathfrak{X})$ which have inverses in $B(\mathfrak{X})$ is an open set in the uniform topology of $B(\mathfrak{X})$, containing with an operator A the sphere $\{B | |A-B| < |A^{-1}|^{-1}\}$. If B is in this sphere, its inverse is given by the series*

$$B^{-1} = A^{-1} \sum_{n=0}^{\infty} [(A-B)A^{-1}]^n.$$

Furthermore, the map $A \to A^{-1}$ of G onto G is a homeomorphism in the uniform operator topology.

Proof. Let $|I-B| < 1$, so that the series $S = \sum_{n=0}^{\infty} (I-B)^n$ converges. Since

$$SB = BS = [I-(I-B)]S = \sum_{n=0}^{\infty}(I-B)^n - \sum_{n=1}^{\infty}(I-B)^n = I,$$

it follows that $\{B \mid |I-B| < 1\} \subset G$. Now let $A \in G$, and let $|A-B| < |A^{-1}|^{-1}$. Then $|I-BA^{-1}| = |(A-B)A^{-1}| < 1$; hence, by what has just been proved, BA^{-1} has an inverse in $B(\mathfrak{X})$, given by the series

$$\sum_{n=0}^{\infty}(I-BA^{-1})^n = \sum_{n=0}^{\infty}[(A-B)A^{-1}]^n.$$

Thus B has an inverse in $B(\mathfrak{X})$, given by the formula as stated in the lemma. This formula shows that

$$|B^{-1}-A^{-1}| = |A^{-1}\sum_{n=0}^{\infty}[(A-B)A^{-1}]^n| \leqq \frac{|A^{-1}|^2|A-B|}{1-|A-B||A^{-1}|},$$

from which it follows that the map $B \to B^{-1}$ of G onto G is a homeomorphism. Q.E.D.

2 Corollary. *Let T, T_1 be in $B(\mathfrak{X})$, λ be in $\varrho(T)$ and $|T-T_1| < |R(\lambda; T)|^{-1}$. Then λ is in $\varrho(T_1)$, and*

$$R(\lambda; T_1) = R(\lambda; T)\sum_{n=0}^{\infty}[(T_1-T)R(\lambda; T)]^n.$$

3 Lemma. *Let T be in $B(\mathfrak{X})$, and let $\varepsilon > 0$. Then there is a $\delta > 0$ such that if T_1 is in $B(\mathfrak{X})$ and $|T_1-T| < \delta$, then $\sigma(T_1) \subseteqq S(\sigma(T), \varepsilon)$ and*

$$|R(\lambda; T_1)-R(\lambda; T)| < \varepsilon, \qquad \lambda \notin S(\sigma(T), \varepsilon).$$

Proof. By Lemma 1,

$$\lim_{\lambda\to\infty}|R(\lambda; T)| = \lim_{\lambda\to\infty}\left|\lambda^{-1}\left(I-\frac{T}{\lambda}\right)^{-1}\right| = 0,$$

and so $|R(\lambda; T)| \leqq N_\varepsilon$ for λ in the complement of $S(\sigma(T), \varepsilon)$. Thus, by Corollary 2, if $\delta_1 = N_\varepsilon^{-1} > |T_1-T|$, we have $\sigma(T_1) \subset S(\sigma(T), \varepsilon)$. Also, by Corollary 2,

$$|R(\lambda; T_1)-R(\lambda; T)| < \frac{N_\varepsilon^2|T_1-T|}{1-|T_1-T|N_\varepsilon} < \varepsilon$$

if $|T_1-T| < \delta_2 = \varepsilon/(N_\varepsilon^2+\varepsilon N_\varepsilon)$. The desired conclusion thus follows by defining δ to be the smaller of δ_1 and δ_2. Q.E.D.

4 Lemma. *Let $T(\mu)$ be an analytic operator valued function defined for $|\mu| < \gamma$, where $\gamma > 0$, and let U be an open set with $\overline{U} \subset \varrho(T(0))$.*

Then there exists a $\delta > 0$ such that if $|\mu| < \delta$, then $\bar{U} \subset \varrho(T(\mu))$ and $R(\lambda; T(\mu))$ is an analytic function of μ for each $\lambda \in U$.

Proof. By Lemma 3, there is a δ_1 such that if $|\mu| < \delta_1$, then $\bar{U} \subset \varrho(T(\mu))$. Let $\delta \leqq \delta_1$ be chosen such that $|T(0)-T(\mu)| < \inf_{\lambda \in U} |R(\lambda; T(0))|^{-1}$ whenever $|\mu| < \delta$. It follows from Corollary 2 that

$$[*] \quad R(\lambda; T(\mu)) = R(\lambda; T(0)) \sum_{n=0}^{\infty} [(R(\mu)-T(0))R(\lambda; T(0))]^n.$$

Since the series converges absolutely and uniformly for $|\mu| < \delta$, and $T(\mu)$ is analytic, it follows from the general theory of analytic functions (cf. III.14) that $R(\lambda; T(\mu))$ is an analytic function of μ, $|\mu| < \delta$, for each $\lambda \in U$. Q.E.D.

5 **Lemma.** *Let T be in $B(\mathfrak{X})$, f be in $\mathscr{F}(T)$, and $\varepsilon > 0$. Then there is a $\delta > 0$ such that if T_1 is in $B(\mathfrak{X})$ and $|T_1 - T| < \delta$, then f is in $\mathscr{F}(T_1)$ and $|f(T)-f(T_1)| < \varepsilon$.*

Proof. Let U be a neighborhood of $\sigma(T)$ on which f is analytic. Let U_1 be a neighborhood of $\sigma(T)$ whose boundary B consists of a finite number of rectifiable Jordan arcs, and such that $U_1 \cup B \subset U$. Then, by Lemma 3, there is a $\delta_1 > 0$ such that $\sigma(T_1) \subset U_1$ for $|T_1 - T| < \delta_1$. Consequently $f \in \mathscr{F}(T_1)$ for $|T_1 - T| < \delta_1$. Also, by Lemma 3, $R(\lambda; T_1)$ is near to $R(\lambda; T)$ uniformly for λ in B if $|T_1 - T|$ is small. Thus, for some positive $\delta \leqq \delta_1$,

$$|f(T_1)-f(T)| = \frac{1}{2\pi} \left| \int_B f(\lambda)\{R(\lambda; T_1) - R(\lambda; T)\} d\lambda \right| < \varepsilon. \quad \text{Q.E.D.}$$

6 **Lemma.** *Let f be in $\mathscr{F}(T(0))$, where $T(\mu)$ is an analytic operator valued function, defined for $|\mu| < \gamma$, where $\gamma > 0$. Then there is a positive δ less than γ such that f is in $\mathscr{F}(T(\mu))$, and such that $f(T(\mu))$ is an analytic operator valued function of μ, for $|\mu| < \delta$.*

Proof. Let B be chosen as in the proof of Lemma 5. By Lemma 4, δ may be found such that the series [*] used in the proof of that lemma converges absolutely and uniformly for λ on B. Thus

$$f(T(\mu)) = \sum_{n=0}^{\infty} \frac{1}{2\pi i} \int_B f(\lambda)[T(\mu) - T(0)]^n [R(\lambda, T(0))]^{n+1} d\lambda,$$

the series converging to an analytic function of μ. Q.E.D.

Lemmas 4 and 6 can be considerably improved in the important case in which $\sigma(T(0))$ contains an isolated point λ_0 such that $E(\lambda_0; T(0))\mathfrak{X}$ is finite-dimensional. To study this situation, we first need a lemma on projections.

7 **Lemma.** *Let E, E_1 be two projections in $\mathfrak{X}$ such that $|E-E_1| < \min(|E|^{-1}, |E_1|^{-1})$. If one of the projections has a finite dimensional range then so does the other and*

$$\dim E\mathfrak{X} = \dim E_1\mathfrak{X}.$$

Proof. Consider the map EE_1E restricted to $E\mathfrak{X}$. Since E is the identity in $E\mathfrak{X}$ and since, by hypothesis, $|EE_1E-E| < 1$, it is seen from Lemma 1 that EE_1E is a one-to-one map of $E\mathfrak{X}$ onto all of itself and thus that $EE_1E\mathfrak{X} = E\mathfrak{X}$. Hence $E\mathfrak{X} \supseteqq EE_1\mathfrak{X} \supseteqq EE_1E\mathfrak{X} = E\mathfrak{X}$, and therefore $E\mathfrak{X} = EE_1\mathfrak{X}$. This shows that $\dim E_1\mathfrak{X} \geqq \dim E\mathfrak{X}$. Similarly, $\dim E\mathfrak{X} \geqq \dim E_1\mathfrak{X}$ and thus $\dim E\mathfrak{X} = \dim E_1\mathfrak{X}$. Q.E.D.

8 **Lemma.** *Let $E(\mu)$ be an analytic projection valued function defined for $|\mu| < \gamma$, where $\gamma > 0$, and let $E(0)\mathfrak{X}$ have the finite dimension m. Then there exists a $\delta > 0$ such that, if $\{x_1, \ldots, x_m\}$ is a basis for $E(0)\mathfrak{X}$, the set $\{E(\mu)x_1, \ldots, E(\mu)x_m\}$ is a basis for $E(\mu)\mathfrak{X}$ when $|\mu| < \delta$.*

Proof. It is readily seen that there exists a δ with $0 < \delta \leqq \gamma$, such that

$$|E(0)-E(\mu)| < \min(|E(0)|^{-1}, |E(\mu)|^{-1}), \qquad |\mu| < \delta.$$

From the proof of Lemma 7, we know that the dimension of $E(\mu)E(0)\mathfrak{X}$ is equal to m. Since the vectors $\{x_1, \ldots, x_m\}$ span $E(0)\mathfrak{X}$, the set $\{E(\mu)x_1, \ldots, E(\mu)x_m\}$ spans the m-dimensional space $E(\mu)E(0)\mathfrak{X}$. Hence these vectors are linearly independent, and form a basis for $E(\mu)\mathfrak{X}$, which is an m-dimensional subspace by Lemma 7. Q.E.D.

9 **Theorem.** *Let $\gamma > 0$ and let $T(\mu)$ be a $B(\mathfrak{X})$ valued function defined and analytic for $|\mu| < \gamma$. Let λ_0 be an isolated point of $\sigma(T(0))$, and suppose that the subspace $E(\lambda_0; T(0))\mathfrak{X}$ has finite dimension m. Let U be an open set with $\bar{U} \cap \sigma(T(0)) = \{\lambda_0\}$. Then there is a positive $\delta < \gamma$, an integer $k \leqq m$, and an integer n, such that for $|\mu| < \delta$ $U \cap \sigma(T(\mu))$ is a finite set $\{\lambda_1(\mu), \ldots, \lambda_k(\mu)\}$. Each function $\lambda_i(\mu)$ depends analytically on the principal value of the fractional power $\mu^{1/n}$ of μ, and satisfies $\lambda_i(0) = \lambda_0$. Moreover, the projections $E(\lambda_i(\mu); T(\mu))$ can be expanded in fractional power Laurent series*

$$E(\lambda_i(\mu); T(\mu)) = \sum_{j=-N}^{\infty} A_{ij}\mu^{j/n}, \qquad |\mu| < \delta,$$

where A_{ij} are operators in $B(\mathfrak{X})$.

Proof. For simplicity of notation we shall write $\sigma(\mu)$ for $U \cap \sigma(T(\mu))$. It is seen from Lemma 6 that, for small μ, $\sigma(\mu)$ is a spectral set of $T(\mu)$. The symbol $E(\mu)$ will be used for $E(\sigma(\mu); T(\mu))$. It follows from Lemma 6 that $E(\mu)$ depends analytically upon μ. If $\{x_1, \ldots, x_m\}$ is a basis for $E(0)\mathfrak{X}$ then, by Lemma 8, there is a positive number δ_1 such that for $|\mu| < \delta_1$ the set $\{E(\mu)x_1, \ldots, E(\mu)x_m\}$ is a basis for $E(\mu)\mathfrak{X}$.

Let the functions t_{ij} be defined by the equations

$$T(\mu)E(\mu)x_i = \sum_{j=1}^{m} t_{ij}(\mu)E(\mu)x_j, \qquad i = 1, \ldots, m.$$

We first show that the t_{ij} are analytic functions if μ is sufficiently small. Let $x_k^* \in \mathfrak{X}^*$ be such that $x_k^*(x_j) = \delta_{kj}$, $k = 1, \ldots, m$. The first of the above equations yields m equations,

$$x_k^*T(\mu)E(\mu)x_i = \sum_{j=1}^{m} t_{ij}(\mu)x_k^*E(\mu)x_j, \qquad k = 1, \ldots, m.$$

Since $x_k^*E(0)x_j = x_k^*x_j = \delta_{kj}$, the determinant $\Delta(\mu)$ of the matrix $(x_k^*E(\mu)x_j)$ equals one for $\mu = 0$, and is therefore bounded away from zero for $|\mu| < \delta_2 < \delta_1$. Consequently, it is possible to solve for the t_{ij}, and obtain in each case the quotient of an analytic function by $\Delta(\mu)$. Since, for $|\mu| < \delta_2$, the function $\Delta(\mu) \neq 0$, its reciprocal, and therefore each t_{ij}, is analytic. Thus the t_{ij} are analytic for $|\mu| < \delta_2$.

If $T_1(\mu)$ denotes the restriction of $T(\mu)$ to the m dimensional space $E(\mu)\mathfrak{X}$ for $|\mu| < \delta_2$, then the spectrum of $T_1(\mu)$ consists of the roots of the equation

$$[*] \qquad d(\lambda, \mu) = \det(\lambda\delta_{ij} - t_{ij}(\mu)) = 0.$$

By Theorem 3.20, $\sigma(\mu) = \sigma(T_1(\mu))$. The function $d(\lambda, \mu)$ is a polynomial of degree m in λ, whose coefficients are analytic functions of μ; for $\mu = 0$, equation [*] has only the m-fold root $\lambda = \lambda_0$. By the Weierstrass preparation theorem (III.14) there exists a positive $\delta \leqq \delta_2$, an integer $k \leqq m$ and an integer n such that if $|\mu| < \delta$, $\mu \neq 0$, $d(\lambda, \mu)$ has exactly k distinct zeros $\lambda_1(\mu), \ldots, \lambda_k(\mu)$, and these zeros are analytic in μ, being given by fractional power series

$$\lambda_j(\mu) = \sum_{p=0}^{\infty} a_{jp}\mu^{p/n}$$

of the principal value of $\mu^{1/n}$.

Let $E_i(\mu) = E(\lambda_i(\mu); T(\mu))$, for $0 < |\mu| < \delta$ and $1 \leqq i \leqq k$. It remains to show that each $E_i(\mu)$ has a Laurent expansion about $\mu = 0$ in terms of $\mu^{1/n}$. Let $q(\lambda) = \prod_{i=2}^{k}(\lambda - \lambda_i(\mu))^m$, $r(\lambda) = 1/q(\lambda)$, and

$$p(\lambda) = \sum_{j=0}^{m} \frac{r^{(j)}(\lambda_1(\mu))}{j!} (\lambda - \lambda_1(\mu))^j.$$

Consider the polynomial

$$s(\lambda) = p(\lambda)q(\lambda).$$

By the Leibnitz rule

$$s^{(j)}(\lambda) = \sum_{i=0}^{j} \binom{j}{i} p^{(i)}(\lambda) q^{(j-i)}(\lambda).$$

The form of q shows that for $0 < |\mu| < \delta$,

$$s^{(j)}(\lambda_i(\mu)) = 0, \qquad j = 0, \ldots, m-1, \qquad i = 2, \ldots, k.$$

Moreover,

$$p^{(j)}(\lambda_1(\mu)) = r^{(j)}(\lambda_1(\mu)), \qquad j = 0, \ldots, m.$$

Thus

$$\begin{aligned} s^{(j)}(\lambda_1(\mu)) &= \sum_{i=0}^{j} \binom{j}{i} q^{(j-i)}(\lambda_1(\mu)) p^{(i)}(\lambda_1(\mu)) \\ &= \sum_{i=0}^{j} \binom{j}{i} q^{(j-i)}(\lambda_1(\mu)) r^{(i)}(\lambda_1(\mu)) \\ &= [q(\lambda)r(\lambda)]^{(j)}(\lambda_1(\mu)), \qquad 0 < |\mu| < \delta. \end{aligned}$$

Since $q(\lambda)r(\lambda) = 1$, we have

$$\begin{aligned} s^{(j)}(\lambda_1(\mu)) &= 0, \qquad 1 \leqq j \leqq m, \\ s(\lambda_1(\mu)) &= 1. \end{aligned}$$

Consequently, by Theorem 1.8

$$E_1(\mu) = s(T(\mu))E(\mu).$$

Since the coefficients of the polynomial s have Laurent expansions in $\mu^{1/n}$ about $\mu = 0$, which contain only a finite number of negative exponents, the projection $E_1(\mu)$ has a Laurent expansion with the same property. The projections $E_i(\mu)$, $i = 2, \ldots, k$ are treated similarly. Q.E.D.

The next theorem gives an extension of Taylor's theorem to functions of an operator.

10 **Theorem.** *Let S and N be commuting operators. Let f be a function analytic in a domain D, including the spectrum $\sigma(S)$ of S and every point within a distance of $\sigma(S)$ not greater than some positive number ε. Suppose that the spectrum $\sigma(N)$ of N lies within the open circle of radius ε about the origin. Then f is analytic on a neighborhood of $\sigma(S+N)$, and*

$$f(S+N) = \sum_{n=0}^{\infty} \frac{f^{(n)}(S)N^n}{n!},$$

the series converging in the uniform topology of operators.

We begin the proof of Theorem 10 by proving the following lemma.

11 **Lemma.** *Let C be a set whose minimum distance from the spectrum $\sigma(T)$ of an operator T is greater than some positive number ε. Then there is a constant K such that*

$$|R(\lambda, T)^n| < K\varepsilon^{-n}, \qquad n \geqq 0, \quad \lambda \in C.$$

Proof. Let the open set U contain $\sigma(T)$ and have a boundary which consists of a finite number of rectifiable Jordan curves. Suppose also that for every λ in C and every α in $U \cup B$ we have $|\lambda - \alpha| > \varepsilon$. Then

$$\begin{aligned} |R(\lambda; T)^n| &= \left| \frac{1}{2\pi i} \int_B (\lambda-\alpha)^{-n} R(\alpha; T) d\alpha \right| \\ &\leqq K\varepsilon^{-n}. \end{aligned}$$

Q.E.D.

Proof of Theorem 10. Let $\delta = \sup_{\lambda \in \sigma(N)} |\lambda|$ so that, by hypothesis, $\delta < \varepsilon$. Choose $\theta < 1$ so that $\delta < \theta\varepsilon < \varepsilon$ and let B be the circle $\{\lambda \mid |\lambda| = \theta\varepsilon\}$. Then

$$|N^n| = \left| \frac{1}{2\pi i} \int_B \lambda^n R(\lambda; N) d\lambda \right| \leqq K(\theta\varepsilon)^{n+1}, \quad n = 0, 1, \ldots.$$

This inequality, together with Lemma 11, shows that the series

$$V = \sum_{n=0}^{\infty} R(\lambda; S)^{n+1} N^n$$

converges uniformly for λ in any set C whose minimum distance from $\sigma(S)$ is greater than ε. Since S and N commute, it is seen, by direct multiplication, that $V(\lambda I - S - N) = (\lambda I - S - N)V = I$. Thus if λ has distance greater than ε from $\sigma(S)$ then λ is in the resolvent set of $S+N$ and

$$R(\lambda; S+N) = \sum_{n=0}^{\infty} R(\lambda; S)^{n+1} N^n.$$

Thus the function f of Theorem 10 is analytic on a neighborhood of $\sigma(S+N)$ as stated.

Now let C denote the union of a finite collection $C_1, \ldots, C_n$ of disjoint closed rectifiable Jordan contours which bound a region D containing every point whose distance from $\sigma(T)$ is less than ε, and which lie, together with D, entirely in the domain of analyticity of f. Suppose, moreover, that the contours C_i are oriented in the positive sense customary in the theory of contour integration. Then

$$\begin{aligned} f(S+N) &= \frac{1}{2\pi i} \int_C f(\lambda) R(\lambda; S+N) d\lambda \\ &= \frac{1}{2\pi i} \sum_{n=0}^{\infty} \int_C f(\lambda)(R(\lambda; S))^{n+1} N^n d\lambda. \end{aligned}$$

On the other hand, it is seen from the resolvent relation

$$R(\lambda_1; S) - R(\lambda_2; S) = (\lambda_2 - \lambda_1) R(\lambda_1; S) R(\lambda_2; S),$$

that $(d/d\lambda)R(\lambda; S) = -(R(\lambda; S))^2$, and, inductively, that

$$\left(\frac{d}{d\lambda}\right)^n R(\lambda; S) = (-1)^n n! (R(\lambda; S))^{n+1}.$$

Hence

$$\int_C f(\lambda)(R(\lambda; S))^{n+1} d\lambda = \frac{(-1)^n}{n!} \int_C f(\lambda) \left(\frac{d}{d\lambda}\right)^n R(\lambda; S) d\lambda.$$

Integrating n times by parts we find that

$$\int_C f(\lambda)(R(\lambda; S))^{n+1} d\lambda = \frac{1}{n!} \int_C \left\{ \left(\frac{d}{d\lambda}\right)^n f(\lambda) \right\} R(\lambda; S) d\lambda,$$

so that

$$f(S+N) = \sum_{n=0}^{\infty} \left\{ \frac{1}{2\pi i} \int_C f^{(n)}(\lambda) R(\lambda; S) d\lambda \right\} N^n$$
$$= \sum_{n=0}^{\infty} \frac{f^n(S) N^n}{n!},$$

and Theorem 10 is proved. Q.E.D.

12 COROLLARY. *If $\sigma(N) = \{0\}$ where N is an operator commuting with an operator S, and if f is a function analytic on a neighborhood of $\sigma(S)$, then f is analytic on a neighborhood of $\sigma(S+N)$, and*

$$f(S+N) = \sum_{n=0}^{\infty} \frac{f^{(n)}(S) N^n}{n!},$$

the series converging in the uniform topology of operators.

We conclude this section with the following lemma which will be needed in Chapter VIII.

13 LEMMA. *If $\lambda \to T(\lambda)$ is an analytic operator valued function defined on a domain D, then the function $\lambda \to T^{-1}(\lambda)$ is defined on an open subset of D and is analytic there. If $T(\lambda)$ is compact for each λ in D and if D is connected, then either $I - T(\lambda)$ has a bounded inverse for no point in D or else this inverse exists except at a countable number of isolated points.*

PROOF. If $\lambda_0 \in D$ is such that $T^{-1}(\lambda_0)$ exists, then it follows from Lemma 6 that for λ sufficiently near λ_0 the inverse $T^{-1}(\lambda)$ exists and is an analytic function of λ in a neighborhood of λ_0. To prove the second part of the lemma we suppose that there is a number λ_0 in D and a sequence $\{\lambda_m\} \subset D$ such that $1 \in \sigma(T(\lambda_m))$, $m \geqq 0$; $\lambda_m \to \lambda_0$; and $\lambda_m \neq \lambda_0$, $m > 0$. It will be shown that $1 \in \sigma(T(\lambda))$ for every $\lambda \in D$. By Theorems 4.5 and 9, for λ sufficiently near λ_0 the points of $\sigma(T(\lambda))$ in some neighborhood of the point $\alpha = 1$ are given by the fractional power series

$$\sum_{j=0}^{\infty} \alpha_j (\lambda - \lambda_0)^{j/n}, \ldots, \sum_{j=0}^{\infty} c_j (\lambda - \lambda_0)^{j/n}.$$

Since $1 \in \sigma(T(\lambda_m))$ for $m \geqq 0$, one of these series assumes the value unity infinitely many times near λ_0 and hence is identically equal to unity. Thus for λ sufficiently near λ_0, $1 \in \sigma(T(\lambda))$. Now let A denote the set of points λ_0 in D with the property that there is a sequence

$\{\lambda_m\} \subset D$ satisfying $1 \in \sigma(T(\lambda_m))$, $m \geqq 0$; $\lambda_m \to \lambda_0$; and $\lambda_m \neq \lambda_0$, $m > 0$. We have shown above that A is open; since it is also closed in D and D is connected, it follows that either $A = \phi$ or $A = D$. This completes the proof of the lemma. Q.E.D.

7. Tauberian Theory

Let $\{f_n\}$ be a sequence of functions in $\mathscr{F}(T)$, where $T \in B(\mathfrak{X})$, and suppose that the sequence $\{f_n(T)\}$ converges in $B(\mathfrak{X})$. If $f \in \mathscr{F}(T)$, it is clear that $\{f(T)f_n(T)\}$ converges. We are interested in the converse problem: under what conditions on f, f_n and T does the convergence of $\{f(T)f_n(T)\}$ imply the convergence of $\{f_n(T)\}$? Theorems describing this situation are in the nature of Tauberian theorems. The prototype of these theorems is the ergodic theorem which arises when $f(\lambda) = 1-\lambda$ and $f_n(\lambda) = n^{-1}\sum_{j=0}^{n-1}\lambda^j$. In this case, one determines conditions under which the convergence of $n^{-1}T^n$ implies the convergence of the averages $n^{-1}(I+\ldots+T^{n-1})$.

We first consider this converse problem for convergence in the uniform operator topology and later discuss the same question when $B(\mathfrak{X})$ has the weak or the strong operator topologies.

Note that if $f(\lambda) \neq 0$ for $\lambda \in \sigma(T)$, then $f(T)$ has a bounded inverse, and hence $f_n(T) = [f(T)]^{-1}f(T)f_n(T)$ converges. However, it is not obvious that the convergence of $\{f_n(T)\}$ can be deduced if f vanishes at certain points of $\sigma(T)$. The next theorem is concerned with this possibility.

1 Theorem. *Let f, f_n be in $\mathscr{F}(T)$, and let $\{f(T)f_n(T)\}$ converge to zero in the uniform operator topology. Let f vanish at a finite set of poles of $R(\lambda; T)$. Suppose that each root λ_0 of f on $\sigma(T)$ has finite order $\alpha(\lambda_0)$, that the sequences $\{f_n^{(m)}(\lambda_0)\}$ converge for $0 \leqq m < \alpha(\lambda_0)$, and that $\lim_{n\to\infty} f_n(\lambda_0) \neq 0$. Then the sequence $\{f_n(T)\}$ converges in the uniform operator topology.*

Proof. Let $\lambda_1, \ldots, \lambda_k$ denote the roots of f on $\sigma(T)$, and let α_i, and ν_i, $i = 1, \ldots, k$, be the order of λ_i as a root of f, and as a pole of $R(\lambda; T)$, respectively.

We first show that $\nu_i \leqq \alpha_i$. Let $g_n = f \cdot f_n$; then $\{g_n(T)\}$ converges to zero in the uniform operator topology. By Corollary 3.23, $\{g_n^{(m)}(\lambda_i)\}$

converges to zero for $0 \leqq m < \nu_i$. If $\alpha_i < \nu_i$, put $m = \alpha_i$. Then since λ_i is a root of f of order α_i we have

$$g_n^{(\alpha_i)}(\lambda_i) = f^{(\alpha_i)}(\lambda_i)f_n(\lambda_i) \to 0.$$

Since $f^{(\alpha_i)}(\lambda_i) \neq 0$, we conclude that $f_n(\lambda_i) \to 0$, contradicting our hypothesis. Hence $\nu_i \leqq \alpha_i$.

Now let σ_1 be the complement in $\sigma(T)$ of the set $\{\lambda_1, \ldots, \lambda_k\}$. By Theorem 3.22,

$$[*] \quad f_n(T) = f_n(T)E(\sigma_1) + \sum_{i=1}^{k} \sum_{j=0}^{\nu_i-1} \frac{f_n^{(j)}(\lambda_i)}{j!}(T-\lambda_i I)^j E(\lambda_i).$$

Since f does not vanish on σ_1, there is an $h \in \mathscr{F}(T)$ such that fh is identically one on σ_1, and identically zero on $\{\lambda_1, \ldots, \lambda_k\}$. Thus $f(T)h(T) = E(\sigma_1)$, so that $f_n(T)E(\sigma_1)$ converges in the uniform topology of operators. It follows from [*] that $\{f_n(T)\}$ will converge in the uniform topology if $\{f_n^{(m)}(\lambda_i)\}$ converges for $m < \nu_i$, $1 \leqq i \leqq k$. Since $\nu_i \leqq \alpha_i$, the theorem is proved. Q.E.D.

2 Corollary. *Let* $|T^n| = o\,(n)$, *and let* $\lambda = 1$ *be a pole of* $R(\lambda; T)$ *of order one. Then* $\{n^{-1}\sum_{j=0}^{n-1} T^j\}$ *converges in the uniform topology to* $E(1)$.

Proof. Let $f_n(\lambda) = n^{-1}(1+\lambda+\ldots+\lambda^{n-1})$, and let $f(\lambda) = 1-\lambda$. Then $f(\lambda)f_n(\lambda) = n^{-1}(1-\lambda^n)$, and $f(T)f_n(T) = n^{-1}(I-T^n)$. Since $|T^n| = o\,(n)$, it follows that $f(T)f_n(T) \to 0$. It follows readily from Theorem 1 that $\{n^{-1}\sum_{j=0}^{n-1} T^j\}$ converges in the uniform operator topology.

Let σ_1 be the complement in $\sigma(T)$ of the set $\{1\}$, and let $h \in \mathscr{F}(T)$ be equal to $(1-\lambda)^{-1}$ in a neighborhood of σ_1, and be identically zero in a neighborhood of $\lambda = 1$. Then $h(T)f(T) = E(\sigma_1)$. Since $f(T)f_n(T) \to 0$, we have

$$f_n(T)E(\sigma_1) = E(\sigma_1)f_n(T) = h(T)f(T)f_n(T) \to 0.$$

From equation [*] of Theorem 1,

$$f_n(T) = f_n(T)E(\sigma_1) + f_n(1)E(1),$$

so that $f_n(T) \to E(1)$ in the uniform topology. Q.E.D.

The central hypothesis in Theorem 1 is the assumption that each zero of f is a pole of $R(\lambda; T)$. This assumption is not required for the analogues of Theorem 1 for the weak and strong topology, which we study next.

3 THEOREM. *Let f, f_n be in $\mathscr{F}(T)$, and let $\{f(T)f_n(T)\}$ converge to zero in the weak operator topology. Suppose that $\{f_n(T)x\}$ is weakly sequentially compact for each x in $\mathfrak{X}$, and that f vanishes at a finite set of points of $\sigma(T)$. If each root λ_0 has finite order $\alpha(\lambda_0)$, if the sequences $\{f_n^{(m)}(\lambda_0)\}$ converge for $0 \leqq m < \alpha(\lambda_0)$, and if $\lim_{n\to\infty} f_n(\lambda_0) \neq 0$, then $\{f_n(T)\}$ converges in the weak operator topology. Moreover,*

$$\mathfrak{X} = \overline{f(T)\mathfrak{X}} \oplus \{x|x \in \mathfrak{X}, f(T)x = 0\}.$$

PROOF. Let $\mathfrak{X}_1 = \overline{f(T)\mathfrak{X}}$, $\mathfrak{X}_2 = \{x|x \in \mathfrak{X}, f(T)x = 0\}$, and $\mathfrak{X}_3^* = \{x^*|x^* \in \mathfrak{X}^*, f(T^*)x^* = 0\}$. If an operator U in $B(\mathfrak{X})$ leaves $\mathfrak{X}_2$ invariant, we let U_2 denote the restriction of U to $\mathfrak{X}_2$.

It will first be shown that, for each non-zero x in $\mathfrak{X}_2$, the sequence $\{f_n(T)x\}$ converges to a non-zero element of $\mathfrak{X}_2$. Since $f(T)$ commutes with $R(\lambda; T)$, $\lambda \in \varrho(T)$, we have $R(\lambda; T)\mathfrak{X}_2 \subseteqq \mathfrak{X}_2$, and, consequently, $R(\lambda; T)_2 = R(\lambda; T_2)$, and $\sigma(T_2) \subseteqq \sigma(T)$. Furthermore, Definition 3.9 shows that $g(T_2) = g(T_2)$, for $g \in \mathscr{F}(T)$. Since $f(T_2) = 0$, it follows from Theorem 3.16 that $\sigma(T_2) \subseteqq \{\lambda_1, \ldots, \lambda_k\}$, the set of zeros of f, and that the order $\nu(\lambda_i)$ of λ_i as a pole of $R(\lambda; T_2)$ does not exceed the order $\alpha(\lambda_i)$ of λ_i as a zero of f. Let g be a function in $\mathscr{F}(T_2)$, such that $g^{(m)}(\lambda_i) = \lim_{n\to\infty} f_n^{(m)}(\lambda_i)$, for $1 \leqq i \leqq k$ and $0 \leqq m < \nu(\lambda_i)$. By Theorem 3.22, $\{f_n(T_2)\}$ converges to $g(T_2)$ in the strong operator topology of $\mathfrak{X}_2$. Since $g(\sigma(T_2))$ does not contain the point 0, the operator $g(T_2)$ has an inverse on the space $\mathfrak{X}_2$. Thus, for each non-zero x in $\mathfrak{X}_2$, $\{f_n(T)x\}$ converges to a non-zero element in $\mathfrak{X}_2$. We can prove, in the same way, that for each non-zero x^* in $\mathfrak{X}_3^*$, $\{f_n(T^*)x^*\}$ converges to a non-zero element in $\mathfrak{X}_3^*$.

We next prove that $\mathfrak{X}_1 \cap \mathfrak{X}_2 = \{0\}$. Since $\{f_n(T)x\}$ is weakly sequentially compact for each x in $\mathfrak{X}$, it follows from II.3.27 and II.3.21, that the sequence $\{|f_n(T)|\}$ is bounded. Since we have assumed that $\{f_n(T)x\}$ converges weakly to zero for $x \in f(T)\mathfrak{X}$, Theorem II.3.6 implies that $\{f_n(T)x\}$ converges weakly to zero for $x \in \mathfrak{X}_1$. But, since $\{f_n(T)x\}$ converges strongly to a non-zero element for each non-zero $x \in \mathfrak{X}_2$, we conclude that $\mathfrak{X}_1 \cap \mathfrak{X}_2 = \{0\}$.

It will next be shown that $\{f_n(T)x\}$ converges in the weak topology for each $x \in \mathfrak{X}$. If not, since $\{f_n(T)x\}$ is weakly sequentially compact, we can extract two sequences $\{f_{n_i}(T)x\}$ and $\{f_{m_i}(T)x\}$, such that

$$\lim_{i\to\infty} f_{n_i}(T)x = y_1, \qquad \lim_{i\to\infty} f_{m_i}(T)x = y_2, \qquad y_1 \neq y_2,$$

both limits being in the weak topology. Since $f(T)f_n(T)$ converges to zero in the weak operator topology, $f(T)(y_1-y_2) = 0$. Thus $y_1-y_2 \in \mathfrak{X}_2$, and $y_1-y_2 \notin \mathfrak{X}_1$. It follows from II.3.13 that there exists an $x_0^* \in \mathfrak{X}^*$ such that $x_0^*(\mathfrak{X}_1) = 0$, and $x_0^*(y_1-y_2) \neq 0$. Since $f(T)\mathfrak{X} \subseteq \mathfrak{X}_1$, we see that $x_0^*(f(T)\mathfrak{X}) = 0$, so that $f(T)^*x_0^* = f(T^*)x_0^* = 0$; i.e., $x_0^* \in \mathfrak{X}_3^*$. It was remarked above that $\{f_n(T^*)x^*\}$ converges strongly for each $x^* \in \mathfrak{X}_3^*$. Thus $\lim_{n\to\infty} x_0^*(f_n(T)x) = \lim_{n\to\infty} (f_n(T^*)x_0^*)(x)$ exists for each $x \in \mathfrak{X}$, and

$$x_0^*(y_1) = \lim_{i\to\infty} x_0^*(f_{n_i}(T)x) = \lim_{i\to\infty} x_0^*(f_{m_i}(T)x) = x_0^*(y_2),$$

contradicting the inequality $x_0^*(y_1-y_2) \neq 0$. This proves that $\{f_n(T)x\}$ converges in the weak topology for each $x \in \mathfrak{X}$.

To prove the final statement of the theorem, let

$$T_0 x = \lim_{n\to\infty} f_n(T)x,$$

the limit being in the weak topology. Then, by II.3.21 and II.3.27, $T_0 \in B(\mathfrak{X})$. Moreover, $T_0x = g(T_2)x$ for x in $\mathfrak{X}_2$. Since $g(T_2)$ has an inverse on $\mathfrak{X}_2$, $g(T_2)\mathfrak{X}_2 = \mathfrak{X}_2$ and $T_0\mathfrak{X} \supseteq \mathfrak{X}_2$. Similarly $T_0^*\mathfrak{X}^* \supseteq \mathfrak{X}_3^*$. But $T_0f(T) = f(T)T_0 = 0$, so $T_0\mathfrak{X} \subseteq \mathfrak{X}_2$ and $T_0\mathfrak{X}_1 = 0$. Thus the operator E, defined by $Ex = g(T_2)^{-1}T_0x$, $x \in \mathfrak{X}$, is a projection whose range is $\mathfrak{X}_2$. Since we know that $\mathfrak{X}_1 \cap \mathfrak{X}_2 = 0$, in order to prove $\mathfrak{X} = \mathfrak{X}_1 \oplus \mathfrak{X}_2$, it is sufficient to show that $Ex = 0$ implies $x \in \mathfrak{X}_1$. If $Ex = 0$, then $T_0x = 0$, so $(T_0^*\mathfrak{X}^*)x = 0$. But $\{x^* | x^*(\mathfrak{X}_1) = 0\} = \mathfrak{X}_3^* \subseteq T_0^*\mathfrak{X}^*$, so $x^*x = 0$ for $x^* \in \mathfrak{X}_3^*$. It follows from II.3.13 and from the definition of $\mathfrak{X}_3^*$, that $x \in \mathfrak{X}_1$; so $(I-E)\mathfrak{X} = \mathfrak{X}_1$. Q.E.D.

The corresponding result for the strong operator topology follows from what we have just proved.

4 **Theorem.** *Let f, f_n be in $\mathscr{F}(T)$, and let $\{f(T)f_n(T)\}$ converge to zero in the strong operator topology. Suppose that $\{f_n(T)x\}$ is weakly sequentially compact for each x in $\mathfrak{X}$, and that f vanishes at a finite set of points of $\sigma(T)$. If each root λ_0 has finite order $\alpha(\lambda_0)$, if the sequences $\{f_n^{(m)}(\lambda_0)\}$ converge for $0 \leqq m < \alpha(\lambda_0)$, and if $\lim_{n\to\infty} f_n(\lambda_0) \neq 0$, then*

$\{f_n(T)\}$ *converges in the strong operator topology. Moreover,*

$$\mathfrak{X} = \overline{f(T)\mathfrak{X}} \oplus \{x|x \in \mathfrak{X}, f(T)x = 0\}.$$

PROOF. All parts of Theorem 4 but the strong convergence of $\{f_n(T)\}$ follow immediately from Theorem 3. The proof of Theorem 3 shows that $\{f_n(T)x\}$ converges for $x \in \mathfrak{X}_2$, and that the sequence $\{|f_n(T)|\}$ is bounded. Since we have assumed that $\{f_n(T)x\}$ converges for $x \in f(T)\mathfrak{X}$, it follows from II.3.6 that $\{f_n(T)x\}$ converges for $x \in \mathfrak{X}_1 = \overline{f(T)\mathfrak{X}}$. Our result now follows from the decomposition $\mathfrak{X} = \mathfrak{X}_1 \oplus \mathfrak{X}_2$. Q.E.D.

We conclude this section with an application of Theorem 4, somewhat different from the ergodic theorem proved in Corollary 2. Further applications will be found in the exercises of Section 8. In the following chapter, Theorems 3 and 4 will form the basis for a comprehensive discussion of ergodic theory.

5 COROLLARY. *Let $\mathfrak{X}$ be reflexive, let λ_n be a sequence in $\varrho(T)$ which converges to zero, and let* $\sup_n |\lambda_n R(\lambda_n; T)| < \infty$. *Then*

$$\mathfrak{X} = \overline{T\mathfrak{X}} \oplus \{x|x \in \mathfrak{X}, Tx = 0\},$$

and the sequence $\{\lambda_n R(\lambda_n; T)\}$ converges in the strong operator topology to the projection E, whose null manifold is $\overline{T\mathfrak{X}}$, and whose range is $\{x|Tx = 0\}$.

PROOF. Let $f_n(\lambda) = \lambda_n(\lambda_n-\lambda)^{-1}$, and let $f(\lambda) = \lambda$. Then $f(T)f_n(T) = \lambda_n TR(\lambda_n; T) = \lambda_n^2 R(\lambda_n; T)-\lambda_n I$; so $f(T)f_n(T) \to 0$ in the uniform operator topology. Since $\mathfrak{X}$ is reflexive, it follows from II.3.28 that for every x in $\mathfrak{X}$ the sequence $\{f_n(T)x\}$ is weakly sequentially compact. As $f_n(0) = 1$ for all n, Theorem 4 shows that $\lim_{n\to\infty} \lambda_n R(\lambda_n; T)$ exists in the strong topology, and $\mathfrak{X}$ has the decomposition above. Examination of the proof of Theorem 3 shows $\lim_{n\to\infty} \lambda_n R(\lambda_n; T) = E$. Q.E.D.

8. Exercises

1 If $T_n, T \in B(\mathfrak{X})$, $T_n \to T$ and $0 \in \varrho(T_n)$, $n = 1, 2, \ldots$, then $0 \in \varrho(T)$ if and only if $\sup_n |T_n^{-1}| < \infty$.

2 Let $g_n \in \mathscr{F}(T)$, $n = 1, 2, \ldots$. Suppose that $g_n(T)$ converges in the uniform operator topology to a compact operator. Let $\lambda \in \sigma(T)$, and $\lim_{n\to\infty} g_n(\lambda) \neq 0$. Show that λ is a pole of $\sigma(T)$, and that $E(\lambda; T)\mathfrak{X}$ has a positive finite dimension. (Hint. See Exercise VII.5.35.)

3 Show that if $g_n \in \mathscr{F}(T)$, $n = 1, 2, \ldots$, and $\{g_n(T)\}$ converges to a projection E in the uniform operator topology, then there exists a spectral subset σ_1 of $\sigma(T)$ such that $E = E(\sigma_1)$. (Hint. See Exercise VII.5.32.)

4 Let $T \in B(\mathfrak{X})$, $T_n \in B(\mathfrak{X})$, $n = 1, 2, \ldots$. Suppose that $T_n \to T$ in the uniform topology of operators. Let $\lambda \in \sigma(T)$ be an isolated point such that $E(\lambda; T)\mathfrak{X}$ is one dimensional. Let U be a neighborhood of λ such that $\bar{U} \cap \sigma(T) = \{\lambda\}$. Show that for n sufficiently large, $U \cap \sigma(T_n)$ contains just one point λ_n, that $\lambda_n \to \lambda$, and that $E(\lambda_n; T_n) \to E(\lambda; T)$ in the uniform operator topology.

5 Let $T(\mu)$ be an analytic function with values in $B(\mathfrak{X})$ defined for $|\mu| < \varepsilon$. Suppose that $\lambda \in \sigma(T(0))$ is an isolated point such that $E(\lambda; T(0))\mathfrak{X}$ is one dimensional. Let U be a neighborhood of λ such that $\bar{U} \cap \sigma(T(0)) = \{\lambda\}$. Show that there exists a positive δ less than ε such that $U \cap \sigma(T(\mu))$ consists of a single point $\lambda(\mu)$ for $|\mu| < \delta$, and that $\lambda(\mu)$ and $E(\lambda(\mu); T(\mu))$ are analytic functions of μ for $|\mu| < \varepsilon$.

6 Let $|T| \leqq 1$, and let T be weakly compact. Then the limit $\lim_{n\to\infty} n^{-1}(I + \ldots + T^{n-1}) = E$ exists in the strong topology of operators. The operator E is a projection of $\mathfrak{X}$ onto $\{x | x \in \mathfrak{X}, Tx = x\}$.

7 Let $\lambda = 0$ be a pole of $R(\lambda; T)$. Suppose that there exist real constants K and ε such that

$$|\lambda R(\lambda; T)| \leqq K \text{ for } 0 < |\lambda| < \varepsilon.$$

Then $\lim_{\lambda\to 0} \lambda R(\lambda; T) = E(\{0\})$ in the uniform topology of operators.

8 Let S be a compact metric space, and let $\alpha : S \to S$ be such that $\varrho(x, y) = \varrho(\alpha x, \alpha y)$ for $x, y \in S$. Show that $\{n^{-1} \sum_{i=0}^{n-1} f(\alpha^i x)\}$ converges for $f \in C(S)$, uniformly for $x \in S$.

9 Let t be an irrational real number, and t_n be the fractional part of nt. If I is a subinterval of $[0, 1)$, let N_k be the number of $\{t_1, \ldots, t_k\}$ which lie in I. Show that $\lim_{k\to\infty} k^{-1}N_k$ exists and is equal to the length of I.

10 Let (S, Σ, μ) be a positive finite measure space. Let $\alpha : S \to S$ be a mapping such that $e \in \Sigma$ if and only if $\alpha(e)$ is in Σ, and $\mu(e) = \mu(\alpha(e))$. If $f \in L_p(S, \Sigma, \mu)$ for $p \geqq 1$, and we put $f_n(S) = n^{-1} \sum_{i=0}^{n-1} f(\alpha^i(s))$ for $n = 1, 2, \ldots$, then $f_n \in L_p(S, \Sigma, \mu)$, and $\lim_{n\to\infty} f_n$ exists.

11 Let $\mathfrak{X}_1$ be a closed subspace of a B-space $\mathfrak{X}$ such that $T_n\mathfrak{X}_1 \subseteqq \mathfrak{X}_1$ for each operator T_n in a sequence $\{T_n, n = 1, 2, \ldots\}$ of commuting operators in $B(\mathfrak{X})$. Let U_n be the restriction of T_n to $\mathfrak{X}_1$ and let $V_n : \mathfrak{X}/\mathfrak{X}_1 \to \mathfrak{X}/\mathfrak{X}_1$ be the map defined by

$$V_n(x+\mathfrak{X}_1) = T_n x+\mathfrak{X}_1, \qquad x \in \mathfrak{X}, \quad n \geqq 1.$$

Let U_n converge weakly to zero and V_n converge weakly to an operator V such that $0 \notin \sigma(V)$. Finally let $\{T_n x\}$ be weakly sequentially compact for each x in $\mathfrak{X}$. Show that (a), the sequence $\{T_n\}$ converges in the weak operator topology, and (b), the space $\mathfrak{X}$ can be written as a direct sum $\mathfrak{X} = \mathfrak{X}_1 \oplus \mathfrak{X}_2$, where $T_n\mathfrak{X}_2 \subseteqq \mathfrak{X}_2$ for all $n \geqq 1$. (Hint. Show first that the sequence $\{T_n\}$ has one and only one accumulation point in the weak operator topology. Compare with the proof of Theorem 7.3.)

12 Assume in addition to the hypothesis of Exercise 11 that $\{U_n\}$ and $\{V_n\}$ converge in the strong operator topology. Show that $\{T_n\}$ converges in the strong operator topology.

13 Let $x_n \in \mathfrak{X}$, $n = 1, 2, \ldots$. Suppose that $|Tx_n - \lambda x_n| \to 0$. Show that $|f(T)x_n - f(\lambda)x_n| \to 0$ for each $f \in \mathscr{F}(T)$.

9. An Operational Calculus for Unbounded Closed Operators

We shall now show that many of the results of the operational calculus for a bounded operator may be extended to the case of a closed operator T with non-empty resolvent set.

We recall that a linear transformation T whose domain is a linear manifold $\mathfrak{D}(T)$ is said to be closed if its graph is closed. Equivalently, if $x_n \in \mathfrak{D}(T)$, $n = 1, 2, \ldots$, $x_n \to x$, and $Tx_n \to y$, then $x \in \mathfrak{D}(T)$ and $Tx = y$. If T is closed and everywhere defined it is in $B(\mathfrak{X})$ (II.2.4); hence we shall suppose that its domain $\mathfrak{D}(T)$ is a proper subset of $\mathfrak{X}$. This important case occurs for many differential operators in various function spaces.

As in the case when T is in $B(\mathfrak{X})$, we define the *resolvent set* $\varrho(T)$ of T to be the set of complex numbers λ such that $(\lambda I - T)^{-1}$ is in

$B(\mathfrak{X})$, and the *spectrum* $\sigma(T)$ of T to be the complement of $\varrho(T)$. As before, (cf. Definition 5.1), the spectrum is divided into three disjoint sets: the point spectrum, continuous spectrum and residual spectrum. It is seen from Lemma 2 below that the spectrum is a closed set. But, in contrast to the case where T is a bounded operator, the spectrum may be a bounded set, an unbounded set, the void set, or even the whole plane. This is observed in Exercise 10.1. We exclude the last possibility, and suppose throughout this section that *$\varrho(T)$ is not void.*

We now show how the development of an operational calculus for T may be based on the calculus already obtained for an operator in $B(\mathfrak{X})$.

1 Definition. By $\mathscr{F}(T)$ we denote the family of all functions f which are analytic on some neighborhood of $\sigma(T)$ and at infinity.

As in the case of a bounded operator (Definition 3.9), the neighborhood need not be connected, and can depend on $f \in \mathscr{F}(T)$.

Let α be a fixed point of $\varrho(T)$, and define

$$A = (T-\alpha I)^{-1} = -R(\alpha; T).$$

Then A defines a one-to-one mapping of $\mathfrak{X}$ onto $\mathfrak{D}(T)$, and

$$TAx = \alpha Ax+x, \qquad x \in \mathfrak{X},$$

$$ATx = \alpha Ax+x, \qquad x \in \mathfrak{D}(T).$$

Our objective is to define an operational calculus for T in terms of that already obtained in Section 3 for the bounded operator A.

If K denotes the complex sphere, with its usual topology, we let $\Phi : K \to K$ be the homeomorphism defined by

$$\mu = \Phi(\lambda) = (\lambda-\alpha)^{-1}, \qquad \Phi(\infty) = 0, \qquad \Phi(\alpha) = \infty.$$

2 Lemma. *If α is in $\varrho(T)$, then $\Phi(\sigma(T)\cup\{\infty\}) = \sigma(A)$, and the relation*

$$\varphi(\mu) = f(\Phi^{-1}(\mu))$$

determines a one-to-one correspondence between f in $\mathscr{F}(T)$ and φ in $\mathscr{F}(A)$.

Proof. Let $\lambda \in \varrho(T)$. Then $0 \neq \mu = (\lambda-\alpha)^{-1}$, and

$$(T - \alpha I)(T - \lambda I)^{-1} = \left[(T - \lambda I) + \frac{I}{\mu}\right] (T - \lambda I)^{-1}$$

$$= I + \frac{(T-\lambda I)^{-1}}{\mu}.$$

But we also have

$$(T - \alpha I)(T - \lambda I)^{-1} = A^{-1}\left[(T - \alpha I) - \frac{I}{\mu}\right]^{-1}$$

$$= \left\{\left[(T - \alpha I) - \frac{I}{\mu}\right]A\right\}^{-1}$$

$$= \mu(\mu I - A)^{-1},$$

which shows that

$$[*] \qquad (T - \lambda I)^{-1} = \mu^2(\mu I - A)^{-1} - \mu I.$$

Thus $\mu \in \varrho(A)$. Conversely, if $\mu \in \varrho(A)$, $\mu \neq 0$, then

$$(\mu I - A)^{-1}A = [A^{-1}(\mu I - A)]^{-1}$$

$$= (\mu A^{-1} - I)^{-1}$$

$$= \frac{I}{\mu}(T - \lambda I)^{-1},$$

showing that $\lambda \in \varrho(T)$. The point $\mu = 0$ is in $\sigma(A)$, since $A^{-1} = T - \alpha I$ is unbounded. The last statement is evident from the definition of Φ. Q.E.D.

3 Definition. For $f \in \mathscr{F}(T)$ we define $f(T) = \varphi(A)$, where $\varphi \in \mathscr{F}(A)$ is given by $\varphi(\mu) = f(\Phi^{-1}(\mu))$.

4 Theorem. *If f is in $\mathscr{F}(T)$ then $f(T)$ is independent of the choice of α in $\varrho(T)$. Let V be an open set containing $\sigma(T)$ whose boundary Γ consists of a finite number of Jordan arcs and such that f is analytic on $V \cup \Gamma$. Let Γ have positive orientation with respect to the (possibly unbounded) set V. Then*

$$f(T) = f(\infty)I + \frac{1}{2\pi i}\int_\Gamma f(\lambda)R(\lambda; T)d\lambda.$$

Proof. It suffices to establish the formula above for $f(T)$ since the integral is independent of α. Given $\alpha \in \varrho(T)$ and the set V we observe first that due to the analyticity of $R(\lambda; T)$, we may assume that $\alpha \notin V \cup \Gamma$, since otherwise, using the Cauchy integral theorem, we can replace V by a set V_1 such that $\alpha \notin V_1 \cup \Gamma_1$ without altering the value of the integral above. Then $U = \Phi^{-1}(V)$ is an open set containing $\sigma(A)$ and whose boundary $C = \Phi^{-1}(\Gamma)$ is positively oriented and consists of a finite number of Jordan arcs. Moreover $\varphi(\mu) =$

$f(\Phi^{-1}(\mu))$ is analytic on $\bar{U}$. Since $\varphi(0) = f(\infty)$ and $0 \in \sigma(A)$, we have from [*], Definition 3.9, and the fact that $d\lambda = -d\mu/\mu^2$

$$\frac{1}{2\pi i}\int_\Gamma f(\lambda)R(\lambda, T)d\lambda = \frac{1}{2\pi i}\int_C \varphi(\mu)[-\mu^{-1}I+R(\mu; A)]d\mu$$
$$= \varphi(A)-\varphi(0)I = f(T)-f(\infty)I. \quad \text{Q.E.D.}$$

Theorem 4, with Theorems 3.10, 3.11, and 3.12, now yields

5 THEOREM. *If f and g are in $\mathscr{F}(T)$, then*
(a) $(f+g)(T) = f(T)+g(T)$;
(b) $(fg)(T) = f(T)g(T)$;
(c) $\sigma(f(T)) = f(\sigma(T)\cup\{\infty\})$;
(d) *if f is in $\mathscr{F}(T)$, g is in $\mathscr{F}(f(T))$, and $F(\xi) = g(f(\xi))$, then F is in $\mathscr{F}(T)$ and $F(T) = g(f(T))$.*

For a comprehensive operational calculus for T, it is important to include a theory of polynomials in T (which will be defined only on proper subsets of $\mathfrak{X}$), and to have rules relating them to the operators $f(T)$. The domain of a polynomial in T is defined in the natural way in the following definition.

6 DEFINITION. For $n = 0, 1, \ldots$ the operator T^n is defined inductively by the relations $T^0 = I$, $T^1 = T$, and

$$\mathfrak{D}(T^n) = \{x | x \in \mathfrak{D}(T^{n-1}), T^{n-1}x \in \mathfrak{D}(T)\}$$
$$T^n x = T(T^{n-1}x), \qquad x \in \mathfrak{D}(T^n).$$

In some of the proofs to follow we shall consider the intersection

$$\bigcap_{n=1}^{\infty} \mathfrak{D}(T^n),$$

which we will denote by the convenient if not quite rigorous symbol $\mathfrak{D}(T^\infty)$. If $P(\lambda) = \sum_{i=0}^n \alpha_i\lambda^i$ is a polynomial of degree n the symbol $P(T)$ will denote the operator $\sum_{i=0}^n \alpha_i T^i$ with domain $\mathfrak{D}(T^n)$.

We now prove an important result.

7 THEOREM. *If T is a closed linear operator with non-void resolvent set, and P is a polynomial, then $P(T)$ is a closed operator.*

PROOF. Suppose that the degree of P is positive. Let $\alpha \in \varrho(T)$, and write

$$P(\lambda) = \sum_{k=0}^{n} b_k(\lambda-\alpha)^{n-k}.$$

Then, under the transformation Φ, $P(\lambda)$ goes into $\mu^{-n}p(\mu)$ where

$$(*) \qquad p(\mu) = \sum_{k=0}^{n} b_k \mu^k.$$

If $x \in \mathfrak{D}(T^n)$, $p(A)x \in \mathfrak{D}(T^n)$, since for any q, A^q maps $\mathfrak{D}(T^n)$ onto $\mathfrak{D}(T^{n+q})$. Also

$$P(T)x = (T-\alpha I)^n p(A)x = p(A)(T-\alpha I)^n x, \qquad x \in \mathfrak{D}(T^n).$$

Now let $x_r \in \mathfrak{D}(T^n)$, $x_r \to x$, $P(T)x_r \to y$. We must show that x is in $\mathfrak{D}(T^n)$, and that $P(T)x = y$. The operator $(T-\alpha I)^n$ is closed, as it is the inverse of the bounded operator A^n. Thus, since $p(A)x_n \to p(A)x$, $p(A)x \in \mathfrak{D}(T^n)$. To see that x is in $\mathfrak{D}(T^n)$, we write

$$p(A)x = b_0 x + b_1 Ax + \ldots + b_n A^n x, \qquad b_0 \neq 0,$$

and note that, since $A\mathfrak{X} = \mathfrak{D}(T)$, all terms except possibly $b_0 x$ are in $\mathfrak{D}(T)$. Thus x is in $\mathfrak{D}(T)$. We proceed by induction. Suppose that $x \in \mathfrak{D}(T^m)$, where $1 \leqq m < n$. Then

$$(T-\alpha I)^m p(A)x = b_0(T-\alpha I)^m x + b_1(T-\alpha I)^{m-1}x + \ldots + b_n A^{n-m}x.$$

By the same argument, $(T-\alpha I)^m x \in \mathfrak{D}(T)$, and hence $x \in \mathfrak{D}(T^{m+1})$. Thus, by induction, $x \in \mathfrak{D}(T^n)$. This completes the proof. Q.E.D.

8 THEOREM. *Let P be a polynomial of degree n, and let the function f in $\mathscr{F}(T)$ have a zero of order m, $0 \leqq m \leqq \infty$, at infinity.*

(a) *If x is in $\mathfrak{D}(T^n)$, then $f(T)x$ is in $\mathfrak{D}(T^{m+n})$, where $m+n = \infty$ if $m = \infty$, and $P(T)f(T)x = f(T)P(T)x$.*

(b) *If $0 \leqq n \leqq m$, and $g(\lambda) = P(\lambda)f(\lambda)$, then g is in $\mathscr{F}(T)$, and $g(T) = P(T)f(T)$.*

PROOF. Let $q = m$, if m is finite, or any positive integer, if $m = \infty$. Let $\alpha \in \varrho(T)$, and $\varphi(\mu) = f(\lambda)$, $(\lambda-\alpha)\mu = 1$. By defining $\beta(\mu) = \mu^{-q}\varphi(\mu)$ when $\mu \neq 0$ and $\beta(0) = \lim_{\mu\to 0} \mu^{-q}\varphi(\mu)$, it is seen that β is in $\mathscr{F}(A)$ and that $A^q\beta(A) = \varphi(A) = f(T)$. Thus $f(T)\mathfrak{X} \subseteq A^q\mathfrak{X} = \mathfrak{D}(T^q)$. Now let $x \in \mathfrak{D}(T^n)$. Then $x = A^n y$ for some y in $\mathfrak{X}$, and $f(T)x = A^{n+q}\beta(A)y \in \mathfrak{D}(T^{n+q})$, proving the first part of (a). To prove the second statement, write $x = A^n y$, and define $p(\mu)$ by formula (*) of the preceding theorem. Then,

$$
\begin{aligned}
P(T)f(T)x &= P(T)A^n\varphi(A)y \\
&= p(A)\varphi(A)y \\
&= \varphi(A)p(A)y \\
&= f(T)p(A)(T-\alpha I)^n x \\
&= f(T)P(T)x.
\end{aligned}
$$

Finally, for (b), let $\gamma(\mu) = g(\lambda)$. Then γ has a removable singularity at $\mu = 0$, and $\gamma(\mu) = \mu^{-n}p(\mu)\varphi(\mu)$. Now $A^n\gamma(A) = p(A)\varphi(A)$, and thus $A^n g(T) = p(A)f(T)$. By operating on both sides by $(T-\alpha I)^n$, we obtain $P(T)f(T) = g(T)$. Q.E.D.

9 THEOREM. *If f is in $\mathscr{F}(T)$, and f has no zeros in $\sigma(T)$, but a zero of finite order n at infinity, then $\{f(T)\}^{-1}$ exists and has domain $\mathfrak{D}(T^n)$.*

PROOF. We let $g(\lambda) = (\lambda-\alpha)^n f(\lambda)$, $\alpha \in \varrho(T)$. Then g and $1/g$ are in $\mathscr{F}(T)$, and $f(T) = A^n g(T)$. Thus $f(T)\mathfrak{X} = \mathfrak{D}(T^n)$, and $[f(T)]^{-1} = [g(T)]^{-1}(T-\alpha I)^n$. Q.E.D.

As a final result, we extend the spectral mapping theorem to polynomials in T.

10 THEOREM. *If P is a polynomial,*

$$P(\sigma(T)) = \sigma(P(T)).$$

PROOF. Let P be of degree n, and suppose $\lambda \notin P(\sigma(T))$. If $g(\xi) = [\lambda - P(\xi)]^{-1}$, g is in $\mathscr{F}(T)$, with no zeros on $\sigma(T)$, and a zero of order n at infinity. Then, by Theorems 8 and 9, $[g(T)]^{-1} = \lambda I - P(T)$ with domain $\mathfrak{D}(T^n)$. Thus $\lambda \notin \sigma(P(T))$, and $\sigma(P(T)) \subseteq P(\sigma(T))$. To prove the converse inclusion, let Q be the polynomial defined by $P(\lambda)-P(\xi) = (\lambda-\xi)Q(\xi)$. If $P(\lambda) \notin \sigma(P(T))$, then $P(\lambda)-P(T) = (\lambda I - T)Q(T)$, has a bounded inverse R, and hence $(\lambda I - T)^{-1} = Q(T)R$ is a closed everywhere defined operator. Thus $\lambda \notin \sigma(T)$. Q.E.D.

Further results on closed operators will be found in the exercises.

10. Exercises

1 Let $\mathfrak{X} = C[0, 1]$, $(Tx)(t) = x'(t)$. Show that T is a closed operator and determine the spectrum when

(a) $\mathfrak{D}(T) = \{x | x' \in C[0, 1], x(0) = 0\}$ (no spectrum).
(b) $\mathfrak{D}(T) = \{x | x' \in C[0, 1]\}$ (no resolvent set).
(c) $\mathfrak{D}(T) = \{x | x' \in C[0, 1], x(0) = x(1)\}$
$(\sigma_p(T) = \{2\pi i n, n = 0, \pm 1, \pm 2, \ldots\}, \sigma_r(T) = \phi, \sigma_c(T) = \phi)$.

2 In the spaces $L_p(-\pi, \pi)$ $1 \leqq p \leqq \infty$ on the unit circle, let T be the differentiation operator $(Tx)(t) = x'(t)$ with domain $\mathfrak{D}(T) = \{x | x$ is absolutely continuous and $x' \in L_p(-\pi, \pi)\}$. Show that T is a closed unbounded operator whose domain is dense for $p < \infty$, whose spectrum is the set $\{\pm in\}$, $n = 0, 1, 2, \ldots$, and that $\sigma(T) = \sigma_p(T)$.

3 In the spaces $L_p(-\infty, \infty)$, $1 \leqq p \leqq \infty$, let T be the operator $(Tx)(t) = x'(t)$ with domain $\mathfrak{D}(T) = \{x | x$ is absolutely continuous on each finite interval, $x' \in L_p(-\infty, \infty)\}$. Show that (a) T is a closed unbounded linear operator whose domain is dense for $p < \infty$, and (b) the spectrum of T is the imaginary axis, and

$$\begin{aligned} R(\lambda; T)(x, t) &= \int_0^\infty e^{-\lambda\xi} x(t+\xi) d\xi, & \mathscr{R}(\lambda) > 0, \\ &= -\int_{-\infty}^0 e^{-\lambda\xi} x(t+\xi) d\xi, & \mathscr{R}(\lambda) < 0. \end{aligned}$$

4 It is known (Bade [1; p. 278]) that a function f is analytic on the imaginary axis and at infinity if and only if it has the representation

$$f(\lambda) = \int_{-\infty}^\infty e^{-\lambda\xi} G(\xi) d\xi$$

with

$$\begin{aligned} G(\xi) &= F_1(\xi), & \xi > 0, \\ &= F_2(\xi), & \xi \leqq 0, \end{aligned}$$

where F_1 and F_2 are entire functions such that for some constant $c > 0$, $|F_i(z)| = O(e^{c|z|})$, $i = 1, 2$, and

$$\begin{aligned} F_1(\xi) &= O(e^{-c\xi}) & \xi \to +\infty, \\ F_2(\xi) &= O(e^{+c\xi}) & \xi \to -\infty. \end{aligned}$$

For the operator T of Exercise 3 show that if $f \in \mathscr{F}(T)$ then

$$f(T)(x, t) = \int_{-\infty}^\infty G(\xi) x(t-\xi) d\xi.$$

5 Let A be a bounded operator and $\lambda_0 \in \sigma_c(A)$ or $\sigma_r(A)$. Then $(A-\lambda_0 I)^{-1}$ is a closed operator and $\sigma((A-\lambda_0 I)^{-1}) = \{\lambda | \lambda = (\lambda_0 - \xi)^{-1}, \xi \in \sigma(A)\}$.

6 Let T be a closed operator with non-empty resolvent set.

Show that if $\lambda \in \varrho(T)$, α is any complex number, and if x is in $\mathfrak{D}(T^n)$, then

$$R(\lambda; T)x = \sum_{i=0}^{n} \frac{(T-\alpha I)^i x}{(\lambda-\alpha)^{i+1}} + \frac{(T-\alpha I)^{n+1}}{(\lambda-\alpha)^{n+1}} R(\lambda; T)x.$$

7 Let T be as in the preceding problem and let f in $\mathscr{F}(T)$ have a zero of order n at $\lambda = \infty$ and no zero on $\sigma(T)$. Show that $[f(T)]^{-1}$ has the domain $\mathfrak{D}(T^n)$ and the representation

$$[f(T)]^{-1}x = \frac{1}{2\pi i}\int_{\Gamma} \frac{(T-\alpha I)^{n+1}}{f(\lambda)(\lambda-\alpha)^{n+1}} R(\lambda, T)x d\lambda$$

where $x \in \mathfrak{D}(T^n)$ and Γ is a suitable contour.

8 Let T satisfy the hypotheses of Exercise 6 and be unbounded. Show that there exists no polynomial of degree n, $n \geqq 1$, not identically zero, such that $P(T)x = 0$ for all x in $\mathfrak{D}(T^n)$.

11. Notes and Remarks

General references. For a general discussion of the theory in and related to this chapter, the reader is referred to the expository articles of Dunford [6] and Taylor [10]. Additional results may be found in the treatises of Hille [1; Chap. 5], Riesz and Sz.-Nagy [1; Chap. 11] and Stone [3; Chap. 4], the latter dealing with Hilbert space theory.

Finite dimensional space. Spectral theory in finite dimensional spaces is a portion of matrix theory. For the many other questions in this theory which we do not discuss the reader should consult MacDuffee [1] or Wedderburn [1]—the former is particularly recommended for its copious references to the literature. The books of Halmos [7], Hamburger and Grimshaw [1], and Schwerdtfeger [1] are close in spirit to the treatment in this section and will be found useful. We will confine our remarks principally to Theorems 1.8—1.10.

A number λ_0 is called (1.2) an *eigenvalue* of the linear operator T if there exists an $x_0 \neq 0$ such that $Tx_0 = \lambda_0 x_0$. The terms "proper value", "characteristic value", "secular value" and "latent-value" or "latent root" are sometimes used by other authors. The latter term is due to Sylvester [2] because such numbers are "latent in a somewhat similar sense as vapour may be said to be latent in water or smoke in a tobacco-leaf." We will not adhere to his terminology. The term "spectrum" is due to Hilbert.

Polynomials of a matrix were used almost from the beginning of the theory, and by 1867 Laguerre [1] had considered infinite power series in a matrix in constructing the exponential function of a matrix. Sylvester [1, 2] constructed arbitrary functions of a matrix with distinct eigenvalues by means of the Lagrange interpolation formula. His method was generalized by Buchheim [1] for the case of multiple eigenvalues, although Buchheim did not express his result as in Theorem 1.8. This form is found first in Giorgi [1]. Even before Sylvester, Frobenius [1; p. 54, 2] had obtained expansions for the resolvent $(\lambda I-T)^{-1}$ in the neighborhood of a pole.

A special case of Theorem 1.9 is due to Weyr [1], and in full generality it was proved by Hensel [1].

In regard to Theorem 1.10, Frobenius [3; p. 11] stated that if f is analytic, then $f(T)$ can be obtained as the sum of the residues of $(\lambda I-T)^{-1}f(\lambda)$ with respect to all the eigenvalues of T. He asserted that this notion had been used in the dissertation of L. Stickelberger (1881), but Frobenius did not develop a precise calculus. The first to make a clear use of this device was Poincaré [1] who employed it where all the roots are distinct. In the case of multiple roots a formula equivalent to Theorem 1.10 was derived by Fantappié [3] on the basis of certain requirements, including the relations in Theorem 1.5, that would be expected for a "reasonable" operational calculus. The formula in Theorem 1.10 was used as a definition for $f(T)$ by Giorgi [1] at the suggestion of É. Cartan, who was undoubtedly familiar with Poincaré's use.

For other related remarks, see MacDuffee [1; Chaps. 9, 10]. An interesting account of various functions of a matrix is given in Rinehart [1].

Functions of an operator. The results of Sections 3 and 4 may be regarded as being a unification of two historical developments. On the one hand, these results are a generalization of those in the theory of matrices, and secondly they are an abstraction of results in the theory of integral equations. Consequently it is hardly possible to give complete and accurate credit of references to many of these notions. For example the resolvent operator, its functional equation and expansion, were employed in both of these theories.

Considerable credit should be conferred upon Hilbert [1] and E. H. Moore [1, 2] for perceiving and establishing this unification. However, it seems fair to state that it was F. Riesz who most fully revealed and developed it along the lines presented here. The reader of his book (F. Riesz [3], particularly Secs. 71—82) will find many of these concepts and results exposed with an approach that is strikingly "modern". Although he dealt principally with compact operators in l_2, he established that the resolvent set is open, that the resolvent operator is analytic and indicated that, at least in the case of a pole, the Cauchy integral theorem could be employed to obtain a projection operator commuting with the operator.

In the case of a bounded or unbounded normal operator in a Hilbert space, many of the results of this section become simpler and can be proved more directly by other methods. With these hypotheses considerable extension is possible. For an elaboration of these remarks, we refer the reader to Volume II of this book or to the books of Stone [3], Halmos [6] or Riesz and Sz.-Nagy [1]. Our remarks will deal only with the B-space case.

The expansion for the resolvent is due to Neumann [1] who established it in potential theory. In a more general context it is due to Hilb [1].

The fact that a closed linear operator on an arbitrary complex B-space has non-void spectrum was proved by Taylor [12]. A special case of the fact that $\max |\sigma(T)| = \lim_{n\to\infty} |T^n|^{1/n}$ was proved by Beurling [1], and the general case by Gelfand [1].

In 1923 Wiener [2] observed that Cauchy's integral theorem and Taylor's theorem remain valid for analytic complex B-space valued functions of a complex variable. It is perhaps surprising that not much application was made of this fact for about a dozen years when a number of researchers independently found it useful. In 1936, Nagumo [1] studied B-algebras from a function-theoretic standpoint and proved, among other things, some theorems due to Riesz for compact operators. Later, Taylor [13] studied certain abstract analytic functions, and Hille [2] applied similar methods in the study of semi-groups. In 1941, the famous paper of Gelfand [1] appeared which, although it partially overlapped with Nagumo's work, developed the ideal theory of B-algebras. In addition, Gelfand used the contour integral to obtain idempotents in B-algebras. Independently, Lorch [6] employed the same device and initiated a study of "spectral sets".

Theorem 3.10 is due to Gelfand [1]. Theorems 3.11, 3.16 and 3.19 are found in Dunford [7], where there is some additional material.

Spectral theory of compact operators. As we have observed, this theory generalizes the work of Fredholm [1] on a type of integral equations. Fredholm's original approach was by means of expansions in determinants which, while intricate, gives detailed representation of the resolvent as the quotient of two entire functions. (See Hellinger and Toeplitz [3; Secs. 9—10] and Chapter XI of the present work for further references.) Of several other methods, we mention the one due to E. Schmidt [2] depending on the possibility (in Hilbert space) of approximating the compact operator by operators with finite dimensional ranges.

Probably the most ingenious and elementary approach is due to F. Riesz [4] and is valid for an arbitrary real or complex B-space. Certain of Riesz's results concerning the adjoint operator were not completely general, but were completed by Hildebrandt [6] and Schauder [6]. For an exposition of this method, see Riesz and Sz.-Nagy [1; Chap. 4] or Zaanen [5; Chap. 11]. A similar treatment is given by Banach [1; Chap. 10].

The line of argument given here is closer to Nagumo [1] and is a special case of some results of Dunford [7; p. 208].

The results of this section have been generalized to locally convex topological linear spaces over the complex numbers by Leray [2] by using a deep theorem on the invariance of domain. Other extensions to spaces more general than B-spaces have been given by Altman [3, 5], Hyers [2], Marinescu [1] and Williamson [3].

The following theorem, called the *Fredholm Alternative*, is a consequence of the theorems in this section.

THEOREM. *Let T be a compact operator in a complex B-space $\mathfrak{X}$, and let λ be a fixed non-zero complex number. Then the nonhomogeneous equations*

$$(N) \qquad (\lambda I - T)x = y,$$

$$(N^*) \qquad (\lambda I - T^*)x^* = y^*$$

have unique solutions for any y in $\mathfrak{X}$ or y^* in $\mathfrak{X}^*$ if and only if the homogeneous equations

$$(H) \qquad (\lambda I - T)x = 0,$$

$$(H^*) \qquad (\lambda I - T^*)x^* = 0$$

have only the zero solutions. Furthermore if one of the homogeneous equations has a non-zero solution, then they both have the same finite number of linearly independent solutions. In this case the equations (N) and (N^*) have solutions if and only if y and y^* are orthogonal to all the solutions of (H^*) and (H), respectively. Moreover the general solution for (N) is found by adding a particular solution of (N) to the general solution of (H).

The results of Fredholm pertaining to the representation of the resolvent or the determinental approach has been discussed by Altman [5], R. Graves [1], Leżánski [1, 2], Michal and Martin [1], Ruston [2, 3, 5, 6], Sikorski [1, 2], and Smithies [1]. See also Grothendieck [3] and Zaanen [5; Chap. 9].

For the purposes of applications it is important to be able to compute the eigenvalues of an operator; this is particularly true for self adjoint compact operators in Hilbert space. We refer the reader to the work of Aronszajn [3, 4] and Collatz [1] for these questions. Also of considerable importance is knowledge concerning the distribution of eigenvalues. In the case of integral operators the reader should consult Hille and Tamarkin [1] and Chang [1]. For related results in abstract spaces see Fan [3, 5], Horn [1], Silberstein [1], Visser and Zaanen [1], Weinberger [1, 2], and Weyl [1].

It is a consequence of Theorem 4.5 that if T is compact and $\sigma(T)$ contains at least one non-zero number, then T has an invariant subspace, that is, there exists a proper closed linear manifold $\mathfrak{X}_0 \subset \mathfrak{X}$ such that $T(\mathfrak{X}_0) \subseteqq \mathfrak{X}_0$. It is an interesting and non-trivial fact that this is true in general. The following theorem was proved by Aronszajn and Smith [1].

THEOREM. *Any compact linear operator in a B-space has a proper invariant subspace.*

Yood [2] has investigated certain properties of an operator that are unchanged upon adding a compact operator, and obtained results related to those in this section. Atkinson [2], Gohberg [1, 2, 3] and Taldykin [1] have also considered similar questions. Using their methods, Kleinecke [1] has proved the following theorem.

THEOREM. *Let $\mathfrak{X}$ be a B-space, and let $\mathfrak{R} \subseteqq B(\mathfrak{X})$ be the intersection of all maximal one-sided ideals in $B(\mathfrak{X})$ containing all uniform limits of operators with finite dimensional range. Then the spectrum of any operator in $\mathfrak{R}$ is a countable set of isolated eigenvalues of finite multiplicity with no limit points except possibly $\lambda = 0$. Further $\mathfrak{R}$ contains any other ideal in $B(\mathfrak{X})$ whose operators have spectra of the nature just specified.*

In certain spaces, e.g. C or L_1, we have seen that there exist operators which are not compact but whose spectra are the same as that of a compact operator. This is not true for normal operators in Hilbert space (see Sz.-Nagy [3; p. 55]). Similarly von Neumann [6; p. 16] has shown that if a (not necessarily bounded) self adjoint operator T in $L_2(-\infty, \infty)$ has a countable set of isolated eigenvalues of finite multiplicity with $\lambda = 0$ as the only limit point of $\sigma(T)$, then T is unitarily equivalent to an integral operator of a certain classical (Carleman) type.

L. M. Graves [6] has extended some of the results of the Riesz theory to mappings of the form $E+T$, where E maps a B-space $\mathfrak{X}$ onto a B-space $\mathfrak{Y}$, and $T : \mathfrak{X} \to \mathfrak{Y}$ is compact. See also L. Schwartz [4] where some of these results are proved for locally convex spaces.

For additional information concerning compact operators, the reader is referred to Atkinson [2, 3, 4], Audin [1], Gohberg [1, 2, 3, 4], Gol'dman and Kračkovskiĭ [1], Hamburger [4], Harazov [1, 2, 3], Keldyš [1], Kračkovskiĭ [1, 2], Kračkovskiĭ and Gol'dman [1, 2, 3], Krein and Krasnosel'skiĭ [1], Livšic [2, 3], Nikol'skiĭ [2], Sz.-Nagy [12, 13] and Zaanen [6, 9].

Perturbation theory. The questions of perturbation theory go back to the work of Lord Rayleigh and E. Schrödinger, but it is Rellich who developed the theory along the lines presented here. (For an expository paper on this theory, see Rellich [1].) The principal result of this section is an extension of a theorem of Rellich [2; I] to a situation admitting elementary divisors. The mode of proof is essentially the same as in Rellich.

In a series of five papers, Rellich [2] considered the nature of the spectrum of the perturbed operator, principally in the neighborhood of an isolated eigenvalue of the unperturbed operator. In his first note, Rellich [2; I] treats bounded self adjoint operators in Hilbert space, and in his third note the hypothesis of boundedness is dropped; in both cases the perturbation depends analytically on a single parameter. Sz.-Nagy [2, 4] gave different proofs for some of these results and extended some to B-spaces; further results in this general setting were obtained by Kato [1, 2] and Wolf [1].

In his second paper Rellich [2; II] discussed continuous perturbation of unbounded self adjoint operators; these results were completed by Heinz [1]. See also Newburgh [1, 2].

Most of these results pertain to the point spectrum. The continuous spectrum is much more difficult to handle, but has been discussed by Friedrichs [1, 2].

Unexpected phenomena can take place under perturbation. For example, Weyl [2] showed that if T is a self adjoint operator (which may have void point spectrum) on a separable Hilbert space, then a self adjoint compact operator K may be found such that $T+K$ has enough eigenvectors to span the entire space. This result was extended to unbounded T by von Neumann [6; p. 11] who showed that the norm of K may be arbitrarily small.

Perturbation theory finds particularly fruitful application to differential equations, and is frequently studied in that context. Particular reference is made to the papers of Titchmarsh [2] and Moser [1].

Theorem 6.10 is due to J. Schwartz [3].

In addition to the papers mentioned above, the reader is referred to Gavurin [1, 2], Hölder [1], Jamison [1, 2], Kato [3, 4], Kleinecke [2, 3], H. P. Kramer [1], V. A. Kramer [1], Krein [8], Lifšic [1, 2, 3, 4], Phillips (6), Rabinovič [1], Rosenbloom [2], Schäfke [3], Schröder [1], Šmulian [17], and Solomyak [1].

Tauberian theory and unbounded operators. The results of Section 7 are found in Dunford [7] and have application in ergodic theory. We will cite some references to the pertinent literature of ergodic theory in the notes to Chapter VIII.

The results in Section 9 are due to Taylor [11], who described the method given here of basing the theory of the unbounded operator on that of the bounded, and also developed a theory for the unbounded operator directly. A somewhat more detailed calculus for a restricted class of unbounded operators was given by Hille [1; Chap. 15, Sec. 1]. A similar program was carried out by Bade [1].

CHAPTER VIII

Applications

1. Semi-groups of Operators

It is well known that the exponential function $f(t) = \exp(ta)$ is the most general continuous real (or complex) function f of the non-negative real variable t which satisfies the functional equations $f(0) = 1$, $f(t+u) = f(t)f(u)$. A corresponding problem is to determine the most general continuous operator valued functions defined on the range $t \geqq 0$, which satisfy the equations

$$\text{(i)} \qquad T(t+u) = T(t)T(u), \qquad T(0) = I, \qquad t, u \geqq 0.$$

An operator valued function $T(t)$ satisfying these equations is called a semi-group of operators. In this section will be found the salient features of the analytical theory of these semi-groups as developed by E. Hille, R. S. Phillips, and K. Yosida. By analogy with the scalar case we could expect that a semi-group $T(t)$ is, in some sense, an exponential function. If we assume that $T(t)$ is continuous in the uniform operator topology, then, as shown in Theorem 2 below, there exists a bounded operator A such that $T(t) = e^{tA}$. If it is merely assumed that $T(t)$ is continuous in the strong operator topology, then the problem becomes more difficult to handle. Still, it will be seen that $T(t)$ can be regarded, in an appropriate sense, as e^{tA}, where now A is an unbounded operator, called the infinitesimal generator of the semi-group. We shall also study the important problem of determining which unbounded closed operators are infinitesimal generators of strongly continuous semi-groups. The solution of this problem to be given in the discussion below will enable us to discuss the "abstract Cauchy problem": to find for a given closed unbounded operator A, a function defined for $t \geqq 0$, with each of its values belonging to the domain of A, which satisfies the equations

$$\text{(ii)} \qquad \frac{dx}{dt} = Ax, \qquad x(0) = x_0$$

where x_0 is a given vector. The theory will be illustrated by solving the abstract Cauchy problem for the special case of an integro-differential operator of the form

$$(Ax)(s) = x''(s) + h(s)x(s) + \int_0^\infty K(s, u)x(u)du.$$

This amounts to solving the ordinary Cauchy problem for a partial integro-differential equation of the form

$$\frac{\partial x}{\partial t} = \frac{\partial^2 x}{\partial s^2} + h(s)x + \int_0^\infty K(s, u)x(u, t)du,$$

$$x(s, 0) = x_0(s),$$

where h, K, and x_0 are given functions. The general theory could be used similarly to solve many other Cauchy problems for partial differential and integro-differential equations.

Throughout this section, $\mathfrak{X}$ will denote a complex B-space, and $\{T(t)\}$ a strongly continuous semi-group of operators in $\mathfrak{X}$; i.e., a family of operators satisfying the conditions of the following definition.

1 Definition. A family $\{T(t)\}$, $0 \leqq t < \infty$, of bounded linear operators in $\mathfrak{X}$ will be called a *strongly continuous semi-group* if

(i) $T(s+t) = T(s)T(t), \quad s, t \geqq 0$;

(ii) $T(0) = I$;

(iii) For each $x \in \mathfrak{X}$, $T(t)x$ is continuous in t on $[0, \infty]$.

If, in addition, the map $t \to T(t)$ is continuous in the uniform operator topology, the family $\{T(t), t \geqq 0\}$ is called a *uniformly continuous semi-group* in $B(\mathfrak{X})$.

It is clear from the operational calculus developed in VII.3 that, for any A in $B(\mathfrak{X})$, e^{tA} is a uniformly continuous semi-group. The following theorem shows that every such semi-group is of this form.

2 Theorem. *Let $\{T(t)\}$ be a uniformly continuous semi-group. Then there exists a bounded operator A such that $T(t) = e^{tA}$ for $t \geqq 0$. The operator A is given by the formula $A = \lim_{h\to 0} (T(h)-I)/h$. For $\mathscr{R}(\lambda)$ sufficiently large, the resolvent of A can be expressed in terms of the semi-group by the formula*

$$R(\lambda; A) = \int_0^\infty e^{-\lambda t}T(t)dt.$$

PROOF. Since $T(0) = I$, it follows by VII.6.5 that there exists an $\varepsilon > 0$ such that $U(t) = \log(T(t))$ is defined and continuous for $0 \leqq t \leqq \varepsilon$. If $nt \leqq \varepsilon$, then

$$U(nt) = \log(T(t)^n) = n \log(T(t)) = nU(t).$$

Thus, $(1/n)U(t) = U(t/n)$ for each t with $0 \leqq t \leqq \varepsilon$, and consequently

$$\frac{m}{n} U(t) = mU(t/n) = U(mt/n)$$

for each rational number m/n with $0 \leqq m/n \leqq 1$, and each t in the interval $0 \leqq t \leqq \varepsilon$. In particular, $(m/n)U(\varepsilon) = U(\varepsilon m/n)$, so that, by continuity, $tU(\varepsilon) = U(\varepsilon t)$ for $0 \leqq t \leqq 1$, and $U(t) = (t/\varepsilon)U(\varepsilon)$ for $0 \leqq t \leqq \varepsilon$. If we put $A = (1/\varepsilon)U(\varepsilon)$, we have

$$[*] \qquad T(t) = e^{tA}$$

for $0 \leqq t \leqq \varepsilon$. If $t > 0$ is arbitrary then $t/n < \varepsilon$ for some sufficiently large integer n, and so we have $e^{tA} = (e^{t/nA})^n = \{T(t/n)\}^n = T(t)$. Thus [*] holds for $0 \leqq t < \infty$. Since $\lim_{h\to 0} (e^{zh}-1)/h \to z$ uniformly in any bounded set of the z-plane, the formula for A follows immediately from VII.3.13.

In the same way we see that $(d/dt)T(t) = AT(t)$ for all $t \geqq 0$, and consequently $T(t)$ is continuous and has a continuous derivative. Since $T(t) = \sum_{n=0}^{\infty}(tA)^n/n!$ it is seen that

$$|T(t)| \leqq e^{t|A|}.$$

Thus, for $\mathscr{R}(\lambda) > |A|$, the integral $\int_0^\infty e^{-\lambda t}T(t)dt$ exists. Since a bounded linear operator commutes with the integral, (III.2.19(c)), we have

$$\begin{aligned}(\lambda I - A)\int_0^s e^{-\lambda t}T(t)dt &= \int_0^s (\lambda I - A)e^{-\lambda t}e^{tA}\,dt \\ &= -\int_0^s \frac{d}{dt}(e^{-\lambda t}e^{tA})dt \\ &= I - e^{-\lambda s}e^{sA}.\end{aligned}$$

Since $\mathscr{R}(\lambda) \geqq |A|$ it is seen that $|e^{-\lambda s}e^{sA}| \leqq e^{(-\lambda+|A|)s} \to 0$ as $s \to \infty$, and thus, by the dominated convergence theorem (III.6.16), we have

$$(\lambda I - A)\int_0^\infty e^{-\lambda t}T(t)dt = I.$$

Thus, since $(\lambda I - A)^{-1}$ exists for $|\lambda| > |A|$ by VII.3.4, we have

$$(\lambda I-A)^{-1} = \int_0^\infty e^{-\lambda t}T(t)dt, \qquad \mathscr{R}(\lambda) > |A|.$$

Q.E.D.

The remaining part of the present section will be devoted primarily to a study of strongly continuous semi-groups. Theorems 10 and 11 to follow show the counterpart of Theorem 2 in this case.

3 Lemma. *Let* $\{T(t)\}$ *be a family of bounded operators defined on a finite closed interval* $[\alpha, \beta]$ *such that* $T(t)x$ *is continuous in* t *for each* $x \in \mathfrak{X}$; *then* $|T(\cdot)|$ *is measurable and bounded on* $[\alpha, \beta]$. *Conversely, if* $\{T(t), 0 \leqq t\}$ *is a semi-group of bounded linear operators in* $\mathfrak{X}$ *and if* $T(\cdot)x$ *is measurable on* $(0, \infty)$ *for each* $x \in \mathfrak{X}$, *then* $T(\cdot)x$ *is continuous at every point in* $(0, \infty)$.

Proof. Since $T(\cdot)x$ is continuous for each x in $\mathfrak{X}$, it is bounded on $[\alpha, \beta]$. Thus the boundedness of $|T(\cdot)|$ follows from the uniform boundedness principle II.1.11. To see that $|T(\cdot)|$ is measurable let $\delta > 0$ and $U = \{t| \, |T(t)| > \delta|\}$. If $t_0 \in U$ we can find an x with $|x| = 1$ and $|T(t)x| > \delta$ for every t in an interval containing t_0. Thus U is open in $[\alpha, \beta]$ and the first statement follows from Theorem III.6.10.

Next we show that if $\{T(t)\}$, $0 \leqq t < \infty$, is a semi-group, if $T(\cdot)x$ is measurable on $(0, \infty)$ for each $x \in \mathfrak{X}$, and if $|T(\cdot)|$ is bounded on each interval $[\delta, 1/\delta]$, $\delta > 0$, then $T(\cdot)x$ is continuous at every point $t_0 > 0$, for any $x \in \mathfrak{X}$. To do this, let $0 < \alpha < \beta < t_0$ and take a $\delta > 0$ such that 2δ is less than 1, α, $t_0-\beta$, and $(t_0+1)^{-1}$. Since $T(t_0)x = T(t)[T(t_0-t)x]$, $t < t_0$, and since the right side is independent of t, it is integrable over $[\alpha, \beta]$. If $|\varepsilon| < \delta$, we have

$$(\beta-\alpha)[T(t_0+\varepsilon)-T(t_0)]x = \int_\alpha^\beta T(t)[T(t_0+\varepsilon-t)-T(t_0-t)]x \, dt.$$

By hypothesis there exists an $M > 0$ such that $|T(t)| < M$ for $t \in [\alpha, \beta]$; further $|[T(t_0+\varepsilon-t)-T(t_0-t)]x|$ is a bounded measurable function for $t \in [\alpha, \beta]$. Hence

$$\begin{aligned}(\beta-\alpha)|[T(t_0+\varepsilon)-T(t_0)]x| &\leqq M \int_\alpha^\beta |[T(t_0+\varepsilon-t)-T(t_0-t)]x|dt \\ &= M \int_{t_0-\beta}^{t_0-\alpha} |[T(s+\varepsilon)-T(s)]x|ds.\end{aligned}$$

It follows from Theorem IV.8.20, or may be verified directly, that as $\varepsilon \to 0$ the last integral approaches zero, and hence $T(\cdot)x$ is continuous at $t = t_0$.

To complete the proof of the second part of the lemma we will show that if $T(\cdot)x$ is measurable on $(0, \infty)$ for each $x \in \mathfrak{X}$, then $|T(\cdot)|$ is bounded on each interval $[\delta, 1/\delta]$, $\delta > 0$. We first show

(i) if $x_0 \in \mathfrak{X}$, then there is a separable closed linear manifold $\mathfrak{X}_0$ of $\mathfrak{X}$ containing x_0 and a null set E_0 of $(0, \infty)$ such that if $t \notin E_0$ then $T(t)x \in \mathfrak{X}_0$ whenever $x \in \mathfrak{X}_0$.

By Lemma III.6.9 there is a null set F_0 such that $\{T(t)x_0 | t \notin F_0\}$ is separable. Thus $\mathfrak{X}_0 = \overline{\text{sp}}\{x_0, T(t)x_0, t \notin F_0\}$ is a separable closed linear subspace of $\mathfrak{X}$, and there is a sequence $\{t_n\}$ with $t_n \notin F_0$, $n = 1, 2, \ldots$, such that the set $\{x_0, T(t_n)x_0, n = 1, 2, \ldots\}$ is fundamental (II.1.4) in $\mathfrak{X}_0$. Now if $t \notin F_0$ and $t+t_n \notin F_0$, $n = 1, 2, \ldots$, then $T(t)x_0$ and $T(t)[T(t_n)x_0] = T(t+t_n)x_0$ belong to $\mathfrak{X}_0$. Let $E_0 = F_0 \cup \bigcup_{n=1}^{\infty}(T - t_n)$, so that E_0 is a null set. It follows from the boundedness of the operator $T(t)$ that if $t \notin E_0$, then $T(t)\mathfrak{X}_0 \subseteqq \mathfrak{X}_0$, which proves (i).

Suppose that there is an interval $[\delta, 1/\delta]$ on which $|T(\cdot)|$ is not bounded. Then there exists $s_n \in [\delta, 1/\delta]$ and $x_n \in \mathfrak{X}$, $|x_n| = 1$ such that $|T(s_n)x_n| > n$, $n = 1, 2, \ldots$. Applying (i) for each n, there is a separable subspace $\mathfrak{X}_n$ containing x_n and a null set E_n with $T(t)\mathfrak{X}_n \subseteqq \mathfrak{X}_n$ for $t \notin E_n$. Define $\mathfrak{X}_\infty = \overline{\text{sp}}\{\mathfrak{X}_n, n = 1, 2, \ldots\}$ and $E_\infty = \bigcup_{n=1}^{\infty} E_n$. It is clear that $\mathfrak{X}_\infty$ is separable, E_∞ is a null set, and $T(t)\mathfrak{X}_\infty \subseteqq \mathfrak{X}_\infty$ for $t \notin E_\infty$. For each t, set

$$|T(t)|' = \sup \{|T(t)x|, x \in \mathfrak{X}_\infty, |x| \leqq 1\};$$

since $x_n \in \mathfrak{X}_\infty$ we have $|T(s_n)|' > n$, $n = 1, 2, \ldots$. Let $\{z_n\}$ be a countable set which is dense on the unit sphere of $\mathfrak{X}_\infty$; since $|T(\cdot)z_n|$ is a measurable real valued function, it follows (III.6.10) that $|T(\cdot)|' = \sup_n |T(\cdot)z_n|$ is also measurable. Also if $t_2 \notin E_\infty$, then for any $x \in \mathfrak{X}_\infty$, $T(t_2)x \in \mathfrak{X}_\infty$ and $|T(t_2)x| \leqq |T(t_2)|'|x|$. Hence we see that

$$\begin{aligned} |T(t_1+t_2)|' &= \sup \{|T(t_1)[T(t_2)x]| \, | x \in \mathfrak{X}_\infty, |x| \leqq 1\} \\ &\leqq \sup \{|T(t_1)y| \, | y \in \mathfrak{X}_\infty, |y| \leqq |T(t_2)|'\} \\ &\leqq |T(t_1)|'|T(t_2)|', \end{aligned}$$

provided $t_2 \notin E_\infty$. We now define ω on $(0, \infty)$ by $\omega(t) = \log |T(t)|'$. From the above we know that ω is a measurable function which never takes the value $+\infty$, that $\omega(t_1+t_2) \leqq \omega(t_1)+\omega(t_2)$ provided at least one of the points $t_i \notin E_\infty$, and that $\omega(s_n) > \log n$ where the s_n are points in $[\delta, 1/\delta]$. That this is a contradiction follows from the statement

(ii) if ω is a measurable extended real valued function on $(0, \infty)$, if $\omega(t) < \infty$ for each $t > 0$, and if $\omega(t_1+t_2) \leqq \omega(t_1)+\omega(t_2)$ for one of the t_i not in a null set E, then ω is bounded above in each finite closed interval.

To see this, let $a > 0$ and let $A = \omega(a)$. Then if $t_1+t_2 = a$ and $t_2 \notin E$, it follows that $A = \omega(a) \leqq \omega(t_1)+\omega(t_2)$. Thus if $F = \{t| 0 < t < a, \omega(t) > A/2\}$ and if $a-F$ is the set of all points of the form $a-t$ with $t \in F$, then $E \cup F \cup (a-F) \supseteqq [0, a]$. Consequently, $\mu(F)+\mu(a-F) \geqq a$ where μ denotes Lebesgue measure. Since $\mu(F) = \mu(a-F)$ we have $\mu(F) \geqq a/2$. This shows that if s_n is a point of a finite interval $[\alpha, \beta]$ at which $\omega(s_n) \geqq 2n$, then the set $\{t|\alpha < t < \beta, \omega(t) \geqq n\}$ has measure at least $\alpha/2$. Hence if ω is not bounded above on $[\alpha, \beta]$ it follows that $\omega(t) = \infty$ on a set with measure at least $\alpha/2$, but this contradicts the hypothesis that $\omega(t) < \infty$ for $t \in (0, \infty)$. This completes the proof of the lemma. Q.E.D.

Our next task is to determine the behavior of $|T(t)|$ as t approaches $+\infty$. For this we will require the following lemma on *subadditive functions*, i.e. functions ω such that $\omega(t_1+t_2) \leqq \omega(t_1)+\omega(t_2)$.

4 LEMMA. *Let ω be a subadditive function defined on $[0, \infty)$, and bounded above on each finite subinterval. Then $\omega_0 = \inf_{t>0} \omega(t)/t$ is finite or equals $-\infty$, and*

$$\omega_0 = \lim_{t\to\infty} \frac{\omega(t)}{t}.$$

PROOF. Given any number $\delta > \omega_0$ there is a point t_0 such that

$$\frac{\omega(t_0)}{t_0} < \delta.$$

For any t we write $t = n(t)t_0+r$, where $n(t)$ is an integer and $0 \leqq r < t_0$. Then

$$\begin{aligned}\frac{\omega(t)}{t} &\leqq \frac{\omega(n(t)t_0)}{t}+\frac{\omega(r)}{t} \\ &\leqq \frac{n(t)\omega(t_0)}{t}+\frac{\omega(r)}{t} \\ &= \frac{\omega(t_0)}{t_0+r/n(t)}+\frac{\omega(r)}{t}.\end{aligned}$$

Thus

$$\limsup_{t\to\infty} \frac{\omega(t)}{t} \leqq \frac{\omega(t_0)}{t_0} < \delta.$$

Since $\omega_0 \leqq \liminf_{t\to\infty} \omega(t)/t$, we see $\lim_{t\to\infty} \omega(t)/t$ exists and equals ω_0.

5 COROLLARY. *The limit* $\omega_0 = \lim_{t\to\infty}(\log|T(t)|)/t$ *exists. For each* $\delta > \omega_0$ *there is a constant* M_δ *such that* $|T(t)| < M_\delta e^{\delta t}$ *for* $t \geqq 0$.

PROOF. Define $\omega(t) = \log |T(t)|$, $t \geqq 0$. Since $\omega(t_1+t_2) = \log |T(t_1+t_2)| \leqq \log (|T(t_1)||T(t_2)|) = \omega(t_1)+\omega(t_2)$, ω is subadditive. The result now follows from Lemmas 3 and 4. Q.E.D.

6 DEFINITION. For $h > 0$ the linear operator A_h is defined by the formula

$$A_h x = \frac{T(h)x - x}{h}, \qquad x \in \mathfrak{X}.$$

Let $\mathfrak{D}(A)$ be the set of all x in $\mathfrak{X}$ for which the limit, $\lim_{h\to 0} A_h x$, exists and define the operator A with domain $\mathfrak{D}(A)$ by the formula

$$Ax = \lim_{h\to 0} A_h x, \qquad x \in \mathfrak{D}(A).$$

The operator A with domain $\mathfrak{D}(A)$ is called the *infinitesimal generator* of the semi-group $T(\cdot)$.

7 LEMMA. (a) *The set* $\mathfrak{D}(A)$ *is a linear manifold and* A *is linear on* $\mathfrak{D}(A)$.

(b) *If* x *is in* $\mathfrak{D}(A)$, *then* $T(t)x$ *is in* $\mathfrak{D}(A)$, $0 \leqq t < \infty$, *and* $(d/dt)T(t)x = AT(t)x = T(t)Ax$.

(c) *If* x *is in* $\mathfrak{D}(A)$, *then* $[T(t)-T(s)]x = \int_s^t T(u)Ax\,du$, *for* $0 \leqq s < t < \infty$.

(d) *If* $t \geqq 0$ *and* g *is a Lebesgue integrable function continuous at* t *then*

$$\lim_{h\to 0} \frac{1}{h} \int_t^{t+h} g(u)T(u)x\,du = g(t)T(t)x.$$

PROOF. Statement (a) is clear from the definitions. To prove (b) let $h > 0$, $t \geqq 0$, and $x \in \mathfrak{D}(A)$. Then $T(t)A_h x = A_h T(t)x$, so that $\lim_{h\to 0} A_h T(t)x = \lim_{h\to 0} T(t)A_h x$, and hence $T(t)x \in \mathfrak{D}(A)$. By definition

$$A(T(t)x) = \lim_{h\to 0} A_h T(t)x.$$

Thus $T(t)Ax = AT(t)x$ for $x \in \mathfrak{D}(A)$. If $t > 0$ and $h > 0$ then

$$\lim_{h\to 0} \left\{\frac{T(t)x - T(t-h)x}{h} - T(t)Ax\right\}$$

$$= \lim_{h\to 0} T(t-h)(A_h x - Ax) + \lim_{h\to 0} [T(t-h) - T(t)]Ax$$

$$= 0$$

by Definition 1 (iii) and Lemma 3. On the other hand,

$$\frac{T(t+h)x - T(t)x}{h} = T(t)A_h x \to T(t)Ax,$$

so that $(d/dt)T(t)x = T(t)Ax$ is established for $x \in \mathfrak{D}(A)$.

Statement (c) follows by applying linear functionals to both sides of (b) and integrating. For (d), see Theorem III.12.8. Q.E.D.

8 Lemma. *The linear manifold $\mathfrak{D}(A)$ is dense in $\mathfrak{X}$ and A is closed on $\mathfrak{D}(A)$.*

Proof. Let x be any vector in $\mathfrak{X}$. Then if $t, h > 0$,

$$A_h \int_0^t T(s)x ds = \frac{1}{h}\int_0^t (T(h+s)x - T(s)x) ds$$

$$= \frac{1}{h}\int_h^{t+h} T(s)x ds - \frac{1}{h}\int_0^t T(s)x ds$$

$$= \frac{1}{h}\int_t^{t+h} T(s)x ds - \frac{1}{h}\int_0^h T(s)x ds \to T(t)x - x$$

as $h \to 0$ by Lemma 7(d). Thus $\int_0^t T(s)x ds \in \mathfrak{D}(A)$. However, $x = \lim_{t\to 0}(1/t)\int_0^t T(s)x ds$, so that $\mathfrak{D}(A)$ is dense in $\mathfrak{X}$. To see that A is closed, suppose $x_n \in \mathfrak{D}(A)$, $n = 1, 2, \ldots$, $\lim_{n\to\infty} x_n = x_0$, and $\lim_{n\to\infty} Ax_n = y_0$. Using Lemma 7(c) we have

$$T(t)x_0 - x_0 = \lim_{n\to\infty} T(t)x_n - x_n$$

$$= \lim_{n\to\infty}\int_0^t T(s)Ax_n ds$$

$$= \int_0^t T(s)y_0 ds,$$

since $\lim_{n\to\infty} T(s)Ax_n = T(s)y_0$ uniformly in $[0, t]$. Thus

$$\lim_{t\to 0} A_t x_0 = \lim_{t\to 0} \frac{1}{t}\int_0^t T(s)y_0 ds = y_0$$

by Lemma 7(d). Consequently $x_0 \in \mathfrak{D}(A)$ and $Ax_0 = y_0$, i.e., A is closed. Q.E.D.

9 Corollary. *A semi-group has a bounded infinitesimal generator if and only if it is uniformly continuous.*

Proof. If $T(\cdot)$ is a uniformly continuous semi-group then, by Theorem 2, it has a bounded infinitesimal generator. Conversely, suppose that $T(\cdot)$ has a bounded generator A. It follows from the preceding lemma that A is everywhere defined. Thus Lemma 3 and the principle of uniform boundedness (II.3.21(ii)) show that

$$\sup_{0\leqq h\leqq 1} |A_h| = K < \infty.$$

By Lemma 3 we have, for each $s \geqq 0$, a constant M_s with

$$|T(t)| \leqq M_s, \qquad 0 \leqq t, \quad |t-s| \leqq 1.$$

Now

$$\begin{aligned} T(t)-T(s) &= T(s)[T(t-s)-I], && t > s,\\ &= -T(t)[T(s-t)-I], && s > t. \end{aligned}$$

Thus

$$|T(t)-T(s)| \leqq M_s \cdot K \cdot |t-s|, \qquad 0 \leqq t, \quad |t-s| < 1,$$

which proves that $T(\cdot)$ is a uniformly continuous semi-group. Q.E.D.

10 Theorem. *Let $T(t)$, $t \geqq 0$, be a strongly continuous semi-group of operators and let $A_h = (T(h)-I)/h$. Then*

$$T(t)x = \lim_{h\to 0} e^{tA_h}x, \qquad x \in \mathfrak{X},$$

uniformly for t in any finite interval.

Proof. We note first that we may write

$$e^{tA_h} = e^{-t/h}e^{(t/h)T(h)} = e^{-t/h}\sum_{n=0}^{\infty}\frac{t^n T(nh)}{n!h^n}.$$

If $\delta > \omega_0$, then by Corollary 5

$$|e^{tA_h}| \leqq e^{-t/h} M_\delta \sum_{n=0}^{\infty} \frac{t^n e^{n\delta h}}{n! h^n}$$
$$= M_\delta \exp\left\{t\left(\frac{e^{\delta h}-1}{h}\right)\right\}.$$

Consequently there is a constant K_t such that

$|e^{sA_h}| \leqq K_t, \quad 0 \leqq s \leqq t, \quad 0 < h \leqq 1$. If $x \in \mathfrak{D}(A)$ and $t \leqq t_0$, then (cf. Lemma 7)

$$|T(t)x - e^{tA_h}x| = \left|\int_0^t \frac{d}{ds}(e^{(t-s)A_h}T(s)x)ds\right|$$
$$= \left|\int_0^t e^{(t-s)A_h}T(s)(Ax - A_h x)ds\right|$$
$$\leqq t_0 K_{t_0} M_\delta e^{\delta t_0}|Ax - A_h x| \to 0$$

as $h \to 0$. Since $\mathfrak{D}(A)$ is dense in $\mathfrak{X}$ by Lemma 8, the conclusion follows from Theorem II.3.6. Q.E.D.

This theorem and the following one together give the analogue of Theorem 2 for strongly continuous semi-groups.

11 THEOREM. *If* $\omega_0 = \lim_{t\to\infty} (\log |T(t)|)/t$ *and* $\mathscr{R}(\lambda) > \omega_0$, *then* $\lambda \in \rho(A)$ *and*

$$R(\lambda; A)x = \int_0^\infty e^{-\lambda t}T(t)x dt, \qquad x \in \mathfrak{X}.$$

PROOF. Suppose that $\omega_0 < \delta < \mathscr{R}(\lambda)$. By Corollary 5 there is a constant M_δ such that $|T(t)| \leqq M_\delta e^{\delta t}$, $t \geqq 0$. Thus the integral $\int_0^\infty e^{-\lambda t}T(t)x dt$ exists if $\mathscr{R}(\lambda) > \omega_0$ and defines a bounded operator. Let

$$R(\lambda)x = \int_0^\infty e^{-\lambda t}T(t)x\, dt, \qquad x \in \mathfrak{X}, \quad \mathscr{R}(\lambda) > \omega_0.$$

Then

$$A_h R(\lambda)x = \frac{1}{h}\int_0^\infty e^{-\lambda t}T(t+h)x dt - \frac{1}{h}\int_0^\infty e^{-\lambda t}T(t)x dt$$
$$= \frac{(e^{\lambda h}-1)}{h}\int_0^\infty e^{-\lambda t}T(t)x dt - \frac{e^{\lambda h}}{h}\int_0^h e^{-\lambda t}T(t)x dt$$
$$\to \lambda R(\lambda)x - x$$

as $h \to 0$ by Lemma 7(d). Thus $R(\lambda)x \in \mathfrak{D}(A)$ and

$$\text{(i)} \qquad (\lambda I - A)R(\lambda)x = x, \qquad x \in \mathfrak{X}.$$

If we can show

$$\text{(ii)} \qquad R(\lambda)(\lambda I - A)x = x, \qquad x \in \mathfrak{D}(A),$$

then $R(\lambda) = R(\lambda; A)$. Thus, in view of (i), it remains to prove $R(\lambda)Ax = AR(\lambda)x$, $x \in \mathfrak{D}(A)$. However, if $x \in \mathfrak{D}(A)$, then $T(t)x \in \mathfrak{D}(A)$ for $t \geqq 0$ and $AT(t)x = T(t)Ax$ by Lemma 7. Consequently, Theorem III.6.20 shows $R(\lambda)x \in \mathfrak{D}(A)$ and

$$A \int_0^\infty e^{-\lambda t} T(t)x dt = \int_0^\infty e^{-\lambda t} T(t)Ax dt. \qquad \text{Q.E.D.}$$

Our next step is to find sufficient conditions on an unbounded operator A that it be the infinitesimal generator of a semi-group.

12 Lemma. *Let A be a closed operator with dense domain whose spectrum lies in the half plane $\mathscr{R}(\lambda) \leqq \omega$. Let $S(t)$, $t \geqq 0$, be strongly continuous in t and satisfy the relations*

$$|S(t)| \leqq Me^{\omega t}, \qquad t \geqq 0,$$

$$R(\lambda; A)x = \int_0^\infty e^{-\lambda t} S(t)x dt, \qquad x \in \mathfrak{X}.$$

Then

$$|R(\lambda; A)^n| \leqq M(\mathscr{R}(\lambda) - \omega)^{-n}, \qquad n = 1, 2, \ldots.$$

Proof. We first establish the formula

$$[*] \qquad R(\lambda; A)^n x = \frac{1}{(n-1)!} \int_0^\infty e^{-\lambda t} t^{n-1} S(t)x dt, \qquad \mathscr{R}(\lambda) > \omega.$$

It follows from the resolvent formula (cf. VII.3.6) that

$$R(\lambda; A) - R(\mu; A) = (\mu - \lambda)R(\lambda; A)R(\mu; A)$$

by letting $\lambda \to \mu$, that

$$\frac{d}{d\lambda} R(\lambda; A) = -R^2(\lambda; A),$$

and inductively, that

$$\frac{d^n}{d\lambda^n} R(\lambda; A) = (-1)^n n! R(\lambda; A)^{n+1}, \qquad \lambda \in \varrho(A).$$

Now suppose $\omega < \delta < \mathscr{R}(\lambda), \mathscr{R}(\mu)$. If

$$f(\lambda, \mu, t) = (e^{-\lambda t} - e^{-\mu t})(\lambda - \mu)^{-1}$$

then

$$|f(\lambda, \mu, t)| = \left| e^{-\mu t} \int_0^t e^{-(\lambda-\mu)s} ds \right|$$
$$\leqq e^{-\mathscr{R}(\mu)t} \int_0^t e^{-\mathscr{R}(\lambda-\mu)s} ds$$
$$< te^{-\delta t}, \qquad t \geqq 0.$$

Consequently,

$$|f(\lambda, \mu, t)t^n S(t)x| \leqq M|x|t^{n+1}e^{-(\delta-\omega)t}.$$

It follows from Corollary III.6.16 that we may let λ approach μ; i.e., we may differentiate

$$\int_0^\infty e^{-\lambda t} t^n S(t)xdt$$

under the integral sign, and obtain

$$\frac{d}{d\lambda}\int_0^\infty e^{-\lambda t} t^n S(t)xdt = -\int_0^\infty e^{-\lambda t} t^{n+1} S(t)xdt, \qquad \mathscr{R}(\lambda) > \omega.$$

Thus, inductively, we find

$$\frac{d^n}{d\lambda^n} R(\lambda; A)x = \int_0^\infty e^{-\lambda t} (-t)^n S(t)xdt.$$

Combining this formula with the one above, we obtain formula [*].

Finally if $\mathscr{R}(\lambda) > \omega$, it is seen from [*] and the assumptions on $S(t)$ that

$$|R(\lambda; A)^n| \leqq \frac{M}{(n-1)!}\int_0^\infty t^{n-1} e^{-(\mathscr{R}(\lambda)-\omega)t} dt$$
$$= \frac{M}{(\mathscr{R}(\lambda)-\omega)^n}. \qquad \text{Q.E.D.}$$

13 Theorem. (*Hille-Yosida-Phillips*) *A necessary and sufficient condition that a closed linear operator A with dense domain be the infinitesimal generator of a strongly continuous semi-group is that there exist real numbers M and ω such that for every real $\lambda > \omega$, λ is in $\varrho(A)$ and*

$$|R(\lambda; A)^n| \leqq M(\lambda-\omega)^{-n}, \qquad n = 1, 2, \ldots.$$

Proof. The necessity of the condition follows from Corollary 5, Theorem 11, and Lemma 12. To prove the sufficiency let $B_\lambda = -\lambda[I-\lambda R(\lambda; A)]$, $\lambda > \omega$. We shall construct the semi-group $T(t)$ as the strong limit as $\lambda \to \infty$ of the semi-groups e^{tB_λ}. Since

$$e^{tB_\lambda} = e^{-\lambda t} \sum_{n=0}^{\infty} \frac{(\lambda^2 t)^n R(\lambda; A)^n}{n!},$$

we have

$$\begin{aligned}|e^{tB_\lambda}| &\leqq Me^{-\lambda t} \sum_{n=0}^{\infty} \frac{(\lambda^2 t)^n}{n!(\lambda-\omega)^n} \\ &< M \sum_{n=0}^{\infty} \frac{(t\omega\lambda)^n}{n!(\lambda-\omega)^n} \\ &= M \exp\,[t\omega\lambda(\lambda-\omega)^{-1}].\end{aligned}$$

If $\omega_1 > \omega$ then for λ sufficiently large $\omega\lambda(\lambda-\omega)^{-1} < \omega_1$. Thus

$$[*] \qquad |e^{tB_\lambda}| < Me^{t\omega_1}$$

for large values of λ.

We next show that $\lim_{\lambda\to\infty} B_\lambda x = Ax$, $x \in \mathfrak{D}(A)$. If $x \in \mathfrak{D}(A)$,

$$|\lambda R(\lambda; A)x - x| = |R(\lambda; A)Ax| \leqq M|Ax|(\lambda-\omega)^{-1} \to 0$$

as $\lambda \to \infty$. Since $|\lambda R(\lambda; A)| \leqq M\lambda(\lambda-\omega)^{-1} < 2M$ for large λ, we conclude (cf. II.3.6) that $\lambda R(\lambda; A)x \to x$, $x \in \mathfrak{X}$ as $\lambda \to \infty$. Therefore $B_\lambda x = \lambda R(\lambda; A)Ax \to Ax$ for x in $\mathfrak{D}(A)$.

Now define $S_\lambda(t) = e^{tB_\lambda}$. For any μ and λ we have $B_\lambda B_\mu = B_\mu B_\lambda$. The series formula $S_\lambda(t) = \sum_{n=0}^{\infty} t^n B_\lambda^n/n!$ shows $B_\mu S_\lambda(t) = S_\lambda(t)B_\mu$. Consequently, if $x \in \mathfrak{D}(A)$ we have (cf. Lemma 7(b))

$$\begin{aligned}S_\lambda(t)x - S_\mu(t)x &= \int_0^t \frac{d}{ds}\,[S_\mu(t-s)S_\lambda(s)x]\,ds \\ &= \int_0^t S_\mu(t-s)(B_\lambda - B_\mu)S_\lambda(s)x\,ds \\ &= \int_0^t S_\mu(t-s)S_\lambda(s)(B_\lambda - B_\mu)x\,ds.\end{aligned}$$

Using [*] we obtain

$$|S_\lambda(t)x - S_\mu(t)x| \leqq M^2 te^{t\omega_1}|B_\lambda x - B_\mu x|,$$

for large values of λ and μ. Thus $S_\lambda(t)x$ converges to a limit uniformly in each finite interval. Since $\mathfrak{D}(A)$ is dense in $\mathfrak{X}$ it follows from the inequality [*] and Theorem II.3.6 that there is a bounded operator $T(t)$ such that $\lim_{\lambda\to\infty} S_\lambda(t)x = T(t)x$, $x \in \mathfrak{X}$. Moreover $|T(t)| \leqq$

$\liminf_{\lambda\to\infty} |S_\lambda(t)| \leqq e^{t\omega_1}$. The continuity in t of $T(t)x$ follows from the uniformity of the convergence. Since $S_\lambda(t)$ is a semi-group it now follows easily that $T(t)$ is a semi-group. Due to the inequality

$$|S_\lambda(t)B_\lambda x - T(t)Ax| \leqq |S_\lambda(t)(B_\lambda x - Ax)| + |(S_\lambda(t) - T(t))Ax|$$
$$\leqq Me^{t\omega_1}|B_\lambda x - Ax| + 2Me^{t\omega_1}|Ax|$$

valid for $x \in \mathfrak{D}(A)$, we may take the limit as $\lambda \to \infty$ (cf. Corollary III.6.16) on both sides of the equation $S_\lambda(t)x - x = \int_0^t S_\lambda(s)B_\lambda x ds$ to obtain

$$T(t)x - x = \int_0^t T(s)Axds.$$

Thus if B is the infinitesimal generator of $T(s)$,

$$Bx = \lim_{t\to 0} \frac{1}{t}\int_0^t T(s)Axds = Ax, \qquad x \in \mathfrak{D}(A).$$

Consequently, $\mathfrak{D}(B) \supseteq \mathfrak{D}(A)$ and B is an extension of A. However, for large λ, $\lambda \in \varrho(A) \cap \varrho(B)$ and the equations $(\lambda I - A)\mathfrak{D}(A) = \mathfrak{X} = (\lambda I - B)\mathfrak{D}(A)$, $(\lambda I - B)\mathfrak{D}(B) = \mathfrak{X}$, imply $\mathfrak{D}(B) = \mathfrak{D}(A)$. Thus $B = A$. Q.E.D.

14 Corollary. *A necessary and sufficient condition that a closed linear operator A with dense domain generate a strongly continuous semi-group $T(t)$ of bounded operators such that $|T(t)| \leqq e^{\omega t}$ for some real number ω is that*

$$[*] \qquad |R(\lambda; A)| \leqq (\lambda - \omega)^{-1}; \qquad \lambda > \omega.$$

Proof. The formula $R(\lambda; A)x = \int_0^\infty e^{-\lambda t}T(t)xdt$, $x \in \mathfrak{X}$, shows the necessity of the inequality [*] above. Clearly [*] implies that $|R(\lambda; A)|^n \leqq (\lambda - \omega)^{-n}$, $\lambda > \omega$, and thus by Theorem 13, (with $M = 1$) A is the generator of a semi-group $T(t)$. It was shown in the proof of Theorem 13 that $|T(t)| \leqq Me^{t\omega_1}$ for each $\omega_1 > \omega$. Thus $|T(t)| \leqq e^{t\omega}$, $t \geqq 0$. Q.E.D.

The next corollary will require the following lemma.

15 Lemma. *Let f be in $L_1(0, \infty)$ and*

$$\int_0^\infty e^{-\lambda t} f(t)dt = 0$$

for $\mathscr{R}(\lambda)$ sufficiently large. Then $f(t) = 0$ almost everywhere.

PROOF. Suppose that the equation holds for $\mathscr{R}(\lambda) \geqq \omega$. Put $f_1(t) = e^{-\omega t}f(t)$, so that $f_1 \in L_1(0, \infty)$ and $\int_0^\infty f_1(t)e^{-\lambda t}\,dt = 0$ for $\mathscr{R}(\lambda) \geqq 0$. We make the change of variable $u = e^{-t}$. Then $t = -\log u$, and $\int_0^\infty e^{-\lambda t} f_1(t)dt = \int_0^1 u^\lambda g(u)du$, where $g(u) = u^{-1}(-\log u)$ is in $L_1[0, 1]$ by Lemma III.10.8. Hence $\int_0^1 u^n g(u)du = 0$, $n = 0, 1, 2, \ldots$. Consequently, by the Weierstrass approximation theorem, and Corollary III.10.6,

$$0 = \int_0^1 h(u)g(u)\,du = \int_0^1 h(u)G(du), \qquad h \in C(0, 1),$$

where $G(E) = \int_E g(u)du$. Since Lebesgue measure is regular so is G and thus, by the Riesz representation theorem (IV.6.3), $\int_E g(u)du = 0$ for every measurable set E. Thus $g(u) = 0$ almost everywhere (III.6.8). Consequently f_1 and thus f vanishes almost everywhere. Q.E.D.

16 COROLLARY. *A necessary and sufficient condition that a closed operator A with dense domain should be the infinitesimal generator of a strongly continuous semi-group is that there exist a strongly continuous family $S(t)$, $t \geqq 0$, of bounded linear operators satisfying $S(0) = I$, $|S(t)| \leqq Me^{\omega t}$ for real numbers M and ω, and such that*

$$R(\lambda; A)x = \int_0^\infty e^{-\lambda t}S(t)x dt, \qquad \lambda > \omega.$$

Then $S(t)$ is the semi-group with infinitesimal generator A.

PROOF. By Lemma 12, $|R(\lambda; A)^n| \leqq M(\lambda-\omega)^{-n}$, $\lambda > \omega$. Thus by Theorem 13, A is the infinitesimal generator of a semi-group $T(t)$ and $|T(t)| \leqq Me^{\omega t}$. By Theorem 11, $R(\lambda; A)x = \int_0^\infty e^{-\lambda t}T(t)x dt$, $x \in \mathfrak{X}$, $\lambda > \omega$. Now if $x^* \in \mathfrak{X}^*$ we have

$$\int_0^\infty e^{-\lambda t}x^*(T(t)x - S(t)x)dt = 0$$

for $\lambda > \omega$. Setting $f(t) = e^{-(\omega+1)t}x^*(T(t)x - S(t)x)$ we have $\int_0^\infty e^{-\lambda t}f(t)dt = 0$ for $\lambda \geqq 0$. It follows from Lemma 15 that $x^*T(t)x = x^*S(t)x$ for almost all t. By continuity this equation holds for all $t \geqq 0$, and thus (cf. II.3.14) $T(t) = S(t)$, $t \geqq 0$. Q.E.D.

We now consider the question of when a strongly continuous semi-group of operators defined on $[0, \infty)$ may be extended to a group $T(t)$ of operators defined on $(-\infty, \infty)$. Such an extension is clearly unique if it exists, and the family $S(t) = T(-t)$, $t \geqq 0$, is a strongly continuous semi-group. Since, if $0 < t < 1$,

$$\frac{S(t)x-x}{t} = \frac{-T(-2)[T(2-t)x-T(2)x]}{-t},$$

it is clear that the infinitesimal generator of $S(t)$ is the closed operator $-A$, $\mathfrak{D}(-A) = \mathfrak{D}(A)$. We shall call the operator A the *infinitesimal generator* of the group and A will be said to *generate* $T(t)$, $-\infty < t < \infty$. The question of whether $T(t)$, $t \geqq 0$, may be extended to a group may be answered in terms of the infinitesimal generator A.

17 Corollary. *A necessary and sufficient condition that a closed linear operator A with dense domain generates a strongly continuous group of bounded operators on $(-\infty, \infty)$ is that there exist real numbers $M > 0$ and $\omega \geqq 0$ such that*

$$[*] \qquad |R(\lambda; A)^n| \leqq M(|\lambda|-\omega)^{-n}, \qquad \lambda > \omega \text{ and } \lambda < -\omega.$$

If A generates $T(t)$, $-\infty < t < \infty$, then $|T(t)| \leqq Me^{\omega|t|}$.

Proof. The necessity of the inequality [*] follows from the remarks above, Theorem 13, and the relations $R(\lambda; -A) = -R(-\lambda; A)$ and $\sigma(-A) = -\sigma(A)$ (cf. VII.9.10). On the other hand if [*] is satisfied, both A and $-A$ satisfy the condition of Theorem 13 and generate semi-groups $T_+(t)$ and $T_-(t)$ respectively. One shows easily that the approximating semi-groups $S_\lambda^+(t)$ and $S_\lambda^-(t)$ (of the proof of Theorem 13) commute, and hence $T_+(t)$ and $T_-(t)$ likewise commute. Thus $W(t) = T_+(t)T_-(t)$ is also a semi-group defined on $[0, \infty)$. However if $x \in \mathfrak{D}(A) = \mathfrak{D}(-A)$,

$$\frac{W(t)x-x}{t} = T_-(t)\left[\frac{T_+(t)x-x}{t}\right] + \frac{T_-(t)x-x}{t}$$

$$\to Ax+(-Ax) = 0$$

as $t \to 0$. Thus $dW(t)x/dt = 0$, and consequently $W(t)x = x$, for $x \in \mathfrak{D}(A)$. Since $\mathfrak{D}(A)$ is dense in $\mathfrak{X}$, it follows that $T_-(t) = T_+(t)^{-1}$. Therefore if we define $T(t) = T_+(t)$, $t \geqq 0$ and $T(t) = T_-(t)$, $t \leqq 0$, $T(t)$ is a strongly continuous group of linear operators with infinitesimal generator A. The inequality $|T(t)| \leqq Me^{\omega|t|}$ is clear. Q.E.D.

Examples. If T is an operator in a space $\mathfrak{X}$ whose elements are functions defined for each s in a set S, we shall, in the following examples, use the notation $(Tx)(s)$ or $T(x, s)$ for the value of Tx at the point s. Of course if $\mathfrak{X}$ is a space, such as an L_p space, whose elements are equivalence classes of functions, then these symbols refer to one of the functions in the equivalence class Tx. Some of the simplest and most important examples of semi-groups of bounded operators arise from the operation of translation $T(t)(x, s) = x(t+s)$ in the spaces $L_p(0, \infty)$ and $C[0, \infty]$. Throughout this discussion $C[0, \infty]$ is the space of all continuous functions on the compact extended non-negative real number system, i.e., the space of all functions x of a non-negative real number for which the limit $x(\infty) = \lim_{s\to\infty} x(s)$ exists. Similarly $C[-\infty, \infty]$ is the space of all continuous functions of a real variable for which both limits $x(\infty)$ and $x(-\infty) = \lim_{s\to-\infty} x(s)$ exist. The infinitesimal generator in these cases is the operator of differentiation $A = d/dt$. We shall verify this and find $\sigma(A)$ for the case of the space $C[0, \infty]$. It is easily seen from the uniform continuity of the functions in $C[0, \infty]$ that $T(t)$ is a strongly continuous semi-group, and moreover $|T(t)| = 1$, $t \geqq 0$. Let $x \in \mathfrak{D}(A)$ and $y = Ax$, then

$$\lim_{h\to 0} \left[\frac{x(t+h)-x(t)}{h} - y(t)\right] = 0$$

uniformly in t. It follows that $y = x' = dx/dt$. On the other hand let x be a function in $C[0, \infty]$ such that $x' \in C[0, \infty]$. Then

$$\left|\frac{x(t+h)-x(t)}{h} - x'(t)\right| \leqq \frac{1}{h}\int_t^{t+h} |x'(s)-x'(t)|ds$$

which approaches zero uniformly in t as $h \to 0$ due to the uniform continuity of x'. Thus

$$\mathfrak{D}(A) = \{x|x' \in C[0, \infty]\}$$

and $A = d/dt$. It follows from the general theory that A is closed. For λ to belong to the resolvent set of A it is required that the differential equation $\lambda y - y' = x$ should have a unique solution in $C[0, \infty]$ for each $x \in C[0, \infty]$. In this case $y = R(\lambda; A)x$. Consideration of the general solution of this differential equation shows $\sigma(A)$ is the half plane $\mathscr{R}(\lambda) \leqq 0$, and

$$R(\lambda; A)(x, t) = \int_0^\infty e^{-\lambda t} x(t+s)ds, \qquad \mathscr{R}(\lambda) > 0.$$

The foregoing discussion applies with minor changes to the case of the space $C[-\infty, \infty]$. Here $T(t)(x, s) = x(t+s)$ defines a strongly continuous group on $(-\infty, \infty)$. Again $A = d/dt$ with domain $\mathfrak{D}(A) = \{x|x' \in C[-\infty, \infty]\}$. In this case $\sigma(A)$ is the imaginary axis and

$$R(\lambda; A)x(t) = \int_0^\infty e^{-\lambda s}x(t+s)ds, \qquad \mathscr{R}(\lambda) > 0$$

$$= -\int_{-\infty}^0 e^{-\lambda s}x(t+s)ds, \qquad \mathscr{R}(\lambda) < 0.$$

The reader will have little difficulty extending the discussion above to the case of the spaces $L_p(0, \infty)$ and $L_p(-\infty, \infty)$, $1 \leqq p < \infty$. As before $A = d/dt$, with $\mathfrak{D}(A)$ consisting of the absolutely continuous functions x in L_p for which $x' \in L_p$. The spectrum and resolvent formulas remain the same as for the spaces C. We shall return to these examples to illustrate the later theory.

As may be expected from the attention we have devoted to Theorem 13, Corollary 14, and Corollary 16, it is important in the present theory to discover when a closed operator A is the generator of a strongly continuous semi-group. Theorem 13 gives necessary and sufficient conditions, but these conditions are often difficult to verify in concrete analytical cases of interest. Consequently, we shall devote the rest of this section to an elucidation of the problem of semi-group generation from the point of view of perturbation theory. Our guiding idea will be the following: if A is a semi-group generator, and if P is not too irregular relative to A, then $A+P$ is a semi-group generator. A precise, though rather special case of this somewhat imprecise principle is the following result, which is a particular case of Theorem 19 below: If A is a semi-group generator, and P is a bounded operator, $A+P$ is a semi-group generator.

The exact range of operators covered by our perturbation principle is singled out in the following definition.

18 Definition. If A is the infinitesimal generator of a strongly continuous semi-group $T(t)$ we denote by $\mathscr{P}(A)$ the class of closed operators P satisfying the conditions

(i) $\mathfrak{D}(P) \supseteqq \mathfrak{D}(A)$;

(ii) For each $t > 0$, there exists a constant $K_t < \infty$ such that $|PT(t)x| \leqq K_t|x|$ for $x \in \mathfrak{D}(A)$;

(iii) The constants K_t in (ii) may be chosen so that $\int_0^1 K_t\, dt$ exists and is finite.

19 Theorem. *Let $T(t)$, $t \geqq 0$, be a strongly continuous semi-group of bounded operators in $\mathfrak{X}$ with infinitesimal generator A. If $P \in \mathscr{P}(A)$ then the operator $A+P$ defined on $\mathfrak{D}(A)$, is closed and is the infinitesimal generator of a semi-group $T(t; A+P)$. Further*

$$T(t; A+P) = \sum_{n=0}^{\infty} S_n(t), \qquad t \geqq 0,$$

where $S_0(t) = T(t)$ and $S_n(t)x = \int_0^t T(t-s)PS_{n-1}(s)x ds$ for $x \in \mathfrak{X}$ and $n \geqq 1$, the series being absolutely convergent, uniformly with respect to t in each finite interval. For each n and x, the function $S_n(t)x$ is continuous for $t \geqq 0$.

We will build up the proof of Theorem 19 through a series of lemmas. Throughout the remainder of this section, we shall use the notations of Definition 18 and Theorem 19.

20 Lemma. *If $P \in \mathscr{P}(A)$, then*

(a) $\mathfrak{D}(P) \supseteqq \cup_{t>0} T(t)\mathfrak{X}$;

(b) *The mapping $x \to PT(t)x$, $x \in \mathfrak{D}(A)$, has a unique extension to a bounded operator (which we shall write $PT(t)$) defined on all of $\mathfrak{X}$;*

(c) *$PT(t)x$ is continuous in t for $t > 0$ and each $x \in \mathfrak{X}$. If $\omega_0 = \lim_{t\to\infty} (\log |T(t)|)/t$, then $\limsup_{t\to\infty} (\log |PT(t)|)/t \leqq \omega_0$;*

(d) *if $\mathscr{R}(\lambda) > \omega_0$, then*

$$PR(\lambda; A)x = \int_0^\infty e^{-\lambda t} PT(t)x\, dt, \qquad x \in \mathfrak{X}.$$

Proof. Since $\mathfrak{D}(A)$ is dense in $\mathfrak{X}$, by Lemma 9, statement (b) clearly follows from (ii) of Definition 18 by virtue of I.6.17.

Let $x_0 \in \mathfrak{X}$, $x_0 = \lim_{n\to\infty} x_n$, where $x_n \in \mathfrak{D}(A)$. Then $T(t)x_n \to T(t)x_0$, and $PT(t)x_n \to \{PT(t)\}x_0$. Since P is closed, $T(t)x_0 \in \mathfrak{D}(P)$ and $P\{T(t)x_0\} = \{PT(t)\}x_0$. This proves (a). To prove (c) let $0 < \delta < t$. The equation $PT(t)x = PT(\delta)T(t-\delta)x$ shows $PT(t)x$ is continuous. Since

$$\log |PT(t)| \leqq \log |PT(\delta)| + \log |T(t-\delta)|,$$

it is seen that

$$\limsup_{t\to\infty} \frac{\log |PT(t)|}{t} \leqq \lim_{t\to\infty} \frac{\log |PT(\delta)|}{t} + \lim_{t\to\infty} \frac{\log |T(t-\delta)|}{t} = \omega_0 .$$

Thus (c) is proved. Statement (d) follows from Theorem III.6.20. Q.E.D.

21 LEMMA. *Let f be a continuous function defined for $t > 0$ with values in $\mathfrak{X}$ and suppose that $\int_0^1 |f(t)|dt < \infty$. If $g(t) = \int_0^t T(t-s)f(s)ds$, then $g(t)$ is in $\mathfrak{D}(P)$,*

$$Pg(t) = \int_0^t PT(t-s)f(s)ds,$$

and g and Pg are continuous functions of t for $t > 0$.

PROOF. The integral defining g exists for each $t \geqq 0$ since $|T(t)|$ is bounded on each finite interval (Lemma 3). For each $s < t$ the vector $T(t-s)f(s)$ is in $\mathfrak{D}(P)$ by Lemma 20. Thus Theorem III.6.20 will show that $g(t) \in \mathfrak{D}(P)$ and also establish the formula given for $Pg(t)$ as soon as it is shown that the function $s \to PT(t-s)f(s)$ is integrable over the interval $[0, t]$. It follows from Lemma 20(b) and the principle of uniform boundedness (II.3.21) that $|PT(\cdot)|$ is bounded on any interval not containing the origin. Let $0 < t_1 < t$ so that the function $s \to |PT(t-s)|$ is bounded and $|f(\cdot)|$ is integrable on the interval $0 \leqq s \leqq t_1$, while $|f(\cdot)|$ is bounded and $s \to |PT(t-s)|$ is, by Lemma 3 and Definition 18(iii), integrable on the interval $t_1 \leqq s \leqq t$.

To see that Pg is continuous for $t > 0$, let $0 < 2\delta < t_0$ and let $M_1 = \sup |PT(s)|$ for $t_0 - 2\delta \leqq s \leqq t_0 + \delta$. Then $|PT(t-s)f(s)| \leqq M_1|f(s)|$ if $|t-t_0| \leqq \delta$. Consequently, by Corollary III.6.16,

$$\lim_{t\to t_0} \int_0^\delta PT(t-s)f(s)ds = \int_0^\delta PT(t_0-s)f(s)ds.$$

Now

$$\int_\delta^t PT(t-s)f(s)ds = \int_0^{t_0} PT(s)f(t-s)\chi_{[0,\, t-\delta]}(s)ds,$$

and if $M_2 = \sup |f(s)|$ for $\delta \leqq s \leqq t_0 + \delta$, the norm of the integral on the right is bounded by $M_2|PT(s)|$. Thus

$$\lim_{t\to t_0} \int_\delta^t PT(t-s)f(s)ds = \int_\delta^{t_0} PT(t_0-s)f(s)ds.$$

Combining this result with the limit above it is seen that Pg is continuous at the arbitrary point $t_0 > 0$. The result just proved, if applied to the case where $P = I$, shows that g is continuous. Q.E.D.

22 Lemma. *Let f be a continuous function defined for $t > 0$ with values in $\mathfrak{X}$ and suppose that $\int_0^\infty e^{-\omega s}|f(s)|ds < \infty$ for some ω. Let $\omega_0 = \lim_{t\to\infty} (\log |T(t)|)/t$ and $F(\lambda) = \int_0^\infty e^{-\lambda s}f(s)ds$. Then, for $\mathscr{R}(\lambda) > \max(\omega, \omega_0)$,*

$$\text{(i)} \qquad R(\lambda; A)F(\lambda) = \int_0^\infty e^{-\lambda t}\int_0^t T(t-s)f(s)dsdt,$$

and

$$\text{(ii)} \qquad PR(\lambda; A)F(\lambda) = \int_0^\infty e^{-\lambda t} P \int_0^t T(t-s)f(s)dsdt.$$

Proof. Suppose that $\mathscr{R}(\lambda) > \max(\omega, \omega_0)$. To prove (i) we have

$$R(\lambda; A)F(\lambda) = \int_0^\infty e^{-\lambda t}T(t)\left\{\int_0^\infty e^{-\lambda s}f(s)ds\right\} dt$$

by Theorem 11. The function $e^{-\lambda(t+s)}T(t)f(s)$ is continuous on the product $(0, \infty)\times(0, \infty)$ and hence measurable on the product. By Tonelli's theorem (III.11.15) we have

$$R(\lambda; A)F(\lambda) = \int_0^\infty e^{-\lambda s}\int_0^\infty e^{-\lambda t}T(t)f(s)dtds.$$

Now for each s we make the substitution $t \to (t-s)$ in the second integral and interchange the order of integration again. Thus

$$R(\lambda; A)F(\lambda) = \int_0^\infty \int_s^\infty e^{-\lambda t}T(t-s)f(s)dtds$$

$$= \int_0^\infty e^{-\lambda t}\int_0^t T(t-s)f(s)dsdt.$$

Statement (ii) now follows from (i), Lemma 21, and Theorem III.6.20. Q.E.D.

The last bit of information we need to be able to prove Theorem 19 concerns integrals of a certain type important in other branches of mathematics as well. We collect this information in the next definition and lemma.

23 Definition. Let F and G be Lebesgue measurable scalar functions defined on $(-\infty, \infty)$. We define the function $F * G$ by putting

$$(F * G)(t) = \int_{-\infty}^\infty F(t-s)G(s)ds$$

for all values of t for which the integral exists. The function $F * G$ is called the *convolution* of F and G.

We remark that if F and G vanish for $s < 0$, then the formula for $F * G$ takes the form

$$(F * G)(t) = \int_0^t F(t-s)G(s)ds.$$

In the next two lemmas are proved certain basic properties of the convolution which will be used in this section and the next.

24 Lemma. (a) *If F and G belong to $L_1(-\infty, \infty)$ (with respect to Lebesgue measure), then $F * G$ is defined for almost all t, is a function in $L_1(-\infty, \infty)$, and $|F * G|_1 \leqq |F|_1|G|_1$.*

(b) *If F is in $L_1(-\infty, \infty)$ and $|G(t)| \leqq M$, then $|(F * G)(t)| \leqq M|F|_1$.*

(c) *Let F and G be defined for $t \geqq 0$ and be Lebesgue integrable over each finite interval. Then $(F * G)(t) = \int_0^t F(t-s)G(s)ds$ is Lebesgue integrable over each finite interval.*

Proof. To prove statement (a) we first show that the function $[s, t] \to G(t-s)$ defined on the plane $(-\infty, \infty)\times(-\infty, \infty)$ is measurable with respect to two-dimensional Lebesgue measure.

If E is a subset of the real axis, let $p(E)$ be the subset of the plane defined by $p(E) = \{[s, t]|s-t \in E\}$. Then $p(\cup_{i=1}^{\infty} E_i) = \cup_{i=1}^{\infty} p(E_i)$, $p(E') = (p(E))'$, and $p(\phi) = \phi$. Since $p(E)$ is open if E is open, it follows immediately that $p(E)$ is a Borel set if E is a Borel set. Now let E be a Borel set of measure zero. By Fubini's theorem (III.11.9), we have

$$\begin{aligned}\int_{-\infty}^{+\infty}\int_{-\infty}^{+\infty} \chi_{p(E)}(s, t)ds\, dt &= \int_{-\infty}^{+\infty}\int_{-\infty}^{+\infty} \chi_E(s-t)dsdt \\ &= \int_{-\infty}^{+\infty}\left\{\int_{-\infty}^{+\infty} \chi_E(s-t)ds\right\} dt \\ &= \int_{-\infty}^{+\infty} 0 \cdot dt = 0.\end{aligned}$$

Thus, $p(E)$ has measure zero if E has measure zero. It follows that if E is in the Lebesgue completion of the σ-field of Borel subsets of the line, $p(E)$ is in the Lebesgue extension of the σ-field of Borel subsets of the plane. Let U be an open set of complex numbers, $E = \{t|G(t) \in U\}$, and $D = \{[s, t]|G(t-s) \in U\}$. Then $D = p(E)$, and now the measurability of the function $G(t-s)$ follows immediately from III.6.10.

Now let $F, G \in L_1$. Then, by Tonelli's theorem (III.11.14),

$$\begin{aligned}\int_{-\infty}^{+\infty} |(F * G)(t)|dt &= \int_{-\infty}^{+\infty} \left| \int_{-\infty}^{+\infty} F(t-s)G(s)ds \right| dt \\ &\leqq \int_{-\infty}^{+\infty} \int_{-\infty}^{+\infty} |F(t-s)G(s)|dsdt \\ &= \int_{-\infty}^{+\infty} \int_{-\infty}^{+\infty} |F(t-s)||G(s)|dtds \\ &= |F|_1 \int_{-\infty}^{+\infty} |G(s)|ds \\ &= |F|_1|G|_1,\end{aligned}$$

from which (a) follows. The proof of (b) is elementary. Statement (c) follows from (a) since

$$\int_0^\beta |(F * G)(t)|dt = \int_0^\beta |(F_1 * G_1)(t)|dt,$$

where $F_1(t) = F(t)$ and $G_1(t) = G(t)$ for $0 < t < \beta$, and $F_1(t) = G_1(t) = 0$ for $t > \beta$ and $t \leqq 0$. Q.E.D.

25 Lemma. (a) *If F and G are Lebesgue measurable functions defined on the real line, then $F * G = G * F$.*

(b) *If F, G and H belong to $L_1(-\infty, +\infty)$, then $(F * G) * H = F * (G * H)$.*

Proof. Part (a) follows immediately from the equations

$$(F * G)(t) = \int_{-\infty}^{+\infty} F(t-s)G(s)ds = \int_{-\infty}^{+\infty} F(s)G(t-s)ds = (G * F)(t).$$

Part (b) follows from the equation

$$\begin{aligned}((F * G) * H)(r) &= \int_{-\infty}^{+\infty} \left\{ \int_{-\infty}^{+\infty} F(s)G(t-s)ds \right\} H(r-t)dt \\ &= \int_{-\infty}^{+\infty} \int_{-\infty}^{+\infty} G(t-s)F(s)H(r-t)ds\,dt \\ &= \int_{-\infty}^{+\infty} \left\{ \int_{-\infty}^{+\infty} G(t-s)H(r-t)dt \right\} F(s)ds \\ &= \int_{-\infty}^{+\infty} \left\{ \int_{-\infty}^{+\infty} G(t)H(r-s-t)dt \right\} F(s)ds \\ &= (F * (G * H))(r),\end{aligned}$$

which holds for almost all r by (a), and by Tonelli's theorem (III.11.14), since

$$\begin{aligned}\int_{-\infty}^{\infty} \int_{-\infty}^{\infty} \int_{-\infty}^{\infty} &|G(t-s)F(s)H(r-t)|drdtds \\ &= |H|_1|G|_1|F|_1 < \infty\end{aligned}$$

and since the measurability of all our functions was proved as the first part of the proof of Lemma 24. Q.E.D.

Finally, we give the proof of Theorem 19.

PROOF OF THEOREM 19. Let $\chi(t) = |T(t)|$ and $\psi(t) = |PT(t)|$. Then χ and ψ are measurable (cf. Lemma 3). If ω is any constant greater than $\omega_0 = \lim_{t\to\infty} \log |T(t)|/t$, then there exists an $M_\omega < \infty$ such that $\chi(t) \leqq M_\omega e^{\omega t}$ for $t \geqq 0$. By Lemma 3 and Definition 18(iii), $\int_0^\beta \psi(t)dt < \infty$ for each $\beta > 0$. Put $\psi^{(1)} = \psi$, and, inductively, $\psi^{(n)} = \psi^{(n-1)} * \psi$. By Lemma 24(c), we see inductively that all the functions $\psi^{(n)}$ are Lebesgue integrable over every finite interval of the positive real axis. Put $\chi^{(0)} = \chi$, $\chi^{(n)} = \chi * \psi^{(n)}$. By Lemma 24(c), $\chi^{(n)}$ is integrable over every finite interval of the positive real axis.

Let $S_0(t) = T(t)$ and define

$$[*] \qquad S_n(t)x = \int_0^t T(t-s)PS_{n-1}(s)xds, \qquad \text{for } x \in \mathfrak{X},$$

inductively. We shall show that this inductive construction is legitimate by proving that

(i) $S_n(t)x \in \mathfrak{D}(P)$ $\qquad x \in \mathfrak{X}, \quad t > 0;$

(ii) $S_n(t)x$ is continuous in t for $t > 0$, $\qquad x \in \mathfrak{X};$

(iii) $|S_n(t)| \leqq \chi^{(n)}(t);$

(iv) $PS_n(t)x$ is continuous in t for $t > 0$, $\qquad x \in \mathfrak{X};$

(v) $|PS_n(t)| \leqq \psi^{(n+1)}(t).$

For $n = 0$, all this is either obvious or follows from Lemmas 20 and 21. Suppose that (i), . . ., (v) are known to hold for $n = m$. Then it is clear from (i), (iv) and (v) that the integral in [*] exists for all $t > 0$, and may be used to define S_{m+1}. Statements (i), (ii) and (iv) for the case $n = m+1$ then follow from Lemma 21. We have

$$\begin{aligned} |S_{m+1}(t)x| &= \left| \int_0^t T(t-s)PS_m(s)xds \right| \\ &\leqq |x| \int_0^t \chi(t-s)\psi^{(m+1)}(s)ds \\ &= |x|\chi^{(m+1)}(t). \end{aligned}$$

By Lemma 21,

$$\begin{aligned} |PS_{m+1}(t)x| &= \left| \int_0^t PT(t-s)PS_m(s)xds \right| \\ &\leqq |x| \int_0^t \psi(t-s)\psi^{(m+1)}(s)ds \\ &= |x|\psi^{(m+2)}(t), \end{aligned}$$

which proves (iii) and (v) for the case $n = m+1$. Consequently, (i),. . ., (v) are proved inductively for all n.

We now obtain an estimate for the series $\sum_{n=0}^{\infty}\chi^{(n)}(t)$ which majorizes $\sum_{n=0}^{\infty}|S_n(t)|$.

By Lemma 20(c), for each $\omega > \omega_0$ there exists a constant $M_\omega < \infty$ such that $\psi(t) = |PT(t)| < M_\omega e^{\omega t}$ for sufficiently large t. On the other hand, we have seen that ψ is integrable over every finite interval of the positive real axis. Thus, if we choose ω_1 sufficiently large,

$$\int_0^\infty e^{-\omega_1 t}\psi(t)dt < \infty.$$

Consequently, by III.6.16,

$$\lim_{\lambda\to\infty}\int_0^\infty e^{-\lambda t}\psi(t)dt = \lim_{\lambda\to\infty}\int_0^\infty e^{-(\lambda-\omega_1)t}\{e^{-\omega_1 t}\psi(t)\}\,dt = 0;$$

so that, if $\omega > \omega_1$ is chosen large enough, we will have

$$\int_0^\infty e^{-\omega t}\psi(t)dt = \gamma < 1.$$

We will now show, inductively, that

$$\chi^{(n)}(t) \leqq M_\omega e^{\omega t}\gamma^n.$$

This is clearly the case if $n = 0$. Suppose that it is true for a given value of n. Then, by Lemma 25(b),

$$\begin{aligned}\chi^{(n+1)}(t) &= \int_0^t \chi^{(n)}(t-s)\psi(s)ds\\ &\leqq M_\omega e^{\omega t}\gamma^n \int_0^t e^{-\omega s}\psi(s)ds\\ &\leqq M_\omega e^{\omega t}\gamma^{n+1}, \qquad\qquad t > 0.\end{aligned}$$

Clearly $\lim_{t\to 0} S_0(t) = I$ in the strong operator topology. Since

$$\chi^{(n)}(t) = \int_0^t \chi(t-s)\psi^{(n)}(s)ds \leqq M_\omega e^{\omega t}\int_0^t e^{-\omega s}\psi^{(n)}(s)ds,$$

it is clear that $\chi^{(n)}(t) \to 0$ as $t \to 0$ for $n \geqq 1$. Thus, since $|S_n(t)| \leqq \chi^{(n)}(t)$, $S_n(t) \to 0$ for $n \geqq 1$. Hence, if we put $S_0(0) = I$, $S_n(0) = 0$ for $n \geqq 0$, $S_n(t)x$ will, for each x in $\mathfrak{X}$, be continuous in t for $t \geqq 0$. Moreover, we will clearly have $|S_n(t)| \leqq M_\omega e^{\omega t}\gamma^n$ for $t \geqq 0$ and $n \geqq 0$.

It now follows that the series $\sum_{n=0}^{\infty} S_n(t)$ converges absolutely, uniformly in each finite interval $[0, \beta]$, and that

$$[**] \qquad \sum_{n=0}^{\infty} |S_n(t)| \leqq (1-\gamma)^{-1} M_\omega e^{\omega t}.$$

We write $S(t) = \sum_{n=0}^{\infty} S_n(t)$, $t \geqq 0$. Since each of the terms of this series is strongly continuous for $t \geqq 0$, it follows that the same is true of $S(t)$. Further $|S(t)| \leqq (1-\gamma)^{-1} M_\omega e^{\omega t}$.

It remains to show that $S(t)$, $t \geqq 0$, is a semi-group of bounded operators with infinitesimal generator $A+P$, $\mathfrak{D}(A+P) = \mathfrak{D}(A)$. To see this note first that, by [**] and III.6.16, we have

$$\int_0^\infty e^{-\lambda s} S(s) x ds = \sum_{n=0}^{\infty} \int_0^\infty e^{-\lambda s} S_n(s) x ds, \qquad x \in \mathfrak{X}, \qquad \mathscr{R}(\lambda) > \omega.$$

By Lemma 22(i)

$$\int_0^\infty e^{-\lambda s} S_n(s) x ds = R(\lambda; A) \int_0^\infty e^{-\lambda s} P S_{n-1}(s) x ds.$$

Now repeated application of Lemma 22(ii) gives

$$\begin{aligned}\int_0^\infty e^{-\lambda s} S_n(s) x ds &= R(\lambda; A) P R(\lambda; A) \int_0^\infty e^{-\lambda s} P S_{n-2}(s) x ds \\ &= \ldots = R(\lambda; A)[PR(\lambda; A)]^n x.\end{aligned}$$

Consequently,

$$\int_0^\infty e^{-\lambda s} S(s) x ds = \sum_{n=0}^{\infty} R(\lambda; A)[PR(\lambda; A)]^n x, \qquad \mathscr{R}(\lambda) > \omega.$$

By Lemma 20(d), and 18(i)

$$|PR(\lambda; A)| \leqq \int_0^\infty e^{-\omega s} \psi(s) ds = \gamma < 1, \qquad \mathscr{R}(\lambda) > \omega.$$

Thus the series $\sum_{n=0}^{\infty} [PR(\lambda; A)]^n$ converges absolutely. Let $\mathfrak{D}(A+P) = \mathfrak{D}(A)$. Recalling that $R(\lambda; A)\mathfrak{X} = \mathfrak{D}(A)$ we have

$$\begin{aligned}&(\lambda I - A - P) R(\lambda; A) \sum_{n=0}^{\infty} PR(\lambda; A) x \\ &= \sum_{n=0}^{\infty} [PR(\lambda; A)]^n x - \sum_{n=1}^{\infty} [PR(\lambda; A)]^n x = x, \qquad x \in \mathfrak{X}.\end{aligned}$$

Moreover, if $x \in \mathfrak{D}(A)$,

$$\begin{aligned}R(\lambda; A)\{\sum_{n=0}^{\infty} [PR(\lambda; A)]^n\}(\lambda I - A - P) x \\ = x + R(\lambda; A) \sum_{n=0}^{\infty} [PR(\lambda; A)]^n P x \\ - R(\lambda; A) \sum_{n=0}^{\infty} [PR(\lambda; A)]^n P x = x.\end{aligned}$$

Thus $\lambda I-A-P$ with domain $\mathfrak{D}(A)$ has the bounded inverse $R(\lambda; A)\sum_{n=0}^{\infty}[PR(\lambda; A)]^n$. Consequently $A+P$ is closed and

$$R(\lambda; A)\sum_{n=0}^{\infty}[PR(\lambda; A)]^n = R(\lambda; A+B).$$

The conclusion of Theorem 19 now follows from Corollary 16. Q.E.D.

Example. In order to illustrate the application of Theorem 19 we consider the space $C[-\infty, \infty]$ and the group of translations $\{T(t)\}$ defined by $T(t)(x, s) = x(t+s)$; this group has the infinitesimal generator $A = d/ds$ (cf. the examples following Corollary 17). Consider the operator $A^2 = d^2/ds^2$ whose domain (cf. VII.9.6) consists of all functions x in $C[-\infty, \infty]$ for which x' and x'' belong to $C[-\infty, \infty]$. That A^2 is closed was shown in Theorem VII.9.7. It is easily seen that $\mathfrak{D}(A^2)$ is dense. Since $\sigma(A)$ has been observed to be the imaginary axis, $\sigma(A^2)$ is the negative real axis (cf. VII.9.10). If $\lambda > 0$, then in view of Theorem VII.9.5

$$R(\lambda; A^2) = -R(\sqrt{\lambda}; A)R(-\sqrt{\lambda}; A).$$

Consequently $|R(\lambda; A^2)| \leqq \lambda^{-1}$, $\lambda > 0$, and it follows from Corollary 14 that A^2 generates a strongly continuous semi-group $T(t; A^2)$ satisfying $|T(t; A^2)| \leqq 1$.

We shall obtain an explicit expression for $T(t; A^2)x$. By Theorem VII.9.4 and Theorem VII.9.5

$$R(\lambda; A^2)(x, s) = \frac{1}{2\pi i}\int_{C_1+C_2}\frac{R(\mu; A)(x, s)d\mu}{(\mu-\sqrt{\lambda})(\mu+\sqrt{\lambda})}, \qquad -\infty < s < \infty,$$

where C_1 and C_2 are small positively oriented circles about the points $\mu = -\sqrt{\lambda}$ and $\mu = \sqrt{\lambda}$ respectively. On making the substitution

$$\begin{aligned} R(\mu; A)(x, s) &= \int_0^{\infty} e^{-\mu t}x(t+s)dt, && \mathscr{R}(\mu) > 0, \\ &= -\int_{-\infty}^{0} e^{-\mu t}x(t+s)dt, && \mathscr{R}(\mu) < 0, \end{aligned}$$

and calculating residues, we obtain the formula

$$R(\lambda; A^2)(x, s) = \int_{-\infty}^{\infty}\frac{e^{-|t|\sqrt{\lambda}}}{2\sqrt{\lambda}}x(s+t)dt, \qquad -\infty < s < \infty.$$

We now substitute

$$\frac{e^{-|t|\sqrt{\lambda}}}{\sqrt{\lambda}} = \int_0^\infty \frac{e^{-\lambda r} e^{-t^2/4r}}{\sqrt{\pi r}}\, dr$$

and invert the order of integration using Tonelli's theorem (III.11.14) obtaining

$$R(\lambda; A^2)(x, s) = \int_0^\infty e^{-\lambda r} \left\{ \frac{1}{2\sqrt{\pi r}} \int_{-\infty}^{\infty} e^{-t^2/4r}\, x(s+t)dt \right\} dr.$$

However, we know from Theorem 11 that

$$R(\lambda; A^2)(x, s) = \int_0^\infty e^{-\lambda r}\, T(r; A^2)(x, s)dr,$$

and thus it follows from Lemma 15 that

$$T(r; A^2)(x, s) = \frac{1}{2\sqrt{\pi r}} \int_{-\infty}^{\infty} e^{-t^2/4r}\, x(s+t)dt, \qquad r > 0.$$

Now let $h \in C[-\infty, \infty]$, and let P be the (unbounded) operator defined as follows:

(a) The domain of P consists of all $x \in C[-\infty, \infty]$ such that x has a continuous derivative in a neighborhood of each point t_0 for which $h(t_0) \neq 0$, and such that $h(t)x'(t)$ is in $C[-\infty, \infty]$;

(b) For $x \in \mathfrak{D}(P)$, $(Px)(t) = h(t)x'(t)$.

The operator P is easily seen to be closed.

We shall now verify the conditions of Definition 18, to show by Theorem 19 that A^2+P with domain $\mathfrak{D}(A^2)$ is the generator of a semi-group. Clearly $\mathfrak{D}(A^2) \subseteqq \mathfrak{D}(P)$. If $x \in \mathfrak{D}(A^2)$ and $t > 0$, then

$$\begin{aligned}
|PT(t; A^2)(x, s)| &\leqq |h| \left| \frac{d}{ds} T(t; A^2)(x, s) \right| \\
&= |h| \left| \frac{-1}{4t\sqrt{\pi t}} \int_{-\infty}^{\infty} (\xi - s) e^{-(\xi-s)^2/4t}\, x(\xi) d\xi \right| \\
&\leqq \frac{|x||h|}{2t\sqrt{\pi t}} \int_0^\infty \xi e^{-\xi^2/4t}\, d\xi = \frac{|x||h|}{\sqrt{\pi t}}.
\end{aligned}$$

Thus conditions (ii) and (iii) are satisfied also. In view of Lemma 7(b), for each $x_0 \in \mathfrak{D}(A^2)$, the function

$$y(s, t) = T(t; A^2+P)(x_0, s)$$

is a solution of the initial value problem

$$(1) \quad \frac{\partial}{\partial t} y(s, t) = \frac{\partial^2}{\partial s^2} y(s, t) + h(s) \frac{\partial}{\partial s} y(s, t),$$

$$(2) \quad \lim_{t \to 0} y(s, t) = x_0(s) \text{ uniformly in } s.$$

It is clear that the same method would establish the existence of a solution to many other initial value problems such as, for example, the one mentioned in the introduction to the chapter (provided that the kernel K defines a bounded linear operator) or the problem expressed by the equations

$$(1') \quad \frac{\partial}{\partial t} y(s, t) = \frac{\partial^2}{\partial s^2} y(s, t) + h_1(s) \frac{\partial}{\partial s} y(s, t) + h_2(s) y(s, t),$$

$$(2') \quad \lim_{t \to 0} y(s, t) = x_0(s) \text{ uniformly in } s,$$

for any $h_1, h_2 \in C[-\infty, \infty]$ and $x_0 \in \mathfrak{D}(A^2)$.

2. Functions of an Infinitesimal Generator

It was shown in Section 9 of Chapter VII how an operational calculus could be constructed for an unbounded closed operator A with non-void resolvent set. Specifically, to each function f analytic on $\sigma(A)$ and at infinity we assigned a bounded operator $f(A)$ such that mapping $f \to f(A)$ was a homomorphism. In this section we suppose A is the infinitesimal generator of a strongly continuous group of operators $T(t)$, $-\infty < t < \infty$, and show in this case how bounded operators may be assigned to each of a larger class of functions. This class will include functions which are analytic on $\sigma(A)$, but not necessarily analytic at infinity. The functions to be considered are bilateral Laplace-Stieltjes transforms. We also discuss the inversion of these operators by limits of polynomials in A.

Throughout this section, we let $T(t)$, $-\infty < t < \infty$, be a strongly continuous group of bounded operators in a complex B-space $\mathfrak{X}$, and A its infinitesimal generator. We recall (cf. 1.17) that there exist positive constants M and ω such that $|T(t)| \leqq Me^{\omega|t|}$, and such that the spectrum of the infinitesimal generator A lies in the vertical strip $-\omega < \mathscr{R}(\lambda) < \omega$. Moreover,

$$R(\lambda; A)x = \int_0^\infty e^{-\lambda t} T(t)x dt, \qquad \mathscr{R}(\lambda) > \omega$$

$$= -\int_{-\infty}^0 e^{-\lambda t} T(t)x dt, \qquad \mathscr{R}(\lambda) < -\omega$$

We begin our analysis with a discussion of bilateral Laplace-Stieltjes transforms.

1 DEFINITION. We denote by $\mathscr{S}(A)$ the family of all finite complex valued measures β defined on the Borel sets in $(-\infty, \infty)$ and such that

$$\int_{-\infty}^\infty e^{(\omega+\varepsilon)|t|} v(\beta; dt) < \infty,$$

where ε is a positive number (which may vary with β). The function

$$f(\lambda) = \int_{-\infty}^\infty e^{-\lambda t} \beta(dt), \quad -(\omega+\varepsilon) < \mathscr{R}(\lambda) < \omega+\varepsilon$$

is called the *bilateral Laplace-Stieltjes transform* of β. We denote by $\mathscr{V}(A)$ the family of bilateral Laplace transforms of measure $\beta \in \mathscr{S}(A)$. If β in $\mathscr{S}(A)$ is continuous with respect to Lebesgue-Stieltjes measure and if F is the function in $L_1(-\infty, \infty)$ given by the Radon-Nikodým theorem (III.10.2) such that $\beta(E) = \int_E F(s)ds$ the function

$$f(\lambda) = \int_{-\infty}^\infty e^{-\lambda t} F(t) dt$$

is called the *bilateral Laplace transform* of F.

We now prove certain basic facts concerning Laplace-Stieltjes transforms.

2 LEMMA. *If f is in $\mathscr{V}(A)$, and $f(\lambda) = \int_{-\infty}^\infty e^{-\lambda t}\beta(dt)$, then f is analytic in a strip $-(\omega+d) < \mathscr{R}(\lambda) < (\omega+d)$. The measures defined for each Borel set E by*

$$\beta^{(n)}(E) = \int_E (-t)^n \beta(dt), \qquad n = 0, 1, 2, \ldots,$$

belong to $\mathscr{S}(A)$ and

$$\frac{d^n f}{d\lambda^n}(\lambda) = f^{(n)}(\lambda) = \int_{-\infty}^\infty e^{-\lambda t} \beta^{(n)}(dt)$$

$$= \int_{-\infty}^\infty (-t)^n e^{-\lambda t} \beta(dt), \qquad -(\omega+\varepsilon) < \mathscr{R}(\lambda) < \omega+\varepsilon.$$

PROOF. For each n and each $\varepsilon_1 < \varepsilon$ there is a constant K such that

$$|t|^n e^{(\omega+\varepsilon_1)|t|} \leqq K e^{(\omega+\varepsilon)|t|}, \qquad -\infty < t < \infty.$$

Thus, since $\beta \in \mathscr{S}(A)$,

$$\int_{-\infty}^{\infty} e^{(\omega+\varepsilon_1)|t|} v(\beta^{(n)}, dt) = \int_{-\infty}^{\infty} e^{(\omega+\varepsilon_1)|t|} |t|^n v(\beta, dt)$$
$$\leqq K \int_{-\infty}^{\infty} e^{(\omega+\varepsilon)|t|} v(\beta, dt) < \infty,$$

and $\beta^{(n)} \in \mathscr{S}(A)$.

Since $|e^{-\mu t} - e^{-\lambda t}||\mu-\lambda|^{-1} \leqq |t| e^{(\omega+\varepsilon)|t|}$ for $|\mathscr{R}(\lambda)| < \omega+\varepsilon$ and $|\mathscr{R}(\lambda)| < \omega+\varepsilon$, we have, by III.6.16 and III.10.6,

$$\lim_{\mu\to\lambda} \frac{f(\lambda)-f(\mu)}{\lambda-\mu} = \lim_{\mu\to\lambda} \int_{-\infty}^{\infty} \frac{e^{-\lambda t} - e^{-\mu t}}{\lambda-\mu} \beta(dt)$$
$$= \int_{-\infty}^{\infty} (-t) e^{-\lambda t} \beta(dt)$$
$$= \int_{-\infty}^{\infty} e^{-\lambda t} \beta^{(1)}(dt), \qquad |\mathscr{R}(\lambda)| < \omega+\varepsilon.$$

Thus f is analytic in $|\mathscr{R}(\lambda)| < \omega+\varepsilon$, and

$$\frac{df}{d\lambda}(\lambda) = f'(\lambda) = \int_{-\infty}^{\infty} e^{-\lambda t} \beta^{(1)}(dt), \qquad |\mathscr{R}(\lambda)| < \omega+\varepsilon.$$

It is clear that an inductive repetition of the argument just given will show that

$$f^{(n)}(\lambda) = \int_{-\infty}^{\infty} e^{-\lambda t} \beta^{(n)}(dt), \qquad |\mathscr{R}(\lambda)| < \omega+\varepsilon.$$

Q.E.D.

3 Definition. If α, β are in $\mathscr{S}(A)$ let $\alpha \times \beta$ be the product measure defined on $(-\infty, \infty)\times(-\infty, \infty)$ (cf. Section III.11). It was shown at the beginning of the proof of Lemma 1.25 that if E is a Borel subset of the real line, the set $P(E) = \{(x, y)|x+y \in E\}$ is a Borel subset of the plane. For each Borel subset E of the real line let $\gamma(E) = (\alpha\times\beta)\{P(E)\}$. We call γ the *convolution* of α and β and write $\gamma = \alpha * \beta$. It is clear that γ is a Borel measure and $\alpha * \beta = \beta * \alpha$. Now let $E_1, \ldots, E_n$ be disjoint Borel sets so that

$$\sum_{i=1}^{n} |\gamma(E_i)| = \sum_{i=1}^{n} |(\alpha\times\beta)(P(E_i))| \leqq \sum_{i=1}^{n} v(\alpha\times\beta, E_i).$$

It follows from this inequality and Lemma III.11.11 that

$$v(\gamma, E) \leqq v(\alpha)\times v(\beta)(P(E)) = v(\alpha) * v(\beta)(E)$$

for every Borel set E on the real line. By III.10.8(b) and Theorem III.11.13, we have

$$[*] \qquad \int_{-\infty}^{\infty} F(r)\gamma(dr) = \int_{-\infty}^{\infty}\int_{-\infty}^{\infty} F(s+t)\alpha(ds)\beta(dt)$$

if F is γ-integrable.

Finally, we remark that by Fubini's theorem

$$\gamma(E) = \int_{-\infty}^{\infty}\int_{-\infty}^{\infty} \chi_E(s+t)\alpha(ds)\beta(dt)$$
$$= \int_{-\infty}^{\infty} \alpha(E-t)\beta(dt).$$

4 **Lemma.** *Let α and β be measures in $\mathscr{S}(A)$ with bilateral Laplace-Stieltjes transforms f and g. Then $\gamma = \alpha * \beta$ is in $\mathscr{S}(A)$ and*

$$\int_{-\infty}^{\infty} e^{-\lambda t}\gamma(dt) = f(\lambda)g(\lambda), \qquad -(\omega+\varepsilon) < \mathscr{R}(\lambda) < \omega+\varepsilon.$$

Proof. Since the variations of γ, α, and β are related by $v(\gamma, E) \leqq v(\alpha) * v(\beta)(E)$,

$$\int_{-\infty}^{\infty} e^{|r|(\omega+\varepsilon)} v(\gamma, dr) \leqq \int_{-\infty}^{\infty}\int_{-\infty}^{\infty} e^{(\omega+\varepsilon)|s+t|} v(\alpha, ds)v(\beta, dt)$$
$$\leqq \left\{\int_{-\infty}^{\infty} e^{(\omega+\varepsilon)|s|} v(\alpha, ds)\right\}\left\{\int_{-\infty}^{\infty} e^{(\omega+\varepsilon)|t|} v(\beta, dt)\right\}$$

by Fubini's theorem, and thus γ is in $\mathscr{S}(A)$. If ε is sufficiently small, it is seen, from formula [*] that, for $|\mathscr{R}(\lambda)| < \omega+\varepsilon$, we have

$$\int_{-\infty}^{\infty} e^{-\lambda t}\gamma(dt) = \int_{-\infty}^{\infty}\int_{-\infty}^{\infty} e^{-\lambda(t+s)}\alpha(dt)\beta(ds)$$
$$= \left\{\int_{-\infty}^{\infty} e^{-\lambda t}\alpha(dt)\right\}\left\{\int_{-\infty}^{\infty} e^{-\lambda s}\beta(ds)\right\}$$
$$= f(\lambda)g(\lambda).$$

Q.E.D.

5 **Definition.** Let $f \in \mathscr{V}(A)$, and

$$f(\lambda) = \int_{-\infty}^{\infty} e^{-\lambda t}\alpha(dt), \qquad -(\omega+\varepsilon) < \mathscr{R}(\lambda) < \omega+\varepsilon$$

where $\alpha \in \mathscr{S}(A)$. If A is the infinitesimal generator of the strongly continuous group $T(t)$ we define the operator $f\{A\}$ by the formula

$$f\{A\}x = \int_{-\infty}^{\infty} T(-t)x\alpha(dt), \qquad x \in \mathfrak{X}.$$

Since $|T(t)| \leqq Me^{\omega|t|}$ the above formula shows that

$$|f\{A\}| \leqq M \int_{-\infty}^{\infty} e^{\omega|t|} v(\alpha; dt).$$

Thus $f\{A\}$ is bounded. We note that for each t_0 the function $f_{t_0}(\lambda) = e^{\lambda t_0}$ is in $\mathscr{V}(A)$, being the transform of the measure which assigns the value one to the point $t = -t_0$ and vanishes on each Borel set not containing $-t_0$. Moreover, $f_{t_0}\{A\} = T(t_0)$. If $|\mathscr{R}(\lambda)| > \omega$ then the function f_α defined by $f_\alpha(\lambda) = (\alpha-\lambda)^{-1}$ is in $\mathscr{V}(A)$ since

$$\begin{aligned} \frac{1}{\alpha-\lambda} &= \int_0^{\infty} e^{-\lambda t} e^{\alpha t}\, dt, & \mathscr{R}(\alpha) < \mathscr{R}(\lambda), \\ &= -\int_{-\infty}^{0} e^{-\lambda t} e^{\alpha t}\, dt, & \mathscr{R}(\alpha) > \mathscr{R}(\lambda). \end{aligned}$$

Moreover,

$$\begin{aligned} f_\alpha\{A\} &= -\int_{-\infty}^{0} e^{\alpha t} T(-t) dt = \int_0^{\infty} e^{-\alpha t} T(t) dt \\ &= R(\alpha; A), & \mathscr{R}(\alpha) > \omega, \end{aligned}$$

and similarly

$$f_\alpha\{A\} = R(\alpha; A), \qquad \mathscr{R}(\alpha) < -\omega.$$

6 THEOREM. *If f and g are in $\mathscr{V}(A)$ then αf, $f+g$ and fg are in $\mathscr{V}(A)$ and*

(a) $(\alpha f)\{A\} = \alpha f\{A\}$;
(b) $(f+g)\{A\} = f\{A\}+g\{A\}$;
(c) $(fg)\{A\} = f\{A\}g\{A\}$.

PROOF. Statements (a) and (b) are clear from the linearity of the formula defining the operator $f\{A\}$. To prove (c) we observe that for each linear functional $x^* \in \mathfrak{X}^*$

$$\begin{aligned} x^* f\{A\}g\{A\}x &= x^* \int_{-\infty}^{\infty} T(-s) g\{A\} x \alpha(ds) \\ &= x^* \int_{-\infty}^{\infty} T(-s) \int_{-\infty}^{\infty} T(-t) x \beta(dt) \alpha(ds) \\ &= \int_{-\infty}^{\infty}\int_{-\infty}^{\infty} x^* T(-s-t) x \beta(dt) \alpha(ds) \\ &= \int_{-\infty}^{\infty} x^* T(-r) x \gamma(dr) \end{aligned}$$

where $\gamma = \alpha * \beta$. By Lemma 4, $(fg)(\lambda) = \int_{-\infty}^{\infty} e^{-\lambda t}\gamma(dt)$. Thus $f\{A\}g\{A\} = (fg)\{A\}$. Q.E.D.

For f in $\mathscr{V}(A)$ and x in $\mathfrak{D}(A^2)$ we shall obtain, in the next two lemmas, a formula for $f\{A\}x$. This formula will enable us to relate the present operational calculus for A to that constructed in Section VII.9 and will be used in the discussion of the inversion of convolution transforms to follow.

It is convenient to begin by making a convention with regard to contour integrals. If c is a real number, then Γ_c will denote the infinite contour consisting of two vertical lines: the line $\lambda = c+i\tau$, $-\infty < \tau < \infty$, directed upward, and the line $\lambda = -c+i\tau$, $-\infty < \tau < \infty$, directed downwards.

7 Lemma. *Let α be a complex number and c be a real number chosen such that $\omega < c < |\mathscr{R}(\alpha)|$. Then, for x in $\mathfrak{D}(A^2)$,*

$$[*] \qquad T(t)x = \frac{1}{2\pi i}\int_{\Gamma_c} \frac{e^{\lambda t} R(\lambda; A)(\alpha I - A)^2 x d\lambda}{(\alpha-\lambda)^2}.$$

Proof. First we note that since $R(\lambda; A) = \int_0^\infty e^{-\lambda t} T(t)dt$, $|R(\lambda; A)| \leqq K\int_0^\infty e^{(\omega-\mathscr{R}(\lambda))t}dt$ for $\mathscr{R}(\lambda) > \omega$, so that $|R(\lambda; A)|$ is uniformly bounded in each half plane $\mathscr{R}(\lambda) > \omega+\varepsilon$. Similarly, $|R(\lambda; A)|$ is uniformly bounded in each half plane $\mathscr{R}(\lambda) < -\omega-\varepsilon$. Thus the integrand in [*] is of the order of $|\lambda|^{-2}$ as $|\lambda| \to \infty$, so that the integral in [*] is well-defined. Let $x \in \mathfrak{D}(A^2)$ and let $B(t)x$ denote the integral in equation [*] above. Suppose first that $t \geqq 0$. Then if $\mathscr{R}(\mu) > c$,

$$\int_0^\infty e^{-\mu t} B(t)x dt = \frac{1}{2\pi i}\int_{\Gamma_c} \frac{R(\lambda; A)(\alpha I - A)^2}{(\alpha-\lambda)^2} x \int_0^\infty e^{(\lambda-\mu)t}\, dtd\mu$$

$$= \frac{1}{2\pi i}\int_{\Gamma_c} \frac{R(\lambda; A)(\alpha I - A)^2 x d\lambda}{(\mu-\lambda)(\alpha-\lambda)^2}.$$

The integrand in the last integral is $O(|\lambda|^{-3})$ as $|\lambda| \to \infty$ in the half plane $\mathscr{R}(\lambda) \geqq c$, and also in the half plane $\mathscr{R}(\lambda) \leqq -c$. Thus, using Cauchy's theorem, we may replace the contour Γ_c by two small negatively oriented circles about $\lambda = \alpha$ and $\lambda = \mu$ respectively and calculate the residues to obtain

$$\int_0^\infty e^{-\mu t} B(t)x\, dt = \frac{R(\mu; A)(\alpha I - A)^2 x}{(\alpha-\mu)^2} - \frac{R(\alpha; A)(\alpha I - A)^2 x}{(\alpha-\mu)^2}$$

$$- \frac{R(\alpha; A)^2(\alpha I - A)^2 x}{\alpha-\mu} = \frac{-x}{\alpha-\mu} - \frac{(\alpha I - A)x}{(\alpha-\mu)^2} + \frac{R(\mu; A)(\alpha I - A)^2 x}{(\alpha-\mu)^2}.$$

Then, successive substitution in the identity

$$R(\mu; A)x = \frac{-x}{\alpha-\mu} + \frac{(\alpha I - A)R(\mu; A)x}{\alpha-\mu}$$

shows that

$$\int_0^\infty e^{-\mu t} B(t)x dt = R(\mu; A)x = \int_0^\infty e^{-\mu t} T(t)x dt.$$

On applying linear functionals to both sides, it follows, by Lemma 1.15, that $B(t)x = T(t)x$, $x \in \mathfrak{D}(A^2)$, $t \geqq 0$.

Applying the same reasoning to the semigroup $T(-t)$, which has the infinitesimal generator $-A$, we see that

$$T(-t)x = \frac{1}{2\pi i}\int_{\Gamma_c} \frac{e^{\lambda t} R(\lambda; -A)(\alpha I - A)^2 x d\lambda}{(\alpha-\lambda)^2}, \qquad t \geqq 0.$$

Substituting $-\lambda$ for λ, and remembering that $R(-\lambda; -A) = -R(\lambda; A)$, we find that

$$T(-t)x = \frac{1}{2\pi i}\int_{\Gamma_c} \frac{e^{-\lambda t} R(\lambda; A)(\alpha I + A)^2 x d\lambda}{(\alpha+\lambda)^2}, \qquad t \geqq 0.$$

If we put $-\alpha$ for α and $-t$ for t, we obtain

$$T(\varepsilon)x = \frac{1}{2\pi i}\int_{\Gamma_c} \frac{e^{\lambda t} R(\lambda; A)(\alpha I - A)^2 x d\lambda}{(\alpha-\lambda)^2}, \qquad t \leqq 0.$$

Thus $T(t)x = B(t)x$ for all real t. Q.E.D.

8 LEMMA. *If f is in $\mathscr{V}(A)$ and $|\mathscr{R}(\alpha)| > c > \omega$, then*

$$f\{A\}x = \frac{1}{2\pi i}\int_{\Gamma_c} \frac{f(\lambda)}{(\alpha-\lambda)^2} R(\lambda; A)(\alpha I - A)^2 x d\lambda, \qquad x \in \mathfrak{D}(A^2),$$

where the number c is such that $\omega < c < |\mathscr{R}(\alpha)|$ and such that f is analytic on the contour Γ_c.

PROOF. Since $f(\lambda) = \int_{-\infty}^{\infty} e^{-\lambda t}\beta(dt)$ for some β in $\mathscr{S}(A)$ by Lemma 7, we have

$$\begin{aligned}
&\frac{1}{2\pi i}\int_{\Gamma_c} \frac{f(\lambda)R(\lambda; A)(\alpha I - A)^2 x d\lambda}{(\alpha-\lambda)^2} \\
&\quad = \frac{1}{2\pi i}\int_{-\infty}^{\infty}\left\{\int_{\Gamma_c} \frac{e^{-\lambda t} R(\lambda; A)(\alpha I - A)^2 x\, d\lambda}{(\alpha-\lambda)^2}\right\}\beta(dt) \\
&\quad = \frac{1}{2\pi i}\int_{-\infty}^{\infty} T(-t)x\beta(dt) = f\{A\}x.
\end{aligned}$$

Fubini's theorem is applicable here since $\int_{-\infty}^{\infty} e^{(\omega+\varepsilon)|t|} v(\beta, dt) < \infty$ for $\varepsilon > 0$, and, as was shown in the first paragraph of the proof of Lemma 7, $|R(\lambda; A)|$ is uniformly bounded on Γ_c. Thus there is a constant $K < \infty$ such that

$$\left| \frac{e^{-\lambda t} R(\lambda; A)(\alpha I - A)^2 x}{(\alpha-\lambda)^2} \right| \leqq K e^{c|t|} (|\lambda|+1)^{-2}. \qquad \text{Q.E.D.}$$

9 **Lemma.** *If C is a closed operator with dense domain and non-void resolvent set, then $\mathfrak{D}(C^n)$ is dense in $\mathfrak{X}$ for each positive integer n.*

Proof. If $\alpha \in \varrho(C)$, then $\mathfrak{D}(C^n) = \mathfrak{D}((\alpha I - C)^n) = (R(\alpha; C))^n \mathfrak{X}$. Thus if $x^*(\mathfrak{D}(C^n)) = 0$, then $[(R(\alpha; C))^n]^* x^* = [R(\alpha; C)^*]^n x^* = 0$. The equation $R(\alpha; C)^* y^* = 0$ implies that $y^*(\mathfrak{D}(C)) = 0$, and since $\mathfrak{D}(C)$ is dense, it follows that $y^* = 0$. Hence $(R(\alpha; C)^*)^n x^* = 0$ implies $x^* = 0$. The desired conclusion follows from the Hahn-Banach theorem. Q.E.D.

10 **Theorem.** *If f is in $\mathscr{V}(A)$ and is analytic at infinity, then $f\{A\}$ is equal to the operator $f(A)$ of Definition* VII.9.3.

Proof. Let f belong to $\mathscr{V}(A)$, so that f is analytic in a strip $|\mathscr{R}(\lambda)| \leqq \omega+\varepsilon$, and let f be analytic at infinity. Then the singularities of f form two bounded sets H_1 and H_2 in the right and left half plane respectively. Choose α such that $\mathscr{R}(\alpha) > \omega+\varepsilon$ and let C_1 and C_2 be two simple closed negatively oriented Jordan curves such that C_1 contains $H_1 \cup \{\alpha\}$, C_2 contains H_2, and C_1 and C_2 do not intersect the lines $\mathscr{R}(\lambda) = \pm(\omega+c)$, $0 < c < d$.

We recall that the operator $g(A)$ is given by the formula (cf. Theorem VII.9.4)

$$g(A) = g(\infty)I + \frac{1}{2\pi i} \int_{C_1+C_2} g(\lambda) R(\lambda; A) d\lambda.$$

Let $g(\lambda) = f(\lambda)(\alpha-\lambda)^{-2}$. Then, if $x \in \mathfrak{D}(A^2)$, it is seen by using the operational calculus of Section VII.9 (cf. Theorem VII.9.5), that

$$\begin{aligned} f(A)x &= f(A) R^2(\alpha; A)(\alpha I - A)^2 x \\ &= \frac{1}{2\pi i} \int_{C_1+C_2} \frac{f(\lambda) R(\lambda; A)(\alpha I - A)^2 x}{(\alpha-\lambda)^2} d\lambda. \end{aligned}$$

Since f is bounded in a neighborhood of infinity, the displayed integrand is $O(|\lambda|^{-2})$ as $|\lambda| \to \infty$ with $|\mathscr{R}(\lambda)| \geqq c$. Thus we may replace the contour C_1+C_2 in the displayed integral by the contour Γ_c. It follows from Lemma 8 that $f(A)x = f\{A\}x$ for $x \in \mathfrak{D}(A^2)$. By Lemma 9, $\mathfrak{D}(A^2)$ is dense in $\mathfrak{X}$, and since the operators $f(A)$ and $f\{A\}$ are bounded, we conclude that $f(A) = f\{A\}$. Q.E.D.

We consider now the problem of inverting an operator of the form $f\{A\}$. The problem is complicated by the fact that the inverse transformation, if it exists, is ordinarily unbounded, and cannot be constructed directly by the operational calculus given so far. However, if a sequence of polynomials λ can be found such that $\lim_{n\to\infty} p_n(\lambda)f(\lambda) = 1$ suitably on $\sigma(A)$, it is to be expected that $\lim_{n\to\infty} p_n(A)f\{A\}x = x$ for every x in $\mathfrak{X}$. Theorem 13 below gives sufficient conditions for the validity of an inversion formula of this type.

11 LEMMA. *Let p be a polynomial in λ of degree m and let f and pf both belong to $\mathscr{V}(A)$. Then $f\{A\}\mathfrak{X} \subseteq \mathfrak{D}(A^m)$ and $p(A)f\{A\}x = (pf)\{A\}x$ for x in $\mathfrak{X}$.*

PROOF. First suppose that x is in $\mathfrak{D}(A^{m+2})$. By Theorem 6, Theorem 10, and Lemma 8,

$$R^m(\alpha; A)(pf)\{A\}x = \frac{1}{2\pi i}\int_{\Gamma_c} \frac{p(\lambda)f(\lambda)R(\lambda; A)(\alpha I - A)^2 x}{(\alpha-\lambda)^{m+2}}\, d\lambda.$$

Writing $p(\lambda) = \sum_{i=0}^m a_i(\alpha-\lambda)^i$ and using Theorem 6, Theorem 10, and Lemma 8 again, we have

$$\begin{aligned} R^m(\alpha; A)(pf)\{A\}x &= \frac{1}{2\pi i}\int_{\Gamma_c}\left\{\sum_{i=0}^m \frac{a_i}{(\alpha-\lambda)^{m+2-i}}\right\} R(\lambda; A)(\alpha I - A)^2 x\, d\lambda \\ &= \left\{\sum_{i=0}^m a_i R^{m-i}(\alpha; A)\right\}\frac{1}{2\pi i}\int_{\Gamma_c}\frac{f(\lambda)R(\lambda; A)(\alpha I - A)^2 x}{(\alpha-\lambda)^2}\, d\lambda \\ &= \{R^m(\alpha; A)\sum_{i=0}^m a_i(\alpha I - A)^i\}p(A)f\{A\}x \\ &= R^m(\alpha; A)p(A)f\{A\}x. \end{aligned}$$

Consequently, $(pf)\{A\}x = p(A)f\{A\}x$ for $x \in \mathfrak{D}(A^{m+2})$. Now let $x \in \mathfrak{X}$ and let $x_n \in \mathfrak{D}(A^{m+2})$ with $x_n \to x$ (cf. Lemma 9). Then $f\{A\}x_n \to f\{A\}x$ and $p(A)f\{A\}x_n = (pf)\{A\}x_n \to (pf)\{A\}x$. Since $p(A)$ is closed on

$\mathfrak{D}(A^m)$ (Theorem VII.9.7), it follows that $f\{A\}x \in \mathfrak{D}(A^m)$ and $p(A)f\{A\}x = (pf)\{A\}x$. Since $\mathfrak{D}(A^m)$ is dense (Lemma 1.8 and Lemma 9) this equation holds for all x in $\mathfrak{X}$. Q.E.D.

12 Definition. A sequence of polynomials $p_n(\lambda)$ will be called an *inverting sequence* for a function $f \in \mathscr{V}(A)$ if

(a) The functions $\{p_n f\}$ belong to $\mathscr{V}(A)$;

(b) $|p_n(\lambda)f(\lambda)| \leqq M$ and $\lim_{n\to\infty} p_n(\lambda)f(\lambda) = 1$ in a strip $|\mathscr{R}(\lambda)| \leqq \omega+\varepsilon$;

(c) $|(p_n f)\{A\}| \leqq M, \quad n = 1, 2, \ldots.$

13 Theorem. *If $\{p_n\}$ is an inverting sequence for a function f in $\mathscr{V}(A)$, then*

$$\lim_{n\to\infty} p_n(A)f\{A\}x = x, \qquad x \in \mathfrak{X}.$$

Proof. First let $x \in \mathfrak{D}(A^2)$. By Lemmas 8 and 11,

$$[*] \quad p_n(A)f\{A\}x = (p_n f)\{A\}x = \frac{1}{2\pi i}\int_{\Gamma_c} \frac{p_n(\lambda)f(\lambda)R(\lambda; A)(\alpha I-A)^2 x}{(\alpha-\lambda)^2}\, d\lambda$$

for $\omega < c < |\mathscr{R}(\alpha)|$ and $c-\omega$ sufficiently small depending on n. We have assumed that $|p_n(\lambda)f(\lambda)| \leqq M$ in a strip $|\mathscr{R}(\lambda)| < \omega+\varepsilon$, where ε is independent of n. In the first paragraph of the proof of Lemma 7 it was seen that $|R(\lambda, A)|$ is uniformly bounded in any half plane $\mathscr{R}(\lambda) > \omega+\varepsilon$ and in any half plane $\mathscr{R}(\lambda) < -(\omega+\varepsilon)$. Hence Cauchy's integral theorem shows that we may take c in formula [*] to be any real constant between ω and $|\mathscr{R}(\alpha)|$, provided that $c < \omega+\varepsilon$. Thus, we need not, and shall not, allow c to depend on n.

Applying Corollary III.6.16 it is seen that

$$\lim_{n\to\infty} p_n(A)f\{A\}x = \frac{1}{2\pi i}\int_{\Gamma_c} \frac{R(\lambda; A)(\alpha I-A)^2 x}{(\alpha-\lambda)^2}\, d\lambda = x,$$

since (Lemma 8) the integral on the right is $g\{A\}x$ where g is identically equal to one. Since the transformations $(p_n f)\{A\}$ are assumed to be uniformly bounded and $\mathfrak{D}(A^2)$ is dense in $\mathfrak{X}$, the theorem now follows from Theorem II.1.18. Q.E.D.

The following corollary shows that when $\mathfrak{X}$ is reflexive, an inverting sequence characterizes the range of $f\{A\}$ in a simple way.

14 Corollary. *Let $\mathfrak{X}$ be reflexive and $\{p_n\}$ be an inverting sequence for a function f in $\mathscr{V}(A)$. Then a vector x is in the range of the*

operator $f\{A\}$ *if and only if* x *is in* $\mathfrak{D}(p_n(A))$ *for each* n *and the sequence* $\{p_n(A)x\}$ *is bounded.*

PROOF. We know from the previous theorem that if $x = f\{A\}y$ then $\lim_{n\to\infty} p_n(A)x = y$ so the sequence $\{p_n(A)y\}$ is bounded. To prove the converse statement, let $\{p_n(A)x\}$ be bounded. Since $\mathfrak{X}$ is reflexive we may select a subsequence of integers $\{n_i\}$ and a vector y such that $x^*p_{n_i}(A)x \to x^*y$ for each $x^* \in \mathfrak{X}^*$ (cf. II.3.28). We shall show that $x = f\{A\}y$.

We have

$$x^*f\{A\}p_{n_i}(A)x = f\{A\}^*x^*p_{n_i}(A)x \to x^*f\{A\}y.$$

However, if $x \in \mathfrak{D}(p_{n_i}(A))$, then $T(t)x \in \mathfrak{D}(p_{n_i}(A))$ for $-\infty < t < +\infty$, by 1.7, and

$$\begin{aligned} f\{A\}p_{n_i}(A)x &= \int_{-\infty}^{\infty} T(-t)p_{n_i}(A)x\alpha(dt) \\ &= p_{n_i}(A)\int_{-\infty}^{\infty} T(-t)x\alpha(dt) = p_{n_i}(A)f\{A\}x \end{aligned}$$

by Theorem III.6.20. Thus $p_{n_i}(A)f\{A\}x \to x$ by Theorem 13, so that $x^*f\{A\}y = x^*x$ for $x^* \in \mathfrak{X}^*$, and $x = f\{A\}y$. Q.E.D.

Examples. Let $\mathfrak{X}$ be one of the spaces $C[-\infty, \infty]$ or $L_p(-\infty, \infty)$, $1 \leq p < \infty$ discussed as examples in Section 1, and let $T(t)$ be the group of translations $T(t)(x, s) = x(t+s)$, with infinitesimal generator $A = d/dt$. We recall that $\sigma(A)$ is the imaginary axis. If $f(\lambda) = \int_{-\infty}^{\infty} e^{-\lambda t}\beta(dt) \in \mathscr{V}(A)$, then the transformation $f\{A\}$ takes the familiar "convolution" form

$$[f\{A\}x](t) = \int_{-\infty}^{\infty} x(t-s)\beta(ds).$$

In particular, if β is continuous with respect to Lebesgue measure, so that $\beta(E) = \int_E F(t)dt$ for some F in $L_1(-\infty, \infty)$, then f is the bilateral Laplace transform of F and

$$[f\{A\}x](t) = \int_{-\infty}^{\infty} F(t-s)x(s)ds.$$

The Stieltjes transform

$$y(t) = \frac{1}{\pi}\int_{-\infty}^{\infty} \operatorname{sech}\frac{(t-s)}{2}\, x(s)ds$$

provides an example for the inversion Theorem 13. In this case,

$$f(\lambda) = \frac{1}{\pi}\int_{-\infty}^{\infty} e^{-\lambda t} \operatorname{sech} \frac{t}{2}\, dt = [\cos \pi\lambda]^{-1}.$$

Since

$$\cos \pi\lambda = \prod_{k=1}^{\infty}\left(1 - \frac{\lambda^2}{\left(k-\frac{1}{2}\right)^2}\right),$$

we choose

$$p_n(\lambda) = \prod_{k=1}^{n}\left(1 - \frac{\lambda^2}{\left(k-\frac{1}{2}\right)^2}\right).$$

The functions $p_n(\lambda)f(\lambda)$ have the representation

$$p_n(\lambda)f(\lambda) = \int_{-\infty}^{\infty} e^{-\lambda t}\, G_n(t)dt$$

where it may be shown that the kernels G_n are positive. Thus

$$|(p_n f)(A)| \leqq \int_{-\infty}^{\infty} G_n(t)dt = (p_n f)(0) = 1, \qquad n = 1, 2, \ldots .$$

Consequently

$$\lim_{n\to\infty} \prod_{k=1}^{n}\left(1 - \frac{\frac{d^2}{dt^2}}{\left(k-\frac{1}{2}\right)^2}\right)\frac{1}{\pi}\int_{-\infty}^{\infty} \operatorname{sech} \frac{(t-s)}{2}\, x(s)ds = x(t)$$

in the norm of any of the spaces $L_p(-\infty, +\infty)$, $1 \leqq p < \infty$, or $C[-\infty, \infty]$. A function $y \in L_p(-\infty, \infty)$, $1 \leqq p < \infty$, has the representation

$$y(t) = \frac{1}{\pi}\int_{-\infty}^{\infty} \operatorname{sech} \frac{(t-s)}{2}\, x(s)ds, \qquad x \in L_p(-\infty, \infty),$$

if and only if the sequence $\{p_n(d/dt)y(t)\}$ is bounded in norm. Reference to analytical details and further applications will be found in the notes at the end of the chapter.

3. Exercises

1 Let $T(t)$ be a semi-group of bounded operators in $\mathfrak{X}$ such that $x^*T(t)x$ is continuous on $[0, \infty)$ for each $x \in \mathfrak{X}$ and $x^* \in \mathfrak{X}^*$. Prove that $T(t)$ is strongly continuous on $[0, \infty)$.

In Exercise 2—9 below $T(t)$ is a strongly continuous semi-group of bounded operators defined on the interval $[0, \infty)$. The operator A is the infinitesimal generator of $T(t)$.

2 Suppose $\lim_{t\to 0} t^{-1}[T(t)x-x] = y$ weakly. Prove that $\lim_{t\to 0} t^{-1}[T(t)x-x] = y$ strongly, and thus $y \in \mathfrak{D}(A)$.

3 Prove that $\cap_{n=1}^{\infty}\mathfrak{D}(A^n)$ is dense in S. (Hint. Let $\mathscr{K}$ denote the class of functions K in $C^\infty(0, \infty)$ each vanishing outside a compact subset of $(0, \infty)$. Prove that if $\mathfrak{Y} = \{y|y = \int_0^\infty K(t)T(t)x, K \in \mathscr{K}, x \in \mathfrak{X}\}$, then $\mathfrak{Y}$ is dense in $\mathfrak{X}$ and $\mathfrak{Y} \subseteqq \cap_{n=1}^{\infty}\mathfrak{D}(A^n)$.)

4 (a) Prove that $\lim_{\lambda\to\infty} \lambda R(\lambda; A)x = x$ if $|\lambda| \to \infty$, $|\arg \lambda| < \pi/2$.

(b) Prove that if $x \in \mathfrak{D}(A^n)$, then

$$\lim_{\lambda\to\infty}\left\{\lambda^{n+1}R(\lambda; A)x - \sum_{k=0}^{n-1}\frac{A^k x}{\lambda^{k+1}}\right\} = A^n x$$

as $|\lambda| \to \infty$, $|\arg \lambda| < \pi/2$.

5 Prove that if $t > 0$ and $c > \omega$, where $\omega = \lim_{t\to\infty}(\log |T(t)|)/t$, then

$$T(t)x = \lim_{r\to\infty}\frac{1}{2\pi}\int_{-r}^{r}\left[1-\frac{|\tau|}{r}\right]e^{(c+i\tau)t}R(c+i\tau; A)x d\tau, \qquad x \in \mathfrak{X},$$

uniformly in each interval $[\varepsilon, 1/\varepsilon]$, $\varepsilon > 0$. Show that at $t = 0$ one obtains the limit $x/2$.

6 (a) Show that

$$[\lambda R(\lambda; A)]^k x = x + \frac{kAx}{\lambda} + \frac{1}{2\pi i}\int_{c-i\infty}^{c+i\infty}\frac{\lambda^k}{\mu^2(\lambda-\mu)^k}R(\mu; A)A^2 x d\mu$$

for $x \in \mathfrak{D}(A^2)$ and $\lambda > c > \omega$, where ω is as in the previous exercise.

(b) Prove that if $t > 0$, then

$$T(t)x = \lim_{k\to\infty}\left[\frac{k}{t}R\left(\frac{k}{t}; A\right)\right]^k x, \qquad x \in \mathfrak{X}.$$

7 Prove that if $t > 0$ and $x \in \mathfrak{D}(A^n)$, then

$$T(t)x = \sum_{k=0}^{n-1} \frac{t^k}{k!} A^k x + \frac{1}{(n-1)!} \int_0^t (t-s)^{n-1} T(s) A^n x \, ds.$$

8 Prove that if $x \in \mathfrak{D}(A^n)$, then

$$\lim_{t \to 0} t^{-n} \left[T(t)x - \sum_{k=0}^{n-1} \frac{t^k}{k!} A^k x \right] = \frac{A^n x}{n!}.$$

9 Defining the differences

$$\Delta_h^n T(t) = h^{-n} \sum_{k=0}^{n} (-1)^{n-k} \binom{n}{k} T(t+kh), \qquad t \geqq 0,$$

show that if $t \geqq s$, then

$$T(t)x = \lim_{h \to 0} \sum_{n=0}^{\infty} \frac{(t-s)^n}{n!} \Delta_h^n T(s)x, \qquad x \in \mathfrak{X},$$

the limit existing uniformly for t in any finite interval.

10 Let E_+^n be the set of points $s = [s_1, \ldots, s_n]$ in Euclidean n-space whose coordinates s_i are non-negative. Let $T(s)$, $s \in E_+^n$ be a family of bounded operators satisfying

$$T(s+t) = T(s)T(t), \quad T(0) = I, \quad \lim_{s \to t} T(s)x = T(t)x, \quad x \in \mathfrak{X}, \quad s, t \in E_+^n.$$

Let $h_i = (0, \ldots, 0, h, 0, \ldots, 0)$ be the vector in E_+^n with $h > 0$ in the i-th place and zeros elsewhere. Let $A(h, i) = h^{-1}[T(h_i) - I]$ and $\mathfrak{D}(A_i)$ be the domain of the operator defined by

$$A_i x = \lim_{h \to 0} A(h, i)x$$

where this limit exists. Prove that $\cap_{i=1}^{n} \mathfrak{D}(A_i)$ is dense in $\mathfrak{X}$ and

$$T(s)x = \lim_{h \to 0} \prod_{i=1}^{n} e^{s_i A(h, i)} x, \qquad x \in \mathfrak{X}.$$

11 (Weierstrass) Let x be a continuous complex valued function defined on a compact subset K of Euclidean n-space. Then x is the uniform limit of polynomials in the n variables. (Hint. Let $\mathfrak{X}$ be the B-space of all functions uniformly continuous on E^n with the uniform norm and suppose $K \subseteqq E_+^n$. Define $T(t)(x, s) = x(t+s)$ for $t \in E_+^n$ and apply Exercise 9.)

12 Let $\mathfrak{X} = L_p(-\pi, \pi)$, $1 \leqq p < \infty$. Show that the family of operators defined by the formula

$$T(t)(x, s) = \frac{1}{2\pi}\int_{-\pi}^{\pi} \frac{(1-e^{-t})x(s+\xi)d\xi}{1-2e^{-t}\cos\xi+e^{-2t}}, \qquad x \in \mathfrak{X},$$

defines a strongly continuous semi-group on $[0, \infty)$. Equivalently,

$$T(t)(x, s) = \sum_{-\infty}^{\infty} e^{-|n|t} x_n e^{ins}$$

if

$$x(s) \sim \sum_{-\infty}^{\infty} x_n e^{ins}, \qquad x_n = \frac{1}{2\pi}\int_{-\pi}^{\pi} x(s)e^{-ins}\, ds.$$

13 Let $\mathfrak{X} = L_p(-\pi, \pi)$, $1 \leq p < \infty$. Show that the family of operators defined by the formula

$$T(t)(x, s) = \frac{1}{2\pi}\int_{-\pi}^{\pi} \theta_3(s-\xi, t)x(\xi)d\xi, \qquad x \in \mathfrak{X},$$

where $\theta_3(s, t) = 1+2\sum_{n=1}^{\infty} e^{-n^2 t}\cos ns$, is a strongly continuous semi-group on $[0, \infty)$. Equivalently, $T(t)(x, s) = \sum_{-\infty}^{\infty} x_n e^{-n^2 t+ins}$ if $x(s) \sim \sum_{-\infty}^{\infty} x_n e^{ins}$. Show that the infinitesimal generator of the semi-group $T(t)$ is the operator A whose domain $\mathfrak{D}(A)$ consists of all periodic functions f of period 2π with absolutely continuous first derivatives and second derivatives f'' lying in L_p, and which is defined by $Af = f''$ for $f \in \mathfrak{D}(A)$.

14 Let $\mathfrak{X} = L_p(-\infty, \infty)$, $1 \leq p < \infty$. Show that the family of operators defined by the formula

$$T(t)(x, s) = \frac{t}{\pi}\int_{-\infty}^{\infty} \frac{x(s-\xi)d\xi}{\xi^2+t^2}, \qquad x \in \mathfrak{X},$$

is a strongly continuous semi-group on $[0, \infty)$.

15 Let $\mathfrak{X} = L_p(-\infty, \infty)$, $1 \leq p < \infty$. Show that the family of operators defined by the formula

$$T(t)(x, s) = \frac{1}{\sqrt{\pi t}}\int_{-\infty}^{\infty} x(s-\xi)e^{\frac{-\xi^2}{t}}\, d\xi, \qquad x \in \mathfrak{X},$$

is a strongly continuous semi-group on $[0, \infty)$. Show that the infinitesimal generator of the semi-group $T(t)$ is the operator A whose domain $\mathfrak{D}(A)$ consists of all functions $f \in L_p$ with absolutely continuous first derivatives and with second derivatives f'' lying in L_p, and which is defined by $Af = f''_4$, $f \in \mathfrak{D}(A)$.

16 Let $\mathfrak{X} = L_p(0, \infty)$, $1 \leq p < \infty$. Let $T(t)(x, s) = x(t+s)$,

$x \in \mathfrak{X}$. Show $T(t)$ is a strongly continuous semi-group whose infinitesimal generator is $A = d/ds$. Show $\sigma(A) = \{\lambda | \mathscr{R}(\lambda) \leqq 0\}$ and $\sigma(T(t)) = \{\lambda | |\lambda| \leqq 1\}$ if $t > 0$.

17 Let A and B be closed linear operators in a complex B-space $\mathfrak{X}$ with non-void resolvent sets.

(a) If $\lambda_0 \in \varrho(A)$ and $R(\lambda_0; A)$ is compact (weakly compact), then $R(\lambda; A)$ is compact (weakly compact) for each $\lambda \in \varrho(A)$.

(b) If $\mathfrak{D}(B) \subseteq \mathfrak{D}(A)$ and if for some $\lambda \in \varrho(A)$, $R(\lambda_0; A)$ is compact (weakly compact), then $R(\mu; B)$ is compact (weakly compact) for every $\mu \in \varrho(B)$.

18 Let A be the infinitesimal generator of a strongly continuous semi-group on $[0, \infty)$. If $B \in B(\mathfrak{X})$ and $R(\lambda; A)$ is compact (weakly compact) for some $\lambda \in \varrho(A)$, then $R(\mu; A+B)$ is compact (weakly compact) for every $\mu \in \varrho(A+B)$.

19 Show that the partial differential equation

$$\frac{\partial x(s, t)}{\partial t} = \frac{\partial x(s, t)}{\partial s} + \int_0^\infty e^{-us} x(u, t) du,$$

$$\lim_{t \to 0} x(s, t) = x_0(s),$$

has a solution for every function $x_0 \in C[0, \infty]$ for which $x_0' \in C[0, \infty]$.

20 Show that the partial differential equation

$$\frac{\partial x(s, t)}{\partial t} = \frac{\partial^2 x(s, t)}{\partial s^2} + e^{-s^2} \frac{\partial}{\partial s} x(s, t) + \int_{-\infty}^{\infty} \frac{x(u, t) du}{1+(s-u)^2},$$

$$\lim_{t \to 0} x(s, t) = x_0(s),$$

has a solution for every function $x_0 \in C[-\infty, \infty]$ for which $x_0', x_0'' \in C[-\infty, \infty]$.

21 Let $\mathfrak{X} = C[-\infty, \infty]$ and $A = d/ds$ with $\mathfrak{D}(A) = \{x | x'(t) \in C[-\infty, \infty]\}$. Show that the closed operator A^3 is not the infinitesimal generator of a strongly continuous semi-group of bounded operators.

4. Ergodic Theory

A fundamental mathematical question in the statistical mechanics of Gibbs and Boltzmann concerns the existence of certain types of time averages. The problem may be formulated in abstract terms as follows: The momentary state of a mechanical system is described by specifying a point in a "phase space" S. The mechanical system is assumed to be governed by the classical Hamiltonian equations and so is subject to a principle of scientific determinism whereby it is known that an initial state x will, after t seconds have elapsed, have passed into a uniquely determined new state y. Since y is uniquely determined by x and t, a function ϕ_t on S to S is defined by the equation $y = \phi_t(x)$. The flow ϕ_t is assumed to have the property that

$$\text{(i)} \qquad \phi_t(\phi_s(x)) = \phi_{t+s}(x),$$

for all points x in phase space and for all times s and t. The identity (i) may be proved for certain mechanical systems and in particular it is easy to verify if the Hamiltonian function is independent of the time. Now any numerical quantity determined by the momentary state of the mechanical system (for instance, the force exerted by the given system, assumed to be a large collection of gas molecules contained in a vessel) will be given as a real function f defined on S. If the initial state of the system is specified by the point x in S, the value of the quantity f at time t will be $f(\phi_t(x))$. The quantity $f(\phi_t(x))$ will ordinarily fluctuate very rapidly with t, as, for instance, in the case where we consider the force exerted by gas molecules on the wall of a container; since the force depends upon the number of molecules recoiling from the wall at any given instant. What is significant, and measurable in the laboratory, is not the quantity $f(\phi_t(x))$ but its average value

$$\text{(ii)} \qquad \frac{1}{T}\int_0^T f(\phi_t(x))dt$$

computed over a certain time interval $0 \leqq t \leqq T$. Ordinarily, the quantity T, which is determined by the inertial character of "macroscopic" instruments, such as pressure gauges, thermometers, etc., is very large compared to the natural rate of evolution of the mechanical system under consideration. This may be expected to be the case, for example, with a gas in a vessel where, in each second, the molecules travel thousands of feet and recoil from the wall millions of times. Thus, from the physical point of view, the time T involved in the experiment is large enough to give a good value for the limit

$$\text{(iii)} \qquad \lim_{T\to\infty} \frac{1}{T}\int_0^T f(\phi_t(x))\,dt,$$

which, in physical theories, is assumed to exist.

Thus, the mathematician is led to the problem of determining whether or not the limit (iii) exists. The next four sections are devoted to a study of this problem and various of its generalizations which arise in the theory of stochastic processes.

It should be observed that this problem is one of interest primarily to the mathematician. To the chemist or physicist the main question, and one which is still in need of a satisfactory discussion, is that of determining when the limit in (iii) is the constant space average

$$\text{(iv)} \qquad \frac{\int_S f(s)\mu(ds)}{\mu(S)}$$

taken with respect to ordinary Lebesgue measure μ in the phase space S. When a mechanical system has this property it is said to be *ergodic.* From the time of Boltzmann various physical assumptions known as *ergodic hypotheses* have been made in order to insure that the system is ergodic. In what follows we shall observe that systems which are metrically transitive, in the sense of G. D. Birkhoff, are ergodic but we shall not endeavor to discuss the difficult and important problem of determining which mechanical systems are metrically transitive.

One key observation in the limit theory to follow is the theorem of Liouville which asserts that, in a conservative mechanical system, the measure μ has the invariance property expressed by the equation

$$\text{(v)} \qquad \mu(\phi_t^{-1}(E)) = \mu(E), \qquad E \in \Sigma,$$

where Σ is the σ-field of all measurable sets in S. If, for each t, we define the linear transformation U_t by the equation

$$\text{(vi)} \qquad (U_t f)(x) = f(\phi_t(x)), \qquad f \in L_p(S, \Sigma, \mu),$$

then equations (i) and (v) show that $\{U_t\}$ is a semi-group of unitary (norm preserving) transformations in $L_p(S, \Sigma, \mu)$. The limit in (iii) may be expressed in terms of U_t as

$$\lim_{T\to\infty} \left\{\frac{1}{T}\int_0^T (U_t f)dt\right\}(x),$$

and we shall show, following J. von Neumann and F. Riesz, that this limit exists in the norm of $L_p(S, \Sigma, \mu)$, and also, following G. D. Birkhoff, that the limit exists for almost all points in the phase space S. Actually, our discussion will treat a semi-group of operators which has a considerably more general form than the one defined by equation (vi).

However, in order to explain the main concepts, it is convenient to avoid certain technical difficulties by first treating the discrete case rather than the continuous case. For example, instead of studying a continuous flow ϕ_t we study the discrete flow ϕ_n, where $\phi_{n+m} = \phi_n\phi_m$. Since $\phi_n = \phi_1^n$, we shall be studying averages of the form

$$\frac{1}{N}\sum_{n=0}^{N-1} f(\phi_1^n(x)), \qquad f \in L_p(S, \Sigma, \mu),$$

where ϕ_1 is a mapping of S into itself.

In probability theory an important class of operators, called *Markov processes*, arises in a natural manner. Let P be a non-negative function defined on $S\times\Sigma$ satisfying the conditions

(α) for each $E \in \Sigma$, $P(\cdot, E)$ is a measurable function on S;

(β) for each $s \in S$, $P(s, \cdot)$ is a countably additive measure on Σ;

(γ) $P(s, S) = 1, \quad s \in S$.

We may regard the number $P(s, E)$ as being the probability that the process sends the point s into the set E after the elapse of one unit of time. With P we may associate the operator T defined by

$$(Tf)(s) = \int_S f(t)P(s, dt),$$

where f belongs to a suitable class of functions on S. The reader will observe that the family of such operators certainly includes those of the form $f(\cdot) \to f(\phi(\cdot))$. It will be seen that the discussion which we give in the sequel includes the operators arising from a Markov process as a special case.

A word about the organization of the following sections. In Section 5, we obtain results in the discrete case concerning convergence in the mean, and in Section 6 the pointwise convergence is treated for the discrete case. These results are applied to obtain, in Section 7, corresponding mean and pointwise convergence theorems for the continuous case. Section 8 is concerned with a certain class of operators T for which the sequence of averages $\{N^{-1} \sum_{n=0}^{N-1} T^n\}$ converges in the uniform topology of operators. Finally, in Section 9, many applications and illustrations of the theory are given in the form of exercises.

5. Mean Ergodic Theorems

In this and the following three sections we shall discuss the behavior of the averages of the iterates of a linear operator and thus attempt to throw some light upon the problems of statistical mechanics and probability theory which were presented in the preceding section. However, it will not be necessary to restrict our attention to operators associated with the flows arising in statistical mechanics or the Markov operators of probability theory. In the present section conditions on an operator T in an arbitrary complex B-space $\mathfrak{X}$ will be given which are necessary and sufficient for the convergence in $\mathfrak{X}$ of the averages

$$A(n) = \frac{1}{n} \sum_{m=0}^{n-1} T^m$$

of the iterates of T. These general conditions will then be interpreted for operators in a Lebesgue space $L_p(S, \Sigma, \mu)$ which have the form that is encountered in statistical mechanics. In the present section only the strong convergence of the averages $A(n)$ is discussed. Questions concerned with the almost everywhere pointwise behavior of $A(n)f$ for a function f in L_p are postponed until the next section.

For simplicity of statement, we will assume throughout the remainder of this chapter that we are dealing with complex B-spaces. In Sections 5, 6, and 7 below, this restriction is not really essential, and the reader will see readily that a slight rewording of the proofs given (if, indeed, any is needed) will establish the results of these sections for real B-spaces also. Only in Section 8 will the restriction to complex B-spaces play any important role.

In the later sections the measure space (S, Σ, μ) will be an arbitrary one but in the mean ergodic theorems as given in Corollary 5,

Theorem 9, and Corollary 10 below, the space (S, Σ, μ) is assumed to be a finite measure space. This is done only to avoid technical complications and the changes needed to give similar results for a general measure space will be found in the exercises.

The symbol $A(n)$ will be used for the averages

$$A(n) = \frac{1}{n} \sum_{m=0}^{n-1} T^m, \qquad n \geqq 1.$$

It is desirable, at times, to emphasize the dependence of $A(n)$ upon T and, on these occasions, we shall use the symbol $A(T, n)$ instead of $A(n)$.

1 THEOREM. *Let the averages $A(n)$ of the iterates of the operator T in the B-space $\mathfrak{X}$ be bounded. Then the set of those points x in $\mathfrak{X}$ for which the sequence $\{A(n)x\}$ is convergent is a closed linear manifold consisting of all vectors x for which the set $\{A(n)x\}$ is weakly sequentially compact and $T^n x/n$ converges to zero.*

PROOF. This theorem is essentially a corollary of Theorem VII.7.4. We shall apply VII.7.4 with $f(\lambda) = 1-\lambda$, $f_n(\lambda) = (\sum_{i=0}^{n-1} \lambda^i)/n$, and with $\mathfrak{X}$ replaced by the set $\mathfrak{X}_0$ of all x in $\mathfrak{X}$ for which the sequence $\{A(n)x\} = \{f_n(T)x\}$ is weakly sequentially compact and $T^n x/n \to 0$. Since $\{A(n)\}$ is bounded, Lemma II.3.30 shows that the set of those x for which $\{A(n)x\}$ is weakly sequentially compact is a closed linear manifold. The identity

$$(*) \qquad \frac{T^n}{n} = A(n) - \frac{n-1}{n} A(n-1)$$

shows that $\{T^n/n\}$ is bounded and hence, by II.1.18, the set of x for which $T^n x/n \to 0$ is a closed linear manifold. Thus $\mathfrak{X}_0$ is a closed linear manifold and, since a continuous linear operator maps weakly convergent sequences into weakly convergent sequences, we have $T\mathfrak{X}_0 \subseteqq \mathfrak{X}_0$. Hence, we may apply VII.7.4 to the restriction of T to the B-space $\mathfrak{X}_0$. The identity $f(T)f_n(T) = (I - T^n)/n$ shows that the hypothesis, $f(T)f_n(T)x \to 0$, of Theorem VII.7.4 holds for x in $\mathfrak{X}_0$. Since f has only the simple root $\lambda = 1$ and since $f_n(1) = 1$, it follows from VII.7.4 that $\{A(n)x\}$ converges for every x in $\mathfrak{X}_0$. Conversely, if, for some x in $\mathfrak{X}$, the sequence $\{A(n)x\}$ is convergent, then it is weakly compact and the identity (*) shows that $T^n x/n \to 0$. Q.E.D.

2 Corollary. *When the strong limit $E = \lim_n A(n)$ exists it is a projection of $\mathfrak{X}$ upon the manifold $\{x | Tx = x\}$ of fixed points of T. The complementary projection has the closure of $(I-T)\mathfrak{X}$ for its range.*

Proof. Since $(I-T)A(n) = (I-T^n)/n$, the identity (*) shows that $TE = E$ and thus that $A(n)E = E$ and $E^2 = E$. Hence E is a projection upon the fixed points. Since $E(I-T) = 0$ the range $E'\mathfrak{X}$ of the complement E' of E contains $(I-T)\mathfrak{X}$. Now let x^* be a linear functional with $x^*(I-T) = 0$. Then $x^* = x^*T = x^*A(n) = x^*E$ and so $x^*E' = 0$. It follows from Corollary II.3.13 that $E'\mathfrak{X}$ is contained in the closure of $(I-T)\mathfrak{X}$. Q.E.D.

3 Corollary. *If the sequence $\{A(n)\}$ is bounded then it converges in the strong operator topology if and only if $T^n x/n$ converges to zero for x in a fundamental set and the sequence $\{A(n)x\}$ is weakly sequentially compact for x in a fundamental set.*

Proof. It was shown at the beginning of the proof of Theorem 1 that the set of x for which $T^n x/n \to 0$ is a closed linear manifold and also that the set of x for which the sequence $\{A(n)x\}$ is weakly sequentially compact is a closed linear manifold. Thus these sets are both $\mathfrak{X}$ and Theorem 1 shows that $\{A(n)x\}$ is convergent for every x in $\mathfrak{X}$. Q.E.D.

4 Corollary. *If $\mathfrak{X}$ is reflexive then the sequence $\{A(n)\}$ converges in the strong operator topology if and only if it is bounded and $\{T^n x/n\}$ converges to zero for each x in a fundamental set.*

Proof. This follows from Corollary 3 and Theorem II.3.28. Q.E.D.

5 Corollary. *Let (S, Σ, μ) be a finite measure space and let T be a linear operator in $L_1(S, \Sigma, \mu)$ which maps essentially bounded functions into essentially bounded functions. Let the averages $A(n)$ be uniformly bounded as operators in $L_\infty(S, \Sigma, \mu)$ as well as in $L_1(S, \Sigma, \mu)$. Then these averages are strongly convergent in $L_1(S, \Sigma, \mu)$ if and only if $T^n f/n$ converges to zero in $L_1(S, \Sigma, \mu)$ for every f in some fundamental set in $L_1(S, \Sigma, \mu)$.*

Proof. If f is the characteristic function of a set in Σ then, since the sequence $\{A(n)f\}$ is bounded as a sequence in $L_\infty(S, \Sigma, \mu)$, it is, by Corollary IV.8.11, weakly sequentially compact as a sequence in $L_1(S, \Sigma, \mu)$. The desired conclusion follows from Corollary 3. Q.E.D.

The preceding general results will now be applied to the case of an operator in $L_p(S, \Sigma, \mu)$ having the special form $Tf = f\varphi$, i.e., the form $(Tf)(s) = f(\varphi(s))$, where f is in $L_p(S, \Sigma, \mu)$, and where φ is a map of S into itself. Throughout the remainder of this section it is assumed that (S, Σ, μ) is a *finite* positive measure space. In order to apply Corollaries 4 and 5 to the operator T we must first discover conditions on φ which are sufficient to insure that the operator T maps $L_p(S, \Sigma, \mu)$ into $L_p(S, \Sigma, \mu)$ and has the corresponding sequence $\{A(n)\}$ bounded. Since the points in $L_p(S, \Sigma, \mu)$ are not functions but classes of equivalent functions, it is seen that T may not be regarded as being defined on $L_p(S, \Sigma, \mu)$ unless $f(\varphi(s)) = g(\varphi(s))$ almost everywhere whenever $f(s) = g(s)$ almost everywhere. The following lemma will give conditions on φ which will insure this and also allow us to infer the μ-measurability of Tf from that of f. This lemma and some of the theorems to follow will apply to functions in a class somewhat more general than $L_p(S, \Sigma, \mu)$. For this reason it is convenient to define the operator T as the mapping in the class of *all* functions with domain S which is given by the equation

$$(Tf)(s) = f(\varphi(s)), \qquad s \in S. \tag{i}$$

Thus if f is a function on S so is Tf and the range of Tf is contained in the range of f.

6 Lemma. *Let $\mathfrak{X} \neq 0$ be a complex B-space, (S, Σ, μ) a finite positive measure space and φ a map of S into itself which satisfies the conditions*

$$\varphi^{-1}(e) \in \Sigma, \qquad \mu(\varphi^{-1}(a)) = 0, \qquad \textit{if } a, e \in \Sigma \textit{ and } \mu(a) = 0. \tag{ii}$$

For every function f on S to $\mathfrak{X}$ let Tf be defined on S by the equation (i). *Then T maps μ-measurable functions into μ-measurable functions and μ-equivalent functions into μ-equivalent functions. Furthermore T is a continuous linear map of the F-space $M(S) = M(S, \Sigma, \mu, X)$ of all $\mathfrak{X}$-valued μ-measurable functions into itself.*

Proof. Since $\mu(S) < \infty$, every μ-measurable function is totally μ-measurable (Definition III.2.10) and thus, as was shown at the beginning of Section IV.11, the space $M(S)$ is an F-space. The condition (ii) shows that T maps μ-equivalent functions into μ-equivalent functions, i.e., $f(\varphi(s)) = g(\varphi(s))$ for μ-almost all s if $f(s) = g(s)$ for μ-almost all s. Thus T may be regarded as a linear map in the F-space $M(S)$ provided that Tf is μ-measurable whenever f is μ-measurable. To see that T has this property we recall (III.5.18) that the Lebesgue extension Σ^* of Σ consists of all sets of the form $e^* = e \cup a$ where e is in Σ and a is a subset of a set b in Σ with $\mu(b) = 0$. Thus, by (ii),

$$\varphi^{-1}(e^*) = \varphi^{-1}(e) \cup \varphi^{-1}(a),$$

where $\varphi^{-1}(e)$ is in Σ and $\varphi^{-1}(a)$ is a subset of the set $\varphi^{-1}(b)$ in Σ with $\mu(\varphi^{-1}(b)) = 0$. Thus $\varphi^{-1}(e^*) \in \Sigma^*$ and hence

$$\text{(iii)} \qquad \varphi^{-1}(\Sigma^*) \subseteq \Sigma^*.$$

Now let f be a μ-measurable function and let G be an open set in $\mathfrak{X}$. Then, by Lemma III.6.9, $f^{-1}(G)$ is in Σ^* and hence, by (iii),

$$(f\varphi)^{-1}(G) = \varphi^{-1}(f^{-1}(G)) \in \Sigma^*.$$

Also, since f is μ-essentially separably valued, conditions (ii) show that Tf is likewise μ-essentially separably valued. Thus, by Lemma III.6.9, Tf is μ-measurable for every μ-measurable function f on S to $\mathfrak{X}$, and T is a linear map of $M(S)$ into itself. It remains to be shown that T is a continuous linear map in $M(S)$. To do this we shall use the closed graph theorem (II.2.4). If $f_n \to f$ and $Tf_n \to g$ in $M(S)$ then, by III.6.13, there is a subsequence $\{f_{n_i}\}$ with $f_{n_i}(s) \to f(s)$ and $f_n(\varphi(s)) \to g(s)$ for every s in S except for those s in a μ-null set e. Thus $f_n(\varphi(s)) \to f(\varphi(s))$ except for s in the set $\varphi^{-1}(e)$. It follows from hypothesis (ii) that $\mu(\varphi^{-1}(e)) = 0$ and so $g(s) = f(\varphi(s))$ for μ-almost all s in S and thus $Tf = g$, which shows that T is closed. The closed graph theorem now yields the continuity of T. Q.E.D.

The next lemma will give an additional restriction on φ which will insure that T maps $L_p(S, \Sigma, \mu, X)$ into itself.

7 LEMMA. *Let $\mathfrak{X} \neq 0$ be a complex B-space, (S, Σ, μ) a finite positive measure space and φ a map of S into itself for which $\varphi^{-1}(\Sigma) \subseteq \Sigma$. Then, for $1 \leqq p < \infty$, the operator T, defined by equation* (i), *maps $L_p(S, \Sigma, \mu, \mathfrak{X})$ into itself if and only if*

$$\text{(iv)} \qquad \sup_{e \in \Sigma} \frac{\mu(\varphi^{-1}(e))}{\mu(e)} = M < \infty.$$

Furthermore, when this condition is satisfied T is a continuous linear map in $L_p(S, \Sigma, \mu, \mathfrak{X})$ whose norm $|T|$ is equal to $M^{1/p}$.

PROOF. First suppose that T maps $L_p = L_p(S, \Sigma, \mu, \mathfrak{X})$ into itself. It will be shown that T is closed and hence (II.2.4) continuous. Since T is defined on L_p, it maps μ-equivalent functions into μ-equivalent functions. Now let $a \neq 0$ be a fixed vector in $\mathfrak{X}$ and let e be a μ-null set in Σ. Then $T(a\chi_e) = a\chi_{\varphi^{-1}(e)} = 0$ in L_p. Thus $\varphi^{-1}(e)$ is a μ-null set in Σ and the conditions (ii) of Lemma 6 are satisfied. Since convergence in L_p implies convergence in measure (III.3.6), Lemma 6 shows that $Tf_n \to Tf$ in measure if $f_n \to f$ in L_p. Thus if $f_n \to f$ in L_p and $Tf_n \to g$ in L_p then $Tf = g$ and T is closed. By the closed graph theorem (II.2.4) T is continuous and thus (II.3.4) bounded. If $a \neq 0$ is a vector in $\mathfrak{X}$ and e is a set in Σ then

$$|a|\mu(\varphi^{-1}(e))^{1/p} = |T(a\chi_e)| \leqq |T||a|\mu(e)^{1/p},$$

$$\mu(\varphi^{-1}(e)) \leqq |T|^p\mu(e),$$

which shows that $M \leqq |T|^p < \infty$. Thus if T maps L_p into L_p it is continuous and (iv) holds.

Conversely, let φ have the property (iv). Let f be a μ-integrable simple function assuming its values $\alpha_1, \ldots, \alpha_n$ on the disjoint sets $e_1, \ldots, e_n$ of positive measure. Then Tf has the values $\alpha_1, \ldots, \alpha_n$ on the sets $\varphi^{-1}(e_1), \ldots, \varphi^{-1}(e_n)$ and

$$\begin{aligned}|Tf| &= \Big[\sum_{i=1}^{n} |\alpha_i|^p\mu(\varphi^{-1}(e_i))\Big]^{1/p} \\ &\leqq M^{1/p}\Big[\sum_{i=1}^{n} |\alpha_i|^p\mu(e_i)\Big]^{1/p} \\ &= M^{1/p}|f|.\end{aligned}$$

Since the μ-integrable simple functions are dense in L_p (III.3.8), this shows that T is continuous on a dense set in L_p and thus has a unique continuous extension $\tilde{T}$ defined on all of L_p and with norm $|\tilde{T}| \leqq M^{1/p}$. Now if $f_n \to f$ in L_p then $f_n \to f$ in μ-measure and Lemma 6 shows that $Tf_n \to Tf$ in measure. Since $Tf_n \to \tilde{T}f$ in L_p we have $(\tilde{T}f)(s) = (Tf)(s)$ for μ-almost all s in S and hence $T = \tilde{T}$ is a continuous linear map in L_p with norm $|T| \leqq M^{1/p}$. On the other hand it is clear from the definition of M that $|T| \geqq M^{1/p}$. Q.E.D.

8. Lemma. *Let (S, Σ, μ) be a finite positive measure space, and φ a map of S into itself for which $\varphi^{-1}(\Sigma) \subseteq \Sigma$. Suppose that there is a constant M for which*

$$\text{(v)} \qquad \frac{1}{n}\sum_{j=0}^{n-1} \mu(\varphi^{-j}(e)) \leqq M\mu(e), \qquad e \in \Sigma, \quad n = 1, 2, \ldots .$$

Then for every p with $1 \leqq p \leqq \infty$ the operator T defined by equation (i) *maps $L_p(S, \Sigma, \mu)$ into itself, and the averages $A(n)$ are uniformly bounded as operators in $L_p(S, \Sigma, \mu)$. Conversely, if the operator T defined by* (i) *maps $L_1(S, \Sigma, \mu)$, into itself, and if the averages $A(n)$ are uniformly bounded as operators in $L_1(S, \Sigma, \mu)$, then there exists a constant M in terms of which* (v) *holds.*

Proof. The symbols $|T|_p$, $|A(n)|_p$ will be used for the norms of T, $A(n)$ respectively when they are considered as operating in $L_p = L_p(S, \Sigma, \mu)$. Likewise for a μ-measurable function f, the symbol $|f|_p$ is used for $(\int_S |f(s)|^p \mu(ds))^{1/p}$ whether or not this is finite. We shall first prove the sufficiency of the condition (v). Using (v), with $n = 2$, it is seen that $\mu(\varphi^{-1}(e)) \leqq (2M-1)\mu(e)$ for every e in Σ, and thus, by Lemma 7, $|T|_p < \infty$ for each p with $1 \leqq p < \infty$. This inequality also shows that $\mu(\varphi^{-1}(e)) = 0$ if $\mu(e) = 0$ and thus

$$(*) \qquad |T|_\infty \leqq 1.$$

Now let f be a μ-integrable simple function which assumes the values $\alpha_1, \ldots, \alpha_k$ on the disjoint sets $e_1, \ldots, e_k$ respectively. Then

$$\frac{1}{n}\sum_{j=0}^{n-1} T^j f = \sum_{i=1}^{k} \alpha_i \frac{1}{n}\sum_{j=0}^{n-1} \chi_{\varphi^{-j}(e_i)},$$

and so, using (v),

$$\left|\frac{1}{n}\sum_{j=0}^{n-1} T^j f\right| \leqq M \sum_{i=1}^{k} |\alpha_i|\mu(e_i) \leqq M|f|_1,$$

and thus

$$(**) \qquad |A(n)|_1 \leqq M.$$

The Riesz convexity theorem (VI.10.11) and the inequalities (*) and (**) now give

$$\log |A(n)|_p \leqq \left(1 - \frac{1}{p}\right) \log |A(n)|_\infty + \frac{1}{p} \log |A(n)|_1$$

$$\leqq \log |A(n)|_1^{1/p} \leqq \log M^{1/p},$$

and so

$$\text{(vi)} \qquad |A(n)|_p \leqq M^{1/p}, \qquad n = 1, 2, \ldots.$$

Conversely, suppose that T maps L_1 into L_1 and that $|A(n)|_1 \leqq M$ for $n \geqq 1$. Then for e in Σ,

$$\frac{1}{n}\sum_{j=0}^{n-1} \mu(\varphi^{-j}(e)) = \int_S \frac{1}{n}\sum_{j=0}^{n-1} \chi_{\varphi^{-j}(e)}(s)\mu(ds)$$
$$= |A(n)\chi_e| \leqq M\mu(e), \qquad n = 1, 2, \ldots,$$

which completes the proof of the lemma. Q.E.D.

Remark. It is worth noting that the final inequality and the inequality (**) taken together show that

$$|A(n)|_1 = \sup_{e \in \Sigma} \frac{1}{n}\sum_{j=0}^{n-1} \mu(\varphi^{-j}(e))/\mu(e).$$

9 Theorem. (*Mean ergodic theorem*) *Let* (S, Σ, μ) *be a finite positive measure space and let* φ *be a mapping of* S *into itself with the properties that* $\varphi^{-1}(\Sigma) \subseteqq \Sigma$ *and*

$$\frac{1}{n}\sum_{j=0}^{n-1} \mu(\varphi^{-j}(e)) \leqq M\mu(e), \qquad e \in \Sigma, \quad n = 1, 2, \ldots,$$

where M *is a constant independent of* e *and* n. *Then, for every* p *with* $1 \leqq p < \infty$, *the transformation* T *defined by the equation* $(Tf)(s) = f(\varphi(s))$ *is a continuous linear map in* $L_p = L_p(S, \Sigma, \mu)$ *and the averages* $A(n)$ *are strongly convergent in* L_p.

Proof. By Lemma 8 the sequence $A(n)$ is bounded. Also if $|f(s)| \leqq K$ for s in S then $|T^n(f, s)|/n \leqq K/n$ and so the theorem follows from Corollaries 4 and 5. Q.E.D.

It should be observed that a *measure preserving transformation* φ (i.e., one for which $\mu(\varphi^{-1}(e)) = \mu(e)$ for every e in Σ) satisfies the hypothesis of the preceding theorem. It has been mentioned earlier that it is exactly this type of transformation which arises in the study of conservative mechanical systems. The transformation φ is *metrically transitive* if, for some e in Σ, $\mu(e\Delta\varphi^{-1}e) = 0$ then either $\mu(e) = 0$ or $\mu(e') = 0$.

10 Corollary. *If, in addition to the conditions of the preceding theorem, it is assumed that φ is measure preserving and metrically transitive then, for each f in $L_1(S, \Sigma, \mu)$, the limit $\lim_n A(n)f$ is almost everywhere equal to the constant space average $\int_S f(s)\mu(ds)/\mu(S)$.*

Proof. Let $g = \lim_n A(n)f$ so that, by Corollary 2, $Tg = g$. Thus, for every Borel set E we have, modulo μ-null sets,

$$\{s|g(s) \in E\} = \{s|g(\varphi s) \in E\},$$

so that $\mu((g^{-1}E)\Delta(\varphi^{-1}g^{-1}E)) = 0$. Hence, it follows from the defition of metric transitivity that either $\mu(g^{-1}E) = 0$ or $\mu(g^{-1}E) = \mu(S)$. For an arbitrary $\varepsilon > 0$ there is, since g is almost separably valued (III.6.10), a sequence $\{E_n\}$ of disjoint Borel sets each of diameter less than ε and such that $g(s) \in \bigcup E_n$ for almost all s in S. Thus for some n we have $\mu(g^{-1}E_n) = \mu(S)$ which means that $|g(s)-g(t)| < \varepsilon$ for all s and t in the complement of a null set. Therefore, for almost all s, $g(s) = g$, a constant. Since φ is measure preserving $\int_S f(s)\mu(ds) = \int_S \{A(n)f(s)\}\mu(ds) = \int_S g(s)\mu(ds) = g\mu(S)$, which completes the proof of the corollary. Q.E.D.

6. Pointwise Ergodic Theorems

The B-space $\mathfrak{X}$ of the preceding section will now be replaced by a space $L_p = L_p(S, \Sigma, \mu)$ where (S, Σ, μ) is an arbitrary positive measure space. For an operator T in L_p and a point f in L_p the symbols $(Tf)(s)$ and $T(f, s)$ will be used for the value at the place s of one of the functions in the equivalence class Tf. As before, the symbol $A(n)$ will denote the average $n^{-1} \sum_{m=0}^{n-1} T^m$ and sometimes, when it is desirable to show the dependence of $A(n)$ upon T, the symbol $A(T, n)$ will be used instead of $A(n)$. As usual the symbol $|f|_p$ will be used for the p norm

$$|f|_p = \left(\int_S |f(s)|^p \mu(ds)\right)^{1/p}, \qquad 1 \leqq p < \infty,$$

$$= \operatorname*{ess\,sup}_{s \in S} |f(s)|, \qquad p = \infty,$$

of a measurable function on S. At times it will be convenient to use the notion of the *p norm of an operator T whose domain is dense in L_1.*

This norm is defined by the equation $|T|_p = \sup |Tf|_p$ where the supremum is taken over all f in the domain of T with $|f|_p \leqq 1$. If $|T|_p < \infty$ and $p < \infty$ then T has a unique continuous extension to L_p and the same symbol T will be used for the extended operator. Most of the results of this section are concerned with an operator T for which $|T|_1 \leqq 1$ and $|T|_\infty \leqq 1$. It follows from the Riesz convexity theorem that $|T|_p \leqq 1$ for $1 \leqq p \leqq \infty$, so that T is a continuous linear map in each space L_p, $1 \leqq p < \infty$. It should be noted that the map $f \to f\varphi$, discussed in Section 5, which arises from a measure preserving transformation $s \to \varphi s$ in S, comes into this category. Another example is that of a Markov process; still another is an operator of the form $f \to a(\cdot)f(\varphi(\cdot))$ where the multiplier a is a measurable function with $|a|_\infty \leqq 1$. Our principal objectives will be to demonstrate the almost everywhere convergence of $\{A(n)f\}$ and to obtain estimates on the function $\sup_n |A(n)f|$.

1 Lemma. (*Hopf*) *Let P be a positive operator in L_∞ with $|P|_\infty \leqq 1$. Let $\{e^k\}$ be a decreasing sequence of characteristic functions with e^k zero for $k > n$. Then there is a sequence $\{g_k, 0 \leqq k\}$ in L_∞ such that if $i \geqq 0$, then*

(i) $0 \leqq g_i \leqq g_i e^{i+1}$;

(ii) $0 \leqq e^{i+2}(g_{i+1} - g_i)$;

(iii) $e^{i+1} = e^{i+1}(g_i + Pg_{i+1} + P^2 g_{i+2} + \ldots)$.

Proof. Let $g_i = 0$ for $i \geqq n$, and, for $i < n$, define g_i by downward induction from the equation

$$g_i = e^{i+1}(1 - Pg_{i+1} - P^2 g_{i+2} - \ldots), \qquad 0 \leqq i < n. \tag{1}$$

Then clearly

$$e^{i+1} = e^{i+1}(g_i + Pg_{i+1} + \ldots), \qquad 0 \leqq i, \tag{2}$$

which proves (iii). We shall next show that

$$g_i + Pg_{i+1} + P^2 g_{i+2} = + \ldots \leqq 1, \qquad 0 \leqq i. \tag{3}$$

This is clear for $i \geqq n$ while for $i < n$ it may be proved by downward induction as follows: since $(1 - e^{i+1})g_i = 0$ we have, by the induction hypothesis,

$$\begin{aligned}(1 - e^{i+1})(g_i + Pg_{i+1} + \ldots) &= (1 - e^{i+1})P(g_{i+1} + Pg_{i+2} + \ldots) \\ &\leqq (1 - e^{i+1})P1 \leqq 1 - e^{i+1},\end{aligned}$$

which, when combined with (2), proves (3). From (1) and (3) it is seen that

$$e^{i+1}g_i = g_i = e^{i+1}[1-P(g_{i+1}+Pg_{i+2}+\ldots)] \geqq 0,$$

which proves (i). To prove (ii) we shall first prove, by induction downwards, that

$$(4) \qquad (g_i+Pg_{i+1}+\ldots)-(g_{i+1}+Pg_{i+2}+\ldots) \geqq 0, \qquad 0 \leqq i.$$

Suppose that (4) holds for $i > j$. Then, by (2) and (3),

$$\begin{aligned}(5) \qquad & e^{j+1}(g_j+Pg_{j+1}+\ldots-g_{j+1}-Pg_{j+2}-\ldots) \\ & = e^{j+1}-e^{j+1}(g_{j+1}+Pg_{j+2}+\ldots) \geqq 0.\end{aligned}$$

Also, since $(1-e^{j+1})g_j = (1-e^{j+1})g_{j+1} = 0$, we see from the induction hypothesis that

$$\begin{aligned}(6) \qquad & (1-e^{j+1})(g_j+Pg_{j+1}+\ldots-g_{j+1}-Pg_{j+2}-\ldots) \\ & = (1-e^{j+1})P(g_{j+1}+Pg_{j+2}+\ldots-g_{j+2}-Pg_{j+3}-\ldots) \geqq 0.\end{aligned}$$

Thus (4) follows by adding (5) and (6). Since $e^{i+2}e^{i+1} = e^{i+2}$ we see from (1) and (4) that

$$e^{i+2}(g_{i+1}-g_i) = e^{i+2}P(g_{i+1}+Pg_{i+2}+\ldots-g_{i+2}-Pg_{i+3}-\ldots) \geqq 0,$$

which proves (ii). Q.E.D.

2 LEMMA. (*Hopf*) *Let P be a positive operator in L_1 with $|P|_1 \leqq 1$. Then, for every real f in L_1 and every positive integer n, we have $\int_E f(s)\mu(ds) \geqq 0$ where $E = \{s| \sup_{1\leqq k\leqq n} A(P, k)(f, s) \geqq 0\}$.*

PROOF. For $i > n$ let E_i be void and for $1 \leqq i \leqq n$ let

$$E_i = \{s|A(i)(f, s) \geqq 0; A(j)(f, s) < 0; j < i\},$$

where $A(n) = A(P, n)$. Let $E^i = E_i \cup E_{i+1} \cup \ldots$ so that $E^1 = E$ and let e_i and e^i be the characteristic functions of E_i and E^i respectively. Then

$$e_i \sum_{j=0}^{i-1} P^j f \geqq 0, \qquad e_i \sum_{j=0}^{m-1} P^j f \leqq 0, \qquad m < i,$$

so that

$$e_i \sum_{j=m}^{i-1} P^j f \geqq 0, \qquad m < i.$$

Upon adding these inequalities we get

$$(1) \qquad \sum_{i>m} \sum_{j=m}^{i-1} e_i P^j f = \sum_{j \geqq m} e^{j+1} P^j f \geqq 0, \qquad 0 \leqq m.$$

The preceding lemma is now applied to the adjoint operator P^* in L_∞ so that there are functions $g_i \in L_\infty$, $0 \leqq i$, with the properties (i), (ii) of that lemma and also with the property

$$(\text{iii}') \qquad e^1 = e^1(g_0 + P^* g_1 + P^{*^2} g_2 + \ldots).$$

Let $g_{-1} = 0$ so that from (1) and (ii) we have

$$0 \leqq \sum_{m \geqq 0} e^{m+1}(g_m - g_{m-1}) \sum_{j \geqq m} e^{j+1} P^j f = \sum_{m \geqq 0} \sum_{j \geqq m} e^{j+1}(g_m - g_{m-1}) P^j f$$

$$= \sum_{j \geqq 0} \sum_{m=0}^{j} (g_m - g_{m-1}) e^{j+1} P^j f = \sum_{j \geqq 0} g_j e^{j+1} P^j f = \sum_{j \geqq 0} g_j P^j f.$$

Consequently

$$0 \leqq \int_S (\sum_{j \geqq 0} g_j P^j f)(s) \mu(ds) = \int_S \sum_{j \geqq 0} (P^{*^j} g_j)(s) f(s) \mu(ds).$$

Since $P^{*^j} g_i \geqq 0$ and $f = f_1$ is negative on $S - E \subseteqq S - E_1$ we see therefore that

$$0 \leqq \int_E \sum_{j \geqq 0} (P^{*^j} g_j)(s) f(s) \mu(ds).$$

Since e^1 is the characteristic function of E it follows from (iii′) that $\sum_{j \geqq 0} (P^{*^j} g_j)(s) = 1$ for S in E. Thus $\int_E f(s) \mu(ds) \geqq 0$. Q.E.D.

3 **Lemma.** *Let A be a bounded subset of L_∞ with $\alpha A \subseteqq A$ for $|\alpha| = 1$. Then* $\sup_{f \in A} |f(\cdot)| = \sup_{f \in A} \mathscr{R} f$.

Proof. The suprema are taken in the lattice ordering of L_∞. Let $g_1 = \sup_{f \in A} |f(\cdot)|$, $g_2 = \sup_{f \in A} \mathscr{R} f$. Then, since $|f(\cdot)| \geqq \mathscr{R} f$, we have $g_1 \geqq g_2$. Hence, to prove the lemma, it will suffice to show that $g_2 \geqq |f(\cdot)|$ for every f in A. If this is false, there is an f in A, an $\varepsilon > 0$, and a measurable set e of positive measure such that $|f(s)| \geqq g_2(s) + \varepsilon$ for s in e. Let N be a positive integer with $4|f|_\infty \pi < N\varepsilon$. Since $0 \leqq \arg f(s) < 2\pi$ there is an integer $i \leqq N$ and a measurable subset e_1 of e with positive measure and such that $2\pi i/N \leqq \arg f(s) < 2\pi(i+1)/N$ for s in e_1. Since $e^{-2\pi i/N} f \in A$ we may and shall assume that $0 \leqq \arg f(s) < 2\pi/N$ for s in e_1. Then

$$|\mathscr{R}f(s)-f(s)| \leqq |f(s) \sin \arg f(s)| \leqq \frac{2\pi|f|_\infty}{N} < \frac{\varepsilon}{2}, \qquad s \in e_1,$$

and hence $\mathscr{R}f \geqq g_2+\varepsilon/2$ for s in e_1 which contradicts the definition of g_2 and proves the lemma. Q.E.D.

4 LEMMA. *Let (S, Σ, μ) be a σ-finite measure space and let T be a bounded linear operator in L_1 whose L_∞ norm is also finite. Then there is a positive linear operator P in L_1 whose L_1 and L_∞ norms do not exceed those of T and which is such that*

$$|(T^n f)(\cdot)| \leqq P^n(|f(\cdot)|), \qquad 1 \leqq n, \quad f \in L_1.$$

PROOF. Using the completeness of the lattice L_∞ (Theorem IV. 8.23), we define Pf for $0 \leqq f \in L_1 \cap L_\infty$ as $\sup \mathscr{R}Tg$ where g varies over all elements in L_∞ with $|g(\cdot)| \leqq f$. Thus, by the preceding lemma,

$$Pf = \sup_{|g(\cdot)| \leqq f} \mathscr{R}Tg = \sup_{|g(\cdot)| \leqq f} |(Tg)(\cdot)|, \qquad 0 \leqq f \in L_\infty.$$

It is clear that $Pf \geqq 0$ if $0 \leqq f \in L_1 \cap L_\infty$. Also, if $|g_1(\cdot)| \leqq f_1 \in L_1 \cap L_\infty$ and $|g_2(\cdot)| \leqq f_2 \in L_1 \cap L_\infty$, then $\mathscr{R}Tg_1 + \mathscr{R}Tg_2 = \mathscr{R}T(g_1+g_2) \leqq P(f_1+f_2)$ from which it follows that $Pf_1+Pf_2 \leqq P(f_1+f_2)$ if $0 \leqq f_1, f_2 \in L_1 \cap L_\infty$. Now let $f_1, f_2, h \in L_1 \cap L_\infty$ with $f_1, f_2 \geqq 0$ and with $|h(\cdot)| \leqq f_1+f_2$ and define $h_2 = h-h_1$ where $h_1(s) = 0$ if $h(s) = 0$ and

$$h_1(s) = \frac{h(s)}{|h(s)|} [|h(s)| \wedge f_1(s)],$$

otherwise. Then clearly $|h_1(\cdot)| \leqq f_1$, $|h_2(\cdot)| \leqq f_2$, so that $\mathscr{R}Th = \mathscr{R}Th_1+\mathscr{R}Th_2 \leqq Pf_1+Pf_2$ which proves that $P(f_1+f_2) \leqq Pf_1+Pf_2$ and establishes the equation

$$P(f_1+f_2) = Pf_1+Pf_2, \qquad 0 \leqq f_1, f_2 \in L_1 \cap L_\infty.$$

If f_1, f_2, g_1, g_2 are non-negative elements of $L_1 \cap L_\infty$ with $f_1-f_2 = g_1-g_2$ then $f_1+g_2 = f_2+g_1$ so that $Pf_1+Pg_2 = Pf_2+Pg_1$ and $Pf_1-Pf_2 = Pg_1-Pg_2$. This shows that Pf may be defined for an arbitrary real element in $L_1 \cap L_\infty$ by the equation $Pf = Pf_1-Pf_2$ where f_1 and f_2 are non-negative functions with $f = f_1-f_2$. It is clear that $P\alpha f = \alpha Pf$ for real scalars α and real functions f in $L_1 \cap L_\infty$. For an arbitrary f in $L_1 \cap L_\infty$ we let $Pf = P\mathscr{R}f + iP\mathscr{J}f$. It is evident that P is additive and that $P\alpha f = \alpha Pf$ for α real. Thus to see that P is linear it suffices to note that $iPf = Pif$, i.e., that $P\mathscr{R}(if) + iP\mathscr{J}(if) = i(P\mathscr{R}f + iP\mathscr{J}f)$ for each f in $L_1 \cap L_\infty$.

To see that $|P|_\infty \leqq |T|_\infty$ we note that $P\mathscr{R}f = \mathscr{R}Pf$ and thus, by Lemma 3,

$$
\begin{aligned}
|(Pf)(\cdot)| &= \sup_{|\alpha|=1} \mathscr{R}P\alpha f = \sup_{|\alpha|=1} P\mathscr{R}\alpha f \\
&\leqq P|f(\cdot)| = \sup_{|g(\cdot)| \leqq |f(\cdot)|} |(Tg)(\cdot)| \leqq |T|_\infty |f|_\infty,
\end{aligned}
$$

for $f \in L_1 \cap L_\infty$, which proves that $|P|_\infty \leqq |T|_\infty$.

To prove that $|P|_1 \leqq |T|_1$ we apply the method used above to the adjoint operator T^*, which is defined in $L_1^* = L_\infty$. In this way we construct an operator P_1 in L_∞ with

$$
\begin{aligned}
[1] \qquad & P_1 \geqq 0, \quad |P_1|_\infty \leqq |T^*|_\infty = |T|_1, \\
& |(T^*f)(\cdot)| \leqq P_1|f(\cdot)| = \sup_{|g(\cdot)| \leqq |f(\cdot)|} |Tg(\cdot)|, \qquad 0 \leqq f \in L_\infty.
\end{aligned}
$$

Let $0 \leqq f \in L_\infty$, $g \in L_1$, and let $\varkappa$ be the natural embedding of L_1 into L_1^{**}. Then

$$
\begin{aligned}
\int_S f(s)(\mathscr{R}Tg)(s)\mu(ds) &= \mathscr{R} \int_S f(s)(Tg)(s)\mu(ds) \\
&\leqq \left| \int_S f(s)(Tg)(s)\mu(ds) \right| = \left| \int_S (T^*f)(s)g(s)\mu(ds) \right| \\
&\leqq \int_S |(T^*f)(s)||g(s)|\mu(ds) \leqq \int_S (P_1 f)(s)|g(s)|\mu(ds) = \nu f,
\end{aligned}
$$

where $\nu = P_1^* \varkappa |g(\cdot)|$. Since (S, Σ, μ) is σ-finite the element ν in L_∞^* may be represented as a bounded additive set function (IV.8.16) whose value on a set e in Σ is $\nu(\chi_e)$ where χ_e is the characteristic function of e. In the preceding inequality we shall replace f by the characteristic function of a set e in Σ and write $\nu(e)$ instead of νf. Thus, by the preceding inequality,

$$
\left| \int_e (Tg)(s)\mu(ds) \right| \leqq \nu(e), \qquad g \in L_1, \quad e \in \Sigma.
$$

The set function $\beta(e) = \int_e (Tg)(s)\mu(ds)$ therefore has its variation $v(\beta, e) \leqq \nu(e)$. Thus, by Lemma III.2.15, $\int_e |(Tg)(s)|\mu(ds) = v(\beta, e) \leqq \nu(e)$, which shows that

$$[2] \qquad \varkappa|(Tg)(\cdot)| \leqq P_1^*\varkappa|g(\cdot)|, \qquad g \in L_1.$$

If h is a non-negative measurable function with $h \leqq f \in L_1 \cap L_\infty$ then h is also in both L_1 and L_∞. Thus an element defined by an expression of the form $\sup_{h \in A} |h(\cdot)|$ will be the same whether the supremum is taken in the complete lattice L_1 or the complete lattice L_∞ provided only that there is an element f in $L_1 \cap L_\infty$ with $|h(\cdot)| \leqq f$ for every h in A. Similarly, if $0 \leqq f \in L_1$, $\nu \in L_1^{**}$, and $0 \leqq \nu \leqq \varkappa f$ then the set function ν is μ-continuous and hence belongs to $\varkappa L_1$. Thus (cf. IV.8.24), for a set A in L_1 for which there exists a non-negative f in L_1 such that $|h(\cdot)| \leqq f$ for h in A, we have $\varkappa \sup_{h \in A} |h(\cdot)| = \sup_{h \in A} \varkappa|h(\cdot)|$. Finally we observe that a bounded set A in L_∞ and the characteristic function χ_e of a set e in Σ satisfy the relation $\sup_{h \in A} \chi_e h = \chi_e \sup_{h \in A} h$.

Using the observations of the preceding paragraph and formula [2] it is seen that for $0 \leqq f \in L_1 \cap L_\infty$ and $e \in \Sigma$ with $\mu(e) < \infty$ we have

$$\begin{aligned}
\varkappa\chi_e Pf &= \varkappa\chi_e \sup_{|g(\cdot)| \leqq f} |Tg(\cdot)| \\
&= \varkappa \sup_{|g(\cdot)| \leqq f} \chi_e|(Tg)(\cdot)| \\
&= \sup_{|g(\cdot)| \leqq f} \varkappa[\chi_e|(Tg)(\cdot)|] \\
&= \sup_{|g(\cdot)| \leqq f} \varkappa|(Tg)(\cdot)| \leqq \sup_{|g(\cdot)| \leqq f} P_1^*\varkappa|g(\cdot)| \\
&= P_1^*\varkappa f.
\end{aligned}$$

Thus, for each set e in Σ with $\mu(e) < \infty$, we have $\int_S \chi_e(s)(Pf)(s)\mu(ds) \leqq |P_1^*\varkappa f|$. Now, from [1], it is seen that $|P_1^*| = |P_1|_\infty \leqq |T|_1$, and so $\int_S (Pf)(s)\mu(ds) \leqq |T|_1|\varkappa f| = |T|_1|f|_1$ for every non-negative f in $L_1 \cap L_\infty$. Since, as shown earlier, $|(Pf)(\cdot)| \leqq P|f(\cdot)|$, we have $\int_S |(Pf)(s)|\mu(ds) \leqq |T|_1|f|_1$ for all f in $L_1 \cap L_\infty$. Thus $|P|_1 \leqq |T|_1$, and P has a unique extension to an operator (which we still call P) defined on all of L_1.

By definition,

$$P|f(\cdot)| = \sup_{|g(\cdot)| \leqq |f(\cdot)|} |(Tg)(\cdot)| \geqq |(Tf)(\cdot)|, \qquad f \in L_1 \cap L_\infty.$$

Since $L_1 \cap L_\infty$ is dense in L_1 and $|P|_1$ has been proved finite, this inequality holds for all $f \in L_1$. It follows by induction that $P^{n+1}|f(\cdot)| \geqq P|(T^n f)(\cdot)| \geqq |(T^{n+1}f)(\cdot)|$. Q.E.D.

5 LEMMA. *Let (S, Σ, μ) be a σ-finite measure space and let T be a linear operator in L_1 with $|T|_\infty \leqq 1$ and $|T|_1 \leqq 1$. Then for $1 \leqq p < \infty$ and f in $L_p(S, \Sigma, \mu)$ we have $\sup_{1 \leqq n \leqq \infty} |A(T, n)(f, s)| < \infty$ for almost all s in S.*

PROOF. We may and shall assume that $f \geqq 0$. Let a' be the complement of the set $a = \{s | f(s) \geqq 1\}$. Since $f \in L_p$ it follows that $\mu(a) < \infty$ and thus f is the sum of a summable function $f\chi_a$ and a bounded function $f\chi_{a'}$. Now, since $|T|_\infty \leqq 1$, we have $|A(n)g|_\infty \leqq |g|_\infty$ for a bounded function g in L_p. Thus in proving the lemma it may be assumed that f is in L_1. In view of Lemma 4 it may also be assumed that T is positive. Let

$$e_\infty = \{s | \sup_{1 \leqq k} A(T, k)(f, s) = \infty\}.$$

Since μ is σ-finite on S it will suffice to prove that $\mu(ee_\infty) = 0$ for every set e in Σ with $\mu(e) < \infty$. Let g be the characteristic function of ee_∞ where $\mu(e) < \infty$. If $\alpha > 0$ and Lemma 2 is applied to the function $f - \alpha g$ it is seen that

$$(*) \qquad \alpha\mu(ee_\infty c) \leqq \int_S f(s)\mu(ds)$$

where $c = \{s | \sup_{1 \leqq k \leqq n} A(T, k)(f - \alpha g, s) \geqq 0\}$. Since $|A(T, k)|_\infty \leqq 1$ it follows that $0 \leqq A(T, k)\alpha g \leqq \alpha$, and thus $c \supseteq e_n$ where $e_n = \{s | \sup_{1 \leqq k \leqq n} A(T, k)(f, s) \geqq \alpha\}$. The inequality (*) then gives

$$\alpha\mu(ee_n) \leqq \int_S f(s)\mu(ds), \qquad n = 1, 2, \ldots,$$

and, since $e_n \to e_\infty$, we have $\alpha\mu(ee_\infty) \leqq \int_S f(s)\mu(ds)$. Since this holds for every $\alpha > 0$ it follows that $\mu(ee_\infty) = 0$. Q.E.D.

6 THEOREM. (*Pointwise ergodic theorem*) *Let (S, Σ, μ) be a positive measure space and let T be a linear operator in $L_1(S, \Sigma, \mu)$ with $|T|_\infty \leqq 1$ and $|T|_1 \leqq 1$. Then, for every p with $1 \leqq p < \infty$ and every function f in $L_p(S, \Sigma, \mu)$, the limit*

$$\lim_{n \to \infty} \frac{1}{n} \sum_{m=0}^{n-1} (T^m f)(s)$$

exists for almost all s in S.

PROOF. Since the functions $T^n f$, $n = 0, 1, \ldots$, all vanish on the complement of a σ-finite set we may and shall assume that (S, Σ, μ) is σ-finite. Since $|T|_\infty \leqq 1$ and $|T|_1 \leqq 1$ it follows from the Riesz convexity theorem that $|T|_p \leqq 1$. For $1 < p < \infty$, L_p is reflexive (IV. 8.2) and thus, from Corollary 5.4, it follows that $A(T, n)f$ converges in L_p for every f in L_p. The decomposition given in Corollary 5.2 then shows that vectors of the form $h = f+(I-T)g$, with $f, g \in L_p$, $Tf = f$, and g bounded are dense in L_p. For such a vector h we have $A(T, n)h = f+(I-T^n)g/n$, and thus, for almost all s in S, we have

$$|A(T, n)(h, s)-f(s)| \leqq 2|g|_\infty/n \to 0.$$

This shows that $A(T, n)h$ converges almost everywhere for every h in a dense set in L_p and, by Lemma 5, $\sup |A(T, n)|(h, s) < \infty$ almost everywhere for every h in L_p. Thus, by Theorem IV.11.2, the sequence $A(T, n)h$ converges almost everywhere for every h in L_p. Since L_p is dense in L_1 we may apply Lemma 5 and Theorem IV.11.2 again to see that the sequence $A(T, n)f$ converges almost everywhere for every f in L_1. Q.E.D.

It is seen from the preceding theorem and from Corollary 5.4 that for $1 < p < \infty$ and for f in L_p the sequence $\{A(n)f\}$ of averages not only converges in L_p but also converges almost everywhere on S. It will next be shown that this sequence of averages is bounded by a function in L_p. The proof of this is based upon the next lemma whose statement refers to the following notation: $A(n) = A(T, n)$, $f^*(s) = \sup |A(n)(f, s)|$, $e^*(\alpha) = \{s | f^*(s) > \alpha\}$, $e(\alpha) = \{s | |f(s)| > \alpha\}$.

7 LEMMA. *Let T be an operator in L_1 with $|T|_\infty \leqq 1$, $|T|_1 \leqq 1$, and let $1 \leqq p < \infty$. Then, for every function f in L_p and every positive number α, we have*

$$\alpha\mu(e^*(2\alpha)) \leqq \int_{e(\alpha)} |f(s)|\mu(ds).$$

PROOF. The lemma will first be proved under the assumption that $\mu(S) < \infty$. If P is the positive operator associated with T as in Lemma 4, then $e^*(\alpha) \subseteqq \{s | \sup_{1 \leqq n} A(P, n)(|f(\cdot)|, s) > \alpha\}$ from which it follows that we may, without loss of generality, assume that T is a positive operator and that f is a non-negative function in L_p. Let

$g(s) = f(s)$ if $f(s) > \alpha$ and $g(s) = 0$ otherwise. Then $f \leqq g+\alpha$, $f-2\alpha \leqq g-\alpha$, and, since $|T|_\infty \leqq 1$, we have

$$A(n)f-2\alpha \leqq A(n)(f-2\alpha) \leqq A(n)(g-\alpha).$$

This shows that $e^*(2\alpha) \subseteq B \subseteq C$, where

$$B = \{s | \sup_{1 \leqq n} A(n)(f-2\alpha, s) \geqq 0\}, \qquad C = \{s | \sup_{1 \leqq n} A(n)(g-\alpha, s) \geqq 0\}.$$

By Lemma 2, $\int_C (g-\alpha)(s)\mu(ds) \geqq 0$, and so

$$\alpha\mu(e^*(2\alpha)) \leqq \alpha\mu(C) \leqq \int_C g(s)\mu(ds).$$

If s is a point such that $g(s) \neq 0$ then $0 < g(s)-\alpha = A(1)(g-\alpha, s)$ and thus s is in C. This means that g vanishes on the complement of C and hence

$$\alpha\mu(e^*(2\alpha)) \leqq \int_S g(s)\mu(ds) = \int_{e(\alpha)} f(s)\mu(ds).$$

This proves the lemma under the additional assumption that $\mu(S) < \infty$. We now allow (S, Σ, μ) to be an arbitrary positive measure space. Since all of the functions $A(n)f$ vanish on the complement of some σ-finite set, we may and shall assume that S is σ-finite. Let $\{S_n\}$ be an increasing sequence of sets whose union is S and with $\mu(S_n) < \infty$. Let $T_n = B_n T B_n$ where B_n is the operation of multiplication by the characteristic function of S_n. Let

$$e^*(m, \alpha) = \{s | \sup_{1 \leqq k \leqq m} A(k)(f, s) > \alpha\},$$

and

$$e^*(n, m, \alpha) = \{s | \sup_{1 \leqq k \leqq m} A(T_n, k)(f, s) > \alpha\}.$$

Then, by what has already been proved,

$$(*) \qquad \alpha\mu(e^*(n, m, 2\alpha)) \leqq \int_{S_n e(\alpha)} f(s)\mu(ds).$$

Since $T_n \to T$ in the strong L_p topology, the function $\sup_{1 \leqq k \leqq m} A(T_n, k)f$ approaches $\sup_{1 \leqq k \leqq m} A(k)f$ in the norm of L_p as $n \to \infty$. On the other hand, since $f \geqq 0$ and since T_n increases with n, the sequence $\{\sup_{1 \leqq k \leqq m} A(T_n,k)(f, s)\}$ is increasing in n. This shows that the sequence $\{e^*(n, m, \alpha)\}$ increases with n and has the limit $e^*(m, \alpha)$. By letting $n \to \infty$ in the inequality (*) one obtains, therefore, the inequality $\alpha\mu(e^*(m, 2\alpha)) \leqq \int_{e(\alpha)} f(s)\mu(ds)$, from which the desired inequality follows by letting $m \to \infty$. Q.E.D.

8 **Theorem.** *Let T be an operator in L_1 with $|T|_\infty \leqq 1$ and $|T|_1 \leqq 1$. For each f in L_p let*

$$f^*(s) = \sup_{1 \leqq n} \left| \frac{1}{n} \sum_{m=0}^{n-1} (T^m f)(s) \right|.$$

Then, if $1 < p < \infty$, we have

$$\int_S f^*(s)^p \mu(ds) \leqq \frac{2^p p}{p-1} \int_S |f(s)|^p \mu(ds),$$

while if $p = 1$ and $\mu(S) < \infty$ we have

$$\int_S f^*(s)\mu(ds) \leqq 2[\mu(S) + \int_S |f(s)| \log^+ |f(s)| \mu(ds)].$$

Note: The symbol $\log^+ a$ is defined for $a > 0$ to be the larger of $\log a$ and 0.

Proof. In view of Lemma 4 we may and shall assume that T and f are both non-negative. The case $1 < p$ will be treated first by using the preceding lemma as follows:

$$\begin{aligned}
\int_S f^*(s)^p \mu(ds) &= p \int_S \left(\int_0^{f^*(s)} a^{p-1}\, da \right) \mu(ds) \\
&= p \int_S \int_0^\infty a^{p-1} \chi_{e^*(a)}(s) \mu(ds) da \\
&= p \int_0^\infty a^{p-1} \mu(e^*(a)) da \\
&\leqq 2p \int_0^\infty a^{p-2} \left\{ \int_{e(a/2)} f(s)\mu(ds) \right\} da \\
&= 2p \int_S \int_0^\infty a^{p-2} f(s) \chi_{e(a/2)}(s) \mu(ds) da \\
&= 2p \int_S f(s) \left\{ \int_0^{2f(s)} a^{p-2}\, da \right\} \mu(ds) \\
&= \frac{2^p p}{p-1} \int_S f(s)^p \mu(ds).
\end{aligned}$$

The case $p = 1$ is treated similarly as follows:

$$\begin{aligned}\int_S f^*(s)\mu(ds) &= \int_S \left\{\int_0^{f^*(s)} da\right\} \mu(ds) \\ &= \int_S \int_0^\infty \chi_{e^*(a)}(s)\mu(ds)da \\ &= \int_0^\infty \mu(e^*(a))da \leqq 2\mu(S) + \int_2^\infty \mu(e^*(a))da,\end{aligned}$$

and

$$\begin{aligned}\int_2^\infty \mu(e^*(a))da &\leqq 2\int_2^\infty a^{-1}\left\{\int_{e(a/2)} f(s)\mu(ds)\right\} da \\ &= 2\int_1^\infty a^{-1}\left\{\int_{e(a)} f(s)\mu(ds)\right\} da \\ &= 2\int_1^\infty \int_S a^{-1} f(s)\chi_{e(a)}(s)\mu(ds)da \\ &= 2\int_S f(s)\left\{\int_1^{\max(1, f(s))} a^{-1} da\right\} \mu(ds) \\ &= 2\int_S f(s)\log^+ f(s)\mu(ds). \qquad \text{Q.E.D.}\end{aligned}$$

9 THEOREM. *Let T_i, $i = 1, \ldots, k$, be linear operators in L_1 with $|T_i|_\infty \leqq 1$ and $|T_i|_1 \leqq 1$, $i = 1, \ldots, k$. Then, for every f in L_p with $p > 1$, the multiple sequence*

$$(1) \qquad (n_1 \ldots n_k)^{-1} \sum_{m_1=0}^{n_1-1} \cdots \sum_{m_k=0}^{n_k-1} (T_k^{m_k} \ldots T_1^{m_1} f)(s)$$

is convergent (as $n_1, \ldots, n_k \to \infty$ independently) almost everywhere on S, as well as in the norm of L_p. Furthermore, this multiple sequence is dominated by a function in L_p.

PROOF. Let $V(n_1, \ldots, n_k) = A(T_k, n_k) \ldots A(T_1, n_1)$, so that $V(n_1, \ldots, n_k)(f, s)$ is the multiple sequence (1). Since $|T_i|_\infty \leqq 1$ and $|T_i|_1 \leqq 1$ it follows from the Riesz convexity theorem (IV.10.11) that $|T_i|_p \leqq 1$ and thus that

$$(2) \qquad |A(T_i, n_i)| \leqq 1, \qquad |V(n_1, \ldots, n_k)| \leqq 1,$$

where here, and throughout the remaining part of the proof, the norm without a subscript refers to the L_p norm. By Corollaries 5.2 and 5.4, there are projection operators E_i with

$$(3) \qquad \lim_n A(T_i, n)f = E_i f, \qquad f \in L_p, \quad i = 1, \ldots, k.$$

From (2) and (3) it follows immediately that

$$(4) \qquad \lim V(n_1, \ldots, n_k)f = E_k \ldots E_1 f, \qquad f \in L_p;$$

for indeed suppose this fact has been established for $k-1$ operators $T_2, \ldots, T_k$ and note that

$$\begin{aligned} &|A(T_k, n_k)\ldots A(T_1, n_1)f - E_k \ldots E_1 f| \\ &\qquad \leqq |A(T_k, n_k)\ldots A(T_2, n_2)\{A(T_1, n_1) - E_1\}f| \\ &\qquad + |\{A(T_k, n_k)\ldots A(T_2, n_2) - E_k \ldots E_2\}E_1 f| \\ &\qquad \leqq |\{A(T_1, n_1) - E_1\}f| + |\{A(T_k, n_k)\ldots A(T_2, n_2) - E_k \ldots E_2\}E_1 f| \end{aligned}$$

approaches zero by the induction hypothesis. By Corollary 5.2, E_i projects L_p onto the manifold $\mathfrak{N}_i$ of those f in L_p for which $T_i f = f$ and the complementary projection $E_i' = I - E_i$ projects L_p onto the closure of the manifold $(I-T_i)L_p$. Thus, if we let $\mathfrak{M}_i$ denote the set of functions of the form $(I-T_i)f$ with f bounded, we have

$$\text{(5)} \qquad \mathfrak{M}_i + \mathfrak{N}_i \text{ is dense in } L_p, \qquad 1 \leqq i \leqq k,$$

a fact which will be needed later. Now let $g = (I-T_1)f \in \mathfrak{M}_1$ with $|f(s)| \leqq K$. Then $A(T_1, n)(g, s) = n^{-1}[f(s) - T_1^n(f, s)]$ and thus $|V(n_1, \ldots, n_k)(g, s)| \leqq 2K/n_1$ for almost all s. This shows that, for almost all s in S,

$$\text{(6)} \qquad \lim V(n_1, \ldots, n_k)(g, s) = 0, \qquad g \in \mathfrak{M}_1.$$

For a function f in $\mathfrak{N}_1$ we have $T_1(f, s) = f(s)$ for almost all s in S and thus $A(T_1, n)(f, s) = f(s)$ for almost all s and all $m = 1, 2, \ldots$. Since the theorem is true for $k = 1$ (Theorem 6), we shall apply induction and assume that it has been proved for the case of $k-1$ transformations $T_2, \ldots, T_k$. The induction hypothesis yields then for f in $\mathfrak{N}_1$ the convergence almost everywhere of the multiple sequence

$$V(n_1, \ldots, n_k)(f, s) = A(T_k, n_k)\ldots A(T_2, n_2)(f, s).$$

This fact combined with (5) and (6) shows that the sequence $V(n_1, \ldots, n_k)(f, s)$ converges almost everywhere on S for every f in a set dense in L_p. It will next be shown that for every f in L_p the sequence (1) is dominated by a function in L_p. To prove this we first use Lemma 4 to find positive operators P_i, $i = 1, \ldots, k$ with $|P_i|_\infty \leqq 1$, $|P_i|_1 \leqq 1$ and such that

$$|A(T_i, n)(f, \cdot)| \leqq A(P_i, n)(|f(\cdot)|), \qquad 1 \leqq i \leqq k, \quad 1 \leqq n.$$

If $g(s) = |f(s)|$ then it follows that $|A(T_1, n_1)(f, s)| \leqq A(P_1, n_1)(g, s)$ for almost all s in S and hence that

$$|V(n_1, \ldots, n_k)(f, s)| \leqq A(P_k, n_k) \cdots A(P_1, n_1)(g, s).$$

Thus, in proving that $V(n_1, \ldots, n_k)f$ is dominated by a function in L_p we may and shall assume that $f \geqq 0$ and that $T_i \geqq 0$, $i = 1, \ldots, k$. By Theorem 8 there are functions $f_1, \ldots, f_k$ in L_p with

$$\begin{aligned} A(T_1, n)f &\leqq f_1, & n_1 &\geqq 1, \\ A(T_2, n_2)A(T_1, n_1)f &\leqq A(T_2, n_2)f_1 \leqq f_2, n_1, & n_2 &\geqq 1, \\ A(T_k, n_k) \cdots A(T_1, n_1)f &\leqq f_k, & n_1, \ldots, n_k &\geqq 1, \end{aligned}$$

which proves that the multiple sequence $V(n_1, \ldots, n_k)f$ is dominated by a function in L_p. We may now deduce the almost everywhere convergence of this sequence from Theorem IV.11.3. In that theorem let A_p be the set of all k-tuples $a = [n_1, \ldots, n_k]$ of integers with $n_i \geqq p$, $i = 1, \ldots, k$, and, for $a = [n_1, \ldots, n_k]$ in A_1, let the operator T_a of Theorem IV.11.3 be replaced by $V(n_1, \ldots, n_k)$. The hypotheses of Theorem IV.11.3 have all been verified and by that theorem we then conclude that the multiple sequence $V(n_1, \ldots, n_k)(f, s)$ converges almost everywhere on S for every f in L_p. Q.E.D.

The preceding theorem is not true if $p = 1$, but there is a k-parameter analogue of Lemma 5. This will be found at the end of the next section.

The convergence theorems of this section have assumed an operator T with $|T|_1 \leqq 1$ and $|T|_\infty \leqq 1$. In some cases, however, they may be applied to prove the convergence almost everywhere of the averages $A(n)(f, s)$ even when $|T|_1 > 1$ and indeed when $|A(n)|_1 \to \infty$. We shall conclude this section with two such theorems whose proofs will use the following lemmas.

10 Lemma. *Let (S, Σ, μ) be a finite positive measure space. A bounded sequence $\{\mu_n\}$ in the B-space of μ-continuous real or complex additive set functions on Σ is weakly sequentially compact provided that*

$$\lim_{\mu(e)=0} \limsup_{n\to\infty} |\mu_n(e)| = 0.$$

Proof. The σ-field Σ is a complete metric space $\Sigma(\mu)$ under the distance function $\varrho(a, b) = \mu(a\Delta b)$ and the functions μ_n are continuous on $\Sigma(\mu)$ (see Section III.7). Thus, for $\varepsilon > 0$, the sets

$$C_k = \{e | e \in \Sigma(\mu), |\mu_n(e)| \leqq \varepsilon, n \geqq k\}$$

are closed in $\Sigma(\mu)$ and, by hypothesis, their union contains a sphere in $\Sigma(\mu)$. Thus (I.6.7, I.6.9) there is a set $a \in \Sigma$, a number $r > 0$, and an integer k such that $e \in C_k$ if $\mu(a \Delta e) < r$. Fix b in Σ with $\mu(b) < r$ and let $b_1 = a+b$, $b_2 = a-b$. Then $b = b_1 - b_2$, $\mu(a \Delta b_1) < r$, and $\mu(a \Delta b_2) < r$. Thus $b_1, b_2 \in C_k$ and hence $|\mu_n(b)| \leqq 2\varepsilon$ for all $n \geqq k$. There is a positive number $r_1 \leqq r$ such that $|\mu_n(b)| \leqq \varepsilon$ for $1 \leqq n \leqq k$ provided that $\mu(b) < r_1$ and this proves that the functions μ_n, $n = 1, 2, \ldots$ are uniformly μ-continuous. The conclusion of the lemma now follows from Theorem IV.9.1.

11 **Lemma.** *A positive linear map P in L_1 is continuous.*

Proof. If not, there are positive elements f_n in L_1 with $|f_n| = 1$, and $|Pf_n| > n^2$. Let $f = \Sigma f_n/n^2$ so that $|Pf| \geqq |P \sum_1^m f_n/n^2| = \sum_1^m |Pf_n|/n^2 \geqq m$ for each $m \geqq 1$, which is a contradiction. Q.E.D.

The final two theorems to be discussed return to the type of operator arising in classical statistical mechanics. They are concerned with a finite positive measure space (S, Σ, μ) and a map φ of S into S. We recall (Lemma 5.6) that if $\varphi^{-1}\Sigma \subseteqq \Sigma$ (i.e., $\varphi^{-1}e \in \Sigma$ for $e \in \Sigma$) and if $\mu(\varphi^{-1}e) = 0$ if $\mu(e) = 0$, then the map $T : f \to f(\varphi)$ is a continuous linear map in the space $M(S)$ of all μ-measurable functions.

12 **Theorem.** *Let (S, Σ, μ) be a finite positive measure space and let φ be a map of S into itself with the properties*

(1) $\quad \varphi^{-1}\Sigma \subseteqq \Sigma$ *and* $\mu(\varphi^{-1}e) = 0$ *if* $\mu(e) = 0$;

(2) $$\lim_{\mu(e)=0} \limsup_{n\to\infty} \frac{1}{n} \sum_{m=0}^{n-1} \mu(\varphi^{-m}e) = 0.$$

Then, for every bounded measurable function f on S, the limit

$$\lim_{n\to\infty} \frac{1}{n} \sum_{m=0}^{n-1} f(\varphi^m s)$$

exists for almost all s in S.

Proof. We shall apply Lemma 10 to the set functions

$$\mu_n(e) = \frac{1}{n} \sum_{m=0}^{n-1} \mu(\varphi^{-m}e), \qquad e \in \Sigma, \quad n \geqq 1.$$

By Lemma III.4.13 these set functions are in the B-space $ca(\Sigma, \mu)$ of μ-continuous additive set functions on Σ. Since $|\mu_n| \leqq |\mu|$ it follows from Lemma 10 and Corollary 5.3 that the limit $m = \lim \mu_n$ exists in $ca(\Sigma, \mu)$. By Corollary 5.2, $m(\varphi^{-1}e) = m(e)$, so that the map $T : f(\cdot) \to f(\varphi(\cdot))$ as an operator in the space $L_1(S, \Sigma, m)$ has its norm $|T|_1 = 1$ (Lemma 5.7). Now let f be a bounded μ-measurable function on S. Then f is m-measurable, m-integrable, and, by Theorem 6, the limit

$$g(s) = \lim_{n\to\infty} \frac{1}{n} \sum_{m=0}^{n-1} f(\varphi^m s)$$

exists for m-almost all points in S. (It should be noted that the only way in which the requirement that f be bounded is used, is to insure the m-integrability of f. In other words the following proof shows the existence of the limit $g(s)$ for μ-almost all s in S provided that f is m-integrable.) Let $e_0 = \{s|(dm/d\mu)(s) \neq 0\}$ so that a subset of e_0 is a μ-null set if and only if it is an m-null set. Thus the limit $g(s)$ exists for μ-almost all points in e_0 and the proof of the existence of $g(s)$ will be complete if it is shown that for μ-almost all points s in e_0' we have $\varphi^m s \in e_0$ for all sufficiently large m. To see this let b be a subset of e_0' which remains in e_0' under all iterates of φ, i.e., $\varphi^m b \subseteq e_0'$ for $m \geqq 0$. Then $b \subseteq \varphi^{-m} e_0'$ and $\mu(b) \leqq \mu_n(e_0') \to m(e_0') = 0$. Thus μ-almost all points in e_0' map, under some iterate of φ, into e_0. It remains to be shown that for μ-almost all s in e_0 we have φs also in e_0; which thus will show that μ-almost all points in e_0' eventually become and remain points of e_0. Let $a \subseteq e_0$ and $\varphi a \subseteq e_0'$. Then $a \subseteq \varphi^{-1} e_0'$ and thus $m(a) \leqq m(\varphi^{-1} e_0') = m(e_0') = 0$. Since $a \subseteq e_0$ we also have $\mu(a) = 0$. Thus $g(s)$ exists for μ-almost all points in S. Q.E.D.

It is natural to ask when the preceding theorem is valid for every μ-integrable function f. An answer is given in the following theorem of C. Ryll-Nardzewski [1; I].

13 THEOREM. *Let (S, Σ, μ) be a finite measure space and φ a map of S into itself with $\varphi^{-1}\Sigma \subseteq \Sigma$ and with $\mu(\varphi^{-1}e) = 0$ for every μ-null set e. Then, for every f in $L_1(S, \Sigma, \mu)$ there is a g in $L_1(S, \Sigma, \mu)$ for which the limit*

$$g(s) = \lim_{n\to\infty} \frac{1}{n} \sum_{m=0}^{n-1} f(\varphi^m s) \tag{1}$$

exists for almost all s in S, if and only if for some constant K,

$$\limsup_{n\to\infty} \frac{1}{n} \sum_{m=0}^{n-1} \mu(\varphi^{-m} e) \leqq K\mu(e), \qquad e \in \Sigma. \tag{2}$$

PROOF. If the limit g exists and if g is in L_1 then, by Lemma 11, the map $f \to g$ is continuous and thus there is a constant K with $|g|_1 \leqq K|f|_1$. Let f be the characteristic function of a set e in Σ; then the convergence in (1) is bounded and hence in the norm of L_1. Thus

$$\int_S \frac{1}{n} \sum_{m=0}^{n-1} f(\varphi^m s)\mu(ds) = \frac{1}{n} \sum_{m=0}^{n-1} \mu(\varphi^{-m} e)$$
$$\to |g| \leqq K|f| = K\mu(e),$$

which proves (2). Now conversely if (2) holds then the set function m in the proof of the preceding theorem satisfies the inequality $m(e) \leqq K\mu(e)$ and thus the μ-integrable function f is also m-integrable. It was observed in the preceding proof that the limit $g(s)$ exists for μ-almost all s and hence it only remains to be shown that g is μ-integrable. It was observed in the preceding proof that $m(\varphi^{-1}e) = m(e)$ for e in Σ and thus the mean ergodic theorem (5.9) shows that g is in $L_1(S, \Sigma, m)$. Since $g\varphi = g$, sets of the form $e = \{s | a < g(s) \leqq b\}$ have $\varphi^{-1}e = e$. Thus, for such sets, $m(e) = \mu(e)$ which proves that $\int g(s)m(ds) = \int g(s)\mu(ds)$. Q.E.D.

7. The Ergodic Theory of Continuous Flows

In Section 4 it was shown how the ergodic hypothesis in classical statistical mechanics has led to questions involving averages of a particular one parameter semi-group of unitary operators. In this section we shall study such questions and show how they may be resolved by a reduction to the study of the averages of the iterates of a single operator. It will not be necessary to restrict the discussion to the type of semi-group arising in classical statistical mechanics as the results to be described are valid in a situation having the same degree of generality enjoyed by the discrete semi-group $\{T^n, 0 \leqq n\}$ discussed in the preceding section.

The basic notion underlying the present discussion is that of a *one parameter semi-group of bounded linear operators in a real or complex B-space.* We recall that such a semi-group is a set $\{T(t), 0 \leqq t\}$ of bounded linear operators in a B-space $\mathfrak{X}$ for which

$$\text{(i)} \qquad T(0) = I, \ T(t+u) = T(t)T(u), \qquad 0 \leqq t, u.$$

In order to be able to assign a meaning to the averages

$$\text{(ii)} \qquad A(\alpha) = \frac{1}{\alpha}\int_0^\alpha T(t)dt, \qquad 0 \leqq \alpha,$$

which are to be our basic concern in this section, it will be desirable to supplement the algebraic condition (i) by a condition concerning the analytical dependence of the semi-group upon the parameter t. We recall that the semi-group $\{T(t), 0 \leqq t\}$ is said to be *strongly continuous* if its dependence upon t is continuous in the strong operator topology, i.e., if $\lim_{t\to u} |T(t)x - T(u)x| = 0$ for each x in $\mathfrak{X}$ and each $u \geqq 0$. The semi-group is said to be *strongly measurable* if, for each x in $\mathfrak{X}$, the function $T(\cdot)x$ is measurable, with respect to Lebesgue measure, on the infinite interval $0 \leqq t$. It was observed in Lemma 1.3 that a strongly measurable semi-group is strongly continuous except possibly at the origin $t = 0$. A related concept is that of strong integrability. The semi-group is said to be *strongly integrable over every finite interval* if, for each x in $\mathfrak{X}$ the function $T(\cdot)x$ is integrable with respect to Lebesgue measure on every finite interval $0 \leqq t \leqq \alpha$. If the semi-group is strongly integrable on every finite interval we may, for each $\alpha \geqq 0$, define a bounded linear operator $A(\alpha)$ called *the average of* $\{T(t), 0 \leqq t\}$ *on the interval* $0 \leqq t \leqq \alpha$. The operator $A(\alpha)$ is defined for $\alpha = 0$ as $A(0) = I$ and for $\alpha > 0$ by the equation $A(\alpha)x = \alpha^{-1}\int_0^\alpha T(t)xdt$. To see that $A(\alpha)$ is a bounded operator we note first that the map $x \to T(\cdot)x$ of $\mathfrak{X}$ into the space L_1 of $\mathfrak{X}$ valued Lebesgue integrable functions on $[0, \alpha]$ is closed: indeed if $x_n \to x$ and $T(\cdot)x_n \to f$ in L_1 then, for some subsequence $\{x_{n_i}\}$, $T(t)x_{n_i} \to f(t)$ for almost all t in $[0, \alpha]$. Since $T(t)x_n \to T(t)x$ for every t we have $T(t)x = f(t)$ for almost all t which proves that the map $x \to T(\cdot)x$ is closed and thus (II.2.4) continuous. It follows immediately that $A(\alpha)$ is a continuous linear operator in the space $\mathfrak{X}$. It is in this sense that the integral appearing in equation (ii) is defined.

This notion of the average $A(\alpha)$ is sufficient for the statement of the first few results for they are concerned with strong limits of the type $\lim_{\alpha\to\infty} A(\alpha)x$. In most of the section, however, the B-space $\mathfrak{X}$ is a space $L_p = L_p(S, \Sigma, \mu)$ where (S, Σ, μ) is a positive measure space and we shall be concerned with the almost everywhere convergence and the almost everywhere boundedness of the averages $A(\alpha)f$ for a particular function f in L_p. Since the point $f_\alpha = A(\alpha)f$ in L_p is a class of equivalent functions and two functions f_α and g_α represent the same point if $f_\alpha(s) = g_\alpha(s)$ except on *a* μ-null set E_α which may vary with α, it is imperative to define more explicitly what is to be meant by the averages $A(\alpha)(f, s)$ of f at a point s in S. To make such a definition we suppose that the semi-group $\{T(t), 0 \leqq t\}$ is strongly integrable on every finite interval and that it operates in L_p. Then, by Lemma 1.3, for each f in L_p the set of all vectors $T(t)f$, $t \geqq 0$, is a separable subset of L_p and hence, by Lemma III.8.5, these vectors all vanish on the complement of a σ-finite subset of S. Thus we may and shall, when defining the average $A(\alpha)(f, s)$, assume that the measure space (S, Σ, μ) is σ-finite. Then, by Theorem III.11.17 there is for each f in L_p, a scalar representation $T(t)(f, s)$ of $T(t)f$ which is measurable in $[t, s]$ (i.e., measurable with respect to the product of Lebesgue measure and μ) and such a measurable representation is unique except on a set of points $[t, s]$ whose product measure is zero. Also Theorem III.11.17 shows that there is a null set $E(f)$ which may depend upon f but which is independent of t and is such that $T(\cdot)(f, s)$ is integrable on every finite t-interval for every s not in $E(f)$. Furthermore, by Theorem III.11.17, for each $\alpha > 0$ the integral

$$\text{(iii)} \qquad A(\alpha)(f, s) = \frac{1}{\alpha}\int_0^\alpha T(t)(f, s)dt, \qquad 0 < \alpha, \quad s \notin E(f),$$

is a scalar representation of the vector $A(\alpha)f$ in L_p. If we had started with another $[t, s]$-measurable representation $K(t, s)$ of $T(t)f$ then, since for almost all s, $K(t, s) = T(t)(f, s)$ for almost all t, it is seen that for almost all s, $\alpha^{-1}\int_0^\alpha K(t, s)dt = A(\alpha)(f, s)$ for all α. Thus for almost all s in S the average $A(\alpha)(f, s)$ as given in equation (iii) is uniquely defined for *all* positive values of α and furthermore furnishes an $[\alpha, s]$-measurable scalar representation of the vector $A(\alpha)f$ as defined in equation (ii).

This uniquely defined average $A(\alpha)(f, s)$ is not only a measurable function of (α, s) but, for every s not in the null set $E(f)$, it is continuous in α on the interval $0 < \alpha < \infty$. This continuity in α shows that, for almost all s in S, the number $f^*(s)$ defined by the equation

$$\text{(iv)} \qquad f^*(s) = \sup_{0<\alpha<\infty} |A(\alpha)(f, s)|$$

is the same as $\sup_{\alpha \in R} |A(\alpha)(f, s)|$ where R is the denumerable set of positive rational numbers. In view of Theorem IV.11.6 it follows that f^* is also the least upper bound of the set $\{|A(\alpha)(f, \cdot)|, 0 < \alpha\}$ when it is regarded as a set of elements in the lattice of real μ-measurable functions. Thus if $\{|A(\alpha)(f, \cdot)|, 0 < \alpha\}$ is a bounded set in this lattice, its supremum

$$\text{(v)} \qquad f^* = \sup_{0<\alpha} |A(\alpha)(f, \cdot)|$$

taken in the lattice ordering is represented by the scalar function (iv) so that no confusion concerning the interpretation of least upper bounds can arise. Similar remarks are valid in the situation discussed in Theorem 10 where averages of the product of k different semi-groups are considered.

We shall, when convenient, continue to use the notation $|U|_p$, which was introduced at the beginning of the last section, for the L_p norm of an operator U. We recall that $|U|_p = \sup |Uf|_p$ where f varies over all elements in the domain of U with $|f|_p \leqq 1$. We shall begin by a discussion of the strong convergence of the averages $A(\alpha)$ as $\alpha \to \infty$ and for this it will not be necessary to restrict the B-space upon which the semi-group acts.

1 **Theorem.** *For $\alpha \geqq 0$ let $A(\alpha)$ be the average on the interval $[0, \alpha]$ of the one parameter semi-group $\{T(t), 0 \leqq t\}$ which is assumed to be strongly integrable on every finite interval. Suppose also that*

$$(1) \lim_{n\to\infty} \frac{T(n)}{n} x = 0, \qquad x \in \mathfrak{X};$$

$$(2) |A(\alpha)| \leqq K, \qquad 0 \leqq \alpha;$$

(3) *for each x in a fundamental set in $\mathfrak{X}$ the set $\{A(\alpha)x, 0 < \alpha\}$ is weakly sequentially compact.*

Then the averages $A(\alpha)$ converge, as $\alpha \to \infty$, in the strong operator topology.

Proof. The proof will be based upon the following identity which will serve to reduce the present theorem to the discrete case already discussed. Let $n = [\alpha]$ so that $\alpha = n+r$ with $0 \leqq r < 1$. Then, for $\alpha \geqq 1$,

$$\begin{aligned} A(\alpha) &= \frac{1}{\alpha}\int_0^n T(t)dt + \frac{1}{\alpha}\int_n^{n+r} T(t)dt \\ &= \frac{1}{\alpha}\sum_{m=0}^{n-1}\int_m^{m+1} T(t)dt + \frac{1}{\alpha}\int_n^{n+r} T(t)dt \\ &= \frac{1}{\alpha}\sum_{m=0}^{n-1}\int_0^1 T(t+m)dt + \frac{1}{\alpha}\int_0^r T(t+n)dt, \end{aligned}$$

and so

$$\text{(vi)} \quad A(\alpha) = \frac{n}{\alpha}\left[\frac{1}{n}\sum_{m=0}^{n-1} T(m)A(1) + \frac{T(n)}{n} rA(r)\right].$$

For a fixed x, $rA(r)x$ is continuous in r and thus the set of all vectors $rA(r)x$ with $0 \leqq r \leqq 1$ is compact. It follows from hypothesis (1) and from Lemma IV.5.4 that

$$\lim_{n\to\infty} \frac{T(n)}{n} rA(r)x = 0$$

uniformly on $0 \leqq r \leqq 1$ and thus hypothesis (3) and the identity (vi) show that the sequence $\{n^{-1}\sum_{m=0}^{n-1} T(m)x\}$ is weakly sequentially compact for each x in a set fundamental in the closure of the range of $A(1)$. Since $T(m) = T^m(1)$, Theorem 4.1 may now be applied to the operator $T(1)$ on the space $\overline{A(1)\mathfrak{X}}$ to give the strong convergence of $A(\alpha)$. Q.E.D.

2 Corollary. *Under the hypothesis of the theorem the limit* $Ex = \lim_\alpha A(\alpha)x$ *projects* $\mathfrak{X}$ *onto the manifold* $\{x|T(t)x = x,\ 0 \leqq t\}$ *of fixed points of the semi-group and the complementary projection* $E' = I-E$ *projects* $\mathfrak{X}$ *onto the closed linear manifold determined by the ranges of all of the operators* $I-T(t)$ *with* $t \geqq 0$.

Proof. Clearly $Ex = x$ if x is a fixed point. Now the identity

$$T(u)A(\alpha) = \frac{1}{\alpha}\int_0^\alpha T(t+u)dt = \frac{1}{\alpha}\left[\int_0^\alpha T(t)dt - \int_0^u T(t)dt + \int_\alpha^{u+\alpha} T(t)dt\right]$$

shows that $|(T(u)-I)A(\alpha)| \leqq 2u\alpha^{-1}$, from which it follows that $T(u)E = E$. Thus every x in the range space of E is a fixed point of the semi-group. Since $E'(T(u)-I) = T(u)-I$, the range of E' contains the union of the ranges of all the operators $T(u)-I$. If x^* is a functional vanishing on the ranges of all the operators $T(u)-I$ then $x^* = x^*T(u) = x^*A(u) = x^*E$ and thus x^* vanishes on the range of E'. The desired conclusion follows from Corollary II.3.13. Q.E.D.

3 Corollary. *In a reflexive space, the averages $A(\alpha)$ are strongly convergent if they are bounded and if $T(n)/n$ converges to zero strongly.*

Proof. This follows from Theorem 1 and Theorem II.3.28. Q.E.D.

4 Corollary. *Let (S, Σ, μ) be a finite positive measure space and $\{T(t), 0 \leqq t\}$ a semi-group of operators in $L_1(S, \Sigma, \mu)$ which is strongly integrable on every finite interval. It is assumed that for some constant K*

$$|A(\alpha)|_1 \leqq K, \qquad |T(t)|_\infty \leqq K, \qquad 0 < \alpha, t,$$

and that $T(n)/n$ converges to zero strongly in $L_1(S, \Sigma, \mu)$. Then the averages converge strongly in $L_1(S, \Sigma, \mu)$.

Proof. If f is in $L_\infty(S, \Sigma, \mu)$ then, for all α and almost all s, $|A(\alpha)(f, s)| \leqq K|f|_\infty$ and it follows (IV.8.9) that the set $\{A(\alpha)f, 0 \leqq \alpha\}$ is weakly sequentially compact in $L_1(S, \Sigma, \mu)$. The conclusion now follows from Theorem 1. Q.E.D.

The remaining part of the section will be concerned exclusively with the case where the semi-group operates in a space $L_p = L_p(S,\Sigma,\mu)$. The basic assumption in the theorems to follow are that the semi-group $\{T(t), 0 \leqq t\}$ is a strongly measurable semi-group in L_1 for which $|T(t)|_1 \leqq 1$ and $|T(t)|_\infty \leqq 1$. There are two immediate conclusions that may be drawn from these assumptions and which will be used repeatedly without further explicit mention of them. The first of these is that $|T(t)|_p \leqq 1$ for all $p \geqq 1$. This follows from the Riesz convexity theorem (VI.10.11). Thus for each $p \geqq 1$ the semi-group $\{T(t), 0 \leqq t\}$ is a semi-group of bounded linear operators in $L_p = L_p(S, \Sigma, \mu)$. The second consequence of the assumptions which we wish to mention here is that, as a semi-group of operators in L_p with $1 \leqq p < \infty$, $\{T(t), 0 \leqq t\}$ is strongly integrable on every finite interval and thus the averages $A(\alpha)$ are defined for $0 \leqq \alpha < \infty$ as linear operators in L_p and have norms $|A(\alpha)|_p \leqq 1$. To see this let $f \in L_1 \cap L_p$ so that $T(t)(f, s)$ is $[t, s]$-measurable and, $|T(t)f|_p \leqq |f|_p$. It follows from Lemma III.11.16 that the L_p-valued function $T(\cdot)f$ is measurable. Since $1 \leqq p < \infty$,

$L_1 \cap L_p$ is dense in L_p, and thus the semi-group is strongly measurable when regarded as operating in L_p. Since $|T(t)|_p \leqq 1$ it is also strongly integrable over every finite interval and therefore $A(\alpha)$ is a bounded operator in L_p with $|A(\alpha)|_p \leqq 1$. In what follows the symbol L_p will always be used for the space $L_p(S, \Sigma, \mu)$.

5 THEOREM. *If $\{T(t), 0 \leqq t\}$ is a strongly measurable semi-group in L_1, with $|T(t)|_1 \leqq 1$ and $|T(t)|_\infty \leqq 1$, then, for every f in L_p with $1 \leqq p < \infty$, the averages $A(\alpha)(f, s)$ converge almost everywhere as $\alpha \to \infty$.*

PROOF. Let $1 \leqq p < \infty$, $f \in L_p$, and let $g(s) = \int_0^1 |T(t)(f, s)|dt$. By Theorem III.11.17 the number $g(s)$ is finite for every s not in a null set $E(f)$ and, for such s, we have $|rA(r)(f, s)| \leqq g(s)$ for every r in the interval $0 \leqq r \leqq 1$. Furthermore, since

$$
\begin{aligned}
|g|_p^p &= \int_S ds \left[\int_0^1 |T(t)(f, s)|dt\right]^p \\
&\leqq \int_S ds \int_0^1 |T(t)(f, s)|^p\, dt = \int_0^1 |T(t)f|_p^p\, dt \leqq |f|_p^p,
\end{aligned}
$$

it is seen that g is in L_p. By Lemma 6.4 there is a positive operator P in L_1 with $|P|_1 \leqq 1$, $|P|_\infty \leqq 1$, and $|T^n(f, \cdot)| \leqq P^n(|f(\cdot)|, \cdot)$, $n = 1, 2, \ldots$. By the Riesz convexity theorem we have also $|P|_p \leqq 1$, and, since $P^n = \sum_0^n P^m - \sum_0^{n-1} P^m$, it is seen from Theorem 5.5 that

$$
\left|\frac{T^n}{n}\, rA(r)(f, s)\right| \leqq \frac{P^n}{n}(g, s) \to 0
$$

almost everywhere on S and uniformly on the interval $0 \leqq r \leqq 1$. The desired conclusion now follows from the identity (vi) and Theorem 5.5. Q.E.D.

6 LEMMA. *Let (S, Σ, μ) be a positive measure space and for $t > 0$ let $T(t)$ be an operator in L_1 with $|T(t)|_1 \leqq 1$ and $|T(t)|_\infty \leqq 1$. It is assumed that either $\{T(t), 0 < t\}$ is a strongly measurable semi-group in L_1 or that the operators $T(t)$ are positive, strongly continuous in t, and satisfy the inequality $T(t)T(v) \geqq T(t+v)$. Let $1 \leqq p < \infty$ and for each u in a set U let f_u be an element of L_p such that the lattice supremum $f = \sup_{u \in U} |f_u(\cdot)|$ is also in L_p. For $\alpha > 0$ let $e(\alpha) = \{s | f(s) > \alpha\}$ and $e^*(\alpha) = \{s | f^*(s) > \alpha\}$ where*

$$f^* = \sup_{u \in U} \sup_{0<\alpha} |A(\alpha)(f_u, \cdot\ )|,$$

and $A(\alpha) = \alpha^{-1} \int_0^\alpha T(t)dt$. *Then*

$$\alpha\mu(e^*(2\alpha)) \leqq \int_{e(\alpha)} f(s)\mu(ds), \qquad \alpha > 0.$$

PROOF. By Theorem III.11.17 there is, for each u in U, a null set $E(u)$ for which the averages $A(\alpha)(f_u, s)$ exist for all non-negative α, all u in U, and all s not in the null set $E(u)$. By Lemma 1.3 the operator $T(t)$ is strongly continuous in t at every point $t > 0$ and thus, since $|T(t)|_p \leqq 1$, it follows that, for each α in the set R of non-negative rational numbers,

$$A(\alpha)f_u = \lim_{n\to\infty} \frac{1}{\alpha n!} \sum_{m=0}^{\alpha n!-1} T\left(\frac{m}{n!}\right) f_u, \qquad u \in U, \quad \alpha \in R.$$

Using Corollary III.6.13, Theorem III.11.17, and the Cantor diagonal process we may find, for each u in U, a sequence $\{n_i\}$ depending upon u and such that

$$A(\alpha)(f_u, s) = \lim_{i\to\infty} \frac{1}{\alpha n_i!} \sum_{m=0}^{\alpha n_i!-1} T\left(\frac{m}{n_i!}\right) (f_u, s)$$

for every α in R and every s not in a null set $E_1(u) \supseteq E(u)$. For u in U and $s \notin E_1(u)$ let

$$f_n^*(u, s) = \sup_{0<k<\infty} \left| \frac{1}{k} \sum_{m=0}^{k-1} T\left(\frac{m}{n!}\right) (f_u, s)\right|$$

so that for $u \in U$, $\alpha \in R$, $s \notin E_1(u)$, and $\varepsilon > 0$ there is an integer $N(u, \alpha, s, \varepsilon)$ with

$$f_n^*(u, s) \geqq |A(\alpha)(f_u, s)| - \varepsilon, \qquad n \geqq N(u, \alpha, s, \varepsilon),$$

and therefore

$$\liminf_{n\to\infty} f_n^*(u, s) \geqq |A(\alpha)(f_u, s)|, \qquad u \in U, \quad \alpha \in R, \quad s \notin E_1(u).$$

Now if $s \notin E_1(u)$ then $s \notin E(u)$ and since, for such s, $A(\alpha)(f_u, s)$ is continuous in α on the interval $0 < \alpha < \infty$, we have

$$\sup_{0<\alpha<\infty} |A(\alpha)(f_u, s)| = \sup_{\alpha \in R} |A(\alpha)(f_u, s)|$$

and thus

$$\liminf_{n\to\infty} f_n^*(u, s) \geqq \sup_{0<\alpha<\infty} |A(\alpha)(f_u, s)|, \qquad u \in U, \quad s \notin E_1(u).$$

Let

$$f^*(u, s) = \sup_{0<\alpha<\infty} |A(\alpha)(f_u, s)|,$$

so that

$$f^* = \sup_u f^*(u, \cdot),$$

and for every u in U

$$\liminf_{n\to\infty} f_n^*(u, s) \geqq f^*(u, s),$$

for almost all s in S. Now let $P(n)$ be the positive operator associated with $T(1/n!)$ as in Lemma 6.4 so that

$$\left|\frac{1}{k}\sum_{m=0}^{k-1} T\left(\frac{m}{n!}\right)(f_u, \cdot)\right| \leqq \frac{1}{k}\sum_{m=0}^{k-1} P(n)^m f, \qquad u \in U, \qquad 1 \leqq k.$$

Let

$$f_n^* = \sup_k \frac{1}{k}\sum_{m=0}^{k-1} P(n)^m f$$

so that $f_n^* \geqq f_n^*(u, \cdot)$ for every u in U and thus

$$\liminf_{n\to\infty} f_n^* \geqq f^*(u, \cdot), \qquad u \in U.$$

It follows that

$$(*) \qquad \liminf_{n\to\infty} f_n^* \geqq \sup_u f^*(u, \cdot) = f^*.$$

Thus, if $e_n^*(\alpha) = \{s | f_n^*(s) > \alpha\}$, we have $e^*(\alpha) \subseteqq \liminf_n e_n^*(\alpha)$. Lemma 6.7 and Fatou's lemma (III.6.19) show that

$$\alpha\mu(e^*(2\alpha)) \leqq \liminf_{n\to\infty} \alpha\mu(e_n^*(2\alpha)) \leqq \int_{e(\alpha)} f(s)\mu(ds). \qquad \text{Q.E.D.}$$

Similarly the inequality $\liminf_n f_n^* \geqq f^*$, together with Fatou's lemma and Theorem 6.8 show that

$$\begin{aligned}\int_S f^*(s)^p \mu(ds) &\leqq \int_S \liminf_{n\to\infty} f_n^*(s)^p \mu(ds)\\ &\leqq \liminf_{n\to\infty} \int_S f_n^*(s)^p \mu(ds)\\ &\leqq \frac{2^p p}{p-1}\int_S f(s)^p \mu(ds), \qquad 1 < p < \infty,\\ &\leqq 2[\mu(S) + \int_S f(s) \log^+ f(s)\mu(ds)], \qquad p = 1.\end{aligned}$$

This proves the following theorem.

7 THEOREM. *Let* $\{T(t), 0 \leqq t\}$ *be a strongly measurable semi-group in* L_1 *with* $|T(t)|_1 \leqq 1$ *and* $|T(t)|_\infty \leqq 1$. *For each* u *in a set* U *let* f_u *be an element of* L_p *such that the lattice supremum* $f = \sup_{u \in U} |f_u(\cdot)|$ *is also in* L_p. *If* $p = 1$ *it is also assumed that* $\mu(S) < \infty$ *and that* $\int_S f(s) \log^+ f(s) \mu(ds) < \infty$. *Then the function*

$$f^* = \sup_{u \in U} \sup_{0 < \alpha < \infty} |A(\alpha)(f_u, \cdot)|$$

is in L_p *and*

$$|f^*|_p \leqq 2\left(\frac{p}{p-1}\right)^{1/p} |f|_p, \qquad 1 < p < \infty,$$

$$\leqq 2[\mu(S) + \int_S f(s) \log^+ f(s) \mu(ds)], \qquad p = 1.$$

The final two theorems in the present section are concerned with generalizations of Theorems 5 and 7 to the case of k-parameter semi-groups. We shall first define the basic concepts associated with such semi-groups and their averages.

8 DEFINITION. *A strongly measurable k-parameter semi-group* of operators in a real or complex B-space $\mathfrak{X}$ is a set $\{T(t_1, \ldots, t_k,)$ $t_1, \ldots, t_k > 0\}$ of bounded linear operators in $\mathfrak{X}$ with the properties

(a) $T(t_1, \ldots, t_k)T(u_1, \ldots, u_k) = T(t_1+u_k, \ldots, t_k+u_k)$;

(b) $T(t_1, \ldots, t_k)x$ is Lebesgue measurable in $[t_1, \ldots, t_k]$ for each x in $\mathfrak{X}$.

9 LEMMA. *If* $T(t_1, \ldots, t_k)$ *is a strongly measurable k-parameter semi-group in a B-space* $\mathfrak{X}$ *then, for each* x *in* $\mathfrak{X}$, $T(t_1, \ldots, t_k)x$ *is continuous on the domain* $t_1, \ldots, t_k > 0$.

PROOF. The case $k = 1$ was proved in Lemma 1.3 and an examination of that proof shows that it is valid also in the general case of Lemma 9. Q.E.D.

If, besides being strongly measurable, the semi-group is strongly integrable on every set determined by inequalities of the form $0 < t_1, \ldots, t_k \leqq \alpha$ then the averages

$$A(\alpha) = \frac{1}{\alpha^k} \int_0^\alpha \cdots \int_0^\alpha T(t_1, \ldots, t_k) dt_1 \ldots dt_k$$

are defined for $\alpha > 0$ in a manner similar to that used in the one parameter case. In particular, $A(\alpha)$ is a bounded linear operator in $\mathfrak{X}$. Also, as in the one parameter case, we see that if $\mathfrak{X}$ is the space $L_p(S, \Sigma, \mu)$ and if $T(t_1, \ldots, t_k)(f, s)$ is a scalar representation of the vector $T(t_1, \ldots, t_k)f$ which is measurable as a function of $t_1, \ldots, t_k$ and s then

$$A(\alpha)(f, s) = \frac{1}{\alpha^k}\int_0^\alpha \cdots \int_0^\alpha T(t_1, \ldots, t_k)(f, s)dt_1 \ldots dt_k$$

is a scalar representation of the vector $A(\alpha)f$ and is, for almost all s in S, continuous in α. This observation shows that the function $f^*(s) = \sup_{0<\alpha} |A(\alpha)(f, s)|$ is the same as the function $\sup_{\alpha\in R} |A(\alpha)(f, s)|$, where R is the set of positive rational numbers. Thus f^* is μ-measurable and, indeed, is the same as the supremum of the set $\{|A(\alpha)(f, \cdot)|, 0 < \alpha\}$ in the lattice of μ-measurable functions on S.

Actually, in the following theorem, which is a generalization of Theorem 7, the situation discussed is somewhat more general than that of a k-parameter semi-group for here we are dealing with the averages of the form

$$\alpha_1^{-1} \ldots \alpha_k^{-1} \int_0^{\alpha_1} \cdots \int_0^{\alpha_k} T_k(t_k) \ldots T_1(t_1)dt_1 \ldots dt_k$$

where $T_1(t_1), \ldots, T_k(t_k)$ are one parameter semi-groups. The interest here lies in the fact that these one parameter semi-groups need not commute, for if they do commute then $T(t_1, \ldots, t_k) = T_k(t_k) \ldots T_1(t_1)$ is a k-parameter semi-group.

10 **Theorem.** *Let the semi-groups* $\{T_i(t), 0 \leqq t\}$, $i = 1, \ldots, k$, *be strongly measurable semi-groups in* L_1 *with* $|T_i(t)|_1 \leqq 1$ *and* $|T_i(t)|_\infty \leqq 1$. *Then, for every* f *in* L_p *with* $1 < p < \infty$, *the functions*

$$(1) \qquad \frac{1}{\alpha_1 \ldots \alpha_k}\int_0^{\alpha_k} \cdots \int_0^{\alpha_1} T_k(t_k) \ldots T_1(t_1)(f, s)dt_1 \ldots dt_k$$

approach a limit almost everywhere on S *as* $\alpha_1 \to \infty, \ldots, \alpha_k \to \infty$ *independently. The limit also exists in the norm of* L_p *and the functions* (1) *are, for* $\alpha_1, \ldots, \alpha_k > 0$, *all dominated by a function in* L_p.

Proof. If $A_i(\alpha)$ is the average of the semi-group $\{T_i(t), 0 \leqq t\}$

on the interval $(0, \alpha)$ then the multiple sequence (1) may be written as $V(\alpha_1, \ldots, \alpha_k)(f, s)$ where

$$(2) \qquad V(\alpha_1, \ldots, \alpha_k) = A_k(\alpha_k) \ldots A_1(\alpha_1).$$

Since $|T_i(t)|_\infty \leqq 1$ and $|T(t)|_1 \leqq 1$ it follows from the Riesz convexity theorem (IV.10.11) that $|T_i(t)|_p \leqq 1$ and thus that

$$(3) \qquad |A_i(\alpha)| \leqq 1, \qquad |V(\alpha_1, \ldots, \alpha_k)| \leqq 1, \qquad 1 \leqq i \leqq k,$$

where here, and throughout the proof, the norm without a subscript refers to the L_p norm.

According to Theorem 1 the limits

$$(4) \qquad E_i f = \lim_{\alpha\to\infty} A_i(\alpha)f, \qquad f \in L_p, \qquad 1 \leqq i \leqq k,$$

exist in the strong L_p topology. From (3) and (4) it follows that

$$(5) \qquad E_k \ldots E_1 f = \lim V(\alpha_1, \ldots, \alpha_k)f, \qquad f \in L_p$$

in the norm of L_p. To see this we use (4) inductively and suppose that (5) has been proved for the product of $k-1$ factors $A_k(\alpha_k), \ldots, A_2(\alpha_2)$. Then

$$\begin{aligned} |V(\alpha_1, \ldots, \alpha_k)f - E_k \ldots E_1 f| &\leqq |A_k(\alpha_k) \ldots A_2(\alpha_2)\{A_1(\alpha_1) - E_1\}f| \\ &\quad + |\{A_k(\alpha_k) \ldots A_2(\alpha_2) - E_k \ldots E_2\}E_1 f| \\ &\leqq |\{A_1(\alpha_1) - E_1\}f| + |\{A_k(\alpha_k) \ldots A_2(\alpha_2) - E_k \ldots E_2\}E_1 f| \end{aligned}$$

approaches zero by the induction hypothesis and establishes (5). It will next be shown that the functions (1) are dominated by a function in L_p which is independent of the parameters $\alpha_1, \ldots, \alpha_k$. The preceding theorem shows that this is the case if $k = 1$. Let us suppose that it has now been established in the case of $k-1$ semi-groups $T_1, \ldots, T_{k-1}$ so that the function

$$g = \sup_{0<\alpha_1, \ldots, \alpha_{k-1}<\infty} |A_{k-1}(\alpha_{k-1}) \ldots A_1(\alpha_1)(f, \cdot)|$$

is in L_p. In the preceding theorem let $u = (\alpha_1, \ldots, \alpha_{k-1})$ be a point in $k-1$ dimensional Euclidean space and let $f_u = A_{k-1}(\alpha_{k-1}) \ldots A_1(\alpha_1)f$. It follows then, from that theorem, that

$$\sup_u \sup_{0<\alpha_k<\infty} |A(\alpha_k)(f_u, \cdot)|,$$

which dominates all the functions (1), is in L_p.

Finally it will be proved that the functions (1) converge almost everywhere on S as $\alpha_1, \ldots, \alpha_k \to \infty$ independently. If $k = 1$ this follows from Theorem 5 and we suppose that it has been established in the case of $k-1$ semi-groups $T_2, \ldots, T_k$. By Corollary 2, functions of the form

$$h = f + \sum_{i=1}^{n} [I - T_1(t_i)]g_i$$

with $T_1(t)f = f$, $0 \leqq t$, and with $g_i \in L_\infty \cap L_p$ are dense in L_p. Thus, since

$$A_1(\alpha)(I - T_1(u)) = \frac{1}{\alpha}\left[\int_0^u T_1(t)dt - \int_\alpha^{u+\alpha} T(t)dt\right],$$

we have, for such a function h,

$$V(\alpha_1, \ldots, \alpha_k)h = A_k(\alpha_k) \ldots A_2(\alpha_2)f + \sum_{i=1}^{n} h_i$$

where $|h_i|_\infty \leqq 2t_i|g_i|_\infty/\alpha_1$. The induction hypothesis then shows that $Vh \to E_k \ldots E_2 f$ almost everywhere on S. To complete the proof we may apply Theorem IV.11.3. We shall let A_p be the set of all elements $a = [\alpha_1, \ldots, \alpha_k]$ with α_i rational and $\alpha_i \geqq p$, $i = 1, \ldots, k$, and let $T_a = V(a) = V(\alpha_1, \ldots, \alpha_k)$. We note that since, for almost all s, $V(\alpha_1, \ldots, \alpha_k)(f, s)$ is continuous for $0 < \alpha_1, \ldots, \alpha_k$ we have

$$\sup_{a, b \in A_p} |V(a)(f, s) - V(b)(f, s)| = \sup_{\alpha_i, \beta_j \geqq p} |V(a)(f, s) - V(b)(f, s)|,$$

where, in the second supremum $\alpha_1, \ldots, \beta_k$ vary over all real numbers greater than or equal to p. Thus, by Theorem IV.11.3,

$$\lim_{p \to \infty} \sup_{\alpha_i, \beta_i \geqq p} |V(a)(f, s) - V(b)(f, s)| = 0,$$

so that $V(a)(f, s)$ converges for almost all s in S as $\alpha_1, \ldots, \alpha_k \to \infty$ independently. Q.E.D.

If $p = 1$ the preceding theorem is no longer true because (even if $k = 1$) the averages need not be bounded by a summable function. However the averages of a k-parameter strongly measurable semi-group $\{T(t_1, \ldots, t_k), t_1, \ldots, t_k > 0\}$ in $L_1(S, \Sigma, \mu)$ with $|T(t_1, \ldots, t_k)|_1 \leqq 1$, $|T(t_1, \ldots, t_k)|_\infty \leqq 1$ do, when operating on a function in $L_1(S, \Sigma, \mu)$, converge almost everywhere. This result, which will be proved by induction on k, is based upon the following key lemma which generalizes Lemma 6 to the case of a k-parameter semi-group.

11 LEMMA. *Let* (S, Σ, μ) *be a positive measure space and let* $\{T(t_1, \ldots, t_k), t_1, \ldots, t_k > 0\}$ *be a strongly measurable semi-group of operators in* $L_1(S, \Sigma, \mu)$ *with* $|T(t_1, \ldots, t_k)|_1 \leqq 1$, $|T(t_1, \ldots, t_k)|_\infty \leqq 1$. *Let* $1 \leqq p < \infty$, $f \in L_p$, *and* $f^*(s) = \sup_{0<\alpha<\infty} |A(\alpha)(f, s)|$ *where*

$$A(\alpha) = \frac{1}{\alpha^k}\int_0^\alpha \cdots \int_0^\alpha T(t_1, \ldots, t_k)dt_1 \ldots dt_k.$$

Then there is an absolute constant c_k, *which is independent of the semi-group and independent of* f, *such that*

$$\mu(e^*(\beta)) \leqq \frac{1}{c_k\beta}\int_{e(c_k\beta)} |f(s)|\mu(ds), \qquad \beta > 0,$$

where $e^*(\beta) = \{s|f^*(s) > \beta\}$ *and* $e(\beta) = \{s||f(s)| > \beta\}$.

The proof of Lemma 11 will be an inductive one based upon its special case $k = 1$ which has been established in Lemma 6. Since the proof is a rather circuitous one, based upon three auxiliary lemmas, it will perhaps be of some help if we make a schematic display of the major logical implications involved in proving Lemma 11 from Lemma 6. To make such a diagram we will use the symbol C_k for Lemma 11 and D_k for its discrete analogue. (D_k is stated explicitly in Lemma 16). Then C_1 is Lemma 6 and D_1 is Lemma 6.7. The symbol CP_k will be used for C_k with the additional hypothesis that the operators in the semi-group are positive and DP_k will stand for the discrete analogue of CP_k. The lemmas designated by CP_k and DP_k are stated explicitly in Lemmas 13 and 14 respectively. In terms of these symbols the logical structure of the proof of Lemma 11 is as follows: $C_1 \Rightarrow CP_k \Rightarrow DP_k \Rightarrow D_k \Rightarrow C_k$. The first implication in this sequence, i.e., the reduction of CP_k to C_1 is obtained by a transformation which reduces a $2k$-parameter semi-group to a k-parameter semi-group. The device is based upon the observation that if a family $\{\varphi_u, 0 < u\}$ of functions form a semi-group under convolution, i.e., $\varphi_u * \varphi_v = \varphi_{u+v}$, and if $\varphi_u(x) = 0$ for $x \leqq 0$, then the transformation

$$S(x_1, \ldots, x_k)$$

$$= \int_0^\infty \cdots \int_0^\infty \varphi_{x_1}(t_1)\varphi_{x_1}(t_2)\varphi_{x_2}(t_3)\varphi_{x_2}(t_4) \cdots \varphi_{x_k}(t_{2k-1})\varphi_{x_k}(t_{2k})T(t_1, \ldots, t_{2k})dt_1 \ldots dt_{2k}$$

reduces the $2k$-parameter semi-group $T(t_1, \ldots, t_{2k})$ to the k-parameter semi-group $S(x_1, \ldots, x_k)$. To be sure that the semi-group $S(x_1, \ldots, x_k)$

has the desired properties the functions φ_u must be chosen with care. The following lemma defines these functions and states the properties we shall use.

12 **Lemma.** *For* $u > 0$ *let* $\varphi_u(x) = u^{-2}\varphi(xu^{-2})$ *where*

$$\varphi(x) = \frac{1}{2}\pi^{-1/2}x^{-3/2}e^{-1/4x}, \qquad x > 0$$

$$= 0, \qquad x \leqq 0.$$

Then $\varphi_u * \varphi_v = \varphi_{u+v}$ *and* $\int_{-\infty}^{\infty}\varphi(x)dx = 1$.

Proof. We have

$$\int_{-\infty}^{\infty}\varphi(x)dx = 2^{-1}\pi^{-1/2}\int_0^{\infty} x^{-3/2}e^{-1/4x}\,dx$$

$$= 2^{-1}\pi^{-1/2}\int_0^{\infty} y^{-1/2}e^{-y/4}\,dy = \pi^{-1/2}\int_0^{\infty} x^{-1/2}e^{-x}\,dx$$

$$= \pi^{-1/2}\Gamma(1/2) = 1.$$

To prove that $\varphi_u * \varphi_v = \varphi_{u+v}$ it will suffice to show that

$$(*) \qquad \int_{-\infty}^{\infty}\varphi(x)e^{-xt}\,dx = e^{-\sqrt{t}}, \qquad t > 0.$$

For, if (*) is known, then

$$\int_{-\infty}^{\infty}\varphi_u(x)e^{-xt}\,dx = \int_{-\infty}^{\infty}\varphi(x)e^{-xu^2t}\,dx = e^{-u\sqrt{t}}, \qquad t > 0,$$

and thus, by Lemma 2.4, we have

$$\int_{-\infty}^{\infty}(\varphi_u * \varphi_v)(x)e^{-xt}\,dx = e^{-(u+v)\sqrt{t}} = \int_{-\infty}^{\infty}\varphi_{u+v}(x)e^{-xt}\,dx, \qquad t > 0,$$

which, in view of Lemma 1.15, proves that $\varphi_u * \varphi_v = \varphi_{u+v}$. To establish the equation (*) let $A = \int_0^{\infty} e^{-(\alpha/u-u)^2}\,du$ for $\alpha > 0$. Then, by placing $u = 1/v$ we find that

$$A = \int_0^{\infty}\frac{\alpha}{u^2}e^{-\left(\frac{\alpha}{u}-u\right)^2}\,du.$$

Upon adding the two formulae for A it is seen that

$$A = \frac{1}{2}\int_0^{\infty}\left(1 + \frac{\alpha}{u^2}\right)e^{-\left(\frac{\alpha}{u}-u\right)^2}\,du,$$

and thus, by putting $\alpha/u - u = v$, that

$$A = \frac{1}{2}\int_{-\infty}^{\infty} e^{-v^2}\,dv = \frac{1}{2}\sqrt{\pi}.$$

Therefore

$$\begin{aligned}\frac{1}{2\sqrt{\pi}}\int_0^\infty \frac{e^{-1/4x}}{x^{3/2}}\, e^{-xt}\,dx &= \frac{e^{-\sqrt{t}}}{2\sqrt{\pi}}\int_0^\infty x^{-3/2}\, e^{-\left(\frac{1}{2\sqrt{x}}-\sqrt{xt}\right)^2}\,dx \\ &= \frac{2e^{-\sqrt{t}}}{\sqrt{\pi}}\int_0^\infty \frac{\sqrt{t}}{2y^2}\, e^{-\left(\frac{\sqrt{t}}{2y}-y\right)^2}\,dy \\ &= \frac{2e^{-\sqrt{t}}}{\sqrt{\pi}}\, A = e^{-\sqrt{t}},\end{aligned}$$

which proves (*) and completes the proof of the lemma. Q.E.D.

We shall now state and prove the lemma referred to as CP_k. For technical reasons occurring later the following lemma is stated for what might be called a positive sub-semi-group rather than for a positive semi-group. The proof of this lemma is the most involved of all the steps in the proof of Lemma 11 as outlined in the diagram: $C_1 \Rightarrow CP_k \Rightarrow DP_k \Rightarrow D_k \Rightarrow C_k$. The proof of the implication $CP_k \Rightarrow DP_k$ which is the proof of Lemma 14 is very similar to the proof of the following lemma but considerably easier in details and thus the reader may prefer to read the proof of Lemma 14 before that of Lemma 13.

13 LEMMA. *Let (S, Σ, μ) be a positive measure space and for $t_1, \ldots, t_k > 0$ let $P(t_1, \ldots, t_k)$ be a positive operator in $L_1(S, \Sigma, \mu)$ with $|P(t_1, \ldots, t_k)|_1 \leqq 1$, $|P(t_1, \ldots, t_k)|_\infty \leqq 1$ and*

$$P(t_1, \ldots, t_k)P(u_1, \ldots, u_k) \geqq P(t_1+u_1, \ldots, t_k+u_k).$$

It is assumed that the operator valued function P is strongly continuous on the domain $t_1, \ldots, t_k > 0$. Let $1 \leqq p < \infty$, $f \in L_p$, and $f^* = \sup_{0<\alpha} A(\alpha)|f(\cdot)|$ where

$$A(\alpha) = \frac{1}{\alpha^k}\int_0^\alpha \cdots \int_0^\alpha P(t_1, \ldots, t_k)\,dt_1 \ldots dt_k.$$

Then there is an absolute constant c_k, which is independent of the operators $P(t_1, \ldots, t_k)$ and independent of f, such that

$$\mu(e^*(\beta)) \leqq \frac{1}{c_k\beta}\int_{e(c_k\beta)} |f(s)|\mu(ds), \qquad \beta > 0,$$

where $e^*(\beta) = \{s|f^*(s) > \beta\}$ and $e(\beta) = \{s||f(s)| > \beta\}$.

Proof. We observe that if $1 \leqq m \leqq k$ and if $T(t_1, \ldots, t_m) = P(t_1, \ldots, t_k)$ then

$$\frac{1}{\alpha^m}\int_0^\alpha \cdots \int_0^\alpha T(t_1, \ldots, t_m)dt_1 \ldots dt_m = \frac{1}{\alpha^k}\int_0^\alpha \cdots \int_0^\alpha P(t_1, \ldots, t_k)dt_1 \ldots dt_k.$$

This shows that if the lemma is known to be true for an integer k it is also true for any integer $m \leqq k$. Thus, to prove the lemma, it will suffice to show that it is true if k is even. If $k = 1$ the lemma has already been proved (Lemma 6). We shall thus suppose that the lemma has been proved for the integer n and shall conclude its validity for the integer $k = 2n$. Let φ_u, φ be the functions defined in Lemma 12 and let

$$S(x_1, \ldots, x_n)$$
$$= \int_0^\infty \cdots \int_0^\infty \varphi_{x_1}(t_1)\varphi_{x_1}(t_2)\varphi_{x_2}(t_3)\varphi_{x_2}(t_4) \ldots \varphi_{x_n}(t_{k-1})\varphi_{x_n}(t_k)P(t_1, \ldots, t_k)dt_1 \ldots dt_k.$$

By Lemma 12 we have

$$\int_0^\infty \cdots \int_0^\infty \varphi_{x_1}(t_1)\varphi_{x_1}(t_2) \ldots \varphi_{x_n}(t_{k-1})\varphi_{x_n}(t_k)dt_1 \ldots dt_k = \left\{\int_0^\infty \varphi(t)dt\right\}^k = 1,$$

and, since $\varphi(t) > 0$, it follows that $S(x_1, \ldots, x_n)$ is a positive operator with $|S(x_1, \ldots, x_n)|_1 \leqq 1$ and $|S(x_1, \ldots, x_n)|_\infty \leqq 1$. Moreover, since $\varphi(x) = 0$ for $x < 0$, we have by Lemma 12, for $0 \leqq f \in L_1(S, \Sigma, \mu)$,

$$S(x_1, \ldots, x_n)S(y_1, \ldots, y_n)f$$
$$= \int_0^\infty \cdots \int_0^\infty \varphi_{x_1}(t_1) \ldots \varphi_{x_n}(t_k)\varphi_{y_1}(u_1) \ldots \varphi_{y_n}(u_k)P(t_1, \ldots, t_k)P(u_1, \ldots, u_k)f dt_1 \ldots du_k$$
$$\geqq \int_{-\infty}^\infty \cdots \int_{-\infty}^\infty \varphi_{x_1}(t_1) \ldots \varphi_{x_n}(t_k)\varphi_{y_1}(u_1) \ldots \varphi_{y_n}(u_k)P(t_1+u_1, \ldots, t_k+u_k)f dt_1 \ldots du_k$$
$$= \int_{-\infty}^\infty \cdots \int_{-\infty}^\infty \varphi_{x_1}(v_1-w_1)\varphi_{y_1}(w_1) \ldots \varphi_{x_n}(v_k-w_k)\varphi_{y_n}(w_k)P(v_1, \ldots, v_k)f dv_1 \ldots dw_k$$
$$= \int_{-\infty}^\infty \cdots \int_{-\infty}^\infty \varphi_{x_1+y_1}(v_1) \ldots \varphi_{x_n+y_n}(v_k)P(v_1, \ldots, v_k)f dv_1 \ldots dv_k$$
$$= S(x_1+y_1, \ldots, x_n+y_n)f.$$

Thus $S(x_1, \ldots, x_n)$ satisfies the hypothesis of Lemma 13 and so by our inductive hypothesis there is a constant c_n such that

$$\mu\{s|f^{**}(s) > \beta\} \leqq \frac{1}{c_n\beta}\int_{e(c_n\beta)} |f(s)|\mu(ds), \qquad \beta > 0,$$

where

$$f^{**} = \sup_{0<\alpha} \frac{1}{\alpha^n}\int_0^\alpha \cdots \int_0^\alpha S(x_1, \ldots, x_n)|f(\cdot)|dx_1 \ldots dx_n.$$

To complete the proof of the lemma it is therefore sufficient to establish the existence of an absolute constant d_k such that $f^{**} \geqq d_k f^*$. For this it will suffice to show that for each $\alpha > 0$ there is an $\alpha_1 = \alpha_1(\alpha) > 0$ such that

$$[*] \quad A(S, \alpha_1)f = \frac{1}{\alpha_1^n}\int_0^{\alpha_1} \cdots \int_0^{\alpha_1} S(x_1, \ldots, x_n)f dx_1 \ldots dx_n$$
$$\geqq d_k \frac{1}{\alpha^k}\int_0^\alpha \cdots \int_0^\alpha P(t_1, \ldots, t_k)f dt_1 \ldots dt_k,$$

for every non-negative function f in L_1. Since

$$A(S, \alpha_1)f$$
$$= 2^{-k}\pi^{-k/2}\int_0^\infty \cdots \int_0^\infty h(\alpha_1, t_1, t_2)h(\alpha_1, t_3, t_4)\ldots h(\alpha_1, t_{k-1}, t_k)P(t_1, \ldots, t_k)f dt_1 \ldots dt_k,$$

where

$$h(\alpha_1, t, v) = \frac{1}{\alpha_1}\int_0^{\alpha_1} \varphi_u(t)\varphi_u(v)du$$
$$= \frac{1}{\alpha_1}(tv)^{-3/2}\int_0^{\alpha_1} u^2 e^{-u^2\left(\frac{1}{4t}+\frac{1}{4v}\right)} du,$$

it is clearly sufficient to show that $h(\alpha_1, t, v) \geqq \delta/\alpha^2$ for all $t, v \leqq \alpha$; for then [*] will follow with $d_k = (\delta/4\pi)^{k/2}$. This will be verified for the function $\alpha_1 = \alpha^{1/2}$, i.e., we shall show that

$$(t_1t_2)^{-3/2}\int_0^{\alpha_1} u^2 e^{-u^2\left(\frac{1}{4t_1}+\frac{1}{4t_2}\right)} du > \delta, \qquad t_1, t_2 \leqq \alpha_1^2.$$

By placing $u = \alpha_1 v$, $t_1 = \alpha_1^2 s_1$, $t_2 = \alpha_1^2 s_2$ this inequality transforms into

$$(s_1s_2)^{-3/2}\int_0^1 v^2 e^{-v^2\left(\frac{1}{4s_1}+\frac{1}{4s_2}\right)} dv > \delta, \qquad 0 < s_1, s_2 \leqq 1.$$

Let $G(a) = \int_0^1 v^2 e^{-v^2 a} dv$ so that we wish to show

$$[**] \qquad (s_1 s_2)^{-3/2} G\left(\frac{1}{4s_1} + \frac{1}{4s_2}\right) > \delta, \qquad 0 < s_1, s_2 \leqq 1.$$

Since G is positive and continuous it suffices to prove [**] in the case where either s_1 or s_2 is near zero. We have

$$G(a) = \int_0^1 v^2 e^{-v^2 a}\, dv = \int_0^1 u^{1/2} e^{-ua}\, du$$

$$= \frac{a^{-3/2}}{2} \int_0^a v^{1/2} e^{-v}\, dv,$$

and so, for some positive k, $G(a) \geqq ka^{-3/2}$ for all sufficiently large values of a. Thus

$$(s_1 s_2)^{-3/2} G\left(\frac{1}{4s_1} + \frac{1}{4s_2}\right) \geqq 8k(s_1+s_2)^{-3/2}$$

if either s_1 or s_2 is near zero. This establishes [**] and completes the proof of the lemma. Q.E.D.

14 Lemma. *Let (S, Σ, μ) be a positive measure space and for every set of k non-negative integers $i_1, \ldots, i_k$ let $P(i_1, \ldots, i_k)$ be a positive operator in $L_1(S, \Sigma, \mu)$ with $|P(i_1, \ldots, i_k)|_1 \leqq 1$, $|P(i_1, \ldots, i_k)|_\infty \leqq 1$, and $P(i_1, \ldots, i_k)P(j_1, \ldots, j_k) \geqq P(i_1+j_1, \ldots, i_k+j_k)$. For f in L_p let $f^* = \sup_n A(n)|f(\cdot)|$ where*

$$A(n) = \frac{1}{n^k} \sum_{i_k=0}^{n-1} \cdots \sum_{i_k=0}^{n-1} P(i_1, \ldots, i_k).$$

Then there is a constant c_k, independent of the operators $P(i_1, \ldots, i_k)$ and the function f, such that

$$\mu(\{s | f^*(s) > \beta\}) \leqq \frac{1}{c_k \beta} \int_{e(c_k\beta)} |f(s)| \mu(ds), \qquad \beta > 0.$$

Proof. For $x_1, \ldots, x_k > 0$ define the operator

$$S(x_1, \ldots, x_k) = e^{-(x_1+\ldots+x_k)} \sum_{i_1=0}^{\infty} \cdots \sum_{i_k=0}^{\infty} \frac{x_1^{i_1} \cdots x_k^{i_1}}{i_1! \ldots i_k!} P(i_1, \ldots, i_k).$$

Since the sum of the coefficients in this series is

$$e^{-(x_1+\ldots+x_k)} \sum_{i_1=0}^{\infty} \cdots \sum_{i_k=0}^{\infty} \frac{x_1^{i_1} \cdots x_k^{i_k}}{i_1! \ldots i_k!} = 1,$$

we see that $|S(x_1, \ldots, x_k)|_1 \leqq 1$, $|S(x_1, \ldots, x_k)|_\infty \leqq 1$, and thus, by the Riesz convexity theorem, that $|S(x_1, \ldots, x_k)|_p \leqq 1$. It is clear also that $S(x_1, \ldots, x_k)$ is a positive operator in L_p. Also

$$\begin{aligned} &S(x_1, \ldots, x_k)S(y_1, \ldots, y_k) \\ &= e^{-(x_1+y_1+\ldots+x_k+y_k)} \sum \frac{x_1^{i_1}\ldots x_k^{i_k} y_1^{j_1} \ldots y_k^{j_k}}{i_1!\ldots i_k!\ldots j_1!\ldots j_k!} P(i_1, \ldots, i_k)\, P(j_1, \ldots, j_k) \\ &\geqq e^{-(x_1+y_1\ldots+x_k+y_k)} \sum \frac{x_1^{i_1}\ldots y_k^{j_k}}{i_1!\ldots j_k!} P(i_1+j_1, \ldots, i_k+j_k), \end{aligned}$$

where Σ is taken over all non-negative integers $i_1, \ldots, i_k, j_1, \ldots, j_k$. Thus

$$\begin{aligned} &S(x_1, \ldots, x_k)S(y_1, \ldots, y_k) \\ &\geqq e^{-(x_1+y_1+\ldots+x_k+y_k)} \sum_{m_1=0}^{\infty} \cdots \sum_{m_k=0}^{\infty} \frac{T(m_1, \ldots, m_k)}{m_1! \ldots m_k!} (x_1+y_1)^{m_1} \ldots (x_k+y_k)^{m_k} \\ &= S(x_1+y_1, \ldots, x_k+y_k), \end{aligned}$$

and so it is seen that the operators $S(x_1, \ldots, x_k)$ satisfy the requirements of the preceding lemma. Consequently, if

$$f^{**} = \sup_{0<\alpha} \frac{1}{\alpha^k} \int_0^\alpha \cdots \int_0^\alpha S(x_1, \ldots, x_k)|f(\cdot)|dx_1 \ldots dx_k,$$

where $f \in L_p$, then

$$\mu(\{s|f^{**}(s) > \beta\}) \leqq \frac{1}{c_k\beta} \int_{e(c_k\beta)} |f(s)|\mu(ds), \qquad \beta > 0.$$

Thus the present lemma will be proved if we establish the existence of an absolute constant d_k with $f^{**} \geqq d_k f^*$. This inequality will follow from the statement that for $0 \leqq f \in L_p$ and for $n \geqq 0$ there is an $\alpha = \alpha(n, f)$ such that

$$[*] \qquad \frac{1}{\alpha^k} \int_0^\alpha \cdots \int_0^\alpha S(x_1, \ldots, x_k) f dx_1 \ldots dx_k \geqq \frac{d_k}{n^k} \sum_{i_1=0}^{n-1} \cdots \sum_{i_k=0}^{n-1} P(i_1, \ldots, i_k) f.$$

Since

$$\frac{1}{\alpha^k}\int_0^\alpha \ldots \int_0^\alpha S(x_1, \ldots, x_k)dx_1 \ldots dx_k$$
$$= \sum_{i_1=0}^{\infty} \ldots \sum_{i_k=0}^{\infty} c_{i_1}(\alpha) \ldots c_{i_k}(\alpha)P(i_1, \ldots, i_k),$$

where $c_m(\alpha) = \alpha^{-1}\int_0^\alpha e^{-x}x^m\,dx/m!$, to prove [*] it suffices to show that there exists a $\delta > 0$ and, for each n, an $\alpha(n) > 0$ such that $c_m(\alpha(n)) \geqq \delta/n$ for $m < n$; for then [*] will hold with $d_k = \delta^k$. This last statement will be verified for $\alpha(n) = n$, i.e., we shall show that, for some $\delta > 0$, $\int_0^n e^{-x}x^m dx/m! \geqq \delta$ for all $m < n$. It thus suffices to show that $\liminf_{m\to\infty} \int_0^m f_m(x)dx > 0$ where $f_m(x) = e^{-x}x^m/m!$. Since $f_m'(x) \geqq 0$ and $f_m''(x) \leqq 0$ on the interval $m-\sqrt{m} \leqq x \leqq m$ we have

$$\int_0^m f_m(x)dx \geqq \int_{m-\sqrt{m}}^m f_m(x)dx \geqq \frac{1}{2}\sqrt{m}\, f_m(m)$$
$$= \frac{1}{2}\frac{e^{-m}\,m^{m+1}\,m^{-1/2}}{m!} \to \frac{e}{\sqrt{8\pi}},$$

by Stirling's formula. Q.E.D.

In proving the implication $DP_k \Rightarrow D_k$ we shall need the following lemma.

15 LEMMA. *For every bounded operator T in L_1, for which $|T|_\infty < \infty$ let $P(T)$ be the positive operator associated with T as in Lemma 6.4. Then $P(T_1)P(T_2) \geqq P(T_1T_2)$.*

PROOF. We recall that, for $0 \leqq f \in L_1 \cap L_\infty$

$$P(T)f = \sup_{|g(\cdot)| \leqq f} |T(g, \cdot)|.$$

Let $|g(\cdot)| \leqq f$ so that, by Lemma 6.4,

$$|T_1T_2(g, \cdot)| \leqq P(T_1)|T_2(g, \cdot)| \leqq P(T_1)P(T_2)|g(\cdot)| \leqq P(T_1)P(T_2)f.$$

Thus $P(T_1T_2)f \leqq P(T_1)P(T_2)f$ for every non-negative function f in $L_1 \cap L_\infty$. A continuity argument shows that this same inequality is valid for all positive f in L_1. Q.E.D.

16 LEMMA. *Let $T_1, \ldots, T_k$ be commuting operators in L_1 with $|T_i|_1$, $|T_i|_\infty \leqq 1$ for $i = 1, \ldots, k$. For $f \in L_p$ let $f^* = \sup_{n \geqq 1} |A(n)(f, \cdot)|$, where*

$$A(n) = \frac{1}{n^k} \sum_{i_1=0}^{n-1} \cdots \sum_{i_k=0}^{n-1} T_1^{i_1} \cdots T_k^{i_k}.$$

Then there is a constant c_k, independent of the operators $T_1, \ldots, T_k$ and the function f, such that

$$\mu(\{s|f^*(s) > \beta\}) \leqq \frac{1}{c_k\beta} \int_{e(c_k\beta)} |f(s)|\mu(ds).$$

PROOF. Using the notation $P(T)$ introduced in Lemma 15 we shall define, for each set $i_1, \ldots, i_k$ on non-negative integers, the operators $P(i_1, \ldots, i_k) = P(T_1^{i_k} \ldots T_k^{i_k})$. Then, by Lemmas 6.4 and 15, the operators $P(i_1, \ldots, i_k)$ satisfy the hypothesis of Lemma 14. Hence, for some constant c_k

$$[*] \qquad \mu(\{f^{**}(s) > \beta\}) \leqq \frac{1}{c_k\beta} \int_{e(c_k\beta)} |f(s)|\mu(ds), \qquad \beta > 0,$$

where

$$f^{**} = \sup_{n \geqq 1} \frac{1}{n^k} \sum_{i_1=0}^{n-1} \cdots \sum_{i_k=0}^{n-1} P(i_1, \ldots, i_k)|f(\cdot)|.$$

But, by Lemma 6.4,

$$f^{**} \geqq \sup_{n \geqq 1} \left| \frac{1}{n^k} \sum_{i_1=0}^{n-1} \cdots \sum_{i_k=0}^{n-1} T_1^{i_1} \cdots T_k^{i_1}(f, \cdot) \right| = f^*.$$

Hence $\{s|f^{**}(s) > \beta\} \supseteqq \{s|f^*(s) > \beta\}$ which, in view of [*], proves the lemma. Q.E.D.

We are now prepared to give the proof of Lemma 11. This lemma will first be proved under the slight additional assumption that the semi-group $T(t_1, \ldots, t_k)$ is defined for all $t_i \geqq 0$. To simplify the notation the proof will be given for the case $k = 2$ but it will be clear that the method is satisfactory for an arbitrary positive integer k. By Theorem III.11.17 there is a μ-null set E for which the averages

$$A(\alpha)(f, s) = \frac{1}{\alpha^2} \int_0^\alpha \int_0^\alpha T(t_1, t_2)(f, s)dt_1\, dt_2$$

exist for all s not in E. Thus, for $s \notin E$, $A(\alpha)(f, s)$ is continuous for $\alpha > 0$. By Lemma 9 the semi-group $\{T(t_1, t_2). t_1, t_2 \geqq 0\}$ is strongly continuous at every point $t_1 > 0$, $t_2 > 0$, and thus, since $|T(t_1, t_2)|_p \leqq 1$, it follows that for each α in the set R of non-negative rational numbers,

$$A(\alpha)f = \lim_{n\to\infty} \frac{1}{(\alpha n!)^2} \sum_{i_1=0}^{\alpha n!-1} \sum_{i_2=0}^{\alpha n!-1} S_1(n)^{i_1} S_2(n)^{i_2} f,$$

where $S_1(n) = T(1/n!, 0)$, $S_2(n) = T(0, 1/n!)$. Using Corollary III.6.13, Theorem III.11.17, and the Cantor diagonal process, we may find a sequence $\{n_j\}$ such that

$$A(\alpha)(f, s) = \lim_{i\to\infty} \frac{1}{(\alpha n_j!)^2} \sum_{i_1=0}^{\alpha n_j!-1} \sum_{i_2=0}^{\alpha n_j!-1} S_1(n_j)^{i_1} S_2(n_j)^{i_2}(f, s)$$

for each α in R and each s not in a null set $E_1 \supseteq E$. For $s \notin E_1$ let

$$f_n^*(s) = \sup_{1\leqq k} \left| \frac{1}{k^2} \sum_{m_1=0}^{k-1} \sum_{m_2=0}^{k-1} S_1(n)^{m_1} S_2(n)^{m_2}(f, s) \right|,$$

so that if $\alpha \in R$, $s \notin E_1$, and $\varepsilon > 0$, there is an integer $N(\alpha, s, \varepsilon)$ with

$$|f_n^*(s)| \geqq |A(\alpha)(f, s)| - \varepsilon, \qquad n \geqq N(\alpha, s, \varepsilon).$$

Therefore

$$\liminf_{n\to\infty} f_n^*(s) \geqq |A(\alpha)(f, s)|, \qquad \alpha \in R, \quad s \notin E_1.$$

Now if $s \notin E_1$ then $s \notin E$ and hence for $s \notin E_1$ the average $A(\alpha)(f, s)$ is continuous in α on the interval $0 < \alpha < \infty$. Thus

$$\liminf_{n\to\infty} f_n^*(s) \geqq f^*(s) = \sup_{0<\alpha} |A(\alpha)(f, s)|, \qquad s \notin E_1.$$

Thus for $\beta > 0$ we have $\liminf_n \chi_n \geqq \chi$ where χ_n and χ are the characteristic functions of $\{s|f_n^*(s) > \beta\}$ and $\{s|f^*(s) > \beta\}$ respectively. Consequently, by Fatou's lemma and Lemma 16, we have

$$\begin{aligned}
\mu(\{s|f^*(s) > \beta\}) &= \int_S \chi(s)\mu(ds) \\
&\leqq \liminf_n \int_S \chi_n(s)\mu(ds) \\
&= \liminf_{n\to\infty} \mu\{s|f_n^*(s) > \beta\} \\
&\leqq \liminf_{n\to\infty} \frac{1}{c_k\beta} \int_{e(c_k\beta)} |f(s)|\mu(ds) \\
&= \frac{1}{c_k\beta} \int_{e(c_k\beta)} |f(s)|\mu(ds).
\end{aligned}$$

This proves Lemma 11 under the additional assumption that the semi-group is defined for all $t_1, \ldots, t_k \geqq 0$. Now suppose that $\{S(t_1, \ldots, t_k), t_1, \ldots, t_k > 0\}$ is a semi-group which satisfies the hypothesis of Lemma 11 and, for $\varepsilon > 0$, let $T(\varepsilon; t_1, \ldots, t_k) = I$ if $t_1 = t_2 = \ldots = t_k = 0$ and otherwise let $T(\varepsilon; t_1, \ldots, t_k) = S(t_1+\varepsilon u_1, \ldots, t_k+\varepsilon u_k)$, where $u_i = t_1+\ldots+t_{i-1}+t_{i+1}+\ldots+t_k$. Then we may apply the result just proved to the semi-group $\{T(\varepsilon; t_1, \ldots, t_k), t_1, \ldots, t_k \geqq 0\}$. For f in L_p we let

$$f^*_\varepsilon = \sup_{0<\alpha} |A(\varepsilon, \alpha)(f, \cdot)|, \qquad f^* = \sup_{0<\alpha} |A(\alpha)(f, \cdot)|,$$

where

$$A(\varepsilon, \alpha) = \frac{1}{\alpha^k}\int_0^\alpha \cdots \int_0^\alpha T(\varepsilon; t_1, \ldots, t_k)dt_1 \ldots dt_k$$

and

$$A(\alpha) = \frac{1}{\alpha^k}\int_0^\alpha \cdots \int_0^\alpha S(t_1, \ldots, t_k)dt_1 \ldots dt_k.$$

Thus it has been proved that

$$[*] \qquad \mu(\{s|f^*_\varepsilon(s) > \beta\}) \leqq \frac{1}{c_k\beta}\int_{e(c_k\beta)} |f(s)|\mu(ds), \qquad \varepsilon, \beta > 0.$$

It is seen from the Lebesgue dominated convergence theorem (III.6.16) and from Lemma 9 that for each f in L_p we have, for $\alpha > 0$, $\lim_{\varepsilon\to 0} A(\varepsilon, \alpha)f = A(\alpha)f$ in the norm of L_p. Thus, by Theorem III.3.6, $A(\varepsilon, \alpha)f \to A(\alpha)f$ in measure and hence, by Corollary III.6.13 and the Cantor diagonal process, we can find a sequence $\varepsilon_i \to 0$ such that $A(\varepsilon_i, \alpha)(f, s) \to A(\alpha)(f, s)$ for μ-almost all s in S and for all α in the set R of positive rational numbers. It follows that $\liminf_{i\to\infty} f^*_{\varepsilon_i}(s) \geqq |A(\alpha)(f, s)|$ for almost all s in S and all α in R. Since, for almost all s in S, $A(\alpha)(f, s)$ is continuous for $0 < \alpha$ we have

$$f^*(s) = \sup_{\alpha\in R} |A(\alpha)(f, s)| \leqq \liminf_{i\to\infty} f^*_{\varepsilon_i}(s)$$

almost everywhere on S. If $e^*(\beta) = \{s|f^*(s) > \beta\}$ and $e^*_{\varepsilon_i}(\beta) = \{s|f^*_{\varepsilon_i}(s) > \beta\}$ then, modulo a null set, $\liminf_{i\to\infty} e^*_{\varepsilon_i}(\beta) \supseteq e^*(\beta)$ and so [*] and Fatou's lemma yield

$$\mu(e^*(\beta)) \leqq \frac{1}{c_k\beta}\int_{e(c_k\beta)} |f(s)|\mu(ds).$$

This completes the proof of Lemma 11 and prepares us for the following basic result on the almost everywhere convergence of the averages of a k-parameter semi-group in $L_1(S, \Sigma, \mu)$.

17 THEOREM. *Let (S, Σ, μ) be a positive measure space and let $\{T(t_1, \ldots, t_k), t_1, \ldots, t_k \geqq 0\}$ be a strongly measurable k-parameter semi-group of operators in $L_1(S, \Sigma, \mu)$ with $|T(t_1, \ldots, t_k)|_1 \leqq 1$ and $|T(t_1, \ldots, t_k)|_\infty \leqq 1$. Then, for every f in $L_1(S, \Sigma, \mu)$, the limit*

$$\lim_{\alpha\to\infty} \frac{1}{\alpha^k}\int_0^\alpha \cdots \int_0^\alpha T(t_1, \ldots, t_k)(f, s)dt_1 \ldots dt_k$$

exists for almost all s in S.

PROOF. Let

$$A(\alpha) = \frac{1}{\alpha^k}\int_0^\alpha \cdots \int_0^\alpha T(t_1, \ldots, t_k)dt_1 \ldots dt_k.$$

For $p > 1$ the set $L_p \cap L_1$ is dense in L_1 and for f in $L_p \cap L_1$ it is seen from Theorem 10 that the limit $\lim_{\alpha\to\infty} A(\alpha)(f, s)$ exists for almost all s in S. In view of Lemma 11 we may apply Theorem IV.11.3, with $A_p =$ the set of all rational numbers $\alpha \geqq p$, to conclude that for every f in L_1

$$\lim_p \sup_{\alpha, \beta\in A_p} |A(\alpha)(f, s) - A(\beta)(f, s)| = 0$$

for almost all s in S. But since, for almost all s, $A(\alpha)(f, s)$ is continuous in α, this means that the limit $\lim_{\alpha\to\infty} A(\alpha)(f, s)$ exists almost everywhere on S. Q.E.D.

8. Uniform Ergodic Theory

In this section we obtain conditions on a bounded linear operator T in a complex B-space which are sufficient to insure that the averages $n^{-1}\sum_{j=0}^{n-1} T^j$ converge in the uniform topology of operators. A special study will be made for certain operators on $C(S)$ and $L_1(S, \Sigma, \mu)$ where a rather complete decomposition of the space S is possible. The results of this section can be advantageously applied to some problems in the theory of Markov processes, although we shall only sketch these applications.

1 LEMMA. *Let T be an operator in a B-space $\mathfrak{X}$ and suppose that $\{n^{-1}T^n\}$ converges to zero in the weak operator topology as n approaches infinity. Then the spectrum of T is a subset of the unit disk $\{z||z| \leqq 1\}$, and any pole λ of T with $|\lambda| = 1$ has order one.*

PROOF. The hypothesis and the uniform boundedness theorem imply that $|T^n/n| \leqq K$, $n = 0, 1, 2, \ldots$, for some K, and so $\limsup_{n\to\infty} |T^n|^{1/n} \leqq 1$. It follows from Lemma VII.3.4 that $|\sigma(T)| \leqq 1$. If λ is a pole of T and $|\lambda| = 1$, then 1 is a pole of $T_1 = T/\lambda$ and $T_1^n/n \to 0$ in the weak operator topology. Hence to prove the second statement it suffices to treat the case that 1 is a pole of T. Suppose that the order of the pole 1 is at least two. It follows from Theorem VII.3.18 that there exists an $x_0 \in E(1; T)\mathfrak{X}$ such that $(I-T)x_0 \neq 0$, but $(I-T)^2x_0 = 0$. Now apply Theorem VII.3.22 to the function f defined by $f(\lambda) = \lambda^n/n$, to obtain

$$\frac{1}{n} T^n x_0 = \frac{1}{n} x_0 + (I-T)x_0.$$

Letting $n \to \infty$, we conclude that $x^*(I-T)x_0 = 0$ for any $x^* \in \mathfrak{X}^*$, which is a contradiction. Hence the only poles of T which lie on the unit circle are simple poles. Q.E.D.

We now give a condition which implies that if $|\sigma(T)| \leqq 1$, then the only spectral points λ with $|\lambda| = 1$ are isolated and the corresponding projections have finite dimensional ranges.

2 LEMMA. *Let T be a bounded linear operator whose spectrum is contained in the unit disk and let $|T^n - K| < 1$ for some positive integer n and some compact operator K. Then every spectral point λ of T with $|\lambda|^n > |T^n - K|$ is isolated and the corresponding projection $E(\lambda; T)$ has a finite dimensional range.*

PROOF. Let n be as in the hypothesis and let ω be a primitive n-th root of unity. It follows from Theorem VII.3.11 that if $\lambda \in \sigma(T)$ and $|\lambda| = 1$, then

$$E(\lambda; T) + E(\lambda\omega; T) + \ldots + E(\lambda\omega^{n-1}; T) = E(\lambda^n; T^n).$$

By Corollary VII.3.21 we have $E(\lambda\omega^p; T)E(\lambda\omega^q; T) = 0$, $p \neq q$, and so if $E(\lambda^n; T^n)$ has a finite dimensional range, so does $E(\lambda; T)$. Also by Theorem VII.3.11, if λ^n is an isolated point of $\sigma(T^n) = [\sigma(T)]^n$, then λ is an isolated point of $\sigma(T)$. Hence it suffices to prove the lemma under the assumption that $n = 1$, which we now make.

Let $V = T-K$. It will first be shown that if $|\mu| > |V|$ and if $(I - R(\mu; V)K)^{-1}$ exists then $R(\mu; T)$ exists and equals $(I-R(\mu; V)K)^{-1}R(\mu; V)$. This follows from the identity

$$R(\mu; V)(\mu I-T) = R(\mu; V)(\mu I-V-K) = I-R(\mu; V)K,$$

and the calculations

$$\begin{aligned}
&[(I-R(\mu; V)K)^{-1}R(\mu; V)][\mu I-T] = I;\\
&[\mu I-T][(I-R(\mu; V)K)^{-1}R(\mu; V)]\\
&= (\mu I-V)R(\mu; V)(\mu I-T)(I-R(\mu; V)K)^{-1}R(\mu; V)\\
&= (\mu I-V)(I-R(\mu; V)K)(I-R(\mu; V)K)^{-1}R(\mu; V) = I.
\end{aligned}$$

Now by VI.5.4 the operator $R(\mu; V)K$ is compact for each μ in the domain $|\mu| > |V|$ and it evidently depends analytically on μ. Since $|R(\mu; V)| \to 0$ as $|\mu| \to \infty$, it follows that for $|\mu|$ sufficiently large $|R(\mu; V)K| < 1$. By Lemma VII.3.4 we conclude that if $|\mu|$ is sufficiently large, then the number 1 is not in the spectrum of $R(\mu; V)K$. It follows from Lemma VII.6.13 that $(I-R(\mu; V)K)^{-1}$ exists and is an analytic function of μ, for $|\mu| > |V|$, except at countably many isolated points in this domain. By the remarks above this shows that $R(\mu; T)$ exists for $|\mu| > |V|$ except at a countable number of isolated points.

Let $\lambda \in \sigma(T)$ and $|\lambda| > |V|$; it remains to show that $E(\lambda; T)$ has finite dimensional range. To do this, let C be a circle with center at λ and radius small enough so that C lies entirely in the domain $|\mu| > |V|$ and does not contain in its interior any point except λ at which $(I-R(\mu; V)K)^{-1}$ fails to exist. For $|\mu|$ large, the Laurent expansion leads to the identity

$$(I-R(\mu; V)K)^{-1} = I+R(\mu; V)K(I-R(\mu; V)K)^{-1},$$

and so by analytic continuation this holds on C as well. Consequently,

$$R(\mu; T) = R(\mu; V)+R(\mu; V)K(I-R(\mu; V)K)^{-1}R(\mu; V).$$

Now, since $R(\mu; V)$ is analytic on C, we have

$$E(\lambda; T) = \frac{1}{2\pi i}\int_C R(\mu; T)d\mu$$
$$= \frac{1}{2\pi i}\int_C R(\mu; V)K(I-R(\mu; V)K)^{-1}R(\mu; V)\, d\mu.$$

Since K is compact, the integrand is a compact operator for each μ on C. By the definition of the integral and Lemma VI.5.3, we conclude that $E(\lambda; T)$ is also a compact operator. But $E(\lambda; T)$ is the identity operator on the subspace $E(\lambda; T)\mathfrak{X}$ and by virtue of Theorem IV.3.5, this implies that $E(\lambda; T)\mathfrak{X}$ is finite dimensional. Q.E.D.

3 THEOREM. *Let T be such that T^n/n converges to zero in the weak topology, and let $|T^n-K| < 1$ for some positive integer n and some compact operator K. Then there are at most a finite number of points $\lambda_1, \ldots, \lambda_q$ of unit modulus in the spectrum of T. Each point λ_k is a simple pole and $E(\lambda_k; T)\mathfrak{X}$ is finite dimensional.*

PROOF. Let $\lambda_k \in \sigma(T)$ and $|\lambda_k| = 1$. By Lemma 2, $E(\lambda_k; T)\mathfrak{X}$ is finite dimensional and so the operator T restricted to this subspace is compact. It follows from Theorems VII.4.5 and VII.3.20 that λ_k is a pole of T. By Lemma 1 it must be a simple pole. Q.E.D.

4 COROLLARY. *If T satisfies the hypothesis of the theorem, then $n^{-1}\sum_{j=0}^{n-1} T^j$ converges to the projection $E(1; T)$ in the uniform topology of operators.*

PROOF. Let $A(n) = n^{-1}\sum_{m=0}^{n-1} T^m$, let σ be that portion of $\sigma(T)$ in the open set $|\lambda| < 1$, and let $\sigma' = \sigma(T)-\sigma$. Then, by Lemma VII.3.13, we have $A(n)E(\sigma) \to 0$. By Theorem VII.3.20 the points $\lambda_1, \ldots, \lambda_q$ are simple poles of the restriction $T_{\sigma'}$ of T to $E(\sigma')\mathfrak{X}$. Thus, since $n^{-1}\sum_{m=0}^{n-1}\lambda_i^m \to 0$ for each $\lambda_i \neq 1$, we see, from Theorem VII.3.22, that $A(n)E(\sigma') \to E(\{1\})$. Since $A(n) = A(n)E(\sigma)+A(n)E(\sigma')$ the corollary is proved. Q.E.D.

We now turn to an examination of certain operators T which are defined in $C(S)$, where S is a compact Hausdorff space. We say that T is *positive* if $(Tf)(s) \geqq 0$, $s \in S$, whenever f is such that $f(s) \geqq 0$, $s \in S$. It will now be seen that if T is positive and satisfies the conditions imposed in Theorem 3, then the points of $\sigma(T)$ of unit modulus are all roots of unity.

5 LEMMA. *Let T be a positive linear operator in $C(S)$ such tha*

T^n/n converges to zero in the weak operator topology and let $|T^n - K| < 1$ for some positive integer n and some compact operator K. Then there exists an integer N such that if $\lambda_1, \ldots, \lambda_q$ are the points of $\sigma(T)$ of unit modulus, then $\lambda_k^N = 1$, $k = 1, \ldots, q$.

PROOF. Let $\lambda \neq 1$ be in $\sigma(T)$ with $|\lambda| = 1$ and let $f \neq 0$ be in the range of $E(\lambda)$ so that $Tf = \lambda f$. Let s_0 be a point of S at which $|f(\cdot)|$ attains its maximum. Since $E(\lambda)C(S)$ is a linear manifold we may suppose that $f(s_0) = 1$. Let δ be the linear functional on $C(S)$ defined by $\delta g = g(s_0)$, $g \in C(S)$. Then $(T^*)^n\delta$ is a non-negative linear functional on $C(S)$ and so by Theorem IV.6.3 there is a non-negative measure π_n in $rca(S)$ such that

$$(T^n g)(s_0) = \int_S g(s)\pi_n(ds), \qquad g \in C(S).$$

Since $\lambda^n f(s_0) = (T^n f)(s_0) = \int_S f(s)\pi_n(ds)$, and $|f(s)| \leqq f(s_0) = 1$ it follows that $\pi_n(S) \geqq 1$, $n = 1, 2, \ldots$.

We now show that for each n, the open set $A_n = \{s \in S | \lambda^{-n} f(s) \neq 1\}$ has π_n-measure zero. Since $T^n f = \lambda^n f$, we have

$$0 = f(s_0) - \lambda^{-n}(T^n f)(s_0) = \int_S \{1 - \lambda^{-n} f(s)\}\pi_n(ds),$$

and, taking real parts,

$$0 = \int_S \{1 - \mathscr{R}(\lambda^{-n} f(s))\}\pi_n(ds).$$

Since $|\mathscr{R}(\lambda^{-n} f(s))| \leqq |f(s)| \leqq 1$, the integrand just written is non-negative. Since π_n is non-negative, it follows from III.2.20(*d*) and III.6.8 that if B_n denotes the set $B_n = \{s \in S | \mathscr{R}(\lambda^{-n} f(s)) \neq 1\}$, then $\pi_n(B_n) = 0$. Since $|f(s)| \leqq f(s_0) = 1$, it is easily seen that $B_n \supseteq A_n$ and hence $\pi_n(A_n) = 0$.

It will now be shown that the complements $A_n' = S - A_n$, $n = 0, 1, 2, \ldots$, cannot be pairwise disjoint. Applying Corollary 4, we define a measure π by the formula

$$\pi = \lim_{n\to\infty} \frac{1}{n} \sum_{j=0}^{n-1} \pi_j = \lim_{n\to\infty} \frac{1}{n} \sum_{j=0}^{n-1} (T^*)^j \delta.$$

Since $\pi_j(S) \geqq 1$, we have $\pi(S) \geqq 1$. Let $A = \cap_{j=0}^{\infty} A_j$; since $A \subseteq A_j$, we have $\pi_j(A) = 0$ for $j = 0, 1, 2, \ldots$ and therefore $\pi(A) = 0$. Let $A' = \cup_{j=0}^{\infty} A_j'$, so that $A' = S - A$. If the sets A_k', $k = 0, 1, 2, \ldots$, are pairwise disjoint then $\pi_j(A_k') = 0$ for $j \neq k$ and so

$$0 \leqq \pi(A'_k) = \lim_{n\to\infty} \frac{1}{n} \sum_{j=0}^{n-1} \pi_j(A'_k) = \lim_{n\to\infty} \frac{1}{n} = 0.$$

Therefore $\pi(A') = \sum_{k=0}^{\infty}\pi(A'_k) = 0$. Since A and A' are complementary sets, we conclude that $0 = \pi(A)+\pi(A') = \pi(S) \geqq 1$. This contradiction proves that the sets $\{A'_k\}$ cannot be pairwise disjoint.

We have shown that there exist distinct integers m and n such that $A'_m A'_n$ contains a point s_1. Consequently $\lambda^{-n}f(s_1) = 1 = \lambda^{-m}f(s_1)$ and so $\lambda^{n-m} = 1$. But λ was any point of $\sigma(T)$ with $|\lambda| = 1$, and since there are only a finite number of such points, the existence of the integer N described in the statement is proved. Q.E.D.

We now collect the results already obtained.

6 Theorem. *Let T be a positive linear operator in $C(S)$ such that T^n/n converges to zero in the weak operator topology and let $|T^n-K| < 1$ for some positive integer n and some compact operator K. Then $\sigma(T)$ may be decomposed into the union of a closed set σ, which lies inside a circle $|z| < \alpha < 1$, and of a finite number of simple poles $e^{2\pi i\theta_k}$, where θ_k is rational, $k = 1, \ldots, q$. If we put $E_k = E(e^{2\pi i\theta_k})$, $E_D = E(\sigma)$ and $D = TE_D$, then each E_k has finite dimensional range. The iterates of T are given by the formula*

$$T^m = \sum_{k=1}^{q} e^{2\pi i m\theta_k} E_k + D^m, \qquad m \geqq 1.$$

Further, there exists a positive number M such that

$$|D^m| \leqq M\alpha^m, \qquad m \geqq 1.$$

Proof. Only the last two statements remain to be proved. Since the points $e^{2\pi i\theta_k}$ are simple poles of T, the expression for T^m is an immediate consequence of Theorem VII.3.22. Finally $D = T_\sigma$ on $E(\sigma)C(S)$ and vanishes on $(I-E(\sigma))C(S)$ so $\sigma(D) = \sigma(T_\sigma)\cup\{0\}$. By Theorem VII.3.20, $\sigma(D) = \sigma\cup\{0\}$ and so $\sigma(D)$ is contained in the disk $|z| < \alpha$ for some $\alpha < 1$. By Lemma VII.3.4 this implies that $\limsup_{m\to\infty} |D^m|^{1/m} < \alpha$, from which it follows that $|D^m| \leqq M\alpha^m$, $m \geqq 1$, for some positive number M. Q.E.D.

7 Theorem. *Let T satisfy the hypothesis of the preceding theorem and let $E_P = \sum_{k=1}^{q} E_k$. Then the subspace $C_P = E_P[C(S)]$ is finite dimensional, and if $C_D = E_D[C(S)]$, then $C(S)$ is the direct sum of C_P and C_D. Both subspaces C_P and C_D are invariant under T, and*

(a) *there exists a positive integer* N *such that* $T^N x = x$, *for* $x \in C_P$;

(b) $T^n x \to 0$ *exponentially fast, for* $x \in C_D$.

Moreover, the subspaces C_P *and* C_D *are uniquely defined by properties* (a) *and* (b), *respectively.*

PROOF. It follows from its definition that E_P is a projection, that $E_P E_D = E_D E_P = 0$, and that $I = E_P + E_D$. Further T commutes with E_P and E_D so this direct sum decomposition is into subspaces invariant under T. Statements (a) and (b) follow from the formula for T^m given in Theorem 6. To prove the final assertion, let $x \in C(S)$ be such that $T^n x \to 0$. Then $x = x_P + x_D$, where $x_P = E_P x \in C_P$ and $x_D = E_D x \in C_D$. It follows that $T^n x_P \to 0$ and $T^n x_D \to 0$; but since $\{T^n x_P\}$ has infinitely many terms equal to x_P, it follows that $x_P = 0$. Hence $C_D = \{x \in C(S) | T^n x \to 0\}$. Similarly if $x \in C(S)$ is such that $T^n x = x$ for some $n \geqq 1$, it follows that $x \in C_P$ and so $C_P = \{x \in C(S) | T^n x = x, \text{ for some } p \geqq 1\}$. Q.E.D.

The conclusions of Theorem 3 are valid for any B-space, and in particular for the space $L_1(S, \Sigma, \mu)$. However, it is not evident that the more complete decompositions obtained in Theorems 6 and 7 for $C(S)$ can be derived for any other spaces. We now prove that for the space $L_1(S, \Sigma, \mu)$ such a derivation is possible due to the fact that its conjugate space may be represented as a space of continuous functions. If T is an operator in the space $L_1(S, \Sigma, \mu)$, we say that T is *positive* if $(Tf)(s) \geqq 0$ μ-almost everywhere on S, whenever $f(s) \geqq 0$ μ-almost everywhere.

8 THEOREM. *Let* (S, Σ, μ) *be a* σ-*finite positive measure space. Suppose that* T *is a positive linear operator in* $L_1(S, \Sigma, \mu)$ *such that* T^n/n *converges to zero in the weak operator topology and let* $|T^n - K| < 1$ *for some positive integer* n *and some compact operator* K. *Then the conclusions of Theorems* 6 *and* 7 *hold if* $C(S)$ *is replaced by* $L_1(S, \Sigma, \mu)$.

PROOF. We consider the adjoint operator T^* in the space $L_\infty(S, \Sigma, \mu)$. It is easily seen that since T is positive, T^* is also positive. Further, if $T^n = K + V$, where K is compact and $|V| < 1$ then $(T^*)^n = K^* + V^*$ and K^* is compact, by Theorem VI.5.2, while $|V^*| < 1$, by Lemma VI.2.2. Now by hypothesis $T^n/n \to 0$ in the weak operator topology. An argument similar to that in the proof of Corollary 4 shows that the sequence $\{T^n/n\}$ converges to zero in the uniform operator topology of $B(L_1)$. By Lemma VI.2.2 this implies that $\{T^{*n}/n\}$ converges in the uniform operator topology of $B(L_\infty)$, and therefore in the weak operator topology of this space. Thus we have shown that the properties assumed for T are also valid for T^*.

Now Theorem V.8.11 asserts that there is a compact Hausdorff space S_1 such that $L_\infty(S, \Sigma, \mu)$ is isometrically isomorphic with the space $C(S_1)$, and that the isomorphism preserves the notion of positivity. Thus it is seen from Theorem 5 that all the points of $\sigma(T^*)$ of unit modulus are roots of unity, and by Lemma VII.3.7 the same holds for $\sigma(T)$. The remaining conclusions in Theorems 6 and 7 are then provided exactly as before, since no special property of the space $C(S)$ was used to establish them. Q.E.D.

We shall conclude this section with a brief indication of how the results already obtained can be applied to the theory of Markov processes. The exposition here is not complete; the reader should refer to the paper of K. Yosida and S. Kakutani [2] for details. Actually the procedure they employ is not exactly the one we will describe but it does not differ in an essential way.

Let (S, Σ, μ) be a finite positive measure space and consider a real valued function P defined on $S \times \Sigma$. The function P is called the *transition probability* function, and we regard the number $P(t, e)$ as giving the probability that, after the elapse of a unit time, the point $t \in S$ will be in the set $e \in \Sigma$. Let $B(S, \Sigma)$ denote the space of bounded Σ-measurable complex valued functions on S with the norm $|f| = \sup_{s \in S} |f(s)|$, $f \in B(S, \Sigma)$. By $ca(S, \Sigma)$ we denote the space of countably additive complex valued measures defined on the σ-field Σ. We assume

$$(\alpha)\quad P(\cdot, e) \in B(S, \Sigma), \qquad e \in \Sigma;$$

$$(\beta)\quad P(t, \cdot) \in ca(S, \Sigma), \qquad t \in S;$$

$$(\gamma)\quad P(t, e) \geqq 0, \qquad t \in S, \quad e \in \Sigma;$$

$$(\delta)\quad P(t, S) = 1, \qquad t \in S.$$

Under these assumptions it is readily seen that the probability that the point $t \in S$ will be in the set $e \in \Sigma$ after an elapse of n unit times, is given by the recurrence

$$P^{(n)}(t, e) = \int_S P^{(n-1)}(t, ds) P(s, e) = \int_S P(t, ds) P^{(n-1)}(s, e), \quad n = 2, 3, \ldots,$$

where $P^{(1)} = P$. The problem of Markov is to investigate the asymptotic behavior of the sequence $\{P^{(n)}(t, e)\}$ and the means $n^{-1} \sum_{j=0}^{n-1} P^{(j)}(t, e)$, where $P^{(0)}(t, e) = \chi_e(t)$.

In this way one is naturally lead to the study of two linear mappings

$$f \to Tf = \int_S f(s)P(\cdot, ds), \qquad f \in B(S, \Sigma),$$

$$\mu \to T'\mu = \int_S \mu(ds)P(s, \cdot), \qquad \mu \in ca(S, \Sigma),$$

which act in the spaces $B(S, \Sigma)$ and $ca(S, \Sigma)$ respectively. It is readily seen that the operator T and its iterates are positive and have norms equal to one. Also the function identically equal to one is left fixed by T, so $1 \in \sigma(T)$. In the study of such processes one makes additional hypotheses on P which will guarantee that there exists an integer n and a compact operator K in $B(S, \Sigma)$ such that $|T^n - K| < 1$.

Recalling Theorem IV.6.18, we note that the space $B(S, \Sigma)$ is isometrically isomorphic with the space $C(S_1)$, where S_1 is some compact Hausdorff space. Since the isomorphism preserves norms and positivity, it is seen that the results of Theorems 6 and 7 are valid for the operator T in the space $B(S, \Sigma)$. It is possible to decompose the transition probability function of P in a way corresponding to the decomposition of the space $B(S, \Sigma)$ established in Theorems 6 and 7. Perhaps somewhat more surprising is the decomposition of the space that can be obtained. Indeed, if g is the dimension of $E(1; T)B(S, \Sigma)$, it is possible to decompose S into g mutually disjoint Borel sets e_i, $i = 1, \ldots, g$, which are called the *ergodic kernels* and are determined up to μ-measure zero, and into the complementary portion $\Delta = S - \bigcup_{i=1}^{g} e_i$, called the *dissipative part* of S. These sets have the property that

$$P(t, e_i) = 1, \qquad t \in e_i,$$

$$\sup_{t \in S} P^{(n)}(t, \Delta) < M\alpha^n, \qquad 0 < \alpha < 1, \qquad n = 1, 2, \ldots.$$

These relations may be interpreted as saying that if $t \in e_i$, then it is transferred by the process T (with probability one) into the same ergodic kernel e_i, and that as time increases the dissipative part is evacuated. Moreover, the kernels e_i cannot be decomposed into smaller sets with the first mentioned property. However, each kernel e_i

can be further split into a finite number of disjoint Borel sets $e_{i1}, \ldots, e_{ik_i}$, where k_i is a divisor of the integer N of Theorem 7, and $\sum_{i=1}^{g} k_i = \dim E_p[B(S, \Sigma)]$. The sets e_{ij}, $j = 1, \ldots, k_i$ are called the *sub-ergodic kernels* belonging to e_i, they are determined up to μ-null sets, and $e_i = \bigcup_{j=1}^{k_i} e_{ij}$. Further, putting $e_{i,k_i+1} = e_{i1}$, we have

$$P(t, e_{i,j+1}) = 1, \qquad t \in e_{ij},$$

so that the process transfers (with probability one) the points of an ergodic kernel cyclically through the sub-ergodic kernels which belong to it.

9. Exercises on Ergodic Theory

1 Show that Theorem 5.9 and Corollary 5.5 remain valid if $p > 1$, even if (S, Σ, μ) is an infinite measure space.

2 Show that if the measure space (S, Σ, μ) in Theorem 5.9 is not finite, the necessary and sufficient condition that the theorem continue to hold in the case $p = 1$ is that for each $\varepsilon > 0$ and each $e \in \Sigma$ such that $\mu(e) < \infty$, there exist an $a \in \Sigma$ such that $\mu(a) < \infty$ and $n^{-1} \sum_{j=0}^{n-1} \mu(\phi^{-j}(e) - a) < \varepsilon$.

3 Let $\{T_t\}$ be a strongly measurable positivity preserving semi-group of operators in L_1 such that $|T(t)|_1 \leqq 1$, $|T(t)|_\infty \leqq 1$, $t > 0$. Show that for each $f \in L_p$, $p \geqq 1$, there exists a measurable function f^* which is finite almost everywhere such that

$$\int_0^\infty |(T_t f)(s)| \beta(t) dt \leqq f^*(s) \int_0^\infty \beta(t) dt$$

for every positive and decreasing function β. Show that if $f \in L_p$, $p > 1$, we can take $f^* \in L_p$.

4 Let $\{T(t_1, \ldots, t_k)\}$ be a strongly measurable positivity preserving k-parameter semi-group of operators in L_1 such that $|T(t_1, \ldots, t_k)|_1 \leqq 1$, $|T(t_1, \ldots, t_k)|_\infty \leqq 1$, $t_i > 0$. Show that for each $f \in L_p$ there exists a measurable function f^* which is finite almost everywhere such that for almost all s

$$\int_0^\infty \cdots \int_0^\infty |T(t_1, \ldots, t_k)(f, s)| \beta(t_1, \ldots, t_k) dt_1 \ldots dt_k \leqq f^*(s) \int_0^\infty \cdots \int_0^\infty \beta(t_1, \ldots, t_k) dt_1 \ldots dt_k$$

provided one of the following conditions is satisfied:

(a) $p \geqq 1$ and $\beta(t_1, \ldots, t_R) = \gamma(t_1^2 + \ldots + t_R^2)$ where γ is a positive and decreasing function, or

(b) $p > 1$ and $\beta(t_1, \ldots, t_R) = \prod_{i=1}^{R} \gamma_i(t_i)$ where $\gamma_i(t_i)$ is positive and decreasing for $t_i > 0$, $i = 1, \ldots, k$.

Show that if $p > 1$ we can take $f^* \in L_p$.

5 Let β be a positive, even, integrable function of the real variable x which is decreasing for $x > 0$. Let $\beta(x) \geqq |\gamma(x)|$, and let γ be measurable and $\int_{-\infty}^{+\infty} \gamma(x)dx = 1$. Then for f in L_p we have

$$\lim_{t\to\infty} t \int_{-\infty}^{+\infty} \gamma(t(x-y))f(y)dy = f(x)$$

for almost all x.

6 Let β be a positive integrable function of the real variables $x_1, \ldots, x_n$ which has the form $\beta(x_1, \ldots, x_n) = \beta_1(x_1^2 + \ldots + x_n^2)$, where β_1 is decreasing. Let $\beta(x_1, \ldots, x_n) \geqq |\gamma(x_1, \ldots, x_n)|$, and let γ be measurable and have $\int_{-\infty}^{+\infty} \ldots \int_{-\infty}^{+\infty} \gamma(x_1, \ldots, x_n)dx_1 \ldots dx_n = 1$. Then for f in L_p we have

$$\lim_{t\to\infty} t^n \int_{-\infty}^{+\infty} \ldots \int_{-\infty}^{+\infty} \gamma(t(x_1-y_1), \ldots, t(x_n-y_n))f(y_1, \ldots, y_n)dy_1 \ldots dy_n$$
$$= f(x_1, \ldots, x_n)$$

for almost all x.

7 (Hardy-Littlewood) Let h be a harmonic function defined in the circle $x^2+y^2 < 1$, and let $1 < p < \infty$. Suppose that $\int_0^{2\pi} |h(re^{i\theta})|^p d\theta \leqq K$, $0 < r < 1$. Let $h^*(\theta) = \max_{0<r<1} h(re^{i\theta})$. Show that there exists an absolute constant C_p such that $\int_0^{2\pi} |h^*(\theta)|^p d\theta \leqq C_p K$. Show that $\lim_{r\to 1} h(re^{i\theta})$ exists almost everywhere. Hint. Cf. Exercise IV.14.60.

8 Let h be a harmonic function defined in the sphere $\sum_{i=1}^{n} x_i^2 < 1$ in Euclidean n-space. Let $1 < p < \infty$, $0 < K < \infty$. Suppose that for each r, $0 < r < 1$, the integral of the function $|h(rx)|^p$ over the surface of the unit sphere is $\leqq K$. Putting $h^*(x) = \sup_{0<r<1} |h(rx)|$ for $|x| = 1$, show that there exists an absolute constant $C_{n,p}$ such that the integral of h^{*p} over the surface of the unit sphere is at most $C_{n,p}K$. Show that $\lim_{r\to 1} h(rx)$ exists for almost all x on the surface of the unit sphere.

9 Let h be an infinitely differentiable function of the $2n$ real variables $x_1, \ldots, x_n, y_1, \ldots, y_n$ defined in the region $x_i^2 + y_i^2 < 1$, $i = 1, \ldots, n$. Suppose that for each $i = 1, \ldots, n$ we have $\partial^2 h/\partial x_i^2 +$

$\partial^2 h/\partial y_i^2 = 0$. Suppose that $p > 1$, and that for $0 < r < 1$

$$\int_0^{2\pi} \cdots \int_0^{2\pi} |f(r\cos\theta_1, r\cos\theta_2, \ldots, r\sin\theta_n)|^p d\theta_1 \ldots d\theta_n \leqq K.$$

Show that $\lim_{r_1\to 1,\ldots,r_n\to 1} f(r_1\cos\theta_1, \ldots, r_n\sin\theta_n)$ exists for almost all $\theta_1, \ldots, \theta_n$.

10 Let h be a harmonic function as in Exercise 7. Show that for almost all θ, $\lim_{z_i\to e^{i\theta}} h(z_i)$ exists, provided that $|\arg(1-z_i e^{-i\theta})| \leqq K < \pi/2$.

Hint. use the method of Exercise 7 with a modified kernel.

11 Let f be a measurable function of the real variable x, let $1 \leqq p \leqq \infty$, and let $f \in L_p$. Show that for almost all x we have

(a) $\displaystyle \lim_{t\to\infty} \sqrt{t/\pi} \int_{-\infty}^{+\infty} e^{-t(x-y)^2} f(y)dy = f(x)$;

(b) $\displaystyle \lim_{t\to\infty} t \int_x^{\infty} e^{t(x-y)} f(y)dy = f(x)$;

(c) $\displaystyle \lim_{t\to\infty} (\pi t)^{-1} \int_{-\infty}^{+\infty} \left(\frac{\sin t(x-y)}{x-y}\right)^2 f(y)dy = f(x)$;

(d) $\displaystyle \lim_{\varepsilon\to 0+} (\varepsilon/\pi) \int_{-\infty}^{+\infty} \frac{1}{\varepsilon^2+(x-y)^2} f(y)dy = f(x)$.

12 Let f be a measurable function of the real variables $x_1, \ldots, x_n$, let $1 \leqq p \leqq \infty$, and let $f \in L_p$. Show that $f(x_1, \ldots, x_n)$ is, for almost all $x_1, \ldots, x_n$, equal to

(a) $\displaystyle \lim_{t\to\infty} \left(\frac{t}{\pi}\right)^{n/2} \int_{-\infty}^{+\infty} \cdots \int_{-\infty}^{+\infty} e^{-t\{(x_1-y_1)^2+\ldots+(x_n-y_n)^2\}} f(y_1, \ldots, y_n)dy_1 \ldots dy_n$;

(b) $\displaystyle \lim_{t\to\infty} \frac{1}{(\pi t)^n} \int_{-\infty}^{+\infty} \cdots \int_{-\infty}^{+\infty} \left\{\frac{\sin t(x_1-y_1)\ldots\sin t(x_n-y_n)}{(x_1-y_1)\ldots(x_n-y_n)}\right\}^2 f(y_1, \ldots, y_n)dy_1 \ldots dy_n$.

13 Show that, by applying Theorems 7.5 and 7.7 *directly* to the semigroup $\{S_t\}$ defined by

$$(S_t f)(x_1, \ldots, x_n)$$
$$= t^{n/2}\pi^{-n/2} \int_{-\infty}^{+\infty} e^{-t\{(x_1-y_1)^2+\ldots+(x_n-y_n)^2\}} f(y_1, \ldots, y_n)dy_1 \ldots dy_n,$$

an inequality strong enough to yield the result (a) of the previous exercise can be obtained. Prove that this implies the Lebesgue theorem which asserts that for an integrable function f,

$$\lim_{h\to 0+} h^{-n}\int_{x_1}^{x_1+h}\cdots\int_{x_n}^{x_n+h} f(y_1,\ldots,y_n)dy_1\ldots dy_n = f(x_1,\ldots,x_n)$$

almost everywhere. Also show that this Lebesgue theorem implies the result (b) of the previous exercise.

14 Let f be as in Exercise 12, and suppose $p>1$. Show that in this case $f(x_1,\ldots,x_n)$ is, for almost all $x_1,\ldots,x_n$, equal to

(a) $\lim \left(\frac{t_1\ldots t_n}{\pi^n}\right)^{\frac{1}{2}}\int_{-\infty}^{+\infty}\cdots\int_{-\infty}^{+\infty}\exp\left(-\sum_{i=1}^{n} t_i(x_i-y_i)^2\right)f(y_1,\ldots,y_n)dy_1\ldots dy_n;$

(b) $\lim \frac{1}{(t_1\ldots t_n)\pi^n}\int_{-\infty}^{+\infty}\cdots\int_{-\infty}^{+\infty}\left(\prod_{i=1}^{n}\frac{\sin t_i(x_i-y_i)}{x_i-y_i}\right)^2 f(y_1,\ldots,y_n)dy_1\ldots dy_n;$

(c) $\lim\,(h_1\ldots h_n)^{-1}\int_{x_1}^{x_1+h_1}\cdots\int_{x_n}^{x_n+h_n} f(y_1,\ldots,y_n)dy_1\ldots dy_n,$

where the limits are taken with respect to $t_1,\ldots,t_n\to\infty$ and $h_1,\ldots,h_n\to 0+$. (Part (c) is a result of Saks, Zygmund and Marcinkiewicz.)

15 Let (S,Σ,μ) be a finite positive measure space, and T an operator in $L_1(S,\Sigma,\mu)$ with $|T|_1\leqq 1$, $|T|_\infty\leqq 1$. Put $f^* = \sup_{1\leqq n}|A(n)(f,\cdot)|$ for each $f\in L_1$. Show that

(a) if $f\in L_1$, $f^*\in L_p(S,\Sigma,\mu)$ for each p, $0<p<1$;

(b) if $\int_S |f(s)|(1+\log^+|f(s)|)^k\mu(ds)<\infty$, then

$$\int_S |f^*(s)|(1+\log^+|f^*(s)|)^{k-1}\mu(ds)<\infty.$$

Establish the corresponding results for a strongly measurable n-parameter semi-group of operators.

16 Show that the convergence almost everywhere in Theorem 6.9 remains valid if (S,Σ,μ) is a finite measure space and $\int_S|f(s)|(1+\log^+|f(s)|)^{k-1}\mu(ds)<\infty$. Establish the corresponding generalization of Theorem 7.10.

17 Show that Exercise 9 remains valid if

$$\int_0^{2\pi}\cdots\int_0^{2\pi}|f(r\cos\theta_1,\ldots,r\sin\theta_n)|(1+\log^+|f(r\cos\theta_1,\ldots,r\sin\theta_n)|)^{n-1}d\theta_1\ldots d\theta_n$$

is finite. Establish a corresponding generalization of Exercise 14.

18 Suppose that $h_1,\ldots,h_n\to 0$ in such a way that $|h_i/h_j|$ remains bounded for $1\leqq i,j\leqq n_0$, and $n_0+1\leqq i,j\leqq n$. Show that the condition

$$\int_{-\infty}^{+\infty} \cdots \int_{-\infty}^{+\infty} |f(x_1, \ldots, x_n)|(1+\log^+|f(x_1, \ldots, x_n)|)dx_1 \ldots dx_n < \infty$$

is then sufficient to guarantee the validity of the results of Exercise 14.

19 Let S be a separable metric space. Let $\mathscr{B}$ be the field of Borel sets of S, and μ a σ-finite measure defined on $\mathscr{B}$, every open subset of S having positive μ-measure. Let $\{T_n\}$ be a sequence of bounded operators in $L_1(S, \mathscr{B}, \mu)$ of the form

$$(T_n f)(s) = \int_S K_n(s, t) f(t) \mu(dt).$$

Suppose that

(i) $K_n(s, t)$ is a bounded uniformly continuous function of s and t for each $n \geqq 1$; $\lim_{s \to s_0} K_n(s, t) = K_n(s_0, t)$ uniformly for $t \in S$, for each $n \geqq 1$.

(ii) If U is an arbitrary neighborhood of s, $K_n(s, t)$ converges uniformly as $n \to \infty$ on $S-U$.

(iii) $\lim_{n\to\infty} (T_n f)(s)$ exists μ-almost everywhere for each $f \in L_1(S, \mathscr{B}, \mu)$.

Show that for each $\nu \in ca(\mathscr{B})$, $\lim_{n\to\infty} \int_S K_n(x, y)\nu(dy)$ exists μ-almost everywhere. (Hint. Decompose ν into its μ-continuous and μ-singular parts. Using the method of Banach's theorem IV.11.2 and the $C(S)$-density of $L_1(S, \mathscr{B}, \mu)$ in $ca(\mathscr{B})$, show that for $\varepsilon > 0$ there exists a $\delta > 0$ such that

$$\mu\left\{ \sup_{1 \leqq n < \infty} \left| \int_S K_n(s, t)\nu(dt) \right| > \varepsilon \right\} < \varepsilon$$

if $v(\nu, S) < \delta$.)

20 Show that if the function β in Exercise 5 is continuous, and γ is Borel-measurable, then, for every regular and finite Borel measure ν, the limit

$$\lim_{t\to\infty} t \int_{-\infty}^{+\infty} \gamma(t(x-y))\nu(dy)$$

exists (Lebesgue) almost everywhere. Establish the corresponding generalization of Exercises 6, 7, 8, 10, 11, 12, and 13.

21 Let (S, Σ, μ) be a measure space. Let A be a closed unbounded operator with dense domain in $L_1(S, \Sigma, \mu)$, such that $R(\lambda; A)$ exists for $\lambda > 0$, and $|R(\lambda; A)|_1 \leqq 1$, $|R(\lambda; A)|_\infty \leqq 1$ for $\lambda > 0$. Show that

$\lim_{\lambda \to \infty} (\lambda R(\lambda; A)f)(s)$ exists almost everywhere for $f \in L_p$, $1 \leqq p < \infty$. Show that if S is a topological space, and if, for each $\lambda > 0$, $R(\lambda; A)f$ is continuous if f is continuous, then for $1 \leqq p < \infty$ and $f \in L_p$ the limit $\lim_{\lambda \to 0} (\lambda R(\lambda; A)f)(s)$ exists almost everywhere.

22 Let T be a transformation in L_1, $T \geqq 0$, $|T|_1 \leqq 1$, and $A(n) = A(T, n)$. Show that if f is real, $f \in L_1$, and $\sup_{0 \leqq n} A(n)f \geqq 0$ almost everywhere, and $\inf_{0 \leqq n} A(n)f \leqq 0$ almost everywhere, then $f = 0$.

23 (Hopf) Let T be a positive linear mapping of $L_1(S, \Sigma, \mu)$ into itself with $|T|_1 \leqq 1$. Let $f \in L_1$, and $g \in L_1$; let $g(s) > 0$ almost everywhere. Show that

$$\sup_{0 \leqq n} \frac{f(s)+(Tf)(s)+ \ldots + (T^n f)(s)}{g(s)+(Tg)(s)+ \ldots +(T^n g)(s)} < \infty$$

almost everywhere. Show from this that

$$\lim_{n \to \infty} \frac{f(s)+\ldots+(T^n f)(s)}{g(s)+\ldots+(T^n g)(s)}$$

exists almost everywhere on the set $\{s | \sum_{n=0}^{\infty} (T^n g)(s) < \infty\}$.

24 (Halmos ergodic theorem) Let (S, Σ, μ) be a positive measure space, $\varphi : S \to S$, and suppose that $\varphi^{-1}e \in \Sigma$ if $e \in \Sigma$. Let ω be a non-negative measurable function defined on S. Suppose that the mapping T defined for each μ-measurable function f by $(Tf)(s) = \omega(s)f(\varphi(s))$ maps $L_1(S, \Sigma, \mu)$ into itself, and that $|T|_1 \leqq 1$. Show that if f is a real function in L_1, there exists no set e of positive measure such that

$$\limsup_{n \to \infty} A(T, n)(f, s) > 0, \qquad \liminf_{n \to \infty} A(T, n)(f, s) < 0,$$

for $s \in e$. Show from this that if $f \in L_1$, $g \in L_1$, and $g(s) > 0$ μ-almost everywhere, then

$$\lim_{n \to \infty} \frac{A(T, n)(f, s)}{A(T, n)(g, s)}$$

exists μ-almost everywhere.

25 (Hurewicz-Oxtoby) Let (S, Σ, μ) be a positive σ-finite measure space, and let $\varphi : S \to S$ have the following properties:

(a) $\varphi^{-1}e \in \Sigma$ if $e \in \Sigma$;

(b) $\mu(e) = 0$ implies $\mu(\varphi^{-1}e) = 0$.

Let $v \in ca(S, \Sigma)$, let $v_{(n)}(e) = \sum_{j=0}^{n} v(\varphi^{-j}e)$, and $\mu_{(n)}(e) = \sum_{j=0}^{n} \mu(\varphi^{-1}e)$. Show that

$$\lim_{n\to\infty} \frac{dv_{(n)}}{d\mu_{(n)}}(s)$$

exists μ-almost everywhere. (Hint. Use the preceding exercise.)

26 Let (S, Σ, μ), φ, ω, and T be as in Exercise 24. Let $g \in L_1$, $g(s) > 0$ almost everywhere, and let f be non-negative and μ-measurable. Show that

$$\lim_{n\to\infty} \frac{A(T, n)(f, s)}{A(T, n)(g, s)}$$

either converges or diverges to $+\infty$ for μ-almost all $s \in S$. Show in particular that if $\omega = 1$, and $\mu(S) < \infty$, then for each non-negative μ-measurable function f the limit $\lim_{n\to\infty} (A(T, n)f)(s)$ exists or diverges to $+\infty$ almost everywhere.

27 Let (S, Σ, μ) be a finite positive measure space, let $0 \leqq f \in L_1(S, \Sigma, \mu)$, and let φ satisfy the conditions of Lemma 5.8. Show that

$$\lim_{n\to\infty} (f(s)f(\varphi s)\ldots f(\varphi^n s))^{1/n}$$

exists μ-almost everywhere.

28 Show that if $\{\mu_n\}$ is a bounded sequence of elements of $ca(S, \Sigma)$ which is such that

$$\lim_{m\to\infty} \limsup_{n\to\infty} |\mu_n(e_m)| = 0$$

for each decreasing sequence $\{e_m\}$ of sets in Σ with $\cap e_m = \varphi$, then $\{\mu_n\}$ is weakly compact.

29 Show that the converse of Theorem 6.12 is also valid.

30 Show that Lemma 6.10 holds as well for finitely additive set functions as for countably additive set functions.

31 (Y. N. Dowker) Let φ be a mapping of a set S into itself, and let Σ be a σ-field of subsets of S such that $\varphi^{-1}e \in \Sigma$ if $e \in \Sigma$. A non-negative element $m \in ca(S, \Sigma)$ is said to be potentially φ-invariant if there exists a non-negative element $\mu \in ca(S, \Sigma)$ such that $\mu(\varphi^{-1}e)$

$= \mu(e)$ and $\mu(e) = 0$ implies $m(e) = 0$. Show that m is potentially invariant if and only if the limit $\tilde{m}(e) = \lim_{n\to\infty} n^{-1} \sum_{i=0}^{n-1} m(\varphi^{-i}e)$ exists for each $e \in \Sigma$, and that $\tilde{m}$ is an element of $ca(S, \Sigma)$ satisfying $\tilde{m}(\varphi^{-1}e) = \tilde{m}(e)$. Hint. Consider the space of all μ-continuous elements of $ca(S, \Sigma)$.

32 (Y. N. Dowker) Let S, Σ, φ, m be as in the preceding exercise. Show that m is potentially invariant if and only if $\lim_{n\to\infty} n^{-1} \sum_{j=1}^{n} f(\varphi^j s)$ exists m-almost everywhere for every $f \in L_\infty(S,\Sigma,m)$.

33 (Dunford-Miller) Let S, Σ, m, φ, be as in Exercise 31. Show that $\lim_{n\to\infty} n^{-1} \sum_{j=0}^{n-1} f(\varphi^i s)$ converges in the mean of $L_1(S, \Sigma, m)$ for every $f \in L_1(S, \Sigma, m)$ only if there exists a constant $K < \infty$ such that

$$\frac{1}{n} \sum_{i=0}^{n-1} m(\varphi^{-i}e) \leqq Km(e), \qquad n = 0, 1, \ldots.$$

34 (Y. N. Dowker) Let S, Σ, φ, m be as in Exercise 31. Show that if for some p, $1 \leqq p < \infty$, the limit $\lim_{n\to\infty} n^{-1} \sum_{j=0}^{n-1} f(\varphi^i s)$ exists in the mean of $L_p(S, \Sigma, m)$ for each $f \in L_p$, then $\lim_{n\to\infty} n^{-1} \sum_{j=0}^{n-1} f(\varphi^i s)$ exists m-almost everywhere for each $f \in L_p$.

35 (Dunford-Miller) Let (S, Σ, μ) be a finite measure space, and, for $x_1, \ldots, x_n > 0$, let $\varphi_{x_1,\ldots,x_n}$ be a mapping of S into itself. Suppose that Λ denotes the Borel-Lebesgue measure of subsets of the set $\{x_1, \ldots, x_n | x_i > 0\}$. Let $\varphi_{x_1,\ldots,x_n}$ have the following properties:

(a) $\{x_1, \ldots, x_n, s | \varphi_{x_1,\ldots,x_n}(s) \in e\}$ is $\Lambda \times \mu$-measurable for each $e \in \Sigma$;

(b) there exists a constant $K < \infty$ such that $t^{-n}\int_0^t \cdots \int_0^t \mu\{s|\varphi_{x_1,\ldots,x_n}(s) \in e\}dx_1 \ldots dx_n \leqq K\mu(e)$ for $0 < t < \infty$.

Show that for each $f \in L_p(S, \Sigma, \mu)$, $1 \leqq p < \infty$,

$$\lim_{t\to\infty} t^{-n} \int_0^t \cdots \int_0^t f(\varphi_{x_1,\ldots,x_n} s)dx_1 \ldots dx_n$$

exists in the mean of L_p and also exists μ-almost everywhere. Hint, Use the "change of measure" method of the last four exercises.

36 (Ryll-Nardzewski) Let (S, Σ, μ) be a positive measure space, and φ a mapping of $S \to S$ such that $\varphi^{-1}e \in \Sigma$ if $e \in \Sigma$ and $\mu(\varphi^{-1}e) = 0$ if $\mu(e) = 0$. Show that the sequence $n^{-1} \sum_{i=0}^{n-1} f(\varphi^i s)$ converges μ-almost everywhere to a function $\tilde{f}$ in L_1 for each $f \in L_1$ if and only if

$$\limsup_{n\to\infty} \frac{1}{n} \sum_{j=0}^{n-1} \mu(\varphi^{-j}e) \leqq K\mu(e)$$

for each set e of finite μ-measure. (Hint. Consider the map $f(s) \to \chi_A(s)f(\varphi s)$ for each $A \in \Sigma$ with $\mu(A) < \infty$.)

37 Let (S, Σ, μ) be a positive measure space, and T a non-negative linear transformation of $L_1(S, \Sigma, \mu)$ into itself. Suppose that $|T|_1 \leqq 1$ and that there exists a constant K such that $|A(T, n)|_\infty \leqq K$. Show that $\lim_{n\to\infty} (A(T, n)f)(s)$ exists μ-almost everywhere for each $f \in L_p$, $1 \leqq p < \infty$.

38 Let S be a compact metric space, and let (S, Σ, μ) be a regular finite measure space. Let φ be a mapping of S into itself such that the set $\{\varphi^i\}$ of mappings is equicontinuous. Show that φ is metrically transitive if and only if $\{\varphi^i x\}$ is dense in S for some point x.

39 (Weyl) Let $[x_1, \ldots, x_n]$ be a point in n-dimensional Euclidean space whose coordinates are independent over the field of rational numbers. Put $x_m^{(n)} = [nx_m]$, where $[y]$ denotes the greatest integer in the real number y. Show that if C is a rectangular subset of the unit cube $\{[x_1, \ldots, x_n] | 0 \leqq x_i < 1\}$, and $v(m)$ denotes the number of indices $k \leqq m$ for which $[x_1^{(k)}, \ldots, x_n^{(k)}] \in C$, then $\lim_{m\to\infty} m^{-1}v(m)$ is the volume of C.

40 Let (S, Σ, μ) be a finite positive measure space, and let (S, Σ^*, v) be the product measure space of (S, Σ, μ) with an infinite number of replicas of itself. Show that the map $\varphi : S \to S$ defined by $\varphi(s_1 \times s_2 \times \ldots) = s_2 \times \ldots$ is metrically transitive.

41 Show that for almost all real numbers x, the digit 7 occurs with limiting frequency 0.1 in the decimal expansion of x, i.e., that $\lim_{n\to\infty} n^{-1}N(n) = 0.1$, where $N(n)$ is the number of 7's occurring among the first n digits of the decimal expansion of x.

42 Let φ be a metrically transitive mapping of a finite positive measure space (S, Σ, μ) into itself and suppose that $\mu(\varphi^{-1}e) = \mu(e)$ for $e \in \Sigma$. Show that if f is non-negative, μ-measurable, but not μ-integrable,

$$\lim_{n\to\infty} \frac{1}{n} \sum_{j=0}^{n-1} f(\varphi^j s) = \infty, \ \mu\text{-almost everywhere.}$$

43 Show that φ is a metrically transitive mapping of a finite positive measure space (S, Σ, μ) into itself if and only if no non-

constant μ-measurable function f can satisfy $f(\varphi s) = f(s)$ μ-almost everywhere, and that in this case, any μ-measurable function g satisfying $g(\varphi s) = \lambda g(s)$ μ-almost everywhere, where $|\lambda| = 1$, satisfies $|g(s)| = 1$ μ-almost everywhere. Show that if $h(\varphi s) = \lambda h(s)$ μ-almost everywhere also, then $h = \alpha g$.

44 (Random ergodic theorem) Let (S, Σ, μ) be a σ-finite measure space. Let $T_0, \ldots, T_9$ be ten operators in L_1, each satisfying $|T_i|_1 \leqq 1$, $|T_i|_\infty \leqq 1$. For each decimal $\alpha = 0.\alpha_1\alpha_2 \ldots$ such that $0 < \alpha < 1$, put

$$A(n, \alpha) = \frac{1}{n} \sum_{j=0}^{n-1} T_{\alpha_j} T_{\alpha_{j-1}} \ldots T_{\alpha_1}.$$

Show that if $f \in L_p(S, \Sigma, \mu)$, $1 \leqq p < \infty$, $\lim_{n\to\infty} A(n, \alpha)f$ exists almost everywhere for (Lebesgue) almost all α. Show that if $p > 1$, $\lim_{n\to\infty} A(n, \alpha)f$ exists in the mean of order p, and that the same holds if $\mu(S) < \infty$ for $p = 1$. (Hint. Consider a suitable operator J of the form

$$(Jf)(s, \alpha) = (T_{\alpha_1} f)(s, \varphi\alpha).)$$

10. Notes and Remarks

Semi-groups of operators. The theory of semi-groups of linear operators is of rather recent origin, the first results in this direction being obtained by M. H. Stone [10], in 1930, who discussed the case of a group of unitary operators in Hilbert space. Since that time a considerable development has taken place, largely due to the efforts of E. Hille, whose treatise [1] gives an extended account of the abstract theory and many applications to concrete mathematical problems. We refer the reader to this book (and its forthcoming revision) for further results as well as for references. In addition, the reader should consult the expository article by R. S. Phillips [9] for some recent developments. We mention here references for the small portion of the theory presented in Section 1, and do not go beyond these results.

The representation of uniformly continuous groups of operators as an exponential was demonstrated independently by Nathan [1; p. 525] and by Nagumo [1; p. 72] and Yosida [7; p. 24]; in fact, Nagumo and Yosida treated the case of a group on a B-algebra. It is seen in Hille [1; p. 162] that if the mapping $t \to T(t)$ is measurable in the uniform operator topology for t in $(0, \infty)$, and if $T(t+s) = T(t)T(s)$, $t, s > 0$, then it is also continuous in this topology for $t > 0$. However, these conditions do not imply the existence of $\lim_{t\to 0} T(t)$ or the differentiability of $T(t)$ for $t > 0$, and so the conclusions of Theorem 2 fail.

It was shown by von Neumann [10] that if $U(t)$, $-\infty < t < \infty$, is a one parameter group of unitary operators on Hilbert space $\mathfrak{H}$ and if $(U(t)x, y)$ is measurable for all $x, y \in \mathfrak{H}$, then $U(t)x$ is continuous on $t\varepsilon(-\infty, \infty)$ for all $x \in \mathfrak{H}$. This was extended by Dunford [12], who showed that if $T(t)$, $t > 0$, is a semi-group of bounded linear operators in a B-space, if $x^*T(\cdot)x$ is measurable on $(0, \infty)$ for any $x^* \in \mathfrak{X}^*$, $x \in \mathfrak{X}$, if $|T(t)| < M$ for t in some interval $(0, a)$ and if $\{T(t)x | t \in (0, \infty)\}$ is separable for each $x \in \mathfrak{X}$, then $T(\cdot)$ is strongly continuous on the right. It follows from Dunford's argument, see also Hille [1; p. 184], that if $T(\cdot)$ is such that $T(\cdot)x$ is measurable on $(0, \infty)$ and if $T(\cdot)$ is bounded on every finite interval of $(0, \infty)$, then $T(\cdot)x$ is continuous on $(0, \infty)$. The fact that this boundedness assumption is unnecessary was proved by Phillips [8] by means of the ingenious device reproduced in the present chapter (cf. 1.3). However, the condition of the measurability of $x^*T(\cdot)x$, $x^* \in \mathfrak{X}^*$, $x \in \mathfrak{X}$, alone is not sufficient to imply the strong continuity. Semi-groups with this weak measurability property have been studied by Feller [1].

The proof of Theorem 10 given here is found in Dunford and Segal [1]. Somewhat more general results of this nature are found in Hille [1; p. 187—9]. The first conditions that a closed linear operator generate a strongly continuous semi-group were given independently by Hille [1; p. 238] and Yosida [8]. The complete characterization of the infinitesimal generator given in Theorem 13 was obtained almost simultaneously by Feller [2], Miyadera [1] and Phillips [6]. For other results in this direction, see Hille [1] and Phillips [10].

The results concerning the perturbation of semi-group generators are due to Phillips [6]. See also Phillips [10].

Functions of an infinitesimal generator. Hille [1; Chap. 15] established an operational calculus for the infinitesimal generator of a semi-group for functions which are Laplace-Stieltjes transforms and are analytic on the spectrum of the generator. A somewhat different treatment was given by Phillips [5]. Bade [1] showed that for the infinitesimal generator of a group, it is possible to extend the class of functions to include polynomials and other functions with similar growth conditions. The treatment in Section 2 follows his work most closely. Theorem 13 generalizes certain theorems of Pollard [1], Widder [2] and Widder and Hirschman [1, 3] on the inversion of convolution transforms by differential operators of infinite order. There is an extensive literature on this problem—particularly on the difficult question of the pointwise convergence of $\{p_n(D)f(D)x(t)\}$ to $x(t)$, where $f(D)$ is a convolution transform and $\{p_n(D)\}$ is an inverting sequence. References to further discussions of this problem may be found in the cited papers and the book of Hirschman and Widder [1]. See also the expository articles of Schoenberg [3] and Widder [3]. The form of Theorem 13 given here is due to Bade [1]. Corollary 14 is a result of Widder and Hirschman [2].

Ergodic theory. Although the development of ergodic theory has taken place principally since 1931, there is considerable literature on the subject. Fortunately, the excellent monograph of Hopf [1] treats the early developments, and the expository articles of Halmos [4], Kakutani [10] and Oxtoby [1] discuss rather completely several different facets of the theory and contain many references to the literature. (See also F. Riesz [18].) Therefore, it is not necessary for us to give an extended account or bibliography of this theory. However, we do wish to make a few observations that may be found of interest, restricting our discussion to the material discussed in the text.

Mean ergodic theorem. The first proof of the mean ergodic theorem was given by von Neumann [11] after it was observed by Koopman [1] that measure preserving mappings in a measure space S give rise to unitary operators in $L_2(S)$. Von Neumann's proof was based on the spectral theory of unitary operators in Hilbert space. Many extensions and generalizations of this ergodic theorem have been given, both to more general B-spaces and to more general operators. The papers of Visser [1], F. Riesz [15, 17], Yosida [4], Kakutani [13], Lorch [8] and Yosida and Kakutani [2] are based on various weak compactness properties. A geometrical proof for operators in Hilbert space, based upon the fact that there is a shortest distance from a point to a convex set, was given by Wiener [3]. Shorter geometrical proofs, valid in uniformly convex spaces, were given by G. Birkhoff [7] and F. Riesz [16, 18]. Another proof, based on the interesting fact that a fixed point for a contraction in Hilbert space is also a fixed point for its adjoint, was given by Riesz and Sz.-Nagy [2]. Theorems for abstract vector lattices were proved by G. Birkhoff [2], Kakutani [7, 8], and F. Riesz [17].

Several generalizations of the mean ergodic theorem to groups or semi-groups of operators more general than the discrete semi-group $\{T^n | n = 0, 1, 2, \ldots\}$ have been presented in the papers of Alaoglu and Birkhoff [1, 2], G. Birkhoff [8], Day [8, 10], Dunford [9, 11], Eberlein [3, 4], and Wiener [3].

Ergodic theorems of the mean type, but in which other methods of summation replace the $(C, 1)$-method ordinarily used, are proved by Cohen [2], Hille [1; Chap. 14] and Phillips [4].

Pointwise ergodic theorem. This theorem was established by G. D. Birkhoff [1], who discussed measure preserving homeomorphisms of manifolds. The case of a finite measure space was treated by Khintchine [1] and the case of an infinite measure space by Stepanoff [1]. Some other generalizations have been given by Doob [1], Dowker [2, 3], Dunford and Miller [1], Halmos [8], Hurewicz [1], Khintchine [2], Oxtoby [2], Riesz [19], Ryll-Nardzewski [1], Wiener [3] and Wiener and Wintner [1].

One extension of the pointwise theorem (Theorem 6.8), called the *dominated* ergodic theorem is due to Wiener [3]. Other proofs of the dominated theorem were given by Yosida and Kakutani [1] and Fukamiya [1]. An important lemma in the proof of either the pointwise or the dominated ergodic theorems is the *maximal* ergodic theorem (Lemma 6.7). A proof of this result was given by Yosida and Kakutani [1], although a similar result was established in the paper of G. D. Birkhoff. The key to the treatment given here, i.e., the generalization of the maximal ergodic theorem to Markov processes (essentially Lemma 6.2) is due to E. Hopf [2]. Other discussions of the maximal ergodic theorem are due to Carathéodory [3], Dowker [2], Hopf [3], Pitt [1], and Riesz [18]. P. Hartman [1] proved the maximal ergodic theorem for the case of a flow.

A pointwise ergodic theorem for an n-parameter group of measure preserving transformations was given by Wiener [3]. The non-commutative case was discussed by Dunford [3] and Zygmund [2]. A generalization to a class of abstract groups was given by Calderón [1].

The theory of Markov processes is a generalization of the theory of point transformations. The ergodic theory of such processes has been treated by Doob [2, 3], Kakutani [16], Yosida [9], Yosida and Kakutani [2], and most recently by E. Hopf [2]. We refer the reader also to the recent treatise of Doob [4].

Uniform ergodic theorem. This result is due to Yosida and Kakutani; see their paper [2], where it is applied to the theory of Markov processes to obtain results previously obtained by Doeblin, Doob, Fréchet, and Kryloff and Bogoliouboff. For a more detailed probabilistic discussion of Markov processes we refer the reader to Doob [4]. It may be remarked that such processes are important in the theory of card shuffling.

REFERENCES

This list includes references for Part II as well as Part I of this volume

Abdelhay, J.

1. *Caractérisation de l'espace de Banach de toutes les suites de nombres réels tendant vers zéro.* C. R. Acad. Sci. Paris 229, 1111–1112 (1949).
2. *On a theorem of representation.* Bull. Amer. Math. Soc. 55, 408–417 (1949).

Abel, N. H.

1. *Untersuchungen über die Reihe:*

$$1+\frac{m}{1}x+\frac{m\cdot(m-1)}{1\cdot 2}x^2+\frac{m\cdot(m-1)\cdot(m-2)}{1\cdot 2\cdot 3}x^3+\ldots \text{ u.s.w.}$$

J. Reine Angew. Math. 1, 311–339 (1826).

Achieser, N. I. (see Ahiezer).

Adams, C. R.

1. *The space of functions of bounded variation and certain general spaces.* Trans. Amer. Math. Soc. 40, 421–438 (1936).

Adams, C. R., and Clarkson, J. A.

1. *On definitions of bounded variation for functions of two variables.* Trans. Amer. Math. Soc. 35, 824–854 (1933).
2. *Properties of functions $f(x,y)$ of bounded variation.* Trans. Amer. Math. Soc. 36, 711–730 (1934). Errata, ibid. 46, 468 (1939).
3. *On convergence in variation.* Bull. Amer. Math. Soc. 40, 413-417 (1934).
4. *The type of certain Borel sets in several Banach spaces.* Trans. Amer. Math. Soc. 45, 322–334 (1939).

Adams, C. R., and Morse, A. P.

1. *On the space (BV).* Trans. Amer. Math. Soc. 42, 194–205 (1937).
2. *Continuous additive functionals on the space (BV) and certain subspaces.* Trans. Amer. Math. Soc. 48, 82–100 (1940).

Agmon, S. (see Mandelbrojt, S.)

Agnew, R. P.

1. *Linear functionals satisfying prescribed conditions.* Duke Math. J. 4, 55–77 (1938).

Agnew, R. P., and Morse, A. P.

1. *Extensions of linear functionals, with applications to limits, integrals, measures, and densities.* Ann. of Math. (2) 39, 20–30 (1938).

Ahiezer, N. I.

1. *Infinite Jacobi matrices and the problem of moments.* Uspehi Mat. Nauk 9, 126–156 (1941).

Ahiezer (Achieser), N. I., and Glazman (Glasmann), I. M.

1. *The theory of linear operators on Hilbert space.* Gosudarstv. Izdat. Tehn.-Teor. Lit., Moscow-Leningrad, 1950. German translation, Akademie-Verlag, Berlin, 1954.

Ahlfors, L. V.

1. *Complex analysis.* McGraw-Hill, New York (1953).

Akilov, G. P.

1. *On the extension of linear operations.* Doklady Akad. Nauk SSSR (N.S.) 57, 643–646 (1947). (Russian) Math. Rev. 9, 241 (1948).
2. *Necessary conditions for the extension of linear operations.* Doklady Akad. Nauk SSSR (N.S.) 59, 417–418 (1948). (Russian) Math. Rev. 9, 358 (1948).

Alaoglu, L.

1. *Weak topologies of normed linear spaces.* Ann. of Math. (2) 41, 252–267 (1940).
2. *Weak convergence of linear functionals* (abstract). Bull. Amer. Math. Soc. 44, 196 (1938).

Alaoglu, L., and Birkhoff, G.

1. *General ergodic theorem.* Proc. Nat. Acad. Sci. U.S.A. 25, 628–630 (1939).
2. *General ergodic theorems.* Ann. of Math. (2) 41, 293–309 (1940).

Albrycht, J.

1. *On a theorem of Saks for abstract polynomials.* Studia Math. 14 (1953), 79–81 (1954).

Alexandroff, A. D.

1. *Additive set functions in abstract spaces,* I–III.
   I. Mat. Sbornik N.S. 8 (50), 307–348 (1940).
   II. ibid. 9 (51), 563–628 (1941).
   III. ibid. 13 (55), 169–238 (1943).

Alexandroff, P., and Hopf, H.

1. *Topologie,* I. J. Springer, Berlin, 1935. Reprinted by Edwards Bros., Ann Arbor.

Alexiewicz, A.

1. *On sequences of operations,* I–IV.
   I. Studia Math. 11, 1–30 (1950).
   II. ibid. 11, 200–236 (1950).
   III. ibid. 12, 84–92 (1951).
   IV. ibid. 12, 93–101 (1951).
2. *Linear operations among bounded measurable functions,* I, II.
   I. Ann. Soc. Polon. Math. 19, 140–161 (1946).
   II. ibid. 19, 161–164 (1946).
3. *On differentiation of vector-valued functions.* Studia Math. 11, 185–196 (1950).
4. *Continuity of vector-valued functions of bounded variation.* Studia Math. 12, 133–142 (1951).
5. *On some theorems of S. Saks.* Studia Math. 13, 18–29 (1953).
6. *A theorem on the structure of linear operations.* Studia Math. 14 (1953), 1–12 (1954).

Alexiewicz, A., and Orlicz, W.

1. *Remarks on Riemann-integration of vector-valued functions.* Studia Math. 12, 125–132 (1951).
2. *On analytic vector-valued functions of a real variable.* Studia Math. 12, 108–111 (1951).
3. *On the differentials in Banach spaces.* Ann. Soc. Polon. Math. 25 (1952), 95–99 (1953).
4. *Analytic operations in real Banach spaces.* Studia Math. 14 (1953), 57–78 (1954).

Altman (Al'tman), M. Š.

1. *On bases in Hilbert space.* Doklady Akad. Nauk SSSR (N.S.) 69, 483–485 (1949). (Russian) Math. Rev. 11, 525 (1950).
2. *On biorthogonal systems.* Doklady Akad. Nauk SSSR (N.S.) 67, 413–416 (1949). (Russian) Math. Rev. 11, 114 (1950).
3. *On linear functional equations in locally convex spaces.* Studia Math. 13, 194–207 (1953).
4. *Mean ergodic theorem in locally convex linear topological spaces.* Studia Math. 13, 190–193 (1953).
5. *The Fredholm theory of linear equations in locally convex topological spaces.* Bull. Acad. Polon. Sci. Cl. III. 2, 267–269 (1954).

Ambrose, W.

1. *Structure theorems for a special class of Banach algebras.* Trans. Amer. Math. Soc. 57, 364–386 (1945).
2. *Measures on locally compact topological groups.* Trans. Amer. Math. Soc. 61, 106–121 (1947).
3. *Direct sum theorem for Haar measures.* Trans. Amer. Math. Soc. 61, 122–127 (1947).
4. *Spectral resolution of groups of unitary operators.* Duke Math. J. 11, 589–595 (1944).

Anzai, H., and Kakutani, S.

1. *Bohr compactifications of a locally compact abelian group,* I, II.
   I. Proc. Imp. Acad. Tokyo 19, 476–480 (1943).
   II. ibid. 19, 533–539 (1943).

Arens, R. F.

1. *Duality on linear spaces.* Duke Math. J. 14, 787–794 (1947).
2. *The space L^ω and convex topological rings.* Bull. Amer. Math. Soc. 52, 931–935 (1946).
3. *Representation of functionals by integrals.* Duke Math. J. 17, 499–506 (1950).
4. *Approximation in, and representation of, certain Banach algebras.* Amer. J. Math. 71, 763–790 (1949).
5. *A topology for spaces of transformations.* Ann. of Math. (2) 47, 480–495 (1946).
6. *On a theorem of Gelfand and Neumark.* Proc. Nat. Acad. Sci. U.S.A. 32, 237–239 (1946).
7. *Representation of Banach *-algebras.* Duke Math. J. 14, 269–282 (1947).
8. *Linear topological division algebras.* Bull. Amer. Math. Soc. 53, 623–630 (1947).
9. *A generalization of normed rings.* Pacific J. Math. 2, 455–471 (1952).

Arens, R. F., and Kaplansky, I.

1. *Topological representation of algebras.* Trans. Amer. Math. Soc. 63, 457–481 (1948).

Arens, R. F., and Kelley, J. L.

1. *Characterizations of the space of continuous functions over a compact Hausdorff space.* Trans. Amer. Math. Soc. 62, 499–508 (1947).

Arnous, E.

1. *Sur les groupes continus de transformations unitaires de l'espace de Hilbert.* Comment. Math. Helv. 19, 50–60 (1946).

Aronszajn, N.

1. *Caractérisation métrique de l'espace de Hilbert, des espaces vectoriels et de certains groupes métriques.* C. R. Acad. Sci. Paris 201, 811–813, 873–875 (1935).
2. *Le correspondant topologique de l'unicité dans la théorie des équations différentielles.* Ann. of Math. (2) 43, 730–738 (1942).
3. *Approximation methods for eigenvalues of completely continuous symmetric operators.* Proceedings of the Symposium on Spectral Theory and Differential Problems, 179–202 (1951). Oklahoma Agricultural and Mechanical College, Stillwater, Oklahoma.
4. *The Rayleigh-Ritz and A. Weinstein methods for approximation of eigenvalues,* I, II. Proc. Nat. Acad. Sci. U.S.A. 34, 474–480, 594–601 (1948).
5. *Sur quelques problèmes concernant les espaces de Minkowski et les espaces vectoriels généraux.* Atti Accad. Naz. Lincei. Rend. Cl. Sci. Fis. Mat. Nat. (6) 26, 374–376 (1937).

Aronszajn, N., and Smith, K. T.

1. *Invariant subspaces of completely continuous operators.* Ann. of Math. (2) 60, 345–350 (1954).

Artemenko, A.

1. *La forme générale d'une fonctionnelle linéaire dans l'espace des fonctions à variation bornée.* Mat. Sbornik N. S. 6 (48), 215-220 (1939). (Russian. French summary) Math. Rev. 1, 239 (1940).
2. *On positive linear functionals in the space of almost periodic functions of H. Bohr.* Comm. Inst. Sci. Math. Méc. Univ. Kharkoff [Zapiski Inst. Mat. Mech.] (4) 16, 111–114 (1940). (Russian. English summary) Math. Rev. 3, 207 (1942).

Arzelà, C.

1. *Intorno alla continuità della somma di infinite funzioni continue.* Rend. dell'Accad. R. delle Sci. dell'Istituto di Bologna, 79–84 (1883–1884).
2. *Funzioni di linee.* Atti della R. Accad. dei Lincei. Rend. Cl. Sci. Fis. Mat. Nat. (4) 5_I, 342–348 (1889).
3. *Sulle funzioni di linee.* Mem. Accad. Sci. Ist. Bologna Cl. Sci. Fis. Mat. (5) 5, 55–74 (1895).
4. *Sulle serie di funzioni,* I, II.
   I. Memorie della R. Accad. delle Sci. dell'Istituto di Bologna Sci. Fis. e Mat. (5) 8, 3–58 (1899).
   II. ibid. (5) 8, 91–134 (1899).
5. *Un' osservazione intorno alle serie di funzioni.* Rend. dell' Accad. R. delle Sci. dell'Istituto di Bologna, 142–159 (1882–1883).

Ascoli, G.

1. *Sugli spazi lineari metrici e le loro varietà lineari.* Ann. Mat. Pura Appl. (4) 10, 33–81, 203–232 (1932).
2. *Le curve limiti di una varietà data di curve.* Atti della R. Accad. dei Lincei Memorie della Cl. Sci. Fis. Mat. Nat. (3) 18, 521–586 (1883–1884).

Atkinson, F. V.

1. *Symmetric linear operators on a Banach space.* Monatsh. Math. 53, 278–297 (1949).
2. *The normal solubility of linear equations in normed spaces.* Mat. Sbornik N. S. 28 (70), 3–14 (1951). (Russian) Math. Rev. 13, 46 (1952).

3. *A spectral problem for completely continuous operators.* Acta Math. Acad. Sci. Hungar. 3, 53–60 (1952). (Russian summary)
4. *On relatively regular operators.* Acta Sci. Math. Szeged 15, 38–56 (1953).
5. *On the second-order linear oscillator.* Univ. Nac. Tucumán. Revista A. 8, 71–87 (1951).

Audin, M.
1. *Sur certaines singularités des transformations linéaires bornées.* C. R. Acad. Sci. Paris 238, 2221–2222 (1954).

Babenko, K. I.
1. *On conjugate functions.* Doklady Akad. Nauk SSSR (N. S.) 62, 157–160 (1948). (Russian) Math. Rev. 10, 249 (1949).

Bade, W. G.
1. *An operational calculus for operators with spectrum in a strip.* Pacific J. Math. 3, 257–290 (1953).
2. *Unbounded spectral operators.* Pacific J. Math. 4, 373–392 (1954).
3. *Weak and strong limits of spectral operators.* Pacific J. Math. 4, 393–413 (1954).
4. *On Boolean algebras of projections and algebras of operators.* Trans. Amer. Math. Soc. 80, 345–360 (1955).

Bade, W. G., and Schwartz, J.
1. *On abstract eigenfunction expansions.* Proc. Nat. Acad. Sci., U.S.A. 42, 519–525 (1956).

Baker, H. F.
1. *On the integration of linear differential equations.* Proc. London Math. Soc. (1) 35, 333–378 (1903).

Banach, S.
1. *Théorie des opérations linéaires.* Monografje Matematyczne, Warsaw, 1932.
2. *Über die Baire'sche Kategorie gewisser Funktionenmengen.* Studia Math. 3, 174–179 (1931).
3. *Sur les opérations dans les ensembles abstraits et leur application aux équations intégrales.* Fund. Math. 3, 133–181 (1922).
4. *Sur les fonctionnelles linéaires,* I, II.
   I. Studia Math. 1, 211–216 (1929).
   II. ibid. 1, 223–239 (1929).
5. *Über homogene Polynome in (L^2).* Studia Math 7, 36–44 (1938).
6. *Teorja operacyj.* Warsaw, 1931. (Polish).
7. *Über metrische Gruppen.* Studia Math. 3, 101–113 (1931).
8. *Sur la convergence presque partout de fonctionnelles linéaires.* Bull. Sci. Math. (2) 50, 27–32, 36–43 (1926).

Banach, S., and Mazur, S.
1. *Zur Theorie der linearen Dimension.* Studia Math. 4, 100–112 (1933).

Banach, S., and Saks, S.
1. *Sur la convergence forte dans les champs L^p.* Studia Math. 2, 51–57 (1930).

Banach, S., and Steinhaus, H.
1. *Sur le principe de la condensation de singularités.* Fund. Math. 9, 50–61 (1927).

Barankin, E. W.
1. *Bounds on characteristic values.* Bull. Amer. Math. Soc. 54, 728–735 (1948).
2. *Bounds for characteristic roots of a matrix.* Bull. Amer. Math. Soc. 51, 767–770 (1945).

Barenblatt, G. I.
1. *On a method of solution of the equation of heat conduction.* Doklady Akad. Nauk SSSR (N. S.) 72, 667–670 (1950). (Russian) Math. Rev. 12, 183 (1951).

Bari (Bary), N. K.
1. *Biorthogonal systems and bases in Hilbert space.* Moskov. Gos. Univ. Učenye Zapiski 148, Matematika 4, 69–107 (1951). (Russian) Math. Rev. 14, 289 (1953).
2. *Sur la stabilité de la propriété d'être un système complet de fonctions.* Doklady Akad. Nauk SSSR (N. S.) 37, 83–87 (1942).

Bargman, V.
1. *Remarks on the determination of a central field of force from the elastic scattering phase shifts.* Phys. Rev. 75, 301–303 (1949).

Barry, J. Y.
1. *On the convergence of ordered sets of projections.* Proc. Amer. Math. Soc. 5, 313–314 (1954).

Bartle, R. G.
1. *Singular points of functional equations.* Trans. Amer. Math. Soc. 75, 366–384 (1953).
2. *On compactness in functional analysis.* Trans. Amer. Math. Soc. 79, 35–57 (1955).
3. *A general bilinear vector integral.* Studia Math. 15, 337–352 (1956).
4. *Implicit functions and solutions of equations in groups.* Math. Zeit. 62, 335–346 (1955).
5. *Newton's method in Banach spaces.* Proc. Amer. Math. Soc. 6, 827–831 (1955).

Bartle, R. G., Dunford, N., and Schwartz, J.
1. *Weak compactness and vector measures.* Canadian J. Math. 7, 289–305 (1955).

Bartle, R. G., and Graves, L. M.
1. *Mappings between function spaces.* Trans. Amer. Math. Soc. 72, 400–413 (1952).

Bassali, W. A. (see Stevenson, A. F.)

Bell, R. P.
1. *Eigenvalues and eigenfunctions for the operator* $\frac{d^2}{dx^2}-|x|$. Philos. Mag. 35, 385–588 (1944).

Bellman, R.
1. *A survey of the theory of the boundedness, stability and asymptotic behavior of solutions of linear and non-linear differential and difference equations.* Office of Naval Research, Washington, D. C., 1949.
2. *Stability theory of differential equations.* McGraw-Hill Co., New York, 1953.

Bennett, A. A.
1. *Newton's method in general analysis.* Proc. Nat. Acad. Sci. U.S.A. 2, 592–598 (1916).

Berezanskiĭ, Yu. M.
1. *On the uniqueness of the determination of Schrödinger's equation from its spectral function.* Doklady Akad. Nauk SSSR (N. S.) 93, 591–594 (1953). (Russian) Math. Rev. 15, 797 (1954).

2. *On hypercomplex systems constructed from a Sturm-Liouville equation on a semi-axis.* Doklady Akad. Nauk SSSR (N. S.) 91, 1245–1248 (1953). (Russian) Math. Rev. 15, 797 (1954).

Berkowitz, J.

1. *On the discreteness of the spectra of Sturm-Liouville operators.* Dissertation, New York University, 1951.

Berri, R.

1. *An investigation of the cone of positive elements in a partially ordered space.* Mat. Sbornik N. S. 23 (65), 419–440 (1948). (Russian) Math. Rev. 10, 380 (1949).

Besicovitch, A. S.

1. *Almost periodic functions.* Cambridge Univ. Press, Cambridge, 1932. Reprinted by Dover Pub. Co., New York.

Bethe, H. A.

1. *Theory of effective image in nuclear scattering.* Phys. Rev. 76, 38–50 (1949).

Beurling, A.

1. *Sur les intégrales de Fourier absolument convergentes et leur application à une transformation fonctionnelle.* Proc. IX Congrès de Math. Scandinaves, Helsingfors, 345–366 (1938).
2. *Un théorème sur les fonctions bornées et uniformément continues sur l'axe réel.* Acta Math. 77, 127–136 (1945).
3. *On the spectral synthesis of bounded functions.* Acta Math. 81, 225–238 (1949).
4. *On two problems concerning linear transformations in Hilbert space.* Acta Math. 81, 239–255 (1949).

Bieberbach, L.

1. *Lehrbuch der Funktionentheorie.* Vol. I, Fourth Ed., 1934; vol. II, Second Ed., 1931, Teubner, Leipzig. Reprinted by Chelsea Pub. Co., New York.

Birkhoff, G. (see also Alaoglu, L.)

1. *Orthogonality in linear metric spaces.* Duke Math. J. 1, 169–172 (1935).
2. *Dependent probabilities and the space (L).* Proc. Nat. Acad. Sci. U.S.A. 24, 154–159 (1938).
3. *Lattice theory.* Amer. Math. Soc. Coll. Publ. 25, New York, 1940.
4. *Integration of functions with values in a Banach space.* Trans. Amer. Math. Soc. 38, 357–378 (1935).
5. *A note on topological groups.* Compositio Math. 3, 427–430 (1936).
6. *On product integration.* J. Math. and Phys. Mass. Inst. Tech. 16, 104–132 (1937).
7. *The mean ergodic theorem.* Duke Math. J. 5, 19–20 (1939).
8. *An ergodic theorem for general semi-groups.* Proc. Nat. Acad. Sci. U.S.A. 25, 625–627 (1939).

Birkhoff, G., and MacLane, S.

1. *A survey of modern algebra.* Macmillan Co., New York, 1941.

Birkhoff, G. D.

1. *Proof of the ergodic theorem.* Proc. Nat. Acad. Sci. U.S.A. 17, 656–660 (1931)
2. *On the asymptotic character of the solutions of certain linear differential operations containing a parameter.* Trans. Amer. Math. Soc. 9, 219–231 (1908).

3. *Boundary value and expansion problems of ordinary linear differential equations*. Trans. Amer. Math. Soc. 9, 373–395 (1908).
4. *Existence and oscillation theorems for a certain boundary value problem*. Trans. Amer. Math. Soc. 10, 259–270 (1909).
5. *Quantum mechanics and asymptotic series*. Bull. Amer. Math. Soc. 39, 681–700 (1933).
6. *Note on the expansion of the Green's function*. Math. Ann. 72, 292–294 (1912).
7. *Note on the expansion problems of ordinary linear differential equations*. Rend. Circ. Mat. Palermo 36, 115–126 (1913).

Birkhoff, G. D., and Kellogg, O. D.
1. *Invariant points in function space*. Trans. Amer. Math. Soc. 23, 96–115 (1922).

Birkhoff, G. D., and Langer, R. E.
1. *The boundary problems and developments associated with a system of ordinary differential equations of the first order*. Proc. Amer. Acad. Arts Sci. (2) 58, 51–128 (1923).

Birman, M. Š.
1. *On the theory of self-adjoint extensions of positive definite operators*. Doklady Akad. Nauk SSSR (N. S.) 91, 189–191 (1953). (Russian) Math. Rev. 15, 326 (1954).

Birnbaum, Z. W., and Orlicz, W.
1. *Über die Verallgemeinerung des Begriffes der zueinander Konjugierten Potenzen*. Studia Math. 3, 1–67 (1931).

Bliss, G. A.
1. *A boundary value problem for a system of ordinary linear differential equations*. Trans. Amer. Math. Soc. 28, 561–589 (1926).

Block, H. D.
1. *Linear transformations on or onto a Banach space*. Proc. Amer. Math. Soc. 3, 126–128 (1952).

Blumenthal, L. M.
1. *Generalized Euclidean space in terms of a quasi-inner product*. Amer. J. Math. 72, 686–698 (1950).

Boas, M. L., Boas, R. P., and Levinson, N.
1. *The growth of solutions of a differential equation*. Duke Math. J. 9, 847–853 (1942).

Boas, R. P., Jr.
1. *Some uniformly convex spaces*. Bull. Amer. Math. Soc. 46, 304–311 (1940).
2. *Expansions of analytic functions*. Trans. Amer. Math. Soc. 48, 467–487 (1940).

Bôcher, M.
1. *On regular singular points of linear differential equations of the second order whose coefficients are not necessarily analytic*. Trans. Amer. Math. Soc. 1, 40–52 (1900).
2. *Green's functions in spaces of one dimension*. Bull. Amer. Math. Soc. 7, 297–299 (1901).
3. *Boundary problems and Green's functions for linear differential and difference equations*. Ann. of Math. (2) 13, 71–88 (1911).
4. *Applications and generalizations of the concept of adjoint system*. Trans. Amer. Math. Soc. 14, 403–420 (1913).
5. *Leçons sur les méthodes de Sturm*. Gauthier-Villars, Paris, 1917.

Bochner, S.

1. *Completely monotone functions in partially ordered space*. Duke Math. J. 9, 519–526 (1942).
2. *Integration von Funktionen, deren Werte die Elemente eines Vectorraumes sind*. Fund. Math. 20, 262–276 (1933).
3. *Additive set functions on groups*. Ann. Math. (2) 40, 769–799 (1939).
4. *Beiträge zur Theorie der fastperiodischen Funktionen*. Math. Ann. 96, 119–147 (1927).
5. *Absolut-additive abstrakte Mengenfunktionen*. Fund. Math. 21, 211–213 (1933).
6. *Vorlesungen über Fouriersche Integrale*. Akad. Verlag., Leipzig, 1932. Reprinted by Chelsea Pub. Co., New York.
7. *Spektraldarstellung linearer Scharen unitärer Operatoren*, S.-B. Preuss. Akad. Wiss., 371–376 (1933).
8. *Inversion formulae and unitary transformations*. Ann. of Math. (2) 35, 111–115 (1934).

Bochner, S., and Fan, K.

1. *Distributive order-preserving operations in partially ordered vector sets*. Ann. of Math. (2) 48, 168–179 (1947).

Bochner, S., and von Neumann, J.

1. *Almost periodic functions in groups*, II. Trans. Amer. Math. Soc. 37, 21–50 (1935).

Bochner, S., and Phillips, R. S.

1. *Additive set functions and vector lattices*. Ann. of Math. (2) 42, 316–324 (1941).
2. *Absolutely convergent Fourier expansions for non-commutative normed rings*. Ann. of Math. (2) 43, 409–418 (1942).

Bochner, S., and Taylor, A. E.

1. *Linear functionals on certain spaces of abstractly-valued functions*. Ann. of Math. (2) 39, 913–944 (1938).

Bohnenblust, H. F.

1. *An axiomatic characterization of L_p-spaces*. Duke Math. J. 6, 627–640 (1940)
2. *A characterization of complex Hilbert spaces*. Portugaliae Math. 3, 103–109 (1942).
3. *Subspaces of $l_{p,n}$ spaces*. Amer. J. Math. 63, 64–72 (1941).
4. *Convex regions and projections in Minkowski spaces*. Ann. of Math. (2) 39, 301–308 (1938).

Bohnenblust, H. F., and Kakutani, S.

1. *Concrete representations of (M)-spaces*. Ann. of Math. (2) 42, 1025–1028 (1941).

Bohnenblust, H. F., and Sobczyk, A.

1. *Extensions of functionals on complex linear spaces*. Bull. Amer. Math. Soc. 44, 91–93 (1938).

Bohr, H.

1. *Zur Theorie der fastperiodischen Funktionen*, I-III.
   I. Acta Math. 45, 29–127 (1925).
   II. ibid. 46, 101–214 (1925).
   III. ibid. 47, 237–281 (1926).

2. *Fastperiodische Funktionen.* Ergebnisse der Math. und ihrer Grenzgebiete I 5, J. Springer, Berlin, 1932. English translation, Chelsea Pub. Co., New York, 1947.
3. *On almost periodic functions and the theory of groups.* Amer. Math. Monthly 56, 595–609 (1949).
4. *A survey of the different proofs of the main theorems in the theory of almost periodic functions.* Proc. International Cong. Math., Cambridge, 1, 339–348 (1950).

Bohr, H., and Følner, E.
1. *On some types of functional spaces. A contribution to the theory of almost periodic functions.* Acta Math. 76, 31–155 (1944).

Bonnesen, T., and Fenchel, W.
1. *Theorie der konvexen Körper.* Ergebnisse der Math. und ihrer Grenzgebiete, III 1, J. Springer, Berlin, 1934.

Bonsall, F. F.
1. *A note on subadditive functionals.* J. London Math. Soc. 29, 125–126 (1954).

Borel, E.
1. *Sur l'équation adjointe et sur certains systèmes d'équations différentielles.* Ann. Sci. École Norm. Sup. (3) 9, 63–90 (1892).

Borg, G.
1. *Über die Stabilität gewisser Klassen von linearen Differentialgleichungen.* Ark. Mat. Astr. Fys. 31A, no. 1 (1944).
2. *Eine Umkehrung der Sturm-Liouvilleschen Eigenwertaufgabe Bestimmung der Differentialgleichung durch die Eigenwerte.* Acta Math. 78, 1–96 (1946).
3. *Inverse problems in the theory of characteristic values of differential systems.* C. R. Dixième Congrès Math. Scandinaves, Copenhagen, 1946.
4. *On the completeness of some sets of functions.* Acta Math. 81, 266–283 (1949).
5. *On a Liapounoff criterion of stability.* Amer. J. Math. 71, 67–70 (1949).
6. *Über die Ableitung der S-Funktion.* Math. Ann. 122, 326–331 (1950–1951).
7. *On the point spectra of* $y''+(\lambda - q(x))\,y = 0$. Amer. J. Math. 73, 122–126 (1951).

Botts, Truman
1. *On convex sets in linear normed spaces.* Bull. Amer. Math. Soc. 48, 150–152 (1942).
2. *Convex sets.* Amer. Math. Monthly 49, 527–531 (1942).

Bourbaki, N.
1. *Sur les espaces de Banach.* C. R. Acad. Sci. Paris 206, 1701–1704 (1938).
2. *Éléments de mathématique, Livre V, Espaces vectoriels topologiques.* Hermann et Cie, Act. Sci. et Ind. 1189, 1229, Paris, 1953, 1955.
3. *Sur certains espaces vectoriels topologiques.* Ann. Inst. Fourier Grenoble 2, 5–16 (1950).
4. *Éléments de mathématique, Livre VI, Intégration.* Hermann et Cie, Act. Sci. et Ind. 1175, Paris, 1952.
5. *Éléments de mathématique, Livre III, Topologie générale.* Hermann et Cie., Act. Sci, et Ind., nos. 858, 916, 1029, 1045, 1084, Paris, 1940–1949.

Bourgin, D. G.
1. *Some properties of Banach spaces.* Amer. J. Math. 64, 597–612 (1942).
2. *Linear topological spaces.* Amer. J. Math. 65, 637–649 (1943).

Brace, J. W.
1. *Transformations on Banach spaces.* Dissertation, Cornell University (1953).
2. *Compactness in the weak topology.* Math. Mag. 28, 125–134 (1955).

Bram, J.
1. *Subnormal operators.* Duke Math. J. 22, 75–94 (1955).

Bray, H. E.
1. *Elementary properties of the Stieltjes integral.* Ann. of Math. (2) 20, 177–186 (1918–1919).

Brauer, A.
1. *Limits for the characteristic roots of a matrix.* Duke Math. J. 13, 387–394 (1946).

Brillouin, L.
1. *La mécanique ondulatoire de Schrödinger; une méthode générale de résolution par approximations successives.* C. R. Acad. Sci. Paris 183, 24–26 (1926).

Brodskiĭ, M. S., and Milman, D. P.
1. *On the center of a convex set.* Doklady Akad. Nauk SSSR (N. S.) 59, 837–840 (1948). (Russian) Math. Rev. 9, 448 (1948).

Brouwer, L. E. J.
1. *Über eineindeutige, stetige Transformationen von Flächen in sich.* Math. Ann. 69, 176–180 (1910).

Browder, F. E.
1. *The Dirichlet problem for linear elliptic equations of arbitrary even order with variable coefficients.* Proc. Nat. Acad. Sci. U.S.A. 38, 230–235 (1952).
2. *The Dirichlet and vibration problems for linear elliptic differential equations of arbitrary order.* Proc. Nat. Acad. Sci. U.S.A. 38, 741–747 (1952).
3. *Assumption of boundary values and the Green's function in the Dirichlet problem for the general linear elliptic equation.* Proc. Nat. Acad. Sci. U.S.A. 39, 179–184 (1953).
4. *Linear parabolic differential equations of arbitrary order; general boundary-value problems for elliptic equations.* Proc. Nat. Acad. Sci. U.S.A. 39, 185–190 (1953).
5. *Strongly elliptic systems of differential equations.* Contributions to the theory of partial differential equations, 15–51. Ann. of Math. Studies, no. 33, Princeton, 1954.
6. *On the eigenfunctions and eigenvalues of the general linear elliptic differential operator.* Proc. Nat. Acad. Sci. 39, 433–439 (1953).
7. *The eigenfunction expansion theorem for the general self-adjoint singular elliptic partial differential operator. I. The analytical foundation.* Proc. Nat. Acad. Sci. U.S.A. 40, 454–459 (1954).
8. *Eigenfunction expansions for singular elliptic operators. II. The Hilbert space argument; parabolic equations on open manifolds.* Proc. Nat. Acad. Sci. U.S.A. 40, 459–463 (1954).

Brown, A.
1. *On a class of operators.* Proc. Amer. Math. Soc. 4, 723–728 (1953).
2. *The unitary equivalence of binormal operators.* Amer. J. Math. 76, 414–434 (1954).

Browne, E. T.
1. *Limits to the characteristic roots of a matrix.* Amer. Math. Monthly 46, 252–265 (1939).

Buchheim, A.

1. *An extension of a theorem of Professor Sylvester's relating to matrices.* Phil. Mag. (5) 22, 173–174 (1886).

Buniakowsky, V.

1. *Sur quelques inégalités concernant les intégrales ordinaires et les intégrales aux différences finies.* Mém. Acad. St. Petersburg (7) 1, no. 9 (1859).

Burgat, P.

1. *Résolutions de problèmes aux limites au moyen de transformations fonctionnelle.* Dissertation, Université de Neuchâtel, Lausanne, 1950.
2. *Résolution de problèmes aux limites au moyen de transformations fonctionnelles.* Z. Angew. Math. Physik 4, 146–152 (1953).

Burkhardt, H.

1. *Sur les fonctions de Green rélatives à une domaine d'une dimension.* Bull. Soc. Math. France 22, 71–75 (1894).

Burnett, D.

1. *The distribution of velocities in a slightly non-uniform gas.* Proc. London Math. Soc. (2) 39, 385–430 (1935).

Burton, L. P.

1. *Oscillation theorems for the solutions of linear, non-homogeneous, second order differential systems.* Pacific J. Math. 2, 281–289 (1952).

Butler, J. B.

1. *Perturbation series for eigenvalues of regular non-symmetric operators.* Technical Report no. 8 to the Office of Ordinance Research, Univ. of California, Berkeley (1955).

Cafiero, F.

1. *Criteri di compattezza per le successioni di funzioni generalmente a variazione limitata,* I, II.
   I. Atti Accad. Naz. Lincei. Rend. Cl. Sci. Fis. *Math. Nat.* (8), 305–311 (1950).
   II. ibid. (8) 8, 450–457 (1950).
2. *Sugli insiemi compatti di funzioni misurabili negli spazi astratti,* Rend. Sem. Mat. Univ. Padova 20, 48–58 (1951).
3. *Sulle famiglie di funzioni additive d'insieme, uniformemente continue.* Atti Accad. Naz. Lincei. Rend. Cl. Sci. Fis. Mat. Nat. (8) 12, 155–162 (1952).
4. *Sul passaggio al limite sotto il segno d'integrale di Stieltjes-Lebesgue negli spazi astratti, con masse variabili con gli integrandi,* I, II.
   I. Atti Accad. Naz. Lincei. Rend. Cl. Sci. Fis. Mat. Nat. (8) 14, 488–494 (1953).
   II. Rend. Sem. Mat. Univ. Padova 22, 223–245 (1953).
5. *Sulle famiglie compatte di funzioni additive di insieme astratto.* Atti del Quarto Congresso dell'Unione Mat. Italiana, Taormin, 1951, vol. II, pp. 30–40. Casa Editrice Perrella, Rome, 1953.

Calderón, A. P.

1. *A general ergodic theorem.* Ann. of Math. (2) 58, 182–191 (1953).

Calderón, A. P., and Zygmund, A.

1. *A note on the interpolation of linear operations.* Studia Math. 12, 194–204 (1951).
2. *On the theorem of Hausdorff-Young and its applications.* Contributions to Fourier Analysis, 166–188. Ann. of Math. Studies, No. 25, Princeton Univ. Press (1950).

3. *A note on the interpolation of sublinear operations.* Amer. J. Math. 78, 282–288 (1956).
4. *On the existence of certain singular integrals.* Acta Math. 88, 85–139 (1952).
5. *On singular integrals.* Amer. J. Math. 78, 289–309 (1956).
6. *Algebras of certain singular operators.* Amer. J. Math. 78, 310–320 (1956).

Calkin, J. W.
1. *Abstract symmetric boundary conditions.* Trans. Amer. Math. Soc. 45, 369–442 (1939).
2. *Two sided ideals and congruences in the ring of bounded operators in Hilbert space.* Ann. of Math. (2) 42, 839–873 (1941).
3. *Symmetric transformations in Hilbert space.* Duke J. Math. 7, 504–508 (1940).

Cameron, R. H.
1. *A "Simpson's Rule" for the numerical evaluation of Wiener's integrals in function space.* Duke Math. J. 18, 111–130 (1951).
2. *The first variation of an indefinite Wiener integral.* Proc. Amer. Math. Soc. 2, 914–924 (1951).
3. *The generalized heat flow equation and a corresponding Poisson formula.* Ann. of Math. (2) 59, 434–462 (1954).
4. *Some examples of Fourier-Wiener transforms of analytic functionals.* Duke Math. J. 12, 485–488 (1945).
5. *The translation pathology of Wiener space.* Duke Math. J. 21, 623–627 (1954).

Cameron, R. H., and Fagan, R. E.
1. *Non- linear transformations of Volterra type in Wiener space.* Trans. Amer. Math. Soc. 75, 552–575 (1953).

Cameron, R. H., and Graves, R. E.
1. *Additive functionals on a space of continuous functions.* I. Trans. Amer. Math. Soc. 70, 160–176 (1951).

Cameron, R. H., and Hatfield, C.
1. *Summability of certain orthogonal developments of non-linear functionals.* Bull. Amer. Math. Soc. 55, 130–145 (1949).
2. *Summability of certain series for unbounded non-linear functionals.* Proc. Amer. Math. Soc. 4, 375–387 (1953).

Cameron, R. H., Lindgren, B. W., and Martin, W. T.
1. *Linearization of certain non-linear functional equations.* Proc. Amer. Math. Soc. 3, 138–143 (1952).

Cameron, R. H., and Martin, W. T.
1. *An expression for the solution of a class of non-linear integral equations.* Amer. J. Math. 66, 281–298 (1944).
2. *The orthogonal development of non-linear functionals in series of Fourier-Hermite functionals.* Ann. of Math. (2) 48, 385–392 (1947).
3. *The transformation of Wiener integrals by non-linear transformations.* Trans. Amer. Math. Soc. 66, 253–283 (1949).
4. *Non-linear integral equations.* Ann. of Math. (2) 51, 629–642 (1950).
5. *Transformations of Wiener integrals under a general class of linear transformations.* Trans. Amer. Math. Soc. 58, 184–219 (1945).
6. *Evaluation of various Wiener integrals by use of Sturm-Liouville differential equations.* Bull. Amer. Math. Soc. 51, 73–90 (1945).

7. *Transformations of Wiener integrals under translations.* Ann. of Math. (2) 45, 386–396 (1944).
8. *The Wiener measure of Hilbert neighborhoods in the space of real continuous functions.* J. Math. Phys. Mass. Inst. Tech. 23, 195–209 (1944).
9. *Fourier-Wiener transforms of analytic functionals.* Duke Math. J. 12, 489–507 (1945).

Cameron, R. H., and Shapiro, J. M.
1. *Non-linear integral equations.* Ann. of Math. (2) 62, 472–497 (1955).

Camp, B. H.
1. *Singular multiple integrals, with applications to series.* Trans. Amer. Math. Soc. 14, 42–64 (1913).

Carathéodory, C.
1. *Vorlesungen über reelle Funktionen.* Second Ed., Teubner, Leipzig, 1927.
2. *Vorlesungen über reelle Funktionen.* Teubner, Berlin and Leipzig, 1918. Reprinted by Chelsea Pub. Co., New York.
3. *Bemerkungen zur Riesz-Fischerschen Satz und zur Ergodentheorie.* Abh. Math. Sem. Hansischen Univ. 14, 351–389 (1941).

Carleman, T.
1. *Sur les équations intégrales singulières à noyau réel et symétrique.* Almquist and Wiksells, Uppsala, 1923.
2. *Zur Theorie der linearen Integralgleichungen.* Math. Zeit. 9, 196–217 (1921).
3. *Über die asymptotische Verteilung der Eigenwerte partiellen Differentialgleichungen.* Ber. Verh. Sachs. Akad. Wiss. Leipzig. Math.-Nat. Kl. 88, 119–132 (1936).

Cartan, H.
1. *Sur la mesure de Haar.* C. R. Acad. Sci. Paris 211, 759–762 (1940).

Cartan, H., and Godement, R.
1. *Théorie de la dualité et analyse harmonique dans les groupes abéliens localement compacts.* Ann. École Norm. Sup. 64, 79–99 (1947).

Cauchy, A.
1. *Oeuvres,* sér. I, t. 12. Gauthier-Villars, Paris, 1900.
2. *Oeuvres,* sér. II, t. 3. Gauthier-Villars, Paris, 1900.

Čech, E.
1. *On bicompact spaces.* Ann. of Math. (2) 38, 823–844 (1937).

Chang, S. H.
1. *On the distribution of the characteristic values and singular values of linear integral equations.* Trans. Amer. Math. Soc. 67, 351–367 (1949).
2. *Generalization of a theorem of Lalesco.* J. London Math. Soc. 22, 185–189 (1947).

Charzyński, Z.
1. *Sur les transformations isométriques des espaces du type (F).* Studia Math. 13, 94–121 (1953).

Chevalley, C.
1. *Theory of Lie groups,* I. Princeton Univ. Press, Princeton, 1946.
2. *Théorie des groupes de Lie,* II. Hermann et Cie., Act. Sci. et Ind. 1152, Paris, 1951.

Chiang, T. P.
1. *A theorem on the normalcy of completely continuous operators.* Acta Sci. Math. Szeged. 14, 188–196 (1952).

Christian, R. R.

1. *On integration with respect to a finitely additive measure whose values lie in a Dedekind complete partially ordered vector space.* Dissert., Yale Univ. (1954).

Clarkson, J. A. (see also Adams, C. R.)

1. *Uniformly convex spaces.* Trans. Amer. Math. Soc. 40, 396–414 (1936).
2. *A characterization of C-spaces.* Ann. of Math. (2) 48, 845–850 (1947).

Clarkson, J. A., and Erdös, P.

1. *Approximation by polynomials.* Duke Math. J. 10, 5–11 (1943).

Clifford, A. H. (see Michal, A. D.)

Coddington, E. A.

1. *On the spectral representation of ordinary self-adjoint differential operators.* Proc. Nat. Acad. Sci. 38, 732–737 (1952).
2. *The spectral representation of ordinary self-adjoint differential operators.* Ann. of Math. (2) 60, 192–211 (1954).
3. *A characterization of ordinary self-adjoint differential systems* (abstract). Bull. Amer. Math. Soc. 58, 42 (1952).
4. *The spectral matrix and Green's function for singular self-adjoint boundary value problems.* Canadian J. Math. 6, 169–185 (1954).

Coddington, E. A., and Levinson, N.

1. *Theory of differential equations.* McGraw-Hill, New York, 1955.
2. *On the nature of the spectrum of singular second order linear differential operators.* Canadian J. Math. 3, 335–338 (1951).
3. *Perturbations of linear systems with constant coefficients possessing periodic solutions.* Contribution to the theory of non-linear oscillations II, 19–35, Princeton, 1952.

Cohen, I. S.

1. *On non-Archimedean normed spaces.* Nederl. Akad. Wetensch., Proc. 51, 693–698 (1948).

Cohen, L. W.

1. *Transformations on spaces of infinitely many dimensions.* Ann. of Math. (2) 37, 326–335 (1936).
2. *On the mean ergodic theorem.* Ann. of Math. (2) 41, 505–509 (1940).

Cohen, L. W., and Dunford, N.

1. *Transformations on sequence spaces.* Duke Math. J. 3, 689–701 (1937).

Collatz, L.

1. *Eigenwertprobleme und ihre numerische Behandlung.* Akademischer Verlag, Leipzig, 1945. Reprinted Chelsea Pub. Co., New York, 1948.

Collins, H. S.

1. *Completeness and compactness in linear topological spaces.* Trans. Amer. Math. Soc. 79, 256–280 (1955).

Cooke, R. G.

1. *Linear operators.* Macmillan, London, 1953.

Cooper, J. L. B.

1. *The spectral analysis of self-adjoint operators.* Quart. J. Math. (Oxford) 16, 31–48 (1945).
2. *Symmetric operators in Hilbert space.* Proc. London Math. Soc. (2) 50, 11–55 (1948).
3. *One-parameter semi-groups of isometric operators in Hilbert space.* Ann. of Math. (2) 48, 827–842 (1947).

Cordes, H. O.

1. *Separation des Variablen in Hilbertschen Räumen*. Math. Ann. 125, 401–434 (1953).
2. *Der Entwicklungssatz nach Produkten bei singulären Eigenwertproblemen partieller Differentialgleichungen, die durch Separation zerfallen*. Nachr. Akad. Wiss. Göttingen. Math.-Phys.-Kl. 1954, 51–69 (1954).

Cotlar, M.

1. *On a theorem of Beurling and Kaplansky*. Pacific J. Math. 4, 459–465 (1954).

Cotlar, M., and Ricabarra, R. A.

1. *On transformations of sets and Koopman's operators*. Revista Unión Mat. Argentina 14, 232–254 (1950). (Spanish) Math. Rev. 12, 719 (1951).

Courant, R., and Hilbert, D.

1. *Methoden der mathematischen physik*. I, II. J. Springer, Berlin, 1924, 1937.
2. *Methods of mathematical physics*. I. Interscience Pub., New York, 1953.

Courant, R., and Lax, A.

1. *Remarks on Cauchy's problem for hyperbolic partial differential equations with constant coefficients in several independent variables*. Comm. Pure Appl. Math. 8, 497–502 (1955).

Cronin, J.

1. *Branch points of solutions of equations in Banach space*. Trans. Amer. Math. Soc. 69, 105–131 (1950).
2. *Branch points of solutions of equations in Banach space*, II. Trans. Amer. Math. Soc. 76, 207–222 (1954).
3. *A definition of degree for certain mappings in Hilbert space*. Amer. J. Math. 73, 763–772 (1951).
4. *Analytic functional mappings*. Ann. of Math. (2) 58, 175–181 (1953).

Daniell, P. J.

1. *A general form of integral*. Ann. of Math. (2) 19, 279–294 (1917–1918).

van Dantzig, D.

1. *Zur topologischen Algebra*, I. Math. Ann. 107, 587–626 (1932).
2. *Einige Sätze über topologische Gruppen*. Jber. Deutsch. Math. Verein. 41, 42–44 (1932).

Davies, R.

1. *Expansions in series of non-orthogonal eigenfunctions*. Industr. Math. 4, 9–16 (1953).

Davis, H. T.

1. *The theory of linear operators*. Principia Press, Bloomington, Indiana, 1936.

Day, M. M.

1. *The space L^p with* $0 < p < 1$. Bull. Amer. Math. Soc. 46, 816–823 (1940).
2. *A property of Banach spaces*. Duke Math. J. 8, 763–770 (1941).
3. *Reflexive Banach spaces not isomorphic to uniformly convex spaces*. Bull. Amer. Math. Soc. 47, 313–317 (1941).
4. *Some more uniformly convex spaces*. Bull. Amer. Math. Soc. 47, 504–517 (1941).
5. *Uniform convexity*, III. Bull. Amer. Math. Soc. 49, 745–750 (1943).
6. *Uniform convexity in factor and conjugate spaces*. Ann. of Math. 45, 375–385 (1944).
7. *Some characterizations of inner- product spaces*. Trans. Amer. Math. Soc. 62, 320–337 (1947).

8. *Means for the bounded functions and ergodicity of the bounded representations of semi-groups.* Trans. Amer. Math. Soc. 69, 276–291 (1950).
9. *Operations in Banach spaces.* Trans. Amer. Math. Soc. 51, 583–608 (1942).
10. *Ergodic theorems for abelian semi-groups.* Trans. Amer. Math. Soc. 51, 399–412 (1942).
11. *Strict convexity and smoothness of normed spaces.* Trans. Amer. Math. Soc. 78, 516–528 (1955).

de Sz.-Nagy, B. (see under Sz.-Nagy)

Devinatz, A., Nussbaum, A. E., and von Neumann, J.
1. *On the permutability of self-adjoint operators.* Ann. of Math. (2) 62, 199–203 (1955).

Dieudonné, J.
1. *Sur le théorème de Hahn-Banach.* Rev. Sci. 79, 642–643 (1941).
2. *Sur la séparation des ensembles convexes dans un espace de Banach.* Rev. Sci. 81, 277–278 (1943).
3. *La dualité dans les espaces vectoriels topologiques.* Ann. Sci. École Norm. Sup. (3) 59, 107–139 (1942).
4. *Sur le théorème de Lebesgue-Nikodým.* Ann. of Math. (2) 42, 547–555 (1941).
5. *Sur le théorème de Lebesgue-Nikodym*, II. Bull. Soc. Math. France 72, 193–239 (1944); errata, ibid., 74, 66–68 (1946).
6. *Complex structures on real Banach spaces.* Proc. Amer. Math. Soc. 3, 162–164 (1952).
7. *Sur les espaces de Köthe.* J. Analyse Math. 1, 81–115 (1951).
8. *Natural homomorphisms in Banach spaces.* Proc. Amer. Math. Soc. 1, 54–59 (1950).
9. *Sur le théorème de Lebesgue-Nikodým*, IV. J. Indian Math. Soc. (N. S.) 15, 77–86 (1951).
10. *Sur le théorème de Lebesgue-Nikodým*, V. Canadian J. Math. 3, 129–139 (1951).
11. *Sur la convergence des suites de mesures de Radon.* Anais Acad. Brasil. Ci. 23, 21–38, 277–282 (1951).
12. *Sur un théorème de Jessen.* Fund. Math. 37, 242–248 (1950).
13. *Recent developments in the theory of locally convex vector spaces.* Bull. Amer. Math. Soc. 59, 495–512 (1953).
14. *Sur le théorème de Lebesgue-Nikodým*, III. Ann. Inst. Fourier Grenoble 23, 25–53 (1947–1948).
15. *Sur un théorème de Šmulian.* Arch. Math. 3, 436–440 (1952).
16. *On biorthogonal systems.* Michigan Math. J. 2, 7–20 (1954).
17. *Sur le produit de composition.* Compositio Math. 12, 17–34 (1954).
18. *Bounded sets in (F)-spaces.* Proc. Amer. Math. Soc. 6, 729–731 (1955).
19. *Sur la bicommutante d'une algèbre d'opérateurs.* Portugaliae Math. 14, 35–38 (1955).
20. *Sur la théorie spectrale.* J. Math. Pures Appl. (9) 35, 175–187 (1956).
21. *Champs de vecteurs non localement triviaux.* Archiv des Math. 7, 6–10 (1956).

Dieudonné, J., and Gomes, A. P.
1. *Sur certains espaces vectoriels topologiques.* C. R. Acad. Sci. Paris 230, 1129–1130 (1950).

Dieudonné, J., and Schwartz, L.
1. *La dualité dans les espaces (F) et (LF).* Ann. Inst. Fourier, Grenoble 1, 61–101 (1950).

Dines, L. L. (see Moskovitz, D.)

Dini, U.

1. *Fonamenti per la teorica delle funzioni di variabili reali.* Pisa (1878).

Dirac, P. A. M.

1. *The principles of quantum mechanics.* Oxford Press, London, 1935.

Ditkin, V. A.

1. *On the structure of ideals in certain normed rings.* Učenye Zapiski Moskov. Gos. Univ. Matematika 30, 83–130 (1939). (Russian, English summary) Math. Rev. 1, 336 (1940).

Dixmier, J.

1. *Les moyennes invariantes dans les semi-groupes et leurs applications.* Acta Sci. Math. Szeged 12 Pars A, 213–227 (1950).
2. *Les fonctionnelles linéaires sur l'ensemble des opérateurs bornés d'un espace de Hilbert.* Ann. of Math. (2) 51, 387–408 (1950).
3. *Sur certains espaces considérés par M. H. Stone.* Summa Brasil. Math. 2, 151–182 (1951).
4. *Sur un théorème de Banach.* Duke Math. J. 15, 1057–1071 (1948).
5. *Les algèbres d'opérateurs dans l'espace hilbertien.* Gauthiers-Villars, Paris, 1957.
6. *Sur une inégalité de E. Heinz.* Math. Ann. 126, 75–78 (1953).
7. *Sur les bases orthonormales dans les espaces préhilbertiens.* Acta Sci. Math. Szeged 15, 29–30 (1953).

Dixon, A. C.

1. *On a class of expansions in oscillating functions.* Proc. London Math. Soc. (2) 3, 83–103 (1905).

Donoghue, W. F., and Smith, K. T.

1. *On the symmetry and bounded closure of locally convex spaces.* Trans. Amer. Math. Soc. 73, 321–344 (1952).

Doob, J. L. (see also Koopman, B. O.)

1. *The law of large numbers for continuous stochastic processes.* Duke Math. J. 6, 290–306 (1940).
2. *Stochastic processes with an integral-valued parameter.* Trans. Amer. Math. Soc. 44, 87–150 (1938).
3. *Asymptotic properties of Markoff transition probabilities.* Trans. Amer. Math. Soc. 63, 393–421 (1948).
4. *Stochastic processes.* Wiley, New York, 1953.

Dorodnicyn, A. A.

1. *Asymptotic laws of distribution of the characteristic values for certain special forms of differential equations of second order.* Uspehi Matem. Nauk (N. S.) 7, no. 6 (52), 3–96 (1952). (Russian) Math. Rev. 14, 877 (1953).

Doubrovsky, V. M. (see Dubrovskiĭ)

Doyle, T. C.

1. *Invariant theory of general ordinary, linear, homogeneous, second order differential boundary problems.* Duke Math. J. 17, 249–261 (1950).

Dowker, Y. N.

1. *Finite and σ-finite invariant measures.* Ann. of Math. (2) 54, 595–608 (1951)
2. *A new proof of the general ergodic theorem.* Acta Sci. Math. Szeged 12 Pars B, 162–166 (1950).
3. *A note on the ergodic theorem.* Bull. Amer. Math. Soc. 55, 379–383 (1949).

Dresden, A.

1. *Solid analytical geometry and determinants.* H. Holt Co., New York, 1930.

Dubrovskiĭ (Doubrovsky), V. M.

1. *On some properties of completely additive set functions and their application to generalization of a theorem of Lebesgue.* Mat. Sbornik N. S. 20 (62), 317–329 (1947). (Russian, English summary) Math. Rev. 9, 19 (1948).
2. *On the basis of a family of completely additive functions of sets and on the properties of uniform additivity and equi-continuity.* Doklady Akad. Nauk SSSR (N. S.) 58, 737–740 (1947). (Russian) Math. Rev. 9, 275 (1948).
3. *On properties of absolute continuity and equi-continuity.* Doklady Akad. Nauk SSSR (N.S.) 63, 483–486 (1948). (Russian) Math. Rev. 10, 361 (1949).
4. *On equi-summable functions and on the properties of uniform additivity and equi-continuity of a family of completely additive set functions.* Izvestiya Akad. Nauk SSSR. Ser. Mat. 13, 341–356 (1949). (Russian) Math. Rev. 11, 90 (1950).
5. *On the property of equicontinuity of a family of completely additive set functions with respect to proper and improper bases.* Doklady Akad. Nauk SSSR (N. S.) 76, 333–336 (1951). (Russian) Math. Rev. 12, 598 (1951).
6. *On a property of a formula of Nikodým.* Doklady Akad. Nauk SSSR (N. S.) 85, 693–696 (1952). (Russian) Math. Rev. 14, 456 (1953).
7. *On certain conditions of compactness.* Izvestiya Akad. Nauk SSSR. Ser. Math. 12, 397–410 (1948). (Russian) Math. Rev. 10, 108 (1949).

Dugundji, J.

1. *An extension of Tietze's theorem.* Pacific J. Math. 1, 353–367 (1951).

Dunford, N. (see also Bartle, R. G., and Cohen, L. W.)

1. *Uniformity in linear spaces.* Trans. Amer. Math. Soc. 44, 305–356 (1938).
2. *Direct decompositions of Banach spaces.* Bol. Soc. Mat. Mexicana 3, 1–12 (1946).
3. *An individual ergodic theorem for non-commutative transformations.* Acta Sci. Math. Szeged 14, 1–4 (1951).
4. *Integration in general analysis.* Trans. Amer. Math. Soc. 37, 441–453 (1935).
5. *On continuous mappings.* Ann. of Math. (2) 41, 639–661 (1940).
6. *Spectral theory.* Bull. Amer. Math. Soc. 49, 637–651 (1943).
7. *Spectral theory* I, *Convergence to projections.* Trans. Amer. Math. Soc. 54, 185–217 (1943).
8. *Integration and linear operations.* Trans. Amer. Math. Soc. 40, 474–494 (1936).
9. *A mean ergodic theorem.* Duke Math. J. 5, 635–646 (1939).
10. *On a theorem of Plessner.* Bull. Amer. Math. Soc. 41, 356–358 (1935).
11. *An ergodic theorem for n-parameter groups.* Proc. Nat. Acad. Sci. U.S.A. 25, 195–196 (1939).
12. *On one parameter groups of linear transformations.* Ann. of Math. (2) 39, 569–573 (1938).
13. *Resolution of the identity for commutative B*-algebras of operators.* Acta Sci. Math. Szeged 12 Pars B, 51–56 (1950).
14. *Spectral theory in abstract spaces and Banach algebras.* Proc. Symposium on Spectral Theory and Differential Problems, 1–65 (1951). Oklahoma Agricultural and Mechanical College, Stillwater, Oklahoma.

15. *Spectral theory.* Proc. Symposium on Spectral Theory etc., 203–208 (1951).
16. *The reduction problem in spectral theory.* Proc. International Congress Math., Cambridge, Mass., 1950, Vol. 2, 115–122.
17. *Spectral theory.* II. *Resolutions of the identity.* Pacific J. Math. 2, 559–614 (1952).
18. *Spectral operators.* Pacific J. Math. 4, 321–354 (1954).

Dunford, N., and Miller, D. S.
1. *On the ergodic theorem.* Trans. Amer. Math. Soc. 60, 538–549 (1946).

Dunford, N., and Morse, A. P.
1. *Remarks on the preceding paper of James A. Clarkson.* Trans. Amer. Math. Soc. 40, 415–420 (1936).

Dunford, N., and Pettis, B. J.
1. *Linear operations on summable functions.* Trans. Amer. Math. Soc. 47, 323–392 (1940).

Dunford, N., and Segal, I. E.
1. *Semi-groups of operators and the Weierstrass theorem.* Bull. Amer. Math. Soc. 52, 911–914 (1946).

Dunford, N., and Schwartz, J.
1. *Convergence almost everywhere of operator averages.* Proc. Nat. Acad. Sci., U.S.A. 41, 229–231 (1955).
2. *Convergence almost everywhere of operator averages.* J. Rational Mech. and Anal. 5, 129–178 (1956).

Dunford, N., and Stone, M. H.
1. *On the representation theorem for Boolean algebras.* Revista Ci., Lima 43, 447–453 (1941).

Dunford, N., and Tamarkin, J. D.
1. *A principle of Jessen and general Fubini theorems.* Duke Math. J. 8, 743–749 (1941).

Dunham, J. L.
1. *The Wentzel-Brillouin-Kramers method of solving the wave equation.* Phys. Rev. 41, 713–720 (1932).
2. *The energy levels of a rotating vibrator.* Phys. Rev. 41, 721–731 (1932).

Dvoretzky, A., and Rogers, C. A.
1. *Absolute and unconditional convergence in normed linear spaces.* Proc. Nat. Acad. Sci. U.S.A. 36, 192–197 (1950).

Eberlein, W. F.
1. *Weak compactness in Banach spaces,* I. Proc. Nat. Acad. Sci. U.S.A. 33, 51–53 (1947).
2. *Closure, convexity, and linearity in Banach spaces.* Ann. of Math. (2) 47, 688–703 (1946).
3. *Abstract ergodic theorems and weak almost periodic functions.* Trans. Amer. Math. Soc. 67, 217–240 (1949).
4. *Abstract ergodic theorems.* Proc. Nat. Acad. Sci. U.S.A. 34, 43–47 (1948).
5. *A note on the spectral theorem.* Bull. Amer. Math. Soc. 52, 328–331 (1946).

Eckart, C.
1. *The penetration of a potential barrier by electrons.* Phys. Rev. 35, 1303–1309 (1930).

Edwards, R. E.
1. *A theory of Radon measures on locally compact spaces.* Acta Math. 89, 133–164 (1953).
2. *Multiplicative norms on Banach algebras.* Proc. Cambridge Philos. Soc. 47, 473–474 (1951).

Eidelheit, M.
1. *Zur Theorie der Konvexen Mengen in linearen normierten Räumen.* Studia Math. 6, 104–111 (1936).

Eidelheit, M., and Mazur, S.
1. *Eine Bemerkung über die Räume vom Typus (F).* Studia Math. 7, 159–161 (1938).

Eilenberg, S.
1. *Banach space methods in topology.* Ann. of Math. (2) 43, 568–579 (1942).

Elconin, V. (see Michal, A. D.)

Elliott, J.
1. *The boundary value problems and semi-groups associated with certain integro-differential operators.* Trans. Amer. Math. Soc. 76, 300–331 (1954).
2. *Eigenfunction expansions associated with singular differential operators.* Trans. Amer. Math. Soc. 78, 406–425 (1955).

Ellis, D.
1. *A modification of the parallelogram law characterization of Hilbert spaces.* Math. Zeit. 59, 94–96 (1953).

Ellis, H. W., and Halperin, I.
1. *Function spaces determined by a levelling length function.* Canadian J. Math. 5, 576–592 (1953).

Erdös, P. (see Clarkson, J. A.)

Erdös, P., and Kac, M.
1. *On certain limit theorems of the theory of probability.* Bull. Amer. Math. Soc. 52, 292–302 (1946).

Esclangon, E.
1. *Nouvelles recherches sur les fonctions quasi-périodiques.* Ann. Obs. Bordeaux 16, 51–226 (1917).

Esser, M.
1. *Analyticity in Hilbert space and self-adjoint transformations.* Amer. J. Math. 69, 825–835 (1947).

Ezrohi, I. A.
1. *The general form of linear operations in spaces with a countable basis.* Doklady Akad. Nauk SSSR (N. S.) 59, 1537–1540 (1948). (Russian) Math. Rev. 9, 448 (1948).

Fagan, R. E. (see Cameron, R. H.)

Fage, M. K.
1. *On symmetry and symmetrizability of influence functions.* Mat. Sbornik N. S. 32 (74), 345–352 (1953). (Russian) Math. Rev. 14, 1088 (1953).
2. *The characteristic function of a one-point boundary problem for an ordinary linear differential equation of second order.* Doklady Akad. Nauk SSSR (N. S.) 96, 929–932 (1954). (Russian) Math. Rev. 16, 362 (1955).

Fan, K. (see also Bochner, S.)
1. *Le prolongement des fonctionnelles continues sur un espace semi-ordonné.* Rev. Sci. 52, 131–139 (1944).

2. *Partially ordered additive groups of continuous functions*. Ann. of Math. (2) 51, 409–427 (1950).
3. *Maximum properties and inequalities for the eigenvalues of completely continuous operators*. Proc. Nat. Acad. Sci. U.S.A. 37, 760–766 (1951).
4. *Les fonctions définies-positives et les fonctions complètement monotones*. Gauthier-Villars, Paris, 1950.
5. *On a theorem of Weyl concerning eigenvalues of linear transformations* I. Proc. Nat. Acad. Sci. U.S.A. 35, 652–655 (1949). II. ibid. 36, 31–35 (1950).

Fantappiè, L.
1. *La teoria dei funzionali analitici, le sue applicazioni e i suoi possibili indirizzi*. Reale Accademia d'Italia, Fondazion Alessandro Volta, Atti dei Convegni. 9, 223–279 (1939). Rome, 1943.
2. *L'analisi funzionale nel campo complesso e i nuovi metodi d'integrazione delle equazioni a derivate parziali*. Rivista Mat. Univ. Parma. 1, 117–120 (1950).
3. *Le calcul des matrices*. C. R. Acad. Sci. Paris 186, 619–621 (1928).

Farnell, A. B.
1. *Limits for the characteristic roots of a matrix*. Bull. Amer. Math. Soc. 50, 789–794 (1944).

Fell, J. M. G., and Kelly, J. L.
1. *An algebra of unbounded operators*. Proc. Nat. Acad. Sci. U.S.A. 38, 592–598 (1952).

Feller, W.
1. *Semi-groups of transformations in general weak topologies*. Ann. of Math. (2) 57, 287–308 (1953).
2. *On the generation of unbounded semi-groups of bounded linear operators*. Ann. of Math. (2) 58, 166–174 (1953).
3. *On positivity preserving semi-groups of transformations on $C[r_1, r_2]$*. Ann. Soc. Polon. Math. 25 (1952), 85–94 (1953).
4. *Diffusion processes in one dimension*. Trans. Amer. Math. Soc. 77, 1–31 (1954).
5. *The parabolic differential equation and the associated semi-group of transformations*. Ann. of Math. (2) 55, 468–519 (1952).
6. *On second order differential operators*. Ann. of Math. (2) 61, 90–105 (1955).
7. *On differential operators and boundary conditions*. Comm. Pure Appl. Math. 8, 203–216 (1955).

Fenchel, W. (see Bonnesen, T.)

Feynman, R. P.
1. *Space-time approach to non-relativistic quantum mechanics*. Rev. Mod. Phys. 20, no. 2, 367–387 (1948).

Fichtenholz, G.
1. *Sur les fonctionnelles linéaires, continues au sens généralisé*. Mat. Sbornik (N. S.) 4 (46), 193–214 (1938).
2. *Sur une classe d'opérations fonctionnelles linéaires*. Mat. Sbornik (N. S.) 4 (46), 215–226 (1938).
3. *Sur les opérations linéaires dans l'espace des fonctions continues*. Bull. Acad. Roy. Belg. 22, 26–33 (1936).

Fichtenholz, G., and Kantorovitch, L. V.
1. *Sur les opérations linéaires dans l'espace des fonctions bornées*. Studia Math. 5, 69–98 (1934).

2. *Quelques théorèmes sur les fonctionnelles linéaires.* Doklady Akad. Nauk SSSR (N. S.) 3, 307–312 (1934). (Russian and French).

Ficken, F. A.

1. *Note on the existence of scalar products in normed linear spaces.* Ann. of Math. (2) 45, 362–366 (1944).

Fischer, C. A.

1. *Necessary and sufficient conditions that a linear transformation be completely continuous.* Bull. Amer. Math. Soc. 27, 10–17 (1920).
2. *Linear functionals of N-spreads.* Ann. of Math. (2) 19, 37–43 (1917–1918).

Fischer, E.

1. *Sur la convergence en moyenne.* C. R. Acad. Sci. Paris 144, 1022–1024 (1907).
2. *Applications d'un théorème sur la convergence en moyenne.* C. R. Acad. Sci. Paris 144, 1148–1151 (1907).

Fishel, B.

1. *The continuous spectra of certain differential equations.* J. London Math. Soc. 27, 175–180 (1952).

Fleischer, I.

1. *Sur les espaces normés non-archimédiens.* Nederl. Akad. Wetensch. Proc. Ser. A. 57, 165–168 (1954).

Følner, E. (see Bohr, H.)

Ford, G. C. (see Mishoe, L. I.)

Fort, M. K., Jr.

1. *Essential and nonessential fixed points.* Amer. J. Math. 72, 315–322 (1950).

Fortet, R.

1. *Remarques sur les espaces uniformément convexes.* C. R. Acad. Sci. Paris 210, 497–499 (1940).
2. *Remarques sur les espaces uniformément convexes.* Bull. Soc. Math. France 69, 23–46 (1941).
3. *Les systèmes d'équations linéaires dans les espaces uniformément convexes.* C. R. Acad. Sci. Paris 211, 422–423 (1940).
4. *Les fonctions aleatoires du type Markoff associées à certaines équations linéaires aux derivées partielles du type parabolique.* J. Math. Pures Appl. 22, 177–243 (1943).

Fréchet, M.

1. *Sur quelques points du calcul fonctionnel.* Rend. Circ. Mat. Palermo 22, 1–74 (1906).
2. *Les espaces abstraits topologiquement affines.* Acta Math. 47, 25–52 (1926).
3. *Les espaces abstraits.* Gauthier-Villars, Paris, 1928.
4. *Sur les ensembles de fonctions et les opérations linéaires.* C. R. Acad. Sci. Paris 144, 1414–1416 (1907).
5. *Sur les opérations linéaires,* I-III.
   I. Trans. Amer. Math. Soc. 5, 493–499 (1904).
   II. ibid. 6, 134–140 (1905).
   III. ibid. 8, 433–446 (1907).
6. *Sur les ensembles compacts de fonctions mesurables.* Fund. Math. 9, 25–32 (1927).
7. *Sur les ensembles compacts de fonctions de carrés sommables.* Acta Sci. Math. Szeged 8, 116–126 (1937).

8. *Sur divers modes de convergence d'une suite de fonctions d'une variable.* Bull. Calcutta Math. Soc. 11, 187–206 (1919–1920).
9. *Essai de géométrie analytique à une infinité de coordonnées.* Nouvelles Ann. de Math. (4) 8, 97–116, 289–317 (1908).
10. *Sur les fonctionnelles bilinéaires.* Trans. Amer. Math. Soc. 16, 215–234 (1915).

Fredholm, I.
1. *Sur une classe d'équations fonctionnelles.* Acta Math. 27, 365–390 (1903).

Freudenthal, H.
1. *Teilweise geordnete Moduln.* Nederl. Akad. Wetensch. Proc. 39, 641–651 (1936).
2. *Einige Sätze über topologische Gruppen.* Ann. of Math. (2) 37, 46–56 (1936).
3. *Über die Friedrichssche Fortsetzung halbbeschränkter Hermitescher Operatoren.* Nederl. Akad. Wetensch. Proc. 39, 832–833 (1936).

Friedman, B., and Mishoe, L. I.
1. *Eigenfunction expansions associated with a non-self adjoint differential equation.* Pacific J. Math. 6, 249–270 (1956).

Friedman, M. D.
1. *Determination of eigenvalues using a generalized Laplace transform.* J. Appl. Phys. 21, 1333–1337 (1950).

Friedrichs, K. O.
1. *Über die Spektralzerlegung eines Integraloperators.* Math. Ann. 115, 249–272 (1938).
2. *On the perturbation of continuous spectra.* Comm. Pure Appl. Math. 1, 361–406 (1948).
3. *Spektraltheorie halbbeschränkter Operatoren,* I-III.
   I. Math. Ann. 109, 465–487 (1934).
   II. ibid. 109, 685–713 (1934).
   III. ibid. 110, 777–779 (1935).
4. *Beiträge zur Theorie der Spektralschar.* Math. Ann. 110, 54–62 (1935).
5. *Über die ausgezeichnete Randbedingung in der Spektraltheorie der halbbeschränkten gewöhnlichen Differentialoperatoren zweiter Ordnung.* Math. Ann. 112, 1–23 (1935).
6. *On differential operators in Hilbert space.* Amer. J. Math. 61, 523–544 (1939).
7. *Spektraltheorie linearer Differentialoperatoren.* Jber. Deutsch. Math. Verein. 45, 181–193 (1935).
8. *Die unitären Invarianten selbstadjugierter Operatoren im Hilbertschen Raum.* Jber. Deutsch. Math. Verein. 45, 79–82 (1935).
9. *The identity of weak and strong extensions of differential operators.* Trans. Amer. Math. Soc. 55, 132–151 (1944).
10. *Criteria for discrete spectra.* Comm. Pure Appl. Math. 3, 439–449 (1950).
11. *Functional analysis and applications.* Inst. Math. Sci., New York Univ., New York, 1956.
12. *Criteria for the discrete character of the spectra of ordinary differential operators.* Courant Anniversary Volume, 145–160, Interscience Pub., 1948.
13. *Spectral representation of linear operators.* Inst. Math. Sci., New York Univ., 1953.
14. *Symmetric hyperbolic linear differential equations.* Comm. Pure Appl. Math. 7, 345–392 (1954).

15. *Differentiability of solutions of linear elliptic differential operators*. Comm. Pure Appl. Math. 6, 299–326 (1953).
16. *Mathematical aspects of the quantum theory of fields*. Interscience Pub., New York and London, 1953.

Frink, O., Jr.
1. *Series expansions in linear vector spaces*. Amer. J. Math. 63, 87–100 (1941).

Frobenius, G.
1. *Über lineare Substitutionen und bilineare Formen*. J. Reine Angew. Math. 84, 1–63 (1878).
2. *Über die schiefe Invariante einer bilinearen oder quadratischen Formen*. J. Reine Angew. Math. 86, 44–71 (1879).
3. *Über die cogredienten Transformationen der bilinearen Formen*. Sitzungsberichte der K. Preuss. Akad. der Wiss. zu Berlin, 7–16 (1896).

Fuchs, L.
1. *Über Relationen, welche für die zwischen je zwei singulären Punkten erstreckten Integrale der Lösungen linearer Differentialgleichungen stattfinden*. J. Reine Angew. Math. 76, 177–213 (1873).

Fuglede, B.
1. *A commutativity theorem for normal operators*. Proc. Nat. Acad. Sci. U.S.A. 36, 35–40 (1950).

Fukamiya, M. (see also Yosida, K.)
1. *On dominated ergodic theorems in L_p ($p \geqq 1$)*. Tôhoku Math. J. 46, 150–153 (1939).
2. *On B^*-algebras*. Proc. Japan Acad. 27, 321–327 (1951).
3. *On a theorem of Gelfand and Neumark and the B^*-algebra*. Kumamoto J. Sci. Ser. A. 1, no. 1, 17–22 (1952).

Fullerton, R. E.
1. *On a semi-group of subsets of a linear space*. Proc. Amer. Math. Soc. 1, 440–442 (1950).
2. *Linear operators with range in a space of differentiable functions*. Duke Math. J. 13, 269–280 (1946).
3. *The representation of linear operators from L^p to L*. Proc. Amer. Math. Soc. 5, 689–696 (1954).
4. *An inequality for linear operators between L^p spaces*. Proc. Amer. Math. Soc. 6, 186–190 (1955).
5. *A characterization of L spaces*. Fund. Math. 38, 127–136 (1951).

Gagaev, B.
1. *On convergence in Banach spaces*. Uspehi Mat. Nauk 3, no. 5 (27), 171–173 (1948). (Russian) Math. Rev. 10, 255 (1949).

Gál, I. S.
1. *Sur la méthode de résonance et sur un théorème concernant des espaces de type (B)*. Ann. Inst. Fourier Grenoble 3, 23–30 (1951).
2. *The principle of condensation of singularities*. Duke Math. J. 20, 27–35 (1953).
3. *On sequences of operations in complete vector spaces*. Amer. Math. Monthly 60, 527–538 (1953).

Galbraith, A. S., and Warschawski, S. E.
1. *The convergence of expansions resulting from a self-adjoint boundary problem*. Duke Math. J. 6, 318–340 (1940).

Gale, D.

1. *Compact sets of functions and function rings.* Proc. Amer. Math. Soc. 1, 303–308 (1950).

Gantmacher, V.

1. *Über schwache totalstetige Operatoren.* Mat. Sbornik N. S. 7 (49), 301–308 (1940).

Gantmacher, V., and Šmulian, V.

1. *Sur les espaces linéaires dont la sphère unitaire est faiblement compacts.* Doklady Akad. Nauk SSSR (N. S.) 17, 91–94 (1937).
2. *Über schwache Kompaktheit in Banachschen Raum.* Mat. Sbornik N. S. 8 (50), 489–492 (1940). (Russian. German summary).

Garabedian, P. R.

1. *The classes L_p and conformal mapping.* Trans. Amer. Math. Soc. 69, 392–415 (1950).

Garabedian, P. R., and Shiffman, M.

1. *On solution of partial differential equations by the Hahn-Banach Theorem.* Trans. Amer. Math. Soc. 76, 288–299 (1954).

Gårding, L.

1. *Linear hyperbolic partial differential equations with constant coefficients.* Acta Math. 85, 2–62 (1950).
2. *Dirichlet's problem for linear elliptic partial differential equations.* Math. Scand. 1, 55–72 (1953).
3. *Le problème de Dirichlet pour les équations aux dérivées partielles élliptiques linéaires dans des domaines bornés.* C. R. Acad. Sci. Paris 233, 1554–1556 (1951).
4. *Dirichlet's problem and the vibration problem for linear elliptic partial differential equations with constant coefficients.* Proc. Symposium Spectral Theory and Differential Problems, 291–301. Oklahoma Agricultural and Mechanical College, Stillwater, Oklahoma, 1951.
5. *L'inégalité de Friedrichs et Lewy pour les équations hyperboliques linéaires d'ordre supérieur.* C. R. Acad. Sci. Paris. 239, 849–850 (1954).
6. *Applications of the theory of direct integrals of Hilbert spaces to some integral and differential operators.* Inst. Fluid Dynamics. Univ. of Maryland, College Park, 1954.

Gavurin, M. K.

1. *On estimates for the characteristic numbers and vectors of a perturbed operator.* Doklady Akad. Nauk SSSR (N. S.) 76, 769–770 (1951). (Russian) Math. Rev. 12, 617 (1951).
2. *On estimates for eigen-values and vectors of a perturbed operator.* Doklady Akad. Nauk SSSR (N. S.) 96, 1093–1095 (1954). (Russian) Math. Rev. 16, 264 (1955).
3. *On the exactness of approximate methods of finding eigen-values of integral operators.* Doklady Akad. Nauk SSSR (N. S.) 97, 13–15 (1954). (Russian) Math. Rev. 16, 264 (1955).

Gelbaum, B. R.

1. *Expansions in Banach spaces.* Duke Math. J. 17, 187–196 (1950).
2. *A nonabsolute basis for Hilbert space.* Proc. Amer. Math. Soc. 2, 720–721 (1951).

Gelfand (Gel'fand), I. M.

1. *Normierte Ringe.* Mat. Sbornik N. S. 9 (51), 3–24 (1941).

2. *Abstrakte Funktionen und lineare Operatoren.* Mat. Sbornik N. S. 4 (46), 235–286 (1938).
3. *Ideale und primäre Ideale in normierten Ringen.* Mat. Sbornik N. S. 9 (51), 41–48 (1941).
4. *Zur Theorie der Charaktere der Abelschen topologischen Gruppen.* Mat. Sbornik N. S. 9 (51), 49–50 (1941).
5. *Über absolut konvergente trigonometrische Reihen und Integrale.* Mat. Sbornik N. S. 9 (51), 51–66 (1941).
6. *Remark on the work of N. K. Bari "Biorthogonal systems and bases in Hilbert space."* Moskov. Gos. Univ. Učenye Zapiski 148, Matematika 4, 224–225 (1951). (Russian) Math. Rev. 14, 289 (1953).

Gelfand, I. M., and Kolmogoroff, A. N.
1. *On rings of continuous functions on a topological space.* Doklady Akad. Nauk SSSR (N. S.) 22, 11–15 (1939).

Gelfand, I. M., and Kostyučenko, A. G.
1. *Expansions in eigenfunctions of differential and other operators.* Doklady Akad. Nauk SSSR (N. S.) 103, 349–352 (1955). (Russian) Math. Rev. 17, 388 (1956).

Gelfand, I. M., and Levitan, B. M.
1. *On the determination of a differential equation from its spectral function.* Izvestiya Akad. Nauk SSSR 15, 309–360 (1951). (Russian) Amer. Math. Soc. Translations (2) 1, 253–304 (1955).
2. *On a simple identity for the characteristic values of a differential operator of second order.* Doklady Akad. Nauk SSSR (N. S.) 88, 593–596 (1953). (Russian) Math. Rev. 15, 33 (1954).

Gelfand (Gel'fand), I. M., and Neumark (Naĭmark), M. A.
1. *On the imbedding of normed rings into the ring of operators in Hilbert space.* Mat. Sbornik N. S. 12 (54), 197–213 (1943).
2. *Normed rings with involution and their representations.* Izvestiya Akad. Nauk SSSR 12, 445–480 (1948). (Russian) Math. Rev. 10, 199 (1949).

Gelfand, I. M., and Raikov, D. A.
1. *On the theory of characters of commutative topological groups.* Doklady Akad. Nauk SSSR (N. S.) 28, 195–198 (1940).
2. *Irreducible unitary representations of locally bicompact groups.* Mat. Sbornik N. S. 13 (55), 301–316 (1943).

Gelfand, I., and Šilov, G.
1. *Über verschiedene Methoden der Einführung der Topologie in die Menge der maximalen Ideale eines normierten Ringes.* Mat. Sbornik N. S. 9 (51), 25–40 (1941). (German, Russian summary).

Gelfand, I. M., and Yaglom, A. M.
1. *Integration in functional spaces and its application in quantum physics.* Uspehi Mat. Nauk (N.S.) 11, 77–114 (1956).

Gillespie, D. C., and Hurwitz, W. A.
1. *On sequences of continuous functions having continuous limits.* Trans. Amer. Math. Soc. 32, 527–543 (1930).

Giorgi, G.
1. *Nuove osservazioni sulle funzioni delle matrici.* Atti Accad. Naz. Lincei. Rend. Cl. Sci. Fis. Mat. Nat. (6) 8, 3–8 (1928).

Glazman, I. M. (see also Ahiezer, N. I.)

1. *On the theory of singular differential operators.* Uspehi Mat. Nauk (N.S.) 5, no. 6 (40), 102–135 (1950). (Russian) Math. Rev. 13, 254 (1952). Amer. Math. Soc. Translation no. 96 (1953).
2. *On the deficiency index of differential operators.* Doklady Akad. Nauk SSSR (N. S.) 64, 151–154 (1949). (Russian) Math. Rev. 10, 538 (1949).
3. *On the spectrum of linear differential operators.* Doklady Akad. Nauk SSSR (N. S.) 80, 153–156 (1951). (Russian) Math. Rev. 13, 654 (1952).
4. *On the character of the spectra of one dimensional singular boundary value problems.* Doklady Akad. Nauk SSSR (N. S.) 87, 5–8 (1952). (Russian) Math. Rev. 14, 1088 (1953).

Glicksberg, I.

1. *The representation of functionals by integrals.* Duke Math. J. 19, 253–261 (1952).

Gödel, K.

1. *The consistency of the continuum hypothesis.* Ann. of Math. Studies, no. 3. Princeton Univ. Press, Princeton, 1940.
2. *Über formal unentscheidbare Sätze der Principia Mathematica und verwandter Systeme,* I. Monatsh. für Math. u. Physik 38, 173–198 (1931).

Godement, R. (see also Cartan, H.)

1. *Théorèmes taubériens et théorie spectrale.* Ann. Sci. École Norm. Sup. (3) 64, 119–138 (1947).
2. *Les fonctions de type positif et la théorie des groupes.* Trans. Amer. Math. Soc. 63, 1–84 (1948).
3. *Sur la théorie des représentations unitaires.* Ann. of Math. (2) 53, 68–124 (1951).
4. *Mémoire sur la théorie des caractères dans les groupes localement compacts unimodulaires.* J. Math. Pures Appl. 30, 1–110 (1951).
5. *Sur une généralisation d'un théorème de Stone.* C. R. Acad. Sci. Paris 218, 901–903 (1944).

Gohberg, I. C.

1. *On linear equations in Hilbert space.* Doklady Akad. Nauk SSSR (N. S.) 76, 9–12 (1951). (Russian) Math. Rev. 13, 46 (1952).
2. *On linear equations in normed spaces.* Doklady Akad. Nauk SSSR (N. S.) 76, 477–480 (1951). (Russian) Math. Rev. 13, 46 (1952).
3. *On linear operators depending analytically on a parameter.* Doklady Akad. Nauk SSSR (N. S.) 78, 629–632 (1951). (Russian) Math. Rev. 13, 46 (1952).
4. *On the index of an unbounded operator.* Mat. Sbornik N. S. 33 (75), 193–198 (1953). (Russian) Math. Rev. 15, 233 (1954).

Gol'dman, M. A. (see also Kračkovskiĭ, S. N.)

Gol'dman, M. A., and Kračkovskiĭ, S. N.

1. *On the null-elements of a linear operator in its Fredholm* region. Doklady Akad. Nauk SSSR (N. S.) 86, 15–17 (1952). (Russian) Math. Rev. 14, 478 (1953).

Goldstine, H. H.

1. *Weakly complete Banach spaces.* Duke Math. J. 4, 125–131 (1938).
2. *The theorem of Hildebrandt.* Studia Math. 7, 157–158 (1938).

Gomes, A. P. (see Dieudonné, J.)

Goodner, D. B.
1. *Projections in normed linear spaces.* Trans. Amer. Math. Soc. 69, 89–108 (1950).

Gowurin, M.
1. *Über die Stieltjessche Integration abstrakter Funktionen.* Fund. Math. 27, 255–268 (1936).

Graff, A. A.
1. *On the theory of systems of linear differential equations in a one-dimensional domain,* I, II.
   I. Mat. Sbornik N. S. 18 (60), 305–328 (1946).
   II. ibid. 21 (63), 143–159 (1947). (Russian. English summary) Math. Rev. 8, 74 (1947) and 8, 186 (1948).

Graves, L. M. (see also Bartle, R. G., and Hildebrandt, T. H.)
1. *Topics in the functional calculus.* Bull. Amer. Math. Soc. 41, 641–662 (1935). Errata, ibid. 42, 381–382 (1936).
2. *The theory of functions of real variables.* McGraw-Hill Company, New York, 1946.
3. *Riemann integration and Taylor's theorem in general analysis.* Trans. Amer. Math. Soc. 29, 163–177 (1927).
4. *Some general approximation theorems.* Ann. of Math. (2) 42, 281–292 (1941).
5. *Some mapping theorems.* Duke Math. J. 17, 111–114 (1950).
6. *A generalization of the Riesz theory of completely continuous transformations.* Trans. Amer. Math. Soc. 79, 141–149 (1955).
7. *Remarks on singular points of functional equations.* Trans. Amer. Math. Soc. 79, 150–157 (1955).

Graves, R. E. (see Cameron, R. H.)

Graves, R. L.
1. *The Fredholm theory in Banach spaces.* Dissertation, Harvard University (1951). Abstract, Bull. Amer. Math. Soc. 58, 479 (1952).

Greco, D.
1. *Sulla convergenza degli sviluppi in serie di autosoluziani associati ad un problema ai limite relativo ad un'equazione differenziale ordinaria del secondo ordine.* Rend. Acc. Sci. Fis. Mat. Napoli (4) 17, 171–189 (1950).

Grinblyum (Grunblum), M. M.
1. *Certains théorèmes sur la base dans un espace du type (B).* Doklady Akad. Nauk SSSR (N. S.) 31, 428–432 (1941).
2. *Biorthogonal systems in Banach space.* Doklady Akad. Nauk SSSR (N. S.) 47, 75–78 (1945).
3. *Sur la théorie des systèmes biorthogonaux.* Doklady Akad. Nauk SSSR (N. S.) 55, 287–290 (1947).
4. *On a property of a basis.* Doklady Akad. Nauk SSSR (N. S.) 59, 9–11 (1948). (Russian) Math. Rev. 10, 307 (1949).
5. *Spectral measure.* Doklady Akad. Nauk SSSR (N. S.) 81, 345–348 (1951). (Russian) Math. Rev. 13, 470 (1952).
6. *Operational integral in a Banach space.* Doklady Akad. Nauk SSSR (N.S.) 71, 5–8 (1950). (Russian) Math. Rev. 11, 601 (1950).

Grosberg, J., and Krein, M.
1. *Sur la décomposition des fonctionnelles en composantes positives.* Doklady Akad. Nauk SSSR (N. S.) 25, 723–726 (1939).

Grosberg, Yu. I.

1. *On linear functionals on the space of functions of bounded variation.* Naukovi Zapiski Kiivkogo Pedinstutu 2, 17–23 (1939). (Ukrainian).

Grothendieck, A.

1. *Critères généraux de compacité dans les espaces vectoriels localement convexes. Pathologie des espaces (LF).* C. R. Acad. Sci. Paris 231, 940–941 (1950).
2. *Critères de compacité dans les espaces fonctionnels généraux.* Amer. J. Math. 74, 168–186 (1952).
3. *Produits tensoriels topologiques et espaces nucléaires.* Memoirs Amer. Math. Soc. no. 16, 1955.
4. *Sur les applications linéaires faiblement compactes d'espaces du type C(K).* Canadian J. Math. 5, 129–173 (1953).
5. *Sur certains espaces de fonctions holomorphes,* I, II.
   I. J. Reine Angew. Math. 192, 35–64 (1953).
   II. ibid. 192, 77–95 (1953).

Grunblum, M. M. (see Grinblyum)

Gurevič, L. A.

1. *On unconditional bases.* Uspehi Mat. Nauk (N. S.) 8, no. 5 (57), 153–156 (1953). (Russian) Math. Rev. 15, 631 (1954).

Haar, A.

1. *Der Massbegriff in der Theorie der kontinuierlichen Gruppen.* Ann. of Math. (2) 34, 147–169 (1933).
2. *Über die Multiplikationstabelle der orthogonalen Funktionensysteme.* Math. Zeit. 41, 769–798 (1930).
3. *Zur Theorie der orthogonalen Funktionensysteme,* I, II.
   I. Math. Ann. 69, 331–371 (1910).
   II. ibid. 71, 38–53 (1911).

Hadamard, J.

1. *Sur les opérations fonctionnelles.* C. R. Acad. Sci. Paris 136, 351–354 (1903).

Hahn, H.

1. *Über die Darstellung gegebener Funktionen durch singuläre Integrale,* II. Denkschriften der K. Akad. Wien. Math.-Naturwiss. Kl. 93, 657–692 (1916).
2. *Über Folgen linearer Operationen.* Monatsh. für Math. und Physik 32, 3–88 (1922).
3. *Über lineare Gleichungssysteme in linearen Räumen.* J. Reine Angew. Math. 157, 214–229 (1927).
4. *Reelle Funktionen.* Akad. Verlag., Leipzig, 1932. Reprinted by Chelsea Pub. Co., New York.
5. *Über die Integrale des Herrn Hellinger und die Orthogonalinvarianten der quadratischen Formen von unendlich vielen Veränderlichen.* Monatsh. für Math. und Physik 23, 161–224 (1912).

Hahn, H., and Rosenthal, A.

1. *Set functions.* Univ. of New Mexico Press, Albuquerque, 1948.

Halmos, P. R.

1. *Normal dilations and extensions of operators.* Summa Brasil. Math. 2, 125–134 (1950).
2. *A nonhomogeneous ergodic theorem.* Trans. Amer. Math. Soc. 66, 284–288 (1949).

3. *Commutativity and spectral properties of normal operators.* Acta Sci. Math. Szeged 12 Pars B, 153–156 (1950).
4. *Measurable transformations.* Bull. Amer. Math. Soc. 55, 1015–1034 (1949).
5. *Measure Theory.* D. Van Nostrand, New York, 1950.
6. *Introduction to Hilbert space and the theory of spectral multiplicity.* Chelsea, New York, 1951.
7. *Finite dimensional vector spaces.* Ann. of Math. Stud. No. 7. Princeton Univ. Press, Princeton, 1942.
8. *An ergodic theorem.* Proc. Nat. Acad. Sci. U.S.A. 32, 156–161 (1946).
9. *Spectra and spectral manifolds.* Ann. Soc. Polon. Math. 25, 43–49 (1952).
10. *Commutators of operators,* I, II.
I. Amer. J. Math. 74, 237–240 (1952).
II. ibid. 76, 191–198 (1954).

Halmos, P. R., and Lumer, G.
1. *Square roots of operators,* II. Proc. Amer. Math. Soc. 5, 589–595 (1954).

Halmos, P. R., Lumer, G., and Schäffer, J. J.
1. *Square roots of operators.* Proc. Amer. Math. Soc. 4, 142–149 (1953).

Halmos, P. R., and von Neumann, J.
1. *Operator methods in classical mechanics,* II. Ann. of Math. (2) 43, 332–350 (1942).

Halperin, I. (see also Ellis, H. W.)
1. *Function spaces.* Canadian J. Math. 5, 273–288 (1953).
2. *Convex sets in linear topological spaces.* Trans. Roy. Soc. Canada Sec. III 47, 1–6 (1953).
3. *Uniform convexity in function spaces.* Duke Math. J. 21, 195–204 (1954).
4. *Reflexivity in the L^λ function spaces.* Duke Math. J. 21, 205–208 (1954).
5. *Closures and adjoints of linear differential operators.* Ann. of Math. (2) 38, 880–919 (1937).

Hamburger, H. L.
1. *Five notes on a generalization of quasi-nilpotent transformations in Hilbert space.* Proc. London Math. Soc. (3) 1, 494–512 (1951).
2. *On a new characterization of self-adjoint differential operators in the Hilbert space L_2.* Proc. Symposium on Spectral Theory and Differential Problems, 229–247 (1951). Oklahoma A. and M. College, Stillwater, Oklahoma.
3. *Remarks on self-adjoint differential operators.* Proc. London Math. Soc. (3) 3, 446–463 (1953).
4. *Über die Zerlegung des Hilbertschen Raumes durch vollstetige lineare Transformationen.* Math. Nachr. 4, 56–69 (1951).
5. *Contributions to the theory of closed Hermitian transformations of deficiency index (m, m).* Ann. of Math. (2) 45, 59–99 (1944).
6. *Contributions to the theory of closed Hermitian transformations of deficiency-index (m, m).* Quart. J. Math. Oxford Ser. 13, 117–128 (1942).
7. *Hermitian transformations of deficiency-index* (1, 1), *Jacobi matrices and undetermined moment problems.* Amer. J. Math. 66, 489–522 (1944).
8. *On a class of Hermiiian transformations containing self-adjoint differential operators.* Ann. of Math. (2) 47, 667–687 (1946).

Hamburger, H. L., and Grimshaw, M. E.
1. *Linear transformations in n-dimensional vector space.* Cambridge Univ. Press, 1951.

Hamel, G.

1. *Über lineare Differentialgleichungen zweiter Ordnung mit periodischen Koeffizienten*. Math. Ann. 73, 371–412 (1913).

Hanson, E. H.

1. *A note on compactness*. Bull. Amer. Math. Soc. 39, 397–400 (1933).

Harazov, D. F.

1. *On a class of linear equations in Hilbert spaces*. Soobščeniya Akad. Nauk Gruzin. SSR 13, 65–72 (1952). (Russian) Math. Rev. 14, 990 (1953).
2. *On a class of linear equations with symmetrizable operators*. Doklady Akad. Nauk SSSR (N. S.) 91, 1023–1026 (1953). (Russian) Math. Rev. 15, 881 (1954).
3. *On the theory of symmetrizable operators with polynomial dependence upon a parameter*. Doklady Akad. Nauk SSSR (N. S.) 91, 1285–1287 (1953). (Russian) Math. Rev. 15, 881 (1954).

Hardy, G. H., and Littlewood, J.

1. *Some properties of fractional integrals*, I, II.
   I. Math. Zeit. 27, 565–606 (1928).
   II. ibid. 34, 403–439 (1932).

Hardy, G. H., Littlewood, J. E., and Pólya, G.

1. *Inequalities*. Cambridge Univ. Press, 1934.

Hartman, P.

1. *On the crgodic theorems*. Amer. J. Math. 69, 193–199 (1947).
2. *On the essential spectra of symmetric operators in Hilbert space*. Amer. J. Math. 75, 229–240 (1953).
3. *The L_2-solutions of linear differential equations of second order*. Duke Math. J. 14, 323–326 (1947).
4. *Unrestricted solution fields of almost separable differential equations*. Trans. Amer. Math. Soc. 63, 560–580 (1948).
5. *On differential equations with non-oscillatory eigenfunctions*. Duke Math. J. 15, 697–709 (1948).
6. *On the linear logarithmico-exponential differential equation of the second order*. Amer. J. Math. 70, 764–779 (1948).
7. *On the spectra of slightly disturbed linear oscillators*. Amer. J. Math. 71, 71–79 (1949).
8. *A characterization of the spectra of the one-dimensional wave equation*. Amer. J. Math. 71, 915–920 (1949).
9. *The number of L_2-solutions of $x''+q(t)x = 0$*. Amer. J. Math. 73, 635–645 (1951).
10. *On bounded Green's kernels for second order linear differential equations*. Amer. J. Math. 73, 646–656 (1951).
11. *On the eigenvalues of differential equations*. Amer. J. Math. 73, 657–662 (1951).
12. *On linear second order differential equations with small coefficients*. Amer. J. Math. 73, 955–962 (1951).
13. *Some examples in the theory of singular boundary value problems*. Amer. J. Math. 74, 107–126 (1952).
14. *On non-oscillatory linear differential equations of second order*. Amer. J. Math. 74, 389–400 (1952).
15. *On the derivatives of solutions of linear second order differential equations*. Amer. J. Math. 75, 173–177 (1953).

16. *On the essential spectra of ordinary differential equations.* Amer. J. Math. 76, 831–838 (1954).

Hartman, P., and Putnam, C.

1. *The least cluster point of the spectrum of boundary value problems.* Amer. J. Math. 70, 847–855 (1948).
2. *The gaps in the essential spectra of wave equations.* Amer. J. Math. 72, 849–862 (1950).

Hartman, P., and Wintner, A.

1. *The (L^2)-space of relative measure.* Proc. Nat. Acad. Sci. U.S.A. 33, 128–132 (1947).
2. *An oscillation theorem for continuous spectra.* Proc. Nat. Acad. Sci. U.S.A. 33, 376–379 (1947).
3. *The asymptotic arcus variation of solutions of real linear differential equations of second order.* Amer. J. Math. 70, 1–10 (1948).
4. *Criteria for the non-degeneracy of the wave equation.* Amer. J. Math. 70, 295–308 (1948).
5. *On the orientation of unilateral spectra.* Amer. J. Math. 70, 309–316 (1948).
6. *On non-conservative linear oscillators of low frequency.* Amer. J. Math. 70, 529–539 (1948).
7. *A criterion for the non-degeneracy of the wave equation.* Amer. J. Math. 71, 206–213 (1949).
8. *On the location of spectra of wave equations.* Amer. J. Math. 71, 214–217 (1949).
9. *On the Laplace-Fourier transcendents.* Amer. J. Math. 71, 367–372 (1949).
10. *Oscillatory and non-oscillatory linear differential equations.* Amer. J. Math. 71, 627–648 (1949).
11. *A separation theorem for continuous spectra.* Amer. J. Math. 71, 650–662 (1949).
12. *On the derivatives of the solutions of the one-dimensional wave equation.* Amer. J. Math. 72, 148–155 (1950).
13. *On the essential spectra of singular eigenvalue problems.* Amer. J. Math. 72, 545–552 (1950).
14. *On an oscillation criterion of Liapounoff.* Amer. J. Math. 73, 885–890 (1951).
15. *On perturbations of the continuous spectrum of the harmonic oscillators.* Amer. J. Math. 74, 79–85 (1952).
16. *An inequality for the amplitudes and arcus in vibration diagrams of time-dependent frequency.* Quart. Appl. Math. 10, 175–176 (1952).
17. *On non-oscillatory linear differential equations.* Amer. J. Math. 75, 717–730 (1953).
18. *On curves defined by binary non-conservative differential systems.* Amer. J. Math. 76, 497–501 (1954).
19. *On the assignment of asymptotic values for the solution of linear differential equations of second order.* Amer. J. Math. 77, 475–483 (1955).
20. *On linear second order differential equations in the unit circle.* Trans. Amer. Math. Soc. 78, 492–500 (1955).
21. *On non-oscillatory linear differential equations with monotone coefficients.* Amer. J. Math. 76, 207–219 (1954).

Hartman, S.

1. *Quelques propriétés ergodiques des fractions continues.* Studia Math. 12, 271–278 (1951).

Hartman, S., Marczewski, E., and Ryll-Nardzewski, C.
1. *Théorèmes ergodiques et leurs applications.* Colloq. Math. 2, 109–123 (1951).
Hartogs, F., and Rosenthal, A.
1. *Über Folgen analytischer Funktionen.* Math. Ann. 104, 606–610 (1931).
Hatfield, C. (see Cameron, R. H.)
Haupt, O.
1. *Über lineare homogene Differentialgleichungen zweiter Ordnung mit periodischen Koeffizienten.* Math. Ann. 79, 278–285 (1918).
Hausdorff, F.
1. *Grundzüge der Mengenlehre.* Verlag von Veit, Leipzig, 1914. Reprinted by Chelsea Pub. Co., New York.
2. *Mengenlehre* (Dritte Auflage). W. de Gruyter, Berlin and Leipzig, 1935. Reprinted by Dover Publications, New York.
3. *Zur Theorie der linearen metrischen Räume.* J. Reine Angew. Math. 167, 294–311 (1932).
Heinz, E.
1. *Beiträge zur Störungstheorie der Spektralzerlegung.* Math. Ann. 123, 415–438 (1951).
2. *Ein v. Neumannscher Satz über beschränkte Operatoren im Hilbertschen Raum.* Nachr. Akad. Wiss. Göttingen. Math.-Phys. Kl. 1952, 5–6 (1952).
3. *Zur Theorie der Hermiteschen Operatoren des Hilbertschen Raumes.* Nachr. Akad. Wiss. Göttingen. Math.-Phys. Kl. 1951, no. 2, 4 pp. (1951).
4. *On an inequality for linear operators in a Hilbert space.* Report on Operator Theory and Group Representations. Pub. no. 387, Nat. Acad. Sci. U.S.A., pp. 27–29 (1955).
5. *Zur Frage der Differenzierbarkeit der S-Funktion.* Math. Ann. 122, 332–333, (1950).
Hellinger, E.
1. *Neue Begründung der Theorie quadratischer Formen von unendlichvielen Veränderlichen.* J. Reine Angew. Math. 136, 210–271 (1909).
Hellinger, E., and Toeplitz, O.
1. *Grundlagen für eine Theorie der unendlichen Matrizen.* Math. Ann. 69, 289–330 (1910).
2. *Grundlagen für eine Theorie der endlichen Matrizen.* Nachr. Akad. Wiss. Göttingen. Math.-Phys. Kl. 1906, 351–355 (1906).
3. *Integralgleichungen und Gleichungen mit unendlichvielen Unbekannten.* Encyklopädie der Mathematischen Wissenschaften II C13, 1335–1616 (1928). Reprinted by Chelsea Pub. Co., New York.
Helly, E.
1. *Über lineare Funktionaloperationen.* S.-B. K. Akad. Wiss. Wien Math.-Naturwiss. Kl. 121, IIa, 265–297 (1912).
2. *Über Systeme linearer Gleichungen mit unendlich vielen Unbekannten.* Monatsh. für Math. u. Phys. 31, 60–91 (1921).
Helson, H.
1. *Spectral synthesis of bounded functions.* Ark. för Mat. 1, 497–502 (1951).
Helson, H., and Quigley, F. D.
1. *Maximal algebras of continuous functions.* Proc. Amer. Math. Soc., 8, 111–114 (1957).

2. *Existence of maximal ideals in algebras of continuous functions.* Proc. Amer. Math. Soc., 8, 115–119 (1957).

Hensel, K.

1. *Über Potenzreihen von Matrizen.* J. Reine Angew. Math. 155, 107–110 (1926).

Herglotz, G.

1. *Über Potenzreihen mit positivem, reellem Teil im Einheitskreis.* S.-B. Sächs. Akad. Wiss. 63, 501–511 (1911).
2. *Über die Integration linearer, partieller Differentialgleichungen mit konstanten Koeffizienten,* I–IV.
   I. S.-B. Sächs. Akad. Wiss. 78, 93–126 (1926).
   II. ibid. 78, 287–318 (1926).
   III. ibid. 80, 69–114 (1928).
   IV. Abh. Math. Sem. Univ. Hamburg 6, 189–197 (1928).

Hewitt, E. (see also Yosida, K.)

1. *Linear functionals on spaces of continuous functions.* Fund. Math. 37, 161–189 (1950).
2. *Integral representation of certain linear functionals.* Ark. för Mat. 2, 269–282 (1952).
3. *Integration on locally compact spaces,* I. Univ. of Washington Pub. in Math. 3, 71–75 (1952).
4. *Certain generalizations of the Weierstrass approximation theorem.* Duke Math. J. 14, 419–427 (1947).
5. *Rings of real-valued continuous functions,* I. Trans. Amer. Math. Soc. 64, 45–99 (1948).
6. *Linear functionals on almost periodic functions.* Trans. Amer. Math. Soc. 74, 303–322 (1953).
7. *A problem concerning finitely additive measures.* Mat. Tidsskr. B 1951, 81–94 (1951).

Heywood, P.

1. *On the asymptotic distribution of eigenvalues.* Proc. London Math. Soc. (3) 4, 456–470 (1954).

Hilb, E.

1. *Über die Auflösung von Gleichungen mit unendlich vielen Unbekannten.* S.-B. Phys.-Med. Soz. Erlangen, 84–89 (1908).
2. *Über Integraldarstellung willkürlicher Funktionen.* Math. Ann. 66, 1–66 (1909).
3. *Über Reihenentwickelung nach den Eigenfunktionen linearer Differentialgleichungen 2ter Ordnung.* Math. Ann. 71, 76–87 (1911).
4. *Über gewöhnliche Differentialgleichungen mit Singularitäten und die dazugehörigen Entwickelungen willkürlicher Funktionen.* Math. Ann. 76, 333–339 (1915).

Hilb, E., and Szász, O.

1. *Allgemeine Reihenentwicklungen.* Encyklopädie der Math. Wiss. II C 11, 1229–1276 (1922).

Hilbert, D. (see also Courant, R.)

1. *Grundzüge einer allgemeinen Theorie der linearen Integralgleichungen,* I–VI.
   I. Nachr. Akad. Wiss. Göttingen. Math.-Phys. Kl., 49–91 (1904).
   II. ibid., 213–259 (1905).
   III. ibid., 307–338 (1905).

IV. ibid., 157–227 (1906).
V. ibid., 439–480 (1906).
VI. ibid., 355–417 (1910).
(Published in book form by Teubner, Leipzig, 1912. Reprinted by Chelsea Pub. Co., New York, 1952).

Hildebrandt, T. H.

1. *On unconditional convergence in normed vector spaces.* Bull. Amer. Math. Soc. 46, 959–962 (1940).
2. *On uniform limitedness of sets of functional operations.* Bull. Amer. Math. Soc. 29, 309–315 (1923).
3. *On bounded functional operations.* Trans. Amer. Math. Soc. 36, 868–875 (1934).
4. *Integration in abstract spaces.* Bull. Amer. Math. Soc. 59, 111–139 (1953).
5. *Lebesgue integration in general analysis.* (Abstract). Bull. Amer. Math. Soc. 33, 646 (1927).
6. *Über vollstetige lineare Transformationen.* Acta Math. 51, 311–318 (1928).
7. *Linear operations on functions of bounded variation.* Bull. Amer. Math. Soc. 44, 75 (1938).
8. *On the moment problem for a finite interval.* Bull. Amer. Math. Soc. 38, 269–270 (1932).
9. *Convergence of sequences of linear operations.* Bull. Amer. Math. Soc. 28, 53–58 (1922).

Hildebrandt, T. H., and Graves, L. M.

1. *Implicit functions and their differentials in general analysis.* Trans. Amer. Math. Soc. 29, 127–153 (1927).

Hildebrandt, T. H., and Schoenberg, I. J.

1. *On linear functional operations and the moment problem for a finite interval in one or several dimensions.* Ann. of Math. (2) 34, 317–328 (1933).

Hilding, S.

1. *On completeness theorems of Paley-Wiener type.* Ann. of Math. (2) 49, 953–954 (1948).
2. *On the closure of disturbed orthonormal sets in Hilbert space.* Ark. Mat. Astr. Fys. 32B, no. 7 (1946), 3 pp.

Hille, E.

1. *Functional analysis and semi-groups.* Amer. Math. Soc. Colloq. Pub. vol. 31, New York (1948).
2. *Notes on linear transformations,* II. *Analyticity of semi-groups.* Ann. of Math. (2) 40, 1–47 (1939).
3. *On the generation of semi-groups and the theory of conjugate functions.* Proc. Roy. Physiog. Soc. Lund 21, 1–13 (1951).
4. *Non-oscillation theorems.* Trans. Amer. Math. Soc. 64, 234–252 (1948).
5. *The abstract Cauchy problem and Cauchy's problem for parabolic differential equations.* J. d'Analyse Math. 3, 81–196 (1953).

Hille, E., and Tamarkin, J. D.

1. *On the characteristic values of linear integral equations.* Acta Math. 57, 1–76 (1931).
2. *On the theory of linear integral equations,* II. Ann. of Math. (2) 35, 445–455 (1934).

Hirschman, I. I., and Widder, D. V. (see also Widder and Hirschman).

1. *The convolution transform.* Princeton Univ. Press, Princeton, 1955.

Hobson, E. W.
1. *The theory of functions of a real variable.* (Two volumes) Second edition, Cambridge Univ. Press, 1921, 1926.
2. *On a general convergence theorem, and the theory of the representation of a function by a series of normal functions.* Proc. London Math. Soc. (2) 6, 349–395 (1908).

Hölder, E.
1. *Über die Vielfachheiten gestörter Eigenwerte.* Math. Ann. 113, 620–628 (1936).
2. *Über einen Mittelwertsatz.* Nachr. Akad. Wiss. Göttingen. Math.-Phys. Kl. 1889, 38–47 (1889).

Holmgren, E.
1. *Über Systeme von linearen partiellen Differentialgleichungen.* Öfvers. Kongl. Vetens.-Akad. Förk. 58, 91–103 (1901).

Hopf, E.
1. *Ergodentheorie.* Ergebnisse der Math. V 2, J. Springer, Berlin, 1937. Reprinted by Chelsea Pub. Co., New York, 1948.
2. *The general temporally discrete Markov process.* J. Rational Mech. and Anal. 3, 13–45 (1954).
3. *Über eine Ungleichung der Ergodentheorie.* S.-B. Math.-Nat. Kl. Bayer. Akad. Wiss., 171–176 (1944).

Horn, A.
1. *On the singular values of a product of completely continuous operators.* Proc. Nat. Acad. Sci. U.S.A. 36, 374–375 (1950).

Hotta, J.
1. *A remark on regularly convex sets.* Kōdai Math. Sem. Rep. 1951, 37–40 (1951).

Hukuhara, M.
1. *Sur l'existence des points invariants d'une transformation dans l'espace fonctionnel.* Jap. J. Math. 20, 1–4 (1950).

Hull, T. E. (see Infeld, L.)

Hulthén, L.
1. *On the Sturm-Liouville Problem connected with a continuous spectrum.* Ark. Mat. Astr. Fys. 35A, no. 25 (1949), 25 pp.

Hurewicz, W.
1. *Ergodic theorems without invariant measure.* Ann. of Math. (2) 45, 192–206 (1944).

Hurewicz, W., and Wallman, H.
1. *Dimension theory.* Princeton Univ. Press, Princeton, 1941.

Hurwitz, W. A. (see Gillespie, D. C.)

Hyers, D. H.
1. *Pseudo-normed linear spaces and abelian groups.* Duke Math. J. 5, 628–634 (1939).
2. *Locally bounded linear topological spaces.* Revista Ci., Lima 41, 555–574 (1939).
3. *Linear topological spaces.* Bull. Amer. Math. Soc. 51, 1–21 (1945).
4. *A note on linear topological spaces.* Bull. Amer. Math. Soc. 44, 76–80 (1938).

Inaba, M.
1. *A theorem on fixed points and its application to the theory of differential equations.* Kumamoto J. Sci. Ser. A. 1, no. 1, 13–16 (1952).

Infeld, L.

1. *On a new treatment of some eigenvalue problems.* Phys. Rev. (2) 59, 737–747 (1941).

Infeld, L., and Hull, T. E.

1. *The factorization method.* Rev. Mod. Phys. 23, 21–68, (1951).

Ingleton, A. W.

1. *The Hahn-Banach theorem for non-Archimedean valued fields.* Proc. Cambridge Philos. Soc. 48, 41–45 (1952).

Ionescu Tulcea, C. T., and Marinescu, G.

1. *Théorie ergodique pour des classes d'opérations non complètement continues.* Ann. of Math. (2) 52, 140–147 (1950).

Iwata, G.

1. *Non-hermitian operator and eigenfunction expansions.* Progress Theoret. Phys. 6, 216–226 (1951).

Iyer, V. G.

1. *On the space of integral functions,* I–III.
   I. J. Indian Math. Soc. (2) 12, 13–30 (1948).
   II. Quart. J. Math. (Oxford) (2) 1, 86–96 (1950).
   III. Proc. Amer. Math. Soc. 3, 874–883 (1952).

Izumi, S.

1. *On the bilinear functionals.* Tôhoku Math. J. 42, 195–209 (1936).
2. *On the compactness of a class of functions.* Proc. Imp. Acad. Tokyo 15, 111–113 (1939).
3. *Lebesgue integral in the abstract space.* Jap. J. Math. 13, 501–513 (1936).
4. *Notes on Banach space,* I. *Differentiation of abstract functions.* Proc. Imp. Acad. Tokyo 18, 127–130 (1942).

Izumi, S., and Sunouchi, G.

1. *Notes on Banach space* (VI): *Abstract integrals and linear operations.* Proc. Imp. Acad. Tokyo 19, 169–173 (1943).

Jackson, D.

1. *Algebraic properties of self-adjoint systems.* Trans. Amer. Math. Soc. 17, 418–424 (1916).

Jacobson, N.

1. *Lectures in abstract algebra.* I. *Basic Concepts.*
   II. *Linear algebras.* D. van Nostrand, New York, 1951, 1953.

James, R. C.

1. *Orthogonality in normed linear spaces.* Duke Math. J. 12, 291–302 (1945).
2. *Orthogonality and linear functionals in normed linear spaces.* Trans. Amer. Math. Soc. 61, 265–292 (1947).
3. *Inner products in normed linear spaces.* Bull. Amer. Math. Soc. 53, 559–566 (1947).
4. *Bases and reflexivity of Banach spaces.* Ann. of Math. (2) 52, 518–527 (1950).
5. *A non-reflexive Banach space isometric with its second conjugate space.* Proc. Nat. Acad. Sci. U.S.A. 37, 174–177 (1951).

Jamison, S. L.

1. *Perturbation of normal operators.* Proc. Amer. Math. Soc. 5, 103–110 (1954).
2. *On analytic normal operators.* Proc. Amer. Math. Soc. 5, 288–290 (1954).

Jerison, M.
1. *Characterizations of certain spaces of continuous functions*. Trans. Amer. Math. Soc. 70, 103–113 (1951).
2. *A property of extreme points of compact convex sets*. Proc. Amer. Math. Soc. 5, 782–783 (1954).

Jessen, B.
1. *The theory of integration in a space of an infinite number of dimensions*. Acta Math. 63, 249–323 (1934).
2. *Bidrag til Integralteorien for Funktioner af unendlig mange Variable*. Copenhagen, 1930.

John, F.
1. *The fundamental solution of linear elliptic differential equations with analytic coefficients*. Comm. Pure Appl. Math. 3, 273–304 (1950).
2. *General properties of solutions of linear elliptic partial differential equations*. Proc. Symposium Spectral Theory and Differential Problems, 113–175. Oklahoma Agricultural and Mechanical College, Stillwater, Oklahoma, 1951.

Jordan, G.
1. *Sur la série de Fourier*. C. R. Acad. Sci. Paris 92, 228–230 (1881).

Jordan, P., and von Neumann, J.
1. *On inner products in linear, metric spaces*. Ann. of Math. (2) 36, 719–723 (1935).

Jost, R. (see also Newton, R. G.)
1. *Bemerkungen zur mathematischen Theorie der Zähler*. Helvetica Phys. Acta 20, 173–182 (1947).

Jost, R., and Kohn, W.
1. *Construction of a potential from a phase shift*. Phys. Rev. 87, 977–992 (1952).

Jost, R., and Pais, A.
1. *On the scattering of a particle by a static potential*. Phys. Rev. 82, 840–851 (1951).

Julia, G.
1. *Sur les racines carrées hermitiennes d'un opérateur hermitien positif donné*. C. R. Acad. Sci. Paris 222, 707–709 (1946).
2. *Remarques sur les racines carrées hermitiennes d'un opérateur hermitien positif borné*. C. R. Acad. Sci. Paris 222, 829–832 (1946).
3. *Sur la representation spectrale des racines hermitiennes d'un opérateur hermitien positif donné*. C. R. Acad. Sci. Paris 222, 1019–1022 (1946).
4. *Sur les racines carrées self-adjoint d'un opérateur self-adjoint positif non borné*. C. R. Acad. Sci. Paris 222, 1061–1063 (1946).
5. *Sur les racines $n^{ièmes}$ hermitiennes d'un opérateur hermitien donné*. C. R. Acad. Sci. Paris 222, 1465–1468 (1946).
6. *Détermination de toutes les racines carrées d'un opérateur hermitien borné quelconque*, I, II.
   I. C. R. Acad. Sci. Paris 227, 792–794 (1948).
   II. ibid. 227, 931–933 (1948).
7. *Introduction mathématique aux théories quantiques*. Paris, 1938.

Kac, M. (see also Erdös, P.)
1. *On distributions of certain Wiener functionals*. Trans. Amer. Math. Soc. 65, 1–13 (1949).

2. *On the average of a certain Wiener functional and a related limit theorem in the calculus of probability.* Trans. Amer. Math. Soc. 59, 401–414 (1946).
3. *On some connections between probability theory and differential and integral equations.* Proc. Second Berkeley Symposium Math. Statistics and Prob., 189–215 (1951).

Kaczmarz, S., and Steinhaus, H.
1. *Theorie der Orthogonalreihen.* Monografje Matematyczne, vol. 6, Warsaw, 1935. Reprinted by Chelsea Pub. Co., New York, 1951.

Kadison, R. V.
1. *A representation theory for commutative topological algebra.* Memoirs Amer. Math. Soc. no. 7, 1951.
2. *Isometries of operator algebras.* Ann. of Math. (2) 54, 325–338 (1951).

Kakutani, S. (see also Anzai, H., Bohnenblust, H. F., and Yosida, K.)
1. *Ein Beweis des Satzes von M. Eidelheit über konvexe Mengen.* Proc. Imp. Acad. Tokyo 13, 93–94 (1937).
2. *Weak topology and regularity of Banach spaces.* Proc. Imp. Acad. Tokyo 15, 169–173 (1939).
3. *Weak topology, bicompact set and the principle of duality.* Proc. Imp. Acad. Tokyo 16, 63–67 (1940).
4. *Two fixed-point theorems concerning bicompact convex sets.* Proc. Imp. Acad. Tokyo 14, 242–245 (1938).
5. *Topological properties of the unit sphere of a Hilbert space.* Proc. Imp. Acad. Tokyo 19, 269–271 (1943).
6. *Some characterizations of Euclidean spaces.* Jap. J. Math. 16, 93–97 (1939).
7. *Mean ergodic theorems in abstract (L)-spaces.* Proc. Imp. Acad. Tokyo 15, 121–123 (1939).
8. *Concrete representation of abstract (L)-spaces and the mean ergodic theorem.* Ann. of Math. (2) 42, 523–537 (1941).
9. *Concrete representation of abstract (M)-spaces. (A characterization of the space of continuous functions.)* Ann. of Math. (2) 42, 994–1024 (1941).
10. *Ergodic theory.* Proc. International Congress Math., Cambridge, Mass. 2, 128–142 (1950).
11. *A proof of Schauder's theorem.* J. Math. Soc. Japan 3, 228–231 (1951).
12. *Über die Metrisation der topologischen Gruppen.* Proc. Imp. Acad. Tokyo 12, 82–84 (1936).
13. *Iteration of linear operations in complex Banach spaces.* Proc. Imp. Acad. Tokyo 14, 295–300 (1938).
14. *Notes on infinite product measure spaces, I, II.*
   I. Proc. Imp. Acad. Tokyo 19, 148–151 (1943).
   II. ibid. 19, 184–188 (1943).
15. *An example concerning uniform boundedness of spectral measures.* Pacific J. Math. 4, 363–372 (1954).
16. *Ergodic theorems and the Markoff process with a stable distribution.* Proc. Imp. Acad. Tokyo 16, 49–54 (1940).
17. *On the uniqueness of Haar's measure.* Proc. Imp. Acad. Tokyo 14, 27–31 (1938).
18. *A proof of the uniqueness of Haar's measure.* Ann. of Math. (2) 49, 225–226 (1948).
19. *On the uniform ergodic theorem concerning real linear operations.* Jap. J. Math. 17, 5–12 (1940).

20. *Some results in the operator-theoretical treatment of the Markoff process.* Proc. Imp. Acad. Tokyo 15, 260–264 (1939).
21. *Simultaneous extension of continuous functions considered as a positive linear operation.* Jap. J. Math. 17, 1–4 (1940).
22. *Weak convergence in uniformly convex spaces.* Tôhoku Math. J. 45, 188–193 (1938).
23. *Rings of analytic functions.* Lectures on functions of a complex variable, pp. 71–83, Ann Arbor, 1955.

Kakutani, S., and Kodaira, K.
1. *Über das Haarsche Mass in der lokal bikompakten Gruppe.* Proc. Imp. Acad. Tokyo 20, 444–450 (1944).

Kakutani, S., and Mackey, G. W.
1. *Two characterizations of real Hilbert space.* Ann. of Math. (2) 45, 50–58 (1944).

Kamke, E.
1. *Mengenlehre.* W. de Gruyter, Berlin and Leipzig, 1928.
2. *Theory of sets.* Translation of Second German edition. Dover Pub. Co., New York.
3. *Neue Herleitung der Oszillationssätze für die linearen selbstadjugierten Randwertaufgaben zweiter Ordnung.* Math. Zeit. 44, 635–658 (1938).

van Kampen, E. R.
1. *Locally bicompact groups and their character groups.* Ann. of Math. (2) 36, 448–463 (1935).

Kantorovitch (Kantorovič), L. V. (see also Fichtenholz, G.)
1. *Lineare halbgeordnete Räume.* Mat. Sbornik N. S. 2 (44), 121–168 (1937).
2. *The method of successive approximations for functional equations.* Acta Math. 71, 63–97 (1939).
3. *Linear operations in semi-ordered spaces,* I. Mat. Sbornik N. S. 7 (49), 210–284 (1940).
4. *Allgemeine Formen gewisser Klassen von linearen Operationen.* Doklady Akad. Nauk SSSR (N. S.) 12, 101–106 (1936).

Kantorovitch (Kantorovič), L. V., and Vulich (Vulih), B. Z.
1. *Sur la représentation des opérations linéaires.* Compositio Math. 5, 119–165 (1938).
2. *Sur un théorème de M. N. Dunford.* Compositio Math. 5, 430–432 (1938).

Kantorovitch (Kantorovič), L. V., Vulih, B. Z., and Pinsker, A. G.
1. *Functional analysis in partially ordered spaces.* Gosudarstr. Izdat. Tehn.-Teor. Lit., Moscow-Leningrad, 1950. (Russian).
2. *Partially ordered groups and linear partially ordered spaces.* Uspehi Matem. Nauk (N. S.) 6, no. 3 (43), 31–98 (1951). (Russian) Math. Rev. 13, 361 (1952).

Kaplansky, I. (see also Arens, R.)
1. *The structure of certain operator algebras.* Trans. Amer. Math. Soc. 70, 219–255 (1951).
2. *The Weierstrass theorem in fields with valuations.* Proc. Amer. Math. Soc. 1, 356–357 (1950).
3. *Lattices of continuous functions,* I, II.
   I. Bull. Amer. Math. Soc. 53, 617–623 (1947).
   II. Amer. J. Math. 70, 626–634 (1948).

4. *Topological rings.* Bull. Amer. Math. Soc. 54, 809–826 (1948).
5. *Primary ideals in group algebras.* Proc. Nat. Acad. Sci. U.S.A. 35, 133–136 (1949).
6. *Products of normal operators.* Duke Math. J. 20, 257–260 (1953).
7. *A theorem on rings of operators.* Pacific J. Math. 1, 227–232 (1951).

Karaseva, T. M.
1. *On the inverse Sturm-Liouville problem for a non-Hermitian operator.* Mat. Sbornik N. S. 32 (74), 477–484 (1953). (Russian) Math. Rev. 14, 1088 (1953).

Karlin, S.
1. *Unconditional convergence in Banach spaces.* Bull. Amer. Math. Soc. 54, 148–152 (1948).
2. *Bases in Banach spaces.* Duke Math. J. 15, 971–985 (1948).

Kato, T.
1. *On the convergence of the perturbation method,* I, II.
I. Progress Theoret. Physics 4, 514–523 (1949).
II. ibid. 5, 96–101, 207–212 (1950).
2. *On the convergence of the perturbation method.* J. Fac. Sci. Univ. Tokyo Sect. I. 6, 145–226 (1951).
3. *Perturbation theory of semi-bounded operators.* Math. Ann. 125, 435–447 (1953).
4. *On the perturbation theory of closed linear operators.* J. Math. Soc. Japan 4, 323–337 (1952).
5. *Notes on some inequalities for linear operators.* Math. Ann. 125, 208–212 (1952).
6. *On some approximate methods concerning the operators T^*T.* Math. Ann. 126, 253–262 (1953).
7. *On the semi-groups generated by Kolmogoroff's differential equations.* J. Math. Soc. Japan 6, 1–15 (1954).
8. *On the upper and lower bounds of eigenvalues.* J. Phys. Soc. Japan 4, 334–339 (1949).

Kay, I.
1. *The inverse scattering problem.* Div. Electromag. Res., Inst. Math. Sci., New York Univ., 1955.

Kay, I., and Moses, H. E.
1. *The determination of the scattering potential from the spectral measure function,* I. Il Nuovo Cimento (10) 2, 917–961 (1955).
2. *The determination of the scattering potential from the spectral measure function,* I–III. Div. of Electromag. Res., Inst. Math. Sci., New York Univ., 1955.

Kaz, I.
1. *On Hilbert spaces generated by positive Hermitian matrix functions.* Zapiski Naučno-Issled. Inst. Mat. Meh. i Har'kov. Mat. Obšč. 22 (1950).

Keldyš, M. V.
1. *On the characteristic values and characteristic functions of certain classes of non-self-adjoint equations.* Doklady Akad. Nauk SSSR (N. S.) 77, 11–14 (1951). (Russian) Math. Rev. 12, 835 (1951).

Kelley, J. L. (see also Arens, R.)
1. *Note on a theorem of Krein and Milman.* J. Osaka Inst. Sci. Tech. Part I. 3, 1–2 (1951).

2. *Banach spaces with the extension property.* Trans. Amer. Math. Soc. 72, 323–326 (1952).
3. *The Tychonoff product theorem implies the axiom of choice.* Fund. Math. 37, 75–76 (1950).
4. *Convergence in topology.* Duke Math. J. 17, 277–283 (1950).
5. *General topology.* D. van Nostrand, New York, 1955.
6. *Commutative operator algebras.* Proc. Nat. Acad. Sci. U.S.A. 38, 598–605 (1952).

Kelley, J. L., and Vaught, R. L.
1. *The positive cone in Banach algebras.* Trans. Amer. Math. Soc. 74, 44–55 (1953).

Kellogg, O. D. (see Birkhoff, G. D.)

Kemble, E. C.
1. *A contribution to the theory of the B.W.K. method.* Phys. Rev. 48, 549–561 (1935).
2. *Note on the Sturm-Liouville eigenvalue-eigenfunction problem with singular endpoints.* Proc. Nat. Acad. Sci. U.S.A. 19, 710–714 (1933).
3. *The fundamental principles of quantum mechanics.* New York, 1937.

Kerner, M.
1. *Abstract differential geometry.* Compositio Math. 4, 308–341 (1937).
2. *Die Differentiale in der allgemeinen Analysis.* Ann. of Math. (2) 34, 564–572 (1933).

Khintchine, A.
1. *Zu Birkhoffs Lösung des Ergodenproblems.* Math. Ann. 107, 485–488 (1933).
2. *Fourierkoeffizienten längs einer Bahn in Phasenraum.* Mat. Sbornik 41, 14–16 (1934).

Kilpi, Y.
1. *Über lineare normale Transformationen im Hilbertschen Raum.* Ann. Acad. Sci. Fennicae. Ser. A 1, no. 154, 38 pp. (1953).

Kinoshita, S.
1. *On essential components of the set of fixed points.* Osaka Math. J. 4, 19–22 (1952).

Klee, V. L., Jr.
1. *The support property of a convex set in a linear normed space.* Duke Math. J. 15, 767–772 (1948).
2. *Dense convex sets.* Duke Math. J. 16, 351–354 (1949).
3. *Convex sets in linear spaces.* Duke Math. J. 18, 443–466 (1951).
4. *Convex sets in linear spaces*, II. Duke Math. J. 18, 875–883 (1951).
5. *Invariant extensions of linear functionals.* Pacific J. Math. 4, 37–46 (1954).
6. *Invariant metrics in groups (Solution of a problem of Banach).* Proc. Amer. Math. Soc. 3, 484–487(1952).
7. *Some characterizations of reflexivity.* Revista Ci., Lima 52, 15–23 (1950).
8. *Convex bodies and periodic homeomorphisms in Hilbert space.* Trans. Amer. Math. Soc. 74, 10–43 (1953).

Kleinecke, D. C.
1. *A generalization of complete continuity.* Technical Report no. 3. to the Office of Ordinance Research, University of California, Berkeley (1954).
2. *Degenerate perturbations.* Technical Report no. 1 to the Office of Ordinance Research, University of California, Berkeley (1953).

3. *Finite perturbations and the essential spectrum.* Technical Report no. 4 to the Office of Ordinance Research, University of California, Berkeley (1954).

Kneser, A.

1. *Untersuchungen über die reellen Nullstellen der Integrale linearer Differentialgleichungen.* Math. Ann. 42, 409–435 (1893).
2. *Untersuchungen über die Darstellung willkürlicher Funktionen in der mathematischen Physik.* Math. Ann. 58, 81–147 (1904).
3. *Beiträge zur Theorie der Sturm-Liouvilleschen Darstellung willkürlicher Funktionen.* Math. Ann. 60, 402–423 (1905).
4. *Die Theorie der Integralgleichungen und die Darstellung willkürlicher Funktionen in der mathematischen Physik,* II. Nachr. Akad. Wiss. Göttingen. Math.-Phys. Kl. 1906, 213–252 (1906).

Knopp, K.

1. *Theory of functions,* I, II. Dover Publications, New York, 1945, 1947.

Kober, H. A.

1. *A theorem on Banach spaces.* Compositio Math. 7, 135–140 (1939).

Kodaira, K. (see also Kakutani, S.)

1. *On ordinary differential equations of any even order and the corresponding eigenfunction expansions.* Amer. J. Math. 72, 502–544 (1950).
2. *On some fundamental theorems in the theory of operators in Hilbert space.* Proc. Imp. Acad. Tokyo 15, 207–210 (1939).
3. *Über die Beziehung zwischen den Massen und den Topologien in einer Gruppe.* Proc. Phys.-Math. Soc. Japan (3) 23, 67–119 (1941).
4. *The eigenvalue problem for ordinary differential equations of the second order and Heisenberg's theory of S-matrices.* Amer. J. Math. 71, 921–945 (1949).
5. *On singular solutions of second order differential operators,* I, II.
   I. Sūgaku 7, 177–191 (1948).
   II. ibid. 2, 113–139 (1948). (Japanese).

Kohn, W. (see Jost, R.)

Kolmogoroff, A. (see also Gelfand, I.)

1. *Zur Normierbarkeit eines allgemeinen topologischen linearen Raumes.* Studia Math. 5, 29–33 (1934).
2. *Über Kompaktheit der Funktionenmengen bei der Konvergenz im Mittel.* Nachr. Ges. Göttingen Math.-Phys. Kl., 60–63 (1931).

Komatuzaki, H.

1. *Sur les projections dans certains espaces du type (B).* Proc. Imp. Acad. Tokyo 16, 274–279 (1940).
2. *Une remarque sur les projections dans certains espaces du type (B).* Proc. Imp. Acad. Tokyo 17, 238–240 (1941).

Koopman, B. O.

1. *Hamiltonian systems and transformations in Hilbert space.* Proc. Nat. Acad. Sci. U.S.A. 17, 315–318 (1931).

Koopman, B. O., and Doob, J. L.

1. *On analytic functions with positive imaginary parts.* Bull. Amer. Math. Soc. 40, 601–605 (1934).

Kostyučenko, A. G. (see Gelfand, I. M.)

Kostyučenko, A., and Skorohod, A.

1. *On a theorem of M. K. Bari.* Uspehi Matem. Nauk (N. S.) 8, no. 5 (57), 165–166 (1953). (Russian) Math. Rev. 15, 632 (1954).

Köthe, G.

1. *Die Teilräume eines linearen Koordinatenraumes.* Math. Ann. 114, 99–125 (1937).
2. *Lösbarkeitsbedingungen für Gleichungen mit unendlich vielen Unbekannten.* J. Reine Angew. Math. 178, 193–213 (1938).
3. *Erweiterung von Linearfunktionen in linearen Räumen.* Math. Ann. 116, 719–732 (1939).
4. *Die Quotienträume eines linearen vollkommenen Räumes.* Math. Z. 51, 17–55 (1947).
5. *Die Stufenräume, eine einfache Klasse linearer vollkommenen Räume.* Math. Z. 51, 317–345 (1948).
6. *Eine axiomatische Kennzeichnung der linearen Räume von Typus ω.* Math. Ann. 120, 634–649 (1949).
7. *Über die Vollständigkeit einer Klasse lokalkonvexer Räume.* Math. Z. 52, 627–630 (1950).
8. *Über zwei Sätze von Banach.* Math. Z. 53, 203–209 (1950).
9. *Neubegründung der Theorie der vollkommenen Räume.* Math. Nachr. 4, 70–80 (1951).
10. *Über zwei Sätze von Banach.* Math. Z. 53, 203–209 (1950).
11. *Funktionalanalysis, Integraltransformationen.* Naturforschung und Medizin in Deutschland, 1939–1946, Band 2, 85–98. Dieterich'sche Verlagsbuchhandlung, Wiesbaden, 1948.

Köthe, G., and Toeplitz, O.

1. *Lineare Räume mit unendlichvielen Koordinaten.* J. Reine Angew. Math. 171, 193–226 (1934).

Kowalewski, G.

1. *Einführung in die Determinantentheorie.* Second ed., W. de Gruyter, Berlin and Leipzig, 1925.

Kozlov, V. Ya.

1. *On bases in the space $L_2[0,1]$.* Mat. Sbornik N. S. 26 (68), 85–102 (1950). (Russian) Math. Rev. 11, 602 (1950).
2. *On a generalization of the concept of basis.* Doklady Akad. Nauk SSSR (N. S.) 73, 643–646 (1950). (Russian) Math. Rev. 12, 110 (1951).

Kračkovskiĭ, S. N. (see also Gol'dman, M. A.)

1. *Canonical representation of null elements of a linear operator in its Fredholm region.* Doklady Akad. Nauk SSSR (N. S.) 88, 201–204 (1953). (Russian) Math. Rev. 14, 1095 (1953).
2. *On properties of a linear operator connected with its generalized Fredholm region.* Doklady Akad. Nauk SSSR (N. S.) 91, 1011–1013 (1953). (Russian) Math. Rev. 15, 437 (1954).
3. *On the extended region of singularity of the operator $T_\lambda = E - \lambda A$.* Doklady Akad. Nauk SSSR (N. S.) 96, 1101–1104 (1954). (Russian) Math. Rev. 16, 263 (1955).

Kračkovskiĭ, S. N., and Gol'dman, M. A.

1. *On the principal part of a completely continuous operator.* Doklady Akad. Nauk SSSR (N.S.) 70, 945–948 (1950). (Russian) Math. Rev. 11, 600 (1950).
2. *Null elements and null functionals of completely continuous operators.* Latvijas PSR Zinãtnu Akad. Vēstis 1950 no. 6 (35), 87–100 (1950). (Russian. Latvian summary) Math. Rev. 13, 251 (1952).

3. *Some properties of a completely continuous operator in Hilbert space.* Latvijas PSR Zinātnu Akad. Vēstis 1950 no. 10 (39), 93–106 (1950). (Russian. Latvian summary) Math. Rev. 15, 440 (1954).

Kračkovskiĭ, S. N., and Vinogradov, A. A.

1. *On a criterion of uniform convexity of a space of type B.* Uspehi Matem. Nauk 7, no. 3 (49) 131–134 (1952). (Russian) Math. Rev. 14, 55 (1953).

Kramer, H. P.

1. *Perturbation of differential operators.* Dissertation, Univ. of California at Berkeley, 1954.

Kramer, V. A.

1. *Investigations in asymptotic perturbation series.* Dissertation, Univ. of California at Berkeley, 1954.
2. *Asymptotic inverse series,* Proc. Amer. Math. Soc. 7, 429–437 (1956).

Kramers, H. A.

1. *Wellenmechanik und halbzahlige Quantisierung.* Zeitschrift für Phys. 39, 828–846 (1926).
2. *Das Eigenwertproblem in eindimensional periodischen Kraftfelde.* Physica 2, 483–490 (1935).

Krasnosel'skiĭ, M. A. (see also Kreĭn, M.)

1. *On certain types of extensions of Hermitian operators.* Ukrain. Mat. Ž. 2, no. 2, 74–83 (1950). (Russian) Math. Rev. 13, 47 (1952).
2. *On self-adjoint extensions of Hermitian operators.* Ukrain. Mat. Ž. 1, no. 1, 21–38 (1949). (Russian) Math. Rev. 13, 954 (1952).
3. *On the deficiency numbers of closed operators.* Doklady Akad. Nauk SSSR (N. S.) 56, 559–561 (1947). (Russian) Math. Rev. 9, 242 (1948).
4. *On the extension of Hermitian operators with a nondense domain of definition.* Doklady Akad. Nauk SSSR (N. S.) 59, 13–16 (1948). (Russian) Math. Rev. 9, 447 (1948).

Krasnosel'skiĭ, M. A., and Rutickiĭ, Ya. B.

1. *On the theory of Orlicz spaces.* Doklady Akad. Nauk SSSR (N. S.) 81, 497–500 (1951). (Russian) Math. Rev. 13, 357 (1952).
2. *Linear integral operators in Orlicz spaces.* Doklady Akad. Nauk SSSR (N. S.) 85, 33–36 (1952). (Russian) Math. Rev. 14, 57 (1953).
3. *Differentiability of non-linear integral operators in Orlicz spaces.* Doklady Akad. Nauk SSSR (N. S.) 89, 601–604 (1953). (Russian) Math. Rev. 15, 137 (1954).

Krein (Kreĭn), M. G.

1. *Sur quelques questions de la géométrie des ensembles convexes situés dans un espace linéaire normé et complet.* Doklady Akad. Nauk SSSR (N. S.) 14, 5–7 (1937).
2. *Sur les opérations linéaires transformant un certain ensemble conique en lui-même.* Doklady Akad. Nauk SSSR (N. S.) 23, 749–752 (1939).
3. *Propriétés fondamentales des ensembles coniques normaux dans l'espace de Banach.* Doklady Akad. Nauk SSSR (N. S.) 28, 13–17 (1940).
4. *Sur la décomposition minimale d'une fonctionnelle linéaire en composantes positives.* Doklady Akad. Nauk SSSR (N. S.) 28, 18–22 (1940).
5. *On positive functionals on almost periodic functions.* Doklady Akad. Nauk SSSR (N. S.) 30, 9–12 (1941).

6. *Sur une généralisation du théorème de Plancherel au cas des intégrales de Fourier sur les groupes topologiques commutatifs*. Doklady Akad. Nauk SSSR (N. S.) 30, 484–488 (1941).
7. *Infinite J-matrices and a matrix moment problem*. Doklady Akad. Nauk SSSR (N. S.) 69, 125–128 (1949). (Russian) Math. Rev. 11, 670 (1950).
8. *On the trace formula in perturbation theory*. Mat. Sbornik N. S. 33 (75), 597–626 (1953). (Russian) Math. Rev. 15, 720 (1954).
9. *The theory of self-adjoint extensions of semi-bounded Hermitian operators and its applications*, I, II.
I. Mat. Sbornik N. S. 20 (62), 431–495 (1947).
II. ibid. 21 (63), 365–404 (1947). (Russian) Math. Rev. 9, 515–516 (1948).
10. *Sur les fonctions de Green non-symétriques oscillatoires des opérateurs differentials ordinaires*. Doklady Akad. Nauk SSSR (N. S.) 25, 643–646 (1939).
11. *On a general method of decomposing Hermite-positive nuclei into elementary products*. Doklady Akad. Nauk SSSR (N. S.) 53, 3–6 (1946).
12. *On Hermitian operators and directing functionals*. Akad. Nauk Ukrain. RSR. Zbirnik Prac. Mat. Inst. 10, 83–106 (1948).
13. *On a one-dimensional singular boundary value problem of even order in the interval* $(0,\infty)$. Doklady Akad. Nauk SSSR (N. S.) 74, 9–12 (1950). (Russian) Math. Rev. 12, 502 (1951).
14. *Solution of the inverse Sturm-Liouville problem*. Doklady Akad. Nauk SSSR (N. S.) 76, 21–24 (1951). (Russian) Math. Rev. 12, 613 (1951).
15. *On certain problems on the maximum and minimum of characteristic values and on the Lyapunov zones of stability*. Akad. Nauk SSSR. Prikl. Mat. Meh. 15, 323–348 (1951). (Russian) Math. Rev. 13, 348 (1952).
16. *On the indeterminate case of the Sturm-Liouville boundary problem in the interval* $(0, \infty)$. Izvestiya Akad. Nauk SSSR Ser. Mat. 16, 293–324 (1952). (Russian) Math. Rev. 14, 558 (1953).
17. *On some cases of effective determination of the density of an inhomogeneous cord from its spectral function*. Doklady Akad. Nauk SSSR (N. S.) 93, 617–620 (1953). (Russian) Math. Rev. 15, 796 (1954).
18. *On a method of effective solution of an inverse boundary problem*. Doklady Akad. Nauk SSSR (N. S.) 94, 987–990 (1954). (Russian) Math. Rev. 16, 38 (1955).
19. *Determination of the density of a nonhomogeneous symmetric cord by its frequency spectrum*. Doklady Akad. Nauk SSSR (N. S.) 76, 345–348 (1951). (Russian) Math. Rev. 13, 43 (1952).

Kreĭn, M., and Krasnosel'skiĭ, M. A.
1. *Stability of the index of an unbounded operator*. Mat. Sbornik N. S. 30 (72), 219–224 (1952). (Russian) Math. Rev. 13, 849 (1952).
2. *Fundamental theorems on the extension of Hermitian operators and some of their applications to the theory of orthogonal polynomials and the moment problem*. Uspehi Matem. Nauk 2, no. 3 (19), 60–107 (1947). (Russian) Math. Rev. 10, 198 (1949).

Kreĭn, M. G., Krasnosel'skiĭ, M. A., and Milman, D. P.
1. *On the deficiency indices of linear operators in Banach spaces and some geometrical questions*. Sbornik Trudov Inst. Akad. Nauk Ukr. SSR 11, 97–112 (1948). (Russian).

Krein, M., and Krein, S.

1. *On an inner characteristic of the set of all continuous functions defined on a bicompact Hausdorff space.* Doklady Akad. Nauk SSSR (N. S.) 27, 427–430 (1940).
2. *Sur l'espace des fonctions continues définies sur un bicompact de Hausdorff et ses sousespaces semiordonnés.* Mat. Sbornik N. S. 13 (55), 1–37 (1943).

Krein, M., and Milman, D.

1. *On extreme points of regularly convex sets.* Studia Math. 9, 133–138 (1940).

Krein, M., Milman, D., and Rutman, M.

1. *A note on basis in Banach space.* Comm. Inst. Sci. Math. Méc. Univ. Kharkoff [Zapiski Inst. Mat. Mech.] (4) 16, 106–110 (1940). (Russian. English summary) Math. Rev. 3, 49 (1942).

Krein, M., and Rutman, M. A.

1. *Linear operators leaving invariant a cone in a Banach space.* Uspehi Matem. Nauk 3, no. 1 (23), 3–95 (1948). Amer. Math. Soc. Translation no. 26 (1950).

Krein, M., and Šmulian, V.

1. *On regularly convex sets in the space conjugate to a Banach space.* Ann. of Math. 41, 556–583 (1940).

Kuller, R. G.

1. *Locally convex topological vector lattices and their representations.* Dissertation, Univ. of Michigan, 1955.

Kunisawa, K.

1. *Some theorems on abstractly-valued functions in an abstract space.* Proc. Imp. Acad. Tokyo 16, 68–72 (1940).

Kuratowski, C.

1. *Sur la propriété de Baire dans les groupes métriques.* Studia Math. 4, 38–40 (1933).

Kürschák, J.

1. *Über Limesbildung und allgemeine Körpertheorie.* J. Reine Angew. Math. 142, 211–253 (1912).

Laasonen, P.

1. *Über die Näherungslösungen der Sturm-Liouvilleschen Eigenwertaufgabe.* Proc. XII Scand. Math. Congress Lund, 1953, 176–182 (1954).

Lagrange, J. L.

1. *Oeuvres,* t. 3. Gauthier-Villars, Paris, 1869.
2. *Oeuvres,* t. 1. Gauthier-Villars, Paris, 1867.

Laguerre, E. N.

1. *Sur le calcul des systèmes linéaires.* Reprinted in *Oeuvres,* t. 1, 221–267 (1898).

Lamson, K. W.

1. *A general implicit function theorem with an application to problems of relative minima.* Amer. J. Math. 42, 243–256 (1920).

Landau, E.

1. *Über einen Konvergenzsatz.* Nachr. Akad. Wiss. Göttingen. Math.-Phys. Kl. IIa 1907, 25–27 (1907).
2. *Über einen Satz von Herrn Esclangon.* Math. Ann. 102, 177–188 (1929).

Langer, R. E. (see also Birkhoff, G. D.)
1. *On the connection formulas and the solutions of the wave equation.* Phys. Rev. 51, 669–676 (1937).
2. *On the wave equation with small quantum numbers.* Phys. Rev. 75, 1573–1578 (1949).
3. *The expansion problem in the theory of ordinary differential systems of the second order.* Trans. Amer. Math. Soc. 31, 868–906 (1929).

LaSalle, J. P.
1. *Pseudo-normed linear spaces.* Duke Math. J. 8, 131–135 (1941).
2. *Application of the pseudo-norm to the study of linear topological spaces.* Revista Ci., Lima 47, 545–563 (1945).
3. *Singular measurable sets and linear functionals.* Math. Mag. 22, 67–72 (1948).

Latshaw, V. V.
1. *The algebra of self-adjoint boundary-value problems.* Bull. Amer. Math. Soc. 39, 969–978 (1933).

Laurikainen, K. V.
1. *Asymptotic eigensolutions of the radical deuteron equations.* Ann. Acad. Sci. Fennicae Ser. A. I, no. 130 (1952), 10 pp.

Lavrentieff, M.
1. *Sur les fonctions d'une variable complexe représentables par des séries de polynomes.* Act. Sci. Ind. 441, Paris, 1936.

Lax, A. (see Courant, R.)

Lax, P. D.
1. *On the existence of Green's function.* Proc. Amer. Math. Soc. 3, 526–531 (1952). Errata, ibid. 3, 993 (1952).
2. *Symmetrizable linear transformations.* Comm. Pure Appl. Math. 7, 633–647 (1954).
3. *Reciprocal extremal problems in function theory.* Comm. Pure Appl. Math. 8, 437–453 (1955).

Lax, P. D., and Milgram, A. N.
1. *Parabolic equations.* Contributions to the theory of partial differential equations, 167–190. Ann. of Math. Studies, no. 33, Princeton, 1954.

Leader, S.
1. *The theory of L^p-spaces for finitely additive set functions.* Ann. Math. (2) 58, 528–543 (1953).

Lebesgue, H.
1. *Sur les intégrales singulières.* Ann. de Toulouse (3) 1, 25–117 (1909).
2. *Leçons sur l'intégration et la recherche des fonctions primitives.* Gauthier-Villars, Paris, 1904. Second edition 1928.

Lefschetz, S.
1. *Algebraic topology.* Amer. Math. Soc. Colloquium Pub. vol. 27, New York, 1942.
2. *Introduction to topology.* Princeton University Press, Princeton, 1949.

Leighton, W.
1. *Bounds for the solutions of a second order linear differential equation.* Proc. Nat. Acad. Sci. U.S.A. 35, 190–193 (1949).
2. *On self-adjoint differential equations of the second order.* Proc. Nat. Acad. Sci. U.S.A. 35, 656–657 (1949).

3. *On the detection of the oscillation of a second order linear differential equation.* Duke Math. J. 17, 57–62 (1950).
4. *On self-adjoint differential equations of second order.* J. London Math. Soc. 27, 33–47 (1952).

Leja, F.
1. *Sur la notion du groupe abstrait topologique.* Fund. Math. 9, 37–44 (1927).

Lengyel, B. A.
1. *On the spectral theorem of self-adjoint operators.* Acta Sci. Math. Szeged 9, 174–186 (1939).
2. *Bounded self-adjoint operators and the problem of moments.* Bull. Amer. Math. Soc. 45, 303–306 (1939).

Lengyel, B. A., and Stone, M. H.
1. *Elementary proof of the spectral theorem.* Ann. of Math. (2) 37, 853–864 (1936).

Leray, J.
1. *La théorie des points fixes et ses applications en analyse.* Proc. International Congress Math., Cambridge, Mass. 2, 202–208 (1950).
2. *Valeurs propres et vecteurs propres d'un endomorphisme complètement continu d'un espace vectoriel à voisinages convexes.* Acta Sci. Math. Szeged 12 Pars B, 177–186 (1950).
3. *Topologie des espaces abstraits de M. Banach.* C. R. Acad. Sci. Paris 200, 1082–1084 (1935).
4. *Lectures on hyperbolic equations with variable coefficients.* Inst. Adv. Studies, Princeton, 1952.

Levi, B.
1. *Sul principio di Dirichlet.* Rend. del Circolo Matem. di Palermo 22, 293–360 (1906).

Levinson, N. (see also Boas, M. L., and Coddington, E. A.)
1. *Gap and density theorems.* Amer. Math. Soc. Colloquium Pub. vol. 26, New York, 1940.
2. *Criteria for the limit-point case for second order linear differential operators.* Časopis Pešt. Mat. Fys. 74, 17–20 (1949).
3. *On the uniqueness of potential in a Schrödinger equation for a given asymptotic phase.* Danske Vid. Selsk. Mat.-Fys. Medd. 25 no. 9, 25 pp. (1949).
4. *The inverse Sturm-Liouville problem.* Mat. Tidsskr. B. 1949, 25–30 (1949).
5. *A simplified proof of the expansion theorem for singular second order linear differential equations.* Duke Math. J. 18, 57–71 (1951).
6. *Addendum to "A simplified proof of the expansion theorem for singular second order differential equations."* Duke Math. J. 18, 719–722 (1951).
7. *The L-closure of eigenfunctions associated with self-adjoint boundary value problems.* Duke Math. J. 19, 23–26 (1952).
8. *Certain relationships between phase shifts and scattering potential.* Phys. Rev. 89, 755–757 (1953).
9. *The expansion theorem for singular self-adjoint linear differential operators.* Ann. of Math. (2) 59, 300–315 (1954).

Levitan, B. M. (see also Gelfand, I. M.)

1. *The application of generalized displacement operators to linear differential equations of the second order.* Uspehi Matem. Nauk (N. S.) 4, no. 1 (29), 3–112 (1949). (Russian) Math. Rev. 11, 116 (1950). Amer. Math. Soc. Translation no. 59 (1951).
2. *Expansions in characteristic functions of differential equations of the second order.* Gosudarstv. Izdat. Tehn.-Teor. Lit., Moscow-Leningrad, 1950.
3. *On a decomposition theorem for characteristic functions of differential equations of second order.* Doklady Akad. Nauk SSSR (N. S.) 71, 605–608 (1950). (Russian) Math. Rev. 11, 720 (1950).
4. *Proof of the theorem on the expansion in eigenfunctions of self-adjoint differential operators.* Doklady Akad. Nauk SSSR (N. S.) 73, 651–654 (1950). (Russian) Math. Rev. 12, 502 (1951).
5. *On a theorem of H. Weyl.* Doklady Akad. Nauk SSSR (N. S.) 82, 673–676 (1952). (Russian) Math. Rev. 13, 844 (1952).
6. *On the completeness of the squares of characteristic functions.* Doklady Akad. Nauk SSSR (N. S.) 83, 349–352 (1952). (Russian) Math. Rev. 14, 171 (1953).
7. *On the asymptotic behavior of a spectral function and on expansion in eigenfunctions of a self-adjoint differential equation of second order,* I, II.
   I. Izvestiya Akad. Nauk SSSR. Ser. Mat. 17, 331–364 (1953).
   II. ibid. 19, 33–58 (1955). (Russian) Math. Rev. 15, 316 (1954) and 16, 1027 (1955).

Lévy, P.

1. *Problèmes concrets d'analyse fonctionnelle. Avec un complément sur les fonctionnelles analytiques par F. Pellegrino.* Gauthier-Villars, Paris. Second edition 484 (1951).
2. *Leçons d'analyse fonctionnelle.* Gauthier-Villars, Paris, 1922.

Leżański, T.

1. *The Fredholm theory of linear equations in Banach spaces.* Studia Math. 13, 244–276 (1953).
2. *Sur les fonctionnelles multiplicatives.* Studia Math. 14 (1953), 13–23 (1954).

Lichtenstein, L.

1. *Zur Analysis der unendlichvielen Variablen,* I. *Entwicklungssätze der Theorie gewöhnlicher linearer Differentialgleichungen zweiter Ordnung.* Rend. Circ. Mat. Palermo 38, 113–166 (1914).

Lidskiĭ, V. B.

1. *On the number of solutions with integrable square of the system of differential equations.* $-y''+P(t)y = \lambda y$. Doklady Akad. Nauk SSSR (N. S.) 95, 217–220 (1954). (Russian) Math. Rev. 15, 957 (1954).

Lifšic, I. M.

1. *On the theory of regular perturbations.* Doklady Akad. Nauk SSSR (N. S.) 48, 79–81 (1945).
2. *On degenerate regular perturbations,* I, II.
   I. Akad. Nauk SSSR Zhurnal Eksper. Teoret. Fiz. 17, 1017–1025 (1947).
   II. ibid. 17, 1076–1089 (1947). (Russian) Math. Rev. 9, 358 (1948).
3. *On the problem of the theory of perturbations connected with quantum statistics.* Uspehi Matem. Nauk (N. S.) 7, no. 1 (47), 171–180 (1952). (Russian) Math. Rev. 14, 185 (1953).

4. *On regular perturbations of an operator with a quasi-continuous spectrum.* Učenye Zapiski Har'kov. Gos. Univ. 28, Zapiski Naučno-Issled. Inst. Mat. Meh. i Har'hov. Mat. Obšč. (4) 20, 77–82 (1950). (Russian) Math. Rev. 14, 565 (1953).

Lindgren, B. W. (see Cameron, R. H.)

Liouville, J.

1. *Sur le développement des fonctions en séries dont les divers termes sont assujeties à satisfaire à une même équation différentielle du sécond ordre contenant un paramètre variable,* I–III.
   I. J. Math. Pures Appl. (1) 1, 253–265 (1836).
   II. ibid. (1) 2, 16–37 (1837).
   III. ibid. (1) 2, 418–436 (1837).
2. *D'un théorème dû à M. Sturm et relatif a une classe de fonctions transcendantes.* J. Math. Pures Appl. (1) 1, 269–277 (1836).

Littlewood, J. (see also Hardy, G. H.)

Littlewood, J., and Paley, R.E.A.C.

1. *Theorems on Fourier series and power series,* I, II.
   I. J. London Math. Soc. 6, 230–233 (1931).
   II. Proc. London Math. Soc. (2) 42, 52–89 (1937).

Livingston, A. E.

1. *The space H^p, $0 < p < 1$, is not normable.* Pacific J. Math. 3, 613–616 (1953).

Livšic, M. S.

1. *Isometric operators with equal deficiency indices, quasi-unitary operators.* Mat. Sbornik N. S. 26 (68), 247–264 (1950). (Russian) Math. Rev. 11, 669 (1950).
2. *On the reduction of a linear non-Hermitian operator to "triangular" form.* Doklady Akad. Nauk SSSR (N. S.) 84, 873–876 (1952). (Russian) Math. Rev. 14, 184 (1953).
3. *On the resolvent of a linear nonsymmetric operator.* Doklady Akad. Nauk SSSR (N. S.) 84, 1131–1134 (1952). (Russian) Math. Rev. 14, 185 (1953).
4. *On spectral decomposition of linear nonselfadjoint operators.* Mat. Sbornik N. S. 34 (76), 145–199 (1954). (Russian) Math. Rev. 16, 48 (1955).
5. *On the theory of self-adjoint systems of differential equations.* Doklady Akad. Nauk SSSR (N. S.) 72, 1013–1016 (1950). (Russian) Math. Rev. 13, 747 (1952).

Livšic, M. S., and Potapov, V. P.

1. *A theorem on the multiplication of characteristic matrix functions.* Doklady Akad. Nauk SSSR (N. S.) 72, 625–628 (1950). (Russian) Math. Rev. 11, 669 (1950).

Loomis, L. H.

1. *An introduction to abstract harmonic analysis.* D. van Nostrand Co., New York, 1953.
2. *Linear functionals and content.* Amer. J. Math. 76, 168–182 (1954).
3. *Abstract congruence and the uniqueness of Haar measure.* Ann. of Math. (2) 46, 348–355 (1945).
4. *Haar measure in uniform structures.* Duke Math. J. 16, 193–208 (1949).
5. *On the representation of σ-complete Boolean algebras.* Bull. Amer. Math. Soc. 53, 757–760 (1947).

Lorch, E. R. (see also Riesz, F.)

1. *Bicontinuous linear transformations in certain vector spaces.* Bull. Amer. Math. Soc. 45, 564–569 (1939).
2. *On a calculus of operators in reflexive vector spaces.* Trans. Amer. Math. Soc. 45, 217–234 (1939).
3. *The Cauchy-Schwarz inequality and self-adjoint spaces.* Ann. of Math. (2) 46, 468–473 (1945).
4. *On certain implications which characterize Hilbert space.* Ann. of Math. (2) 49, 523–532 (1948).
5. *Return to the self-adjoint transformation.* Acta Sci. Math. Szeged 12 Pars B, 137–144 (1950).
6. *The spectrum of linear transformation.* Trans. Amer. Math. Soc. 52, 238–248 (1942).
7. *The integral representation of weakly almost-periodic transformations in reflexive vector spaces.* Trans. Amer. Math. Soc. 49, 18–40 (1941).
8. *Means of iterated transformations in reflexive vector spaces.* Bull. Amer. Math. Soc. 45, 945–957 (1939).
9. *The structure of normed abelian rings.* Bull. Amer. Math. Soc. 50, 447–463 (1944).
10. *The theory of analytic functions in normed abelian vector rings.* Trans. Amer. Math. Soc. 54, 414–425 (1943).
11. *Functions of self-adjoint transformations in Hilbert space.* Acta Sci. Math. Szeged 7, 136–146 (1934).
12. *Differentiable inequalities and the theory of convex bodies.* Trans. Amer. Math. Soc. 71, 243–266 (1951).
13. *Su certe estensioni del concetto di volume.* Rend. Acc. Naz. Lincei (8) 16, 25–29 (1954).
14. *On the volume of smooth convex bodies in Hilbert space.* Math. Zeit. 61, 391–407 (1955).

Lorentz, G. G.

1. *On the theory of spaces Λ.* Pacific J. Math. 1, 411–429 (1951).
2. *Some new functional spaces.* Ann. of Math. (2) 51, 37–55 (1950).
3. *Operations in linear metric spaces.* Duke Math. J. 15, 755–761 (1948).
4. *Funktionale und Operationen in den Räumen der Zahlenfolgen.* Doklady Akad. Nauk SSSR (N. S.) 1, 81–85 (1935).

Lovaglia, A. R.

1. *Locally uniformly convex Banach spaces.* Trans. Amer. Math. Soc. 78, 225–238 (1955).

Löwig, H.

1. *Komplexe euklidische Räume von beliebiger endlicher oder unendlicher Dimensionszahl.* Acta Sci. Math. Szeged 7, 1–33 (1934).

Löwner, K.

1. *Grundzüge einer Inhaltslehre im Hilbertschen Räume.* Ann. of Math. (2) 40, 816–833 (1939).

Lukomskiĭ, T. I.

1. *On the theory of matrix representations of unbounded self-adjoint operators.* Doklady Akad. Nauk SSSR (N. S.) 70, 377–379 (1950). (Russian) Math. Rev. 11, 669 (1950).

Lumer, G. (see Halmos, P. R.)
Ma, S. T.
1. *On a general condition of Heisenberg for the S-matrix.* Phys. Rev. 71, 195–200 (1947).
Maak, W.
1. *Fastperiodische Funktionen.* Springer, Berlin, 1950.
MacDuffee, C. C.
1. *The theory of matrices.* Ergebnisse der Math. und ihrer Grenzgebiete, vol. 2, Berlin, 1933. Reprinted Chelsea Pub. Co., New York, 1946.
McEwen, W. H.
1. *Spectral theory and its application to differential eigenvalue problems.* Amer. Math. Monthly 60, 223–233 (1953).
Macintyre, A. J., and Rogosinski, W. W.
1. *Extremum problems in the theory of analytic functions.* Acta Math. 82, 275–325 (1950).
Mackey, G. W. (see also Kakutani, S.)
1. *On convex topological linear spaces.* Trans. Amer. Math. Soc. 60, 519–537 (1946).
2. *On infinite dimensional linear spaces.* Trans. Amer. Math. Soc. 57, 155–207 (1945).
3. *Note on a theorem of Murray.* Bull. Amer. Math. Soc. 52, 322–325 (1946).
4. *Commutative Banach algebras.* Mimeographed lecture notes, Harvard University, 1952.
5. *Functions on locally compact groups.* Bull. Amer. Math. Soc. 56, 385–412 (1950).
MacPhail, M. S.
1. *Absolute and unconditional convergence.* Bull. Amer. Math. Soc. 53, 121–123 (1947).
McShane, E. J.
1. *Linear functionals on certain Banach spaces.* Proc. Amer. Math. Soc. 1, 402–408 (1950).
2. *Integration.* Princeton University Press, Princeton, 1944.
3. *Order-preserving maps and integration processes.* Ann. of Math. Studies, no. 31, Princeton Univ. Press, 1953.
4. *Images of sets satisfying the condition of Baire.* Ann. of Math. (2), 51, 380–386 (1950).
Maddaus, I., Jr.
1. *On types of "weak" convergence in linear normed spaces.* Ann. of Math. (2) 42, 229–246 (1941).
2. *On completely continuous linear transformations.* Bull. Amer. Math. Soc. 44, 279–282 (1938).
Maeda, F. (see also Ogasawara, T.)
1. *Unitary equivalence of self-adjoint operators and constants of motion.* J. Sci. Hiroshima Univ. A. 6, 283–290 (1936).
Maharam, D.
1. *The representation of abstract measure functions.* Trans. Amer. Math. Soc. 65, 279–330 (1949).
2. *The representation of abstract integrals.* Trans. Amer. Math. Soc. 75, 154–184 (1953).

3. *On kernel representation of linear operators.* Trans. Amer. Math. Soc. 79, 229–255 (1955).

Makai, E.

1. *Asymptotische Abschätzung der Eigenwerte gewisser Differentialgleichungen zweiter Ordnung.* Ann. Scuola Norm. Sup. Pisa (2) 10, 123–126 (1941).

Mandelbrojt, S.

1. *Un théorème de fermeture.* C. R. Acad. Sci. Paris 231, 16–18 (1950).
2. *Théorèmes généraux de fermeture.* C. R. Acad. Sci. Paris 232, 284–286 (1951).
3. *Théorèmes d'approximation et problèmes des moments.* C. R. Acad. Sci. Paris 232, 1054–1056 (1951).
4. *General theorems of closure.* Rice Inst. Pamphlet, Houston, 1951.
5. *Quelques théorèmes d'unicité.* Proc. International Cong. Math., Cambridge, Mass. 1, 349–355 (1950).
6. *Théorèmes généraux de fermeture.* J. Analyse Math. 1, 180–208 (1951). (Hebrew summary).
7. *Quelques nouveaux théorèmes de fermeture.* Ann. Soc. Polon. Math. 25 (1952), 241–251 (1953).

Mandelbrojt, S., and Agmon, S.

1. *Une généralisation du théorème tauberien de Wiener.* C. R. Acad. Sci. Paris 228, 1394–1396 (1949). (cf. Math. Rev. 11, 99 (1950)).
2. *Une généralisation du théorème tauberien de Wiener.* Acta Sci. Math. Szeged 12, Pars B, 167–176 (1950).

Marčenko, V. A.

1. *Concerning the theory of a differential operator of second order.* Doklady Akad. Nauk SSSR (N. S.) 72, 457–460 (1950). (Russian) Math. Rev. 12, 183 (1951).
2. *On the transformation formulas generated by a linear differential operator of second order.* Doklady Akad. Nauk SSSR (N. S.) 74, 657–660 (1950). (Russian) Math. Rev. 12, 707 (1951).

Marcinkiewicz, J.

1. *Sur les multiplicateurs des séries de Fourier.* Studia Math. 8, 78–91 (1939).

Marinescu, G. (see also Ionescu Tulcea, C. T.)

1. *Opérations relativement complètement continues.* Acad. Repub. Pop. Române. Stud. Cerc. Mat. 2, 107–194 (1951). (Romanian. Russian and French summaries) Math. Rev. 16, 487 (1955).

Markouchevitch, A.

1. *Sur les bases (au sens large) dans les espaces linéaires.* Doklady Akad. Nauk SSSR (N. S.) 41, 227–229 (1943).
2. *Sur la meilleure approximation.* Doklady Akad. Nauk SSSR (N. S.) 44, 262–264 (1944).
3. *Sur une généralisation d'un théorème de Menchoff.* Mat. Sbornik N. S. 15 (57), 433–436 (1944). (Russian. French summary) Math. Rev. 6, 274 (1945).

Markov (Markoff), A.

1. *Quelques théorèmes sur les ensembles abéliens.* Doklady Akad. Nauk SSSR (N. S.) 10, 311–314 (1936).
2. *On mean values and exterior densities.* Mat. Sbornik N. S. 4 (46), 165–191 (1938).

Martin, R. S. (see Michal, A. D.)
Martin, W. T. (see Cameron, R. H.)
Maruyama, G.
1. *Notes on Wiener integrals.* Kōdai Math. Sem. Rep. 1950, 41–44 (1950).
Masani, P. R.
1. *Multiplicative Riemann integration in normed rings.* Trans. Amer. Math. Soc. 61, 147–192 (1947).
Maslow, A.
1. *A note on Birkhoff's product-integral.* Leningrad State Univ. Annals [Uchenye Zapiski] 83 [Math. Ser. 12], 42–56 (1941). (Russian. English summary). Math. Rev. 7, 455 (1946).
Mautner, F. I.
1. *On eigenfunction expansions.* Proc. Nat. Acad. U.S.A. 39, 49–53 (1953).
Mazur, S. (see also Banach, S., and Eidelheit, M.)
1. *Über konvexe Mengen in linearen normierte Räumen.* Studia Math. 4, 70–84 (1933).
2. *Über die kleinste konvexe Menge, die eine gegebene kompakte Menge enthält.* Studia Math. 2, 7–9 (1930).
3. *Über schwache Konvergenz in den Räumen* (L^p). Studia Math. 4, 128–133 (1933).
4. *Une remarque sur l'homéomorphie des champs fonctionnels.* Studia Math. 1, 83–85 (1929).
5. *Sur les anneaux linéaires.* C. R. Acad. Sci. Paris 207, 1025–1027 (1938).
Mazur, S., and Orlicz, W.
1. *Über Folgen linearer Operationen.* Studia Math. 4, 152–157 (1933).
2. *Grundlegende Eigenschaften der polynomischen Operationen*, I, II.
I. Studia Math. 5, 50–68 (1934).
II. ibid. 5, 179–189 (1934).
3. *Sur les espaces métriques linéaires*, I, II.
I. Studia Math. 10, 184–208 (1948).
II. ibid. 13, 137–179 (1953).
Mazur, S., and Ulam, S.
1. *Sur les transformations isométriques d'espace vectoriels normés.* C. R. Acad. Sci. Paris 194, 946–948 (1932).
Medvedev, Yu. T.
1. *Two criteria of compactness of families of functions.* Doklady Akad. Nauk SSSR (N. S.) 90, 337–340 (1953). (Russian) Math. Rev. 14, 1072 (1953).
Mergelyan, S. N.
1. *On the representation of functions by series of polynomials on closed sets.* Doklady Akad. Nauk SSSR (N. S.) 78, 405–408 (1951). (Russian) Math. Rev. 13, 23 (1952). Amer. Math. Soc. Translation, no. 85.
2. *Uniform approximations to functions of a complex variable.* Uspehi Matem. Nauk (N. S.) 7, no. 2 (48), 31–122 (1952). (Russian) Math. Rev. 14, 547 (1953). Amer. Math. Soc. Translation, no. 101.
Michael, E.
1. *Transformations from a linear space with weak topology.* Proc. Amer. Math. Soc. 3, 671–676 (1952).
2. *Locally multiplicatively-convex topological algebras.* Memoirs Amer. Math. Soc. no. 11, 1952.

Michal, A. D.
1. *General differential geometries and related topics.* Bull. Amer. Math. Soc. 45, 529–563 (1939).
Michal, A. D., and Clifford, A. H.
1. *Fonctions analytiques implicites dans des espaces vectoriels abstraits.* C. R. Acad. Sci. Paris 197, 735–737 (1933).
Michal, A. D., and Elconin, V.
1. *Completely integrable differential equations in abstract spaces.* Acta Math. 68, 71–107 (1937).
Michal, A. D., and Martin, R. S.
1. *Some expansions in vector space.* J. Math. Pures et Appl. (9) 13, 69–91 (1934).
Michlin (Mihlin), C. G.
1. *On the convergencc of the Fredholm series.* Doklady Akad. Nauk SSSR (N. S.) 42, 373–376 (1944).
Mikusinski, J. G.
1. *Sur certains espaces abstraits.* Fund. Math. 36, 125–130 (1949).
Miller, D. S. (see Dunford, N.)
Miller, K. S.
1. *A Sturm-Liouville problem associated with iterative methods.* Ann. of Math. (2) 53, 520–530 (1951).
2. *Construction of the Green's function of a linear differential system.* Math. Mag. 26, 1–8 (1952).
3. *Self-adjoint differential systems.* Quart. J. Math. Oxford (2) 3, 175–178 (1952).
Miller, K. S., and Schiffer, M. M.
1. *On the Green's function of ordinary differential systems.* Proc. Amer. Math. Soc. 3, 433–441 (1952).
2. *Monotonic properties of the Green's function.* Proc. Amer. Math. Soc. 3, 948–956 (1952).
Milman (Mil'man), D. P. (see also Brodskiĭ, M. S., and Kreĭn, M.)
1. *On some criteria for the regularity of spaces of the type (B).* Doklady Akad. Nauk SSSR (N. S.) 20, 234 (1938).
2. *Characteristics of extremal points of regularly convex sets.* Doklady Akad. Nauk SSSR (N. S.) 57, 119–122 (1947). (Russian) Math. Rev. 9, 192 (1948).
3. *Accessible points of a functional compact set.* Doklady Akad. Nauk SSSR (N. S.) 59, 1045–1048 (1948). (Russian) Math. Rev. 9, 449 (1948).
4. *Isometry and extremal points.* Doklady Akad. Nauk SSSR (N. S.) 59, 1241–1244 (1948). (Russian) Math. Rev. 9, 516 (1948).
5. *Multimetric spaces. Analysis of the invariant subsets of a multinormed bicompact space under a semigroup of nonincreasing operators.* Doklady Akad. Nauk SSSR (N. S.) 67, 27–30 (1949). (Russian) Math. Rev. 11, 117 (1950).
6. *Extremal points and centers of convex bicompacts.* Uspehi Matem. Nauk (N. S.) 4, no. 5 (33), 179–181 (1949). (Russian) Math. Rev. 11, 117 (1950).
7. *The facial structure of a convex bicompact space and integral decompositions of means.* Doklady Akad. Nauk SSSR (N. S.) 83, 357–360 (1952). (Russian) Math. Rev. 13, 848 (1952).

8. *Sur une classification des points du spectre d'un opérateur linéaire*. Doklady Akad. Nauk SSSR (N. S.) 33, 279–287 (1941).

Milman (Mil'man), D. P., and Rutman, M. A.

1. *On a more precise theorem about the completeness of the system of extremal points of regularly convex sets*. Doklady Akad. Nauk SSSR (N. S.) 60, 25–27 (1948). (Russian) Math. Rev. 9, 448 (1948).

Milne, W. E.

1. *The behavior of a boundary value problem as the interval becomes infinite*. Trans. Amer. Math. Soc. 30, 797–802 (1928).
2. *On the degree of convergence of expansions in an infinite interval*. Trans. Amer. Math. Soc. 31, 906–918 (1929).
3. *The numerical determination of characteristic numbers*. Phys. Rev. 35, 863–867 (1930).

Mimura, Y. (see also Yosida, K.)

1. *Über Funktionen von Funktionaloperatoren in einem Hilbertschen Raum*. Jap. J. Math. 13, 119–128 (1936).

Minkowski, H.

1. *Gesammelte Abhandlungen*. Vol. II. Teubner, Berlin, 1911.

Miranda, C.

1. *Próblemi di esistenza in analisi funzionale*. Litografia Tacchi, Pisa, 1949.
2. *Sul principio di Dirichlet per le funzioni armoniche*. Atti Accad. Naz. Lincei. Rend. Cl. Sci. Fis. Mat. Nat. (8) 3, 55–59 (1947).

Mirkil, H.

1. *The work of Šilov on commutative semi-simple Banach algebras*. Technical Report, Contract 218(00), Office of Naval Research.

Mishoe, L. I. (see also Friedman, B.)

1. *On the expansion of an arbitrary function in terms of the eigenfunctions of a non-self-adjoint differential system*. Dissert., New York University, 1953.

Mishoe, L. I., and Ford, G. C.

1. *Studies in the eigenfunction series associated with a non-self-adjoint differential system*. Tech. Report Nat. Sci. Foundation, 1955.
2. *On the uniform convergence of a certain eigenfunction series*. Pacific J. Math. 6, 271–278 (1956).

Miyadera, I.

1. *Generation of a strongly continuous semi-group of operators*. Tôhoku Math. J. 4 (2), 109–114 (1952).

Mohr, E.

1. *Die Konstruktion der Greenschen Funktion im erweiterten Sinne*. J. Reine Angew. Math. 189, 129–140 (1951).

Molčanov, A. M.

1. *On conditions for discreteness of the spectrum of self-adjoint differential equations of second order*. Trudy Moskov. Mat. Obšč. 2, 169–199 (1953). (Russian) Math. Rev. 15, 224–5 (1954).

Monna, A. F.

1. *On a linear P-adic space*. Nederl. Akad. Wetensch. Verslagen, Afd. Natuurkunde 52, 74–82 (1943). (Dutch, English summary).
2. *On weak and strong convergence in a P-adic Banach space*. Nederl. Akad. Wetensch. Verslagen, Afd. Natuurkunde 52, 207–211 (1943). (Dutch. English summary).

3. *On non-Archimedean linear spaces.* Nederl. Akad. Wetensch. Verslagen, Afd. Natuurkunde 52, 308–321 (1943). (Dutch. English summary).
4. *Linear functional equations in non-Archimedean Banach spaces.* Nederl. Akad. Wetensch. Verslagen, Afd. Natuurkunde 52, 654–661 (1943). (Dutch. English summary).
5. *On ordered groups and linear spaces.* Nederl. Akad. Wetensch. Verslagen, Afd. Natuurkunde 53, 178–182 (1944). (Dutch. English summary).
6. *On the integral of a function whose values are elements of a non-Archimedean valued field.* Nederl. Akad. Wetensch. Verslagen, Afd. Natuurkunde 53, 385–399 (1944). (Dutch. English summary).
7. *Sur les espaces linéaires normés.* I–VI.
   I. Nederl. Akad. Wetensch., Proc. 49, 1045–1055 (1946).
   II. ibid. 1056–1062 (1946).
   III. ibid. 1134–1141 (1946).
   IV. ibid. 1142–1152 (1946).
   V. ibid. 51, 197–210 (1948).
   VI. ibid. 52, 151–160 (1949).
8. *Espaces linéaires à une infinité dénombrable de coordonnée.* Nederl. Akad. Wetensch., Proc. 53, 1548–1559 (1950).
9. *Sur une classe d'espaces linéaires normés.* Nederl. Akad. Wetensch. Proc. 55, 513–525 (1952).

Montroll, E. W.
1. *Markoff chains, Wiener integrals, and quantum theory.* Comm. Pure Appl. Math. 5, 415–453 (1952).

Moore, E. H.
1. *Introduction to a form of general analysis.* The New Haven Math. Colloquium of the Amer. Math. Soc., 1906.
2. *General analysis,* I, II. Mem. Amer. Philos. Soc., Philadelphia, 1935, 1939.

Moore, R. L.
1. *Foundations of point set theory.* Amer. Math. Soc. Colloquium Publications, vol. 13, New York, 1932.

Morse, A. P. (see also Adams, C. R., Agnew, R. P., and Dunford, N.)
1. *A theory of covering and differentiation.* Trans. Amer. Math. Soc. 55, 205–235 (1944).

Morse, M.
1. *Bilinear functionals over $C \times C$.* Acta Sci. Math. Szeged 12, Pars. B, 41–48 (1950).

Morse, M., and Transue, W.
1. *Functionals of bounded Fréchet variation.* Canadian J. Math. 1, 153–165 (1949).
2. *Functionals F bilinear over the product $A \times B$ of two pseudo-normed vector spaces,* I, II.
   I. Ann. of Math. (2) 50, 777–815 (1949).
   II. ibid. (2) 51, 576–614 (1950).
3. *Integral representations of bilinear functionals.* Proc. Nat. Acad. Sci. U.S.A. 35, 136–143 (1949).
4. *The generalized Fréchet variation and Riesz-Young-Hausdorff type theorems.* Rend. Circ. Mat. Palermo (2) 2, 5–35 (1953).

Moser, J.
1. *Störungstheorie des kontinuierlichen Spektrums für gewöhnliche Differentialgleichungen zweiter Ordnung.* Math. Ann. 125, 366–393 (1953).

Moses, H. E. (see Kay, I.)

Moskovitz, D., and Dines, L. L.
1. *Convexity in a linear space with an inner product.* Duke Math. J. 5, 520–534 (1939).
2. *On the supporting-plane property of a convex body.* Bull. Amer. Math. Soc. 46, 482–489 (1940).

Munroe, M. E.
1. *Absolute and unconditional convergence in Banach spaces.* Duke Math. J. 13, 351–365 (1946).
2. *Introduction to measure and integration.* Addison Wesley, Cambridge, Mass., 1953.
3. *A note on weak differentiability of Pettis integrals.* Bull. Amer. Math. Soc. 52, 167–174 (1946).
4. *A second note on weak differentiability of Pettis integrals.* Bull. Amer. Math. Soc. 52, 668–670 (1946).

Müntz, Ch. H.
1. *Über den Approximationssatz von Weierstrass.* Math. Abhandlungen H. A. Schwarz gewidmet. Berlin, 303–312 (1914).

Murray, F. J.
1. *On complementary manifolds and projections in spaces L_p and l_p.* Trans. Amer. Math. Soc. 41, 138–152 (1937).
2. *Quasi-complements and closed projections in reflexive Banach spaces.* Trans. Amer. Math. Soc. 58, 77–95 (1945).
3. *The analysis of linear transformations.* Bull. Amer. Math. Soc. 48, 76–93 (1942).
4. *Linear transformations between Hilbert spaces and the application of this theory to linear partial differential equations.* Trans. Amer. Math. Soc. 37, 301–338 (1935).
5. *Linear transformations in L_p, $p > 1$.* Trans. Amer. Math. Soc. 39, 83–100 (1936).
6. *Bilinear transformations in Hilbert space.* Trans. Amer. Math. Soc. 45, 474–507 (1939).
7. *An introduction to linear transformations in Hilbert space.* Ann. of Math. Studies, no. 4, Princeton, 1941.

Murray, F. J., and von Neumann, J.
1. *On rings of operators*, I, II, IV.
   I. Ann. of Math. (2) 37, 116–229 (1936).
   II. Trans. Amer. Math. Soc. 41, 208–248 (1937).
   IV. Ann. of Math. (2) 44, 716–808 (1943).

Myers, S. B.
1. *Equicontinuous sets of mappings.* Ann. of Math. (2) 47, 496–502 (1946).
2. *Banach spaces of continuous functions.* Ann. of Math. (2) 49, 132–148 (1948).
3. *Spaces of continuous functions.* Bull. Amer. Math. Soc. 55, 402–407 (1949).
4. *Normed linear spaces of continuous functions.* Bull. Amer. Math. Soc. 56, 233–241 (1950).

Nachbin, L.

1. *On the axiom of the nonconvergent sequences in some linear topological space.* Revista Unión Mat. Argentina 12, 129–150 (1947). (Spanish. French summary).
2. *A characterization of the normed vector ordered spaces of continuous functions over a compact space.* Amer. J. Math. 71, 701–705 (1949).
3. *A theorem of the Hahn-Banach type for linear transformations.* Trans. Amer. Math. Soc. 68, 28–46 (1950).

Nagata, J.

1. *On lattices of functions on topological spaces and of functions on uniform spaces.* Osaka Math. J. 1, 166–181 (1949).

Nagumo, M.

1. *Einige analytische Untersuchungen in linearen metrischen Ringen.* Jap. J. Math. 13, 61–80 (1936).
2. *Degree of mapping in convex linear topological spaces.* Amer. J. Math. 73, 497–511 (1951).
3. *Characterisierung der allgemeinen euklidischen Räume durch eine Postulate für Schwerpunkte.* Jap. J. Math. 12, 123–128 (1936).

Nagy, B. von Sz. — (see under Sz.-Nagy).

Naĭmark, M. A. (see under Neumark).

Nakamura, M.

1. *Notes on Banach space,* X. *Vitali-Hahn-Saks' theorem and K-spaces.* Tôhoku Math. J. (2) 1, 101–108 (1949).
2. *Notes on Banach space,* XI. *Banach lattices with positive bases.* Tôhoku Math. J. (2) 2, 135–141 (1950).
3. *Complete continuities of linear operators.* Proc. Japan Acad. 27, 544–547 (1951).

Nakamura, M., and Sunouchi, S.

1. *Note on Banach spaces* (IV). *On a decomposition of additive set functions.* Proc. Imp. Acad. Tokyo 18, 333–335 (1942).

Nakamura, M., and Umegaki, H.

1. *A remark on theorems of Stone and Bochner.* Proc. Japan Acad. 27, 506–507 (1951).

Nakano, H.

1. *Topology and linear topological spaces.* Maruzen Co., Tokyo, 1951.
2. *Modulared semi-ordered linear spaces.* Maruzen Co., Tokyo, 1950.
3. *Riesz-Fischerscher Satz im normierten teilweise geordneten Modul.* Proc. Imp. Acad. Tokyo 18, 350–353 (1942).
4. *Über Erweiterungen von allgemein teilweise geordneten Moduln,* I. Proc. Imp. Acad. Tokyo 18, 626–630 (1942).
5. *Über Erweiterungen von allgemein teilweise geordneten Moduln,* II. Proc. Imp. Acad. Tokyo 19, 138–143 (1943).
6. *Modulared linear spaces.* J. Fac. Sci. Univ. Tokyo. Sect. I. 6, 85–131 (1950).
7. *Zur Eigenwerttheorie normaler Operatoren.* Proc. Phys.-Math. Soc. Japan (3) 21, 315–339 (1939).
8. *Über Abelsche Ringe von Projektionsoperatoren.* Proc. Phys.-Math. Soc. Japan (3) 21, 357–375 (1939).
9. *Unitärinvariante hypermaximale normale Operatoren.* Ann. of Math. (2) 42, 657–664 (1941).

10. *Unitärinvarianten in allgemeinen Euklidischen Raum.* Math. Ann. 118, 112–133 (1941).
11. *Funktionen mehrerer hypermaximaler normaler Operatoren.* Proc. Phys.-Math. Soc. Japan (3) 21, 713–728 (1939).
12. *Modern spectral theory.* Maruzen Co., Tokyo, 1950.
13. *Spectral theory in the Hilbert space.* Japan Soc. for Promotion of Sci., Tokyo, 1953.
14. *Über normierte teilweise geordnete Moduln.* Proc. Imp. Acad. Tokyo 17, 311–317 (1941).
15. *Stetige lineare Funktionale auf dem teilweise geordnete Modul.* J. Fac. Sci. Imp. Univ. Tokyo, 4, 201–382 (1942).
16. *Über ein lineare Funktional auf dem teilweise geordneten Modul.* Proc. Imp. Acad. Tokyo 18, 548–552 (1942).
17. *Über den Beweis des Stoneschen Satzes.* Ann. of Math. (2) 42, 665–667 (1941).
18. *Reduction of Bochner's theorem to Stone's theorem.* Ann. of Math. (2) 49, 279–280 (1948).

Nakayama, T. (see Yosida, K.).

Nathan, D. S.

1. *One-parameter groups of transformations in abstract vector spaces.* Duke Math. J. 1, 518–526 (1935).

Neĭgauz, M. G.

1. *On the determination of the asymptotic behavior of a function $q(x)$ by the properties of the operator $-y'' + q(x)y$.* Doklady Akad. Nauk SSSR (N. S.) 102, 25–28 (1955). (Russian) Math. Rev. 17, 370 (1956).

Neĭmark, F. A.

1. *Extension of a Hermitian operator to one permutable with a given Hermitian operator.* Doklady Akad. Nauk SSSR (N. S.) 66, 9–12 (1949). (Russian) Math. Rev. 11, 371 (1950).

Nemyčkiĭ, V. (see Niemytzki).

Neumann, C.

1. *Untersuchungen über das logarithmische und Newtonsche Potential.* Teubner Leipzig, 1877.

von Neumann, J. (see also Bochner, S., Devinatz, A., Halmos, P. R., Jordan, P., and Murray, F. J.).

1. *On complete topological spaces.* Trans. Amer. Math. Soc. 37, 1–20 (1935).
2. *Zur Algebra der Funktionaloperationen und Theorie der Normalen Operatoren.* Math. Ann. 102, 370–427 (1929–1930).
3. *Eine Spektraltheorie für allgemeine Operatoren eines unitären Raumes.* Math. Nachr. 4, 258–281 (1951).
4. *Functional operators,* I. Annals of Math. Studies, no. 21, Princeton University Press, Princeton, 1950.
5. *On a certain topology for rings of operators.* Ann. of Math. (2) 37, 111–115 (1936).
6. *Charakterisierung des Spektrums eines Integraloperators.* Act. Sci. et Ind. 229, Paris, 1935.
7. *Allgemeine Eigenwerttheorie Hermitescher Funktionaloperatoren.* Math. Ann. 102, 49–131 (1929–1930).

8. *Mathematische Begründung der Quantenmechanik.* Nachr. Gesell. Wiss. Göttingen. Math.-Phys. Kl., 1–57 (1927).
9. *Almost periodic functions in a group,* I. Trans. Amer. Math. Soc. 36, 445–492 (1934).
10. *Über einen Satz von Herrn M. H. Stone.* Ann. of Math. (2) 33, 567–573 (1932).
11. *Proof of the quasi-ergodic hypothesis.* Proc. Nat. Acad. Sci. U.S.A. 18, 70–82 (1932).
12. *Zum Haarschen Mass in topologischen Gruppen.* Comp. Math. 1, 106–114 (1934).
13. *On rings of operators,* III. Ann. of Math. (2) 41, 94–161 (1940).
14. *On some algebraical properties of operator rings.* Ann. of Math. (2) 44, 709–715 (1943).
15. *On rings of operators. Reduction theory.* Ann. of Math. (2) 50, 401–485 (1949).
16. *Über adjungierte Funktionaloperatoren.* Ann. of Math. (2) 33, 294–310 (1932).
17. *The uniqueness of Haar's measure.* Mat. Sbornik N. S. 1 (43), 721–734 (1936).
18. *Über Funktionen von Funktionaloperatoren.* Ann. of Math. (2) 32, 191–226 (1931).
19. *Einige Sätze über messbare Abbildungen.* Ann. of Math. (2) 33, 574–586 (1932).
20. *Zur Operatorenmethode in der klassischen Mechanik.* Ann. of Math. (2) 33, 587–642, 789–791 (1932).
21. *Algebraische Repräsentanten der Funktionen "bis auf eine Menge vom Masse Null."* J. Reine Angew. Math. 165, 109–115 (1931).
22. *On an algebraic generalization of the quantum mechanical formalism,* I. Mat. Sbornik 1 (43), 415–482 (1936).
23. *Mathematische Grundlagen der Quantenmechanik.* J. Springer, Berlin, 1932. English translation, Princeton Univ. Press, Princeton, 1955.
24. *Approximative properties of matrices of high finite order.* Port. Math. 3, 1–62 (1942).

Neumark (Naĭmark), M. A. (see also Gelfand, I.)

1. *Positive definite operator functions on a commutative group.* Izvestiya Akad. Nauk SSSR 7, 237–244 (1943).
2. *Rings of operators in Hilbert space.* Uspehi Mat. Nauk (N. S.) 4, no. 4 (32) 83–147 (1949). (Russian) Math. Rev. 11, 186 (1950). German translation in "Sowjetische Arbeiten zur Funktional-analysis," Verlag Kultur und Fortschritt, Berlin, 1954.
3. *On a representation of additive operator set functions.* Doklady Akad. Nauk SSSR (N. S.) 41, 359–361 (1943).
4. *On the deficiency index of linear differential operators.* Doklady Akad. Nauk SSSR (N. S.) 82, 517–520 (1952). (Russian) Math. Rev. 14, 277 (1953).
5. *Linear differential operators.* Gosudarstr. Izdat. Tehn.-Teo. Lit., Moscow, 1954.
6. *On the square of a closed symmetric operator.* Doklady Akad. Nauk SSSR (N. S.) 26, 806–870 (1940). ibid. 28, 207–208 (1940).

7. *Self-adjoint extensions of the second kind of a symmetric operator*. Izvestiya Akad. Nauk SSSR 4, 53–104 (1940). (Russian. English summary) Math. Rev. 2, 104 (1941).
8. *Spectral functions of a symmetric operator*. Izvestiya Akad. Nauk SSSR (N. S.) 4, 277–318 (1940). (Russian. English summary) Math. Rev. 2, 105 (1941).
9. *On the spectrum of singular non-self-adjoint differential operators of the second order*. Doklady Akad. Nauk SSSR (N. S.) 85, 41–44 (1952). (Russian) Math. Rev. 14, 473 (1953).
10. *Investigation of the spectrum and expansion in eigenfunctions of singular non-self-adjoint differential operators of the second order*. Uspehi Matem. Nauk (N. S.) 8, no. 4 (56), 174–175 (1953). (Russian) Math. Rev. 15, 530 (1954).
11. *On expansion in characteristic functions of non-self-adjoint singular differential operators of the second order*. Doklady Akad. Nauk SSSR (N. S.) 89, 213–216 (1953). (Russian) Math. Rev. 15, 33 (1954).
12. *Investigation of the spectrum and the expansion in eigenfunctions of a non-self-adjoint operator of the second order on a semi-axis*. Trudy Moskov. Mat. Obšč. 3, 181–270 (1954). (Russian) Math. Rev. 15, 959 (1954).

Newton, R. G., and Jost, R.
1. *The construction of potentials from the S-matrix for systems of differential equations*. Il Nuovo Cimento (10) 1, 590–622 (1955).

Nicolescu, M.
1. *On the criterion of compactness of A. Kolmogorov*. Acad. Repub. Pop. Române. Bul. Sti. Ser. Mat. Fiz. Chim. 2, 407–415 (1950). (Roumanian. Russian and French summaries). Math. Rev. 13, 357 (1952).

Niemytzki (Nemyčkiĭ), V.
1. *The method of fixed points in analysis*. Uspehi Matem. Nauk 1, 141–174 (1936). (Russian).
2. *Problems of the qualitative theory of differential equations*. Vestnik Moskov. Univ. Ser. Fiz.-Mat. Estest. Nauk 1952, no. 8, 19–39 (1952). (Russian) Math. Rev. 14, 753 (1953).

Nikodým, O. M.
1. *Remarques sur les intégrales de Stieltjes en connexion avec celles de MM. Radon et Fréchet*. Ann. Soc. Polon. Math. 18, 12–24 (1945).
2. *Sur les fonctionnelles linéaires*. C. R. Acad. Sci. Paris 229, 16–18, 169–171, 288–289 (1949).
3. *Remarques sur la pseudo-topologie et sur les fonctionnelles linéaires*. C. R. Acad. Sci. Paris 229, 863–865 (1949).
4. *Un nouvel appareil mathématique pour la théorie des quanta*. Ann. Inst. H. Poincaré 11, 49–112 (1949).
5. *Sur les familles bornées de fonctions parfaitement additives d'ensemble abstrait*. Monatsh. für Math. u. Phys. 40, 418–426 (1933).
6. *Sur les suites convergentes de fonctions parfaitement additives d'ensemble abstrait*. Monatsh. für Math. u. Phys. 40, 427–432 (1933).
7. *Sur les fonctions d'ensembles*. Comptes Rendus du I Congrès des Math. des Pays Slaves, Warsaw, 304–313 (1929).
8. *Sur une généralisation des intégrales de M. J. Radon*. Fund. Math. 15, 131–179 (1930).

9. *Contribution à la théorie des fonctionnelles linéaires en connexion avec la théorie de la mesure des ensembles abstraits.* Mathematica, Cluj, 5, 130–141 (1931).
10. *Sur les opérateurs normaux maximaux dans l'espace hilbertien séparable et complet*, I, II.
I. C. R. Acad. Sci. Paris 238, 1373–1375 (1954).
II. ibid. 238, 1467–1469 (1954).

Nikol'skiĭ, V. N.
1. *Best approximation and basis in a Fréchet space.* Doklady Akad. Nauk SSSR (N. S.) 59, 639–642 (1948). (Russian) Math. Rev. 10, 128 (1949).
2. *Linear equations in normed linear spaces.* Izvestiya Akad. Nauk SSSR 7, 147–166 (1943). (Russian. English summary) Math. Rev. 5, 187 (1944).

Nikovič, I. A.
1. *On the Fredholm series.* Doklady Akad. Nauk SSSR (N. S.) 59, 423–425 (1948). (Russian) Math. Rev. 9, 592 (1948).

Nirenberg, L.
1. *Remarks on strongly elliptic partial differential equations.* Comm. Pure Appl. Math. 8, 648–674 (1955).

Nussbaum, A. E. (see Devinatz, A.)

Nyman, B.
1. *On the one-dimensional translation group and semi-group in certain function spaces.* Dissertation, University of Uppsala (1950). Math. Rev. 12, 108 (1951).

Ogasawara, T.
1. *Compact metric Boolean algebras and vector lattices.* J. Sci. Hirosima Univ. Ser. A. 11, 125–128 (1942).
2. *On Fréchet lattices*, I. J. Sci. Hirosima Univ. Ser. A. 12, 235–248 (1943). (Japanese) Math. Rev. 10, 544 (1949).
3. *Remarks on a vector lattice with a metric function.* J. Sci. Hiroshima Univ. Ser. A. 13, 317–325 (1944). (Japanese) Math. Rev. 10, 544 (1949).
4. *Commutativity of Archimedean semi-ordered groups.* J. Sci. Hirosima Univ. Ser. A. 12, 249–254 (1943). (Japanese) Math. Rev. 10, 544 (1949).
5. *Theory of vector lattices*, I, II.
I. J. Sci. Hirosima Univ. Ser. A. 12, 17–35 (1942).
II. ibid. 12, 217–234 (1943). (Japanese) Math. Rev. 10, 545 (1949).
6. *Some general theorems and convergence theorems in vector lattices.* J. Sci. Hiroshima Univ. Ser. A. 14, 14–25 (1949).
7. *On the integral representation of unbounded self-adjoint transformations.* J. Sci. Hiroshima Univ. Ser. A. 6, 279–281 (1936).

Ogasawara, T., and Maeda, F.
1. *Representation of vector lattices.* J. Sci. Hiroshima Univ. Ser. A. 12, 17–35 (1942). (Japanese) Math. Rev. 10, 544 (1949).
2. *Remarks on representation of vector lattices.* J. Sci. Hiroshima Univ. Ser. A. 12, 217–234 (1934). (Japanese) Math. Rev. 10, 594 (1949).

Ōhira, K.
1. *On a certain complete, separable and metric space.* Mem. Fac. Sci. Kyūsyū Univ. A. 6, 9–15 (1951).
2. *On some characterizations of abstract Euclidean spaces by properties of orthogonality.* Kumamoto J. Sci. Ser. A. 1, no. 1, 23–26 (1952).

O'Neill, B.
1. *Essential sets and fixed points.* Amer. J. Math. 75, 497–509 (1953).
Ono, T.
1. *A generalization of the Hahn-Banach theorem.* Nagoya Math. J. 6, 171–176 (1953).
Orihara, M.
1. *On the regular vector lattice.* Proc. Imp. Acad. Tokyo 18, 525–529 (1942).
Orlicz, W. (see also Alexiewicz, A., and Mazur, S.)
1. *Über unbedingte Konvergenz in Funktionräumen,* I, II.
I. Studia Math. 4, 33–37 (1933).
II. ibid. 4, 41–47 (1933).
2. *Über konjugierte Exponentenfolgen.* Studia Math. 3, 200–211 (1931).
3. *Über eine gewisse Klasse von Räumen von Typus B.* Bull. Int. Acad. Polon. Sci. Sér. A., 207–220 (1932).
4. *Ein Satz über die Erweiterung von linearen Operationen.* Studia Math. 5, 127–140 (1934).
5. *Sur les opérations linéaires dans l'espace des fonctions bornées.* Studia Math. 10, 60–89 (1948).
6. *Linear operations in Saks spaces* (I). Studia Math. 11, 237–272 (1950).
7. *Über Folgen linearer Operationen, die von einem Parameter abhängen.* Studia Math. 5, 160–170 (1934).
8. *Beiträge zur Theorie der Orthogonalentwicklungen,* I–VI.
I. Studia Math. 1, 1–39 (1929).
II. ibid. 1, 241–255 (1929).
III. Bull. Int. Acad. Polon. Sci. Sér. A. 8–9, 229–238 (1932).
IV. Studia Math. 5, 1–14 (1934).
V. ibid. 6, 20–38 (1936).
VI. ibid. 8, 141–147 (1939).
Orlov, S. A.
1. *On the deficiency index of linear differential operators.* Doklady Akad. Nauk SSSR (N. S.) 92, 483–486 (1953). (Russian) Math. Rev. 15, 802 (1954).
Owchar, M.
1. *Wiener integrals of multiple variations.* Proc. Amer. Math. Soc. 3, 459–470 (1952).
Oxtoby, J. C.
1. *Ergodic sets.* Bull. Amer. Math. Soc. 58, 116–136 (1952).
2. *On the ergodic theorem of Hurewicz.* Ann. of Math. (2) 49, 872–884 (1948).
3. *Invariant measures in groups which are not locally compact.* Trans. Amer. Math. Soc. 60, 215–237 (1946).
4. *The category and Borel class of certain subsets of L_p.* Bull. Amer. Math. Soc. 43, 245–248 (1937).
Oxtoby, J. C., and Ulam, S.
1. *On the existence of a measure invariant under a transformation.* Ann. of Math. (2) 40, 560–566 (1939).
Pais, A. (see Jost, R.)
Paley, R. E. A. C. (see also Littlewood, J.)
1. *A proof of a theorem on bilinear forms.* J. London Math. Soc. 6, 226–230 (1931).

2. *A note on bilinear forms.* Bull. Amer. Math. Soc. 39, 259–260 (1933).
3. *Some theorems on orthogonal functions.* Studia Math. 3, 226–238 (1931).

Paley, R. E. A. C., and Wiener, N.
1. *Fourier transforms in the complex domain.* Amer. Math. Soc. Colloquium Pub. no. 19, New York, 1934.

Paley, R. E. A. C., Wiener, N., and Zygmund, A.
1. *Notes on random functions.* Math. Zeit. 37, 647–668 (1933).

Parker, W. V.
1. *Limits to the characteristic roots of a matrix.* Duke Math. J. 10, 479–482 (1943).

Pauli, W.
1. *Meson theory of nuclear forces.* Interscience Pub., New York, 1946.

Peano, G.
1. *Integrazione per serie delle equazioni differenziali lineari.* Atti R. Acc. Sci. Torino 22, 293–302 (1887). German translation in Math. Ann. 32, 450–456 (1888).

Peck, J. E. L.
1. *An ergodic theorem for a noncommutative semi-group of linear operators.* Proc. Amer. Math. Soc. 2, 414–421 (1951).

Peter, F., and Weyl, H.
1. *Die Vollständigkeit der primitiven Darstellungen einer geschlossenen kontinuierlichen Gruppe.* Math. Ann. 97, 737–755 (1927).

Pettis, B. J. (see also Dunford, N.)
1. *A note on regular Banach spaces.* Bull. Amer. Math. Soc. 44, 420–428 (1938).
2. *A proof that every uniformly convex space is reflexive.* Duke Math. J. 5, 249–253 (1939).
3. *Remarks on a theorem of E. J. McShane.* Proc. Amer. Math. Soc. 2, 166–171 (1951).
4. *On integration in vector spaces.* Trans. Amer. Math. Soc. 44, 277–304 (1938).
5. *On continuity and openness of homomorphisms in topological groups.* Ann. of Math. (2) 52, 293–308 (1950).
6. *Absolutely continuous functions in vector spaces.* (Abstract) Bull. Amer. Math. Soc. 45, 677 (1939).
7. *Differentiation in Banach spaces.* Duke Math. J. 5, 254–269 (1939).
8. *Linear functionals and completely additive set functions.* Duke Math. J. 4, 552–565 (1938).

Phillips, R. S. (see also Bochner, S.)
1. *On weakly compact subsets of a Banach space.* Amer. J. Math. 65, 108–136 (1943).
2. *A characterization of Euclidean spaces.* Bull. Amer. Math. Soc. 46, 930–933 (1940).
3. *On linear transformations.* Trans. Amer. Math. Soc. 48, 516–541 (1940).
4. *A note on ergodic theory.* Proc. Amer. Math. Soc. 2, 662–669 (1951).
5. *Spectral theory for semi-groups of linear operators.* Trans. Amer. Math. Soc. 71, 393–415 (1951).
6. *Perturbation theory for semi-groups of linear operators.* Trans. Amer. Math. Soc. 74, 199–221 (1953).
7. *Integration in a convex linear topological space.* Trans. Amer. Math. Soc. 47, 114–145 (1940).

8. *On one parameter semi-groups of linear transformations.* Proc. Amer. Math. Soc. 2, 234–237 (1951).
9. *Semi-groups of operators.* Bull. Amer. Math. Soc. 61, 16–33 (1955).
10. *An inversion formula for Laplace transforms and semi-groups of linear operators.* Ann. of Math. (2) 59, 325–356 (1954).
11. *The adjoint semi-group.* Pacific J. Math. 5, 269–283 (1955).
12. *A decomposition of additive set functions.* Bull. Amer. Math. Soc. 46, 274–277 (1940).
13. *Linear ordinary differential operators of the second order.* Div. Electromag. Res., Inst. Math. Sci., New York Univ., 1952.

Picone, M.

1. *Sui valori eccezionali di un parametro da cui dipende un'equazione differenziale del secondo ordine.* Ann. R. Scuola Norm. Sup. Pisa (1) 11, 1–141 (1910).

Pierce, R.

1. *Cones and the decomposition of functionals.* Math. Mag. 24, 117–122 (1951).

Pincherle, S.

1. *Funktionaloperationen und -Gleichungen.* Encyklopädie der Math. Wiss. II A 11, 761–817 (1905). French translation: *Équations et opérations fonctionnelles.* Enc. des sciences math., II, Vol. 5, fasc. 1, no. 26, 1–86.

Pinsker, A. (see also Kantorovitch, L. V.)

1. *On a class of operations in K-spaces.* Doklady Akad. Nauk SSSR (N. S.) 36, 227–230 (1942).
2. *On normed K-spaces.* Doklady Akad. Nauk SSSR (N. S.) 33, 12–14 (1941).
3. *Universal K-spaces.* Doklady Akad. Nauk SSSR (N. S.) 49, 8–11 (1945).
4. *On the decomposition of K-spaces into elementary spaces.* Doklady Akad. Nauk SSSR (N. S.) 49, 168–171 (1945).
5. *On separable K-spaces.* Doklady Akad. Nauk SSSR (N. S.) 49, 318–319 (1945).
6. *Completely linear functionals in K-spaces.* Doklady Akad. Nauk SSSR (N. S.) 55, 299–302 (1947).
7. *On concrete representations of linear semi-ordered spaces.* Doklady Akad. Nauk SSSR (N. S.) 55, 379–381 (1947).

Pitt, H. R.

1. *Some generalizations of the ergodic theorem.* Proc. Cambridge Phil. Soc. 38, 325–343 (1942).

Plancherel, M.

1. *Integraldarstellungen willkürlicher Funktionen.* Math. Ann. 67, 519–534 (1909).

Plessner, A. I.

1. *Über halbunitäre Operatoren.* Doklady Akad. Nauk SSSR (N. S.) 25, 710–712 (1939).
2. *Spectral theory of linear operators*, I. Uspehi Mat. Nauk 9, 3–125 (1941). (Russian) Math. Rev. 3, 210 (1942).

Plessner, A. I., and Rohlin, V. A.

1. *Spectral theory of linear operators*, II. Uspehi Mat. Nauk (N. S.) 1 (11), no. 1, 71–191 (1946). (Russian) Math. Rev. 9, 43 (1948).

Poincaré, H.

1. *Sur les groupes continus.* Cambridge Phil. Trans. 18, 220–255 (1899). Reprinted in Oeuvres 3, 173–212.

2. *Sur les équations de la physique mathématique.* Rend. Circ. Mat. Palermo 8, 57–156 (1894).

Pollard, H.

1. *Integral transforms.* Duke Math. J. 13, 307–330 (1946).

Pólya, G.

1. *Remark on Weyl's note "Inequalities between the two kinds of eigenvalues of a linear transformation".* Proc. Nat. Acad. Sci. U.S.A. 36, 49–51 (1950).

Pontrjagin, L.

1. *Topological groups.* [translated by E. Lehmer] Princeton University Press, Princeton, 1939.

Poole, E. G. C.

1. *Introduction to the theory of linear differential equations.* Oxford Univ. Press, 1936.

Potapov, V. P. (see Livšic, M. S.)

Potter, R. L.

1. *On self-adjoint differential equations of the second order.* Pacific J. Math. 3, 467–491 (1953).

Povzner, A.

1. *On some applications of a class of Hilbert spaces of functions.* Doklady Akad. Nauk SSSR (N. S.) 74, 13–16 (1950). (Russian) Math. Rev. 12, 343 (1951).
2. *On the spectra of bounded functions.* Doklady Akad. Nauk SSSR (N. S.) 57, 755–758 (1947). (Russian) Math. Rev. 9, 236 (1948).
3. *On the spectra of bounded functions and the Laplace transform.* Doklady Akad. Nauk SSSR (N. S.) 57, 871–874 (1947). (Russian) Math. Rev. 9, 236 (1948).
4. *On some general inversion theorems of Plancherel type.* Doklady Akad. Nauk SSSR (N. S.) 57, 123–125 (1947). (Russian) Math. Rev. 9, 193 (1948).
5. *Sur les équations du type de Sturm-Liouville et des fonctions positives.* Doklady Akad. Nauk SSSR (N. S.) 43, 367–371 (1944).
6. *On differential equations of Sturm-Liouville type on a half-axis.* Mat. Sbornik N. S. 23 (65), 3–52 (1948). (Russian) Math. Rev. 10, 299 (1949). Amer. Math. Soc. Translation no. 5, 1950.
7. *On the method of directing functionals of M. G. Krein.* Zapiski Naučno-Issled. Inst. Mat. Meh. i Har'kov. Mat. Obšč. (4) 20, 43–52 (1950).
8. *On the differentiation of the spectral function of the Schrödinger equation.* Doklady Akad. Nauk SSSR (N. S.) 79, 193–196 (1951). (Russian) Math. Rev. 13, 241 (1952).

Price, G. B.

1. *The theory of integration.* Trans. Amer. Math. Soc. 47, 1–50 (1940).

Prüfer, H.

1. *Neue Herleitung der Sturm-Liouvilleschen Reihenentwickelung stetiger Funktionen.* Math. Ann. 95, 499–518 (1926).

Pták, V.

1. *On complete topological linear spaces.* Cehoslovack. Mat. Ž. 3 (78), 301–364 (1953). (Russian. English summary) Math. Rev. 16, 262 (1955).
2. *On a theorem of W. F. Eberlein.* Studia Math. 14, 276–284 (1954).
3. *Weak compactness in convex topological linear spaces.* Cehoslovack. Mat. Ž. 4 (79), 175–186 (1954). (Russian summary).

Putnam, C. R. (see also Hartman, P.)

1. *On normal operators in Hilbert space.* Amer. J. Math. 73, 357–362 (1951).
2. *On commutators of bounded matrices.* Amer. J. Math. 73, 127–131 (1951).
3. *On the spectra of commutators.* Proc. Amer. Math. Soc. 5, 929–931 (1954).
4. *An application of spectral theory to a singular calculus of variations problem.* Amer. J. Math. 70, 780–803 (1948).
5. *The cluster spectra of bounded potentials.* Amer. J. Math. 71, 612–620 (1949).
6. *An oscillation criterion involving a minimum principle.* Duke Math. J. 16, 633–636 (1949).
7. *On the spectra of certain boundary value problems.* Amer. J. Math. 71, 109–111 (1949).
8. *On isolated eigenfunctions associated with bounded potentials.* Amer. J. Math. 72, 135–147 (1950).
9. *The comparison of spectra belonging to potentials with a bounded difference.* Duke Math. J. 18, 267–273 (1951).
10. *On the least eigenvalue of Hill's equation.* Quart. Appl. Math. 9, 310–314 (1951).
11. *The spectra of quantum-mechanical operators.* Amer. J. Math. 74, 377–388 (1952).
12. *On the unboundedness of the essential spectrum.* Amer. J. Math. 74, 578–585 (1952).
13. *A sufficient condition for an infinite discrete spectrum.* Quart. Appl. Math. 11, 484–486 (1953).
14. *On the gap in the spectrum of the Hill equation.* Quart. Appl. Math. 11, 496–498 (1953).
15. *Integrable potentials and half-line spectra.* Proc. Amer. Math. Soc. 6, 243–246 (1955).
16. *On the continuous spectra of singular boundary value problems.* Canadian J. Math. 6, 420–426 (1954).
17. *Note on a limit-point criterion.* J. London Math. Soc. 29, 126–128 (1954).
18. *Necessary and sufficient conditions for the existence of negative spectra.* Quart. Appl. Math. 13, 335–337 (1955).

Quigley, F. D. (see Helson, H.)

Rabinovič, Yu. L.

1. *On the continuous dependence upon a parameter of the spectrum of a symmetric linear integral operator.* Moskov. Gos. Univ. Učenye Zapiski 148, Matematika 4, 181–191 (1951). (Russian) Math. Rev. 14, 289 (1953).

Radon, J.

1. *Über lineare Funktionaltransformationen und Funktionalgleichungen.* S.-B. Akad. Wiss. Wien 128, 1083–1121 (1919).
2. *Theorie und Anwendungen der absolut additiven Mengenfunktionen.* S.-B. Akad. Wiss. Wien 122, 1295–1438 (1913).

Raikov (Raĭkov), D. A. (see also Gelfand, I. M.)

1. *Harmonic analysis on commutative groups with the Haar measure and the theory of characters.* Trav. Inst. Math. Steklov 14, 5–86 (1945). (Russian) Math. Rev. 8, 133 (1947). [German translation in "Sowjetische Arbeiten zur Funktional-analysis", Verlag Kultur und Fortschritt, Berlin, 1954.]

2. *A new proof of the uniqueness of the Haar's measure.* Doklady Akad. Nauk SSSR (N. S.) 34, 211–214 (1942).
3. *Positive definite functions on commutative groups with an invariant measure.* Doklady Akad. Nauk SSSR (N. S.) 28, 296–300 (1940).

Ramaswami, V.
1. *Normed algebras, isomorphism and the associative postulate.* J. Indian Math. Soc. (N. S.) 14, 47–64 (1950).

Rapoport, I. M.
1. *On singular boundary problems for ordinary linear differential equations.* Doklady Akad. Nauk SSSR (N. S.) 79, 21–24 (1951). (Russian) Math. Rev. 13, 558 (1952).
2. *On estimation of eigenvalues of Hermitian operators.* Doklady Akad. Nauk SSSR (N. S.) 103, 199–202 (1955). (Russian) Math. Rev. 17, 388 (1956).

Reid, W. T.
1. *Symmetrizable completely continuous linear transformations in Hilbert space.* Duke Math. J. 18, 41–56 (1951).
2. *Expansion problems associated with a system of linear integral equations.* Trans. Amer. Math. Soc. 33, 475–485 (1931).
3. *A new class of self-adjoint boundary value problems.* Trans. Amer. Math. Soc. 52, 381–425 (1942).

Reiter, H. J.
1. *Investigations in harmonic analysis.* Trans. Amer. Math. Soc. 73, 401–427 (1952).
2. *On a certain class of ideals in the L^1-algebra of a locally compact abelian group.* Trans. Amer. Math. Soc. 75, 505–509 (1953).

Rellich, F.
1. *Störungstheorie der Spektralzerlegung.* Proc. Internat. Congress of Math., Cambridge, Mass. 1, 606–613 (1950).
2. *Störungstheorie der Spektralzerlegung,* I–V.
   I. Math. Ann. 113, 600–619 (1936).
   II. ibid. 113, 677–685 (1936).
   III. ibid. 116, 555–570 (1939).
   IV. ibid. 117, 356–382 (1940–1941).
   V. ibid. 118, 462–484 (1941–1943).
3. *Spektraltheorie in nichtseparabeln Räumen.* Math. Ann. 110, 342–356 (1935).
4. *Die zulässigen Randbedingungen bei den singulären Eigenwertproblemen der mathematischen Physik.* Math. Zeit. 49, 702–723 (1944).
5. *Die Eindeutigkeitssatz für die Lösungen quantenmechanische Vertauschungsrelationen.* Nachr. Akad. Wiss. Göttingen. Math.-Phys. Kl. 1946, 107–115 (1946).
6. *Halbbeschränkte gewöhnliche Differentialoperatoren zweiter Ordnung.* Math. Ann. 122, 243–368 (1951).
7. *Spectral theory of a second order ordinary differential equation.* Inst. Math. Sci., New York University, 1951.

Rey Pastor, J.
1. *Functional analysis and the general theory of functions.* Reale Accademia d'Italia, Fondezione Allesandro Volta. Atti dei Convegni. 9, 339–372 (1939), Rome 1943 (Spanish).

Rickart, C. E.

1. *Integration in a convex linear topological space.* Trans. Amer. Math. Soc. 52, 498–521 (1942).
2. *An abstract Radon-Nikodým theorem.* Trans. Amer. Math. Soc. 56, 50–66 (1944).
3. *Decomposition of additive set functions.* Duke Math. J. 10, 653–665 (1943).
4. *The singular elements of a Banach algebra.* Duke Math. J. 14, 1063–1077 (1947).
5. *Isomorphic groups of linear transformations.* Amer. J. Math. 72, 451–464 (1950).
6. *Banach algebras with an adjoint operation.* Ann. of Math. (2) 47, 528–550 (1946).
7. *The uniqueness of norm problem in Banach algebras.* Ann. of Math. (2) 51, 615–628 (1950).
8. *Representation of certain Banach algebras on Hilbert space.* Duke Math. J. 18, 27–39 (1951).
9. *On spectral permanence for certain Banach algebras.* Proc. Amer. Math Soc. 4, 191–196 (1953).
10. *Theory of Banach algebras* (forthcoming book).

Riesz, F.

1. *Stetigkeitsbegriff und abstrakte Mengenlehre.* Atti del IV Congresso Intern. dei Matem., Bologna. 2, 18–24 (1908).
2. *Untersuchungen über Systeme integrierbarer Funktionen.* Math. Ann. 69, 449–497 (1910).
3. *Sur certains systèmes singuliers d'équations integrales.* Ann. École Norm. Sup. (3) 28, 33–62 (1911).
4. *Über lineare Funktionalgleichungen.* Acta Math. 41, 71–98 (1918).
5. *Sur les systèmes orthogonaux de fonctions.* C. R. Acad. Sci. Paris 144, 615–619 (1907).
6. *Les systèmes d'équations linéaires à une infinité d'inconnues.* Gauthier-Villars, Paris, 1913.
7. *Sur les opérations fonctionnelles linéaires.* C. R. Acad. Sci. Paris 149, 974–977 (1909).
8. *Zur Theorie des Hilbertschen Raumes.* Acta Sci. Math. Szeged 7, 34–38 (1934).
9. *Sur une espèce de géométrie analytiques des systèmes de fonctions sommables.* C. R. Acad. Sci. Paris 144, 1409–1411 (1907).
10. *Démonstration nouvelle d'un théorème concernant les opérations fonctionnelles linéaires.* Ann. l'École Norm. Sup. (3) 31, 9–14 (1914).
11. *Sur la répresentation des opérations fonctionnelles linéaires par des intégrales de Stieltjes.* Proc. Roy. Physiog. Soc. Lund 21, no. 16, 145–151 (1952).
12. *Sur les suites de fonctions mesurables.* C. R. Acad. Sci. Paris 148, 1303–1305 (1909).
13. *Sur la convergence en moyenne*, I, II.
    I. Acta Sci. Math. Szeged 4, 58–64 (1928–1929).
    II. ibid. 4, 182–185 (1928–1929).
14. *Über die linearen Transformationen des komplexen Hilbertschen Raumes.* Acta Sci. Math. Szeged 5, 23–54 (1930–1932).
15. *Some mean ergodic theorems.* J. London Math. Soc. 13, 274–278 (1938).

16. *Another proof of the mean ergodic theorem.* Acta Sci. Math. Szeged 10, 75–76 (1941–1943).
17. *Sur la théorie ergodique des espaces abstraits.* Acta Sci. Math. Szeged 10, 1–20 (1941–1943).
18. *Sur la théorie ergodique.* Comment. Math. Helv. 17, 221–239 (1945).
19. *On a recent generalization of G. D. Birkhoff's ergodic theorem.* Acta Sci. Math. Szeged 11, 193–200 (1946–1948).
20. *Über quadratische Formen von unendlich vielen Veränderlichen.* Nachr. Akad. Wiss. Göttingen Math.-Phys. Kl. 190–195 (1910).
21. *Sur les fonctions des transformations hermitiennes dans l'espace de Hilbert.* Acta Sci. Math. Szeged 7, 147–159 (1935).
22. *Über Sätze von Stone und Bochner.* Acta Sci. Math. Szeged 6, 184–198 (1933).
23. *Sur quelques notions fondamentales dans la théorie générale des opérations linéaires.* Ann. of Math. (2) 41, 174–206 (1940).

Riesz, F., and Lorch, E. R.
1. *The integral representation of unbounded self-adjoint transformations in Hilbert space.* Trans. Amer. Math. Soc. 39, 331–340 (1936).

Riesz, F., and Sz.–Nagy, B.
1. *Leçons d'analyse fonctionnelle.* Akadémiai Kiadó, Budapest, 1952.
2. *Über Kontraktionen des Hilbertschen Raumes.* Acta Sci. Math. Szeged 10, 202–205 (1941–1943).

Riesz, M.
1. *Sur les maxima des formes bilinéaires et sur les fonctionnelles linéaires.* Acta Math. 49, 465–497 (1926).
2. *Sur les ensembles compacts de fonctions sommables.* Acta Sci. Math. Szeged 6, 136–142 (1933).
3. *Sur les fonctions conjugées.* Math. Zeit. 27, 218–244 (1927).
4. *L'intégrale de Riemann-Liouville et le problème de Cauchy.* Acta Math. 81, 1–223 (1949).

Rinehart, R. F.
1. *The equivalence of definitions of a metric function.* Amer. Math. Monthly 62, 395–414 (1955).

Riss, J.
1. *Transformation de Fourier des distributions.* C. R. Acad. Sci. Paris 229, 12–14 (1949).

Roberts, B. D.
1. *On the geometry of abstract vector spaces.* Tôhoku Math. J. 39, 42–59 (1934).

Robison, G. B.
1. *Invariant integrals over a class of Banach spaces.* Pacific J. Math. 4, 123–150 (1954).

Rogers, C. A. (see Dvoretzky, A.)

Rogosinski, W. W. (see Macintyre, A. J.)

Rogosinski, W. W., and Shapiro, H. S.
1. *On certain extremum problems for analytic functions.* Acta Math. 90, 287–318 (1953).

Rohlin, V. A. (see also Plessner, A. I.)
1. *On endomorphism of compact commutative groups.* Izvestiya Akad. Nauk SSSR Ser. Mat. 13, 329–340 (1949). (Russian) Math. Rev. 11, 40 (1950).

2. *Selected topics from the metric theory of dynamical systems.* Uspehi Matem. Nauk (N. S.) 4, no. 2 (30), 57–128 (1949). (Russian) Math. Rev. 11, 40 (1950).
3. *On the decomposition of a dynamical system into transitive components.* Mat. Sbornik N. S. 25 (67), 235–249 (1949). (Russian) Math. Rev. 11, 373 (1950).

Rosenblatt, M.
1. *On a class of Markoff processes.* Trans. Amer. Math. Soc. 71, 120–135 (1951).

Rosenbloom, P. C.
1. *Elements of mathematical logic.* Dover Publications, New York, 1950.
2. *Perturbations of linear operators in Banach spaces.* Arch. Math. 6, 89–101 (1955).

Rosenthal, A. (see Hahn, H., and Hartogs, F.)

Rosser, J. B.
1. *Logic for mathematicians.* McGraw-Hill Co., New York, 1953.

Rota, G. C.
1. *Extension theory of ordinary linear differential operators.* Dissertation, Yale University, 1956.

Rothe, E. H.
1. *Zur Theorie der topologischen Ordnung und der Vektorfelder in Banachschen Räumen.* Composito Math. 5, 177–196 (1937–1938).
2. *Topological proofs of uniqueness theorems in the theory of differential and integral equations.* Bull. Amer. Math. Soc. 45, 606–613 (1939).
3. *Critical points and gradient fields of scalars in Hilbert space.* Acta Math. 85, 73–98 (1951).
4. *Gradient mappings.* Bull. Amer. Math. Soc. 59, 5–19 (1953).
5. *Completely continuous scalars and variational methods.* Ann. of Math. (2) 47, 580–592 (1946).
6. *Gradient mappings and extrema in Banach spaces.* Duke Math. J. 15, 421–431 (1948).
7. *Mapping degree in Banach spaces and spectral theory.* Math. Z. 63, 195–218 (1955).

Rubin, H., and Stone, M. H.
1. *Postulates for generalizations of Hilbert space.* Proc. Amer. Math. Soc. 4, 611–616 (1953).

Rudin, W.
1. *Analyticity, and the maximum modulus principle.* Duke Math. J. 20, 449–457 (1953).

Ruston, A. F.
1. *A note on convexity in Banach spaces.* Proc. Cambridge Philos. Soc. 45, 157–159 (1949).
2. *On the Fredholm theory of integral equations for operators belonging to the trace class of a general Banach space.* Proc. London Math. Soc. (2) 53, 109–124 (1951).
3. *Direct products of Banach spaces and linear functional equations.* Proc. London Math. Soc. (3) 1, 327–384 (1951).
4. *A short proof of a theorem on reflexive spaces.* Proc. Cambridge Philos. Soc. 45, 674 (1949).

5. *Formulae of Fredholm type for compact linear operations on a general Banach space.* Proc. London Math. Soc (3) 3, 368–377 (1953).
6. *Operators with a Fredholm theory.* J. London Math. Soc. 29, 318–326 (1954).

Rutickiĭ, Ya. B. (see Krasnosel'skiĭ, M. A.)

Rutman, M. (see Krein, M., and Milman, D. P.)

Rutovitz, D.
1. *On the L_p-convergence of eigenfunction expansions.* Quart. J. Math. Oxford (2) 7, 24–38 (1956).

Ryll-Nardzewski, C.
1. *On the ergodic theorems*, I, II.
   I. *Generalized ergodic theorems.* Studia Math. 12, 65–73 (1951).
   II. *Ergodic theory of continued fractions.* ibid. 12, 74–79 (1951).

Saks, S. (see also Banach, S.)
1. *Theory of the integral.* Second ed. Monografje Matematyczne, vol. 7, Warsaw, 1937. Reprinted Stechert-Hafner Pub. Co., New York.
2. *On some functionals*, I, II.
   I. Trans. Amer. Math. Soc. 35, 549–556 (1933).
   II. ibid. 41, 160–170 (1937).
3. *Addition to the note on some functionals.* Trans. Amer. Math. Soc. 35, 967–974 (1933).
4. *Integration in abstract metric spaces.* Duke Math. J. 4, 408–411 (1938).
5. *Sur les fonctionnelles de M. Banach et leurs applications aux développements de fonctions.* Fund. Math. 10, 189–196 (1928).

Saks, S., and Tamarkin, J. D.
1. *On a theorem of Hahn-Steinhaus.* Ann. of Math. (2) 34, 595–601 (1933).

Salem, R.
1. *Sur une extension du théorème de convexité de M. Marcel Riesz.* Colloq. Math. 1, 6–8 (1947).

Salem, R., and Zygmund, A.
1. *A convexity theorem.* Proc. Nat. Acad. Sci. U.S.A. 34, 443–447 (1948).

San Juan, R.
1. *Generalization of a theorem of Steinhaus on linear functionals.* Las Ciencias. Madrid 17, no. 2, 205–208 (1952). (Spanish) Math. Rev. 14, 657 (1953).

Sargent, W. L. C.
1. *On linear functionals in spaces of conditionally integrable functions.* Quart. J. Math., Oxford. Ser. (2) 1, 288–298 (1950).
2. *On some theorems of Hahn, Banach and Steinhaus.* J. London Math. Soc. 28, 438–451 (1953).

Schaerf, H. M.
1. *Sur l'unicité des mesures invariantes.* C. R. Acad. Sci. Paris 229, 1053–1055 (1949). Errata, 230, 795 (1950).
2. *Sur l'unicité de la mesure de Haar.* C. R. Acad. Sci. Paris 229, 1112–1113 (1949).

Schäffer, J. J. (see also Halmos, P. R.)
1. *On unitary dilations of contractions.* Proc. Amer. Math. Soc. 6, 322 (1955).
2. *On some problems concerning operators in Hilbert space.* Anais Acad. Brasil. Ci. 25, 87–90 (1953).

Schäfke, F. W.
1. *Über einige unendliche lineare Gleichungssysteme.* Math. Nachr. 3, 40–58 (1949).

2. *Das Kriterium von Paley und Wiener in Banachschen Raum.* Math. Nachr. 3, 59–61 (1949).
3. *Über Eigenwertprobleme mit zwei Parametern.* Math. Nachr. 6, 109–124 (1951).

Schatten, R.
1. *A theory of cross-spaces.* Ann. of Math. Studies, No. 26, Princeton University Press, Princeton, 1950.

Schauder, J.
1. *Zur Theorie stetiger Abbildungen in Funktionalräumen.* Math. Z. 26, 47–65, 417–431 (1927).
2. *Der Fixpunktsatz in Funktionalräumen.* Studia Math. 2, 171–180 (1930).
3. *Eine Eigenschaft des Haarschen Orthogonalsystemes.* Math. Z. 28, 317–320 (1928).
4. *Invarianz des Gebietes in Funktionalräumen.* Studia Math. 1, 123–139 (1929).
5. *Über den Zusammenhang zwischen der Eindeutigkeit und Lösbarkeit partieller Differentialgleichungen zweiter Ordnung vom elliptischen Typus.* Math. Ann. 106, 661–721 (1932).
6. *Über lineare, vollstetige Funktionaloperationen.* Studia Math. 2, 183–196 (1930).
7. *Über die Umkehrung linearer, stetiger Funktionaloperationen.* Studia Math. 2, 1–6 (1930).

Schiffer, M. M. (see Miller, K. S.)

Schmeidler, W.
1. *Integralgleichungen mit Anwendungen in Physik und Technik.* Akademische Verlagsgesellschaft, Leipzig, 1950.

Schmidt, E.
1. *Über die Auflösung linearer Gleichungen mit unendlich vielen Unbekannten.* Rend. del Circolo Matematico di Palermo. 25, 53–77 (1908).
2. *Auflösung der allgemeinen linearen Integralgleichung.* Math. Ann. 64, 161–174 (1907).

Schoenberg, I. J. (see also Hildebrandt, T. H.)
1. *A remark on M. M. Day's characterization of innerproduct spaces and a conjecture of L. M. Blumenthal.* Proc. Amer. Math. Soc. 3, 961–964 (1952).
2. *On local convexity in Hilbert space.* Bull. Amer. Math. Soc. 48, 432–436 (1942).
3. *On smoothing operations and their generating functions.* Bull. Amer. Math. Soc. 59, 199–230 (1953).

Schreiber, M.
1. *Generealized spectral resolution for operators in Hilbert space.* Dissertation, University of Chicago, 1955.

Schreier, J.
1. *Ein Gegenbeispiel zur Theorie der schwachen Konvergenz.* Studia Math. 2, 2, 58–62 (1930).

Schreier, O.
1. *Abstrakte kontinuierliche Gruppen.* Abhand. Math. Sem. Hamburgischen Univ. 4, 15–32 (1926).

Schröder, J.
1. *Fehlerabschätzungen zur Störungsrechnung bei linearen Eigenwertproblemen mit Operatoren eines Hilbertschen Raumes.* Math. Nachr. 10, 113–128 (1953).

Schrödinger, E.

1. *Quantisierung als Eigenwertproblem.* Annalen der Physik (4) 80, 437–490 (1926).
2. *Verwasschene Eigenwertsspectra.* S.-B. Preussischen Akad. Wiss., 1929, 668–682 (1929).

Schur, A.

1. *Zur Entwickelung willkürlicher Funktionen nach Lösungen von Systemen linearer Differentialgleichungen.* Math. Ann. 82, 213–239 (1921).

Schur, I.

1. *Über die charakteristischen Wurzeln einer linearen Substitution mit einer Anwendung auf die Theorie der Integralgleichungen.* Math. Ann. 66, 488–510 (1909).

Schur, J.

1. *Über lineare Transformationen in der Theorie der unendlichen Reihen.* J. Reine Angew. Math. 151, 79–111 (1920).

Schwartz, H. M.

1. *Sequences of Stieltjes integrals,* I–III.
   I. Bull. Amer. Math. Soc. 47, 947–955 (1941).
   II. Duke Math. J. 10, 13–22 (1943).
   III. ibid. 10, 595–610 (1943).

Schwartz, J. (see also Bade, W. G., Bartle, R. G., and Dunford, N.)

1. *A note on the space* L_p^*. Proc. Amer. Math. Soc. 2, 270–275 (1951).
2. *Perturbations of spectral operators, and applications,* I. Pacific J. Math. 4, 415–458 (1954).
3. *Two perturbation formulae.* Comm. Pure Appl. Math. 8, 371–376 (1955).

Schwartz, L. (see also Dieudonné, J.)

1. *Analyse et synthèse harmoniques dans les espaces de distributions.* Canadian J. Math. 3, 503–512 (1951).
2. *Sur une propriété de synthèse spectrale dans les groupes non compacts.* C. R. Acad. Sci. Paris 227, 424–426 (1948).
3. *Théorie générale des fonctions moyennepériodiques.* Ann. of Math. (2) 48, 857–929 (1947).
4. *Homomorphismes et applications complètement continues.* C. R. Acad. Sci. Paris 236, 2472–2473 (1953).
5. *Théorie des distributions,* I, II. Act. Sci. Ind., 1091, 1122, Hermann et Cie., Paris (1951).

Schwarz, H. A.

1. *Gesammelte Mathematische Abhandlungen,* Band I. J. Springer, Berlin, 1890.

Schwerdtfeger, H.

1. *Les fonctions de matrices.* Act. Sci. et Ind., 649. Hermann et Cie., Paris, 1938.

Sears, D. B.

1. *On the solutions of a linear second order differential equation which are of integrable square.* J. London Math. Soc. 24, 207–215 (1949).
2. *Note on the uniqueness of Green's functions associated with certain differential equations.* Canadian J. Math. 2, 314–325 (1950).
3. *On the spectrum of a certain differential equation.* J. London Math. Soc. 26, 205–210 (1951).

4. *An expansion in eigenfunctions.* Proc. London Math. Soc. (2) 53, 396–421 (1951).
5. *Some properties of a differential equation.* J. London Math. Soc. 27, 180–188 (1952).
6. *Integral transforms and eigenfunction theory.* Quart. J. Math. Oxford (2) 5, 47–58 (1954).
7. *Some properties of a differential equation.* J. London Math. Soc. 29, 354–366 (1954).

Sears, D. B., and Titchmarsh, E. C.
1. *Some eigenfunction formulae.* Quart. J. Math. Oxford (2) 1, 165–175 (1950).

Sebastião e Silva, J.
1. *Integration and derivation in Banach spaces.* Univ. Lisboa. Revista Fac. Ci. A. Ci. Mat. (2) 1, 117–166 (1950). Errata 401–402 (1951). (Portuguese) Math. Rev. 13, 45 (1952).
2. *Analytic functions and functional analysis.* Portugaliae Math. 9, 1–130 (1950). (Portuguese. French summary) Math. Rev. 11, 524 (1950).
3. *Sui fondamenti della teoria dei funzionali analitici.* Portugaliae Math. 12, 1–47 (1953).

Segal, I. E. (see also Dunford, N.)
1. *Postulates for general quantum mechanics.* Ann. of Math. (2) 48, 930–948 (1947).
2. *The group algebra of a locally compact group.* Trans. Amer. Math. Soc. 61, 69–105 (1947).
3. *The span of the translations of a function in a Lebesgue space.* Proc. Nat. Acad. Sci. U.S.A. 30, 165–169 (1944).
4. *An extension of Plancherel's formula to separable unimodular groups.* Ann. of Math. (2) 52, 272–292 (1950).
5. *Decompositions of operator algebras,* I, II. Memoirs Amer. Math. Soc. no. 9, 1951.
6. *Invariant measures on locally compact spaces.* J. Indian Math. Soc. (N. S.) 13, 105–130 (1949).

Seidel, Ph. L.
1. *Note über eine Eigenschaft der Reihen, welche discontinuierliche Functionen darstellen.* Abhandlungen der Bayerischen Akad. der Wiss. München 5, 379–393 (1847–1848).

Seifert, G.
1. *A third order boundary value problem arising in aeroelastic wing theory.* Quart. Appl. Math. 9, 210–218 (1951).
2. *A third order irregular boundary value problem and the associated series.* Pacific J. Math. 2, 395–406 (1952).

Seitz, F.
1. *The modern theory of solids.* New York, 1940.

Shah, S. M.
1. *Note on eigenfunction expansions.* J. London Math. Soc. 27, 58–64 (1952).

Shapiro, H. S. (see also Rogosinski, W. W.)
1. *Extremal problems for polynomials and power series.* Dissertation, Mass. Inst. Tech., 1952.

2. *Applications of normed linear spaces to function-theoretic extremal problems.* Lectures of functions of a complex variable, 399–404. Univ. of Michigan Press, Ann Arbor, 1955.

Shapiro, J. M. (see Cameron, R. H.)

Shiffman, M. (see Garabedian, P. R.)

Shirota, T.

1. *A generalization of a theorem of I. Kaplansky.* Osaka Math. J. 4, 121–132 (1952).

Shohat, J. A., and Tamarkin, J. D.

1. *The problem of moments.* Math. Surveys, no. 1, Amer. Math. Soc., New York, 1943.

Sikorski, R.

1. *On multiplication of determinants in Banach spaces.* Bull. Acad. Polon. Sci. Cl. III. 1, 219–221 (1953).
2. *On Leżański's determinants of linear equations in Banach spaces.* Studia Math. 14 (1953), 24–48 (1954).

Silberstein, J P. O.

1. *On eigenvalues and singular values of compact linear operators in Hilbert space.* Proc. Cambridge Philos. Soc. 49, 201–212 (1953).

Šilov, G. (see also Gelfand, I.)

1. *Ideals and subrings of the ring of continuous functions.* Doklady Akad. Nauk SSSR (N. S.) 22, 7–10 (1939).
2. *Sur la théorie des idéaux dans les anneaux normés de fonctions.* Doklady Akad. Nauk SSSR (N. S.) 27, 900–903 (1940).
3. *On normed rings possessing one generator.* Mat. Sbornik N. S. 21 (63), 25–47 (1947). (Russian. English summary) Math. Rev. 9, 445 (1948).
4. *On regular normed rings.* Trav. Inst. Math. Steklov 21, 118 pp. (1947). (Russian. English summary) Math. Rev. 9, 596 (1948).
5. *On the extension of maximal ideals.* Doklady Akad. Nauk SSSR (N.S.) 29, 83–84 (1940).

da Silvas Dias, C. L.

1. *Topological vector spaces and their application in analytic functional spaces.* Bol. Soc. Mat. São Paulo 5 (1950), 1–58 (1952). (Portuguese) Math. Rev. 13, 249 (1952).

Silverman, R. J.

1. *Invariant linear functions.* Trans. Amer. Math. Soc. 81, 411–424 (1956).

Šin, D.

1. *On quasi-differential operators in Hilbert space.* Doklady Akad. Nauk SSSR (N. S.) 18, 523–526 (1938).
2. *Über die Lösungen einer quasi-Differentialgleichung der n-ten Ordnung.* Mat. Sbornik N. S. 7 (49), 479–532 (1940).
3. *Quasi-differential operators in Hilbert space.* Mat. Sbornik (N. S.) 13 (55), 39–70 (1943). (Russian) Math. Rev. 6, 179 (1945).

Singer, I. M., and Wermer, J.

1. *Derivations on commutative normed algebras.* Math. Ann. 129, 260–264 (1955).

Širohov, M. F.

1. *Functions of elements of partial ordered spaces.* Doklady Akad. Nauk SSSR (N. S.) 74, 1057–1060 (1950). (Russian) Math. Rev. 12, 341 (1951).

Sirvint, G. (or J.)
1. *Sur les transformations intégrales de l'espace L.* Doklady Akad. Nauk SSSR (N. S.) 18, 255–257 (1938).
2. *Schwache Kompaktheit in den Banachschen Räumen.* Doklady Akad. Nauk SSSR (N. S.) 28, 199–202 (1940).
3. *Weak compactness in Banach spaces.* Studia Math. 11, 71–94 (1950).

Skorohod, A. (see Kostyučenko, A.)

Slobodyanskiĭ, M. G.
1. *On estimates for the eigenvalues of a self-adjoint operator.* Akad. Nauk SSSR. Prikl. Mat. Meh. 19, 295–314 (1955). (Russian) Math. Rev. 17, 286 (1956).

Smiley, M. F.
1. *A remark on S. Kakutani's characterization of (L)-spaces.* Ann. of Math. (2) 43, 528–529 (1942).

Smith, K. T. (see also Aronszajn, N., and Donoghue, W. F.)
1. *Sur le théorème spectral.* C. R. Acad. Sci. Paris 234, 1024–1025 (1952).

Smithies, F.
1. *The Fredholm theory of integral equations.* Duke Math. J. 8, 107–130 (1941).
2. *A note on completely continuous transformations.* Ann. of Math. (2) 38, 626–630 (1937).

Šmulian, V. L. (Šmul'yan, Yu. L.)(see also Gantmacher, V., and Krein, M.)
1. *Sur les ensembles régulièrement fermés et faiblement compacts dans l'espace du type (B).* Doklady Akad. Nauk SSSR (N. S.) 18, 405–407 (1938).
2. *Linear topological spaces and their connection with the Banach spaces.* Doklady Akad. Nauk SSSR (N. S.) 22, 471–473 (1939).
3. *Sur les topologies différentes dans l'espace de Banach.* Doklady Akad. Nauk SSSR (N. S.) 23, 331–334 (1939).
4. *On some geometrical properties of the sphere in a space of the type (B).* Doklady Akad. Nauk SSSR (N. S.) 24, 648–652 (1939).
5. *On the principle of inclusion in the space of the type (B).* Mat. Sbornik N. S. 5 (47), 317–328 (1939). (Russian. English summary) Math. Rev. 1, 335 (1940).
6. *On some geometrical properties of the unit sphere in the space of the type (B).* Mat. Sbornik N. S. 6 (48), 77–94 (1939). (Russian. English summary) Math. Rev. 1, 242 (1940).
7. *Sur la dérivabilité de la norme dans l'espace de Banach.* Doklady Akad. Nauk SSSR (N. S.) 27, 643–648 (1940).
8. *Über lineare topologische Räume.* Mat. Sbornik N. S. 7 (49), 425–448 (1940).
9. *Sur la structure de la sphère unitaire dans l'espace de Banach.* Mat. Sbornik N. S. 9 (51), 545–561 (1941).
10. *Sur les espaces linéaires topologiques,* II. Mat. Sbornik N. S. 9 (51), 727–730 (1941).
11. *Sur quelques propriétés géométriques de la sphère dans les espaces linéaires semi-ordonnés de Banach.* Doklady Akad. Nauk SSSR (N. S.) 30, 394–398 (1941).
12. *On some problems of the functional analysis.* Doklady Akad. Nauk SSSR (N. S.) 38, 157–159 (1943).
13. *Sur les ensembles compacts et faiblement compacts dans l'espace du type (B).* Mat. Sbornik N. S. 12 (54), 91–95 (1943).

14. *On compact sets in the space of measurable functions.* Mat. Sbornik N. S. 15 (57), 343–346 (1944). (Russian. English summary) Math. Rev. 6, 276 (1945).
15. *Isometric operators with infinite deficiency indices and their orthogonal extensions.* Doklady Akad. Nauk SSSR (N. S.) 87, 11–14 (1952). (Russian) Math. Rev. 14, 882 (1953).
16. *Operators with degenerate characteristic functions.* Doklady Akad. Nauk SSSR (N. S.) 93, 985–988 (1953). (Russian) Math. Rev. 15, 803 (1954).
17. *Completely continuous perturbations of operators.* Doklady Akad. Nauk SSSR (N. S.) 101, 35–38 (1955). (Russian) Math. Rev. 16, 933 (1955).

Šnol, E.
1. *Behavior of eigenfunctions and the spectrum of Sturm-Liouville operators.* Uspehi Mat. Nauk (N. S.) 9, no. 4 (62), 113–132 (1954). (Russian) Math. Rev. 16, 824 (1955).

Sobczyk, A. (see also Bohnenblust, H. F.)
1. *Projections in Minkowski and Banach spaces.* Duke Math. J. 8, 78–106 (1941).
2. *Projection of the space (m) on its subspace (c_0).* Bull. Amer. Math. Soc. 47, 938–947 (1941).
3. *On the extension of linear transformations.* Trans. Amer. Math. Soc. 55, 153–169 (1944).

Sobolev, S. L.
1. *The equations of mathematical physics.* 2d ed. Gosudarstv. Izdat. Tehn.-Teor. Lit., Moscow-Leningrad, 1950 (Russian).
2. *On a theorem of functional analysis.* Mat. Sbornik N. S. 4 (46), 471–497 (1938).

Solomyak, M. Z.
1. *On characteristic values and characteristic vectors of a perturbed operator.* Doklady Akad. Nauk SSSR (N. S.) 90, 29–32 (1953). (Russian) Math. Rev. 15, 136 (1954).

Sonine, N.
1. *Recherches sur les fonctions cylindriques et le developpement des fonctions continues en séries.* Math. Ann. 16, 1–80 (1880).

Soukhomlinoff, G. A.
1. *Über Fortsetzung von linearen Funktionalen in linearen komplexen Räumen und linearen Quaternionräumen.* Mat. Sbornik N. S. 3 (45), 353–358 (1938). (Russian. German summary). Zbl. f. Math. 19, 169 (1938–1939).

Sparre Andersen, E., and Jessen, B.
1. *Some limit theorems on integrals in an abstract set.* Danske Vid. Selsk. Math.-Fys. Medd. 22, No. 14 (1946).
2. *On the introduction of measures in infinite product sets.* Danske Vid. Selsk. Math.-Fys. Medd. 25, No. 4 (1948).

Spragens, W. H.
1. *On series of Walsh eigenfunctions.* Proc. Amer. Math. Soc. 2, 202–204 (1951).

Šreĭder, Yu. A.
1. *The structure of maximal ideals in rings of measures with convolution.* Mat. Sbornik N. S. 27 (69), 297–318 (1950). (Russian) Math. Rev. 12, 420 (1951). Amer. Math. Soc. Translation no. 81 (1953).

Staševskaya, V. V.

1. *On inverse problems of spectral analysis for a class of differential equations.* Doklady Akad. Nauk SSSR (N. S.) 93, 409–411 (1953). (Russian) Math. Rev. 15, 873 (1954).

Steinhaus, H. (see also Banach, S., and Kacmarz, S.)

1. *Sur les développements orthogonaux.* Bull. Int. Acad. Polon. Sci. Sér.A. 11–39 (1926).
2. *Additive und stetige Funktionaloperationen.* Math. Z. 5, 186–221 (1919).

Stekloff, W.

1. *Sur les expressions asymptotiques de certaines fonctions définies par des équations différentielles linéaires du deuxième ordre, et leurs applications au problème du développement d'une fonction arbitraire en séries procédant suivant les dites fonctions.* Comm. Soc. Math. Kharkow (2) 10 (2–6), 97–199 (1907–1909). Rev. Sem. Publ. Math. 21, 117 (1913).

Stepanoff, W.

1. *Sur une extension du théorème ergodique.* Compositio Math. 3, 239–253 (1936).

Steinberg, H.

1. *Diffusion processes with absorption.* Thesis, Yale University, 1954.

Stevenson, A. F., and Bassali, W. A.

1. *On the possible forms of differential equation which can be factorized by the Schrödinger-Infeld method.* Canadian J. Math. 4, 385–395 (1952).

Stewart, F. M.

1. *Integration in noncommutative systems.* Trans. Amer. Math. Soc. 68, 76–104 (1950).

Stieltjes, T. J.

1. *Recherches sur les fractions continues.* Ann. Fac. Sci. Toulouse (1) 8, J. 1–22 (1894).

Stokes, G. G.

1. *On the critical values of the sums of periodic series.* Trans. Cambridge Phil. Soc. 8, 533–583 (1849).

Stone, M. H. (see also Dunford, N., Lengyel, B., and Rubin, H.)

1. *Applications of the theory of Boolean rings to general topology.* Trans. Amer. Math. Soc. 41, 375–481 (1937).
2. *Convexity.* Mimeographed lecture notes, The University of Chicago, 1946.
3. *Linear transformations in Hilbert space and their applications to analysis.* Amer. Math. Soc. Colloquium Pub. vol. 15, New York, 1932.
4. *The generalized Weierstrass approximation theorem.* Math. Mag. 21, 167–184, 237–254 (1947–1948).
5. *On the compactification of topological spaces.* Ann. de la Soc. Polonaise de Math. 21, 153–160 (1948).
6. *Notes on integration*, I–IV.
   I. Proc. Nat. Acad. Sci. U.S.A. 34, 336–342 (1948).
   II. ibid. 34, 447–455 (1948).
   III. ibid. 34, 483–490 (1948).
   IV. ibid. 35, 50–58 (1949).
7. *A general theory of spectra*, I, II.
   I. Proc. Nat. Acad. Sci. U.S.A. 26, 280–283 (1940).
   II. ibid. 27, 83–87 (1941).

8. *Boundedness properties in function-lattices.* Canadian J. Math. 1, 176–186 (1949).
9. *The theory of representations for Boolean algebras.* Trans. Amer. Math. Soc. 40, 37–111 (1936).
10. *Linear transformations in Hilbert space,* I–III.
I. Proc. Nat. Acad. Sci. U.S.A. 15, 198–200 (1929).
II. ibid. 15, 423–425 (1929).
III. ibid. 16, 172–175 (1930).
11. *On the theorem of Gelfand-Mazur.* Ann. Soc. Polon. Math. 25 (1952), 238–240 (1953).
12. *On the foundations of harmonic analysis.* Proc. Roy. Physiog. Soc. Lund 21, no. 17, 152–172 (1952).
13. *On unbounded operators in Hilbert space.* J. Indian Math. Soc. (N. S.) 15 (1951), 155–192 (1952).
14. *On a theorem of Pólya.* J. Indian Math. Soc. (N. S.) 12, 1–7 (1948).
15. *The algebraization of harmonic analysis.* Math. Student 17, 81–92 (1949).
16. *On one-parameter unitary groups in Hilbert space.* Ann. of Math. (2) 33, 643–648 (1932).
17. *Certain integrals analogous to Fourier integrals.* Math. Zeit. 28, 654–676 (1928).
18. *An unusual type of expansion problem.* Trans. Amer. Math. Soc. 26, 335–355 (1924).
19. *A comparison of the series of Fourier and Birkhoff.* Trans. Amer. Math. Soc. 28, 695–761 (1926).
20. *Irregular differential systems of order two and the related expansion problem.* Trans. Amer. Math. Soc. 29, 23–53 (1927).
21. *The expansion problems associated with regular differential systems of the second order.* Trans. Amer. Math. Soc. 29, 826–844 (1927).

Štraus, A. V.
1. *On the theory of the generalized resolvent of a symmetric operator.* Doklady Akad. Nauk SSSR (N. S.) 78, 217–220 (1951). (Russian) Math. Rev. 12, 837 (1951).
2. *On characteristic properties of generalized resolvents.* Doklady Akad. Nauk SSSR (N. S.) 82, 209–212 (1952). (Russian) Math. Rev. 13, 755 (1952).
3. *On the theory of Hermitian operators.* Doklady Akad. Nauk SSSR (N. S.) 67, 611–614 (1949). (Russian) Math. Rev. 11, 186 (1950).
4. *On generalized resolvents of a symmetric operator.* Doklady Akad. Nauk SSSR (N. S.) 71, 241–244 (1950). (Russian) Math. Rev. 11, 600 (1950).
5. *Generalized resolvents of symmetric operators.* Izvestiya Akad. Nauk SSSR Ser. Mat. 18, 51–86 (1954). (Russian) Math. Rev. 16, 48 (1955).

Strutt, M. J. O.
1. *Lamésche, Mathieusche und verwandte Funktionen in Physik und Technik.* Ergebnisse der Math., I 3, J. Springer, Berlin, 1932.
2. *Reelle Eigenwerte verallgemeinerter Hillscher Eigenwertaufgaben 2. Ordnung.* Math. Zeit. 49, 593–643 (1943–1944).

Sturm, C.
1. *Sur les équations différentielles du second ordre.* J. Math. Pures Appl. (1) 1, 106–136 (1836).
2. *Sur une classe d'équations à différences partielles.* J. Math. Pures Appl. (1) 1, 373–444 (1836).

Sunouchi, G. (see also Izumi, S.)
1. *On the sequence of additive set functions.* J. Math. Soc. Japan 3, 290–295 (1951).

Sunouchi, H.
1. *On integral representations of bilinear functionals.* Proc. Japan Acad. 27, 159–161 (1951).

Sunouchi, S. (see Nakamura, M.)

Sylvester, J. J.
1. *On the equation to the secular inequalities in the planetary theory.* Phil. Mag. 16, 267–269 (1883). Reprinted in *Collected Papers* 4, 110–111.
2. *Sur les puissances et les racines de substitutions linéaires.* C. R. Acad. Sci. Paris 94, 55–59 (1882). Reprinted in *Collected Papers* 3, 562–564.

Szász, O. (see also Hilb, E.)
1. *Über die Approximation stetiger Funktionen durch lineare Aggregate von Potenzen.* Math. Ann. 77, 482–496 (1915–1916).

Sz.-Nagy, B. von (see also Riesz, F.)
1. *Sur les lattis linéaires de dimension finie.* Comm. Math. Helv. 17, 209–213 (1945).
2. *Perturbations des transformations linéaires fermées.* Acta Sci. Math. Szeged 14, 125–137 (1951).
3. *Spektraldarstellung linearer Transformationen des Hilbertschen Raumes.* Ergebnisse der Math., V 5, J. Springer, Berlin, 1942. Reprinted Edwards Bros., Ann Arbor, Mich., 1947.
4. *Perturbations des transformations autoadjointes dans l'espace de Hilbert.* Comment. Math. Helv. 19, 347–366 (1946–1947).
5. *On the set of positive functions in L_2.* Ann. of Math. (2) 39, 1–13 (1938).
6. *On semi-groups of self-adjoint transformations in Hilbert space.* Proc. Nat. Acad. Sci. U.S.A. 24, 559–560 (1938).
7. *On uniformly bounded linear transformations in Hilbert space.* Acta Sci. Math. Szeged 11, 152–157 (1947).
8. *Sur les contractions de l'espace de Hilbert.* Acta Sci. Math. Szeged 15, 87–92 (1953).
9. *A moment problem for self-adjoint operators.* Acta Math. Acad. Sci. Hungar. 3, 285–293 (1952).
10. *Transformations de l'espace de Hilbert, fonctions de type positif sur un groupe.* Acta Sci. Math. Szeged 15, 104–114 (1954).
11. *Prolongements des transformations de l'espace de Hilbert qui sortent de cet espace.* Akad. Kiadó, Budapest, 1955. (Appendix for Riesz and Sz.-Nagy [1]).
12. *On a spectral problem of Atkinson.* Acta Math. Acad. Sci. Hungar. 3, 61–66 (1952). (Russian summary).
13. *On the stability of the index of unbounded linear transformations.* Acta Math. Acad. Sci. Hungar. 3, 49–52 (1952). (Russian summary).
14. *Über messbare Darstellungen Liescher Gruppen.* Math. Ann. 112, 286–296 (1936).
15. *Expansion theorems of Paley-Wiener type.* Duke Math. J. 14, 975–978 (1947).

Tagamlitzki, Y.

1. *Sur quelques applications de la théorie générale des espaces vectoriels partiellement ordonnés.* Annuaire [Godišnik] Fac. Sci. Phys. Math., Univ. Sofia, Livre 1, Partie II. 45, 263–286 (1949). (Bulgarian. French summary) Math. Rev. 12, 420 (1951).
2. *Zur Geometrie des Kegels in den Hilbertschen Räumen.* Annuaire [Godišnik] Fac. Sci. Phys. Math., Univ. Sofia, Livre 1, Partie II. 47, 85–107 (1952). (Bulgarian. Russian and German summaries) Math. Rev. 15, 135 (1954).

Takahashi, T.

1. *On the compactness of the function-set by the convergence in mean of general type.* Studia Math. 5, 141–150 (1934).

Taldykin, A. T.

1. *On linear equations in Hilbert space.* Mat. Sbornik N. S. 29 (71), 529–550 (1951). Errata 30 (72), 463 (1952). (Russian) Math. Rev. 13, 564 (1952).

Tamarkin, J. D. (see also Dunford, N., Hille, E., Saks, S., Shohat, J. A., and Stone, M. H.)

1. *On the compactness of the space L_p.* Bull. Amer. Math. Soc. 38, 79–84 (1932).
2. *Sur quelques points de la théorie des équations différentielles linéaires ordinaires et sur la généralisation de la série de Fourier.* Rend. Circ. Mat. Palermo 24, 345–382 (1912).
3. *Some general problems of the theory of ordinary linear differential equations and expansions of an arbitrary function in a series of fundamental functions.* Math. Zeit. 27, 1–54 (1927).

Tamarkin, J. D., and Zygmund, A.

1. *Proof of a theorem of Thorin.* Bull. Amer. Math. Soc. 50, 279–282 (1944).

Taylor, A. E. (see also Bochner, S.)

1. *The extension of linear functionals.* Duke Math. J. 5, 538–547 (1939).
2. *The weak topologies of Banach spaces.* Revista Ci., Lima 42, 355–366 (1940); 43, 465–474 (1941); 44, 45–63 (1942).
3. *The weak topologies of Banach spaces.* Proc. Nat. Acad. Sci. U.S.A. 25, 438–440 (1939).
4. *On certain Banach spaces whose elements are analytic functions.* Actas Acad. Ci. Lima 12, 31–43 (1949).
5. *Weak convergence in the space H^p.* Duke Math. J. 17, 409–418 (1950).
6. *New proofs of some theorems of Hardy by Banach space methods.* Math. Mag. 23, 115–124 (1950).
7. *Banach spaces of functions analytic in the unit circle,* I, II.
   I. Studia Math. 11, 145–170 (1950).
   II. ibid. 12, 25–50 (1951).
8. *Conjugations of complex linear spaces.* Univ. California Publ. Math. (N. S.) 2, 85–102 (1944).
9. *Spectral theory of unbounded closed operators.* Proc. Symposium on Spectral Theory and Differential Problems, 267–275 (1951). Oklahoma Agricultural and Mechanical College, Stillwater, Oklahoma.
10. *Analysis in complex Banach spaces.* Bull. Amer. Math. Soc. 49, 652–669 (1943).
11. *Spectral theory of closed distributive operators.* Acta Math. 84, 189–224 (1951).
12. *The resolvent of a closed transformation.* Bull. Amer. Math. Soc. 44, 70–74 (1938).

13. *Linear operations which depend analytically upon a parameter.* Ann. of Math. (2) 39, 574–593 (1938).
14. *A note on unconditional convergence.* Studia Math. 8, 148–153 (1939).

Teichmüller, O.
1. *Braucht der Algebraiker das Auswahlaxiom?* Deutsche Math. 4, 567–577 (1939).
2. *Operatoren im Wachsschen Raum.* J. Reine Angew. Math. 174, 73–124 (1935).

Temple, G.
1. *The computation of characteristic numbers and characteristic functions.* Proc. London Math. Soc. (2) 29, 257–280 (1929).

Thomas, J.
1. *Untersuchungen über das Eigenwertproblem*
$$\frac{d}{dx}\left((f(x)\,\frac{dy}{dx}\right) + \lambda g(x)y = 0, \qquad \int_a^b A(x)y(x)dx = \int_a^b B(x)y(x)dx = 0.$$
Math. Nachr. 6, 229–261 (1951).

Thorin, G. O.
1. *Convexity theorems.* Comm. Sém. Math. Univ. Lund no. 9, 1948.
2. *An extension of a convexity theorem due to M. Riesz.* Comm. Sém. Math. Univ. Lund no. 4, 1939.

Tingley, A. J.
1. *A generalization of the Poisson formula for the solution of the heat flow equation.* Dissertation, Univ. of Minnesota, 1952.

Titchmarsh, E. C. (see also Sears, D. B.)
1. *The theory of functions.* Clarendon Press, Oxford, 1932.
2. *Some theorems on perturbation theory,* I–IV.
   I. Proc. Roy. Soc. London, Ser. A., 200, 34–46 (1949).
   II. ibid. 201, 473–479 (1950).
   III. ibid. 207, 321–328 (1951).
   IV. ibid. 210, 30–47 (1951).
3. *Introduction to the theory of Fourier integrals.* Oxford Univ. Press, 1937.
4. *Weber's integral theorem.* Proc. London Math. Soc. (2) 22, 15–28 (1923).
5. *On expansion in eigenvalues,* I–VIII.
   I. Proc. London Math. Soc. 14, 274–278 (1939).
   II. Quart. J. Math. Oxford 11, 129–140 (1940).
   III. ibid. 11, 141–145 (1940).
   IV. ibid. 12, 33–50 (1941).
   V. ibid. 12, 89–107 (1941).
   VI. ibid. 12, 154–166 (1941).
   VII. ibid. 16, 103–114 (1945).
   VIII. ibid. 16, 115–128 (1945).
6. *An eigenfunction problem occurring in quantum mechanics.* Quart. J. Math. Oxford 13, 1–10 (1942).
7. *On the eigenvalues of differential equations.* J. London Math. Soc. 19, 66–68 (1944).
8. *On the discreteness of the spectrum associated with certain differential equations.* Ann. Mat. Pura Appl. (4) 28, 141–147 (1949).
9. *On the uniqueness of Green's function associated with a second order differential operator.* Canadian J. Math. 1, 191–198 (1949).

10. *Eigenfunction problems with periodic potentials.* Proc. Roy. Soc. London, Ser. A., 203, 501–514 (1950).
11. *On the discreteness of spectra of differential equations.* Acta Sci. Math. Szeged 12 Pars B, 16–18 (1950).
12. *On the summability of eigenfunction expansions.* Quart. J. Math. Oxford (2) 2, 250–268 (1951).
13. *Travaux récents sur la théorie des fonctions charactéristiques.* Bull. Soc. Roy. Sci. Liège 20, 543–561 (1951).
14. *On the convergence of eigenfunction expansions.* Quart. J. Math. Oxford (2) 3, 139–144 (1952).
15. *Some properties of eigenfunction expansions.* Quart. J. Math. Oxford (2) 5, 59–70 (1954).
16. *Eigenfunction expansions associated with second-order differential equations.* Oxford Univ. Press, London, 1946.

Titov, N. S.
1. *On different kinds of convergence of elements and linear operators in Banach spaces.* Doklady Akad. Nauk SSSR (N. S.) 52, 569–572 (1946).
2. *Concerning various forms of convergence of elements or linear operators in Banach spaces.* Uspehi Matem. Nauk 1, no. 5–6 (15–16) 228–229 (1946). (Russian) Math. Rev. 10, 307 (1949).

Toeplitz, O. (see also Hellinger, E., and Köthe, G.)
1. *Die linearen vollkommenen Räume der Funktiontheorie.* Comment. Math. Helv. 23, 222–242 (1949).
2. *Über allgemeine lineare Mittelbildungen.* Prace Math.-Fiz. 22, 113–119 (1911).

Tomita, M.
1. *On the regularly convex hull of a set in a conjugate Banach space.* Math. J. Okayama Univ. 3, 143–145 (1954).

Tornheim, L.
1. *Normed fields over the real and complex fields.* Michigan Math. J. 1, 61–68 (1952).

Transue, W. (see Morse, M.)

Tseng, Y. Y.
1. *On generalized biorthogonal expansions in metric and unitary spaces.* Proc. Nat. Acad. Sci. U.S.A. 28, 170–175 (1942).
2. *Generalized inverses of unbounded operators between two unitary spaces.* Doklady Akad. Nauk SSSR (N. S.) 67, 431–434 (1949). (Russian) Math. Rev. 11, 115 (1950).
3. *Properties and classification of generalized inverses of closed operators.* Doklady Akad. Nauk SSSR (N. S.) 67, 607–610 (1949). (Russian) Math. Rev. 11, 115 (1950).

Tsuji, M.
1. *On the integral representation of unitary and self-adjoint operators in Hilbert space.* Jap. J. Math. 19, 287–297 (1948).
2. *On the compactness of space L^p $(p>0)$ and its application to integral equations.* Kōdai Math. Sem. Rep., 33–36 (1951).

Tukey, J. W.
1. *Some notes on the separation of convex sets.* Portugaliae Math. 3, 95–102 (1942).

Tulajkov, A.

1. *Zur Kompaktheit im Raum L_p für $p=1$.* Nachr. Ges. Wiss. Göttingen, Math.-Phys. Kl. 167–170 (1933).

Tychonoff, A.

1. *Ein Fixpunktsatz.* Math. Ann. 111, 767–776 (1935).

Udin, A. I.

1. *Some geometric questions in the theory of linear semi-ordered spaces.* Leningrad State Univ. Annals [Učenye Zapiski] Math. Ser. 10, 64–83 (1940). (Russian) Math. Rev. 2, 314 (1941).

Ulam, S. (see Mazur, S., and Oxtoby, J. C.)

Umegaki, H. (see Nakamura, M.)

van Dantzig, D. (see under Dantzig)

van Kampen, E. R. (see under Kampen)

van der Waerden, B. L. (see under Waerden)

Vaught, R. L. (see Kelley, J. L.)

Veblen, O.

1. *Invariants of quadratic differential forms.* Cambridge Univ. Press, London, 1933.

Veress, P.

1. *Über kompakte Funktionenmengen und Bairesche Klassen.* Fund. Math. 7, 244–249 (1925).
2. *Über Funktionenmengen.* Acta Math. Sci. Szeged 3, 177–192 (1927).

Vidav, I.

1. *Über eine Vermutung von Kaplansky.* Math. Z. 62, 330 (1955).

Vinogradov, A. A. (see Kračkovskiĭ, S. N.)

Vinokurov, V. G.

1. *On biorthogonal systems spanning a given subspace.* Doklady Akad. Nauk SSSR (N. S.) 85, 685–687 (1952). (Russian) Math. Rev. 14, 183 (1953).

Višik, M. I.

1. *Linear extensions of operators and boundary conditions.* Doklady Akad. Nauk SSSR (N. S.) 65, 433–436 (1949). (Russian) Math. Rev. 11, 38–39 (1950).
2. *On linear boundary problems for differential equations.* Doklady Akad. Nauk SSSR (N. S.) 65, 785–788 (1949). (Russian) Math. Rev. 11, 39 (1950).
3. *On a general form of solvable boundary problems for homogeneous and non-homogeneous elliptic differential operators.* Doklady Akad. Nauk SSSR (N. S.) 82, 181–184 (1952). (Russian) Math. Rev. 14, 279 (1953).
4. *The method of orthogonal and direct decomposition in the theory of elliptic differential equations.* Mat. Sbornik N. S. 25 (67), 189–234 (1949). (Russian) Math. Rev. 11, 520 (1950).
5. *On strongly elliptic systems of differential equations.* Mat. Sbornik N. S. 29 (71), 615–676 (1951). (Russian) Math. Rev. 14, 174 (1953).

Visser, C.

1. *On the iteration of linear operations in a Hilbert space.* Neder. Akad. Wetensch. Proc. 41, 487–495 (1938).
2. *Note on linear operators.* Neder. Akad. Wetensch. Proc. 40, 270–272 (1937).

Visser, C., and Zaanen, A. C.

1. *On the eigenvalues of compact linear transformations.* Nederl. Akad. Wetensch. Proc. Ser. A. 55, 71–78 (1952).

Vitali, G.

1. *Sulle funzioni integrali*. Atti R. Accad. delle Sci. di Torino 40, 753–766 (1905).
2. *Sull'integrazione per serie*. Rend. del Circolo Mat. di Palermo 23, 137–155 (1907).

Volterra, V.

1. *Theory of functionals*. Blackie and Sons, London and Glasgow, 1930.

von Neumann, J. (see under Neumann)

von Sz.-Nagy, B. (see under Sz.-Nagy)

Vulich (Vulih), B. Z. (see also Kantorovitch, L. V.)

1. *Une definition du produit dans les espaces semi-ordonnés linéaires*. Doklady Akad. Nauk SSSR (N. S.) 26, 850–854 (1940).
2. *Sur les propriétés du produit et de l'élément inverse dans les espaces semi-ordonnés linéaires*. Doklady Akad. Nauk SSSR (N. S.) 26, 855–859 (1940).
3. *Linear spaces with given convergence*. Leningrad State University Annals [Učenye Zapiski] Math. Ser. 10, 40–63 (1940). (Russian) Math. Rev. 1, 313 (1940).
4. *Sur l'intégrale de Stieltjes de fonctions, dont les valeurs appartiennent à un espace semi-ordonné*. Leningrad State Univ. Annals [Učenye Zapiski] Math. Ser. 12, 3–29 (1941). (Russian. French summary) Math. Rev. 8, 30 (1947).
5. *Sur les fonctionnelles linéaires dans les espaces semi-ordonnés linéaires*. Doklady Akad. Nauk SSSR (N. S.) 52, 95–98 (1946).
6. *Sur les opérations linéaires multiplicatives*. Doklady Akad. Nauk SSSR (N. S.) 52, 383–386 (1946).
7. *Sur quelques opérations non-linéaires dans les espaces semi-ordonnés linéaires*. Doklady Akad. Nauk SSSR (N. S.) 52, 475–478 (1946).
8. *Concrete representations of linear partially ordered spaces*. Doklady Akad. Nauk SSSR (N. S.) 58, 733–736 (1947). (Russian) Math. Rev. 9, 290 (1948).
9. *The product in linear partially ordered spaces and its application to the theory of operations*, I. Mat. Sbornik N. S. 22 (64), 27–78 (1948). (Russian) Math. Rev. 10, 46 (1949).
10. *The product in linear partially ordered spaces and its application to the theory of operations*, II. Mat. Sbornik N. S. 22 (64), 267–317 (1948). (Russian) Math. Rev. 10, 46 (1949).
11. *On the concrete representation of partially ordered lineals*. Doklady Akad. Nauk SSSR (N. S.) 78, 189–192 (1951). (Russian) Math. Rev. 13, 140 (1952).
12. *Sur les formes générales de certaines opérations linéaires*. Mat. Sbornik N. S. 2 (44), 275–305 (1937).
13. *Sur les opérations linéaires dans l'espace des fonctions sommables*. Mathematica, Cluj. 13, 40–54 (1937).
14. *On a generalized notion of convergence in a Banach space*. Ann. of Math. (2) 38, 156–174 (1937).
15. *Some questions of the theory of linear partially ordered sets*. Izvestiya Akad. Nauk SSSR Ser. Mat. 17, 365–388 (1953). (Russian) Math. Rev. 14, 328 (1954).

van der Waerden, B. L.

1. *Moderne Algebra*, I, II. Springer, Berlin, 1930, 1931.

Wallach, S.
1. *On the location of spectra of differential equations*. Amer. J. Math. 70, 833–841 (1948).
2. *The spectra of periodic potentials*. Amer. J. Math. 70, 842–848 (1948).

Walsh, J. L.
1. *On the convergence of the Sturm-Liouville series*. Ann. of Math. (2) 24, 109–120 (1923).
2. *Über die Entwicklung einer analytischen Funktion nach Polynomen*. Math. Ann. 96, 430–436 (1926).
3. *Über die Entwicklung einer Funktion einer komplexen Veränderlichen nach Polynomen*. Math. Ann. 96, 437–450 (1926).
4. *Interpolation and approximation by rational functions in the complex domain*. Amer. Math. Soc. Colloquium Publication, vol. 20, New York, 1935.

Walters, S. S.
1. *The space H^p with $0 < p < 1$*. Proc. Amer. Math. Soc. 1, 800–805 (1950).
2. *Remarks on the space H^p*. Pacific J. Math. 1, 455–471 (1951).

Ward, L. E.
1. *A third order irregular boundary value problem and the associated series*. Trans. Amer. Math. Soc. 34, 417–434 (1932).

Warschawski, S. E. (see Galbraith, A. S.)

Wassilkoff, D.
1. *Partially ordered linear systems, Banach spaces and systems of functions*. Doklady Akad. Nauk SSSR 35, 135–137 (1942).
2. *Classification of orderings of linear systems*. Doklady Akad. Nauk SSSR (N. S.) 39, 167–169 (1943).
3. *On the theory of partially ordered linear systems and linear spaces*. Ann. of Math. 44, 580–609 (1943). [see Math. Rev. 5, 186 (1944)].
4. *Orderings of abstract sets and linear systems*. Izvestiya Akad. Nauk SSSR Ser. Mat. 7, 203–236 (1943). (Russian. English summary) Math. Rev. 6, 130 (1945).

Ważewski, T.
1. *Sur l'évaluation du domaine d'existence des fonctions implicites dans le cas des espaces abstraits*. Fund. Math. 37, 5–24 (1950).

Wecken, F. J.
1. *Zur Theorie linearer Operatoren*. Math. Ann. 110, 722–725 (1935).
2. *Unitärinvarianten selbstadjugierter Operatoren*. Math. Ann. 116, 422–455 (1939).

Wedderburn, J. H. M.
1. *Lectures on matrices*. Amer. Math. Soc. Colloquium Pub. 17, New York, 1934.

Wehausen, J. V.
1. *Transformations in linear topological spaces*. Duke Math. J. 4, 157–169 (1938).
2. *Transformations in metric spaces and ordinary differential equations*. Bull. Amer. Math. Soc. 51, 113–119 (1945).

Weierstrass, K.
1. *Mathematische Werke*, Band 1. Mayer und Müller, Berlin, 1894.
2. *Mathematische Werke*, Band 3, Mayer und Müller, Berlin, 1903.

Weil, A.

1. *L'intégration dans les groupes topologiques et ses applications.* Hermann et Cie., Act. Sci. et Ind. 869, Paris, 1940.
2. *Sur les fonctions presque périodiques de von Neumann.* C. R. Acad. Sci. Paris 200, 38–40 (1935).
3. *Sur les groupes topologiques et les groupes mesures.* C. R. Acad. Sci. Paris 202, 1147–1149 (1936).

Weinberger, H. F.

1. *An optimum problem in the Weinstein method for eigenvalues.* Pacific J. Math. 2, 413–418 (1952).
2. *Error estimation in the Weinstein method for eigenvalues.* Proc. Amer. Math. Soc. 3, 643–646 (1952).
3. *An extension of the classical Sturm-Liouville theory.* Duke Math. J. 22, 1–14 (1955).

Weinstein, A.

1. *Quantitative methods in Sturm-Liouville theory.* Proc. Symposium on Spectral Theory and Differential Problems (1951). Oklahoma Agricultural and Mechanical College, Stillwater, Oklahoma.

Wentzel, G.

1. *Eine Verallgemeinerung der Quantenbedingungen für die Zwecke der Wellenmechanik.* Zeit. für Physik 38, 518–529 (1926).

Wermer, J. (see also Singer, I. M.)

1. *The existence of invariant subspaces.* Duke Math. J. 19, 615–622 (1952).
2. *Invariant subspaces of bounded operators.* Proc. XII Scand. Math. Congress, Lund (1953).
3. *Commuting spectral measures on Hilbert space.* Pacific J. Math. 4, 355–361 (1954).
4. *On invariant subspaces of normal operators.* Proc. Amer. Math. Soc. 3, 270–277 (1952).
5. *On restrictions of operators.* Proc. Amer. Math. Soc. 4, 860–865 (1953).
6. *On algebras of continuous functions.* Proc. Amer. Math. Soc. 4, 866–869 (1953).
7. *On a class of normed rings.* Arkiv. för Mat. 2, 537–551 (1953).
8. *Ideals in a class of commutative Banach algebras.* Duke Math. J. 20, 273–278 (1953).
9. *Algebras with two generators.* Amer. J. Math. 76, 853–859 (1954).
10. *Subalgebras of the algebra of all continuous complex-valued functions on the circle.* Amer. J. Math. 78, 225–242 (1956).
11. *Polynomial approximaton on an arc in C^3.* Ann. of Math. (2) 62, 269–270 (1955).

Westfall, J.

1. *Zur Theorie der Integralgleichungen.* Dissertation, Göttingen, 1905.

Weyl, H. (see also Peter, F.)

1. *Inequalities between the two kinds of eigenvalues of a linear transformation.* Proc. Nat. Acad. Sci. U.S.A. 35, 408–411 (1949).
2. *Über beschränkte quadratische Formen, deren Differenz vollstetig ist.* Rend. Circ. Mat. Palermo 27, 373–392 (1909).
3. *Raum, Zeit, Materie.* Vierte Aufl., J. Springer, Berlin, 1921.
4. *Ramifications, old and new, of the eigenvalue problem.* Bull. Amer. Math. Soc. 56, 115–139 (1950).

5. *Über gewöhnliche Differentialgleichungen mit Singularitäten und die zugehörigen Entwicklungen willkürlicher Funktionen.* Math. Ann. 68, 220–269 (1910).
6. *Almost periodic invariant vector sets in a metric vector space.* Amer. J. Math. 71, 178–205 (1949).
7. *Über gewöhnliche lineare Differentialgleichungen mit singulären Stellen und ihre Eigenfunktionen.* Nachr. Akad. Wiss. Göttingen. Math.-Phys. Kl. 1909, 37–64 (1909).
8. *Über gewöhnliche Differentialgleichungen mit singulären Stellen und ihre Eigenfunktionen.* Nachr. Akad. Wiss. Göttingen. Math.-Phys. Kl. 1910, 442–467 (1910).
9. *The method of orthogonal projection in potential theory.* Duke Math. J. 7, 411–444 (1940).

Weyr, E.
1. *Note sur la théorie des quantités complexes formées avec n unités principales.* Bull. Sci. Math. (2) 11, 205–215 (1887).

Whitney, H.
1. *On ideals of differentiable functions.* Amer. J. Math. 70, 635–658 (1948).

Whyburn, G. T.
1. *Analytic topology.* Amer. Math. Soc. Colloq. Pub. vol. 28, New York, 1942.
2. *Open mappings on locally compact spaces.* Mem. Amer. Math. Soc. no. 1, New York, 1950.

Whyburn, W. M.
1. *Differential equations with general boundary conditions.* Bull. Amer. Math. Soc. 48, 692–704 (1942).

Widder, D. V. (see also Hirschman, I. I.)
1. *The Laplace transform.* Princeton Univ. Press, Princeton, 1941.
2. *Inversion formulas for convolution transforms.* Duke Math. J. 14, 217–249 (1947).
3. *The convolution transform.* Bull. Amer. Math. Soc. 60, 444–456 (1954).

Widder, D. V., and Hirschman, I. I.
1. *The inversion of a general class of convolution transforms.* Trans. Amer. Math. Soc. 66, 135–201 (1949).
2. *A representation theory for a general class of convolution transforms.* Trans. Amer. Math. Soc. 67, 69–97 (1949).
3. *Convolution transforms with complex kernels.* Pacific J. Math. 1, 211–225 (1951).

Wiegmann, N. A.
1. *A note on infinite normal matrices.* Duke Math. J. 16, 535–538 (1949).

Wielandt, H.
1. *Eigenwerttheorie.* Naturforschung und Medizin in Deutschland 1939–1946, Band 2, 85–98. Dieterich'sche Verlagsbuchhandlung, Wiesbaden, 1948.
2. *Über die unbeschränktheit der Operatoren der Quantenmechanik.* Math. Ann. 121, 21 (1949).

Wiener, N. (see also Paley, R. E. A. C.)
1. *Limit in terms of continuous transformation.* Bull. de la Soc. Math. de France 50, 119–134 (1922).
2. *Note on a paper of M. Banach.* Fund. Math. 4, 136–143 (1923).
3. *The ergodic theorem.* Duke Math. J. 5, 1–18 (1939).

4. *The Fourier integral and certain of its applications*. Cambridge Univ. Press, 1933. Reprinted by Dover Pub., New York.
5. *Tauberian theorems*. Ann. of Math. (2) 33, 1–100 (1932).
6. *Generalized harmonic analysis*. Acta Math. 55, 117–285 (1930).
7. *The average value of a functional*. Proc. London Math. Soc. (2) 22, 454–467 (1924).
8. *Differential space*. J. Math. Phys. Mass. Inst. Tech. 2, 131–174 (1923).

Wiener, N., and Wintner, A.
1. *Harmonic analysis and ergodic theory*. Amer. J. Math. 63, 415–426 (1941).

Wilansky, A.
1. *The basis in Banach space*. Duke Math. J. 18, 795–798 (1951).
2. *An application of Banach linear functionals to summability*. Trans. Amer. Math. Soc. 67, 59–68 (1949).

Wilder, C. E.
1. *Expansion problems of ordinary linear differential equations with auxiliary conditions at more than two points*. Trans. Amer. Math. Soc. 18, 415–442 (1917).
2. *Problems in the theory of ordinary linear differential equations with auxiliary conditions at more than two points*. Trans. Amer. Math. Soc. 19, 157–186 (1918).

Wilder, R. L.
1. *Introduction to the foundations of mathematics*. Wiley, New York, 1952.

Wilkins, J. E., Jr.
1. *Definitely self-conjugate adjoint integral equations*. Duke Math. J. 11, 155–166 (1944).

Williamson, J. H.
1. *Spectral representation of linear transformations in ω*. Proc. Cambridge Philos. Soc. 47, 461–472 (1951).
2. *Linear transformations in arbitrary linear spaces*. J. London Math. Soc. 28, 203–210 (1953).
3. *Compact linear operators in linear topological spaces*. J. London Math. Soc. 29, 149–156 (1954).
4. *On topologising the field $C(t)$*. Proc. Amer. Math. Soc. 5, 729–734 (1954).

Windau, W.
1. *Über lineare Differentialgleichungen vierter Ordnung mit singuläritaten und die dazugehörigen Darstellungen willkürlicher Funktionen*. Math. Ann. 83, 256–279 (1921).

Wintner, A. (see also Hartman, P. and Wiener, N.)
1. *Spektraltheorie der unendlichen Matrizen*. Hirzel, Leipzig, 1929.
2. *Zur Theorie der beschränkten Bilinearformen*. Math. Z. 30, 228–289 (1929).
3. *The unboundedness of quantum-mechanical matrices*. Physical Rev. 71, 738–739 (1947).
4. *(L_2)-connections between the kinetic and potential energies of linear systems*. Amer. J. Math. 69, 5–13 (1947).
5. *On the Laplace-Fourier transcendents occurring in mathematical physics*. Amer. J. Math. 69, 87–98 (1947).
6. *Asymptotic integrations of the adiabatic oscillator*. Amer. J. Math. 69, 251–272 (1947).
7. *Stability and high frequency*. J. Appl. Physics 18, 941–942 (1947).

8. *Stability and spectrum in the wave mechanics of lattices.* Phys. Rev. 72, 81–82 (1947).
9. *On the normalization of characteristic differentials in continuous spectra.* Phys. Rev. 72, 516–517 (1947).
10. *On the location of continuous spectra.* Amer. J. Math. 70, 22–30 (1948).
11. *Asymptotic integrations of the adiabatic oscillator in its hyperbolic range.* Duke Math. J. 15, 55–67 (1948).
12. *On Dirac's theory of continuous spectra.* Phys. Rev. 73, 781–785 (1948).
13. *A new criterion for non-oscillatory differential equations.* Quart. Appl. Math. 6, 183–185 (1948).
14. *A criterion of oscillatory stability.* Quart. Appl. Math. 7, 115–117 (1949).
15. *A priori Laplace transformation of linear differential equations.* Amer. J. Math. 71, 587–594 (1949).
16. *On almost free linear motions.* Amer. J. Math. 71, 595–602 (1949).
17. *On the smallness of isolated eigenfunctions.* Amer. J. Math. 71, 603–611 (1949).
18. *A criterion for the non-existence of L_2-solutions of a non-oscillatory differential equation.* J. London Math. Soc. 25, 347–351 (1950).
19. *On the non-existence of conjugate points.* Amer. J. Math. 73, 368–380 (1951).
20. *On linear instability.* Quart. Appl. Math. 13, 192–195 (1955).

Wirtinger, W.

1. *Beiträge zu Riemann's Integrationsmethode für hyperbolische Differentialgleichungen, und deren Anwendungen auf Schwingungsprobleme.* Math. Ann. 48, 365–389 (1897).

Wittich, H.

1. *Über das Anwachsen der Lösungen linearer Differentialgleichungen.* Math. Ann. 124, 277–288 (1952).

Wolf, F.

1. *Analytic perturbation of operators in Banach spaces.* Math. Ann. 124, 317–333 (1952).
2. *Simplicity of spectra in general operators.* (Abstract) Bull. Amer. Math. Soc. 60, 345 (1954).

Wolfson, K.

1. *On the spectrum of a boundary value problem with two singular endpoints.* Amer. J. Math. 72, 713–719 (1950).
2. *On the separation of spectra.* Proc. Amer. Math. Soc. 4, 408–409 (1953).

Wright, F. B.

1. *Absolute valued algebras.* Proc. Nat. Acad. Sci. U.S.A. 39, 330–332 (1953).

Yaglom, A. M. (see Gelfand, I. M.)

Yamabe, H.

1. *On an extension of Helly's theorem.* Osaka Math. J. 2, 15–17 (1950).

Yood, B.

1. *Banach algebras of continuous functions.* Amer. J. Math. 73, 30–42 (1951).
2. *Properties of linear transformations preserved under addition of a completely continuous transformation.* Duke Math. J. 18, 599–612 (1951).
3. *On fixed points for semi-groups of linear operators.* Proc. Amer. Math. Soc. 2, 225–233 (1951).
4. *Transformations between Banach spaces in the uniform topology.* Ann. of Math. (2) 50, 486–503 (1949).

5. *Additive groups and linear manifolds of transformations between Banach spaces.* Amer. J. Math. 71, 663–677 (1949).
6. *Difference algebras of linear transformations on a Banach space.* Pacific J. Math. 4, 615–636 (1954).

Yosida, K.

1. *On vector lattices with a unit.* Proc. Imp. Acad. Tokyo 17, 121–124 (1941).
2. *Vector lattices and additive set functions.* Proc. Imp. Acad. Tokyo 17, 228–232 (1941).
3. *On the unitary equivalence in general Euclidean space.* Proc. Japan Acad. 22, 242–245 (1946).
4. *Mean ergodic theorem in Banach spaces.* Proc. Imp. Acad. Tokyo 14, 292–294 (1938).
5. *Ergodic theorems of Birkhoff-Khintchine's type.* Jap. J. Math. 17, 31–36 (1940).
6. *An abstract treatment of the individual ergodic theorem.* Proc. Imp. Acad. Tokyo 16, 280–284 (1940).
7. *On the group embedded in the metrical complete ring.* Jap. J. Math. 13, 7–26 (1936).
8. *On the differentiability and the representation of one-parameter semi-groups of linear operators.* J. Math. Soc. Japan 1, 15–21 (1948).
9. *The Markoff process with a stable distribution.* Proc. Imp. Acad. Tokyo 16, 43–48 (1940).
10. *On Titchmarsh-Kodaira's formula concerning Weyl-Stone's eigenfunction expansion.* Nagoya Math. J. 1, 49–58 (1950). Errata, ibid. 6, 187–188 (1953).
11. *On the theory of spectra.* Proc. Imp. Acad. Tokyo 16, 378–383 (1940).
12. *Normed rings and spectral theorems,* I–VI.
    I. Proc. Imp. Acad. Tokyo 19, 356–359 (1943).
    II. ibid. 19, 466–470 (1943).
    III. ibid. 20, 71–73 (1944).
    IV. ibid. 20, 183–185 (1944).
    V. ibid. 20, 269–273 (1944).
    VI. ibid, 20, 451–453 (1944).

Yosida, K. and Fukamiya, M.

1. *On regularly convex sets.* Proc. Imp. Acad. Tokyo 17, 49–52 (1941).

Yosida, K., and Hewitt, E.

1. *Finitely additive measures.* Trans. Amer. Math. Soc. 72, 46–66 (1952).

Yosida, K., and Kakutani, S.

1. *Birkhoff's ergodic theorem and the maximal ergodic theorem.* Proc. Imp. Acad. Tokyo 15, 165–168 (1939).
2. *Operator-theoretical treatment of Markoff process and mean ergodic theorem.* Ann. of Math. (2) 42, 188–228 (1941).

Yosida, K., Mimura, Y., and Kakutani, S.

1. *Integral operator with bounded kernel.* Proc. Imp. Acad. Tokyo 14, 359–362 (1938).

Yosida, K., and Nakayama, T.

1. *On the semi-ordered ring and its application to the spectral theorem,* I, II.
    I. Proc. Imp. Acad. Tokyo 18, 555–560 (1942).
    II. ibid. 19, 144–147 (1943).

Young, L. C.
1. *On an inequality of Marcel Riesz*. Ann. of Math. (2) 40, 567–574 (1939).

Zaanen, A. C. (see also Visser, C.)
1. *On a certain class of Banach spaces*. Ann. of Math. (2) 47, 654–666 (1946).
2. *Integral transformations and their resolvents in Orlicz and Lebesgue spaces*. Compositio Math. 10, 56–94 (1952).
3. *Normalisable transformations in Hilbert space and systems of linear integral equations*. Acta Math. 83, 197–248 (1950).
4. *Note on a certain class of Banach spaces*. Nederl. Akad. Wetensch. Proc. 52, 488–499 (1949).
5. *Linear analysis*. P. Noordhoff, Groningen, and Interscience Pub., New York, 1953.
6. *Characterization of a certain class of linear transformations in an arbitrary Banach space*. Nederl. Akad. Wetensch. Proc. Ser. A. 54, 87–93 (1951).
7. *Über vollstetige symmetrische und symmetrisierbare Operatoren*. Nieuw Arch. Wiskunde (2) 22, 57–80 (1943).
8. *On the theory of linear integral equations*, I. Nederl. Akad. Wetensch. Proc. 49, 194–204 (1946).
9. *On linear functional equations*. Nieuw Arch. Wiskunde (2) 22, 269–282 (1948).

Zalcwasser, Z.
1. *Sur une propriété du champ des fonctions continues*. Studia Math. 2, 63–67 (1930).

Zermelo, E.
1. *Beweis, dass jede Menge wohlgeordnet werden kann*. Math. Ann. 59, 514–516 (1904).
2. *Neuer Beweis für die Möglichkeit einer Wohlordnung*. Math. Ann. 65, 107–128 (1908).

Zimmerberg, H. J.
1. *On normalizable transformations in Hilbert space*. Acta Math. 86, 85–88 (1951).
2. *Definite integral systems*. Duke Math. J. 15, 371–388 (1948).

Zorn, M.
1. *A remark on method in transfinite algebra*. Bull. Amer. Math. Soc. 41, 667–670 (1935).

Zwahlen, R.
1. *Ein "neues" Eigenwertproblem*. Actes Soc. Helv. Sci. Nat. 133, 60–65 (1954).

Zygmund, A. (see also Calderón, A. P., Paley, R. E. A. C., Salem, R., and Tamarkin, J. D.)
1. *Trigonometrical Series*. Monografje Matematyczne, Warsaw, 1935. Reprinted Dover and Chelsea Pub. Co., New York.
2. *An individual ergodic theorem for non-commutative transformations*. Acta Sci. Math. Szeged 14, 103–110 (1951).
3. *On a theorem of Paley*. Proc. Cambridge Phil. Soc. 34, 125–133 (1938).
4. *On the convergence and summability of power series on the circle of convergence* (I). Fund. Math. 30, 171–196 (1938).

NOTATION INDEX

AUTHOR INDEX

SUBJECT INDEX

Section numbers are followed by page numbers in parentheses.

D

E

G

H

I

J

K

L

M

P

Q

R

S

T

Y

Z